Grundlehren der mathematischen Wissenschaften 319

A Series of Comprehensive Studies in Mathematics

Springer
Berlin
Heidelberg
New York
Barcelona
Hong Kong
London
Milan
Paris
Singapore
Tokyo

Martin R. Bridson
André Haefliger

Metric Spaces of Non-Positive Curvature

With 84 Figures

Springer

Martin R. Bridson
University of Oxfod
Mathematical Institute
24-29 St. Giles'
Oxford OX1 3LB
Great Britain
email: bridson@maths.ox.ac.uk

André Haefliger
Université de Genève
Section de Mathématiques
2-4 Rue du Lièvre, CP-240
CH-1211 Genève 24
Switzerland
email: Andre.Haefliger@math.unige.ch

Library of Congress Cataloging-in-Publication Data
Bridson, Martin R., 1964-
Metric spaces of non-positive curvature / Martin R. Bridson, André Haefliger.
p. cm. - (Grundlehren der mathematischen Wissenschaften, ISSN 0072-7830; 319)
Includes bibliographical references and index.
ISBN 3-540-64324-9 (hardcover: alk. paper)
1. Metric spaces. 2. Geometry, Differential. I. Haefliger, André. II. Title. III. Series.
QA611.28.B75 1999 514'.32-dc21 99-38163 CIP

Mathematics Subject Classification (1991): 57C23, 20F32, 57Mxx, 53C70

ISSN 0072-7830
ISBN 3-540-64324-9 Springer-Verlag Berlin Heidelberg New York

Printed in Germany

Cover design: MetaDesign plus GmbH, Berlin
Typesetting: Typeset by the authors and reformatted by Frank Herweg, Hirschberg-Leutershausen, using a Springer TEX macro-package

Printed on acid-free paper SPIN: 10483755 41/3143/ko-5 4 3 2 1 0

For Julie and Minouche

Introduction

The purpose of this book is to describe the global properties of complete simply-connected spaces that are non-positively curved in the sense of A. D. Alexandrov and to examine the structure of groups that act properly on such spaces by isometries. Thus the central objects of study are metric spaces in which every pair of points can be joined by an arc isometric to a compact interval of the real line and in which every triangle satisfies the CAT(0) *inequality*. This inequality encapsulates the concept of non-positive curvature in Riemannian geometry and allows one to reflect the same concept faithfully in a much wider setting — that of geodesic metric spaces. Because the CAT(0) condition captures the essence of non-positive curvature so well, spaces that satisfy this condition display many of the elegant features inherent in the geometry of non-positively curved manifolds. There is therefore a great deal to be said about the global structure of CAT(0) spaces, and also about the structure of groups that act on them by isometries — such is the theme of this book.

The origins of our study lie in the fundamental work of A. D. Alexandrov[1]. He gave several equivalent definitions of what it means for a metric space to have curvature bounded above by a real number κ. Let us begin by explaining one of Alexandrov's definitions; this formulation has been given prominence by M. Gromov, who termed it the CAT(κ) inequality. (The initial A is in honour of Alexandrov, and the initials C and T are in honour of E. Cartan and A. Toponogov, each of whom made an important contribution to the understanding of curvature via inequalities for the distance function.)

Given a real number κ, let M_κ^2 denote the following space: if $\kappa < 0$ then M_κ^2 is real hyperbolic space $\mathbb{H}^2$ with the distance function scaled by a factor of $1/\sqrt{-\kappa}$; if $\kappa = 0$ then M_κ^2 is the Euclidean plane; if $\kappa > 0$ then M_κ^2 is the 2-sphere $\mathbb{S}^2$ with the metric scaled by a factor $1/\sqrt{\kappa}$. Alexandrov pointed out that one could define curvature bounds on a space by comparing triangles in that space to triangles in M_κ^2. A natural class of spaces in which to study triangles is the following. A metric space X is called a *geodesic space* if every pair of points $x, y \in X$ can be joined by a continuous path of length $d(x, y)$; the image of such a path is called a geodesic segment. In general there may be many geodesic segments joining x to y, but nevertheless it is convenient to use the notation $[x, y]$ for a choice of such a segment. A geodesic triangle Δ in X consists of three points $x, y, z \in X$ and three geodesic segments $[x, y]$, $[y, z]$, $[z, x]$. A *comparison triangle* for Δ in M_κ^2 is a geodesic triangle $\overline{\Delta}$ in M_κ^2 with vertices $\bar{x}, \bar{y}, \bar{z}$

[1] In 1957 Alexandrov wrote an article summarizing his ideas [Ale57]

such that $d(x, y) = d(\overline{x}, \overline{y})$, $d(y, z) = d(\overline{y}, \overline{z})$ and $d(z, x) = d(\overline{z}, \overline{x})$. (If $\kappa \leq 0$ then such a $\overline{\Delta}$ always exists; if $\kappa > 0$ then it exists provided the perimeter of Δ is less than $2\pi/\sqrt{\kappa}$; in both cases it is unique up to an isometry of M_κ^2.) The point $\overline{p} \in [\overline{x}, \overline{y}]$ is called *a comparison point* in $\overline{\Delta}$ for $p \in [x, y]$ if $d(x, p) = d(\overline{x}, \overline{p})$. Comparison points on $[y, z]$ and $[z, x]$ are defined similarly. A geodesic space X is said to satisfy the CAT(κ) inequality (more briefly, *X is a* CAT(κ) *space*) if, for all geodesic triangles Δ in X,

$$d(p, q) \leq d(\overline{p}, \overline{q})$$

for all comparison points $\overline{p}, \overline{q} \in \overline{\Delta} \subseteq M_\kappa^2$.

Alexandrov defines a metric space to be of *curvature* $\leq \kappa$ if each point of the space has a neighbourhood which, equipped with the induced metric, is a CAT(κ) space. He and the Russian school which he founded have made an extensive study of the local properties of such spaces. A complete Riemannian manifold has curvature $\leq \kappa$ in the above sense if and only if all of its sectional curvatures are $\leq \kappa$. The main point of making the above definition, though, is that there are many examples of spaces other than Riemannian manifolds whose curvature is bounded above. An interesting class of non-positively curved polyhedral complexes is provided by the buildings of Euclidean (or affine) type which arose in the work of Bruhat and Tits on algebraic groups. Many other examples will be described in the course of this book.

In recent years, CAT(-1) and CAT(0) spaces have come to play an important role both in the study of groups from a geometrical viewpoint and in the proofs of certain rigidity theorems in geometry. This is due in large part to the influence of Mikhael Gromov. Of particular importance are the lectures which Gromov gave in February 1981 at Collège de France in Paris. In these lectures (an excellent account of which was written by Viktor Schroeder [BGS95]) Gromov explained the main features of the global geometry of manifolds of non-positive curvature, essentially by basing his account on the CAT(0) inequality. In the present book we shall pursue this approach further in order to describe the global properties of CAT(0) spaces and the structure of groups which act on them by isometries. Two particular features of our treatment are that we give very detailed proofs of the basic theorems, and we describe many examples.

We have divided our book into three parts: Part I is an introduction to the geometry of geodesic spaces, in Part II the basic theory of spaces with upper curvature bounds is developed, and more specialized topics are covered in Part III. We shall now outline the contents of each part. Before doing so, we should emphasize that many of the chapters can be read independently, and we therefore suggest that if you are particularly interested in the material from a certain chapter, then you should turn directly to that chapter. (References are given in the text whenever material from earlier chapters[2] is needed.)

---

[2] We shall write (7.11) to direct readers to item 7.11 in the part of the book that they are reading, and (I.7.11) to direct readers to item 7.11 in Part I. The chapters in Part III are labelled by letters and subdivided into smaller sections, giving rise to references of the form (III.Γ.7.11).

In Part I we examine such basic concepts as distance (metric spaces), geodesics, the length of a curve, length (inner) metrics, and the notion of the (upper) angle between two geodesics issuing from the same point in a metric space. (This concept of angle, which is due to Alexandrov, plays an essential role throughout the book.) Part I also contains various examples of geodesic spaces. Of these, the most important are the model spaces M_κ^n, which we introduce in Chapter I.2 and study further in I.6. One can describe M_κ^n as the complete, simply connected, Riemannian n-manifold of constant sectional curvature κ. However, in keeping with the spirit of this book, we shall define M_κ^n directly as a metric space and deduce the desired properties of the space and its group of isometries directly from this definition.

We shall augment the supply of basic examples in Part I by describing several methods for constructing new examples of geodesic metric spaces out of more familiar ones: products, gluing, cones, spherical joins, quotients, induced path metrics and limits. Most of these constructions are due to Alexandrov and the Russian school. In Chapter I.7 we shall describe the general properties of geometric complexes, as established by Bridson in his thesis. And in the final chapter of Part I we shall turn our attention to groups: after gathering some basic facts about group actions, we shall describe some of the basic ideas in geometric group theory.

In Part II we set about our main task — exploring the geometry of CAT(κ) spaces. We shall give several different formulations of the CAT(κ) condition, all due to Alexandrov, and prove that they are equivalent. One quickly sees that CAT(κ) spaces enjoy significant properties. For example, one can see almost immediately that in a complete CAT(0) space angles exist in a strong sense, the distance function is convex, every bounded set has a unique circumcentre, one has orthogonal projections onto closed convex subsets, etc. Early in Part II we shall also examine how CAT(κ) spaces behave with regard to the basic constructions introduced in Chapter I.5.

Following these basic considerations, we turn our attention to a richer circle of ideas based on a key observation of Alexandrov: when considering a triangle Δ in a complete CAT(0) space X, if one gets any non-trivial equality in the CAT(0) condition, then Δ spans an isometrically embedded Euclidean triangle in X. This observation leads quickly to results concerning the existence of flat polygons and flat strips, and thence a product decomposition theorem.

Much of the force and elegance of the theory of non-positively curved spaces rests on the fact that there is a local-to-global theorem which allows one to use local information about the space to make deductions about the global geometry of its universal cover and about the structure of groups which act by isometries on the universal cover. More precisely, we have the following generalization of the Cartan-Hadamard theorem: for $\kappa \leq 0$, a complete simply-connected geodesic space satisfies the CAT(κ) inequality locally if and only if it satisfies the CAT(κ) inequality globally. (In Chapter II.4, following a proof of Alexander and Bishop, we shall actually prove a more general statement concerning metric spaces whose metrics are locally convex.)

A more concise account of much of the material presented in Chapters II.1-II.4 and II.8-II.9 of the present book can be found in the first two chapters of Ballmann's lecture notes [Ba95].

Even if one were ultimately interested only in CAT(0) spaces, there are aspects of the subject that force one to consider geodesic metric spaces satisfying the CAT(κ) inequality for arbitrary κ. An important link between CAT(0) spaces and CAT(1) spaces is provided by a theorem of Berestovskii, which shows that the Euclidean cone C_0Y over a geodesic space Y is a CAT(0) space if and only if Y is a CAT(1) space. (A similar statement holds with regard to the κ-cone $C_\kappa Y$, where κ is arbitrary.) This theorem is used in Chapter II.5 to establish the *link condition*, a necessary and sufficient condition (highlighted by Gromov) which translates questions concerning the existence of CAT(0) metrics on polyhedral complexes into questions concerning the structure of links of vertices. The importance of the link condition is that in many circumstances (particularly in dimension two) it provides a practical method for deciding if a given complex supports a metric of non-positive curvature. Thus we are able to construct interesting examples. Two-dimensional complexes are a particularly rich source of examples, partly because the link condition is easier to check than in higher dimensions, but also because the connections between group theory and geometry are closest in dimension two, and in dimension two any subcomplex of a non-positively curved complex is itself non-positively curved.

In Chapter II.6 we begin our study of groups that act by isometries on CAT(0) spaces. First we establish basic properties of individual isometries and groups of isometries. Individual isometries are divided into three classes according to the behaviour of their displacement functions. If the displacement function is constant then the isometry is called a Clifford translation. The Clifford translations of a CAT(0) space X form a pre-Hilbert space H, which is a generalization of the Euclidean de Rham factor in Riemannian geometry: if X is complete then there is an isometric splitting $X = X' \times H$. We also show that the group of isometries of a compact non-positively curved space is a topological group with finitely many connected components, the component of the identity being a torus.

In the early nineteen seventies, Gromoll-Wolf and Lawson-Yau proved several striking theorems concerning the structure of those groups that are the fundamental groups of compact non-positively curved Riemannian manifolds, including the Flat Torus Theorem, the Solvable Subgroup Theorem and the Splitting Theorem. In Chapters II.6 and II.7 we generalize these results to the case of groups that act properly and cocompactly by isometries on CAT(0) spaces. These generalizations have a variety of applications to group theory and topology.

In Chapters II.8 and II.9 we explore the geometry at infinity in CAT(0) spaces. Associated to any complete CAT(0) space one has a boundary at infinity ∂X, which can be constructed as the set of equivalence classes of geodesic rays in X, two rays being considered equivalent if their images are a bounded distance apart. There is a natural topology on $\overline{X} = X \cup \partial X$ called the cone topology. If X is complete and locally compact, $\overline{X}$ is compact. If X is a Riemannian manifold, $\overline{X}$ is homeomorphic to a closed ball, but for more general CAT(0) spaces the topology of ∂X can be rather complicated. An alternative construction of $\overline{X}$ is obtained by taking the closure of X in the Banach space $\mathcal{C}_*X$ of continuous functions on X modulo additive constants, where X is embedded in $\mathcal{C}_*X$ by the map that assigns to $x \in X$ the class of the function $y \mapsto d(x, y)$. In this description of $\overline{X}$ the points of ∂X emerge as classes of Busemann functions, and we are led to examine the geometry of horoballs in CAT(0) spaces.

There is a natural metric $\angle$ on the set ∂X: given $\xi, \mu \in \partial X$, one takes the supremum over all points $p \in X$ of the angle between the geodesics issuing from p in the classes ξ and μ. The topology on ∂X associated to this metric is in general weaker than the cone topology. (For instance if X is a CAT(-1) space, one gets the discrete topology.) We shall explain two significant facts concerning $\angle$: first, if X is a complete CAT(0) space then $(\partial X, \angle)$ is a CAT(1) space; secondly, the length metric associated to $\angle$, called the *Tits metric*, encodes the geometry of flat subspaces in X, in particular it determines how X can split as a product.

In Chapter III.H we shall revisit the study of boundaries in the context of Gromov's δ-hyperbolic spaces. In the context of CAT(0) spaces, the δ-hyperbolic condition is closely related to the idea of a visibility space, which was introduced in the context of smooth manifolds by Eberlein and O'Neill. Intuitively speaking, visibility spaces are "negatively curved on the large scale". In Chapter II.9 we shall see that if a proper CAT(0) space X admits a cocompact group of isometries, then X is a visibility space if and only if it does not contain an isometrically embedded copy of the Euclidean plane.

The main purpose of the remaining three chapters in Part II is to provide examples of CAT(0) spaces: in Chapter II.11 we describe various gluing techniques that allow one to build new examples out of more classical ones; in Chapter II.10 we describe elements of the geometry of symmetric spaces of non-compact type in terms of the metric approach to curvature developed in earlier chapters; and in Chapter II.12 we introduce simple complexes of groups as a forerunner to the general theory of complexes of groups developed in Chapter III.$\mathcal{C}$.

Complexes of groups were introduced by Haefliger to describe group actions on simply-connected polyhedral complexes in terms of suitable local data on the quotient. They are a natural generalization of the concept of a graph of groups, which is due to Bass and Serre. In order to work effectively with polyhedral complexes in this context, one needs a combinatorial description of them; the appropriate object to focus on is the partially ordered set of cells in the first barycentric subdivision of the complex, which provides the motivating example for objects that we call *scwols* (small categories without loops).

Associated to any action of a group on a scwol there is a complex of groups over the quotient scwol. If a complex of groups arises from such an action, it is said to be *developable*. In contrast to the one-dimensional case (graphs of groups), complexes of groups are not developable in general. However, if a complex of groups is non-positively curved, in a suitable sense, then it is developable.

The foundations of the theory of complexes of groups are laid out in Chapter III.$\mathcal{C}$. The developability theorem for non-positively curved complexes of groups is proved in Chapter III.$\mathcal{G}$, where it is treated in the more general context of groupoids of local isometries.

There are two other chapters in Part III. In the first, Chapter III.H, we describe elements of Gromov's theory of δ-hyperbolic metric spaces and discuss the relationship between non-positive curvature and isoperimetric equalities. In the second, Chapter III.Γ, we shall delve more deeply into the nature of groups that act properly and cocompactly by isometries on CAT(0) spaces. In particular, we shall analyse the algorithmic properties of such groups and explore the diverse nature of their subgroups. We shall also show that many theorems concerning groups of isometries of CAT(0) spaces can be extended to larger classes of groups — hyperbolic and semihyperbolic groups. The result is a substantial (but not comprehensive) account of the role which non-positive curvature plays in geometric group theory.

Having talked at some length about what this book contains, we should say a few words about what it does not contain. First we should point out that besides defining what it means for a metric space to have curvature bounded above, Alexandrov also defined what it means for a metric space to have curvature bounded below by a real number κ. (He did so essentially by imposing the reverse of the CAT(κ) inequality.) The theory of spaces with lower curvature bounds, particularly their local properties, has been developed extensively by Alexandrov and the Russian school, and such spaces play an important role in the study of collapsing for Riemannian manifolds. We shall not consider the theory of such spaces at all in this book, instead we refer the reader to the excellent survey article of Burago, Gromov and Perel'man [BGP92].

We should also make it clear that our treatment of the theory of non-positively curved spaces is by no means exhaustive; the study of such spaces continues to be a highly active field of research, encompassing many topics that we do not cover in this book. In particular, we do not discuss the conformal structure on the boundary of a CAT(-1) space, nor do we discuss the construction of Patterson measures at infinity, the geodesic flow in singular spaces of non-positive curvature, the theory of harmonic maps into CAT(0) spaces, rigidity theorems etc.

It is our intention that the present book should be able to serve as an introductory text. Although we shall arrive at non-trivial results, our lines of reasoning will be elementary, and we have written with the intention of making the material accessible to students whose background encompasses little more than a reasonable course in topology and an acquaintance with the basic concepts of group theory. Thus, for example, we expect the reader to understand what a manifold is and to be familiar with the definition of the fundamental group of a space, but a nodding acquaintance with the notion of a Riemannian metric will be quite sufficient for a complete understanding of this book. In any case, all such knowledge will be much less important than an enthusiasm for direct geometric arguments.

Acknowledgements: We thank the many colleagues whose comments helped to improve the content and exposition of the material presented in this book. In particular we thank Dick Bishop, Marc Burger, Mike Davis, Thomas Delzant, David Epstein,

Pierre de la Harpe, Panos Papasoglu, Frédéric Paulin, John Roe, Ralph Strebel and Dani Wise.

We offer our heartfelt gratitude to Felice Ronga for his invaluable assistance in preparing this book for publication. We are particularly grateful for the long days that he spent transforming our rough sketches into the many figures that accompany the text and for his help in solving the many word processing problems we encountered.

We thank the Swiss National Science Foundation for its financial support and we thank the mathematics department at the University of Geneva for providing us with the facilities and equipment needed to prepare this book.

The first author thanks the Engineering and Physical Sciences Research Council of Great Britain for the Advanced Fellowship which currently supports his research. The National Science Foundation of America, the Alfred P. Sloane Foundation, and the Frank Buckley Foundation have also provided financial support for Bridson's work during the course of this project, and it is with pleasure that he takes this opportunity to thank each of them. Above all, he thanks his wife Julie Lynch Bridson for her constant love and support, and he thanks Minouche and André Haefliger for welcoming him so warmly into their home.

The second author expresses his deep gratitude to his wife Minouche for her unconditional support during his career, and for her dedication and interest during the preparation of this book.

Geneva, March 1999 MRB, AH

Table of Contents

Part I. Geodesic Metric Spaces

This part of the book is an introduction to the geometry of geodesic spaces. The ideas that you will find here are elementary, and we have written with the intention of making all of the material accessible to first year graduate students.

Our treatment begins with such basic concepts as distance, length, geodesic, and angle. In Chapter I.2 we introduce the model spaces M_κ^n (as metric spaces) and establish basic facts about their geometry and their isometry groups; further aspects of their geometry are explained in Chapter I.6. In the intervening chapters we present various examples of geodesic spaces, establish basic facts about length metrics, and describe methods for manufacturing new spaces out of more familiar ones. Chapter I.7 is an introduction to the geometry of metric polyhedral complexes, and Chapter I.8 is an introduction to some of the basic ideas in geometric group theory.

Each chapter can be read independently (modulo references to earlier definitions).

Chapter I.1 Basic Concepts

The fundamental concept with which we shall be concerned throughout this book is that of *distance*. We shall explore the geometry of spaces whose distance functions possess various properties that compare favourably to those of the distance function in Euclidean space. If such properties are identified and articulated clearly, then by proceeding in simple steps from these defining properties one can recover much of the elegant structure of classical geometry, now adapted to a wider context.

Metric Spaces

We begin by recalling the basic properties normally required of distance functions.

1.1 Definitions. Let X be a set. A *pseudometric* on X is a real-valued function $d : X \times X \to \mathbb{R}$ satisfying the following properties, for all $x, y, z \in X$:

$$\begin{aligned} \textit{Positivity:} \quad & d(x, y) \geq 0 \text{ and } d(x, x) = 0. \\ \textit{Symmetry:} \quad & d(x, y) = d(y, x). \\ \textit{Triangle Inequality:} \quad & d(x, y) \leq d(x, z) + d(z, y). \end{aligned}$$

A pseudometric is called a *metric* if it is *positive definite*, i.e.

$$d(x, y) > 0 \text{ if } x \neq y.$$

We shall often refer to $d(x, y)$ as the *distance* between the points x and y. A *metric space* is a pair (X, d), where X is a set and d is a metric on X. A metric space is said to be *complete* if every Cauchy sequence in it converges. If Y is a subset of X, then the restriction of d to $Y \times Y$ is called the *induced metric* on Y. Given $x \in X$ and $r > 0$, the open ball of radius r about x (i.e. the set $\{y \mid d(x, y) < r\}$) shall be denoted $B(x, r)$, and the closed ball $\{y \mid d(x, y) \leq r\}$ shall be denoted $\overline{B}(x, r)$. (Note that $\overline{B}(x, r)$ may be strictly larger than the closure of $B(x, r)$.) Associated to the metric d one has the topology with basis the set of open balls $B(x, r)$. The metric space is said to be *proper* if, in this topology, for every $x \in X$ and every $r > 0$, the closed ball $\overline{B}(x, r)$ is compact.

An *isometry* from one metric space (X, d) to another (X', d') is a bijection $f : X \to X'$ such that $d'(f(x), f(y)) = d(x, y)$ for all $x, y \in X$. If such a map exists then

(X, d) and (X', d') are said to be *isometric*. The group of all isometries from a metric space (X, d) to itself will be denoted $\mathrm{Isom}(X, d)$ or, more briefly, $\mathrm{Isom}(X)$.

1.2 First Examples. Perhaps the easiest example of a metric space is obtained by taking the intuitive notion of distance $d(x, y) = |x-y|$ on the real line. More generally, we can consider the usual Euclidean metric on $\mathbb{R}^n$, the set of n–tuples of real numbers $x = (x_1, \dots, x_n)$. The distance $d(x, y)$ between two points $x = (x_1, \dots, x_n)$ and $y = (y_1, \dots, y_n)$ is given by

$$d(x, y) := \left(\sum_{i=1}^{n} |x_i - y_i|^2 \right)^{\frac{1}{2}}.$$

Throughout this book we shall use the symbol $\mathbb{E}^n$ *to denote the metric space* $(\mathbb{R}^n, d)$.

The Euclidean scalar product of two vectors $x, y \in \mathbb{R}^n$ is the number

$$(x \mid y) := \sum_{i=1}^{n} x_i y_i;$$

and the Euclidean norm of x is $\|x\| := (x \mid x)^{1/2}$. A useful way to view the metric on $\mathbb{E}^n$ is to note that $d(x, y) = \|x - y\|$. The triangle inequality for $\mathbb{E}^n$ then follows from the Cauchy-Schwarz inequality, $|(x \mid y)| \leq \|x\|.\|y\|$ (see 4.1).

$\mathbb{E}^n$ is the most familiar of the model spaces M^n_κ, whose central role in this book was described in the introduction. Another familiar model space is $\mathbb{S}^n$, the n-dimensional sphere, which may be described as follows. Let $S^n = \{x \in \mathbb{E}^{n+1} \mid \|x\| = 1\}$. This set inherits two obvious metrics from $\mathbb{E}^{n+1}$. The first, and least useful, is the induced metric. The second, and most natural, is obtained by defining the distance d between two points x and y on the sphere to be the angle at 0 between the line segments joining 0 to x and 0 to y respectively. We shall denote the metric space (S^n, d) by $\mathbb{S}^n$. It requires some thought to verify the triangle inequality for $\mathbb{S}^n$. We shall return to this point in Chapter 2, where we consider the basic properties of the model spaces M^n_κ.

Of course, not all metric spaces enjoy as much structure as the preceding classical examples. Indeed, given any set one can define a metric on it by defining the distance between every pair of distinct points to be 1. Such pathological examples render the theory of general metric spaces rather dull, and it is clear that we must restrict our notion of distance further if we wish to obtain a class of spaces whose geometry is in any way comparable to the preceding classical examples. For the most part, we shall consider only spaces in which every pair of points can be joined by a geodesic, which is defined as follows.

Geodesics

1.3 Definitions. Let (X, d) be a metric space. A *geodesic path* joining $x \in X$ to $y \in X$ (or, more briefly, a *geodesic* from x to y) is a map c from a closed interval $[0, l] \subset \mathbb{R}$ to X such that $c(0) = x$, $c(l) = y$ and $d(c(t), c(t')) = |t - t'|$ for all $t, t' \in [0, l]$ (in particular, $l = d(x, y)$). If $c(0) = x$, then c is said to *issue* from x. The image α of c is called a *geodesic segment* with endpoints x and y. (There is a 1–1 correspondence between geodesic paths in X and pairs (α, x), where α is a geodesic segment in X and x is an endpoint of α.)

Let $I \subseteq \mathbb{R}$ be an interval. A map $c : I \to X$ is said to be a *linearly reparameterized geodesic* or a *constant speed geodesic*, if there exists a constant λ such that $d(c(t), c(t')) = \lambda\, |t - t'|$ for all $t, t' \in I$. Under the same hypotheses, we say that c parameterizes its image *proportional to arc length.*

A *geodesic ray* in X is a map $c : [0, \infty) \to X$ such that $d(c(t), c(t')) = |t - t'|$ for all $t, t' \geq 0$. A *geodesic line* in X is a map $c : \mathbb{R} \to X$ such that $d(c(t), c(t')) = |t - t'|$ for all $t, t' \in \mathbb{R}$ (according to the context, we may also refer to the image of c as a geodesic line).

A *local geodesic* in X is a map c from an interval $I \subseteq \mathbb{R}$ to X with the property that for every $t \in I$ there exists $\varepsilon > 0$ such that $d(c(t'), c(t'')) = |t' - t''|$ for all $t', t'' \in I$ with $|t - t'| + |t - t''| \leq \varepsilon$.

(X, d) is said to be a *geodesic metric space* (or, more briefly, a *geodesic space*) if every two points in X are joined by a geodesic. We say that (X, d) is *uniquely geodesic* if there is exactly one geodesic joining x to y, for all $x, y \in X$.

Given $r > 0$, a metric space (X, d) is said to be *r-geodesic* if for every pair of points $x, y \in X$ with $d(x, y) < r$ there is a geodesic joining x to y. And X is said to be *r-uniquely geodesic* if there is a unique geodesic segment joining each such pair of points x and y.

A subset C of a metric space (X, d) is said to be *convex* if every pair of points $x, y \in C$ can be joined by a geodesic in X and the image of every such geodesic is contained in C. If this condition holds for all points $x, y \in C$ with $d(x, y) < r$, then C is said to be *r-convex*.

Notation. Henceforth, when referring to a generic metric space, we shall follow the common practice of writing X instead of (X, d), except in cases where ambiguity may arise.

1.4 Remarks

(1) If for every pair of points x and y in a complete metric space X there exists a point $m \in X$ such that $d(x, m) = d(y, m) = \frac{1}{2}d(x, y)$, then X is a geodesic space. If such a midpoint exists for all points $x, y \in X$ with $d(x, y) < r$ then X is r-geodesic.

(2) We emphasize that the paths which are commonly called geodesics in differential geometry need not be geodesics in the above sense; in general they will only be local geodesics. Thus, for example, a unit speed parameterization of an arc of a great circle on $\mathbb{S}^2$ is a geodesic (in the sense of this book) only if it has length at most π (cf. 2.3).

(3) According to the definition of convexity that we have adopted, a geodesic segment in a space that is not uniquely geodesic need not be convex.

1.5 Example. The most familiar example of a uniquely geodesic metric space is $\mathbb{E}^n$: the unique geodesic segment joining two points x and y is the line segment between them, namely the set of points $\{(1-t)x + ty \mid 0 \leq t \leq 1\}$. A subset C is convex in the sense of the above definition if and only if it is convex in the linear sense, i.e. if the linear segment joining each pair of points of C is entirely contained in C. A subset $X \subseteq \mathbb{E}^n$, equipped with the induced metric, is a geodesic space if and only if it is convex. More generally, when endowed with the induced metric, a subset of a uniquely geodesic metric space will be geodesic if and only if it is convex. For instance a circle in $\mathbb{E}^2$ with the induced metric is not a geodesic space, but a round disc is.

From the point of view of this book, the most fundamental examples of geodesic spaces are the model spaces of constant curvature M^n_κ; these are the subject of Chapter 2. Later in Part I we shall present three other major classes of geodesic metric spaces: normed vector spaces, complete Riemannian manifolds, and polyhedral complexes of piecewise constant curvature. In the last two cases the existence of geodesic paths is not so obvious; determining when such spaces are uniquely geodesic is also a non-trivial matter. The case of normed vector spaces is much easier.

Let V be a real vector space. Recall that a *norm*, denoted $v \mapsto \|v\|$, is a map $V \to \mathbb{R}$ such that $d(v, w) := \|v - w\|$ defines a metric and $\|\lambda v\| = |\lambda|.\|v\|$ for all $v \in V$ and $\lambda \in \mathbb{R}$. The triangle inequality for the metric associated to the norm is equivalent to the statement that $\|v + w\| \leq \|v\| + \|w\|$ for all $v, w \in V$. For the purposes of this book, we define a *normed vector space* to be a *real* vector space V together with a choice of norm. V is called a *Banach space* if the metric associated to this norm is complete.

We claim that every normed space V, equipped with the metric $d(v, w) = \|v - w\|$, is a geodesic metric space. Indeed, it is easy to see that $t \mapsto (1-t)v + tw$ defines a path $[0, 1] \to X$ which is a linearly reparameterized geodesic from v to w. We shall denote the image of this path $[v, w]$.

1.6 Proposition. *Every normed vector space V is a geodesic space. It is uniquely geodesic if and only if the unit ball in V is strictly convex (in the sense that if u_1 and u_2 are distinct vectors of norm* 1, *then* $\|(1-t)u_1 + tu_2\| < 1$ *for all* $t \in (0, 1)$*).*

Proof. In the light of our previous observation that $[v, w]$ is a geodesic segment, we see that V is uniquely geodesic if and only if, given any $v, v', v'' \in V$, the equality $d(v, v') + d(v', v'') = d(v, v'')$ implies that $v' \in [v, v'']$. Thus, writing v_1 in place of $v' - v$ and v_2 in place of $v'' - v'$, we see that V is uniquely geodesic if and only if $\|v_1 + v_2\| < \|v_1\| + \|v_2\|$ whenever v_1 and v_2 are linearly independent.

If we write $v_i = a_i u_i$, where $a_i = \|v_i\|$, and let $t = a_1/(a_1 + a_2)$, then

$$\begin{aligned} v_1 + v_2 &= (a_1 + a_2)\left(\frac{a_1}{a_1 + a_2}u_1 + \frac{a_2}{a_1 + a_2}u_2\right) \\ &= (a_1 + a_2)(tu_1 + (1-t)u_2). \end{aligned}$$

Hence $\|v_1 + v_2\| < \|v_1\| + \|v_2\|$ if and only if $\|tu_1 + (1-t)u_2\| < 1$. □

1.7 Example (The ℓ^1 and ℓ^∞ norms on $\mathbb{R}^n$).

The ℓ^1 norm of a vector $x = (x_1, \dots, x_n) \in \mathbb{R}^n$ is defined to be $\sum |x_i|$, and the ℓ^∞ norm of x is defined to be $\max_i |x_i|$. It is obvious that the maps $\mathbb{R}^n \to \mathbb{R}$ thus defined are indeed norms. We denote the associated metrics by d_1 and d_∞ respectively. If $n \geq 2$ then neither the unit ball in $(\mathbb{R}^n, d_1)$ nor that in $(\mathbb{R}^n, d_\infty)$ is strictly convex, so by (1.6) these spaces are not uniquely geodesic.

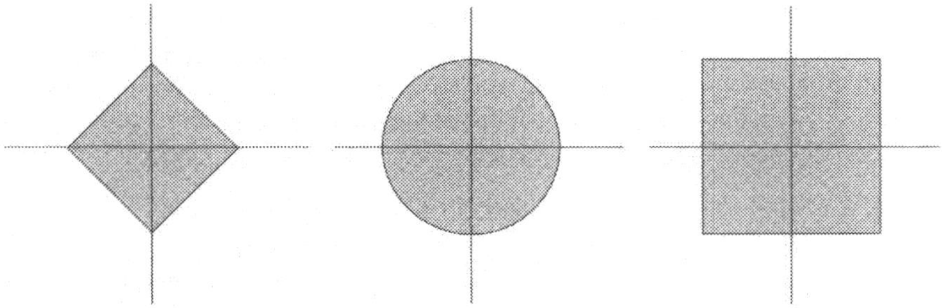

Fig. 1.1 The unit ball in $\mathbb{R}^2$ with the norms $\ell_1, \ell_2, \ell_\infty$

In Chapter 4 we shall use (1.6) to show that the metric space associated to the ℓ^p norm on $\mathbb{R}^n$ is uniquely geodesic if $1 < p < \infty$.

1.8 Exercise. Give a complete description of the geodesics in $(\mathbb{R}^2, d_1)$ and $(\mathbb{R}^2, d_\infty)$.

Perhaps the easiest examples of geodesic metric spaces which are not manifolds are yielded by the following construction.

1.9 Metric Graphs. Intuitively speaking, metric graphs are the spaces that one obtains by taking a connected graph (i.e., a connected 1-dimensional CW-complex), metrizing the individual edges of the graph as bounded intervals of the real line, and then defining the distance between two points to be the infimum of the lengths of paths joining them, where "length" is measured using the chosen metrics on the edges. It does not take long to realize that if one is not careful about the way in which the metrics on the edges are chosen, then various unpleasant pathologies can arise. Before considering these, we give a more precise formulation of the above construction.

A combinatorial graph $\mathcal{G}$ consists of two (possibly infinite) sets $\mathcal{V}$ (the vertices) and $\mathcal{E}$ (the edges) together with two maps $\partial_0 : \mathcal{E} \to \mathcal{V}$ and $\partial_1 : \mathcal{E} \to \mathcal{V}$ (the endpoint maps). We assume that $\mathcal{V}$ is the union of the images of ∂_0 and ∂_1. (A combinatorial graph in our sense is not a graph in the sense of Serre [Ser77].)

One associates to $\mathcal{G}$ the set $X_{\mathcal{G}}$ (more briefly, X) that is obtained by taking the quotient of $\mathcal{E} \times [0, 1]$ by the equivalence relation generated by $(e, i) \sim (e', i')$ if $\partial_i(e) = \partial_{i'}(e')$, where $e, e' \in \mathcal{E}$ and $i, i' \in \{0, 1\}$. Let $p : \mathcal{E} \times [0, 1] \to X$ be the quotient map. We identify $\mathcal{V}$ with the image in X of $\mathcal{E} \times \{0, 1\}$. For each $e \in \mathcal{E}$, let $f_e : [0, 1] \to X$ denote the map that sends $t \in [0, 1]$ to $p(e, t)$. Note that f_e is injective on $(0, 1)$. If $f_e(0) = f_e(1)$, the edge e is a called a *loop*.

To define a metric on X, one first specifies a map

$$\lambda : \mathcal{E} \to (0, \infty)$$

associating a *length* $\lambda(e)$ to each edge e. A *piecewise linear path* is a map $c : [0, 1] \to X$ for which there is a partition $0 = t_0 \leq t_1 \leq \cdots \leq t_n = 1$ such that each $c|_{[t_i, t_{i+1}]}$ is of the form $f_{e_i} \circ c_i$, where $e_i \in \mathcal{E}$ and c_i is an affine map from $[t_i, t_{i+1}]$ into $[0, 1]$. We say that c joins x to y if $c(0) = x$ and $c(1) = y$. The *length of* c is defined to be $l(c) = \sum_{i=0}^{n-1} l(c_i)$, where $l(c_i) = \lambda(e_i)\,|c_i(t_i) - c_{i+1}(t_{i+1})|$. We assume that X is connected, i.e. any two points are joined by such a path.

We define a pseudometric $d : X \times X \to [0, \infty]$ by setting $d(x, y)$ equal to the infimum of the length of piecewise linear paths joining x to y. The space X with its pseudometric d is called a *metric graph*. For any edge e, the distance between $p(e, 1/2)$ and $\partial_i(e)$ is $\lambda(e)/2$.

If the graph has only one edge e and this is a loop, then the corresponding metric graph is isometric to a circle of length $\lambda(e)$. Let us consider some other simple examples.

(i) Suppose that the set of vertices $\mathcal{V} = \{v_n\}_n$ is indexed by the non-negative integers, and that the set of edges $\{e_n\}_n$ is indexed by the positive integers. Suppose $\partial_0(e_n) = v_{n-1}$ and $\partial_1(e_n) = v_n$. In this case, if $\lambda(e_n) = 1$ for each n, then X is isometric to $[0, \infty)$; if $\lambda(e_n) = 1/2^n$, then X is isometric to the interval $[0, 2)$ (which is not complete); in general X is isometric to $[0, b)$ where $b = \sum \lambda(e_n)$.

(ii) Suppose that $\mathcal{V}$ has two elements v_0 and v_1, that the set of edges $\mathcal{E} = \{e_n\}_n$ is indexed by the positive integers, and that $\partial_i(e_n) = v_i$. If $\lambda(e_n) = 1/n$, then d is not a metric, because $d(v_0, v_1) = 0$. If $\lambda(e_n) = 1 + 1/n$, then d is a metric, $d(v_0, v_1) = 1$, the metric space X is complete, but there is no geodesic joining v_0 to v_1.

Coming back to the general situation, we leave the reader to check the following facts.

(1) The pseudometric d is actually a metric if for each vertex v the set $\{\lambda(e) \mid e \in \mathcal{E}, \text{ and } \partial_0(e) = v \text{ or } \partial_1(e) = v\}$ is bounded away from 0.

(2) If the set of edge lengths $\lambda(e)$ is finite, then X with the metric d is a complete geodesic space.

A combinatorial graph $\mathcal{G}$ is called a *tree* if the corresponding metric graph X where all the edges have length one is connected and simply connected.

(3) If the graph $\mathcal{G}$ is a tree, then the pseudometric on $X_{\mathcal{G}}$ associated to *any* length function $\lambda : \mathcal{E} \to (0, \infty)$ makes $X_{\mathcal{G}}$ a geodesic metric space.

Cayley Graphs. Many interesting examples of metric graphs are given by the following construction.

The *Cayley graph* [3] $\mathcal{C}_{\mathcal{A}}(\Gamma)$ of a group Γ with respect to a generating set $\mathcal{A}$ is the metric graph whose vertices are in 1-1 correspondence with the elements of Γ and which has an edge (labelled a) of length one joining γ to γa for each $\gamma \in \Gamma$ and $a \in \mathcal{A}$. In the notation of (1.9), $\mathcal{V} = \Gamma$, $\mathcal{E} = \{(\gamma, a) \mid \gamma \in \Gamma, a \in \mathcal{A}\}$, $\partial_0(\gamma, a) = \gamma$, $\partial_1(\gamma, a) = \gamma a$, and $\lambda : \mathcal{E} \to [0, \infty)$ is the constant function 1.

Cayley graphs will play a significant role in Chapter 8. A simple example is the Cayley graph of the free abelian group $\mathbb{Z}^2$ with basis $\{x, y\}$: this can be identified with the integer lattice in the Euclidean plane.

Angles

We now turn to Alexandrov's definition [Ale51 and Ale57a] of the angle between geodesics issuing from a common point in an arbitrary metric space. For this we shall need the following tool for comparing the geometry of an arbitrary metric space to that of the Euclidean plane. (This method of comparison was used extensively by A.D. Alexandrov and it plays a central role in Part II of this book. For historical references see [Rin61].)

1.10 Definition. Let X be a metric space. A *comparison triangle in* $\mathbb{E}^2$ for a triple of points (p, q, r) in X is a triangle in the Euclidean plane with vertices $\overline{p}, \overline{q}, \overline{r}$ such that $d(p, q) = d(\overline{p}, \overline{q})$, $d(q, r) = d(\overline{q}, \overline{r})$ and $d(p, r) = d(\overline{p}, \overline{r})$. Such a triangle is unique up to isometry, and shall be denoted $\overline{\Delta}(p, q, r)$. The interior angle of $\overline{\Delta}(p, q, r)$ at $\overline{p}$ is called the *comparison angle* between q and r at p and is denoted $\overline{\angle}_p(q, r)$. (The comparison angle is well-defined provided q and r are both distinct from p.)

1.11 The Law of Cosines in $\mathbb{E}^n$. We shall make frequent use of the following standard fact from Euclidean geometry; this is commonly called the *cosine rule* or the *law of cosines*.

In a Euclidean triangle with distinct vertices A, B, C, with sides of length $a = d(B, C)$, $b = d(A, C)$, $c = d(A, B)$, and with interior angle γ at the vertex C (opposite the side of length c), the following identity holds:

$$c^2 = a^2 + b^2 - 2ab \cos \gamma.$$

In particular, for fixed a and b, the length c is a strictly increasing function of γ.

The following concept is what Alexandrov calls the *upper angle* between geodesics in a metric space.

[3] Cayley graphs were introduced by Arthur Cayley in 1878 [Cay1878].

1.12 Definition of Angle. Let X be a metric space and let $c : [0, a] \to X$ and $c' : [0, a'] \to X$ be two geodesic paths with $c(0) = c'(0)$. Given $t \in (0, a]$ and $t' \in (0, a']$, we consider the comparison triangle $\overline{\Delta}(c(0), c(t), c'(t'))$, and the comparison angle $\overline{\angle}_{c(0)}(c(t), c'(t'))$. The (Alexandrov) *angle* or the *upper angle* between the geodesic paths c and c' is the number $\angle_{c,c'} \in [0, \pi]$ defined by:

$$\angle(c, c') := \limsup_{t,t' \to 0} \overline{\angle}_{c(0)}(c(t), c'(t')) = \lim_{\varepsilon \to 0} \sup_{0<t,t'<\varepsilon} \overline{\angle}_{c(0)}(c(t), c'(t')).$$

If the limit $\lim_{t,t' \to 0} \overline{\angle}_{c(0)}(c(t), c'(t'))$ exists, then we say the *angle exists in the strict sense.*

One can express $\angle(c, c')$ purely in terms of the distance function by noting that

$$\cos(\overline{\angle}_{c(0)}(c(t), c'(t'))) = \frac{1}{2tt'}\left(t^2 + t'^2 - d(c(t), c'(t'))^2\right).$$

The angle between two geodesic segments which have a common endpoint is defined to be the angle between the unique geodesics which issue from this point and whose images are the given segments. If X is uniquely geodesic, $p \neq x$ and $p \neq y$, then the angle between the geodesic segments $[p, x]$ and $[p, y]$ may be denoted $\angle_p(x, y)$.

Note that in $\mathbb{E}^n$, the Alexandrov angle is equal to the usual Euclidean angle.

1.13 Remarks and Examples.

(1) The angle between c and c' depends only on the germs of these paths at 0: if $c'' : [0, l] \to X$ is any geodesic path for which there exists $\varepsilon > 0$ such that $c''|_{[0,\varepsilon]} = c'|_{[0,\varepsilon]}$, then the angle between c and c'' is the same as that between c and c'.

(2) The angle between the incoming and outgoing germs of a geodesic at any interior point along its image is π. In other words, if $c : [a, b] \to X$ is a geodesic path with $a < 0 < b$, and if we define $c' : [0, -a] \to X$ and $c'' : [0, b] \to X$ by $c'(t) = c(-t)$ and $c''(t) = c(t)$, then $\angle(c', c'') = \pi$.

(3) In a metric tree, the angle between two geodesic segments which have a common endpoint is either 0 or π.

(4) In the above definition of angle, it is important that one takes a limsup; in general the limit $\lim_{t,t' \to 0} \overline{\angle}_{c(0)}(c(t), c'(t'))$ does not exist. For instance, we shall see in Chapter 4 that such limits exist in a normed vector space if and only if the norm derives from a scalar product. In contrast, in the metric spaces which are of primary concern in this book, namely those with curvature bounded above, this limit always exists (II.3.1).

(5) Consider $(\mathbb{R}^2, d_\infty)$, where $d_\infty((x, y), (x', y')) := \max\{|x - x'|, |y - y'|\}$. For every integer $n > 1$, the map $t \mapsto (t, [t(1 - t)]^n)$ defines a geodesic path $[0, 1/n] \to (\mathbb{R}^2, d_\infty)$. These geodesics all issue from a common point and their germs are pairwise disjoint, but the angle between any two of them is zero.

This last example illustrates the fact that the angle between distinct geodesics issuing from the same point may be 0, even if their germs are distinct. Thus, in general, $(c, c') \mapsto \angle(c, c')$ does not define a metric on the set of (germs of) geodesics issuing from a point. However, this function clearly satisfies the axioms of symmetry and positivity, so the following proposition, due to Alexandrov [Ale51], shows that it is a pseudometric. If one were to take lim inf instead of lim sup in the definition of Alexandrov angle, then this property would fail.

1.14 Proposition. *Let X be a metric space and let c, c' and c'' be three geodesic paths in X issuing from the same point p. Then,*

$$\angle(c', c'') \leq \angle(c, c') + \angle(c, c'').$$

Proof. We argue by contradiction. If this inequality were not true, then there would exist $\delta > 0$ such that $\angle(c', c'') > \angle(c, c') + \angle(c, c'') + 3\delta$. Using the definition of the lim sup, we could then find an $\varepsilon > 0$ such that:

(i) $\qquad \angle_p(c(t), c'(t')) < \angle(c, c') + \delta \quad$ for all $t, t' < \varepsilon$,

(ii) $\qquad \angle_p(c(t), c''(t'')) < \angle(c, c'') + \delta \quad$ for all $t, t'' < \varepsilon$,

(iii) $\qquad \angle_p(c'(t'), c''(t'')) > \angle(c', c'') - \delta \quad$ for some $t', t'' < \varepsilon$.

Let t' and t'' be as in (iii). Consider a triangle in $\mathbb{E}^2$ with vertices $0, x', x''$ such that $d(0, x') = t'$, $d(0, x'') = t''$, and such that the angle α at the vertex 0 satisfies:

$$\overline{\angle}_p(c'(t'), c''(t'')) > \alpha > \angle(c', c'') - \delta.$$

In particular $\alpha < \pi$, so the triangle is non-degenerate. Notice that the left-most inequality implies that $d(x', x'') < d(c'(t'), c''(t''))$. The right-most inequality implies that $\alpha > \angle(c, c') + \angle(c, c'') + 2\delta$, and hence enables us to choose a point $x \in [x', x'']$ such that the angle α' (resp. α'') between the Euclidean segments $[0, x']$ and $[0, x]$ (resp. $[0, x'']$ and $[0, x]$) is bigger than $\angle(c, c') + \delta$ (resp. $\angle(c, c'') + \delta$).

Let $t = d(0, x)$. Because $t \leq \max\{t', t''\} < \varepsilon$, condition (i) implies that

$$\overline{\angle}_p(c(t), c'(t')) < \angle(c, c') + \delta < \alpha',$$

and hence $d(c(t), c'(t')) < d(x, x')$. Similarly, $d(c(t), c''(t'')) < d(x, x'')$. Thus

$$d(c'(t'), c''(t'')) > d(x', x'') = d(x, x') + d(x, x'') > d(c(t), c'(t')) + d(c(t), c''(t'')),$$

which contradicts the triangle inequality in X. □

1.15 Exercise. Generalize definition 1.12 to give a notion of (upper) angle between arbitrary continuous paths that issue from the same point in an arbitrary metric space and do not return to that point. Then adapt the proof of the preceding proposition so that it applies to this more general situation.

The Strong Upper Angle

Following Alexandrov ([Ale51] and [Ale57a]), we give an alternative description of the angle between geodesics issuing from a common point. Let X be a metric space and let $c : [0,a] \to X$ and $c' : [0,a'] \to X$ be two geodesic paths with $c(0) = c'(0) = p$. The *strong upper angle* γ between c and c' is the number $\gamma(c,c') \in [0,\pi]$ defined by:

$$\gamma(c,c') := \limsup_{s\to 0;\, t\in(0,a']} \overline{\angle}_p(c(s),c'(t)) = \lim_{\varepsilon\to 0} \sup_{s\in(0,\varepsilon];\, t\in(0,a']} \overline{\angle}_p(c(s),c'(t)).$$

Alexandrov proved that this notion of strong upper angle is equivalent to the notion of angle given in (1.12).

1.16 Proposition. *Let X be a metric space. For all geodesics $c : [0,a] \to X$ and $c' : [0,a'] \to X$ with $c(0) = c'(0)$ one has $\angle(c,c') = \gamma(c,c')$.*

Proof. Let $p = c(0) = c'(0)$. It is clear from the definition that $\angle(c,c') \le \gamma(c,c')$. In order to establish the reverse inequality it suffices to show that for each fixed $t \in (0,a']$ we have $\limsup_{s\to 0} \overline{\angle}_p(c(s),c'(t)) \le \angle(c,c')$. For this we need a technical lemma:

1.17 Lemma. *Let $u_{s,t} = d(c(s),c'(t))$ and let $\gamma_{s,t} = \overline{\angle}_p(c(s),c'(t))$. Then*

$$\frac{t-u_{s,t}}{s} \le \cos\gamma_{s,t} \le \frac{t-u_{s,t}}{s} + \frac{s}{2t}.$$

Proof of the lemma. From the definition of $\gamma_{s,t}$ we have

$$\cos\gamma_{s,t} = \frac{s^2+t^2-u_{s,t}^2}{2st} = \frac{t-u_{s,t}}{s} + \frac{s^2-(u_{s,t}-t)^2}{2st}.$$

And by the triangle inequality $|u_{s,t}-t| \le s$, whence $0 \le \frac{s^2-(u_{s,t}-t)^2}{2st} \le \frac{s}{2t}$. □

From the lemma we have $\liminf_{s\to 0}\cos\gamma_{s,t} = \liminf_{s\to 0}\frac{t-u_{s,t}}{s}$ for each fixed $t \in (0,a']$. If $t' < t$ then by the triangle inequality $t - t' \ge u_{s,t} - u_{s,t'}$, and hence $t' - u_{s,t'} \le t - u_{s,t}$. Therefore

$$\liminf_{s,t'\to 0}\cos\gamma_{s,t'} \le \liminf_{s\to 0}\cos\gamma_{s,t}.$$

The left hand side of this inequality is equal to $\cos\angle(c,c')$, so since cos is decreasing on $[0,\pi]$ we have $\limsup_{s\to 0}\overline{\angle}_p(c(s),c'(t)) \le \angle(c,c')$ for each $t \in (0,a']$, as required. □

The Length of a Curve

The last of the fundamental concepts which we introduce in this first chapter is the notion of *length* for curves in an arbitrary metric space. We begin by fixing some terminology. Let X be a metric space. For us, a *curve* or a *path* in X is a *continuous* map c from a compact interval $[a, b] \subset \mathbb{R}$ to X. We say that c joins the point $c(a)$ to the point $c(b)$. If $c_1 : [a_1, b_1] \to X$ and $c_2 : [a_2, b_2] \to X$ are two paths such that $c_1(b_1) = c_2(a_2)$, their *concatenation* is the path $c : [a_1, b_1 + b_2 - a_2] \to X$ defined by $c(t) = c_1(t)$ if $t \in [a_1, b_1]$ and $c(t) = c_2(t + a_2 - b_1)$ if $t \in [b_1, b_1 + b_2 - a_2]$. More generally, the concatenation of a finite sequence of paths $c_i : [a_i, b_i] \to X$, with $c_i(b_i) = c_{i+1}(a_{i+1})$ for $i = 1, 2, \dots, n-1$, is defined inductively by concatenating $c_1, \dots, c_{n-1}$ and then concatenating the result with c_n.

1.18 Definition of Length. Let X be a metric space. The *length* $l(c)$ of a curve $c : [a, b] \to X$ is

$$l(c) = \sup_{a=t_0 \le t_1 \le \cdots \le t_n = b} \sum_{i=0}^{n-1} d(c(t_i), c(t_{i+1})),$$

where the supremum is taken over all possible partitions (no bound on n) with $a = t_0 \le t_1 \le \cdots \le t_n = b$.

The length of c is either a non-negative number or it is infinite. The curve c is said to be *rectifiable* if its length is finite.

1.19 Example. An easy example of a non-rectifiable path in the metric space $X = [0, 1]$ can be constructed as follows. Let $0 = t_0 < t_1 < \cdots < t_n < \dots$ be an infinite sequence of real numbers in $[0, 1]$ such that $\lim_{n\to\infty} t_n = 1$. Let $c : [0, 1] \to X$ be any path such that $c(0) = 0$ and $c(t_n) = \sum_{k=1}^{n} (-1)^{k+1}/k$. This is not a rectifiable path, because its length is bounded below by the sum of the harmonic series.

Let Y be the graph of c in $[0, 1] \times [0, 1]$, with the induced metric. Y is a compact, path-connected metric space, but there is no rectifiable path in Y joining $(0, 0)$ to the point $(1, \ln 2)$.

We note some basic properties of length.

1.20 Proposition. *Let (X, d) be a metric space and let $c : [a, b] \to X$ be a path.*

(1) $l(c) \ge d(c(a), c(b))$, *and* $l(c) = 0$ *if and only if c is a constant map.*

(2) *If ϕ is a weakly monotonic map from an interval $[a', b']$ onto $[a, b]$, then* $l(c) = l(c \circ \phi)$.

(3) Additivity: *If c is the concatenation of two paths c_1 and c_2, then* $l(c) = l(c_1) + l(c_2)$.

(4) *The reverse path $\overline{c} : [a, b] \to X$ defined by $\overline{c}(t) = c(b + a - t)$ satisfies* $l(\overline{c}) = l(c)$.

(5) *If c is rectifiable of length l, then the function* $\lambda : [a, b] \to [0, l]$ *defined by* $\lambda(t) = l(c|_{[a,t]})$ *is a continuous weakly monotonic function.*

(6) Reparameterization by arc length: *If c and* λ *are as in (5), then there is a unique path* $\tilde{c} : [0, l] \to X$ *such that*

$$\tilde{c} \circ \lambda = c \text{ and } l(\tilde{c}|_{[0,t]}) = t.$$

(7) Lower semicontinuity: *Let* (c_n) *be a sequence of paths* $[a, b] \to X$ *converging uniformly to a path c. If c is rectifiable, then for every* $\varepsilon > 0$ *there exists an integer* $N(\varepsilon)$ *such that*

$$l(c) \leq l(c_n) + \varepsilon$$

whenever $n > N(\varepsilon)$.

Proof. Properties (1), (2), (3) and (4) are immediate. Property (3) reduces the proof of (5) to showing that, given $\varepsilon > 0$, one can partition $[a, b]$ into finitely many subintervals so that the length of c restricted to each of these subintervals is at most ε. To see that this can be done, we first use the uniform continuity of the map $c : [a, b] \to X$ to choose $\delta > 0$ such that $d(c(t), c(t')) < \varepsilon/2$ for all $t, t' \in [a, b]$ with $|t - t'| < \delta$. Since $l(c)$ is finite, we can find a partition $a = t_0 < t_1 < \cdots < t_k = b$ such that

$$\sum_{i=0}^{k-1} d(c(t_i), c(t_{i+1}) > l(c) - \varepsilon/2.$$

Taking a refinement of this partition if necessary, we may assume that $|t_i - t_{i+1}| < \delta$ for $i = 0, \ldots, k-1$, and hence $d(c(t_i), c(t_{i+1})) < \varepsilon/2$. But $l(c|_{[t_i,t_{i+1}]}) \geq d(c(t_i), c(t_{i+1}))$, and $l(c) = \sum l(c|_{[t_i,t_{i+1}]})$ by (3). Hence

$$l(c) = \sum_{i=0}^{k-1} l(c|_{[t_i,t_{i+1}]}) \geq \sum_{i=0}^{k-1} d(c(t_i), c(t_{i+1}) > l(c) - \varepsilon/2,$$

with each summand in the first sum no less than the corresponding summand in the second sum. Hence, for all i we have $l(c|_{[t_i,t_{i+1}]}) - d(c(t_i), c(t_{i+1}) \leq \varepsilon/2$, and in particular $l(c|_{[t_i,t_{i+1}]}) < \varepsilon$.

(6) follows from (5) and (2). To check (7), we again choose $a = t_0 < t_1 < \cdots < t_k = b$ such that

$$l(c) \leq \sum_{i=0}^{k-1} d(c(t_i), c(t_{i+1})) + \varepsilon/2.$$

Then we choose $N(\varepsilon)$ big enough to ensure that $d(c(t), c_n(t)) < \varepsilon/4k$ for all $n > N(\varepsilon)$ and all $t \in [a, b]$. By the triangle inequality, $d(c(t_i), c(t_{i+1})) \leq 2\varepsilon/4k + d(c_n(t_i), c_n(t_{i+1}))$. Hence

$$l(c) \leq k\frac{\varepsilon}{2k} + \sum_{i=0}^{k-1} d(c_n(t_i), c_n(t_{i+1})) + \frac{\varepsilon}{2} \leq \varepsilon + l(c_n).$$

□

1.21 Definition. A path $c : [a, b] \to X$ is said to be *parameterized proportional to arc length* if the map λ defined in 1.20(5) is linear.

1.22 Remark. It follows from 1.20(2) and (5) that every path in X of length l has the same image as a path $[0, 1] \to X$ of length l which is parameterized proportional to arc length. It also follows from 1.20(5) that if a path joining x to y in X has length $d(x, y)$ then its reparameterization by arc length is a geodesic path joining x to y.

An easy but useful observation is that for any (continuous) path $c : [0, 1] \to X$,

$$l(c) = \sup_{n>0} \sum_{i=0}^{n-1} d(c(i/n), c((i+1)/n)).$$

1.23 Exercice. Show that if a sequence of paths $c_n : [a, b] \to X$ converges uniformly to a path c and $\limsup_n l(c_n)$ is finite, then c is rectifiable.

Find a sequence of rectifiable paths $c_n : [0, 1] \to [0, 1]$ converging uniformly to a non-rectifiable path c (see 1.19).

1.24 Constructing Metrics Using Chains. In certain situations one encounters a numerical measure of separation between the points of a set X that does not satisfy the triangle inequality, say $\rho : X \times X \to [0, \infty)$ with $\rho(x, x) = 0$ and $\rho(x, y) = \rho(y, x)$ for all $x, y \in X$. In such situations one can obtain a pseudometric on X by defining

$$d_\rho(x, y) := \inf \sum_{i=0}^{n-1} d(x_i, x_{i+1}),$$

where the infimum is taken over all chains $x = x_0, \dots, x_n = y$, no bound on n (cf. (5.19) and (7.4)).

One way in which examples can arise is when one has a metric space (X, d) and a function $f : [0, \infty) \to [0, \infty)$ with $f(0) = 0$; one then considers $\rho := f \circ d$.

Chapter I.2 The Model Spaces M_κ^n

In this chapter we shall construct the metric spaces M_κ^n, which play a central role in later chapters, serving as standard models to which one can profitably compare more general geodesic spaces. One way to describe M_κ^n is as the complete, simply connected, Riemannian n-manifold of constant sectional curvature $\kappa \in \mathbb{R}$. However, in keeping with the spirit of this book, we shall first define and study the M_κ^n purely as metric spaces (cf. [Iv92]), and defer consideration of their Riemannian structure until Chapter 6.

For each integer n, the spaces M_κ^n fall into three qualitatively distinct classes, according to whether κ is zero, positive or negative. To simplify the notation, we concentrate first on the cases $\mathbb{E}^n = M_0^n$, $\mathbb{S}^n = M_1^n$ and $\mathbb{H}^n = M_{-1}^n$ (we shall explain at the end of the chapter how each M_κ^n can be obtained from one of these special cases simply by scaling the metric). We shall give a concrete model for each of the spaces under consideration, describe its metric explicitly and verify the triangle inequality. In each case the triangle inequality is seen to be intimately connected with (the appropriate form of) the law of cosines. We give an explicit description of the geodesics in each case, and also of the hyperplanes; the latter play an important role in our description of $\mathrm{Isom}(M_\kappa^n)$.

In Chapter 6 we shall give alternative models for the metric spaces M_κ^n and we shall describe their natural Riemannian metric.

Euclidean n-Space $\mathbb{E}^n$

In 1.2 we introduced the notation $\mathbb{E}^n$ for the metric space obtained by equipping the vector space $\mathbb{R}^n$ with the metric associated to the norm arising from the Euclidean scalar product $(x \mid y) = \sum_{i=1}^n x_i y_i$, where $x = (x_1, \dots, x_n)$ and $y = (y_1, \dots, y_n)$. As we noted earlier, $\mathbb{E}^n$ is a uniquely geodesic space; the triangle inequality is a consequence of the Cauchy-Schwarz inequality, and the geodesic segments in $\mathbb{E}^n$ are the subsets of the form $[x, y] = \{ty + (1-t)x \mid 0 \le t \le 1\}$.

By definition, a hyperplane H in $\mathbb{E}^n$ is an affine subspace of dimension $n-1$. Given a point $P \in H$ and a unit vector u orthogonal to H (u is unique up to sign), H may be written as $H = \{Q \in \mathbb{E}^n \mid (Q - P \mid u) = 0\}$. Every hyperplane arises in the following way: given two distinct points $A, B \in \mathbb{E}^n$, the set of points equidistant from

A and B is called the *hyperplane bisector* of A and B; it is a hyperplane that contains the midpoint P of the geodesic segment $[A, B]$ and is orthogonal to the vector $A - B$.

Associated to each hyperplane there is an isometry r_H of $\mathbb{E}^n$, called *the reflection through H*. If P is a point on H and u is a unit vector orthogonal to H, then for all $A \in \mathbb{E}^n$ one has $r_H(A) = A - 2(A - P \mid u)u$. The set of fixed points of r_H is H. If $A \in \mathbb{E}^n$ does not belong to H, then H is the hyperplane bisector of A and $r_H(A)$. Conversely, if H is the hyperplane bisector of A and B, then $r_H(A) = B$.

The n-Sphere $\mathbb{S}^n$

The n-dimensional sphere $\mathbb{S}^n$ is the set $\{x = (x_1, \dots, x_{n+1}) \in \mathbb{R}^{n+1} \mid (x \mid x) = 1\}$, where $(\cdot \mid \cdot)$ denotes the Euclidean scalar product. It is endowed with the following metric:

2.1 Proposition (Definition of the Metric). *Let $d : \mathbb{S}^n \times \mathbb{S}^n \to \mathbb{R}$ be the function that assigns to each pair $(A, B) \in \mathbb{S}^n \times \mathbb{S}^n$ the unique real number $d(A, B) \in [0, \pi]$ such that*

$$\cos d(A, B) = (A \mid B).$$

Then d is a metric.

Proof. Clearly $d(A, B) = d(B, A) \geq 0$, and $d(A, B) = 0$ if and only if $A = B$. The triangle inequality is a special case of (1.14), because $d(A, B)$ is the angle between the segments $[O, A]$ and $[O, B]$ in $\mathbb{E}^{n+1}$. □

In order to harmonize our treatment of $\mathbb{S}^n$ with that of $\mathbb{H}^n$, we give a second proof of the triangle inequality based on the spherical law of cosines (which can also be deduced from (1.14)). But first we describe what we mean by a spherical triangle in $\mathbb{S}^n$ and give meaning to the notion of vertex angle. We must be careful about our use of language here, because we do not wish to presuppose that d is a metric, and therefore we should not use the definitions of geodesic, angle and triangle from Chapter 1.

It is convenient to use the vector space structure on $\mathbb{E}^{n+1}$, which contains $\mathbb{S}^n$. A great circle in $\mathbb{S}^n$ is, by definition, the intersection of $\mathbb{S}^n$ with a 2-dimensional vector subspace of $\mathbb{E}^{n+1}$. There is a natural way to parameterize arcs of great circles: given a point $A \in \mathbb{S}^n$, a unit vector $u \in \mathbb{E}^{n+1}$ with $(u \mid A) = 0$ and a number $a \in [0, \pi]$, consider the path $c : [0, a] \to \mathbb{S}^n$ given by $c(t) = (\cos t)A + (\sin t)u$. Note that $d(c(t), c(t')) = |t - t'|$ for all $t, t' \in [0, a]$. The image of c is contained in the great circle where the vector subspace spanned by A and u intersects $\mathbb{S}^n$. We shall refer to the image of c as *a minimal great arc* or, more accurately, the great arc with initial vector u joining A to $c(a)$. If $a = \pi$, then for any choice of u one has $c(\pi) = -A$. On the other hand, if $d(A, B) < \pi$ then there is a unique minimal great arc joining A to B. If $B \neq A$ then the initial vector u of this arc is the unit vector in the direction of $B - (A \mid B)A$.

By definition, the *spherical angle* between two minimal great arcs issuing from a point of $\mathbb{S}^n$, with initial vectors u and v say, is the unique number $\alpha \in [0, \pi]$ such that $\cos \alpha = (u \mid v)$. A *spherical triangle* Δ in $\mathbb{S}^n$ consists of a choice of three distinct points (its vertices) $A, B, C \in \mathbb{S}^n$, and three minimal great arcs (its sides), one joining each pair of vertices. The *vertex angle* at C is defined to be the spherical angle between the sides of Δ joining C to A and C to B.

2.2 The Spherical Law of Cosines. *Let Δ be a spherical triangle with vertices A, B, C. Let $a = d(B, C)$, $b = d(C, A)$ and $c = d(A, B)$ (spherical distances). Let γ denote the vertex angle at C. Then,*

$$\cos c = \cos a \, \cos b + \sin a \, \sin b \cos \gamma.$$

Proof. Let u and v be the initial vectors of the sides of Δ joining C to A and B respectively. By definition, $\cos \gamma = (u \mid v)$. And

$$\begin{aligned} \cos c &= (A \mid B) \\ &= \big((\cos b)C + (\sin b)u \mid (\cos a)C + (\sin a)v\big) \\ &= \cos a \, \cos b \, (C \mid C) + \sin a \, \sin b \, (u \mid v) \\ &= \cos a \, \cos b + \sin a \, \sin b \, \cos \gamma, \end{aligned}$$

as required. □

2.3 The Triangle Inequality and Geodesics. *For all $A, B, C \in \mathbb{S}^n$,*

$$d(A, B) \leq d(A, C) + d(C, B),$$

with equality if and only if C lies on a minimal great arc joining A to B. Thus:

(1) *$(\mathbb{S}^n, d)$ is a geodesic metric space.*

(2) *The geodesic segments in $\mathbb{S}^n$ are the minimal great arcs.*

(3) *If $d(A, B) < \pi$ then there is a unique geodesic joining A to B.*

(4) *Any open (resp. closed) ball of radius $r \leq \pi/2$ (resp. $< \pi/2$) in $\mathbb{S}^n$ is convex.*

Proof. Let $a = d(C, B)$, $b = d(A, C)$, $c = d(A, B)$. We assume that A, B and C are distinct (the case where they are not is trivial). Consider a spherical triangle Δ with vertices A, B, C. Let γ be the vertex angle at C. The cosine function is strictly decreasing on $[0, \pi]$, so for fixed a and b, as γ increases from 0 to π, the function $\gamma \mapsto \cos a \cos b + \sin a \sin b \cos \gamma$ decreases from $\cos(b - a)$ to $\cos(b + a)$. So by the law of cosines we have $\cos c \geq \cos(b + a)$, and hence $c \leq b + a$, with equality if and only if $\gamma = \pi$ and $b + a \leq \pi$. The conditions for equality hold if and only if C belongs to a minimal great arc joining A to B.

The convexity assertion in (3) follows from the fact, noted above, that if $d(A, B) < \pi$ then there is a unique minimal great arc joining A to B. This arc is the intersection of $\mathbb{S}^n$ with the positive cone in $\mathbb{E}^{n+1}$ spanned by A and B, and hence consists of

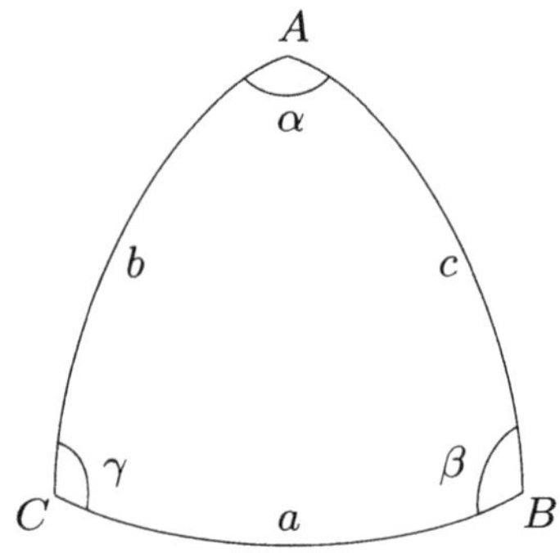

Fig. 2.1 A spherical triangle in $\mathbb{S}^n$

points of the form $\lambda A + \mu B$ with $\lambda + \mu \geq 1$. Suppose that A and B both lie in the closed ball $\overline{B}(P, r) \subseteq \mathbb{S}^n$ of radius $r < \pi/2$ about P. By definition, $C \in \overline{B}(P, r)$ if and only if $(C \mid P) \geq \cos r$. But if $\lambda + \mu \geq 1$ then $(\lambda A + \mu B \mid P) = \lambda(A \mid P) + \mu(B \mid P) \geq (\lambda + \mu) \cos r \geq \cos r$. Hence the minimal great arc from A to B is contained in $\overline{B}(P, r)$. □

Hyperplanes in $\mathbb{S}^n$. A hyperplane H in $\mathbb{S}^n$ is, by definition, the intersection of $\mathbb{S}^n$ with an n-dimensional vector subspace of $\mathbb{E}^{n+1}$. Note that H, with the induced metric from $\mathbb{S}^n$, is isometric to $\mathbb{S}^{n-1}$. We define the reflection r_H through H to be the isometry of $\mathbb{S}^n$ obtained by restricting to $\mathbb{S}^n$ the Euclidean reflection through the hyperplane of $\mathbb{E}^{n+1}$ spanned by H. Given two distinct points $A, B \in \mathbb{S}^n$, the set of points in $\mathbb{S}^n$ that are equidistant from A and B is a hyperplane, called the hyperplane bisector for A and B; it is the intersection of $\mathbb{S}^n$ with the vector subspace of $\mathbb{E}^{n+1}$ orthogonal to the vector $A - B$. If $A \in \mathbb{S}^n$ does not belong to H, then H is the hyperplane bisector of A and $r_H(A)$. Conversely, if H is the hyperplane bisector of A and B, then $r_H(A) = B$.

Hyperbolic n-Space $\mathbb{H}^n$

When approaching hyperbolic geometry from a Riemannian viewpoint, it is common to begin by introducing the Poincaré model for $\mathbb{H}^n$ (see Chapter 6). However, from the purely metric viewpoint which we are adopting, it is more natural to begin with the *hyperboloid model*. One benefit of this approach is that it brings into sharp focus the fact that $\mathbb{S}^n$ is obtained from $\mathbb{E}^{n+1}$ in essentially the same way as $\mathbb{H}^n$ is obtained from $\mathbb{E}^{n,1}$, where $\mathbb{E}^{n,1}$ denotes the vector space $\mathbb{R}^{n+1}$ endowed with the symmetric bilinear form that associates to vectors $u = (u_1, \dots, u_{n+1})$ and $v = (v_1, \dots, v_{n+1})$ the real number $\langle u \mid v \rangle$ defined by

$$\langle u \mid v \rangle = -u_{n+1} v_{n+1} + \sum_{i=1}^{n} u_i v_i .$$

The quadratic form associated to $\langle \cdot \mid \cdot \rangle$ is nondegenerate of type $(n, 1)$.

The orthogonal complement of $v \in \mathbb{E}^{n,1}$ with respect to this quadratic form is, by definition, the n-dimensional vector subspace $v^\perp \subseteq \mathbb{E}^{n,1}$ consisting of vectors u such that $\langle u \mid v\rangle = 0$. If $\langle v \mid v\rangle < 0$, then the restriction of the given quadratic form to $v^\perp$ is positive definite. If $v \neq 0$ but $\langle v \mid v\rangle = 0$, then $v^\perp$ is tangent to what is called the *light cone* $\{x \mid \langle x \mid x\rangle = 0\}$; in this case the given quadratic form restricts to a degenerate form of rank $n-1$ on $v^\perp$. If $\langle v \mid v\rangle > 0$, then the restriction to $v^\perp$ of the given quadratic form is non-degenerate of type $(n-1, 1)$.

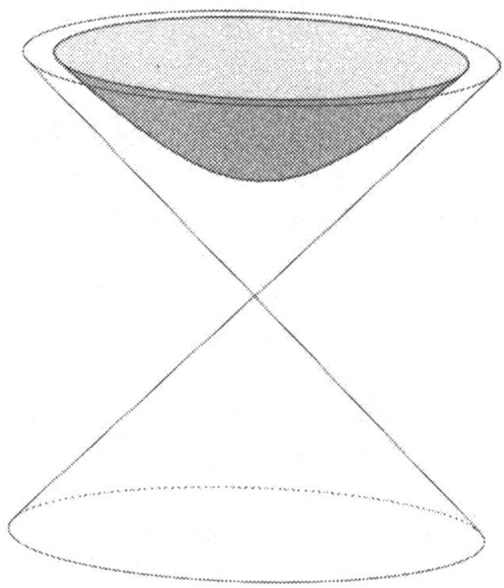

Fig. 2.2 The hyperbolic plane $\mathbb{H}^2$

2.4 Definition. (Real) hyperbolic n-space $\mathbb{H}^n$ is

$$\{u = (u_1, \dots, u_{n+1}) \in \mathbb{E}^{n,1} \mid \langle u \mid u\rangle = -1,\ u_{n+1} > 0\}.$$

In other words, $\mathbb{H}^n$ is the upper sheet of the hyperboloid $\{u \in \mathbb{E}^{n,1} \mid \langle u \mid u\rangle = -1\}$.

Note that if $u \in \mathbb{H}^n$ then $u_{n+1} \geq 1$, with equality if and only if $u_i = 0$ for all $i = 1, \dots, n$.

2.5 Remark. We shall need the observation that for all $u, v \in \mathbb{H}^n$ one has $\langle u \mid v\rangle \leq -1$, with equality if and only if $u = v$. Indeed,

$$\begin{aligned}
\langle u \mid v\rangle &= \left(\sum_{i=1}^{n} u_i v_i\right) - u_{n+1}v_{n+1}\\
&\leq \left(\sum_{i=1}^{n} u_i^2\right)^{1/2}\left(\sum_{i=1}^{n} v_i^2\right)^{1/2} - u_{n+1}v_{n+1}\\
&= (u_{n+1}^2 - 1)^{1/2}(v_{n+1}^2 - 1)^{1/2} - u_{n+1}v_{n+1},
\end{aligned}$$

and one checks easily that this last expression is ≤ -1 for any numbers $u_{n+1}, v_{n+1} \geq 1$, with equality if and only if $u_{n+1} = v_{n+1}$.

As an immediate consequence we see that $\langle u - v \mid u - v \rangle \geq 0$, for all $u, v \in \mathbb{H}^n$, with equality if and only if $u = v$.

2.6 Proposition (Definition of the Metric). *Let $d : \mathbb{H}^n \times \mathbb{H}^n \to \mathbb{R}$ be the function that assigns to each pair $(A, B) \in \mathbb{H}^n \times \mathbb{H}^n$ the unique non-negative number $d(A, B) \geq 0$ such that*

$$\cosh d(A, B) = -\langle A \mid B \rangle.$$

Then d is a metric.

Proof. The content of Remark 2.5 is that d is well-defined and positive definite. d is clearly symmetric, so it only remains to verify the triangle inequality. As in the case of $\mathbb{S}^n$, we shall deduce this from the appropriate form of the law of cosines. □

Again as for $\mathbb{S}^n$, in order to state the law of cosines we must first define primitive notions of segment, angle and triangle in $\mathbb{H}^n$. Eventually (2.8) we shall see that the geodesic lines in $\mathbb{H}^n$ are the intersections of $\mathbb{H}^n$ with 2-dimensional subspaces of $\mathbb{E}^{n,1}$. Thus a hyperbolic segment should be a compact subarc of such a line of intersection. More precisely, given $A \in \mathbb{H}^n$ and a unit vector $u \in A^\perp \subseteq \mathbb{E}^{n,1}$ (i.e., $\langle u \mid u \rangle = 1$ and $\langle A \mid u \rangle = 0$), we consider the path $c : \mathbb{R} \to \mathbb{H}^n$ defined by $t \mapsto (\cosh t)A + (\sinh t)u$. Note that $d(c(t), c(t')) = |t - t'|$ for all $t, t' \in \mathbb{R}$. Given $a \geq 0$, we define the image under c of the interval $[0, a]$ to be the *hyperbolic segment* joining A to $c(a)$, and denote it $[A, c(a)]$. Given $B \in \mathbb{H}^n$ distinct from A, let u be the unit vector in the direction of $(B + \langle A \mid B \rangle A)$. This is the unique unit vector $u \in A^\perp$ such that $B = (\cosh a)A + (\sinh a)u$, where $a = d(A, B)$. We shall refer to u as the initial vector of the hyperbolic segment $[A, B]$.

By definition, the *hyperbolic angle* between two hyperbolic segments issuing from a point of $\mathbb{H}^n$, with initial vectors u and v, say, is the unique number $\alpha \in [0, \pi]$ such that $\cos \alpha = \langle u \mid v \rangle$. A *hyperbolic triangle* Δ in $\mathbb{H}^n$ consists of a choice of three distinct points (its vertices) $A, B, C \in \mathbb{H}^n$, and the three hyperbolic segments joining them (its sides). The vertex angle at C is defined to be the hyperbolic angle between the segments $[C, A]$ and $[C, B]$.

2.7 The Hyperbolic Law of Cosines. *Let Δ be a hyperbolic triangle with vertices A, B, C. Let $a = d(B, C), b = d(C, A)$ and $c = d(A, B)$. Let γ denote the vertex angle at C. Then*

$$\cosh c = \cosh a \, \cosh b - \sinh a \, \sinh b \, \cos \gamma.$$

Proof. Let u and v be the initial vectors of the hyperbolic segments joining C to A and B respectively. By definition, $\cos \gamma = \langle u \mid v \rangle$. And

$$\begin{aligned}\cosh c &= -\langle A \mid B\rangle \\ &= -\langle(\cosh b)C + (\sinh b)u \mid (\cosh a)C + (\sinh a)v\rangle \\ &= -\cosh a\,\cosh b\,\langle C \mid C\rangle - \sinh a\,\sinh b\,\langle u \mid v\rangle \\ &= \cosh a\,\cosh b - \sinh a\,\sinh b\,\cos\gamma,\end{aligned}$$

as required. □

There are many other useful trigonometric formulae in hyperbolic geometry (see, for example, [Thu97], section 2.4).

2.8 Corollary (The Triangle Inequality and Geodesics).
For all $A, B, C \in \mathbb{H}^n$,

$$d(A, B) \le d(A, C) + d(C, B),$$

with equality if and only if C lies on the hyperbolic segment joining A to B. Thus:

(1) *$\mathbb{H}^n$ is a uniquely geodesic metric space.*

(2) *The unique geodesic segment joining A to B is the hyperbolic segment* $[A, B]$.

(3) *If the intersection of a 2-dimensional vector subspace of $\mathbb{E}^{n,1}$ with $\mathbb{H}^n$ is non-empty, then it is a geodesic line, and all geodesic lines in $\mathbb{H}^n$ arise in this way.*

(4) *All the balls in $\mathbb{H}^n$ are convex.*

Proof. Let $a = d(C, B)$, $b = d(A, C)$, $c = d(A, B)$. We assume that A, B and C are distinct (the case where they are not is trivial). Consider a hyperbolic triangle Δ with vertices A, B, C. Let γ be the vertex angle at C. For fixed a and b, as γ increases from 0 to π, the function $\gamma \mapsto \cosh a \cosh b - \sinh a \sinh b \cos\gamma$ strictly increases from $\cosh(b - a)$ to $\cosh(b + a)$. So by the law of cosines we have $\cosh c \le \cosh(b + a)$, and hence $c \le a + b$, with equality if and only if $\gamma = \pi$. But $\gamma = \pi$ if and only if $C \in [A, B]$. Statements (1) to (3) are now immediate from the definitions.

The proof that the balls are convex is similar to the one given in (2.3) (one notes that a point C in $[A, B]$ is of the form $\lambda A + \mu B$, with $\lambda + \mu \le 1$ and $\lambda, \mu \ge 0$). □

Henceforth we shall always assume that $\mathbb{S}^n$ and $\mathbb{H}^n$ are metrized as in (2.2) and (2.8).

Exercise. Consider a geodesic triangle in $\mathbb{H}^n$ with vertex angles α, β and $\pi/2$. Let d be the length of the side opposite the angle β. Prove that $\cos\beta = \cosh d\, \sin\alpha$.

Hyperplanes in $\mathbb{H}^n$. A *hyperplane H* in $\mathbb{H}^n$ is, by definition, a non-empty intersection of $\mathbb{H}^n$ with an n-dimensional vector subspace of $\mathbb{E}^{n,1}$. With the induced metric, H is isometric to $\mathbb{H}^{n-1}$. Given two distinct points $A, B \in \mathbb{H}^n$, the set of points in $\mathbb{H}^n$ that are equidistant from A and B is a hyperplane, called the *hyperplane bisector* of A and B; it is the intersection of $\mathbb{H}^n$ with the vector subspace $(A - B)^\perp \subset \mathbb{E}^{n,1}$.

The reflection r_H through H is the isometry of $\mathbb{H}^n$ which is given by $X \mapsto X - 2\langle X \mid u\rangle u$, where u is a unit vector orthogonal to the vector space spanned by H (it is unique up to sign). The set of fixed points of r_H is H. If $A \in \mathbb{H}^n$ does not belong to H, then H is the hyperplane bisector of A and $r_H(A)$. Conversely, if H is the hyperplane bisector of A and B, then $r_H(A) = B$.

2.9 Proposition. *The spherical (respectively, hyperbolic) angle between two geodesic segments $[C, A]$ and $[C, B]$ in $\mathbb{S}^n$ (respectively, $\mathbb{H}^n$) is equal to the Alexandrov angle between them.*

Proof. This proof was communicated to us by Ralph Strebel. We give the details in the hyperbolic case; the spherical case in entirely similar.

Let $a = d(A, C)$ and $b = d(B, C)$, and let γ denote the hyperbolic angle between $[C, A]$ and $[C, B]$. Given numbers s, t with $0 < s \leq a$ and $0 < t \leq b$, one considers the points $A_s \in [C, A]$ and $B_t \in [C, B]$ that are a distance s and t from C respectively. Let $c_{s,t} = d(A_s, B_t)$ and let $\gamma_{s,t}$ be the angle at the vertex $\overline{C}$ in the comparison triangle $\overline{\Delta}(A_s, B_t, C) \subseteq \mathbb{E}^2$ (notation as in 1.10). The Euclidean law of cosines yields

$$c_{s,t}^2 = s^2 + t^2 - 2st\cos\gamma_{s,t}.$$

The hyperbolic law of cosines relates $c_{s,t}$ to s and t:

$$\cosh c_{s,t} = \cosh s\,\cosh t - \sinh s\,\sinh t\,\cos\gamma. \tag{2.9-1}$$

But the inverse of $\cosh|_{[0,\infty]}$ is not differentiable at 0, so the behaviour of

$$\cos\gamma_{s,t} = \frac{s^2 + t^2 - c_{s,t}^2}{2st}$$

cannot be gleaned immediately from (2.9-1). To circumvent this difficulty, we introduce an auxiliary function $h\colon \mathbb{R} \to \mathbb{R}$, given by the power series

$$h(x) = \sum_{i=1}^{\infty} \frac{1}{(2i)!} x^i.$$

This power series converges on all of $\mathbb{R}$ and it is related to cosh by the formula $\cosh x - 1 = h(x^2)$. Since $h(0) = 0$ and $h'(0) = 1/2$, the function h is invertible in a neighbourhood of 0; this local inverse is given by a power series of the form

$$h^{-1}(x) = 2x + \sum_{i=2}^{\infty} a_i x^i.$$

We continue our analysis of $\cosh c_{s,t}$. Using (2.9-1) we can rewrite $h(c_{s,t}^2) = \cosh c_{s,t} - 1$ as

$$\begin{aligned} &\cosh c\,\cosh t - 1 - \sinh s\,\sinh t\,\cos\gamma \\ &\quad = (\cosh s - 1)\cosh t + (\cosh t - 1) - \sinh s\,\sinh t\,\cos\gamma \\ &\quad = h(s^2)\,\cosh t + h(t^2) - \sinh s\,\sinh t\,\cos\gamma. \end{aligned}$$

This last expression defines an analytic function $g: \mathbb{R}^2 \to \mathbb{R}$ with the following three properties:

$$g(0,0) = 0, \quad g(s,0) = h(s^2), \quad g(0,t) = h(t^2).$$

Moreover the coefficient of st in the power series representation of g is $-\cos\gamma$.

The first of these properties implies that the composition $f := h^{-1} \circ g$, which is equal to $c^2_{s,t}$ for small positive s and t, is defined in a small disc around the origin. It has there an absolutely convergent power series representation that we can write in the form

$$f(s,t) = \sum_{j=1}^{\infty} f_{j,0} s^j + \sum_{k=1}^{\infty} f_{0,k} t^k - st\Big(\sum_{j=1}^{\infty}\sum_{k=1}^{\infty} f_{j,k} s^{j-1} t^{k-1}\Big).$$

By the second and third properties of g stated above, the first and second sums in this expression for f are equal to s^2 and t^2, respectively, so if s and t are small and positive then the term in parenthesis equals $(s^2 + t^2 - c^2_{s,t})/(st)$. Finally, by combining the power series expression of h^{-1} given above with our remark about the st coefficient of g, we see that $f_{1,1} = 2\cos\gamma$. Hence

$$\cos\gamma_{s,t} = \frac{s^2 + t^2 - c^2_{s,t}}{2st} = \cos\gamma + \sum_{k>1 \text{ or } j>1} f_{j,k} s^{j-1} t^{k-1}$$

tends to $\cos\gamma$ when $t, s \to 0_+$. □

Remark. This proof shows that the angle between two geodesic segments issuing from the same point in $\mathbb{H}^n$ exists in the strict sense (as defined in 1.12).

The Model Spaces M^n_κ

2.10 Definition. Given a real number κ, we denote by M^n_κ the following metric spaces:

(1) if $\kappa = 0$ then M^n_0 is Euclidean space $\mathbb{E}^n$;

(2) if $\kappa > 0$ then M^n_κ is obtained from the sphere $\mathbb{S}^n$ by multiplying the distance function by the constant $1/\sqrt{\kappa}$;

(3) if $\kappa < 0$ then M^n_κ is obtained from hyperbolic space $\mathbb{H}^n$ by multiplying the distance function by $1/\sqrt{-\kappa}$.

The following is an immediate consequence of our previous results:

2.11 Proposition. *M^n_κ is a geodesic metric space. If $\kappa \le 0$, then M^n_κ is uniquely geodesic and all balls in M^n_κ are convex. If $\kappa > 0$, then there is a unique geodesic segment joining $x, y \in M^n_\kappa$ if and only if $d(x,y) < \pi/\sqrt{\kappa}$. If $\kappa > 0$, closed balls in M^n_κ of radius $< \pi/(2\sqrt{\kappa})$ are convex.*

When working with M^n_κ, one constantly finds it necessary to phrase hypotheses and conclusions in terms of the quantity $\pi/\sqrt{\kappa}$, which made its first appearance in the preceding proposition. For this reason, we introduce the following device.

2.12 The Diameter of M^n_κ. Henceforth we shall write D_κ to denote the *diameter of* M^n_κ. More precisely, $D_\kappa := \pi/\sqrt{\kappa}$ for $\kappa > 0$ and $D_\kappa := \infty$ for $\kappa \leq 0$.

2.13 The Law of Cosines in M^n_κ. *Given a geodesic triangle in M^n_κ with sides of positive length a, b, c and angle γ at the vertex opposite to the side of length c,*

$$\text{for } \kappa = 0 : \qquad c^2 = a^2 + b^2 - 2ab\cos(\gamma)$$

$$\text{for } \kappa < 0 : \quad \cosh(\sqrt{-\kappa}\,c) = \cosh(\sqrt{-\kappa}\,a)\cosh(\sqrt{-\kappa}\,b) - \sinh(\sqrt{-\kappa}\,a)\sinh(\sqrt{-\kappa}\,b)\cos(\gamma)$$

$$\text{for } \kappa > 0 : \quad \cos(\sqrt{\kappa}\,c) = \cos(\sqrt{\kappa}\,a)\cos(\sqrt{\kappa}\,b) + \sin(\sqrt{\kappa}\,a)\sin(\sqrt{\kappa}\,b)\cos(\gamma).$$

In particular, fixing a, b and κ, one sees that c is a strictly increasing function of γ, varying from $|b-a|$ to $a+b$ as γ varies from 0 to π.

Note that one can pass from the formula for $\kappa > 0$ to the formula for $-\kappa$ by replacing $\sqrt{\kappa}$ with $i\sqrt{\kappa} = \sqrt{-1\kappa}$, since for any real t one has $\cos it = \cosh t$ and $\sin it = i\sinh t$.

Proof. These formulae are obtained from the law of cosines for $\mathbb{S}^n$ and $\mathbb{H}^n$ by rescaling. □

Alexandrov's Lemma

In Chapter 1 we introduced the notion of a comparison triangle to enable us to compare triples of points in an arbitrary metric space to corresponding triples of points in the Euclidean plane. In Part II we shall have occasion to compare metric spaces not only to $\mathbb{E}^2$, but also to M^2_κ for $\kappa \neq 0$. In order to do so we shall need the following:

2.14 Lemma (Existence of Comparison Triangles in M^2_κ). *Let κ be a real number, and let p, q, r be three points in a metric space X; if $\kappa > 0$ assume that $d(p,q) + d(q,r) + d(r,p) < 2D_\kappa$. Then, there exist points $\overline{p}, \overline{q}, \overline{r} \in M^2_\kappa$ such that $d(p,q) = d(\overline{p},\overline{q})$, $d(q,r) = d(\overline{q},\overline{r})$ and $d(r,p) = d(\overline{r},\overline{p})$.*

The triangle $\Delta(\overline{p},\overline{q},\overline{r}) \subseteq M^2_\kappa$ with vertices $\overline{p},\overline{q},\overline{r}$ is called a comparison triangle for the triple (p,q,r); it is unique up to an isometry of M^2_κ. If $\Delta \subseteq X$ is a geodesic triangle with vertices p, q, r, then $\Delta(\overline{p},\overline{q},\overline{r})$ is also said to be a comparison triangle for Δ.

Proof. Let $a = d(p,q)$, $b = d(p,r)$, $c = d(q,r)$. We may assume that $a \leq b \leq c$. By the triangle inequality, $c \leq a + b$, so in particular if $\kappa > 0$ then $c < \pi/\sqrt{\kappa}$. Hence we may solve to find a unique $\gamma \in [0,\pi]$ which is related to a, b, c by the equation in the statement of the law of cosines for M^2_κ (2.12). We fix a point $\overline{p} \in M^2_\kappa$ and construct two geodesic segments $[\overline{p},\overline{q}]$ and $[\overline{p},\overline{r}]$, of lengths a and b respectively, meeting at an angle γ. By the law of cosines, $d(\overline{q},\overline{r}) = c$.

The asserted uniqueness of $\Delta(\overline{p},\overline{q},\overline{r})$ is a special case of 2.15. □

It follows from Proposition 2.9 that the definition of Alexandrov angle (1.12) would remain the same if we were to use comparison triangles in M^2_κ rather $\mathbb{E}^2$. However, the way in which the limiting angle is obtained would be different: instead of $\overline{\angle}$ one would consider the following quantity.

2.15 Definition. Let X be a metric space and let κ be a real number. Let $p, q, r \in X$ be three distinct points with $d(p,q) + d(q,r) + d(r,p) < 2D_\kappa$. The *$\kappa$-comparison angle* between q and r at p, denoted $\angle^{(\kappa)}_p(q,r)$, is the angle at $\overline{p}$ in a comparison triangle $\Delta(\overline{p},\overline{q},\overline{r}) \subseteq M^2_\kappa$ for (p,q,r). (In particular $\angle^{(0)}_p(q,r) = \overline{\angle}_p(q,r)$.)

Later, particularly in Chapter (II.4), we shall need to bound the geometry of large geodesic triangles that are constructed from smaller triangles in a controlled manner. The following technical lemma is invaluable in this regard. The reader may find it helpful to view this lemma in the following light: take two (sufficiently small) geodesic triangles in M^2_κ that have one edge in common, with the sum of the interior angles at one end of this edge, C say, at least π; delete this common edge; imagine the remaining four edges as rigid bars, joined by hinges at the vertices; imagine straightening this arrangement so as to decrease the interior angle at C to π. The main content of the following lemma is that during such a straightening process the angles at the vertices other than C increase.

We shall need the following terminology: given distinct points $x, y \in M^2_\kappa$, with $d(x,y) < D_\kappa$, there is (up to reparameterization) a unique local geodesic ('line') $\mathbb{R} \to M^2_\kappa$ passing through x and y; the image of this line separates M^2_κ into two connected components; two points $z, w \in M^2_k$ are said to lie on *opposite sides of the line* if they are in different connected components of its complement.

2.16 Alexandrov's Lemma. *Consider four distinct points $A, B, B', C \in M^2_\kappa$; if $\kappa > 0$, assume that $d(C,B) + d(C,B') + d(A;B) + d(A,B') < 2D_\kappa$. Suppose that B and B' lie on opposite sides of the line through A and C.*

Consider the geodesic triangles $\Delta = \Delta(A,B,C)$ and $\Delta' = \Delta(A,B',C)$. Let α, β, γ (resp. α', β', γ') be the angles of Δ (resp. Δ') at the vertices A, B, C (resp. A, B', C). Assume that $\gamma + \gamma' \geq \pi$. Then,

(1) $d(B,C) + d(B',C) \leq d(B,A) + d(B',A)$.

Let $\overline{\Delta}$ be a triangle in M^2_κ with vertices $\overline{A}, \overline{B}, \overline{B}'$ such that $d(\overline{A},\overline{B}) = d(A,B)$, $d(\overline{A},\overline{B}') = d(A,B')$ and $d(\overline{B},\overline{B}') = d(B,C) + d(C,B') < D_\kappa$. Let $\overline{C}$ be the point

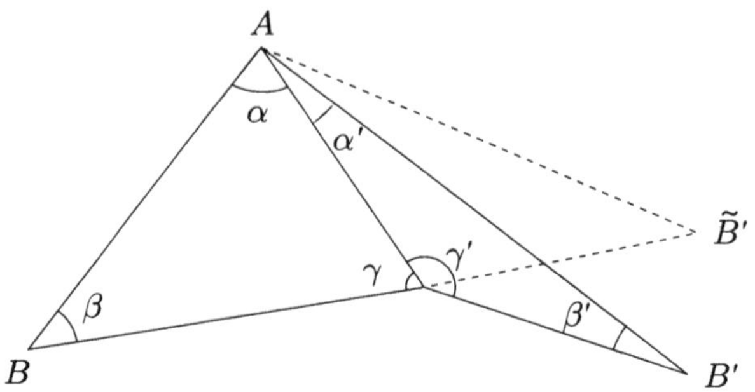

Fig. 2.3 Alexandrov's Lemma

of $[\overline{B}, \overline{B}']$ *with* $d(\overline{B}, \overline{C}) = d(B, C)$. *Let* $\overline{\alpha}, \overline{\beta}, \overline{\beta}'$ *be the angles of* $\overline{\Delta}$ *at the vertices* $\overline{A}, \overline{B}, \overline{B}'$. *Then,*

(2) $\overline{\alpha} \geq \alpha + \alpha'$, $\overline{\beta} \geq \beta$, $\overline{\beta}' \geq \beta'$ *and* $d(A, C) \leq d(\overline{A}, \overline{C})$*; any one equality implies the others, and occurs if and only if* $\gamma + \gamma' = \pi$.

Proof. Let $\tilde{B}' \in M^2_\kappa$ be the unique point such that $d(\tilde{B}', C) = d(B', C)$ and C lies on the geodesic segment $[\tilde{B}', B]$. Because $\gamma + \gamma' \geq \pi$, the angle at C between $[C, A]$ and $[C, \tilde{B}']$ is no greater than the angle between $[C, A]$ and $[C, B']$. Hence, by the law of cosines, $d(\tilde{B}', A) \leq d(B', A)$. Therefore, $d(B, A) + d(B', A) \geq d(B, A) + d(\tilde{B}', A) \geq d(B, \tilde{B}') = d(B, C) + d(C, \tilde{B}')$. This proves (1).

As in (1), we have $d(\overline{A}, \overline{B}') \geq d(A, \tilde{B}')$, and by the triangle inequality, $d(\overline{B}, \overline{B}') \geq d(B, B')$; in each case equality is strict unless $C \in [B, B']$, that is, $\gamma + \gamma' = \pi$. Applying the law of cosines to each of these inequalities we get $\overline{\alpha} \geq \alpha + \alpha'$ and $\overline{\beta} \geq \beta$. Exchanging the roles of β and β' we get $\overline{\beta}' \geq \beta'$. A further application of the law of cosines shows $d(A, C) \leq d(\overline{A}, \overline{C})$. All of these inequalities are strict except when $\gamma + \gamma' = \pi$, in which case $\overline{\Delta}$ is isometric to the union of Δ and Δ' with the common side $[A, C]$ deleted, so equality holds everywhere. □

The Isometry Groups Isom(M^n_κ)

Our next goal is to describe the isometry groups of the model spaces M^n_k. A key observation in this regard is that each of these groups is generated by reflections in hyperplanes. If one rescales the metric on a space then one does not alter its group of isometries, so there are essentially only three cases to consider $\mathbb{E}^n$, $\mathbb{S}^n$ and $\mathbb{H}^n$.

The following proposition shows in particular that the group of isometries acts transitively on M^n_κ.

2.17 Proposition. *Given any positive integer* k *and* $2k$ *points* $A_1, \dots, A_k$ *and* $B_1, \dots, B_k$ *in* M^n_κ *such that* $d(A_i, A_j) = d(B_i, B_j)$ *for all* $i, j \in \{1, \dots, k\}$*, there exists an isometry mapping* A_i *to* B_i*, for all* $i \in \{1, \dots, k\}$*. Moreover, one can obtain such an isometry by composing* k *or fewer reflections through hyperplanes.*

Proof. Arguing by induction on k (the case $k = 0$ is vacuous) we may assume that there exists an isometry ϕ such that $\phi(A_i) = B_i$ for all $i \in \{1, \dots, k-1\}$, and ϕ is the product of $k - 1$ or fewer reflections. If $\phi(A_k) = B_k$ then we are done. If not, then let H denote the hyperplane bisector of $\phi(A_k)$ and B_k. As usual, we let r_H denote the reflection through H. Note that $B_i \in H$ for $i = 1, \dots, k-1$, because $d(B_i, \phi(A_k)) = d(\phi(A_i), \phi(A_k)) = d(A_i, A_k) = d(B_i, B_k)$. Hence $r_H \circ \phi$, which is a product of at most k hyperplane reflections, sends A_i to B_i for all $i \in \{1, \dots, k\}$. □

2.18 Proposition. *Let ϕ be an isometry of M_κ^n.*

(1) *If ϕ is not the identity, then the set of points which it fixes is contained in a hyperplane.*

(2) *If ϕ acts as the identity on a hyperplane H, then ϕ is either the identity or the reflection r_H through H.*

(3) *ϕ can be written as the composition of $n + 1$ or fewer reflections through hyperplanes.*

Proof.

(1) If ϕ is not the identity, then there is a point A such that $\phi(A) \neq A$. We claim that the set of points fixed by ϕ is contained in the hyperplane bisector of A and $\phi(A)$. Indeed, if B is fixed by ϕ then $d(A, B) = d(\phi(A), \phi(B)) = d(\phi(A), B)$, and hence $B \in H$.

(2) Suppose ϕ acts as the identity on a hyperplane H. Given any point A such that $\phi(A) \neq A$, the preceding argument shows that H must be contained in the hyperplane bisector of A and $\phi(A)$. But no hyperplane is properly contained in any other, so H is the hyperplane bisector of A and $\phi(A)$, and hence $r_H(A) = \phi(A)$. But A was arbitrary, so $\phi = r_H$.

(3) Fix a collection of $n + 1$ points $A_0, \dots, A_n$ that is not contained in any hyperplane. Given an isometry ϕ, let $B_0 = \phi(A_0), \dots, B_n = \phi(A_n)$. According to the preceding proposition, there is an isometry ϕ' which maps each A_i to B_i and which can be written as the composition of at most $n + 1$ reflections through hyperplanes. The isometry $(\phi')^{-1} \circ \phi$ fixes each of the A_i, and hence, by part (1), it is the identity. □

We require a base hyperplane $H_0 \subset M_\kappa^n$. In $\mathbb{E}^n$ we take $H_0 = \{0\} \times \mathbb{E}^{n-1}$; in $\mathbb{S}^n$ we take $H_0 = \mathbb{S}^n \cap (\{0\} \times \mathbb{E}^n)$; in $\mathbb{H}^n$ we take $H_0 = \mathbb{H}^n \cap (\{0\} \times \mathbb{E}^{n-1,1})$. In each case there is a natural identification of H_0 with M_κ^{n-1}, and we use this to define what we mean by *a hyperplane in* H_0. (In the light of (2.20) this definition will become obsolete.)

2.19 Exercise. Let H be a hyperplane in M_κ^n and suppose that $H_0 \neq H$. If $H_0 \cap H$ is not empty then it is a hyperplane in H_0. Conversely, every hyperplane in H_0 is the intersection of H_0 with some hyperplane in M_κ^n. (Because we defined hyperplanes in $\mathbb{S}^n$ (resp. $\mathbb{H}^n$) as intersections of the sphere (resp. hyperboloid) in $\mathbb{R}^{n+1}$ with vector subspaces, this is just an exercise in linear algebra.)

A set of $n+1$ points in M^n_κ is said to be in *general position* if it is not contained in any hyperplane. It is obvious that such sets exist.

2.20 Proposition. *Let S_1 and S_2 be subsets of M^n_κ and let $f : S_1 \to S_2$ be an isometry. Then,*

(1) *there exists $\psi \in \text{Isom}(M^n_\kappa)$ such that $\psi|_{S_1} = f$;*

(2) *the restriction of ψ to the intersection of all hyperplanes containing S_1 is unique.*

Proof. We first consider the case where $S_1 = S_2 = H$ is a hyperplane. Because every hyperplane is the bisector of some pair of points, the action of $\text{Isom}(M^n_\kappa)$ obviously sends hyperplanes to hyperplanes. And it follows from (2.17) that the action of $\text{Isom}(M^n_\kappa)$ on the set of hyperplanes in M^n_κ is transitive. Hence there is no loss of generality in assuming that $H = H_0$.

Let $\Sigma \subset H_0 = M^{n-1}_\kappa$ be a set of n points in general position and let $f : H_0 \to H_0$ be an isometry. (2.17) gives $\psi \in \text{Isom}(M^n_\kappa)$ such that $\psi|_\Sigma = f|_\Sigma$. In particular Σ is contained in the hyperplane $\psi^{-1}(H_0)$. Since Σ is assumed not to lie in any hyperplane of H_0, it follows that $\psi^{-1}(H_0) = H_0$ (see 2.19). But now, applying (2.18) to $H_0 = M^{n-1}_\kappa$, since the fixed points of $\psi^{-1}f \in \text{Isom}(H_0)$ are not contained in any hyperplane of H_0, we have $\psi^{-1}f = \text{id}_{H_0}$, i.e. $f = \psi|_{H_0}$.

We now consider the general case $f : S_1 \to S_2$. If S_1 and S_2 are contained in hyperplanes H_1 and H_2, then we can replace them by $S_i' = \psi_i(S_i) \subset H_0$, where $\psi_i : H_i \to H_0$ is an isometry. Then, by induction on the dimension of M^n_κ (the case $n = 1$ is trivial) we can assume that any isometry $S_1' \to S_2'$ is the restriction of an isometry of H_0, and the first step of the proof then applies.

Thus we may assume that S_1 is not contained in any hyperplane of M^n_κ. It follows that some finite subset $S_0 \subset S_1$ is not contained in any hyperplane of M^n_κ. By (2.17), there exists $\psi_0 \in \text{Isom}(M^n_\kappa)$ such that $\psi_0|_{S_0} = f|_{S_0}$. If $\psi_0|_{S_1} = f$ then we are done. If not, then there would exist $x \in S_1$ such that $\psi_0(x) \neq f(x)$, and by applying (2.17) again we would get $\psi_1 \in \text{Isom}(M^n_\kappa)$ such that $\psi_1|_{S_0} = f|_{S_0}$ and $\psi_1(x) = f(x)$. But this cannot happen, because we would have a non-trivial element $\psi_1^{-1}\psi_0 \in \text{Isom}(M^n_\kappa)$ fixing a set S_0 that was not contained in any hyperplane, contradicting (2.18). This completes the proof of (1).

Suppose that ψ and ψ' are both such that $\psi|_{S_1} = \psi'|_{S_1}$. Then $\psi^{-1}\psi'$ and its inverse send any hyperplane containing S_1 to another such hyperplane, and therefore restricts to an isometry on the intersection of all such hyperplanes. Let I denote this intersection. By making iterated use of (2.19) and (1) we see that I is isometric to M^m_κ for some $m \leq n$. If S_1 were contained in some $(m-1)$-dimensional hyperplane $P \subset I$, then by making further use of (2.19) we could construct a hyperplane in M^n_κ such that $I \cap H = P$, contradicting the fact that I is the intersection of all hyperplanes containing S. Therefore S_1 is not contained in any hyperplane of $I \cong M^m_\kappa$, and (2.18) implies that $\psi^{-1}\psi'$ restricts to the identity on I. □

2.21 Definition. Let $m \leq n$ be non-negative integers. A subset $P \subset M^n_\kappa$ is called an m–plane if it is isometric to M^m_κ.

2.22 Corollary. *If $m < n$ then every m–plane in M^n_κ is the intersection of $(n-m)$ hyperplanes. Every subset of M^n_κ is contained in a unique m–plane of minimal dimension.*

Proof. Fix $m < n$. There certainly exists an m–plane $P_0 \subset M^n_\kappa$ that is the intersection of $(n-m)$ hyperplanes. The preceding proposition shows that any other m–plane is the image of P_0 by an isometry of M^n_κ, and any such isometry takes hyperplanes to hyperplanes.

The unique m–plane of minimal dimension containing S is the intersection of all hyperplanes containing S. □

2.23 The Groups Isom(M_κ). Using (2.16) we can determine the full group of isometries of $\mathbb{E}^n$, $\mathbb{S}^n$ and $\mathbb{H}^n$. In the case of $\mathbb{E}^n$ and $\mathbb{S}^n$, a more pedestrian account (from a different viewpoint) will be given in Chapter 4. A more explicit description of the individual isometries of $\mathbb{H}^n$ will be given in Chapter 6.

Let $D(n)$ denote the group of displacements of $\mathbb{E}^n$, that is, the group of affine isomorphisms of $\mathbb{E}^n$ that preserve the distance. The subgroup of $D(n)$ fixing the origin $0 \in \mathbb{E}^n$ is, by definition, the subgroup of $\mathrm{GL}(n, \mathbb{R})$ that consists of matrices which preserve the Euclidean metric. (The action of $\mathrm{GL}(n, \mathbb{R})$ is the usual linear action; a matrix $A = [a_{i,j}]$ acts thus: $A(x) = (y_1, \dots, y_n)$ where $y_i = \sum_j a_{i,j}x_j$.) A simple calculation shows that the action of A preserves the Euclidean norm if and only if $A \in O(n)$, where $O(n)$ denotes the group of orthogonal matrices, i.e., those real (n, n)-matrices A which satisfy $A\,{}^tA = I$, where ${}^tA$ is the transpose of A and I is the identity matrix.

$D(n)$ also contains an abelian subgroup consisting of translations $\tau_a : x \mapsto x+a$. This subgroup obviously acts transitively, so any element of $D(n)$ can be written uniquely as the composition of an orthogonal linear transformation and a translation. For each orthogonal transformation ϕ we have $\phi\tau_a\phi^{-1} = \tau_{\phi(a)}$, and hence the group of translations is normal in $D(n)$. The quotient of $D(n)$ by this normal subgroup is isomorphic to $O(n)$ and $D(n)$ is naturally isomorphic to the semi-direct product $\mathbb{R}^n \rtimes O(n)$ (cf. 4.13). Finally, we observe that $D(n)$ contains all reflections through hyperplanes in $\mathbb{E}^n$, because $O(n)$ contains all reflections in hyperplanes through 0, and $D(n)$ (indeed its translation subgroup $\mathbb{R}^n$) acts transitively on $\mathbb{E}^n$. Therefore, by part (3) of the preceding proposition, $D(n)$ is the whole of Isom($\mathbb{E}^n$).

We have already noted that the natural action of $O(n+1)$ on $\mathbb{E}^{n+1}$ preserves the Euclidean norm; hence it preserves the n-sphere $\mathbb{S}^n$. A trivial calculation shows that the induced action of $O(n+1)$ on $\mathbb{S}^n$ is by isometries. The action is faithful, so we obtain an injective map from $O(n+1)$ to Isom($\mathbb{S}^n$). But, as noted above, $O(n+1)$ contains all reflections through hyperplanes in $\mathbb{E}^{n+1}$ that contain 0, and the restriction of such reflections to $\mathbb{S}^n$ are, by definition, the hyperplane reflections of $\mathbb{S}^n$. We have shown that such reflections generate Isom($\mathbb{S}^n$), so the natural map from $O(n+1)$ to Isom($\mathbb{S}^n$) is actually an isomorphism.

Consider the group $\mathrm{GL}(n+1, \mathbb{R})$ (thought of as matrices, as above) with the usual linear action on $\mathbb{E}^{n,1} (= \mathbb{R}^{n+1})$. Let $O(n, 1)$ denote the subgroup formed by those matrices which leave invariant the bilinear form $\langle \cdot \mid \cdot \rangle$. A simple calculation shows that $O(n, 1)$ consists of those $(n+1, n+1)$-matrices A such that ${}^tAJA = J$, where J is the diagonal matrix with entries $(1, \dots, 1, -1)$ in the diagonal. While $O(n, 1)$ preserves the hyperboloid $\{x \in \mathbb{E}^{n,1} \mid \langle x \mid x \rangle = -1\}$, certain of its elements interchange the two sheets. Let $O(n, 1)_+ \subseteq O(n, 1)$ be the subgroup of index two that preserves the upper sheet $\mathbb{H}^n$. This subgroup consists of those matrices in $O(n, 1)$ whose bottom right hand entry is positive. It is clear from the definition that $O(n, 1)_+$ acts by isometries on $\mathbb{H}^n$. A direct calculation shows that the stabilizer of the point $(0, \dots, 0, 1)$ is naturally isomorphic to $O(n)$. The group $O(n, 1)_+$ contains all the reflections through hyperplanes of $\mathbb{H}^n$, so by the preceding proposition, it is equal to the full group $\mathrm{Isom}(\mathbb{H}^n)$ of isometries of $\mathbb{H}^n$.

We summarize the preceding discussion:

2.24 Theorem.
(1) $\mathrm{Isom}(\mathbb{E}^n) \cong \mathbb{R}^n \rtimes O(n)$.
(2) $\mathrm{Isom}(\mathbb{S}^n) \cong O(n+1)$.
(3) $\mathrm{Isom}(\mathbb{H}^n) \cong O(n, 1)_+$.
In all three cases, the stabilizer of a point is isomorphic to $O(n)$.

Approximate Midpoints

The following result will be needed several times in Part II.

2.25 Lemma. *For every* $\kappa \in \mathbb{R}$, $l < D_\kappa$ *and* $\varepsilon > 0$, *there exists a constant* $\delta(\kappa, l, \varepsilon) > 0$ *such that for all* $x, y \in M^2_\kappa$ *with* $d(x, y) \leq l$, *if* $d(x, m')$ *and* $d(m', y)$ *are both less than* $\frac{1}{2}d(x, y) + \delta$, *then* $d(m', m) < \varepsilon$, *where* m *is the midpoint of* $[x, y]$.

This result is a consequence of the transitivity properties of $\mathrm{Isom}(M^n_\kappa)$ and the following general observation.

2.26 Lemma. *Let* X *be a proper geodesic space. If there is a unique geodesic segment* $[x, y]$ *joining* $x, y \in X$, *then for each* $\varepsilon > 0$ *there exists* $\eta > 0$ *such that* $[d(x, z) + d(z, y) < d(x, y) + \eta]$ *implies* $d(z, [x, y]) < \varepsilon$.

Proof. Let $S_\varepsilon = \{p \in X \mid d(p, [x, y]) = \varepsilon\}$. Because X is geodesic, this set is empty only if every point of X is a distance less than ε from $[x, y]$. We consider the case where it is non-empty. Because X is proper, S_ε is compact. Let η be the minimum value attained by $f(z) = d(x, z) + d(z, y) - d(x, y)$ on S_ε. Note that $\eta > 0$ because $[x, y]$ is unique.

If $d(u, [x, y]) \geq \varepsilon$ then there exists $v \in S_\varepsilon$ such that $d(x, u) = d(x, v) + d(v, u)$. Therefore $d(x, u) + d(u, y) = d(x, v) + [d(v, u) + d(u, y)] \geq d(x, v) + d(v, y)$. Whence $d(x, u) + d(u, y) \geq d(x, y) + f(v) \geq d(x, y) + \eta$. □

Proof of (2.25). Because the action of $\mathrm{Isom}(M_\kappa^2)$ is transitive on pairs of equidistant points, we may assume that $[x, y]$ is the initial segment of a fixed geodesic segment $[x, y_0]$ of length l. Let η_0 be a constant such that if $d(x, z) + d(z, y_0) < d(x, y_0) + \eta_0$ then $d(z, [x, y_0]) < \varepsilon/2$.

We have $d(x, m') + d(m', y_0) \leq d(x, m') + d(m', y) + d(y, y_0)$, which is at most $d(x, y) + d(y, y_0) + 2\delta = d(x, y_0) + 2\delta$. So if $2\delta < \eta_0$, then $d(m', p) < \varepsilon/2$ for some $p \in [x, y_0]$. Then, $d(p, x) \leq d(p, m') + d(m', x) \leq \varepsilon/2 + \delta + \frac{1}{2}d(x, y)$ and similarly $d(p, y) \leq \varepsilon/2 + \delta + \frac{1}{2}d(x, y)$. But p lies on the geodesic segment $[x, y_0]$ so therefore $d(p, m) \leq \varepsilon/2 + \delta$. Thus it suffices to take $\delta = \min\{\varepsilon/3, \eta_0/2\}$. □

Chapter I.3 Length Spaces

In this section we consider metric spaces in which the distance between two points is given by the infimum of the lengths of curves which join them — such a space is called a *length space*. In this context, it is natural to allow metrics for which the distance between two points may be infinite. A convenient way to describe this is to introduce the notation $[0, \infty]$ for the ordered set obtained by adjoining the symbol ∞ to the set of non-negative reals and decreeing that $\infty > a$ for all real numbers a. We also make the convention that $a + \infty = \infty$ for all $a \in [0, \infty]$. Having made this convention, we can define a (generalized) metric on a set X to be a map $d : X \times X \to [0, \infty]$ satisfying the axioms stated in (1.1).

Henceforth we shall allow metrics and pseudometrics to take the value ∞.

Length Metrics

3.1 Definition. Let (X, d) be a metric space. d is said to be a *length metric* (otherwise known as an *inner metric*) if the distance between every pair of points $x, y \in X$ is equal to the infimum of the length of rectifiable curves joining them. (If there are no such curves then $d(x, y) = \infty$.) If d is a length metric then (X, d) is called a *length space*.

A complete metric space X is a length space if and only if it has approximate midpoints in the sense that for all $x, y \in X$ and $\varepsilon > 0$ there exists $z \in X$ such that $\max\{d(x, z), d(z, y)\} \leq \varepsilon + d(x, y)/2$.

An arbitrary metric space gives rise to a length space in an obvious way:

3.2 Proposition. *Let (X, d) be a metric space, and let $\bar{d} : X \times X \to [0, \infty]$ be the map which assigns to each pair of points $x, y \in X$ the infimum of the lengths of rectifiable curves which join them. (If there are no such curves then $\bar{d}(x, y) = \infty$.)*

(1) *$\bar{d}$ is a metric.*

(2) *$\bar{d}(x, y) \geq d(x, y)$ for all $x, y \in X$.*

(3) *If $c : [a, b] \to X$ is continuous with respect to the topology induced by $\bar{d}$, then it is continuous with respect to the topology induced by d. (The converse is false in general.)*

(4) *If a map $c : [a, b] \to X$ is a rectifiable curve in (X, d), then it is a continuous and rectifiable curve in $(X, \bar{d})$.*

(5) *The length of a curve $c : [a, b] \to X$ in $(X, \bar{d})$ is the same as its length in (X, d).*

(6) $\bar{\bar{d}} = \bar{d}$.

Proof. Properties (1) and (2) are immediate from the definition of length. (2) implies (3), and (6) is a consequence of (4) and (5). Property (4) is a consequence of 1.20(5), so it only remains to prove (5). Let $c : [a, b] \to X$ be a path which has length $\bar{l}(c)$ with respect to the metric $\bar{d}$, and length $l(c)$ with respect to the metric d. On the one hand we have that $\bar{l}(c) \geq l(c)$, by (2), and on the other hand

$$\begin{aligned}\bar{l}(c) &= \sup_{a=t_0\leq t_1\leq\cdots\leq t_k=b} \sum_{i=0}^{k-1} \bar{d}(c(t_i), c(t_{i+1})) \\ &\leq \sup_{a=t_0\leq t_1\leq\cdots\leq t_k=b} \sum_{i=0}^{k-1} l(c|_{[t_{i-1},t_i]}) = l(c).\end{aligned}$$

Hence $\bar{l}(c) = l(c)$. □

3.3 Definition. Let (X, d) be a metric space. The map $\bar{d}$ defined in (3.2) is called *the length metric* (or inner metric) associated to d, and $(X, \bar{d})$ is called the length space associated to (X, d). Note that $\bar{d} = d$ if an only if (X, d) is a length space.

The *induced length metric* on a subset $Y \subseteq X$ is the length metric associated to the restriction of d to $Y \times Y$ (which in general will not be the same as the restriction to $Y \times Y$ of $\bar{d}$).

3.4 Examples. Let (X, d) be a metric space. The identity map on X induces a continuous map $(X, \bar{d}) \to (X, d)$, but this is not a homeomorphism in general. For instance, the metric space (Y, d) considered in (1.19) is homeomorphic to $[0, 1]$, but the associated length space is homeomorphic to the *disjoint* union of $[0, 1)$ and $\{1\}$. More generally, the length space associated to a metric space (X, d) is connected if and only if every pair of points in (X, d) can be joined by a rectifiable curve.

As another example, we can consider the set of rational numbers $\mathbb{Q}$ with the usual metric d induced from $\mathbb{R}$. In the associated length metric $\bar{d}$, the distance between every pair of distinct points of $\mathbb{Q}$ is infinite, hence $\bar{d}$ induces the discrete topology on $\mathbb{Q}$.

3.5 Example. In the Euclidean plane we consider the complement X of an open sector of angle $\alpha < \pi$, namely $X = \{x \in \mathbb{R}^2 \mid (x \mid e_1) \leq \cos(\alpha/2)\,\|x\|\}$, where $e_1 = (0, 1)$, and $(\cdot \mid \cdot)$ is the Euclidean scalar product on $\mathbb{R}^2$, with associated norm $\|\cdot\|$.

We consider X as a length space, with the induced length metric from $\mathbb{E}^2$. This is a uniquely geodesic space: if the geodesic in $\mathbb{E}^2$ which joins $x \in X$ to $x' \in X$ lies entirely in X then this is the unique geodesic segment joining x to x' in X; otherwise, the unique geodesic from x to x' in X is the concatenation of the Euclidean geodesics joining x to 0 and 0 to x'.

3.6 Exercises

(1) Prove that the induced length metric that $\mathbb{S}^n$ inherits from $\mathbb{E}^{n+1}$ coincides with the metric defined in 2.1. Prove that for all $\kappa \neq 0$, all $x \in M^n_\kappa$ and all $r > 0$ (assuming $r < \pi/\sqrt{\kappa}$ if $\kappa > 0$), the induced length metric on $\{y \in M^n_\kappa \mid d(x, y) = r\} \subseteq M^n_\kappa$ makes it isometric to $M^{n-1}_{\frac{1}{\kappa}\sin^2(r\sqrt{\kappa})}$. (If $\kappa < 0$ then $\sin(r\sqrt{\kappa}) = \sin(ir\sqrt{-\kappa}) = i\sinh(r\sqrt{-\kappa})$.)

(2) Give an example of a subset $Y \subseteq \mathbb{E}^2$ such that the distance between each pair of points in the induced length metric is finite but the topology which this metric induces on Y does not coincide with the topology given by the restriction of the Euclidean metric.

Recall that any metric space admits a *completion*. In other words, every metric space X can be embedded isometrically as a dense subspace of a complete metric space X', which is unique up to isometry. One can obtain X' as the set of Cauchy sequences in X modulo the equivalence relation: $(x_n) \sim (y_n)$ if and only $\forall \varepsilon > 0\ \exists N_\varepsilon > 0$ such that $d(x_n, y_n) < \varepsilon$ for all $n > N_\varepsilon$. Each $x \in X$ is identified with the class of the stationary sequence $x_n = x$, and the metric on X is extended to X' by: $d([x_n], [y_n]) = \lim_n d(x_n, y_n)$.

(3) Prove that if X is a length space then its completion X' is a length space. (The converse is obviously false, e.g. (3.4).)

(4) Prove that there exists a geodesic metric space which is locally compact but whose completion is neither geodesic nor locally compact.

(Hint: Consider the induced path metric on the following subset of the Euclidean plane: $(0, 1] \times \{0\} \cup (0, 1] \times \{1\} \cup \bigcup_{n=1}^{\infty} \{1/n\} \times [0, 1]$.)

(5) Let d_1 and d_2 be two metrics on a space X which are Lipschitz equivalent, i.e. there exists a positive constant C such that

$$\frac{1}{C} d_2(x, y) \leq d_1(x, y) \leq C d_2(x, y),$$

for each $x, y \in X$. Show that the same inequality is then satisfied by the length metrics associated to d_1 and d_2.

(6) If one has a metric space (X, d) and a positive function $f : [0, \infty) \to [0, \infty)$, then one can apply the construction of (1.24) with $\rho := f \circ d$. Show that if X is a length space with $f(a + b) \geq f(a) + f(b)$ for all $a, b \geq 0$ and $\lim_{t\to 0} f(t)/t = 1$, then $d_\rho = d$.

The Hopf-Rinow Theorem

In general, a length space need not be a geodesic space, even if the distance between every pair of points is finite. For example, the induced metric on the Euclidean plane minus the origin is a length metric, but there is no geodesic joining $x \in \mathbb{R}^2 - \{0\}$ to $-x$. The space in this example is not complete, and in fact if one restricts to complete locally compact spaces, then a length space in which the distance between each pair of points is finite must be a geodesic metric space. This was known to Cohn-Vossen [CV35a] and proved earlier in case of surfaces by Hopf and Rinow [HoRi32] (see also the Appendix in [Rh52]). For a more general statement of the Hopf-Rinow Theorem, see Ballmann [Ba95, p.13-14].

3.7 Proposition (Hopf-Rinow Theorem). *Let X be a length space. If X is complete and locally compact, then*

(1) *every closed bounded subset of X is compact;*

(2) *X is a geodesic space.*

Proof. We follow the treatment of [GrLP81]. For (1) it suffices to prove that closed balls about a fixed point $a \in X$ are compact. Given $r > 0$, we denote by $\overline{B}(r) = \{x \in X \mid d(a, x) \leq r\}$ the closed ball with centre a and radius r. Consider the set of non-negative numbers ρ such that $\overline{B}(\rho)$ is compact. This is an interval containing 0; we claim that it is both open and closed. Because X is assumed to be locally compact, the interval contains a neighbourhood of 0. To see that it contains a neighbourhood of each of its other points, we fix $\rho \geq 0$ such that $\overline{B}(\rho)$ is compact, and use the local compactness of X to cover $\overline{B}(\rho)$ with finitely many balls $B(x_i, \varepsilon_i)$ such that each $\overline{B}(x_i, \varepsilon_i)$ is compact. There is a strictly positive lower bound, 2δ say, on the distance from any point in $\overline{B}(\rho)$ to the closed set $X \smallsetminus \bigcup B(x_i, \varepsilon_i)$, and hence $\overline{B}(\rho + \delta)$ is a closed subset of the compact set $\bigcup \overline{B}(x_i, \varepsilon_i)$.

It remains to prove that if $\overline{B}(r)$ is compact for all $r < \rho$ then $\overline{B}(\rho)$ is compact. It suffices to show that every sequence of points $x_n \in \overline{B}(\rho)$ such that $d(a, x_n)$ converges to ρ has a convergent subsequence. We fix such a sequence (x_n).

Let ε_p be a sequence of positive numbers tending to 0. For each p and each n, we can find a point y_n^p such that $d(a, y_n^p) \leq \rho - \varepsilon_p/2$ and $d(y_n^p, x_n) \leq \varepsilon_p$ (to see this one chooses a path c of length smaller than $d(a, x_n) + \varepsilon_p/2$ joining a to x_n, and then chooses a convenient point y_n^p on this path). For each p, the points y_n^p are contained in the compact ball $\overline{B}(\rho - \varepsilon_p/2)$, hence we can extract from $(y_n^1)_{n \in \mathbb{N}}$ a convergent subsequence $(y_{n_k^1}^1)_{k \in \mathbb{N}}$; from the sequence $(y_{n_k^1}^2)_{k \in \mathbb{N}}$ we can then extract a convergent subsequence, and so on. Eventually, by a diagonal process, we obtain a sequence of integers $(n_k)_{k \in \mathbb{N}}$ such that the sequence $(y_{n_k}^p)_{k \in \mathbb{N}}$ converges for all p. We claim that the corresponding sequence $(x_{n_k})_{k \in \mathbb{N}}$ is Cauchy. Indeed, for a given $\varepsilon > 0$, we can choose p such that $\varepsilon_p < \varepsilon/3$ and then use the fact that the sequence $y_{n_k}^p$ is convergent (hence Cauchy) to see that for large k, k' we have $d(y_{n_k}^p, y_{n_{k'}}^p) < \varepsilon/3$, and hence $d(x_{n_k}, x_{n_{k'}}) < d(x_{n_k}, y_{n_k}^p) + d(y_{n_k}^p, y_{n_{k'}}^p) + d(y_{n_{k'}}^p, x_{n_{k'}}) < \varepsilon_p + \varepsilon/3 + \varepsilon_p < \varepsilon$.

We shall now prove (2). Let a and b be distinct points of X. For every integer $n > 1$, there is a path $c_n : [0, 1] \to X$, parameterized proportional to its arc length, such that $l(c_n) < d(a, b) + 1/n$. Such a family of paths $\{c_n\}$ is equicontinuous; indeed for all $t, t' \in [0, 1]$ we have:

$$|t - t'| = \frac{l(c_n|_{[t,t']})}{l(c_n)} \geq \frac{d(c_n(t), c_n(t'))}{d(a, b) + 1};$$

hence $d(c_n(t), c_n(t')) < \varepsilon$ if $|t - t'| < \frac{\varepsilon}{d(a,b)+1}$. The image of each path c_n is contained in the compact set $\overline{B}(2\, d(a, b))$. By the Arzelà-Ascoli theorem (see below) there is a subsequence of the $(c_n)_{n\in\mathbb{N}}$ converging uniformly to a path $c : [0, 1] \to X$. Finally, by the lower semicontinuity of length (1.18(7)), we have

$$l(c) \leq \liminf l(c_{k_n}) = d(a, b).$$

But $l(c) \geq d(a, b)$, so in fact $l(c) = d(a, b)$, and therefore c is a linearly reparameterized geodesic joining a to b. □

3.8 Corollary. *A length space is proper if and only if it is complete and locally compact.*

The argument used to prove part (2) of the preceding proposition shows, more generally, that if two points in a proper metric space are joined by a rectifiable curve, then among all curves joining them there is one of minimal length (see [Hi1900]).

3.9 Remark. The terminology 'proper metric space', introduced in (1.1), arises from the fact that a metric space (X, d) is proper if and only if, given a base point $x_0 \in X$, the function $x \to d(x, x_0)$ is a proper map from X to $\mathbb{R}$ in the usual topological sense, i.e., the inverse image of every compact set is compact.

Because we shall have rather frequent need of it, we include a proof of the Arzelà-Ascoli theorem. Recall that a sequence of maps f_n from one metric space to another is said to be equicontinuous if for every $\varepsilon > 0$ there exists $\delta > 0$ such that for every $n \in \mathbb{N}$, if $d(x, y) < \delta$ then $d(f_n(x), f_n(y)) < \varepsilon$.

3.10 Lemma (Arzelà-Ascoli). *If X is a compact metric space and Y is a separable metric space, then every sequence of equicontinuous maps $f_n : Y \to X$ has a subsequence that converges (uniformly on compact subsets) to a continuous map $f : Y \to X$.*

Proof. We first fix a countable dense set $Q = \{q_1, q_2, \dots\}$ in Y. Then we use the compactness of X to choose a subsequence $f_{n(1)}$ such that the sequence of points $f_{n(1)}(q_1)$ converges in X as $n \to \infty$; we call the limit point $f(q_1)$. We then pass to a further subsequence $f_{n(2)}$ to ensure that $f_{n(2)}(q_2)$ converges to some point $f(q_2)$, as $n \to \infty$. Proceeding by recursion on k, we pass to further subsequences $f_{n(k)}$ so that as $n(k)$ tends to infinity $f_{n(k)}(q_j)$ converges to $f(q_j)$ for all $j \leq k$. The diagonal subsequence $f_{n(n)}$ has the property that $\lim_{n\to\infty} f_{n(n)}(q) = f(q)$ for all $q \in Q$.

By equicontinuity, for every $\varepsilon > 0$ there exists $\delta > 0$ such that if $d(y, y') < \delta$ then $d(f_n(y), f_n(y')) < \varepsilon$ for all n. So taking the limit, $d(f(q), f(q')) \leq \varepsilon$ for all $q, q' \in Q$ with $d(q, q') < \delta$. Since X is compact (hence complete) it follows that f has a unique continuous extension $Y \to X$, which again satisfies this inequality.

It remains to show that the convergence of $f_{n(n)}$ to f is uniform on compact subsets. Given $\varepsilon > 0$ we choose δ as above. Given a compact subset $C \subseteq Y$ we fix $N > 0$ so that for every $y \in C$ there exists $j(y) < N$ with $d(y, q_{j(y)}) < \delta$, then we fix M sufficiently large so that $d(f_{n(n)}(q_j), f(q_j)) < \varepsilon$ for all $n > M$ and all $j < N$. Then, for all $y \in C$ and all $n > M$ we have:

$$\begin{aligned} &d(f(y), f_{n(n)}(y)) \\ &\quad \leq d(f(y), f(q_{j(y)})) + d(f(q_{j(y)}), f_{n(n)}(q_{j(y)})) + d(f_{n(n)}(q_{j(y)}), f_{n(n)}(y)) \\ &\quad\quad \leq \varepsilon + \varepsilon + \varepsilon. \end{aligned}$$

Thus $f_{n(n)} \to f$ uniformly on compact sets as $n \to \infty$. □

3.11 Corollary. *If X is a compact metric space and if $c_n : [0, 1] \to X$ is a sequence of linearly reparameterized geodesics, then there exists a linearly reparameterized geodesic $c : [0, 1] \to X$ and a subsequence $c_{n(i)}$ such that $c_{n(i)} \to c$ uniformly as $n(i) \to \infty$.*

Proof. For every $n \in \mathbb{N}$ and all $t, t' \in [0, 1]$ we have $d(c_n(t), c_n(t')) \leq D|t - t'|$, where D is the diameter of X. Thus the c_n are equicontinuous. A uniform limit of geodesics is a geodesic. □

Later we shall need the following variation on (3.11). It is important to note that here it is not necessary to pass to a subsequence in order to guarantee convergence.

3.12 Lemma. *Let x and y be points in a proper geodesic metric space X. Suppose that there is a unique geodesic segment joining x to y in X; let $c : [0, 1] \to X$ be a linear parameterization of this segment. Let $c_n : [0, 1] \to X$ be linearly reparameterized geodesics in X, and suppose that the sequences of points $c_n(0)$ and $c_n(1)$ converge to x and y respectively. Then, $c_n \to c$ uniformly.*

Proof. We fix $R > 0$ so that the image of each of the paths c_n lies in the (compact) closed ball of radius R about x. If the sequence c_n did not converge to c pointwise, then there would exist $\varepsilon > 0$, $t_0 \in (0, 1)$ and an infinite subsequence c_{n_i} such that $d(c_{n_i}(t_0), c(t_0)) \geq \varepsilon$ for all n_i. The previous corollary would then yield a subsequence of the c_{n_i} converging uniformly to a linearly reparameterized geodesic $c' : [0, 1] \to X$ joining $x = \lim c_{n_i}(0)$ to $y = \lim c_{n_i}(1)$. But since $d(c'(t_0), c(t_0)) \geq \epsilon$, this would contradict the uniqueness of c.

Thus $c_n \to c$ pointwise. Using the fact that c and c_n are geodesic it is easy to see that the convergence must be uniform. □

Let X be a uniquely geodesic space. Let $c(x, y)$ denote the linear reparameterization $[0, 1] \to X$ of the geodesic segment joining x to y. Geodesics in X are said to vary continuously with their endpoints if $c(x_n, y_n) \to c(x, y)$ uniformly whenever $x_n \to x$ and $y_n \to y$.

3.13 Corollary. *If a proper metric space X is uniquely geodesic, then geodesics in X vary continuously with their endpoints.*

The following exercise shows that the hypothesis of properness in the above corollary is necessary.

3.14 Exercise. Let Y_n be the subspace of $\mathbb{S}^2$ obtained by taking a sector bounded by two arcs of great circles, each of length π, which meet at the north and south poles, at an angle $\pi/4$, and then removing the open ball of radius $1/n$ about the north pole. Y_n with the induced length metric is a topological disc whose boundary consists of three geodesic segments, two of which have length $\pi - \frac{1}{n}$; let α_n be one of the latter segments. Let X be the space obtained by taking the interval $[0, \pi]$ and, for every integer $n > 2$, attaching Y_n to an initial segment of $[0, \pi]$ by an isometry $\alpha_n \to [0, \pi - \frac{1}{n}]$; endow X with the unique length metric that restricts to the given metric on $[0, \pi]$ and on each of the Y_n.

Show that X is complete, uniquely geodesic, and admits a (Lipschitz-1) contraction to a point, but that geodesic segments in X do not vary continuously with their endpoints.

We close this section by noting a further application of the Arzelà-Ascoli theorem in the spirit of the Hopf-Rinow theorem.

3.15 Definition. By a *loop* in a metric space X we mean a continuous map $c : \mathbb{S}^1 \to X$. Such a loop is called a *closed local geodesic* if there exists a constant λ such that $d(c(\theta), c(\theta')) = \lambda\, d(\theta, \theta')$ for all sufficiently close $\theta, \theta' \in \mathbb{S}^1$ (where the circle $\mathbb{S}^1$ is equipped with its usual length metric).

Recall that two loops $c, c' : \mathbb{S}^1 \to X$ are said to be *homotopic in* X, and we write $c \simeq c'$, if there exists a continuous map $F : \mathbb{S}^1 \times [0, 1] \to X$ such that $F(\theta, 0) = c(\theta)$ and $F(\theta, 1) = c'(\theta)$ for all $\theta \in \mathbb{S}^1$. A space X is said to be *semi-locally simply-connected* if every $x \in X$ has a neighbourhood such that each loop in that neighbourhood is homotopic in X to a constant map.

3.16 Proposition. *If X is a compact, semi-locally simply-connected, geodesic space, then every loop $c : \mathbb{S}^1 \to X$ is homotopic either to a constant path or to a closed local geodesic.*

Proof. The fact that X is compact and semi-locally simply-connected implies that there exists $r > 0$ such that every closed loop of length less than r is homotopic to a constant map. So if c is not homotopic to a constant map then $\ell = \inf\{l(c') \mid c' \simeq c\}$

is strictly positive. We choose a sequence of loops parameterized by arc length, $c_n : \mathbb{S}^1 \to X$ which are homotopic to c whose lengths tend to ℓ. By the Arzelà-Ascoli theorem (3.10), this sequence has a convergent subsequence; we must show that the limit c_∞, which is obviously a closed local geodesic, is homotopic to c. For this, we choose n sufficiently large so that $d(c_n(\theta), c_\infty(\theta)) < r/4$ for all $\theta \in S^1$, then we fix $\theta_0, \dots, \theta_m$ so that $d(c_n(\theta_i), c_n(\theta_{i+1}) < r/4$ and $d(c_\infty(\theta_i), c_\infty(\theta_{i+1})) < r/4$ for all i (indices mod m). Let p_i be a path of length $< r/4$ joining $c_\infty(\theta_i)$ to $c_n(\theta_i)$. For each i, one obtains a loop of length less than r by concatenating p_i, $c_n|_{[\theta_i,\theta_{i+1}]}$, then p_{i+1} and $c_\infty|_{[\theta_i,\theta_{i+1}]}$ with reversed orientation. A loop of such length is null-homotopic. It follows that c_∞ is homotopic to c_n and hence c. □

Riemannian Manifolds as Metric Spaces

We now turn our attention to the study of Riemannian manifolds from the metric viewpoint. Recall that a Riemannian manifold is a differentiable manifold X together with an assignment of a scalar product to the tangent space T_xX to X at each point x, such that these scalar products vary continuously with x. Such an assignment of scalar products is called a *Riemannian metric* on X. More explicitly, in the classical notation, if $(x_1, \dots, x_n)$ are local coordinates (of class C^1) in an open set U of X then, at each $x \in U$, the scalar product on the tangent space T_xX is given by a formula of the form

$$ds^2 = \sum_{i,j=1}^{n} g_{ij}(x)\, dx_i\, dx_j$$

where the $g_{ij}(x)$ are continuous functions on U. This means that, at each point $x \in U$, if $v = \sum v_i(x)\partial/\partial x_i$ and $w = \sum w_i(x)\partial/\partial x_i$ are two vectors in T_xX, then their scalar product is $(v|w) = \sum_{i,j=1}^{n} g_{ij}(x)v_i(x)w_j(x)$.

If $t \mapsto c(t)$ is a differentiable path in X, then one writes $\dot{c}(t) \in T_{c(t)}X$ to denote its velocity vector at time t, and $|\dot{c}(t)|$ to denote its norm with respect to the given scalar product on $T_{c(t)}X$.

3.17 Definitions. If $c(t)$ and $c'(t)$ are two continuously differentiable paths such that $c(0) = c'(0)$, then the *Riemannian angle* between them at $c(0)$ is the angle between the vectors $\dot{c}(0), \dot{c}'(0) \in T_{c(t)}X$, namely the unique $\alpha \in [0, \pi]$ such that

$$\cos\alpha = \frac{(\dot{c}(0) \mid \dot{c}'(0))}{|\dot{c}(0)|\, |\dot{c}'(0)|}.$$

The *Riemannian length* $l_R(c)$ of a piecewise differentiable path $c : [a, b] \to X$ is defined by:

$$l_R(c) = \int_a^b |\dot{c}(t)|\, dt.$$

3.18 Proposition (The Distance Function on a Riemannian Manifold). *Let X be a connected Riemannian manifold. Given $x, y \in X$, let $d(x, y)$ be the infimum of the*

Riemannian length of piecewise continuously differentiable paths $c : [0, 1] \to X$ *such that* $c(0) = x$ *and* $c(1) = y$.

(1) *d is a metric on X.*

(2) *The topology on X defined by this distance is the same as the given (manifold) topology on X.*

(3) *(X, d) is a length space.*

3.19 Example. The simplest example of a Riemannian manifold is $\mathbb{R}^n$ with the Riemannian metric

$$ds^2 = \sum_{i=1}^{n} dx_i^2.$$

The associated metric is the Euclidean metric, i.e., the associated length space is $\mathbb{E}^n$.

We shall often abuse terminology to the extent of referring to the length space (X, d) defined in (3.18) as a Riemannian manifold.

Proof of 3.18. It is clear that d is a pseudometric. To check that it is positive definite, we fix $p \in X$ and consider a local coordinate system $(x_1, \dots, x_n)$ defined in a neighbourhood V of p such that the map $q \mapsto (x_1(q), \dots, x_n(q))$ is a diffeomorphism of V onto an open set of $\mathbb{R}^n$ containing the closed unit ball centered at the origin, and $x_i(p) = 0$ for $i = 1, \dots, n$. Let U be the set of points $q \in V$ such that $\sum_{i=1}^n x_i(q)^2 < 1$. Suppose that in these local coordinates the Riemannian metric is given by the expression

$$ds^2 = \sum_{i,j=1}^{n} g_{ij}(q) dx_i dx_j.$$

We want to compare it with the Euclidean metric on U given by

$$ds_E^2 = \sum_{i}^{n} dx_i^2.$$

Given $q \in U$ and $v \in T_q X$, we denote by $d_E(p, q)$ the Euclidean distance $\sum_{i=1}^n x_i(q)^2$ and by $|v|_E$ (resp. $|v|$) the norm of v with respect to the metric ds_E^2 (resp. ds^2).

Let m (resp. M) be the infimum (resp. the supremum) of $|v|$ over all $v \in T_q X$ with $|v|_E = 1$ and $q \in U$. By compactness, we have $m > 0$ and $M < \infty$. And for all $v \in T_q(U)$, we have

$$m|v|_E \le |v| \le M|v|_E.$$

Therefore, for each piecewise differentiable curve c in U, we have

$$m\, l_E(c) \le l_R(c) \le M l_E(c)$$

where $l_E(c)$ denotes the length with respect to the Riemannian metric ds_E^2.

In particular, if c joins p to q, then $l_R(c) \geq m\, d_E(p, q)$, hence $d(p, q) \geq m\, d_E(p, q)$. On the other hand, if $q \notin U$, then $d(p, q) \geq m$. This shows that the pseudometric d is positive definite. Thus we have established (1).

The balls $B_E(p, r) := \{q \in U \mid d_E(p, q) < r\}$ form a fundamental system of neighbourhoods of p for the given topology on the manifold X. And for $r < m$, we have

$$B_E(p, r/M) \subseteq B(p, r) \subseteq B_E(p, r/m).$$

Thus the given topology on X agrees with the topology associated to the metric d.

To prove (3), let $\bar{d}$ be the length metric associated to d. Given $p, q \in X$ and $\varepsilon > 0$, we can find a piecewise differentiable curve c joining p to q and such that $l_R(c) < d(p, q)+\varepsilon$. And from the definition of length (1.18) it is clear that $l(c) \leq l_R(c)$. Hence

$$d(p, q) \leq \bar{d}(p, q) \leq l(c) \leq l_R(c) < d(p, q) + \varepsilon.$$

As ε is arbitrary, we conclude that $d = \bar{d}$. □ □

By combining the preceding result with the Hopf-Rinow Theorem (3.7) we obtain:

3.20 Corollary. *Every complete, connected, Riemannian manifold is a geodesic metric space.*

3.21 Remarks. One can prove that the Riemannian length of any piecewise continuously differentiable curve is equal to its length in the metric constructed in (3.18). Indeed a similar statement is true for curves in any smooth manifold endowed with a continuous Finsler metric (see [BusM41], [Rin61]). Recall that a *Finsler metric* on a smooth manifold X is an assignment to each tangent space T_xX of a norm; this norm varies continuously with x. Proposition 3.18 is also true for Finsler manifolds (with the same proof).

In the same vein, we note that the Riemannian angle between two geodesics issuing from the same point of a smooth enough Riemannian manifold is equal to the Alexandrov angle between them. This will be proved in (II.1.A).

3.22 Remark. A *Riemannian isometry* of a Riemannian manifold X is, by definition, a diffeomorphism of X whose differential preserves the scalar product on each tangent space. A Riemannian isometry is obviously an isometry of the associated length space (X, d). The converse is also true if the Riemannian metric is of class C^2 (see, for example, [Hel78]).

If X is a Riemannian manifold and $Y \subseteq X$ is a smoothly embedded submanifold, then the restriction to T_xY of the given scalar product on T_xX gives a scalar product on the tangent space to Y at each point $x \in Y$. Thus Y inherits a Riemannian structure from X, and we can consider the associated length space, as defined in (3.18). We note a consequence of (3.18):

3.23 Proposition. *Let X be a Riemannian manifold and let $Y \subseteq X$ be a smoothly embedded submanifold. Then, the induced length metric on Y (as defined in 3.3) coincides with the length metric associated to the Riemannian structure which Y inherits from X.*

A similar observation holds for Finsler metrics.

Length Metrics on Covering Spaces

A continuous map $p : Y \to X$ between topological spaces is said to be a *covering map* if it is surjective and every point $x \in X$ has an open neighbourhood U such that $p^{-1}U$ is a disjoint union of sets $\widetilde{U}_\alpha$ such that p restricts to a homeomorphism of each $\widetilde{U}_\alpha$ onto U. Whenever one has a covering $p : Y \to X$, the covering space Y inherits all of the local structure of the base X. For example, a covering space of a manifold is again a manifold; given a covering of a complete Riemannian manifold, one can use the covering map to pull back a complete Riemannian metric; if the base manifold is non-positively curved, then the pull-back metric on the covering space will also be non-positively curved. Similarly, since measuring the length of a curve in a metric space is a purely local process, there is a natural way to pull back the definition of length to covering spaces. The purpose of this paragraph is to make this last idea precise. We also give a criterion for recognizing covering spaces; this will be needed in Chapter II.4.

3.24 Definition. Let X be a length space and let $\widetilde{X}$ be a topological space. Let $p : \widetilde{X} \to X$ be a continuous map that is a *local homeomorphism* (i.e., every point $\tilde{x} \in \widetilde{X}$ has an open neighbourhood U such that p maps U homeomorphically onto an open set of X)[4]. Given a path $c : [0, 1] \to \widetilde{X}$, we define its *length* $l(c)$ to be the length of the curve $p \circ c$ in X. We define a pseudometric on $\widetilde{X}$ by setting the distance $\tilde{d}(\tilde{x}, \tilde{y})$ between two points $\tilde{x}, \tilde{y} \in \widetilde{X}$ equal to the infimum of the length of paths in $\widetilde{X}$ joining them. (One checks easily that provided $\widetilde{X}$ is Hausdorff, this is indeed a metric; it may be infinite if $\widetilde{X}$ is not connected or if the given length metric on X is infinite.) The metric $\tilde{d}$ is called the *metric induced on $\widetilde{X}$ by p*.

3.25 Proposition. *Let X be a length space, and let $\widetilde{X}$ be a topological space that is Hausdorff. Let $p : \widetilde{X} \to X$ be a continuous map that is a local homeomorphism, and let $\tilde{d}$ be the metric induced on $\widetilde{X}$ by p.*

(1) *If one endows $\widetilde{X}$ with the metric $\tilde{d}$, then p becomes a local isometry.*

(2) *$\tilde{d}$ is a length metric.*

(3) *$\tilde{d}$ is the unique metric on $\widetilde{X}$ that satisfies properties (1) and (2).*

[4] Such a continuous map $p : \widetilde{X} \to X$ will also be called an *étale map*.

Proof. For (1), we must show that, given $\tilde{x} \in \widetilde{X}$ with image $p(\tilde{x}) = x$, if $r > 0$ is sufficiently small then the restriction of p to $B(\tilde{x}, r)$ is an isometry onto $B(x, r)$. Let $\widetilde{U}$ be an open neighbourhood of $\tilde{x}$ such that the restriction of p to $\widetilde{U}$ is a homeomorphism onto its image $p(\widetilde{U}) = U$, and let $s : U \to \widetilde{U}$ be the inverse of $p|_{\widetilde{U}}$. Let $r > 0$ be small enough so that $B(x, 2r) \subseteq U$. We claim that s restricted to $B(x, r)$ is an isometry onto $B(\tilde{x}, r)$. Indeed, given $y, z \in B(x, r)$ and $\varepsilon \in (0, r)$, since X is a length space, there exists a path c joining y to z in $B(x, 2r)$ of length smaller than $d(y, z) + \varepsilon$; its image under s is a path joining $s(y)$ to $s(z)$, hence $\tilde{d}(s(y), s(z)) < d(y, z) + \varepsilon$ for arbitrarily small $\varepsilon > 0$. But, by construction, p does not increase distances, therefore $\tilde{d}(s(y), s(z)) = d(y, z)$. It remains to prove that $s : B(x, r) \to B(\tilde{x}, r)$ is surjective. But this is clear, because $p \circ s$ restricts to the identity on $B(x, r)$ and p does not increase distances. This completes the proof of (1).

Because p is a local isometry, the length of a curve in $\widetilde{X}$ is the same as the length of its image under p; assertion (2) follows immediately.

If $\tilde{d}'$ is any metric on $\widetilde{X}$ that satisfies (1), then the identity map id : $(\widetilde{X}, \tilde{d}) \to (\widetilde{X}, \tilde{d}')$ is a local isometry, in particular it preserves the length of curves. So if $\tilde{d}'$ is assumed to be a length metric then id : $(\widetilde{X}, \tilde{d}) \to (\widetilde{X}, \tilde{d}')$ must be an isometry. □

3.26 Definition. A metric space X is said to be *locally uniquely geodesic* if for each point $x \in X$ there is an $r > 0$ such that every pair of points $y, z \in B(x, r)$ can be joined by a unique geodesic in X and this geodesic lies in $B(x, r)$.

3.27 Remarks.

(1) It is a classical theorem in differential geometry that a Riemannian manifold of class C^2, when considered as a metric space (see 3.18), is locally uniquely geodesic.

(2) If X is locally uniquely geodesic and proper then geodesics vary continuously with their endpoints locally. That is to say, with the notation of the above definition, geodesics in $B(x, r)$ will vary continuously with their endpoints (cf. 3.13).

In the following proof we shall use the fact that if $\tilde{X}$ is Hausdorff and if $p : \tilde{X} \to X$ is a local homeomorphism, then p has the property that 'lifts are unique': if f and f' are two continuous maps of a connected space Y into $\widetilde{X}$ such that $p \circ f = p \circ f'$, and if $f(y) = f'(y)$ for some point $y \in Y$, then $f = f'$ on the whole of Y. To see that p has this property, one simply notes that the set of points in Y at which f and f' coincide is both open and closed.

3.28 Proposition. *Let $p : \widetilde{X} \to X$ be a map of length spaces such that*

(1) *X is connected,*

(2) *p is a local homeomorphism,*

(3) *the length of every path in $\widetilde{X}$ is not bigger than the length of its image under p,*

(4) *X is locally uniquely geodesic and geodesics in X vary continuously with their endpoints locally, and*

(5) *$\widetilde{X}$ is complete.*

Then, p is a covering map.

Note that conditions (2) and (3) are satisfied if p is a local isometry.

Proof. We first prove that given a rectifiable curve $c : [0, 1] \to X$ and a point $\tilde{x} \in \widetilde{X}$ such that $p(\tilde{x}) = c(0)$, there is a unique path $\tilde{c} : [0, 1] \to \widetilde{X}$ that is a lift of c at $\tilde{x}$, in the sense that $\tilde{c}(0) = \tilde{x}$ and $p(\tilde{c}(t)) = c(t)$ for all $t \in [0, 1]$. Suppose that such a lift has been constructed over an interval $[0, a)$. Let $t_n \in [0, a)$ be a sequence of points converging to a. By (3), we have $d(\tilde{c}(t_n), \tilde{c}(t_m)) \leq l(\tilde{c}|_{[t_n,t_m]}) \leq l(c|_{[t_n,t_m]})$. As the sequence of numbers $l(c|_{[0,t_n]})$ is Cauchy, $\tilde{c}(t_n)$ is a Cauchy sequence in $\widetilde{X}$, and hence converges to a unique point, which has the properties required of $\tilde{c}(a)$. Thus the given lifting can be extended to the closed interval $[0, a]$; this shows that the maximal subinterval of $[0, 1]$ that contains 0 and on which a lifting $\tilde{c}$ exists is closed; it is also open by condition (2), hence it is the whole interval.

As X is (path) connected, the argument of the preceding paragraph implies that the restriction of p to each connected component of $\widetilde{X}$ is a surjection onto X. What remains to be proved is that for every $x \in X$ there is a neighbourhood U of x such that the restriction of p to each component of $p^{-1}U$ is a homeomorphism onto its image. We shall show that it suffices to let the role of U be played by any ball $B(x, r)$ which is uniquely geodesic, the geodesic segment joining x to each point $y \in B(x, r)$ depending continuously of its endpoints.

Given such a ball, we fix a point $\tilde{x} \in p^{-1}(x)$, and for each point $y \in B(x, r)$ we denote by $c_y : [0, 1] \to B(x, r)$ the linearly reparameterized geodesic joining x to y, and we let $\tilde{c}_y$ denote the unique lifting of c_y with $\tilde{c}_y(0) = \tilde{x}$. Let $s_{\tilde{x}} : B(x, r) \to \widetilde{X}$ denote the map $y \mapsto \tilde{c}_y(1)$. We claim that $s_{\tilde{x}}$ is a homeomorphism onto an open set of $\widetilde{X}$. By (2), it is sufficient to check that it is continuous at y.

Because p is a local homeomorphism, we can cover the image of c_y with a finite number of balls $B_k \subseteq B(x, r)$, so that $c_y([(k-1)/n, k/n]) \subseteq B_k$ for $k = 1, \dots, n$, and there exist continuous maps $s^k : B_k \to \widetilde{X}$ with $s^k \circ p$ equal to the identity on B_k and $s^k(c_y(t)) = s_{\tilde{x}}(c_y(t))$ for all $t \in [(k-1)/n, k/n]$. As geodesic segments in $B(x, r)$ depend continuously of their endpoints, if $\delta > 0$ is small enough then for all z with $d(y, z) < \delta$, we have $c_z([(k-1)/n, k/n]) \subseteq B_k$. And we may define a continuous map $\tilde{s} : B(y, \delta) \times [0, 1] \to \widetilde{X}$ by: $\tilde{s}(z, t) = s^k(c_z(t))$ if $t \in [(k-1)/n, k/n]$; to see that this map is well-defined and continuous, one observes that on the connected set $B(y, \delta) \times \{t_{k/n}\}$ the definitions using s_{k-1} and s^k agree at $(y, t_{k/n})$ and hence everywhere (because hypothesis (2) ensures that p has the property that lifts are unique). Since $t \mapsto \tilde{s}(c_z(t))$ is a lifting of c_z at $\tilde{x}$, it must coincide with $\tilde{c}_z$. Therefore the restriction of $s_{\tilde{x}}$ to $B(y, \delta)$ agrees with the continuous map $z \mapsto \tilde{s}(z, 1)$; in particular $s_{\tilde{x}}$ is continuous at y.

We have shown that $p^{-1}(B(x, r))$ is the union of the open sets $s_{\tilde{x}}(B(x, r))$ where $\tilde{x} \in p^{-1}(x)$, and p restricted to each of thee sets is a homeomorphism onto $B(x, r)$. Finally we observe that these sets must be disjoint; for if $\tilde{y} \in s_{\tilde{x}}(B(x, r)) \cap s_{\tilde{x}'}(B(x, r))$, then the lifts of $\tilde{c}_{p(\tilde{y})}$ beginning at $\tilde{x}$ and $\tilde{x}'$ both end at $\tilde{y}$, hence they must coincide and $\tilde{x} = \tilde{x}'$. Therefore p is a covering map. $\square$

3.29 Remark. An examination of the preceding proof shows that one can obtain the same conclusion under the following alternative hypotheses.

(1′) p is surjective,

(2′) p is a local homeomorphism,

(3′) the length of every path in $\widetilde{X}$ is not bigger than the length of its image under p,

(4′) for every $x \in X$ there exists $r > 0$ such that, for every $y \in B(x, r)$, there is a unique constant speed geodesic $c_y : [0, 1] \to B(x, r)$ joining x to y, and c_y varies continuously with y, and

(5′) for every $\tilde{y} \in p^{-1}(y)$ there is a continuous lifting $\tilde{c}_y : [0, 1] \to \widetilde{X}$ of c_y with $\tilde{c}_y(1) = \tilde{y}$.

3.30 Exercises

(1) Give an example of a compact geodesic space that shows that condition (4′) is strictly weaker than condition (4). (Hint: Consider a cone with a small vertex angle.)

(2) By adapting example (3.14), show that the continuity requirement in conditions (4) and (4′) is necessary, even for locally contractible spaces (if one does not assume that the space is proper).

Manifolds of Constant Curvature

3.31 Definition. By definition, an n-dimensional *manifold of constant curvature* κ is a length space X that is locally isometric to M^n_κ. In other words, for every point $x \in X$ there is an $\varepsilon > 0$ and an isometry ϕ from $B(x, \varepsilon)$ onto a ball $B(\phi(x), \varepsilon) \subset M^n_\kappa$.

3.32 Theorem. *Let X be a complete, connected, n-dimensional manifold of constant curvature* κ. *When endowed with the induced length metric (3.24), the universal covering of X is isometric to* M^n_κ.

Proof. The following proof is due to C. Ehresmann [Ehr54]. In the first part of the proof we do not assume that X is complete.

By definition, a *chart* $\phi : U \to M^n_\kappa$ is an isometry from an open set $U \subseteq X$ onto an open set $\phi(U) \subseteq M^n_\kappa$. If $\phi' : U' \to M^n_\kappa$ is another chart and if $U \cap U'$ is connected, then by (2.20) there is a unique isometry $g \in \mathrm{Isom}(M^n_\kappa)$ such that ϕ and $g \circ \phi'$ are equal on $U \cap U'$.

Consider the set of all pairs (ϕ, x), where $\phi : U \to M^n_\kappa$ is a chart and $x \in U$. We say that two such pairs (ϕ, x) and (ϕ', x') are equivalent if $x = x'$ and if the restrictions of ϕ and ϕ' to a small neighbourhood of x coincide. This is indeed an equivalence relation and the equivalence class of (ϕ, x) is called the *germ* of ϕ at x. Let $\hat{X}$ be the set of all equivalence classes, i.e. the set of all germs of charts. Let $\hat{p} : \hat{X} \to X$ and $\hat{D} : \hat{X} \to M^n_\kappa$ be the maps that send the germ of ϕ at x to x and $\phi(x)$ respectively.

Notice that there is a natural action of $G = \mathrm{Isom}(M^n_\kappa)$ on $\hat{X}$: if $\hat{x}$ is the germ of ϕ at x and if $g \in G$, then $g.\hat{x}$ is the germ of $g \circ \phi$ at x. The map $\hat{D} : \hat{X} \to M^n_\kappa$ is G-equivariant: $g.(\hat{D}(\hat{x})) = \hat{D}(g.\hat{x})$, and according to (2.20), given $\hat{x}, \hat{x}' \in \hat{p}^{-1}(x)$, there is a unique $g \in G$ such that $g.\hat{x} = \hat{x}'$.

There is a natural topology on $\hat{X}$, called the *germ topology*, with respect to which $\hat{p}$ is a covering and $\hat{D}$ is a local homeomorphism. The basic open sets defining this topology are U_ϕ, where $\phi : U \to M^n_\kappa$ is a chart and $U_\phi \subseteq \hat{X}$ is the set of germs of ϕ at the various points of U. The restriction of $\hat{p}$ (resp. $\hat{D}$) to U_ϕ is a homeomorphism onto U (resp. $\phi(U)$). Moreover, if U is connected then $\hat{p}^{-1}(U)$ is the disjoint union of the open sets $U_{g\circ\phi}$, where $g \in G$. Thus $\hat{p} : \hat{X} \to X$ is a covering map (and in particular $\hat{X}$ is Hausdorff). Indeed $\hat{p} : \hat{X} \to X$ is a *Galois covering with Galois group* G (meaning that G acts by homeomorphisms on $\hat{X}$ preserving the fibres of $\hat{p}$ and acting simply transitively on each fibre).

Choose a base point $x_0 \in X$ and a chart ϕ defined at x_0. Let $\hat{x}_0 \in \hat{X}$ be the germ of ϕ at x_0, and let $\tilde{X}$ be the connected component of $\hat{X}$ containing $\hat{x}_0$. Let $p : \tilde{X} \to X$ and $D : \tilde{X} \to M^n_\kappa$ be the restrictions of $\hat{p}$ and $\hat{D}$ to $\tilde{X}$. Let $\Gamma \subset G$ be the subgroup of G that leaves $\tilde{X}$ invariant. Then $p : \tilde{X} \to X$ is a Galois covering with Galois group Γ and the map D is a local homeomorphism which is Γ-equivariant. If we endow $\tilde{X}$ with the unique length metric $\tilde{d}$ such that p is a local isometry (3.24), then D becomes a local isometry.

Now we assume that X is complete. Then $(\tilde{X}, \tilde{d})$ is also complete, and (3.28) tells us that $D : \tilde{X} \to M^n_\kappa$ is a covering. As M^n_κ is simply connected and $\tilde{X}$ connected, D must be a homeomorphism. Thus $\tilde{X}$ is simply connected (hence the universal covering of X), and D is an isometry. □

3.33 Remark. The map $D : \tilde{X} \to M^n_\kappa$ constructed in the proof of (3.29) is called the *developing map* of X, and the group $\Gamma \subset \text{Isom}(M^n_\kappa)$ (which is naturally isomorphic to $\pi_1(X, x_0)$) is called the *holonomy group* of X with respect to the germ (ϕ, x_0).

An analogous theorem with the same proof is valid for spaces with a (G, Y)-geometric structure — see III.$\mathcal{G}$.1.11 where we consider the more general case of orbifolds.

Chapter I.4 Normed Spaces

In this chapter we return to the study of normed spaces, which were introduced briefly in Chapter 1.

Hilbert Spaces

4.1 Definition. A *scalar product* (or inner-product) on a real vector space V is a symmetric bilinear map $V \times V \to \mathbb{R}$, written $(v, w) \mapsto (v \mid w)$, with the property that $(v \mid v) > 0$ for all $v \neq 0$. Associated to a scalar product one has a norm $\|v\| := (v \mid v)^{1/2}$, and hence a metric. A *pre-Hilbert space* (or inner-product space) is a real vector space V equipped with a scalar product; it is called a *Hilbert space* if the associated metric is complete.

In order to see that the above expression really does define a norm, we must verify the triangle inequality. This is a consequence of the *Cauchy–Schwarz inequality*: $|(v \mid w)| \leq \|v\|.\|w\|$. The validity of this inequality is clear if $w = 0$, and for the general case one expands the right-most term of the expression $0 \leq \|v - \lambda w\|^2 = (v - \lambda w \mid v - \lambda w)$ and sets the scalar λ equal to $(v \mid w)/(w \mid w)$.

Using the Cauchy–Schwarz inequality, we deduce the triangle inequality:

$$\begin{aligned}\|v + w\|^2 &= (v + w \mid v + w) \\ &= \|v\|^2 + (v \mid w) + (w \mid v) + \|w\|^2 \leq \|v\|^2 + 2\|v\|.\|w\| + \|w\|^2.\end{aligned}$$

4.2 Remark. The ℓ^2 norm on $\mathbb{R}^n$ is the norm associated to the Euclidean scalar product $(x \mid y) := \sum_i^n x_i y_i$, where $x = (x_1, \dots, x_n)$ and $y = (y_1, \dots, y_n)$. The associated metric space is $\mathbb{E}^n$. The existence of orthonormal bases (as constructed by the Gram-Schmidt process, for example) implies that *every* n-dimensional pre-Hilbert space is isometric to $\mathbb{E}^n$.

A trivial calculation with the scalar product shows that the unit ball in a pre-Hilbert space is strictly convex, so by (1.6) we have:

4.3 Lemma. *Every pre-Hilbert space is uniquely geodesic.*

Not every uniquely geodesic normed space is a pre-Hilbert space. Indeed, P. Jordan and J. von Neumann [JvN35] proved the following metric characterization of pre-Hilbert spaces.

4.4 Proposition (The Parallelogram Law). *A norm $\|\cdot\|$ on a real vector space V arises from a scalar product if and only if the norm satisfies the parallelogram law: for all $v, w \in V$,*

$$\|v - w\|^2 + \|v + w\|^2 = 2(\|v\|^2 + \|w\|^2).$$

If a norm does satisfy this condition, then one recovers the scalar product by setting

$$(v \mid w) = \frac{1}{4}(\|v + w\|^2 - \|v - w\|^2).$$

Proof. If the norm derives from a scalar product then

$$\|v + w\|^2 - \|v - w\|^2 = (v + w \mid v + w) - (v - w \mid v - w) = 4(v \mid w).$$

Conversely, we show that if the norm satisfies the parallelogram law then the following formula defines a scalar product:

$$(v \mid w) = \frac{1}{4}(\|v + w\|^2 - \|v - w\|^2).$$

It is clear that $(v \mid w) = (w \mid v)$ and $(v \mid v) = \|v\|^2 \geq 0$, so it suffices to check that when w is fixed, $(v \mid w)$ is a linear function of v. Applying the parallelogram law to the pairs of vectors $(v' + w, v'')$ and $(v' - w, v'')$, we get

$$\|v' + v'' + w\|^2 + \|v' - v'' + w\|^2 = 2\|v' + w\|^2 + 2\|v''\|^2$$
$$\|v' + v'' - w\|^2 + \|v' - v'' - w\|^2 = 2\|v' - w\|^2 + 2\|v''\|^2.$$

Subtracting the second equality from the first and using the definition of the scalar product, we find

$$(v' + v'' \mid w) + (v' - v'' \mid w) = 2(v' \mid w).$$

In particular if $v' = v''$, since it is clear that $(0 \mid w) = 0$, we have

$$(2v' \mid w) = 2(v' \mid w),$$

and hence the above formula can be rewritten

$$(v' + v'' \mid w) + (v' - v'' \mid w) = (2v' \mid w).$$

Replacing v' by $1/2(v' + v'')$ and v'' by $(1/2(v' - v'')$, we get

$$(v' \mid w) + (v'' \mid w) = (v' + v'' \mid w) \tag{$*$}$$

for all $v', v'', w \in V$.

We still have to check that $(\lambda v \mid w) = \lambda(v \mid w)$ for any real number λ. The equality (*) implies that this is true for any rational number, hence by continuity for any real number. $\square$

There are important qualitative differences between the geometry of those norms that satisfy the parallelogram law and those which do not. We note one important difference concerning the notion of angle (1.12). For further differences and characterizations of pre-Hilbert spaces see [Rob82].

Recall that the angle $\angle(c, c')$ between two geodesics $c, c' : [0, \varepsilon] \to X$ issuing from a point p in a metric space X, is $\limsup_{t,t' \to 0} \overline{\angle}_p(c(t), c(t'))$, where $\overline{\angle}_p(c(t), c(t'))$ is the vertex angle at $\overline{p}$ in the comparison triangle $\overline{\Delta}(p, c(t), c'(t')) \subset \mathbb{E}^2$.

4.5 Proposition. *A normed vector space V is a pre-Hilbert space if and only if the limit $\lim_{t,t' \to 0} \overline{\angle}_0(c(t), c'(t'))$ exists for all pairs of geodesic rays c, c' issuing from $0 \in V$.*

Proof. All linear segments in V are geodesics (1.6) and therefore every vector subspace of V is isometrically embedded. If V is a pre-Hilbert space, then every geodesic ray $c : [0, \infty) \to V$ issuing from $0 \in V$ is of the form $t \mapsto tu$ (cf. (4.3)). By restricting our attention to the subspace spanned by any two such rays, we reduce to the case $V = \mathbb{E}^2$, where it is clear that the stated limit exists.

Conversely, suppose that the above limit exists for all pairs of rays issuing from the origin in the normed vector space V. Given any two linearly independent unit vectors $u, u' \in V$, we consider the rays $c, c' : [0, \infty) \to V$ defined by $c(t) = tu$ and $c'(t) = tu'$. We claim that if $\lim_{t,t' \to 0} \overline{\angle}_0(c(t), c'(t'))$ exists and is equal to α, then $\alpha = \overline{\angle}_0(c(t), c'(t'))$ for all $t > 0$ and $t' > 0$. Indeed, $\alpha = \lim_{s \to 0} \overline{\angle}_0(c(st), c'(st'))$, and by the law of cosines, $\cos \overline{\angle}_0(c(st), c'(st')) = \frac{1}{2s^2tt'}(s^2t^2 + s^2t'^2 - \|stu - st'u'\|^2)$, which is independent of s.

To complete the proof we must show that V satisfies the the parallelogram law. It is sufficient to check it for two linearly independent vectors v and w. Applying the argument of the preceding paragraph to the normalized vectors $u = v/\|v\|$ and $u' = (v + w)/\|v + w\|$, we see that the angles at $\overline{0}$ in the comparison triangles $\overline{\Delta}(0, v, v + w)$ and $\overline{\Delta}(0, v, 1/2\,(v + w))$ are equal. The parallelogram law for v and w follows easily from this observation, by the law of cosines. $\square$

4.6 The Hilbert Spaces $\ell^2(S)$. We first consider $\ell^2(\mathbb{Z})$, the set of bi-infinite sequences of real numbers $x = (x_n)$ such that $\sum |x_n|^2$ is finite. Let $\|x\|_2 := (\sum |x_n|^2)^{1/2}$. We claim that, when equipped with the term-wise operations of addition and scalar multiplication, $\ell^2(\mathbb{Z})$ is a real vector space. That it is closed under scalar multiplication is clear. And one sees that if $x = (x_n)$ and $y = (y_n)$ lie in $\ell^2(\mathbb{Z})$ then so too does $x + y = (x_n + y_n)$ by observing that the triangle inequality in $\mathbb{E}^n$ gives a bound on the partial sums of $x + y$:

$$\left(\sum_{i=-n}^{n} |x_i + y_i|^2\right)^{1/2} \leq \left(\sum_{i=-n}^{n} |x_i|^2\right)^{1/2} + \left(\sum_{i=-n}^{n} |y_i|^2\right)^{1/2} \leq \|x\|_2 + \|y\|_2.$$

We define a scalar product on $\ell^2(\mathbb{Z})$, with associated norm $\| \cdot \|_2$, by: $(x \mid y) := \sum_i x_i y_i$. The convergence of this sum is assured by the Cauchy–Schwarz inequality for the Euclidean scalar product:

$$\left(\sum_{i=-n}^{n} |x_i y_i| \right)^2 \leq \sum_{i=-n}^{n} |x_i|^2 \sum_{i=-n}^{n} |y_i|^2 \leq \|x\|_2^2 \, \|y\|_2^2.$$

Now let S be any set. We consider functions $x : S \to \mathbb{R}$. By definition, $\sum_S x(s)$ is the supremum of the sums $\sum_C x(s)$ taken over the finite subsets $C \subset S$. If $\sum_S x(s)$ is finite, then $x(s) = 0$ for all but countably many $s \in S$ (see [Dieu63] for example).

Let $\ell^2(S)$ be the set of functions x such that $\sum_S |x(s)|^2$ is finite. Arguing exactly as in the case $S = \mathbb{Z}$, we see that $\ell^2(S)$ is closed under the operations of pointwise addition and scalar multiplication, and that when equipped with the scalar product $(x \mid y) := \sum_S x(s)y(s)$ it is a pre-Hilbert space. The associated norm is denoted $\| \cdot \|_2$.

4.7 Proposition. *Let S be a set. The metric associated to the norm $\| \cdot \|_2$ defined above is complete, and hence $\ell^2(S)$ is a Hilbert space.*

Proof. Let d denote the metric associated to the norm $\| \cdot \|_2$. Let $(x^{(m)})$ be a Cauchy sequence in $\ell^2(S)$. Given $\varepsilon > 0$, we fix N such that for all $m, m' > N$,

$$d(x^{(m)}, x^{(m')}) = \left(\sum_S |x^{(m)}(s) - x^{(m')}(s)|^2 \right)^{1/2} < \varepsilon.$$

Then $|x^{(m)}(s) - x^{(m')}(s)| < \varepsilon$ for all $s \in S$ if $m, m' > N$, and hence for fixed s the sequence of numbers $(x^{(m)}(s))$ is Cauchy, with limit $x^\infty(s)$, say. This defines a function $x^\infty : S \to \mathbb{R}$; we claim that $x^\infty \in \ell^2(S)$. From the above inequality we have that, for all finite subset $C \subset S$

$$\sum_{s \in C} |x^{(m)}(s) - x^{(m')}(s)|^2 < \varepsilon^2.$$

Fixing m and letting $m' \to \infty$ in this expression we get

$$\sum_{s \in C} |x^{(m)}(s) - x^\infty(s)|^2 \leq \varepsilon^2.$$

Hence $x^{(m)} - x^\infty \in \ell^2(S)$, and therefore $x^\infty = x^{(m)} - (x^{(m)} - x^\infty) \in \ell^2(S)$. Since ε and C were arbitrary, the displayed inequality also shows that $d(x^{(m)}, x^\infty) \to 0$ as $m \to \infty$. □

Recall that a metric space X is said to satisfy a given condition *locally* (e.g., "X is locally compact") if for every $x \in X$ there exists $\varepsilon > 0$ such that the closed ball of radius ε is (when equipped with the induced metric) a metric space which satisfies the specified property (e.g., $\overline{B}(x, \varepsilon)$ is compact).

4.8 Example. Let S be an infinite set. Let $\delta^s(\varepsilon)$ denote the sequence whose only non-zero term is in position m, and this term is ε. The infinite set $\{\delta^s(\varepsilon) \mid s \in S\}$ is obviously contained in the closed ball of radius ε about $0 \in \ell^2(S)$ but does not have an accumulation point, thus $\ell^2(S)$ is not a locally compact space.

4.9 Exercise. Let d be the metric associated to any norm on $\mathbb{R}^n$, and let d_2 denote the Euclidean metric on $\mathbb{R}^n$. Show that there exist constants $M \geq m \geq 0$ such that $m\, d(x, y) \leq d_2(x, y) \leq M\, d(x, y)$ for all $x, y \in \mathbb{R}^n$. Hence deduce that every finite dimensional normed space is proper (and hence complete).

Further to the preceding example and exercise, we note that a classical result of F. Riesz states that a normed vector space is locally compact if and only if it is finite dimensional (see [La68, p.37] or [Dieu63, V.9]). We prove a particular case.

4.10 Lemma. *A pre-Hilbert space is locally compact if and only if it is finite dimensional.*

Proof. Suppose that V is an infinite dimensional pre-Hilbert space, with scalar product $(v \mid w)$. Let $u_1, u_2, \dots$ be a linearly independent set of vectors. From this we can construct an infinite sequence of orthonormal vectors, i.e., a sequence $v_1, v_2, \dots$ such that $(v_i \mid v_j) = \delta_{i,j}$ (Kronecker's delta). In order to do so, we invoke the Gram-Schmidt procedure: $v_1 := u_1/\|u_1\|$ and, inductively, $w_n := u_n - \sum_{i=1}^{n-1}(u_n \mid v_i)$ and $v_n := w_n/\|w_n\|$.

Now, given any $\varepsilon > 0$, we claim that the closed ball of radius ε is not compact. Indeed, the vectors εv_n form an infinite sequence contained in this ball, and the sequence does not have an accumulation point, because if $n \neq m$ then $d(\varepsilon v_n, \varepsilon v_m)^2 = (\varepsilon v_n - \varepsilon v_m \mid \varepsilon v_n - \varepsilon v_m) = 2\varepsilon^2$. □

4.11 Example: The Space of Finite Sequences. We have noted that all finite dimensional pre-Hilbert spaces are actually Hilbert spaces. An easy example of a pre-Hilbert space which is not a Hilbert space is provided by the space C_0 consisting of those bi-infinite sequences of real numbers (x_n) all but finitely many of whose entries are zero; this is a subspace of $\ell^2(\mathbb{Z})$, and one endows it with the induced scalar product.

If $x \in \ell^2(\mathbb{Z})$ is a sequence all of whose entries are non-zero, then one can obtain a Cauchy sequence $(x^{(m)})$ in C_0 by defining $x_n^{(m)} = x_n$ if $m > |n|$ and $x_n^{(m)} = 0$ otherwise. This Cauchy sequence does not have a limit in C_0. Notice that this argument shows that C_0 is dense in $\ell^2(\mathbb{Z})$.

Isometries of Normed Spaces

Let V be a normed vector space. Given any $a \in V$, the map $V \to V$ given by $x \mapsto x+a$ defines an isometry of V; we shall refer to this isometry as *translation by a*. The abelian subgroup of Isom(X) formed by such translations is clearly isomorphic to the underlying additive group of V. We introduce the notation $\mathcal{O}(V)$ for the subgroup of Isom(X) consisting of those linear transformations of V that preserve its norm.

Recall that, given two groups G and H, and an action of H on G (i.e., a homomorphism $H \to Aut(G)$, written $h \mapsto \varphi_h$) one forms the *semidirect product* $G \rtimes H$ by endowing the set $G \times H$ with the operation $(g, h) \cdot (g', h') = (g\varphi_h(g'), hh')$. For the semidirect product referred to in part (2) of the following proposition, the action of $\mathcal{O}(V)$ on V is the obvious linear action.

4.12 Theorem. *Let V be a normed vector space.*

(1) *Every isometry of V can be expressed uniquely as a linear transformation followed by a translation.*

(2) $\mathrm{Isom}(V) \cong V \rtimes \mathcal{O}(V)$.

Proof. The uniqueness assertion in (1) is clear, since the image of 0 under the given isometry determines the translation whose existence is asserted. By composing an arbitrary isometry φ of V with the translation $x \mapsto x - \varphi(0)$, we reduce the proof of (1) to showing that an isometry φ which preserves the origin is linear, i.e., $\varphi(\lambda x) = \lambda\varphi(x)$ and $\varphi(x+y) = \varphi(x)+\varphi(y)$ for all $x, y \in \mathbb{R}^n$ and all $\lambda \in \mathbb{R}$. Mazur and Ulam [MaUl32] proved that any isometry of a normed space that preserves the origin is linear (see [Bana32, XI.2] for a clearer proof). We shall give the proof only in the simple case where V is uniquely geodesic.

The first of these equalities follows easily from the fact that the only geodesic lines through $0 \in V$ are the 1-dimensional subspaces of V, hence φ must map each such line isometrically onto another such line. The second equality follows from the fact that φ, as an isometry, must send the midpoint of a line segment to the midpoint of the image segment (because such segments are the unique geodesics in V). The midpoint of the unique geodesic segment joining x to y is $(x + y)/2$, hence $\varphi(x + y) = 2\varphi((x + y)/2) = 2\varphi(x) + \varphi(y))/2) = \varphi(x) + \varphi(y)$.

The decomposition in (1) allows us to express any $\varphi \in \mathrm{Isom}(V)$ as the product of a linear transformation A_φ and a translation $x \mapsto x + \varphi$. We claim that the map from Isom(V) to $V \rtimes \mathcal{O}(V)$ given by $\varphi \mapsto (\varphi(0), A_\varphi)$ is an isomorphism. The uniqueness assertion in (1) assures that this map is well-defined and injective. It is clearly surjective, so it only remains to check that it is a homomorphism. Given $\phi, \psi \in \mathrm{Isom}(V)$, the action of $\phi\psi$ on V is:

$$\begin{aligned}\phi\psi(x) &= \phi(A_\psi(x) + \psi(0)) \\ &= A_\phi(A_\psi(x) + \psi(0)) + \phi(0) = A_\phi A_\psi(x) + [A_\phi\psi(0) + \phi(0)].\end{aligned}$$

So the image of $\phi\psi$ in $V \rtimes \mathcal{O}(V)$ is $(\phi(0)+A_\phi\psi(0), A_\phi A_\psi) = (\phi(0), A_\phi)\cdot(\psi(0), A_\psi)$. □

Of course, not all linear transformations of a normed real vector space are isometries. As we noted earlier, in the case of Euclidean space an n-by-n matrix A acts as an isometry if and only if $A^t A = I$, that is $A \in O(n)$. Thus we have an alternative proof of 2.24(1).

4.13 Corollary. $\mathrm{Isom}(\mathbb{E}^n) \cong \mathbb{R}^n \rtimes O(n)$.

$\mathbb{S}^n$ is the set of vectors in $\mathbb{R}^{n+1}$ of Euclidean norm 1. The distance between two points $A, B \in \mathbb{S}^n$ is the Euclidean angle between $[0, A]$ and $[0, B]$ at $0 \in \mathbb{R}^{n+1}$. Since this angle is arcsin of one half the Euclidean distance from A to B, any isometry of $\mathbb{E}^{n+1}$ which fixes 0 must restrict to an isometry of $\mathbb{S}^n$. In particular this is the case for the action of $O(n)$. We obtain an alternative proof of 2.24(2) by showing that these are the only isometries of $\mathbb{S}^n$:

4.14 Proposition. $\mathrm{Isom}(\mathbb{S}^n) = O(n+1)$.

Proof. We must show that every isometry of $\mathbb{S}^n \subseteq \mathbb{E}^{n+1}$ can be extended to an isometry of $\mathbb{E}^{n+1}$. Given an isometry φ of $\mathbb{S}^n$, the desired extension $\tilde{\varphi}$ to $\mathbb{E}^{n+1}$ is given by writing each $x \in \mathbb{E}^{n+1}$ in polar coordinates $t_x v_x$, where v_x is a unit vector and $t_x \geq 0$ is a number; one defines $\tilde{\varphi}(tv) = t\varphi(v)$. Hence $t_{\tilde{\varphi}(x)} = t_x$ for all $x \in \mathbb{R}^{n+1}$. Furthermore, since φ is an isometry of $\mathbb{S}^n$, if we write $\alpha(x, y)$ for the angle at the origin between the line segments $[0, x]$ and $[0, y]$, then for all $x, y \in \mathbb{R}^{n+1}$ we have $\alpha(\tilde{\varphi}(x), \tilde{\varphi}(y)) = \alpha(\varphi(v_x), \varphi(v_y)) = \alpha(v_x, v_y) = \alpha(x, y)$. But then, applying the law of cosines to the Euclidean triangle with vertices $0, x, y$, we can express the Euclidean distance between x and y in terms of the φ-invariant quantities t_x, t_y and $\alpha(x, y)$:

$$d(x, y)^2 = t_x^2 + t_y^2 - 2t_x t_y \cos\alpha(x, y).$$

□

4.15 Exercises

(1) Show that every linear transformation of a pre-Hilbert space V which preserves orthogonality is a homothety, i.e., there exists a constant $\lambda > 0$ such that $d(\phi(v), \phi(w)) = \lambda d(v, w)$ for all $v, w \in V$.

(2) Prove the Mazur-Ulam theorem [MaUl32] for finite dimensional spaces.

ℓ^p Spaces

In this paragraph we describe an important class of uniquely geodesic normed spaces that are not Hilbert spaces. Given $p \geq 1$ and a vector $x = (x_1, \dots, x_n)$ in $\mathbb{R}^n$, the ℓ^p norm of x is defined by:

$$\|x\|_p := \left(\sum_{i=1}^{n} |x_i|^p\right)^{1/p}.$$

The associated metric is denoted d_p. It is obvious that $\|\cdot\|_1$ is a norm, but less obvious for $p > 1$.

4.16 Proposition. *For every $n \in \mathbb{N}$ and every real number $p > 1$, the map $x \mapsto \|x\|_p$ defines a norm on $\mathbb{R}^n$. In particular, for any two non-zero vectors $x, y \in \mathbb{R}^n$, we have*

$$\|x+y\|_p \leq \|x\|_p + \|y\|_p.$$

Furthermore, one has equality in this expression if and only if x is a positive multiple of y, hence for all $1 < p < \infty$, *the metric space* $(\mathbb{R}^n, d_p)$ *associated to the* ℓ^p *norm is uniquely geodesic.*

Proof. To check the triangle inequality, one considers the real number $q > 1$ defined by

$$1/p + 1/q = 1$$

We claim that for any real numbers $a \geq 0, b \geq 0$, one has

$$a^{1/p} b^{1/q} \leq a/p + b/q \tag{1}$$

with equality if and only if $a = b$. The non-trivial case is when $a > 0$ and $b > 0$. Then, since the function $\log : [0, \infty) \to \mathbb{R}$ is strictly concave, we have

$$\log(a/p + b/q) \geq 1/p \,\log a + 1/q \,\log b$$

with equality if and only if $a = b$. Composing both sides of this inequality with the strictly increasing function exp, we get $a/p + b/q \geq a^{1/p} b^{1/q}$, with equality if and only if $a = b$.

Recall that, given two vectors $x = (x_1, \ldots, x_n)$ and $y = (y_1, \ldots, y_n)$ in $\mathbb{R}^n$, we have the Hölder inequality

$$\sum_{i=1}^{n} |x_i|\,|y_i| \leq \|x\|_p \|y\|_q. \tag{2}$$

We shall verify this inequality for the non-trivial case $x \neq 0, y \neq 0$. Applying inequality (1) to $a_i = \frac{|x_i|^p}{\|x\|_p^p}$ and $b_i = \frac{|y_i|^q}{\|y\|_q^q}$, we have

$$\frac{|x_i|}{\|x\|_p} \frac{|y_i|}{\|y\|_q} \leq 1/p \frac{|x_i|^p}{\|x\|_p^p} + 1/q \frac{|y_i|^q}{\|y\|_q^q}.$$

Summing over $i = 1, \ldots, n$, we then get the Hölder inequality and see that equality holds if and only if $a_i = b_i$ for each i.

We are now ready to prove the triangle inequality for d_p. One has for each $i = 1, \ldots, n$

$$|x_i + y_i|^p \leq |x_i|\,|x_i + y_i|^{p-1} + |y_i|\,|x_i + y_i|^{p-1}. \tag{3-i}$$

From the Hölder inequality we get

$$\sum_{i=1}^{n} |x_i|\,|x_i + y_i|^{p-1} \leq \left(\sum_{i=1}^{n} |x_i|^p\right)^{1/p} \left(\sum_{i=1}^{n} |x_i + y_i|^{(p-1)q}\right)^{1/q} \tag{4}$$

$$\sum_{i=1}^{n} |y_i|\,|x_i + y_i|^{p-1} \leq \left(\sum_{i=1}^{n} (|y_i|^p\right)^{1/p} \left(\sum_{i=1}^{n} |x_i + y_i|^{(p-1)q}\right)^{1/q} \tag{5}$$

hence, as $(p-1)q = p$,

$$\sum_{i=1}^{n} |x_i + y_i|^p \leq \left(\sum_{i=1}^{n} |x_i|^p\right)^{1/p} + \left(\sum_{i=1}^{n} |y_i|^p\right)^{1/p} \left(\sum_{i=1}^{n} |x_i + y_i|^p\right)^{1/q}$$

which leads to the triangle inequality.

In the above triangle inequality, equality can occur if and only equality occurs simultaneously in (3-i), (4) and (5). Equality in (3-i) implies that x_i and y_i have the same sign and that if x and y are non-zero then $x + y$ is also non-zero. Equality in (4) and (5) implies that

$$\frac{|x_i|^p}{\|x\|_p^p} = \frac{|x_i + y_i|^q}{\|x_i + y_i\|_q^q} = \frac{|y_i|^p}{\|y\|_p^p}$$

for each $i = 1, \ldots, n$, hence x has to be a positive multiple of y. Hence, by (1.6), $(\mathbb{R}^n, d_p)$ is uniquely geodesic for all $1 < p < \infty$. □

4.17 The Banach Spaces $\ell^p(S)$. Let S be a set. Let $\ell^p(S)$ denote the set of maps $x : S \to R$ such that $\sum_S |x(s)|^p$ is finite (cf. 4.6). Define $\|x\|_p = (\sum_S |x(s)|^p)^{1/p}$. We claim that, when equipped with the term-wise operations of addition and scalar multiplication, $\ell^p(S)$ is a real vector space. That it is closed under scalar multiplication is clear. In order to see that if $x, y \in \ell^p(S)$ then $x + y \in \ell^p(S)$, one uses the triangle inequality in $(\mathbb{R}^n, d_p)$ to bound the sums over finite subsets $C \subset S$:

$$\left(\sum_{s\in C} |x(s) + y(s)|^p\right)^{1/p} \leq \left(\sum_{s\in C} |x(s)|^p\right)^{1/p} + \left(\sum_{s\in C} |y(s)|^p\right)^{1/p} \leq \|x\|_p + \|y\|_p.$$

This establishes the triangle inequality showing $\|\cdot\|$ to be a norm on $\ell^p(S)$. By arguing as in the preceding proposition one see that the associated metric space is uniquely geodesic if and only if $p > 1$.

4.18 Proposition. *For every set S and every real number $1 \leq p < \infty$, the metric space associated to $\ell^p(S)$ is complete. In other words, $\ell^p(S)$ is a Banach space.*

Proof. Modulo obvious changes of notation, the proof of Proposition 4.7 applies *verbatim*. □

Chapter I.5 Some Basic Constructions

In this section we describe some basic constructions that allow one to manufacture interesting new metric spaces out of more familiar ones. We consider whether such properties as being a geodesic space are preserved by these constructions. Each of the constructions which we consider here will play a significant role in Part II.

Products

5.1 Definition. The *product* of two metric spaces X_1 and X_2 is the set $X_1 \times X_2$ endowed with the metric:

$$d((x_1, x_2), (y_1, y_2))^2 = d(x_1, y_1)^2 + d(x_2, y_2)^2.$$

For example, $\mathbb{E}^n \times \mathbb{E}^m$ is isometric to $\mathbb{E}^{n+m}$.

In the proof of the next proposition we shall need the following characterization of geodesic segments.

5.2 Exercise. Let X be a metric space. A continuous path $c : I \to X$ is a linearly reparameterized geodesic if and only if $d(c(s), c(t)) = 2\, d(c(s), c((s+t)/2))$ for all $s, t \in I$; in other words $c((s+t)/2)$ is a midpoint of $c(s)$ and $c(t)$.

5.3 Proposition. *Let X be the product of the metric spaces X_1 and X_2.*

(0) *X is complete if and only if both X_1 and X_2 are complete.*

(1) *X is a length space if and only if both X_1 and X_2 are length spaces.*

(2) *X is a geodesic space if and only if both X_1 and X_2 are geodesic spaces.*

(3) *Let $I \subseteq \mathbb{R}$ be a compact interval. The map $c : I \to X$ given by $t \mapsto (c_1(t), c_2(t))$ is a linearly reparameterized geodesic if and only if the maps c_1 and c_2 are both linearly reparameterized geodesics.*

(4) *An isometry $\gamma \in \mathrm{Isom}(X)$ decomposes as a product (γ_1, γ_2), with $\gamma_1 \in \mathrm{Isom}(X_1)$ and $\gamma_2 \in \mathrm{Isom}(X_2)$, if and only if, for every $x_1 \in X_1$ there exists a point denoted $\gamma_1(x_1) \in X_1$ such that $\gamma(\{x_1\} \times X_2) = \{\gamma_1(x_1)\} \times X_2$.*

Proof. Assertion (0) is obvious. We first prove (1). The projection of X onto X_1 is distance-decreasing and hence length-decreasing, and it restricts to an isometry on slices of the form $X_1 \times \{x_2\}$. From this observation it follows easily that if X is a length space then so too is X_1 (and similarly X_2). For the converse, given $(x_1, x_2), (y_1, y_2) \in X$ and $\varepsilon > 0$, we fix paths $c_1 : [0, 1] \to X_1$ and $c_2 : [0, 1] \to X_2$ joining x_1 to y_1 and x_2 to y_2 respectively, chosen so that $l(c_1)^2 < d(x_1, y_1)^2 + \varepsilon^2/2$ and $l(c_2)^2 < d(x_2, y_2)^2 + \varepsilon^2/2$, and parameterized proportional to arc length. Consider the path $c : [0, 1] \to X$ defined by $c(t) = (c_1(t), c_2(t))$. This path joins (x_1, x_2) to (y_1, y_2) and, according to (1.20), has length $\sup_{n>0} \sum_{i=0}^{n-1} d(c(i/n), c((i+1)/n))$. Because c_1 and c_2 are parameterized proportional to arc length, for all $n > i \geq 0$ we have that

$$\begin{aligned} n^2\, d(c(i/n), c((i+1)/n))^2 & \\ &= n^2\, d(c_1(i/n), c_1((i+1)/n))^2 + n^2\, d(c_2(i/n), c_2((i+1)/n))^2 \\ &\leq l(c_1)^2 + l(c_2)^2 \\ &< d(x_1, y_1)^2 + \varepsilon^2/2 + d(x_2, y_2)^2 + \varepsilon^2/2 \\ &= d((x_1, x_2), (y_1, y_2))^2 + \varepsilon^2. \end{aligned}$$

Hence, $n\, d(c(i/n), c((i+1)/n)) < d((x_1, x_2), (y_1, y_2)) + \varepsilon$ for all $n > i \geq 0$, so $l(c) < d((x_1, x_2), (y_1, y_2)) + \varepsilon$. Since ε was arbitrary, this shows that X is a length space.

Assertion (2) follows immediately from (3). A trivial calculation shows that if c_1 and c_2 are linearly reparameterized geodesics then so too is c. For the converse we apply criterion (5.2). To this end, given $t, s \in I$, we let $x = (x_1, x_2) := (c_1(t), c_2(t))$ and $y = (y_1, y_2) := (c_1(s), c_2(s))$, and we denote the midpoint $(c_1((t+s)/2), c_2((t+s)/2))$ by $m = (m_1, m_2)$. We must show that

$$d(m_1, y_1) = \frac{1}{2}\, d(x_1, y_1).$$

If two numbers $a, b \in [0, 1]$ satisfy $a + b = 1$, then $a^2 + b^2 \geq 1/2$, with equality if and only if $a = b = 1/2$. Combining this with the triangle inequality, we have

$$\frac{1}{2}\, d(x_1, y_1)^2 \leq d(x_1, m_1)^2 + d(m_1, y_1)^2,$$

with equality if and only if $\frac{1}{2} d(x_1, y_1) = d(x_1, m_1) = d(m_1, y_1)$. Adding this to the similar inequality for x_2, m_2, y_2, we obtain:

$$\frac{1}{2}\, \{d(x_1, y_1)^2 + d'(x_2, y_2)^2\} \leq d(x_1, m_1)^2 + d(x_2, m_2)^2 + d(m_1, y_1)^2 + d(m_2, y_2)^2.$$

The left hand side of this inequality is equal to $\frac{1}{2} d(x, y)^2$ and the right hand side is equal to $d(x, m)^2 + d(y, m)^2$. But since c is a linearly reparameterized geodesic, $d(x, m) = d(y, m) = \frac{1}{2} d(x, y)$. Thus all of the above inequalities must actually hold with equality, and in particular $d(m_1, y_1) = \frac{1}{2} d(x_1, y_1)$.

The necessity of the condition given in (4) is clear. In order to see that it is sufficient, we assume that γ is an isometry of $X_1 \times X_2$ mapping $\{x_1\} \times X_2$ onto

$\{\gamma_1(x_1)\} \times X_2$ for each $x_1 \in X_1$. Notice that the map $x_1 \mapsto \gamma_1(x_1)$ must be an isometry of X_1. Indeed its surjectivity follows from that of γ, and it is distance preserving because for all $x_1, y_1 \in X_1$

$$d(x_1, y_1) = \inf\{d\big((x_1, x_2), (y_1, y_2)\big) \mid x_2, y_2 \in X_2\}.$$

If we now show that the map $x_2 \mapsto \gamma_2(x_2)$ of X_2 to X_2 defined by $(\gamma_1(x_1), \gamma_2(x_2)) = \gamma(x_1, x_2)$ is independent of x_1, then we can apply the preceding argument to see that it too must be an isometry.

Suppose that γ maps (x_1, x_2) to $(\gamma_1(x_1), \gamma_2(x_2))$ and (y_1, x_2) to $(\gamma_1(y_1), \gamma_2'(x_2))$. We have to check that $\gamma_2(x_2) = \gamma_2'(x_2)$. But,

$$\begin{aligned} d(x_1, y_1)^2 &= d((x_1, x_2), (y_1, x_2))^2 \\ &= d(\gamma(x_1, x_2), \gamma(y_1, x_2))^2 \\ &= d(\gamma_1(x_1), \gamma_1(y_1))^2 + d(\gamma_2(x_2), \gamma_2'(x_2))^2 \\ &= d(x_1, y_1)^2 + d(\gamma_2(x_2), \gamma_2'(x_2))^2. \end{aligned}$$

Hence $d(\gamma_2(x_2), \gamma_2'(x_2)) = 0$, as required. □

5.4 Remark. Given any finite number of metric spaces $(X_1, d_1), \dots, (X_n, d_n)$, one can consider their product $\prod_{i=1}^n X_i$ with the metric in which the distance from $x = (x_1, \dots, x_n)$ to $y = (y_1, \dots, y_n)$ is given by the formula

$$d(x, y)^2 = \sum_{i=1}^{n} d_i(x_i, y_i)^2.$$

If $1 \leq m < n$ then the natural bijection $\prod_{i=1}^m X_i \times \prod_{j=m+1}^n X_j \to \prod_{k=1}^n X_k$ is an isometry. And by induction on n (or directly) one can prove the analogue of (5.3) in this more general setting.

5.5 Exercises

(1) One might describe the space constructed in (5.4) as the ℓ^2-product of the spaces (X_i, d_i). More generally, for any $1 \leq p \leq \infty$, one can define an ℓ^p-metric on $\prod_{i=1}^n X_i$ by the formula

$$d(x, y)^p = \sum_{i=1}^{n} d_i(x_i, y_i)^p.$$

Using the properties of the ℓ^p-norm in $\mathbb{R}^n$, prove that d is a metric and that the analogues of (5.3) parts (0) to (3) are valid if $1 < p < \infty$.

(2) Consider now the product $\prod_{i \in \mathbb{N}} X_i$ of countably many metric spaces (X_i, d_i). Prove that the function which associates to $x = (x_i),\ y = (y_i) \in \prod_{i \in \mathbb{N}} X_i$ the number $d(x, y) \in [0, \infty]$ defined by

$$d(x, y) = \sum_{i=1}^{\infty} d_i(x_i, y_i)^2$$

is a metric on $\prod_{i\in\mathbb{N}} X_i$. In order to get a metric with values in $[0,\infty)$ one must choose a sequence of basepoints $x_i \in X_i$ and consider only those sequences (y_i) that are a finite distance from (x_i). This subspace is called *the ℓ^2-product of the pointed spaces* $(X_i; x_i)$. Which parts of (5.3) have valid analogues in this setting?

Check that if $X_i = \mathbb{R}$ for each i, and we choose $0 \in X_i$ as the basepoint for each i, then then ℓ^2-product is isometric to $\ell^2(\mathbb{N})$.

κ-Cones

The purpose of this section is to describe an important construction which is essentially due to Berestovskii [Ber83] (see the article by Alexandrov, Berestovskii and Nikolaev [AleBN86]). In order to offset the rather technical appearance of this construction, we first recall the law of cosines in the model space M^n_κ (see 2.12); this indicates the origins of the definition that follows. We also draw the reader's attention to (5.8).

Let γ be the angle between two geodesic segments $[x_0, x_1]$ and $[x_0, x_2]$ in M^n_κ. Let $a = d(x_0, x_1)$, $b = d(x_0, x_2)$ and $c = d(x_1, x_2)$. If $\kappa > 0$ then assume that $a + b + c < 2D_\kappa$. With this notation:

for $\kappa = 0$:
$$c^2 = a^2 + b^2 - 2ab\cos(\gamma)$$

for $\kappa < 0$:
$$\cosh(\sqrt{-\kappa}c) = \cosh(\sqrt{-\kappa}a)\cosh(\sqrt{-\kappa}b) - \sinh(\sqrt{-\kappa}a)\sinh(\sqrt{-\kappa}b)\cos(\gamma)$$

for $\kappa > 0$:
$$\cos(\sqrt{\kappa}c) = \cos(\sqrt{\kappa}a)\cos(\sqrt{\kappa}b) + \sin(\sqrt{\kappa}a)\sin(\sqrt{\kappa}b)\cos(\gamma).$$

5.6 Definition (The κ-Cone over a Metric Space). Given a metric space Y and a real number κ, the *κ-cone* $X = C_\kappa Y$ over Y is the metric space defined as follows. If $\kappa \le 0$ then, as a set, X is the quotient of $[0,\infty) \times Y$ by the equivalence relation given by: $(t, y) \sim (t', y')$ if $(t = t' = 0)$ or $(t = t' > 0$ and $y = y')$. If $\kappa > 0$ then X is the quotient of $[0, D_\kappa/2] \times Y$ by the same relation. The equivalence class of (t, y) is denoted ty. The class of $(0, y)$ is denoted 0 and is called the *vertex of the cone*, or the *cone point*.

Let $d_\pi(y, y') := \min\{\pi, d(y, y')\}$. The distance between two points $x = ty$ and $x' = t'y'$ in X is defined so that $d(x, x') = t$ if $x' = 0$ and so that $\cos(\angle_0^{(\kappa)}(x, x')) = d_\pi(y, y')$ if $t, t' > 0$. This is achieved by defining:
for $\kappa = 0$:
$$d(x, x')^2 = t^2 + t'^2 - 2tt'\cos(d_\pi(y, y')),$$
for $\kappa < 0$:
$$\cosh(\sqrt{-\kappa}d(x, x')) = \cosh(\sqrt{-\kappa}t)\cosh(\sqrt{-\kappa}t') - \sinh(\sqrt{-\kappa}t)\sinh(\sqrt{-\kappa}t')\cos(d_\pi(y, y')),$$

and for $\kappa > 0$:

$$d(x, x') \leq D_\kappa \quad \text{and}$$
$$\cos(\sqrt{\kappa}d(x, x')) = \cos(\sqrt{\kappa}t)\cos(\sqrt{\kappa}t') + \sin(\sqrt{\kappa}t)\sin(\sqrt{\kappa}t')\cos(d_\pi(y, y')).$$

For $\kappa = 0$, the κ-cone over Y is also called the Euclidean cone over Y. We shall prove in Proposition 5.9(1) that these formulae do indeed define a metric on X (it is only the triangle inequality that is not obvious).

5.7 Remarks. In the notation of the preceding definition: if $t, t' > 0$, then $d(x, x') = t + t'$ if and only if $d(y, y') \geq \pi$. This observation has two consequences. First, if one replaces the given metric d on Y by the metric d_π, then the induced distance function on $X = C_\kappa Y$ remains unaltered. Notice that one can recover the metric d_π on Y from the distance function on X, because if $x = ty,\ x' = t'y'$ with $t, t' > 0$, then $d_\pi(y, y') = \angle_0^{(\kappa)}(x, x')$. It follows that, since the length metrics $\overline{d}$ and $\overline{d}_\pi$ agree on Y, one can recover the metric on Y from the distance function on X under the added hypothesis that Y is a length space.

Secondly, a path $c : [-t_1, t_2] \to C_\kappa Y$, where $t_1, t_2 > 0$ and $c(0) = 0$, is a geodesic if and only if there exist $y_1, y_2 \in Y$ such that $d(y_1, y_2) \geq \pi$, $c(-s) = sy_1$ for $s \in [0, t_1]$ and $c(s) = sy_2$ for $s \in [0, t_2]$.

The above definition is motivated in part by the following example.

5.8 Proposition. *If Y is the sphere $\mathbb{S}^{n-1}$, then $X = C_\kappa Y$ is isometric to M^n_κ if $\kappa \leq 0$, and a closed ball of radius $D_\kappa/2$ in M^n_κ if $\kappa > 0$.*

Proof. Let o be a point in M^n_κ, identify $\mathbb{S}^{n-1}$ to the unit sphere in the tangent space $T_oM^n_\kappa$ and let $\exp_o : T_o(M^n_\kappa) \to M^n_\kappa$ be the exponential map (see 6.16). Then the map $C_\kappa\mathbb{S}^{n-1} \to M^n_\kappa$ which sends the class of (t, y) to $\exp_o(ty)$ is an isometry onto M^n_κ if $\kappa \leq 0$, and an isometry onto the closed hemisphere centered at o if $\kappa > 0$. This follows from the definition of the metric on $C_\kappa\mathbb{S}^{n-1}$ and the law of cosines in M^n_κ. □

5.9 Proposition.

(1) *The formulae in (5.6) define a metric on $X = C_\kappa Y$.*

(2) *Y is complete if and only if X is complete.*

Proof. (1) Consider three points $x_i = t_iy_i$, $i = 1, 2, 3$ in X. We want to prove that $d(x_1, x_3) \leq d(x_1, x_2) + d(x_2, x_3)$. This is easy to check if one of the t_i is 0, because in that case the assertion follows from the triangle inequality in M^2_κ.

Assume that $t_i > 0$ for $i = 1, 2, 3$. We consider two cases.

Case 1: $d(y_1, y_2) + d(y_2, y_3) < \pi$.

It follows from the triangle inequality in Y that $d(y_1, y_3) < \pi$. Consider three points $\overline{y}_1, \overline{y}_2, \overline{y}_3$ in $\mathbb{S}^2$ such that $d(y_i, y_j) = d(\overline{y}_i, \overline{y}_j)$, for $i, j \in \{1, 2, 3\}$. As in (5.8), the subcone $C_\kappa\{y_1, y_2, y_2\}$ is isometric to the subcone $C_\kappa\{\overline{y}_1, \overline{y}_2, \overline{y}_3\} \subseteq C_\kappa(\mathbb{S}^2) \subseteq M^3_\kappa$, so we may appeal to the triangle inequality in M^3_κ.

Case 2: $d(y_1, y_2) + d(y_2, y_3) \geq \pi$.

Consider three points $\overline{y}_1, \overline{y}_2, \overline{y}_3$ occurring in the given order on the circle $\mathbb{S}^1$ such that $d(\overline{y}_1, \overline{y}_2) = d_\pi(y_1, y_2)$ and $d(\overline{y}_2, \overline{y}_3) = d_\pi(y_2, y_3)$. We identify $C_\kappa\mathbb{S}^1$ to M^2_κ if $\kappa \leq 0$ and to a closed hemisphere in M^2_κ if $\kappa > 0$. If $\overline{x}_i = t_i\overline{y}_i$ for $i = 1, 2, 3$, then $d(x_1, x_2) = d(\overline{x}_1, \overline{x}_2)$ and $d(x_2, x_3) = d(\overline{x}_2, \overline{x}_3)$ and $d(x_1, x_3) \leq t_1 + t_3$. As $d(\overline{y}_1, \overline{y}_2) + d(\overline{y}_2, \overline{y}_3) \geq \pi$, by Alexandrov's lemma (2.16) we have $t_1 + t_3 \leq d(\overline{x}_1, \overline{x}_2) + d(\overline{x}_2, \overline{x}_3)$. Therefore, $d(x_1, x_3) \leq d(x_1, x_2) + d(x_2, x_3)$.

(2) Assume that X is complete. Given a Cauchy sequence (y_n) in Y, let $x_n = t_0 y_n$, where $t_0 > 0$ and $t_0 \leq D_\kappa$ if $\kappa > 0$. From the definition of the metric on X, one sees immediately that (x_n) is a Cauchy sequence in X, and hence converges to some point a distance t_0 from 0, let us say $t_0 y$. Then, $y = \lim y_n$.

Conversely, assume that Y is complete. Let $x_n = t_n y_n$ be a Cauchy sequence in X which does not converge to $0 \in X$. Since it is Cauchy, this sequence is contained in a ball of finite radius about $0 \in X$, in other words the t_n are bounded. Passing to a subsequence, we may assume that they converge, to $t \neq 0$ say. For t_n, t_m sufficiently close to t, one can estimate $d(y_n, y_m)$ in terms of $d(x_n, x_m)$, thus (y_n) is Cauchy, and hence converges, to y say. It follows that a subsequence of the Cauchy sequence $(t_n y_n)$ converges to ty. □

5.10 Proposition (Characterization of Geodesics). *Let $x_1 = t_1 y_1$ and $x_2 = t_2 y_2$ be elements of $C_\kappa Y$.*

(1) *If $t_1, t_2 > 0$ and $d(y_1, y_2) < \pi$, then there is a bijection between the set of geodesic segments joining y_1 to y_2 in Y and the set of geodesic segments joining x_1 to x_2 in X.*

(2) *In all other cases, there is a geodesic segment in X joining x_1 to x_2; this segment is unique, except possibly in the case where $\kappa > 0$ and $d(x_1, x_2) = D_\kappa$.*

(3) *Any geodesic segment joining x_1 to x_2 is contained in the closed ball of radius $\max\{t_1, t_2\}$ about the vertex $0 \in C_\kappa Y$.*

Proof. (1) Consider a geodesic segment $[y_1, y_2] \subseteq Y$ joining y_1 to y_2. The subcone $C_\kappa[y_1, y_2] \subseteq C_\kappa Y$ is isometric to a sector in M^2_κ which is convex, hence there is a geodesic segment joining x_1 to x_2 which is contained in $C_\kappa[y_1, y_2]$.

For the converse, we consider a geodesic segment $[x_1, x_2] \subseteq X$ joining $x_1 = t_1 y_1$ to $x_2 = t_2 y_2$; let $x = ty$ be an arbitrary point of $[x_1, x_2]$. Notice that $t > 0$, for otherwise $d(x_1, x_2) = t_1 + t_2$, and this would imply $d(y_1, y_2) \geq \pi$, contrary to the hypothesis. Therefore the projection $ty \mapsto y$ of $[x_1, x_2]$ into Y is well-defined; we want to prove that the image of $[x_1, x_2]$ under this projection is a geodesic segment joining y_1 to y_2 in Y. For this it is sufficient to prove that $d(y_1, y) + d(y, y_2) = d(y_1, y_2)$.

To check this equality, we consider in M^2_κ comparison triangles $\overline{\Delta}_1 = \Delta(\overline{0}, \overline{x}, \overline{x}_1)$ for $(0, x, x_1)$ and $\overline{\Delta}_2 = \Delta(\overline{0}, \overline{x}, \overline{x}_2)$ for $(0, x, x_2)$, and we assume that these triangles are arranged so that $\bar{x}_1$ and $\bar{x}_2$ are on opposite sides of the common segment $[\bar{0}, \bar{x}]$ (cf. Alexandrov's lemma). Note that the vertex angle of $\overline{\Delta}_1$ at $\bar{0}$ is equal to $d(y_1, y)$ and the vertex angle of $\overline{\Delta}_2$ at $\bar{0}$ is $d(y, y_2)$. As $d(\bar{x}_1, \bar{x}) + d(\bar{x}, \bar{x}_2) = d(x_1, x_2) < t_1 + t_2 = d(\bar{0}, \bar{x}_1) + d(\bar{0}, \bar{x}_2)$, we have $d(y_1, y) + d(y, y_2) < \pi$. Let $\tilde{\Delta}(\tilde{0}, \tilde{x}_1, \tilde{x}_2)$ be a comparison triangle in M^2_κ for $(0, x_1, x_2)$. The angle at $\tilde{0}$ is $d(y_1, y_2)$. By the law of cosines 2.13, since $d(\bar{x}_1, \bar{x}_2) \leq d(\tilde{x}_1, \tilde{x}_2)$, we have $d(y_1, y) + d(y, y_2) \leq d(y_1, y_2)$. The reverse inequality is simply the triangle equality in Y, hence $d(y_1, y_2) = d(y_1, y) + d(y, y_2)$.

(2) If we are not in case (1), then $d(x_1, x_2) = t_1 + t_2$ and the path $c : [0, t_1 + t_2] \to X$ that sends $t \in [0, t_1]$ to $(t_1 - t)y_1$ and $t \in [t_1, t_2]$ to $(t - t_1)y_2$ is a geodesic path joining x_1 to x_2. We must prove that this is the only geodesic segment joining x_1 to x_2, if $t_1 + t_2 < \pi/\kappa$. If $t_1 = 0$ or $t_2 = 0$, it is easy to see that this is true, so we need only consider the case where $t_1, t_2 > 0$ and $d(y_1, y_2) \geq \pi$.

It is sufficient to prove that if $x = ty$ satisfies $d(x_1, x) + d(x, x_2) = d(x_1, x_2)$, and $d(x_1, x) < t_1$, then $y = y_1$. As in the proof of (1), we construct the two comparison triangles $\overline{\Delta}_1$ and $\overline{\Delta}_2$ in M^2_κ. The vertex angles at $\bar{0}$ of $\overline{\Delta}_1$ and $\overline{\Delta}_2$ are respectively $d_\pi(y_1, y)$ and $d_\pi(y, y_2)$, and so the sum of these two angles is not less than π. But $d(\bar{x}_1, \bar{0}) + d(\bar{0}, \bar{x}_2) = d(x_1, x_2) = d(\bar{x}_1, \bar{x}) + d(\bar{x}, \bar{x}_2)$ (and this sum is smaller than $2D_\kappa$ if $\kappa > 0$). Such an equality is only possible if both of the triangles $\overline{\Delta}_1$ and $\overline{\Delta}_2$ are degenerate, in which case $x = tx_1$.

Part (3) follows from the convexity of balls in M^2_κ (see the beginning of the proof of (1)). □

As with the other constructions considered in this section, it is natural to ask whether the cone over a geodesic space is again a geodesic space; conversely, one might ask whether knowing that the cone over a space is geodesic, one can deduce that the space itself is geodesic. In the latter case, one sees immediately that the answer is no, because as we noted earlier, the isometry type of $C_\kappa Y$ is not changed if one truncates the metric on Y at any value $\geq \pi$. However, one can obtain a positive result by taking account of this phenomenon.

Recall that a subset Y of a metric space X is said to be *convex* it every pair of points in Y can be joined by a geodesic segment and every such geodesic segment is contained in Y.

5.11 Corollary. *Let X be the κ-cone $C_\kappa Y$ over a metric space Y. Then the following conditions are equivalent:*

(1) *X is a geodesic space;*

(2) *any ball in X centred at the vertex of the cone is convex;*

(3) *there is an open ball centred at the vertex of the cone which is convex;*

(4) *Y is π-geodesic.*

The equivalence of (1) to (4) remains valid if one replaces "geodesic" by "uniquely geodesic" and "convex" by "convex and uniquely geodesic".

Proof. Part (3) of the preceding proposition shows that (1) $\Longrightarrow$ (2) $\Longrightarrow$ (3). If we know that a ball of radius $2t > 0$ centred at the vertex of the cone is convex, then

$y_1, y_2 \in Y$ with $d(y_1, y_2) < \pi$, then any points of the form $x_1 = ty_1$ and $x_2 = ty_2$ can be joined by a geodesic in X and hence, by part (1) of the preceding proposition, there is a geodesic joining y_1 to y_2 in Y. Thus (3) $\Longrightarrow$ (4). Part (1) of the proposition also shows that the ball of radius t about the cone point is uniquely geodesic if and only Y is uniquely π-geodesic.

Assume that Y is π-convex. Consider two points $x_1 = t_1y_1$ and $x_2 = t_2y_2$ of X. If $t_1 > 0$, $t_2 > 0$ and $d(y_1, y_2) < \pi$, then according to part (1) of the preceding proposition, each geodesic segment joining y_1 to y_2 corresponds to a geodesic joining x_1 to x_2. And in all other cases, $d(x_1, x_2) = t_1 + t_2$ and part (2) of the preceding applies. Thus (4) $\Longrightarrow$ (1). □

5.12 Exercises

(1) Show that if Y is a length space then $C_\kappa Y$ is a length space.

(2) Let (Y, d) and (Y', d') be length spaces. Prove that every isometry $f : C_\kappa Y \to C_\kappa Y'$ that sends the cone point to the cone point, induces an isometry $(Y, d) \to (Y, d')$.

Spherical Joins

The product construction allows one to obtain $\mathbb{E}^{n+m}$ from the pair of spaces $(\mathbb{E}^n, \mathbb{E}^m)$. In this paragraph, we describe a method for combining pairs of metric spaces which when applied to the pair $(\mathbb{S}^n, \mathbb{S}^m)$ yields $\mathbb{S}^{n+m+1}$. This construction, which is called the *spherical join*, is due to Berestovskii [Ber83] (see also [CD93]).

5.13 Definition. Let (Y_1, d^1) and (Y_2, d^2) be two metric spaces. As a set, their *spherical join* $Y_1 * Y_2$ is $[0, \pi/2] \times Y_1 \times Y_2$ modulo the equivalence relation which identifies (θ, y_1, y_2) to (θ', y_1', y_2') whenever $[(\theta = \theta' = 0)$ and $y_1 = y_1']$ or $[(\theta = \theta' = \pi/2)$ and $y_2 = y_2']$ or $[\theta = \theta' \notin \{0, \pi/2\}$ and $y_1 = y_1', y_2 = y_2']$. The equivalence class of (θ, y_1, y_2) will normally be denoted $(\cos\theta\; y_1 + \sin\theta\; y_2)$. Sometimes we shall denote the class of $(0, y_1, y_2)$ (resp. $(\pi/2, y_1, y_2)$) simply by y_1 (resp. y_2), thus implicitly identifying Y_1 and Y_2 to subsets of $Y_1 * Y_2$.

We define a metric d on $Y_1 * Y_2$ by requiring that the distance between the points $y = (\cos\theta\; y_1 + \sin\theta\; y_2)$ and $y' = (\cos\theta' y_1' + \sin\theta' y_2')$ be at most π, and that d satisfy the formula

$$\cos(d(y, y')) = \cos\theta\ \cos\theta'\ \cos(d^1_\pi(y_1, y_1')) + \sin\theta\ \sin\theta'\ \cos(d^2_\pi(y_2, y_2')).$$

5.14 Remarks

(1) If one equips Y_1 and Y_2 with the truncated metrics d^1_π and d^2_π, then the natural inclusion of each space into $Y_1 * Y_2$ is an isometry onto its image.

(2) If Y_1 is a single point then $Y_1 * Y_2$ is naturally isometric to the cone $C_1 Y_2$.

The fact that the formula in (5.13) does indeed define a metric on $Y_1 * Y_2$ is implicit in the following proposition.

5.15 Proposition. *For any metric spaces Y_1 and Y_2, there is a natural isometry of $C_0(Y_1 * Y_2)$ onto $C_0Y_1 \times C_0Y_2$.*

Proof. Consider the map $\Phi : C_0(Y_1 * Y_2) \to C_0Y_1 \times C_0Y_2$ defined by

$$\Phi(t(\cos\theta\ y_1 + \sin\theta\ y_2)) = (t\cos\theta\ y_1,\ t\sin\theta\ y_2).$$

We claim that Φ is an isometry. Given two points $x = t(\cos\theta\ y_1 + \sin\theta\ y_2)$ and $x' = t'(\cos\theta' y_1' + \sin\theta' y_2')$ in $C_0(Y_1 * Y_2)$,

$$\begin{aligned} &d(x, x')^2 \\ &\quad = t^2 + t'^2 - 2tt'(\cos\theta\ \cos\theta' \cos(d_\pi^1(y_1, y_1')) + \sin\theta\ \sin\theta' \cos(d_\pi^2(y_2, y_2'))). \end{aligned}$$

On the other hand,

$$\begin{aligned} (\Phi(x), \Phi(x'))^2 = {} & t^2\cos^2\theta\ + t'^2\cos^2\theta' - 2tt'\cos\theta\ \cos\theta' \cos(d_\pi^1(y_1, y_1')) \\ & + t^2\sin^2\theta\ + t'^2\sin^2\theta' - 2tt'\sin\theta\ \sin\theta' \cos(d_\pi^2(y_2, y_2')). \end{aligned}$$

These expressions are obviously equal. □

5.16 Corollary. *There is a natural isometry from $\mathbb{S}^n * \mathbb{S}^m$ to $\mathbb{S}^{n+m+1}$.*

Proof. $Y = \mathbb{S}^{n+m+1}$ is the only geodesic space with $C_0Y = \mathbb{E}^{n+m+2} = C_0\mathbb{S}^n \times C_0\mathbb{S}^m$. □

5.17 Exercise. Let X and Y be non-empty metric spaces. Prove that $X * Y$ is path-connected and give an example where it is not locally connected. Prove that if X is path-connected and Y is non-empty then $X * Y$ is simply-connected.

Quotient Metrics and Gluing

A natural way to construct interesting new metric spaces is to take a disjoint collection of known metric spaces and glue them together. The purpose of this section is to give a precise meaning to this idea. The two basic ingredients of the discussion are the notions of *disjoint union* and *quotient pseudometric*.

5.18 The Disjoint Union of Metric Spaces. Let $(X_\lambda, d_\lambda)_{\lambda\in\Lambda}$ be a family of metric spaces. Their disjoint union $X = \coprod_\lambda X_\lambda$ is the metric space whose underlying set is the disjoint union of the sets X_λ (i.e. the set of pairs (x, λ) with $\lambda \in \Lambda$ and $x \in X_\lambda$) and whose distance function d is defined by $d((x, \lambda), (x', \lambda)) = d_\lambda(x, x')$ and $d((x, \lambda), (x', \lambda')) = \infty$ if $\lambda \neq \lambda'$.

Most of the time we shall simply declare that the spaces X_λ are disjoint, meaning that we implicitly identify each X_λ with a subset of the disjoint union by the map $x \mapsto (x, \lambda)$.

5.19 Quotient Pseudometrics. Let X be a set with an equivalence relation $\sim$, let $\overline{X} = X/{\sim}$ be the set of equivalence classes and let $p : X \to \overline{X}$ be the natural projection. Associated to any metric d on X there is a pseudometric $\overline{d}$ on $\overline{X}$ defined by the formula:

$$\overline{d}(\overline{x}, \overline{y}) = \inf \sum_{i=1}^{n} d(x_i, y_i),$$

where the infimum is taken over all sequences $C = (x_1, y_1, x_2, y_2, \dots, x_n, y_n)$ of points of X such that $x_1 \in \overline{x}$, $y_n \in \overline{y}$, and $y_i \sim x_{i+1}$ for $i = 1, \dots, n-1$. Such a sequence will be called an *n-chain* joining $\overline{x}$ to $\overline{y}$ and its length $l(C)$ is defined to be $\sum_{i=1}^n d(x_i, y_i)$. Using the triangle inequality for d, we see that there is no loss of generality in assuming $y_i \neq x_{i+1}$.

It is obvious that $\overline{d}$ is symmetric and satisfies the triangle inequality, but in general $\overline{d}$ is only a pseudometric rather than a metric (for instance if there is an equivalence class which is dense in X then $\overline{d}$ is identically zero). We call $\overline{d}$ *the quotient pseudometric* associated to the relation $\sim$. Note that $\overline{d}(\overline{x}, \overline{y}) \leq d(x, y)$ for all $x, y \in X$.

5.20 Lemma. *Let (X, d) be a length space, let $\sim$ be an equivalence relation on X and let $\overline{d}$ be the quotient pseudometric on $\overline{X} = X/{\sim}$. If $\overline{d}$ is a metric then $(\overline{X}, \overline{d})$ is a length space.*

Proof. Suppose that $\overline{d}(\overline{x}, \overline{y})$ is a finite positive number a. Given $\varepsilon > 0$, one can find an n-chain joining $\overline{x}$ to $\overline{y}$ such that $\sum_{i=1}^n d(x_i, y_i) < a + \varepsilon/2$. Because d is a length metric, for each $i = 1, \dots, n$ there exists a continuous path $c_i : [0, 1] \to X$ of length smaller than $d(x_i, y_i) + \varepsilon/n$ such that $c_i(0) = x_i$ and $c_i(1) = y_i$. Because $\overline{d}(\overline{x}, \overline{y}) \leq d(x, y)$ for all $x, y \in X$, the length of the path $\overline{c}$ in X which is the concatenation of the paths $p \circ c_i$ is a curve in $\overline{X}$ of length smaller than $a + \varepsilon$ joining $\overline{x}$ to $\overline{y}$. □

5.21 Examples

(1) If X is a closed interval in $\mathbb{R}$ of length a and $\sim$ is the equivalence relation identifying the extremities of this interval, then the quotient $(\overline{X}, \overline{d})$ is isometric to a circle of length a.

(2) Let $X = \coprod_{\mathbb{Z}} I_n$ where each I_n is isometric to the the unit interval $[0, 1]$. Let $\sim$ be the equivalence relation that identifies the initial point of I_n with the terminal point of I_{n-1} for each $n \in \mathbb{Z}$. The quotient $(\overline{X}, \overline{d})$ is isometric to $\mathbb{R}$.

(3) *(Metric graphs.)* We use the notation of (1.9). The metric graph associated to a combinatorial graph $\mathcal{G}$ with edges $\mathcal{E}$, vertices $\mathcal{V}$, endpoint maps $\partial_0, \partial_1 : \mathcal{E} \to \mathcal{V}$, and length function $\lambda : \mathcal{E} \to (0, \infty)$ is the quotient of $\mathcal{V} \cup \bigcup_{e \in \mathcal{E}} ([0, \lambda(e)] \times \{e\})$ by the equivalence relation generated by $(0, e) \sim \partial_0(e)$ and $(\lambda(e), e) \sim \partial_1(e)$.

(4) Let X be a sector in the Euclidean plane (resp. the hyperbolic plane) which is the convex hull of two geodesic rays c and c' with $c(0) = c'(0) = x_0$ that form an angle $\alpha \leq \pi$. On X we consider the equivalence relation generated by $c(t) \sim c'(t)$, $\forall t \in [0, \infty)$. Then the set $\overline{X}$ of equivalence classes with the quotient metric is isometric to the Euclidean cone (resp. the hyperbolic cone) over a circle of length α.

(5) Let X be a length space and let $(U_\lambda : \lambda \in \Lambda)$ be a family of open subsets such that $X = \bigcup_\Lambda U_\lambda$. Let $i_\lambda : U_\lambda \hookrightarrow X$ be the natural inclusion and let $i : \coprod_\Lambda U_\lambda \to X$ be the induced map on the disjoint union. Then X is naturally isometric to the quotient of $\coprod_\Lambda U_\lambda$ by the equivalence relation $[x \sim x'$ iff $i(x) = i(x')]$. (It is easy to see that the natural map from the quotient to X preserves the length of paths.)

(6) Given an action of a group Γ by isometries on a metric space X one can define an equivalence relation by $x \sim y$ if and only if there exists $\gamma \in \Gamma$ such that $\gamma.x = y$. The quotient space $\overline{X}$ is the set of Γ-orbits and the quotient pseudometric is given by the formula: $\overline{d}(\overline{x}, \overline{y}) = \inf\{d(x, y) \mid x \in \overline{x}, y \in \overline{y}\}$. In other words, one need only consider 1-chains. Indeed any n-chain $C = (x_1, y_1, x_2, y_2, \ldots, x_n, y_n)$ with $n > 1$ and $y_1 = \gamma.x_2$ can be replaced by the $(n-1)$-chain $(x_1, \gamma.y_2, \gamma.x_3, \gamma.y_3, \ldots, \gamma.x_n, \gamma.y_n)$ whose length is not bigger than $l(C)$.

If the action is free and proper (see 8.2), then the projection $p : X \to \Gamma\backslash X = \overline{X}$ is a covering and a local isometry.

(7) In (3.24) we described how, given a length space (X, d) and a local homeomorphism $p : Y \to X$, one can construct an induced length metric $\tilde{d}$ on Y. The quotient of Y by the equivalence relation $[y \sim y'$ iff $p(y) = p(y')]$ is naturally isometric to X. In symbols, $\overline{\tilde{d}} = d$.

5.22 Exercises

(1) In the following examples, describe the geodesics and the balls in the corresponding quotient metric space.

(i) Consider a circle S in $\mathbb{E}^2$ and the equivalence relation on $\mathbb{E}^2$ generated by $[x \sim y$ if $x, y \in S]$.

(ii) Let S be as above but now consider the equivalence relation generated by $[x \sim y$ if x and y are antipodal points on $S]$.

(2) Let Γ be a group acting by isometries on a metric space X. Assume there is a closed convex set F whose translates by Γ cover X and assume also that for each $x \in F$ there exists an $\varepsilon > 0$ such that $\{\gamma \in \Gamma | \gamma.F \cap B(x, \varepsilon) \neq \emptyset\}$ is finite.

Consider on X the equivalence relation $\sim$ whose classes are the orbits of Γ. Let $\sim_F$ be the restriction of $\sim$ to F. Show that the natural bijection from $F/{\sim_F}$ to $X/{\sim} = \Gamma\backslash X$ is an isometry (when each is endowed with its respective quotient metric). (Hint: Show that any path in X is contained in the union of a finite number of translates of F.)

As an example, let Γ be the subgroup of $\text{Isom}(\mathbb{E}^2)$ generated by two linearly independent translations $x \mapsto x + a$ and $x \mapsto x + b$. Consider the parallelogram $F = \{ta + sb | t, s \in [0, 1]\}$. Then $\Gamma\backslash\mathbb{E}^2$ is isometric to the quotient of F by the equivalence relation generated by $[ta \sim ta+b,\ sb \sim b+sb]$, which is a torus locally isometric to $\mathbb{E}^2$.

(3) Let $\sim$ be an equivalence relation defined on a metric space Y and let $\sim_\kappa$ be the equivalence relation on its κ-cone $X = C_\kappa Y$ defined by $[ty \sim_\kappa t'y'$ iff $y \sim y'$ and $t = t']$. Let $\overline{Y}$ and $\overline{X}$ be the quotients of Y and X by these equivalence relations. Prove that $C_\kappa \overline{Y}$ is naturally isometric to $\overline{X}$.

(4) For $i = 1, 2$, let $Y_i \subseteq X_i$ be a subspace. Let $f : Y_1 \to Y_2$ be a bi-Lipschitz homeomorphism and let X be the quotient of the disjoint union of X_1 and X_2 by the equivalence relation generated by $y \sim f(y)$ for all $y \in Y_1$. Show that the natural map $X_i \to X$ is bi-Lipschitz. Show also that if each X_i is proper and each $Y_i \subset X_i$ is closed, then X is proper.

Gluing Along Isometric Subspaces

The most obvious way of gluing metric spaces is by attaching them along isometric subspaces.

5.23 Definition. Let $(X_\lambda, d_\lambda)_{\lambda\in\Lambda}$ be a a family of metric spaces with closed subspaces $A_\lambda \subset X_\lambda$. Let A be a metric space and suppose that for each $\lambda \in \Lambda$ we have an isometry $i_\lambda : A \to A_\lambda$. Let X denote the quotient of the disjoint union $\coprod_\Lambda X_\lambda$ by the equivalence relation generated by $[i_\lambda(a) \sim i_{\lambda'}(a)\ \forall a \in A,\ \lambda, \lambda' \in \Lambda]$. We identify each X_λ with its image in X and write

$$X = \bigsqcup_A X_\lambda.$$

X is called the *gluing* (or *amalgamation*) of the X_λ along A.

5.24 Lemma. *Let $X = \bigsqcup_A X_\lambda$. In the quotient pseudometric on X, the distance between $x \in X_\lambda$ and $y \in X_{\lambda'}$ is given by the formula:*

$$\begin{aligned} d(x,y) &= d_\lambda(x,y) && \text{if } \lambda = \lambda' \\ d(x,y) &= \inf_{a\in A}\{d_\lambda(x, i_\lambda(a)) + d_{\lambda'}(i_{\lambda'}(a), y)\} && \text{if } \lambda \neq \lambda'. \end{aligned}$$

And:

(1) *d is a metric on X.*

(2) *If Λ is finite and each X_λ is proper, then X is proper.*

(3) *If each X_λ is a geodesic space and A is proper, then X is a geodesic space.*

(4) *If each of the X_λ is a length space then X is a length space, moreover d is the unique length metric such that the induced length metric on each $X_\lambda \subset X$ is d_λ.*

Proof. To see that d is the quotient pseudometric note that, by the triangle inequality in X_λ, any m-chain C in $\coprod_\Lambda X_\lambda$ joining $x_\lambda \in X_\lambda$ to $x_{\lambda'} \in X_{\lambda'}$ can be replaced by a 2-chain whose length is no greater than $l(C)$. And if $\lambda = \lambda'$ then (again by the triangle inequality) one can actually take a 1-chain.

The content of (1) is that d is positive definite. This follows immediately from the fact that A_λ is closed in X_λ for every $\lambda \in \Lambda$.

In order to establish (2) one simply observes that, since $d_\lambda = d$ on $X_\lambda \times X_\lambda$, the intersection of X_λ with any closed bounded set $C \subseteq X$ is closed and bounded in X_λ. For then if each (X_λ, d_λ) is proper and Λ finite, C is a finite union of compact sets, hence it is compact.

An easy compactness argument shows that if A is proper, then for every $x \in X_\lambda$, $y \in X_{\lambda'}$ with $\lambda \neq \lambda'$, there exists $a \in A$ such that $d(x, y) = d(x, i_\lambda(a)) + d(i_{\lambda'}(a), y)$. Part (3) follows immediately from this observation. We leave the (easy) proof of (4) as an exercise. □

5.25 Exercises. Let (X_λ, d_λ) and X be as in (5.23). Prove the following statements.

(1) If each X_λ is complete, then X is complete.

(2) If each X_λ is locally compact and Λ is finite, then X is locally compact.

(3) Give examples to show that, for $\Lambda = \{1, 2\}$, it may happen that X_1 and X_2 are complete (resp. locally compact) geodesic spaces but X is not a geodesic space. (Hint: In the complete case, you may wish to consider a metric graph with countably many edges emanating from a single vertex, the n-th having length $1 + \frac{1}{n}$, and then double this space along the set of free endpoints. In the locally compact case, you may wish to consider a bounded region in the Euclidean plane.)

(4) Is the metric constructed in (5.24) determined by the fact that $d|_{X_\lambda \times X_\lambda} = d_\lambda$? What about the associated length metric?

5.26 Successive Gluing. Let $(X_j, d_j), j = 1, 2 \dots, n$ be a sequence of metric spaces; assume that an isometry f_2 from a closed subspace A_2 of X_2 onto a closed subspace $f_2(A_2)$ of X_1 is given and form the amalgamation $Y_2 := X_1 \sqcup_{A_2} X_2$ as in (5.23); assume that an isometry f_3 of a closed subspace A_3 of X_3 onto a closed subspace $f_3(A_3)$ of Y_2 is given and form the metric space $Y_3 := (X_1 \sqcup_{A_2} X_2) \sqcup_{A_3} X_3$; proceeding inductively, suppose that for every $j \geq 2$ an isometry f_j of a closed subspace A_j of X_j onto a closed subspace of Y_{j-1} is given and define $Y_j = Y_{j-1} \sqcup_{A_j} X_j$. We say that Y_n is obtained from the sequence $X_1, \dots, X_n$ by *successive gluing*. Note that Y_j is isometrically embedded in Y_{j+1}. Because of this, one can glue an infinite sequence of spaces in the same manner.

Note also that, for each i, the natural inclusion $p_i : X_i \to Y_n$ is an isometric embedding and that Y_n is the quotient of the disjoint union $\coprod_{i=1}^n X_i$ by the equivalence relation which identifies two points if they have the same projection in Y_n.

When one pictures ways of gluing spaces together, one quickly thinks of natural processes that do not conform to the simple templates described in (5.23) and (5.26) — see (5.21) or (5.29) for instance, and Chapters 7 and II.11 for many more examples. Often one wants to take a more local approach to gluing, identifying subspaces by means of (not necessarily injective) local isometries, for example. The following lemma is very useful in this regard (see II.11.4).

5.27 Lemma. *Let X be a metric space and let $(\overline{X}, \overline{d})$ be the quotient of X by an equivalence relation $\sim$. Let $\overline{x} \in \overline{X}$ and suppose that there exists $\varepsilon(\overline{x}) = \varepsilon > 0$ such that:*

(i) *for all $x, x' \in \overline{x}$ and $z \in B(x, \varepsilon)$, $z' \in B(x', \varepsilon)$ with $z \sim z'$, we have $d(x, z) = d(x', z')$; and*

(ii) *$X' = \bigcup_{x \in \overline{x}} B(x, \varepsilon)$ is a union of equivalence classes.*

Then,

(1) *for every $y \in X$ with $\overline{d}(\overline{x}, \overline{y}) < \varepsilon$, there exists $x \in \overline{x}$ such that $d(x, y) = \overline{d}(\overline{x}, \overline{y})$,*

(2) *for every $x \in \overline{x}$ and $y \in B(x, \varepsilon)$ we have $d(x, y) = \overline{d}(\overline{x}, \overline{y})$;*

(3) *let d' and $\sim'$ be the restrictions of d and $\sim$ to X' and let $(\overline{X}', \overline{d}')$ be the quotient of X' by the equivalence relation $\sim'$: then the inclusion of X' into X induces an isometry of $B(\overline{x}', \varepsilon/2)$ onto $B(\overline{x}, \varepsilon/2)$, where $\overline{x}'$ is the $\sim'$ equivalence class of $x \in \overline{x}$.*

Proof. We first prove (1). If $d(\overline{x}, \overline{y}) < \varepsilon$, then the definition of $\overline{d}$ yields an m-chain $C = (x_1, y_1, \ldots, x_m, y_m)$ joining $x_1 \in \overline{x}$ to $y = y_m$ with $l(C) < \varepsilon$. It is sufficient to prove that if $m > 1$ then there is an $(m-1)$-chain C' joining an element of $\overline{x}$ to y with $l(C') \leq l(C)$. As $y_1 \in B(x_1, \varepsilon)$ and $y_1 \sim x_2$, using (ii) we can find $x_1' \in \overline{x}$ such that $x_2 \in B(x_1', \varepsilon)$. By (i) we have $d(x_1', x_2) = d(x_1, y_1)$, hence $d(x_1', y_2) \leq d(x_1', x_2) + d(x_2, y_2) = d(x_1, y_1) + d(x_2, y_2)$. Thus we can replace C by the $(m-1)$-chain $C' = (x_2', y_2, \ldots, x_m, y_m)$ and $l(C') \leq l(C)$.

To prove (2), note that by the first part of the proof we have $d(\overline{y}, \overline{x}) = d(y, x') < \varepsilon$ for some $x' \in \overline{x}$, and by (i) (with $z = z' = y$) we have $d(x, y) = d(y, x')$.

Finally (ii) implies that the inclusion $X' \to X$ induces an injection $\overline{X}' \to \overline{X}$; and (1) and (2) imply that for each $\delta \leq \varepsilon$ the ball $B(\overline{x}', \delta)$ is mapped bijectively onto the ball $B(\overline{x}, \delta)$. To measure the distance between two points of $B(\overline{x}', \varepsilon/2)$ we can use chains of length $< \varepsilon$, and such a chain is contained in X': this proves (3). □

5.28 Corollary. *If for each $\overline{x} \in \overline{X}$ there exists $\varepsilon(\overline{x}) > 0$ as in (5.27), then $\overline{d}$ is a metric, and for all $x, y \in X$ with $x \in \overline{x}$ and $d(x, y) < \varepsilon(\overline{x})$ we have $d(x, y) = d(\overline{x}, \overline{y})$.*

5.29 Exercises

(1) Let P be a convex polygon in $\mathbb{H}^2$ with $4g$ sides. Proceeding clockwise around P, we orient its edges and label them (in order) $a_1, b_1, a_1', b_1', \ldots, a_g, b_g, a_g', b_g'$. Suppose that $l(a_i) = l(a_i')$ and $l(b_i) = l(b_i')$ for all i, where l denotes length. Let j_i (resp. h_i) be an isometry from a_i to a_i' (resp. b_i to b_i') that reverses orientation. Let $\sim$ be the equivalence relation on P generated by: $[x \sim j_i(x)\ \forall x \in a_i]$ and $[x \sim h_i(x)\ \forall x \in b_i]$, for $i = 1, \ldots, g$. Prove that the quotient pseudometric on $S = P/\!\sim$ is a metric (using Lemma 5.27) and that S is a compact surface (of genus g). Show that if the sum of the angles at the vertices of P is 2π, then S is locally isometric to the hyperbolic plane. Deduce from this and (3.32) that $\Gamma = \pi_1 S$ acts by isometries on $\mathbb{H}^2$ and that the quotient metric on $\Gamma\backslash\mathbb{H}^2$ (in the sense of 5.21(6)) is isometric to S.

(2) *(Gluing with a tube.)* Let X_1 and X_2 be two disjoint metric spaces, let C be a circle of length ℓ and let $i_1 : C \to X_1$ and $i_2 : C \to X_2$ be two local closed

geodesics. Let X be the quotient of the disjoint union of $X_1 \cup (C \times [0,1]) \cup X_2$ by the equivalence relation generated by $i_1(t) \sim (t,0)$ and $i_2(t) \sim (t,1)$. Show that the quotient pseudometric is a metric. (Hint: Use Lemma 5.27 and successive gluing.)

Give a necessary condition on ℓ so that the natural inclusions of X_1 and X_2 into X are isometric embeddings.

Limits of Metric Spaces

In this final section of the chapter we shall explore some notions of limit for sequences of metric spaces, and consider spaces whose points are themselves metric spaces. The idea of taking the limit of a sequence of spaces has played a central role in many recent advances in geometry and geometric group theory. Hausdorff introduced this idea in the case where the sequence of spaces concerned is embedded in a fixed ambient space, but the range of recent applications rests upon the idea of taking limits in the absence of any obvious ambient space, and this innovation is due to Gromov.

The first part of this section is based on [BriS94].

Gromov-Hausdorff Convergence

We begin by considering the classical construction of Hausdorff, partly for completeness but also because the basic structure of the argument in the following proof serves as a blueprint for future proofs. Two particular features to note are: first, compact sets are approximated by finite sets in a uniform way; secondly, a diagonal sequence argument is used to construct a limit object as (the closure of) an increasing union of finite sets.

5.30 Definition. Let X be a metric space and let $V_\varepsilon(A)$ denote the ε-neighbourhood of a subset $A \subset X$. The Hausdorff distance between $A, B \subset X$ is defined by:

$$d_H(A,B) = \inf\{\varepsilon \mid A \subseteq V_\varepsilon(B) \text{ and } B \subseteq V_\varepsilon(A)\}.$$

5.31 Lemma. *Let X be a compact metric space and let $\mathcal{C}X$ be the set of closed subspaces of X. Then, $(\mathcal{C}X, d_H)$ is a compact metric space.*

Proof. The only nontrivial point to check is that $\mathcal{C}X$ is compact.

Consider a sequence C_i in $\mathcal{C}X$. We must exhibit a convergent subsequence. First notice that given any $\varepsilon > 0$ there exists an integer $M(\varepsilon)$ such that, in its induced metric from X, every $A \in \mathcal{C}X$ can be covered by $M(\varepsilon)$ open balls of radius ε. Indeed, because X is compact one can cover it with $M(\varepsilon)$ balls of radius $\varepsilon/2$, then for each such ball which intersects A one chooses a point in the intersection and takes the ball of radius ε about that point. In this way, for every positive integer n and every C_i, by taking duplicates if necessary, we may assume that C_i is covered by precisely $M(1/n)$ balls of radius $1/n$. We wish to retain the centres of the balls from the n-th stage of this construction at subsequent stages. Thus we define $N(1/n) = \sum_{m \leq n} M(1/m)$ and cover C_i with $N(1/n)$ balls of radius $1/n$ with centres $x_n(i,j) \in C_i$ for $j = 1, \dots, N(1/n)$, where $x_{n+1}(i,j) = x_n(i,j)$ if $j \leq N(1/n)$. This allows us to drop the subscript n from $x_n(i,j)$. Let $\Sigma(i,n) := \{x(i,j) : j \leq N(1/n)\}$ and note that $d_H(C_i, \Sigma(i,n)) \leq 1/n$ for all i.

$$\begin{aligned}
C_1 &\ni x(1,1), x(1,2), \dots, x(1,j), \dots \\
C_2 &\ni x(2,1), x(2,2), \dots, x(2,j), \dots \\
&\ \ \vdots \\
C_i &\ni x(i,1), x(i,2), \dots, x(i,j), \dots \\
&\ \ \vdots
\end{aligned}$$

Because X is compact, we may pass to a subsequence of the C_i in order to assume that the sequence $x(i,1)$ converges in X, to $x(\omega, 1)$ say. Let C_i^1 denote this subsequence. Inductively, we may pass to further subsequences C_i^k in order to assume that for $j = 1, \dots, k$ each of the sequences $(x(i,j))_i$ converges in X to $x(\omega, j)$. Let C_ω be the closure in X of $\{x(\omega,j) \mid j \in \mathbb{N}\}$. We claim that the *diagonal sequence* C_k^k converges to C_ω in $\mathcal{C}X$. To simplify the notation, we write $C_{k'}$ in place of C_k^k.

Let $n > 0$ be an integer. Given any integer $l > 0$, since $d_H(C_i, \Sigma(i,n)) \leq 1/n$ for all i, in particular $x(i,l)$ is a distance at most $1/n$ from $\Sigma(i,n)$, and hence $x(\omega, l)$ is a distance at most $1/n$ from $\Sigma(\omega, n) := \{x(\omega,j) : j \leq N(1/n)\}$. Thus $d_H(C_\omega, \Sigma(\omega, n)) \leq 1/n$. And if k' is large enough to ensure that $d(x(k',j), x(\omega,j)) \leq 1/n$ for all $j \leq N(1/n)$, then:

$$\begin{aligned}
&d_H(C_{k'}, C_\omega) \\
&\qquad \leq d_H(C_{k'}, \Sigma(k',n)) + d_H(\Sigma(k',n), \Sigma(\omega,n)) + d_H(\Sigma(\omega,n), C_\omega) \leq 3/n.
\end{aligned}$$

□

One can rephrase Hausdorff convergence in terms of convergence of sequences of points:

5.32 Lemma. *C_n converges to $C \in \mathcal{C}X$ if and only if:*

(1) *for all $x \in C$ there exists a sequence $x_n \in C_n$ such that $x_n \to x$ in X, and*

(2) *every sequence $y_{n(i)} \in C_{n(i)}$ with $n(i) \to \infty$ has a convergent subsequence whose limit point is an element of C.*

Proof. Exercise. □

We wish to consider what it means for a sequence of compact metric spaces to converge to a limit space when there is no obvious ambient space containing the sequence. For this we need the following definition.

5.33 Definition. A subset S of a metric space X is said to be ε*-dense* if every point of X lies in the ε-neighbourhood of A. (Under the same circumstances, S is called an ε*-net* in X.)

An ε*-relation* between two (pseudo)metric spaces X_1 and X_2 is a subset $R \subseteq X_1 \times X_2$ such that:

(1) for $i = 1, 2$, the projection of R to X_i is ε-dense, and

(2) if $(x_1, x_2), (x_1', x_2') \in R$ then $|d_{X_1}(x_1, x_1') - d_{X_2}(x_2, x_2')| < \varepsilon$. The relation is said to be *surjective* if its projection onto each X_i is surjective. If there exists an ε-relation between X_1 and X_2 then we write $X_1 \sim_\varepsilon X_2$, and if there is a surjective ε-relation then we write $X_1 \simeq_\varepsilon X_2$.

We define[5] the *Gromov-Hausdorff* distance between X_1 and X_2 to be:

$$D_H(X_1, X_2) := \inf\{\varepsilon \mid X_1 \simeq_\varepsilon X_2\}.$$

If there exists no ε such that $X_1 \simeq_\varepsilon X_2$, then $D_H(X_1, X_2)$ is infinite.

Sometimes, instead of writing "$(x, y) \in R$" we shall write "x is related to y" or "x corresponds to y".

5.34 Lemma.

(1) *If* $X_1 \sim_\varepsilon X_2$ *then* $X_1 \simeq_{3\varepsilon} X_2$.

(2) D_H *satisfies the triangle inequality and hence defines a pseudometric (which may take the value* ∞*) on any set of metric spaces.*

Proof. Exercise. □

Terminology: We say that a sequence of (pseudo)metric spaces X_n converges to X in the *Gromov-Hausdorff metric* (or, *in Gromov-Hausdorff space*), and write $X_n \to X$, if and only if $D_H(X_n, X) \to 0$ as $n \to \infty$.

5.35 Remarks and Exercises

(1) The graph of $f : X_1 \to X_2$ is a 0-relation if and only if f is an isometry.

(2) The following alternative notion of distance between metric spaces is the one used by Gromov in [Gro81b]. Given two metric spaces X_1 and X_2, consider all metrics d on the disjoint union $X_1 \coprod X_2$ that restrict to the given metrics on X_1 and X_2, and define

$$D'_H(X_1, X_2) := \inf_d d_H(X_1, X_2),$$

where d_H is as in definition (5.30).

[5] Some authors prefer to use $\sim_\varepsilon$ instead of $\simeq_\varepsilon$ in this definition; the difference is not significant, 5.34(1).

(3) If X_1 be a space with only one point and X_2 is a space consisting of two points a distance one apart, then $D_H(X_1, X_2) = 1$ and $D'_H(X_1, X_2) = 1/2$.

(4) Let X_1 and X_2 be metric spaces. Show that $D'_H(X_1, X_2) < \varepsilon$ if and only if there is a metric space Y and isometric embeddings $j_i : X_i \hookrightarrow Y$ such that $d_H(j_1(X_1), j_2(X_2)) < \varepsilon$. (Hint: Given Y and $\eta > 0$, perturb $j_1(X_1)$ in $Y \times [0, \eta]$ to make it disjoint from $j_2(X_2)$.)

(5) Let X_1 and X_2 be metric spaces. Prove that $2D'_H(X_1, X_2) = D_H(X_1, X_2)$.

(Hint: To see that $2D'_H(X_1, X_2) \leq D_H(X_1, X_2)$, given a surjective ε-relation $R \subset X_1 \times X_2$, let Z be the disjoint union of a family of copies of $[0, \varepsilon]$ indexed by R and consider a quotient of $X_1 \coprod X_2 \coprod Z$.)

The Gromov-Hausdorff distance between a metric space and any dense subset of it is zero, and hence limits of sequences of spaces are not unique in general. However:

5.36 Proposition. *Let A and B be compact metric spaces. A and B are isometric if and only if $D_H(A, B) = 0$.*

Proof. We shall show that if $D_H(A, B) = 0$ then A and B are isometric, the other implication is trivial. Let $\{a_n\}_n$ be a countable dense subset of A and let R_m be a surjective $(1/m)$-relation between A and B. We choose $b_{m,n} \in B$ so that $(a_n, b_{m,n}) \in R_m$. By passing to a subsequence of $(b_{m,1})_m$ we may assume that $b_{m,1} \to b_1$ in B. By passing to a further subsequence we may assume that $b_{m,2} \to b_2$, and so on. Thus for all n, n' and infinitely many m we have that $|d_A(a_n, a_{n'}) - d_B(b_{m,n}, b_{m,n'})| < 1/m$, and hence $d_A(a_n, a_{n'}) = d_B(b_n, b_{n'})$. The desired isometry $A \to B$ is the unique continuous extension of $a_n \mapsto b_n$. □

(5.36) does not extend to proper metric spaces:

5.37 Exercise. Consider the following two metric graphs: each is constructed by attaching a segment to integer points of the real line; in the first case the segment attached at m has length $|\sin m|$ and in the second case it has length $|\sin(m + \frac{1}{2})|$. Prove that the Gromov-Hausdorff distance between these spaces is zero but that they are not isometric. (Hint: Given $\varepsilon > 0$, let $\delta > 0$ be such that $|x - y| < \delta$ implies $|\sin x - \sin y| < \varepsilon$. Note that there exist integers r and s such that $|(r - \frac{1}{2}) - 2s\pi| < \delta$.)

5.38 Proposition. *Let X be a complete space that is a Gromov-Hausdorff limit of a sequence X_n.*

(1) *If each X_n is a length space, then X is a length space.*

(2) *If each X_n is proper, then X is proper.*

(3) *If each X_n is a proper geodesic space, then so too is X.*

Proof. In order to prove (1), since X is complete it is enough to show that X has approximate midpoints. Suppose that $x, y \in X$ and $\varepsilon > 0$ are given.

If n is sufficiently large then there is a surjective ε-relation $R_n \subseteq X \times X_n$. Choose $x_n, y_n \in X_n$ such that $(x, x_n) \in R_n$ and $(y, y_n) \in R_n$, and let $z_n \in X_n$ be such that $\max\{d(x_n, z_n), d(z_n, y_n)\} < d(x_n, y_n)/2 + \varepsilon$. Choose $z \in X$ with $(z, z_n) \in R_n$. Then,

$$d(x, z) < d(x_n, z_n) + \varepsilon < \frac{1}{2} d(x_n, y_n) + \varepsilon + \varepsilon < \frac{1}{2} d(x, y) + \frac{5}{2}\varepsilon.$$

Similarly $d(y, z) \leq d(x, y) + 5\varepsilon/2$. This proves (1).

Each closed ball in X is the limit of balls in the X_n, and if a complete space is the limit of compact spaces then it is compact (5.40). This proves (2). Part (3) is immediate from (1), (2) and the Hopf-Rinow theorem. □

The interested reader should be able to think of many more properties that are preserved under the taking of Gromov-Hausdorff limits. We should mention one property that is not preserved: if the spaces concerned are not proper, then a complete limit of geodesic spaces need not be geodesic. To see this, consider the metric graph X with two vertices and edges $\{e_m \mid m \in \mathbb{Z}_+\}$, where e_m has length $(1 + 1/m)$. This is not a geodesic space, but it is the limit of the geodesic spaces X_n, where X_n differs from X only in that its n-th edge has length 1.

In the proof of (5.31) we relied on the fact that closed subspaces of a fixed compact metric space are uniformly compact in the following sense.

5.39 Definition. A family $(C_\lambda)_{\lambda \in \Lambda}$ of metric spaces is said to be *uniformly compact* if there is a uniform bound on their diameters and for every $\varepsilon > 0$ there exists an integer $N(\varepsilon)$ such that each C_λ can be covered by $N(\varepsilon)$ balls of radius ε.

5.40 Exercise. If a sequence of compact spaces converges in Gromov-Hausdorff space, then it is uniformly compact and the completion of the limit is compact. (Hint: Recall that a metric space is compact if and only if it is complete and totally bounded[6].)

The following was an ingredient in Gromov's proof of his polynomial growth theorem (see [Gro81b] and (8.37)).

5.41 Theorem (Gromov). *If a sequence of compact metric spaces C_i is uniformly compact, then it has a subsequence that converges in the Gromov-Hausdorff metric.*

Proof. We follow the proof of (5.31). Thus we first construct a sequence of points $(x(i, j))_j$ in each C_i, with the property that for all $n \in \mathbb{N}$ the balls of radius $1/n$ about the first $N(1/n)$ terms in the sequence cover C_i. Let $S(i, n) = \{x(i, 1), \dots, x(i, N(i/n)\}$.

[6] A metric space is *totally bounded* if, for every $\varepsilon > 0$, it is the union of finitely many balls $B(x, \varepsilon)$.

$$\begin{aligned}
C_1 &\ni x(1,1), x(1,2), \ldots, x(1,j), \ldots \\
C_2 &\ni x(2,1), x(2,2), \ldots, x(2,j), \ldots \\
&\vdots \\
C_i &\ni x(i,1), x(i,2), \ldots, x(i,j), \ldots \\
&\vdots
\end{aligned}$$

Let d_i denote the metric on C_i. By passing to a subsequence C_i^2 of the C_i, we may assume that the numbers $d_i(x(i,1), x(i,2))$ tend to a limit, $\delta(1,2)$ say, as $i \to \infty$. By passing to increasingly rarer subsequences C_i^k, given any integer k we may assume that $d_i(x(i,j), x(i,j')) \to \delta(j,j')$ as $i \to \infty$ for all $1 \leq j < j' \leq k$. We define $\delta : \mathbb{N} \times \mathbb{N} \to [0,\infty)$ recursively in this manner, and claim that the diagonal subsequence C_k^k is convergent.

Let I be the set of i indexing this diagonal subsequence of spaces.

We shall construct the desired limit as the completion of a countable space. To this end, we take a countable set $\hat{C}_\infty = \{x_1, x_2, \ldots\}$ and define a pseudometric on $\hat{C}_\infty$ by setting $d_\infty(x_j, x_k) = \delta(j,k)$. We then take equivalence classes under the relation that identifies points which are a distance zero apart, and define C_∞ to be the metric completion of this space. It is convenient to continue to write x_j for the image of x_j in C_∞.

Fix an integer $n > 0$. Because $S(i,n)$ is a $(1/n)$-net for C_i, given any $l > N(1/n)$ and $i \in \mathbb{N}$ we have $d_i(x(i,l), x(i,j)) < 1/n$ for some $j \leq N(1/n)$. If we fix l and let the index i tend to infinity through the diagonal indexing set I, then the same choice of j must recur infinitely often, and for this value we have $d_\infty(x_j, x_l) \leq 1/n$. Thus $S(\infty, n) := \{x_1, \ldots, x_{N(1/n)}\}$ is $(1/n)$-dense in $\hat{C}_\infty$ and hence in C_∞.

If $i \in I$ is sufficiently large, then by definition $R_n := \{(x(i,j), x_j\} \subset S(i,n) \times S(\infty,n)\}$ is a surjective $(1/n)$-relation. Moreover $S(i,n)$ is $(1/n)$-dense in C_i and $S(\infty,n)$ is $(1/n)$-dense in C_∞, therefore R_n is an $(1/n)$-relation. Since $n > 0$ is arbitrary, this proves the desired convergence.

Finally we note that C_∞ is totally bounded (indeed we have described a finite $(1/n)$-covering for every $n > 0$), so since it is complete by construction, C_∞ is compact. $\square$

In [Gro81b] Gromov established the above compactness criterion by a different argument, realizing the C_i as compact subspaces of a fixed compact space.

5.42 Remarks

(1) Let d denote the usual metric on $\mathbb{H}^m$, let B_n be the closed ball of radius n about a fixed point and consider the sequence of metric spaces $X_n = (B_n, \frac{1}{n}d)$. We claim that $\{X_n \mid n \in \mathbb{N}\}$ is *not* uniformly compact. To see this, we note that the volume $V(n)$ of a ball of radius n in $\mathbb{H}^m$ grows like e^n, so if one could cover X_n by $N(\varepsilon)$ balls of radius $\varepsilon > 0$, then one could cover (B_n, d) by $N(\varepsilon)$ balls of radius $n\varepsilon$, and hence $V(n) \leq N(\varepsilon).V(\varepsilon n)$. If we fix ε and let $n \to \infty$, this inequality becomes absurd.

It follows from (5.40) that the sequence X_n does not have a subsequence that converges in Gromov-Hausdorff space.

(2) By way of contrast, note that in $\mathbb{E}^m$ the rescaled balls $(B_n, \frac{1}{n}d)$ are all isometric; in particular they are uniformly compact. The previous inequality regarding volumes is of course not absurd in this case, because $V(n) = c\,n^m$, where c is a constant depending on m. Polynomial growth is the real key here: in [Pan83] Pierre Pansu proved some remarkable results concerning uniform compactness and convergence of nilpotent groups.

(3) There are many applications of Gromov's compactness criterion (5.41) in differential geometry, but to expose even a part of this literature would take us well beyond the scope of this book. For an introduction to convergence theorems in Riemannian geometry see [GrLP81], [BlW97] and [Pet96]. We mention one result to give a flavour of some of the applications: *For all $n \in \mathbb{N}$, all $D > 0$, all $v_0 > 0$ and all $k \in \mathbb{R}$, the class $\mathcal{M}^n(k, v_0, D)$ of n-dimensional Riemannian manifolds with volume $\geq v_0$, diameter $\leq D$ and sectional curvature $\leq k$ is precompact in Gromov-Hausdorff space (and in the $C^{1,\alpha}$ topology*[7] *for every $\alpha < 1$). Moreover, $\mathcal{M}^n(k, v_0, D)$ contains only finitely many different manifolds up to diffeomorphism.* In the form stated, this result is due to Grove, Peters and Wu [GPW90]. There are similar results by a number of authors, beginning with the work of Cheeger and Gromov [ChGr86].

Convergence of Pointed Spaces

Gromov-Hausdorff convergence works well in contexts where one wishes to consider sequences of *compact* metric spaces, but it is a less satisfactory concept for convergence for non-compact spaces. One obvious disadvantage is that the distance between a compact space and an unbounded space is always infinite. Thus, for example, Gromov-Hausdorff convergence is insufficient to capture the intuitive notion that as the radius of a sphere of constant curvature tends to infinity, to an observer standing at the north pole, the sphere looks increasing like Euclidean space.

The key here is that the intuitive sense of convergence comes from observations taken at a fixed point.

5.43 Definition. Consider a sequence of metric spaces X_n with basepoints $x_n \in X_n$. The sequence of pointed spaces (X_n, x_n) is said to converge to (X, x) if for every $r > 0$ the sequence of closed balls $\overline{B}(x_n, r)$ (with induced metrics) converges to $\overline{B}(x, r) \subseteq X$ in the Gromov-Hausdorff metric.

Under the same circumstances we say that X is a pointed Gromov-Hausdorff limit of the X_n and that (X_n, x_n) converges to (X, x) in the pointed Gromov-Hausdorff sense.

Just as (unpointed) Gromov-Hausdorff convergence is most suitable for sequences of compact metric spaces, so pointed convergence is most suitable for the study of proper spaces. In particular:

[7] For the definition of the $C^{1,\alpha}$ topology see [Pet96].

5.44 Lemma. *If a complete space is the (pointed) Gromov-Hausdorff limit of a sequence of proper spaces, then it is itself proper.*

And Gromov's pre-compactness criterion becomes:

5.45 Theorem. *Let (X_n, x_n) be a sequence of pointed metric spaces. If for every $r > 0$ and $\varepsilon > 0$ there exists an integer $N(r, \varepsilon)$ such that $B(x_n, r)$ can be covered by $N(r, \varepsilon)$ ε-balls, then a subsequence of (X_n, x_n) converges in the pointed Gromov-Hausdorff sense.*

We leave the reader to formulate and prove the analogue of (5.38) for pointed convergence.

5.46 Remark (Related notions of convergence). There are a number of situations (notably in low-dimensional topology and geometric group theory) where one has a natural degeneration of metric structure in which proper spaces approximate spaces that are not proper. Such a situation arises, for example, in the study of degenerations of hyperbolic structures [Sha91]. In light of (5.44), we need a weaker definition of convergence to cover such cases. A suitable notion was introduced by Paulin [Pau88] (see also, Bestvina [Bes88]). The idea of this generalization is that finite subsets in the limit space should be approximated by finite subsets of the limiting sequence (cf. (II.3.10)). There is a natural extension of this situation where one has a fixed group acting on a sequence of spaces and one wishes to take an equivariant limit. Such equivariant limits have played an important role in the study of 3-manifolds (see [Sha91]) and many recent developments in geometric group theory, e.g. [Sela97]. (See section 4 of [BriS94] for further remarks.)

Ultralimits and Asymptotic Cones

Many of the key arguments in the preceding section are characterized by the fact that one takes repeated subsequences to obtain the desired limiting space. A considerable clarification of the argument would result if one could extract a convergent subsequence all at once; non-principal ultrafilters provide a tool for doing so. They also provide a means of constructing a limiting object (ultralimit) in cases where no Gromov-Hausdorff limit of a sequence of spaces exists. For example, given a finitely generated group Γ with a word metric d, it is natural to consider the sequence of metric spaces $X_n = (\Gamma, \frac{1}{n}d)$; the identity element serves as a basepoint. One expects limit points of this sequence to contain information about the asymptotic properties of the group Γ. If Γ has polynomial growth then X_n converges in the pointed Gromov-Hausdorff topology ([Gro81b], [Pan83]). In general though, X_n will not even contain a convergent subsequence. One remedies this by passing to ultralimits (5.50), the significance of which were brought to the fore in this context by van den Dries and Wilkie [DW84].

In order to motivate the definition of ultrafilters and ultralimits, let us pursue remark (5.46) and reflect that one might define a metric space (Y, d) to be a limit of a sequence of metric spaces (Y_n, d_n) if for every finite set of points $\{p_1, \dots, p_i\} \subset Y$ and every $\varepsilon > 0$, for infinitely many n one can find subsets $\{p_1^n, \dots, p_i^n\} \subset Y_n$ such that $|d_n(p_j^n, p_k^n) - d(p_j, p_k)| < \varepsilon$ for all $1 \le j, k \le i$. This is too weak a notion of limit for most purposes (but not all (II.3.10)). In order to obtain a stricter notion of convergence one should replace "for infinitely many n" by "for almost all n", and one should also require that the points of Y account for all sequences of points in the approximating sequence that might be said to converge. In order to quantify "almost all" we need a measure:

5.47 Definition. A *non-principal ultrafilter on* $\mathbb{N}$ is a finitely additive probability measure ω such that all subsets $S \subset \mathbb{N}$ are ω-measurable, $\omega(S) \in \{0, 1\}$ and $\omega(S) = 0$ if S is finite.

5.48 Exercise (Existence of Non-Principal Ultrafilters). Let N be a set and let $\mathcal{P}N$ denote the set of its subsets. A *filter for N* is a map $\mu : \mathcal{P}N \to \{0, 1\}$ such that: $\mu(\emptyset) = 0$, $\mu(N) = 1$; if $S \subset T$ then $\mu(S) \le \mu(T)$; and if $\mu(S) = \mu(T) = 1$ then $\mu(S \cap T) = 1$. And μ is called an *ultrafilter* if in addition $\mu(S) + \mu(N \smallsetminus S) = 1$ for all $S \in \mathcal{P}N$. (One thinks of μ as labelling subsets S as large if $\mu(S) = 1$ and small if $\mu(S) = 0$.)

There is a natural ordering on the set of filters for any set N, namely $\mu \preceq \mu'$ if $\mu(S) \le \mu'(S)$ for all $S \in \mathcal{P}N$. Given $S \in \mathcal{P}N$, one obtains an ultrafilter by defining $\mu_S(T) = 1$ if and only if $S \subset T$. Such an ultrafilter is called principal.

(1) Prove that a filter is $\preceq$-maximal if and only if it is an ultrafilter.

(2) If N is an infinite set then the map $\mu_f : \mathcal{P}N \to \{0, 1\}$ which assigns the value 1 to a set if and only if its complement is finite is a filter. Prove that an ultrafilter ω on N is not principal if and only if $\mu_f \preceq \omega$.

(3) By applying Zorn's lemma to the filters μ with $\mu_f \preceq \mu$, deduce that there exist non-principal ultrafilters in the sense of (5.47).

Non-principal ultrafilters pick out convergent subsequences:

5.49 Lemma. *Let ω be a non-principal ultrafilter on $\mathbb{N}$. For every bounded sequence of real numbers a_n there exists a unique point $l \in \mathbb{R}$ such that $\omega\{n : |a_n - l| < \varepsilon\} = 1$ for every $\varepsilon > 0$. One writes $l = \lim_\omega a_n$.*

Proof. Exercise. □

One should be aware that $\lim_\omega a_n$ depends very much on the choice of ω. For example, given two convergent sequences (b_n) and (c_n), if one defines a_n to be b_n if n is even and c_n if n is odd, then $\lim_\omega a_n$ will equal $\lim b_n$ if the even integers have ω-measure one, and $\lim c_n$ if the odd integers have ω-measure one. A similar dependence on the choice of ω is built into the following definition.

5.50 Definition (ω-Limits and Asymptotic Cones). Let ω be a non-principal ultrafilter on $\mathbb{N}$. Let (X_n, d_n) be a sequence of metric spaces with basepoints p_n and let X_∞ denote the set of sequences (x_n), where $x_n \in X_n$ and $d_n(x_n, p_n)$ is bounded independently of n. Consider the equivalence relation $[(x_n) \sim (y_n)$ iff $\lim_\omega d_n(x_n, y_n) = 0]$, and let X_ω denote the set of equivalence classes. Endow X_ω with the metric $d_\omega((x_n), (y_n)) = \lim_\omega d_n(x_n, y_n)$. One writes $(X_\omega, d_\omega) = \lim_\omega(X_n, d_n)$ (if the choice of basepoints is not important).

One may be interested in the case of a fixed metric space X with a basepoint and a metric d, and $X_n = (X, \frac{1}{n}d)$. In this case X_ω is called an *asymptotic cone*[8] of X and is denoted $\text{Cone}_\omega(X)$.

5.51 Remark. One can construct metric spaces X (subsets of the line even) and non-principal ultrafilters $\omega \neq \omega'$ so that $\text{Cone}_\omega(X)$ and $\text{Cone}_{\omega'}(X)$ are not homeomorphic. Simon Thomas and Boban Velickov recently showed that this can happen even when X is the Cayley graph of a finitely generated group.

Ultralimits are related to Gromov-Hausdorff limits by the following exercise.

5.52 Exercise. Suppose that the pointed space (X_∞, p_∞) is the Gromov-Hausdorff limit of a sequence of proper spaces (X_n, p_n). Show that for every non-principal ultrafilter ω on $\mathbb{N}$, the ultralimit (X_ω, p_ω) is isometric to (X_∞, p_∞). (See (5.55) with regard to the need for properness.)

5.53 Lemma. *The ultralimits of all sequences of metric spaces are complete.*

Proof. We sketch the proof and leave the details to the reader. Let $X_\omega = \lim_\omega X_n$. Let d_n denote the metric on X_n. Let $(x^j_\omega)_j$ be a Cauchy sequence in X_ω and represent each entry x^j_ω by a sequence $(x^j_i)_i$ where $x^j_i \in X_i$. Let $A_0 = \mathbb{N}$ and for $k \in \mathbb{N}$ inductively define a strictly decreasing sequence of subsets $A_k \subset A_{k-1}$ with $\omega(A_k) = 1$ so that for all i and all j between 1 and k we have $|d_\omega(x^j_\omega, x^k_\omega) - d_i(x^j_i, x^k_i)| < 1/2^k$. Set $y_i = x^k_i$ for all $i \in A_{k-1} \smallsetminus A_k$ and prove that the sequence (y_i) defines a point $y_\omega \in X_\omega$ such that $x^j_\omega \to y_\omega$ as $j \to \infty$. $\square$

5.54 Exercise. Show that any ultralimit of length spaces is a length space, and any ultralimit of geodesic spaces is a geodesic space.

5.55 Remarks. If X is a proper metric space, then the map which sends each $x \in X$ to the constant sequence at x defines an isometry from X to the ultralimit of the constant sequence $X_n = X$. However, this is not true of complete metric spaces in general. Indeed, if X is a countably infinite set endowed with the discrete metric, $d(x, y) = 1$ for all $x \neq y$, then for any non-principal ultrafilter ω, the ultralimit of the constant sequence $X_n = X$ is an uncountable set with the discrete metric.

[8] Some authors, e.g. [KlL97], find it useful to allow sequences of scaling factors other than $(1/n)$ in the definition of an asymptotic cone.

Asymptotic cones and ultralimits have played a significant role in recent proofs of rigidity theorems (e.g. [KlL97]). This follows the influence of Gromov [Gro93]. The asymptotic cone of a finitely generated group was introduced by van den Dries and Wilkie [DW84] as a device for streamlining Gromov's proof of his polynomial growth theorem. See [Gro81b], [Dru97], [Paps96] and [Bri99a] for results connecting the geometry and topology of asymptotic cones of groups to algebraic properties of groups.

Chapter I.6 More on the Geometry of M_κ^n

In this chapter we return to the study of the model spaces M_κ^n. We begin by describing alternative constructions of $\mathbb{H}^n = M_{-1}^n$ attributed to Klein and Poincaré[9]. In each case we describe the metric, geodesics, hyperplanes and isometries explicitly. In the case of the Poincaré model, this leads us naturally to a discussion of the Möbius group of $\mathbb{S}^n$ and of the one point compactification of $\mathbb{E}^n$. We also give an explicit description of how one passes between the various models of hyperbolic space. In the final paragraph we explain how the metric on M_κ^n can be derived from a Riemannian metric and give explicit formulae for the Riemannian metric.

The Klein Model for $\mathbb{H}^n$

In Chapter 2 we constructed $\mathbb{H}^n$ as a subset of $\mathbb{E}^{n,1} = \mathbb{R}^{n+1}$. This subset does not contain $0 \in \mathbb{R}^{n+1}$, and each 1-dimensional subspace of $\mathbb{R}^{n+1}$ intersects it in at most one point. Therefore, the natural projection from $\mathbb{R}^{n+1} \setminus \{0\}$ to the real projective space $\mathbb{P}^n$ of dimension n restricts to an injection on $\mathbb{H}^n$. One way to obtain the Klein model for $\mathbb{H}^n$ is to simply transport the geometry of $\mathbb{H}^n \subset \mathbb{E}^{n,1}$ by the map p. We shall explain this in some detail.

Let $\mathbb{P}^n$ denote real projective n-space, i.e. the set of 1-dimensional vector subspaces of $\mathbb{R}^{n+1}$. Let $p : \mathbb{R}^{n+1} \setminus \{0\} \to \mathbb{P}^n$ be the natural projection, which associates to each point $X \in \mathbb{R}^{n+1} \setminus \{0\}$ the line passing through 0 and X. Let $x = p(X)$. The coordinates of X are called homogeneous coordinates for x; thus "the" homogeneous coordinates of x are well defined up to multiplication by any non-zero real number. The projective transformations of $\mathbb{P}^n$ are, by definition, the transformations of $\mathbb{P}^n$ induced by the linear transformations of $\mathbb{R}^{n+1}$; they form a group $\mathrm{PGL}(n+1, \mathbb{R})$ naturally isomorphic to the quotient of the group $\mathrm{GL}(n+1, \mathbb{R})$ by the normal subgroup formed by non-zero multiples of the identity matrix. A projective line (more generally a projective p-plane) is a subset of $\mathbb{P}^n$ formed by the 1-dimensional subspaces contained in a 2-dimensional (resp. a $(p+1)$-dimensional) subspace of $\mathbb{R}^{n+1}$.

The projection p sends the set of points $X \neq 0$ of $\mathbb{E}^{n,1} = \mathbb{R}^{n+1}$ such that $\langle X \mid X \rangle = 0$ onto the set of points in $\mathbb{P}^n$ whose homogeneous coordinates satisfy the

[9] The basic properties of real hyperbolic spaces were first established by Lobachevski in the nineteenth century. See [Mil82] for an account of the subsequent developments of the subject.

quadratic equation $\sum_{i=1}^n X_i^2 - X_{n+1}^2 = 0$; the set of such points is a quadric, and shall be denoted Q. The map p restricts to a bijection from $\mathbb{H}^n$ to the interior of the quadric Q, i.e. the connected component of $\mathbb{P}^n - Q$ which is the set of points whose homogeneous coordinates satisfy $\sum_{i=1}^n X_i^2 - X_{n+1}^2 < 0$. Hence p induces an isomorphism from the group $O(n, 1)_+$ of isometries of $\mathbb{H}^n$ to the group of projective transformations of $\mathbb{P}^n$ that preserve the interior of the quadric Q. The geodesic lines in $\mathbb{H}^n$ are the intersections of $\mathbb{H}^n$ with the vector subspaces of $\mathbb{E}^{n,1}$ of dimension 2; they are mapped by p onto the non-empty intersections of the interior of Q with the projective lines of $\mathbb{P}^n$.

The projective form of the Klein model for hyperbolic n-space is the interior of the quadric Q endowed with the unique metric such that $d(p(X), p(Y)) = d(X, Y)$ for all $X, Y \in \mathbb{H}^n \subseteq \mathbb{E}^{n,1}$. This metric admits an elegant description in terms of projective geometry: given two distinct points x and y in the interior of Q, let x_∞ and y_∞ be the points of intersection of Q with the projective line through x and y, arranged so that x_∞, x, y, y_∞ occur in order on the projective line through x and y.

6.1 Lemma. *The distance between two points x, y in the projective Klein model is given by the formula*

$$d(x, y) = \frac{1}{2} \log(x, y, x_\infty, y_\infty),$$

where $(x, y, x_\infty, y_\infty)$ is the cross ratio of the four aligned point x, y, x_∞, y_∞.

Proof. Recall that the *cross ratio* (x, y, v, w) of four aligned distinct points $x, y, v, w \in \mathbb{P}^n$ is defined as follows. Let X and Y be points of $\mathbb{R}^{n+1} - \{0\}$ projecting by p to x and y. Then there are pairs of real numbers (λ, μ) and (λ', μ') such that $p(\lambda X + \mu Y) = v$ and $p(\lambda' X + \mu' Y) = w$. By definition, $(x, y, v, w) = \lambda\mu'/\mu\lambda' \in [0, \infty]$. This definition is independent of the various choices. An alternative description of (x, y, v, w) is to say that it is the cross ratio (X, Y, V, W) (in the sense of 6.4(3) below) of any four aligned points $X, Y, V, W \in \mathbb{R}^{n+1} - \{0\}$ such that $p(X) = x$, $p(Y) = y$, $p(V) = v$, $p(W) = w$. If h is a projective transformation, then $(h(x), h(y), h(v), h(w)) = (x, y, v, w)$.

To prove the lemma, we use the invariance of the cross ratio under projective transformations and the fact that $\mathrm{Isom}(\mathbb{H}^n)$ acts transitively on pairs of points which are a fixed distance apart. This enables us to reduce to the case where x and y are the images under p of points $X = (X_1, \dots, X_{n+1})$ and $Y = (Y_1, \dots, Y_{n+1})$, all of whose coordinates are zero except $X_{n+1} = 1$, and $Y_1 = \sinh r$ and $Y_{n+1} = \cosh r$, where $r = d(x, y)$. Then $X_\infty = (\cosh r + \sinh r)X - Y$ and $Y_\infty = (\cosh r - \sinh r)X - Y$ are sent by p to x_∞ and y_∞ respectively. One calculates that $(x, y, x_\infty, y_\infty) = e^{2r}$. $\square$

One can obtain a more specific parameterization of the Klein model by replacing the interior of the quadric $Q \subseteq \mathbb{P}^n$ with the open unit ball $B^n = \{x \in \mathbb{R}^n \ : \ \|x\| < 1\}$, where $\|x\|$ denotes the Euclidean norm of x. This is essentially done by identifying B^n with the disc $B^n \times \{1\} \subset \mathbb{R}^{n+1}$ and noting that there is exactly one point of this disc on each line in $\mathbb{R}^{n+1}$ that represents a point in the interior of the quadric Q.

The following proposition describes this construction more precisely.

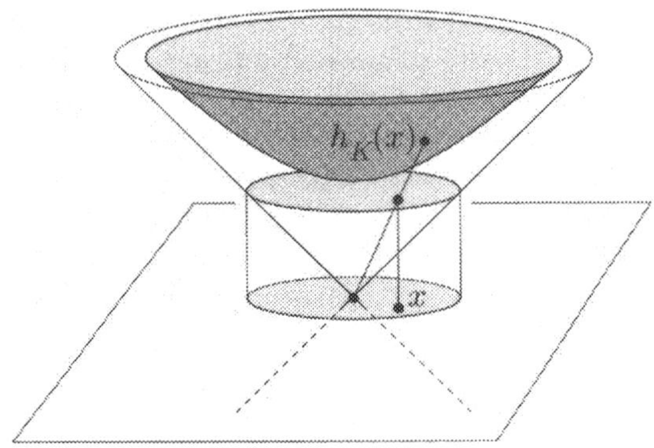

Fig. 6.1 The hyperboloid model and the Klein model for $\mathbb{H}^n$

6.2 Proposition (The Metric for the Klein Model). *Let h_K be the homeomorphism from B^n to $\mathbb{H}^n$ that sends the point $x = (x_1, \ldots, x_n) \in B^n$ to $\frac{1}{\sqrt{1-\|x\|^2}}(x_1, \ldots, x_n, 1) \in \mathbb{H}^n$. (Note that this image point is the intersection of $\mathbb{H}^n$ with the line through 0 and $(x, 1) \in \mathbb{R}^n \times \mathbb{R}^1 = \mathbb{E}^{n,1}$.) The ball B^n, equipped with the pull-back by h_K of the hyperbolic metric of $\mathbb{H}^n$ is the Klein model for hyperbolic n-space. The distance $d(x, y)$ between two points x, y in this model is given by the formula:*

$$\cosh d(x, y) = \frac{1 - (x \mid y)}{\sqrt{1 - \|x\|^2}\sqrt{1 - \|y\|^2}},$$

where $(x \mid y)$ is the Euclidean scalar product and $\| \cdot \|$ is the Euclidean norm.

Given two distinct points $x, y \in B^n$, the unique hyperbolic geodesic line containing x and y is the intersection of B^n with the affine line in $\mathbb{R}^n$ through x and y. Let x_∞ and y_∞ be the two intersection points of this line with the boundary $\mathbb{S}^{n-1}$ of B^n, arranged so that x_∞, x, y, y_∞ occur in order on the line through x and y. The hyperbolic distance $d(x, y)$ between x and y is given by

$$d(x, y) = \frac{1}{2} \log(x, y, x_\infty, y_\infty) = \frac{1}{2} \log \left(\frac{\|x - y_\infty\|.\|y - x_\infty\|}{\|x - x_\infty\|.\|y - y_\infty\|} \right).$$

Proof. The first formula for $d(x, y)$ is a straightforward computation. Arguing as we did in (6.1), one can use the projective invariance of the cross ratio and the fact that $\mathrm{Isom}(\mathbb{H}^2)$ acts transitively on pairs a fixed distance apart, to reduce the verification of the second formula to the case $x = (0, \ldots, 0)$ and $y = (\frac{\sinh r}{\cosh r}, 0, \ldots, 0)$, where $r = d(x, y)$. In this case $x_\infty = (-1, 0, \ldots, 0)$ and $y_\infty = (1, 0, \ldots, 0)$, and $(x, y, x_\infty, y_\infty)$ is e^{2r}. □

A great advantage of the Klein model is that the geodesic lines are very easy to describe: they are the non-empty intersection of B^n with Euclidean lines. However, the angle between two geodesic segments in the Klein model that issue from the same point is *not* equal to the Euclidean angle observed by simply regarding these segments in $B^n \subset \mathbb{E}^n$. Thus considerations of angles in the Klein model do not conform to Euclidean intuition.

For instance, orthogonality can be understood as follows. We can consider the projective space $\mathbb{P}^n$ as the union of its affine part $\mathbb{E}^n$ and the hyperplane at infinity. A hyperplane in the Klein model is a non-empty intersection of B^n with a hyperplane H of $\mathbb{P}^n$. The pole of this hyperplane is, by definition, the point of $x \in \mathbb{P}^n \setminus B^n$ such that the lines which issue from x and are tangent to the boundary $\mathbb{S}^{n-1}$ of B^n cut $\mathbb{S}^{n-1}$ along $H \cap \mathbb{S}^{n-1}$.

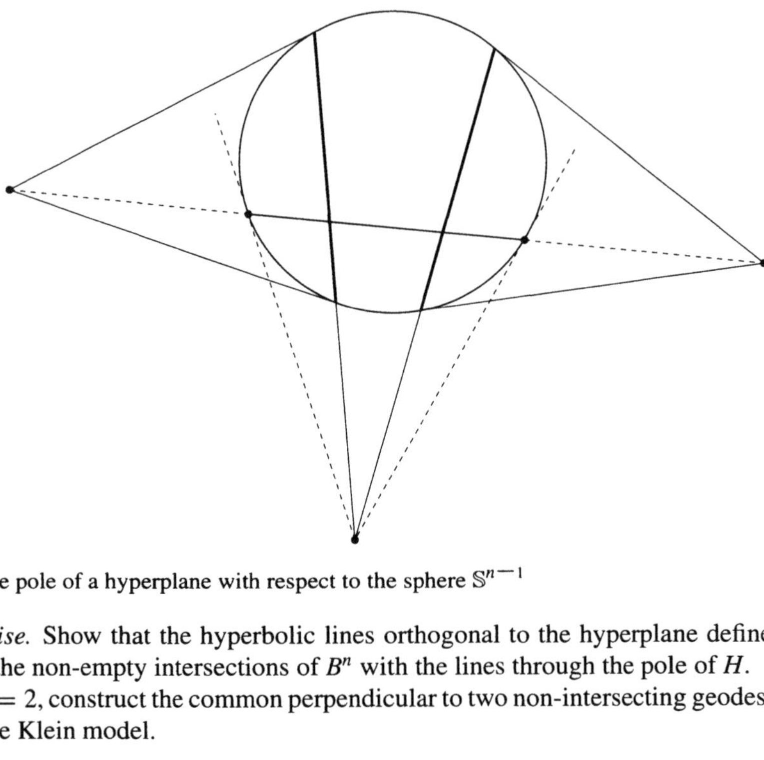

Fig. 6.2 The pole of a hyperplane with respect to the sphere $\mathbb{S}^{n-1}$

6.3 Exercise. Show that the hyperbolic lines orthogonal to the hyperplane defined by H are the non-empty intersections of B^n with the lines through the pole of H.

For $n = 2$, construct the common perpendicular to two non-intersecting geodesic lines in the Klein model.

The Möbius Group

6.4 Definitions.

(1) Let $\widehat{\mathbb{E}^n} = \mathbb{E}^n \cup \{\infty\}$ be the one point compactification of $\mathbb{E}^n$. There is a natural homeomorphism from $\mathbb{S}^n \subset \mathbb{E}^{n+1}$ to $\widehat{\mathbb{E}^n}$, given by *stereographic projection* p_N from the north pole $N = (0, \dots, 0, 1) \in \mathbb{S}^n$: identify $\mathbb{E}^n$ with the hyperplane $x_{n+1} = 0$ in $\mathbb{E}^{n+1}$ via the map $(x_1, \dots, x_n) \mapsto (x_1, \dots, x_n, 0)$; the stereographic projection $p_N : \mathbb{S}^n \to \widehat{\mathbb{E}^n}$ maps the north pole N to ∞ and maps each $x \in \mathbb{S}^n \setminus \{N\}$ to the point of intersection of the hyperplane $\mathbb{E}^n$ with the line through N and x.

(2) A *circle* in $\widehat{\mathbb{E}^n}$ is either a Euclidean circle in $\mathbb{E}^n$ or the union of a straight line in $\mathbb{E}^n$ and $\{\infty\}$. More generally, a *k-sphere* in $\widehat{\mathbb{E}^n}$ is either a k-dimensional sphere in $\mathbb{E}^n$ or the union of a k-dimensional affine subspace of $\mathbb{E}^n$ and $\{\infty\}$.

(3) The *cross ratio* of four distinct points x, y, v, w situated on a circle of $\widehat{\mathbb{E}^n}$ is the real number (x, y, v, w) defined by the formula

$$(x, y, v, w) = \frac{\|x - w\|.\|y - v\|}{\|x - v\|.\|y - w\|},$$

where $\|\cdot\|$ is the Euclidean norm; by convention, $\frac{x}{\infty} = 0$ and $\frac{\infty}{\infty} = 1$, and $\|x - \infty\| = \infty$ for all $x \in \mathbb{E}^n$.

(4) An *inversion* (or reflection) I_S of $\widehat{\mathbb{E}^n}$ with respect to an $(n-1)$-sphere S is defined as follows. If S is a sphere contained in $\mathbb{E}^n$ with centre a and radius $r > 0$, then I_S exchanges a and ∞ and is defined otherwise by

$$I_S(x) = \frac{r^2}{\|x - a\|^2}(x - a) + a.$$

In other words, I_S sends each point $x \neq a$ to a point on the geodesic ray from a that passes through x; this point is characterized by the fact that the product of its distance from a with the distance from x to a is equal to r^2. In particular, the points of S are precisely those left fixed by I_S. If S is the union of $\{\infty\}$ and a hyperplane $H \subseteq \mathbb{E}^n$, then I_S fixes ∞ and its restriction to $\mathbb{E}^n$ is the Euclidean reflection r_H.

We leave the reader to check the following properties of inversions.

6.5 Proposition. *Let I_S be an inversion of $\widehat{\mathbb{E}^n}$ with respect to a sphere S. Then,*

(1) *I_S maps spheres of $\widehat{\mathbb{E}^n}$ to spheres of $\widehat{\mathbb{E}^n}$;*

(2) *I_S preserves the cross ratio;*

(3) *I_S preserves the Euclidean angles between intersecting spheres;*

(4) *the stereographic projection p_N from $\mathbb{S}^n$ onto $\widehat{\mathbb{E}^n}$ is the restriction to $\mathbb{S}^n$ of the inversion of $\widehat{\mathbb{E}^{n+1}}$ with respect to the sphere whose center is the north pole N and whose radius is $\sqrt{2}$.*

From this proposition it follows that the stereographic projection p_N maps spheres to spheres and preserves cross ratios and angles. We define *an inversion of $\mathbb{S}^n$ with respect to an $(n-1)$-sphere $S \subset \mathbb{S}^n$* to be the restriction to $\mathbb{S}^n$ of the inversion of $\widehat{\mathbb{E}^{n+1}}$ with respect to the unique n-sphere that is orthogonal to $\mathbb{S}^n$ and contains S. Note that stereographic projection p_N conjugates such an inversion of $\mathbb{S}^n$ to the inversion of $\widehat{\mathbb{E}^n}$ with respect to the $(n-1)$-sphere $p_N(S)$.

6.6 Definition. The Möbius group Möb(n) (resp. Möb($\mathbb{S}^n$)) is the group of transformations of $\widehat{\mathbb{E}^n}$ (resp. of $\mathbb{S}^n$) generated by inversions in $(n-1)$-spheres.

Exercise. Show that the subgroup of the Möbius group Möb(n) fixing the point ∞ is the group of similarities of $\mathbb{E}^n$ (extended by the identity on $\{\infty\}$). For instance any

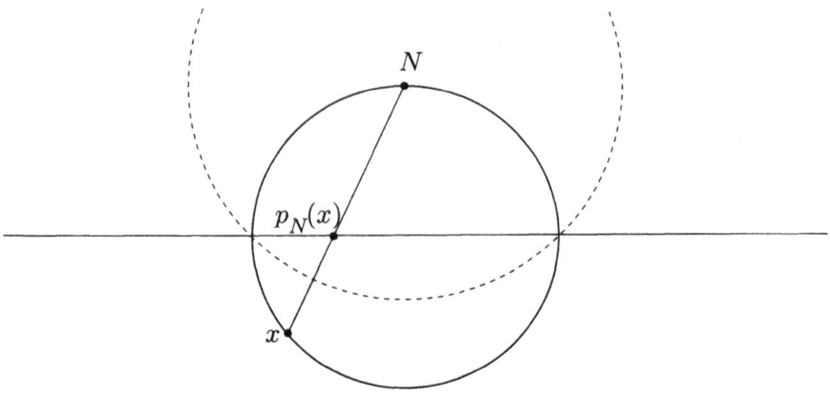

Fig. 6.3 The stereographic projection viewed as the restriction of an inversion

homothety fixing the origin is the product of two inversions with respect to spheres centred at the origin.

The Poincaré Ball Model for $\mathbb{H}^n$

As with the Klein model, in the Poincaré ball model the points of hyperbolic space are represented by the points of the unit ball B^n in $\mathbb{E}^n$, but the geometry imposed upon the ball is quite different. In particular, we shall see that the geodesic lines in the Poincaré model are the intersection of B^n with those Euclidean lines and circles which meet the boundary of B^n orthogonally. A great advantage of the Poincaré model is that it is conformally correct in the sense that the angle between two geodesic paths issuing from a point is equal to the Euclidean angle between these paths.

6.7 The Poincaré Ball Model. We define the Poincaré metric on B^n by pulling back the metric from $\mathbb{H}^n \subseteq \mathbb{E}^{n,1}$ via the homeomorphism $h_P : B^n \to \mathbb{H}^n$ given by

$$h_P(x) = \left(\frac{2x}{1-\|x\|^2}, \frac{1+\|x\|^2}{1-\|x\|^2}\right) \in \mathbb{R}^n \times \mathbb{R} = \mathbb{E}^{n,1},$$

where $\|x\|$ is the Euclidean norm of x. This map associates to x the point of intersection of $\mathbb{H}^n$ with the line through the points $(0, -1)$ and $(x, 1)$ in $\mathbb{R}^n \times \mathbb{R} = \mathbb{E}^{n,1}$.

More explicitly, the hyperbolic distance $d(x, y)$ between $x, y \in B^n$ in the Poincaré model is given by the formula

$$\cosh d(x,y) = \frac{(1+\|x\|^2)(1+\|y\|^2) - 4(x \mid y)}{(1-\|x\|^2)(1-\|y\|^2)} = 1 + 2\frac{\|x-y\|^2}{(1-\|x\|^2)(1-\|y\|^2)},$$

where $(x \mid y)$ is the Euclidean scalar product.

6.8 Proposition. *In the Poincaré ball model for hyperbolic n-space, the unique geodesic segment joining distinct points $x, y \in B^n$ is the arc which they bound on the unique Euclidean circle (or line) that passes through them and intersects the boundary of B^n orthogonally. If x_∞ and y_∞ are the points of intersection of this circle with $\mathbb{S}^{n-1}$, arranged so that x_∞, x, y, y_∞ occur in order on the circle, then the hyperbolic distance between x and y is given by the formula*

$$d(x, y) = \log (x, y, x_\infty, y_\infty).$$

Proof. We shall deduce the stated properties of the Poincaré model from those of the Klein model by examining the following explicit description of the isometry $h := (h_K)^{-1}h_P : B^n \to B^n$ between the two models (see figure 6.4). Briefly, this isometry is obtained as the composition of two projections, the first stereographic, the second orthogonal.

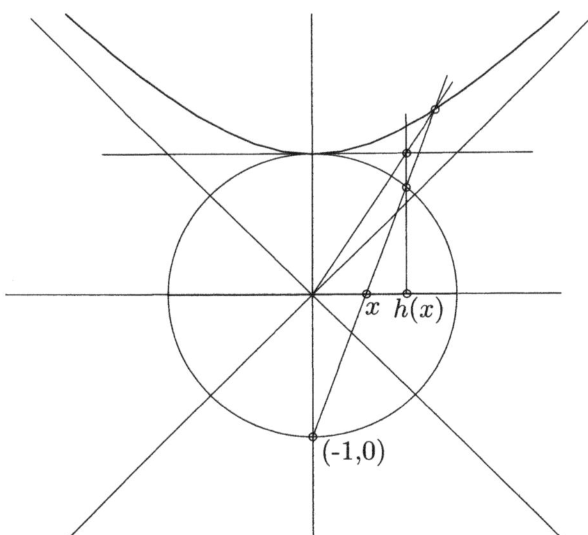

Fig. 6.4 The map $h := (h_K)^{-1}h_P$ from the Poincaré ball to the Klein model

Consider B^n as the intersection of the unit ball B^{n+1} in $\mathbb{E}^{n+1} = \mathbb{R}^n \times \mathbb{R}$ with the hyperplane $\mathbb{R}^n \times \{0\}$. Let $p_S : B^n \to \mathbb{S}^n \subseteq \mathbb{R}^n \times \mathbb{R}$ be the stereographic projection from the south pole of $\mathbb{S}^n$, sending B^n onto the upper hemisphere of $\mathbb{S}^n$; namely, $p_S : x \mapsto (\frac{2x}{1+\|x\|^2}, \frac{1-\|x\|^2}{1+\|x\|^2})$. In other words, p_S is the restriction to B^n of the inversion of $\widehat{\mathbb{E}}^{n+1}$ in the sphere with centre $(0, \dots, 0, -1)$ and radius $\sqrt{2}$. The isometry $h = (h_K)^{-1}h_P : x \mapsto \frac{2x}{1+\|x\|^2}$ is obtained by post-composing the stereographic projection p_S with the orthogonal projection from the upper hemisphere of $\mathbb{S}^n \subseteq \mathbb{R}^n \times \{0\}$ onto $B^n \subseteq \mathbb{R}^n \times \{0\}$.

We claim that the isometry h sends the set of arcs of those circles and lines in B^n that are orthogonal to the boundary $\mathbb{S}^{n-1}$ onto the set of straight line segments in B^n. Indeed p_S sends the former set of arcs onto the set of arcs in the upper hemisphere of $\mathbb{S}^n$ that are contained in circles orthogonal to the equatorial sphere $\mathbb{S}^{n-1}$; and the orthogonal projection from the upper hemisphere of $\mathbb{S}^n$ onto B^n then sends the set of such circular arcs onto the set of straight line segments in B^n. Thus, since the geodesic segment joining two distinct points in the Klein model is a straight line segment, the geodesic arc joining two distinct points $x, y \in B^n$ in the Poincaré model is the arc bounded by x and y on the unique Euclidean circle (or line) that passes through them and meets $\mathbb{S}^{n-1}$, the boundary of B^n, orthogonally. Let x_∞ and y_∞ be the points of intersection of this circle with $\mathbb{S}^{n-1}$, arranged so that x_∞, x, y, y_∞ occur in order on the circle.

Because p_S is the restriction of an inversion of $\widehat{\mathbb{E}^{n+1}}$ the cross ratios $(x, y, x_\infty, y_\infty)$ and $(p_S(x), p_S(y), x_\infty, y_\infty)$ are equal (6.6(4)). The points $p_S(x), p_S(y), x_\infty, y_\infty$ lie on a semicircle orthogonal to $\mathbb{R}^n \times \{0\}$, so it is easy to calculate the cross ratio of their projections onto $\mathbb{R}^n \times \{0\}$:

6.9 Exercise. Let $p, q \in \mathbb{E}^n$ be distinct points on a semicircular arc with endpoints a and b. Let p' and q' denote the orthogonal projections of p and q onto the line segment $[a, b]$. Prove that $(p, q, a, b) = (p', q', a, b)^{1/2}$.

Hence $(h(x), h(y), x_\infty, y_\infty) = (p_S(x), p_S(y), x_\infty, y_\infty)^{1/2} = (x, y, x_\infty, y_\infty)^{1/2}$. Therefore, $\log(x, y, x_\infty, y_\infty)$ is equal to $\frac{1}{2}\log(h(x), h(y), x_\infty, y_\infty)$, which is the hyperbolic distance in the Klein model between $h(x)$ and $h(y)$, by (6.2). Since h is an isometry, this establishes the desired formula for $d(x, y)$ in the Poincaré model. □

In Chapter 2 we described how reflections through hyperplanes generate $\operatorname{Isom}(\mathbb{H}^n)$. In the Klein model for hyperbolic space (based on B^n), the hyperplanes are simply the intersections of B^n with Euclidean hyperplanes. The next proposition describes the hyperplanes in the Poincaré ball model of hyperbolic space, and the subsequent corollary records what this description tells us about the structure of the isometry group of the model.

6.10 Proposition. *Let $h_P : B^n \to \mathbb{H}^n$ be the isometry from the Poincaré ball to $\mathbb{H}^n$ defined in (6.7). This map sends the set of intersections of B^n with $(n-1)$-dimensional spheres $S \subseteq \widehat{\mathbb{E}^n}$ orthogonal to $\mathbb{S}^{n-1}$ bijectively onto the set of hyperplanes $H \subseteq \mathbb{H}^n$. Moreover, h_P conjugates inversion in S to reflection through $h_P(S)$, i.e., $h_P I_S h_P^{-1} = r_{h_P(S)}$.*

We proved in (2.18) that reflections through hyperplanes generate $\operatorname{Isom}(\mathbb{H}^n)$, hence:

6.11 Corollary. *The group of isometries of the Poincaré ball model for $\mathbb{H}^n$ is generated by the restrictions to B^n of inversions of $\widehat{\mathbb{E}^n}$ in spheres orthogonal to $\mathbb{S}^{n-1}$. Thus* $\operatorname{Isom}(\mathbb{H}^n)$ *is naturally isomorphic to the Möbius group* $\text{Möb}(\mathbb{S}^{n-1})$.

Proof of 6.10. One way to prove this proposition is to calculate an explicit formula for $h_P^{-1} r_H h_P$, the conjugate of a hyperplane reflection. We indicate two other, more instructive, proofs.

Let S be an $(n-1)$-sphere orthogonal to $\mathbb{S}^{n-1}$, and consider the restriction to B^n of the inversion I_S. According to (6.5), I_S sends the set of circles orthogonal to $\mathbb{S}^{n-1}$ to itself and preserves cross ratios. Hence, by the the preceding proposition, I_S is an isometry of the Poincaré ball. Its fixed point set is $S \cap B^n$. But, according to (2.18), reflections in hyperplanes are the only isometries of $\mathbb{H}^n$ whose fixed point sets separate $\mathbb{H}^n$ into two connected components; and hyperplanes can be characterized as those subsets of $\mathbb{H}^n$ which arise as such fixed point sets. Since the fixed point set of the isometry $h_P I_S h_P^{-1}$ has this separation property, and its fixed point set is $h_P(S)$, we deduce that $h_P I_S h_P^{-1}$ is the reflection in the hyperplane $h_P(S)$.

It remains to be shown that every hyperplane in $\mathbb{H}^n$ arises as $h_P(S)$ for some S. Since every hyperplane is the bisector of some pair of points in $\mathbb{H}^n$, it suffices to show that the set of points in the Poincaré model that are equidistant from a fixed pair of points x_0 and y_0 form an $(n-1)$-sphere orthogonal to $\mathbb{S}^{n-1}$. According to the first formula in (6.7), the set of such points is determined by the equation:

$$\|x_0 - z\|^2 = k\|y_0 - z\|,$$

where $k = (1 - \|x\|^2)/(1 - \|y\|^2)$. It is an ancient observation of Apollonius that this is the equation of a sphere (if $k \neq 1$) or a Euclidean hyperplane (if $k = 1$). Moreover, since we know that the geodesic lines in the Poincaré model are the intersection with B^n of (Euclidean) lines and circles orthogonal to $\mathbb{S}^{n-1}$, and since we also know that the connected components of the complement of a hyperplane in $\mathbb{H}^n$ are geodesically convex, the above bisecting sphere/hyperplane must meet $\mathbb{S}^{n-1}$ orthogonally, as was to be proved. □

A different approach to proving the proposition is to first factor h_P as $h_K h$, where $h = (h_K)^{-1} h_P$ is the map from the Poincaré to the Klein model considered in (6.8). As in that proof, we write $h = \pi \circ p_S$, where p_S is a stereographic projection sending B^n onto the upper hemisphere of the unit sphere $\mathbb{S}^n_+ \subseteq \mathbb{R}^n \times \mathbb{R}$ and π is orthogonal projection onto $B^n = B^n \times \{0\}$. The map p_S sends the set of intersections of B^n with $(n-1)$-spheres orthogonal to $\mathbb{S}^{n-1}$ onto the set of half-spheres in $\mathbb{S}^n_+$ that are orthogonal to the equator $\mathbb{S}^{n-1}$. The orthogonal projection π sends the set of such half-spheres onto the set of intersections of B^n with affine hyperplanes of $\mathbb{E}^n$. The latter is precisely the set of hyperbolic hyperplanes in the Klein model. To finish the proof, one notes that if S is an $(n-1)$-sphere orthogonal to $\mathbb{S}^{n-1}$, then $h_P I_S h_P^{-1}$ is an isometry that fixes the hyperplane $h_P(S)$; it is not the identity, so it must be a reflection through this hyperplane.

The Poincaré Half-Space Model for $\mathbb{H}^n$

6.12 Definitions. Let H^n denote the upper half-space $\{(x_1, \ldots, x_n) \in \mathbb{E}^n \mid x_n > 0\} \subseteq \widehat{\mathbb{E}^n}$. The *Cayley transform* is the restriction to B^n of the inversion $I_S : \widehat{\mathbb{E}^n} \to \widehat{\mathbb{E}^n}$, where S is the sphere with centre $(0, \ldots, 0, -1)$ and radius $\sqrt{2}$.

The Cayley transform gives a homeomorphism from the Poincaré ball model for $\mathbb{H}^n$ to H^n. We define the *Poincaré half-space model* for $\mathbb{H}^n$ to be the half-space H^n endowed with the unique metric with respect to which this homeomorphism is an isometry.

The *boundary* ∂H^n of the half-space model H^n is the set of points $\{(x_1, \ldots, x_n) \in \mathbb{E}^n \mid x_n = 0\} \cup \{\infty\} \subset \widehat{\mathbb{E}^n}$. One can extend the Cayley transform by considering the restriction of the above inversion $I_S : \widehat{\mathbb{E}^n} \to \widehat{\mathbb{E}^n}$ to the closure of B^n; the resulting map sends $\mathbb{S}^{n-1}$, the boundary of B^n, bijectively onto ∂H^n.

Applying the considerations of (6.5) to the Cayley transform, one sees immediately that:

6.13 Proposition. *Propositions 6.8 and 6.10 and Corollary 6.11 remain valid if one replaces the Poincaré ball model B^n by the half-space model H^n and $\mathbb{S}^{n-1}$ by ∂H^n.*

In Chapter II.8 we shall describe a natural compactification $\overline{\mathbb{H}}^n$ of $\mathbb{H}^n$ that depends only on the metric (and not on any specific model). The subspace formed by the ideal points of this compactification is homeomorphic to $\mathbb{S}^{n-1}$, and there are natural bijections from it to the sphere bounding the ball model in $\mathbb{E}^n$, and to ∂H^n. We shall see that isometries of $\mathbb{H}^n$ extend uniquely to homeomorphisms of $\overline{\mathbb{H}}^n$; the induced action of $\text{Isom}(H^n)$ on $\partial \mathbb{H}^n$ is faithful and transitive.

We have seen that the generators of the isometry group of the Poincaré ball model are restrictions to B^n of inversions of $\widehat{\mathbb{E}^n}$. Thus they extend to continuous homeomorphisms of the closed ball $B^n \cup \mathbb{S}^{n-1}$. Similarly, there is a natural extension of isometries of the half-space model to homeomorphisms of ∂H^n. With respect to these actions, the above bijections from $\partial \mathbb{H}^n$ to $\mathbb{S}^{n-1}$ and ∂H^n are equivariant.

The study of individual isometries of $\mathbb{H}^n$ is well-documented and we shall not repeat it here. However, it is worth recording some important examples: for the action of $\text{Isom}(\mathbb{H}^n)$ on the Poincaré ball model, the subgroup fixing $0 \in B^n \subseteq \mathbb{E}^n$ consists of the restrictions to B^n of the usual linear action of $O(n)$ on $\mathbb{E}^n$; for the action of $\text{Isom}(\mathbb{H}^n)$ on the Poincaré half-space model H^n, the subgroup fixing the point $\{\infty\} \in \partial H^n$ is generated by the restrictions to H^n of Euclidean translations, homotheties and reflections of $\mathbb{E}^n$ preserving H^n — it is isomorphic to $\text{Isom}(\mathbb{E}^{n-1}) \rtimes \mathbb{R}$, where the left hand factor consists of those isometries of H^n that preserve subsets of the form $\mathbb{R}^{n-1} \times \{a\}$ and the action of the right hand factor is conjugation by Euclidean homotheties $a \mapsto e^t a$.

Since the actions of Isom($\mathbb{H}^n$) on $\mathbb{H}^n$ and $\partial\mathbb{H}^n$ are both transitive, these examples serve to describe the isotropy subgroups of arbitrary points in $\overline{\mathbb{H}}^n$, up to conjugation in Isom($\mathbb{H}^n$).

Isometries of $\mathbb{H}^2$

We conclude our considerations of the standard models for hyperbolic space with a classical description of Isom($\mathbb{H}^2$). This requires that we introduce a complex coordinate in the Poincaré model: identify $\mathbb{E}^2$ with the complex plane $\mathbb{C}$, the unit ball B^2 with the disc $\{z \in \mathbb{C} \mid |z| \leq 1\}$ and the upper half-plane H^2 with $\{z \in \mathbb{C} \mid \Im z > 0\}$, where $\Im z$ is the imaginary part of z. Given two distinct points z and w in B^n or H^n, let c be the unique circle in the Riemann sphere $\mathbb{C} \cup \{\infty\} = \widehat{\mathbb{E}^2}$ that passes through z and w and meets the boundary of the model orthogonally. Let z_∞ and w_∞ be the two points at which the circle c cuts the boundary of the model, labelled so that z_∞, z, w, w_∞ occur in order on c. The hyperbolic distance between z and w is

$$d(z, w) = \log{(z, w, z_\infty, w_\infty)},$$

where $(z, w, z_\infty, w_\infty) = \frac{(z-w_\infty)(w-z_\infty)}{(z-z_\infty)(w-w_\infty)}$ is real and greater than one.

In (6.12) we related B^n to H^n by the Cayley transform; in the present context this is given by the formula $z \mapsto \frac{1-i\bar{z}}{\bar{z}-1}$. (We warn the reader that the term Cayley transform is also commonly used to describe the orientation preserving isometry $H^2 \to B^2$ defined by $z \mapsto \frac{z-i}{z+i}$.)

6.14 Proposition. *The group of isometries of the Poincaré half-space model for $\mathbb{H}^2$ is precisely the group of maps $H^2 \to H^2$ obtained by letting the group* GL(2, $\mathbb{R}$) *act thus:*
if $ad - bc > 0$ then the action of $\begin{pmatrix} a & b \\ c & d \end{pmatrix}$ is

$$z \mapsto \frac{az+b}{cz+d};$$

if $ad - bc < 0$ then the action of $\begin{pmatrix} a & b \\ c & d \end{pmatrix}$ is

$$z \mapsto \frac{a\bar{z}+b}{c\bar{z}+d}.$$

The kernel of this action consists of all scalar multiples of the identity matrix. Thus Isom($\mathbb{H}^2$) $\cong$ PGL(2, $\mathbb{R}$).

The group of isometries of the ball model for $\mathbb{H}^2$ is precisely the group of maps of the form:

$$z \mapsto \frac{pz+\overline{q}}{qz+\overline{p}}, \qquad z \mapsto \frac{p\bar{z}+\overline{q}}{q\bar{z}+\overline{p}},$$

where $p, q \in \mathbb{C}$ and $|p|^2 - |q|^2 = 1$.

Proof. A direct calculation shows that the given maps are indeed isometries of the half-space model H^2, and the map $\mathrm{GL}(2,\mathbb{R}) \to \mathrm{Isom}(H^2)$ described by the given formulae is a homomorphism with kernel the group of scalar multiples of the identity. To see that the given action is transitive on H^2, it suffices to consider $z \mapsto az$ with $a \geq 0$ and $z \mapsto z + b$.

The Cayley transform conjugates the given maps of H^2 to those of B^2, hence the latter group of maps acts transitively on B^2. This group also contains $O(2)$ (in the form of the maps $z \mapsto pz$ and $z \mapsto p\bar{z}$, with $|p| = 1$), and since this is the stabilizer of 0 in $\mathrm{Isom}(B^2)$, it follows that the given maps are the whole of $\mathrm{Isom}(\mathbb{H}^2)$. □

M^n_κ as a Riemannian Manifold

In Chapter 2 we defined the metric spaces M^n_κ. The main purpose of this section is to exhibit M^n_κ as the length space associated to a Riemannian manifold. Initially, we defined the set M^n_κ to be a certain submanifold of a Euclidean space. We shall give an explicit description of a Riemannian metric on this submanifold. We shall then prove that M^n_κ is the associated length space. We do so by combining some facts about $\mathrm{Isom}(M^n_\kappa)$ with an explicit calculation of the Riemannian length of certain curves. We shall also give explicit formulae for the Riemannian metrics on the Poincaré models of hyperbolic space.

We begin by recalling once more that a Riemannian metric on a differentiable manifold X is an assignment to each tangent space T_xX of a positive definite bilinear form (i.e., a scalar product) such that these forms vary continuously with x. By relaxing the requirement that these bilinear forms must be positive definite one obtains the notion of a pseudo-riemannian metric.

The Riemannian Metric on M^2_κ

As we pointed out in Chapter 3, $\mathbb{E}^n$ has a natural Riemannian metric, the Euclidean Riemannian metric $ds^2 = \sum_{i=1}^n dx_i^2$ (see Chapter 3 for notation). More explicitly, one identifies the tangent space $T_x\mathbb{E}^n$ of $\mathbb{E}^n$ at x with the set of pairs (x, v), where v is a vector in $\mathbb{E}^n$; then one endows $T_x\mathbb{R}^n$ with the unique vector space structure and scalar product so that the projection $(x, v) \mapsto v$ is a linear isometry of $T_x\mathbb{E}^n$ onto $\mathbb{E}^n$.

$\mathbb{S}^n$ is the differentiable submanifold of $\mathbb{E}^{n+1}$ consisting of vectors of Euclidean norm one. As such it inherits an induced Riemannian metric: the tangent space $T_x\mathbb{S}^n$ to $\mathbb{S}^n$ at $x \in \mathbb{S}^n$ is the vector subspace of $T_x\mathbb{E}^{n+1}$ consisting of pairs (x, v) with $(x \mid v) = 0$ (Euclidean scalar product), and the restriction to this subspace of the given scalar product on $T_x\mathbb{E}^{n+1}$ defines a Riemannian metric on $\mathbb{S}^n$. In other words, for every $x \in \mathbb{S}^n$, the map $(x, v) \mapsto v$ gives a bijection from $T_x\mathbb{S}^n$ to the n-dimensional subspace $x^\perp \subset \mathbb{E}^{n+1}$ orthogonal to x, and the Riemannian metric on $\mathbb{S}^n$ is the unique assignment of scalar products that makes all of these maps isometries.

In an entirely analogous way, one can obtain a Riemannian metric on $\mathbb{H}^n$ by regarding it as a differentiable submanifold of $\mathbb{E}^{n,1}$. The latter has a natural pseudo-riemannian metric $ds^2 = (\sum_{i=1}^n dx_i^2) - dx_{n+1}^2$; the desired Riemannian metric on $\mathbb{H}^n$ is obtained by restricting the given bilinear form on $T_x\mathbb{E}^{n,1}$ to $T_x\mathbb{H}^n$ for each $x \in \mathbb{H}^n$ (these restrictions are positive definite). More explicitly: the tangent space $T_x\mathbb{H}^n$ to $\mathbb{H}^n$ at $x \in \mathbb{H}^n$ is the vector subspace of $T_x\mathbb{E}^{n,1}$ consisting of pairs (x, v) such that $v \in x^\perp$ (that is, $\langle x \mid v\rangle=0$); for every $x \in \mathbb{H}^n$, the map $(x, v) \mapsto v$ gives a bijection from $T_x\mathbb{H}^n$ to the n-dimensional subspace $x^\perp \subset \mathbb{E}^{n,1}$ orthogonal to x; the desired Riemannian metric on $\mathbb{H}^n$ is the unique assignment of scalar products that makes all of these maps isometries. (The restriction of $\langle \cdot \mid \cdot\rangle$ to $x^\perp$ is positive definite for all $x \in \mathbb{H}^n$.)

The natural Riemannian metric on M^n_κ is obtained from the preceding special cases by scaling. There are two equivalent ways of doing this. If $\kappa > 0$, then one can either identify M^n_κ with $\mathbb{S}^n$ and multiply each of the scalar products defining the Riemannian metric on the latter by a factor of $1/\kappa$, or else one can view M^n_κ as the Riemannian submanifold $\{x \in \mathbb{E}^{n+1} \mid (x \mid x) = 1/\kappa\} \subseteq \mathbb{E}^{n+1}$ and give it the induced Riemannian metric. Similarly, if $\kappa < 0$, then one can either identify M^n_κ with $\mathbb{H}^n$ and multiply each of the scalar products defining the Riemannian metric on the latter by a factor of $1/-\kappa$, or else one can view M^n_κ as the Riemannian submanifold $\{x \in \mathbb{E}^{n+1} \mid \langle x \mid x\rangle = 1/\kappa\} \subseteq \mathbb{E}^{n,1}$ and give it the induced (pseudo-)Riemannian metric.

We should be absolutely clear about the logical structure of what we are doing at this point. We have introduced a Riemannian metric on the manifold M^n_κ; this allows us to speak, for example, of the Riemannian length of curves in M^n_κ and of Riemannian isometries from M^n_κ to itself or other manifolds. But *a priori* the Riemannian length of a curve may be different from its length in the metric space M^n_κ, and we may not assume that a Riemannian isometry from M^n_κ to itself (i.e., a map whose derivative preserves the Riemannian metric) is an isometry of the metric space M^n_κ, nor vice versa.

6.15 Lemma. *Given $\kappa \in \mathbb{R}$, suppose that M^n_κ is equipped with the Riemannian metric described above. Let d denote the metric on M^n_κ defined in (2.10).*

(1) *The action of* $\mathrm{Isom}(M^n_\kappa, d)$ *on M^n_κ is by Riemannian isometries.*

(2) *For all $x \in M^n_\kappa$ and $r > 0$ (with $r < D_\kappa = \pi/\sqrt{\kappa}$ if $\kappa > 0$), if $\kappa \neq 0$ then the induced Riemannian metric on $\{y \in M^n_\kappa \mid d(x, y) = r\} \subseteq M^n_\kappa$ makes it isometric, as a Riemannian manifold, to a sphere of radius $\frac{1}{\sqrt{|\kappa|}}\,|\sin(r\sqrt{\kappa})|$ in $\mathbb{E}^n$. (If $\kappa < 0$ then $\sin(r\sqrt{\kappa}) = \sin(ir\sqrt{-\kappa}) = i\sinh(r\sqrt{-\kappa})$.)*

Proof. Because the metric and Riemannian metric on M^n_κ were defined by compatible scaling processes, it is enough to consider the cases $M^n_\kappa = \mathbb{E}^n, \mathbb{S}^n, \mathbb{H}^n$. It is obvious that the group $D(n)$ of displacements of $\mathbb{E}^n$, the orthogonal group $O(n+1)$ acting as usual on $\mathbb{S}^n \subseteq \mathbb{E}^{n+1}$, and the group $O(n, 1)_+$ acting linearly on $\mathbb{H}^n \subseteq \mathbb{E}^{n,1}$, all preserve the Riemannian metric. Thus (1) is a consequence of (2.21).

Having proved (1), it suffices to verify assertion (2) for a particular choice of x, because $\text{Isom}(M^n_\kappa)$ acts transitively. In the case of $\mathbb{S}^n$, we let $x = (1, 0, \dots, 0)$. By definition, $\{y \in \mathbb{S}^n \mid d_{\mathbb{S}^n}(x, y) = r\} = \{y \in \mathbb{E}^{n+1} \mid \|y\| = 1,\ \cos r = (y \mid x)\} = \{(y_1, \dots, y_{n+1}) \in \mathbb{E}^{n+1} \mid y_1 = \cos r, \sum_{i>1} y_i^2 = \sin^2 r\}$, which is a Euclidean sphere of radius $\sin r$. The Riemannian metric which this set inherits as a submanifold of $\mathbb{S}^n$ is, by definition, the same as that which it inherits from the ambient manifold $\mathbb{E}^n$. This proves (2) in the case of $\mathbb{S}^n$.

The case of $\mathbb{H}^n \subseteq \mathbb{E}^{n,1}$ is analogous. Let $x = (0, \dots, 0, 1) \in \mathbb{H}^n \subseteq \mathbb{E}^{n,1}$. Then, $\{y \in \mathbb{H}^n \mid d_{\mathbb{H}^n}(x, y) = r\} = \{y \in \mathbb{E}^{n,1} \mid \langle y \mid y\rangle = -1,\ \cosh r = -\langle y \mid x\rangle\} = \{(y_1, \dots, y_{n+1}) \in \mathbb{E}^{n,1} \mid y_{n+1} = \cosh r, \sum_{i=1}^n y_i^2 = \sinh^2 r\}$. This is a sphere of radius $\sinh r$ in the affine subspace $\mathbb{E}^n \times \{\cosh r\} \subseteq \mathbb{E}^{n,1}$, and the projection of this subspace onto its first factor gives a (Riemannian) isometry to $\mathbb{E}^n$. □

6.16 The Exponential Map. We have constructed a Riemannian metric on M^n_κ, which gives a norm on each tangent space $T_xM^n_\kappa$. Each non-zero vector $v \in T_xM^n_\kappa$ can be written uniquely as ru, where r is a positive number and u is a vector of norm one. We call (r, u) the *polar coordinates* of v. Given a point $x \in T_xM^n_\kappa$, the *exponential map at* x is the map $\exp_x$ from $T_xM^n_\kappa$ to M^n_κ that sends the origin to x and sends the vector with polar coordinates (r, u) to the endpoint of the geodesic segment issuing from x with initial vector u and length r (as defined in Chapter 2). If $\kappa \le 0$ then the domain of $\exp_x$ is the whole of $T_xM^n_\kappa$; if $\kappa > 0$ then $\exp_x$ is only defined on the open ball of radius $D_\kappa = \pi/\sqrt{\kappa}$.

More explicitly, for the case $\kappa = -1$, if one identifies unit vectors in $T_x\mathbb{H}^n$ to unit vectors in $\mathbb{E}^{n,1}$, as described above, then

$$\exp_x(ru) = (\cosh r)x + (\sinh r)u.$$

Similarly, for the case $\kappa = 1$, if one identifies unit vectors in $T_x\mathbb{S}^n$ to unit vectors in $\mathbb{E}^{n+1}$, as described above, then

$$\exp_x(ru) = (\cos r)x + (\sin r)u.$$

In (2.11) we proved that for each pair of points $x, y \in M^n_\kappa$ (assuming that $d(x, y) < D_\kappa$), there is a unique geodesic joining x to y in M^n_κ. Hence $\exp_x$ is a bijection — onto M^n_κ if $\kappa \le 0$, and onto $B(x, \pi/\sqrt{\kappa})$ if $\kappa > 0$ — and the explicit expression above shows that $\exp_x$ is a diffeomorphism. In particular, one can use $\exp_x$ to pull the Riemannian metric on M^n_κ back to $T_xM^n_\kappa$, where the availability of polar coordinates facilitates an elegant and instructive description of it, (6.17(2)). This process is rather like using planar charts to describe the surface of the Earth.

Notation. The metric arising from the given scalar product on $T_xM^n_\kappa$ makes it isometric to Euclidean space, and hence there is an associated Riemannian metric on $T_xM^n_\kappa$. Let du^2 denote the induced Riemannian metric on the unit sphere in $T_xM^n_\kappa$.

6.17 Proposition.

(1) *For every $\kappa \in \mathbb{R}$, the metric on M^n_κ that was defined in (2.10) is equal to the length metric associated the Riemannian metric given above.*

(2) *For every $x \in M^n_\kappa$, the Riemannian metric on $T_xM^n_\kappa$ that is obtained by pulling back the given Riemannian metric on M^n_κ via the exponential map* $\exp_x$, *is given in polar coordinates (r, u) by the formula*

$$ds^2 = dr^2 - \frac{1}{\kappa}\sinh^2(\sqrt{-\kappa}\, r)\, du^2 \qquad \textit{if } \kappa < 0;$$

$$ds^2 = dr^2 + \frac{1}{\kappa}\sin^2(\sqrt{\kappa}\, r)\, du^2 \qquad \textit{if } \kappa > 0;$$

$$ds^2 = dr^2 + r^2\, du^2 \qquad \textit{if } \kappa = 0.$$

Proof. We first prove (2). The case $\kappa = 0$ is clear, because the exponential map is an isometry and the Euclidean Riemannian metric is given in polar coordinates by the formula $ds^2 = dr^2 + r^2du^2$. The given formulae for $\kappa \neq 0$ follow from 6.15(2) and the fact that the smooth curve $r \mapsto \exp_x(ru)$ is orthogonal to spheres centred at x.

We shall deduce (1) from (2) in the case $\kappa = -1$; the other cases are entirely similar. Given $x, y \in \mathbb{H}^n$ with $d(x, y) = r$, there is a unique unit vector $u_0 \in T_x\mathbb{H}^n$ such that $y = \exp_x(ru_0)$. Consider a piecewise differentiable curve $c : [0, 1] \to \mathbb{H}^n$ joining x to y. By writing c in the form $c(t) = \exp_x(r(t)u(t))$ we obtain a formula for its Riemannian length:

$$l_R(c) = \int_0^1 \sqrt{|\dot r(t)|^2 + (\sinh r(t))^2|\dot u(t)|^2}\, dt \geq \int_0^1 |\dot r(t)|\, dt \geq \left|\int_0^1 \dot r(t)dt\right| = r.$$

One obtains equality in this expression if and only if $\dot u(t) = 0$ and $\dot r(t) \geq 0$ for all $t \in [0, 1]$. These conditions are satisfied by the path, $t \mapsto \exp_x(tru_0)$ (and by no other, up to reparameterization). Thus the Riemannian distance between x and y is $r = d(x, y)$. □

6.18 Proposition. *The Riemannian metrics for the Poincaré ball and half-space models of hyperbolic space are:*

(1) *for the ball model $B^n \subseteq \mathbb{E}^n$,*

$$ds^2 = \frac{4ds_E^2}{(1 - \|x\|^2)^2}$$

(2) *and for the half-space model $\{x \in \mathbb{E}^n \mid x_n > 0\}$,*

$$ds^2 = \frac{ds_E^2}{x_n^2},$$

where $\|x\|$ is the Euclidean norm of x and $ds_E^2 = \sum_{i=1}^n dx_i^2$ is the Euclidean Riemannian metric.

Proof. The formula for the ball model can be calculated directly as the pull-back of the Riemannian metric on $\mathbb{H}^n$ by the isometry $h_P : B^n \to \mathbb{H}^n \subseteq \mathbb{E}^{n,1}$ defined prior to (6.8). Alternatively, one can verify that the isometries of the Poincaré ball preserve the given Riemannian metric; the action is transitive, so in order to prove the proposition it then suffices to calculate the Riemannian length of curves emanating from $O \in B^n$; this can be done using polar coordinates on B^n, as in the preceding proposition.

The Riemannian metric on the half-space model H^n can be deduced from that on B^n by using Cayley transform $B^n \to H^n$. □

6.19 Exercises

(1) Prove that the action of $\text{Isom}(M_\kappa^n)$ on the Riemannian manifold M_κ^n induces a simply transitive action on the associated bundle of n-frames, i.e. the space of $(n+1)$-tuples $(x, u_1, \dots, u_n)$, where $x \in M_\kappa^n$ and $u_1, \dots, u_n \in T_x M_\kappa^n$ are mutually orthogonal unit vectors. (cf. 2.15.)

(2) Using the Poincaré ball model, construct in the hyperbolic plane for each $n > 4$ a regular n-gon with vertex right angles, and more generally regular n-gons with vertex angles equal to $\alpha < (n-2)\pi/n$.

Given numbers $\alpha_1, \dots, \alpha_n \in (0, \pi)$ with $\sum_{i=1}^n \alpha_i < (n-2)\pi$, construct in $\mathbb{H}^2$ n-gons with vertex angles $\alpha_1, \dots, \alpha_n$.

(3) The map $(x_1, \dots, x_n) \mapsto (x_1, \dots, x_{n-1}, 1)$ reduces the length of any rectifiable path that is contained in the subspace of the upper half-space model consisting of points with $x_n < 1$.

(4) Fix $l > 0$. Let $S \subset \mathbb{E}^2$ and $S' \subset \mathbb{H}^2$ be circles of length l with centres c and c' respectively. Let $p, q \in S$ and $p', q' \in S'$ be points such that $\angle_c(p, q) = \angle_{c'}(p', q') > 0$ and suppose that the tangent lines to S at p and q (resp. to S' at p' and q') intersect at r (resp. r'). Show that $d_{\mathbb{H}}(r', p') > d_{\mathbb{E}}(r, p)$ and $d_{\mathbb{H}}(p', q') < d_{\mathbb{E}}(p, q)$.

(Hint: Consider the upper half-space model of hyperbolic space. When endowed with the induced path metric, the horosphere $x_3 = 1$ is isometric to $\mathbb{E}^2$; regard S as lying on this horosphere. Identify $\mathbb{H}^2$ with the unique 2-plane in $\mathbb{H}^3$ that contains S and take $p = p'$ and $q = q'$; note that the hyperbolic segment $[p, r']$ is the intersection of this 2-plane with the vertical 2-plane containing the Euclidean segment $[p, r]$. Argue that r is the image of r' under the projection map described in (3).)

Chapter I.7 M_κ–Polyhedral Complexes

The simplest examples of geodesic metric spaces that are not manifolds are provided by metric graphs, which we introduced in (1.9). In this section we shall study their higher dimensional analogues, M_κ–polyhedral complexes. Roughly speaking, an M_κ–polyhedral complex is a space that one gets by taking the disjoint union of a family of convex polyhedra from M_κ^n and gluing them along isometric faces (see (7.37)). The complex is endowed with the quotient metric (5.19). The main result in this chapter is the following theorem from Bridson's thesis [Bri91] (see (7.19)).

Theorem. *If an M_κ–polyhedral complex has only finitely many isometry types of cells, then it is a complete geodesic metric space.*

Metric Simplicial Complexes

We begin with an informal description of an M_κ-simplicial complex. There are technical reasons why simplicial complexes are easier to work with than general polyhedral complexes, so since any polyhedral complex can be made simplicial by subdivision (7.49), it makes sense to begin with the simplicial case.

There are two equivalent ways to view a metric simplicial complex. The first is to imagine building the complex from a disjoint union of geodesic simplices in M_κ by gluing simplices together along isometric faces. (One restricts the gluing so that each simplex injects into the quotient.) The quotient pseudometric (7.4) is called the *intrinsic pseudometric* of the complex. This is the approach to M_κ-simplicial complexes that we shall emphasize. An alternative approach is described in the appendix to this chapter. This second approach is more appropriate in situations where one wishes to use geometry to study a problem that has been encoded in the combinatorics of an abstract simplicial complex.

Whichever approach one takes, the issues that one faces are the same. One hopes to obtain a complete geodesic metric space, but as we have already seen in the 1-dimensional case (1.9), in order to do so one must add additional hypotheses. Under very mild hypotheses one can prove that the intrinsic pseudometric is a metric and the complex is a length space (7.10). Under the same hypotheses we prove that each point in the complex has a neighbourhood that is isometric to a neighbourhood of the vertex in the κ-cone over the space of directions at that point (7.16).

If the complex under consideration is locally compact, then one can deduce the existence of geodesics under mild hypotheses using the Hopf-Rinow Theorem (3.7). However we do not wish to assume that the complexes under consideration are locally compact. (This is particularly important for the applications to complexes of groups in Part III.) Instead, following [Bri91], we consider complexes that have only *finitely many isometry types of simplices*. (Note that whenever one has a cocompact group action on a complex, a natural thing to do is to metrize the simplices of the quotient and extend equivariantly; this gives a complex with only finitely many isometry types of cells.)

7.1 Geodesic Simplices in M^n_κ. Let $n \leq m$ be positive integers. An *n-plane*[10] in M^m_κ is, by definition, a subspace isometric to M^n_κ. We say that $(n+1)$ points in M^m_κ are *in general position* if they are not contained in any $(n-1)$-plane.

A *geodesic n-simplex* $S \subset M^m_\kappa$ is the convex hull of $(n+1)$ points in general position; the points are called the *vertices* of S. If $\kappa > 0$, then the set of vertices of S is required to lie in an open hemisphere, i.e. in an open ball of radius $D_\kappa/2$.

A *face* $T \subseteq S$ is the convex hull of a non-empty subset of the vertices of S; if $T \neq S$ then T is called a *proper* face. Note that a face of a geodesic n-simplex is a geodesic n'-simplex for some $n' \leq n$. The *interior* of S is the set of points that do not lie in any proper face.

7.2 Definition of an M_κ-Simplicial Complex. Let $(S_\lambda : \lambda \in \Lambda)$ be a family of geodesic simplices $S_\lambda \subset M^{n_\lambda}_\kappa$. Let $X = \bigcup_{\lambda\in\Lambda}(S_\lambda \times \{\lambda\})$ denote their disjoint union, let $\sim$ be an equivalence relation on X and let $K = X/\!\sim$. Let $p : X \to K$ be the natural projection and define $p_\lambda : S_\lambda \to K$ by $p_\lambda(x) := p(x, \lambda)$.

K is called an M_κ–simplicial complex if

(1) for every $\lambda \in \Lambda$, the map p_λ is injective, and

(2) if $p_\lambda(S_\lambda) \cap p_{\lambda'}(S_{\lambda'}) \neq \emptyset$ then there is an isometry $h_{\lambda,\lambda'}$ from a face $T_\lambda \subset S_\lambda$ onto a face $T_{\lambda'} \subset S_{\lambda'}$ such that $p(x, \lambda) = p(x', \lambda')$ if and only if $x' = h_{\lambda,\lambda'}(x)$.

The set of isometry classes of the faces of the geodesic simplices S_λ is denoted Shapes(K).

Terminology. M_0 (resp. M_1 , M_{-1}) simplicial complexes are often called piecewise Euclidean (resp. spherical, hyperbolic) complexes. M_κ–simplicial complexes are also called metric simplicial complexes, or simplicial complexes of piecewise-constant curvature.

7.3 Definitions (Simplices, Stars and Segments). Let K be as in (7.2). A subset $S \subset K$ is called an *m-simplex* if it is the image under some p_λ of an m-dimensional face of S_λ. If T is a subset of S and $p_\lambda^{-1}(T)$ is a face of $p_\lambda^{-1}(S)$, then T is called a *face* of S. The *interior* of S is the image under p_λ of the interior of $p_\lambda^{-1}(S)$. The *support* of a point $x \in K$ is the unique simplex supp(x) that contains x in its interior.

[10] Any such subspace is an intersection of hyperplanes.

We define a metric d_S on S by $d_S(p_\lambda(x), p_\lambda(y)) = d_{S_\lambda}(x, y)$. Condition 7.2(2) ensures that this is independent of λ. Observe that if T is a face of S then d_T is the restriction of d_S. A bijection $K_1 \to K_2$ between M_κ–simplicial complexes is called a *simplicial isometry*[11] if it maps each simplex of K_1 isometrically onto a simplex of K_2.

Let $x \in K$. The *(closed) star* of x, denoted $St(x)$, is the union of the simplices of K that contain x. The *open star* of x, denoted $st(x)$, is the union of the interiors of the simplices of K that contain x.

The image under p_λ of a geodesic segment $[x_\lambda, y_\lambda] \subseteq T_\lambda \subseteq S_\lambda$ is called a *segment* in the simplex $T = p_\lambda(T_\lambda)$ and is denoted $[x, y]$, where $x = p_\lambda(x_\lambda)$ and $y = p_\lambda(y_\lambda)$. The *length* of $[x, y]$ is defined to be $d_T(x, y)$. (Condition 7.2(2) ensures that the notation $[x, y]$ and the definition of length are unambiguous.)

K comes equipped with the quotient pseudometric described in (5.19) — the definition was in terms of chains. In the present setting, we can use the extra structure inherent in the definition of K to restrict the sort of chains that have to be considered. The following combinatorial device is useful in this regard.

7.4 Definition (m-Strings and the Intrinsic Pseudometric). An *m–string in K from x to y* is a sequence $\Sigma = (x_0, x_1, \dots, x_m)$ of points of K such that $x = x_0$, $y = x_m$, and for each $i = 0, \dots, m-1$, there exists a simplex $S(i)$ containing x_i and x_{i+1}. We call m the *size* of Σ, and define the *length* of Σ to be

$$l(\Sigma) := \sum_{i=0}^{m-1} d_{S(i)}(x_i, x_{i+1}).$$

Every m–string determines a path in K, given by the concatenation of the segments $[x_i, x_{i+1}]$. We denote this path $P(\Sigma)$. When the integer m is not important, we may refer to Σ simply as a *string*.

The *intrinsic pseudometric* on K is defined by:

$$d(x, y) := \inf\{l(\Sigma) \mid \Sigma \text{ a string from } x \text{ to } y\}.$$

If there is no string from x to y, then $d(x, y) := \infty$.

We now have two pseudometrics on K, the quotient pseudometric (defined in terms of m–chains (5.19)), and the intrinsic pseudometric (defined in terms of m–strings).

7.5 Lemma. *For any M_κ–simplicial complex K, the intrinsic pseudometric and the quotient pseudometric coincide. More precisely, there is an m–chain joining x to y in K if and only if there is an m–string of the same length joining x to y.*

[11] sometimes called an isometric isomorphism

Proof. We have $p : \coprod S_\lambda \to K$. Every m–chain $C = (\hat{x}_1, \hat{y}_1, \dots, \hat{x}_m, \hat{y}_m)$ projects to an m–string of the same length $(p(\hat{x}_1), p(\hat{x}_2), \dots, p(\hat{x}_m), p(\hat{y}_m))$. Conversely, to each m–string $\Sigma = (x_0, x_1, \dots, x_m)$ we can associate an m–chain $C_\Sigma = (\hat{x}_1, \hat{y}_1, \dots, \hat{x}_m, \hat{y}_m)$ such that for each i there exists $\lambda \in \Lambda$ with $\hat{x}_i = (x_i', \lambda)$, $\hat{y}_i = (y_i', \lambda)$, and $p_\lambda[x_i', y_i'] = [x_i, x_{i+1}]$. It is clear that C_Σ projects to Σ and that $l(\Sigma) = l(C_\Sigma)$. □

Until further notice, K will denote a fixed M_κ–simplicial complex and d will denote the intrinsic pseudometric on K.

7.6 Remark. If S is a simplex in K then $d_S(x, y) \geq d(x, y)$ for all $x, y \in S$, but one does not get equality in general. For instance, consider the 1–complex that has three vertices and three 1-simplices of lengths 1, 1 and 3 connecting the vertices pairwise. If S is the simplex of length 3, then the restriction of d to S does not coincide with d_S.

If Shapes(K) is finite, then one can always subdivide so as to arrange that $d = d_S$ for all simplices S.

The following simple example shows that the intrinsic pseudometric on K will not be a metric in general.

7.7 Example. Consider the metric graph that has two vertices and countably many edges e_n connecting these vertices, where e_n has length $1/n$. Subdivide the edges to make it simplicial. The distance between the two original vertices in the intrinsic pseudometric is zero.

The pathology in this example derives from the fact that there are points at which the following quantity is zero.

7.8 Definition. Let $x \in K$ and let

$$\varepsilon(x) := \inf \{\varepsilon(x, S) \mid S \subset K \text{ a simplex containing } x\},$$

where $\varepsilon(x, S) := \inf \{d_S(x, T) \mid T \text{ a face of } S \text{ and } x \notin T\}$. (If $S = \{x\}$ then we define $\varepsilon(x, S)$ to be ∞.)

7.9 Lemma *Fix $x \in K$. If $y \in K$ is such that $d(x, y) < \varepsilon(x)$, then any simplex S which contains y also contains x, and $d(x, y) = d_S(x, y)$.*

Proof. It is enough to show that if $\Sigma = (x_0, \dots, x_m)$ is an m–string of length $l(\Sigma) < \varepsilon(x)$ from $x = x_0$ to $y = x_m$, with $m \geq 2$, then $\Sigma' = (x_0, x_2, \dots, x_m)$ is an $(m-1)$–string with $l(\Sigma') \leq l(\Sigma)$.

From the definition of an m–string, we know that there is a simplex $S(1)$ containing both x_1 and x_2, and since $l(\Sigma) < \varepsilon(x)$, the point $x_0 = x$ also belongs to $S(1)$. As

$d_{S(1)}(x_0, x_2) \leq d_{S(1)}(x_0, x_1) + d_{S(1)}(x_1, x_2)$, we deduce that Σ' is an $(m-1)$–string of length $\leq l(\Sigma)$. $\square$

7.10 Corollary. *If $\varepsilon(x) > 0$ for every $x \in K$, then the intrinsic pseudometric is a metric and (K, d) is a length space.*

Proof. The lemma shows that $d(x, y) > 0$ if $x \neq y$. It is then obvious from the definition that d is a length metric. $\square$

The following simple example shows that we must impose extra hypotheses in order to ensure that d is a *complete* length metric.

7.11 Example. Let L be the quotient of the disjoint union $\bigcup_{\mathbb{N}}([0, 1/2^n] \times \{n\})$ by the equivalence relation that identifies the terminal point of the n-th interval to the initial point of the $(n+1)$-th interval. Then $\varepsilon(x) > 0$ for every $x \in L$, and L is isometric to $[0, 2)$.

In this example there are infinitely many simplices in a ball of finite radius. If one has only finitely many simplices in any ball then (K, d) will be a proper length space, hence complete and geodesic (3.7). (For an alternative proof see Ballmann [Ba90].)

The following result is the first of a number in which the set of isometry classes Shapes(K) plays a prominent role. In order to work with this set we need a concrete realization of it.

7.12 Definition (Model Simplices)**.** We represent each of the isometry classes in Shapes(K) by a fixed geodesic simplex $\overline{S} \subset M^n_\kappa$ that is isometric to the simplices in that class. $\overline{S}$ is called a *model simplex.* If S is an n-simplex in K then (S, d_S) is isometric to a unique model simplex $\overline{S}$. We associate to S an isometry $f_S : \overline{S} \to S$ called the *characteristic map* of S.

Note that $d(f_S(\overline{x}), f_S(\overline{y})) \leq d_{\overline{S}}(\overline{x}, \overline{y})$ for all $\overline{x}, \overline{y} \in \overline{S}$ (compare with 7.6).

7.13 Theorem. *If* Shapes(K) *is finite, then (K, d) is a complete length space.*

Proof. First we prove that d is a length metric. According to (7.9), it suffices to show that $\varepsilon(x) > 0$ for every $x \in K$. Let $\overline{S} \in$ Shapes(K) and let $\overline{x}$ be a point in the interior of $\overline{S}$. Define $\varepsilon(\overline{x}) > 0$ to be the minimum of the finite set of numbers $d(j(y), \overline{T})$, where $\overline{T}$ is a face of some $\overline{S}' \in$ Shapes(K) and j ranges through the isometries of $\overline{S}$ onto those faces of $\overline{S}'$ whose interiors are disjoint from $\overline{T}$. It follows immediately from (7.9) that if $x \in K$ is such that $f_S(\overline{x}) = x$, then $\varepsilon(x) \geq \varepsilon(\overline{x})$. In particular $\varepsilon(x) > 0$.

In order to prove that K is complete, we must show that every Cauchy sequence (x_n) in K contains a convergent subsequence. Let S_n be a simplex of K containing x_n. Because Shapes(K) is finite, we can pass to a subsequence such that $\overline{S}_n = \overline{S}_0$ for every $n \in \mathbb{N}$. Let $\overline{x}_n = f_{S_n}^{-1}(x_n)$. By passing to a further subsequence, we may assume that $(\overline{x}_n)$ converges in $\overline{S}_0$ to a point $\overline{x}$.

Let $\overline{T} = \text{supp}(\overline{x})$. Fix $\varepsilon_0 > 0$ such that $\varepsilon_0 < \varepsilon(\overline{x})$ and $d(\overline{x}, j(\overline{x})) > 2\varepsilon_0$ for every $j \in \text{Isom}(\overline{T})$ with $\overline{x} \neq j(\overline{x})$. Note that if $\overline{S} = \overline{S}' = \overline{S}_0$ then either $f_S(\overline{x}) = f_{S'}(\overline{x})$ or else $d(f_S(\overline{x}), f_{S'}(\overline{x})) > \varepsilon_0$.

We fix an integer N such that $d(\overline{x}_n, \overline{x}) < \varepsilon_0/3$ and $d(x_n, x_N) < \varepsilon_0/3$ for all $n \geq N$. Then, $d(f_{S_n}(\overline{x}), f_{S_N}(\overline{x})) \leq d(f_{S_n}(\overline{x}), x_n) + d(x_n, x_N) + d(x_N, f_{S_N}(\overline{x}))$. But $d(f_{S_n}(\overline{x}), x_n) = d(f_{S_n}(\overline{x}), f_{S_n}(\overline{x}_n))$ is no greater than $d(\overline{x}, \overline{x}_n)$ (see the last sentence of (7.12)), and similarly $d(f_{S_N}(\overline{x}), x_N) \leq d(\overline{x}, \overline{x}_N)$. Therefore $d(f_{S_n}(\overline{x}), f_{S_N}(\overline{x})) < \varepsilon_0$ if $n \geq N$, and hence $f_{S_n}(\overline{x}) = f_{S_N}(\overline{x})$. So if we let $x = f_{S_N}(\overline{x})$, then for all $n \geq N$ we have $d(x_n, x) = d(f_{S_n}(\overline{x}_n), f_{S_n}(\overline{x})) \leq d(\overline{x}_n, \overline{x})$, which goes to zero as $n \to \infty$. □

In (7.19) we shall see that if Shapes(K) is finite then K is actually a complete *geodesic* space. In the appendix (7A.4) we consider a weaker hypothesis which is enough to ensure that K is a complete length space but not enough to ensure the existence of geodesics.

Geometric Links and Cone Neighbourhoods

In this section we study the local structure of M_κ–simplicial complexes. We shall show that if $\varepsilon(x) > 0$ (in the notation of (7.8)) then the geometry of K in a neighbourhood of x is entirely determined by the *infinitesimal* geometry at x (see 7.16). The infinitesimal geometry at $x \in K$ is encoded in the space of directions of geodesics issuing from that point, which is called the *geometric link* Lk(x, K).

Let us first reflect on how to describe the neighbourhood of a point in a geodesic simplex in terms of the tangent space at that point.

7.14 The Link of a Point in a Geodesic Simplex. Let x be a point of a geodesic m-simplex S in M^n_κ, where $m \geq 1$. The *geometric link* Lk(x, S) of x in S is the set of unit vectors at x that point into S, i.e. the set of initial vectors of geodesic segments joining x to the points of S. The scalar product on the tangent space $T_x M^n_\kappa$ induces a length metric on the set of unit vectors making it isometric to $\mathbb{S}^{n-1}$. Thus we may identify Lk(x, S) with a subset of $\mathbb{S}^{n-1}$. If x is a vertex of S then this subset is a geodesic simplex; if x lies in the interior of S then Lk(x, S) is isometric to $\mathbb{S}^{n-1}$.

An equivalent way to view Lk(x, S) is as the set of $\sim$ equivalence classes of geodesic segments $[x, y]$ in S, where $y \in S \smallsetminus \{x\}$ and $[x, y_1] \sim [x, y_2]$ if $[x, y_1] \subseteq [x, y_2]$. The distance between the classes of $[x, y]$ and $[x, y']$ is the angle $\angle^{(\kappa)}_x(y, y')$. Motivated by this, in what follows we shall write $\angle_S(u, u')$ to denote the distance between points $u, u' \in$ Lk(x, S).

If $\varepsilon > 0$ is less that the distance from x to the union of those faces of S that do not contain x, then the ε-neighbourhood of x in S is naturally isometric to the ε-neighbourhood of the vertex in the κ-cone $C_\kappa(\text{Lk}(x, S))$: the isometry sends $tu \in C_\kappa(\text{Lk}(x, S))$ to the point a distance t from x along the geodesic segment in the direction u. (See 5.8)

The geometric link of a point in an M_κ–complex is assembled from the links of the points in the model simplices by the natural identifications.

7.15 The Link Lk(x, K) of a Point in K. Fix $x \in K$. Given $y, y' \in st(x) \setminus \{x\}$, we say that the segments $[x, y]$ and $[x, y']$ define the same *direction* at x if one of them is contained in the other. (Note that every equivalence class is contained in a simplex.) The *geometric link of x in K* (usually just called the link) is the set $\text{Lk}(x, K)$ of directions at x. If x lies in the simplex S, then the *link of x in S*, denoted $\text{Lk}(x, S)$ is the subset of $\text{Lk}(x, K)$ consisting of those directions which are contained in S.

Suppose $S = p_\lambda(S_\lambda)$ (notation of (7.2)). The *distance* (angle) between two directions $u, u' \in \text{Lk}(x, S)$ is defined to be $\angle_{S_\lambda}(u_\lambda, u'_\lambda)$, where $u_\lambda, u'_\lambda \in \text{Lk}(x_\lambda, S_\lambda)$ are the directions of the unique geodesic segments in S_λ that project onto geodesic segments in the classes u and u' respectively. Note that p_λ induces a canonical isometry $\text{Lk}(x_\lambda, S_\lambda) \to \text{Lk}(x, S)$ and that if T is a face of S then $\text{Lk}(x, T)$ is a subspace of $\text{Lk}(x, S)$.

We construct a pseudometric on $\text{Lk}(x, K)$ in strict analogy with (7.4). An *m–string* in $\text{Lk}(x, K)$ joining u to v is a sequence $\tilde{\Sigma} = (u_0, \dots, u_m)$ of points in $\text{Lk}(x, K)$ such that $u_0 = u$, $u_m = v$, and for each i there is a simplex $S(i)$ such that u_i and u_{i+1} are contained in $\text{Lk}(x, S(i))$. The *length* of $\tilde{\Sigma}$ is defined to be

$$l(\tilde{\Sigma}) := \sum_{i=0}^{m-1} \angle_{S(i)}(u_i, u_{i+1}),$$

and the *intrinsic pseudometric* on $\text{Lk}(x, K)$ is the function:

$$d(u, v) := \inf\{l(\tilde{\Sigma}) \mid \tilde{\Sigma} \text{ a string joining } u \text{ to } v\}.$$

If there is no string connecting u and v, then $d(u, v) := \infty$.

The intrinsic pseudometric on $\text{Lk}(x, K)$ coincides with the quotient metric associated to the projection $\coprod_\Lambda \text{Lk}(x_\lambda, S_\lambda) \to \text{Lk}(x, K)$ induced by $p : \coprod_\Lambda S_\lambda \to K$ where, by definition, $\text{Lk}(x_\lambda, S_\lambda) = \emptyset$ if $x \notin p(S_\lambda)$ and otherwise $x = p_\lambda(x_\lambda)$. Note that if $x \in K$ is a vertex, then $\text{Lk}(x, K)$ is an M_1–simplicial complex whose model simplices $\text{Shapes}(\text{Lk}(x, K))$ form a subset of $\{\text{Lk}(\overline{x}, \overline{S}) \mid \overline{x} \text{ a vertex of } \overline{S} \in \text{Shapes}(K)\}$.

7.16 Theorem. *Let K be an M_κ-simplicial complex, and let $x \in K$. If the number $\varepsilon(x)$ defined in (7.8) is strictly positive, then $B(x, \varepsilon(x)/2)$ is naturally isometric to the open ball of radius $\varepsilon(x)/2$ about the cone point in $C_\kappa(\text{Lk}(x, K))$.*

This is the first of a number of results in which we shall need the following device for examining m–strings.

7.17 Definition (The Development of an m–String)**.** We fix a base point $\overline{x}$ in M^2_κ, a base ray $c : [0, D_\kappa] \to M^2_\kappa$ with $c(0) = \overline{x}$, and an orientation of M^2_κ. (The point of the orientation is that it provides a uniform way of ordering pairs of directions that are not antipodal.)

Let x be a point of K and let $\Sigma = (x_0, \dots, x_m)$ be an m–string in $st(x)$ that avoids x (in the sense that x does lie on any of the segments $[x_i, x_{i+1}]$). Let $S(i)$ be a simplex containing $[x_i, x_{i+1}]$ and x. Let $\overline{\Sigma} = (\overline{x}_0, \dots, \overline{x}_m)$ be the sequence of points in M_κ^2 such that $\overline{x}_0$ lies in the image of c, and for each i we have $d(\overline{x}, \overline{x}_i) = d_{S(i)}(x, x_i)$ and $d(\overline{x}_{i-1}, \overline{x}_i) = d_{S(i)}(x_{i-1}, x_i)$, and if $\overline{x}, \overline{x}_i, \overline{x}_{i+1}$ are not aligned, then the initial vectors $[\overline{x}, \overline{x}_i]$ and $[\overline{x}, \overline{x}_{i+1}]$ occur in the order of the given orientation of M_κ^2. The sequence $\overline{\Sigma}$ is uniquely determined by these conditions and is called *the development* of the m–string Σ in M_κ^2.

An important point to note is that the sum of the distances $d(\overline{x}_{i-1}, \overline{x}_i)$ is equal to $l(\Sigma)$.

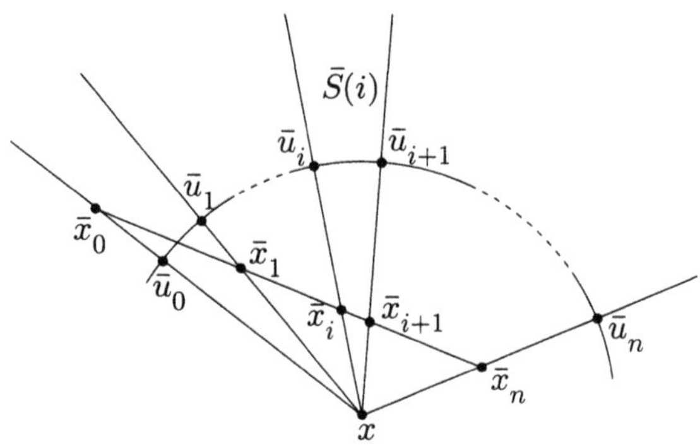

Fig. 7.1 The development of Σ

The Proof of Theorem 7.16. We identify $B(x, \varepsilon(x))$ with the ball of radius $\varepsilon(x)$ about the cone point in $C_\kappa\big(\mathrm{Lk}(x, K)\big)$ by writing tu to denote the point a distance t along a segment issuing from x in the direction $u \in \mathrm{Lk}(x, K)$. Let d_c denote the pseudometric on $C_\kappa\big(\mathrm{Lk}(x, K)\big)$ (now transferred to $B(x, \varepsilon(x))$). What we must prove is that for any $y, y' \in B(x, \varepsilon(x)/2)$ we have $d(y, y') = d_c(y, y')$. (Lemma 7.9 ensures that the above identification is well-defined, i.e. if $d(x, y) < \varepsilon(x)$ then there is a unique segment joining x to y.)

Let $y = tu$ and $y' = t'u'$. Lemma 7.9 tells us that $t = d(x, y) = d_c(x, y)$ and therefore we restrict attention to the case $t, t' > 0$. Note that $d(y, y') \leq t + t'$ and $d_c(y, y') \leq t + t'$.

Claim 1: If $d(u, u') < \pi$ then $d(y, y') \leq d_c(y, y') < t + t'$.

By definition, $d(u, u')$ is the supremum of the lengths of the strings in $\mathrm{Lk}(x, K)$ connecting u to u'. Thus there exists an m–string $\tilde{\Sigma} = (u_0, \dots, u_m)$ with $u = u_0$ and $u' = u_m$ such that $d(u, u') \leq l(\tilde{\Sigma}) < \pi$. Let $S(i)$ be a simplex such that $u_i, u_{i+1} \in \mathrm{Lk}(x, S(i))$.

From $\tilde{\Sigma}$ we shall construct a corresponding m–string Σ joining y to y' by first constructing its development. To this end, we fix a basepoint $\overline{x} \in M_\kappa^2$ and choose $\overline{x}_0^{\tilde{\Sigma}}, \overline{x}_m^{\tilde{\Sigma}} \in M_\kappa^2$ such that $d(\overline{x}, \overline{x}_0^{\tilde{\Sigma}}) = t$, $d(\overline{x}, \overline{x}_m^{\tilde{\Sigma}}) = t'$ and $\angle_{\overline{x}}^{(k)}(\overline{x}_0^{\tilde{\Sigma}}, \overline{x}_m^{\tilde{\Sigma}}) = l(\tilde{\Sigma})$. Then we choose points $\overline{x}_1, \dots, \overline{x}_{m-1}$ on the geodesic segment $[\overline{x}_0^{\tilde{\Sigma}}, \overline{x}_m^{\tilde{\Sigma}}]$ such that

$\angle_{\bar{x}}^{(k)}(\bar{x}_i, \bar{x}_{i+1}) = d_{\mathrm{Lk}(x,S(i))}(u_i, u_{i+1})$. Let $t_i = d(\bar{x}, \bar{x}_i)$ and let $x_i = t_i u_i$. Note that $[x_i, x_{i+1}] \subset S(i)$ and $d_{S(i)}(x_i, x_{i+1}) = d(\bar{x}_i, \bar{x}_{i+1})$.

The m–string $\Sigma = (x_0, \ldots, x_m)$ joins $y = x_0$ to $y' = x_m$ in $B(x, \varepsilon(x))$, and therefore $d(y, y') \leq l(\Sigma) = \sum_i d_i(x_i, x_{i+1}) = d(\bar{x}_0^{\tilde{\Sigma}}, \bar{x}_m^{\tilde{\Sigma}}) < t + t'$. But $d_c(y, y') = \inf_{\tilde{\Sigma}} d(\bar{x}_0^{\tilde{\Sigma}}, \bar{x}_m^{\tilde{\Sigma}})$. Thus Claim 1 is proved.

If $d(y, y') = t + t'$, then $d(u, u') \geq \pi$ by Claim 1, and therefore $d_c(y, y') = t + t'$. In the light of this observation, the theorem follows from Claim 1 and:

Claim 2: If $d(y, y') < t + t'$, then $d(u, u') < \pi$ and $d_c(y, y') \leq d(y, y')$.

Let $\Sigma = (x_0, \ldots, x_m)$ be an m–string joining $y = x_0$ to $y' = x_m$ in $B(x, \varepsilon)$, and suppose $l(\Sigma) < t + t'$. The triangle inequality tells us that x is not contained in the path determined by Σ. Thus x_i can be written uniquely as $x_i = t_i u_i$, and we have an m–string $\tilde{\Sigma} = (u_0, \ldots, u_m)$ joining u to u' in $\mathrm{Lk}(x, K)$.

Let $\overline{\Sigma} = (\bar{x}_0, \ldots, \bar{x}_m)$ be the development of Σ in M_κ^2. If $l(\tilde{\Sigma})$ were greater than π, then for some i with $0 < i < m$ there would exist $\bar{x}_i' \in [\bar{x}_i, \bar{x}_{i+1}]$ such that $\angle_{\bar{x}}^{(\kappa)}(\bar{x}_0, \bar{x}_i') = \pi$; let $t_i' = d(\bar{x}, \bar{x}_i')$. Then, for $\overline{\Sigma}' := (\bar{x}_0, \ldots, \bar{x}_i, \bar{x}_i')$ and $\overline{\Sigma}'' := (\bar{x}_i', \bar{x}_{i+1}, \ldots, \bar{x}_m)$ we would have $l(\overline{\Sigma}) = l(\overline{\Sigma}') + l(\overline{\Sigma}'') \geq (t_0 + t_i') + (t_m - t_i') = t + t'$, which is a contradiction. Thus $\pi > l(\tilde{\Sigma}) = \angle_{\bar{x}}^{(\kappa)}(\bar{x}_0, \bar{x}_m) \geq d(u_0, u_m) = d(u, u')$. Since $d(\bar{x}_0, \bar{x}_m)$ and $d_c(y, y')$ are obtained by applying the law of cosines to the M_κ-triangles with vertex angles $\angle_{\bar{x}}^{(\kappa)}(\bar{x}_0, \bar{x}_m)$ and $d(u, u')$ respectively (see 5.6), we deduce that $l(\Sigma) \geq d(\bar{x}_0, \bar{x}_m) \geq d_c(y, y')$. Therefore $d(y, y') = \inf l(\Sigma) \geq d_c(y, y')$. □

The following map was implicit in the arguments of the preceding proof.

7.18 Definition (Radial Projection). The map $St(x) \setminus \{x\} \to \mathrm{Lk}(x, K)$ that associates to each point y the direction of $[x, y]$ is called *radial projection* from x. Thus the m-string $\tilde{\Sigma}$ considered in the preceding proof was the radial projection of the m-string Σ. If x is a vertex then this map takes open simplices to open simplices and sends segments in $st(x)$ to segments in $\mathrm{Lk}(x, K)$.

The Existence of Geodesics

In this section we shall prove the theorem of Bridson [Bri91] mentioned in the introduction.

7.19 Theorem. *Let K be a connected M_κ–simplicial complex. If* Shapes(K) *is finite, then K is a complete geodesic space.*

The outline of the proof of this theorem is as follows. We shall define a useful subset of the set of m–strings called *taut strings*. This class is large enough to ensure that the infimum in the definition of the intrinsic pseudometric on K can be taken over taut strings rather than all strings (7.24). On the other hand, taut strings are rather efficient in the sense that their size is bounded above by a linear function of their length (7.30); this will be proved by an induction on the dimension of K. Thus the proof of Theorem 7.19 is reduced to showing that if one fixes m then there is a shortest m–string among all those with specified endpoints. This is proved by a simple "quasi-compactness" argument (7.27).

The proof of Theorem 7.19 will also provide us with a characterization of the geodesics in M_κ–simplicial complexes that will be useful in Chapter II.11 when we come to construct polyhedral complexes using the gluing techniques introduced in (5.19).

Taut Strings

Before defining what it means for an m–string to be taut, we must make some observations about small subcomplexes of K. Suppose that S and S' are closed simplices in K which have non-empty intersection, and consider $L = S \cup S'$ equipped with its intrinsic metric, i.e. the metric such that $d(x, y) = d_S(x, y)$ if $x, y \in S$, $d(x, y) = d_{S'}(x, y)$ if $x, y \in S'$, and $d(x, y) = \inf_{z \in S \cap S'}(d_S(x, z) + d_{S'}(z, y))$ if $x \in S$ and $y \in S'$. Then L is a geodesic metric space, and the minimal m–string associated to any geodesic segment has size at most 2 (cf. 5.24).

7.20 Definition of a Taut String. An m–string $\Sigma = (x_0, x_1, \dots, x_m)$ in K is *taut* if it satisfies the following two conditions for $i = 1, \dots, m-1$:

(1) there is no simplex containing $\{x_{i-1}, x_i, x_{i+1}\}$;

(2) if $x_{i-1}, x_i \in S(i)$ and $x_i, x_{i+1} \in S(i+1)$ then the concatenation of the line segments $[x_{i-1}, x_i]$ and $[x_i, x_{i+1}]$ is a geodesic segment in $L = S(i) \cup S(i+1)$.

Notice that if a string is taut then only its first and last entries can lie in the interior of a top dimensional simplex of K.

7.21 Lemma. *For a fixed integer m, if Σ is an m–string from x to y in K, and if Σ is of minimal length among all m–strings joining these points, then $P(\Sigma)$ (the path in K determined by Σ) is the path determined by some taut n–string with $n \leq m$.*

Proof. Let $\Sigma = (x_0, x_1, \dots, x_m)$. Suppose that for some i there exists a simplex S containing x_{i-1}, x_i and x_{i+1}, and let Σ^i denote the $(m-1)$–string obtained from Σ by deleting the entry x_i. The triangle inequality for d_S gives $d_S(x_{i-1}, x_{i+1}) \leq d_S(x_{i-1}, x_i) + d_S(x_i, x_{i+1})$, with equality if and only if x_i lies on the line segment $[x_{i-1}, x_{i+1}]$. Thus $l(\Sigma^i) \leq l(\Sigma)$. But $l(\Sigma)$ is minimal, so in fact $l(\Sigma^i) = l(\Sigma)$, and hence x_i must lie on the line segment $[x_{i-1}, x_{i+1}]$. This implies that Σ^i determines the same path as Σ. We can repeat this procedure until no simplex of K contains three successive entries of the resulting string — the first condition for tautness.

We now show that Σ satisfies the second condition for tautness. Let S and S' be any simplices containing $\{x_{i-1}, x_i\}$ and $\{x_i, x_{i+1}\}$ respectively. Every geodesic segment in the complex $L = S \cup S'$ can be expressed as the concatenation of one or two line segments. So if the concatenation of the line segments $[x_{i-1}, x_i]$ and $[x_i, x_{i+1}]$ were not a geodesic segment in L, then we could replace x_i by some $x_i' \in S \cap S'$ to obtain an m–string from x_0 to x_m which would be strictly shorter than Σ, contradicting the minimality of $l(\Sigma)$. □

7.22 Example. Figure 7.2 illustrates the fact that a taut string is not necessarily a local geodesic. Here K is a planar 2–complex with three 2–simplices, and the string (a, b, c) is taut but not a local geodesic.

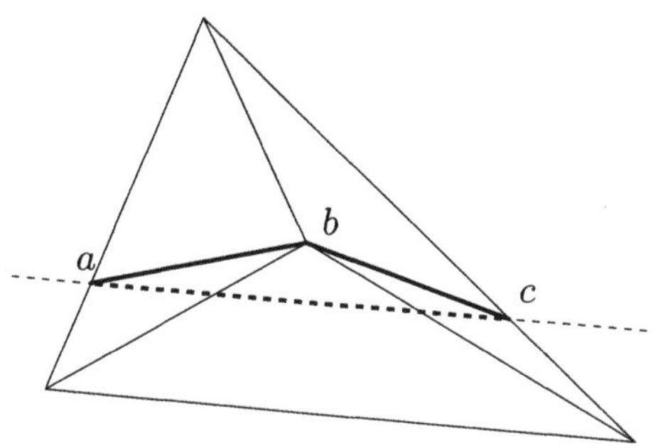

Fig. 7.2 A taut string which is not a local geodesic

7.23 Lemma. *Let x be a vertex of K and let $\Sigma = (x_0, \dots, x_m)$ be a taut string in $st(x)$, the open star of x. Suppose that $x_i \neq x$ for each $i = 0, \dots, m$. If $\kappa > 0$, assume in addition that for each i, $d_{S(i)}(x, x_i) < D_\kappa/2$, where $S(i)$ is a simplex containing $[x_i, x_{i+1}]$.*

Then, the image $\tilde{\Sigma}$ of Σ under radial projection (as defined in (7.18)) is a taut string in $\mathrm{Lk}(x, K)$ of length $< \pi$.

Proof. The radial projection of Σ into $\mathrm{Lk}(x, K)$ is $\tilde{\Sigma} = (u_0, \dots, u_m)$, where u_i is the direction determined by the segment $[x, x_i]$. Let $\overline{\Sigma} = (\overline{x}_0, \dots, \overline{x}_m)$ be the development of Σ in M_κ^2 with respect to the base point $\overline{x} \in M_\kappa^2$ (see 7.17); if $\kappa > 0$, we have $d(\overline{x}, \overline{x}_i) < D_\kappa/2$. We claim that the concatenation of the geodesic segments $[\overline{x}_0, \overline{x}_1], \dots, [\overline{x}_{m-1}, \overline{x}_m]$ is the geodesic segment $[\overline{x}_0, \overline{x}_m]$. As $l(\tilde{\Sigma}) = \angle_{\overline{x}}(\overline{x}_0, \overline{x}_m)$, it will follow that $l(\tilde{\Sigma}) < \pi$.

Since every local geodesic in $B(\overline{x}, D_\kappa/2)$ is a geodesic, it is enough to show that the concatenation of $[\overline{x}_{i-1}, \overline{x}_i]$ and $[\overline{x}_i, \overline{x}_{i+1}]$ is $[\overline{x}_{i-1}, \overline{x}_{i+1}]$. If this were not the case, then arbitrarily close to $\overline{x}_i$ on the ray issuing from $\overline{x}$ and passing through $\overline{x}_i$ there would be a point $\overline{x}_i'$ such that $d(\overline{x}_{i-1}, \overline{x}_i') + d(\overline{x}_i', \overline{x}_{i+1}) < d(\overline{x}_{i-1}, \overline{x}_i) + d(\overline{x}_i, \overline{x}_{i+1})$. Let $S(i)$ and $S(i+1)$ be simplices containing $[x_{i-1}, x_i]$ and $[x_i, x_{i+1}]$ respectively. Let y_i be a point of $St(x) \smallsetminus st(x)$ such that $x_i \in [x, y_i]$; note that $y_i \in S(i) \cap S(i+1)$. As x_i is in the open star of x, we have $x_i \neq y_i$, and if $\overline{x}_i'$ is chosen close enough to $\overline{x}_i$, there is a point $x_i' \in [x, y_i]$ such that $d(x, x_i') = d(\overline{x}, \overline{x}_i')$. As $x_i' \in S(i) \cap S(i+1)$ and $d(x_{i-1}, x_i') = d(\overline{x}_{i-1}, \overline{x}_i')$ and $d(x_i, x_{i+1}') = d(\overline{x}_i, \overline{x}_{i+1}')$, we would have $d(x_{i-1}, x_i) + d(x_i, x_{i+1}) > d(x_{i-1}, x_i') + d(x_i, x_{i+1}')$, contrary to the hypothesis that Σ is taut.

It remains to show that $\tilde{\Sigma}$ is taut. The first condition for tautness is obvious. To establish the second condition, for each i we consider the map from $S(i) \cup S(i+1)$

(with its intrinsic metric) to $C_\kappa\big(\mathrm{Lk}(x, S(i) \cup S(i+1))\big)$, as in (7.16). The argument of the previous paragraph shows that the image under this map of the concatenation of $[x_{i-1}, x_i]$ and $[x_i, x_{i+1}]$ is a geodesic in $C_\kappa\big(\mathrm{Lk}(x, S(i) \cup S(i+1))\big)$. Therefore, by (5.10), the concatenation of $[u_{i-1}, u_i]$ and $[u_i, u_{i+1}]$ is a geodesic in $\rho S(i) \cup \rho S(i+1)$. $\square$

The Main Argument

In this subsection we present the proof of Theorem 7.19, as outlined above. In the first step (7.24), we prove that the distance between two points is the infimum of the length of taut strings joining them. In the second step (7.28), we prove that for every $\ell > 0$, there is an integer N depending only on the finite set Shapes(K) such that the size of taut strings of length $< \ell$ is bounded by N.

7.24 Proposition. *Let K be an M_κ–simplicial complex with* Shapes(K) *finite. Then for any two points $x, y \in K$, we have*

$$d(x, y) = \inf\{l(\Sigma) : \Sigma \textit{ a taut string from } x \textit{ to } y\}.$$

The following lemma is taken from the thesis of Gabor Moussong [Mou88] who used m–strings to prove the existence of geodesics in locally compact complexes.

7.25 Lemma. *If L is a finite M_κ–simplicial complex, and if the points x and y can be joined by an m–string in L, where m is a fixed integer, then there is a shortest m–string from x to y in L.*

Proof. Let $X \subset L^{m+1}$ denote the set of m–strings from x to y in L. We show that X is closed and hence compact. The length function l on m–strings is continuous on X, and therefore attains a minimum if X is closed.

Notice that for any $w, z \in L$, the set $st(w) \cap st(z)$ is empty if and only if there is no closed simplex of L containing both w and z. Thus, if $\mathbf{z} = (z_0, z_1, \dots, z_m)$ is *not* an m–string, then for some i between 0 and $m-1$ we have $st(z_i) \cap st(z_{i+1}) = \emptyset$. Hence $\mathbf{z}$ has a neighbourhood disjoint from X, namely $(L \times \dots L \times st(z_i) \times st(z_{i+1}) \times L \times \dots \times L)$. Thus X is closed. $\square$

7.26 Model Complexes (Quasi-Compactness). Given K with Shapes(K) finite, and a positive integer m, one can build a finite set of *model subcomplexes*, that is, connected complexes obtained by taking at most m (not necessarily distinct) simplices from Shapes(K) and identifying faces by isometries. Any subcomplex K_0 of K which can be expressed as the union of at most m closed simplices must be simplicially isometric to one of these models, i.e. there is a bijection from K_0 to the model which respects the simplicial structure and restricts to an isometry on each simplex S equipped with its intrinsic metric d_S. The existence of this finite set of models allows us to pass from the case of finite complexes to the case of interest, complexes with Shapes(K) finite.

Notice that K_0 with the induced metric from K is not in general isometric to the model with its intrinsic metric. However, since the length of a string is defined in terms of the local metrics d_S, a given m–string in K_0 and the corresponding m–string in the model have the same length. This is the key to the following lemma.

7.27 Lemma. *If K is an M_κ–simplicial complex with* Shapes(K) *finite, and two points x and y can be joined by an m–string in K, where m is a fixed integer, then there is a shortest m–string from x to y in K.*

Proof. For any fixed pair of elements x and y there are only finitely many bipointed models, $(K'; x', y')$, for $(K_0; x, y)$ as K_0 runs over all subcomplexes of K which contain both x and y and can be expressed as the union of at most m closed simplices. Thus, any m–string from x to y in K corresponds to an m–string of the same length from x' to y' in one of the finitely many models under consideration, and *vice versa.* The present lemma now follows by application of (7.25) to each of the models. □

Proof of 7.24. Apply (7.21) and (7.27). □

We are now in a position to prove the required lower bound on the length of taut m–strings. The idea is to localize the problem to the star of a single vertex and then use (7.23) to reduce to a lower-dimensional complex. The proof which we present here is a refinement of the original proof [Bri91], and appears in some unpublished seminar notes of R. L. Bishop.

7.28 Theorem. *Let K be an M_κ-simplicial complex with* Shapes(K) *finite. Then, for every $\ell > 0$ there exists an integer $N > 0$, depending only on* Shapes(K), *such that for every taut m–string in K of length at most ℓ we have $m \leq N$.*

Proof. We proceed by induction on the dimension of K. In order to clarify the induction process, we articulate an intermediate conclusion and use the following scheme of proof:

Φ_n : The statement of the theorem is true for complexes of dimension $\leq n$.

Ψ_n : Suppose that K is of dimension n. Then there exists a constant R, depending only on the finite set Shapes(K), with the following property: if x is a vertex of K and $\Sigma = (x_0, \dots, x_m)$ is a taut m–string contained in $st(x)$, and if $[x, x_i]$ has length less than $D_\kappa/2$ for each i, then $m \leq R$.

Assertions Φ_1 and Φ_1 are trivial. We shall prove: $\Phi_{n-1} \Rightarrow \Psi_n \Rightarrow \Phi_n$ for all $n \geq 2$.

$\Phi_{n-1} \Rightarrow \Psi_n$:

Let K, Σ and x be as in statement Ψ_n.

If the path determined by Σ were to pass through x, then x would have to be an entry in Σ. The first condition for tautness would then imply that Σ has size at most 2. Thus we may take $R \geq 3$ and restrict our attention to the case where $\Sigma = (x_0, x_1, \dots, x_m)$ with $m \geq 2$ and the path $P(\Sigma)$ determined by Σ does not pass through x. We can then radially project Σ into $\text{Lk}(x, K)$, as in (7.18).

According to (7.23), the image of Σ under this projection is a taut m–string $\tilde{\Sigma}$ in $\text{Lk}(x, K)$ of length smaller than π.

Now, by applying condition Φ_{n-1} to the spherical complex $\text{Lk}(x, K)$, we obtain a constant $N(x)$, depending only on Shapes(Lk(x, K)), such that any taut string in $\text{Lk}(x, K)$ of length at most π has size at most $N(x)$.

There are only finitely many possibilities for Shapes(Lk(x, K)) as x ranges over the vertices of K, and the set of these possibilities depends only on Shapes(K). So setting R equal to the minimum of the constants $N(x)$ as x ranges over the set of vertices of K finishes the proof of the present implication.

$\Psi_n \Rightarrow \Phi_n$:

If Φ_n were false, then in some n-dimensional metric simplicial complex K with only finitely many isometry types of simplices there would be taut chains of bounded length and arbitrarily large size, and hence taut strings of arbitrarily small length and arbitrarily large size. But because Shapes(K) is finite, there is a constant η such that any set of diameter at most η is contained in the open star of some vertex $x \in K$ intersected with $B(x, D_\kappa/2)$ (one simply chooses η small enough so that each $S \in$ Shapes(K) is the union of the subsets $S \smallsetminus V_\eta(S \smallsetminus st(x))$, where x runs over the vertices of S and V_η denotes the η-neighbourhood). Thus we obtain a contradiction to Ψ_n. □

The Proof of Theorem 7.19. Given $x, y \in K$, we can apply the preceding result to obtain a bound on the size of taut strings joining x to y. Then, by (7.28), one need only take the infimum in the definition of $d(x, y)$ over m–strings with m fixed. According to (7.26) there is a shortest such m–string joining x to y. □

We amplify two aspects of our proof of the existence of geodesics.

7.29 Corollary. *If K is an M_κ-simplicial complex with* Shapes(K) *finite, then every local geodesic of finite length in K is the concatenation of a finite number of segments (each contained in a simplex).*

7.30 Corollary. *If K is an M_κ-simplicial complex with* Shapes(K) *finite, then there exists a constant $\alpha > 0$ such that every taut m–string in K has length at least $\alpha m - 1$.*

Proof. It is clear from the definition that a substring of a taut string is itself taut. And according to (7.28) there exists an integer N such that if $m \geq N$ then the length of every taut m–string in K is at least 1. Let $\alpha = 1/N$ and let μ be the integer part of αm. Given a taut m–string $\Sigma = (x_0, \ldots, x_m)$, we can view it as the concatenation of the $\mu + 1$ substrings $\Sigma_1 = (x_0, \ldots, x_N)$, $\Sigma_2 = (x_N, \ldots, x_{2N})$, $\ldots$, $\Sigma_{\mu+1} = (x_{\mu N}, \ldots, x_m)$. Each of the first μ of these strings has length at least 1. Hence the length of Σ is at least $\mu = [\alpha m] > \alpha m - 1$. □

This technical observation has the following consequence relating the metric structure of K to its combinatorial structure. The *1-skeleton* of K, denoted $K^{(1)}$, is the graph whose vertices are the vertices of K and whose edges are the 1-simplices of K.

7.31 Proposition. *Let K be an M_κ-simplicial complex with* Shapes(K) *finite. Let $K_1^{(1)}$ denote the 1-skeleton of K, considered as a metric graph with all edge lengths 1; let d_1 be the associated metric. Then the natural injection $j : K_1^{(1)} \hookrightarrow K$ is a quasi-isometry*[12].

Proof. Because Shapes(K) is finite, every point of K is within a bounded distance of the image of j. And if we set k equal to the length of the longest edge of any simplex in Shapes(K), then clearly $d(j(u), j(v)) \leq k\, d_1(u, v)$ for all vertices $u, v \in K^{(1)}$.

Let the constant α be as in (7.30). We fix a taut m–string $\Sigma = (x_0, \ldots, x_m)$ determining a path in K that is a geodesic joining the vertices $j(u)$ and $j(v)$. Let $y_0 = j(u)$ and, inductively for $i = 1, \ldots, m-1$, define y_i to be a vertex of K that lies in $St(y_{i-1}) \cap St(x_{i+1})$. Let $y_m = j(v)$. This gives a sequence of vertices connecting $j(u)$ to $j(v)$, and since each successive pair of these vertices cobounds an edge in $K^{(1)}$, we deduce that $d_1(u, v) \leq m$. But from (7.28) we have $d(j(u), j(v)) \geq \alpha m - 1$. Hence $\alpha\, d_1(u, v) - 1 \leq d(j(u), j(v)) \leq k\, d_1(u, v)$. □

Cubical Complexes

We turn our attention briefly to cubical complexes. These are more rigid objects than M_κ-simplicial complexes and in many ways they are easier to work with. Moreover, there exist many interesting examples (see II.5).

The unit n-cube I^n is the n-fold product $[0, 1]^n$; it is isometric to a cube in $\mathbb{E}^n$ with edges of length one. By convention, I^0 is a point.

The faces of the 1-cube $[0, 1]$ are the subsets $\{0\}$, $\{1\}$ and $[0, 1]$, the first two have dimension 0 and the last has dimension 1. A *face* of I^n is a subset S of I^n which is a product $S_1 \times \cdots \times S_n$ of faces of $[0, 1]$; the *dimension* of S is the sum of the dimensions of the S_i. (Each k-dimensional face is isometric to I^k.)

We define a *cubical complex*, by mimicking the definition of an M_κ–simplicial complex, using unit cubes instead of geodesic simplices.

[12] See I.8.14 for the definition of quasi-isometry

7.32 Definition of a Cubical Complex. A cubical complex K is the quotient of a disjoint union of cubes $X = \coprod_\Lambda I^{n_\lambda}$ by an equivalence relation $\sim$. The restrictions $p_\lambda : I^{n_\lambda} \to K$ of the natural projection $p : X \to K = X/\sim$ are required to satisfy:

(1) for every $\lambda \in \Lambda$ the map p_λ is injective;

(2) if $p_\lambda(I^{n_\lambda}) \cap p_{\lambda'}(I^{n_{\lambda'}}) \neq \emptyset$ then there is an isometry $h_{\lambda,\lambda'}$ from a face $T_\lambda \subset I^{n_\lambda}$ onto a face $T_{\lambda'} \subset I^{n_{\lambda'}}$ such that $p_\lambda(x) = p_{\lambda'}(x')$ if and only if $x' = h_{\lambda,\lambda'}(x)$.

7.33 Basic Structure. Let K be as in (7.32). A subset $C \subset K$ is called an m-dimensional cube if it is the image under some p_λ of an m-dimensional face of I^{n_λ}. The notions of segments, m-strings, and so on, are defined as in the simplicial case, as is the pseudometric on K. In contrast to the simplicial case, the intrinsic pseudometric on a cubical complex is always a metric because the number $\varepsilon(x)$ defined in (7.8) is positive for every $x \in K$. If K is finite dimensional then d is also complete (a simplified version of the argument given in (7.9) applies). In fact, if K is finite dimensional, then it is a complete geodesic space (see (7.43)).

If K is not finite dimensional then it might not be complete. For instance, let H be an infinite dimensional Hilbert space, fix an orthogonal basis, and consider the cubes spanned by the finite subsets of the basis elements; these form a cubical complex that is not complete.

M_κ-Polyhedral Complexes

We began this chapter by studying simplicial complexes rather than more general polyhedral complexes, partly because simplicial complexes are easier to work with and partly because we did not wish to obscure the main ideas by the troublesome technicalities which must accompany a precise definition of polyhedra in general. Now we shall consider the general case.

Intuitively speaking, an M_κ-polyhedral space is a cell complex whose cells are endowed with local metrics making each isometric to the convex hull of a finite number of points in M_κ^n; these local metrics agree on the intersection of cells. (A precise definition is given in (7.37).) One can measure the length of paths using these local metrics, and the intrinsic pseudometric on the complex is obtained by setting the distance between two points equal to the infimum of the lengths of paths connecting them. As in the simplicial case, we are interested in the question of whether this infimum is attained. We also wish to establish that polyhedral complexes have the sort of local cone structure described in (7.16).

In order to establish the existence of geodesics, we shall describe a canonical process of subdivision that transforms any M_κ-polyhedral complex K into an isometric M_κ-simplicial complex called the *second barycentric subdivision of K*.

Recall that a subset of a geodesic space is said to be convex if every geodesic segment whose endpoints lie in the subset is entirely contained in the subset. The *convex hull* of a subset P of a geodesic space is the intersection of all convex sets containing P.

7.34 Convex M_κ-Polyhedral Cells. Fix $\kappa \in \mathbb{R}$. By definition, a *convex M_κ-polyhedral cell* $C \subset M_\kappa^n$ is the convex hull of a finite set of points $P \subset M_\kappa^n$; if $\kappa > 0$, then P (hence C) is required to lie in an open ball of radius $D_\kappa/2$. The *dimension* of C is the dimension of the smallest[13] m-plane containing it. The *interior* of C is the interior of C as a subset of this m-plane.

Let H be a hyperplane in M_κ^n. If C lies in one of the closed half-spaces bounded by H, and if $H \cap C \neq \emptyset$, then $F = H \cap C$ is called a *face* of C; if $F \neq C$ then it is called a *proper face*. The *dimension of a face F* is the dimension of the smallest m-plane containing it. The *interior* of F is the interior of F in this plane. The 0-dimensional faces of C are called its *vertices*. The *support* of $x \in C$, denoted supp(x), is the unique face containing x in its interior.

The following discussion is parallel to (7.14) and (7.15). Given $x \in C$, the directions at x pointing into C form a space $\text{Lk}(x, C)$. The distance (angle) between two directions $u, u' \in \text{Lk}(x, C)$ is denoted $\angle_C(u, u')$. If C has dimension m, then $\text{Lk}(x, C)$ can be identified with a subset of $\mathbb{S}^{m-1}$; if x is an interior point then $\text{Lk}(x, C)$ is the whole of $\mathbb{S}^{m-1}$; if x is a vertex of C then $\text{Lk}(x, C)$ is a convex polyhedral cell in $\mathbb{S}^{m-1}$.

7.35 Exercise. Let C be a convex M_κ–polyhedral cell of dimension m, let $x \in C$ and suppose that supp(x) has dimension $k > 0$. Prove that $\text{Lk}(x, C)$ is isometric to the spherical join (5.13) of $\mathbb{S}^{k-1}$ and a convex M_1–polyhedral cell of dimension $m-k-1$.

7.36 Proposition. *Let C be a convex M_κ–polyhedral cell. Then:*

(1) *Each face of C is a convex M_κ–polyhedral cell.*

(2) *The intersection of any two faces of C is again a face.*

(3) *C has only finitely many faces.*

(4) *C is the convex hull of its set of vertices.*

(5) *If C is the convex hull of a set P, then there is a unique minimal subset of P with convex hull C, and this is the vertex set of C.*

(6) *If $f : C \to C_1$ is an isometry from C to a convex M_κ–polyhedral cell C_1, and if F is a face of C, then $f(F)$ is a face of C_1.*

Proof. Suppose that C is the convex hull of the finite set P. Let F be a face of C and suppose that $F = C \cap H$ where H is a hyperplane. Because the open half-spaces defined by H are convex, $F = H \cap C$ is the convex hull of $H \cap P$. This proves (1).

The content of (2) is that the intersection of two closed half-spaces X_1 and X_2, bounded by hyperplanes H_1 and H_2 say, is contained in a half-space bounded by a hyperplane that intersects $X_1 \cap X_2$ in $H_1 \cap H_2$. Part (3) follows immediately from (1), and (4) and (5) are proved by induction on the dimension of faces.

Part (6) is a consequence of the fact that any isometry between subsets of M_κ^n is the restriction of an element of $\text{Isom}(M_\kappa^n)$ (see (2.20)). Any such global isometry

[13] i.e. the intersection of all hyperplanes containing C

obviously takes the intersections of C with hyperplanes to intersections of C_1 with hyperplanes. □

For more details, see Eggleston's book on convexity [Eg77]. (Note that because convex polyhedral cells in M^n_κ correspond to convex polyhedral cones in $\mathbb{R}^{n+1}$, one can convert most questions about such cells into questions concerning the convex geometry of Euclidean space.) Further references include [Berg77], [Bro88] and [Grü67].

7.37 Definition of an M_κ-Polyhedral Complex. Let $(C_\lambda : \lambda \in \Lambda)$ be a family of convex M_κ–polyhedral cells and let $X = \bigcup_{\lambda\in\Lambda}(C_\lambda \times \{\lambda\})$ denote their disjoint union. Let $\sim$ be an equivalence relation on X and let $K = X/\!\sim$. Let $p : X \to K$ be the natural projection and define $p_\lambda : C_\lambda \to K$ by $p_\lambda(x) := p(x, \lambda)$.

K is called an M_κ–polyhedral complex if:

(1) for every $\lambda \in \Lambda$, the restriction of p_λ to the interior of each face of C_λ is injective;

(2) for all $\lambda_1, \lambda_2 \in \Lambda$ and $x_1 \in C_{\lambda_1}$, $x_2 \in C_{\lambda_2}$, if $p_{\lambda_1}(x_1) = p_{\lambda_2}(x_2)$ then there is an isometry $h : \text{supp}(x_1) \to \text{supp}(x_2)$ such that $p_{\lambda_1}(y) = p_{\lambda_2}(h(y))$ for all $y \in \text{supp}(x_1)$.

The set of isometry classes of the faces of the cells C_λ is denoted Shapes(K).

The auxiliary definitions and terminology needed to describe the geometry of M_κ–polyhedral complexes are essentially the same as in the simplicial case:

7.38 Cells, Stars, Links and the Intrinsic Pseudometric. Let K be as in (7.37). A subset $C \subset K$ is called an *n-cell* if it is the image $p_\lambda(F)$ of some n-dimensional face $F \subset C_\lambda$; the interior of C is the image under p_λ of the interior of F.

The *intrinsic pseudometric* on K is the quotient pseudometric d associated to the projection $p : \coprod_\lambda C_\lambda \to K$. As in (7.5) one shows that it is equivalent to define $d(x, y)$ to be the infimum of the lengths of piecewise geodesic paths in K joining x to y: a *piecewise geodesic path* is a map $c : [a, b] \to K$ such that there is subdivision $a = t_0 \leq t_1 \leq \cdots \leq t_k = b$ and geodesic paths $c_i : [t_{i-1}, t_i] \to C_{\lambda_i}$ such that for each $t \in [t_{i-1}, t_i]$ we have $c(t) = p_{\lambda_i}(c_i(t))$; the length $l(c)$ of c is $l(c) := \sum_{i=1}^k l(c_i)$. (7.37(2) ensures that $l(c)$ is independent of the choice of subdivision.)

Fix $x \in K$. The *open star* of x, denoted $st(x)$, is the union of the interiors of the cells that contain x. The *geometric link* Lk(x, K) of x in K is the space of directions at x endowed with the quotient metric associated to the projection $\coprod_\Lambda \text{Lk}(x_\lambda, C_\lambda) \to \text{Lk}(x, K)$ induced by $\coprod_\Lambda C_\lambda \to K$ (compare with (7.15)). As in the simplicial case, one can give a more explicit description of the metric on Lk(x, K), except that now one has to use piecewise geodesic paths in place of m-strings (as with the pseudometric).

As in (7.8), for each $x \in K$ one defines $\varepsilon(x)$ to be the distance from x to the closure of $st(x)$ minus $st(x)$, where distance is measured in the local metrics $d_C(p_\lambda(x), p_\lambda(y)) = d_{C_\lambda}(x, y)$ on the cells $C \subset K$. By following the proof of (7.16) we obtain:

7.39 Theorem. *Let K be an M_κ-polyhedral complex, and let $x \in K$. If $\varepsilon(x) > 0$ then $B(x, \varepsilon(x)/2)$ is naturally isometric to the open ball of radius $\varepsilon(x)/2$ about the cone point in $C_\kappa(\mathrm{Lk}(x, K))$.*

7.40 Examples. We use the notation of (7.37).

(1) K is a metric graph if and only if all of its cells have dimension at most 1.

(2) K is an M_κ-simplicial complex if and only if each of the cells C_λ is a geodesic simplex, each of the maps p_λ is injective, and the intersection of any two cells in K is empty or a single face.

(3) K is a cubical complex if and only if each of the cells C_λ is isometric to a cube I^{n_λ}, each of the maps p_λ is injective, and the intersection of any two cells in K is empty or a single face.

(4) There are many interesting polyhedral complexes all of whose cells are cubes, but they do not all satisfy the conditions of (3). We shall use the term *cubed complex* to describe this larger class of complexes, except that in the 2-dimensional case we shall use the term *squared complex.*

A simple example of a cubed complex that is not a cubical complex is the n-dimensional torus obtained by forming the quotient of I^n by the equivalence relation that associates to each point in the $(n-1)$-dimensional faces its orthogonal projection on the opposite face.

7.41 Exercises

(1) Describe all of the isometrically distinct squared complexes that arise as the quotient of a single square I^2 by an equivalence relation.

(2) Describe all of the isometrically distinct polyhedral complexes that one can obtain as the quotient of a single equilateral triangle and describe the links of the vertices in each case.

One of the complexes that you should consider is called the dunce hat, which can be described as follows: if the vertices of the triangle are x_0, x_1, x_2, then the dunce hat is the space that you get by identifying $[x_0, x_1]$ with $[x_1, x_2]$ by the isometry mapping x_0 to x_2, and by identifying $[x_0, x_1]$ with $[x_2, x_0]$ by the isometry mapping x_0 to x_2. The dunce hat has one vertex, the link of which has two 0-cells, with a loop based at each and a third edge connecting them.

Show that the dunce hat is simply-connected. Then prove that it is contractible (this is harder).

Barycentric Subdivision

We shall now explain how to subdivide an M_κ-polyhedral complex so as to obtain an isometric M_κ-simplicial complex. The approach is rather obvious: one subdivides each of the cells of K, or equivalently $\coprod_\Lambda C_\lambda$ (in the notation of (7.37)), into geodesic simplices. However there are some technical difficulties: the process of subdivision must be sufficiently canonical to ensure that the subdivisions of all cells agree on faces of intersection, and one has to arrange for the conditions of (7.40(2)) to hold.

7.42 The Barycentre of a Convex Polyhedral Cell. Let $C \subset M^n_\kappa$ be a convex polyhedral cell. We shall define a point b_C in the interior of C that is fixed by all isometries of C.

First we define a barycentre for finite sets in M^n_κ. In the case $\kappa = 0$, we identify $\mathbb{E}^n$ with the hyperplane in $\mathbb{E}^{n+1}$ formed by the points $v = (x_0, \ldots, x_n)$ with $\sum_{i=0}^n x_i = 1$. Given k such points $v_1, \ldots, v_k$, we define their barycentre to be $\frac{1}{k}(\sum_{j=1}^k v_j)$.

In the case $\kappa = 1$, we identify $\mathbb{S}^n$ with the unit sphere about $0 \in \mathbb{E}^{n+1}$. We define the barycentre of a finite set of points $\{v_1, \ldots, v_k\}$ contained in an open hemisphere of $\mathbb{S}^n$ to be $\sum_{j=1}^k v_j / || \sum_{j=1}^k v_j ||$.

In the case $\kappa = -1$, we identify $\mathbb{H}^n$ with the upper sheet of the hyperboloid in $\mathbb{E}^{n,1}$. The barycentre of a finite subset $\{v_1, \ldots, v_k\}$ in $\mathbb{H}^n$ is the unique point on the hyperboloid that lies on the half-line that issues from 0 and passes through $\sum_{j=1}^k v_j$.

The barycentre of a finite subset $V \subset M^n_\kappa$ (assuming that V is contained in an open ball of radius $D_\kappa/2$) is defined by rescaling the metric on $\mathbb{E}^n$, $\mathbb{S}^n$ or $\mathbb{H}^n$.

The *barycentre* of a convex polyhedral cell $C \subset M^n_\kappa$ is defined to be the barycentre of its set of vertices; it is denoted b_C.

7.43 Lemma. *Let C be a convex M_κ-polyhedral cell. The barycentre of C lies in the interior of C and is fixed by any isometry of C.*

Proof. It is clear from the definition that b_C lies in the convex hull of the vertex set of C and not in the convex hull of any proper subset of it. Thus b_C is in the interior of C (by 7.36(5)). The invariance of b_C under isometries follows from the invariance of the vertex set (7.36(6)). □

7.44 Barycentric Subdivision of a Convex Cell. Let $C \subset M_\kappa$ be a convex polyhedral cell. The *first barycentric subdivision of* C, denoted C', is the M_κ-simplicial complex defined as follows. There is one geodesic simplex in C' corresponding to each strictly ascending sequence of faces $F_0 \subset F_1 \cdots \subset F_k$ of C; the simplex is the convex hull of the barycentres of the F_i. Note that the intersection in C of two such simplices is again such a simplex. The natural map from the disjoint union of these geodesic simplices to C imposes on C the structure of an M_κ-simplicial complex — this is C'.

Notice that if $x_1, x_2 \in C$ belong to the same simplex of C' and F_1 and F_2 are the minimal faces of C containing these points, then $F_1 \subseteq F_2$ or $F_2 \subseteq F_1$.

As an immediate consequence of (7.36(6)) we have:

7.45 Lemma. *Any isometry $C_1 \to C_2$ between convex M_κ-polyhedral cells defines a simplicial isometry $C_1' \to C_2'$.*

7.46 Lemma. *Let $K = \coprod C_\lambda/\!\sim$ be an M_κ–polyhedral complex, and let $p_\lambda : C_\lambda \to K$ be the natural projection. Then:*

(1) *The restriction of p_λ to each simplex of the barycentric subdivision C'_λ is injective.*

(2) *Let $\lambda_1, \lambda_2 \in \Lambda$, let S_1 and S_2 be simplices of C'_{λ_1} and C'_{λ_2} and suppose that p_{λ_1} and p_{λ_2} are injective. If $\{x \in S_1 \mid p_{\lambda_1}(x) \in p_{\lambda_2}(S_2)\}$ is not empty, then it is a face of S_1.*

Proof. (1) If $p_\lambda(x) = p_\lambda(y)$, then there is an isometry $h : \text{supp}(x) \to \text{supp}(y)$ such that $y = h(x)$. Therefore the faces $\text{supp}(x)$ and $\text{supp}(y)$ have the same dimension. If x and y are in the same simplex of C'_λ, then by the last sentence of (7.44), $\text{supp}(x) = \text{supp}(y)$ and h is the identity.

(2) It follows from the second condition in the definition of a polyhedral complex that the set described in (2) is a union of faces. It therefore suffices to show that it is convex.

Let $x_1, y_1 \in C_{\lambda_1}$ and $x_2, y_2 \in C_{\lambda_2}$ be such that $p_{\lambda_1}(x_1) = p_{\lambda_2}(x_2)$ and $p_{\lambda_1}(y_1) = p_{\lambda_2}(y_2)$. Then $\text{supp}(x_1)$ and $\text{supp}(x_2)$ (resp. $\text{supp}(y_1)$ and $\text{supp}(y_2)$) have the same dimension. According to the last sentence of (7.44), if x_1, y_1 belong to the same simplex S_1 of the barycentric subdivision of C_{λ_1} then we may assume that $\text{supp}(x_1) \subseteq \text{supp}(y_1)$, in which case $\text{supp}(x_2) \subseteq \text{supp}(y_2)$. Condition (2) in the definition of a polyhedral complex yields an isometry $h : \text{supp}(y_1) \to \text{supp}(y_2)$ such that $h(y_1) = y_2$ and $p_{\lambda_2}(h(x)) = p_{\lambda_1}(x)$ for each $x \in \text{supp}(y_1)$. As p_{λ_2} is injective, we also have $h(x_1) = x_2$, hence $p_{\lambda_1}([x_1, y_1]) = p_{\lambda_2}([x_2, y_2])$, showing that the set in (2) is convex. □

7.47 Barycentric Subdivision of M_κ-Polyhedral Complexes. Let K be an M_κ–polyhedral complex. Let $p : \coprod_\Lambda C_\lambda \to K$ be as in (7.37). For each cell C_λ, we index the simplices of the barycentric subdivision C'_λ by a set I_λ; so C'_λ is the M_κ–simplicial complex associated to $\coprod_{I_\lambda} S_i \to C_\lambda$. Let $\Lambda' = \coprod_\Lambda I_\lambda$. By composing the natural maps $\coprod_{I_\lambda} S_i \to C_\lambda$ and $p : \coprod_\Lambda C_\lambda \to K$ we get a projection $p' : \coprod_{i\in\Lambda'} S_i \to K$. Let K' be the quotient of $\coprod_{i\in\Lambda'} S_i$ by the equivalence relation $[x \sim y \text{ iff } p'(x) = p'(y)]$.

K' is called the *first barycentric subdivision* of K. There is a natural identification of sets $K \to K'$.

7.48 Lemma. *With the notation of (7.47):*

(1) *K' is an M_κ–polyhedral complex.*

(2) *The restriction of p' to each simplex S_i is an injection.*

(3) *The natural map $K \to K'$ is an isometry.*

Proof. For (1) We must check the conditions of (7.37): the first is obviously satisfied and the second follows from (7.45). Part (2) is a restatement of (7.46(1)), and (3) is an easy consequence of the definitions. □

In general K' will not be an M_κ–simplicial complex because although the cells of K' are geodesic simplices (in their local metrics), the intersection of two cells is in general a union of faces rather than a single face. To remedy this, we take the *second barycentric subdivision* of K, which is by definition the first barycentric subdivision of K'.

7.49 Proposition. *If K is an M_κ-polyhedral complex, then the second barycentric subdivision K'' of K is an M_κ-simplicial complex and the natural map $K \to K''$ is an isometry.*

Proof. This follows immediately from (7.46(2)) and (7.48). □

7.50 Theorem. *Let K be an M_κ-polyhedral complex. If* Shapes(K) *is finite, then K is a complete geodesic metric space.*

Proof. Shapes(K'') is the set of isometry classes of the simplices in the second barycentric subdivisions of the model cells in Shapes(K). In particular, since Shapes(K) is finite, Shapes(K'') is finite, and the theorem follows from (7.49) and (7.19). □

We end this section with an example to indicate how Theorem 7.50 together with our earlier results on the nature of geodesics can be combined to construct new examples of complete geodesic spaces.

7.51 Example. Let K_1 and K_2 be 2–dimensional M_0–polyhedral (i.e. piecewise Euclidean) complexes with Shapes(K_1) and Shapes(K_2) finite. For $i = 1, 2$, let $c_i : S_i \to K_i$ be a closed local geodesic, where S_i is a circle of length $l(c_i)$. By scaling the metrics on K_1 and K_2 we can arrange that $l(c_1) = l(c_2) = 2\pi$. Consider the metric space obtained by gluing a cylinder $\mathbb{S}^1 \times [0, 1]$ to the disjoint union of K_1 and K_2 by attaching the ends of the cylinder to the K_i by the maps c_i.

According to (7.29), each c_i is the concatenation of a finite number of segments (in the sense of (7.3)). It follows that we may subdivide the cells of K_1 and K_2 into smaller polyhedra so that the curves c_i run in the 1-skeleton. We can then triangulate the cylinder $\mathbb{S}^1 \times [0, 1]$ so that its attaching maps c_i map cells isometrically to cells. The resulting quotient complex $K_1 \cup (\mathbb{S}^1 \times [0, 1]) \cup K_2/\sim$ is then a Euclidean polyhedral complex.

We shall see many more specific examples of this construction in Chapter II.11.

7.52 Exercise. Show that the subdivision in (7.51) is possible even if K_1 and K_2 are not 2-dimensional.

More on the Geometry of Geodesics

In this section we gather a number of technical results concerning geodesics in M_κ–polyhedral complexes K with Shapes(K) finite. First we note that although we have shown that for every $x \in K$ there is a unique geodesic joining x to y whenever

$y \in B(x, \varepsilon(x))$, it does not follow that K is locally uniquely geodesic. Indeed we shall now show (7.55) that requiring K to be locally uniquely geodesic is a surprisingly stringent condition.

7.53 Definition. The *injectivity radius* of a geodesic space X is the supremum of the set of non-negative numbers r such that any two points in X a distance $\leq r$ apart are joined by a unique geodesic. We denote this number *injrad*(X).

An easy compactness argument shows that if a compact geodesic space X is locally uniquely geodesic then *injrad*$(X) > 0$. We shall show that M_κ–polyhedral complexes with Shapes(K) finite share this property. For this we need the following lemma.

7.54 Lemma. *If K is an M_κ–simplicial complex with* Shapes(K) *finite, then there is a constant $\varepsilon_0 > 0$ such that for every $x \in K$ there exists $y \in K$ with $B(x, \varepsilon_0) \subseteq B(y, \varepsilon(y)/4)$, where $\varepsilon(y)$ is as defined in (7.8).*

Proof. Let $K^{(n)}$ denote the n–skeleton of K (that is, the union of the closed simplices of dimension at most n). We shall construct a constant ε_0 so that, for all n, if $x \in K^{(n)}$ lies in the $8\varepsilon_0$–neighbourhood of $K^{(n-1)}$ then there exists $y \in K^{(n-1)}$ with $B(x, 4\varepsilon_0) \subseteq B(y, \varepsilon(y)/4)$.

Let Σ denote the disjoint union of the model simplices $\overline{S} \in$ Shapes(K). Consider the map $\eta : \Sigma \to (0, \infty)$ defined on $s \in \overline{S}$ by:

$$\eta(s) = \min\{d(s, F) \mid F \text{ a face of } \overline{S} \text{ and } s \notin F\}.$$

Note that while η is not a continuous map, its restriction to the interior of each face is continuous.

Let η_0 be $1/8$ of the minimum value attained by η on $\Sigma^{(0)}$, the 0–skeleton of Σ. Then, inductively for $n \leq D = \dim(K)$, we define η_n to be $1/8$ of the minimum value attained by η on the compact set (on which η is continuous) obtained by deleting the open η_{n-1}–neighbourhood of $\Sigma^{(n-1)}$ from $\Sigma^{(n)}$. Notice that $8\eta_n \leq \eta_{n-1}$. Let $\varepsilon_0 = \eta_D$.

Arguing by induction on n we prove that if $x \in K^{(n)} \smallsetminus K^{(n-1)}$ then either $8\eta_n < \varepsilon(x)$ or else there exists $y \in K^{(n-1)}$ with $B(x, 4\eta_n) \subseteq B(y, \varepsilon(y)/4)$. The case $n = 0$ follows from Lemma 7.9. Consider $x \in K^{(n)}$. If x lies in $K^{(n)}$ minus the open η_{n-1}–neighbourhood of $K^{(n-1)}$ then $\varepsilon(x) \geq 8\eta_n$. On the other hand, if there exists $z \in K^{(n-1)}$ such that $d(x, z) \leq \eta_{n-1}$, then $B(x, 8\eta_n) \subseteq B(x, \eta_{n-1}) \subseteq B(z, 2\eta_{n-1})$. But then, by induction, either $\varepsilon(z) \geq 8\eta_{n-1}$ and hence $B(x, 2\eta_n) \subseteq B(z, \varepsilon(z)/4)$, or else there exists $y \in K^{(n-2)}$ such that $B(x, 4\eta_n) \subset B(z, 4\eta_{n-1}) \subseteq B(y, \varepsilon(y)/4)$. □

Recall that a metric space is said to be *r-uniquely geodesic* if every pair of points a distance less than r apart can be joined by exactly one geodesic segment.

7.55 Proposition. *Let K be an M_κ–polyhedral complex with* Shapes(K) *finite. The following conditions are equivalent:*

(1) *K is locally uniquely geodesic;*

(2) *K has positive injectivity radius;*

(3) $\mathrm{Lk}(x, K)$ *is π-uniquely geodesic for every $x \in K$.*

Proof. Subdividing if necessary, we may assume that K is simplicial.

That (2) implies (1) is trivial. If (1) holds then, by (7.39), for every $x \in K$ a neighbourhood of the cone point in $C_\kappa(\mathrm{Lk}(x, K))$ is uniquely geodesic. By (5.11), this implies that $\mathrm{Lk}(x, K)$ is π-uniquely geodesic, thus (3) holds.

A second application of (5.11) shows that if $\mathrm{Lk}(x, K)$ is π-uniquely geodesic then the whole of $C_\kappa(\mathrm{Lk}(x, P))$ is uniquely geodesic. So by (7.39), $B(x, \varepsilon(x)/4)$ is uniquely geodesic. Hence, by the preceding lemma, there is a constant ε_0 such that for every $x \in X$ the ball $B(x, \varepsilon_0)$ is contained in a uniquely geodesic subspace of K. In particular, *injrad*$(K) \geq \varepsilon_0$. Thus (3) implies (2). □

For future reference we note a result related to (7.55).

7.56 Lemma. *Let K be an M_κ-polyhedral complex with* $\mathrm{Shapes}(K)$ *finite. If the points x and y lie in the same open cell of K, then for sufficiently small $\varepsilon > 0$ there exists an isometry from $B(x, \varepsilon)$ to $B(y, \varepsilon)$ that restricts to an isometry from $B(x, \varepsilon) \cap C$ to $B(y, \varepsilon) \cap C$ for every closed cell C containing x.*

Proof. If we take $\varepsilon < \frac{1}{2}\min\{\varepsilon(x), \varepsilon(y)\}$, then the metrics induced on $B(x, \varepsilon)$ and $B(y, \varepsilon)$ are the length metrics given by measuring the length of paths in these balls using the given local metrics d_B on the individual cells B that contain x and y. Thus it is enough to exhibit a bijection from $B(x, \varepsilon)$ to $B(y, \varepsilon)$ that restricts to an isometry from $B(x, \varepsilon) \cap B$ to $B(y, \varepsilon) \cap B$, where these intersections are endowed with the restriction of the local metric d_B.

This observation allows us to argue in the models $S \in \mathrm{Shapes}(K)$ and hence in M_κ^n. More specifically, it is enough to show that given two points x' and y' in M_κ^n (with $d(x', y') \leq \pi/\sqrt{\kappa}$ if $\kappa > 0$), there is a canonical isometry of M_κ that interchanges x and y and has the property that it sends every hyperplane containing x and y to itself. (We need this isometry to be canonical so as to ensure that the local isometries induced on each $B(x, \varepsilon) \cap B$ agree on common faces.) The reflection of M_κ^n in the hyperplane bisector $H(x', y')$ of x' and y' has the desired property. □

Arguments Using Finite Models

Proposition 7.55 is an example of how, in many respects, complexes with $\mathrm{Shapes}(K)$ finite behave as if they were compact (or cocompact) spaces. We shall now present two more examples of this phenomenon. Our main interest in the first result lies with the corollary, but in the next chapter we shall have need of the stronger, more technical statement of the proposition.

7.57 Proposition. *Let x and y be points in an M_κ–polyhedral complex K. Suppose that* $\mathrm{Shapes}(K)$ *is finite. Suppose that there is a unique geodesic segment joining x*

to y in K and let $c : [0,1] \to K$ be a linear parameterization of this segment. Let $c_n : [0,1] \to K$, $n = 1, 2, \dots$ be linearly reparameterized geodesics in K, and suppose that the sequences of points $c_n(0)$ and $c_n(1)$ converge to x and y respectively. Then, $c_n \to c$ uniformly.

7.58 Corollary. *Let K be an M_κ–polyhedral complex with* Shapes(K) *finite. If K is uniquely geodesic, then geodesics in K vary continuously with their endpoints.*

Proof of 7.57. Subdividing, we may assume that K is an M_κ–simplicial complex K with Shapes(K) finite. Suppose $x_n \to x$ and $y_n \to y$. Let $c_n : [0,1] \to K$ be a linearly reparameterized geodesic joining x_n to y_n and let $c : [0,1] \to P$ be the linearly reparameterized geodesic joining x to y. There is a uniform bound on $d(x, x_n) + l(c_n) + d(y_n, y)$, so by (7.30) there exists an integer N such that image of each of the paths c_n lies in a connected subcomplex $K_n \subset K$ that can be expressed as the union of at most N closed cells, and which contains geodesic segments from x to x_n and from y to y_n. We fix a definite choice of K_n for each positive integer n.

Because Shapes(K) is finite, there are only finitely many equivalence classes of such complexes K_n modulo the relation: $K_n \sim K_m$ if there is a map $K_n \to K_m$ that fixes x, y and is a simplicial isometry (i.e., a homeomorphism whose restriction to each closed simplex in K_n is a local isometry onto a closed simplex in K_m). Such a map is an isometry between K_n and K_m equipped with their intrinsic metrics (not their induced metrics from K). We shall call $\sim$ equivalence classes models, denoted μ.

This notion of a model carries a little more information than that considered in (7.26) and (7.27). A model μ in the present sense consists not only of a (model) compact M_κ–complex L_μ with two distinguished points x^μ and y^μ, it also comes equipped with a specified family of simplicial isometries $\phi_n : L_\mu \to K_n \subseteq K$, one for each $K_n \in \mu$, such that $\phi_n(x^\mu) = x$ and $\phi_n(y^\mu) = y$. An important point to note is that since the maps ϕ_n are simplicial isometries, they preserve the length of paths. Thus, when considered as maps to K (as they will be henceforth) they do not increase distances.

Because there are only finitely many models, after deleting a finite number of terms if necessary, we may decompose (c_n) into finitely many infinite subsequences so that $K_i \sim K_j$ for all c_i and c_j in any subsequence. We focus our attention on one such subsequence, and to simplify the notation we write μ for the corresponding model and c_n for the paths comprising the subsequence.

Let $\tilde{c}_n = \phi_n^{-1} c_n : [0,1] \to L_\mu$. Because ϕ_n is length-preserving, $\tilde{c}_n$ is a linearly reparameterized geodesic of length $l(c_n)$. Let $\tilde{x}_n = \tilde{c}_n(0)$ and let $\tilde{y}_n = \tilde{c}_n(1)$. Notice that by our initial specification of K_n, there exists a geodesic segment of length $d(x, x_n)$ connecting $\tilde{x}$ to $\tilde{x}_n$ in L_μ, and a geodesic segment of length $d(y, y_n)$ connecting $\tilde{y}$ to $\tilde{y}_n$. Hence $d(\tilde{x}, \tilde{y}) \leq d(x, x_n) + l(c_n) + d(y, y_n)$, which approaches $d(x, y)$ as $n \to \infty$. Since ϕ_n does not increase distance, we deduce that $d(\tilde{x}, \tilde{y}) = d(x, y)$. It follows that since c is the *unique* reparameterized geodesic $[0,1] \to K$ joining x to y, and since ϕ_n is injective, there is a unique linearly reparameterized geodesic $\tilde{c} : [0,1] \to L_\mu$ joining $\tilde{x}$ to $\tilde{y}$ in L_μ; furthermore, $c = \phi_n \circ \tilde{c}$ for all n.

We can now appeal to (3.11) to see that the sequence of paths $\tilde{c}_n : [0,1] \to L_\mu$ converges uniformly to $\tilde{c}$. But $d(c(t), c_n(t)) = d(\phi_n\tilde{c}(t), \phi_n\tilde{c}_n(t)) \le d(\tilde{c}(t), \tilde{c}_n(t))$. Hence $c_n \to c$ uniformly. $\square$

Given an M_κ–polyhedral complex K, model cells $\overline{S}, \overline{S}' \in \text{Shapes}(K)$ and points $\overline{x} \in \overline{S}$ and $\overline{y} \in \overline{S}'$, we consider the sets of points in K that lie above $\overline{x}$ and $\overline{y}$:

$$X_K = \{x \in K \mid \exists \text{ isometry } f \text{ from a face of } \overline{S} \text{ to a closed cell in } K \text{ with } f(\overline{x}) = x\},$$

$$Y_K = \{y \in K \mid \exists \text{ isometry } g \text{ from a face of } \overline{S}' \text{ to a closed cell in } K \text{ with } g(\overline{y}) = y\}.$$

7.59 Proposition. *Let K be an M_κ-polyhedral complex. If* Shapes(K) *is finite then for all $\overline{S}, \overline{S}' \in$* Shapes($K$) *and all $\overline{x} \in \overline{S}$ and $\overline{y} \in \overline{S}'$, the set of numbers $\{d(x,y) \mid x \in X_K, y \in Y_K\}$ is discrete.*

Proof. The number of points of X_K or Y_K that lie in any given closed cell of K is bounded by the maximum of the orders of the isometry groups of the model cells $\overline{S} \in \text{Shapes}(K)$. Hence, for each finite subcomplex $L \subset P$ the set of numbers $\{d_L(x,y) \mid x \in X_K \cap L, y \in Y_K \cap L\}$ is finite, where L denotes the intrinsic metric on L (not the induced metric from K).

Let $\Sigma_N = \bigcup\{d_L(x,y) \mid x \in X_K \cap L, y \in Y_K \cap L\}$, where the union is taken only over those subcomplexes $L \subset K$ which can be expressed as the union of at most N closed cells. For each integer N there are only finitely many such subcomplexes up to simplicial isometry, so the final sentence of the previous paragraph implies that Σ_N is finite. But according to (7.28), for every $\ell > 0$ there exists an integer N such that every geodesic in K of length $\ell > 0$ is contained in a subcomplex which is the union of at most N cells. In particular, $\{d(x,y) \mid x \in X_K, y \in Y_K\} \cap [0,\ell) \subset \Sigma_N$. $\square$

Alternative Hypotheses

We close this chapter with a couple of exercises and examples to illustrate both the usefulness and deficiencies of alternatives to the hypothesis that Shapes(K) is finite. Further examples and exercises are contained in the appendix.

Let us first consider how one might weaken the hypothesis that Shapes(K) is finite so as to accommodate infinite dimensional complexes. An obvious hypothesis to try is that Shapes(K) contains only finitely many model simplices in each dimension. In this generality we have:

7.60 Exercise. Let K be a metric simplicial complex. If Shapes(K) contains only finitely many model simplices in each dimension then the intrinsic pseudometric on K is actually a metric.

On the other hand, we do not necessarily obtain geodesic metric spaces in this generality.

7.61 Example. Let Δ^n denote the standard Euclidean simplex of dimension n (i.e. the convex hull of the standard basis vectors in $\mathbb{R}^{n+1}$. For every positive integer n we have $\Delta^n \subset \Delta^{n+1}$ via the inclusion $(x_1, \cdots, x_{n+1}) \mapsto (x_1, \cdots, x_{n+1}, 0)$ of $\mathbb{R}^{n+1}$ into $\mathbb{R}^{n+2}$. Let Δ^∞ be the union of Δ^n for n in $\mathbb{N} \setminus \{0\}$ with the obvious piecewise Euclidean structure. We consider the complex K obtained by joining two copies $\Delta^\infty_+, \Delta^\infty_-$ of Δ^∞ along the subcomplex opposite the vertex $P = (1, 0, \dots)$ (that is, we join them along $\Delta^n_\pm \cap \{(x_1, \cdots, x_{n+1}) \in \mathbb{R}^{n+1} \mid x_1 = 0\}$ for every n in $\mathbb{N} \setminus \{0\}$). We claim that there is no geodesic joining P_+ and P_- in K. Indeed, the distance from P to the opposite face of Δ^n decreases to the limit 1 as $n \to \infty$, and the distance between P_+ and P_- in K is 2, which is strictly less than the length of any path joining P_+ to P_- in K.

On the other hand, it is easy to verify the following:

7.62 Exercise. Let K be a metric simplicial complex. If for every vertex $x \in K$ each ball of finite radius about K (in the intrinsic pseudometric) contains only cells which are modelled on a finite subset of Shapes(K), then (K, d) is a complete geodesic metric space.

Appendix: Metrizing Abstract Simplicial Complexes

In this appendix we describe an alternative approach to metric simplicial complexes. This approach is often appropriate in situations where one wishes to use geometry to study a problem that has been encoded in the combinatorics of an abstract simplicial complex.

We shall recall the definition of an abstract simplicial complex, define the affine realization of such a complex, and discuss related notions such as barycentric coordinates. We shall then recast the definition of a metric simplicial complex in this language.

In (7A.13) we generalize (7.13) by giving a weaker criterion for the intrinsic metric on a metric simplicial complex to be complete. We shall also define join and cone constructions for simplicial complexes and explain how they are related to the analogous metric constructions given in Chapter 5. Finally, we give a criterion (7A.15) for spaces to have the homotopy type of a finite simplicial complex.

Abstract Simplicial Complexes

7A.1 Definitions. An *abstract simplicial complex K* consists of a non-empty set V (its set of vertices) together with a collection $\mathcal{S}$ of non-empty finite subsets of V, such that $\{v\} \in \mathcal{S}$ for all $v \in V$, and if $S \in \mathcal{S}$ then every non-empty subset T of S is also in $\mathcal{S}$. The elements of $\mathcal{S}$ are called the *simplices* of K. A simplex $S \in \mathcal{S}$ is called an *n-simplex* if S has cardinality $n+1$ and n is called the dimension of S. The elements of S are called its *vertices* and the non-empty subsets of S are called its faces. The dimension $\dim K$ of K is the supremum of the dimensions of its simplices.

A simplicial complex K_0 is called a subcomplex of K if its set of vertices V_0 is a subset of V and each simplex of K_0 is also a simplex of K. To each n-simplex S of K is associated a subcomplex $\underline{S}$ of K of dimension n whose set of simplices is the set of faces of S.

Consider the equivalence relation on V generated by: $u \sim v$ if $\{u, v\} \in \mathcal{S}$. The complex K is said to be *connected* if V consists of a single $\sim$ equivalence class.

Let K_1 and K_2 be be abstract simplicial complexes with vertex sets V_1 and V_2 respectively. A *simplicial map* from K_1 to K_2 is a map $f : V_1 \to V_2$ that sends each simplex of K_1 to a simplex of K_2.

7A.2 Definition of Cones, Joins and Links. Let K_1 and K_2 be abstract simplicial complexes whose vertex sets V_1, V_2 are disjoint. The *simplicial join* of K_1 and K_2, denoted $K_1 * K_2$, is a simplicial complex with vertex set $V_1 \cup V_2$; a subset of $V_1 \cup V_2$ is a simplex of $K_1 * K_2$ if and only if it is a simplex of K_1, a simplex of K_2, or the union of a simplex of K_1 and a simplex of K_2. Note that

$$\dim(K_1 * K_2) = \dim K_1 + \dim K_2 + 1.$$

For example, the join of a p-simplex and a q-simplex is a $(p+q+1)$-simplex.

If K is a simplicial complex with vertex set V, then the *simplicial cone* over K, denoted $C(K)$, is the simplicial join of K and a complex with only one vertex v_0. The vertex v_0 is called the *cone vertex* of $C(K)$ and $\dim C(K) = \dim K + 1$.

Given a simplex S in an abstract simplicial complex K, the *link* of S in K, denoted $lk(S, K)$, is the subcomplex of K consisting of those simplices T such that $T \cap S = \emptyset$ and $T \cup S$ is a simplex of K.

Note that the subcomplex of K whose simplices are the faces of simplices containing S is isomorphic to the join $\underline{S} * lk(S, K)$.

7A.3 Affine Realization. Let K be an abstract simplicial complex with vertex set V. Let W denote the real vector space with basis V. The *affine realization* $|S|$ of a simplex S of K is the affine simplex in W that has vertex set S; in other words, $|S|$ consists of those vectors $x = \sum_{v\in S} x_v\, v$ with $x_v \in [0, 1]$ and $\sum_{v\in S} x_v = 1$. We give $|S|$ the topology and affine structure induced from the finite dimensional subspace of W spanned by the vertices of S; this subspace is isomorphic to $\mathbb{R}^{n+1}$, where n is the dimension of S. Given two points x and y in $|S|$, the affine segment with extremities x and y is denoted $[x, y]$.

We define the *affine realization* (or simply the *realization*) of K to be the subset $|K|$ of W which is the union of the affine realizations of the simplices of K. (It is convenient to abuse terminology to the extent of writing S and K to mean the realization of S and K, and we shall do so frequently.) The realization of S is called a *closed simplex* in K. The coordinates x_v of a point x of $K \subset W$ are called its *barycentric coordinates*.

The *interior* of a simplex S is the set of points in its affine realization whose coordinates x_v, $v \in S$, lie in $(0, 1)$. Given a point $x \in K$, the *open star* $st(x)$ of x is defined to be the union of $\{x\}$ and the interior of those closed simplices S which contain x; the *closed star* $St(x)$ of x is the union of the closed simplices which contain x. (If v is a vertex, then its open star $st(v)$ is the set of points x of K with barycentric coordinates $x_v > 0$.)

Let K_1 and K_2 be abstract simplicial complexes with vertex sets V_1 and V_2 and let W_1 and W_2 be the real vector spaces with bases V_1 and V_2 respectively. If $f : V_1 \to V_2$ gives a simplicial map from K_1 to K_2, then the linear map $W_1 \to W_2$ sending $v \in V_1$ to $f(v) \in V_2$ maps $|K_1|$ to $|K_2|$ and its restriction to each $|S|$ is an affine map onto $|f(S)|$. This map, which we denote $|f| : |K_1| \to |K_2|$ will be called the *(affine) realization* of the simplicial map f.

7A.4 The Barycentric Subdivision. Let K be an abstract simplicial complex with vertex set V. The barycentric subdivision K' of K is the abstract simplicial complex whose vertices are the simplices S_i of K, and whose n-simplices are the sets $\{S_0, \dots, S_n\}$ where $S_0 \subset S_1 \subset \cdots \subset S_n$.

There is a natural bijection from the affine realization of K' to the affine realization of K: this map is affine on each simplex of K' and sends the vertex of K' corresponding to the n-simplex S of K to the barycentre of the affine realization of S (namely the point with barycentric coordinates $x_v = 0$ if $v \notin S$ and $x_v = 1/(n+1)$ if $v \in S$) and is affine on each simplex of K'.

7A.5 Topology on the Affine Realization. For the moment we do not consider any topology on the affine realization of K, but mention in passing that a useful topology for the purposes of homotopy theory is the *weak topology*, which is characterized by the property that a subspace of K is closed if and only if its intersection with each closed simplex is closed.

An alternative topology on K can be obtained by introducing on W the scalar product such that the basis V is orthonormal and considering the associated metric. The restriction of this metric to K is called the *metric of barycentric coordinates*. If K is not locally finite, then the topology associated to this metric is not the weak topology. Indeed if a vertex v lies in infinitely many simplices, then any set containing exactly one point ($\neq v$) in the interior of each simplex of $St(v)$ is closed in the weak topology, but in general it is not closed with respect to the metric of barycentric coordinates because v might be an accumulation point of this set.

On the other hand, an important theorem of Dowker [Dow52] shows that the identity map of K is a homotopy equivalence from K equipped with the weak topology to K equipped with the topology induced by the metric of barycentric coordinates.

7A.6 Barycentric Coordinates for Geodesic Simplices in M^m_κ. For a geodesic n-simplex $S \subset M^m_\kappa$ with vertices $v_0, \dots, v_n$ one has natural barycentric coordinates, which are defined as follows (compare with 7.42).

$\kappa = 0$: Identify $\mathbb{E}^m$ with the hyperplane in $\mathbb{E}^{m+1}$ formed by the points $y = (y_0, \dots, y_n)$ with $\sum_{i=0}^m y_i = 1$. Any point $x \in S \subset \mathbb{E}^m$ can be written uniquely as $x = \sum_{i=0}^{n} x_{v_i} v_i$ with $x_{v_i} \in [0, 1]$; the reals numbers x_{v_i} are the barycentric coordinates of x.

$\kappa > 0$: Identify M_κ^m with the sphere of radius $1/\sqrt{\kappa}$ centred at the origin in $\mathbb{E}^{m+1}$. The n-simplex S_0 in $\mathbb{E}^{m+1}$ spanned by the vertices $v_0, \dots, v_n$ does not contain 0 and therefore is mapped bijectively onto the simplex S by the radial projection from 0 to M_κ^m. The barycentric coordinates of points $x \in S$ are defined to be the barycentric coordinates of the corresponding points in S_0.

$\kappa < 0$: Identify M_κ^m with the upper sheet of the hyperboloid $\sum_{i=1}^{m} x_i^2 - x_{m+1}^2 = 1/\sqrt{\kappa}$ in $\mathbb{E}^{m,1}$ and proceed as in the case $\kappa > 0$.

Barycentric coordinates are preserved by isometries of M_κ^m. If S is an n-simplex of an M_κ-simplicial complex, then the barycentric coordinates of a point $x \in S$ are defined to be the barycentric coordinates of the unique point $\overline{x} \in \overline{S}$ that is mapped to x by the characteristic map $f_S : \overline{S} \to S$ (see 7.12). The *barycentre* of a simplex S is the point whose barycentric coordinates are all equal.

7A.7 Affine and Projective Maps. Given geodesic n-simplices $S \subset M_\kappa^m$ with vertices $v_0, \dots, v_n$ and $S' \subset M_{\kappa'}^{m'}$ with vertices $v_0', \dots, v_n'$, there is a unique map from S to S' sending v_i to v_i' and preserving the barycentric coordinates. Similarly, given an n-simplex $T = \{u_0, \dots, u_n\}$ in an abstract simplicial complex, there is a unique map from S to the affine realization of T sending v_i to u_i for all i and preserving the barycentric coordinates. Such a map will be called an *affine map*.

A projective map from S to S' (or $|T|$) is one for which there exist positive real numbers λ_i such that the point of S with barycentric coordinates $(x_0, \dots, x_n)$ is sent to the point with barycentric coordinates

$$\frac{\lambda_0 x_0}{\sum_{i=0}^{n} \lambda_i x_i}, \dots, \frac{\lambda_n x_n}{\sum_{i=0}^{n} \lambda_i x_i}.$$

The image under a projective map of a geodesic segment in S is a geodesic segment in S', but in general a linear parameterization of this segment is not mapped to a linear parameterization of its image.

7A.8 Subdivision. Given a geodesic n-simplex $S \subset M_\kappa^m$ with vertex set $V = \{v_0, \dots, v_n\}$ and a point $v \in S \smallsetminus V$, one can subdivide S into smaller n-simplices: the n-simplices $S_{V'}$ in this subdivision correspond to subsets $V' \subset V$ of n vertices whose convex hull does not contain v; $S_{V'}$ is the convex hull of $V' \cup \{v\}$. With respect to the barycentric coordinates of S and $S_{V'}$, the inclusion $S_{V'} \hookrightarrow S$ is projective.

Metrizing Simplicial Complexes

Let K be an M_κ-simplicial complex as defined in (7.2). Associated to K one has an abstract simplicial complex, whose vertex set V is the set of vertices of K; a subset $S \subseteq V$ is an (abstract) simplex if and only if S is the vertex set of a simplex in K. There is a canonical bijection from K onto the affine realization of this abstract simplicial complex; this map extends the natural bijection between the sets of vertices and is affine on each simplex with respect to the barycentric coordinates as defined in (7A.6).

Conversely, when presented with an abstract simplicial complex K, one can turn it into a geometric object (an M_κ-complex) as follows: metrize the simplices of the realization of K by choosing an affine isomorphism to a geodesic simplex in M^n_κ; choose these isomorphisms so that the induced metrics coincide on faces of intersection. This construction leads to a well-defined notion of a piecewise linear (PL) path in K and the intrinsic pseudometric on K is the infimum of the lengths of PL paths joining them.

7A.9 Alternative Definition of an M_κ–Simplicial Complex. Let κ be a real number. An *M_κ–simplicial complex* consists of the following information:

(1) an abstract simplicial complex K;

(2) a set Shapes(K) of geodesic simplices $\overline{S}_i \subset M^{n_i}_\kappa$;

(3) for every closed simplex S in the realization of K, an affine isomorphism $f_S : \overline{S} \to S$, where $\overline{S} \in$ Shapes(K); if S' is a face of S, then $f_S^{-1} \circ f_{S'}$ is required to be an isometry from $\overline{S}'$ onto a face $\overline{S}$.

A *piecewise linear* (PL) path in K is a path that is the concatenation of a finite number of affine segments. The length of a segment I in a simplex $S \subset K$ is defined to be the length of $f_S^{-1}(I)$. The intrinsic metric on K is defined by setting $d(x, y)$ equal to the infimum of the length of PL-paths joining x to y.

Note that the set of PL paths in K depends only on the affine structure of K and not on the chosen maps f_S. It is easy to see that the above definition of the intrinsic metric d agrees with that given in (7.4) and therefore (K, d) is a length space; it is a complete geodesic space if Shapes(K) is finite (7.19). Moreover, the characterization of geodesics in (7.29) shows that if Shapes(K) is finite then $d(x, y)$ is equal to the length of the shortest PL path connecting them (i.e. the infimum in the definition of d is attained).

7A.10 Regular M_κ-Simplicial Complexes. An M_κ-simplicial complex K in which all of the model simplices $\overline{S} \in$ Shapes(K) have the same edge lengths is called *regular*. We describe the cases $\kappa = 0$ and $\kappa = 1$.

In the notation of (7A.3), let W be the vector space with basis the vertex set of K, endowed with the metric associated to the scalar product for which this basis is orthonormal. Each n-simplex S of K, with the induced metric d_S, is isometric to a regular n-simplex in $\mathbb{E}^n$ with edge lengths $\sqrt{2}$. In this way K becomes a piecewise Euclidean complex. If we wish, we may rescale the metric so that each edge has length one; this is called the *standard Euclidean realization* of K.

Consider the unit sphere in W (the set of vectors of norm one) with the induced length metric. Radial projection from the origin $0 \in W$ onto the unit sphere restricts to an injection on K, and this projection identifies each n-simplex S of K with a spherical n-simplex whose edge lengths are all $\pi/2$. In this case, K becomes a piecewise spherical complex in a natural way, called the *all-right spherical realization* of K.

If K is of finite dimension n, then in the above examples Shapes(K) $= \{\Delta^i | i = 0, 1, \ldots, n\}$, where Δ^i is the Euclidean (resp. spherical) i-simplex in $\mathbb{E}^{i+1}$ (resp. $\mathbb{S}^i$) spanned by the vectors of the standard basis of $\mathbb{R}^{i+1}$.

7A.11 Exercises

(1) Prove that the identity map of any abstract simplicial complex induces a bi-Lipschitz homeomorphism between any two regular M_κ-simplicial complexes associated to K (where κ is not fixed).

(2) Let K and d be as in (7A.5). Prove that if Shapes(K) is finite then the canonical map from K to the standard Euclidean realization of the underlying simplicial which is the identity on the set of vertices and affine on each simplex is a homeomorphism.

(3) Let K_1 and K_2 be abstract simplicial complexes. Show that the all-right spherical realization of $K_1 * K_2$ is the spherical join (as defined in 5.13) of the all-right realizations of K_1 and K_2.

(4) Let K be an M_κ-simplicial complex and fix a vertex $v \in K$. In (7.15) we defined the geometric link $\text{Lk}(v, K)$ of v in K; it is a spherical complex. On the other hand, the subcomplex $St(v) \smallsetminus st(v)$ is endowed with an induced M_κ-simplicial structure from K. Prove that the map which associates to each $x \in St(v) \smallsetminus st(v)$ the initial vector of $[v, x] \in \text{Lk}(v, K)$ is a map that is not affine in general, but is projective in the barycentric coordinates.

A Criterion for Completeness

7A.12 Definition. Given any geodesic n-simplex S in M_κ^m, there is an affine map ϕ_S of S onto the standard n-simplex Δ^n in $\mathbb{E}^{n+1}$ spanned by the basis vectors. Note that ϕ_S is unique modulo the action of Isom(Δ^n). The *distortion* of S is the smallest number λ such that $\frac{1}{\lambda} d(\phi_S(x), \phi_S(y)) \leq d(x, y) \leq \lambda d(\phi(x), \phi(y))$ for every $x, y \in S$.

The *distortion* of an M_κ-simplicial complex K is the supremum of the distortions of the model simplices $\overline{S} \in$ Shapes(K).

7A.13 Theorem. *Let K be a finite dimensional M_κ-simplicial complex. If the distortion of K is finite, then K is a complete length space.*

Proof. Equip K with its intrinsic pseudometric d. It follows from (7.10) that (K, d) is a metric space and it is clear that it is a length space. (However it need not be a geodesic space, 7A.14.)

Let K_0 be the affine realization of the simplicial complex associated to K and let $\phi : K \to K_0$ be the canonical map which is affine on each simplex. We view K_0 as a regular M_0-simplicial complex. For each model n-simplex $\overline{S} \in$ Shapes(K) we have a bi-Lipschitz map $\phi_{\overline{S}} : \overline{S} \to \Delta^n$. To say that K has finite distortion means that there is a uniform bound, λ say, on the Lipschitz constants of these maps. It follows that the length of each PL path in K is at most λ times the length of the same path in K_0, and vice versa. Since the pseudometric on each of the complexes is defined by taking the infimum of the lengths of such paths, it follows that the map $\phi : K \to K_0$ is a bi-Lipschitz homeomorphism (with Lipschitz constant λ). In particular, since K_0 is complete (K, d) is complete. □

7A.14 Example. Consider the metric graph with two vertices y, z and countably many edges joining them, the n-th of which has length $(1 + \frac{1}{n})$. By introducing a new vertex in the middle of each edge we make this graph simplicial. It has bounded distortion, but it is not a geodesic space: $d(y, z) = 1$ whereas every path from y to z has length strictly larger than 1.

Spaces with the Homotopy Type of Simplicial Complexes

We close with a result which shows that many geodesic spaces have the homotopy type of locally compact complexes. There are much more general results of this type (see [Bor66], [Hu(S)65], [West77]), but the simple result given below indicates the ideas that we wish to exemplify and is sufficient for the purposes of Part II.

Let X be a topological space. A collection of subsets of a space X is said to be *locally finite* if each point $x \in X$ has a neighbourhood that meets only finitely many of the given sets. An abstract simplicial complex is said to be locally finite if each vertex belongs to only finitely many simplices.

7A.15 Lemma. *Let X be a geodesic space and suppose that there exists $\varepsilon > 0$ such that X can be covered by a locally finite collection of open balls $\mathcal{U} = \{B(x_i, \varepsilon) \mid i \in I\}$ such that $B(x_i, \varepsilon)$ and $B(x_i, 3\varepsilon)$ are convex and uniquely-geodesic, and suppose that geodesics in these balls vary continuously with their endpoints. Then X is homotopy equivalent to the geometric realization of a locally finite simplicial complex K (namely the nerve of the covering $\mathcal{U}$). Moreover, if X is compact then K is finite.*

Proof. Consider the abstract simplicial complex K whose vertex set $\{v_i\}_{i\in I}$ is indexed by the elements of I and which has an r-simplex $\{v_{i_0}, \dots, v_{i_r}\}$ for each non-empty intersection $\bigcap_{j=0}^{r} B(x_{i_j}, \varepsilon) \neq \emptyset$. This complex is called the *nerve* of the covering $\mathcal{U}$. We consider the affine realization of this complex, which we also denote K, with the weak topology (7A.5).

We shall define maps $f : X \to K$ and $g : K \to X$ such that fg is homotopic to the identity of K and gf is homotopic to the identity of X. In the second case it suffices to prove that for every $x \in X$ there exists $i \in I$ such that x has a neighbourhood V with $V \cup gf(V) \subset B(x_i, 3\varepsilon)$, for then we obtain a homotopy h_t from $h_0 = gf$ to $h_1 = id_X$ by defining $h_t(x)$, for each $t \in [0, 1]$, to be the point a distance $t\, d(x, gf(x)$ from $gf(x)$ on the unique geodesic segment $[x, gf(x)]$; this segment is contained in $B(x_i, 3\varepsilon)$ and is assumed to vary continuously with its endpoints.

The desired map $f : X \to K$ is obtained by using a partition of the unity subordinate to the covering $\mathcal{U}$. For each $i \in I$, let $f_i : X \to \mathbb{R}$ be the function which is identically zero outside of the ball $B(x_i, \varepsilon)$ and which maps $x \in B(x, \varepsilon_i)$ to $\varepsilon - d(x, x_i)$. Then $f(x)$ is defined to be the point whose j-th barycentric coordinate is

$$\frac{f_j(x)}{\sum_{i \in I} f_i}.$$

Note that f maps $B(x_i, \varepsilon)$ to the open star of v_i.

The desired map $g : K \to X$ is defined inductively on the skeleta of K. For each vertex v of K, let $St'(v)$ be the set of points $x \in K$ whose barycentric coordinate x_v is not smaller than any of the other barycentric coordinates of x. (This subset of K corresponds to the closed star of the vertex v in the barycentric subdivision K' under the bijection from K' to K described in 7A.4.) First, for each $i \in I$ we map v_i to x_i. Proceeding by induction, we assume that a continuous map g has been constructed on the r-skeleton $K^{(r)}$ of K and that, for each vertex v_i, this map sends $K^{(r)} \cap St'(v_i)$ to $B(x_i, \varepsilon)$. We extend g across each $(r+1)$-simplex $S = \{v_{i_0}, \dots, v_{i_{r+1}}\}$ by mapping the barycentre b_S of S to a point in the intersection $\bigcap_{j=0}^{r+1} B(x_{i_j}, \varepsilon)$, and then mapping the affine segment $[b_S, y]$ affinely onto the geodesic segment $[g(b_S), g(y)]$ for each point y in the boundary of S. This extension of g is well-defined and continuous because for each y there is a vertex v_i of S such that both $g(b_S)$ and $g(y)$ lie in $B(x_i, \varepsilon)$, where geodesics are unique and vary continuously with their endpoints. By construction, if v is a vertex of a simplex S of K, then g maps S into the ball $B(x_i, 3\varepsilon)$, and hence g maps $St(v_i)$ to $B(x_i, 3\varepsilon)$.

It follows that gf maps $B(x_i, \varepsilon)$ to $B(x_i, 3\varepsilon)$ and the argument at the end of the second paragraph shows that gf is homotopic to the identity.

To see that fg is homotopic to the identity, note that if S is a simplex of K with vertices $\{v_{i_0}, \dots, v_{i_r}\}$, then $g(S) \subset \bigcup_{j=0}^{r} B(x_{i_j}, \varepsilon)$, and hence $fg(S)$ is contained in the union of the open stars in K of the vertices of S. This completes the proof, because if L is the realization of any locally finite simplicial complex, then by a standard argument, any map $F : L \to L$ that sends each simplex S into the union of the open stars of the vertices of S is homotopic to the identity. (One proves this by proceeding one simplex at a time using the obvious "straight line" homotopies, [Spa66, 3.3.11].) □

Chapter I.8 Group Actions and Quasi-Isometries

In this chapter we study group actions on metric and topological spaces. Following some general remarks, we shall describe how to construct a group presentation for an arbitrary group Γ acting by homeomorphisms on a simply connected topological space X. If X is a simply connected length space and Γ is acting properly and cocompactly by isometries, then this construction gives a finite presentation for Γ. In order to obtain a more satisfactory description of the relationship between a length space and any group which acts properly and cocompactly by isometries on it, one should regard the group itself as a metric object; in the second part of this chapter we shall explore this idea. The key notion in this regard is *quasi-isometry*, an equivalence relation among metric spaces that equates spaces which look the same on the large scale (8.14).

Group Actions on Metric Spaces

First we need some vocabulary.

8.1 Terminology for group actions. An *action* of a group Γ on a topological space X is a homomorphism $\Phi : \Gamma \to \text{Homeo}(X)$, where Homeo($X$) is the group of self-homeomorphisms of X. We shall usually suppress all mention of Φ and write $\gamma.x$ for the image of $x \in X$ under $\Phi(\gamma)$, and $\gamma.Y$ for the image of a subset $Y \subseteq X$. We shall write $\Gamma.Y$ to denote $\bigcup_{\gamma\in\Gamma} \gamma.Y$.

An action Φ is said to be *faithful* (or effective) if $\ker \Phi = \{1\}$. An action is said to be *free* if, for every $x \in X$ and every $\gamma \in \Gamma \smallsetminus \{1\}$, one has $\gamma.x \neq x$. And it is said to be *cocompact* if there exists a compact set $K \subseteq X$ such that $X = \Gamma.K$. Given an action of a group Γ on a space X, we write Γ_x for the stabilizer (*isotropy subgroup*) of $x \in X$, that is $\Gamma_x = \{\gamma \in \Gamma \mid \gamma.x = x\}$.

If X is a metric space, then one says that Γ is *acting by isometries* on X if the image of Φ is contained in the subgroup $\text{Isom}(X) \subset \text{Homeo}(X)$.

8.2 Definition of a Proper Action. Let Γ be a group acting by isometries on a metric space X. The action is said to be *proper* (alternatively, "Γ acts properly on X") if for each $x \in X$ there exists a number $r > 0$ such that the set $\{\gamma \in \Gamma \mid \gamma.B(x, r) \cap B(x, r) \neq \emptyset\}$ is finite.

8.3 Remarks

(1) It is more usual to define an action of a group Γ on a topological space X to be proper if for every compact subset $K \subset X$ the set of elements $\{\gamma \in \Gamma \mid \gamma.K \cap K \neq \emptyset\}$ is finite. The definition that we have adopted is equivalent to the standard definition in the case of actions by groups of isometries on proper metric spaces, but in general it is more restrictive. In fact our definition implies that every compact subset $K \subset X$ has an open neighbourhood U such that $\{\gamma \in \Gamma \mid \gamma.U \cap U \neq \emptyset\}$ is finite.

To see this, cover K with finitely many open balls $B(x_i, r_i)$, where each x_i is in K and each of the sets $S(i) = \{\gamma \in \Gamma \mid B(x_i, 2r_i) \cap \gamma.B(x_i, 2r_i) \neq \emptyset\}$ is finite. Let U be the union of the balls $B(x_i, r_i)$. If there were infinitely many distinct elements $\gamma_n \in \Gamma$ such that $\gamma_n.U \cap U \neq \emptyset$, then for some fixed indices i_0 and i_1 there would be infinitely many $\gamma_n \in \Gamma$ such that $\gamma_n.B(x_{i_0}, r_{i_0}) \cap B(x_{i_1}, r_{i_1}) \neq \emptyset$. But then we would have infinitely many elements $\gamma_n^{-1}\gamma_m \in S(i_0)$, which is a contradiction.

(2) Let X be a length space. If one endows the universal covering of X with the induced length metric, then the action of $\pi_1 X$ by deck transformations on $\widetilde{X}$ is a proper action (see 3.22(1)). (This statement would not remain true if one were to replace "$\exists\, r > 0$" by "$\forall\, r > 0$" in the definition of properness.)

(3) In (8.2) we do not assume that the group Γ acts faithfully on X. However, if the action of Γ on X is proper then every isotropy subgroup Γ_x will be finite.

8.4 Exercises

(1) Let X be a length space. If there is a group that acts properly and cocompactly by isometries on X, then X is complete and locally compact, so by the Hopf-Rinow Theorem it is a proper geodesic space.

(2) Show that the action of a group Γ on a locally compact space X is proper in the sense of (8.3(1)) if and only if:

(i) $\Gamma\backslash X$ with the quotient topology is Hausdorff,

(ii) for every $x \in X$, the isotropy subgroup Γ_x is finite, and

(iii) there is a Γ_x-invariant neighbourhood U of x with $\Gamma_x = \{\gamma \in \Gamma \mid \gamma.U \cap U \neq \emptyset\}$.

(3) If a group Γ acts properly by isometries on a metric space X, then the pseudo-metric on $\Gamma\backslash X$ that the construction of (5.19) associates to the equivalence relation $x \sim \gamma.x$ is the metric described in (8.5(2)) and the metric topology is the quotient topology.

We gather some basic facts about proper actions on metric spaces.

8.5 Proposition. *Suppose that the group Γ acts properly by isometries on the metric space X. Then:*

(1) *For each $x \in X$, there exists $\varepsilon > 0$ such that if $\gamma.B(x, \varepsilon) \cap B(x, \varepsilon) \neq \emptyset$ then $\gamma \in \Gamma_x$.*

(2) *The distance between orbits in X defines a metric on the space $\Gamma\backslash X$ of Γ-orbits.*

(3) *If the action is proper and free, then the natural projection $p : X \to \Gamma\backslash X$ is a covering map and a local isometry.*

(4) *If a subspace Y of X is invariant under the action of a subgroup $H \subseteq \Gamma$, then the action of H on Y is proper.*

(5) *If the action of Γ is cocompact then there are only finitely many conjugacy classes of isotropy subgroups in Γ.*

Proof. Let $x \in X$. By properness, there exists $r > 0$ such that the orbit $\Gamma.x$ of x meets the ball $B(x, r)$ in only a finite number of points. To prove (1), choose ε small enough to ensure that $B(x, 2\varepsilon) \cap \Gamma.x = \{x\}$.

The distance $d(\Gamma.x, \Gamma.y) = d(x, \Gamma.y)$ between orbits is obviously a pseudometric, and it follows from (1) that $d(\Gamma.x, \Gamma.y) = d(x, \Gamma.y)$ is strictly positive, thus (2) is true. (3) follows easily from (1), and (4) is immediate from the definitions. To prove (5), we fix a compact set K whose translates by Γ cover X, and cover K with finitely many balls $B(x_i, r_i)$ such that each of the sets $\{\gamma \in \Gamma \mid \gamma.B(x_i, r_i) \cap B(x_i, r_i) \neq \emptyset\}$ is finite; let Σ be the union of these sets. For each $x \in X$ there exists $\gamma \in \Gamma$ such that $x \in \gamma.K$. Since $\gamma^{-1}\Gamma_x\gamma = \Gamma_{\gamma^{-1}.x}$, we have $\gamma^{-1}\Gamma_x\gamma \subset \Sigma$. □

In the following proof we need some elementary facts about lifting maps to covering spaces. Such material is covered in any introductory course on algebraic topology, e.g. [Mass91].

8.6 Proposition. *Let X be a length space that is connected and simply connected, and suppose that the group $\Gamma \subset$ Isom(X) acts freely and properly on X. Consider $\Gamma\backslash X$ with the metric described in (8.5(2)): $d(\Gamma.x, \Gamma.y) := \inf_{\gamma\in\Gamma} d(x, \gamma.y)$. Let $N(\Gamma)$ be the normalizer of Γ in* Isom(X).

Then, there is a natural isomorphism $N(\Gamma)/\Gamma \cong$ Isom$(\Gamma\backslash X)$, induced by the map $\nu \mapsto \overline{\nu}$, where $\nu \in N(\Gamma)$ and $\overline{\nu} \in$ Isom$(\Gamma\backslash X)$ is the map $\Gamma.x \mapsto \Gamma.(\nu(x))$.

Proof. By definition, $N(\Gamma) = \{\nu \in \text{Isom}(X) \mid \nu\gamma\nu^{-1} \in \Gamma\ \forall\gamma \in \Gamma\}$. So if $\nu \in N(\Gamma)$ then $\nu.\Gamma = \Gamma.\nu$, and hence $\nu.(\gamma.x) \in \Gamma.(\nu.x)$ for every $x \in X$ and $\gamma \in \Gamma$. Therefore the map $\overline{\nu} : \Gamma.x \mapsto \Gamma.(\nu.x)$ is well-defined. $\overline{\nu}$ is obviously an isometry. If $\overline{\nu}$ is the identity map, then for any choice of basepoint $x_0 \in X$, we have $\nu.x_0 = \gamma.x_0$, for some $\gamma \in \Gamma$. But this implies that $\nu = \gamma$, because both ν and γ are liftings to X of the identity map on $\Gamma\backslash X$ and these lifts coincide at one point x_0, so they must coincide everywhere, because X is connected. Thus $\nu \mapsto \overline{\nu}$ defines an injective homomorphism $N(\Gamma)/\Gamma \to$ Isom$(\Gamma\backslash X)$.

It remains to prove that this map is surjective. Let $\overline{\mu}$ be an isometry of $\Gamma\backslash X$, and let $p : X \to \Gamma\backslash X$ be the natural projection (which is a covering map, by (8.5(3)). As X is connected and simply connected, the composition $\overline{\mu} \circ p$ lifts to a continuous bijection $\mu : X \to X$ projecting to $\overline{\mu}$. This map μ is a local isometry, because p is locally an isometry. It follows that μ is an isometry, because it preserves the lengths of the curves in X, and X is assumed to be a length space. Moreover, for each $\gamma \in \Gamma$, the isometry $\mu^{-1}\gamma\mu$ projects to the identity of $\Gamma\backslash X$, hence it is an element of Γ, since X is connected. □

The Group of Isometries of a Compact Metric Space

Let Y be a compact metric space. One defines a metric on the group $\mathrm{Isom}(Y)$ by

$$d(\alpha, \alpha') := \sup_{y \in Y} d(\alpha.y, \alpha'.y).$$

8.7 Proposition. *The metric on* $\mathrm{Isom}(Y)$*, defined above, is invariant by left and right translations. With the topology induced by this metric,* $\mathrm{Isom}(Y)$ *is a compact topological group.*

Proof. The invariance of the metric by left and right translations is obvious, as is the fact that $\alpha \mapsto \alpha^{-1}$ preserves distances from the identity in $\mathrm{Isom}(Y)$. The composition $(\alpha, \beta) \mapsto \alpha\beta$ is continuous because $d(\alpha\beta, \alpha'\beta') \leq d(\alpha\beta, \alpha\beta') + d(\alpha\beta', \alpha'\beta') = d(\beta, \beta') + d(\alpha, \alpha')$, hence $\mathrm{Isom}(Y)$ is a topological group.

To see that $\mathrm{Isom}(Y)$ is compact, one can apply Arzelà-Ascoli theorem (3.10): the set of isometries of Y is equicontinuous, so every sequence has a subsequence converging uniformly to a map from Y to Y; the limit map is obviously an isometry. □

Presenting Groups of Homeomorphisms

In this paragraph we construct a presentation for an arbitrary group Γ acting by homeomorphisms on a simply connected topological space X. If X is a simply connected length space and Γ is acting properly and cocompactly by isometries, then the construction which we shall describe gives a finite presentation for Γ. Our proof relies on a simple observation concerning the topology of group presentations (8.9).

8.8 Free Groups and Presentations. We write $F(\mathcal{A})$ to denote the free group on a set $\mathcal{A}$. The elements of $F(\mathcal{A})$ are equivalence classes of words over the alphabet [14] $\mathcal{A}^{\pm 1}$: a *word* is a finite sequence $a_1 \dots a_n$ where $a_i \in \mathcal{A}^{\pm 1}$; one may insert or delete a subword of the form aa^{-1}, and two words are said to be equivalent if one can pass from one to the other by a finite sequence of such deletions and insertions. A word $a_1 \dots a_n$ is said to be *reduced* if each $a_i \neq a_{i-1}^{-1}$. There is a unique reduced word in each equivalence class and therefore we may regard the elements of $F(\mathcal{A})$ as reduced words.

The group operation on $F(\mathcal{A})$ is given by concatenation of words — the empty word is the identity.

Let G be a group and let $S \subset G$ be a subset. We write $\langle\langle S \rangle\rangle$ to denote the smallest normal subgroup of G that contains S (i.e. the *normal closure* of S).

A *presentation* for a group Γ consists of a set $\mathcal{A}$, an epimorphism $\pi : F(\mathcal{A}) \twoheadrightarrow \Gamma$, and a subset $\mathcal{R} \subset F(\mathcal{A})$ such that $\langle\langle \mathcal{R} \rangle\rangle = \ker \pi$. One normally suppresses mention

[14] Given a set $\mathcal{A}$, the elements of $\mathcal{A}^{-1}$ are by definition the symbols a^{-1} where $a \in \mathcal{A}$. The notation $\mathcal{A}^{\pm 1}$ is used to denote the disjoint union of $\mathcal{A}$ and $\mathcal{A}^{-1}$.

of π and writes $G = \langle \mathcal{A} \mid \mathcal{R} \rangle$. The presentation is called *finite* if both of the sets $\mathcal{A}$ and $\mathcal{R}$ are finite, and Γ is said to be *finitely presentable* (or finitely presented) if it admits such a presentation.

The Cayley graph $\mathcal{C}_{\mathcal{A}}(\Gamma)$ of a group Γ with a specified generating set $\mathcal{A}$ was defined in (1.9). Directed edges in $\mathcal{C}_{\mathcal{A}}(\Gamma)$ are labelled by the generators and their inverses, and hence there is a 1–1 correspondence between words in the alphabet $\mathcal{A}^{\pm 1}$ and edge paths issuing from each vertex of $\mathcal{C}_{\mathcal{A}}(\Gamma)$. An edge path will be a loop if and only if the word labelling it is equal to the identity in Γ. The action of Γ on itself by left multiplication extends to a free action on $\mathcal{C}_{\mathcal{A}}(\Gamma)$: the action of γ_0 sends the edge labelled a emanating from a vertex γ to the edge labelled a emanating from the vertex $\gamma_0\gamma$.

Basic definitions concerning 2-complexes (adapted to the needs of this chapter) are given in the appendix.

8.9 Lemma. *Let Γ be a group with generating set $\mathcal{A}$ and let $\mathcal{R}$ be a subset of the kernel of the natural map $F(\mathcal{A}) \to \Gamma$. Consider the 2-complex that one obtains by attaching 2-cells to all of the edge-loops in the Cayley graph $\mathcal{C}_{\mathcal{A}}(\Gamma)$ that are labelled by reduced words $r \in \mathcal{R}$. This 2-complex is simply-connected if and only if $\langle\langle \mathcal{R} \rangle\rangle = \ker(F(\mathcal{A}) \twoheadrightarrow \Gamma)$.*

Proof. The Cayley graph $\mathcal{C}_{\mathcal{A}}(F(\mathcal{A}))$ of $F(\mathcal{A})$ is a tree. $\mathcal{C}_{\mathcal{A}}(\Gamma)$ is the quotient of this tree by the (free) action of $N := \ker(F(\mathcal{A}) \twoheadrightarrow \Gamma)$. Thus there is a natural identification $\pi_1(\mathcal{C}_{\mathcal{A}}(\Gamma), 1) = N$, and a word in the generators $\mathcal{A}^{\pm 1}$ defines an element of N if and only if it is the label on an edge-loop in $\mathcal{C}_{\mathcal{A}}(\Gamma)$ that begins and ends at the identity vertex.

Let $u \in F(\mathcal{A})$ be a reduced word and consider the vertex of $\mathcal{C}_{\mathcal{A}}(\Gamma)$ that is reached from the vertex 1 by following an edge path labelled u. If we attach the boundary of a 2-cell to $\mathcal{C}_{\mathcal{A}}(\Gamma)$ by means of the edge loop labelled r that begins at this vertex, then by the Seifert-van Kampen theorem [Mass91], the fundamental group of the resulting 2-complex will be the quotient of N by $\langle\langle u^{-1}ru \rangle\rangle$. More generally, if for every reduced word $r \in \mathcal{R}$ and every vertex v of $\mathcal{C}_{\mathcal{A}}(\Gamma)$, we attach a 2-cell along the loop labelled r that begins at v, then the fundamental group of the resulting 2-complex will be the quotient of N by the normal closure of $\mathcal{R}$. In particular, this 2-complex will be simply-connected if and only if N is equal to the normal closure of $\mathcal{R}$. $\square$

In the following proposition we do not assume that Γ is finitely generated, nor do we make any assumption concerning the discreteness of the action.

8.10 Theorem. *Let X be a topological space, let Γ be a group acting on X by homeomorphisms, and let $U \subset X$ be an open subset such that $X = \Gamma.U$.*

(1) *If X is connected, then the set $S = \{\gamma \in \Gamma \mid \gamma.U \cap U \neq \emptyset\}$ generates Γ.*

(2) *Let $\mathcal{A}_S$ be a set of symbols a_s indexed by S. If X and U are both path-connected and X is simply connected, then $\Gamma = \langle \mathcal{A}_S \mid \mathcal{R} \rangle$, where*

$$\mathcal{R} = \{a_{s_1} a_{s_2} a_{s_3}^{-1} \mid s_i \in S;\ U \cap s_1.U \cap s_3.U \neq \emptyset;\ s_1 s_2 = s_3 \text{ in } \Gamma\}.$$

Proof. First we prove (1). Let $H \subset \Gamma$ be the subgroup of Γ generated by S, let $V = H.U$ and let $V' = (\Gamma \setminus H).U$. If $V \cap V' \neq \emptyset$, then there exist $h \in H$, $h' \in \Gamma \setminus H$ such that $h^{-1}h'.U \cap U \neq \emptyset$ and hence $h' \in HS \subset H$, contrary to assumption. Thus the open sets V and V' are disjoint. V is non-empty and X is connected, therefore $V' = \emptyset$ and $H = \Gamma$.

We now prove (2). Let K be the combinatorial 2-complex obtained from the Cayley graph $\mathcal{C}_{\mathcal{A}_S}(\Gamma)$ by attaching a 2-cell to each of the edge-loops labelled by the words $r \in \mathcal{R}$. According to the previous lemma, it is enough to show that K is simply connected.

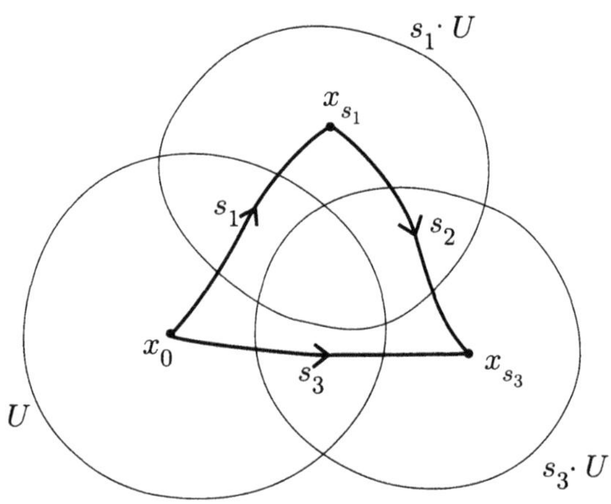

Fig. 8.1 The relations in Theorem 8.10

Fix $x_0 \in U$. For each $s \in S$ we choose a point $x_s \in U \cap s.U$ and then (using the fact that U is path-connected) we choose a path from x_0 to x_s in U, and a path from x_s to $s.x_0$ in $s.U$; call the concatenation of these paths c_s. Let $p : \mathcal{C}_{\mathcal{A}_S}(\Gamma) \to X$ be the Γ-equivariant map that sends 1 to x_0 and sends the edge labelled a_s emanating from 1 to the path c_s. Because X is assumed to be simply connected, we can extend this to a continuous Γ-equivariant map $p : K \to X$.

Let D be the standard 2-disc. To complete the proof of the proposition, we must show that every continuous map $\ell : \partial D \to \mathcal{C}_{\mathcal{A}_S}(\Gamma)$ can be continuously extended to a map $D \to K$. It is enough to consider locally injective edge loops. Because X is simply-connected, the map $p\ell : \partial D \to X$ extends to a map $\phi : D \to X$. Because D is compact and U is open, there is a finite triangulation $\mathcal{T}$ of D with the property that for every vertex v of $\mathcal{T}$ there exists $\gamma_v \in \Gamma$ such that ϕ maps all of the triangles incident at v into $\gamma_v.U$.

Let t be a triangle of $\mathcal{T}$ with vertices v_1, v_2, v_3. We have

$$\begin{aligned}\phi(t) &\subseteq \gamma_{v_1}.U \cap \gamma_{v_2}.U \cap \gamma_{v_3}.U \\ &= \gamma_{v_1}.\left(U \cap \gamma_{v_1}^{-1}\gamma_{v_2}.U \cap \gamma_{v_1}^{-1}\gamma_{v_3}.U\right) \\ &= \gamma_{v_2}.\left(\gamma_{v_2}^{-1}\gamma_{v_1}.U \cap U \cap \gamma_{v_2}^{-1}\gamma_{v_3}.U\right) \neq \emptyset.\end{aligned}$$

It follows that $s_1 := \gamma_{v_1}^{-1}\gamma_{v_2}$, $s_2 := \gamma_{v_2}^{-1}\gamma_{v_3}$, and $s_3 := \gamma_{v_1}^{-1}\gamma_{v_3}$ are elements of S, and $a_{s_1}a_{s_2}a_{s_3}^{-1} \in \mathcal{R}$. We can therefore extend the map $v \mapsto \gamma_v$ to a map $\hat{\phi}$ from the 1-skeleton of $\mathcal{T}$ to $\mathcal{C}_{\mathcal{A}_S}(\Gamma)$ by sending the edge connecting v_i and v_{i+1} to the edge labelled s_i emanating from γ_{v_i} (indices mod 3). If $v_i \in \partial D$, then we choose the γ_{v_i} so that the restriction of $\hat{\phi}$ to ∂D is a reparameterization of ℓ.

Finally, since $\hat{\phi}$ sends the boundary of each triangle in $\mathcal{T}$ to a circuit in $\mathcal{C}_{\mathcal{A}_S}(\Gamma)$ labelled by an element of $\mathcal{R}$, and since each such circuit is the boundary of a 2-cell in K, we may extend $\hat{\phi}$ to a continuous map $D \to K$. □

8.11 Corollary. *A group is finitely presented if and only if it acts properly and cocompactly by isometries on a simply-connected geodesic space.*

Proof. Given a group Γ acting properly and cocompactly by isometries on a simply-connected geodesic space X, one chooses a compact set $C \subset X$ such that $\Gamma.C = X$ and an open ball $B(x_0, R)$ containing C, then one applies the theorem with $U = B(x_0, R)$. Because X is proper (8.4) and the action of Γ is proper, the set of generators S described in (8.10) is finite and hence so is the set $\mathcal{R}$.

Conversely, given a finite presentation $\langle \mathcal{A} \mid \mathcal{R} \rangle$ of a group, one can metrize the simply-connected 2-complex constructed in the proof of (8.9) as a piecewise Euclidean complex in which all of the edges have length 1 and all of the 2-cells are regular polygons. The natural action of Γ on its Cayley graph extends in an obvious way to an action of Γ by isometries on this complex; the action is proper and cocompact (cf. Appendix). □

8.12 Remarks

(1) Theorem 8.10 is due to Murray Macbeath [Mac64]. See page 31 of [Ser77] for an alternative treatment and additional references.

(2) In 8.10(2) it is necessary to require that X is simply-connected. To see this, let Γ be the group of order four generated by a rotation ϕ of the circle $\mathbb{S}^1$ through an angle $\pi/2$ and let U be an open arc of length π on the circle. Then the set S in (8.10) is $\{\phi, \phi^{-1}\}$ and the set $\mathcal{R}$ is empty, thus $\langle S \mid \mathcal{R} \rangle$ is not a presentation of Γ.

To obtain further examples we consider groups Γ that are finitely generated but not finitely presentable. There are uncountably many such groups [Rot95]; some examples are given in section (III.Γ.5). Let $\mathcal{A}$ be a finite generating set for Γ and consider the action of Γ on the Cayley graph $\mathcal{C}_{\mathcal{A}}(\Gamma)$. If we take U to be the open ball of radius 1 about the identity then $S = \mathcal{A}^{\pm 1}$, in the notation of (8.10), and one cannot present Γ with finitely many relations on these generators.

(3) An action of a group Γ on a metric space X is said to be cobounded if there exists a bounded set $B \subset X$ such that $\Gamma.B = X$. (8.11) does not remain true if one replaces the hypothesis 'cocompactly' by 'coboundedly'. For example, consider $\ell_2(\mathbb{Z})$ and let δ_i denote the sequence whose only non-zero term is a 1 in the i-th place. Define $\tau_i : \ell_2(\mathbb{Z}) \to \ell_2(\mathbb{Z})$ by $\sigma \mapsto \sigma + \delta_i$. The subgroup $T \subset \mathrm{Isom}(\ell_2(\mathbb{Z}))$ generated by $\{\delta_i \mid i \in \mathbb{N}\}$ is not finitely generated, but its action is proper and cobounded.

8.13 Exercise. Consider the ℓ_2 metric on the space $\prod_{\mathbb{N}} \mathbb{S}^1$ defined in (5.5(2)). Let T be as above. Prove that $T\backslash\ell_2(\mathbb{Z})$ is isometric to the subspace of $\prod_{\mathbb{N}} \mathbb{S}^1$ consisting of sequences a finite distance from the constant sequence 0.

Quasi-Isometries

Our next goal is to explain the comments that we made at the beginning of this chapter concerning the relationship between length spaces and the geometry of groups which act properly and cocompactly by isometries on them.

One of the main themes of this book is the large scale geometry of metric spaces. In this context one needs a language that will lend precision to observations such as the following: if one places a dot at each integer point along a line in the Euclidean plane, then the line and the set of dots become indistinguishable when viewed from afar, whereas the line and the plane remain visibly distinct. One makes this observation precise by saying that the set of dots is *quasi-isometric* to the line whereas the line is not quasi-isometric to the plane.

8.14 Definition of Quasi-Isometry. Let (X_1, d_1) and (X_2, d_2) be metric spaces. A (not necessarily continuous) map $f : X_1 \to X_2$ is called a (λ, ε)*-quasi-isometric embedding* if there exist constants $\lambda \geq 1$ and $\varepsilon \geq 0$ such that for all $x, y \in X_1$

$$\frac{1}{\lambda} d_1(x, y) - \varepsilon \leq d_2(f(x), f(y)) \leq \lambda d_1(x, y) + \varepsilon.$$

If, in addition, there exists a constant $C \geq 0$ such that every point of X_2 lies in the C–neighbourhood of the image of f, then f is called a (λ, ε)*-quasi-isometry*. When such a map exists, X_1 and X_2 are said to be *quasi-isometric*.

8.15 Remark. Some authors refer to the maps defined above as *coarse* quasi-isometries. In general the term "coarse" is used to describe properties that are insensitive to all finite perturbations — when studying such properties there is little reason to care about whether maps are continuous or not.

8.16 Exercises

(1) Show that if there exists a (λ, ε)–quasi-isometry $f : X_1 \to X_2$ then there exists a (λ', ε')-quasi-isometry $f' : X_2 \to X_1$ (for some λ' and ε') and a constant $k \geq 0$ such that $d(ff'(x'), x') \leq k$ and $d(f'f(x), x) \leq k$ for all $x' \in X_2$ and all $x \in X_1$. Such a map f' is called a *quasi-inverse* for f.

(2) Prove that the composition of a (λ, ε)-quasi-isometric embedding and a (λ', ε')-quasi-isometric embedding is a (μ, ν)-quasi-isometric embedding with $\mu = \lambda\lambda'$ and $\nu = \lambda'\varepsilon + \varepsilon'$. Deduce that the composition of any two quasi-isometries is a quasi-isometry.

(3) Let X a be metric space. We consider two maps $f, g : X \to X$ to be equivalent, and write $f \sim g$, if $\sup_x d(f(x), g(x))$ is finite. Let $[f]$ denote the $\sim$ equivalence class

of f. The *quasi-isometry group of* X, denoted $\mathcal{QI}(X)$, is the set of $\sim$ equivalence classes of quasi-isometries $X \to X$. Show that composition of maps induces a group structure on X, and that any quasi-isometry $\phi : X \to X'$ induces an isomorphism $\phi_* : \mathcal{QI}(X) \to \mathcal{QI}(X')$.

(4) Give an example of an unbounded metric space X for which the natural map $\mathrm{Isom}(X) \to \mathcal{QI}(X)$ is an isomorphism; give an example where $\mathrm{Isom}(X)$ is trivial but $\mathcal{QI}(X)$ is infinite; and give an example where $\mathcal{QI}(X)$ is trivial but $\mathrm{Isom}(X)$ is infinite. (In the first two cases it suffices to consider discrete subspaces of $\mathbb{R}$ — but there are more interesting examples.)

8.17 Examples

(1) A metric space is quasi-isometric to a one-point space if and only it has finite diameter. More generally, the inclusion $Y \hookrightarrow X$ of a subset Y of a metric space X is a quasi-isometry if and only if Y is *quasi-dense* in X, i.e. there exists a constant $C > 0$ such that every point of X lies in the C–neighbourhood of some point of Y. For example, the natural inclusion $\mathbb{Z} \hookrightarrow \mathbb{R}$ is a quasi-isometry.

(2) *Every finitely generated group is a metric space, well-defined up to quasi-isometry.*[15] Given a group Γ with generating set $\mathcal{A}$, the first step in realizing the geometry of the group is to give Γ the *word metric* associated to $\mathcal{A}$: this is the metric obtained by defining $d_{\mathcal{A}}(\gamma_1, \gamma_2)$ to be the shortest word in the pre-image of $\gamma^{-1}\gamma_2$ under the natural projection $F(\mathcal{A}) \twoheadrightarrow \Gamma$. The action of Γ on itself by left multiplication gives an embedding $\Gamma \to \mathrm{Isom}(\Gamma, d_{\mathcal{A}})$. (The action of $\gamma_0 \in \Gamma$ by right multiplication $\gamma \mapsto \gamma\gamma_0$ is an isometry only if γ_0 lies in the centre of Γ.)

The word metrics associated to different finite generating sets $\mathcal{A}$ and $\mathcal{A}'$ of Γ are Lipschitz equivalent, i.e. there exists $\lambda > 0$ such that $\frac{1}{\lambda} d_{\mathcal{A}}(\gamma_1, \gamma_2) \leq d_{\mathcal{A}'}(\gamma_1, \gamma_2) \leq \lambda\, d_{\mathcal{A}}(\gamma_1, \gamma_2)$ for all $\gamma_1, \gamma_2 \in \Gamma$. One sees this by expressing the elements of $\mathcal{A}$ as words in the generators $\mathcal{A}'$ and vice versa — the constant λ is the length of the longest word in the dictionary of translation.

If two metrics d and d' on a set X are Lipschitz equivalent then the identity map $\mathrm{id} : (X, d) \to (X, d')$ is a quasi-isometry. Thus statements such as "the finitely generated group Γ is quasi-isometric to the metric space Y" or "the finitely generated groups Γ_1 and Γ_2 are quasi-isometric" are unambiguous, and we shall speak like this often.

(3) The Cayley graph $\mathcal{C}_{\mathcal{A}}(\Gamma)$ of a group Γ with respect to a generating set $\mathcal{A}$ was defined in (1.10); it is a metric graph with edges of length one. The induced metric on its vertex set Γ is exactly the word metric $d_{\mathcal{A}}$ described in (2). As a special case of (1) we see that $(\Gamma, d_{\mathcal{A}})$ is quasi-isometric to $\mathcal{C}_{\mathcal{A}}(\Gamma)$ and that the Cayley graphs associated to different finite generating sets for Γ are quasi-isometric.

[15] Mikhael Gromov has been extremely successful in promoting the study of groups as geometric objects [Gro84, 87, 93]. This approach to group theory (which we shall consider at some length in Part III) lacked prominence for much of this century but was inherent in the seminal work of Max Dehn [Dehn87]. See [ChaM82] for historical details up to 1980.

One can also think of the inclusion $\Gamma \hookrightarrow \mathcal{C}_{\mathcal{A}}(\Gamma)$ as being given by the natural action of Γ on $\mathcal{C}_{\mathcal{A}}(\Gamma)$ — it is $\gamma \mapsto \gamma.1$. From this point of view, the fact that $\Gamma \hookrightarrow \mathcal{C}_{\mathcal{A}}(\Gamma)$ is a quasi-isometry provides a simple illustration of (8.19).

Quasi-Isometries Arising from Group Actions

8.18 Lemma. *Let (X, d) be a metric space. Let Γ be a group with finite generating set $\mathcal{A}$ and associated word metric $d_{\mathcal{A}}$. If Γ acts by isometries on X, then for every choice of basepoint $x_0 \in X$ there exists a constant $\mu > 0$ such that $d(\gamma.x_0, \gamma'.x_0) \leq \mu\, d_{\mathcal{A}}(\gamma, \gamma')$ for all $\gamma, \gamma' \in \Gamma$.*

Proof. Let $\mu = \max\{d(x_0, a.x_0) \mid a \in \mathcal{A} \cup \mathcal{A}^{-1}\}$. If $d_{\mathcal{A}}(\gamma, \gamma') = n$ then $\gamma^{-1}\gamma' = a_1 a_2 \dots a_n$ for some $a_j \in \mathcal{A} \cup \mathcal{A}^{-1}$. Let $g_i = a_1 a_2 \dots a_i$. By the triangle inequality, $d(\gamma.x_0, \gamma'.x_0) = d(x_0, \gamma^{-1}\gamma'.x_0) \leq d(x_0, g_1.x_0) + d(g_1.x_0, g_2.x_0) + \dots + d(g_{n-1}.x_0, \gamma^{-1}\gamma'.x_0)$. And for each i we have $d(g_{i-1}.x_0, g_i.x_0) = d(x_0, g_{i-1}^{-1} g_i.x_0) = d(x_0, a_i.x_0) \leq \mu$. $\square$

The following result was discovered by the Russian school in the nineteen fifties (see [Ef53],[Sv55]). It was rediscovered by John Milnor some years later [Mil68, Lemma 2].

8.19 Proposition (The Švarc-Milnor Lemma). *Let X be a length space. If Γ acts properly and cocompactly by isometries on X, then Γ is finitely generated and for any choice of basepoint $x_0 \in X$, the map $\gamma \mapsto \gamma.x_0$ is a quasi-isometry.*

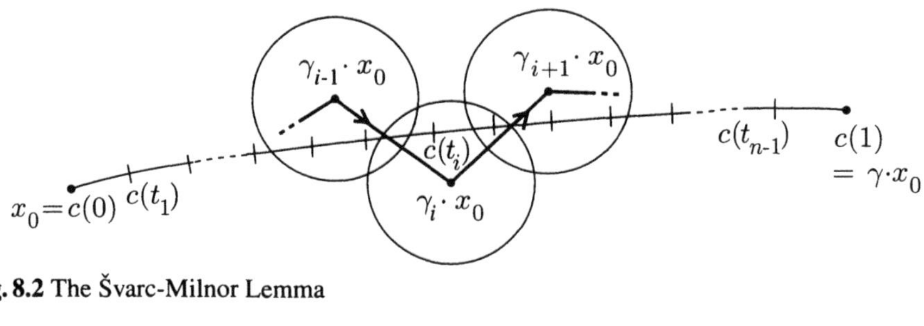

Fig. 8.2 The Švarc-Milnor Lemma

Proof. Let $C \subset X$ be a compact set with $\Gamma.C = X$. We choose $x_0 \in X$ and $D > 0$ such that $C \subset B(x_0, D/3)$ and let $\mathcal{A} = \{\gamma \in \Gamma \mid \gamma.B(x_0, D) \cap B(x_0, D) \neq \emptyset\}$. Because X is proper (8.4) and the action of Γ is proper, $\mathcal{A}$ is finite.

In (8.10) we showed that $\mathcal{A}$ generates Γ. (A second proof is given below.) Let $d_{\mathcal{A}}$ be the word metric on Γ associated to $\mathcal{A}$. Lemma 8.18 yields a constant μ such that $d(\gamma.x_0, \gamma'.x_0) \leq \mu\, d_{\mathcal{A}}(\gamma, \gamma')$ for all $\gamma, \gamma' \in \Gamma$, so it only remains to bound $d_{\mathcal{A}}(\gamma, \gamma')$ in terms of $d(\gamma.x_0, \gamma'.x_0)$. Because both metrics are Γ-invariant, it is enough to compare $d_{\mathcal{A}}(1, \gamma)$ and $d(x_0, \gamma.x_0)$.

Given $\gamma \in \Gamma$ and a path $c : [0, 1] \to X$ of finite length with $c(0) = x_0$ and $c(1) = \gamma.x_0$, we can choose a partition $0 = t_0 < t_1 < \ldots < t_n = 1$ of $[0, 1]$ such that $d(c(t_i), c(t_{i+1})) \leq D/3$ for all i. For each t_i we choose an element $\gamma_i \in \Gamma$ such that $d(c(t_i), \gamma_i.x_0) \leq D/3$; choose $\gamma_0 = 1$ and $\gamma_n = \gamma$. Then, for $i = 1, \ldots, n$ we have $d(\gamma_i.x_0, \gamma_{i-1}.x_0) \leq D$ and hence $a_i := \gamma_{i-1}^{-1}\gamma_i \in \mathcal{A}$.

$$\gamma = \gamma_0(\gamma_0^{-1}\gamma_1)\ldots(\gamma_{n-2}^{-1}\gamma_{n-1})(\gamma_{n-1}^{-1}\gamma_n) = a_1 \ldots a_{n-1}a_n.$$

Because X is a length space, we can choose the curve c considered above to have length less than $d(x_0, \gamma.x_0) + 1$. If we take as coarse a partition $0 = t_0 < t_1 < \ldots < t_n = 1$ as is possible with $d(c(t_i), c(t_{i+1})) \leq D/3$, then $n \leq (d(x_0, \gamma.x_0)+1)\,(3/D)+1$. Since γ can be expressed as a word of length n, we get $d_{\mathcal{A}}(1, \gamma) \leq (d(x_0, \gamma.x_0) + 1)\,(3/D) + 1$. □

8.20 Exercises

(1) Let $\varphi : \Gamma_1 \to \Gamma_2$ be a homomorphism between finitely generated groups. Show that if φ is a quasi-isometric embedding then $\ker(\phi)$ is finite, and that ϕ is a quasi-isometry if and only if $\ker(\phi)$ and $\Gamma_2/\mathrm{im}(\varphi)$ are both finite.

(2) Let T_n denote the connected metric tree in which every vertex has valence n and every edge has length 1. Prove that if $n, m \geq 3$ then T_n is quasi-isometric to T_m.

(Hint: Rather than doing this directly, you can show that every finitely generated free group occurs as a subgroup of finite index in the free group of rank 2; the Cayley graph of the free group of rank r is T_{2r}, so the case where n and m are even then follows from (1) and the fact that finitely generated groups are quasi-isometric to their Cayley graphs. In the case where n is odd, consider the Cayley graph of $G_{2,n} = \mathbb{Z}_2 * \mathbb{Z}_n$; the kernel of the abelianization map $G_{2,n} \to \mathbb{Z}_2 \times \mathbb{Z}_n$ is free and has finite index.)

8.21 Remark (Commensurability versus Quasi-Isometry). Two groups are said to be *commensurable* if they contain isomorphic subgroups of finite index. Commensurable groups are quasi-isometric (8.20(1)), but quasi-isometric groups need not be commensurable, as we shall now discuss.

First we note that if one uses the Švarc-Milnor lemma to show that finitely generated groups are quasi-isometric by getting them to act cocompactly by isometries on the same length space, then in general it can be difficult to decide if the groups are commensurable. A setting in which one can decide quite easily is that of semi-direct products of the form $\mathbb{Z}^2 \rtimes_\phi \mathbb{Z}$, which were considered in [BriG96].

Each such group acts properly and cocompactly by isometries on one of the 3-dimensional geometries $\mathbb{E}^3$, Nil or Sol (see [Sco83] or [Thu97]). These three possibilities are mutually exclusive and determine the quasi-isometry type of the group. $\mathbb{Z}^2 \rtimes_\phi \mathbb{Z}$ acts properly and cocompactly by isometries on $\mathbb{E}^3$ if $\phi \in \mathrm{GL}(2, \mathbb{Z})$ has finite order; it acts properly and cocompactly by isometries on Nil if ϕ has infinite order and its eigenvalues have absolute value 1; and it acts properly and cocompactly by isometries on Sol if ϕ has an eigenvalue λ with $|\lambda| > 1$.

In each of the cases $\mathbb{E}^3$ and Nil, all of the groups concerned have isomorphic subgroups of finite index. But in the case of Sol, the groups $\mathbb{Z}^2 \rtimes_\phi \mathbb{Z}$ and $\mathbb{Z}^2 \rtimes_{\phi'} \mathbb{Z}$ will have isomorphic subgroups of finite index if and only if the corresponding eigenvalues λ, λ' have a common power (see [BriG96]). For arbitrary $n > 0$, the groups $\mathbb{Z}^n \rtimes \mathbb{Z}$ have not been classified up to quasi-isometry (although there has been some progress [BriG96]).

Interesting examples of quasi-isometric groups that are not commensurable also arise in the study of cocompact arithmetic lattices in Lie groups. In this setting there are many interesting invariants of commensurability — see [NeRe92] for example. Further examples of groups that are quasi-isometric but not commensurable are described in (III.Γ.4.23).

Quasi-Geodesics

Let Γ and X be as in the (8.19). One would like to know to what extent the information encoded in the geometry of geodesics in X is transmitted to Γ by the quasi-isometry $X \to \Gamma$ given by (8.19). This leads us to the following definition.

8.22 Definition of Quasi-Geodesics. A (λ, ε)-*quasi-geodesic* in a metric space X is a (λ, ε)-quasi-isometric embedding $c : I \to X$, where I is an interval of the real line (bounded or unbounded) or else the intersection of $\mathbb{Z}$ with such an interval. More explicitly,

$$\frac{1}{\lambda}\,|t - t'| - \varepsilon \le d(c(t), c(t')) \le \lambda\,|t - t'| + \varepsilon$$

for all $t, t' \in I$. If $I = [a, b]$ then $c(a)$ and $c(b)$ are called the endpoints of c. If $I = [0, \infty)$ then c is called a *quasi-geodesic ray*.

Quasi-geodesics will play an important role in Chapter III.H. In particular we shall see that quasi-geodesics in hyperbolic spaces such as $\mathbb{H}^n$ follow geodesics closely. In general geodesic spaces this is far from true, as the following example indicates.

8.23 Exercise. Prove that the map $c : [0, \infty) \to \mathbb{E}^2$ given in *polar* coordinates by $t \mapsto (t, \log(1 + t))$ is a quasi-geodesic ray.

Some Invariants of Quasi-Isometry

In the next few sections we shall describe some invariants of quasi-isometry. We begin with a result that illustrates the fact that quasi-isometries preserve more algebraic structure than one might at first expect. We shall see further examples of this phenomenon in Part III.

8.24 Proposition. *Let Γ_1 and Γ_2 be groups with finite generating sets $\mathcal{A}_1$ and $\mathcal{A}_2$. If Γ_1 and Γ_2 are quasi-isometric and Γ_2 has a finite presentation $\langle \mathcal{A}_2 \mid \mathcal{R}_2 \rangle$, then Γ_1 has a finite presentation $\langle \mathcal{A}_1 \mid \mathcal{R}_1 \rangle$.*

Proof. The strategy of the proof is as follows. We shall build combinatorial 2-complexes K_1 and K_2 by attaching a locally-finite collection of 2-cells to the Cayley graphs of Γ_1 and Γ_2; it will be clear that K_2 is simply connected and our goal will be to prove that K_1 is simply connected. Because Γ_1 and Γ_2 are quasi-isometric, there is a quasi-isometry $f : K_1 \to K_2$ with quasi-inverse $f' : K_2 \to K_1$. In order to show that K_1 is simply connected, we use f to map loops in K_1 to loops into K_2, we choose a filling (i.e. homotopy disc) in K_2 and map it back to K_1 using the quasi-inverse $f' : K_2 \to K_1$; a suitable approximation to the resulting (non-continuous) map of a disc into K_1 yields a genuine filling for the original loop in K_1.

Let $i \in \{1, 2\}$. We write $\mathcal{C}_i$ to denote the Cayley graph $\mathcal{C}_{\mathcal{A}_i}(\Gamma_i)$. Let ρ be the length of the longest word in $\mathcal{R}_2$. According to (8.9), the 2-complex K_2 obtained by attaching 2-cells to $\mathcal{C}_2$ along all edge-loops of length $\leq \rho$ is simply connected.

Let $f : \Gamma_1 \to \Gamma_2$ and $f' : \Gamma_2 \to \Gamma_1$ be (λ, ε)-quasi-isometries and let $\mu > 0$ be such that $d(f'f(\gamma), \gamma) \leq \mu$ for every $\gamma \in \Gamma_1$. Let $m = \max\{\lambda, \varepsilon, \mu, \rho\}$ and let $M = 3(3m^2 + 5m + 1)$. Let K_1 be the complex obtained by attaching a 2-cell to $\mathcal{C}_1$ along each edge loop of length $\leq M$. Let ℓ be an edge-loop in $\mathcal{C}_1$ that visits the vertices $g_1, \dots, g_n$ in that order. We view ℓ as a map $\ell : \partial D \to \mathcal{C}_1$, where D is a 2-dimensional disc. We will be done if we can show that ℓ has a continuous extension $\hat{\ell} : D \to K_1$ (see 8.9).

Let $v_1, \dots, v_n$ be the inverse images of the g_i, arranged in cyclic order around ∂D, and let $\phi : \partial D \to \mathcal{C}_2$ be a map that sends the edge (subarc) bounded by $\{v_i, v_{i+1}\}$ onto a geodesic in $\mathcal{C}_2$ connecting $f(g_i)$ to $f(g_{i+1})$, where the indices are taken mod n. Because K_2 is simply connected, we can extend ϕ to a continuous map $\hat{\phi} : D \to K_2$. We associate to each point $x \in D$ an element $\gamma_x \in \Gamma_2$ such that either $\hat{\phi}(x) = \gamma_x$ or else γ_x is a vertex of the edge or open 2-cell in which $\hat{\phi}(x)$ lies. We specify that $\gamma_{v_i} = f(g_i)$. Notice that because $\hat{\phi}$ is continuous, $d(\gamma_{x_1}, \gamma_{x_2}) \leq \rho$ for all sufficiently close $x_1, x_2 \in D$. Notice also that $d(\phi(x), \gamma_x) \leq 1/2$ for every $x \in \partial D$.

We fix a triangulation $\mathcal{T}$ of D such that $d(\gamma_{x_1}, \gamma_{x_2}) \leq \rho$ if x_1, x_2 are adjacent vertices. We require that $v_1, \dots, v_n$ be vertices of $\mathcal{T}$.

Define $\hat{\ell}|_{\partial D} = \ell$ and $\hat{\ell}(x) = f'(\gamma_x)$ for every vertex x of $\mathcal{T}$ that lies in the interior of D. We claim that $\hat{\ell}$ sends each pair of adjacent vertices $x_1, x_2 \in \mathcal{T}$ to elements of Γ_1 that are a distance at most $M/3$ apart in $\mathcal{C}_1$. Once this claim is proved, we can extend $\hat{\ell}$ across the edges of $\mathcal{T}$ be sending each edge to a geodesic in $\mathcal{C}_1$, and since every circuit of length $\leq M$ in $\mathcal{C}_1$ bounds a disc in K_1 (by construction), we can then extend $\hat{\ell}$ continuously across the faces of $\mathcal{T}$.

We must show that $d(\hat{\ell}(x_1), \hat{\ell}(x_2)) \leq M/3$. This is obvious except in the case where x_1 is in the interior of D and $x_2 \in \partial D$, with x_2 between v_i and v_{i+1} say. In that case,

$$\begin{aligned} d(\hat{\ell}(x_1), \hat{\ell}(x_2)) = &\ d(f'(\gamma_{x_1}), \ell(x_2)) \\ &\leq d(f'(\gamma_{x_1}), f'(\gamma_{x_2})) + d(f'(\gamma_{x_2}), f'\phi(x_2)) + d(f'\phi(v_i), f'\phi(x_2)) \\ &\qquad + d(f'\phi(v_i), \ell(v_i)) + d(\ell(v_i), \ell(x_2)). \\ &\leq (\lambda\rho + \varepsilon) + (\lambda/2 + \varepsilon) + [\lambda\, d(\phi(v_i), \phi(v_{i+1})) + \varepsilon] + d(f'f(g_i), g_i) + 1 \\ &\leq (\lambda\rho + \varepsilon) + (\lambda/2 + \varepsilon) + [\lambda(\lambda + \varepsilon) + \varepsilon] + \mu + 1 \\ &\leq 3m^2 + 5m + 1 \leq M/3. \end{aligned}$$

□

8.25 Remarks. Let X and Y be metric spaces. X is said to be a *quasi-retract* of Y if there exist constants $\lambda, \varepsilon, C > 0$, and maps $f : X \to Y$, and $f' : Y \to X$ such that $d(f'(y_1), f'(y_2)) \leq \lambda d(y_1, y_2) + \varepsilon$ for all $y_1, y_2 \in Y$, and $d(f(x_1), f(x_2)) \leq \lambda d(x_1, x_2) + \varepsilon$ and $d(f'f(x_1), x_1) \leq C$ for all $x_1, x_2 \in X$. The preceding proof actually shows that any quasi-retract of a finitely presented group is finitely presented.

Juan Alonso [Alo90] pointed out that by keeping track of the number of faces in the triangulations that occur in the proof of (8.24), one can deduce that having a solvable word problem is an invariant of quasi-isometry among finitely presented groups.

8.26 Exercise. Let Γ be a group with a finite generating set $\mathcal{A}$. Show that if Γ is finitely presentable, and if $\Gamma = \langle \mathcal{A} \mid \mathcal{R} \rangle$ is any presentation for Γ, then there is a finite subset $\mathcal{R}' \subset \mathcal{R}$ such that $\langle\langle \mathcal{R}' \rangle\rangle = \ker(F(\mathcal{A}) \twoheadrightarrow \Gamma)$.

The Ends of a Space

Recall that a map $f : X \to Y$ between topological spaces is said to be proper if $f^{-1}(C) \subseteq X$ is compact whenever $C \subseteq Y$ is compact. The following definition is due to Freudenthal [Fr31].

8.27 Definition of Ends. Let X be a topological space. A *ray* in X is a map $r : [0, \infty) \to X$. Let $r_1, r_2 : [0, \infty) \to X$ be proper rays. r_1 and r_2 are said to *converge to the same end* if for every compact $C \subseteq X$ there exists $N \in \mathbb{N}$ such that $r_1[N, \infty)$ and $r_2[N, \infty)$ are contained in the same path component of $X \smallsetminus C$. This defines an equivalence relation on continuous proper rays; the equivalence class of r is denoted *end*(r) and the set of equivalence classes is denoted Ends(X). If the cardinality of Ends(X) is m, then X is said to have m ends.

Convergence of ends is defined by: *end*$(r_n) \to$ *end*(r) if and only if for every compact set $C \subset X$ there exists a sequence of integers N_n such that $r_n[N_n, \infty)$ and $r[N_n, \infty)$ lie in the same path component of $X \smallsetminus C$ whenever n is sufficiently large. We define a topology on Ends(X) by describing its closed sets. A subset $B \subset$ Ends(X) is defined to be closed if it satisfies the following condition: if *end*$(r_n) \in B$ for all $n \in \mathbb{N}$, then *end*$(r_n) \to$ *end*(r) implies *end*$(r) \in B$.

Let X be a metric space. By definition, a *k-path* connecting x to y is a finite sequence of points $x = x_1, \dots, x_n = y$ in X such that $d(x_i, x_{i+1}) \leq k$ for $i = 1, \dots, n-1$.

8.28 Lemma. *Let X be a proper geodesic space and let $k > 0$. Let r_1 and r_2 be proper rays in X. Let $\mathcal{G}_{x_0}(X)$ denote the set of geodesic rays issuing from $x_0 \in X$. Then:*

(1) *end(r_1) = end(r_2) if and only if for every $R > 0$ there exists $T > 0$ such that $r_1(t)$ can be connected to $r_2(t)$ by a k-path in $X \smallsetminus B(x_0, R)$ whenever $t > T$.*

(2) *The natural map $\mathcal{G}_{x_0}(X) \to$ Ends(X) is surjective.*

(3) *Let $r \in \mathcal{G}_{x_0}(X)$ and let $n > 0$ be an integer. Let $\tilde{V}_n \subset \mathcal{G}_{x_0}(X)$ be the set of proper rays r' such that $r(n, \infty)$ and $r'(n, \infty)$ lie in the same path component of $X \smallsetminus B(x_0, n)$. The sets $V_n = \{end(r') \mid \tilde{V}_n\}$ form a fundamental system of neighbourhoods for end(r) in* Ends(X).

Proof. Every compact subset of X is contained in an open ball about x_0 and vice versa, so one may replace compact sets by open balls $B(x_0, R)$ in the definition of Ends(X). Part (3) follows immediately from (2) and this observation. Part (1) also follows from this observation, because if $x_1, \dots, x_n$ is a *k-path* connecting x_1 to x_n in $X \smallsetminus B(x_0, R + k)$, then the concatenation of any choice of geodesics $[x_i, x_{i+1}]$ gives a continuous path from x_1 to x_n in $X \smallsetminus B(x_0, R)$.

It remains to prove (2). Let $r : [0, \infty) \to X$ be a proper ray. Let $c_n : [0, d_n] \to X$ be a geodesic joining x_0 to $r(n)$; extend c_n to be constant on $[d_n, \infty)$. Because X is proper, the Arzelà-Ascoli theorem (3.10) furnishes us with a subsequence of the c_n converging to a geodesic ray $c : [0, \infty) \to X$, and it is clear that *end*(c) = *end*(r). ☐

8.29 Proposition. *If X_1 and X_2 are proper geodesic spaces, every quasi-isometry $f : X_1 \to X_2$ induces a homeomorphism $f_{\mathcal{E}}$:* Ends$(X_1) \to$ Ends(X_2).

The map $\mathcal{QI}(X_1) \to$ Homeo(Ends(X_1)) *given by $f \mapsto f_{\mathcal{E}}$ is a homomorphism.*

Proof. Let r be a geodesic ray in X_1 and let $f_*(r)$ be a ray in X_2 obtained by concatenating some choice of geodesic segments $[fr(n), fr(n+1)]$, $n \in \mathbb{N}$. Because f is a (λ, ε)-quasi-isometry, this is a proper ray. It is clear that *end*$(f_*(r))$ is independent of the choice of the geodesic segments $[fr(n), fr(n+1)]$.

Define $f_{\mathcal{E}}$: Ends$(X_1) \to$ Ends(X_2) by *end*$(r) \mapsto$ *end*$(f_*(r))$ for every geodesic ray r in X_1. The image under f of any k-path in X_1 is a $(\lambda k + \varepsilon)$-path in X_2, so 8.28(1) ensures that $f_{\mathcal{E}}$ is well-defined on equivalence classes, and that it is continuous. 8.28(2) ensures that $f_{\mathcal{E}}$ is defined on the whole of Ends(X_1).

It is clear that if $f' : X_2 \to X_1$ and $f : X_1 \to X_2$ are quasi-isometries then $f'_{\mathcal{E}} f_{\mathcal{E}} = (f'f)_{\mathcal{E}}$, and if $f' : X_2 \to X_1$ is a quasi-inverse for f, then $f'_{\mathcal{E}} f_{\mathcal{E}} = (f'f)_{\mathcal{E}}$ is the identity map on Ends(X_1). ☐

In the light of 8.17(2) and the preceding result, the following definition of Ends(Γ) is unambiguous.

8.30 Definition (The Ends of a Group). Let Γ be a group and let $\mathcal{C}$ be its Cayley graph with respect to a finite generating set. We define $\text{Ends}(\Gamma) := \text{Ends}(\mathcal{C})$.

8.31 Exercise. Let T_n be a metric tree in which all vertices have valence $n \geq 3$ and all edges have length one. Construct a homeomorphism from $\text{Ends}(T_n)$ to the Cantor set.

(Hint: The Cantor set may be thought of as the subspace of $[0, 1]$ consisting of those numbers whose ternary expansion consists only of zeros and twos. Embed T_3 in the plane (topologically), fix a base point, and encode the trajectory of each geodesic ray as a sequence of zeros and twos describing when the ray turns left and right; regard this sequence as a ternary expansion. Use (8.20) and (8.29) to pass from $n = 3$ to the general case.)

Amalgamated free products and HNN extensions, which appear in part (5) of the following theorem, will be discussed in more detail in section (III.Γ.6).

8.32 Theorem. *Let Γ be a finitely generated group.*

(1) *Γ has $0, 1, 2$ or infinitely many ends.*

(2) *Γ has 0 ends if and only if it is finite.*

(3) *Γ has 2 ends if and only if it contains $\mathbb{Z}$ as a subgroup of finite index.*

(4) *$\text{Ends}(\Gamma)$ is compact. If it is infinite then it is uncountable and each of its points is an accumulation point.*

(5) *Γ has infinitely many ends if and only if Γ can be expressed as an amalgamated free product $A *_C B$ or HNN extension $A*_C$ with C finite, $|A/C| \geq 3$ and $|B/C| \geq 2$.*

This result shows, for example, that if a geodesic space X has three ends, or if the set of ends is countable, then X does not admit a proper cocompact action by a group of isometries.

The first four parts of this theorem are due to Hopf [Ho43]. Part (5) is due to Stallings [St68]; its proof is beyond the scope of the ideas presented here. Part (2) is trivial. We shall prove parts (1) and (4). We leave (3) as an exercise (8.34).

Proof of 8.32(1) and (4) We fix a finite generating set for Γ and work with the corresponding Cayley graph $\mathcal{C}$. The action of Γ on itself by left multiplication extends to an action by isometries on $\mathcal{C}$, giving a homomorphism $\Gamma \to \text{Homeo}(\text{Ends}(\mathcal{C}))$ as in(8.29). Let H be the kernel of this map. If $\text{Ends}(\mathcal{C})$ is finite, then H has finite index in Γ.

We prove (1) by arguing the contrapositive. Suppose that $\mathcal{C}$ has finitely many ends, and let e_0, e_1, e_2 be three distinct ones. We fix two geodesic rays $r_1, r_2 : [0, \infty) \to X$ with $r_1(0) = r_2(0)$ equal to the identity vertex 1, and $end(r_i) = e_i$. Because H has finite index in Γ, there is a constant μ such that every vertex of $\mathcal{C}$ lies in the μ-neighbourhood of H. It follows that there is a proper ray $r_0 : [0, \infty) \to \mathcal{C}$ with $end(r_0) = e_0$, $d(r_0(n), 1) \geq n$, and $r_0(n) \in H$ for every $n \in \mathbb{N}$. Let $\gamma_n = r_0(n)$.

We fix $\rho > 0$ such that $r_1([\rho, \infty))$, $r_2([\rho, \infty))$, and $r_0([\rho, \infty))$ lie in different path components of $\mathcal{C} \setminus B(1, \rho)$. If $t, t' > 2\rho$, then $d(r_1(t), r_2(t')) > 2\rho$, because any path joining $r_1(t)$ to $r_2(t')$ must pass through $B(1, \rho)$.

Since H acts trivially on Ends($\mathcal{C}$), we have $end(\gamma_n.r_i) = end(r_i)$ for $i = 1, 2$. Let $n > 3\rho$. Then $\gamma_n.r_i(0) = \gamma_n$ lies in a different path component to $r_i([\rho, \infty))$, so $\gamma_n.r_i$ must pass through $B(1, \rho)$. Thus $\gamma_n.r_1(t) \in B(1, \rho)$ and $\gamma_n.r_2(t') \in B(1, \rho)$ for some $t, t' > 2\rho$. But γ_n is an isometry, so this implies $d(r_1(t), r_2(t')) < 2\rho$, which is a contradiction.

We now prove (4). Given a sequence of ends e_n, we choose geodesic rays $r_n : [0, \infty) \to \mathcal{C}$ with $r_n(0) = 1$ and $end(r_n) = e_n$. The Arzelà-Ascoli theorem (or a simple finiteness argument) shows that a subsequence of these rays converges on compact subsets to some geodesic ray r; it follows that the corresponding sequence of ends converges to $end(r)$. Thus Ends($\mathcal{C}$) is sequentially compact, and since it satisfies the first axiom of countability (8.28(3), it is compact. It is clear that Ends($\mathcal{C}$) is Hausdorff, and a compact Hausdorff space in which every point is an accumulation point is uncountable (exercise). Thus it suffices to prove that if Ends($\mathcal{C}$) is infinite then every $e \in$ Ends($\mathcal{C}$) is an accumulation point.

Given $e_1, e_2 \in$ Ends($\mathcal{C}$), let $D = D(e_1, e_2)$ be the maximum integer such that $r_1([D, \infty))$ and $r_2([D, \infty))$ lie in the same path component of $\mathcal{C} \setminus B(1, D)$ for some (hence all) geodesic rays with $end(r_i) = e_i$ and $r_i(0) = 1$. Note that $e_n \to e$ if and only if $D(e_n, e) \to \infty$ as $n \to \infty$.

Fix $e_0 \in$ Ends($\mathcal{C}$) and a geodesic ray r_0 with $r_0(0) = 1$ and $end(r_0) = e_0$. Let $\gamma_n = r_0(n)$. We shall construct a sequence of ends e^m such that $e^m \neq e_0$ and $D(e^m, e_0) \to \infty$ as $m \to \infty$. Let e_1, e_2 be distinct ends, neither of which is e_0 and let r_i be a geodesic ray with $r_i(0) = 1$ and $end(r_i) = e_i$ for $i = 1, 2$. Let $M = \max D(e_i, e_j)$, $i, j = 0, 1, 2$. If we fix $\rho > M$, then by arguing as in (1) we see that for large n at most one of the rays $\gamma_n.r_i$ can pass through $B(1, \rho)$. To each of the other two rays $\gamma_n.r_j$ we apply the Arzelà-Ascoli theorem as in (8.28) to construct a geodesic ray r_j' with $r_j'(0) = 1$ and $e_j' := end(r_j') = end(\gamma_n.r_j)$. Since r_0 and r_j' have terminal segments in the same path component of $\mathcal{C} \setminus B(1, \rho)$, we have $D(e_0, e_j') > \rho$. The e_j' are distinct, so at least one of them is not equal to e_0: rename this end e^1.

We now repeat the above argument with e_1 replaced by e^1 and call the ray constructed in the argument e^2. We replace e^1 be e^2 and repeat again. At each iteration of this process the integer M is increased. Thus we obtain the desired sequence of ends e^m with $e^m \neq e_0$ and $D(e^m, e_0) \to \infty$ as $m \to \infty$. □

We note a consequence of 8.32(5):

8.33 Corollary. *If two finitely generated groups are quasi-isometric and one splits as an amalgamated free product or HNN extension of the type described in 8.32(5), then so does the other.*

8.34 Exercise. Prove 8.32(3). (Hint: Choose a ball B about the identity that separates the ends of the Cayley graph $\mathcal{C}$, and choose $\gamma \in \Gamma \setminus B$ whose action on Ends($\mathcal{C}$) is trivial. Check that γ has infinite order by considering $\gamma^n.r$, where r is a geodesic ray

r starting at γ that does not pass through B. Finally, argue that the sequences $n \mapsto \gamma^n$ and $n \mapsto \gamma^{-n}$ tend to different ends of C, and that any point of C is within a bounded distance of $\langle \gamma \rangle$.)

Growth and Rigidity

The material that follows is of a different nature to what went before: greater background is assumed, non-trivial facts are quoted without proof, and some of the exercises are hard. The material in this section is included for the information of the reader; our purpose in presenting it is to indicate that there are many situations in geometry where it is useful to employ techniques involving quasi-isometries of groups.

There is an extensive literature on the growth of groups. Much of it focuses on the search for closed formulae which describe the number of vertices in the ball of radius n about the identity in the Cayley graph of a group (see [GrH97]), but the coarser aspects of the theory are also interesting.

8.35 Definition. Let Γ be a group with generating set $\mathcal{A}$. Let $\beta_{\mathcal{A}}(n)$ be the number of vertices in the closed ball of radius n about $1 \in \mathcal{C}_{\mathcal{A}}(\Gamma)$. The *growth function* of Γ with respect to $\mathcal{A}$ is $n \mapsto \beta_{\mathcal{A}}(n)$.

One says that Γ has *polynomial growth of degree* $\leq d$ if there exists a constant k such that $\beta_{\mathcal{A}}(n) \leq kn^d$ for all $n \in \mathbb{N}$.

8.36 Exercises

(1) Prove that having polynomial growth of degree $\leq d$ is an invariant of quasi-isometry. Prove that a group has growth of degree ≤ 1 if and only if it is finite or contains $\mathbb{Z}$ as a subgroup of finite index (cf. 8.40).

(2) Show that if Γ_1 and Γ_2 are quasi-isometric then with respect to any finite generating sets $\mathcal{A}_1$ for Γ_1 and $\mathcal{A}_2$ for Γ_2 one has $\beta_{\mathcal{A}_1}(n) \approx \beta_{\mathcal{A}_2}(n)$, where $\approx$ is the equivalence relation on functions $\mathbb{N} \to \mathbb{R}$ defined as follows: $f \approx g$ if and only if $f \preccurlyeq g$ and $g \preccurlyeq f$, where $f \preccurlyeq g$ if and only if there exists a constant $k > 0$ such that $f(n) \leq k\,g(kn + k) + k$ for all $n \in \mathbb{N}$. (This observation allows one to drop the subscript $\mathcal{A}$ or to write $\beta_\Gamma(n)$ when discussing the $\approx$ properties of $\beta_{\mathcal{A}}(n)$.)

(3) Prove that if Γ contains a non-abelian free group, then the growth function of Γ satisfies $2^n \preccurlyeq \beta_\Gamma(n)$. More generally, prove that if $H \subset \Gamma$ is finitely generated then $\beta_H(n) \preccurlyeq \beta_\Gamma(n)$.

(4) The growth of $\mathbb{Z}^m$ with respect to any finite generating set is polynomial of degree m.

(5) If $n \geq 2$, then any group Γ which acts properly and cocompactly by isometries on $\mathbb{H}^n$ has exponential growth. (Hint: One way to see this is to show that the number of disjoint balls of radius 1 that one can fit into a ball or radius r in $\mathbb{H}^n$ grows as an

exponential function of r, and this property is inherited by the Cayley graph of Γ. One can also argue that Γ must have a non-abelian free subgroup.)

8.37 Examples

(1) R. Grigorchuk [Gri83] has constructed finitely generated groups Γ for which $2^{n^{\alpha_1}} \preccurlyeq \beta(n) \preccurlyeq 2^{n^{\alpha_2}}$, where $0 < \alpha_1 < \alpha_2 < 1$.

(2) Recall that the lower central series of a group Γ is (Γ_n), where $\Gamma_0 = \Gamma$ and $\Gamma_n = [\Gamma, \Gamma_{n-1}]$ is the subgroup generated by n-fold commutators. Γ is said to be *nilpotent* if its lower central series terminates in a finite number of steps, i.e. $\Gamma_c = \{1\}$ for some $c \in \mathbb{N}$. All finitely generated nilpotent groups have polynomial growth — see [Dix60], [Mil68], [Wo64]. Guivarc'h [Gui70, 73], Bass [Bass72] and others calculated the degree of growth: $\beta(n) \approx n^d$ where

$$d = \sum_{i=0}^{c} (i+1)\, \mathrm{rk}_{\mathbb{Q}} \Gamma_i/\Gamma_{i+1}.$$

Gromov proved the following remarkable converse to (8.37(2)) — see [Gro81b] and [DW84].

8.38 Theorem. *Let Γ be a finitely generated group. If Γ has polynomial growth then Γ contains a nilpotent subgroup of finite index.*

In order to prove this theorem one must use the hypothesis on growth to construct a subgroup of finite index $\Delta \subset \Gamma$ for which there is a surjection $\Delta \twoheadrightarrow \mathbb{Z}$. Once one has constructed such a map, there is an obvious induction on the degree of polynomial growth: the degree of polynomial growth of Δ is the same as that of Γ, and it is not hard to show that the kernel of any map $\Delta \twoheadrightarrow \mathbb{Z}$ is finitely generated and the degree of its polynomial growth is less than that of Γ; by induction the kernel has a nilpotent subgroup of finite index and hence so does Γ.

In order to construct a map to $\mathbb{Z}$, Gromov proves that a subsequence of the metric spaces $X_n = (\Gamma, \frac{1}{n}d)$ converges (in the pointed Gromov-Hausdorff sense (5.44)) to a complete, connected, locally compact, finite-dimensional metric space X_∞. A classical theorem of Montgomery and Zippin [MoZ55] implies that the isometry group of X_∞ is a Lie group G with finitely many connected components. Using this theorem, and classical results about the subgroups of Lie groups, Gromov analyses the action of Γ on X_∞ to produce the desired map $\Delta \twoheadrightarrow \mathbb{Z}$.

8.39 Corollary. *If a finitely generated group Γ is quasi-isometric to a nilpotent group then Γ contains a nilpotent subgroup of finite index.*

A group Γ is called *polycyclic* if it has a sequence of subgroups $\{1\} = \Gamma_0 \subset \Gamma_1 \subset \cdots \subset \Gamma_n = \Gamma$ such that each Γ_i is normal in Γ_{i+1} and Γ_{i+1}/Γ_i is a cyclic group. The number of factors Γ_{i+1}/Γ_i isomorphic to $\mathbb{Z}$ is called the *Hirsch length* of

Γ. In [BriG96] Bridson and Gersten proved that the Hirsch length is an invariant of quasi-isometry for polycyclic groups. [16]

Every finitely generated nilpotent group N is polycyclic. Arguing by induction on the Hirsch length, one can show that the degree of polynomial growth of N is equal to the Hirsch length if and only if N does not contain a subgroup of the form $\mathbb{Z} \rtimes_\phi \mathbb{Z}$ where $\phi \in \mathrm{GL}(n, \mathbb{Z})$ is an element of infinite order, cf. (II.7.16) and (III.Γ.4.17). And N contains such a subgroup if and only if it is not virtually abelian. By combining these facts with Gromov's Theorem (8.38) one gets the following result.

8.40 Theorem. *If a finitely generated group is quasi-isometric to $\mathbb{Z}^n$ then it contains $\mathbb{Z}^n$ as a subgroup of finite index.*

This theorem is contained in a more general result of P. Pansu concerning quasi-isometries between nilpotent groups [Pan83]. The proof that we have sketched is based on that of Bridson and Gersten [BriG96].

8.41 Quasi-Isometric Rigidity of Lattices. In recent years a number of results have been proved concerning the quasi-isometric rigidity of lattices in semi-simple Lie groups [Es98], [EsF97], [KlL97], [Schw95]. See [Fa97] for a survey and further references. There are rigidity results of two types in these works: in the first type of result one shows that non-uniform lattices in certain Lie groups (e.g. $\mathrm{SO}(n, 1)$ in [Schw95]) are quasi-isometric if and only if they are commensurable; in the second type of result one shows that if an abstract finitely generated group G is quasi-isometric to a group in the given class of lattices (e.g. irreducible lattices in higher rank groups in [KlL97]), then there is a short exact sequence $1 \to F \to G \to \Gamma \to 1$, where F is finite and Γ is a lattice of the given type.

Both types of rigidity results have also been proved for classes of groups other than lattices (e.g. [FaMo98] and [KaL95].)

Quasi-Isometries of the Model Spaces

We consider which of the model spaces are quasi-isometric.

8.42 Proposition.

(1) *M^n_κ is quasi-isometric to a point if and only if $\kappa > 0$.*

(2) *$\mathbb{E}^n$ is quasi-isometric to $\mathbb{E}^m$ if and only if $n = m$.*

(3) *$\mathbb{H}^n$ is quasi-isometric to $\mathbb{H}^m$ if and only if $n = m$.*

(4) *$\mathbb{E}^n$ is not quasi-isometric to $\mathbb{H}^m$ if $n, m \geq 2$.*

[16] More generally, Gersten [Ger93a] proved that cohomological dimension is an invariant of quasi-isometry among groups that have a subgroup of finite index with a compact Eilenberg-Mac Lane space.

Remark. It is clear that $\mathbb{H}^n$ is quasi-isometric to M^n_κ for all $\kappa < 0$.

Proof. In the light of the considerations of growth described in the Exercises (8.36), the only non-trivial point that remains is to prove that $\mathbb{H}^n$ and $\mathbb{H}^m$ are not quasi-isometric if $n \neq m$. This is explained properly in (III.H.3) in terms of the boundaries at infinity of the spaces, but it can also be proved using the elementary argument outlined in the following exercise.

8.43 Exercise (Coarse Filling).

(1) Let k be a positive constant. A map between metric spaces $f : X \to Y$ is said to be k-continuous if for every $x \in X$ there exists $\nu > 0$ such that $d(f(x_1), f(x_2)) \leq k$ whenever $d(x, x_1) < \nu$ and $d(x, x_2) < \nu$.

Let $\mathbb{D}^{n+1}$ be a closed unit ball in $\mathbb{E}^n$ and let $S^n = \partial\mathbb{D}^{n+1}$. Show that the following property (but not the choice of constants) is an invariant of quasi-isometry of metric spaces X. Fix $n \in \mathbb{N}$.

(Φ_n) For all sufficiently large $k > 0$ there exist constants $\lambda, \lambda' > 0$ such that for every $R > 0$ and every $x \in X$, every k-continuous map $S^n \to X \smallsetminus B(x, R)$ extends to a λk-continuous map $\mathbb{D}^{n+1} \to X \smallsetminus B(x, \lambda' R)$.

(2) Show that $\mathbb{H}^n$ has property Φ_{n-i} for every $i \geq 2$, but does not have property Φ_{n-1}.

This exercise provides a simple illustration of the fact that one can use coarse metric analogues of the basic tools of algebraic topology (in this case higher homotopy groups) to describe the large scale geometry of a space (in this case dimension). For more sophisticated applications of this approach, in particular the development of cohomology theories that are invariant under quasi-isometry — see [Roe96], [Ger95] and the survey article of Bloch and Weinberger [BlWe97].

8.44 Exercises

(1) Prove that the natural homomorphism $\mathrm{GL}(n, \mathbb{R}) \to \mathcal{QI}(\mathbb{E}^n)$ is injective.

(2) Prove that the image of the natural homomorphism $\mathrm{Isom}(\mathbb{E}^n) \to \mathcal{QI}(\mathbb{E}^n)$ coincides with that of $O(n)$.

(3) Prove that the natural map $\mathrm{Isom}(\mathbb{H}^n) \to \mathcal{QI}(\mathbb{H}^n)$ is injective if $n \geq 2$.

(4) Prove that the map in (3) is not surjective. (Hint: One can construct quasi-isometries of $\mathbb{H}^n$ as follows. Given any C^1-diffeomorphism f of $\mathbb{S}^{n-1} = \partial\mathbb{H}^n$, send 0 to itself and for each geodesic ray $c : [0, \infty) \to \mathbb{H}^n$ with $c(0) = 0$, send $c(t)$ to $c'(t)$, where $c'(0) = 0$ and $c'(\infty) = f(c(\infty))$.)

The conclusion of exercise (3) is valid for other irreducible symmetric spaces of non-compact type, but the conclusion of (4) is not. (See the references listed in (8.41).)

Approximation by Metric Graphs

We close our discussion of quasi-isometries with a result that allows one to replace (perhaps very complicated) length spaces by simple metric graphs in the same quasi-isometry class. This can be useful when studying quasi-isometry invariants because, for example, it allows one to use induction on the length of paths (cf. III.H.2).

8.45 Proposition (Approximation by Graphs). *There exist universal constants α and β such that there is an (α, β)-quasi-isometry from any length space to a metric graph all of whose edges have length one. (Every vertex of the graph lies in the image of this map.)*

Proof. Let (X, d_X) be a length space. Consider a subset $V \subset X$ with the property that $d_X(u, v) \geq 1/3$ for all pairs of distinct points $u, v \in V$. Using Zorn's lemma, we may assume that V is not properly contained in any subset of X with the same property. In particular every point of X is a distance at most 1/3 from V. Let $\mathcal{G}$ be the graph with vertex set V which has an edge joining two vertices u, v if and only if $d_X(u, v) \leq 1$. We consider the metric graph $(\mathcal{G}, d_{\mathcal{G}})$ obtained by setting all edge lengths equal to 1. We claim that the map $X \to \mathcal{G}$ sending x to a choice of closest point in V is the desired quasi-isometry.

Since $(V, d_{\mathcal{G}}) \hookrightarrow (\mathcal{G}, d_{\mathcal{G}})$ and $(V, d_X) \hookrightarrow (X, d_X)$ are obviously quasi-isometries with universal parameters, it suffices to show that $id : (V, d_{\mathcal{G}}) \to (V, d_X)$ is a $(3, 1)$-quasi-isometry. By construction this map does not increase distances. Conversely, given $u, v \in V$ and $\eta > 0$ we can find a path of length $d_X(u, v) + \eta$ connecting u to v in X. Along the image of this path we choose N equally spaced points p_i, beginning with u and ending with v, where N is the least integer greater than $3(d_X(u, v)+\eta)+1$. For each of these points p_i we choose $u_i \in V$ with $d_X(p_i, u_i) \leq 1/3$. For each i we have $d_X(u_i, u_{i+1}) \leq 1$, and hence $u_i = u_{i+1}$ or else u_i and u_{i+1} are connected by an edge in $\mathcal{G}$. It follows that $d_{\mathcal{G}}(u, v) \leq N - 1 \leq 3(d_X(u, v) + \eta) + 1$ and hence (since $\eta > 0$ is arbitrary) $id : (V, d_{\mathcal{G}}) \to (V, d_X)$ is a $(3, 1)$-quasi-isometry. □

8.46 Exercises

(1) Calculate the universal constants in the above proposition.

(2) (*Quasi-length spaces*). k-paths were defined prior to (8.28). The length of a k-path $x_1, \dots, x_n$ is defined to be $\sum d(x_i, x_{i+1})$. X said to be *coarsely connected* if there is a constant k such that every pair of points in X can be joined by a k-path, and X is called a *quasi-length space* if there exist constants $k, \lambda, \varepsilon > 0$ such that for every pair of points $x, y \in X$ can be connected by a k-path of length at most $\lambda\, d(x, y) + \varepsilon$.

Prove that the class of quasi-length spaces is closed under quasi-isometry and generalize (8.45) to include such spaces.

Prove that any metric space which admits a cocompact action by a finitely generated group is coarsely connected, but that it need not be a quasi-length space. (Hint: Truncate the metric on a nice space by defining $d'(x, y) = \max\{d(x, y), 1\}$.)

Appendix: Combinatorial 2-Complexes

Combinatorial complexes are topological objects with a specified combinatorial structure. They are defined by a recursion on dimension; the definition of an open cell is defined by a simultaneous recursion. If K_1 and K_2 are combinatorial complexes, then a continuous map $K_1 \to K_2$ is said to be combinatorial if its restriction to each open cell of K_1 is a homeomorphism onto an open cell of K_2.

A combinatorial complex of dimension 0 is simply a set with the discrete topology; each point is an open cell. Having defined $(n-1)$-dimensional combinatorial complexes and their open cells, one constructs n-dimensional combinatorial complexes as follows.

Take the disjoint union of an $(n-1)$-dimensional combinatorial complex $K^{(n-1)}$ and a family $(e_\lambda \mid \lambda \in \Lambda)$ of copies of closed n-dimensional discs. Suppose that for each $\lambda \in \Lambda$ a homeomorphism is given from ∂e_λ (a sphere) to an $(n-1)$-dimensional combinatorial complex S_λ, and that a combinatorial map $S_\lambda \to K^{(n-1)}$ is also given; let $\phi_\lambda : \partial e_\lambda \to K^{(n-1)}$ be the composition of these maps. Define K to be the quotient of $K^{(n-1)} \cup \coprod_\Lambda e_\lambda$ by the equivalence relation generated by $t \sim \phi_\lambda(t)$ for all $\lambda \in \Lambda$ and all $t \in \partial e_\lambda$. Then K, with the quotient topology, is an n-dimensional combinatorial complex whose open cells are the (images of) open cells in $K^{(n-1)}$ and the interiors of the e_λ.

8A.1 Attaching 2-Cells. We are interested primarily in 2-dimensional combinatorial complexes. Let K be such a complex with 2-cells $(e_\lambda \mid \lambda \in \Lambda)$. Let X denote the 1-skeleton of K. The attaching maps $\phi_\lambda : \partial e_\lambda \to K^{(1)}$ are combinatorial loops (i.e. edge loops) in X, and thus one describes K as *the 2-complex obtained by attaching the 2-cells e_λ to X along the loops ϕ_λ*.

There is an obvious way to endow any combinatorial 2-complex K with a piecewise Euclidean structure: one metrizes the 1-skeleton as a metric graph with edges of length one, and one metrizes each e_λ as a regular Euclidean polygon with sides of length one. We write $K^{\mathbb{E}}$ to denote the piecewise Euclidean complex obtained in this way.

8A.2 The 2-Complex Associated to a Group Presentation. Associated to any group presentation $\langle \mathcal{A} \mid \mathcal{R} \rangle$ one has a 2-complex[17] $K = K(\mathcal{A} : \mathcal{R})$ that is compact if and only if the presentation is finite. K has one vertex and it has one edge ε_a (oriented and labelled a) for each generator $a \in \mathcal{A}$; thus edge loops in the 1-skeleton of K are in 1–1 correspondence with words in the alphabet $\mathcal{A}^{\pm 1}$: the letter a^{-1} corresponds to traversing the edge ε_a in the direction opposite to its orientation, and the word $w = a_1 \dots a_n$ corresponds to the loop that is the concatenation of the directed edges $a_1, \dots, a_n$; one says that w *labels* this loop. The 2-cells e_r of K are indexed by the relations $r \in \mathcal{R}$; if $r = a_1 \dots a_n$ then e_r is attached along the loop labelled $a_1 \dots a_n$. The map that sends the homotopy class of ε_a to $a \in \Gamma$ gives an isomorphism $\pi_1 K(\mathcal{A} : \mathcal{R}) \cong \Gamma$ (by the Seifert-van Kampen theorem).

There is a natural Γ-equivariant identification of the Cayley graph $\mathcal{C}_{\mathcal{A}}(\Gamma)$ with the 1-skeleton of the universal cover of $K(\mathcal{A} : \mathcal{R})$: fix a base vertex $v_0 \in \widetilde{K}(\mathcal{A} : \mathcal{R})$, identify $\gamma.v_0$ with γ, and identify the edge of $\mathcal{C}_{\mathcal{A}}(\Gamma)$ labelled a issuing from γ with the (directed) edge at $\gamma.v_0$ in the pre-image of ε_a.

8A.3 Exercises

(1) Prove that $K_1 = K(a, b : aba^{-1}b^{-1})$ and $K_2 = K(a, b, c : abc, bac)$ are tori but that $K_1^{\mathbb{E}}$ and $K_2^{\mathbb{E}}$ are not isometric.

(2) Following Whitehead [Wh36], we define the *star graph* of a presentation $\langle \mathcal{A} \mid \mathcal{R} \rangle$ to be the graph with vertices $\{(a, \eta) \mid a \in \mathcal{A}, \eta = \pm 1\}$ that has one edge joining (a, η) to (a', η') for each occurrence of the subwords $(a^{\eta} a'^{-\eta'})$ and $(a'^{-\eta'} a^{-\eta})$ among the relators $r \in \mathcal{R}$ (the terminal and initial letters of a word are to be counted as being consecutive, thus $a'^{\eta'} \dots a^{-\eta}$ contributes to the count for the subword $a^{-\eta} a'^{\eta'}$).

Show that the link of the vertex in $K(\mathcal{A} : \mathcal{R})$ is the star graph of $\langle \mathcal{A} \mid \mathcal{R} \rangle$.

(3) Using (2), prove that $K(a_1, \dots, a_n : a_1 \dots a_n a_1^{-1} \dots a_n^{-1})$ is a closed surface (of genus $n/2$) if and only if n is even.

(4) Using (2), prove that if $\mathcal{A}$ is finite and w is a reduced word in which a and a^{-1} both occur exactly once, for every $a \in \mathcal{A}$, then $K = K(\mathcal{A} : w)$ can be obtained by gluing a number of closed surfaces together at a point.

8A.4 Homotopy and van Kampen Diagrams. Let K be a combinatorial complex with basepoint $x_0 \in K^{(0)}$. We consider edge paths in K; such a path is the concatenation of a finite number of directed edges $\varepsilon_i : [0, 1] \to K^{(1)}$ with $\varepsilon_i(1) = \varepsilon_{i+1}(0)$, or else it is the constant path at a vertex. An edge loop is by definition an edge path

[17] Sometimes called the Cayley complex.

that begins and ends at the same vertex. Given an edge path c we write $\bar{c}$ to denote the same path with reversed orientation (thus, for example, if $\varepsilon : [0, 1] \to K$ is a directed edge then $\bar{\varepsilon} : [0, 1] \to K$ is the inverse $\bar{\varepsilon}(0) = \varepsilon(1)$ and $\bar{\varepsilon}(1) = \varepsilon(0)$).

Two edge paths in K are said to be related by an *elementary homotopy* if one of the following conditions holds:

(i) *(Backtracking)* c' is obtained from c by inserting or deleting a subpath of the form $\varepsilon\bar{\varepsilon}$, where ε is a directed edge;

(ii) *(Pushing across a 2-cell)* c and c' can be expressed as concatenations $c = \alpha u \beta$ and $c' = \alpha u' \beta$, where $u^{-1}u'$ is the attaching loop of some 2-cell of K. (Note that u or u' may be the empty path.)

Two edge paths are said to be *combinatorially equivalent* if they are related by a finite sequence of elementary homotopies. The set of equivalence classes of paths that begin and end at x_0 is denote $\pi_1^c(K, x_0)$. Concatenation of paths induces a group structure on $\pi_1^c(K, x_0)$, where the class of the constant path $[x_0]$ serves as the identity. If c is equivalent to the constant path, then we define Area(c) to be the minimal number of elementary homotopies of type (ii) that one must apply in conjunction with elementary homotopies of type (i) to reduce c to the constant path. It is a standard exercise using the Seifert-van Kampen theorem to show that the natural map $\pi_1^c(K, x_0) \to \pi_1(K, x_0)$ is an isomorphism (see [Mass91], for example).

A *van Kampen diagram* in K is a combinatorial map $\Delta \to K$ where Δ is a connected, simply connected, planar 2-complex. The boundary cycle of Δ, when oriented, gives an edge loop c in $K^{(1)}$ that is null-homotopic. Δ is called a van Kampen diagram for c. A simple induction shows that c can be reduced to the constant loop by applying elementary homotopies of type (i) together with Area(Δ) elementary homotopies of type (ii), where Area(Δ) is the number of 2-cells in Δ. Thus Area(c) $\leq$ Area(Δ). Conversely, if c can be reduced to the constant path $[x_0]$ by elementary homotopies, then arguing by induction of the total number of elementary homotopies required, one can show that there is a van Kampen diagram Δ for c with Area(Δ) = Area(c). This important observation is due to van Kampen [K32b] (cf. Strebel's appendix to [GhH90]).

If K is the standard 2-complex of a finite presentation $K = K(\mathcal{A} : \mathcal{R})$, edge paths c in K are in natural bijection with words in the generators $\mathcal{A}^{\pm 1}$, and two loops are related by a homotopy of type (ii) if and only if the corresponding words have the form $w = \alpha u \beta$ and $w' = \alpha u' \beta$, where $r = u'u^{-1} \in \mathcal{R}^{\pm 1}$. This means precisely that there is an equality $w' = (\alpha r \alpha^{-1})w$ in the free group $F(\mathcal{A})$. It follows that if an edge loop in K is labelled by the word w, then Area(c), as defined above, is the least integer N_w for which there is an equality of the form

$$w = \prod_{i=1}^{N_w} x_i r_i x_i^{-1}$$

in $F(\mathcal{A})$, where the $x_i \in F(\mathcal{A})$ are arbitrary and $r_i \in \mathcal{R}^{\pm 1}$. Thus it is natural to define Area(w) = N_w. The function $f(n) = \max\{\text{Area}(w) : |w| \leq n,\ w = 1 \text{ in } \pi_1 K\}$ is called the *Dehn function* of $\langle \mathcal{A} \mid \mathcal{R} \rangle$. Dehn functions are widely studied in connection with questions concerning the complexity of the word problem in finitely presented groups (cf. Section III.Γ.5).

The Dehn functions associated to different finite presentations of the same group (or more generally quasi-isometric groups) are $\simeq$ equivalent in the following sense (see [Alo90]): given functions $g, h : \mathbb{N} \to [0, \infty)$, one writes $g \preceq h$ if there is a constant $K > 0$ such that $g(n) \leq K\, h(Kn) + Kn + K$, and one writes $g \simeq h$ if $g \preceq h$ and $h \preceq g$.

The following picture shows a van Kampen diagram over $\langle a, b, c \mid [a, b]\rangle$ for the word $a^2bcb^{-1}a^{-1}bac^{-1}ba^{-2}b^{-2}$. The reader might prove as an exercise that no van Kampen diagram for this word is homeomorphic to a disc.

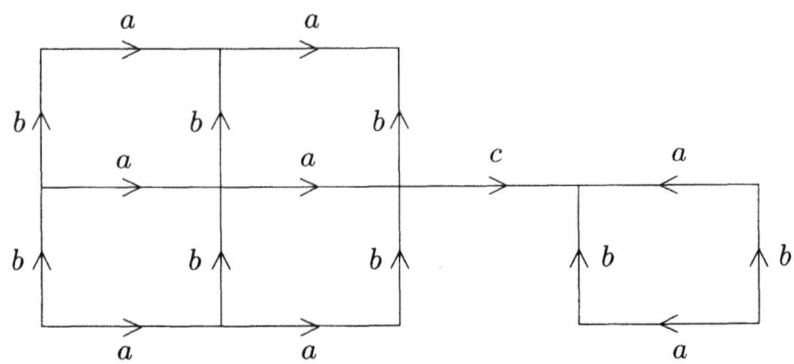

Fig. 8.3 A van Kampen diagram over $\langle a, b, c \mid [a, b]\rangle$

Part II. CAT(κ) Spaces

In Part I we assembled basic facts about geometric notions such as length, angle, geodesic etc., and presented various constructions of geodesic metric spaces. The most important of the examples which we considered are the model spaces M_κ^n. The central role which these spaces play in the scheme of this book was explained in the introduction: we seek to elucidate the structure of metric spaces by comparing them to M_κ^n; if favourable comparisons can be drawn then one can deduce much about the structure of the spaces at hand. We are now in a position to set about this task.

In Part II we shall study the basic properties of spaces whose curvature is bounded from above by a real number κ. Roughly speaking, a space has curvature $\leq \kappa$ if every point of the space has a neighbourhood in which geodesic triangles are no fatter than their comparison triangles in M_κ^2. In Chapter 1 we give several precise formulations of this idea (all due to A.D. Alexandrov) and prove that they are equivalent. In subsequent chapters we develop the theory of spaces which satisfy these conditions, concentrating mainly (but not exclusively) on the case of non-positive curvature. We punctuate our discussion of the general theory with chapters devoted to various classes of examples.

Unless further qualification is made, in all that follows κ will denote an arbitrary real number. The diameter of M_κ^2 will be denoted D_κ (thus D_κ is equal to $\pi/\sqrt{\kappa}$ if $\kappa > 0$, and ∞ otherwise).

Chapter II.1 Definitions and Characterizations of CAT(κ) Spaces

In this chapter we present the various definitions of a CAT(κ) space and prove that they are equivalent.

The CAT(κ) Inequality

Recall that a geodesic segment joining two points p and q of a metric space X is the image of a path of length $d(p, q)$ joining p to q. We shall often write $[p, q]$ to denote a definite choice of geodesic segment, but immediately offset the dangers of this notation by pointing out that in general such a segment is not determined by its endpoints, i.e., without further assumptions on X there may be many geodesic segments joining p to q.

By definition, a *geodesic triangle* Δ in a metric space X consists of three points $p, q, r \in X$, its *vertices*, and a choice of three geodesic segments $[p, q], [q, r], [r, p]$ joining them, its *sides*. Such a geodesic triangle will be denoted $\Delta([p, q], [q, r], [r, p])$ or (less accurately if X is not uniquely geodesic) $\Delta(p, q, r)$. If a point $x \in X$ lies in the union of $[p, q]$, $[q, r]$ and $[r, p]$, then we write $x \in \Delta$.

Recall from (I.2.13) that a triangle $\overline{\Delta} = \Delta(\overline{p}, \overline{q}, \overline{r})$ in M_κ^2 is called a comparison triangle for $\Delta = \Delta([p, q], [q, r], [r, p])$ if $d(\overline{p}, \overline{q}) = d(p, q)$, $d(\overline{q}, \overline{r}) = d(q, r)$ and $d(\overline{p}, \overline{r}) = d(p, r)$. Such a triangle $\overline{\Delta} \subseteq M_\kappa^2$ always exists if the perimeter $d(p, q) + d(q, r) + d(r, p)$ of Δ is less than than twice the diameter D_κ of M_κ^2, and it is unique up to isometry (see I.2.13). We will write $\overline{\Delta} = \overline{\Delta}(p, q, r)$ or $\Delta(\overline{p}, \overline{q}, \overline{r})$, according to whether a specific choice of $\overline{p}, \overline{q}, \overline{r}$ is required. A point $\overline{x} \in [\overline{q}, \overline{r}]$ is called *a comparison point* for $x \in [q, r]$ if $d(q, x) = d(\overline{q}, \overline{x})$. Comparison points on $[\overline{p}, \overline{q}]$ and $[\overline{p}, \overline{r}]$ are defined in the same way. If $p \neq q$ and $p \neq r$, the *angle* of Δ at p is the Alexandrov angle between the geodesic segments $[p, q]$ and $[p, r]$ issuing from p, as defined in (I.1.12).

1.1 Definition of a CAT(κ) **Space.** Let X be a metric space and let κ be a real number. Let Δ be a geodesic triangle in X with perimeter less than $2D_\kappa$. Let $\overline{\Delta} \subset M_\kappa^2$ be a comparison triangle for $\overline{\Delta}$. Then, Δ is said to satisfy the CAT(κ) *inequality* if for all $x, y \in \Delta$ and all comparison points $\overline{x}, \overline{y} \in \overline{\Delta}$,

$$d(x, y) \leq d(\overline{x}, \overline{y}).$$

If $\kappa \leq 0$, then X is called a CAT(κ) space (more briefly, "X is CAT(κ)") if X is a geodesic space all of whose geodesic triangles satisfy the CAT(κ) inequality.

If $\kappa > 0$, then X is called a CAT(κ) space if X is D_κ-geodesic and all geodesic triangles in X of perimeter less than $2D_\kappa$ satisfy the CAT(κ) inequality. (In this definition we admit the possibility that the metric on X may take infinite values. And we remind the reader that to say X is D_κ-geodesic means that all pairs of points a distance less than D_κ apart are joined by a geodesic.)

Note that in our definition of a CAT(κ) space we do not require that X be complete. Complete CAT(0) spaces are often called *Hadamard spaces.*

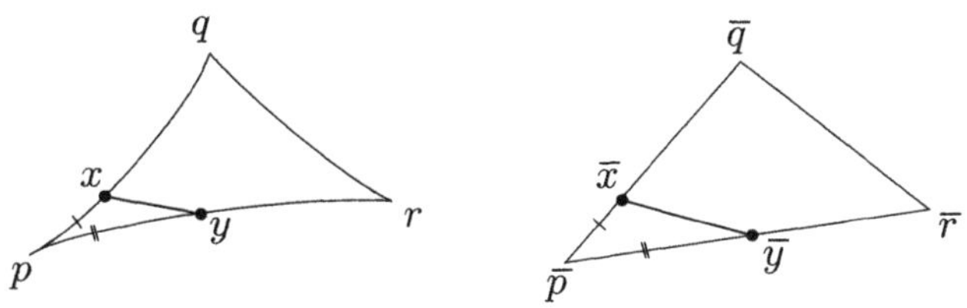

Fig. 1.1 The CAT(κ) inequality

1.2 Definition. A metric space X is said to be of *curvature* $\leq \kappa$ if it is locally a CAT(κ) space, i.e. for every $x \in X$ there exists $r_x > 0$ such that the ball $B(x, r_x)$, endowed with the induced metric, is a CAT(κ) space.

If X is of curvature ≤ 0 then we say that it is *non-positively curved.*

The definition given above was introduced by A.D. Alexandrov [Ale51]. It provides a good notion of an upper bound on curvature in an arbitrary metric space. Classical comparison theorems in differential geometry show that if a Riemannian manifold is sufficiently smooth (for example if it is C^3) then it has curvature $\leq \kappa$ in the above sense if and only if all of its sectional curvatures are $\leq \kappa$ (see the Appendix to this chapter).

The terminology "CAT(κ)" was coined by M. Gromov [Gro87, p.119]. The initials are in honour of E. Cartan, A.D. Alexandrov and V.A. Toponogov, each of whom considered similar conditions in varying degrees of generality. In Chapter 4 we shall prove that if a complete simply connected length space has curvature ≤ 0 then it is a CAT(0) space.

1.3 Remark. Let X be a geodesic space. If each pair of geodesics $c_1 : [0, a_1] \to X$ and $c_2 : [0, a_2] \to X$ with $c_1(0) = c_2(0)$ satisfy the inequality $d(c_1(ta_1), c_2(ta_2)) \leq t\, d(c_1(a_1), c_2(a_2))$ for all $t \in [0, 1]$, then one says that *the metric on X is convex.* It is easy to see that the metric on a CAT(0) space is convex (cf. 2.2). In general having a convex metric is a weaker property than being CAT(0) (cf. 1.18). There are, however, several important classes of spaces in which convexity of the metric is equivalent to the CAT(0) condition, including Riemannian manifolds (1A.8) and M_κ-polyhedral complexes (5.4).

1.4 Proposition. *Let X be a* CAT(κ) *space.*

(1) *There is a unique geodesic segment joining each pair of points $x, y \in X$ (provided $d(x, y) < D_\kappa$ if $\kappa > 0$), and this geodesic segment varies continuously with its endpoints.*

(2) *Every local geodesic in X of length at most D_κ is a geodesic.*

(3) *The balls in X of radius smaller than $D_\kappa/2$ are convex (i.e., any two points in such a ball are joined by a unique geodesic segment and this segment is contained in the ball).*

(4) *The balls in X of radius less than D_κ are contractible.*

(5) *(Approximate midpoints are close to midpoints.) For every $\lambda < D_\kappa$ and $\varepsilon > 0$ there exists $\delta = \delta(\kappa, \lambda, \varepsilon)$ such that if m is the midpoint of a geodesic segment $[x, y] \subset X$ with $d(x, y) \leq \lambda$ and if*

$$\max\{d(x, m'), d(y, m')\} \leq \frac{1}{2}d(x, y) + \delta,$$

then $d(m, m') < \varepsilon$.

Proof. (1) Consider $p, q \in X$ with $d(p, q) < D_\kappa$. Let $[p, q]$ and $[p, q]'$ be geodesic segments joining p to q, let $r \in [p, q]$ and $r' \in [p, q]'$ be such that $d(p, r) = d(p, r')$. Let $[p, r]$ and $[r, q]$ be the two geodesic segments whose concatenation is $[p, q]$. Any comparison triangle in M_κ^2 for $\Delta([p, q]', [p, r], [r, q])$ is degenerate and the comparison points for r and r' are the same. The CAT(κ) inequality implies that $d(r, r') = 0$, hence $r = r'$.

If X is proper then it follows immediately from (I.3.12) that $[p, q]$ depends continuously on p and q. We consider the general case. First note that given any positive number $\ell < D_\kappa$, there is a constant $C = C(\ell, \kappa)$ such that if $c, c' : [0, 1] \to M_\kappa^2$ are two linearly parameterized geodesic segments of length at most ℓ and if $c(0) = c'(0)$, then $d(c(t), c'(t)) \leq Cd(c(1), c'(1))$ for all $t \in [0, 1]$ (see 3.20). For $\kappa \leq 0$ we even have $d(c(t), c'(t)) \leq t\, d(c(1), c'(1))$.

Let p_n and q_n be sequences of points converging to p and q respectively. We assume that $d(p_n, q_n)$ and $d(p, q_n)$ are smaller than $\ell < D_k$. Let c, c_n, c_n' be linear parameterizations $[0, 1] \to X$ of the geodesic segments $[p, q]$, $[p_n, q_n]$ and $[p, q_n]$ respectively. Applying the CAT(κ) inequality, we have $d(c(t), c_n(t)) \leq d(c(t), c_n'(t)) + d(c_n'(t), c_n(t) \leq C(d(q, q_n) + d(p, p_n))$, where $C = C(\ell, \kappa)$ is as above. Therefore c_n converges uniformly to c.

To prove (2), we fix a local geodesic $c : [0, \lambda] \to X$ of length $\lambda \leq D_\kappa$ and consider $S = \{t \in [0, \lambda] \mid c|_{[0,t]}$ is a geodesic$\}$. This set is obviously closed, and it contains a neighbourhood of 0 by hypothesis; we must prove that it is open. Suppose $t_0 \in S$ and $0 < t_0 < \lambda$. Because c is a local geodesic, there exists a positive number $\varepsilon < \lambda - t_0$ such that $c|_{[t_0-\varepsilon, t_0+\varepsilon]}$ is geodesic. Thus $c([0, t_0 + \varepsilon])$ forms two sides of the geodesic triangle $\Delta = \Delta(c(0), c(t_0), c(t_0 + \varepsilon))$. The comparison triangle $\overline{\Delta} \subset M_\kappa^2$ must be degenerate: if it were not then the CAT(κ) inequality applied to points $x \in [c(0), c(t_0)]$ and $y \in [c(t_0), c(t_0 + \varepsilon)]$ close to $c(t_0)$ would contradict the fact that

$c|_{[t_0-\varepsilon,t_0+\varepsilon]}$ is geodesic. Thus $c|_{[0,t_0+\varepsilon]}$ has length $d(c(0), c(t_0+\varepsilon))$, so $(t_0, t_0+\varepsilon) \subseteq S$ and we are done.

Assertion (3) follows from the CAT(κ) inequality and the fact that balls in M_κ^2 of radius $< D_\kappa/2$ are convex (I.2.12).

(4) follows easily from (1): if $B = \overline{B}(x, \ell)$, where $\ell < D_\kappa$, then the map $B \times [0, 1] \to X$ which associates to (y, t) the point a distance $t\,d(x, y)$ from y along the geodesic segment $[x, y]$ is a continuous retraction of B to x.

In (I.2.25) we proved that approximate midpoints are close to midpoints in M_κ^2, and having noted this, (5) follows immediately from the CAT(κ) inequality for $\Delta(x, y, m')$. □

1.5 Corollary. *For $\kappa \le 0$, any* CAT(κ) *space is contractible, in particular it is simply connected and all of its higher homotopy groups are trivial.*

1.6 Exercise. Let X be a metric space whose metric is convex in the sense of (1.3). Prove that X satisfies statements (1) to (4) of (1.4) with $D_\kappa = \infty$.

Characterizations of CAT(κ) Spaces

The CAT(κ) inequality can be reformulated in a number of different ways, all due to Alexandrov. We note four such reformulations immediately, and in what follows we shall pass freely between these and definition (1.1), using whichever formulation is best suited to the situation at hand and referring to that formulation as the CAT(κ) inequality.

1.7 Proposition. *Fix $\kappa \in \mathbb{R}$. Let X be a metric space that is D_κ-geodesic. The following conditions are equivalent (when $\kappa > 0$ we assume that the perimeter of each geodesic triangle considered is smaller than $2D_\kappa$):*

(1) *X is a* CAT(κ) *space.*

(2) *For every geodesic triangle $\Delta([p, q], [q, r], [r, p])$ in X and every point $x \in [q, r]$, the following inequality is satisfied by the comparison point $\overline{x} \in [\overline{q}, \overline{r}] \subset \overline{\Delta}(p, q, r) \subset M_\kappa^2$:*

$$d(p, x) \le d(\overline{p}, \overline{x}).$$

(3) *For every geodesic triangle $\Delta([p, q], [q, r], [r, p])$ in X and every pair of points $x \in [p, q]$, $y \in [p, r]$ with $x \ne p$ and $y \ne p$, the angles at the vertices corresponding to p in the comparison triangles $\overline{\Delta}(p, q, r) \subset M_\kappa^2$ and $\overline{\Delta}(p, x, y) \subset M_\kappa^2$ satisfy:*

$$\angle_p^{(\kappa)}(x, y) \le \angle_p^{(\kappa)}(q, r).$$

(4) *The Alexandrov angle (as defined in I.1.12) between the sides of any geodesic triangle in X with distinct vertices is no greater than the angle between the corresponding sides of its comparison triangle in M_κ^2.*

(5) *For every geodesic triangle $\Delta([p,q],[p,r],[q,r])$ in X with $p \neq q$ and $p \neq r$, if γ denotes the Alexandrov angle between $[p,q]$ and $[p,r]$ at p and if $\Delta(\hat{p},\hat{q},\hat{r}) \subset M_\kappa^2$ is a geodesic triangle with $d(\hat{p},\hat{q}) = d(p,q)$, $d(\hat{p},\hat{r}) = d(p,r)$ and $\angle_{\hat{p}}(\hat{q},\hat{r}) = \gamma$, then $d(q,r) \geq d(\hat{q},\hat{r})$.*

Proof. It is clear that (1) implies (2) and that (4) is equivalent to (5). And (4) follows immediately from (3) and the observation that one can use comparison triangles in M_κ^2 rather than $\mathbb{E}^2$ when defining the Alexandrov angle (I.2.15). We shall prove that (1) and (3) are equivalent, that (2) implies (3), and that (4) implies (2).

Let p, q, r, x, y be as in (3). We write $\bar{z}$ to denote comparison points in $\overline{\Delta} = \overline{\Delta}(p,q,r) \subset M_\kappa^2$, and $\bar{z}'$ to denote comparison points in $\overline{\Delta} = \overline{\Delta}(p,x,y) \subset M_\kappa^2$. Consider the vertex angles $\bar{\alpha} = \angle_p^{(\kappa)}(q,r)$ and $\bar{\alpha}' = \angle_p^{(\kappa)}(x,y)$ at $\bar{p}$ and $\bar{p}'$. According to the law of cosines (I.2.13), the inequality $d(\bar{x},\bar{y}) \geq d(x,y) = d(\bar{x}',\bar{y}')$ is valid if and only if $\bar{\alpha}' \leq \bar{\alpha}$. Thus (1) and (3) are equivalent.

Maintaining the notation of the previous paragraph, we shall now show that (2) implies (3). Let $\Delta(\bar{p}'',\bar{x}'',\bar{r}'') \subset M_\kappa^2$ be a comparison triangle for $\Delta(p,x,r)$. Let $\bar{\alpha}''$ denote the vertex angle at $\bar{p}''$. By (2), we have $d(x,y) \leq d(\bar{x}'',\bar{y}'')$, where $\bar{y}'' \in [\bar{p}'',\bar{r}'']$ is the comparison point for y. But $d(\bar{x}',\bar{y}') = d(x,y)$, so $\bar{\alpha}' \leq \bar{\alpha}''$. Again by (2), $d(\bar{x}'',\bar{r}'') = d(x,r) \leq d(\bar{x},\bar{r})$, hence $\bar{\alpha}'' \leq \bar{\alpha}$. Thus $\bar{\alpha}' \leq \bar{\alpha}$.

Finally we prove that (4) implies (2). Given a geodesic triangle Δ in X, say $\Delta = \Delta([p,q],[q,r],[r,p])$, and a point $x \in [q,r]$ distinct from q and r, we fix a geodesic segment $[x,p]$ joining x to p in X. Let γ and γ' be the angles at x which $[x,p]$ makes with the subsegments of $[q,r]$ joining x to q, and x to r, respectively. Let β be the angle at q between the geodesic segments $[q,p]$ and $[q,r]$. Let $\Delta(\bar{p},\bar{q},\bar{r})$ be a comparison triangle for $\Delta(p,q,r)$ in M_κ^2 and let $\bar{\beta}$ be the angle at the vertex $\bar{q}$.

Consider comparison triangles $\Delta(\tilde{p},\tilde{x},\tilde{q})$ and $\Delta'(\tilde{p},\tilde{x},\tilde{r})$ in M_κ^2 for the geodesic triangles $\Delta(p,x,q)$ and $\Delta(p,x,r)$ respectively. We choose these comparison triangles in such a way that they share the edge $[\tilde{x},\tilde{p}]$ and $\tilde{q}$ and $\tilde{r}$ lie on opposite sides of the line which passes through $\tilde{p}$ and $\tilde{x}$. Let $\tilde{\gamma} = \angle_{\tilde{x}}(\tilde{p},\tilde{q})$, $\tilde{\gamma}' = \angle_{\tilde{x}}(\tilde{p},\tilde{r})$ and $\tilde{\beta} = \angle_{\tilde{q}}(\tilde{p},\tilde{x})$. By I.1.13(2) we have $\gamma+\gamma' \geq \pi$, hence, by (4), $\tilde{\gamma}+\tilde{\gamma}' \geq \pi$. Alexandrov's lemma implies that $\tilde{\beta} \leq \bar{\beta}$. Therefore, using the law of cosines, we have $d(\bar{p},\bar{x}) \geq d(\tilde{p},\tilde{x}) = d(p,x)$. □

1.8 Other Notions of Angle. Let X be a metric space and suppose that we have a map A which associates a number $A(c,c') \in [0,\pi]$ to each pair of geodesics c, c' in X that have a common initial point. One might regard A as a reasonable notion of angle if for each triple of geodesics c, c' and c'' issuing from a common point we have:

(1) $A(c,c') = A(c',c)$;

(2) $A(c,c'') \leq A(c,c') + A(c',c'')$;

(3) if c is the restriction of c' to an initial segment of its domain then $A(c,c') = 0$;

(4) if $c : [-a,a] \to X$ is a geodesic and $c_-, c_+ : [0,a] \to X$ are defined by $c_-(t) = c(-t)$ and $c_+(t) = c(t)$, then $A(c_-,c_+) = \pi$.

The Alexandrov angle satisfies these conditions (see I.1.13 and 14). The Riemannian angle also satisfies these conditions. And a trivial example is obtained by defining $A(c, c') = \pi$ unless c and c' have a common initial segment.

If in (1.7) one replaces the Alexandrov angle by a function A satisfying the above conditions, then the implications (5) $\iff$ (4) $\implies$ (2) remain valid. This observation provides us with a useful tool for proving that certain spaces are CAT(κ) (see (1A.2) and (10.10)).

1.9 Exercises.

(1) Let X be a geodesic metric space. Prove that the following conditions are equivalent:

(a) X is a CAT(κ) space.

(b) (cf. 1.7(2)) For every geodesic triangle $\Delta([p, q], [q, r], [r, p])$ in X (with perimeter smaller than $2D_\kappa$ if $\kappa > 0$), the point $m \in [q, r]$ with $d(q, m) = d(r, m)$ and its comparison point $\overline{m} \in [\overline{q}, \overline{r}] \subset \overline{\Delta}(p, q, r) \subset M_\kappa^2$ satisfy:

$$d(p, m) \leq d(\overline{p}, \overline{m}).$$

Show further that in the case $\kappa = 0$ the above conditions are equivalent to:

(c) **The CN inequality**[18] **of Bruhat and Tits** [BruT72]. For all $p, q, r \in X$ and all $m \in X$ with $d(q, m) = d(r, m) = d(q, r)/2$, one has:

$$d(p, q)^2 + d(p, r)^2 \geq 2\, d(m, p)^2 + \frac{1}{2}\, d(q, r)^2.$$

(In $\mathbb{E}^2$ one gets equality by a simple calculation with the scalar product.)

(2) Condition (1.7(5)) can be reformulated more analytically using the law of cosines. In the case $\kappa = 0$, if we let $a = d(p, q)$, $b = d(p, r)$, $c = d(q, r)$ and write γ for the Alexandrov angle at p between $[p, q]$ and $p, r]$, then the required condition is:

$$c^2 \geq a^2 + b^2 - 2ab \cos \gamma.$$

Find the equivalent reformulations for $\kappa < 0$ and $\kappa > 0$.

(3) Let X be a CAT(κ) space. If $p, x, y \in X$ are such that $d(x, p) + d(p, y) < D_\kappa$, then the geodesic segment $[x, y]$ is the union of $[x, p]$ and $[p, y]$ if and only if $\angle_p(x, y) = \pi$.

The CAT(κ) 4-Point Condition

All of the reformulations of the CAT(κ) condition given above concern the geometry of triangles. There is also a useful reformulation concerning the geometry of quadrilaterals.

---

[18] Courbure négative

1.10 Definition. A *subembedding in M^2_κ* of a 4-tuple of points (x_1, y_1, x_2, y_2) from a pseudometric space X is a 4-tuple of points $(\overline{x}_1, \overline{y}_1, \overline{x}_2, \overline{y}_2)$ in M^2_κ such that $d(\overline{x}_i, \overline{y}_j) = d(x_i, y_j)$ for $i, j \in \{1, 2\}$, and $d(x_1, x_2) \leq d(\overline{x}_1, \overline{x}_2)$ and $d(y_1, y_2) \leq d(\overline{y}_1, \overline{y}_2)$.

X is said to satisfy the CAT(κ) 4-point condition if every 4-tuple of points (x_1, y_1, x_2, y_2) with $d(x_1, y_1,) + d(y_1, x_2) + d(x_2, y_2) + d(y_2, x_1) < 2D_\kappa$ has a subembedding in M^2_κ.

Note that every subspace of a CAT(κ) space X satisfies the CAT(κ) 4-point condition. This condition first appears in the work of Reshetnyak [Resh68]. It was used extensively by Korevaar and Schoen in their work on harmonic maps [KS93], and also by Nikolaev [Ni95].

Recall that a pair of points x, y in a metric space X is said to have *approximate midpoints* if for every $\delta > 0$ there exists $m' \in X$ such that $\max\{d(x, m'), d(y, m')\} < \frac{1}{2}d(x, y)+\delta$. If X is complete and every pair of points in X has approximate midpoints, then X is a length space.

1.11 Proposition. *Let X be a complete metric space. The following conditions are equivalent:*

(1) *X is a* CAT(κ) *space.*

(2) *X satisfies the* CAT(κ) *4-point condition and every pair of points $x, y \in X$ with $d(x, y) < D_\kappa$ has approximate midpoints.*

Proof. We first prove that (1) implies (2). The existence of geodesics in X ensures the existence of approximate midpoints. In order to show that X satisfies the CAT(κ) 4-point condition we must construct a subembedding in M^2_κ of each 4-tuple (x_1, y_1, x_2, y_2) from X with $d(x_1, y_1,) + d(y_1, x_2) + d(x_2, y_2) + d(y_2, x_1) < 2D_\kappa$.

Given such a 4-tuple, we consider the quadrilateral $Q \subset M^2_\kappa$ formed by comparison triangles $\Delta(\tilde{x}_1, \tilde{x}_2, \tilde{y}_1) = \overline{\Delta}(x_1, x_2, y_1)$ and $\Delta(\tilde{x}_1, \tilde{x}_2, \tilde{y}_2) = \overline{\Delta}(x_1, x_2, y_2)$ with a common edge $[\tilde{x}_1, \tilde{x}_2]$ and with $\tilde{y}_1$ and $\tilde{y}_2$ on opposite sides of the line containing $[\tilde{x}_1, \tilde{x}_2]$.

If Q is convex then the diagonals $[\tilde{x}_1, \tilde{x}_2]$ and $[\tilde{y}_1, \tilde{y}_2]$ intersect in some point $\tilde{z}$. Let $z \in [x_1, x_2]$ be such that $d(x_1, z) = d(\tilde{x}_1, \tilde{z})$. From the CAT($\kappa$) inequality (and triangle inequality) we have:

$$d(y_1, y_2) \leq d(y_1, z) + d(z, y_2) \leq d(\tilde{y}_1, \tilde{z}) + d(\tilde{z}, \tilde{y}_2) = d(\tilde{y}_1, \tilde{y}_2).$$

By construction, $d(x_1, x_2) = d(\tilde{x}_1, \tilde{x}_2)$, so if Q is convex then $(\tilde{x}_1, \tilde{y}_1, \tilde{x}_2, \tilde{y}_2)$ is a subembedding of (x_1, y_1, x_2, y_2) in M^2_κ.

If Q is not convex then one of the $\tilde{x}_i$, say $\tilde{x}_2$, is in the interior of the convex hull of the other three vertices of Q. By applying Alexandrov's lemma (I.2.16) we obtain a quadrilateral in M^2_κ with vertices $\overline{x}_1, \overline{y}_1, \overline{x}_2, \overline{y}_2$ where $d(\overline{x}_i, \overline{y}_j) = d(x_i, y_j)$ for $i, j = 1, 2$ and $\overline{x}_2 \in [\overline{y}_1, \overline{y}_2]$ is such that $d(\overline{x}_1, \overline{x}_2) > d(\tilde{x}_1, \tilde{x}_2) = d(x_1, x_2)$. And $d(\overline{y}_1, \overline{y}_2) = d(\overline{y}_1, \overline{x}_2) + d(\overline{x}_2, \overline{y}_2) = d(y_1, x_2) + d(x_2, y_2) \geq d(y_1, y_2)$. Therefore $(\overline{x}_1, \overline{y}_1, \overline{x}_2, \overline{y}_2)$ is a subembedding of (x_1, y_1, x_2, y_2).

We shall now show that (2) implies (1). As a first step, we prove that (2) implies the existence of midpoints for pairs of points less than D_κ apart; as X is complete, the existence of geodesics follows immediately. We fix $x, y \in X$ with $d(x, y) < D_\kappa$ and consider a sequence of approximate midpoints m_i with $\max\{d(x, m_i), d(y, m_i)\} \le \frac{1}{2}d(x, y) + \frac{1}{i}$. If we can prove that this is a Cauchy sequence then the limit will be a midpoint for x and y. We fix an arbitrary $\varepsilon > 0$ and fix λ between $d(x, y)$ and D_κ.

According to (I.2.25), there exists $\delta = \delta(\kappa, \lambda, \varepsilon)$ such that in M^2_κ, if $d(p, q) \le l$ and $\max\{d(p, m'), d(q, m')\} \le \frac{1}{2}d(p, q) + \delta$, then $d(m, m') < \varepsilon$, where m is the midpoint of $[p, q]$.

If $(\overline{x}, \overline{m}_i, \overline{y}, \overline{m}_j)$ is a subembedding of (x, m_i, y, m_j), then by definition $d(\overline{m}_i, \overline{m}_j) \ge d(m_i, m_j)$. Also, $d(\overline{x}, \overline{y}) \le d(\overline{x}, \overline{m}_i) + d(\overline{m}_i, \overline{y}) = d(x, m_i) + d(m_i, y) \le d(x, y) + 1/i$. Thus, for i and j sufficiently large we have $d(\overline{x}, \overline{y}) \le \lambda$. And if $\max\{1/i, 1/j\} < \delta(\kappa, \lambda, \varepsilon)$ then $d(\overline{m}_i, m) < \varepsilon$ and $d(\overline{m}_j, m) < \varepsilon$, where m is the midpoint of $[\overline{x}, \overline{y}]$. Since $d(\overline{m}_i, \overline{m}_j) \ge d(m_i, m_j)$, we have proved that m_i is a Cauchy sequence, as required.

It remains to prove that triangles in X satisfy the CAT(κ) inequality. Consider a triangle $\Delta(x, y, z) \subset X$ of perimeter less than $2D_\kappa$ and let m be a point on $[x, y]$. Let $(\overline{z}, \overline{x}, \overline{m}, \overline{y})$ be a subembedding of (z, x, m, y) in M^2_κ. As $d(x, y) \le d(\overline{x}, \overline{y}) \le d(\overline{x}, \overline{m}) + d(\overline{m}, \overline{y}) = d(x, y)$, we see that $\Delta(\overline{x}, \overline{y}, \overline{z})$ is a comparison triangle for $\Delta(x, y, z)$ and $\overline{m}$ is the comparison point for $m \in [\overline{x}, \overline{y}]$. By the definition of subembedding, $d(z, m) \le d(\overline{z}, \overline{m})$. Thus $\Delta(x, y, z)$ satisfies the CAT(κ) inequality (as characterized in 1.7(2)). □

CAT(κ) Implies CAT(κ') if $\kappa \le \kappa'$

The following basic theorem shows in particular that real hyperbolic space $\mathbb{H}^n$ is a CAT(0) space.

1.12 Theorem.

(1) *If X is a* CAT(κ) *space, then it is a* CAT(κ') *space for every $\kappa' \ge \kappa$.*

(2) *If X is a* CAT(κ') *space for every $\kappa' > \kappa$, then it is a* CAT(κ) *space.*

Proof. We first prove (2). Given $x, y \in X$ with $d(x, y) < D_\kappa$, we have $d(x, y) < D_{\kappa'}$ for all $\kappa' > \kappa$ sufficiently close to κ. Thus if X is $D_{\kappa'}$-geodesic for all $\kappa' > \kappa$, then it is D_κ-geodesic.

Given a geodesic triangle $\Delta = \Delta(p, q, r)$ in X with perimeter $< 2D_\kappa$, we choose $\kappa' > \kappa$ sufficiently close to κ to ensure that the perimeter of Δ is less than $2D_{\kappa'}$. Let $a = d(p, q)$, $b = d(p, r)$ and $c = d(q, r)$. Let γ be the Alexandrov angle at p between $[p, q]$ and $[p, r]$. In the light of the law of cosines for M^2_κ, if $\kappa \ge 0$ then characterization (1.7(5)) of the CAT(κ') inequality becomes:

$$\cos(c\sqrt{\kappa'}) \le \cos(a\sqrt{\kappa'})\cos(b\sqrt{\kappa'}) + \sin(a\sqrt{\kappa'})\sin(b\sqrt{\kappa'})\cos(\gamma).$$

If $\kappa > 0$, then passing to the limit we get the same inequality with κ' replaced by κ. If $\kappa = 0$, then in the limit we get $c^2 \le a^2 + b^2 - 2ab\cos\gamma$. Thus we obtain the CAT(κ) inequality (in the guise of 1.7(5)). The case $\kappa < 0$ is similar.

We now prove (1). If $\kappa' \geq \kappa$, then since X is assumed to be D_κ-geodesic it is *a fortiori* $D_{\kappa'}$-geodesic.

Given a geodesic triangle in X of perimeter less than D_κ we compare its comparison triangles $\Delta_\kappa \subset M_\kappa^2$ and $\Delta_{\kappa'} \subset M_{\kappa'}^2$. It is sufficient to prove that the angles at the vertices of Δ_κ are not smaller than the angles at the corresponding vertices of $\Delta_{\kappa'}$. Using the law of cosines in M_κ^2, we reduce to proving the following lemma. □

1.13 Lemma. *Fix $\kappa' > \kappa$ and let Δ (resp. Δ') be a geodesic triangle in M_κ^2 (resp. $M_{\kappa'}^2$) with vertices A, B, C (resp. A', B', C') and opposite sides of length a, b, c (resp. a, b, c'), where the side opposite A has length a etc.. Suppose $a+b+c < 2D_{\kappa'}$, and suppose that the angles at the vertices C and C' are equal and lie in $(0, \pi)$. Then $c' < c$.*

Proof. We introduce polar coordinates (r, θ) in M_κ^2 and $M_{\kappa'}^2$ (in open balls of radius $D_{\kappa'} = \pi/\sqrt{\kappa'}$ if $\kappa' > 0$), see (I.6.16). Consider the map h sending the point of $M_{\kappa'}^2$ with polar coordinates (r, θ) to the point with the same coordinates in M_κ^2. We take C and C' to be the centres of these polar coordinates (so $h(C') = C$). Because the derivative of h at the origin is an isometry, we may assume $h(A') = A$ and $h(B') = B$. The image of h is the ball centred at C with radius $D_{\kappa'}$, hence it contains the triangle Δ.

We wish to show that h increases the length of every path that is not radial in the polar coordinates. For this we must show that if v is a vector in the tangent space $T_x M_{\kappa'}^2$ at a point x in the domain of h, then the norm of its image $T_x h(v)$ by the differential $T_x h$ of h at x is no smaller than the norm of v, and if v is not radial then $\|T_x h(v)\| > \|v\|$.

Recall (I.6.17) that in the polar coordinates on M_κ^2 the Riemannian metric is given by

$$ds^2 = dr^2 + f(\kappa, r)^2 d\theta^2,$$

where the function $\kappa \mapsto f(\kappa, r)$ is defined on the interval $(-\infty, (\pi/r)^2)$ by

$$f(\kappa, r) = \begin{cases} \frac{1}{\sqrt{-\kappa}} \sinh(\sqrt{-\kappa}\, r) & \text{if } \kappa < 0 \\ r & \text{if } \kappa = 0 \\ \frac{1}{\sqrt{\kappa}} \sin(\sqrt{\kappa}\, r) & \text{if } \kappa > 0. \end{cases}$$

It is easy to show that the function $\kappa \mapsto f(\kappa, r)$ is continuous and strictly decreasing.

Let $v = v_r \frac{\partial}{\partial r}(r, \theta) + v_\theta \frac{\partial}{\partial \theta}(r, \theta)$ be a tangent vector at a point x with polar coordinates (r, θ) in the domain of h; the square of its norm is $v_r^2 + f(\kappa', r)^2 v_\theta^2$. The image of v under the derivative of h has the same expression in the polar coordinates on M_κ^2 and the square of its norm is $v_r^2 + f(\kappa, r)^2 v_\theta^2$, hence $\|T_x h(v)\|^2 \geq \|v\|^2$ with equality only if $v_\theta = 0$. Thus $T_x h$ strictly increases the norm of vectors which are not radial, and so h strictly increases the length of paths which are not radial. It follows that the segment $[A, B] \subset \Delta$, which has length c, is the image under h of a path from A' to B' of length less than c; but $c' = d(A', B')$, so $c' < c$. □

Simple Examples of CAT(κ) Spaces

Pre-Hilbert spaces are obviously CAT(0). Conversely we have:

1.14 Proposition. *If a normed real vector space V is* CAT(κ) *for some $\kappa \in \mathbb{R}$, then it is a pre-Hilbert space.*

Proof. If V is not a pre-Hilbert space then, as in (I.4.5), there exist $u, v \in V$ such that $\lim_{t\to 0} \overline{\angle}_0(tu, tv) = \lim_{t\to 0} \angle_0^{(\kappa)}(tu, tv)$ does not exist, whereas 1.7(3) implies that if V were CAT(κ) then this limit would exist. $\square$

1.15 Examples

(1) When endowed with the induced metric, a convex subset of Euclidean space $\mathbb{E}^n$ is CAT(0). More generally, a subset of a CAT(κ) space, equipped with the induced metric, is CAT(κ) if and only if it is D_κ-convex. Here, it is important to distinguish between the induced metric and the induced length metric: many non-convex subsets of $\mathbb{E}^n$ are CAT(0) spaces when endowed with the induced length metric.

(2) Let X be the planar set which is the complement of the quadrant $\{(x, y) \mid x > 0, y > 0\}$, endowed with the induced length metric from $\mathbb{E}^2$. In (I.3.5) we described the geodesics in this space. It follows immediately from this description and Alexandrov's lemma (I.2.16) that X satisfies the angle criterion 1.7(4) for a CAT(0) space.

A similar argument applies to the complement of any sector in the plane, and a local version of the same argument shows that the complement of any polygon in M_κ^2 has curvature $\leq \kappa$.

(3) If X_1 and X_2 are CAT(0) spaces, then their product $X_1 \times X_2$ (as defined in (I.5.1)) is also a CAT(0) space. This is easily seen using the characterization of CAT(0) spaces given in (1.9(1c)).

(4) A metric simplicial graph is CAT(κ) if and only if it does not contain any essential loops of length less than $2D_\kappa$; in other words, every locally injective path of length less than $2D_\kappa$ is injective.

(5) An $\mathbb{R}$–*tree* is a metric space T such that:

(i) there is a *unique* geodesic segment (denoted $[x, y]$) joining each pair of points $x, y \in T$;

(ii) if $[y, x] \cap [x, z] = \{x\}$, then $[y, x] \cup [x, z] = [y, z]$.

$\mathbb{R}$–trees are CAT(κ) spaces for every κ. Indeed in such spaces every geodesic triangle with distinct vertices x_1, x_2, x_3 is degenerate in the sense that there exists $v \in T$ such that $[x_i, x_j] = [x_i, v] \cup [v, x_j]$ if $i \neq j$. In particular, the angle at each vertex is either 0 or π, and a vertex angle of π occurs if and only if one vertex lies on the geodesic segment joining the other two vertices. Thus (1.7(4)) holds for all $\kappa \in \mathbb{R}$. Conversely, any metric space which is CAT(κ) for all κ is an $\mathbb{R}$–tree.

Simply connected metric simplicial graphs, as defined in (I.1.10), are the easiest examples of $\mathbb{R}$-trees. In general an $\mathbb{R}$-tree cannot be made simplicial. For example, consider the set $[0, \infty) \times [0, \infty)$ with the distance $d(x, y)$ between two points $x = (x_1, x_2)$ and $y = (y_1, y_2)$ defined by $d(x, y) = x_1 + y_1 + |x_2 - y_2|$ if $x_2 \neq y_2$ and $d(x, y) = |x_1 - y_1|$ if $x_2 = y_2$.

The asymptotic cone of $\mathbb{H}^n$ provides another non-simplicial example. In this case, the complement of every point in the $\mathbb{R}$-tree has infinitely many connected components.

The apparently trivial nature of the geometry of $\mathbb{R}$-trees is deceptive and belies their importance in low-dimensional topology and the delicacy of questions concerning their isometry groups. We refer the reader to [Sha91] for a survey to 1991 and [Pau96] for an account of recent developments in this area.

1.16 Exercises.

(1) Let X be the closed subset of $\mathbb{E}^3$ which is the complement of the octant $\{(x, y, z) \mid x > 0, y > 0, z > 0\}$. Show that X, endowed with the induced length metric, is *not* a CAT(0) space (or indeed a CAT(κ) space for any κ).

(2) Let $\kappa \geq 0$. Prove that $X_1 \times X_2$ is a CAT(κ) space if and only if both X_1 and X_2 are CAT(κ) spaces.

(3) Give a second proof of (1.14) by showing that if a normed space satisfies the CAT(κ) inequality then the norm satisfies the parallelogram law.

(4) Let C be a geodesic segment in M_κ^2. Consider $M_\kappa^2 \smallsetminus C$ equipped with the induced length metric and let X denote its metric completion. Prove using Alexandrov's lemma that X has curvature $\leq \kappa$ but that it is only a CAT(κ) space if $\kappa > 0$ and C has length D_κ.

(5) Prove that when endowed with the induced length metric, the complement of an open horoball in $\mathbb{H}^2$ is a CAT(-1) space. (Hint: Use criterion 1.7(4).)

Historical Remarks

Prior to the work of A.D. Alexandrov, H. Busemann had introduced a weaker notion of curvature in metric spaces [Bus48,55]. The book of W. Rinow [Rin61] provides good historical references for the period to 1960, and the report of K. Menger [Men52] explains the importance of the early work of Wald [Wa36]. We close this chapter with two remarks concerning the work of Alexandrov and Busemann.

1.17 The Relative Excess of a Triangle. Following Alexandrov, given $\kappa \in \mathbb{R}$ and a geodesic triangle $\Delta = \Delta([p, q], [q, r], [r, p])$ in a metric space X with perimeter smaller than Δ_k, we define the *relative excess* $\delta_\kappa(\Delta)$ to be the difference of the sum of the vertex angles of Δ and the sum of the corresponding angles for a comparison triangle in M_κ^2.

It follows immediately from (1.7(4)) that $\delta_\kappa(\Delta) \leq 0$ for any geodesic triangle Δ in a CAT(κ) space. Alexandrov proved the non-trivial fact that the converse is also true: in a geodesic space X, if the relative excess δ_κ of every geodesic triangle is non-positive, then X is a CAT(κ) space. The proof of this result requires a rather long argument (see Alexandrov [Ale57a]).

Alexandrov shows also that a locally geodesic metric space X is of curvature $\leq \kappa$ if and only if, for each point $x \in X$ and each geodesic triangle Δ in a neighbourhood of x, the inequality $\delta_0(\Delta)/A \leq \kappa$ holds, where A is the area of a comparison triangle for Δ in $\mathbb{E}^2$.

1.18 Busemann's Approach to Curvature. Busemann [Bus48] made an extensive study of spaces X whose metric is *locally convex* in the sense that every point has a neighbourhood on which the induced metric is convex in the sense (1.3). (Equivalently, every point $x \in X$ has an open neighbourhood which is a geodesic space and for every geodesic triangle $\Delta = \Delta([p,q],[q,r],[r,p])$ in U the distance between the midpoints of $[p,q]$ and $[p,r]$ is not bigger than the distance between the comparison points in a comparison triangle $\overline{\Delta} \subset \mathbb{E}^2$.) The metric on any space of non-positive curvature is locally convex, but not conversely. For example, if V is a normed space whose unit ball is strictly convex in the sense of (I.1.4), then the metric on V is convex, whereas V is non-positively curved if and only if it is a pre-Hilbert space (cf. 1.14).

More generally one might define a geodesic space X to be of curvature $\leq k$ (in a weak sense) if every point of X has a neighbourhood such that given any three points p, q, r in this neighbourhood (with $d(p,q) = d(q,r) + d(r,p) < D_\kappa$) and geodesics $[p,q]$ and $[p,r]$, the midpoints $x \in [p,q]$ and $y \in [p,r]$ and their comparison points in $\Delta(\overline{p},\overline{q},\overline{r}) \subset M_\kappa^2$ satisfy the inequality $d(x,y) \leq d(\overline{x},\overline{y})$. We shall see in the Appendix that this condition is equivalent to Alexandrov's definition of curvature $\leq k$ if X is a sufficiently smooth Riemannian manifold.

Appendix: The Curvature of Riemannian Manifolds

The purpose of this appendix is to relate Alexandrov's definition of upper curvature bounds (1.2) to the classical notion of sectional curvature. In what follows M will be a smooth Riemannian manifold of class C^3 and dimension n equipped with its associated distance function d, as in (I.3.18). We begin with a technical criterion for deciding when a Riemannian manifold is of curvature $\leq \kappa$.

1A.1 Lemma. *Fix $\kappa \in \mathbb{R}$, $x \in M$ and $o \in M_\kappa^n$. Suppose that for a suitable positive number $\varepsilon < D_\kappa$ there exists a diffeomorphism e from $B(o,\varepsilon)$ onto an open set $U \subseteq M$ such that $e(o) = x$ and*

(a) *for all $y \in B(o,\varepsilon)$ and $v \in T_yM_\kappa^n$, we have $|T_ye(v)| \geq |v|$, where $T_ye : T_yM_\kappa^n \to T_{e(y)}M$ denotes the differential of e at y;*

(b) *$|T_ye(v)| = |v|$ if v is tangent to the geodesic joining o to y; in particular $T_oe : T_oM_\kappa^n \to T_xM$ is an isometry.*

Then:

(1) $U = B(x, \varepsilon)$. *Indeed, for each* $y \in B(o, \varepsilon)$, *the geodesic segment* $[o, y]$ *is mapped isometrically by* e *onto a geodesic segment joining* $x = e(0)$ *to* $e(y)$ *in* M, *and this segment is unique;*

(2) *for all* $y, z \in B(o, \varepsilon/2)$, *we have* $d(e(y), e(z)) \geq d(y, z)$.

Proof. As in (I.3.17), the Riemannian length of a curve c will be denoted $l_R(c)$. The hypotheses of the lemma imply that for any piecewise C^1 curve $c : [0, 1] \to B(o, \varepsilon)$, we have $l_R(e \circ c) \geq l_R(c)$ with equality if c is the geodesic path joining o to y. This together with the fact that any piecewise C^1 curve in M joining x to a point outside of U has length at least ε, implies that $d(o, y) = d(x, e(y))$, that $U = B(x, \varepsilon)$, and that the image under e of the geodesic segment $[o, y]$ is the unique geodesic segment joining x to $e(y)$.

If $y, z \in B(o, \varepsilon/2)$, then $d(e(y), e(z)) < \varepsilon$ is the infimum of the lengths $l_R(c)$ of piecewise C^1 curves c of length $< \varepsilon$ joining $e(y)$ to $e(z)$. Any such curve is the image under e of a piecewise C^1 curve joining y to z whose length (by (a)) is not bigger than $l_R(c)$. Hence $d(e(y), e(z)) \leq d(y, z)$. This proves (2). □

1A.2 Proposition. *Let* M *be a Riemannian manifold and let* $\kappa \in \mathbb{R}$. *Let* $p \in M$ *and suppose that* $\varepsilon > 0$ *is such that for each point* $x \in B(p, \varepsilon/2)$ *one can find a map* $e_x : B(o, \varepsilon) \to B(x, \varepsilon)$ *satisfying the hypotheses of (1A.1). Then* $B(p, \varepsilon/2)$ *is a CAT(κ) space.*

Proof. Let $B = B(p, \varepsilon/2)$. Lemma 1A.1 implies that $B \subseteq M$ is convex and uniquely geodesic. It also implies that the Riemannian angles at the vertices of any geodesic triangle $\Delta(x, y, z) \subset B$ with distinct vertices are no greater than the corresponding angles in a comparison triangle $\overline{\Delta} = \Delta(\overline{x}, \overline{y}, \overline{z}) \subset M_\kappa^2$. As remarked in (1.8), this implies that B satisfies condition (2) of (1.7). □

1A.3 Jacobi Fields and Sectional Curvature. We recall some classical constructions and facts from Riemannian geometry; these are proved in any textbook on the subject (e.g. [Mil63] or [ChEb75]). We consider a Riemannian manifold M that is smooth enough (class C^3 suffices) to ensure that the constant speed local geodesics $t \mapsto c(t)$ in M are the solutions of the differential equation

$$\frac{D}{dt}\dot{c}(t) = 0,$$

where $\frac{D}{dt}$ denotes the covariant derivative applied to smooth vector fields along the path c, and $\dot{c}(t) \in T_{c(t)}M$ is the velocity vector of c at t.

For each vector $v \in T_xM$ whose norm $|v|$ is small enough, there is a unique constant speed local geodesic $c_v : [0, 1] \to M$ such that $c_v(0) = x$ and $\dot{c}_v(0) = v$; the point $c_v(1)$ is denoted $\exp(v)$. We shall need the following classical fact: every point of M has a neighbourhood U for which one can find $\varepsilon > 0$ such that exp is defined on $T^\varepsilon U = \cup_{x\in U} T^\varepsilon_x M$, where $T^\varepsilon_x = \{v \in T_xM|\ |v| < \varepsilon\}$; the restriction $\exp_x$ of exp to each $T^\varepsilon_x M$ is a diffeomorphism onto an open subset of M and for each $x \in U$ and $v \in T^\varepsilon_x M$, the path c_v is the unique constant speed geodesic $[0, 1] \to X$ joining x to $\exp(v)$.

Given $x \in U$, a non-zero vector $u \in T^\varepsilon_x M$ and $v \in T_xM$, there is a one-parameter family of constant speed geodesics c_{u+sv}, where s varies over a small neighbourhood of $0 \in \mathbb{R}$. Associated to this family of geodesics one has the vector field $J(t) = \frac{d}{ds} c_{t(u+sv)}(t)|_{s=0}$ along the geodesic path $c = c_u$. This satisfies the second order differential equation

$$\frac{D^2}{dt^2} J(t) = (-R(J(t), \dot{c}(t))) \left(\dot{c}(t)\right),$$

with initial conditions $J(0) = 0$ and $\frac{D}{dt}J(0) = v = \lim_{t\to 0} J(t)/t$, where $R(\ ,\ ) \in \mathrm{End}(TM)$ is the curvature tensor. Any vector field satisfying such an equation is called a *Jacobi vector field* along the geodesic path c. If $J(0) = 0$ and $\frac{D}{dt}$ is orthogonal to $\dot{c}(0)$, then $J(t)$ is orthogonal to $\dot{c}(t)$ for every t.

If $J(t) \neq 0$ is orthogonal to $\dot{c}(t)$, the *sectional curvature* of M along the 2-plane in $T_{c(t)}M$ spanned by the orthogonal vectors $J(t)$ and $\dot{c}(t)$ is by definition the number

$$K(t) = \frac{(R(J(t), \dot{c}(t))\dot{c}(t) \mid J(t))}{|J(t)|^2\ |\dot{c}(t)|^2}.$$

Given $\kappa \in \mathbb{R}$, if the sectional curvature of M along every 2-plane in TM is $\leq \kappa$, then one says *the sectional curvature of M is $\leq \kappa$.*

The following lemma is due to H. Rauch [Rau51].

1A.4 Lemma. *Let J be a Jacobi vector field along a unit speed geodesic $c : [0, \varepsilon] \to M$. If $J(t)$ is orthogonal to $\dot{c}(t)$ for all t, then*

$$|J|''(t) \geq K(t)\, |J|(t)$$

at each point where $J(t) \neq 0$.

If $K(t) \leq \kappa$, $J(0) = 0$ and $|\frac{D}{dt}J|(0) = 1$, then for every $t < D_\kappa$ in the domain of c we have

$$|J|(t) \geq j_\kappa(t),$$

where j_κ is the solution of the differential equation $j''_\kappa(t) = -\kappa j_\kappa(t)$ with initial conditions $j_\kappa(0) = 0, j'_\kappa(0) = 1$.

Proof. For any vector fields $X(t)$ and $Y(t)$ along c, we have $\frac{d}{dt}(X(t) \mid Y(t)) = (\frac{D}{dt}X(t) \mid Y(t)) + (X(t) \mid \frac{D}{dt}Y(t))$. Thus we may write

$$|J|'(t) = \frac{d}{dt} (J(t) \mid J(t))^{1/2} = \left(\frac{D}{dt}J(t) \mid J(t)\right)\ |J(t)|^{-1},$$

hence

$$\begin{aligned}|J|''(t) = &\left(\frac{D^2}{dt^2}J(t) \mid J(t)\right) |J(t)|^{-1} + \left(\frac{D}{dt}J(t) \mid \frac{D}{dt}J(t)\right) |J(t)|^{-1} \\ &- \left(\frac{D}{dt}J(t) \mid J(t)\right)^2 |J(t)|^{-3} \\ = &- (R(J(t), \dot{c}(t))\dot{c}(t) \mid J(t)) \, |J(t)|^{-3} \\ &+ \left(|\frac{D}{dt}J(t)|^2 |J(t)|^2 - (\frac{D}{dt}J(t) \mid J(t))^2\right) |J(t)|^{-3}.\end{aligned}$$

Therefore by the Cauchy-Schwarz inequality

$$|J|''(t) \geq K(t)\, |J|(t).$$

Now assume $K(t) \leq \kappa$, $J(0) = 0$ and $|\frac{D}{dt}J|(0) = 1$. The solution of the differential equation $j''_\kappa(t) = -\kappa j_\kappa(t)$ with initial conditions $j_\kappa(0) = 0, j'_\kappa(0) = 1$ is $j_\kappa(t) = \frac{1}{\sqrt{\kappa}} \sin(\sqrt{\kappa} t)$ if $\kappa > 0$; it is $j_k(t) = t$ if $\kappa = 0$; and $j_\kappa(t) = \frac{1}{\sqrt{-\kappa}} \sinh(\sqrt{-\kappa} t)$ if $\kappa < 0$.

The first part of the lemma implies that

$$\left(|J|'(t) j_\kappa(t) - |J|(t) j'_\kappa(t)\right)' \geq 0,$$

and hence $|J|'(t) j_\kappa(t) - |J|(t) j'_\kappa(t) \geq 0$ (because this function of t vanishes at $t = 0$). In other words, as $j_\kappa(t) > 0$ for $0 < t < D_\kappa$,

$$(*) \qquad \frac{|J|'(t)}{|J|(t)} \geq \frac{j'_\kappa(t)}{j_\kappa(t)}$$

for $0 < t < D_\kappa$.

By de l'Hôpital's rule,

$$\lim_{t \to 0} \frac{|J|(t)}{j_\kappa(t)} = \frac{|J|'(0)}{j'_\kappa(0)} = \frac{\lim_{t \to 0} |J(t)|/t}{1} = 1.$$

By integrating $(*)$ we get $|J(t)| \geq j_\kappa(t)$. □

1A.5 Lemma. *Let M be as in (1A.3). If M has sectional curvature $\leq \kappa$, then for every $p \in M$ there exists a neighbourhood V of p and a positive number $\varepsilon > 0$ such that for all $x \in V$ one can find a map $e : B(o, \varepsilon) \to M$ satisfying the hypotheses of Lemma 1A.1.*

Proof. Given a compact neighbourhood V of p one can find a positive number ε such that, for all $x \in V$, the exponential map $\exp_x : T_x^\varepsilon M \to B(x, \varepsilon)$ is well defined and is a diffeomorphism. Fix $o \in M_\kappa^n$ and $x \in V$ and identify $T_o M_\kappa^n$ with $T_x M$ by means of a linear isometry. Define $e = \exp_x \circ \exp_o^{-1}$. This map satisfies condition (b) of Lemma 1A.1.

Let $u, v \in T_o M_\kappa^n$ be such that $|u| = 1, |v| = 1$ and $(v \mid u) = 0$. Let $J_\kappa(t)$ (resp. $J(t)$) be the Jacobi field that is associated to the one-parameter family of geodesics $c_{t(u+sv)}(t) = \exp_o(t(u + sv))$ (resp. $\exp_x(t(u + sv))$ as in (1A.3), where s varies over a small neighbourhood of $0 \in \mathbb{R}$. By construction the differential of e

maps $J_\kappa(t)$ to $J(t)$. It follows immediately from the description of the Riemannian metric on M_κ^2 in terms of polar coordinates (cf. I.6.17), that $j_\kappa(t) = |J_\kappa|(t)$, and hence the second part of Lemma 1A.4 implies that e satisfies condition (a) of (1A.1). □

The following theorem is due to Alexandrov [Ale51]. The fact that non-positive sectional curvature implies non-positive curvature in the sense of (1.2) was proved by E. Cartan [Car28].

1A.6 Theorem. *A smooth Riemannian manifold M is of curvature $\leq \kappa$ in the sense of Alexandrov (1.2) if and only if the sectional curvature of M is $\leq \kappa$.*

Proof. (1A.5) and (1A.2) together imply that if the sectional curvature of M is bounded above by κ, then M is of curvature $\leq \kappa$ in the sense of Alexandrov.

To prove the converse we use the following classical fact [Car28]. Let $x \in M$ and let u, v be orthogonal unit vectors in T_xM. For t small enough, we have unique geodesics $t \mapsto c_u(t)$ and $c_v(t)$ issuing from x with $\dot{c}_u(0) = u$ and $\dot{c}_v(0) = v$. Let K be the sectional curvature of M along the 2-plane in T_xM spanned by u and v. Let $d(\varepsilon) = d(c_u(\varepsilon), c_v(\varepsilon))$. Then[19]

$$d(\varepsilon)^2 = 2\varepsilon^2 - \frac{K}{6}\varepsilon^4 + O(\varepsilon^5).$$

Consider a geodesic triangle in M_κ^2 with two orthogonal sides of length ε; let $c(\varepsilon)$ be the length of the third side. As M is assumed to be of curvature $\leq \kappa$, we have $d(\varepsilon) \geq c(\varepsilon)$ (see (1.7(5)) and (1A.7)). If $\kappa = 0$, then $d(\varepsilon)^2 \geq c(\varepsilon)^2 = 2\varepsilon^2$, hence $K \leq 0$. If $\kappa < 0$, then using the law of cosines we can express $c(\varepsilon)$ as $\cosh(\sqrt{-\kappa}\, c(\varepsilon)) = \cosh^2(\sqrt{-\kappa}\,\varepsilon)$. The inequality $d(\varepsilon) \geq c(\varepsilon)$ implies $\cosh(\sqrt{-\kappa}\, d(\varepsilon)) \geq \cosh(\sqrt{-\kappa}\, c(\varepsilon))$, hence

$$1 - \kappa\varepsilon^2 + \frac{1}{6}(\kappa K + \kappa^2)\,\varepsilon^4 + O(\varepsilon^5) \geq 1 - \kappa\varepsilon^2 + \frac{1}{3}\kappa^2\varepsilon^4 + O(\varepsilon^5),$$

which implies $K \leq \kappa$. The case $\kappa > 0$ is treated similarly. □

The "if" implication implies the following result.

1A.7 Corollary. *Let M be a smooth Riemannian manifold and let $c, c' : [0, \varepsilon] \to M$ be geodesics issuing from a common point $x = c(0) = c'(0)$. The limit as $t, t' \to 0$ of the comparison angle $\overline{\angle}_x(c(t), c'(t'))$ exists and is equal to the Riemannian angle between the velocity vectors $\dot{c}(0)$ and $\dot{c}'(0)$. In other words, the Riemannian angle between c and c' is equal to the Alexandrov angle between them.*

Proof. As M is locally compact, the point x is contained in a neighbourhood where the sectional curvature of M is bounded by some number κ. Hence x has a convex neighbourhood that is a CAT(κ) space. It follows for 1.7(3) that the angle $\angle_x^{(\kappa)}(c(t), c'(t'))$

[19] A proof in modern language can be found in unpublished lecture notes of Wolfang Meyer on "Toponogov's Theorem and Applications".

at $\bar{x}$ in a comparison triangle $\overline{\Delta}(x, c(t), c('t')) \subset M_\kappa^2$ is a non-increasing function of $t, t' > 0$. Hence the desired limit exists — call it α. It follows from (I.2.9) that $\angle_x(c(t), c'(t'))$ also converges to α as $t, t' \to 0$. To compute α we can take $t = t'$; observe that $2\sin(\overline{\angle}(c(t), c'(t))/2) = d(c(t), c'(t))/t$. Let $\exp_x$ be the exponential map defined on a neighbourhood of $0 \in T_xM$. There are unit vectors $u, u' \in T_xM$ such that $c(t) = \exp_x(tu)$ and $c'(t) = \exp_x(tu')$; we have to prove that α is the angle between u and u', namely that $2\sin(\alpha/2) = ||u - u'||$. As the differential of $\exp_x$ at 0 is an isometry on the tangent spaces, the argument used in the proof of (I.3.18) shows that if t is small enough, then $\lim_{t\to 0} d(c(t), c'(t))/||tu - tu'|| = 1$, hence $2\sin(\alpha/2) = \lim_{t\to 0} d(c(t), c'(t))/t = ||u - u'||$. □

1A.8 Remark. The proof of (1A.6) shows that if a smooth Riemannian manifold M is of curvature $\leq \kappa$ in the weak sense of (1.18), then its sectional curvature is bounded above by κ, and hence M is of curvature $\leq \kappa$ in the sense of Alexandrov (1.2).

Chapter II.2 Convexity and its Consequences

In this chapter we concentrate mainly on CAT(0) spaces, establishing basic properties that we shall appeal to repeatedly in subsequent chapters. We shall also describe the extent to which each of the results presented can be extended to the case of CAT(κ) spaces with $\kappa > 0$.

The most basic of the properties defining the nature of CAT(0) spaces is the convexity of the metric (2.2). From this property alone one can deduce a great deal about the geometry of CAT(0) spaces (cf. [Bus55]). We shall also examine orthogonal projections onto complete convex subspaces (2.4), and we shall see that the existence of unique centres for bounded subsets implies that every compact group of isometries of a complete CAT(0) space must have a fixed point (2.8).

The third section of this chapter begins with the observation (from [Ale51]) that when considering a triangle Δ in a CAT(0) space X, if one gets any non-trivial equality in the CAT(0) condition, then Δ spans an isometrically embedded Euclidean triangle in X. (An analogous result holds for CAT(κ) spaces in general (2.10).) This insight leads quickly to results concerning the global structure of CAT(0) spaces (e.g. 2.14) which illustrate the subtlety of Alexandrov's definition (1.1).

Convexity of the Metric

We recall the definition of a convex function.

2.1 Definition. A function $f : I \to \mathbb{R}$ defined on an interval I (not necessarily closed or compact) is said to be *convex* if, for any $t, t' \in I$ and $s \in [0, 1]$,

$$f((s-1)t + st') \leq (s-1)f(t) + sf(t').$$

A function $f : X \to \mathbb{R}$ defined on a geodesic metric space is *convex* if, for any geodesic path $c : I \to X$ parameterized proportional to arc length, the function $t \mapsto f(c(t))$ defined on the interval I is convex. Equivalently, for each geodesic path $c : [0, 1] \to X$ parameterized proportional to arc length, we have

$$f(c(s)) \leq (s-1)f(c(0)) + sf(c(1))$$

for each $s \in [0, 1]$.

2.2 Proposition. *If X is a* CAT(0) *space, then the distance function $d : X \times X \to \mathbb{R}$ is convex, i.e. given any pair of geodesics $c : [0, 1] \to X$ and $c' : [0, 1] \to X$, parameterized proportional to arc length, the following inequality holds for all $t \in [0, 1]$:*

$$d(c(t), c'(t)) \leq (1 - t)\, d(c(0), c'(0)) + t\, d(c(1), c'(1)).$$

Proof. We first assume that $c(0) = c'(0)$ and consider a comparison triangle $\overline{\Delta} \subset \mathbb{E}^2$ for $\Delta(c(0), c(1), c'(1))$. Given $t \in [0, 1]$, elementary Euclidean geometry tells us that $d(\overline{c(t)}, \overline{c'(t)}) = t\, d(\overline{c(1)}, \overline{c'(1)}) = t\, d(c(1), c'(1))$. And by the CAT(0) inequality, $d(c(t), c'(t)) \leq d(\overline{c(t)}, \overline{c'(t)})$. Hence we obtain $d(c(t), c'(t)) \leq t\, d(c(1), c'(1))$.

In the general case, we introduce the linearly reparameterized geodesic $c'' : [0, 1] \to X$ with $c''(0) = c(0)$ and $c''(1) = c'(1)$. By applying the preceding special case, first to c and c'' and then to c' and c'' with reversed orientation, we obtain: $d(c(t), c''(t)) \leq t\, d(c(1), c''(1))$ and $d(c''(t), c'(t)) \leq (1 - t)\, d(c''(0), c'(0))$. Hence,

$$d(c(t), c'(t)) \leq d(c(t), c''(t)) + d(c''(t), c'(t)) \leq t\, d(c(1), c'(1)) + (1 - t)\, d(c(0), c'(0)),$$

as required. □

2.3 Exercises.

(1) Let X be a CAT(κ) space and fix $x_0 \in X$. Prove that the restriction of $x \mapsto d(x, x_0)$ to the open ball of radius $D_\kappa/2$ about x_0 is convex.

(2) Let X be a CAT(0) space. Prove that for all points $p, q, r \in X$, if $c : [0, a] \to X$ and $c' : [0, b] \to X$ are the unique geodesics joining q to p and r to p, respectively, then $d(c(t), c'(t)) \leq d(q, r)$ for all $t < \min\{a, b\}$.

Convex Subspaces and Projection

In subsequent sections we shall make frequent use of orthogonal projections onto complete, convex subsets of CAT(0) spaces. *Orthogonal projection* (or simply 'projection') is the name given to the map $\pi : X \to C$ constructed in the following proposition. (A careful inspection of the proof of this proposition shows that, modulo the usual restrictions on scale, one can also project onto convex subsets of CAT(κ) spaces when $\kappa > 0$, see 2.6(1).)

2.4 Proposition. *Let X be a* CAT(0) *space, and let C be a convex subset which is complete in the induced metric. Then,*

(1) *for every $x \in X$, there exists a unique point $\pi(x) \in X$ such that $d(x, \pi(x)) = d(x, C) := \inf_{y \in C} d(x, y)$;*

(2) *if x' belongs to the geodesic segment $[x, \pi(x)]$, then $\pi(x') = \pi(x)$;*

(3) *given $x \notin C$ and $y \in C$, if $y \neq \pi(x)$ then*[20] $\angle_{\pi(x)}(x, y) \geq \pi/2$;

(4) *the map $x \mapsto \pi(x)$ is a retraction of X onto C which does not increase distances; the map $H : X \times [0, 1] \to X$ associating to (x, t) the point a distance $t\, d(x, \pi(x))$ from x on the geodesic segment $[x, \pi(x)]$ is a continuous homotopy from the identity map of X to π.*

Proof. To show the existence of $\pi(x)$, we consider a sequence of points $y_n \in C$ such that $d(y_n, x)$ tends to $d(x, C)$. We claim that this is a Cauchy sequence. Once we have proved that this is the case, we can take $\pi(x)$ to be the limit point, whose existence is guaranteed by the completeness of C. Moreover, the fact that every such sequence (y_n) is Cauchy also establishes the uniqueness of $\pi(x)$, because if there were a second point $\pi(x)' \in C$ with $d(\pi(x)', x) = d(x, C)$, then the sequence whose terms were alternately $\pi(x)$ and $\pi(x)'$ would satisfy the definition of (y_n), but would not be Cauchy.

Let $D = d(x, C)$ and let $\varepsilon > 0$ be small compared to D. By hypothesis, there exists $N > 0$ such that $d(y_n, x) < D + \varepsilon$ whenever $n > N$. We fix $n, m > N$ and consider a comparison triangle $\overline{\Delta}(x, y_n, y_m) \subset \mathbb{E}^2$. We then draw two circles about $\bar{x}$, one of radius D and one of radius $D + \varepsilon$. An elementary calculation in Euclidean geometry shows that any line segment which is entirely contained in the closed annular region bounded by these two circles can have length at most $2\sqrt{2\varepsilon D + \varepsilon^2}$. The line segment $[\bar{y}_n, \bar{y}_m]$ must be contained in this annular region, for if it were not then there would exist $\bar{z} \in [\bar{y}_n, \bar{y}_m]$ with $d(\bar{z}, \bar{x}) < D$. But then, by the CAT(0) inequality, the corresponding point $z \in [y_n, y_m] \subset C$ would satisfy $d(x, z) < D$, contradicting the definition of D. Hence, if $n, m > N$, then $d(y_n, y_m) < 2\sqrt{2\varepsilon D + \varepsilon^2}$. Thus the sequence (y_n) is Cauchy.

(2) follows from the triangle inequality.

For (3), one observes that if $\angle_{\pi(x)}(x, y)$ were less that $\pi/2$, then one could find points $x' \in [\pi(x), x]$ and $y' \in [\pi(x), y]$ distinct from $\pi(x)$ such that, in the comparison triangle $\overline{\Delta}(x', \pi(x), y')$ the angle at $\overline{\pi(x)}$ would be $< \pi/2$. This, together with the CAT(0) inequality, would imply that for some point $p \in [\pi(x), y'] \subset C$ we would have $d(x', p) < d(x', \pi(x))$. But by (2), $d(x', \pi(x)) = d(x', C)$.

(4) We claim that if $x_1, x_2 \in X$ do not belong to C and if $\pi(x_1) \neq \pi(x_2)$, then $d(x_1, x_2) \geq d(H(x_1, t), H(x_2, t))$ for all $t \in [0, 1]$. To this end, we write $x_i(t) = H(x_i, t)$ and consider the quadrilateral in $\mathbb{E}^2$ which is obtained by adjoining comparison triangles $\overline{\Delta}(x_1, \pi(x_1), \pi(x_2))$ and $\overline{\Delta}(x_1, \pi(x_2), x_2)$ along the edge $[\bar{x}_1, \overline{\pi(x_2)}]$ with $\overline{x_2}$ and $\overline{\pi(x_1)}$ on different sides. It follows from (3), together with (1.7(4)), that the angles at $\overline{\pi(x_1)}$ and $\overline{\pi(x_2)}$ are not less than $\pi/2$. Hence $d(x_1, x_2) = d(\bar{x}_1, \bar{x}_2) \geq d(\overline{x_1(t)}, \overline{x_2(t)}) \geq d(x_1(t), x_2(t))$. □

The first part of the following corollary implies in particular that for any $r > 0$ the closed r-neighbourhood of a convex set in a complete CAT(0) space is itself convex.

[20] $\angle_{\pi(x)}(x, y)$ is the Alexandrov angle between the geodesic segments $[\pi(x), x)]$ and $[\pi(x), y)]$

2.5 Corollary. *Let C be a complete convex subset in a* CAT(0) *space X. Let d_C be the distance function to C, namely $d_C(x) = d(x, C)$. Then:*

(1) *d_C is a convex function;*

(2) *for all $x, y \in X$, we have $|d_C(x) - d_C(y)| \leq d(x, y)$;*

(3) *the restriction of d_C to a sphere with centre x and radius $r \leq d_C(x)$ attains its infimum at a unique point y and $d_C(x) = d_C(y) + r$.*

Proof. Let π be the projection of X onto C.

(1) Let $c : [0, 1] \to X$ be a linear parameterization of a geodesic segment and let $c' : [0, 1] \to C$ be a linear parameterization of the geodesic segment $[\pi(c(0)), \pi(c(1))]$. By the convexity of the distance function we have

$$\begin{aligned} d_C(c(t)) &\leq d(c(t), c'(t)) \\ &\leq (1-t)d(c(0), c'(0)) + td(c(1), c'(1)) \\ &= (1-t)d_C(c(0)) + td_C(c(1)). \end{aligned}$$

For (2), note $d_C(x) \leq d(x, \pi(y)) \leq d(x, y) + d(y, \pi(y)) = d(x, y) + d_C(y)$.

(3) Let y be the point on $[x, \pi(x)]$ such that $d(x, y) = r$. Then $d_C(x) = d_C(y) + r$ by (2.4(2)). If y' is such that $d(x, y') = r$ and $d_C(y') \leq d_C(y)$, then

$$d(x, \pi(y')) \leq d(x, y') + d(y', \pi(y')) = r + d_C(y) = d(x, \pi(x)),$$

hence $\pi(y') = \pi(x)$ and $y' = y$. □

2.6 Exercises.

(1) Let X be a CAT(κ) space and let $C \subseteq X$ be a complete subset which is D_κ-convex. If we replace X by $V = \{x \in X \mid d_C(x) < D_\kappa/2\}$ then parts (1) to (3) of (2.4) remain valid, as does (2.5). Moreover, the map H described in (2.4(4)) is a homotopy from id_V to π.

(2) Let X be a complete CAT(κ) space and let $C \subseteq X$ be a closed D_κ-convex subspace. Endow $Y = X \setminus C$ with the induced path metric, let $\overline{Y}$ be its completion, and define $\mathrm{Bd}(Y) = \overline{Y} \setminus Y$. Show that any geodesic in $\overline{Y}$ with endpoints in Bd(Y) is entirely contained in Bd(Y). (Hint: Y is open in $\overline{Y}$ and there is a natural map $\phi : \overline{Y} \to X$; if a local geodesic in $\overline{Y}$ met Bd(Y) only at its endpoints, then its image under ϕ would be a local geodesic of the same length with endpoints in C.)

The Centre of a Bounded Set

Early in this century Elie Cartan [Car28] proved that in a complete, simply connected manifold M of non-positive curvature, given any finite subset $\{x_1, \dots, x_n\}$, the function $x \mapsto \sum_i d(x, x_i)^2$ has a unique minimum which can usefully be regarded as the "centre" of the subset. Using this idea, he proved the existence of a fixed point for the action of any compact group of isometries of M. Using essentially the same idea (but with a different notion of centre) Bruhat and Tits proved a fixed-point theorem for group actions on Euclidean buildings [BruT72]. In this section we explain how this idea applies to complete CAT(0) spaces, and we also explain how to construct a centre for suitably small bounded sets in complete CAT(κ) spaces with $\kappa > 0$.

Given a bounded subset Y of a metric space X, the *radius* of Y is by definition the infimum of the positive numbers r such that $Y \subseteq B(x, r)$ for some $x \in X$.

2.7 Proposition. *Let X be a complete* CAT(κ) *space. If $Y \subseteq X$ is a bounded set of radius $r_Y < D_\kappa/2$, then there exists a unique point $c_Y \in X$, called the centre[21] of Y, such that $Y \subseteq \overline{B}(c_Y, r_Y)$.*

Proof. Let (x_n) be a sequence of points in X with the property that $Y \subseteq B(x_n, r_n)$ and $r_n \to r_Y$ as $n \to \infty$. We shall prove that (x_n) is a Cauchy sequence; the idea of the proof is very similar to that of (2.4). Since X is complete, this sequence will have a limit point, and such a limit has the property required of c_Y. The fact that every such sequence (x_n) is Cauchy also establishes the uniqueness of c_Y, as in (2.4).

We fix a basepoint $O \in M_\kappa^2$. Given $\varepsilon > 0$ we choose numbers $R \in (r_Y, D_\kappa/2)$ and $R' < r_Y$ such that any geodesic segment which is entirely contained in the annular region $A = B(O, R) \smallsetminus B(O, R')$ has length less than ε.

For sufficiently large n, n' we have $r_n, r_{n'} < R$. Let m be the midpoint of the unique geodesic segment joining x_n to $x_{n'}$. For each $y \in Y$, we consider a comparison triangle $\overline{\Delta}_y = \Delta(O, \overline{x_n}, \overline{x_{n'}}) \subset M_\kappa^2$ for $\Delta(y, x_n, x_{n'})$. If it were the case that for every $y \in Y$ the midpoint of $[\overline{x_n}, \overline{x_{n'}}] \subset \overline{\Delta}(y)$ belonged to $B(O, R')$, then by the CAT(κ) inequality we would have $Y \subset B(m, R')$, contradicting the fact that $R' < r_Y$. Therefore, there exists $y \in Y$ such that the midpoint of $[\overline{x_n}, \overline{x_{n'}}] \subset \overline{\Delta}_y$ lies in the annulus A. This implies that at least half of $[\overline{x_n}, \overline{x_{n'}}]$ lies in A, and hence $[\overline{x_n}, \overline{x_{n'}}]$ has length less than 2ε. □

For an alternative proof in the case $\kappa = 0$, see [Bro88, p. 157]. Finer results of a similar nature can be found in the papers of U. Lang and V. Schroeder [LS97a].

2.8 Corollary.

(1) *If X is a complete* CAT(0) *space and Γ is a finite group of isometries of X or, more generally, a group of isometries with a bounded orbit, then the fixed-point set of Γ is a non-empty convex subspace of X.*

(2) *If a group Γ acts properly and cocompactly by isometries on a* CAT(0) *space, then Γ contains only finitely many conjugacy classes of finite subgroups.*

Proof. In order to see that the fixed point set of Γ is non-empty, one simply applies the preceding proposition with the role of Y played by a bounded orbit of Γ; since Y is Γ–invariant, so is its centre. If an isometry fixes $p, q \in X$, then it must fix the unique geodesic segment $[p, q]$ pointwise. Hence the fixed-point set of Γ is convex.

Part (2) follows immediately from (1) and I.8.5(5). □

[21] more precisely, circumcentre.

Flat Subspaces

Most of the results that we have presented so far follow fairly directly from the definition of a CAT(κ) space. In this paragraph we begin to strike a richer vein of ideas, based on the observation (from [Ale51]) that when considering a triangle Δ of perimeter at most $2D_\kappa$ in a CAT(κ) space X, if one gets any non-trivial equality in the CAT(κ) condition then the natural map $\overline{\Delta} \to \Delta$ from any comparison triangle $\overline{\Delta} \subset M_\kappa^2$ extends to give an isometric embedding of the convex hull of $\overline{\Delta}$ into X (see (2.9) and (2.10)).

Flat Triangles

By definition, the (closed) convex hull of a subset A of a geodesic space X is the intersection of all (closed) convex subspaces of X containing A. The convex hull of a geodesic triangle $\Delta(p, q, r) \subset X$ coincides with the convex hull of its vertex set $\{p, q, r\}$; in a general CAT(0), for example in the complex hyperbolic plane $\mathbb{C}\mathrm{H}^2$ the convex hull of three points in general position is 4–dimensional (see (10.12)).

2.9 Proposition (Flat Triangle Lemma). *Let Δ be a geodesic triangle in a* CAT(0) *space X. If one of the vertex angles of Δ is equal to the corresponding vertex angle in a comparison triangle $\overline{\Delta} \subset \mathbb{E}^2$ for Δ, then "Δ is flat", more precisely, the convex hull of Δ in X is isometric to the convex hull of $\overline{\Delta}$ in $\mathbb{E}^2$.*

Proof. Let $\Delta = \Delta(p, q, q')$ and suppose that $\angle_p(q, q') = \overline{\angle}_p(q, q')$. Let r be any point in the interior of the segment $[q, q']$. The first step of the proof is to show that our hypothesis on $\angle_p(q, r)$ implies that for all such r we have equality in the CAT(0) condition, i.e. $d(\overline{p}, \overline{r}) = d(p, r)$. Let $\Delta' = \Delta(p, q, r)$, and let $\Delta'' = \Delta(p, q', r)$. We consider comparison triangles $\widetilde{\Delta}' = \Delta(\tilde{p}, \tilde{q}, \tilde{r})$ for Δ', and $\widetilde{\Delta}'' = \Delta(\tilde{p}, \tilde{q}', \tilde{r})$ for Δ'' in $\mathbb{E}^2$, and assume that these are arranged with a common side $[\tilde{p}, \tilde{r}]$ so that $\tilde{q}$ and $\tilde{q}'$ are not on the same side of the line through $[\tilde{p}, \tilde{r}]$. Let $\overline{r}$ be the comparison point in $\overline{\Delta}$ for r. As the sum of the angles at $\tilde{r}$ of $\tilde{\Delta}'$ and $\tilde{\Delta}''$ is not less than π, we can apply Alexandrov's lemma (I.2.16). We have

$$\angle_p(q, q') \le \angle_p(q, r) + \angle_p(q', r) \le \angle_{\tilde{p}}(\tilde{q}, \tilde{r}) + \angle_{\tilde{p}}(\tilde{q}', \tilde{r}) \le \angle_{\overline{p}}(\overline{q}, \overline{q}'),$$

where the second inequality follows from the CAT(0) condition (1.7(4)). By assumption $\angle_p(q, q') = \angle_{\overline{p}}(\overline{q}, \overline{q}')$, hence we have equality everywhere, in particular $\angle_{\tilde{p}}(\tilde{q}, \tilde{r}) + \angle_{\tilde{p}}(\tilde{q}', \tilde{r}) = \angle_{\overline{p}}(\overline{q}, \overline{q}')$. Therefore, by Alexandrov's lemma again, $d(p, r) = d(\tilde{p}, \tilde{r}) = d(\overline{p}, \overline{r})$, as desired.

Let j be the map from the convex hull $C(\overline{\Delta})$ of $\overline{\Delta} \subset \mathbb{E}^2$ to X which, for every $\overline{r} \in [\overline{q}, \overline{q}']$, sends the geodesic segment $[\overline{p}, \overline{r}]$ isometrically onto the geodesic segment $[p, r]$. We claim that j is an isometry onto its image; it then follows that the unique geodesic joining any two points of the image of j will be contained in the image, so j maps $C(\overline{\Delta})$ onto $C(\Delta)$. Consider two points $\overline{x} \in [\overline{p}, \overline{r}]$ and $\overline{x}' \in [\overline{p}, \overline{r}']$ in $C(\overline{\Delta})$, where $\overline{r}$ and $\overline{r}'$ lie on $[\overline{q}, \overline{q}']$, and $\overline{r}$ is between $\overline{q}$ and $\overline{r}'$. Let $x = j(\overline{x})$, $x' = j(\overline{x}')$, $r = j(\overline{r})$, $r' = j(\overline{r}')$, and let $\delta_1, \delta_2, \delta_3$ be the angles $\angle_p(q, r)$, $\angle_p(r, r')$, $\angle_p(r', q')$, respectively; $\overline{\delta}_1, \overline{\delta}_2, \overline{\delta}_3$ will denote the corresponding angles in $\overline{\Delta}$. The first step of the proof shows that $\Delta(\overline{p}, \overline{q}, \overline{r})$ is a comparison triangle for $\Delta(p, q, r)$, hence $\delta_1 \leq \overline{\delta}_1$. Similarly, $\delta_2 \leq \overline{\delta}_2$ and $\delta_3 \leq \overline{\delta}_3$. But $\angle_p(q, q') \leq \delta_1 + \delta_2 + \delta_3 \leq \overline{\delta}_1 + \overline{\delta}_2 + \overline{\delta}_3 = \angle_{\overline{p}}(\overline{q}, \overline{q}') = \angle_p(q, q')$, so in fact $\delta_2 = \overline{\delta}_2$. Hence $d(x, x') = d(\overline{x}, \overline{x}')$, as required. □

2.10 Exercises.

(1) Let $\Delta(p, q, r)$ be a geodesic triangle in a CAT(0) space, and let $\Delta(\overline{p}, \overline{q}, \overline{r}) \subset \mathbb{E}^2$ be its comparison triangle. Use (2.9) to prove that if there exists a point $\overline{x}$ in the interior of $[\overline{p}, \overline{r}]$ and a point $\overline{y} \in [\overline{p}, \overline{q}]$ with $\overline{y} \neq \overline{p}$ such that $d(x, y) = d(\overline{x}, \overline{y})$, then the triangle $\Delta(p, q, r)$ is flat.

(2) Generalize Proposition 2.9 and (1) to arbitrary CAT(κ) spaces.

Flat Polygons

2.11 The Flat Quadrilateral Theorem. *Consider four points p, q, r, s in a* CAT(0) *space X. Let $\alpha = \angle_p(q, s)$, $\beta = \angle_q(p, r)$, $\gamma = \angle_r(q, s)$ and $\delta = \angle_s(r, p)$. If $\alpha + \beta + \gamma + \delta \geq 2\pi$, then this sum is equal to 2π and the convex hull of the four points p, q, r, s is isometric to the convex hull of a convex quadrilateral in $\mathbb{E}^2$.*

Proof. The method of proof is similar to that of the preceding proposition. Let $\Delta_1 = \Delta(p, q, s)$ and $\Delta_2 = \Delta(r, q, s)$. To begin, we construct a quadrilateral in $\mathbb{E}^2$ by joining comparison triangles $\overline{\Delta}_1 = \overline{\Delta}(p, q, s)$ and $\overline{\Delta}_2 = \overline{\Delta}(r, q, s)$ along the edge $[\overline{q}, \overline{s}]$ so that $\overline{p}$ and $\overline{r}$ lie on opposite sides of the line that passes through $\overline{q}$ and $\overline{s}$.

We denote the angles of $\overline{\Delta}_1$ at the vertices $\overline{p}, \overline{q}, \overline{s}$ by $\overline{\alpha}, \overline{\beta}_1, \overline{\delta}_1$, respectively, and those of $\overline{\Delta}_2$ at the vertices $\overline{r}, \overline{q}, \overline{s}$ by $\overline{\gamma}, \overline{\beta}_2, \overline{\delta}_2$. From (1.7(4)) and (I.1.14) we have

$$\alpha \leq \overline{\alpha},\ \beta \leq \overline{\beta}_1 + \overline{\beta}_2,\ \delta \leq \overline{\delta}_1 + \overline{\delta}_2,\ \gamma \leq \overline{\gamma}.$$

We are assuming that $\alpha + \beta + \gamma + \delta \geq 2\pi$, and the sum of the angles of a Euclidean quadrilateral is 2π, so we conclude that in the above expression we have equality everywhere. In particular $\overline{\beta}_1 + \overline{\beta}_2 = \beta \leq \pi$ and $\overline{\delta}_1 + \overline{\delta}_2 = \delta \leq \pi$; therefore the Euclidean quadrilateral Q with vertices $\{\overline{p}, \overline{q}, \overline{r}, \overline{s}\}$ is convex. The preceding proposition then implies that Δ_1 and Δ_2 bound flat triangles. Let j be the map from the convex hull C of Q to X whose restrictions, j_1 and j_2, to the convex hulls $C(\overline{\Delta}_1)$ and $C(\overline{\Delta}_2)$ of the triangles $\overline{\Delta}_1$ and $\overline{\Delta}_2$ are isometries onto the convex hulls of Δ_1 and Δ_2. To check that j is an isometry, we have to prove that, for all $\overline{x}_1 \in C(\overline{\Delta}_1)$ and $\overline{x}_2 \in C(\overline{\Delta}_2)$, if $j(\overline{x}_1) = x_1$ and $j(\overline{x}_2) = x_2$, then $d(x_1, x_2) = d(\overline{x}_1, \overline{x}_2)$. This will follow once we have proved that the angle $\mu = \angle_q(x_1, x_2)$ is equal to the angle $\overline{\mu} = \angle_{\overline{q}}(\overline{x}_1, \overline{x}_2)$.

To check this equality, we let $\beta_1' = \angle_q(p, x_1)$, $\beta_1'' = \angle_q(x_1, s)$, $\beta_2'' = \angle_q(s, x_2)$ and $\beta_2' = \angle_q(x_2, r)$. As above, one sees that each of these angles is equal to the corresponding angle in the comparison figure. Hence

$$\beta \leq \beta_1' + \mu + \beta_2' \leq \beta_1' + \beta_1'' + \beta_2'' + \beta_2' = \overline{\beta}_1 + \overline{\beta}_2 = \beta.$$

And since $\beta_1' = \overline{\beta}_1'$ and $\beta_2' = \overline{\beta}_2'$, we deduce $\mu = \overline{\mu}$. □

2.12 Exercises.

(1) Let X be a CAT(0) space and consider n distinct points $\{p_i\}_{i=1}^n$ in X. Let $\alpha_i = \angle_{p_i}(p_{i-1}, p_{i+1})$, where indices are taken mod n. Prove that if $\sum \alpha_i \geq (n-2)\pi$ then the convex hull in X of $\{p_i\}_{i=1}^n$ is isometric to a convex n–gon in $\mathbb{E}^2$.

(2) **(The Sandwich Lemma)** Let X be a CAT(0) space. We write $d_C(x) = \inf\{d(x, c) \mid c \in C\}$ to denote the distance of a point from a closed subspace $C \subset X$. Let C_1 and C_2 be two complete, convex subspaces of X. Prove that if the restriction of d_{C_1} to C_2 is constant, equal to a say, and the restriction of d_{C_2} to C_1 is constant, then the convex hull of $C_1 \cup C_2$ is isometric to $C_1 \times [0, a]$.

(3) With the hypothesis of (2), let $p : X \to C_1$ be orthogonal projection. Let C' be an arbitrary subspace of X. Prove that if the restriction of p to C' is an isometry onto C_1 then the restriction of $d_{C_1}(x)$ to C' is constant, equal to b say, and there is a unique isometry j from $C' \times [0, b]$ to the convex hull of $C_1 \cup C'$ such that $j(x, 0) = x$ and $j(x, b) = p(x)$ for all $x \in C'$.

Flat Strips

Recall that a *geodesic line* in a metric space X is a map $c : \mathbb{R} \to X$ such that $d(c(t), c(t')) = |t - t'|$ for all $t, t' \in \mathbb{R}$. Two geodesic lines c, c' in X are said to be *asymptotic* if there exists a constant K such that $d(c(t), c'(t)) \leq K$ for all $t \in \mathbb{R}$. We remind the reader that if a function $\mathbb{R} \to \mathbb{R}$ is convex and bounded, then it is constant.

2.13 The Flat Strip Theorem. *Let X be a* CAT(0) *space, and let $c : \mathbb{R} \to X$ and $c' : \mathbb{R} \to X$ be geodesic lines in X. If c and c' are asymptotic, then the convex hull of $c(\mathbb{R}) \cup c'(\mathbb{R})$ is isometric to a flat strip $\mathbb{R} \times [0, D] \subset \mathbb{E}^2$.*

Proof. Let π be the projection of X onto the closed convex subspace $c(\mathbb{R})$ (cf. (2.4)). By reparameterizing if necessary, we may assume that $c(0) = \pi(c'(0))$, i.e., that $c(0)$ is the point on $c(\mathbb{R})$ closest to $c'(0)$.

The function $t \mapsto d(c(t), c'(t))$ is convex, non-negative and bounded, hence constant, equal to D say. Similarly, for all $a \in \mathbb{R}$, the function $t \mapsto (c(t+a), c'(t))$ is constant. In particular $d(c(a+t), c'(t)) = d(c(a), c'(0)) \geq d(c(0), c'(0))$, and hence $\pi(c'(t)) = c(t)$ for all t. This same inequality shows that the projection π' onto $c'(\mathbb{R})$ maps $c(t)$ to $c'(t)$.

Given $t < t'$, we consider the quadrilateral in X which is the union of the geodesic segments $[c(t), c(t')]$, $[c(t'), c'(t')]$, $[c'(t'), c'(t)]$ and $[c'(t), c(t)]$. According to (2.4), all of the angles of this quadrilateral are at least $\pi/2$. By (2.11), the convex hull of $\{c(t), c'(t), c(t'), c'(t')\}$ is isometric to a Euclidean rectangle, and therefore the map $j : \mathbb{R} \times [0, D] \to X$ which sends (t, s) to the point on the geodesic segment $[c(t), c'(t)]$ a distance s from $c(t)$ is an isometry onto the convex hull of $c(\mathbb{R}) \cup c'(\mathbb{R})$. □

In view of the Flat Strip Theorem, the terms *parallel* and *asymptotic* are synonymous when used to describe geodesic lines in a CAT(0) space. Both are in common use.

2.14 A Product Decomposition Theorem. *Let X be a* CAT(0) *space and let $c : \mathbb{R} \to X$ be a geodesic line.*

(1) *The union of the images of all geodesic lines $c' : \mathbb{R} \to X$ parallel to c is a convex subspace X_c of X.*

(2) *Let p be the restriction to X_c of the projection from X to the complete convex subspace $c(\mathbb{R})$. Let $X_c^0 = p^{-1}(c(0))$. Then, X_c^0 is convex (in particular it is a* CAT(0) *space) and X_c is canonically isometric to the product $X_c^0 \times \mathbb{R}$.*

Proof. Given two points $x_1, x_2 \in X_c$ we fix geodesic lines c_1 and c_2 parallel to c such that x_1 lies in the image of c_1 and x_2 lies in the image of c_2. Because c_1 is parallel to c_2, we can apply the Flat Strip Theorem and deduce that the convex hull of $c_1(\mathbb{R}) \cup c_2(\mathbb{R})$ is isometric to a flat strip; in particular, it is the union of images of geodesic lines parallel to c. This proves that X_c is convex, hence it is a CAT(0) space.

In order to prove (2), we restrict our attention to those geodesic lines c' which are parallel to c and for which $p(c'(0)) = c(0)$. Every $x \in X_c$ lies in the image of a unique such geodesic line, which we denote c_x.

Let $j : X_c^0 \times \mathbb{R} \to X_c$ be the bijection defined by $j(x, t) = c_x(t)$. Using the Flat Strip Theorem, it is clear that this map is an isometry provided $d(x, x') = d(c_x(\mathbb{R}), c_{x'}(\mathbb{R}))$ for all $x, x' \in X_c^0$. Thus the following lemma completes the proof of the theorem. □

2.15 Lemma. *Consider three geodesic lines $c_i : \mathbb{R} \to X$, $i = 1, 2, 3$ in a metric space X. Suppose that the union of each pair of these lines is isometric to the union of two parallel lines in $\mathbb{E}^2$. Let $p_{i,j}$ be the map that assigns to each point of $c_j(\mathbb{R})$ the unique closest point on $c_i(\mathbb{R})$. Then $p_{1,3} \circ p_{3,2} \circ p_{2,1} = p_{1,1}$, the identity of $c_1(\mathbb{R})$.*

Proof. This proof was simplified by use of an idea suggested to us by Phil Bowers and Kim Ruane (cf. [BoRu96b]). If $p_{1,3} \circ p_{3,2} \circ p_{2,1}$ were not the identity on $c_1(\mathbb{R})$, then it would be a translation by a non-zero real number, b say, so $p_{1,3} \circ p_{3,2} \circ p_{2,1}(c_1(t)) = c_1(t + b)$ for all $t \in \mathbb{R}$.

If we reparameterize c_2 and c_3 so that $p_{2,1}(c_1(0)) = c_2(0)$ and $p_{3,2}(c_2(0)) = c_3(0)$, then $p_{1,3}(c_3(0)) = c_1(b)$. Let $a_1 = d(c_1(\mathbb{R}), c_2(\mathbb{R}))$, $a_2 = d(c_2(\mathbb{R}), c_3(\mathbb{R}))$, $a_3 = d(c_1(\mathbb{R}), c_3(\mathbb{R}))$, and let $a = a_1 + a_2 + a_3$. By hypothesis, $d(c_1(t), c_2(t + s)) = \sqrt{a_1^2 + s^2}$, $d(c_2(t), c_3(t + s)) = \sqrt{a_2^2 + s^2}$ and $d(c_3(t), c_1(t + s)) = \sqrt{a_3^2 + (s - b)^2}$. Therefore, for all s,

$$\begin{aligned} d(c_1(0), c_1(as + b)) &\\ &\leq d(c_1(0), c_2(a_1 s)) + d(c_2(a_1 s), c_3((a_1 + a_2)s)) \\ &\quad + d(c_3((a_1 + a_2)s), c_1((a_1 + a_2 + a_3)s + b)) \\ &= a_1\sqrt{1 + s^2} + a_2\sqrt{1 + s^2} + a_3\sqrt{1 + s^2} = a\sqrt{1 + s^2}. \end{aligned}$$

But since c_1 is a geodesic, $d(c_1(0), c_1(as + b)) = |as + b|$, so we have $(as + b)^2 \leq a^2(1 + s^2)$ for all $s \in \mathbb{R}$, which is impossible if $b \neq 0$. □

Chapter II.3 Angles, Limits, Cones and Joins

In this chapter we examine how upper curvature bounds influence the behaviour of angles, limits of sequences of spaces, and the cone and join constructions described in (I.5). We then use the results concerning limits and cones to describe the space of directions at a point in a CAT(κ) space.

Angles in CAT(κ) Spaces

In CAT(κ) spaces, angles exist in the following strong sense.

3.1 Proposition. *Let X be a CAT(κ) space and let $c : [0, a] \to X$ and $c' : [0, a'] \to X$ be two geodesic paths issuing from the same point $c(0) = c'(0)$. Then the κ-comparison angle $\angle_{c(0)}^{(\kappa)}(c(t), c'(t'))$ is a non-decreasing function of both $t, t' \geq 0$, and the Alexandrov angle $\angle(c, c')$ is equal to $\lim_{t,t'\to 0} \angle_{c(0)}^{(\kappa)}(c(t), c'(t')) = \lim_{t\to 0} \angle_{c(0)}^{(\kappa)}(c(t), c'(t))$. Hence, in the light of (I.2.9),*

$$\angle(c, c') = \lim_{t\to 0} 2 \arcsin \frac{1}{2t} d(c(t), c'(t)).$$

Proof. Immediate from 1.7(3) and the fact that one can take comparison triangles in M_κ^2 instead of $\mathbb{E}^2$ in the definition (I.1.12) of the Alexandrov angle (see I.2.9). □

3.2 Notation for Angles. For the convenience of the reader we recall the following notation. Let p, x, y be points of a metric space X such that $p \neq x, p \neq y$.

$\angle_p^{(\kappa)}(x, y)$ denotes the comparison angle in M_κ^2 (see I.2.15);

$\overline{\angle}_p(x, y) = \angle_p^{(0)}(x, y)$ denotes the comparison angle in $\mathbb{E}^2$ (see I.1.12);

if there are unique geodesic segments $[p, x]$ and $[p, y]$, then we write $\angle_p(x, y)$ to denote the (Alexandrov) angle between these segments (see I.1.12).

Recall that a real-valued function f on a topological space Y is said to be *upper semicontinuous* if $f(y) \geq \limsup_{n\to\infty} f(y_n)$ whenever $y_n \to y$ in Y.

3.3 Proposition. *Let X be a CAT(κ) space. For all points $p, x, y \in X$ with $\max\{d(p, x), d(p, y)\} < D_\kappa$,*

(1) *the function* $(p, x, y) \mapsto \angle_p(x, y)$ *is upper semicontinuous, and*

(2) *for fixed* $p \in X$, *the function* $(x, y) \mapsto \angle_p(x, y)$ *is continuous.*

Proof. (1) Let (x_n), (y_n) and (p_n) be sequences of points converging to x, y and p respectively. Let c, c', c_n and c'_n be linear parameterizations $[0, 1] \to X$ of the geodesic segments $[p, x]$, $[p, y]$, $[p_n, x_n]$ and $[p_n, y_n]$ respectively. For $t \in (0, 1]$, let $\alpha(t) = \angle_p^{(\kappa)}(c(t), c'(t))$ and let $\alpha_n(t) = \angle_{p_n}^{(\kappa)}(c_n(t), c'_n(t))$. According to (3.1), $\alpha(t)$ and $\alpha_n(t)$ are non-decreasing functions of t and $\alpha := \angle_p(x, y) = \lim_{t\to 0} \alpha(t)$ and $\alpha_n := \angle_{p_n}(x_n, y_n) = \lim_{t\to 0} \alpha_n(t)$. And for fixed t we have $\alpha_n(t) \to \alpha(t)$ as $n \to \infty$. Given $\varepsilon > 0$, let $T > 0$ be such that $\alpha(t) - \varepsilon/2 \leq \alpha$ for all $t \in (0, T]$. Then for n big enough, $\alpha_n(T) \leq \alpha(T) + \varepsilon/2$, therefore $\alpha_n \leq \alpha_n(T) \leq \alpha(T) + \varepsilon/2 \leq \alpha + \varepsilon$. Thus $\limsup \alpha_n \leq \alpha$ as required.

(2) We keep the above notations, but we assume that $p_n = p$ for all n. Let $\beta_n = \angle_p(x, x_n)$ and $\gamma_n = \angle_p(y, y_n)$. By (1.7(4)), $\beta_n \to 0$ and $\gamma_n \to 0$ as $n \to \infty$. By the triangle inequality for Alexandrov angles, $|\alpha - \alpha_n| \leq \beta_n + \gamma_n$. Hence $\lim_{n\to\infty} \alpha_n = \alpha$. □

3.4 Remark. With regard to part (1) of the preceding proposition, we note that in general it will not be true that $p \mapsto \angle_p(x, y)$ is a continuous function. For example, consider the CAT(0) space X obtained by endowing the subset $\{(x_1, 0) \mid x_1 \in \mathbb{R}\} \cup \{(x_1, x_2) \mid x_1 \geq 0, x_2 \in \mathbb{R}\}$ of the plane with the induced length metric from $\mathbb{E}^2$. Let $p = (0, 0)$ and let $p_n = (-1/n, 0)$. Given any $x = (x_1, x_2)$ and $y = (y_1, y_2) \in X$ such that $x_1 \geq 0$ and $y_1 \geq 0$, the angle which x and y subtend at p is equal to the usual Euclidean angle, but $\angle_{p_n}(x, y) = 0$ for all n.

The following addendum to (3.3) will be useful later.

3.5 Proposition. *Let X be a* CAT(κ) *space, let* $c : [0, \varepsilon] \to X$ *be a geodesic segment issuing from* $p = c(0)$ *and let* y *be a point of* X *distinct from* p *(with* ε *and* $d(p, y)$ *less than* D_κ *if* $\kappa > 0$*). Then,*

$$\lim_{s\to 0} \overline{\angle}_p(c(s), y) = \angle_p(c(\varepsilon), y).$$

Proof. As $s \mapsto \angle_p^{(\kappa)}(c(s), y)$ is non-decreasing, $\gamma := \lim_{s\to 0} \angle_p^{(\kappa)}(c(s), y)$ exists. By (I.2.9), we have $\gamma = \lim_{s\to 0} \overline{\angle}_p(c(s), y)$. This last expression is, by definition, the strong upper angle between $[p, y]$ and c, which we showed in (I.1.16) to be equal to the Alexandrov angle. □

3.6 Corollary (First Variation Formula). *With the notation of the preceding proposition,*

$$\lim_{s\to 0} \frac{d(c(0), y) - d(c(s), y)}{s}$$

exists and is equal to $\cos \angle_p(c(s), y)$.

Proof. This follows from the preceding proposition and the Euclidean law of cosines. □

4-Point Limits of CAT(κ) Spaces

We now turn our attention to limits of sequences of CAT(κ) spaces. We want a notion of convergence for metric spaces that is strict enough to ensure that if a complete geodesic space is a limit of CAT(κ) spaces then it is itself a CAT(κ) space. On the other hand, we would like a notion of convergence that is weak enough to allow a wide range of applications. We would also like a definition that allows us to construct a "tangent space" at each point in a CAT(κ) space X as a limit of the sequence (X, d_n), where $d_n(x, y) = n\, d(x, y)$ (see 3.19). Since the local structure of a CAT(κ) space can differ wildly from one point to another, this seems a lot to ask, but in our discussion of ultralimits (I.5) we saw that a single sequence of spaces can have a wide variety of "limits" if the notion of limit is defined in terms of approximations to finite configurations of points. Since the CAT(κ) inequality can be characterized by a condition on 4-point configurations (1.10), we are led to the following definition (cf. [Ni95]).

3.7 Definition of 4-Point Limits. A metric space (X, d) is *a 4-point limit* of a sequence of metric spaces (X_n, d_n) if, for every 4-tuple of points (x_1, x_2, x_3, x_4) from X and every $\varepsilon > 0$, there exist infinitely many integers n such that there is a 4-tuple $(x_1(n), x_2(n), x_3(n), x_4(n))$ from X_n with $|d(x_i, x_j) - d_n(x_i(n), x_j(n))| < \varepsilon$ for $1 \le i, j \le 4$.

3.8 Remark. The unrestrictive nature of 4-point limits means that if (X, d) is a limit of a sequence of metric spaces (X_n, d_n) in most reasonable senses (for instance, a Gromov-Hausdorff limit or an ultralimit) then it is a 4-point limit of the sequence. A single sequence (X_n, d_n) may have a wide variety of 4-point limits. For example, given a metric space Y, every subspace $X \subset Y$ is a 4-point limit of the constant sequence $X_n = Y$.

The final sentence of the preceding remark shows that if one wishes to deduce that a 4-point limit of a sequence of CAT(κ) spaces is CAT(κ) then one must impose an hypothesis on the limit to ensure the existence of geodesics. The most satisfactory way of doing this is to require that the space be complete and have approximate midpoints (cf. 1.11).

3.9 Theorem (Limits of CAT(κ) Spaces). *Let (X, d) be a complete metric space. Let (X_n, d_n) be a sequence of* CAT(κ_n) *spaces. Suppose that (X, d) is a 4-point limit of the sequence (X_n, d_n) and that $\kappa = \lim_{n\to\infty} \kappa_n$. Suppose also that every pair of points $x, y \in X$ with $d(x, y) < D_\kappa$ has approximate midpoints. Then X is a* CAT(κ) *space.*

Proof. According to (1.12), it suffices to show that X is a CAT(κ') space for every $\kappa' > \kappa$, and according to (1.11) for this it suffices to show that X satisfies the CAT(κ') 4-point condition. Fix $\kappa' > \kappa$. For n sufficiently large we have $\kappa_n < \kappa'$, so by (1.12) we may assume that all of the X_n are CAT(κ') spaces.

Let (x_1, y_1, x_2, y_2) be a 4-tuple of points from X with $d(x_1, y_1,) + d(y_1, x_2) + d(x_2, y_2) + d(y_2, x_1) < 2D_{\kappa'}$. According to the definition of a 4-point limit, there is a sequence of integers n_i and 4-tuples $(x_1(n_i), y_1(n_i), x_2(n_i), y_2(n_i))$ from X_{n_i} such that, for $j, k \in \{0, 1\}$, as $n_i \to \infty$ we have $d(x_j(n_i), x_k(n_i)) \to d(x_j, x_k)$ and $d(y_j(n_i), y_k(n_i)) \to d(y_j, y_k)$, and $d(x_j(n), y_k(n)) \to d(x_j, y_k)$.

Each X_{n_i} is a CAT(κ') space, so the 4-tuple $(x_1(n_i), y_1(n_i), x_2(n_i), y_2(n_i))$ has a subembedding $(\overline{x}_1(n_i), \overline{y}_1(n_i), \overline{x}_2(n_i), \overline{y}_2(n_i))$ in $M^2_{\kappa'}$. We may assume that all of the points $\overline{x}_1(n_i)$ are equal, so these 4-tuples are all contained in a compact ball in $M^2_{\kappa'}$. Passing to a subsequence if necessary, we may then assume that the sequences $\overline{x}_2(n_i)$ and $\overline{y}_k(n_i)$ converge, to $\overline{x}_2$ and $\overline{y}_k$ say. Clearly $(\overline{x}_1, \overline{y}_1, \overline{x}_2, \overline{y}_2)$ is a subembedding of (x_1, y_1, x_2, y_2). Thus we have shown that X satisfies the CAT(κ') 4-point condition, and we conclude that X is CAT(κ') for every $\kappa' > \kappa$. □

We articulate some special cases of (3.9) that are of particular interest.

3.10 Corollary. *Fix $\kappa \in \mathbb{R}$. Let (X, d) be a complete metric space and let (X_n, d_n) be a sequence of* CAT(κ) *spaces.*

(1) *If (X, d) is a (pointed or unpointed) Gromov-Hausdorff limit of (X_n, d_n), then (X, d) is a* CAT(κ) *space.*

(2) *If (X, d) is an ultralimit of (X_n, d_n), then (X, d) is a* CAT(κ) *space.*

Let (Y, δ) be a CAT(κ) *space and let ω be a non-principal ultrafilter on $\mathbb{N}$. Let $Y_n = Y$, let $\delta_n = n.\delta$ and let $\delta'_n = \frac{1}{n}\delta$.*

(3) $\lim_\omega(Y_n, \delta_n)$ *is a* CAT(0) *space.*

(4) *If $\kappa = 0$, then* $\text{Cone}_\omega Y = \lim_\omega(Y_n, \delta'_n)$ *is a* CAT(0) *space.*

(5) *If $\kappa < 0$, then* $\text{Cone}_\omega Y$ *is an $\mathbb{R}$-tree.*

Proof. In each case the existence of approximate midpoints follows easily from the hypothesis. For (3), (4) and (5) we recall that an ultralimit of geodesic spaces is always complete and geodesic, and that if (Y, d) is CAT(κ) then $(Y, \lambda.d)$ is CAT($\lambda^{-2}\kappa$). □

Metric Completion

For the most part, in this book we work with spaces that are assumed to be complete and we only resort to the additional hypothesis of local compactness when it appears unavoidable. One advantage of this approach is that any CAT(κ) space can be realized as a dense, convex subset of a *complete* CAT(κ) space, namely its metric completion.

3.11 Corollary. *The metric completion (X', d') of a* CAT(κ) *space (X, d) is a* CAT(κ) *space.*

Proof. It is clear that (X', d') is a 4-point limit of the constant sequence $(X_n, d_n) = (X, d)$ and that it has approximate midpoints. It is complete by hypothesis, so we can apply (3.9). □

3.12 Remark. We saw in (I.3.6) that the completion of a geodesic space need not be geodesic, so our appeal to (3.9) in the above proof masked the fact that the curvature hypothesis was being used to prove the existence of geodesics.

3.13 Exercises

(1) Give an example of a space X such that X and its completion X' are locally compact geodesic metric spaces and X is non-positively curved but X' is not.

(2) Let $U \subset M_\kappa^2$ be a bounded subset that is open and convex. Let $c : [0, l] \to M_\kappa^2$ be an arc-length parameterization of the (rectifiable) boundary curve of U. Let $P_n \subset M_\kappa^2$ be the polygonal disc bounded by the concatenation of the geodesic segments $[c(il/n), c((i+1)l/n)]$, $i = 0, \dots, n-1$. Let $X_n = M_\kappa^2 \setminus P_n$. Prove that $M \setminus U$ is the (4-point) limit of the sequence X_n (where all of the spaces considered are endowed with the induced path metric from M_κ^2).

(3) Fix $\kappa \leq 0$, consider an open convex subset $U \subset M_\kappa^2$, let $\overline{U}$ be its closure and let $\mathrm{Bd}(U) = \overline{U} \setminus U$. Prove the following facts and use (2) and 1.15(2) to deduce that $M_\kappa^2 \setminus U$ has curvature $\leq \kappa$. (Hint: One can reduce to the case where U is bounded by restricting attention to a suitable ball in M_κ^2.)

(i) Fix $p \in U$. Show that for all $x \in M_\kappa^2 \setminus U$ the geodesic segment $[x, p]$ intersects $\mathrm{Bd}(U)$ in exactly one point, which we denote x'.

(ii) Show that $x \mapsto x'$ is a continuous map and that $\mathrm{Bd}(U)$ is a 1-manifold.

(iii) Show that if U is bounded and c is a simple closed curve enclosing U, then the length of $\mathrm{Bd}(U)$ (which is homeomorphic to a circle) is less than that of c. (Hint: Consider the orthogonal projection onto $\overline{U}$.)

(iv) Show that if U is unbounded then each component of its boundary is the image of a map $c : \mathbb{R} \to M_\kappa^2$ whose restriction to each compact subinterval is rectifiable. In the case $\kappa = 0$, prove that if $\mathrm{Bd}(U)$ is not connected then it consists of two parallel lines.

Cones and Spherical Joins

The purpose of this paragraph is to prove the following important theorem of Berestovskii [Ber83] (see also [AleBN86]). This result provides a basic connection between CAT(1) spaces and CAT(κ) spaces in general. The first evidence of the importance of this link will be seen in the next paragraph, where we shall combine (3.14) with results on limits in order to prove that the 'tangent space' at any point of a CAT(κ) space has a natural CAT(0) metric. We shall make further use of Berestovskii's theorem in Chapter 5, when we discuss curvature in polyhedral complexes.

The definition of a κ-cone was given in Chapter I.5.

3.14 Theorem (Berestovskii). *Let Y be a metric space. The κ-cone $X = C_\kappa Y$ over Y is a* CAT(κ) *space if and only if Y is a* CAT(1) *space.*

Proof. First we assume that Y is a CAT(1) space and prove that X is a CAT(κ) space. By I.5.10, we know that any two points $x_1, x_2 \in X$ with $d(x_1, x_2) < D_\kappa$ are joined by a unique geodesic segment $[x_1, x_2]$. We have to check that the CAT(κ) inequality is satisfied for any geodesic triangle with vertices $x_i = t_i y_i,\ i = 1, 2, 3$ (assuming that its perimeter is less than $2D_\kappa$). If one of the t_i is zero, then the triangle $[x_1, x_2] \cup [x_2, x_3] \cup [x_3, x_1]$, with its induced metric, is isometric to its comparison triangle in M^2_κ, so we may assume each $t_i > 0$.

Following Berestovskii [Be83], we consider three cases:

(a) $d(y_1, y_2) + d(y_2, y_3) + d(y_3, y_1) < 2\pi$.

(b) $d(y_1, y_2) + d(y_2, y_3) + d(y_3, y_1) \geq 2\pi$ but $d(y_i, y_j) < \pi$ for all $i, j = 1, 2, 3$.

(c) One of the $d(y_i, y_j)$ is $\geq \pi$.

Case (a). Let $\Delta = \Delta([y_1, y_2], [y_2, y_3], [y_3, y_1]) \subset Y$. We fix a comparison triangle $\overline{\Delta}$ in $M^2_1 = \mathbb{S}^2$ with vertices $\bar{y}_1, \bar{y}_2, \bar{y}_3$. The comparison map $\overline{\Delta} \to \Delta$ extends naturally to a bijection from $C_\kappa \overline{\Delta} \subset M^3_\kappa$ to $C_\kappa \Delta$. Let $x = ty$ be a point on the segment $[x_2, x_3]$ and let $\bar{y}$ be the comparison point for y in $[\bar{y}_2, \bar{y}_3]$. The triangle $\Delta(\bar{x}_1, \bar{x}_2, \bar{x}_3) \subset M^3_\kappa$, with $\bar{x}_i = t_i \bar{y}_i$, can be viewed as a κ-comparison triangle for $\Delta(x_1, x_2, x_3)$, with $\bar{x} = t\bar{y}$ as the comparison point for x. As the CAT(1) inequality holds for $\overline{\Delta}$, we have $d(y_1, y) \leq d(\bar{y}_1, \bar{y})$. From this we see (using the formula I.5.6 defining $d(x_1, x)$) that $d(x_1, x) \leq d(\bar{x}_1, \bar{x})$.

Case (b). Let $\Delta(\tilde{0}, \tilde{x}_1, \tilde{x}_2)$ and $\Delta(\tilde{0}, \tilde{x}_1, \tilde{x}_3)$ be comparison triangles in M^2_κ for $\Delta(0, x_1, x_2)$ and $\Delta(0, x_1, x_3)$, respectively, chosen so that $\tilde{x}_2$ and $\tilde{x}_3$ are on different sides of $[\tilde{x}_1, \tilde{0}]$ (where, as usual, 0 denotes the cone point of X). From the definition of the metric on $C_\kappa Y$ we have $\angle_{\tilde{0}}(\tilde{x}_1, \tilde{x}_2) = d(y_1, y_2)$, $\angle_{\tilde{0}}(\tilde{x}_1, \tilde{x}_3) = d(y_1, y_3)$, and $\angle_{\tilde{x}_1}(\tilde{0}, \tilde{x}_2) = \angle_{x_1}(0, x_2)$ and $\angle_{\tilde{x}_1}(\tilde{0}, \tilde{x}_3) = \angle_{x_1}(0, x_3)$. And by hypothesis we have $d(y_1, y_2) + d(y_1, y_3) > \pi$, hence

$$\angle_{\tilde{0}}(\tilde{x}_2, \tilde{x}_3) = 2\pi - \angle_{\tilde{0}}(\tilde{x}_2, \tilde{x}_1) - \angle_{\tilde{0}}(\tilde{x}_3, \tilde{x}_1) \leq d(y_2, y_3) = \angle_0(x_2, x_3)$$

and $d(\tilde{x}_2, \tilde{x}_3) \leq d(x_2, x_3)$. Therefore, in a comparison triangle $\overline{\Delta} = \Delta(\bar{x}_1, \bar{x}_2, \bar{x}_3)$ in M^2_κ for $\Delta(x_1, x_2, x_3)$, we have:

$$\begin{aligned}\angle_{\bar{x}_1}(\bar{x}_2, \bar{x}_3) \geq \angle_{\tilde{x}_1}(\tilde{x}_2, \tilde{x}_3) &= \angle_{\tilde{x}_1}(\tilde{x}_2, \tilde{0}) + \angle_{\tilde{x}_1}(\tilde{0}, \tilde{x}_3) \\ &= \angle_{x_1}(x_2, 0) + \angle_{x_1}(0, x_3) \geq \angle_{x_1}(x_2, x_3)\end{aligned}$$

Hence condition (1.7(4)) is satisfied.

Case (c). Suppose $d(y_1, y_3) \geq \pi$. Then the geodesic segment $[x_1, x_3]$ is the concatenation of $[x_1, 0]$ and $[0, x_3]$. Let $\overline{\Delta}_1 = \Delta(\bar{0}, \bar{x}_1, \bar{x}_2)$ and $\overline{\Delta}_3 = \Delta(\bar{0}, \bar{x}_3, \bar{x}_2)$ be comparison triangles in M^2_κ for the triangles $\Delta_1 = \Delta(0, x_1, x_2)$ and $\Delta_3 = \Delta(0, x_3, x_2)$, chosen so that the vertices $\bar{x}_1$ and $\bar{x}_3$ lie on different sides of the common segment $[0, x_2]$. (In the case $\kappa > 0$ such comparison triangles exist because the perimeter of $\Delta(x_1, x_2, x_3)$ is assumed to be smaller than $2D_\kappa$). The sum of the angles at $\bar{0}$ in $\overline{\Delta}_1$ and $\overline{\Delta}_3$ is $d_\pi(y_1, y_2) + d_\pi(y_2, y_3) \geq \pi$. When we straighten the union of these two triangles to get a comparison triangle for $\Delta(x_1, x_2, x_3)$, Alexandrov's lemma (I.2.16) ensures that the angles do not decrease, and therefore $\Delta(x_1, x_2, x_3)$ satisfies the CAT(κ) inequality.

It remains to be proved that if X is a CAT(κ) space then Y is a CAT(1) space. From I.5.10(1) we know that pairs of points a distance less than π apart in Y are joined by a unique geodesic segment. Let $\Delta([y_1, y_2], [y_2, y_3], [y_3, y_1])$ be a triangle in Y of perimeter less than 2π and let $\overline{\Delta} = \Delta(\overline{y}_1, \overline{y}_2, \overline{y}_3)$ be a comparison triangle in M_1^2. Given $y \in [y_2, y_3]$, let $\overline{y} \in [\overline{y}_2, \overline{y}_3]$ be its comparison point. On the subcone $C_\kappa\Delta$, we consider three points $x_i = \varepsilon y_i$, $i = 1, 2, 3$, where ε is positive (and small enough to ensure that the perimeter of the triangle with vertices x_1, x_2, x_3 is less than $2D_\kappa$). The cone $C_\kappa\overline{\Delta}$ is a subcone of $C_\kappa M_1^2 \subseteq M_\kappa^3$ and the points $\overline{x}_i = \varepsilon\overline{y}_i$, $i = 1, 2, 3$, are the vertices of a comparison triangle $\overline{\Delta}'$ for the geodesic triangle Δ' with vertices x_1, x_2, x_3. If $x = ty \in [x_2, x_3]$, then its comparison point is $\overline{x} = t\overline{y}$. By hypothesis $d(x_1, x) \leq d(\overline{x}_1, \overline{x})$. Hence $d(y_1, y) \leq d(\overline{y}_1, \overline{y})$, by the definition of the metric on $C_\kappa Y$. □

3.15 Corollary. *The join $Y_1 * Y_2$ of two metric spaces Y_1 and Y_2 is a* CAT(1) *space if and only if Y_1 and Y_2 are* CAT(1) *spaces.*

Proof. This follows immediately from the theorem, because we saw in (I.5.15) that $C_0(Y_1 * Y_2)$ is isometric to $C_0Y_1 \times C_0Y_2$, and $C_0Y_1 \times C_0Y_2$ is a CAT(0) space if and only if C_0Y_1 and C_0Y_2 are CAT(0) spaces (1.16(2)). □

3.16 Corollary. *Let Y be a metric space. The following conditions are equivalent.*

(1) *$C_\kappa Y$ is a* CAT(κ) *space.*

(2) *$C_\kappa Y$ has curvature $\leq \kappa$.*

(3) *A neighbourhood of the cone point $0 \in C_\kappa Y$ is a* CAT(κ) *space.*

Proof. Trivially, (1) $\Longrightarrow$ (2) $\Longrightarrow$ (3). And (3) $\Longrightarrow$ (1) is a consequence of the preceding theorem. □

3.17 Example. If S is isometric to a circle of length a, then its cone $C_\kappa S$ is a CAT(κ) space if and only if $a \geq 2\pi$.

The Space of Directions

3.18 Definition (The Space of Directions and the Tangent Cone). Let X be a metric space. Two non-trivial geodesics c and c' issuing from a point $p \in X$ are said to define the *same direction* at p if the Alexandrov angle between them is zero. The triangle inequality for angles (I.1.14) implies that [$c \sim c'$ iff $\angle(c, c') = 0$] is an equivalence relation on the set of non-trivial geodesics issuing from p, and $\angle$ induces a metric on the set of equivalence classes. The resulting metric space is called the *space of directions* at p and is denoted $S_p(X)$. (Note that two geodesics segments issuing from p may have the same direction but intersect only at p.)

$C_0S_p(X)$, the Euclidean cone over $S_p(X)$, is called the *tangent cone* at p.

If X is a Riemannian manifold, then $S_p(X)$ is isometric to the unit sphere in the tangent space of X at p and $C_0S_p(X)$ is the tangent space itself. If X is a polyhedral complex, then $S_p(X)$ is isometric to $\mathrm{Lk}(p, X)$, the geometric link of p. In a Riemannian manifold with non-empty boundary, if p is a point on the boundary then $S_p(X)$ may not be complete; for instance if X is a closed round ball in $\mathbb{E}^n$ and p belongs to the boundary of this ball, then $S_p(X)$ is isometric to an open hemisphere in $\mathbb{S}^{n-1}$.

The following theorem is due to I. G. Nikolaev [Ni95].

3.19 Theorem. *Let $\kappa \in \mathbb{R}$. If a metric space X has curvature $\leq \kappa$, then the completion of the space of directions at each point $p \in X$ is a* CAT(1) *space, and the completion of the tangent cone at p is a* CAT(0) *space.*

Proof. In the light of Berestovskii's Theorem (3.14), it suffices to prove that the completion of $C_0(S_p(X))$ is a CAT(0) space. Following the outline of a proof by B. Kleiner and B. Leeb [KlL97], we shall prove this by showing that $C_0(S_p(X))$ satisfies the CAT(0) 4-point condition and has approximate midpoints (cf. (1.11) and (3.9)). In the light of (3.16), it suffices to establish these conditions in a neighbourhood of the cone point.

Since $S_p(X)$ depends only on a neighbourhood of p, we may assume that X is a CAT(κ) space of diameter less than $D_\kappa/2$, in which case there is a unique geodesic segment $[p, x]$ for every $x \in X$.

Let $j : X \to C_0S_p(X)$ be the map that sends p to the cone point 0 and sends each $x \neq p$ to the point a distance $d(x, p)$ from 0 that projects to the class of $[p, x]$ in $S_p(X)$. And for each $t \in [0, 1]$, let tx denote the point of X a distance $td(p, x)$ from p on the geodesic segment $[p, x]$.

For each $\varepsilon \in (0, 1]$, define d_ε to be the following pseudometric on X:

$$d_\varepsilon(x, y) = \frac{1}{\varepsilon}\, d(\varepsilon x, \varepsilon y).$$

Note that d_ε satisfies the CAT($\varepsilon^2\kappa$) 4-point condition (and hence the CAT(0) 4-point condition if $\kappa \leq 0$).

Fix $x, y \in X$ distinct from p, and let $\gamma_\varepsilon = \overline{\angle}_p(\varepsilon x, \varepsilon y)$. By the Euclidean law of cosines,

$$d_\varepsilon(x, y)^2 = |x|^2 + |y|^2 - 2|x|.|y| \cos \gamma_\varepsilon,$$

where $|x| = d(p, x)$ and $|y| = d(p, y)$. And from (3.1) we have $\angle_p(x, y) = \lim_{\varepsilon\to 0} \gamma_\varepsilon$.

By letting $\varepsilon \to 0$ in the above formula, we see that $d_0(x, y) := \lim_{\varepsilon\to 0} d_\varepsilon(x, y)$ exists for all $x, y \in X$, and that $j : (X, d_0) \to C_0S_p(X)$ is a distance-preserving map of pseudometric spaces. Moreover, (X, d_0) satisfies the CAT(0) 4-point condition, because it satisfies the CAT($\varepsilon^2\kappa$) 4-point condition for every ε. The image of j contains a neighbourhood of the cone point in $C_0S_p(X)$, so this neighbourhood also satisfies the CAT(0) 4-point condition. In order to complete the proof of the theorem, it only remains to establish the existence of approximate midpoints for pairs of points $j(x), j(y) \in C_0S_p(X)$.

We first consider the case $\kappa \leq 0$. Let m_ε be the midpoint of the segment $[\varepsilon x, \varepsilon y]$. We claim that when ε is small, $\frac{1}{\varepsilon}j(m_\varepsilon)$ is an approximate midpoint for $j(x), j(y)$. To see this, note that by the convexity of the distance function on X, we have $d_0(x, y) \leq d(x, y)$ for any $x, y \in X$, and hence

$$\begin{aligned} d(j(x), \frac{1}{\varepsilon} j(m_\varepsilon)) &= \frac{1}{\varepsilon} d(\varepsilon j(x), j(m_\varepsilon)) = \frac{1}{\varepsilon} d_0(\varepsilon x, m_\varepsilon) \\ &\leq \frac{1}{\varepsilon} d(\varepsilon x, m_\varepsilon) = \frac{1}{2\varepsilon} d(\varepsilon x, \varepsilon y) = \frac{1}{2} d_\varepsilon(x, y). \end{aligned}$$

Similarly $d(j(y), \frac{1}{\varepsilon} j(m_\varepsilon)) \leq \frac{1}{2} d_\varepsilon(x, y)$. And as $d(j(x), j(y)) = \lim_{\varepsilon \to 0} d_\varepsilon(x, y)$, we are done.

In the case $\kappa > 0$, the argument is entirely similar except that one compensates for the invalidity of the inequality $d_0(x, y) \leq d(x, y)$ by noting that for any $x, y \in B(p, r)$ and $\varepsilon > 0$, we have $d_\varepsilon(x, y) \leq C(r) d(x, y)$, where $C(r) \to 1$ as $r \to 0$. The existence of this constant is a consequence of the following lemma and the CAT(κ) inequality for $\Delta(p, x, y)$. □

3.20 Lemma. *For each $\kappa > 0$ there is a function $C : [0, D_\kappa) \to \mathbb{R}$ such that $\lim_{r \to 0} C(r) = 1$ and for all $p \in M_\kappa^2$ and all $x, y \in B(p, r)$*

$$d(\varepsilon x, \varepsilon y) \leq \varepsilon C(r)\, d(x, y),$$

where εx denotes the point a distance $\varepsilon\, d(p, x)$ from p on the geodesic $[p, x]$.

Proof. Regard $B(p, r) \subset M_\kappa^2$ as the ball of (Euclidean) radius r in $\mathbb{R}^2$ with the Riemannian metric g_κ expressed in polar coordinates (ρ, θ) by $ds^2 = d\rho^2 + \frac{1}{\kappa} \sin^2(\sqrt{\kappa}\rho) d\theta^2$. The associated metric on $B(p, r)$ is denoted d_κ. Because εx is also the point a distance $\varepsilon d(p, x)$ from p along the Euclidean geodesic from p to x, we have

$$d_{\mathbb{E}^2}(\varepsilon x, \varepsilon y) = \varepsilon\, d_{\mathbb{E}^2}(x, y). \tag{1}$$

And by the compactness of $\overline{B(p, r)}$, there is a constant $C(r) > 1$, which goes to 1 as $r \to 0$, such that the norm of any tangent vector v in the Euclidean metric is related to the norm $\| \ \|_\kappa$ for g_κ by:

$$\frac{1}{\sqrt{C(r)}} \|v\|_\kappa \leq \|v\|_{\mathbb{E}^2} \leq \sqrt{C(r)}\, \|v\|_\kappa .$$

This gives a corresponding Lipschitz relation between the lengths of curves in the two metrics, and hence

$$\frac{1}{\sqrt{C(r)}} d_\kappa(x, y) \leq d_{\mathbb{E}^2}(x, y) \leq \sqrt{C(r)}\, d_\kappa(x, y)$$

for all $x, y \in B(p, r)$. And this, together with (1), proves the lemma. □

3.21 Remark. The above proof applies more generally to any Riemannian metric for which one has normal coordinates.

3.22 Exercise. Let (X, d) be a space of curvature $\leq \kappa$. For each positive integer n, define $d_n(x, y) := n\, d(x, y)$. Prove that the tangent cone $C_0 S(X)$ of X at p (and its completion) is a 4-point limit of the sequence of metric spaces (X, d_n).

Chapter II.4 The Cartan-Hadamard Theorem

Requiring a Riemannian manifold to have non-positive sectional curvature is a restriction on the infinitesimal geometry of the space. Much of the power and elegance of the theory of such manifolds stems from the fact that one can use this infinitesimal condition to make deductions about the global geometry and topology of the manifold. The result which underpins this passage from the local to the global context is a fundamental theorem that is due to Hadamard [H1898] in the case of surfaces and to Cartan [Car28] in the case of arbitrary Riemannian manifolds of non-positive curvature. The main purpose of this chapter is to show that the Cartan-Hadamard Theorem can be generalized to the context of complete geodesic metric spaces. We shall see in subsequent chapters that this generalization is of fundamental importance in the study of complete (1-connected) metric spaces of curvature $\leq \kappa$, where $\kappa \leq 0$. Related results concerning non-simply connected spaces and CAT(κ) spaces with $\kappa > 0$ will also be presented in this chapter. For a more complete treatment of the case $\kappa > 0$, see Bowditch [Bow95c].

Local-to-Global

The proof of the Cartan-Hadamard theorem that we shall give is very close to that given by S. Alexander and R.L. Bishop [AB90]. The main argument is best viewed in the general setting of spaces in which the metric is locally convex in the sense of (1.18). We remind the reader that the metric on a space X is said to be *convex* if X is a geodesic space and all geodesics $c_1 : [0, a_1] \to X$ and $c_2 : [0, a_2] \to X$ with $c_1(0) = c_2(0)$ satisfy the inequality $d(c_1(ta_1), c_2(ta_2)) \leq t d(c_1(a_1), c_2(a_2))$ for all $t \in [0, 1]$. The metric on a space is said to be *locally convex* if every point has a neighbourhood in which the induced metric is convex. If the metric on X is locally convex then in particular X is locally contractible, and therefore has a universal covering $p : \widetilde{X} \to X$. In Chapter I.3 we showed that there is a unique length metric on $\widetilde{X}$ making p a local isometry — this is called the induced length metric.

4.1 The Cartan-Hadamard Theorem. *Let X be a complete connected metric space.*

(1) *If the metric on X is locally convex, then the induced length metric on the universal covering $\widetilde{X}$ is (globally) convex. (In particular there is a unique*

geodesic segment joining each pair of points in $\widetilde{X}$ and geodesic segments in $\widetilde{X}$ vary continuously with their endpoints.)

(2) *If X is of curvature $\leq \kappa$, where $\kappa \leq 0$, then $\widetilde{X}$ (with the induced length metric) is a* CAT(κ) *space.*

This theorem is a variation on a result of M. Gromov ([Gro87], p.119), a detailed proof of which was given by W. Ballmann in the locally compact case [Ba90]. S. Alexander and R.L. Bishop [AB90] proved (4.1) under the additional hypothesis that $\widetilde{X}$ is a geodesic metric space.

4.2 Remarks

(1) The Cartan-Hadamard Theorem is of considerable interest from a purely topological viewpoint, because it provides a tool for showing that the universal coverings of many compact metric spaces are contractible.

(2) We emphasize that in (4.1) we do not assume that X is a geodesic space. Thus, although there exist (non-simply connected) complete length spaces of non-positive curvature which are not geodesic spaces,[22] (4.1) shows that the universal covering of such a space (with the induced length metric) is always a geodesic space.

The second part of (4.1) is deduced from the first by a patchwork process that will be described in the next section. In this section we shall prove 4.1(1). The main work is contained in the following three lemmas.

In what follows, it is convenient to use the term "(local) geodesic" in place of "linearly reparameterized (local) geodesic" and we shall do so freely throughout this chapter.

4.3 Lemma. *Let X be a metric space. Suppose that the metric on X is locally complete and locally convex. Let $c : [0, 1] \to X$ be a local geodesic joining x to y. Let $\varepsilon > 0$ be small enough so that for every $t \in [0, 1]$ the induced metric on the closed ball $\overline{B}(c(t), 2\varepsilon)$ is complete and convex. Then:*

(1) *For all $\overline{x}, \overline{y} \in X$ with $d(x, \overline{x}) < \varepsilon$ and $d(y, \overline{y}) < \varepsilon$, there is exactly one*[23] *local geodesic $\overline{c} : [0, 1] \to X$ joining $\overline{x}$ to $\overline{y}$ such that $t \mapsto d(c(t), \overline{c}(t))$ is a convex function.*

(2) *Moreover,*

$$l(\overline{c}) \leq l(c) + d(x, \overline{x}) + d(y, \overline{y}).$$

Proof. We first prove that if $\overline{c}$ exists then it is unique. Note that if $\overline{c}$ does exist, then the convexity of $t \mapsto d(c(t), \overline{c}(t))$ implies that $d(c(t), \overline{c}(t)) \leq \varepsilon$ for every $t \in [0, 1]$. Let $c', c'' : [0, 1] \to X$ be local geodesics such that $d(c(t), c'(t)) < \varepsilon$ and $d(c(t), c''(t)) <$

[22] For instance the graph with two vertices joined by countably many edges the n-th of which has length $1 + \frac{1}{n}$.

[23] There may be many local geodesics joining $\overline{x}$ to $\overline{y}$, but only one has this property.

ε for all $t \in [0, 1]$. Because the metric on each of the balls $\overline{B}(c(t), 2\varepsilon)$ is convex, the function $t \mapsto d(c'(t), c''(t))$ is locally convex, hence convex. In particular, if $c'(0) = c''(0)$ and $c'(1) = c''(1)$, then $c' = c''$.

Next we prove that (1) implies (2). To this end, we continue to consider c' and c'' with $d(c(t), c'(t)) < \varepsilon$ and $d(c(t), c''(t)) < \varepsilon$ for all $t \in [0, 1]$, and we now assume that $c'(0) = c''(0)$. By convexity, $d(c'(t), c''(t)) \leq t\, d(c('1), c''(1))$. Combining this with the fact that $l(c'|_{[0,t]}) = t\, l(c')$ and $l(c''|_{[0,t]}) = d(c''(0), c''(t))$ for small $t > 0$, we have:

$$\begin{aligned} t\, l(c'') &= d(c''(0), c''(t)) = d(c'(0), c''(t)) \\ &\leq d(c'(0), c'(t)) + d(c'(t), c''(t)) \\ &\leq t\, l(c') + t\, d(c'(1), c''(1)), \end{aligned}$$

and hence $l(c'') \leq l(c') + d(c'(1), c''(1))$.

Let $\overline{c}'$ be the unique local geodesic from $\overline{x}$ to y that satisfies the conditions of the lemma. By applying the argument of the preceding paragraph with $c' = \overline{c}'$ and $c'' = \overline{c}$ we get $l(\overline{c}) \leq l(\overline{c}') + d(y, \overline{y})$. And by applying the argument with $c'(t) := c(1-t)$ and $c''(t) := \overline{c}'(1-t)$ we get $l(\overline{c}') \leq l(c) + d(x, \overline{x})$. Thus $l(\overline{c}) \leq l(c) + d(x, \overline{x}) + d(y, \overline{y})$.

It remains to prove the existence of $\overline{c}$. Let $A > 0$ and consider the following statement:

P(A) *For all $a, b \in [0, 1]$ with $0 < b - a \leq A$ and all $\overline{p}, \overline{q} \in X$ with $d(c(a), \overline{p}) < \varepsilon$ and $d(c(b), \overline{q}) < \varepsilon$, there is a local geodesic $\overline{c} : [a, b] \to X$ such that $\overline{c}(a) = \overline{p}$, $\overline{c}(b) = \overline{q}$ and $d(c(t), \overline{c}(t)) < \varepsilon$ for all $t \in [a, b]$.*

If $A < \varepsilon / l(c)$ then $P(A)$ is clearly true, so it suffices to prove:

Claim. *If $P(A)$ is true then $P(3A/2)$ is also true.*

Let $a, b \in [0, 1]$ be such that $0 < b - a < 3A/2$. Divide $[a, b]$ into three equal parts with endpoints $a < a_1 < b_1 < b$. Let $\overline{p}, \overline{q} \in X$ be such that $d(c(a), \overline{p}) < \varepsilon$ and $d(c(b), \overline{q}) < \varepsilon$. Proceeding recursively, we shall construct Cauchy sequences (p_n) and (q_n) in the ε-neighbourhoods of $c(a_1)$ and $c(b_1)$ respectively (figure 4.1). We shall then construct a local geodesic from a to b, as required in $P(3A/2)$, by taking the union of a local geodesic joining $\overline{p}$ to the limit of the sequence (q_n) and a local geodesic joining the limit of the sequence (p_n) to $\overline{q}$.

To this end, we define $p_0 := c(a_1)$ and $q_0 := c(b_1)$. Then, assuming that p_{n-1} and q_{n-1} have been defined, we use $P(A)$ to construct local geodesics $c_n : [a, b_1] \to X$ and $c'_n : [a_1, b] \to X$ joining $\overline{p}$ to q_{n-1} and p_{n-1} to $\overline{q}$ with $d(c(t), c_n(t)) < \varepsilon$ for $t \in [a, b_1]$ and $d(c(t), c'_n(t)) < \varepsilon$ for $t \in [a_1, b]$. Let $p_n := c_n(a_1)$ and let $q_n := c'_n(b_1)$. By convexity we have $d(p_0, p_1) < \varepsilon/2$ and $d(q_0, q_1) < \varepsilon/2$. More generally, the convexity of the metric in the balls $B(c(t), \varepsilon)$ tells us that on $[a, b_1]$ the function $t \mapsto d(c_n(t), c_{n+1}(t))$ is locally convex, hence convex, and therefore $d(p_n, p_{n+1}) \leq d(q_{n-1}, q_n)/2$. Similarly, $d(q_n, q_{n+1}) \leq d(p_{n-1}, p_n)/2$. It follows that

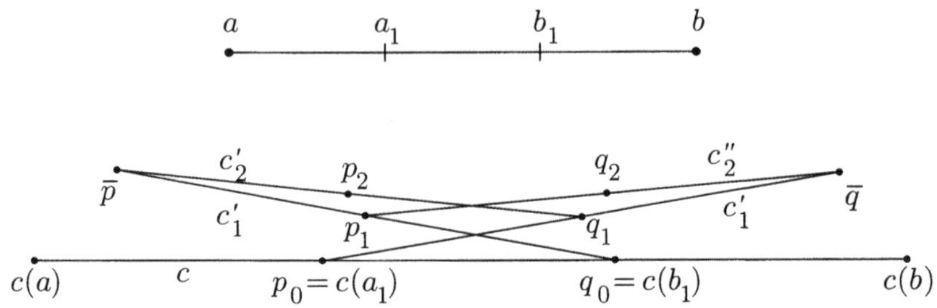

Fig. 4.1 Constructing $\overline{c}$

$d(p_n, p_{n+1}) < \varepsilon/2^{n+1}$ and $d(q_n, q_{n+1}) < \varepsilon/2^{n+1}$ for all $n \in \mathbb{N}$, and therefore (p_n) and (q_n) are Cauchy sequences in $B(p_0, \varepsilon)$ and $B(q_0, \varepsilon)$ respectively.

Now, the function $t \mapsto d(c_n(t), c_{n+1}(t))$ is convex and bounded by $d(q_{n-1}, q_n) < \varepsilon/2^n$. Thus $(c_n(t))$ is a Cauchy sequence in the complete ball $\overline{B}(c(t), \varepsilon)$ for each $t \in [a, b_1]$. Similarly, $(c'_n(t'))$ is a Cauchy sequence in $\overline{B}(c'(t'), \varepsilon)$ for each $t' \in [a_1, b]$. It follows that the local geodesics c_n and c'_n converge uniformly to local geodesics whose restrictions to the common interval $[a_1, b_1]$ coincide. The union of these two local geodesics gives the local geodesic $\overline{c} : [a, b] \to X$ required to complete the proof of the Claim. □

An Exponential Map

In the case of 1–connected Riemannian manifolds of non-positive curvature, the Cartan–Hadamard Theorem follows easily from the fact that the exponential map from the tangent space at each point is a covering map. In the more general context of (4.1) one uses the following analogue of the exponential map.

4.4 Definition. Let X be a metric space and let $x_0 \in X$. We define $\widetilde{X}_{x_0}$ to be the set of all (linearly reparameterized) *local geodesics* $c : [0, 1] \to X$ issuing from x_0, together with the constant map $\tilde{x}_0$ at x_0. We equip $\widetilde{X}_{x_0}$ with the metric

$$d(c, c') := \sup\{d(c(t), c'(t)) \mid t \in [0, 1]\}.$$

And we define $\exp : \widetilde{X}_{x_0} \to X$ to be the map $c \mapsto c(1)$.

4.5 Lemma. *Let X be a metric space and suppose that the metric on X is locally complete and locally convex.*

(1) *$\widetilde{X}_{x_0}$ is contractible (in particular it is simply connected).*

(2) *$\exp : \widetilde{X}_{x_0} \to X$ is a local isometry.*

(3) *There is a unique local geodesic joining $\tilde{x}_0$ to each point of $\widetilde{X}_{x_0}$.*

Proof. (1) There is a natural homotopy $\widetilde{X}_{x_0} \times [0,1] \to \widetilde{X}_{x_0}$ from the identity map of $\widetilde{X}_{x_0}$ to the constant map at $\tilde{x}_0$ given by $(c, s) \mapsto r_s(c)$, where $r_s(c) : [0,1] \to X$ is the path $t \mapsto c(st)$.

(2) Part (1) of Lemma 4.3 implies that for every $c \in \widetilde{X}_{x_0}$ there exists $\varepsilon > 0$ such that the restriction of exp to the ball $B(c, \varepsilon)$ is an isometry onto $B(c(1), \varepsilon)$.

(3) Because exp is a local isometry, a continuous path in $\widetilde{X}_{x_0}$ is a local geodesic if and only if its image under exp is a local geodesic in X. In particular, $\tilde{c} \mapsto \exp \circ \tilde{c}$ is a bijection from the set of local geodesics $\tilde{c} : [0,1] \to \widetilde{X}_{x_0}$ issuing from $\tilde{x}_0$ to the set of local geodesics $c : [0,1] \to X$ issuing from x_0. Thus for each $c \in \widetilde{X}_{x_0}$, the path $\tilde{c} : [0,1] \to \widetilde{X}_{x_0}$ defined by $\tilde{c}(s) = r_s(c)$ is the unique local geodesic joining $\tilde{x}_0$ to c in $\widetilde{X}_{x_0}$. □

4.6 Lemma. *Let X be a metric space and let $x_0 \in X$. If the metric on X is complete and locally convex, then the metric on $\widetilde{X}_{x_0}$ is complete.*

Proof. Let (c_n) be a Cauchy sequence in $\widetilde{X}_{x_0}$. Because X is complete, for every $t \in [0,1]$ the Cauchy sequence $(c_n(t))$ converges in X, to $c(t)$ say. Part (2) of Lemma 4.3 shows that the lengths $l(c_n)$ are uniformly bounded, and hence the pointwise limit $t \mapsto c(t)$ of the curves c_n is a local geodesic. □

4.7 Corollary. *Let X be a connected metric space and let $x_0 \in X$. If the metric on X is complete and locally convex, then:*

(1) $\exp : \widetilde{X}_{x_0} \to X$ *is a universal covering map (in particular it is surjective);*

(2) *there is a unique local geodesic joining each pair of points in $\widetilde{X}_{x_0}$.*

Proof. Assertion (1) follows immediately from (I.3.28), the preceding lemma, and the fact that $\widetilde{X}_{x_0}$ is simply connected (4.5(1)).

We claim that every path in X is homotopic (rel endpoints) to a unique local geodesic. Since $x_0 \in X$ is arbitrary, it suffices to consider continuous paths $c : [0,1] \to X$ issuing from x_0. And since $\widetilde{X}_{x_0}$ is a universal covering, the set of paths in X that are homotopic to c (rel endpoints) is in 1-1 correspondence with the set of paths in $\widetilde{X}_{x_0}$ that issue from $\tilde{x}_0$ and have the same endpoint as the lifting of c that issues from $\tilde{x}_0$. By (4.5(3)), the latter set of paths contains a unique local geodesic. This proves the claim.

Let $p, q \in \widetilde{X}_{x_0}$. Because $\widetilde{X}_{x_0}$ is simply connected, there is only one homotopy class of continuous paths joining p to q. The projection $\exp : \widetilde{X}_{x_0} \to X$ sends this class of paths bijectively onto a single homotopy class (rel endpoints) of paths in X, and the argument of the preceding paragraph shows that the latter class contains a unique local geodesic. □

The Proof of Theorem 4.1(1)

Let X be as in 4.1(1) and fix $x_0 \in X$. We have shown that $\exp : \widetilde{X}_{x_0} \to X$ is a covering map. In (I.3.25) we showed that (under suitable hypotheses) the induced length metric on a covering space is the only length metric that makes the covering map a local isometry. Thus (4.5(1)) implies that the length metric which X induces on $\widetilde{X}_{x_0}$ is the same as the length metric associated to the metric defined in (4.4). According to (4.7(2)), there is a unique local geodesic joining each pair of points in $\widetilde{X}_{x_0}$. And (4.3(1)) shows that these local geodesics must vary continuously with their endpoints. To complete the proof of Theorem 4.1(1), we apply the following lemma with $Y = \widetilde{X}_{x_0}$.

4.8 Lemma. *Let Y be a simply connected length space whose metric is complete and locally convex. Suppose that for every pair of points $p, q \in Y$ there is a unique local geodesic $c_{p,q} : [0, 1] \to Y$ joining p to q. If these geodesics vary continuously with their endpoints then:*

(1) *each $c_{p,q}$ is a geodesic;*

(2) *the metric on Y is convex.*

Proof. (1) It suffices to show that for every rectifiable curve $\gamma : [0, 1] \to Y$ and every $t \in [0, 1]$ we have $l(c_{\gamma(0),\gamma(t)}) \leq l(\gamma|_{[0,t]})$. Because the metric on Y is locally convex, for sufficiently small t the local geodesic $c_{\gamma(0),\gamma(t)}$ is actually a geodesic. Thus the subset of $[0, 1]$ consisting of those t' for which the above inequality holds for all $t \leq t'$ is non-empty. It is clear that this subset is closed. We claim that it is also open, and hence is the whole of $[0, 1]$. Indeed if t_0 is such that $l(c_{\gamma(0),\gamma(t)}) \leq l(\gamma|_{[0,t]})$ for all $t \leq t_0$, then (4.3(2)) tells us that when ε is sufficiently small,

$$\begin{aligned} l(c_{\gamma(0),\gamma(t_0+\varepsilon)}) &\leq l(c_{\gamma(0),\gamma(t_0)}) + l(\gamma|_{[t_0,t_0+\varepsilon]}) \\ &\leq l(\gamma|_{[0,t_0]}) + l(\gamma|_{[t_0,t_0+\varepsilon]}) = l(\gamma|_{[0,t_0+\varepsilon]}), \end{aligned}$$

as required.

(2) We have shown that Y is a complete uniquely geodesic space in which geodesics vary continuously with their endpoints. In order to prove that the metric on Y is convex, it suffices to show that $d(c_{p,q_0}(1/2), c_{p,q_1}(1/2)) \leq (1/2)d(q_0, q_1)$ for each pair of geodesics $c_{p,q_0}, c_{p,q_1} : [0, 1] \to Y$. Suppose that q_0 and q_1 are joined by the geodesic $s \mapsto q_s$. Because the metric on Y is locally convex, we know from (4.3(1)) that when s and s' are sufficiently close $d(c_{p,q_s}(1/2), c_{p,q_{s'}}(1/2)) \leq (1/2)d(q_s, q_{s'})$. Partition $[0, 1]$ finely enough $0 = s_0 < \cdots < s_n = 1$ so that the above inequality holds with $\{s, s'\} = \{s_i, s_{i+1}\}$ for $i = 0, \ldots, n-1$. By adding the resulting inequalities we get $d(c_{p,q_0}(1/2), c_{p,q_1}(1/2)) \leq (1/2)d(q_0, q_1)$. □

Alexandrov's Patchwork

4.9 Proposition (Alexandrov's Patchwork). *Let X be a metric space of curvature $\le \kappa$ and suppose that there is a unique geodesic joining each pair of points that are a distance less than D_κ apart. If these geodesics vary continuously with their endpoints, then X is a* CAT(κ) *space.*

This proposition is an immediate consequence of the following two lemmas and the characterization of CAT(κ) space in terms of angles (1.7(4)). The idea of the proof is portrayed in figure 4.2.

4.10 Gluing Lemma for Triangles. *Fix $\kappa \in \mathbb{R}$. Let X be a metric space in which every pair of points a distance less than D_κ apart can be joined by a geodesic. Let $\Delta = \Delta([p, q_1], [p, q_2], [q_1, q_2])$ be a geodesic triangle in X that has perimeter less than $2D_\kappa$. Suppose that the vertices of Δ are distinct. Consider $r \in [q_1, q_2]$ with $r \neq q_1$ and $r \neq q_2$. Let $[p, r]$ be a geodesic segment joining p to r.*

Let $\overline{\Delta}_i$ be a comparison triangle in M_κ^2 for $\Delta_i = \Delta_i([p, q_i], [p, r], [q_i, r])$. If for $i = 1, 2$, the vertex angles of $\overline{\Delta}_i$ are no smaller than the corresponding angles of Δ_i, then the same is true for the angles of any comparison triangle for Δ in M_κ^2.

Proof. Let $\Delta_1(\overline{p}, \overline{q}_1, \overline{r})$ and $\Delta_2(\overline{p}, \overline{q}_2, \overline{r})$ be comparison triangles in M_κ^2 for Δ_1 and Δ_2, glued along $[\overline{p}, \overline{r}]$ so that $\overline{q}_1$ and $\overline{q}_2$ do not lie on the same side of the line through $\overline{p}$ and $\overline{r}$. The sum of the angles of Δ_1 and Δ_2 at r is at least π, so the sum of the comparison angles is as well. Therefore we can apply Alexandrov's Lemma (I.2.16). □

4.11 Lemma. *Fix $\kappa \in \mathbb{R}$. Let X be a metric space of curvature $\le \kappa$. Let $q : [0, 1] \to X$ be a linearly reparameterized geodesic joining two distinct points $q_0 = q(0)$ and $q_1 = q(1)$ and let p be a point of X which is not in the image of q. Assume that for each $s \in [0, 1]$ there is a linearly reparameterized geodesic $c_s : [0, 1] \to X$ joining p to $q(s)$, and assume that the function $s \mapsto c_s$ is continuous. Then the angles at p, q_0 and q_1 in the geodesic triangle Δ with sides $c_0([0, 1])$, $c_1([0, 1])$ and $q([0, 1])$) are no greater than the corresponding angles in any comparison triangle $\overline{\Delta} \subseteq M_\kappa^2$. (If $\kappa > 0$, assume that the perimeter of Δ is smaller than $2D_\kappa$.)*

Proof. By hypothesis, every point of X has a neighbourhood which (with the induced metric) is a CAT(κ) space, and the map $c : [0, 1] \times [0, 1] \to X$ given by $c : (s, t) \mapsto c_s(t)$ is continuous. Hence there exist partitions $0 = s_0 < s_1 < \cdots < s_k = 1$ and $0 = t_0 < t_1 < \cdots < t_k = 1$ such that for all i and j there is an open ball $U_{i,j} \subset X$ of radius $< D_\kappa/2$ which is a CAT(κ) space and which contains $c([s_{i-1}, s_i] \times [t_{j-1}, t_j])$.

Consider the sequence of adjoining triangles $\Delta_1, \dots, \Delta_k$ where Δ_i is the union of the geodesic segments $c_{s_{i-1}}([0, 1])$, $c_{s_i}([0, 1])$ and $q([s_{i-1}, s_i])$. Repeated use of the Gluing Lemma (4.10) reduces the present proposition to the assertion that the angles at the vertices of each Δ_i are no greater than the corresponding angles in comparison triangles $\overline{\Delta}_i \subset M_\kappa^2$. In order to prove this assertion, we subdivide each Δ_i and make further use of the Gluing Lemma (see figure 4.2). Specifically, for each i we consider the sequence of $2k-1$ adjoining geodesic triangles $\Delta_i^1, \Delta_i^2, \widetilde{\Delta}_i^2, \ldots, \Delta_i^k, \widetilde{\Delta}_i^k$, where Δ_i^1 is the geodesic triangle in $U_{1,1}$ with vertices $p, c_{s_{i-1}}(t_1), c_{s_i}(t_1)$, Δ_i^j the geodesic triangle in $U_{i,j}$ with vertices $c_{s_{i-1}}(t_{j-1}), c_{s_i}(t_{j-1}), c_{s_{i-1}}(t_j)$ and $\widetilde{\Delta}_i^j$ the geodesic triangle in $U_{i,j}$ with vertices $c_{s_{i-1}}(t_j), c_{s_i}(t_{j-1}), c_{s_i}(t_j)$. Since $U_{i,j}$ is a CAT(κ) space, each vertex angle in these small triangles is no greater than the corresponding angle in a comparison triangle in M_κ^2. By making repeated use of the gluing lemma for triangles, we see that the vertex angles of Δ_i also satisfy the desired inequality. □

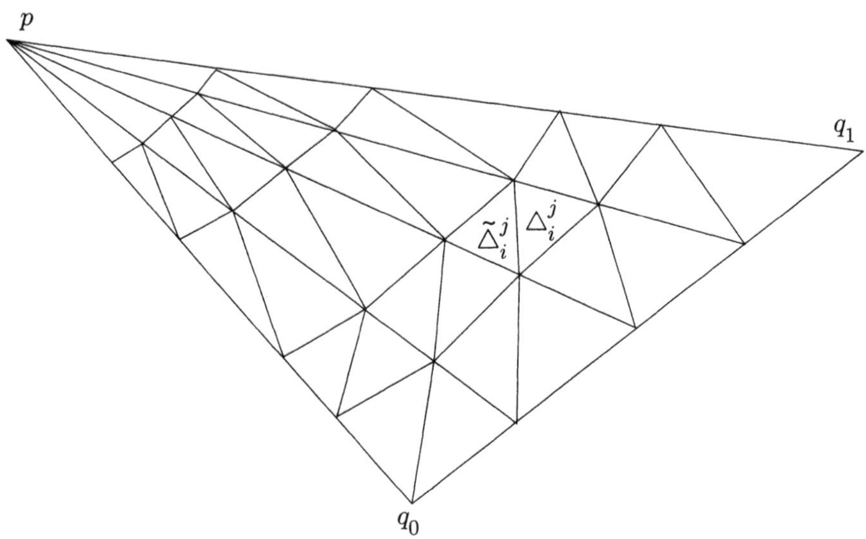

Fig. 4.2 Alexandrov's Patchwork

4.12 Corollary. *Fix $\kappa \in \mathbb{R}$. Let X be a metric space of curvature $\leq \kappa$. Suppose that all closed balls in X of radius less than $D_\kappa/2$ are compact. Then X is a* CAT(κ) *space if and only if every pair of points a distance $< D_\kappa$ apart can be joined by a unique geodesic segment.*

Proof. Immediate from (4.9) and (I.3.13). □

Local Isometries and π_1-Injectivity

We point out some topological consequences of the Cartan-Hadamard Theorem.

4.13 Theorem. *Let X be a complete, non-positively curved metric space, and fix a basepoint $x_0 \in X$.*

(1) *Every homotopy class $\gamma \in \pi_1(X, x_0)$ contains a unique local geodesic c_γ : $[0, 1] \to X$. (The uniqueness implies that a non-constant locally geodesic loop cannot be homotopic to a constant loop.)*

(2) *Every non-trivial element of $\pi_1(X, x_0)$ has infinite order.*

Proof. Consider the universal covering $p : \widetilde{X} \to X$ endowed with the induced length metric. Fix $\tilde{x}_0 \in \widetilde{X}$ with $p(\tilde{x}_0) = x_0$. The fundamental group $\pi_1(X, x_0)$ acts on $\widetilde{X}$ by deck transformations; the action is free and by isometries. Each loop $c : [0, 1] \to X$ in the homotopy class $\gamma \in \pi_1(X, x_0)$ lifts uniquely to a path $\tilde{c} : [0, 1] \to \widetilde{X}$ such that $\tilde{c}(0) = \tilde{x}_0$ and $\tilde{c}(1) = \gamma.\tilde{x}_0$. Because p is a local isometry, $\tilde{c}$ will be a local geodesic if and only if c is a local geodesic.

The Cartan-Hadamard Theorem tells us that $\widetilde{X}$ is a CAT(0) space. Hence there is a unique local geodesic $\tilde{c}_\gamma : [0, 1] \to X$ joining $\tilde{x}_0$ to $\gamma.\tilde{x}_0$. This proves (1).

Because $\widetilde{X}$ is a CAT(0) space, any isometry of finite order will have a fixed point (2.8). Thus (2) follows from the fact that the action of $\pi_1(X, x_0)$ is free. □

Recall that a map $f : Y \to X$ between metric spaces is said to be locally an isometric embedding if for every $y \in Y$ there is an $\varepsilon > 0$ such that the restriction of f to $B(y, \varepsilon)$ is an isometry onto its image.

4.14 Proposition. *Let X and Y be complete connected metric spaces. Suppose that X is non-positively curved and that Y is locally a length space. If there is a map $f : Y \to X$ that is locally an isometric embedding, then Y is non-positively curved and:*

(1) *For every $y_0 \in Y$, the homomorphism $f_* : \pi_1(Y, y_0) \to \pi_1(X, f(y_0))$ induced by f is injective.*

(2) *Consider the universal coverings $\widetilde{X}$ and $\widetilde{Y}$ with their induced length metrics. Every continuous lifting $\tilde{f} : \widetilde{Y} \to \widetilde{X}$ of f is an isometric embedding.*

Proof. First we prove that Y is non-positively curved. Each point $y \in Y$ has a neighbourhood $B(y, \varepsilon)$ that is isometric to its image in X. Shrinking ε if necessary, we may assume that the image of $B(y, \varepsilon)$ under f is contained in a neighbourhood of $f(y)$ that is CAT(0) in the induced metric. Fix $p, q \in B(y, \varepsilon/3)$. Because Y is locally a length space, there is a sequence of paths joining p to q in $\overline{B}(y, 3\varepsilon/4)$ whose lengths converge to $d(p, q)$. Parameterize these paths by arc length. It follows from (1.4(5)) that the points halfway along the images of these paths form a Cauchy sequence in the complete space $f(\overline{B}(y, 3\varepsilon/4))$. The limit $f(m)$ of this sequence is a midpoint for $f(p)$ and $f(q)$ in $f(\overline{B}(y, 3\varepsilon/4))$. Because we are in a CAT(0) neighbourhood of $f(y)$, the point $f(m)$ must lie in the open ball of radius $\varepsilon/3$ about $f(y)$. Thus we obtain a midpoint m between p and q in $B(y, \varepsilon/3)$. Since Y is complete, we deduce that $B(y, \varepsilon/3)$ is a geodesic space. And since geodesic triangles in $B(y, \varepsilon/3)$ are isometric to geodesic triangles in a CAT(0) neighbourhood of $f(y)$, it follows that $B(y, \varepsilon/3)$ is a CAT(0) space. Thus Y is non-positively curved.

Y is complete and non-positively curved, so by (4.13) every non-trivial element $\gamma \in \pi_1(Y, y_0)$ is represented by a local geodesic $c_\gamma : [0, 1] \to Y$. The image of c_γ

under f is again a local geodesic (of the same length), so it is not homotopic to a constant map (4.13) and therefore $f_*(\gamma) \neq 1$.

Let $u : \widetilde{X} \to X$ and $v : \widetilde{Y} \to Y$ be the universal covering maps. Let $\tilde{f} : \widetilde{Y} \to \widetilde{X}$ be a lifting of f. By the Cartan-Hadamard Theorem, both $\widetilde{X}$ and $\widetilde{Y}$ are CAT(0) spaces, so the distance between each pair of points is the length of the unique local geodesic joining them. But if c is a local geodesic in $\widetilde{Y}$, then $\tilde{f} \circ c$ is a local geodesic of the same length in $\widetilde{X}$, because $\tilde{f} \circ c$ is a lift of $f \circ v \circ c$, and u, v and f all carry local geodesics to local geodesics of the same length. □

Injectivity Radius and Systole

We now turn our attention to the study of spaces which contain closed geodesics (isometrically embedded circles).

4.15 Definition. Recall (I.7.53) that the *injectivity radius, injrad*(X), of a geodesic space X is the supremum of the non-negative numbers r such that if $d(x, y) \leq r$ then there is a unique geodesic segment joining x to y in X. If X contains any isometrically embedded circles then the injectivity radius of X is obviously bounded above by one half of the infimum of the lengths of such circles. (An isometrically embedded circle of length ℓ is, by definition, the image of an isometric embedding $M^1_\rho \to X$, where $\sqrt{\rho} = 2\pi/\ell$.) If X contains such circles then *Sys*(X), the *systole of* X, is defined to be the infimum of their lengths.

We urge the reader to consider the many phenomena which may cause X not to contain any isometrically embedded circles.

The following result, which appears in [Gro87], complements Proposition 4.9. We follow a proof of R. Charney and M. Davis [CD93]. Recall that a metric space X is said to be *cocompact* if there exists a compact subset $K \subseteq X$ such that $X = \bigcup\{\gamma.K \mid \gamma \in \mathrm{Isom}(X)\}$.

4.16 Proposition. *A cocompact, proper, geodesic space X of curvature $\leq \kappa$ fails to be a* CAT(κ) *space if and only if it contains an isometrically embedded circle of length $< 2D_\kappa$. Moreover, if it contains such a circle then it contains a circle of length* $Sys(X) = 2\,injrad(X)$.

Proof. The sufficiency of the condition in the first assertion is clear; we shall prove its necessity. Let $r = injrad(X)$. As X is assumed to be cocompact and locally CAT(κ) we have $r > 0$. Suppose that X is not a CAT(κ) space. Then, by (4.12), we have $r < D_\kappa$.

We shall use the term (non-degenerate) *digon* to mean the union of two distinct geodesic segments $[x, y]$ and $[x, y]'$, called its sides, joining two points x and y in X. We shall refer to x and y as the vertices of the digon. The first step of the proof is to show that there exists a digon of perimeter $2r$ and the second step is to show that such a digon is an isometrically embedded circle.

By (I.3.13), geodesics of length $< r$ emanating from a fixed point in X vary continuously with their terminal endpoint. Thus we may apply (4.11) to any geodesic triangle Δ in X whose perimeter is less than $2r$ and conclude that the angles at the vertices of Δ are no bigger than the corresponding angles in a comparison triangle in M_κ^2.

By the definition of r, there is a sequence of digons in X whose perimeter tends to $2r$. Because X is proper and cocompact, we can translate these digons into a compact subspace and extract a subsequence $D(n)$ converging to the union of two geodesic segments $[x, y]$ and $[x, y]'$ of length r; we have to show that these two segments are distinct. If this were not the case, then for n big enough the digon $D(n) = [x_n, y_n] \cup [x_n, y_n]'$ would be very narrow, consequently if m is the midpoint of $[x_n, y_n]$ and m' is the midpoint of $[x_n, y_n]'$, the geodesic triangles Δ_1 and Δ_2 with vertices x_n, m, m' and y_n, m, m', respectively, would each have perimeter less than $2r$. The, by (4.11), the vertex angles in Δ_1 and Δ_2 would be less than the corresponding angles in their comparison triangles in M_κ^2. But the comparison triangle for $\Delta = [x_n, m] \cup [m, y_n] \cup [x_n, y_n]'$ is degenerate, so the angles at the vertices corresponding to x_n and y_n are zero. Alexandrov's lemma shows that these angles are not smaller than the angles at the vertices corresponding to x_n and y_n in comparison triangles for Δ_1 and Δ_2 respectively. Therefore these latter comparison triangles are also degenerate, so $[m, x_n] = [m', x_n]$ and $[m, y_n] = [m', y_n]$, contradicting the hypothesis that $[x_n, y_n]$ is distinct from $[x_n, y_n]'$. Thus $D = [x, y] \cup [x, y]'$ is a (non-degenerate) digon.

We say that a point $z \in [x, y]$ is *opposite* the point $z' \in [x, y]'$ if $d(z, x) + d(z', x) = r$. We claim that $d(z, z') = r$. If not, then the geodesic triangles with vertices z, z', x and z, z', y, respectively, would be of perimeter smaller than $2r$ and we could apply the same argument as above to conclude that the geodesic segments $[x, y]$ and $[x, y]'$ were the same. It follows that D is an isometric image of a circle. □

We saw in Chapter I.7 that in many respects M_κ-polyhedral complexes with only finitely many isometry types of cells behave like cocompact spaces. Here is another important instance of this phenomenon.

4.17 Proposition. *Let K be an M_κ-polyhedral complex with* Shapes(K) *finite and suppose that K has curvature $\leq \kappa$. If K is not a* CAT(κ) *space, then K contains an isometrically embedded circle of length $Sys(K) = 2\, injrad(K)$. If $\kappa > 0$ and K is not a* CAT(κ) *space then $Sys(K) < 2\pi/\sqrt{\kappa}$.*

Proof. Since K is locally CAT(κ) it is *a fortiori* locally geodesic. So by (I.7.55), we have $r = injrad(K) > 0$. In order to apply the arguments of (4.16) it is enough to establish the existence of a non-degenerate digon whose sides have length r. In the present setting, as on previous occasions, we shall use the hypothesis that Shapes(K) is finite to reduce to consideration of the compact case.

First we choose a sequence of (non-degenerate) digons $D(n)$ with endpoints x_n and y_n such that $d(x_n, y_n)$ approaches r. Since the lengths of the sides of the digons $D(n)$ are uniformly bounded, (I.7.30) yields an integer N such that each $D(n)$ is contained in a subcomplex L_n of K that can be expressed as the union of at most N closed cells. Because Shapes(K) is finite, there are only finitely many possibilities for L_n up to isometric isomorphism, so by passing to a further subsequence we may assume that all of the L_n are isometrically isomorphic. In other words there exists a finite M_κ-polyhedral complex L with Shapes(L) $\subseteq$ Shapes(K) such that for every positive integer n there is a bijective map $\phi_n : L \to L_n$ that sends cells to cells and restricts to an isometry on the individual cells equipped with their local metrics.

Each of the maps ϕ_n is a length-preserving map of geodesic spaces, so $d(x_n, y_n) = d(\phi_n^{-1}(x_n), \phi_n^{-1}(y_n))$ and $\phi_n^{-1}D(n) \subset L$ is a geodesic digon with sides of length $d(x_n, y_n)$. Since L is compact, we may pass to a subsequence and assume that these digons converge in L to a geodesic digon $\widetilde{D}$ with endpoints $\tilde{x}$ and $\tilde{y}$. The argument given in (4.16) shows that the width of the digons $D(n)$ is bounded away from zero, so since $\phi_n : \phi_n^{-1}D(n) \to D(n)$ does not increase distances, $\widetilde{D}$ is non-degenerate. However, we are not quite done at this stage, because in general the sides of $\phi_n\widetilde{D}$ will not be geodesics in K: we have $d(\phi_n(\tilde{x}), \phi_n(\tilde{y})) \leq d(\tilde{x}, \tilde{y}) = r$ for all n, but we do not get equality in general. However we do know that

$$\begin{aligned} d(\phi_n(\tilde{x}), \phi_n(\tilde{y})) - r &\leq d(\phi_n(\tilde{x}), x_n) + d(x_n, y_n) + d(y_n, \phi_n(\tilde{y})) - r \\ &\leq d(\tilde{x}, \phi_n^{-1}(x_n)) + d(\phi_n^{-1}(y_n), \tilde{y}) + d(x_n, y_n) - r, \end{aligned}$$

which tends to zero as $n \to \infty$.

Let $\overline{x} \in \overline{S}$ and $\overline{y} \in \overline{S}'$ be points in the model cells $\overline{S}, \overline{S}' \in$ Shapes(L) $\subseteq$ Shapes(K) such that $f_S(\overline{x}) = \tilde{x}$ and $f_{B'}(\overline{y}) = \tilde{y}$ for some closed cells $S \ni \tilde{x}$ and $S' \ni \tilde{y}$ in L. By construction, for all n we have $\phi_n(\tilde{x}) \in X_K$ and $\phi_n(\tilde{y}) \in Y_K$, where X_K and Y_K are as in (I.7.59). We proved in (I.7.59) that the set of numbers $\{d(x, y) \mid x \in X_K, y \in Y_K\}$ is discrete. Hence for all sufficiently large n we have $d(\phi_n(\tilde{x}), \phi_n(\tilde{y})) = r$, and $\phi_n(\widetilde{D})$ is a non-degenerate geodesic digon with sides of length r. □

Chapter II.5 M_κ-Polyhedral Complexes of Bounded Curvature

In this chapter we return to the study of metric polyhedral complexes, which we introduced in Chapter I.7. Our focus now is on complexes whose curvature is bounded from above. The first important point to be made is that in this context one can reformulate the CAT(κ) condition in a number of useful ways. In particular, Gromov's *link condition* (5.1) enables one to reduce the question of whether or not a complex supports a metric of curvature $\leq \kappa$ to a question about the geometry of the links in the complex. This opens the way to arguments that proceed by induction on dimension, as we saw in (I.7). If the cells of the complex are sufficiently regular then the link condition can be interpreted as a purely combinatorial condition; in particular this is the case for cubical complexes and many 2-dimensional complexes.

This chapter is organized as follows. We begin by establishing the equivalence of various characterizations of what it means for a polyhedral space to have curvature bounded above. We then address the question of which complexes have the property that their local geodesics can be prolonged indefinitely, and we describe some general results of a topological nature. There then follows a discussion of flag complexes and all-right spherical complexes. Two important results arising from this discussion are: there is no topological obstruction to the existence of CAT(1) metrics on simplicial complexes (5.19), and there is a purely combinatorial criterion for deciding whether or not a cubical complex is non-positively curved (5.20). In the fourth section we describe some constructions of cubical complexes and outline Davis's construction of aspherical manifolds that are not covered by Euclidean space. The remaining sections deal with two-dimensional complexes. Two-dimensional complexes provide a rich source of examples of non-positively curved spaces, and their usefulness is enhanced by the close link between geometry and group theory in dimension two (cf. 5.45). Two-dimensional complexes also enjoy some important properties which are not shared by their higher dimensional cousins (see 5.27).

Further explicit examples of non-positively curved polyhedral complexes will be described in Chapters II.12 and III.Γ, and the techniques described in Chapters II.11, II.12 and III.*C* provide means of constructing many other examples.

Characterizations of Curvature $\leq \kappa$

We begin by describing criteria for recognizing when an M_κ-polyhedral complex is a CAT(κ) space. In the course of our discussion we shall need to call on a number of facts from Chapter I.7 concerning M_κ-polyhedral complexes with only finitely many isometry types of cells. Most of these facts can be proved much more easily in the special case where K is locally compact, as the interested reader can verify.

The following criterion for an M_κ-complex to have curvature $\leq \kappa$ is due to M. Gromov [Gro87, p.120]. A proof in the locally compact case was given by Ballmann [Ba90] and in the general case by Bridson [Bri91].

5.1 Definition. An M_κ-polyhedral complex satisfies *the link condition* if for every vertex $v \in K$ the link complex $\text{Lk}(v, K)$ is a CAT(1) space.

5.2 Theorem. *An M_κ-polyhedral complex K, with* Shapes(K) *finite, has curvature $\leq \kappa$ if and only if it satisfies the link condition.*

Proof. Let v be a vertex of K. According to (I.7.39), there exists $\varepsilon(v) > 0$ such that $B(v, \varepsilon(v))$ is convex and isometric to the $\varepsilon(v)$ neighbourhood of the cone point in $C_\kappa(\text{Lk}(v, K))$. Thus, in the light of Berestovskii's theorem (3.14), we see that K satisfies the link condition if and only if every vertex has a neighbourhood that is a CAT(κ) space. To complete the proof of the theorem one simply notes that, given any $x \in K$, if v is a vertex of supp(x) and $\eta > 0$ is sufficiently small, then $B(x, \eta)$ is isometric to $B(x', \eta)$ for some $x' \in B(v, \varepsilon(v))$ (see I.7.56). $\square$

5.3 Remark. In the above theorem one can weaken the hypotheses and require only that the number $\varepsilon(x)$ defined in (I.7.39) is positive for every $x \in K$.

5.4 Theorem. *Let K be an M_κ-polyhedral complex with* Shapes(K) *finite. If $\kappa \leq 0$ then the following conditions are equivalent:*

(1) *K is a* CAT(κ) *space;*

(2) *K is uniquely geodesic;*

(3) *K satisfies the link condition and contains no isometrically embedded circles;*

(4) *K is simply connected and satisfies the link condition.*

If $\kappa > 0$ then the following conditions are equivalent:

(5) *K is a* CAT(κ) *space;*

(6) *K is $(\pi/\sqrt{\kappa})$-uniquely geodesic;*

(7) *K satisfies the link condition and contains no isometrically embedded circles of length less than $2\pi/\sqrt{\kappa}$.*

Proof. The implications (1) $\Longrightarrow$ (2) and (5) $\Longrightarrow$ (6) are clear. (4.17) together with (5.2) shows that (3) is equivalent to (1) and that (5) is equivalent to (7). As a consequence (I.7.1), the Cartan-Hadamard Theorem and (5.2), we have (4) $\Longrightarrow$ (3).

We argue by induction on the dimension of K in order to see that (2) $\Longrightarrow$ (4) and (6) $\Longrightarrow$ (7). For complexes of dimension 1 these implications are clear. If K satisfies (2) or (6) then in particular its injectivity radius is positive, so by (I.7.55) $\text{Lk}(x, K)$ is uniquely π-geodesic for every $x \in K$. Hence, by induction, $\text{Lk}(x, K)$ is a CAT(1) space. (5.2) completes the proof of (6) $\Longrightarrow$ (7). To complete the proof of (2) $\Longrightarrow$ (4), we need the additional fact that if (2) holds then by (I.7.58) geodesics vary continuously with their endpoints in K, so K is contractible (hence simply connected). Indeed, for fixed $x_0 \in K$, we obtain a contraction $K \times [0, 1] \to K$ by sending (x, t) to the point a distance $t\, d(x_0, x)$ from x along the unique geodesic from x to x_0. □

The same arguments are enough to prove a local version of the above theorem:

5.5 Theorem. *Let K be an M_κ-polyhedral complex with* Shapes(K) *finite. The following conditions are equivalent:*

(1) *K has curvature $\leq \kappa$;*

(2) *K satisfies the link condition;*

(3) *K is locally uniquely geodesic;*

(4) *K has positive injectivity radius.*

The 2-Dimensional Case.

Let K be a 2-dimensional M_κ-complex. The link of a vertex $v \in K$ is a metric graph whose vertices correspond to 1-cells incident at v and whose edges correspond to corners of the 2-cells S incident at v; the length of an edge is the vertex angle at the corresponding corner of the model simplex $\overline{S} \in \text{Shapes}(K)$. A metric graph is a CAT(κ) space if and only if every locally injective loop in the graph has length at least $2\pi/\sqrt{\kappa}$, therefore:

5.6 Lemma. *A 2-dimensional M_κ–complex K satisfies the link condition if and only if for each vertex $v \in K$ every injective loop in* $\text{Lk}(v, K)$ *has length at least 2π.*

Extending Geodesics

Many important results concerning the global geometry of simply connected Riemannian manifolds X of non-positive curvature rely in an essential way upon the fact that if the manifold is complete then one can extend any geodesic path in X to a complete geodesic line. When seeking to generalize such results to CAT(0) spaces, it is important to understand which spaces share this extension property. We shall address this question in the case of polyhedral complexes and topological manifolds.

It is useful to phrase the extension property locally.

5.7 Definition. A geodesic metric space X is said to have the *geodesic extension property* if for every local geodesic $c : [a, b] \to X$, with $a \neq b$, there exists $\varepsilon > 0$ and a local geodesic $c' : [a, b + \varepsilon] \to X$ such that $c'|_{[a,b]} = c$.

5.8 Lemma. *Let X be a complete geodesic metric space.*

(1) *X has the geodesic extension property if and only if every local geodesic $c : [a, b] \to X$ with $a \neq b$ can be extended indefinitely, i.e. there exists a local isometry $\tilde{c} : \mathbb{R} \to X$ such that $\tilde{c}|_{[a,b]} = c$.*

(2) *If X is a* CAT(0) *space, then X has the geodesic extension property if and only if every non-constant geodesic can be extended to a geodesic line $\mathbb{R} \to X$.*

Proof. For (1), one simply notes that if X has the geodesic extension property then, for every local geodesic $c : [a, b] \to X$, the set of numbers $t \geq 0$ such that $c : [a, b] \to X$ can be extended to a local isometry $[a - t, b + t] \to X$ is open; the completeness of X assures that it is also closed, and hence the whole of $[0, \infty)$.

Assertion (2) is an immediate consequence of the fact that in a CAT(0) space every local geodesic is in fact a geodesic (1.4(2)). $\square$

In the case of polyhedral complexes, the geodesic extension property can be rephrased in terms of the combinatorial structure of the complex.

5.9 Definition. Let K be an M_κ–polyhedral complex. A closed n-cell B in K is said to be a *free face* if it is contained in the boundary of exactly one cell B' of higher dimension and the intersection of the interior of B' with some small neighbourhood of an interior point of B is connected. (The second clause in the preceding sentence is necessary in order to avoid suggesting, for example, that if both ends of a 1-cell are attached to a single vertex in a graph then that vertex is a free face.)

5.10 Proposition. *Let K be an M_κ–polyhedral complex of curvature $\leq \kappa$ with* Shapes(K) *finite. K has the geodesic extension property if and only if it has no free faces.*

Proof. By (I.7.39), a small neighbourhood of each point $x \in K$ is isometric to the κ-cone on $\mathrm{Lk}(x, K)$. It follows from (I.5.7) that in any cone $C_\kappa Y$, a local geodesic from ty incident at the cone point 0 can be extended past 0 if and only if there exists $y' \in Y$ with $d(y, y') \geq \pi$. Therefore K has the geodesic extension property if and only if for every $x \in K$ and every $u \in \mathrm{Lk}(x, K)$ there exists $u' \in \mathrm{Lk}(x, K)$ with $d(u, u') \geq \pi$. We call u' an antipode for u.

Suppose that K has a free face of dimension n. If x is an interior point of this face, then $\mathrm{Lk}(x, K)$ is isometric to a closed hemisphere in $\mathbb{S}^n$. In particular, some points of $\mathrm{Lk}(x, K)$ do not have antipodes, and K does not have the geodesic extension property.

Suppose now that K does not have any free faces. We shall prove that every point in every link $\mathrm{Lk}(x, K)$ has an antipode. We argue by induction on the dimension of K; the 1-dimensional case is trivial. Given $x \in K$ and $u \in \mathrm{Lk}(x, K)$, we choose

$u' \in \mathrm{Lk}(x, K)$ with $u' \neq u$ (note u must exist, for otherwise x would be a free face of dimension 0). If u' is in a different connected component to u then they are an infinite distance (greater than π!) apart and we are done. If not then we connect u to u' by a geodesic in $\mathrm{Lk}(x, K)$. Since K has no free faces, neither does $\mathrm{Lk}(x, K)$, so using our inductive hypothesis we may extend the geodesic $[u, u']$. Indeed, because $\mathrm{Lk}(x, K)$ is complete (7.19), we may extend $[u, u']$ to a local geodesic $c : \mathbb{R} \to \mathrm{Lk}(x, K)$ with $c(0) = u$.

By hypothesis, $\mathrm{Lk}(x, K)$ is a CAT(1) space. In a CAT(1) space any local geodesic of length π is actually a geodesic (1.10), thus $c(\pi)$ is the desired antipodal point for u. $\square$

5.11 Exercise. Give an example to show that the curvature hypothesis in the preceding proposition is necessary.

Here is another important class of spaces with the geodesic extension property.

5.12 Proposition. *If a complete metric space X has curvature $\leq \kappa$ and is homeomorphic to a finite dimensional manifold, then it has the geodesic extension property.*[24]

Proof. Let n be the dimension of X. Suppose that a certain local geodesic c cannot be extended past $x \in X$. We fix a small CAT(κ) neighbourhood B of x and a point $y \in B$ in the image of c. Because the unique geodesic from y to x cannot be extended locally, no geodesic to y from any point of B passes through x. Thus the geodesic retraction of B to y restricts to give a retraction of $B \setminus \{x\}$ within itself. It follows that the local homology group $H_n(B, B \setminus \{x\})$ is trivial. On the other hand, we may choose a homeomorphism from the n-dimensional disc D^n onto a neighbourhood V of x in B, and by excision, $H_n(B, B \setminus \{x\}) = H_n(V, V \setminus \{x\}) = H_n(D^n, D^n \setminus \{0\}) = \mathbb{Z}$ (see [Spa66]). $\square$

At this point, when we have temporarily found the need to use some non-trivial topology, we note a result that relates complexes to more general spaces of bounded curvature.[25]

5.13 Proposition. *Let X be a compact metric space. If there exists $\kappa \in \mathbb{R}$ such that X has curvature $\leq \kappa$, then X has the homotopy type of a finite simplicial complex. (In fact, X is a compact ANR.)*

Proof. Every point of X has a neighbourhood which is CAT(κ) in the induced metric. Since X is compact, it follows that there exists $\varepsilon \in (0, \pi/6\sqrt{\kappa})$ such that $B(x, 3\varepsilon)$ is CAT(κ) for every $x \in X$. According to (1.4), geodesics in $B(x, 3\varepsilon)$ are unique and vary continuously with their endpoints, and $B(x, \varepsilon)$ is convex. Lemma I.7A.19

---

[24] Experts will notice that all that is actually required in the proof of (5.12) is that X is a homology manifold.

[25] An equivariant version of this result is proved in a recent preprint of Pedro Ontaneda entitled " Cocompact CAT(0) spaces are almost extendible".

implies that X is homotopy equivalent to the nerve of any covering of X by balls of radius ε.

The parenthetical assertion in the statement of the proposition is a special case of [Hu(S), III.4.1]. (A deep theorem of Jim West [West77] states that every compact ANR has the homotopy type of a finite simplicial complex, so this parenthetical assertion is stronger than the main assertion of the proposition.) □

5.14 Remark. The preceding theorem does *not* say that every compact non-positively curved space is homotopy equivalent to a compact *non-positively curved* simplicial complex. In fact this is not true even for compact Riemannian manifolds of non-positive curvature, as was recently shown by M. Davis, B. Okun and F. Zheng, and (independently) by B. Leeb.

Flag Complexes

In the introduction to this chapter we alluded to the usefulness of having a purely combinatorial criterion for checking whether complexes satisfy the link condition and we indicated that such a criterion exists for cubed complexes. The purpose of this section is to describe this condition.

5.15 Definition. Let L be an abstract simplicial complex and let $\mathcal{V}$ be its set of vertices. L is called a *flag complex* (alternatively, *"L satisfies the no triangles condition"*) if every finite subset of $\mathcal{V}$ that is pairwise joined by edges spans a simplex. In other words, if $\{v_i, v_j\}$ is a simplex of L for all $i, j \in \{1, \dots, n\}$, then $\{v_1, \dots, v_n\}$ is a simplex of L.

5.16 Remarks

(1) If L is a flag complex, then so is the link of each vertex in L.

(2) The *girth* of a simplicial graph is the least number of edges in any reduced circuit in the graph. A simplicial graph L is a flag complex if and only if its girth is at least 4.

(3) The barycentric subdivision of every simplicial complex is a flag complex.

(4) (*No triangles condition*). Let L be a simplicial complex. Then L is a flag complex if and only if the statement $\Phi_n(L)$ is false for all $n > 0$, where the statement Φ_n is defined inductively as follows: $\Phi_0(L)$ is the statement that there exist vertices $v_0, v_1, v_2 \in L$ such that $\{v_i, v_j\}$ is a simplex for all $i, j \in \{0, 1, 2\}$ but $\{v_0, v_1, v_2\}$ is not a simplex of L; and Φ_{n+1} is the statement that there exists a vertex $v_0 \in L$ such that $\Phi_n(\mathrm{Lk}(v_0, L))$.

(5) The simplicial join of two flag complexes is again a flag complex.

5.17 Definition. An *all-right* spherical complex is an M_1–simplicial complex each of whose edges has length $\pi/2$.

The link of every vertex in an all-right spherical complex is itself an all-right spherical complex.

The following theorem is due to M. Gromov [Gro87].

5.18 Theorem. *Let L be a finite dimensional all-right spherical complex. L is* CAT(1) *if and only if it is a flag complex.*

Proof. First we observe that if there exist vertices $v_0, v_1, v_2 \in L$ that are pairwise joined by edges but do not span a simplex, then $[v_0, v_1] \cup [v_1, v_2] \cup [v_2, v_0]$ is a locally geodesic loop in L. To see this, note that in $\text{Lk}(v_1, L)$, which is an all-right complex, the vertices corresponding to the edges $[v_0, v_1]$ and $[v_1, v_2]$ are not joined by an edge, so the distance between them is at least π and therefore $[v_0, v_1] \cup [v_1, v_2] \cup [v_2, v_0]$ is geodesic in a neighbourhood of v_1 (by 5.4).

If L is CAT(1), then so is the link of every vertex $v \in L$, the link of every vertex $v' \in \text{Lk}(v, L)$, the link of every vertex $v'' \in \text{Lk}(v', \text{Lk}(v, L))$, etc.. In particular, none of these successive links can contain a geodesic circle of length less than 2π. Each of these links is an all-right spherical complex. Thus, applying the argument of the preceding paragraph to these successive links, we deduce that the statement $\Phi_n(L)$ defined in (5.16(4)) is false for all $n > 0$, and therefore L is a flag complex.

It remains to prove that if L is a flag complex then it is CAT(1). Proceeding by induction on the dimension of L, we may assume that $\text{Lk}(v, L)$ is CAT(1) for every vertex $v \in L$, and therefore it suffices (by 5.4(7)) to prove that any curve $\ell \subset L$ which is isometric to a circle must have length at least 2π.

We consider the geometry of geodesic circles in arbitrary all-right complexes K. We claim that if v is a vertex of K and ℓ is a geodesic circle, then each connected component ℓ' of $\ell \cap B(v, \pi/2)$ must have length π. To see this, consider the development of ℓ' in $\mathbb{S}^2$ (Fig. I.7.1): this is a local geodesic (with the same length as ℓ') in an open hemisphere of $\mathbb{S}^2$ (the ball of radius $\pi/2$ about the image of v). Since the endpoints of this local geodesic lie on the boundary of the hemisphere, it must have length π.

Now suppose that the length of ℓ is less than 2π. In this case ℓ cannot contain two disjoint arcs of length $\geq \pi$ and therefore cannot meet two disjoint balls of the form $B(v, \pi/2)$. Thus the set of vertices $v \in K$ such that ℓ meets $B(v, \pi/2)$ is pairwise joined by edges in K. If K were a flag complex, this set of vertices would have to span a simplex S and ℓ would be contained in S, which is clearly impossible. Thus we conclude that all-right flag complexes do not contain isometrically embedded circles of length less than 2π. $\square$

The following corollary was proved previously by Berestovskii [Ber86]. It shows that there is no topological obstruction to the existence of a CAT(1) metric on a finite dimensional complex.

5.19 Corollary. *The barycentric subdivision of every finite dimensional simplicial complex K supports a piecewise spherical* CAT(1) *metric.*

Proof. Consider the barycentric subdivision K' of the given complex — this is a flag complex. The canonical all-right spherical metric on K' makes it CAT(1). □

We refer the reader to (I.7.40) for an explanation of the terminology "cubed" versus "cubical". The following theorem is due to M. Gromov [Gro87].

5.20 Theorem. *A finite dimensional cubed complex has non-positive curvature if and only if the link of each of its vertices is a flag complex.*

Proof. According to (5.2), a cubed complex has non-positive curvature if and only if the link of each vertex is CAT(1). Such a link is an all-right spherical complex, so we can apply Theorem 5.18. □

5.21 Moussong's Lemma. There is a useful generalization of (5.18) due to Gabor Moussong [Mou88]. A spherical simplicial complex L is called a *metric flag complex* (see [Da99]) if it satisfies the following condition: if a set of vertices $\{v_0, \dots, v_k\}$ is pairwise joined by edges $e_{i,j}$ in L and there exists a spherical k-simplex in $\mathbb{S}^k$ whose edges lengths are $l(e_{i,j})$, then $\{v_0, \dots, v_k\}$ is the vertex set of a k-simplex in L.

Moussong's Lemma asserts that if all the edges of a spherical simplicial complex L have length at least $\pi/2$, then L is CAT(1) if and only if it is a metric flag complex.

This lemma has been used by Moussong [Mou88] (resp. M. Davis [Da98]) to prove that the appropriate geometric realization of any Coxeter system (resp. building) is CAT(0). See also [CD95a].

Constructions with Cubical Complexes

The following construction is essentially due to Mike Davis.

5.22 Proposition. *Given any finite dimensional simplicial complex L, one can construct a cubical complex K such that the link of every vertex in K is isomorphic to L. The group of isometries of K acts transitively on the set of vertices of K, and if L has a finite number m of vertices, then K has 2^m vertices.*

Proof. Let V be the vertex set of L and let E be a Euclidean vector space with orthonormal basis $\{e_s \mid s \in V\}$. We shall construct K as a subset of the following union of cubes $\mathcal{C} = \bigcup_T \{\sum_{s \in T} x_s e_s \mid x_s \in [0, 1]\}$, where T varies over the finite subsets of V. A face of $\mathcal{C}$ belongs to K if and only if it is parallel to a face spanned by basis vectors $e_{s_0}, \dots, e_{s_k}$, where $s_0, \dots, s_k$ are the vertices of a k-simplex in L. Note that the vertices of K are just the vertices of $\mathcal{C}$. The link in K of the vertex at the origin is isomorphic to L.

Let r_s be the reflection in the hyperplane that passes through $e_s/2$ and is orthogonal to e_s, and let G be the subgroup of Isom(E) generated by these reflections. G is the direct sum of the cyclic groups of order two generated by the r_s (so if V is finite then $|G| = 2^{|V|}$). The action of G leaves K invariant and acts simply transitively on the set of vertices of K. Since we already know that the link of one vertex is isomorphic to L, this proves the proposition. □

According to (5.20), the cubical complex K constructed above will be non-positively curved if and only if L is a flag complex. Thus, for example, if L is the barycentric subdivision of any simplicial complex, then the universal cover $\widetilde{K}$ of the complex constructed above will be a CAT(0) space, and in particular it will be contractible.

5.23 A Compact Aspherical Manifold Whose Universal Cover is Not a Ball. In 1983 Mike Davis solved an important open question in the study of topological manifolds. Until that time it was unknown whether or not there existed closed aspherical manifolds whose universal cover was not homeomorphic to Euclidean space. Davis constructed the first such examples [Da83]. We shall sketch his construction, proceeding in several steps from the construction in (5.22).

Step 1: Let L, K and G be as in (5.22) but assume now that L is a finite complex. Let F be the intersection of K with the cube $\{\sum_s x_s e_s \in E \mid x_s \in [0, 1/2]\}$. Note that F is a cubical complex (with edges of length $1/2$). Note also that F is a strict fundamental domain for the action of G on K, i.e. each G-orbit meets F in exactly one point. Given a $(k-1)$-simplex σ of L with vertices $s_1, \dots, s_k$, we write G_σ to denote the subgroup of G generated by the reflections $r_{s_1}, \dots, r_{s_k}$ and F^σ to denote the intersection of F with the affine subspace defined by the equations $x_{s_1} = 1/2, \dots, x_{s_k} = 1/2$. We also write $e_\sigma = \sum_{i=1}^k e_{s_i}$. Note that G_σ is the stabilizer of F^σ in G.

We endow G with the discrete topology. Thus $G \times F$ is the disjoint union of copies of F that are indexed by the elements of G. We identify K with the quotient of $G \times F$ by the equivalence relation $[(g, x) \sim (gh, x)$ if $x \in F^\sigma$ and $h \in G_\sigma]$, where σ ranges over the simplices of L. We write this identification $(g, x) \mapsto g.x$.

An important point to note is that F can be identified with the simplicial cone CL' over the barycentric subdivision L' of L. Indeed one obtains a bijection $CL' \to F$ that is affine on each simplex by sending the cone vertex of CL' to the origin of E and the barycentre of each simplex σ of L to $\frac{1}{2}e_\sigma \in F$. This isomorphism identifies F^σ with the subcomplex L^σ of $L' \subset CL'$ that is the intersection of the stars in L' of the vertices $s_i \in \sigma$.

Step 2: Let M be a compact piecewise-linear[26] n-manifold with non-empty boundary ∂M and let L be a triangulation of ∂M (compatible with the given PL-structure). We mimic the construction of K in Step 1, replacing $G \times F$ by $G \times M$. Explicitly, we define N to be the quotient of $G \times M$ by the equivalence relation $[(g, x) \sim (gh, x)$ if $x \in L^\sigma \subset \partial M$ and $h \in G_\sigma]$, where σ ranges over the simplices of L.

[26] henceforth abbreviated to PL

We claim that N is a compact n-manifold without boundary. It is sufficient to check that the equivalence class of a point of the form $(1, x) \in G \times \partial M$ has a neighbourhood homeomorphic to an n-ball. Because ∂M is a PL-manifold of dimension $n-1$, if $\sigma \subset L$ is a $(k-1)$-simplex then L^σ is PL-homeomorphic to an $(n-k)$-ball. The star in L' of the barycentre of σ is the simplicial join (see I.7A.2) of L^σ and the boundary of the simplex σ. It follows that for each point x in the interior of L^σ, there is a PL-homeomorphism j from a neighbourhood U of x in M to a neighbourhood of 0 in $(\mathbb{R}^+)^k \times \mathbb{R}^{n-k}$, with $j(U \cap L) \subset \partial(\mathbb{R}^+)^k \times \mathbb{R}^{n-k}$: the map j sends x to 0 and sends $U \cap L^{s_i}$, for each vertex s_i of σ, into the subspace consisting of vectors whose i-th coordinate is 0.

Consider the homomorphism $\alpha : G_\sigma \to \text{Isom}(\mathbb{E}^n)$ sending each r_{s_i} to the reflection in the i-th coordinate hyperplane of $\mathbb{R}^n$. The map sending $(g, y) \in G_\sigma \times U$ to $\alpha(g).j(y)$ induces a homeomorphism from a neighbourhood in N of the equivalence class of $(1, x)$ onto a neighbourhood of $0 \in \mathbb{R}^k \times \mathbb{R}^{n-k}$.

Step 3: Let K be as in Step 1 and let M be as in Step 2, and assume now that the manifold M is contractible. In this case there is a homotopy equivalence $f_0 : CL \to M$ which restricts to the identity on $L = \partial M$ and hence on L'. The map $(g, x) \mapsto (g, f_0(x))$ is compatible with the equivalence relations on $G \times CL' = G \times F$ and $G \times M$ and thus we obtain a G-equivariant homotopy equivalence $f : K \to N$.

If the triangulation L of ∂M is a flag complex (this can always be achieved by passing to a barycentric subdivision), then by (5.20) the cubical complex K is non-positively curved. In particular, its universal cover is contractible, so all of its higher homotopy groups are trivial, i.e. K is aspherical. Since N is homotopy equivalent to K, it too is aspherical (so its universal cover is contractible).

In dimensions 4 and above it is possible to construct compact contractible n-manifolds whose boundary is not simply connected (the first examples were given by B. Mazur [Maz61] and V. Poenaru [Poe60]). In this case Davis shows that the universal covering of N, though contractible, is not simply connected at infinity [27] and hence is not homeomorphic to $\mathbb{R}^n$.

In the above construction N is not endowed with any geometry — we merely related its topology to a space K where we could use geometry (specifically, the Cartan-Hadamard theorem) to prove that the space was aspherical. Remarkably, if the dimension of M (and hence N) is at least 5, then Ancel, Davis and Guibault [AnDG97] prove that N can always be metrized as a cubical complex of non-positive curvature.

[27] This means that $\widetilde{N}$ cannot be exhausted by a sequence of compact sets C such that any loop in the complement of C is contractible in $\widetilde{N} \smallsetminus C$.

Two-Dimensional Complexes

We now focus our attention on 2-dimensional M_κ–complexes of curvature $\leq \kappa$. There are three main reasons for doing so. First of all, in the 2-dimensional case it is easy to check if an explicitly described M_κ–complex has curvature $\leq \kappa$. Secondly, there are many natural and interesting examples. And thirdly, 2-dimensional complexes enjoy some important properties that fail in higher dimensions (for example Theorem 5.27). We refer the reader to [Hag93], [BaBr94,95], [BaBu96], [BuM97], [BriW99], [Sw98] and [Wi96a] for examples of recent results in this active area.

5.24 Checking the Link Condition. As we noted earlier, it is easy to check whether or not a given 2-dimensional M_κ–complex K_0 has curvature $\leq \kappa$, because the link condition for such a complex admits a simple reformulation:

(5.6). *A 2-dimensional M_κ–complex K_0 satisfies the link condition if and only if for each vertex $v \in K_0$ every injective loop in* $\mathrm{Lk}(v, K_0)$ *has length at least 2π.*

Thus, if we label each of the corners of the 2-cells of K_0 with the angle α_i at that corner, then the link condition for K_0 is equivalent to a system of simultaneous inequalities of the form $\alpha_{i_1} + \cdots + \alpha_{i_m} \geq 2\pi$ — there is one inequality for each injective loop in the link of each vertex. Moreover, if $\kappa \leq 0$ then for each n-sided face of K_0 we know that the sum of the vertex angles of that face is bounded by the sum of the angles in a Euclidean n-gon: $\alpha_{i'_1} + \cdots + \alpha_{i'_m} \leq (n-2)\pi$. If $\kappa < 0$ then each of this last set of inequalities will be sharp.

Changing perspective, we may view the above system of inequalities as giving a necessary (but not sufficient) condition for a given combinatorial 2-complex to admit the structure of an M_κ–complex of non-positive curvature. For suppose that we are given a connected, combinatorial 2-complex K such that the attaching map of each 2-dimensional cell is a closed locally injective loop in the graph $K^{(1)}$; we say that the cell is *n-sided* if this loop crosses exactly n edges. A *corner* is defined to be a pair of successive 1-cells in the boundary of a 2-cell (this corresponds to an edge in the link of the vertex at which the 1-cells are incident). To each corner we assign a real variable α_i and consider the simultaneous system of inequalities requiring that the sum of the α_i around any injective loop in the link graph of each vertex of K is at least 2π, and the sum of the α_i around each n-sided face of K is at most $(n-2)\pi$. If this system of inequalities does not admit a solution with each $\alpha_i \in (0, \pi]$, then K does not support an M_κ-polyhedral structure of non-positive curvature (cf. [Ger87]).

There is a partial converse: if there is a solution such that all of the corners of each face have the same angle, then one can obtain a metric of non-positive curvature by metrizing all of the 2-cells as regular Euclidean n-gons of side length 1. However, the existence of a less symmetric solution to the above system of inequalities does not guarantee the existence an M_κ–complex structure of non-positive curvature (although it does imply that K is aspherical [Ger87], [Pri88]). The key point to observe is that while a solution appears to allow one to assign appropriate angles to the corners of K, in general one cannot assign edge lengths in a consistent way that is compatible with metrics on the 2-cells that have the desired angles. Readers wishing to understand this difficulty should consider the 2-complex with one vertex, two directed 1-cells labelled a and b respectively, and one 2-cell, whose boundary traces out the path a (in the positive direction), then b twice, then a in the reverse direction, then b in the reverse direction. The fundamental group of this complex is solvable but not virtually abelian, and therefore it is not the fundamental group of any compact non-positively curved space (cf. 7.5).

We note some situations in which the link condition is immediate. (These are analogous to the small cancellation conditions of combinatorial group theory, see [LyS77].)

5.25 Proposition. *Let K be a combinatorial 2-complex and suppose that there is an integer N such that K contains no n-sided faces with $n > N$. Let p and q be positive integers such that every face of K has at least p sides and every simple closed loop in the link graph of every vertex of K has combinatorial length at least q.*

(1) *If $(p, q) \in \{(3, 6),\ (4, 4),\ (6, 3)\}$ then K can be metrized as a piecewise Euclidean complex of non-positive curvature.*

(2) *If $(p, q) \in \{(3, 7),\ (4, 5),\ (5, 4),\ (6, 3)\}$ then K can be metrized as a piecewise hyperbolic complex of curvature ≤ -1.*

Proof. For (1) it suffices to metrize each n-sided 2-cell of K as a regular Euclidean n-gon with sides of length 1.

For every $\varepsilon > 0$ and $N \in \mathbb{N}$ there exists $\delta(\varepsilon, N) > 0$ such that if $n \leq N$ then the vertex angles in a regular n-gon in $\mathbb{H}^2$ with sides of length $\leq \delta$ are at least $[\pi(n-2)/n] - \varepsilon$. For (2) it suffices to metrize each n-sided 2-cell of K as a regular hyperbolic n-gon with sides of length $\delta(\pi/21, N)$. □

5.26 Exercise. Let K be a piecewise Euclidean 2-complex each of whose faces is metrized as a regular n-gon, where n may vary from face to face but is uniformly bounded. Prove that if every injective loop in the link of each vertex of K has length $> 2\pi$, then K can be metrized as an M_κ-complex of curvature $\leq \kappa$ for all $\kappa < 0$.

Subcomplexes and Subgroups in Dimension 2

This section is dedicated to the proof of the following theorem.

5.27 Theorem. *If K is a 2-dimensional M_κ–complex of curvature $\leq \kappa$, then every finitely presented subgroup of $\pi_1(K)$ is the fundamental group of a compact, 2-dimensional, M_κ–complex of curvature $\leq \kappa$ with the geodesic extension property.*

This theorem exemplifies a principle which can be applied to any class of combinatorial 2-complexes that is closed under passage to subcomplexes and covering spaces. This principle is founded upon the concept of a *tower*, which was developed in the context of 3-manifolds by C.D. Papakyriakopoulos [Papa57] and adapted to the setting of combinatorial complexes by Jim Howie [How81]. Roughly speaking, Papakyriakopoulos's insight was that a large class of maps can be expressed as the composition of a series of inclusions and covering maps, and this enables one to exploit the fact that by lifting a map to a proper covering space one can often simplify it (in some measurable sense). The relevance of this idea in the present context was brought to our attention by Peter Shalen.

5.28 Definition. Given connected CW-complexes Y and Z, with Y compact, a map $g : Y \to Z$ will be called an *admissible tower* if it can be written

$$g = i_0 \circ p_1 \circ i_1 \circ \cdots \circ p_h \circ i_h$$

where each $p_r : Z_r \to Y_{r-1}$ is a connected covering of a compact complex, Z_r has the induced cell structure, and each $i_r : Y_r \to Z_r$ is the inclusion map of a compact connected subcomplex. $Y_h = Y$ and $Z_0 = Z$.

A map of CW complexes is said to be *combinatorial* if it sends open cells homeomorphically onto open cells. (Note that a tower is a combinatorial map.) Let X be a compact CW-complex. By a *tower lift* of a combinatorial map $f : X \to Z$ we will mean a decomposition $f = g \circ f'$, where $f' : X \to Y$ is a combinatorial map and $g : Y \to Z$ is an admissible tower. In this setting it is convenient to write $f : X \xrightarrow{f'} Y \xrightarrow{g} Z$.

The following result is a special case of Lemma 3.1 in [How81].

5.29 Lemma. *If X and Z are connected CW-complexes and X is compact, then every combinatorial map $f : X \to Z$ has a maximal tower lift $f : X \xrightarrow{f'} Y \xrightarrow{g} Z$ such that $f'_* : \pi_1 X \to \pi_1 Y$ is surjective.*

Proof. Let Y_0 denote the image of f. We measure the complexity of f by means of the non-negative integer $c(f) := d(X) - d(Y_0)$, where $d(K)$ denotes the number of 0-cells of a compact complex K.

Let $i_0 : Y_0 \to Z$ be the inclusion map and consider $f : X \xrightarrow{f_0} Y_0 \xrightarrow{i_0} Z$. If the map $\pi_1 X \to \pi_1 Y$ induced by f_0 is not a surjection then by elementary covering space theory (see [Mass91] for example) there exists a proper connected covering $p_1 : Z_1 \to Y_0$ such that $f = i_0 \circ p_1 \circ i_1 \circ f_1$, where $f_1 : X \to Z_1$ is a lifting of f_0 with image Y_1 and $i_1 : Y_1 \to Z_1$ is the inclusion map.

We claim that since p_1 is a non-trivial connected covering, $c(f_1) < c(f_0)$. Note first that $(p_1 \circ i_1)$, which is surjective, is not injective (for otherwise its inverse would give a section of p_1). Hence there is a non-trivial deck transformation τ of Z_1 such that $\tau.Y_1 \cap Y_1 \neq \emptyset$. This intersection is the union of closed cells, hence there exist 0-cells $v \in Y_1$ such that $\tau.v = v$. Since $p_1(\tau.v) = p_1(v)$, we have $d(Y_1) > d(Y_0)$ and hence $c(f_1) < c(f_0)$.

If $(f_1)_* : \pi_1 X \to \pi_1 Y$ is not surjective then there exists a proper connected covering $p_2 : Z_2 \to Y_1$ such that $f_1 = i_1 \circ p_2 \circ i_2 \circ f_1$, where $f_2 : X \to Z_2$ is a lifting of f_1 with image Y_2 and $i_2 : Y_2 \to Z_2$ is the inclusion map. And $c(f_2) < c(f_1) < c(f_0)$. Proceeding in this manner, since $c(f)$ is a non-negative integer, after a finite number of steps we obtain a tower lift $f : X \xrightarrow{f_n} Y_n \to Z$ of f such that f_n induces a surjection on π_1. $\square$

5.30 Proposition. *Let $\mathbb{K}$ be a class of (not necessarily compact) combinatorial complexes that is closed under the operations of passing to finite subcomplexes and to connected covers. Let Γ be a finitely presented group. If there exists an injective homomorphism of groups $\phi : \Gamma \to \pi_1 K$ for some $K \in \mathbb{K}$, then there exists a compact complex $K' \in \mathbb{K}$, of dimension at most 2, such that $\Gamma = \pi_1 K'$.*

Proof. Our goal is to construct a combinatorial map of complexes $f : X \to K$ where X is compact, $\pi_1 X = \Gamma$, the image of f is contained in 2-skeleton of K and $f_* : \pi_1 X \to \pi_1 K$ is an injection (that represents ϕ). The preceding lemma implies that such a map f has a tower lift $f : XX \xrightarrow{f'} K'X \xrightarrow{g} K$ with $f'_* : \pi_1 X \to \pi_1 K'$ surjective. Since $f_* = g_* \circ f'_*$ is injective, it then follows that f'_* is an isomorphism. The compact complex K' is obtained from K by repeatedly taking subcomplexes and coverings, hence $K' \in \mathbb{K}$; and because the image of f is contained in the 2-skeleton of K, the complex K' is at most 2-dimensional. Thus it suffices to construct X and the combinatorial map $f : X \to K$ representing ϕ.

To say that Γ is finitely presented means that there is a surjection $F(\mathcal{A}) \to \Gamma$ from the free group on a finite set $\mathcal{A} = \{a_1, \dots, a_m\}$ such that the kernel N of this surjection is the normal closure of a finite subset $R \subset F(\mathcal{A})$. Without loss of generality we may assume that the number of generators m is chosen to be minimal, in which case each a_i determines a non-trivial element of Γ, which we again denote a_i.

Let X_0 be a graph with one 0-cell x_0, and one 1-cell for each of the generators a_i, oriented and labelled a_i. (This labelling identifies $\pi_1 X_0$ with the free group F.) We fix a basepoint $k_0 \in K$ and for each generator $a_i \in \Gamma$ we choose a locally injective loop in the 1-skeleton of K that represents $\phi(a_i) \in \pi_1(K, k_0)$. We then define $\hat{f} : X_0 \to K$ by sending the oriented edge labelled a_i to a monotone parameterization of this loop. We introduce a new combinatorial structure on X_0 by decreeing that each point in $\hat{f}^{-1}(K^{(0)})$ is a 0-cell; $\hat{f}$ then becomes a combinatorial map. Note that $\hat{f}_* : F = \pi_1 X_0 \to \pi_1 K$ is the composition of $F \to \Gamma$ and $\phi : \Gamma \to \pi_1 K$, and so has image $\phi(\Gamma)$ and kernel N.

Let S be an oriented circle with basepoint. Each of the defining relators $r \in R$ determines an edge path $\alpha_r : S \to X_0$ which begins at the vertex $x_0 \in X_0$ and then proceeds to cross (in order) the oriented 1-cells whose labels are the letters of the word $r \in F$.

Because the word r in the generators a_i represents the identity in Γ, the loop $\hat{f} \circ \alpha_r$ is null-homotopic in K. So by van Kampen's lemma [K33b, Lemma 1] there exists a simply connected, planar, 2-complex Δ_r, a map $\beta_r : S \to \Delta_r$ that is a monotone parameterization of the boundary cycle of Δ_r, and a combinatorial map $f_r : \Delta_r \to K^{(2)}$ such that $f_r \circ \beta_r = \hat{f} \circ \alpha_r$ (cf. I.8A).

Let X be the combinatorial 2-complex obtained by taking the quotient of $X_0 \cup \coprod_r \Delta_r$ by the equivalence relation generated by $\alpha_r(x) \sim \beta_r(x)$ for all $x \in S$ and all $r \in R$. Let $p : X_0 \to X$ be the quotient map. Let $f : X \to K$ be the combinatorial map induced by $\hat{f} : X_0 \to K$ and $f_r : \Delta_r \to K$. By the Seifert-van Kampen theorem, $p_* : \pi_1 X_0 \to \pi_1 X$ is onto. And by construction $r \in \ker p_*$ for all $r \in R$, so $N \subseteq \ker p_*$. But $\hat{f}_* = f_* \circ p_*$ has kernel N, so $f_* : \pi_1 X \to \pi_1 K$ is an isomorphism onto the image of $\hat{f}_*$, which is $\phi(\Gamma)$. □

In order to apply the preceding proposition we need the following easy observation.

5.31 Lemma. *Fix $\kappa \in \mathbb{R}$ and let $\mathbb{K}(\kappa)$ denote the class of connected M_κ–complexes K of dimension at most 2 that satisfy the link condition and for which* Shapes(K) *is finite. If $K \in \mathbb{K}(\kappa)$ then, when equipped with the induced length metric, every connected subcomplex of K is in $\mathbb{K}(\kappa)$ and every connected cover of K is in $\mathbb{K}(\kappa)$.*

Proof. The asserted closure property of $\mathbb{K}(\kappa)$ with respect to covers is not special to dimension 2. Indeed, if K is an n-dimensional M_κ–complex with Shapes(K) finite and $\widehat{K}$ is a connected cover of K (with the induced path metric) then $\widehat{K}$ is an M_κ–complex with Shapes($\widehat{K}$) = Shapes(K). And since the covering map $\widehat{K} \to K$ is a local isometry, K has curvature $\leq \kappa$ (i.e., satisfies the link condition) if and only if $\widehat{K}$ does.

The asserted closure property of $\mathbb{K}(\kappa)$ with respect to subcomplexes clearly fails above dimension 2. But if K is 2-dimensional and v is a vertex of the subcomplex $L \subset K$ then Lk(v, L) is (locally isometric to) a subgraph of Lk(v, K). So if Lk(v, K) contains no closed injective loops of length less than 2π, then neither does Lk(v, L). The absence of such loops is equivalent to the link condition. □

5.32 Remark. One of the central problems in low-dimensional homotopy theory is the Whitehead conjecture, which asserts that if the second homotopy group of a 2-dimensional CW complex is trivial, then so is the second homotopy group of each of its connected subcomplexes. The preceding result, in conjunction with the Cartan-Hadamard theorem, shows that M_κ–complexes of non-positive curvature satisfy the Whitehead conjecture.

We now turn our attention to showing that if a compact 2-dimensional M_κ–complex satisfies the link condition then it collapses onto a subcomplex with the geodesic extension property. When combined with the preceding two results, this completes the proof of Theorem 5.27.

We recall Whitehead's notion of an elementary collapse (which is the starting point for simple-homotopy theory [Coh75]). Recall that a cell e of a combinatorial complex K is, by definition, a *free face* if it lies in the boundary of exactly one cell e' of higher dimension and the intersection of the interior of e' with some small neighbourhood of an interior point of e is connected. We define an *elementary collapse* of K

to be any subcomplex obtained by removing a free face and the interior of the unique higher dimensional cell in whose boundary the free face lies. We say that K collapses onto a subcomplex L if there is a sequence of complexes $K = K_0, \dots, K_n = L$ such that each K_{i+1} is an elementary collapse of K_i.

It is clear that an elementary collapse does not alter the fundamental group of a complex, so the following lemma completes the proof of Theorem 5.27.

5.33 Lemma. *Let K be a compact two dimensional M_κ–complex. If K satisfies the link condition, then it collapses onto a subcomplex which (when equipped with its intrinsic metric) has the geodesic extension property.*

Proof. According to (5.10) an M_κ–complex which satisfies the link condition has the geodesic extension property if and only if it has no free faces. K has only finitely many cells, so after finitely many elementary collapses we obtain the desired subcomplex. □

5.34 Exercises

(1) How unique is the subcomplex yielded by the preceding lemma?

(2) If K is non-positively curved, compact, simply connected and two dimensional, then it collapses to a point.

Knot and Link Groups

We began our discussion of 2-complexes by asserting that there are many interesting examples. In Chapter 12 we shall describe how to construct examples using complexes of groups and in Chapter III.Γ we shall give explicit examples that are of interest in group theory. In the present section we shall describe a construction adapted from classical considerations in 3-manifold topology, in particular Dehn's presentation for the fundamental group of the complement of a knot in $\mathbb{S}^3$. The geometric construction here is similar to earlier work of Weinbaum [We71] and others who studied these groups from the point of view of small cancellation theory (see [LyS77, p.270]). The adaptation to the non-positively curved case was first noticed by I. Aitchison and (independently) by D. Wise [Wi96a].

5.35 Theorem. *If $\mathcal{K} \subset \mathbb{R}^3$ is an alternating link then $\pi_1(\mathbb{R}^3 - \mathcal{K})$ is the fundamental group of a compact 2-dimensional piecewise-Euclidean 2-complex of non-positive curvature.*

5.36 Remark. This result is not the strongest of its ilk, but the degree of generality chosen allows a straightforward and instructive proof. In fact, all link groups are fundamental groups of non-positively curved spaces: this is a consequence of Thurston's geometrization theorem for Haken manifolds [Mor84], [Thu82] and the fact that link complements are irreducible [Papa57]; cf. [Bri98b] and [Le95]. In addition to (5.35), we shall cover hyperbolic knots (Theorem 11.27) and torus knots (11.15).

For clarity of exposition we shall consider only knots; the adaptation to the case of links is entirely straightforward.

5.37 Definitions. We consider smooth embeddings $f : S^1 \hookrightarrow \mathbb{R}^3$. Two such embeddings f_0 and f_1 are said to be equivalent if there is a smooth embedding $F : S^1 \times [0,1] \to \mathbb{R}^3 \times [0,1]$ such that $F(x,t) = (f(x,t),t)$ for all $x \in S^1$ and $t \in [0,1]$, where $f(x,0) = f_0(x)$ and $f(x,1) = f_1(x)$. A (tame) *knot* is, by definition, an equivalence class of smooth embeddings $S^1 \hookrightarrow \mathbb{R}^3$. It is convenient to work with a definite choice of representative for the class under consideration, or even just the image of this representative, which we write $\mathcal{K} \subset \mathbb{R}^3$. We may refer to $\mathcal{K}$ itself as 'the knot'.

Let $\pi : \mathbb{R}^3 \to \mathbb{R}^2 = \mathbb{R}^2 \times \{0\}$ be the obvious projection. Every knot has a representative $f : S^1 \to \mathcal{K} \subset \mathbb{R}^3$ such that $\pi \circ f$ has a nowhere zero derivative, all of its self-intersections are transverse, and for every $x \in \operatorname{im}(\pi \circ f)$ the cardinality of $\mathcal{K} \cap \pi^{-1}(x)$ is at most two. $\Pi = \pi \circ f$ is called a (regular) *projection* of $\mathcal{K}$. Let Λ denote the image of Π. The points $x \in \Lambda$ for which $\mathcal{K} \cap \pi^{-1}(x)$ has cardinality two are called its *double points*. If x is a double point, then the elements of $\mathcal{K} \cap \pi^{-1}(x)$ are called the *overpass* and *underpass* at x, according to which has the greater third coordinate. Π is called an *alternating* projection if when traversing $\mathcal{K}$ in a monotone manner one encounters underpasses and overpasses alternately. By definition, an *alternating knot* is one that admits an alternating projection.

Π is called a *prime* projection if it does not admit a *reducing circle*, i.e, an embedded circle $S \subset \mathbb{R}^2$ such that S intersects Λ in exactly two points, neither of which is a double point, and both components of $\mathbb{R}^2 \smallsetminus S$ contain a double point of Λ. Let Σ be a smoothly embedded 2-sphere in $\mathbb{R}^3$ that intersects $\mathcal{K}$ transversally at exactly two points p_1 and p_2. Let $\mathcal{I}$ and $\mathcal{O}$ be the connected components of $\mathbb{R}^3 \smallsetminus \Sigma$. Let $[p_1, p_2]$ be an arc joining p_1 to p_2 on Σ. Let $\mathcal{K}_1$ and $\mathcal{K}_2$ be the knots obtained by smoothing $[p_1,p_2] \cup (\mathcal{K} \cap \mathcal{O})$ and $[p_1,p_2] \cup (\mathcal{K} \cap \mathcal{I})$ respectively. In this situation, $\mathcal{K}$ is said to be the *connected sum* of $\mathcal{K}_1$ and $\mathcal{K}_2$, written $\mathcal{K} = \mathcal{K}_1 \# \mathcal{K}_2$.

5.38 Remark. The above definition of connected sum can be viewed as an operation $\# : \{knots\} \times \{knots\} \to \{knots\}$. This is well-defined (in particular it is independent of the choices of $\mathcal{K}$, $\mathcal{K}_1$ and $\mathcal{K}_2$ within their equivalence classes) and makes the set of knots into an abelian monoid (see [BurZ85], [Rol90]).

5.39 Lemma. *Every alternating knot $\mathcal{K}$ is the connected sum of finitely many knots each of which has an alternating prime projection.*

Proof. Let $\Pi = \pi \circ f$ be an arbitrary alternating projection of $\mathcal{K}$, with image Λ. If Π is not prime, then there exists a reducing circle $S \subset \mathbb{R}^2$. Let $\overline{p_1}$ and $\overline{p_2}$ be the points at which S intersects Λ, and let Λ_1 and Λ_2 be the components into which S separates Λ. Let $p_1 = \pi^{-1}(\overline{p_1}) \cap \mathcal{K}$ and $p_2 = \pi^{-1}(\overline{p_2}) \cap \mathcal{K}$. By closing the cylinder

$S \times \mathbb{R}$ with horizontal discs well above and below $\mathcal{K}$ we obtain an embedded 2-sphere $\Sigma \subset \mathbb{R}^3$ that is transverse to $\mathcal{K}$ and meets it only at p_1 and p_2. Σ separates $\mathcal{K}$ into two components, parameterized by the restriction of $f : S^1 \to \mathcal{K}$ to the connected components of $S^1 \setminus \Pi^{-1}\{\overline{p_1}, \overline{p_2}\}$; we call these two components of the circle C_1 and C_2. Thus $\mathcal{K} = \mathcal{K}_1 \# \mathcal{K}_2$, where (interchanging the indices as necessary) $\mathcal{K}_i$ has projection Π_i, which begins at p_i, then describes Λ_i as parameterized by $\Pi|_{C_i}$, then follows the arc of the reducing circle S joining p_i to $p_{i\pm 1}$.

In order to complete the proof we must check that the projection Π_i is alternating. The double points of Π_i are precisely those of $\Pi|_{C_i}$, so since $\Pi = \pi \circ f$ is alternating it only remains to prove that if the first double point along Π is encountered as an overpass, then the last is encountered as an underpass, and vice versa. But this is easily seen by combining the fact that underpasses and overpasses alternate along $f(C_i)$ with the observation that the number of underpasses on $f(C_i)$ is equal to the number of overpasses, because there is one of each for each double point in $\Pi(C_i)$. □

5.40 Corollary. *If $\mathcal{K} \subset \mathbb{R}^3$ is an alternating knot then the fundamental group of $(\mathbb{R}^3 \setminus \mathcal{K})$ is an amalgamated free product of the form*

$$G_0 *_{C_1} G_1 *_{C_2} \cdots *_{C_n} G_n,$$

where each G_i is the fundamental group of the complement of a knot with a prime alternating projection and each C_i is infinite cyclic.

Proof. This follows from (5.39) upon application of the Seifert-van Kampen theorem (see [BurZ85]). □

5.41 The Dehn Complex of a Projection.

We maintain the notation and conventions established in (5.37). In particular, Λ is the image of the regular projection Π of the knot $\mathcal{K}$. We need an additional definition: fix a neighbourhood V_x about each double point x of Λ, small enough so that V_x intersects Λ only in two small arcs meeting transversally at x; a connected component of $(\Lambda \cap V_x) \setminus \{x\}$ is called a *germ*. Thus each edge of the graph Λ has two germs, one at each end. We label germs o or u according to whether they are the image of an arc in $\mathcal{K}$ going through an overpass or an underpass. When embellished with this extra information, Λ is called a *knot diagram*.

The Dehn complex $\mathcal{D}(\Pi)$ of the projection Π is a 2-complex with two vertices v_+, v_- (which are thought of as lying, respectively, above and below the plane of projection). The 1-cells of $\mathcal{D}(\Pi)$ are in 1–1 correspondence with the connected components $A_0, A_1, \ldots, A_n$ of $\mathbb{R}^2 \setminus \Lambda$, the i-th 1-cell being oriented from v_+ to v_- and labelled A_i; the 2-cells of $\mathcal{D}(\Pi)$ are in 1–1 correspondence with the double points of Λ. They are attached as follows: given a double point $x \in \Lambda$, let $A_{i_x(1)}, A_{i_x(2)}, A_{i_x(3)}, A_{i_x(4)}$ be the (not necessarily distinct) components of $\mathbb{R}^2 \setminus \Lambda$ that one encounters as one proceeds anticlockwise around x, beginning at a germ labelled o; for each x, a 2-cell is attached to the 1-skeleton of $\mathcal{D}(\Pi)$ by a locally injective map from its boundary to the edge-path labelled $A_{i_x(1)}A^{-1}_{i_x(2)}A_{i_x(3)}A^{-1}_{i_x(4)}$, where A_i^{-1} is A_i traversed in the opposite direction.

Choosing v_+ as basepoint and writing a_i for the homotopy class of the loop $A_iA_0^{-1}$, we see that the fundamental group of $\mathcal{D}(\Pi)$ has presentation

$$\mathcal{P}_\Pi = \langle a_0, \ldots, a_n \mid a_0 = 1,\, a_{i_x(1)}a^{-1}_{i_x(2)}a_{i_x(3)}a^{-1}_{i_x(4)} = 1 \text{ for every double point } x\rangle.$$

This is the well-known Dehn presentation of the fundamental group of $\mathbb{R}^3 \smallsetminus \mathcal{K}$. The following exercise is standard.

5.42 Exercise. Show that $\mathcal{P}_\Pi$ is indeed a presentation of $\pi_1(\mathbb{R}^3 \smallsetminus \mathcal{K})$.

(Hint: Isotop $\mathcal{K}$ to lie in the union of $\mathbb{R}^2 \times \{0\}$ and the spheres of some small radius r about the double points of the projection. Place a ball of radius $2r$ about each double point and apply the Seifert-van Kampen theorem.)

Henceforth we view $\mathcal{D}(\Pi)$ as a piecewise Euclidean complex with the intrinsic metric obtained by metrizing each 2-cell as a unit square.

5.43 Proposition. $\mathcal{D}(\Pi)$ *is non-positively curved if and only if* Π *is a prime alternating projection.*

Proof. We first describe the links $\text{Lk}(v_\pm, \mathcal{D}(\Pi))$. Each edge in one of these link graphs has length $\pi/2$, so our concern will be to determine when they contain circuits of length less than 4. The vertex set of $\text{Lk}(v_\pm, \mathcal{D}(\Pi))$ is in 1–1 correspondence with the 1-cells of $\mathcal{D}(\Pi)$, from which it inherits a labelling $A_0, \ldots, A_n$. The edges of $\text{Lk}(v_+, \mathcal{D}(\Pi))$ (resp. $\text{Lk}(v_-, \mathcal{D}(\Pi))$) are in 1–1 correspondence with pairs (x, g), where x is a double point of Λ and g is a germ incident at x labelled u (resp. o). The edge corresponding to (x, g) joins the vertices labelled by the regions which meet along g.

If the projection Π is not prime, then there exists a reducing circle S. The two components of $S \smallsetminus \Lambda$ lie in distinct connected components of $\mathbb{R}^2 \smallsetminus \Lambda$, A_i and A_j say. These components meet along two distinct edges, which means that they meet along four distinct germs. At least two of these germs must carry the same label, o or u, and therefore give rise to two edges joining the vertices labelled A_i and A_j in the corresponding link graph — whence a circuit of length two.

To say that the projection Π is not alternating means precisely that the germs at either end of some edge e in Λ carry the same label, u say. These two germs give rise to two distinct edges in $\text{Lk}(v_+, \mathcal{D}(\Pi))$ connecting the vertices labelled by the two components of $\mathbb{R}^2 \smallsetminus \Lambda$ that abut along e.

If Π is alternating then one can divide the components of $\mathbb{R}^2 \smallsetminus \Lambda$ into two types, let us call them black and white: a bounded component is said to be white (respectively, black) if the anticlockwise orientation of its boundary orients each of the edges in its boundary from the germ labelled o to the germ labelled u (respectively, u to o); the unbounded region A_0 is coloured according to the opposite convention. Note that if two regions abut along an edge of Λ then they have opposite colours. If we give the vertices of $\mathrm{Lk}(v_\pm, \mathcal{D}(\Pi))$ the induced colouring, then the preceding sentence implies that the endpoints of each edge in $\mathrm{Lk}(v_\pm, \mathcal{D}(\Pi))$ are coloured differently. It follows that all closed circuits in $\mathrm{Lk}(v_\pm, \mathcal{D}(\Pi))$ are of even length, so to check the link condition we need only rule out circuits of length two.

There are two edges connecting the vertices of $\mathrm{Lk}(v_\pm, \mathcal{D}(\Pi))$ labelled A_i and A_j if and only if the regions A_i and A_j meet across two germs of edges in Λ that carry the same label. If Π is alternating, the germs at either end of an edge carry different labels, whence A_i and A_j meet across two distinct edges. But then, by connecting the midpoints of these edges with arcs in the interiors of A_i and A_j we obtain a reducing circle showing that Π is not prime. □

Proof of Theorem 5.35. According to Corollary 5.40, the fundamental group Γ of the complement of an alternating knot $\mathcal{K}$ is of the form $G_0 *_{C_1} G_1 *_{C_1} \cdots *_{C_n} G_n$, where each G_i is the fundamental group of a knot with a prime alternating projection, and each C_i is infinite cyclic. The preceding proposition shows that each G_i is the fundamental group of a compact, piecewise Euclidean 2-complex of non-positive curvature. It follows from (11.17 and I.7.29) that Γ is also the fundamental group of such a complex (but not necessarily a complex built from squares). □

5.44 Exercises

(1) Consider the projection of the trivial knot given in polar coordinates by the equation $r = (1/\sqrt{2}) + \cos\theta$. Show that the Dehn complex of this projection is a M'obius strip. Which projections of the trivial knot have non-positively curved Dehn complexes?

(2) The Hopf link consists of two unknotted circles in $\mathbb{R}^3$ linked in such a way that they cannot be separated but have a planar projection with only two crossings (figure 5.1)? What is the Dehn complex of this minimal projection of the Hopf link?

(3) The Borromean rings is a link with three components such that if one removes any one of the components then the other two can be unlinked (fig. 5.1). What is the Dehn complex of the projection shown in (fig. 5.1).

From Group Presentations to Negatively Curved 2-Complexes

In [Rips82] Rips introduced a construction that associates to any finite group presentation a short exact sequence $1 \to N \to \Gamma \to G \to 1$, where N is a finitely generated group, Γ is a finitely presented group satisfying an arbitrarily strict small cancellation condition (cf. 5.25) and G is the group defined by the original presentation. We shall modify Rips's construction so as to arrange that G is the fundamental group of a negatively curved 2-complex. An earlier modification of this sort was given by Daniel Wise [Wi98a].

5.45 Theorem. *There is an algorithm that associates to every finite presentation a short exact sequence*

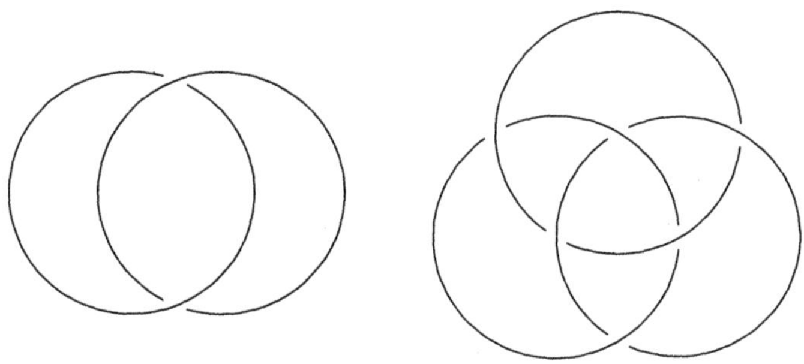

Fig. 5.1 Hopf link and Borromean rings

$$1 \to N \to \pi_1 K \to G \to 1,$$

where G is the group given by the presentation, N is a finitely generated group and K is a compact, negatively curved, piecewise hyperbolic 2-complex.

Proof. Given a finite presentation $\langle x_1, \ldots, x_n \mid r_1, \ldots, r_m \rangle$ of G, let $\rho = \max |r_i| + 2$ and let M be the least integer such that $M^2 \geq (10nM + 5m) + M$.

The idea of the proof is to expand the given presentation by adding a finite number of new generators a_j; one uses these new generators to 'unwrap' the relations in G, replacing $r_i = 1$ with a relation of the form $r_i = v_i$, where v_i is a word in the generators a_j; then one forces the subgroup generated by the a_j to be normal by adding extra relations of the form $x_i^{\varepsilon} a_j x_i^{-\varepsilon} = u_{i,j,\varepsilon}$, where the words $u_{i,j,\varepsilon}$ involve only the letters a_k. The art of the construction comes in choosing the words v_i and $u_{i,j,\varepsilon}$. We shall choose them in a way that allows us to metrize the 2-cells of the standard 2-complex K of the presentation (I.8A) as right-angled hyperbolic pentagons with the words r_i completely contained in the interior of one side. We then arrange for the 2-complex to have negative curvature.

K is defined as follows:

- K has one 0-cell.
- K has $(M + n)$ 1-cells, labelled $x_1, \ldots, x_n, a_1, \ldots, a_M$. (The loops labelled a_i generate the normal subgroup N in the statement of the theorem.)
- K has $(m + 2nM)$ 2-cells, which are of two types:
- the 2-cells of the first type are in 1-1 correspondence with the original relators r_i;
- the 2-cells of the second type are in 1-1 correspondence with triples $\tau = (x_i, a_j, \varepsilon)$, where $\varepsilon = \pm 1$, $i = 1, \ldots, n$ and $j = 1, \ldots, M$.
- The attaching map of the 2-cell corresponding to r_i traces out the path labelled

$$\alpha_{i,0} r_i \alpha_{i,0}^{\rho - |r_i| - 1} \alpha_{i,1}^{\rho} \alpha_{i,2}^{\rho} \alpha_{i,3}^{\rho} \alpha_{i,4}^{\rho},$$

where each $\alpha_{i,j} \in \{a_1, \ldots, a_M\}$ — the exact choice of $a_{i,j}$ is deferred.

- The 2-cell corresponding to $\tau = (x_i, a_j, \varepsilon)$ is attached by a path labelled

$$\beta_{0,\tau} x_i^{\varepsilon} a_j x_i^{-\varepsilon} \beta_{0,\tau}^{\rho-4} \beta_{1,\tau}^{\rho} \beta_{2,\tau}^{\rho} \beta_{3,\tau}^{\rho} \beta_{4,\tau}^{\rho},$$

  where each $\beta_{i,\tau} \in \{a_1, \dots, a_M\}$ — again, the exact choice of $\beta_{i,\tau}$ is deferred.
- Let L be the length of each side in a regular right-angled hyperbolic pentagon. Each 1-cell in K is metrized so that it has length L/ρ.
- Each 2-cell is metrized as a regular right-angled hyperbolic pentagon, so that the arcs of the boundary that are labelled $\alpha_{i,j}^{\rho}$ (in the case of 2-cells of the first type) or β_{τ}^{ρ} (in the case of 2-cells of the second type) become sides of the pentagon.

We claim that we may choose the $\alpha_{i,j}$ and the $\beta_{i,\tau}$ so as to ensure that the piecewise hyperbolic complex K described above satisfies the link condition (5.1). How could it fail to do so? Well, the vertex angles at the corners of the 2-cells are all either $\pi/2$ or π (we call the former 'sharp corners') and at corners with a vertex angle π we either have one edge labelled x_i and one labelled a_j, or else we have both the incoming and outgoing edge labelled by the same a_i. Each 2-cell has only five sharp corners and at each of these corners we have arranged that the labels on both the incoming and outgoing edge are drawn from $\{a_1, \dots, a_M\}$.

In order to ensure that there are no circuits of length less than π in the link of the vertex of K, it suffices to arrange that no ordered pair of labels $(a_i, a_{i'})$ occurs at more than one sharp corner, and that at no sharp corner do we have $a_i = a_{i'}$. This amounts to choosing a family of $(m+2nM)$ words $\alpha_{i,0}\alpha_{i,1}\alpha_{i,2}\alpha_{i,3}\alpha_{i,4}$ and $\beta_{0,\tau}\beta_{1,\tau}\beta_{2,\tau}\beta_{3,\tau}\beta_{4,\tau}$ with the following properties: no 2-letter subword $a_j a_{j'}$ appears more than once; there is no subword of the form $a_i a_i$; and for all j, j' there is at most one word of the form $a_j * * * a_{j'}$. We shall call a family *good* if it has these properties.

We seek a good family words. To this end, following Wise [98a], we list all of the words $a_i a_{i'}$ with $i < i'$ in lexicographical order:

$$(a_1a_2), \dots, (a_1a_M), (a_2, a_3), \dots, (a_2a_M), (a_3a_4), \dots\dots\dots, (a_{M-1}a_M).$$

We then concatenate the words in this list to form one long word W_M. By construction, no 2-letter subword $a_j a_{j'}$ appears more than once in W_M and there is no subword of the form $a_i a_i$. Moreover, for each fixed j, j', the word W_M has at most one subword of length 5 that is of the form $a_j * * * a_{j'}$.

W_M has length $M^2 - M$, and we chose M so that $M^2 - M \geq 5(m + 2nM)$. Thus we can divide W_M into $(m + 2nM)$ disjoint subwords of length 5:

$$(a_1a_2a_1a_3a_1), (a_4a_1a_5a_1a_6), \dots .$$

This is the *good* family of words that we were seeking. □

5.46 Remarks

(1) We emphasize that K was obtained from the given presentation in an algorithmic way: the integer M depended in a simple way on the number of generators and

relators; the metric depended only on ρ, which is an invariant of the presentation; and the "good" family of words used to attach the 2-cells of K were derived from M by a simple algorithm.

(2) Notice that all of the 1-cells of K are geodesic circles and that they are the shortest homotopically non-trivial loops in K.

5.47 Exercise (Incoherence). Robert Bieri [Bi76a] showed that if H and Γ are finitely presented groups of cohomological dimension at most two, and if $H \subset \Gamma$ is normal, then H is either free or else it has finite index in Γ. Prove that the group N constructed in (5.45) is not free, and hence deduce that if G is infinite then N is not finitely presented.

Groups that contain subgroups which are finitely generated but not finitely presented are called *incoherent*. (See [Wi98a] for further results in this direction.)

Chapter II.6 Isometries of CAT(0) Spaces

In Chapters 2 and 6 of Part I we described the isometry groups of the most classical examples of CAT(0) spaces, Euclidean space and real hyperbolic space. Already in these basic examples there is much to be said about the structure of the isometry group of the space, both with regard to individual isometries and with regard to questions concerning the subgroup structure of the full group of isometries. More generally, the study of isometries of non-positively curved manifolds is well-developed and rather elegant. In this chapter we shall study isometries of arbitrary CAT(0) spaces X.

We begin with a study of individual isometries $\gamma \in \mathrm{Isom}(X)$. We divide isometries into three types: elliptic, hyperbolic and parabolic, in analogy with the classification of isometries of real hyperbolic space discussed in Chapter I.6. We develop some basic properties concerning the structure of the set of points moved the minimal distance by a given isometry γ.

In the second part of this chapter we consider groups of isometries. After noting some general facts, we focus on the subgroup of $\mathrm{Isom}(X)$ consisting of those isometries which move every element of X the same distance — Clifford translations. These translations play an important role in the description of the full isometry groups of compact non-positively curved spaces (6.17). In the last section of this chapter we shall prove a theorem (6.21) that relates direct product decompositions of groups which act properly and cocompactly by isometries on CAT(0) spaces to splittings of the spaces themselves.

We refer to Chapter I.8 for basic facts and definitions concerning group actions. In particular we remind readers that in the case of spaces which are not locally compact, the definition of proper group action that we shall be using is non-standard.

Individual Isometries

In order to analyze the behaviour of isometries of a CAT(0) space, we first divide the isometries into three disjoint classes. Membership of these classes is defined in terms of the behaviour of the displacement function of an isometry. This function is defined as follows.

The Displacement Function

6.1 Definitions. Let X be a metric space and let γ be an isometry of X. The *displacement function* of γ is the function $d_\gamma : X \to \mathbb{R}_+$ defined by $d_\gamma(x) = d(\gamma.x, x)$. The *translation length* of γ is the number $|\gamma| := \inf\{d_\gamma(x) \mid x \in X\}$. The set of points where d_γ attains this infimum will be denoted $\text{Min}(\gamma)$. More generally, if Γ is a group acting by isometries on X, then $\text{Min}(\Gamma) := \bigcap_{\gamma\in\Gamma} \text{Min}(\gamma)$.

An isometry γ is called *semi-simple* if $\text{Min}(\gamma)$ is non-empty. An action of a group by isometries of X is called semi-simple if all of its elements are semi-simple.

6.2 Proposition. *Let X be a metric space and let γ be an isometry of X. Let Γ be a group acting by isometries on X.*

(1) $\text{Min}(\gamma)$ *is γ-invariant and* $\text{Min}(\Gamma)$ *is Γ-invariant.*

(2) *If α is an isometry of X, then $|\gamma| = |\alpha\gamma\alpha^{-1}|$, and* $\text{Min}(\alpha\gamma\alpha^{-1}) = \alpha.\text{Min}(\gamma)$. *In particular, if α commutes with γ, then it leaves* $\text{Min}(\gamma)$ *invariant. If N is a normal subgroup of Γ, then* $\text{Min}(N)$ *is Γ-invariant.*

(3) *If X is a* CAT(0) *space, then the displacement function d_γ is convex, and hence* $\text{Min}(\gamma)$ *is a closed convex set.*

(4) *If $C \subset X$ is non-empty, complete, convex, and γ-invariant, then $|\gamma| = |\gamma|_C|$ and γ is semi-simple if and only if $\gamma|_C$ is semi-simple. Thus* $\text{Min}(\gamma)$ *is non-empty if and only if $C \cap$* $\text{Min}(\gamma)$ *is non-empty.*

Proof. The proofs of (1) and (2) are easy, and (3) follows from the convexity of the distance function in a CAT(0) space (2.2). To prove (4), consider the projection p of X onto C. By (2.4) we have $p(\gamma.x) = \gamma.p(x)$ and $d(\gamma.x, x) \geq d(\gamma.p(x), p(x))$, for all $x \in X$, hence $p(\text{Min}(\gamma)) = \text{Min}(\gamma) \cap C = \text{Min}(\gamma|_C)$. ☐

6.3 Definition. Let X be a metric space. An isometry γ of X is called

(1) *elliptic* if γ has a fixed point,

(2) *hyperbolic* (or axial) if d_γ attains a strictly positive minimum,

(3) *parabolic* if d_γ does not attain its minimum, in other words if $\text{Min}(\gamma)$ is empty.

Every isometry is in one of the above disjoint classes and it is semi-simple if and only if it is elliptic or hyperbolic. If two isometries of X are conjugate in Isom(X), then they have the same translation length and lie in the same class.

6.4 Examples

(1) Consider the half-space model $\{x = (x_1, \dots, x_n) \in \mathbb{R}^n \mid x_n > 0\}$ for hyperbolic space $\mathbb{H}^n$. If $\lambda > 0$ and $\lambda \neq 1$, the map $x \mapsto \lambda x$ is a hyperbolic isometry γ with $|\gamma| = |\log \lambda|$. In this case $\text{Min}(\gamma)$ is a single geodesic line, namely $\{x \in \mathbb{R}^n \mid x_n > 0,\ x_i = 0 \text{ if } i < n\}$. Every hyperbolic isometry of $\mathbb{H}^n$ with the translation length $|\log \lambda|$ is conjugate in Isom($\mathbb{H}^n$) to the composition of γ and an orthogonal transformation that fixes the axis of γ pointwise.

An orientation preserving isometry of $\mathbb{H}^2$ is parabolic if and only if it is conjugate in $\text{Isom}(\mathbb{H}^2)$ to an isometry of the form $\gamma : (x_1, x_2) \mapsto (x_1 + 1, x_2)$. Note that $|\gamma| = 0$ and that the level sets of d_γ are the subsets $x_2 = \text{const}$.

The half-space model for $\mathbb{H}^2$ can be described as the space of complex numbers z whose imaginary part $\mathcal{I}m(z)$ is positive. In (I.6.14) we showed that the full isometry group for this model can be identified with $\text{PGL}(2, \mathbb{R})$, the quotient of the group of real $(2, 2)$-matrices by the subgroup formed by scalar multiples of the identity. It is easy to show that the isometry determined by an element of $\text{GL}(2, \mathbb{R})$ is semi-simple if and only if this element is semi-simple in the classical sense, i.e. it is conjugate over $\mathbb{C}$ to a diagonal matrix. More generally, in Chapter 10 we shall describe the action of $\text{GL}(n, \mathbb{R})$ by isometries on the symmetric space $P(n, \mathbb{R})$ of positive definite (n, n)-matrices and prove that an element of $\text{GL}(n, \mathbb{R})$ acts as a semi-simple isometry if and only if it is a semi-simple matrix in the classical sense.

(2) Let γ be an isometry of $\mathbb{E}^n$. We have seen (I.2.23 and I.4.13) that γ is of the form $x \mapsto Ax + b$, where $A \in O(n)$ and $b \in \mathbb{R}^n$. If γ does not fix a point then $-b$ is not in the image of $A - I$, so $A - I$ is not invertible. Thus $Av = v$ for some $v \in \mathbb{R}^n \setminus \{0\}$. Let V_1 be the vector subspace generated by v and let V_2 be its orthogonal complement. Then $\mathbb{E}^n$ is isometric to the product $V_1 \times V_2$, and because γ maps each line that is parallel to V_1 to another such line, it follows from I.5.3(4) that γ splits as the product of a translation γ_1 of V_1 and an isometry γ_2 of V_2. Because V_2 is isometric to $\mathbb{E}^{n-1}$, arguing by induction on n it is easy to prove the following:

6.5 Proposition. *Every isometry of $\mathbb{E}^n$ is semi-simple. More precisely, an isometry γ of $\mathbb{E}^n$ is either elliptic, or else it is hyperbolic and there is an integer k with $0 < k \leq n$ such that* $\text{Min}(\gamma)$ *is an affine subspace E_1 of dimension k. And in the second case, if E_2 is an affine subspace which is an orthogonal complement of E_1 (i.e. $\mathbb{E}^n = E_1 \times E_2$), then γ is the product of a non-trivial translation γ_1 of E_1 and an elliptic isometry γ_2 of E_2.*

In contrast to the finite dimensional case, infinite dimensional Hilbert spaces do admit parabolic isometries (see Example 8.28).

6.6 Exercises

(1) Let X be a metric space and let γ be an isometry of X. Show that

$$\lim_{n\to\infty} \frac{1}{n} d(x, \gamma^n.x)$$

exists for all $x \in X$. Show that $\lim_{n\to\infty} \frac{1}{n} d(x, \gamma^n.x)$ is independent of x, and if γ is semi-simple then $|\gamma| = \lim_{n\to\infty} \frac{1}{n} d(x, \gamma^n.x)$.

(Hint: $n \mapsto d(x, \gamma^n.x)$ is a subadditive function. A function $f : \mathbb{N} \to \mathbb{N}$ is subadditive if $f(m + n) \leq f(m) + f(n)$ for all m and n. It is a classical fact that $\lim_{n\to\infty} f(n)/n$ exists. Indeed for fixed $d > 0$, any integer n can be written uniquely as $n = qd + r$ with $0 \leq r < d$. The subadditivity condition implies that

$f(n)/n \leq f(d)/d + f(r)/n$, hence $limsup_{n\to\infty} f(n)/n \leq f(d)/d$ for any d. Therefore $\limsup_{n\to\infty} f(n)/n \leq \liminf_{d\to\infty} f(d)/d$.)

(2) Let K be a connected M_κ-simplicial complex with Shapes(K) finite (in the sense of Chapter I.7). Show that every simplicial isometry of K is semi-simple. Show further that if a group Γ acts by cellular isometries on K, then the set of translation lengths $\{|\gamma| : \gamma \in \Gamma\}$ is a discrete subset of the real line. (Hint: Given an isometry γ, there exists a model simplex $\overline{S} \in$ Shapes(K), a sequence of points $x_n \in \overline{S}$ and simplices $\phi_n(\overline{S}) \subset K$ such that $d_\gamma(\phi_n(x_n)) \to |\gamma|$ as $n \to \infty$. Combine this observation with (I.7.59).)

(3) Recall that an $\mathbb{R}$–tree is a geodesic space that is CAT(κ) for all $\kappa \in \mathbb{R}$. Prove that every isometry γ of an $\mathbb{R}$–tree is semi-simple. (Hint: Let $x \in X$ and let m be the midpoint of $[x, \gamma.x]$. Prove that $d(m, \gamma.m) = |\gamma|$.)

Semi-Simple Isometries

In this paragraph we note some basic facts about semi-simple isometries. First we give an easy characterization of elliptic isometries.

6.7 Proposition. *Let X be a complete* CAT(0) *space, and let γ be an isometry of X. Then, γ is elliptic if and only if γ has a bounded orbit. And if γ^n is elliptic for some integer $n \neq 0$, then γ is elliptic.*

Proof. The first assertion is a special case of (2.7). If γ^n is elliptic, fixing $x \in X$ say, then the orbit of x under γ is finite, hence bounded. □

The following result concerning the structure of Min(γ) when γ is hyperbolic will be used repeatedly both in the remainder of this chapter and in subsequent chapters.

6.8 Theorem. *Let X be a* CAT(0) *space.*

(1) *An isometry γ of X is hyperbolic if and only if there exists a geodesic line $c : \mathbb{R} \to X$ which is translated non-trivially by γ, namely $\gamma.c(t) = c(t+a)$, for some $a > 0$. The set $c(\mathbb{R})$ is called an axis of γ. For any such axis, the number a is actually equal to $|\gamma|$.*

(2) *If X is complete and γ^m is hyperbolic for some integer $m \neq 0$, then γ is hyperbolic.*

Let γ be a hyperbolic isometry of X.

(3) *The axes of γ are parallel to each other and their union is* Min(γ).

(4) Min(γ) *is isometric to a product $Y \times \mathbb{R}$, and the restriction of γ to* Min(γ) *is of the form $(y, t) \mapsto (y, t + |\gamma|)$, where $y \in Y, t \in \mathbb{R}$.*

(5) *Every isometry α that commutes with γ leaves* Min(γ) $= Y \times \mathbb{R}$ *invariant, and its restriction to $Y \times \mathbb{R}$ is of the form (α', α''), where α' is an isometry of Y and α'' a translation of $\mathbb{R}$.*

Proof. The "if" direction in (1) is a special case of 6.2(4). Conversely, we claim that if γ is hyperbolic then every point of $\text{Min}(\gamma)$ lies on an axis of γ, namely the union of the geodesic segments $[\gamma^n.x, \gamma^{n+1}.x]_{n\in\mathbb{Z}}$. Since local geodesics in CAT(0) spaces are geodesics, it suffices to show that $[x, \gamma^2.x]$ is the concatenation of $[x, \gamma.x]$ and $[\gamma.x, \gamma^2.x]$, and this is equivalent to showing that $d(m, \gamma.m) = d(m, \gamma.x) + d(\gamma.x, \gamma.m) = 2\,d(x, m)$, where m is the midpoint of the geodesic segment $[x, \gamma.x]$. As $\text{Min}(\gamma)$ is convex, it contains m, hence $d(m, \gamma.m) = d(x, \gamma.x)$. But $d(x, \gamma.x) = 2\,d(x, m)$, so $d(m, \gamma.m) = 2\,d(x, m)$, as required.

To complete the proof of (3) we must show that if $c, c' : \mathbb{R} \to X$ are axes for γ then they are parallel. Since $\gamma.c(t) = c(t + |\gamma|)$ and $\gamma.c'(t) = c(t + |\gamma|)$, the convex function $t \mapsto d(c(t), c'(t))$ is periodic of period $|\gamma|$. In particular it is bounded and therefore constant.

For (4), one notes that $\text{Min}(\gamma)$ is a convex subspace of X, and hence is itself a CAT(0) space. In the light of (3), we can apply the Decomposition Theorem (2.14) to the set $\text{Min}(\gamma)$ to obtain a product decomposition $\text{Min}(\gamma) = Y \times \mathbb{R}$, where each $\{y\} \times \mathbb{R}$ is an axis for γ, so in particular $\gamma.(y, t) = (y, t + |\gamma|)$.

(5) Let α be an isometry of X that commutes with γ. It leaves $\text{Min}(\gamma)$ invariant (6.2(2)). Moreover, α takes axes of γ to axes of γ. Hence (see I.5.3) α preserves the product decomposition $\text{Min}(\gamma) = Y \times \mathbb{R}$ and splits as (α', α''). Because $\alpha'' \in \text{Isom}(\mathbb{R})$ commutes with the translation defined by γ, it must be a translation.

Finally we prove (2). If γ^m is hyperbolic, then $\text{Min}(\gamma^m)$ splits as $Y \times \mathbb{R}$ where γ^m is the identity on Y and a non-trivial translation on $\mathbb{R}$. As γ commutes with γ^m, by (5) we see that the restriction of γ to $\text{Min}(\gamma^m) = Y \times \mathbb{R}$ splits as (γ', γ''), where γ'^m is the identity of Y and γ'' a non-trivial translation of $\mathbb{R}$. But Y is a complete CAT(0) space (because it is closed and convex in X), so the periodic isometry γ' must have a fixed point in Y. The product of any such fixed point with $\mathbb{R}$ yields an axis for γ, so by (1) we are done. □

In relation to part (5) of the preceding proposition, we note the following fact concerning isometries of product spaces.

6.9 Proposition. *Let X be a metric space which splits as a product $X' \times X''$, and let $\gamma = (\gamma', \gamma'')$ be an isometry preserving this decomposition. Then, γ is semi-simple if and only if γ' and γ'' are semi-simple. Moreover,* $\text{Min}(\gamma) = \text{Min}(\gamma') \times \text{Min}(\gamma'')$.

Proof. γ is semi-simple if and only if $\text{Min}(\gamma) \neq \emptyset$, so it is enough to show $\text{Min}(\gamma) = \text{Min}(\gamma') \times \text{Min}(\gamma'')$.

For any $x', y' \in X'$ and $x'', y'' \in X''$, we have

$$d(\gamma'(x'), x') < d(\gamma'(y'), y') \iff d(\gamma(x', x''), (x', x'')) < d(\gamma(y', x''), (y', x''))$$
$$d(\gamma''(x''), x'') < d(\gamma''(y''), y'') \iff d(\gamma(x', x''), (x', x'')) < d(\gamma(x', y''), (x', y'')).$$

Thus (from $\Leftarrow$), we have $\text{Min}(\gamma) \subseteq \text{Min}(\gamma') \times \text{Min}(\gamma'')$ and (from $\Rightarrow$) $\text{Min}(\gamma) \supseteq \text{Min}(\gamma') \times \text{Min}(\gamma'')$. □

On the General Structure of Groups of Isometries

In this section we gather some general facts about groups acting properly by isometries. The first result does not involve any hypothesis of curvature.

6.10 Proposition. *Suppose that the group Γ acts properly by isometries on the metric space X. Then:*

(1) *If a subspace X' of X is invariant under the action of a subgroup Γ' of Γ, then the action of Γ' on X' is proper.*

(2) *If the action of Γ is cocompact then every element of Γ is a semi-simple isometry of X.*

(3) *If the action of Γ is cocompact then the set of translation distances $\{|\gamma| \mid \gamma \in \Gamma\}$ is a discrete subset of $\mathbb{R}$.*

(4) *Assume that X splits isometrically as a product $X' \times X''$ and that each element γ of Γ splits as (γ', γ''). Let N be a normal subgroup of Γ formed by elements of the form $\gamma = (\mathrm{id}_{X'}, \gamma'')$, and assume that there is a compact subset K in X'' such that $X'' = \bigcup_{\gamma\in N} \gamma''.K$. Then the induced action of Γ/N on X' is proper. (In this action the coset $(\gamma', \gamma'')N$ acts as γ'.)*

Proof. (1) is immediate from the definitions.

(2) Let K be a compact set whose translates by the action of Γ cover X. Fix $\gamma \in \Gamma$ and consider its displacement function d_γ. Let (x_n) be a sequence of points in X such that $d_\gamma(x_n) \to |\gamma|$ as $n \to \infty$. We choose elements $\gamma_n \in \Gamma$ such that $y_n := \gamma_n.x_n \in K$. Note that $d(\gamma_n\gamma\gamma_n^{-1}.y_n, y_n) = d(\gamma.x_n, x_n)$ tends to $|\gamma|$, as $n \to \infty$. Hence, for every point $x \in K$, the sequence $d(\gamma_n\gamma\gamma_n^{-1}.x, x)$ remains bounded. Because the action of Γ is proper, there are an infinite number of integers n such that $\gamma_n\gamma\gamma_n^{-1}$ is equal to the same element $\overline{\gamma}$ of Γ. Passing to a subsequence if necessary, we assume that $\overline{\gamma} = \gamma_n\gamma\gamma_n^{-1}$ for every positive integer n. Because K is compact, we may pass to a further subsequence in order to assume that (y_n) converges to some point $y \in K$. Then, for every positive integer n, the function d_γ assumes its minimum at $\gamma_n^{-1}.y$, because $d_\gamma(\gamma_n^{-1}.y) = d(\gamma\gamma_n^{-1}.y, \gamma_n^{-1}.y) = d(\overline{\gamma}.y, y) = \lim d(\overline{\gamma}.y_n, y_n) = \lim d(\gamma.x_n, x_n) = |\gamma|$.

(3) Let $K \subset X$ be a compact set whose translates by Γ cover X. In order to obtain a contradiction, let us suppose that there is a sequence of elements $\gamma_n \in \Gamma$ and a number $a \geq 0$ such that $|\gamma_n| \neq |\gamma_m|$ for all $m \neq n$, and $|\gamma_n| \to a$ as $n \to \infty$. According to (2), we can choose points $x_n \in X$ such that $d(\gamma_n.x_n, x_n) = |\gamma_n|$. Replacing γ_n by a suitable conjugate if necessary, we may assume that $x_n \in K$ for all n. Since K is compact, it is contained in a bounded set, say $B(x, r)$. But then, for all n sufficiently large, $\gamma_n.B(x, r+a+1) \cap B(x, r) \neq \emptyset$, which contradicts the properness of the action (see I.8.3(1)).

(4) Let $x' \in X'$ and choose a point $x = (x', x'') \in X$ such that $x'' \in K$. As the action of Γ is proper, there exists $\varepsilon > 0$ such that $\{\gamma \in \Gamma \mid \gamma.(B(x', \varepsilon)\times K)\cap(B(x', \varepsilon)\times K) \neq \emptyset\}$ is finite (see I.8.3(1)). If $\gamma' \in \Gamma/N$ is such that $\gamma'.B(x', \varepsilon) \cap B(x', \varepsilon) \neq \emptyset$, then there exists $\gamma = (\gamma', \gamma'')$ in the coset γ' such that $\gamma''.x'' \in K$. But this implies that

$\gamma.(B(x', \varepsilon) \times K) \cap (B(x', \varepsilon) \times K) \neq \emptyset$. Thus the number of such γ' is finite and the action of Γ/N on X' is proper. $\square$

6.11 Remark. Parts (2) and (3) of the preceding proposition obviously remain valid under the weaker hypothesis that Γ is a subgroup of a group that acts properly and cocompactly by isometries. In section 5 of Chapter III.Γ we shall see that in the context of CAT(0) spaces, many groups that cannot act properly and cocompactly by isometries can nevertheless be embedded in groups that do.

The following result places severe restrictions on the way in which central extensions can act by isometries on CAT(0) spaces. Some topological consequences of this will be discussed in Chapter 7. We emphasize that in the following result we assume nothing beyond what is stated about the group actions involved, in particular the action of the group Γ is not required to be proper or semi-simple.

6.12 Theorem. *Let X be a* CAT(0) *space and let Γ be a finitely generated group acting by isometries on X. If Γ contains a central subgroup $A \cong \mathbb{Z}^n$ that acts faithfully by hyperbolic isometries (apart from the identity element), then there exists a subgroup of finite index $H \subset \Gamma$ which contains A as a direct factor.*

Proof. Fix $\alpha \in A$, a non-trivial element. According to 6.8(5), the action of Γ leaves $\text{Min}(\alpha) = Y \times \mathbb{R}$ invariant, and the restriction of each $\gamma \in \Gamma$ to $Y \times \mathbb{R}$ is of the form (γ', γ''), where γ' is an isometry of Y and γ'' is a translation of $\mathbb{R}$. The map $\gamma \mapsto \gamma''$ defines a homomorphism from Γ to a finitely generated group of translations of $\mathbb{R}$. Such a group of translations is isomorphic to $\mathbb{Z}^m$, for some m, so we have a surjective homomorphism $\psi : \Gamma \to \mathbb{Z}^m$. The image under ψ of A is non-trivial, because $\psi(\alpha)$ is non-trivial.

We compose ψ with the projection of $\mathbb{Z}^m$ onto a suitable direct summand so as to obtain a homomorphism $\phi : \Gamma \to \mathbb{Z}$ such that $\phi(A)$ is non-trivial. We then choose $a \in A$ so that $\phi(a)$ generates $\phi(A)$. Let $H_0 = \phi^{-1}(\phi(A))$ and note that H_0 has finite index in Γ. The map $\phi(a) \mapsto a$ splits $\phi|_{H_0}$, giving $H_0 = \ker\phi \times \langle a \rangle$ (since a is central) and $A = A' \times \langle a \rangle$, where $A' = A \cap \ker\phi$. By induction on m (the rank of A) we may assume that A' is a direct factor of a subgroup of finite index $H' \subset \ker\phi$. Let $H = H' \times \langle a \rangle$. $\square$

6.13 Remark. The first paragraph of the above proof establishes the following: if Γ is any group of isometries of a CAT(0) space X such that every homomorphism $\Gamma \to \mathbb{R}$ is trivial, then the centre of Γ contains no hyperbolic elements.

Clifford Translations and the Euclidean de Rham Factor

A simply connected complete Riemannian manifold of non-positive curvature admits a canonical splitting $M = N \times \mathbb{E}^n$ such that N cannot be further decomposed as a Riemannian product with a Euclidean factor; the factor $\mathbb{E}^n$ is called the Euclidean de Rham factor. In this section we construct a similar splitting for arbitrary CAT(0) spaces.

6.14 Definition. An isometry γ of a metric space X is called a *Clifford translation* if d_γ is a constant function, i.e., $\text{Min}(\gamma) = X$.

The pre-Hilbert space H constructed in the following theorem is the analogue for CAT(0) spaces of the Euclidean de Rham factor in Riemannian geometry.

6.15 Theorem. *Let X be a* CAT(0) *space.*

(1) *If γ is a non-trivial Clifford translation of X, then X splits as a product $X = Y \times \mathbb{R}$, and $\gamma(y, t) = (y, t + |\gamma|)$ for all $y \in Y$ and all $t \in \mathbb{R}$.*

(2) *If X splits as a product $X' \times X''$, every Clifford translation of X preserves this splitting and is the product of a Clifford translation of X' and a Clifford translation of X''.*

(3) *The Clifford translations form an abelian subgroup of the group of isometries of X; let H denote this subgroup.*

(4) *The group H is naturally a pre-Hilbert space, where the norm of a Clifford translation γ is equal to its translation length $|\gamma|$.*

(5) *If X is complete, then H is a Hilbert space.*

(6) *If the space H of Clifford translations is complete (in particular if it is finite dimensional or X is complete), then X admits a splitting $Y \times H$ such that, for every $y \in Y$, the subspace $\{y\} \times H$ is $H.y$, the orbit of y under the group H. Furthermore, every isometry of X preserves this splitting.*

Proof. Part (1) is a special case of 6.8(4). To prove (2), we consider the foliation of X by the axes of a non-trivial Clifford translation γ. If these axes project to points in X' then γ is of the form (id, γ''). Otherwise, the orthogonal projection of these axes give a foliation of X' by geodesic lines that are parallel; if $\gamma(x_0', x_0'') = (x_1', x_1'')$ then x_0' and x_1' lie on the same line. We obtain a Clifford translation γ' of X' by translating each of these parallel geodesic lines a distance $d(x_0', x_1')$ in the appropriate direction. (The product decomposition theorem (2.14) ensures that this is an isometry.) In the same way we obtain a Clifford translation γ'' of X'' with $\gamma''(x_0'') = \gamma''(x_1'')$. Now, (γ', γ'') is a Clifford translation of X whose action on the γ-axis through (x_0', x_0'') is the same as that of γ, and therefore by (1) we have $\gamma = (\gamma', \gamma'')$.

For (3) we consider two Clifford translations α and β of X. If α is non-trivial, then X splits as a product $Y \times \mathbb{R}$, and α is the product of the identity on Y and a translation τ_α of $\mathbb{R}$. By (2), β is the product of a Clifford translation β' of Y and a translation

τ_β of $\mathbb{R}$. Thus $\alpha\beta = \beta\alpha$ is the Clifford translation $(\beta', \tau_\alpha\tau_\beta)$ on $X = Y \times \mathbb{R}$. (Notice that if β' is non-trivial, then by applying (1) to the action of β' on Y we get that X splits as $Y' \times \mathbb{E}^2$, where α and β project to the identity on Y' and translations of $\mathbb{E}^2$.)

(4) We first define the scalar multiplication giving the vector space structure on the abelian group H of Clifford translations. Each non-trivial Clifford translation γ defines a splitting $Y \times \mathbb{R}$ of X such that $\gamma(y, t) = (y, t + |\gamma|)$. Given any $\lambda \in \mathbb{R}$, the map $(y, t) \mapsto (y, t + \lambda|\gamma|)$ is again a Clifford translation of X, which we denote $\lambda \cdot \gamma$. We define $\lambda \cdot \gamma$ to be the product of γ by the scalar λ, and claim that this defines a vector space structure on H. The only non-trivial point to check is that $\lambda \cdot (\alpha\beta)$ is the composition of $\lambda \cdot \alpha$ and $\lambda \cdot \beta$, for all $\alpha, \beta \in H$ and $\lambda \in \mathbb{R}$. But this follows easily from the parenthetical remark at the end of the preceding paragraph.

With the above definition of scalar multiplication, it is clear that the map which assigns to each $\gamma \in H$ its translation length $|\gamma|$ is a norm (the triangle inequality being an immediate consequence of the triangle inequality in X). In order to complete the proof of (4) we must show that this norm satisfies the parallelogram law. But the parallelogram law only involves comparing two vectors, and so we may again use the final sentence of the proof of (3) to reduce to the case $X = \mathbb{E}^2$, where the result is clear.

(5) For any $x \in X$ and $\alpha \in H$, the map $\alpha \mapsto \alpha.x$ is an isometry of H onto $H.x$, because $d(\alpha.x, \beta.x) = d(x, \alpha^{-1}\beta.x) = |\alpha^{-1}\beta|$. Thus, if α_n is a Cauchy sequence in H, then $\alpha_n.x$ is a Cauchy sequence in X. If we assume that X is complete, then this sequence must have a limit point, which we denote by $\alpha.x$. The map $x \mapsto \alpha.x$ is an isometry of X, moreover it is a Clifford translation, because for all $x \in X$ we have $d(x, \alpha.x) = \lim d(x, \alpha_n(x)) = \lim |\alpha_n|$. And by construction, $|\alpha^{-1}\alpha_n| = d(\alpha.x, \alpha_n.x) \to 0$, as $n \to \infty$. This proves (5).

(6) Suppose that H is complete. Choose a base point $x_0 \in X$. Let p denote the projection of X onto the closed convex set $H.x_0$, and let Y be $p^{-1}(x_0)$. We claim that the map $(y, \alpha) \mapsto \alpha.y$ is an isometry from $Y \times H$ onto X. This map is surjective because, given $x \in X$, its projection $p(x) \in H.x_0$ is of the form $\alpha.x_0$; we have $p(\alpha^{-1}.x) = \alpha^{-1}.p(x) = x_0$, hence $y := \alpha^{-1}.x$ belongs to Y and $x = \alpha.y$. Consider $y_1, y_2 \in Y$ and $\alpha_1, \alpha_2 \in H$ and let $\alpha = \alpha_1^{-1}\alpha_2$; we have to check that $d(\alpha_1.y_1, \alpha_2.y_2)^2 = d(y_1, y_2)^2 + |\alpha|^2$. We can assume that α is not the identity, otherwise the assertion is obvious. Henceforth we work with the splitting $Y' \times \mathbb{R}$ of X associated to α, as in (1). Without loss of generality, we may assume that x_0 corresponds to the point $(x_0, 0)$; thus Y is contained in $Y' = Y' \times \{0\}$, and y_1, y_2 correspond to the points $(y_1, 0)$ and $(y_2, 0)$. Therefore, $d(\alpha_1.y_1, \alpha_2.y_2)^2 = d(y_1, \alpha.y_2)^2 = d((y_1, 0), (y_2, |\alpha|))^2 = d(y_1, y_2)^2 + |\alpha|^2$, as required.

Finally, let γ be an isometry of X. For each $x \in X$, $H.x$ is the union of all axes of Clifford translations passing through x. As γ carries an axis of a Clifford translation α onto an axis of the Clifford translation $\gamma\alpha\gamma^{-1}$, we have $\gamma(H.x) = H(\gamma.x)$. Therefore, by I.5.3(4), γ preserves the splitting $X = Y \times H$. □

The next result indicates how Clifford translations may arise.

6.16 Lemma. *Let X be a complete* CAT(0) *space in which every geodesic can be extended to geodesic line. If a group Γ acts cocompactly by isometries on X and $\alpha \in$* Isom(X) *commutes with Γ, then α is a Clifford translation.*

Proof. The displacement function d_α is Γ-invariant, hence bounded because the action of Γ is cocompact. The geodesic segment connecting any pair of distinct points $x, y \in X$ can be extended to a geodesic line $c : \mathbb{R} \to X$. The function $t \mapsto d(c(t), \alpha.c(t))$ is convex and bounded, hence constant, so $d_\alpha(x) = d_\alpha(y)$ for all $x, y \in X$. □

The Group of Isometries of a Compact Metric Space of Non-Positive Curvature

Let Y be a compact metric space. In I.8.7 we showed that Isom(Y) equipped with the metric

$$d(\alpha, \alpha') = \sup_{y \in Y} d(\alpha.y, \alpha'.y)$$

is a compact topological group. In this section we show that one can say considerably more if Y is non-positively curved.

If Y is compact and non-positively curved, then by the Cartan-Hadamard theorem (4.1), its universal covering is a proper CAT(0) space X. And we can think of Y as the quotient $\Gamma\backslash X$ of X, where Γ is the fundamental group of Y acting freely and properly by isometries (as deck transformations) on X. In the following proof we shall need the fact (proved in I.8.6) that the group of isometries Isom(Y) of Y is naturally isomorphic to the quotient $N(\Gamma)/\Gamma$, where $N(\Gamma)$ is the normalizer of Γ in Isom(X).

Recall (5.7) that a complete geodesic space Y is said to have the *geodesic extension property* if every local isometry from a non-trivial compact interval of the real line into Y can be extended to a local isometry of the real line into Y. If Y has this property then so does any covering space of it. Examples of non-positively curved spaces which enjoy this property include closed topological manifolds and metric simplicial complexes with no free faces (5.10).

6.17 Theorem. *Let Y be a compact connected metric space of non-positive curvature, the quotient of the* CAT(0) *space X by a group Γ acting properly and freely on X by isometries. (Thus $\Gamma = \pi_1 Y$ and $X = \widetilde{Y}$.) Suppose that Y has the geodesic extension property. Then:*

(1) *The group* Isom(Y)*, equipped with the metric described above is a compact topological group with a finite number of connected components.*

(2) *The connected component of the identity in* Isom(Y) *is isomorphic to a torus — the quotient by a lattice Λ of a finite dimensional Euclidean vector space H_0, which is a subspace of the Clifford translations of X.*

(3) *In fact,* H_0 *is isomorphic to the centralizer* $C(\Gamma)$ *of* Γ *in* $\mathrm{Isom}(X)$ *and* $\Lambda = C(\Gamma) \cap \Gamma$ *is the centre of* Γ.

(4) *If the centre of* $\Gamma = \pi_1 Y$ *is trivial, then* $\mathrm{Isom}(Y)$ *is a finite group. In general, the centre of* Γ *is a free abelian group whose rank is smaller than or equal to the maximal dimension of flat subspaces of* X.

Proof. We showed in (6.16) that the centralizer $C(\Gamma)$ of Γ in $\mathrm{Isom}(X)$ consists entirely of Clifford translations.

We wish to describe the component of the identity in $\mathrm{Isom}(Y)$. Because Y is compact and the covering map $p : X \to Y$ is a local isometry, we can choose $\varepsilon > 0$ such that the restriction of p to every closed ball of radius ε in X is an isometry onto its image. Let $\overline{\nu}$ be an isometry of Y whose distance to the identity (in the metric defined above) is equal to $a \leq \varepsilon$. Then, there is a Clifford translation ν of X of norm a which is a lifting of $\overline{\nu}$: the map ν sends each point $x \in X$ to the unique point of the ball of radius ε and centre x whose image under p is $\overline{\nu}(p(x))$. The map ν is clearly length-preserving, and hence is an isometry of X. Moreover it commutes with every element of Γ, and hence is a Clifford translation. Indeed, for every $x \in X$, $d(x, \nu(x)) = d(p(x), \overline{\nu}(p(x))$ is equal to a.

The group of Clifford translations of X which centralize Γ form a Hilbert subspace H_0 of the Hilbert space H of all Clifford translations of X (cf. 6.15). As X is locally compact, H is finite dimensional. The argument of the previous paragraph shows that the natural homomorphism p_* from $C(\Gamma) = H_0$ to $\mathrm{Isom}(Y)$ induced by p sends the ball of radius ε centred at $0 \in H_0$ isometrically onto the ball of radius ε centred at the identity in $\mathrm{Isom}(Y)$. A first consequence of this observation is that the identity has a connected open neighbourhood in $\mathrm{Isom}(Y)$. In light of the fact that $\mathrm{Isom}(Y)$ is a compact topological group, this establishes (1). A second consequence of this observation is that the homomorphism p_* is a covering map from H_0 onto the connected component of the identity in $\mathrm{Isom}(Y)$; its kernel, $C(\Gamma) \cap \Gamma$, must be a lattice in $C(\Gamma) = H_0$. This establishes parts (2) and (3) of the theorem. (4) follows immediately from (3) and the fact that H_0 is isometrically embedded in X (cf. 6.15). □

6.18 Remarks

(1) In the preceding theorem, the hypothesis that Y has the geodesic extension property is essential. (Consider for example the case of a compact ball in $\mathbb{E}^n$.)

(2) The preceding proof, like that of (I.8.6), is not valid for proper actions by groups which have torsion. Let X be a CAT(0) space and suppose that $\Gamma \subset \mathrm{Isom}(X)$ acts properly and cocompactly. Let $N(\Gamma)$ denote the normalizer of Γ in $\mathrm{Isom}(X)$. As in (I.8.6), there is a natural homomorphism from $N(\Gamma)/\Gamma$ to the isometry group of the compact orbit space $\Gamma\backslash X$, but in general this map is neither injective nor surjective (Exercise: why not?). Note however that the image of $N(\Gamma)/\Gamma$ in $\mathrm{Isom}(\Gamma\backslash X)$ is closed, so with the induced topology $N(\Gamma)/\Gamma$ becomes a compact topological group (cf. I.8.7).

For future reference we note an obvious consequence of (6.17).

6.19 Corollary. *Let X be a* CAT(0) *space with the geodesic extension property and suppose that $\Gamma \subset \text{Isom}(X)$ acts properly and cocompactly. If Γ has a torsion-free subgroup of finite index with trivial centre, then the centralizer of Γ in* Isom(X) *is trivial.*

Lemma 6.16 can be viewed as a special case of the following lemma which will be needed in the proof of the Splitting Theorem (6.21) as well various points in later chapters.

6.20 Lemma. *If X is a complete* CAT(0) *space with the geodesic extension property, and Γ is a group which acts cocompactly on X by isometries, then X contains no Γ-invariant, closed, convex subsets other than X and the empty set.*

Proof. Let C be a non-empty Γ-invariant, closed, convex subset of X. Let π denote the orthogonal projection of X onto C. Since C is Γ-invariant, it contains a Γ orbit, and since the action of Γ is cocompact there is a constant K such that every point of X is within a distance K of C. This implies that in fact C itself is the whole of X, for if there were a point $x \in X \smallsetminus C$ then one could extend the geodesic $[\pi(x), x]$ to a geodesic ray $c : [0, \infty) \to X$ such that $d(c(t), C) = t$ for all t (because $t \mapsto d(c(t), C)$ is convex and $d(c(t), C) = t$ for small t). □

A Splitting Theorem

Splitting theorems of the following type were proved in the framework of Riemannian manifolds of non-positive curvature by D. Gromoll and J. Wolf [GW71] and B. Lawson and S. Yau [LaY72] (see also Schroeder [Sch85]). A similar theorem for CAT(0) spaces was proved by Claire Baribaud in her diplôme work [Bar93]. A splitting theorem of a slightly different nature will be proved in Chapter 9.

6.21 Theorem. *Let X be a* CAT(0) *space with the geodesic extension property. Suppose that $\Gamma = \Gamma_1 \times \Gamma_2$ acts properly and cocompactly by isometries on X and suppose that Γ satisfies one of the following conditions:*

(1) *Γ has finite centre and Γ_1 is torsion free; or*

(2) *the abelianization of Γ_1 is finite.*

Then, X splits as a product of metric spaces $X_1 \times X_2$ and Γ preserves the splitting; the action of $\Gamma = \Gamma_1 \times \Gamma_2$ on $X = X_1 \times X_2$ is the product action and $\Gamma\backslash X = \Gamma_1\backslash X_1 \times \Gamma_2\backslash X_2$.

6.22 Corollary. *Let Y be a compact geodesic space of non-positive curvature that has the geodesic extension property. Assume that the fundamental group of Y splits as a product $\Gamma = \Gamma_1 \times \Gamma_2$ and that Γ has trivial center. Then Y splits as a product $Y_1 \times Y_2$ such that the fundamental group of Y_i is Γ_i, $i = 1, 2$.*

We shall deduce Theorem 6.21 from the following more technical result.

6.23 Proposition. *Let X be a* CAT(0) *space with the geodesic extension property. Suppose that $\Gamma = \Gamma_1 \times \Gamma_2$ acts properly and cocompactly on X. If the action of Γ_1 on the closed convex hull C of one of its orbits is cocompact, then X splits as a product of metric spaces $X_1 \times X_2$ and Γ preserves the splitting. Moreover, the subspaces of the form $X_1 \times \{x_2\}$ are precisely the closed convex hulls of the Γ_1-orbits*

Proof. We shall present the proof of the proposition in several steps. Let Σ be the set of closed, convex, non-empty, Γ_1-invariant subspaces of X, and let $\mathcal{N}$ be the subset of Σ consisting of those subspaces which are minimal with respect to inclusion. Note that since a non-empty intersection of subspaces in Σ is again in Σ, the members of $\mathcal{N}$ are disjoint. Note also that each member of $\mathcal{N}$ is the closed convex hull of some Γ_1-orbit $C(\Gamma_1.x)$.

Claim 1. $\mathcal{N}$ is non-empty.

Let C be as in the proposition and let K be a compact set such that $\Gamma_1.K = C$. Every Γ_1-invariant closed, convex, non-empty subspace of C intersects K, so the intersection of a decreasing sequence of such subspaces has a non-empty intersection with K; such an intersection is again in Σ. Therefore we may apply Zorn's lemma to deduce the existence of a minimal, closed, convex, Γ_1-invariant subset in C.

Given $C_1, C_2 \in \mathcal{N}$, let $p_i : X \to C_i$ denote the projection of X onto C_i and let $d = d(C_1, C_2) := \inf\{d(x_1, x_2) \mid x_1 \in C_1, x_2 \in C_2\}$.

Claim 2. There is a unique isometry j of $C_1 \times [0, d]$ onto the convex hull of $C_1 \cup C_2$ such that $j(x, 0) = x$ and $j(x, d) = p_2(x)$.

The function $d_{C_1} : x \mapsto d(x, p_1(x))$, which is convex and Γ_1-invariant, must be constant on C_2, because if there were points $x, y \in C_2$ such that $d_{C_1}(x) < d_{C_1}(y)$ then $\{z \in C_2 \mid d_{C_1}(z) \leq d_{C_1}(x)\}$ would be a closed, convex, Γ_1-invariant proper subspace of C_2, contradicting the minimality of C_2. Apply 2.12 (2).

Claim 3. For every $x \in X$ there exists a unique $C_x \in \mathcal{N}$ such that $x \in C_x$. In fact, $C_x = C(\Gamma_1.x)$, the closed convex hull of the Γ_1-orbit of x.

The uniqueness of C_x is obvious because distinct elements of $\mathcal{N}$ are disjoint. And if $x \in C_x$ then $\Gamma_1.x \subset C_x$ and hence $C(\Gamma_1.x) = C_x$, because C_x is minimal.

Let $X' = \bigcup\{C_i \mid C_i \in \mathcal{N}\}$. To establish the existence of C_x it is enough to show that X' is a closed, convex, Γ-invariant subspace of X (see 6.20).

We first prove that X' is Γ-invariant. Each $C \in \mathcal{N}$ is Γ_1-invariant. Given $C = C(\Gamma_1.x) \in \mathcal{N}$ and $\gamma_2 \in \Gamma_2$, we have $\gamma_2.C = C(\gamma_2\Gamma_1.x) = C(\Gamma_1.\gamma_2 x)$, a closed, convex, Γ_1-invariant subspace of X. And if $C' \subset C(\Gamma_1.\gamma_2 x)$ were a smaller such subspace then $\gamma_2^{-1}C' \subset C$ would contradict the minimality of C. Thus $\gamma_2.C \in \mathcal{N}$.

Claim 2 implies that X' is convex.

To prove that X' is closed, we consider a point $x \in X$ and a sequence of points $x_n \in X'$ converging to x. Passing to a subsequence we may assume that $d(x_n, x_{n+1}) \leq 1/2^n$. Let C_n be the element of $\mathcal{N}$ containing x_n, let $p_n : X \to C_n$ be the orthogonal projection, and let P_n be the restriction to C_1 of the composition

$p_n p_{n-1} \dots p_2$. Note that for any $z \in C_1$, the sequence $P_n(z)$ is a Cauchy sequence because $d(P_n(z), P_{n+k}(z)) \leq \sum_{i=n}^{n+k-1} d(C_i, C_{i+1}) \leq \sum_{i=n}^{n+k-1} d(x_i, x_{i+1}) \leq 1/2^{n-1}$. Since X is complete, the sequence of Γ_1-equivariant maps $P_n : C_1 \to X$ converges. The limit $P : C_1 \to X$ is a Γ_1-equivariant isometry of C_1 onto its image, which is therefore closed, convex and Γ_1-invariant. The minimality of C_1 implies that $P(C_1)$ is also minimal, hence $P(C_1) \in \mathcal{N}$.

It remains to show that $x \in P(C_1)$. Let $z_n \in C_1$ be such that $P_n(z_n) = x_n$. Then, $d(P(z_n), x) \leq d(P(z_n), x_n) + d(x_n, x) = d(P(z_n), P_n(z_n)) + d(x_n, x) \leq 1/2^{n-1} + d(x_n, x)$, which tends to 0 as $n \to \infty$. Thus x lies in the closure of $P(C_1)$, hence in $P(C_1)$.

Claim 4. If $C_1, C_2, C_3 \in \mathcal{N}$, and if $p_i : X \to C_i$ denotes projection onto C_i, then $p_1 = p_1 p_2$ on C_3.

If each C_i is a single point (the case where Γ_1 is finite) then the claim is trivial. If not, then each point $x \in C_3$ lies on a non-trivial geodesic path in C_3. By hypothesis this can be extended to a geodesic line $c(\mathbb{R})$ in X. It follows from Claim 2 that $c(\mathbb{R})$ cannot intersect any member of $\mathcal{N}$ other than C_3 and hence, by Claim 3, $c(\mathbb{R}) \subset C_3$. The present claim then follows by application of (2.15) to the parallel geodesic lines $c(\mathbb{R})$, $p_1 c(\mathbb{R})$ and $p_2 c(\mathbb{R})$.

Fix $X_1 \in \mathcal{N}$. Let $p : X \to X_1$ be orthogonal projection. It is Γ_1-equivariant. Let X_2 denote the metric space $(\mathcal{N}, d)$, where $d(C, C') = \inf\{d(x, x') \mid x \in C, x' \in C'\}$. Given $x \in X$, let $C_x \in \mathcal{N}$ be as in Claim 3.

By restriction to X_1 we get an action of Γ_1 on X_1. The action of Γ_1 on X_2 is trivial. There is an isometric action of Γ_2 on X_2 by $\gamma_2.C_x = C_{\gamma_2.x}$. According to Claims 2 and 4, $x \mapsto p(\gamma_2.x)$ defines an action of Γ_2 on X_1 commuting with the action of Γ_1, and by Claim 2 this action is by isometries. Thus we obtain an action of Γ on $X_1 \times X_2$ by isometries.

Claim 5. The map $X \to X_1 \times X_2$ given by $x \mapsto (p(x), C_x)$ is a Γ-equivariant isometry.

The map is Γ-equivariant by construction.

Given $x, x' \in X$, let p' be projection onto $C_{x'}$. By applying Claim 2 twice and then applying Claim 4, we see that $X \to X_1 \times X_2$ is an isometry:

$$\begin{aligned} d(x, x')^2 &= d(x, p'(x'))^2 + d(C_x, C_{x'})^2 \\ &= d(p(x), pp'(x'))^2 + d(C_x, C_{x'})^2 \\ &= d(p(x), p(x'))^2 + d(C_x, C_{x'})^2. \end{aligned}$$

It is clear from the construction of $X_1 \times X_2$ that the action of Γ on it has the properties stated in the proposition. □

We now turn to the proof of Theorem 6.21.

Proof of Theorem 6.21(1). We first prove that hypothesis (6.21(1)) is sufficient to guarantee that the hypothesis of Proposition 6.23 holds. This follows from the next lemma, which can be found in [GW71] and [ChEb75].

6.24 Lemma. *Let X be a proper* CAT(0) *space. Let $\Gamma = \Gamma_1 \times \Gamma_2$ be a group acting properly on X by isometries, and suppose that the centre of Γ_2 is finite. Let $C \subset X$ be a closed, convex, Γ_1-invariant subspace and suppose that there exists a compact subset $K \subset X$ such that $C \subset \Gamma.K$. Then there is a compact subset $K' \subset C$ such that $C = \Gamma_1.K'$.*

Proof. Assuming that there is no compact subspace K' such that $C(\Gamma_1.K) = \Gamma_1.K'$, we shall construct below a sequence of distinct elements $\beta_n \in \Gamma_2$ such that, for any $\beta \in \Gamma_2$, the family of displacement functions $d_{\beta_n\beta\beta_n^{-1}}$ is uniformly bounded on K. As Γ acts properly on X, this implies that the subset $\{\beta_n\beta\beta_n^{-1}\}_n \subset \Gamma_2$ is finite. Hence we may pass to a subsequence and assume that all of the $\beta_n\beta\beta_n^{-1}$ are equal. Applying this argument as β ranges over a finite generating set for Γ_2 (cf. I.8.10), we obtain a sequence of distinct elements β_i such that $\beta_m\beta\beta_m^{-1} = \beta_n\beta\beta_n^{-1}$ for every $\beta \in \Gamma_2$ and all $n, m \in \mathbb{N}$. But then we would have an infinite number of elements $\beta_n\beta_1^{-1}$ in the centre of Γ_2, which is supposed to be finite.

To construct the sequence (β_n), first note that if the desired K' does not exist then one can find a sequence of points x_n in C such that $d(x_n, \Gamma_1.K) \to \infty$. Choose $\beta_n \in \Gamma_2$ so that $\beta_n(x_n) \in \Gamma_1.K$. Then $d(\beta_n^{-1}(K), K) \geq d(\beta_n^{-1}(K), C(\Gamma_1.K)) \geq d(x_n, C(\Gamma_1.K)) - \text{diam}(K)$. Hence, after passing to a subsequence, we may assume that all of the elements β_n are distinct.

It remains to check that, for any $\beta \in \Gamma_2$, the family of displacement functions $d_{\beta_n\beta\beta_n^{-1}}$ is uniformly bounded on K. Observe that the displacement function d_β (which is convex) is constant on the orbits of Γ_1 because β commutes with Γ_1; if r is the maximum of d_β on K, then the set of points where d_β is not bigger than r is a closed, convex, Γ_1-invariant subspace containing $\Gamma_1.K$ and hence C. Therefore d_β is bounded by r on $C(\Gamma_1.K)$. As $d_{\beta_n\beta\beta_n^{-1}}(\beta_n.x_n) = d_\beta(x_n) \leq r$, for any $x \in K$ we have

$$d_{\beta_n\beta\beta_n^{-1}}(x) \leq r + 2\,\text{diam}(K).$$

Indeed if $\alpha_n \in \Gamma_1$ is such that $\beta_n\alpha_n.x_n \in K$, then

$$\begin{aligned}
&d(\beta_n\beta\beta_n^{-1}.x, x)\\
&\leq d(\beta_n\beta\beta_n^{-1}.x,\ \beta_n\beta\beta_n^{-1}.(\beta_n\alpha_n.x_n)) + d(\beta_n\beta\alpha_n.x_n,\ \alpha_n\beta_n.x_n) + d(\alpha_n\beta_n.x_n,\ x)\\
&= d(\beta_n\beta\beta_n^{-1}.x,\ \beta_n\beta\beta_n^{-1}.(\beta_n\alpha_n.x_n)) + d(\alpha_n\beta_n\beta.x_n,\ \alpha_n\beta_n.x_n) + d(\alpha_n\beta_n.x_n,\ x)\\
&\leq \text{diam}(K) + r + \text{diam}(K).
\end{aligned}$$

□

Applying 6.23, we get a splitting $X_1 \times X_2$ of X preserved by the action of Γ, the action of Γ_1 being trivial on X_2. It remains to show that the action of Γ_2 on the first factor X_1 is also trivial. To see this we apply Corollary 6.18. As Γ_1 is assumed to be torsion free and as it acts properly on X_1, it acts freely on X_1. The centre of Γ_1 is assumed to be finite and torsion free, hence trivial, so by (6.18) the centralizer of Γ_1 in $\text{Isom}(X_1)$ is trivial. As each element of Γ_2 commutes with Γ_1, it follows that the action of Γ_2 on X_1 is trivial. □

Proof of Theorem 6.21(2). We wish to apply (6.23). In order to do so, we shall prove by induction on the maximum dimension of flats in X that Γ_1 acts cocompactly on the closed convex hull of each Γ_1-orbit in X. If the centre of Γ is finite then we can apply (6.24). As we remarked in (6.13), because Γ_1 has finite abelianization it has no elements of infinite order in its centre, so if the centre of Γ (which is a finitely generated abelian group) is infinite then there exists an element of infinite order γ in the centre of Γ_2. Such a γ is a Clifford translation (6.16), so we get a Γ-invariant splitting $X = Y \times \mathbb{R}$, where γ acts trivially on Y and Γ acts by translations on the second factor $\mathbb{R}$. Because the abelianization of Γ_1 is finite, its action on the second factor is trivial; in particular, every Γ_1-orbit in X is contained in some slice $Y \times \{t\}$. Applying our inductive hypothesis to the action of $\Gamma/\langle \gamma \rangle = \Gamma_1 \times \Gamma_2/\langle \gamma \rangle$ on Y we deduce that the action of Γ_1 on the closed convex hull of each of its orbits in Y (and hence X) is cocompact.

In order to complete the proof we must show that the action of Γ_2 on X_1 arising from (6.23) is trivial. For this it suffices to show that for every $\gamma_2 \in \Gamma_2$ there exists a Γ_1-orbit $\Gamma_1.x$ in X with closed convex hull $C(\Gamma_1.x)$ such that, if $p : X \to C(\Gamma_1.x)$ is the orthogonal projection then $p\gamma_2$ restricts to the identity on $C(\Gamma_1.x)$. If γ_2 is elliptic then its action on the closed convex hull of any orbit in $\mathrm{Min}(\gamma_2)$ is trivial, so we choose $x \in \mathrm{Min}(\gamma_2)$. If γ_2 is hyperbolic, then we consider $\mathrm{Min}(\gamma_2)$ with its splitting $Y' \times \mathbb{R}$ as in (6.8(4)). (Recall that the action of γ_2 on Y' is trivial.) As Γ_1 commutes with γ_2, it preserves $\mathrm{Min}(\gamma_2)$ and its splitting. As the abelianization of Γ_1 is finite, its action on the second factor is trivial, so each Γ_1-orbit in $\mathrm{Min}(\gamma_2)$ is contained in a slice $Y' \times \{t\}$. If p is the projection onto such an orbit, then $p\gamma_2$ is the identity, because the action of γ_2 on Y' is trivial. □

6.25 Exercises

(1) Let Γ be the fundamental group of a closed surface of positive genus. Construct a proper cocompact action of $\Gamma \times \mathbb{Z}$ on $\mathbb{H}^2 \times \mathbb{R}$ with the property that Γ does not act cocompactly on the convex hull of any of its orbits. (Hint: Observe that there exist non-trivial homomorphisms $\Gamma \to \mathbb{Z}$ via which Γ acts on $\mathbb{R}$. Combine such an action with a cocompact action of Γ on $\mathbb{H}^2$ and consider the diagonal action on $\mathbb{H}^2 \times \mathbb{R}$. Extend this to a cocompact action of $\Gamma \times \mathbb{Z}$.)

(2) Show that the hypotheses of (6.21(1)) can be weakened to: the centre of Γ is finite and Γ_1 modulo its centre has a torsion-free subgroup of finite index.

(3) Let X be a proper CAT(0) space with the geodesic extension property. Let Γ be a group acting by isometries on X such that the closure in $\overline{X}$ of the Γ-orbit of a point of X contains the whole of ∂X (notation of Chapter 8). Suppose that no point of ∂X is fixed by all the elements of Γ. If Γ splits as a product $\Gamma_1 \times \Gamma_2$, prove there is a splitting $X = X_1 \times X_2$ such that if $\gamma = (\gamma_1, \gamma_2)$ and $x = (x_1, x_2)$, then $\gamma.(x_1, x_2) = (\gamma_1.x_1, \gamma_2.x_2)$. (cf. Schroeder [Sch85].)

Chapter II.7 The Flat Torus Theorem

This is the first of a number of chapters in which we study the subgroup structure of groups Γ that act properly by semi-simple isometries on CAT(0) spaces X. In this chapter our focus will be on the abelian subgroups of Γ. The Flat Torus Theorem (7.1) shows that the structure of such subgroups is faithfully reflected in the geometry of the flat subspaces in X. One important consequence of this fact is the Solvable Subgroup Theorem (7.8): if Γ acts properly and cocompactly by isometries on a CAT(0) space, then every solvable subgroup of Γ is finitely generated and virtually abelian. In addition to algebraic results of this kind, we shall also present some topological consequences of the Flat Torus Theorem.

Both the Flat Torus Theorem and the Solvable Subgroup Theorem were discovered in the setting of smooth manifolds by Gromoll and Wolf [GW71] and, independently, by Lawson and Yau [LaY72]. Our proofs are quite different to the original ones.

Throughout this chapter we shall work primarily with proper actions by semi-simple isometries rather than cocompact actions. Working in this generality has a number of advantages: besides the obvious benefit of affording more general results, one can exploit the fact that when restricting an action to a subgroup one again obtains an action of the same type. (In particular this facilitates induction arguments.)

The Flat Torus Theorem

Recall that, given a group of isometries Γ, we write $\mathrm{Min}(\Gamma)$ for the set of points which are moved the minimal distance $|\gamma|$ by every $\gamma \in \Gamma$.

7.1 Flat Torus Theorem. *Let A be a free abelian group of rank n acting properly by semi-simple isometries on a* CAT(0) *space X. Then:*

(1) $\mathrm{Min}(A) = \bigcap_{\alpha\in A} \mathrm{Min}(\alpha)$ *is non-empty and splits as a product $Y \times \mathbb{E}^n$.*

(2) *Every element $\alpha \in A$ leaves* $\mathrm{Min}(A)$ *invariant and respects the product decomposition; α acts as the identity on the first factor Y and as a translation on the second factor $\mathbb{E}^n$.*

(3) *The quotient of each n-flat $\{y\} \times \mathbb{E}^n$ by the action of A is an n-torus.*

(4) *If an isometry of* X *normalizes* A*, then it leaves* $\mathrm{Min}(A)$ *invariant and preserves the product decomposition.*

(5) *If a subgroup* $\Gamma \subset \mathrm{Isom}(X)$ *normalizes* A*, then a subgroup of finite index in* Γ *centralizes* A*. Moreover, if* Γ *is finitely generated, then* Γ *has a subgroup of finite index that contains* A *as a direct factor.*

Proof. We shall prove parts (1), (2) and (3) by induction on the rank of A. As the action of A is proper and by semi-simple isometries, each non-trivial element of A is a hyperbolic isometry. Suppose that $A \cong \mathbb{Z}^n$ and choose generators $\alpha_1, \dots, \alpha_n$. We have seen (6.8) that $\mathrm{Min}(\alpha_1)$ splits as $Z \times \mathbb{E}^1$ where α_1 acts trivially on the first factor and acts as a translation of amplitude $|\alpha_1|$ on the second factor. Every $\alpha \in A$ commutes with α_1 and therefore preserves the subspace $Z \times \mathbb{E}^1$ with its decomposition, acting by translation on the factor $\mathbb{E}^1$ (see (6.8)).

We claim that the subgroup $N \subset A$ formed by the elements $\alpha \in A$ which act trivially on the factor Z is simply the subgroup of A generated by α_1. To see this, note that N, which is free abelian, acts properly on each line $\{y\} \times \mathbb{E}^1$ by translation, hence it is cyclic; and since $\alpha_1 \in N$ is primitive in A, it must generate N.

The free abelian group $A_0 = A/N$ is of rank $n-1$. Its action on Z is proper (6.10(4)) and by semi-simple isometries (6.9). As Z is a convex subspace of X, it is a CAT(0) space, so by induction $\mathrm{Min}(A_0) \subset Z$ splits as $Y \times \mathbb{E}^{n-1}$, where A_0 acts trivially on Y and acts by translations on $\mathbb{E}^{n-1}$ with quotient an $(n-1)$-torus. Thus $\mathrm{Min}(A) = Y \times \mathbb{E}^{n-1} \times \mathbb{E}^1 = Y \times \mathbb{E}^n$, and (1), (2) and (3) hold.

If an isometry γ normalizes A, then it obviously preserves $\mathrm{Min}(A)$; we claim that it also preserves its product decomposition. Indeed γ maps A-orbits onto A-orbits, and therefore maps the convex hull of each A-orbit onto the convex hull of an A-orbit; and it follows from (3) that the convex hull of the A-orbits of points of $\mathrm{Min}(A)$ are the n-flats $\{y\} \times \mathbb{E}^n$, $y \in Y$. This proves (4).

It also follows from (3) that for any number r there are only finitely many elements $\alpha \in A$ with translation length $|\alpha| = r$. But $|\gamma\alpha\gamma^{-1}| = |\alpha|$ for all $\gamma \in \mathrm{Isom}(X)$, so if $\Gamma \subset \mathrm{Isom}(X)$ normalizes A, then the image of the homomorphism $\Gamma \to \mathrm{Aut}(A)$ given by conjugation must be finite (because the number of possible images for each basis element $\alpha_i \in A$ is finite). The kernel of this homomorphism is a subgroup of finite index in Γ and its elements centralize A. If Γ is finitely generated then (6.12) tells us that there exists a further subgroup of finite index which contains A as a direct factor. This proves (5). $\square$

We generalize the Flat Torus Theorem to the case of virtually abelian groups. (A group is said to *virtually* satisfy a property if it has a subgroup of finite index that satisfies that property.)

7.2 Corollary. *Let* Γ *be a finitely generated group which acts properly by semi-simple isometries on a complete* CAT(0) *space* X*. Suppose that* Γ *contains a subgroup of finite index that is free abelian of rank* n*. Then:*

(1) X *contains a* Γ*-invariant closed convex subspace isometric to a product* $Y \times \mathbb{E}^n$.

(2) *The action of Γ preserves the product structure on $Y \times \mathbb{E}^n$, acting as the identity on the first factor, and acting cocompactly on the second.*

(3) *Any isometry of X which normalizes Γ preserves $Y \times \mathbb{E}^n$ and its splitting.*

Proof. Let $A_0 \cong \mathbb{Z}^n$ be a subgroup of finite index in Γ. Because Γ is finitely generated, it contains only finitely many subgroups of index $|\Gamma/A_0|$. Let A be the intersection of all of these subgroups. Note that A is a characteristic subgroup of finite index in Γ, that is, $\varphi(A) = A$ for all $\varphi \in \text{Aut}(\Gamma)$. In particular, if $\alpha \in \text{Isom}(X)$ normalizes Γ, then it also normalizes A. As A is of finite index in A_0, it is also isomorphic to $\mathbb{Z}^n$.

According to (7.1), Min(A) splits as $Z \times \mathbb{E}^n$, where Z is a closed convex subspace of X; A acts trivially on the first factor, and cocompactly by translations on the second factor. Because $A \subset \Gamma$ is normal, Γ acts by isometries of Min(A), preserving the splitting (7.1(4)). Thus we obtain an induced action of the finite group Γ/A on the complete CAT(0) space Z. According to (2.8), the fixed point set for this action is a non-empty, closed, convex subset of Z; call this subset Y.

By construction $Y \times \mathbb{E}^n$ is Γ-invariant and the action of Γ on the first factor is trivial. Since the action of A on each flat $\{y\} \times \mathbb{E}^n$ is cocompact, the action of Γ on each $\{y\} \times \mathbb{E}^n$ is also cocompact.

To prove (3), one first notes that $Y \times \mathbb{E}^n$ is preserved by any isometry of X that normalizes Γ, because (as we observed above) the normalizer of Γ in Isom(X) is contained in the normalizer of A. Assertion (3) then follows, as in (7.1), from the fact that the flats $\{y\} \times \mathbb{E}^n$ are precisely those subsets of $Y \times \mathbb{E}^n$ that arise as convex hulls of Γ-orbits. □

7.3 Remarks

(1) The invariant subspace $Y \times \mathbb{E}^n$ constructed in the preceding corollary contains Min(Γ), but in general it is not equal to Min(Γ). For example, consider the standard action of the Klein bottle group $\Gamma = \langle \alpha, \beta \mid \alpha^{-1}\beta^{-1}\alpha = \beta \rangle$ on the Euclidean plane: the action of α is a translation of norm 1 followed by a reflection about an axis of this translation; β acts as a Clifford translation whose axis is orthogonal to that of α.

In this case, the invariant subspace yielded by Corollary 7.2 is the whole of the plane, while Min(Γ) is empty.

(2) (Bieberbach Theorem) A finitely generated group Γ acts properly and cocompactly by isometries on $\mathbb{E}^n$ if and only if Γ contains a subgroup $A \cong \mathbb{Z}^n$ of finite index. In the light of the preceding corollary and (6.5), in order to prove the 'if' direction in this statement it suffices to get Γ to act properly by isometries on $\mathbb{E}^d$ for some d. To produce such an action, one fixes a proper action of A by translations on $\mathbb{E}^n$ and considers the quotient X of $\Gamma \times \mathbb{E}^n$ by the relation $[(\gamma a, x) \sim (\gamma, a.x)$ for all $x \in \mathbb{E}^n$ and $a \in A]$. Let $m = |\Gamma/A|$. There is a natural identification of $\mathbb{E}^{mn}$ with the set of sections of the projection $X \to \Gamma/A$, and the natural action of Γ on the set of sections gives a proper action by isometries on $\mathbb{E}^{mn}$.

See section 4.2 of [Thu97] for an elegant account of the discrete isometry groups of $\mathbb{E}^n$.

Cocompact Actions and the Solvable Subgroup Theorem

The main goal of this paragraph is to prove that if a group Γ acts properly and cocompactly by isometries on a CAT(0) space then all of its solvable subgroups are finitely generated and virtually abelian. First we show that abelian subgroups of Γ are finitely generated and satisfy the ascending chain condition. For this we need the following geometric observation. Recall that a metric space X is said to be cocompact if there exists a compact subset $K \subset X$ whose translates by Isom(X) cover X.

7.4 Lemma. *If a proper* CAT(0) *space X is cocompact, then there is a bound on the dimension of isometrically embedded flat subspaces $\mathbb{E}^m \hookrightarrow X$.*

Proof. We fix $x_0 \in X$ and choose r sufficiently large to ensure that the translates of the ball $B(x_0, r)$ by Isom(X) cover X. As $\overline{B}(x_0, 2r)$ is compact, it can be covered by a finite number N of balls B_i of radius $r/2$. If there is a flat subspace of dimension p in X, then we can translate it by the action of Isom(X) so as to obtain a convex subspace $D_p \subset B(x_0, 2r)$ isometric to a Euclidean ball of dimension p and radius r. An orthonormal frame at the centre of D_p gives $2p$ points on the boundary of D_p, each pair of which are a distance at least $\sqrt{2}r$ apart. Each of these points is contained in a different ball B_i of our chosen covering of $\overline{B}(x_0, 2r)$. Therefore $2p < N$. □

If A is an abelian group, then its *rank* $\text{rk}_{\mathbb{Q}}A$ is, by definition, the dimension of the $\mathbb{Q}$-vector space $A \otimes \mathbb{Q}$. In other words, $\text{rk}_{\mathbb{Q}}A$ is the greatest integer n such that A contains a subgroup isomorphic to $\mathbb{Z}^n$.

7.5 Theorem (Ascending Chain Condition). *Let $H_1 \subseteq H_2 \subseteq \cdots$ be an ascending chain of virtually abelian subgroups in a group Γ. If Γ acts properly and cocompactly by isometries on a* CAT(0) *space, then $H_n = H_{n+1}$ for sufficiently large n.*

7.6 Corollary. *If a group Γ acts properly and cocompactly by isometries on a* CAT(0) *space, then every abelian subgroup of Γ is finitely generated.*

Proof of the corollary. If $A \subset \Gamma$ is not finitely generated then there exists a sequence of elements $a_1, a_2, \dots$ such that, for all $n > 0$, the subgroup H_n generated by $\{a_1, \dots, a_n\}$ does not contain a_{n+1}. Thus we obtain a strictly ascending chain of subgroups $H_1 \subset H_2 \subset \cdots$. □

Proof of the theorem. The proof of the corollary shows that if there were a strictly ascending chain of virtually abelian subgroups in Γ then there would be a strictly ascending chain of finitely generated virtually abelian subgroups. Thus we may assume that the H_i are finitely generated.

Let X be a CAT(0) space on which Γ acts properly and cocompactly by isometries. The existence of a proper cocompact group action implies that X is proper (I.8.4) and that each element of Γ acts on X as a semi-simple isometry (6.10). Lemma 7.4 gives a bound on the rank of flat subspaces in X, so it follows from the Flat Torus

Theorem (7.1) that there is a bound on the rank of the abelian subgroups in Γ. Thus, without loss of generality, we may assume that there is an integer n such that each H_i contains a finite-index subgroup $A_i \subset H_i$ isomorphic to $\mathbb{Z}^n$.

If $n = 0$, the groups H_i are finite, and by (2.8) there is a bound on the orders of finite subgroups in Γ.

Consider the case $n > 0$. We claim that A_1 has finite index in each H_i. To see this, consider the action of A_1 on H_i by left translation and the induced action of A_1 on the finite set H_i/A_i. The stabilizer $B_i \subset A_1$ of the coset A_i is contained in A_i. Since B_i has finite index in A_1, it has rank n, therefore it has finite index in A_i, and hence in H_i.

To complete the proof we must show that the index of A_1 in H_i is bounded independent of i. By the Flat Torus Theorem (7.1), the convex subspace $\text{Min}(A_1)$ splits isometrically as $Y \times \mathbb{E}^n$, and A_1 acts trivially on Y and cocompactly by translations on $\mathbb{E}^n$. According to (7.2), for each i there is a convex subspace Y_i of Y such that H_i preserves $Y_i \times \mathbb{E}^n$ with its product structure and acts trivially on the factor Y_i.

As the action of Γ on X is proper and cocompact, for each $R > 0$, there is an integer N_R such that, for every $x \in X$, the set $\{\gamma \in \Gamma \mid \gamma.x \in B(x, R)\}$ has cardinality less than N_R. We can choose R big enough so that in the action of A_1 on the second factor of $Y \times \mathbb{E}^n$, the orbit of any ball of radius R covers $\mathbb{E}^n$. Then, given $h \in H_i$ and $x \in Y_i \times \{0\}$, we can find an element $a \in A_1$ such that $d(x, ah(x)) < R$. It follows that the index of A_1 in each H_i is bounded by N_R, which is independent of i. $\square$

7.7 Remarks

(1) If X is a complete simply-connected Riemannian manifold of non-positive curvature, then every abelian subgroup of $\text{Isom}(X)$ either contains an element of infinite order or else it is finite. Indeed, one can easily show that the set of points fixed by any group of isometries of X is a complete geodesically convex submanifold, and this allows one to argue by induction on the dimension of the fixed point sets of the elements of Γ (see [ChEb75] and [BaGS85]).

A similar induction can be used to show that any abelian subgroup of $\text{Isom}(X)$ which acts properly by semi-simple isometries is finitely generated. The analogous statements are not true for proper CAT(0) spaces in general (see (7.11) and (7.15)).

(2) In the preceding theorem one can replace the hypothesis that the action of Γ on X is cocompact by the following assumptions: (i) the action of Γ is proper and by semi-simple isometries; (ii) there is a bound on the dimension of flat subspaces in X; (iii) the set of translation numbers $\{\, |\gamma| \mid \gamma \in \Gamma\}$ is discrete at zero; and (iv) there is a bound on the order of finite subgroups in Γ.

The examples given in (7.11) and (7.15) show that each of these conditions is necessary. Torsion-free groups acting by cellular isometries on M_κ-complexes K with $\text{Shapes}(K)$ finite satisfy all of these conditions.

Recall that the n-th derived subgroup of a group G is defined recursively by $G^{(0)} = G$ and $G^{(n+1)} = [G^{(n)}, G^{(n)}]$, where the square brackets denote the commutator subgroup. G is said to be *solvable* of derived length n if n is the least integer such that $G^{(n)} = \{1\}$.

7.8 Solvable Subgroup Theorem. *If the group Γ acts properly and cocompactly by isometries on a* CAT(0) *space X, then every virtually solvable subgroup $S \subset \Gamma$ is finitely generated and contains an abelian subgroup of finite index. (And the action of S on X is as described in Corollary 7.2.)*

Proof. Any countable group can be written as an ascending union of its finitely generated subgroups, so in the light of the preceding result it is enough to consider the case where S is finitely generated and solvable. Arguing by induction on the derived length of S, we may assume that the commutator subgroup $S^{(1)} = [S, S]$ is finitely generated and virtually abelian. As in the proof of (7.2), we choose a free abelian subgroup A that has finite index in $S^{(1)}$ and is characteristic in $S^{(1)}$, hence normal in S.

By applying (7.1(5)) we obtain a subgroup S_0 of finite index in S that contains A as a direct factor. The commutator subgroup of S_0 has trivial intersection with A, and hence injects into the finite group $S^{(1)}/A$. The following lemma completes the proof. □

7.9 Lemma. *If a finitely generated group Γ has a finite commutator subgroup, then Γ has an abelian subgroup of finite index.*

Proof. The action of Γ by conjugation on $[\Gamma, \Gamma]$ gives a homomorphism from Γ to the finite group $\mathrm{Aut}([\Gamma, \Gamma])$. Let $\Gamma_1 \subset \Gamma$ denote the kernel of this map and notice that $g^n h = h g^n [g, h]^n$ for all $g, h \in \Gamma_1$ and all $n \in \mathbb{Z}$ (because $[\Gamma, \Gamma] \cap \Gamma_1$ is central). Setting n equal to the order of $[\Gamma, \Gamma]$, we see that g^n is central in Γ_1 for every $g \in \Gamma_1$. But this implies that the finitely generated abelian group $\Gamma_1/Z(\Gamma_1)$ is torsion, hence finite. Thus Γ_1 (and hence Γ) is virtually abelian. □

7.10 Remarks

(1) The hypotheses of the Solvable Subgroup Theorem can be relaxed as indicated in (7.7(2)). In particular, if Γ acts freely by semi-simple isometries on an M_κ-polyhedral complex K with Shapes(K) finite, then every solvable subgroup of Γ is finitely generated and virtually abelian.

(2) Let H be a group that acts properly and cocompactly by isometries on a CAT(0) space and let Γ be a non-uniform, irreducible, lattice in a semi-simple Lie group of real rank at least two that has no compact factors. Such lattices have solvable subgroups that are not virtually abelian (see [Mar90]), so by the Solvable Subgroup Theorem, the kernel of any homomorphism $\phi : \Gamma \to H$ must be infinite. The Kazhdan-Margulis Finiteness Theorem ([Zim84] 8.1.2) states that every normal subgroup of Γ is either finite or of finite index. Thus the image of ϕ must be finite.

Proper Actions That Are Not Cocompact

Our first goal in this section is to examine how (7.5) can fail when the action is not cocompact. We begin with a construction which shows that if G is a countable abelian group all of whose elements have finite order, then G admits a proper action on a proper CAT(0) space (a metric simplicial tree in fact). We shall then describe how this construction can be modified to yield proper actions by infinitely generated abelian groups such as $\mathbb{Q}$. In each case the action has the property that Min(G) is empty.

7.11 Example: Infinite Torsion Groups. Let G be a group which is the union of a strictly increasing sequence of finite subgroups $\{1\} = G_0 \subset G_1 \subset G_2 \subset \dots$. Note that G can be taken to be any countably infinite abelian group that has no elements of infinite order. (The reader may find it helpful to keep in mind the example $G_n \cong (\mathbb{Z}/2\mathbb{Z})^n$.)

Let X be the quotient of $G \times [0, \infty)$ by the equivalence relation which identifies (g, t) with (g', t') if $g^{-1}g' \in G_i$ and $t = t' \geq i$. Let $[g, t]$ denote the equivalence class of (g, t). We endow X with the unique length metric such that each of the maps $t \mapsto [g, t]$ from $[0, \infty)$ into X is an isometry. With this metric, X becomes a metric simplicial tree whose vertices are the points $[g, i]$, where $g \in G$, $i \in \mathbb{Z}$.

The action of G on $G \times [0, \infty)$ by left translation on the first factor and the identity on the second factor induces an action of G by isometries on X. This action is proper. Indeed, if we denote by $I_g(i)$ the interval which is the union of points $[g, t]$ with $i - 3/4 \leq t \leq i + 3/4$, then $g'(I_g(i)) \cap I_g(i) \neq \emptyset$ if and only if g' belongs to the finite subgroup G_i. Each element $g \in G_i$ is an elliptic isometry, fixing in particular those points $[1, t]$ with $t \geq i$. Finally, we note that the map $[g, t] \mapsto t$ induces an isometry from the quotient space $G \backslash X$ onto $[0, \infty)$.

7.12 Exercises

(1) Let X be a metric simplicial tree with all edges of length 1. Suppose that each vertex of X has valence at most m. Let T_m be the regular tree in which each vertex has valence m and each edge has length 1. Prove that there exists an injective homomorphism $\phi : \text{Isom}(X) \to \text{Isom}(T_n)$ and a ϕ-equivariant isometric embedding $X \hookrightarrow T_n$.

(2) Deduce that there exist proper CAT(0) spaces Y such that Isom(Y) contains subgroups that act properly and cocompactly on Y and also contains abelian subgroups that act properly on Y by semi-simple isometries but are not finitely generated. (Hint: The free group of rank m acts freely and cocompactly by isometries on T_{2m}. The construction of (7.11) with $G_n \cong (\mathbb{Z}/r\mathbb{Z})^n$ produces a tree of bounded valence.)

We shall need the following (easy) fact in the course of the next example.

(3) Consider two actions $\Phi : G \to \text{Isom}(X)$ and $\Psi : G \to \text{Isom}(Y)$ of an abstract group G on metric spaces X and Y. Suppose that the induced action of $G/\ker\Phi$ on X is proper and the restricted action $\Psi|_{\ker\Phi}$ is proper. Show that the diagonal action $g \mapsto (\Phi(g), \Psi(g))$ of G on $X \times Y$ is proper.

7.13 Example: Infinitely Generated Abelian Groups. We describe a proper action of $\mathbb{Q}$, the additive group of rational numbers, by semi-simple isometries on a proper 2-dimensional CAT(0) space (the product of a simplicial tree and a line in fact).

$\mathbb{Q}$ acts by translations on the line $\mathbb{E}^1$ in the obvious way (the action is not proper of course). $\mathbb{Q}/\mathbb{Z}$ is a countable abelian group with no elements of infinite order, so by the preceding example it acts properly on a simplicial tree X. According to the preceding exercise, the diagonal action of $\mathbb{Q}$ on $X \times \mathbb{E}^1$ (via the obvious homomorphism $\mathbb{Q} \to \mathbb{Q}/\mathbb{Z} \times \mathbb{Q}$) is a proper action. This action is by semi-simple isometries. Note that $X \times \mathbb{E}^1$ is proper.

In exactly the same way, we can combine the natural action of $\mathbb{Q}^n$ by translations on $\mathbb{E}^n$ with a proper action of the torsion group $\mathbb{Q}^n/\mathbb{Z}^n$ on a tree to obtain a proper action of $\mathbb{Q}^n$ by semi-simple isometries on a proper CAT(0) space of dimension $n+1$ (the product of $\mathbb{E}^n$ and a tree in fact). The following lemma shows that the dimension of this example is optimal.

7.14 Lemma. *Suppose that the group Γ acts properly by semi-simple isometries on a proper* CAT(0) *space X and suppose that Γ contains a normal subgroup $A \cong \mathbb{Z}^n$. If X does not contain an isometrically embedded copy of* $[0, \infty) \times \mathbb{E}^n$*, then A has finite index in Γ.*

Proof. By the Flat Torus Theorem, Min(A) splits as $Y \times \mathbb{E}^n$ and Γ preserves Min(A) and its splitting. Let $N \subset \Gamma$ be the subgroup consisting of elements which act trivially on Y. Note that N contains A. Since the action of N on each of the flats $\{y\} \times \mathbb{E}^n$ is proper and the action of A is cocompact, A must have finite index in N.

Y (with the metric induced from X) is a proper CAT(0) space, so it is either bounded or else it contains a geodesic ray; we are assuming that X does not contain a copy of $[0, \infty) \times \mathbb{E}^n$, so Y must be bounded. Therefore, since the action of Γ/N on Y is proper (6.10(4)), Γ/N and hence Γ/A must be finite. □

The following exercise gives a more explicit description of a proper action of $\mathbb{Q}$ that can be constructed by the above method, and describes the quotient space (which is a piecewise Euclidean 2-complex of finite area).

7.15 Exercises

(1) **A Non-Positively Curved 2-complex K with $\pi_1 K = \mathbb{Q}$**

The 1-skeleton of K consists of a half-line $[0, \infty)$ with vertices v_n at the integer points, together with countably many oriented 1-cells $\{\varepsilon_n\}_{n \in \mathbb{N}}$, where ε_n has length $1/n!$ and has both of its endpoints attached to v_n. For every non-negative integer n we attach a 2-cell e_n to K, where e_n is metrized as a rectangle with sides of length $1/n!$ and 1. We orient the boundary of e_n and describe the attaching map reading around the boundary in the positive direction: the side of length $1/n!$ traces across ε_n in the direction of its orientation, beginning at v_n, then a side of length 1 runs along $[0, \infty)$, joining v_n to v_{n+1}, then the side of length $1/n!$ traces $(n+1)$ times around ε_{n+1} in the direction opposite to its orientation, and finally the remaining side of length 1 runs back along $[0, \infty)$ to v_n.

Note that the 2-cell e_n has area $1/n!$, and hence K has total area $e = 2 \cdot 71 \ldots$.

(a) Prove that $\pi_1 K = \mathbb{Q}$.

(b) Prove that when endowed with the induced length metric, the universal cover of K is a proper CAT(0) space and the action of $\pi_1 K = \mathbb{Q}$ on it by deck transformations is a proper action by semi-simple isometries.

(c) Prove that the action of $\mathbb{Q}$ is precisely that which one obtains by applying the construction of (7.13) with $\mathbb{Q}/\mathbb{Z}$ expressed as the union of the finite cyclic groups $G_n = \langle x_n \mid x_n{}^{n!} = 1\rangle$, where $G_n \hookrightarrow G_{n+1}$ by $x_n \mapsto x_{n+1}^{n+1}$.

(2) In contrast, prove that if a group G acts freely on a non-positively curved M_κ-complex K by cellular isometries, and if Shapes(K) is finite, then every abelian subgroup of G is finitely generated. (This is a matter of verifying the remarks that we made in 7.7(2). You will need to use 6.6(2).)

Polycyclic Groups

One would like to weaken the hypotheses of the Solvable Subgroup Theorem so as to allow proper actions which are not cocompact. The preceding examples show that in order to do so one must contend with the possible presence of infinitely generated abelian groups. On the other hand, in some situations (e.g., when one is considering actions on Riemannian manifolds) one knows that the groups at hand do not contain infinitely generated abelian subgroups. If the abelian subgroups of a solvable group are finitely generated (as the case with finitely generated nilpotent groups for example) then the group is *polycyclic* (see [Seg85]). Polycyclic groups were defined following (I.8.39).

7.16 Theorem. *A polycyclic group* Γ *acts properly by semi-simple isometries on a complete* CAT(0) *space if and only if* Γ *is virtually abelian.*

Proof. For the "if" assertion, see (7.3(2)). To prove the "only if" assertion we argue by induction on the Hirsch length of Γ. This reduces us to consideration of the case $\Gamma = H \rtimes_\phi \mathbb{Z}$ with H finitely generated and virtually abelian. As in the proof of (7.2), we can choose a free abelian subgroup $A \subset H$ that has finite index and is characteristic in H (hence normal in Γ). According to 7.1(5), the abelian group $A \times \ker\phi$ has finite index in Γ. □

The preceding proof relied on an appeal to 7.1(5) which is underpinned by the following observation: the translation numbers for the elements of a free abelian group A acting properly by semi-simple isometries on a CAT(0) space X can be seen by looking at the action of A on a flat subspace $\mathbb{E}^n \hookrightarrow X$ which has quotient an n-torus. We close this section with a lemma and some exercises that hint at the wider utility of this observation.

7.17 Lemma. *Let X be a complete* CAT(0) *space, and let $\Gamma \cong \mathbb{Z}^n$ be a group which acts properly on X by semi-simple isometries.*

(1) *The translation length function $\gamma \mapsto |\gamma|$ is the restriction to Γ of a norm on $\Gamma \otimes \mathbb{R} \cong \mathbb{R}^n$, and this norm satisfies the parallelogram law (i.e., arises from a scalar product).*

(2) *If $S \subseteq \Gamma$ is a subset with the property that there does not exist any scalar product on $\Gamma \otimes \mathbb{R} \cong \mathbb{R}^n$ such that $(s \mid s) = (s' \mid s')$ for all $s, s' \in S$, then S does not lie in a single conjugacy class of the full isometry group* Isom(X).

(3) *Let G be an abstract group with a free abelian subgroup $G' \cong \mathbb{Z}^n$. If $S \subseteq G'$ is a subset with the property that there does not exist any scalar product on $G' \otimes \mathbb{R} \cong \mathbb{R}^n$ such that $(s \mid s) = (s' \mid s')$ for all $s, s' \in S$, and if S lies in a single conjugacy class of G, then there does not exist a proper action of G by semi-simple isometries on any complete* CAT(0) *space.*

Proof. According to the Flat Torus Theorem (7.1), Γ acts by Clifford translations on the complete CAT(0) space $\text{Min}(\Gamma) = Y \times \mathbb{E}^n$. The map of Γ into $H_{\text{Min}(\Gamma)}$, the group of Clifford transformations of $\text{Min}(\Gamma)$, extends in the obvious (linear) way to give an injection $\Gamma \otimes \mathbb{R} \to H_{\text{Min}(\Gamma)}$. The norm $\alpha \mapsto |\alpha|$ on $H_{\text{Min}(\Gamma)}$ makes it a pre-Hilbert space (6.15). This proves (1).

Part (2) follows immediately from (1) and the fact that translation length is preserved under conjugation. (3) follows immediately from (2). □

It is easy to construct sets S as in part (3) of the preceding lemma. Any infinite subset of $\mathbb{Z}^n$ has the appropriate property, as do many finite sets, for example any set which meets a coset of a cyclic subgroup in more than two points.

7.18 Exercises

(1) Let X be a complete CAT(0) space. Prove that if an abelian subgroup $\Gamma \subset \text{Isom}(X)$ acts properly by semi-simple isometries, then the intersection of Γ with each conjugacy class of Isom(X) is finite.

(2) Let F be the free group on $\{a, b, c\}$ and let ϕ be the automorphism of F given by $\phi(a) = a, \phi(b) = ba, \phi(c) = ca^2$. Prove that the group $\Gamma = F \rtimes_\phi \mathbb{Z}$, which has presentation $\langle a, b, c, t \mid tat^{-1} = a, tbt^{-1} = ba, tct^{-1} = ca^2\rangle$, cannot act properly by semi-simple isometries on any complete CAT(0) space. (Hint: Note that a, t generate a free abelian subgroup of Γ and apply part (3) of the above lemma to the set $S = \{t, at, a^2t\}$.)

The example in (7.18(2)) is due to S.M. Gersten [Ger94]. He used the above example to show that the outer automorphism group of any free group of rank at least 4 cannot act properly and cocompactly by isometries on any complete CAT(0) space. (A different argument shows that $\text{Out}(F_3)$ does not admit such an action either [BriV95].)

Actions That Are Not Proper

The following version of the Flat Torus Theorem can be applied to actions which are not proper.

7.19 Lemma. *If Γ is an abelian group acting by isometries on a complete* CAT(0) *space, then the set of elements of Γ that act as elliptic isometries is a subgroup.*

Proof. Because they commute, if $\gamma, \gamma' \in \Gamma$ are semi-simple then $\text{Min}(\gamma)$ and $\text{Min}(\gamma')$ intersect (6.2). If γ and γ' are elliptic then $\gamma\gamma'$ fixes $\text{Min}(\gamma) \cap \text{Min}(\gamma')$ pointwise. □

7.20 Theorem. *Let Γ be a finitely generated abelian group acting by semi-simple isometries on a complete* CAT(0) *space X. (The action of Γ need not be proper.)*

(1) $\text{Min}(\Gamma) := \bigcap_{\gamma\in\Gamma} \text{Min}(\gamma)$ *is non-empty and splits as a product $Y \times \mathbb{E}^n$.*

(2) *Every $\gamma \in \Gamma$ leaves* $\text{Min}(\Gamma)$ *invariant and respects the product structure; γ acts as the identity on the first factor Y and acts by translation on the second factor $\mathbb{E}^n$; moreover, $n \leq \text{rk}_{\mathbb{Q}}\Gamma$ and the action of Γ on $\mathbb{E}^n$ is cocompact.*

(3) *The n–flats $\{y\} \times \mathbb{E}^n$ are precisely those subsets of* $\text{Min}(\Gamma)$ *which arise as the convex hull of a Γ–orbit.*

(4) *If an isometry $\alpha \in \text{Isom}(X)$ normalizes Γ, then α leaves* $\text{Min}(\Gamma)$ *invariant and preserves its splitting $Y \times \mathbb{E}^n$.*

Proof. Let $\mathcal{E}$ denote the subgroup of Γ consisting of those elements which act as elliptic isometries. An easy induction argument on the number of elements needed to generate $\mathcal{E}$ shows that $\text{Min}(\mathcal{E})$ is non-empty. By (6.2), $\text{Min}(\mathcal{E})$ is convex and Γ-invariant. The action of $\Gamma/\mathcal{E}$ on $\text{Min}(\mathcal{E})$ is by hyperbolic isometries (except for the identity element). From this point one can proceed by induction on $\text{rk}_{\mathbb{Q}}\Gamma$ to prove parts (1), (2) and (3) of the theorem. The argument is essentially the same as in (7.1), so we omit the details. Part (4) is also proved as in (7.1). □

7.21 Exercise. Let Γ and X and n be as in the above theorem. Show that if a convex subspace $Z \subset X$ is Γ-invariant and isometric to $\mathbb{E}^r$ for some r, then $r \geq n$. Show further that if $r = n$ then $Z \subset \text{Min}(\Gamma)$.

Some Applications to Topology

In this section we gather a number of results which illustrate the great extent to which the structure of the fundamental group determines the topology of a non-positively curved space. Most of the results which we present are applications of the Flat Torus Theorem.

Recall that a *strong deformation retraction* of topological space Y onto a subspace Z is a continuous map $F : [0, 1] \times Y \to Y$ such that $F|_{\{0\}\times Y} = \mathrm{id}_Y$, $\mathrm{im}\, F|_{\{1\}\times Y} = Z$, and $F(t, z) = z$ for all $t \in [0, 1]$ and $z \in Z$. Recall also that a *flat manifold* is the quotient of Euclidean space $\mathbb{E}^n$ by a group acting freely and properly by isometries.

7.22 Theorem. *Let Y be a compact non-positively curved geodesic space, and suppose that its fundamental group $\Gamma = \pi_1 Y$ is virtually solvable. Then, Γ is virtually abelian, and*

(1) *there is a strong deformation retraction of Y onto a compact subspace that is isometric to a flat manifold;*

(2) *if Y has the geodesic extension property, then Y itself is isometric to a flat manifold.*

Proof. According to (7.8), Γ is virtually abelian. Corollary 7.2 yields a Γ-invariant copy of Euclidean space E in the universal cover $\widetilde{Y}$ of Y, such that Γ acts properly by isometries on this flat subspace. Since the action of Γ on $\widetilde{Y}$ is free, $Z := \Gamma\backslash E$ is a flat manifold.

$E \subset \widetilde{Y}$ is closed, complete and convex; consider the orthogonal projection $\pi : \widetilde{Y} \to E$, as defined in (2.4). Because E is Γ-invariant, the projection map is Γ-equivariant. Indeed there is a Γ-equivariant retraction $R : [0, 1] \times \widetilde{Y} \to \widetilde{Y}$ that sends (t, x) to the point a distance $t\, d(x, \pi(x))$ from x on the geodesic segment $[x, \pi(x)]$.

Let $p : \widetilde{Y} \to Y$ denote the universal covering map (so $Z = p(E)$). Because the retraction R is Γ-equivariant, the map $[0, 1] \times Y \to Z$ given by $(t, p(x)) = p(R(t, y))$ is well-defined. This is the desired strong deformation retraction of Y onto Z.

The key point to observe for the second part of the theorem is that under the additional hypothesis that Y has the geodesic extension property, the Euclidean subspace E must actually be the whole of $\widetilde{Y}$. This follows from (6.20) and the fact that $\widetilde{Y}$ inherits the geodesic extension property from Y (because this property is local and $\widetilde{Y}$ is locally isometric to Y). □

Combining Theorem 7.8 with Propositions (5.10) and (5.12) we obtain:

7.23 Corollary. *Let Y be a compact geodesic space of non-positive curvature, and suppose that $\pi_1 Y$ is virtually solvable.*

(1) *If Y is homeomorphic to a topological n-manifold (or indeed an homology manifold) then the universal cover of Y is isometric to $\mathbb{E}^n$.*

(2) *If Y is isometric to an M_κ-complex with no free faces, then the complex is actually a flat manifold.*

Our next result is of a more combinatorial nature. Recall that the girth of a graph is the combinatorial length of the shortest injective loop that it contains.

7.24 Proposition. *Let K be a finite, connected, 2-dimensional simplicial complex with no free faces (i.e., every edge of K lies in the boundary of at least two 2-*

simplices). Suppose that the for each vertex $v \in K$ the girth of the link of v is at least 6. Suppose further that the fundamental group of K is solvable.

Then, K is homeomorphic to either a torus or a Klein bottle, and exactly six 2-simplices meet at each vertex of K.

Proof. We metrize K as a piecewise Euclidean complex with each 2-simplex isometric to an equilateral triangle of side length 1. Metrized thus, K satisfies the link condition (5.1), and hence is non-positively curved. The preceding result implies that K is isometric to a flat manifold, hence it is homeomorphic to either a torus or a Klein bottle. The fact that the metric is non-singular at the vertices means that exactly six 2-simplices meet there. □

One of the aspects of the study of non-positively curved spaces that we shall not explore in this book is the theory of super-rigidity and the influence that ergodic theory has had on the study of lattices in Lie groups, à la Margulis. (Basic references for this material include [Mar90] and [Zim84]. See also [BuM96], [GrS92] and [GroP91]).

The following consequence of the Flat Torus Theorem provides a very simple example of a result from this circle of ideas.

7.25 Proposition. *Let H be a group that acts properly by semi-simple isometries on a* CAT(0) *space X and let Γ be an irreducible lattice in a semi-simple Lie group of real rank $n \geq 2$ that has no compact factors. If X does not contain an isometrically embedded copy of $\mathbb{E}^n$, then the image of every homomorphism $\Gamma \to H$ is finite.*

Proof. The Flat Torus Theorem implies that H does not contain a subgroup isomorphic to $\mathbb{Z}^n$, whereas Γ does. Thus the kernel of any homomorphism $\phi : \Gamma \to H$ must be infinite. If a normal subgroup of Γ is not finite, then it is of finite index ([Zim84] 8.1.2), therefore the image of ϕ must be finite. □

Low-Dimensional Topology

In this section we present some applications of Theorems 6.12 and 7.1 to the study of manifolds in dimensions two and three. In the course of the discussion we shall require a number of standard facts about 3-manifolds and surface automorphisms. We refer to [Bir76] for facts about the mapping class group, to [Rol91] and [Hem76] for basic facts about 3-manifolds, and to [Sco83] for facts about the geometry of 3-manifolds.

Mapping Class Groups. Let Σ be an oriented surface of finite type, i.e., a compact, connected, oriented, 2-dimensional manifold with a finite number of points and open discs deleted. (The surface may be closed, i.e. the set of punctures and deleted discs may be empty.) A closed curve $c : S^1 \to \Sigma$ is said to be simple if it has no self-intersections; it is said to be separating if the complement of its image is not connected.

We consider the group $\mathcal{M}(\Sigma)$ of isotopy classes of those orientation preserving self-homeomorphisms of Σ which send each puncture to itself and restrict to the identity on the boundary of Σ. (We are only allowing isotopies through homeomorphisms which restrict to the identity on $\partial\Sigma$.) Associated to each simple closed curve c in Σ, there is an element of $\mathcal{M}(\Sigma)$ defined as follows. One considers a product neighbourhood of the image of c, identified with $(\mathbb{R}/\mathbb{Z}) \times [0, 1]$. The *Dehn twist* associated to c is the map T_c defined on the given product neighbourhood by (using the above coordinates) $(\theta, t) \mapsto (\theta + t, t)$; one extends T_c to be the identity on the remainder of Σ. The class of $\mathcal{M}(\Sigma)$ thus defined is clearly independent of the choice of product neighbourhood.

We shall restrict our attention to surfaces Σ which are either closed of genus at least 3, or else have genus at least two and at least one boundary component or two punctures. With this restriction, we have the following result, which is a consequence of Theorem 6.12 and an unpublished result of G. Mess[28] concerning centralizers in the mapping class group. This consequence was first observed by Kapovich and Leeb [KaL95] (from a different point of view).

7.26 Theorem. *If Σ is as above, then the mapping class group $\mathcal{M}(\Sigma)$ cannot act properly by semi-simple isometries on any complete* CAT(0) *space.*

Proof. The restrictions we imposed on Σ ensure that there is a separating simple closed curve c on Σ with the property that the closure of one of the components of $\Sigma \smallsetminus c$ is a sub-surface of genus 2 with no punctures and only one boundary component, c itself, and the Dehn twist in c is non-trivial in $\mathcal{M}(\Sigma)$. Let Σ_2 denote the sub-surface of genus 2. The subgroup of $\mathcal{M}(\Sigma)$ given by homeomorphisms which (up to isotopy) are the identity on the complement of Σ_2 is naturally isomorphic to $\mathcal{M}(\Sigma_2)$, thus it will suffice to prove that $\mathcal{M}(\Sigma_2)$ cannot act properly by semi-simple isometries on any CAT(0) space. Notice that the mapping class of the Dehn twist T_c is central in $\mathcal{M}(\Sigma_2)$.

Let Σ_2' denote a closed surface of genus 2 with a single puncture p, and let $\overline{\Sigma}_2$ denote the corresponding closed (unpunctured) surface. There is an obvious inclusion of Σ_2 into Σ_2' sending c to the boundary of a neighbourhood D of the puncture in Σ_2', and an even more obvious inclusion of Σ_2' into $\overline{\Sigma}_2$. These inclusions induce surjective maps from $\mathcal{M}(\Sigma_2)$ to $\mathcal{M}(\Sigma_2')$ and from $\mathcal{M}(\Sigma_2')$ to $\mathcal{M}(\overline{\Sigma}_2)$. (For the first map, one extends homeomorphisms of Σ_2 to be the identity on D.) One checks easily that both maps are surjective homomorphisms, well-defined on isotopy classes. The kernel of the first homomorphism is the infinite cyclic subgroup generated by the mapping class of the Dehn twist T_c. The kernel of the second homomorphism is the subgroup, isomorphic to $\pi_1(\overline{\Sigma}_2)$, which consists of mapping classes represented by homeomorphisms which are the identity off a neighbourhood of some essential closed curve based at p, and which drag p around that curve; a more algebraic way of describing these mapping classes is to say that they are precisely those elements of $\mathcal{M}(\Sigma_2')$ which induce an *inner* automorphism of $\pi_1(\overline{\Sigma}_2, p)$. (See [Bir76]

[28] MSRI preprint #05708, 1990

for more details.) Thus the kernel K of the composition of the natural homomorphisms $\mathcal{M}(\Sigma_2) \to \mathcal{M}(\Sigma_2') \to \mathcal{M}(\overline{\Sigma}_2)$ is a central extension of $\pi_1(\overline{\Sigma}_2)$ with infinite cyclic centre.

Using the above geometric description of the mapping classes in the kernel of $\mathcal{M}(\Sigma_2') \to \mathcal{M}(\overline{\Sigma}_2)$, one sees that K is the fundamental group of the unit tangent bundle of $\overline{\Sigma}_2$ (where the surface is endowed with a metric of constant negative curvature)[29]. In particular, K is (isomorphic to) a cocompact lattice in the Lie group $\widetilde{\text{PSL}(2,\mathbb{R})}$, the universal covering group of PSL(2, $\mathbb{R}$). Now, $\widetilde{\text{PSL}(2,\mathbb{R})}$ does not contain the fundamental group of any closed surface of genus ≥ 2 (see [Sco83]), so since every finite index subgroup of $\pi_1(\overline{\Sigma}_2)$ is the fundamental group of such a surface, the central extension $1 \to \langle T_c \rangle \to K \to \pi_1(\overline{\Sigma}_2) \to 1$ cannot be split even after passing to a subgroup of finite index in $\pi_1(\overline{\Sigma}_2)$. Thus, by (6.12), K (and hence $\mathcal{M}(\Sigma)$) cannot act properly by semi-simple isometries on any CAT(0) space. □

3-Manifolds

There has been a good deal of work recently on the question of which 3-manifolds admit metrics of non-positive curvature (see [Le95], [Bri98b] and references therein). We shall not describe that body of work. Instead, we present a single consequence of our earlier results in order to exemplify the great extent to which the structure of the fundamental group determines the topology of a 3-manifold, particularly in the presence of non-positive curvature.

Building on a sequence of works by various authors (including Tukia, Scott and Mess), Casson and Jungreis [CasJ94] and Gabai [Gab93] independently proved the following deep theorem: if a 3-manifold M is compact, irreducible and $\pi_1 M$ contains a normal subgroup isomorphic to $\mathbb{Z}$, then M is a Seifert fibre space (i.e., it can be foliated by circles). We prove a related (but much easier) result.

7.27 Theorem. *Let M be a closed 3-manifold which is non-positively curved (i.e., locally* CAT(0)*). If $\pi_1 M$ contains a normal subgroup isomorphic to $\mathbb{Z}$, then:*

(1) *M can be foliated by circles.*

(2) *A finite-sheeted covering of M is homeomorphic to a product $\Sigma \times S^1$, where Σ is a closed surface of positive genus.*

(3) *M supports a Riemannian metric of non-positive sectional curvature.*

Proof. Consider the universal covering $p : \widetilde{M} \to M$, equipped with the length metric making p a local isometry. The action of $\Gamma := \pi_1 M$ on $\widetilde{M}$ by deck transformations is an action by hyperbolic isometries. Suppose that $\gamma \in \Gamma$ generates an infinite cyclic normal subgroup $C \subset \Gamma$. Because C is normal, $\text{Min}(\gamma)$ is Γ-invariant, and hence is the whole of the universal cover (see (6.20) or (6.16)). The splitting $\widetilde{M} = Y \times \mathbb{R}$ given by (6.8) is preserved by the action of Γ; the group C acts by translations on the second factor and acts trivially on the first factor. Hence the foliation of $\widetilde{M}$ by

[29] This is Mess's unpublished observation.

the lines $\{y\} \times \mathbb{R}$ descends to a foliation of M by circles. Since $Y \times \mathbb{R}$ is a simply connected 3-manifold, Y is a simply connected 2-manifold (cf. [Bing59]). Since Y is contractible (it is a complete CAT(0) space) it must be homeomorphic to $\mathbb{R}^2$.

Let $\Gamma' \subset \Gamma$ be a subgroup of finite index that contains C as a direct factor (6.12), say $\Gamma' = C \times K$. The action of $K = \Gamma'/C$ on Y is proper and cocompact (6.10). Since K is a subgroup of Γ (which acts freely on the CAT(0) space $\widetilde{M}$) it is torsion-free (2.8). Thus the action of K on Y is free, $K \backslash Y$ is a closed surface Σ, and the quotient of $\widetilde{M}$ by the action of $\Gamma' = C \times K$ is homeomorphic to $\Sigma \times S^1$.

If Σ has genus at least two, then it follows from Kerckhoff's solution to the Nielsen Realization problem [Ker83] that there is a metric of constant negative curvature on Y that is Γ/C invariant; the resulting product metric on $\Sigma \times \mathbb{R}$ descends to the desired Riemannian metric of non-positive curvature on M. If Σ has genus 1 then Γ is virtually abelian and (7.2) implies that M admits a flat metric. □

Chapter II.8 The Boundary at Infinity of a CAT(0) Space

In this chapter we study the geometry at infinity of CAT(0) spaces. If X is a simply connected complete Riemannian n-manifold of non-positive curvature, then the exponential map from each point $x \in X$ is a diffeomorphism onto X. At an intuitive level, one might describe this by saying that, as in our own space, the field of vision of an observer at any point in X extends indefinitely through spheres of increasing radius. One obtains a natural compactification $\overline{X}$ of X by attaching to X the inverse limit of these spheres. $\overline{X}$ is homeomorphic to a closed n-ball; the ideal points $\partial X = \overline{X} \smallsetminus X$ correspond to geodesic rays issuing from an arbitrary basepoint in X and are referred to as *points at infinity*.

We shall generalize this construction to the case of complete CAT(0) spaces X. We shall give two constructions of ∂X, the first follows the *visual* description given above (and is due to Eberlein and O'Neill [EbON73]) and the second (described by Gromov in [BaGS85]) arises from a natural embedding of X into the space of continuous functions on X. These constructions are equivalent (Theorem 8.13). In general ∂X is not a sphere. If X is not locally compact then in general $\overline{X}$ and ∂X will not even be compact. We call $\overline{X}$ the *bordification* of X.

Isometries of X extend uniquely to homeomorphisms of $\overline{X}$ (see (8.9)). In the last section of this chapter we shall characterize parabolic isometries of complete CAT(0) spaces in terms of their fixed points at infinity.

Asymptotic Rays and the Boundary ∂X

8.1 Definition. Let X be a metric space. Two geodesic rays $c, c' : [0, \infty) \to X$ are said to be *asymptotic* if there exists a constant K such that $d(c(t), c'(t)) \leq K$ for all $t \geq 0$. The set ∂X of *boundary points* of X (which we shall also call the *points at infinity*) is the set of equivalence classes of geodesic rays — two geodesic rays being equivalent if and only if they are asymptotic. The union $X \cup \partial X$ will be denoted $\overline{X}$. The equivalence class of a geodesic ray c will be denoted $c(\infty)$. A typical point of ∂X will often be denoted ξ.

Notice that the images of two asymptotic geodesic rays under any isometry γ of X are again asymptotic geodesic rays, and hence γ extends to give a bijection of $\overline{X}$, which we shall continue to denote by γ.

8.2 Proposition. *If X is a complete* CAT(0) *space and $c : [0, \infty) \to X$ is a geodesic ray issuing from x, then for every point $x' \in X$ there is a unique geodesic ray c' which issues from x' and is asymptotic to c.*

The uniqueness assertion in this proposition follows immediately from the convexity of the distance function in CAT(0) spaces (2.2). For if two asymptotic rays c' and c'' issue from the same point, then $d(c'(t), c''(t))$ is a bounded, non-negative, convex function defined for all $t \geq 0$; it vanishes at 0 and hence is identically zero. In order to prove the asserted existence of c' we shall need the following lemma.

8.3 Lemma. *Given $\varepsilon > 0$, $a > 0$ and $s > 0$, there exists a constant $T = T(\varepsilon, a, s) > 0$ such that the following is true: if x and x' are points of a* CAT(0) *space X with $d(x, x') = a$, if c is a geodesic ray issuing from x, and if σ_t is the geodesic joining x' to $c(t)$ with $\sigma_t(0) = x'$, then $d(\sigma_t(s), \sigma_{t+t'}(s)) < \varepsilon$ for all $t \geq T$ and all $t' > 0$.*

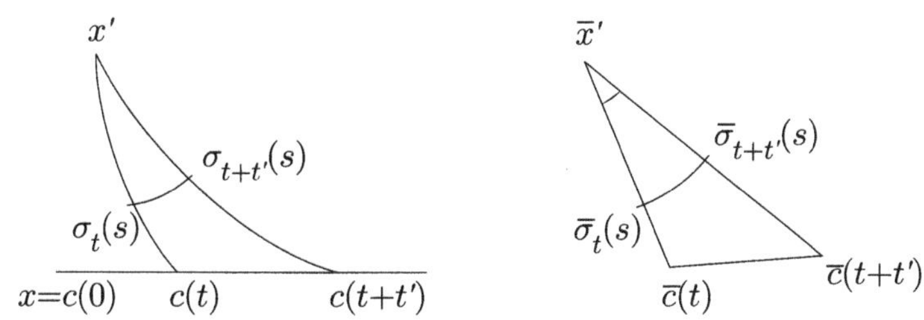

Fig. 8.1 The construction of asymptotic rays.

Proof. By the triangle inequality,

$$|t - a| \leq d(x', c(t)) \leq t + a,$$

$$|t + t' - a| \leq d(x', c(t + t')) \leq t + t' + a.$$

Let α be the comparison angle $\overline{\angle}_{x'}(c(t), c(t + t'))$. By the law of cosines,

$$\begin{aligned}\cos(\alpha) &= \frac{d(x', c(t))^2 + d(x', c(t + t'))^2 - t'^2}{2d(x', c(t))\, d(x', c(t + t'))} \\ &\geq \frac{(t - a)^2 + (t + t' - a)^2 - t'^2}{2(t + a)(t + t' + a)} \\ &= \frac{(t - a)(t + t' - a)}{(t + a)(t + t' + a)}.\end{aligned}$$

Thus $\cos\alpha \to 1$ as $t \to \infty$, and hence $\alpha \to 0$, where the convergence is uniform in $t' > 0$.

On the other hand, if $s \leq \min\{d(x', c(t)), d(x', c(t + t'))\}$ then by the CAT(0) inequality in the comparison triangle for $\Delta(x, c(t), c(t + t'))$,

$$d(\sigma_t(s), \sigma_{t+t'}(s)) \leq d(\overline{\sigma_t(s)}, \overline{\sigma_{t+t'}(s)}) = 2s\sin(\alpha/2).$$

Thus for sufficiently large t we have $d(\sigma_t(s), \sigma_{t+t'}(s)) < \varepsilon$. □

Proof of (8.2). It follows from the preceding lemma that for every $s \geq 0$ the sequence $(\sigma_n(s))$ (which is defined for sufficiently large n) is Cauchy, and hence converges in X to a point which we shall call $c'(s)$. As the pointwise limit of geodesics, $s \mapsto c'(s)$ must be a geodesic ray — this is the desired ray c' issuing from x'. One can see that c' is asymptotic to c by checking that $d(c(s), c'(s)) \leq d(x, x')$ for all $s > 0$. Indeed by considering the comparison triangle $\overline{\Delta}(x, x', c(n))$ one sees that

$$d(\sigma_n(s), c(s)) \leq d(\overline{\sigma_n(s)}, \overline{c(s)}) \leq d(x, x'),$$

for all $n > 0$, where the first inequality comes from the CAT(0) condition and the second from elementary Euclidean geometry. □

Let $c : [0, \infty) \to \mathbb{E}^n$ be a geodesic ray. The asymptotic ray which issues from $x \in \mathbb{E}^n$ and is asymptotic to c is the unique ray parallel to c. The following exercises describe the asymptotic rays in $\mathbb{H}^n$.

8.4 Exercises

(1) The Poincaré metric on the open ball $B^n \subset \mathbb{R}^n$ was defined in (I.6.7). The image of each geodesic ray $c : [0, \infty) \to B^n$ is an arc of a circle that is orthogonal to the unit sphere about the origin in $\mathbb{E}^n$. The closure of this arc has one endpoint on the sphere and $c(t)$ converges to this point as $t \to \infty$.

Show that geodesic rays in this model of hyperbolic space are asymptotic if and only if the closures of their images intersect the sphere at the same point.

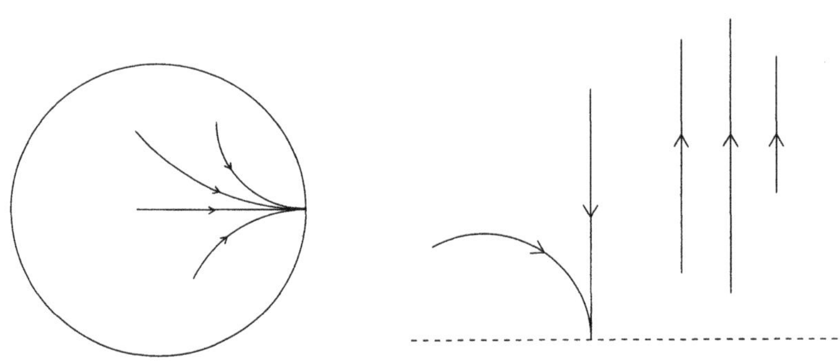

Fig. 8.2 Asymptotic rays in the Poincaré model

(2) B^n can also be viewed as the set of points in the Klein model for hyperbolic space. In this model the images of geodesic rays are half-open affine intervals whose closures have one point on the bounding sphere (I.6.2). Show that, as in (1), geodesic rays in the Klein model are asymptotic if and only if the closures of their images intersect the bounding sphere at the same point.

(3) The geodesic lines in the hyperboloid model of hyperbolic space $\mathbb{H}^n \subset \mathbb{E}^{n,1}$ are intersections with 2-dimensional vector subspaces in $\mathbb{E}^{n,1}$. Show that if two

geodesic rays in $\mathbb{H}^n$ are asymptotic, then the intersection of the corresponding 2-dimensional subspaces of $\mathbb{E}^{n,1}$ is a line in the light cone.

(4) Give an example of a non-complete convex subspace of the Euclidean plane for which the conclusion of (8.2) is not valid.

The Cone Topology on $\overline{X} = X \cup \partial X$

Throughout this section we assume that X is a complete CAT(0) space. Let $\overline{X} = X \cup \partial X$. We wish to define a topology on $\overline{X}$ such that the induced topology on X is the original metric topology.

8.5 Definition of $\overline{X}$ as an Inverse Limit. We fix a point $x_0 \in X$ and consider the system of closed balls $\overline{B}(x_0, r)$ centred at x_0. The projection p_r of X onto $\overline{B}(x_0, r)$ is well-defined because $\overline{B}(x_0, r)$ is a complete convex subset of X (see 2.4). If $x \notin B(x_0, r)$ then $p_r(x)$ is the point of $[x_0, x]$ a distance r from x_0. For $r' \geq r$, we have $p_r \circ p_{r'} = p_r$. And for $r \in [0, r']$ and all $x' \in X$ with $d(x', x_0) = r'$, the path $r \mapsto p_r(x')$ is the unique geodesic segment joining x_0 to x'.

The projections $p_r|_{\overline{B}(x_0,r')} : \overline{B}(x_0, r') \to \overline{B}(x_0, r)$, where $r' \geq r$, form an inverse system, and we consider $\varprojlim \overline{B}(x_0, r)$ with the inverse limit topology [Spa66]. A point in this space is a map $c : [0, \infty) \to X$ such that if $r' \geq r$ then $p_r(c(r')) = c(r)$. Such maps are of two types: either $c(r') \neq c(r)$ for all $r' \neq r$ (in which case c is a geodesic ray issuing from x_0), or else there is a minimum $r_0 \geq 0$ such that $c(r) = c(r_0)$ for all $r \geq r_0$ (in which case the restriction of c to $[0, r_0]$ is the geodesic segment joining x_0 to $c(r_0)$, and the restriction of c to $[r_0, \infty)$ is a constant map). Thus $\varprojlim \overline{B}(x_0, r)$ may be viewed as a subspace of the set of maps $[0, \infty) \to X$. (The inverse limit topology coincides with the topology of uniform convergence on compact subsets.)

There is a natural bijection $\phi(x_0) : \overline{X} \to \varprojlim \overline{B}(x_0, r)$, which associates to $\xi \in \partial X$ the geodesic ray that issues from x_0 in the class of ξ, and which associates to $x \in X$ the map $c_x : [0, \infty) \to X$ whose restriction to $[0, d(x_0, x)]$ is the geodesic segment joining x_0 to x and whose restriction to $[d(x_0, x), \infty)$ is the constant map at x. Let $\mathcal{T}(x_0)$ be the topology on $\overline{X}$ for which $\phi(x_0)$ is a homeomorphism. The inclusion $X \hookrightarrow \overline{X}$ gives a homeomorphism from X onto a dense open set of $\overline{X}$.

Notice that for every $x \in X$ and $\xi \in \partial X$ there is a natural path with compact image $[x, \xi]$ joining x to ξ in $\overline{X}$; this is the union of ξ and the image of $[0, \infty)$ by the geodesic ray c with $c(0) = x$ and $c(\infty) = \xi$. Notice also that a sequence of points (x_n) in $X \subset \overline{X}$ converges to a point $\xi \in \partial X$ if and only if the geodesics joining x_0 to x_n converge (uniformly on compact subsets) to the geodesic ray that issues from x_0 and belongs to the class of ξ.

8.6 Definition of the Cone Topology. Let X be a complete CAT(0) space. The *cone topology* on $\overline{X}$ is the topology $\mathcal{T}(x_0)$ defined above. (This is independent of the choice of basepoint $x_0 \in X$ (see 8.8).)

The topology on ∂X induced by the cone topology on $\overline{X}$ will also be referred to as the cone topology. When equipped with this topology, ∂X is sometimes called the *visual boundary* of X. However, we shall normally refer to it simply as *the boundary*. Note that ∂X is a closed subspace of $\overline{X}$ and that $\overline{X}$ is compact if X is proper.

In order to work effectively with the cone topology, we need to understand an explicit neighbourhood basis for it. A basic neighbourhood of a point at infinity has the following form: given a geodesic ray c and positive numbers $r > 0$, $\varepsilon > 0$, let

$$U(c, r, \varepsilon) = \{x \in \overline{X} \mid d(x, c(0)) > r, d(p_r(x), c(r)) < \varepsilon\},$$

where p_r is the natural projection of $\overline{X} = \varprojlim \overline{B}(c(0), r)$ onto $\overline{B}(c(0), r)$.

8.7 Exercise. Prove that one obtains a basis for the topology $\mathcal{T}(x_0)$ on $\overline{X}$ by taking the set of all open balls $B(x, r) \subset X$, together with the collection of all sets of the form $U(c, r, \varepsilon)$, where c is a geodesic ray with $c(0) = x_0$.

8.8 Proposition. *For all $x_0, x_0' \in X$, the topologies $\mathcal{T}(x_0)$ and $\mathcal{T}(x_0')$ are the same.*

Proof. It is enough to prove that the projection $p_{r_0} : \overline{X} \to \overline{B}(x_0, r_0)$ is continuous when $\overline{X}$ is endowed with the topology $\mathcal{T}(x_0')$. The continuity at points $y \in X$ is obvious. Given $\xi \in \partial X$, let c_0 and c_0' be the geodesic rays issuing from x_0 and x_0' respectively such that $c_0(\infty) = c_0'(\infty) = \xi$. Let $x = c_0(r_0) = p_{r_0}(\xi)$ and let ε be an arbitrary positive number. Let R be a number bigger than $r_0 + d(x_0, x_0')$ and also bigger than the number $T(\varepsilon/3, d(x_0, x_0'), r_0)$ given by (8.3). We shall argue that $p_{r_0} U(c_0', R, \varepsilon/3) \subset B(x, \varepsilon)$. To see this, given $y \in \overline{X}$, let c_y be the geodesic segment or the geodesic ray joining x_0' to y. We claim that if $y \in U(c', R, \varepsilon/3)$ then

$$\begin{aligned} d(p_{r_0}(y), x) &\leq d(p_{r_0}(y), p_{r_0}(c_y(R))) + d(p_{r_0}(c_y(R)), p_{r_0}(c_0'(R))) + d(p_{r_0}(c_0'(R)), x) \\ &< \varepsilon/3 + \varepsilon/3 + \varepsilon/3. \end{aligned}$$

The bound on the second term comes from the definition of $U(c', R, \varepsilon/3)$, which implies that $d(c_y(R), c'(R)) < \varepsilon/3$. The desired bounds on the first and third terms are obtained by applying (8.3) with x_0 in the rôle of x', and x_0' in the rôle of x; the rôle of c is played in the first case by c_y and in the second case by c_0'. □

8.9 Corollary. *Let γ be an isometry of a complete* CAT(0) *space X. The natural extension of γ to $\overline{X}$ is a homeomorphism.*

Proof. The action of γ on ∂X is induced by its action on geodesic rays: $t \mapsto c(t)$ is sent to $t \mapsto \gamma.c(t)$. An equivalent description of the action of γ on $\overline{X} = X \cup \partial X$ can be obtained by fixing $x_0 \in X$ and noting that γ conjugates the projection $X \to \overline{B}(x_0, r)$ to the projection $X \to \overline{B}(\gamma(x_0), r)$, and therefore induces a homeomorphism $\overline{\gamma}$ from $\varprojlim \overline{B}(x_0, r)$ to $\varprojlim \overline{B}(y_0, r)$. Modulo the natural identifications $\phi(x_0)$ and $\phi(\gamma(x_0))$ described in (8.5), this homeomorphism $\overline{\gamma} : \overline{X} \to \overline{X}$ is the natural extension of γ described above. □

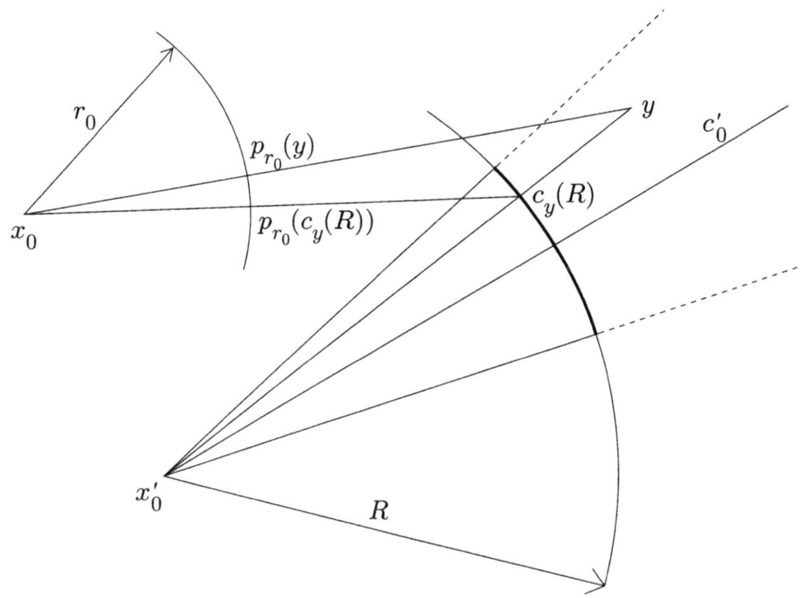

Fig. 8.3 The cone topology in independent of base point

8.10 Remark. If X is a proper CAT(0) space, then the natural map $c \mapsto \text{end}(c)$ from the set of geodesic rays to Ends(X) (as defined in I.8.27) induces a continuous surjection $\partial X \to \text{Ends}(X)$. If the asymptotic classes of c and c' are in the same path component of ∂X, then $\text{end}(c) = \text{end}(c')$.

8.11 Examples of ∂X

(1) If X is a complete n–dimensional Riemannian manifold of non-positive sectional curvature, then ∂X is homeomorphic to $\mathbb{S}^{n-1}$, the $(n-1)$–sphere. Indeed, given a base point $x \in X$, one obtains a homeomorphism by considering the map $T^1_x \to \partial X$ which associates to each unit vector u tangent to X at x the class of the geodesic ray c which issues from x with velocity vector u.

(2) As a special case of (1), consider the Poincaré ball model (I.6.8) for real hyperbolic n-space. In (8.4) we noted that two geodesic rays c and c' in $\mathbb{H}^n$ in this model are asymptotic if and only if $c(t)$ and $c'(t)$ converge to the same point of $\mathbb{R}^n$ as $t \to \infty$. Thus the visual boundary $\partial\mathbb{H}^n$ is naturally identified with the sphere of Euclidean radius 1 centred at $0 \in \mathbb{R}^n$, and $\overline{\mathbb{H}}^n$ is homeomorphic to the unit ball $\overline{B}(0,1)$ in $\mathbb{E}^n$.

We noted in (8.4) that geodesic rays in the Klein model for $\mathbb{H}^n$ also define the same point in the bounding sphere. It follows that the natural transformation $h_K^{-1}h_P$ between the models (see I.6.2 and I.6.7) extends continuously to the bounding sphere as the identity map.

(3) In [DaJ91a] M. Davis and T. Janusciewicz used Gromov's hyperbolization technique to produce interesting examples of contractible topological manifolds X

in all dimensions $n \geq 5$. Each of their examples supports a complete CAT(0) metric d such that (X, d) admits a group of isometries which acts freely and cocompactly. However, in contrast to (1), one can construct these examples so as to ensure that ∂X is not a sphere (see (5.23)). If X is a complete CAT(0) 3-manifold, then one can use a theorem of Rolfsen ([Rol68]) to show that ∂X is homeomorphic to a 2-sphere and $\overline{X}$ is a 3-ball.

(4) If X_0 is a closed convex subspace of a complete CAT(0) space X, then ∂X_0 is a closed subspace of ∂X. For instance, consider the Klein model of hyperbolic space and let F be any closed set in $\partial \mathbb{H}^n$ identified to the unit sphere in $\mathbb{E}^n$, as in example (2). Consider the convex hull $\widehat{F}$ of F, that is, the intersection with B^n of the closed affine half-spaces containing F. (This is the same as the intersection of the closed hyperbolic half-spaces whose boundary contains F.) The boundary at infinity of $\widehat{F}$ is F.

(5) Let X be an $\mathbb{R}$–tree (1.15(5)) and fix a basepoint $x \in X$. The boundary of X is homeomorphic to the projective limit as $r \to \infty$ of the spheres $S_r(x)$. Each such sphere is totally disconnected, hence ∂X is also totally disconnected. If X is locally compact, then each sphere in X contains only a finite number of points, hence ∂X is a compact space. If X is an infinite simplicial $\mathbb{R}$–tree in which every vertex has valence at least three, then ∂X is a Cantor set (see (I.8.31)).

(6) Let $X = X_1 \times X_2$ be the product of two complete CAT(0) spaces X_1 and X_2. If $\xi_1 \in \partial X_1$ and $\xi_2 \in \partial X_2$ are represented by the geodesic rays c_1 and c_2 and if $\theta \in [0, \pi/2]$, we shall denote by $(\cos\theta)\,\xi_1 + (\sin\theta)\,\xi_2$ the point of ∂X represented by the geodesic ray $t \mapsto (c_1(t\cos\theta)), c_2(t\sin\theta))$. The boundary ∂X is naturally homeomorphic to the spherical join $\partial X_1 * \partial X_2$, that is, the quotient of the product $\partial X_1 \times [0, \pi/2] \times \partial X_2$ by the equivalence relation which identifies (ξ_1, θ, ξ_2) to (ξ_1', θ, ξ_2') if and only if $[\theta = 0$ and $\xi_2 = \xi_2']$ or $[\theta = \pi/2$ and $\xi_1 = \xi_1']$.

(7) If in the previous example we take X_1 to be a metric simplicial tree in which all the vertices have the same valence ≥ 3 and we take $X_2 = \mathbb{R}$, then the boundary of $X_1 \times X_2$ is the suspension of a Cantor set — in particular it is connected but not locally connected[30]. In contrast, Swarup [Sw96], building on work of Bowditch and others, has shown that if X is compact and has curvature $\leq \kappa < 0$ and if X is not quasi-isometric to $\mathbb{R}$, then the boundary of its universal covering is connected if and only if it is locally connected.

(8) The Menger and Sierpinski curves can be obtained as the visual boundaries of certain CAT(0) polyhedral complexes: see (12.34(4)) and [Ben92].

[30] Further examples are given in "CAT(0) groups with non-locally connected boundary", by M. Mihalik and K. Ruane, to appear in Proc. London Math. Soc.

Horofunctions and Busemann Functions

Let X be a complete CAT(0) space. We have explained how one can attach to X a boundary at infinity ∂X, whose points correspond to classes of asymptotic geodesic rays in X. This boundary was attached to X, intuitively speaking, by fixing a point $x_0 \in X$ and attaching to each geodesic ray issuing from x_0 an endpoint at infinity. Since the homeomorphism type of the space $\overline{X} = X \cup \partial X$ obtained by this construction is independent of the choice of x_0, one expects that there might be a more functorial way of constructing the boundary. Our first goal in this section is to describe such a construction.

Following Gromov [BaGS87], we shall describe a natural embedding of an arbitrary metric space X into a certain function space. This embedding has the property that when X is complete and CAT(0), the closure of X is naturally homeomorphic to $\overline{X}$. In this construction, the ideal points of $\overline{X}$ appear as equivalence classes of *Busemann functions* (see 8.16).

8.12 The Space $\widehat{X}$. Let X be any metric space. We denote by $\mathcal{C}(X)$ the space of continuous functions on X equipped with the topology of uniform convergence on *bounded* subsets (so in particular if X is proper then this is the more familiar topology of uniform convergence on compact subsets). Let $\mathcal{C}_*(X)$ denote the quotient of $\mathcal{C}(X)$ by the 1-dimensional subspace of constant functions, and let $\overline{f}$ denote the image in $\mathcal{C}_*(X)$ of $f \in \mathcal{C}(X)$. Notice that $\overline{f}_n \to \overline{f}$ in $\mathcal{C}_*(X)$ if and only if there exist constants $a_n \in \mathbb{R}$ such that $f_n + a_n \to f$ uniformly on bounded subsets; equivalently, given a base point $x_0 \in X$, the sequence of functions $z \mapsto f_n(z) - f_n(x_0)$ converges to the function $z \mapsto f(z) - f(z_0)$ uniformly on all balls $B(x_0, r)$. In fact, given a base point $x_0 \in X$, the space $\mathcal{C}_*(X)$ is homeomorphic to the subspace of $\mathcal{C}(X)$ of continuous functions f on X such that $f(x_0) = 0$.

It is also useful to observe that, given $x, y \in X$, the value of $f(x) - f(y)$ is an invariant of $\overline{f}$.

There is a natural embedding $\iota : X \hookrightarrow \mathcal{C}_*(X)$ obtained by associating to each $x \in X$ the equivalence class $\overline{d}_x$ of the function $d_x : y \mapsto d(x, y)$. We shall denote by $\widehat{X}$ the closure of $\iota(X)$ in $\mathcal{C}_*(X)$. The ultimate goal of this section is to prove the following theorem.

8.13 Theorem. *Let X be a complete* CAT(0) *space. The natural inclusion $\iota : X \to \mathcal{C}_*(X)$ extends uniquely to a homeomorphism $\overline{X} \to \widehat{X}$.*

It will be convenient to identify X with $\iota(X)$, and we shall do so freely, unless there is a danger of ambiguity.

8.14 Definition. $h \in \mathcal{C}(X)$ is said to be a *horofunction* (centred at $\overline{h}$) if $\overline{h} \in \widehat{X} \smallsetminus X$. The sublevel sets $h^{-1}(-\infty, r] \subset X$ are called (closed) *horoballs*, and the sets $h^{-1}(r)$ are called *horospheres* centred at $\overline{h}$.

8.15 Exercises

(1) Let X be a metric space. Prove that the map $\iota : X \to C_*(X)$ is a homeomorphism onto its image.

(2) Prove that if X is proper then $\widehat{X}$ and $\widehat{X} \smallsetminus X$ are compact. (Hint: Given a sequence $\overline{f}_n$ in $C_*(X)$ and $x_0 \in X$, there exist functions $f_n \in \overline{f}_n$ such that $f_n(x_0) = 0$.)

(3) Show that if $\overline{h} \in \widehat{X}$ then $h(x) - h(y) \leq d(x, y)$ for all $x, y \in X$. Show that if X is CAT(0) then $h \circ c : [a, b] \to \mathbb{R}$ is convex for every geodesic segment $c : [a, b] \to X$.

The following general observation explains the connection between horofunctions and sequences of points tending to infinity.

8.16 Lemma. *Let X be a complete metric space. If the sequence $\overline{d}_{x_n}$ converges to $\overline{h} \in \widehat{X} \smallsetminus X$, then only finitely many of the points x_n lie in any bounded subset of X.*

Proof. Fix $x_0 \in X$ and $h \in \overline{h}$ with $h(x_0) = 0$. If there is an infinite number of points x_n contained in some ball $B(x_0, r)$, passing to a subsequence we can assume that $d(x_0, x_n)$ converges as $n \to \infty$. By hypothesis the sequence of functions $x \mapsto d(x_n, x) - d(x_n, x_0)$ converges uniformly to h on $B(x_0, r)$. Therefore, given $\varepsilon > 0$ we can find $N > 0$ such that $|h(x) - d(x_n, x) + d(x_n, x_0)| < \varepsilon$ for all $n > N$ and $x \in B(x_0, r)$, and $|d(x_n, x_0) - d(x_{n'}, x_0)| < \varepsilon$, for all $n, n' > N$. In particular $|h(x_n) - d(x_n, x_n) + d(x_n, x_0)| < \varepsilon$ and $|h(x_n) - d(x_n, x_{n'}) + d(x_{n'}, x_0)| < \varepsilon$. Hence $d(x_n, x_{n'}) < 2\varepsilon$. Thus the sequence (x_n) is Cauchy and converges to some point $z \in X$. Therefore $\overline{h} = \overline{d}_z$, contrary to hypothesis. □

Typical examples of horofunctions are the Busemann functions b_c associated to geodesic rays c. We shall prove that in complete CAT(0) spaces every horofunction is a Busemann function.

8.17 Definition of a Busemann Function. Let X be a metric space and let $c : [0, \infty) \to X$ be a geodesic ray. The function $b_c : X \to \mathbb{R}$ defined by

$$b_c(x) = \lim_{t \to \infty} (d(x, c(t)) - t)$$

is called the *Busemann function* associated to the geodesic ray c. (On occasion b_c may be denoted $b_{x,\xi}$, where $x = c(0)$ and $\xi = c(\infty)$.)

The first part of the following lemma implies that the limit in the above definition does indeed exist. Examples of Busemann functions in CAT(0) spaces are given at the end of this section.

8.18 Lemma. *Let X be a metric space and let $c : [0, \infty) \to X$ be a geodesic ray.*

(1) *For each $x \in X$, the function $[0, \infty) \to \mathbb{R}$ given by $t \mapsto d(x, c(t)) - t$ is non-increasing and is bounded below by $-d(x, c(0))$.*

(2) *If X is a* CAT(0) *space, then as $t \to \infty$ the functions $d_{c(t)}(x) = d(x, c(t)) - t$ converge to b_c uniformly on bounded subsets.*

Proof. By the triangle inequality, if $0 \le t' < t$ then $d(x, c(t)) \le d(x, c(t')) + t - t'$, and hence $d(x, c(t)) - t \le d(x, c(t')) - t'$. But we also have $t - t' \le d(x, c(t')) + d(x, c(t))$, so taking $t' = 0$ we get $-d(x, c(0)) \le d(x, c(t)) - t$. This proves (1).

Part (2) is an immediate consequence of (8.3). □

Our next goal is to establish a natural correspondence between horofunctions and equivalence classes of Busemann functions.

8.19 Proposition. *Let X be a complete* CAT(0) *space and let x_n be a sequence of points in X. Then $\overline{d}_{x_n}$ converges to a point of $\widehat{X} \smallsetminus X$ if and only if x_n converges in $\overline{X}$ to a point of ∂X, that is, for fixed x_0 the sequence of geodesic segments $[x_0, x_n]$ converges to a geodesic ray $[x_0, \xi]$.*

We defer the proof of this proposition for a moment and move directly to:

8.20 Corollary (The Ideal Points $\widehat{X} \smallsetminus X$). *If X is a complete* CAT(0) *space, then the Busemann functions associated to asymptotic rays in X are equal up to addition of a constant, and for every $x_0 \in X$ the map $i : \partial X \to \widehat{X} \smallsetminus X$ defined by $i(\xi) = \overline{b_{x_0,\xi}}$ is a bijection.*

Proof. We first prove that i is surjective. Given $\overline{h} \in \widehat{X} \smallsetminus X$ choose $x_n \in X$ with $\overline{d}_{x_n} \to \overline{h}$. By the proposition, there exists $\xi \in \partial X$ such that $x_n \to \xi$ in $\overline{X}$. Let $c : [0, \infty) \to X$ be the geodesic ray with $c(0) = x_0$ and $c(\infty) = \xi$, and let $y_{2n} = x_n$ and $y_{2n+1} = c(n)$ for all n. Then $y_n \to \xi$, so by the proposition $\overline{d_{y_n}}$ converges to a point of $\widehat{X} \smallsetminus X$, and since $\widehat{X}$ is Hausdorff this point must be $\overline{h}$. Thus $\overline{d_{c(n)}} \to \overline{h}$, so by definition (8.17) $\overline{h} = \overline{b}_{x_0,\xi}$.

Given any $x_0' \in X$, in the preceding paragraph one can replace c by $c' : [0, \infty) \to X$, where $c'(0) = x_0'$ and $c'(\infty) = \xi$ to deduce that $\overline{b}_{x_0',\xi} = \overline{h} = \overline{b}_{x_0,\xi}$. Thus the Busemann functions of asymptotic rays differ only by an additive constant.

Given distinct points $\xi, \xi' \in \partial X$, let c and c' be the geodesic rays in X with $c(0) = c'(0) = x_0$, $c(\infty) = \xi$ and $c'(\infty) = \xi'$. The sequence z_n defined by $z_{2n} = c(n)$ and $z_{2n+1} = c'(n)$ does not converge in $\overline{X}$, so by the proposition, the sequence $\overline{d_{z_n}}$ does not converge in $\widehat{X}$, hence $\overline{b}_{x_0,\xi} \neq \overline{b}_{x_0,\xi'}$. Thus i is injective. □

In order to prove (8.19), we shall need the following lemma.

8.21 Lemma. *Let X be a* CAT(0) *space.*

(1) *Given a ball $B(x_0, \rho)$ and $\varepsilon > 0$, there exists $r > 0$ such that, for all $z \in B(x_0, \rho)$ and all $x \in X$ with $d(x, x_0) \ge r$,*

$$d(z, y) + d(x, y) - d(x, z) \le \varepsilon,$$

where y is the point of $[x_0, x]$ a distance r from x_0.

(2) *Consider $x_0, y, x \in X$ with $y \in [x_0, x]$ and $d(x_0, y) = \rho > 0$. Then,*

$$d(x, z) - d(x, y) \geq \frac{d(y, z)^2}{2\rho}$$

for all z such that $d(x_0, z) = \rho$.

Proof. (1) We first note that if a segment $[p, q]$ of length t is contained in an infinite strip in $\mathbb{E}^2$ whose boundary is the union of two parallel lines a distance ρ apart, and if t' is the length of its orthogonal projection onto the side of the strip, then

$$|t - t'| \leq \rho^2/t.$$

Indeed, if q' is the orthogonal projection of q onto the straight line through p parallel to the sides of the strip and q'' is the orthogonal projection of q' on $[p, q]$, then $t' = d(p, q')$ and $|t - t'| \leq d(q, q'')$; obviously $d(q, q')/t = d(q, q'')/d(q, q')$ and $d(q, q') \leq \rho$, so we have $|t - t'| \leq \rho^2/t$.

Choose r big enough so that $\rho^2/(r - \rho) < \varepsilon/2$. If $X = \mathbb{E}^2$ and z' is the orthogonal projection of z onto the line through x and x_0, then using the above inequality twice we get: $d(z, x) - d(z', x) < \varepsilon/2$ and $d(z, y) - d(z', y) < \varepsilon/2$. But $d(z', x) = d(z', y) + d(y, x)$, hence $d(z, y) + d(y, x) < d(z, x) + \varepsilon$. In the general case, one applies the CAT(0) inequality to the triangle $\Delta(x_0, z, x)$ and considers the comparison point for $y \in [x_0, x]$.

(2) We first check the inequality in the case $X = \mathbb{E}^2$. We have $d(z, x) \geq d(x, y) + d(y, z')$, where z' is the projection of z on the line through x_0 and x. But $d(z', y)/d(y, z) = d(y, z)/2\rho$, whence the desired inequality for $X = \mathbb{E}^2$. The inequality in the general case is obtained by applying the CAT(0) inequality to the triangle $\Delta(x_0, z, x)$ and considering the comparison point for y. □

Proof of (8.19). Choose a basepoint $x_0 \in X$ and let $\tilde{d}_x$ be the function on X defined by $\tilde{d}_x(z) = d(x, z) - d(x, x_0)$.

Assume first that x_n converges to $\xi \in \partial X$. We want to prove that the sequence $\tilde{d}_{x_n}$ converges uniformly on any ball $B(x_0, \rho)$. Given $\varepsilon > 0$, choose $r > 0$ as in (8.21(1)). Choose N big enough so that $d(x_n, x_0) > r$ for $n > N$. If y_n is the point of $[x_0, x_n]$ at distance r from x_0, then by making two applications of (8.21(1)) we see that for all $z \in B(x_0, \rho)$

$$\begin{aligned} |\tilde{d}_{x_n}(z) - \tilde{d}_{x_m}(z)| &= |d(x_n, z) - d(x_n, y_n) - d(x_m, z) + d(x_m, y_m)| \\ &\leq |d(y_n, z) - d(y_m, z)| + 2\varepsilon \\ &\leq d(y_n, y_m) + 2\varepsilon. \end{aligned}$$

By hypothesis $d(y_n, y_m) \to 0$ as $m, n \to \infty$. Thus we have uniform convergence on the ball $B(x_0, \rho)$.

Conversely, assume $(\tilde{d}_{x_n})$ converges uniformly on bounded sets but (x_n) is unbounded (cf. 8.16). We need to prove that the geodesic segments $[x_0, x_n]$ converge to a geodesic ray issuing from x_0. Given $\rho > 0$ and $\varepsilon > 0$, we fix an integer N such that $d(x_n, x_0) > \rho$ and $|d(x_n, z) - d(x_n, x_0) - d(x_m, z) + d(x_m, x_0)| < \varepsilon$ for all $m, n > N$ and $z \in B(x_0, \rho)$. Let y_n be the point on the geodesic segment $[x_0, x_n]$ at distance ρ from

x_0. In the above inequality, replacing z by y_n, we get $|d(x_m, y_n) - d(x_m, y_m)| < \varepsilon$, because $d(x_0, y_n) = d(x_0, y_m) = \rho$. (8.21(2)) implies that for $n, m > N$ we have $d(y_n, y_m)^2/2\rho \leq d(x_m, y_n) - d(x_m, y_m) < \varepsilon$, hence y_n is a Cauchy sequence. □

Proof of Theorem 8.13. Because X is dense in $\overline{X}$, there is at most one continuous extension to $\overline{X}$ of $\iota : X \hookrightarrow \widehat{X}$. We showed in (8.20) that the map $i : \partial X \to \widehat{X} \smallsetminus X$ given by $i(c(\infty)) = \overline{b}_c$ is a bijection, so it only remains to show that if we extend ι by i then the resulting bijection $i : \overline{X} \to \widehat{X}$ and its inverse are continuous on the complement of X.

We choose a base point x_0 in X and identify $\mathcal{C}_*(X)$ to the subspace $\mathcal{C}_0(X)$ of $\mathcal{C}(X)$ consisting of those functions f such that $f(x_0) = 0$; thus $\widehat{X}$ is identified to a subspace of $\mathcal{C}(X)$. With this identification, the map $i : \overline{X} \to \widehat{X}$ associates to $x \in X$ the function $\tilde{d}_x(z) := d(x, z) - d(x, x_0)$, and associates to $\xi \in \partial X$ the Busemann function b_c of the geodesic ray c with $c(0) = x_0$ and $c(\infty) = \xi$.

A fundamental system of neighbourhoods for a point $x \in X$ is given by the balls $B(x, r)$. A fundamental system of neighbourhoods for b_c in $\widehat{X}$ is the collection of sets

$$V(b_c, \rho, \varepsilon) = \{f \in \widehat{X} \mid |f(z) - b_c(z)| < \varepsilon,\ \forall z \in B(x_0, \rho)\},$$

where $\rho > 0$ and $\varepsilon > 0$. A fundamental system of neighbourhoods for $c(\infty)$ in $\overline{X}$ was described in (8.6):

$$U(c, r, \delta) = \{x \in \overline{X} \mid d(x, x_0) > r,\ d(p_r(x), c(r)) < \delta\}$$

where $r > 0$, $\delta > 0$, and $p_r : \overline{X} \to B(x_0, r)$ is the map whose restriction to X is the projection on $B(x_0, r)$ and which maps $c(\infty) \in \partial X$ onto $c(r)$.

By letting $\rho, \varepsilon, r, \delta$ take all positive rational values, we see that each point of $\overline{X}$ and $\widehat{X}$ has a countable fundamental system of neighbourhoods. (i.e. as topological spaces $\overline{X}$ and $\widehat{X}$ are first countable). It is therefore sufficient to prove that both i and i^{-1} carry convergent sequences to convergent sequences. In other words, writing $\tilde{\delta}_u$ to denote $\tilde{d}_u$ or $b_{x_0,u}$ (according to whether $u \in X$ or $u \in \partial X$), what we must show is that a sequence x_n in $\overline{X}$ converges to $\xi \in \partial X$ if and only if $\tilde{\delta}_{x_n} \to b_{x_0,\xi}$ uniformly on bounded subsets. But this was done in (8.18) and (8.19). □

Characterizations of Horofunctions

Let $\mathcal{B}$ be the set of functions h on X satisfying the following three conditions:

(i) h is convex;

(ii) $|h(x) - h(y)| \leq d(x, y)$ for all $x, y \in X$;

(iii) For any $x_0 \in X$ and $r > 0$, the function h attains its minimum on the sphere $S_r(x_0)$ at a unique point y and $h(y) = h(x_0) - r$.

Notice that if $h \in \mathcal{B}$ then $h' \in \mathcal{B}$ whenever $\overline{h} = \overline{h}'$.

8.22 Proposition. *Let X be a complete* CAT(0) *space. For functions $h : X \to \mathbb{R}$, the following conditions are equivalent:*

(1) *h is a horofunction;*

(2) *$h \in \mathcal{B}$;*

(3) *h is convex, for every $t \in \mathbb{R}$ the set $h^{-1}(-\infty, t]$ is non-empty, and for each $x \in X$ the map $c_x : [0, \infty) \to X$ associating to $t > 0$ the projection of x onto $h^{-1}(-\infty, h(x) - t]$ is a geodesic ray.*

Proof. (1) $\implies$ (2): Let h be a horofunction. After modification by an additive constant we may assume that h is the Busemann function b_c associated to a geodesic ray issuing from x. Let x_n be a sequence of points tending to $c(\infty)$ along the geodesic ray c, and let $\tilde{d}_{x_n}$ be the function defined by $\tilde{d}_{x_n}(z) = d(x_n, z) - d(x_n, x_0)$. Then $b_c = \lim_{n\to\infty} \tilde{d}_{x_n}$. The convexity of the metric on X (2.2) ensures that each $\tilde{d}_{x_n}$ satisfies conditions (i) and (iii) in the definition of $\mathcal{B}$, and (ii) is simply the triangle inequality for $\tilde{d}_{x_n}$. Each of these is a closed condition, so the limit b_c also satisfies them.

(2) $\implies$ (3): Consider a function $h \in \mathcal{B}$, a point $x \in X$, and a number $t > 0$. Let $c_x(t)$ be the unique point of $S_t(x)$ such that $h(c_x(t)) = h(x) - t$ (condition (iii)). This point is the projection of x onto the closed convex subset $A_t = h^{-1}((-\infty, h(x) - t])$. Let $r : [0, t] \to X$ be the geodesic segment joining x to $c_x(t)$. Because h is convex, we have that $h(x) - h(r(s)) \leq s$ for all $s \in [0, t]$. Condition (iii) applied to the sphere $S_s(x)$ implies that $c_x(s) = r(s)$, hence $t \mapsto c_x(t)$ is a geodesic ray.

(3) $\implies$ (1): Let b be the Busemann function associated to the geodesic ray c_x. For convenience (replacing h by $h - h(x)$ if necessary) we assume that $h(x) = 0$. We want to prove that $b = h$. For $r > 0$ let π_r be the projection of X onto the closed convex set $A_r = h^{-1}(-\infty, -r]$. Note that $d(x, \pi_r(x)) = r$.

Given $y \in X$, for each (sufficiently large) $r > 0$, we have $h(y) = d(y, \pi_r(y)) - r$. Consider the Euclidean quadrilateral obtained by joining the comparison triangles $\overline{\Delta}(x, y, \pi_r(x))$ and $\overline{\Delta}(y, \pi_r(y), \pi_r(x))$ along the edge $[\overline{\pi_r(x)}, \overline{y}]$, so that $\overline{\pi_r(y)}$ and $\overline{x}$ lie on opposite sides of the line through $\overline{y}$ and $\overline{\pi_r(x)}$ (see Figure 8.3). (The degenerate case $\pi_r(y) = \pi_r(x)$ is easily dealt with.)

Let φ_r be the angle of the triangle $\overline{\Delta}(x, y, \pi_r(x))$ at the vertex $\overline{\pi_r(x)}$ and let $\alpha_r, \beta_r, \psi_r$ be the angles of the triangle $\overline{\Delta}(y, \pi_r(x), \pi_r(y))$ at the vertices $\overline{\pi_r(x)}, \overline{\pi_r(y)}$ and $\overline{y}$ respectively. Because $d(\pi_r(x), \pi_r(y)) \leq d(x, y)$ (cf. 2.4), and $d(x, \pi_r(x))$ and $d(y, \pi_r(y))$ both tend to ∞ as $r \to \infty$, the angles φ_r and ψ_r tend to 0 as $r \to \infty$. From (1.7) and (2.4) we have

$$\varphi_r + \alpha_r \geq \angle_{\pi_r(x)}(x, y) + \angle_{\pi_r(x)}(y, \pi_r(y)) \geq \angle_{\pi_r(x)}(x, \pi_r(y)) \geq \pi/2.$$

As $\beta_r \geq \pi/2$, we have $\lim_{r\to\infty} \beta_r = \lim_{r\to\infty} \alpha_r = \pi/2$. And therefore $b(y) = \lim_{r\to\infty}(d(y, \pi_r(x)) - r) = \lim_{r\to\infty} d(y, \pi_r(y)) - r = h(y)$. □

8.23 Exercises

(1) Prove that the Busemann functions associated to asymptotic rays in a CAT(0) space X differ only by an additive constant even if X is not complete. (Hint: Use (2.15).)

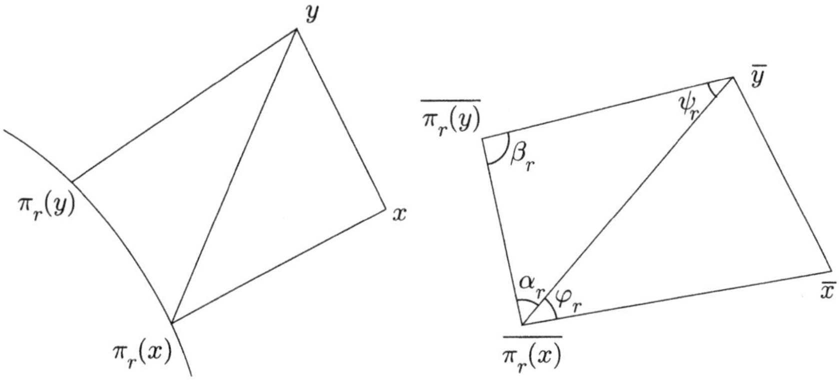

Fig. 8.4 Characterizing horofunctions

(2) Let c be a geodesic ray in $\mathbb{E}^n$. Prove that $h_c(x) = -r < 0$ if and only if $c(r)$ is the image of x under orthogonal projection to $\text{im}(c)$.

(3) Prove that in $\mathbb{H}^n$ the assertion of (2) is false.

(4) Let X be a complete CAT(0) space and let $c : [0, \infty) \to X$ be a geodesic ray. Fix a geodesic $\overline{c} : \mathbb{R} \to \mathbb{E}^2$ in the Euclidean plane. For every $x \in X$ and every $n > 0$ consider a comparison triangle $\overline{\Delta}(x, c(0), c(n))$ with $\overline{c(0)} = \overline{c}(0)$ and $\overline{c(n)} = \overline{c}(n)$. Let $\overline{c}(r_n)$ be the foot of the perpendicular from $\overline{x}$ to $\overline{c}(\mathbb{R})$. Prove that the sequence r_n converges to a point r such that $b_c(x) = -r$.

8.24 Examples of Busemann Functions

(1) Consider $\mathbb{R}^n$ equipped with the Euclidean metric, and let $c : [0, \infty) \to \mathbb{R}^n$ be a geodesic ray of the form $c(t) = a + te$, where e is a vector of unit length and $a \in \mathbb{R}^n$ is fixed. The associated Busemann function b_c is defined by

$$b_c(x) = (a - x \mid e)$$

where $(\cdot \mid \cdot)$ is the Euclidean scalar product. Its horospheres (i.e., the sets where b_c is constant) are precisely those hyperplanes which are orthogonal to the direction determined by e.

(2) Consider the Poincaré disc model for real hyperbolic space $\mathbb{H}^n$. The Poisson kernel is the function $P : \mathbb{H}^n \times \partial\mathbb{H}^n \to \mathbb{R}$ defined by $P(x, \xi) = \frac{1-|x|^2}{|\xi - x|^2}$, where $|\cdot|$ denotes the Euclidean norm. The Busemann function $b_{x,\xi}$ associated to the geodesic ray c with $c(0) = x$ and $c(\infty) = \xi$ is

$$b_{x,\xi} = -\log P(x, \xi).$$

(3) Let $X = X_1 \times X_2$ be the product of two CAT(0) spaces X_1 and X_2. Let c_1 and c_2 be geodesic rays in X_1 and X_2 respectively, and let $\xi_1 = c_1(\infty) \in \partial X_1$ and $\xi_2 = c_2(\infty) \in \partial X_2$. If $\theta \in [0, \pi/2]$, then $c(t) = (c_1(t\cos\theta), c_2(t\sin\theta))$ is a geodesic

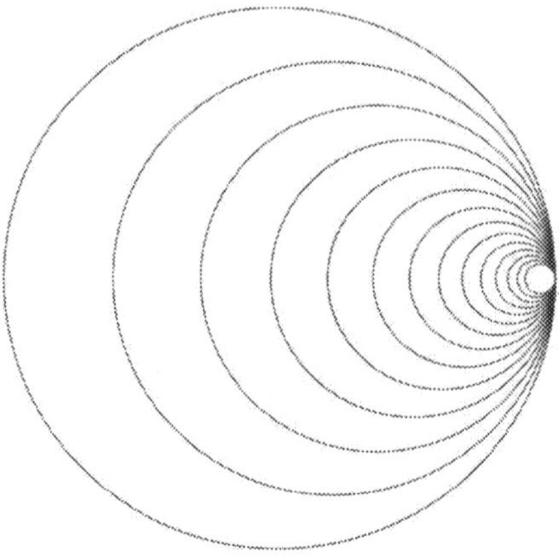

Fig. 8.5 Horospheres in the Poincaré model

ray in X and $c(\infty)$ is denoted $(cos\theta)\xi_1 + (\sin\theta)\xi_2$ (see (8.9(5)). If b_1, b_2, b are the Busemann functions associated to c_1, c_2, c respectively, then

$$b = (\cos\theta)b_1 + (\sin\theta)b_2.$$

Indeed, since $\lim_{t\to\infty}(d(x, c(t)) + t)/2t = 1$,

$$\begin{aligned} b(x) &= \lim_{t\to\infty}(d(x, c(t)) - t) = \lim_{t\to\infty}\frac{1}{2t}(d(x, c(t))^2 - t^2) \\ &= \lim_{t\to\infty}\frac{1}{2t}\left[d(x_1, c_1(t\cos\theta))^2 - t^2\cos^2\theta + d(x_2, c_2(t\sin\theta))^2 - t^2\sin^2\theta\right] \\ &= (\cos\theta)\, b_1(x_1) + (\sin\theta)\, b_2(x_2). \end{aligned}$$

(4) Let X be the tree constructed in (7.11). The boundary ∂X consists of exactly one point and (up to sign) the natural projection onto $[0, \infty)$ is a Busemann function.

Further examples of Busemann functions and horospheres are described in (10.69).

Parabolic Isometries

Recall from (6.1) that an isometry γ of a metric space X is said to be *parabolic* if $d(\gamma.x_0, x_0) > \inf\{d(\gamma.x, x) \mid x \in X\}$ for every $x_0 \in X$. In this section we consider the action of such isometries on CAT(0) spaces X, and more particularly on $\overline{X}$.

To say that an isometry γ of a CAT(0) space X is parabolic means precisely that if one takes a sequence of points $x_n \in X$ such that the sequence of numbers $d_\gamma(x_n)$ approximates $|\gamma|$, then the sequence x_n does not converge in X. But if X is proper, then such a sequence must have a convergent subsequence in $\overline{X}$, which is compact. Thus one expects to find fixed points of parabolic isometries at infinity.

8.25 Proposition. *Let γ be an isometry of a proper* CAT(0) *space X. If γ is parabolic, then it fixes at least one point $u \in \partial X$ and leaves invariant all of the horoballs centred at u.*

Proof. Consider the displacement function d_γ of γ. It is a γ-invariant convex function which does not attain its infimum in X. Thus the proposition is a consequence of the following lemma.

8.26 Lemma. *Let X be a proper* CAT(0) *space and let $f : X \to \mathbb{R}$ be a continuous convex function which does not attain its infimum $a \in \mathbb{R} \cup \{-\infty\}$. If γ is an isometry of X such that $f(\gamma.x) = f(x)$ for all $x \in X$, then γ fixes at least one point $u \in \partial X$, and also fixes all of the horofunctions centred at u.*

8.27 Remark. Notice that, in general, when an isometry γ fixes a point $u \in \partial X$, the action of γ may permute the set of horofunctions centred at u. On the other hand, if γ leaves invariant a horofunction centred at u then it must fix u and all of the horofunctions centred at u.

Proof of Lemma 8.26 For every $t > a$, we let $A_t = \{x \in X \mid f(x) \leq t\}$. This is a closed, convex, non-empty subspace of X. Obviously $A_t \subseteq A_{t'}$ if $t < t'$, and by our hypothesis on f we have $\bigcap_{t>a} A_t = \emptyset$. Since X is proper, for every closed ball B in X we have $B \cap A_t = \emptyset$ when t is close enough to a.

Let π_t be the projection of X onto A_t and let f_t be the function $x \mapsto d(x, \pi_t(x))$. Choose a base point $x_0 \in X$, and let $h_t(x) = f_t(x) - f_t(x_0)$. We claim that we can find a sequence of numbers t_n tending to a such that the corresponding sequence of functions $h_{t_n} : x \mapsto f_{t_n}(x) - f_{t_n}(x_0)$ converges uniformly on compact sets to some function h. Indeed, the family of functions $h_t : x \mapsto f_t(x) - f_t(x_0)$ is uniformly bounded on balls and equicontinuous because $|h_t(x) - h_t(y)| \leq d(x, y)$ (8.22), and hence, since X is proper, we can apply the Arzelà-Ascoli theorem (I.3.9) to obtain the desired convergent sequence of functions h_{t_n}. By Corollary 2.5, the limit h of the sequence of functions (h_{t_n}) satisfies the list of conditions preceding (8.22) and hence h is a horofunction. Since each of the functions h_{t_n} is invariant by γ, so too is h. □

The following example shows that (8.25) does not remain true without the hypothesis that X is proper.

8.28 Example. Let H be the Hilbert space $\ell^2(\mathbb{Z})$, as defined in Chapter I.4. (We write points in H as $x = (x_n)$.) The boundary at infinity ∂H can be identified with the set of equivalence classes of $H \setminus \{0\}$ modulo the relation generated by scalar multiplication by positive numbers: $x \sim y$ if there exists $\lambda > 0$ such that $x_n = \lambda y_n$ for all $n \in \mathbb{Z}$.

Translations of H act trivially on ∂H. One example of an isometry that does not act trivially on ∂H is the shift map σ. By definition, $\sigma.x$ is the sequence whose n-th entry is x_{n+1}. The only element of H fixed by σ is 0, because if $x_n = x_{n+1} \neq 0$ for all $n \in \mathbb{Z}$, then $(x_n)_{n \in \mathbb{Z}}$ is not an element of H. Let γ be the isometry $x \mapsto \sigma.x + \delta$, where δ is the sequence whose only non-zero entry is $\delta_0 = 1$. Arguing as above, one sees that γ does not fix any point in H, so it is not elliptic. But γ does not fix any point in ∂H either, so it can't be hyperbolic because a hyperbolic isometry must fix at least two points at infinity, namely the endpoints of its axes. Hence γ is a parabolic isometry that does not fix any point of ∂H.

8.29 Exercise. Calculate the translation length $|\gamma|$ of the isometry γ described in the preceding example, and prove directly that d_γ does not attain this value.

8.30 Example. Let $X = X_1 \times X_2$ be the product of two complete CAT(0) spaces. Let γ be an isometry of X that splits as $\gamma = (\gamma_1, \gamma_2)$. If γ_i fixes $\xi_i \in \partial X_i$, for $i = 1, 2$, then γ fixes the whole segment $(\cos\theta)\xi_1 + (\sin\theta)\xi_2 \in \partial X$ for $\theta \in [0, \pi/2]$ (in the notation of 8.11(6)). If γ_1 is parabolic and preserves the horofunctions of X_1 centered at ξ_1, then γ is parabolic (6.9), but it preserves the horofunction centred at $(\cos\theta)\xi_1 + (\sin\theta)\xi_2$ if and only if $\theta = 0$ or if γ_2 preserves the horofunctions centred at ξ_2.

Chapter II.9 The Tits Metric and Visibility Spaces

Let X be a complete CAT(0) space. In the preceding chapter we constructed a boundary at infinity ∂X and studied the cone topology on it. (This topology makes ∂X compact if X is proper.) $\partial\mathbb{H}^n$ and $\partial\mathbb{E}^n$ are homeomorphic in the cone topology, but at an intuitive level they appear quite different when viewed from within the space. Consider how the apparent distance between two points at infinity changes as one moves around in $\mathbb{E}^n$ and $\mathbb{H}^n$: in $\mathbb{E}^n$ the angle subtended at the eye of an observer by the geodesic rays going to two fixed points at infinity does not depend on where the observer is standing; in $\mathbb{H}^n$ the angle depends very much on where the observer is standing, and by standing in the right place he can make the angle π. Thus by recording the view of ∂X from various points inside X one obtains a metric structure that discriminates between the boundaries of $\mathbb{E}^n$ and $\mathbb{H}^n$. In this chapter we shall consider the same metric structure in the context of complete CAT(0) spaces.

More precisely, we shall define an *angular metric* on the boundary of a complete CAT(0) space X. The associated length metric is called the *Tits metric*, and the corresponding length space is called the *Tits boundary of* X, written $\partial_T X$. The natural map $\partial_T X \to \partial X$ is continuous (9.7), but in general it is not a homeomorphism. $\partial_T\mathbb{H}^n$ is a discrete space. In contrast, $\partial_T\mathbb{E}^n$ is isometric to $\mathbb{S}^{n-1}$ and the natural map $\partial_T\mathbb{E}^n \to \partial\mathbb{E}^n$ is a homeomorphism.

This contrast in behaviour is indicative of the fact that the Tits boundary encodes the geometry of flats in a complete CAT(0) space X (see (9.21)). Moreover, if X has the geodesic extension property, then the existence of product compositions for X is determined by $\partial_T X$ (see (9.24)).

$\mathbb{H}^n$ is a *visibility space* (9.28), which means that every pair of distinct points in $\partial\mathbb{H}^n$ can be connected by a geodesic line in $\mathbb{H}^n$. The concept of visibility is closely related to Gromov's δ-hyperbolic condition (see III.H). In the final section of this chapter we shall prove that a complete cocompact CAT(0) space X contains an isometrically embedded copy of $\mathbb{E}^2$ if and only if X is not a visibility space (The Flat Plane Theorem (9.33)).

Angles in $\overline{X}$

Most of the results in this chapter concern the behaviour of angles in the bordification of a complete CAT(0) space.

9.1 Definition. Let X be a complete CAT(0) space. Given $x, y \in X$ and $\xi, \eta \in \partial X$, we shall use the symbol $\angle_x(\xi, \eta)$ to denote the angle at x between the unique geodesic rays which issue from x and lie in the classes ξ and η respectively, and we write $\angle_x(y, \xi)$ to denote the angle at x between the geodesic segment $[x, y]$ and the geodesic ray that issues from x and is in the class ξ.

The continuity properties of angles in $\overline{X}$ are much the same as those of angles in X (see (3.3)).

9.2 Proposition (Continuity Properties of Angles). *Let X be a complete metric* CAT(0) *space.*

(1) *For fixed $p \in X$, the function $(x, x') \mapsto \angle_p(x, x')$, which takes values in $[0, \pi]$, is continuous at all points $(x, x') \in \overline{X} \times \overline{X}$ with $x \neq p$ and $x' \neq p$.*

(2) *The function $(p, x, x') \mapsto \angle_p(x, x')$ is upper semicontinuous at points $(p, x, x') \in X \times \overline{X} \times \overline{X}$ with $x \neq p$ and $x' \neq p$.*

Proof. Suppose $x_n \to x \in \overline{X}$, $x'_n \to x' \in \overline{X}$ and $p_n \to p \in X$, where $x \neq p$ and $x' \neq p'$. Let c and c' be the geodesic paths or rays joining p to x and x' respectively, and let c_n and c'_n be the geodesic paths or rays joining p_n to x_n and x'_n respectively. We fix $t > 0$ so that for all sufficiently large n the points $c_n(t)$ and $c'_n(t)$ are defined and different from x_n and x'_n. By definition $\angle_{p_n}(c_n(t), c'_n(t)) = \angle_{p_n}(x_n, x'_n)$ and $\angle_p(c(t), c(t)) = \angle_p(x, x)$. And from the convexity of the metric on X (2.2) and the definition of the topology on $\overline{X}$ we have $c_n(t) \to c(t)$ and $c'_n(t) \to c'(t)$.

Thus, by replacing x_n and x'_n by $c_n(t)$ and $c'_n(t)$, we can reduce the present proposition to the corresponding properties for angles in X (see (3.3)). $\square$

With the notion of asymptotic rays in hand, one can consider triangles which have vertices in $\overline{X}$ rather than just X. Many basic facts about triangles in CAT(0) spaces can be extended to this context. The following proposition provides a useful illustration of this fact (cf. (2.9)).

9.3 Proposition (Triangles with One Vertex at Infinity). *Let Δ be a triangle in a* CAT(0) *space X with one vertex at infinity; thus Δ consists of two asymptotic rays, c and c' say, together with the geodesic segment joining $c(0)$ to $c'(0)$. Let $x = c(0)$, $x' = c'(0)$ and $\xi = c(\infty) = c'(\infty)$. Let $\gamma = \angle_x(x', \xi)$ and $\gamma' = \angle_{x'}(x, \xi)$. Then:*

(1) $\gamma + \gamma' \leq \pi$;

(2) *$\gamma + \gamma' = \pi$ if and only if the convex hull of Δ is isometric to the convex hull of a triangle in $\mathbb{E}^2$ with one vertex at infinity and interior angles γ and γ'.*

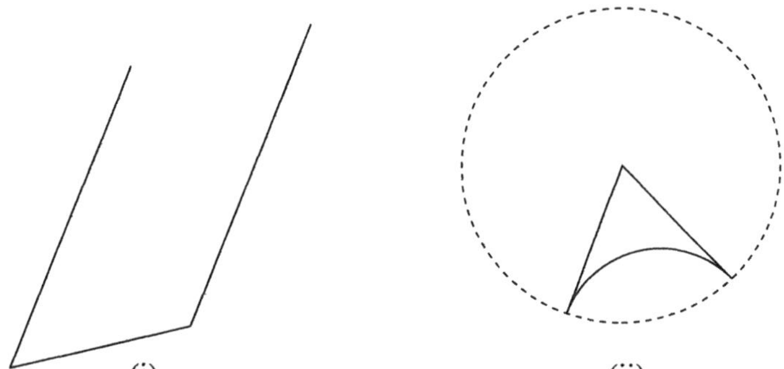

(i) (ii)

Fig. 9.1 (i) A triangle in $\mathbb{E}^2$ with one vertex at infinity. (ii) A triangle in $\mathbb{H}^2$ with two vertices at infinity

Proof. Let Δ_t denote the geodesic triangle in X with vertices x, x' and $c'(t)$. Let γ_t and γ_t' be the angles at the vertices x and x'. Let $\overline{\gamma_t}$ and $\overline{\gamma_t}'$ be the corresponding angles in a comparison triangle $\overline{\Delta}_t$. Let α_t denote the angle at x between c and the geodesic segment joining x to $c'(t)$.

With this notation, $\gamma \le \gamma_t + \alpha_t$, $\gamma_t \le \overline{\gamma_t}$, $\gamma' = \gamma_t' \le \overline{\gamma_t}'$ and $\overline{\gamma_t} + \overline{\gamma_t}' \le \pi$. Hence,

$$\gamma + \gamma' \le \gamma_t + \alpha_t + \gamma_t' \le \overline{\gamma_t} + \overline{\gamma_t}' + \alpha_t \le \pi + \alpha_t.$$

But it follows from (8.3) that $\lim_{t\to\infty} \alpha_t = 0$, hence $\gamma + \gamma' \le \pi$.

We also have that $\lim_{t\to\infty} \gamma_t = \gamma$ because, by the triangle inequality for angles (I.1.14), $|\gamma_t - \alpha_t| \le \gamma \le \gamma_t + \alpha_t$.

Suppose now that $\gamma + \gamma' = \pi$. We claim that $\gamma' = \gamma_t' = \overline{\gamma_t}'$ for all $t > 0$. Since $\overline{\gamma_t}' \ge \gamma_t' = \gamma'$ and $\overline{\gamma_t}'$ is a non-decreasing function of t, it suffices to show that $\lim_{t\to\infty} \overline{\gamma_t}' \le \gamma'$. But $\gamma_t + \gamma_t' \le \gamma_t + \overline{\gamma_t}' \le \overline{\gamma_t} + \overline{\gamma_t}' \le \pi$, so passing to the limit as $t \to \infty$, we get $\pi = \gamma + \gamma' \le \gamma + \lim_{t\to\infty} \overline{\gamma_t}' \le \pi$. Thus $\overline{\gamma_t}' = \gamma_t'$, and (2.9) implies that the triangles Δ_t are flat for all $t > 0$.

Let $\overline{\Delta}$ be a Euclidean triangle with one vertex at infinity, consisting of two asymptotic geodesic rays $\bar{c}$ and $\bar{c}'$ issuing from points $\bar{x}$ and $\bar{x}'$ which are a distance $d(x, x')$ apart, such that the angles of $\overline{\Delta}$ at $\bar{x}$ and $\bar{x}'$ are γ and γ' respectively. For all $t > 0$, the Euclidean triangle $\overline{\Delta}_t = \Delta(\bar{x}, \bar{x}', \bar{c}'(t))$ is a comparison triangle for Δ_t. As t varies, the isometries from the convex hull of $\overline{\Delta}_t$ onto the convex hull of Δ_t, established in the previous paragraph fit together, to give an isometry from the convex hull of $\overline{\Delta}$ onto the convex hull of Δ. $\square$

The Angular Metric

Intuitively speaking, in order to obtain a true measure of the separation of points at infinity in a CAT(0) space, one should view them from all points of the space. This motivates the following definition.

9.4 Definition. Let X be a complete CAT(0) space. The angle $\angle(\xi, \eta)$ between $\xi, \eta \in \partial X$ is defined to be:

$$\angle(\xi, \eta) = \sup_{x \in X} \angle_x(\xi, \eta).$$

9.5 Proposition (Properties of the Angular Metric). *Let X be a complete* CAT(0) *space.*

(1) *The function $(\xi, \xi') \mapsto \angle(\xi, \xi')$ defines a metric on ∂X called the angular metric. The extension to ∂X of any isometry of X is an isometry of ∂X with the angular metric. (We show in (9.9) that this metric is complete.)*

(2) *The function $(\xi, \xi') \mapsto \angle(\xi, \xi')$ is lower semicontinuous with respect to the cone topology: for every $\varepsilon > 0$, there exist neighbourhoods U of ξ and U' of ξ' such that $\angle(\eta, \eta') > \angle(\xi, \xi') - \varepsilon$ for all $\eta \in U$ and $\eta' \in U'$.*

(3) *If X is proper and cocompact then for all $\xi, \xi' \in \partial X$ there exist $y \in X$ and $\eta, \eta' \in \partial X$ such that $\angle_y(\eta, \eta') = \angle(\xi, \xi')$.*

Proof. (1) The triangle inequality for $\angle$ follows from the triangle inequality for Alexandrov angles (I.1.13). We have to check that if $\xi \neq \xi'$ then $\angle(\xi, \xi') > 0$. Let c and c' be geodesic rays issuing from the same point $x \in X$ and such that $c(\infty) = \xi$ and $c'(\infty) = \xi'$. We choose $t > 0$ large enough to ensure that $d(c(t), c'(t)) > 0$. Let c'' be the geodesic ray issuing from $c(t)$ that is in the class ξ'. If $\angle_{c(t)}(\xi, \xi')$ were zero, then $\angle_{c(t)}(x, \xi')$ would be π, and the concatenation of $[x, c(t)]$ and $c''|_{[t,\infty)}$ would be a geodesic ray distinct from c', contradicting the fact that c' is the only geodesic ray in the class ξ' that issues from x.

(2) For fixed $p \in X$, the function $(\xi, \xi') \mapsto \angle_p(\xi, \xi')$ is continuous (9.2). The supremum of a bounded family of continuous functions is lower semicontinuous.

(3) Choose a sequence of points x_n in X such that $\lim_{n\to\infty} \angle_{x_n}(\xi, \xi') = \angle(\xi, \xi')$. As X is assumed to be cocompact, there exists a compact subset $C \subset X$ and a sequence of isometries $\gamma_n \in \text{Isom}(X)$ such that $\gamma_n.x_n = y_n \in C$. Clearly $\angle_{y_n}(\gamma_n.\xi, \gamma_n.\xi_n') = \angle_{x_n}(\xi, \xi')$ and $\angle(\gamma_n.\xi, \gamma_n.\xi') = \angle(\xi, \xi')$. Passing to a subsequence we may assume that y_n converges to a point $y \in C$. Furthermore, since we are assuming that X is proper, ∂X is compact, so by passing to a further subsequence we may assume that $\gamma_n.\xi$ and $\gamma_n.\xi'$ converge in ∂X, to points η and η' say. The lower semicontinuity of $\angle(\gamma_n.\xi, \gamma_n.\xi')$ yields the third inequality given below, and the upper semicontinuity of $\angle_{y_n}(\gamma_n.\xi, \gamma_n.\xi')$ (see (9.2)) yields the first (the remaining inequality and the two equalities hold by definition):

$$\begin{aligned}\angle(\xi, \xi') &= \lim_{n\to\infty} \angle_{y_n}(\gamma_n.\xi, \gamma_n.\xi') \leq \angle_y(\eta, \eta') \leq \angle(\eta, \eta') \\ &\leq \liminf_{n\to\infty} \angle(\gamma_n.\xi, \gamma_n.\xi') = \angle(\xi, \xi').\end{aligned}$$

Equality holds throughout. $\square$

9.6 Examples

(1) If $X = \mathbb{E}^n$ then $\angle_p(\xi, \xi') = \angle_q(\xi, \xi')$ for all $p, q \in \mathbb{E}^n$ and all $\xi, \xi' \in \partial X$, hence $\angle_p(\xi, \xi') = \angle(\xi, \xi')$. It follows that the map which associates to each $\xi \in \partial X$ the point of intersection of $[0, \xi]$ with the unit sphere about $0 \in \mathbb{E}^n$ gives an isometry from $\partial\mathbb{E}^n$ to $\mathbb{S}^{n-1}$.

(2) If X is real hyperbolic space $\mathbb{H}^n$, then $\angle(\xi, \xi') = \pi$ for any distinct points $\xi, \xi' \in \partial\mathbb{H}^n$, because there is geodesic line $c : \mathbb{R} \to \mathbb{H}^n$ such $c(\infty) = \xi$ and $c(-\infty) = \xi'$. We shall see later that the same is true for any complete CAT(−1) space X (cf. 9.13).

(3) If X is the Euclidean cone C_0Y on a metric space Y, then ∂X with its angular metric is isometric to Y with the truncated metric $\max\{d, \pi\}$. (This generalizes (1).)

Our next goal is to prove:

9.7 Proposition. *Let X be a complete* CAT(0) *space.*

(1) *The identity map from ∂X equipped with the angular metric to ∂X with the cone topology is continuous. (In general it is not a homeomorphism, cf. 9.6(2)).*

(2) *∂X with the angular metric is a complete metric space.*

We postpone the proof of this proposition for a moment, because it will be clearer if we first articulate a reformulation of the definition of $\angle$. In fact we take this opportunity to list several such reformulations. (The proof of their equivalence intertwines, so it is natural to prove them all at once.) The list of corollaries given below indicates the utility of these different descriptions of $\angle$. Recall that for $x, y \in X$ distinct from $p \in X$, the symbol $\overline{\angle}_p(x, y)$ denotes the angle at the vertex corresponding to p in a comparison triangle $\overline{\Delta}(p, x, y) \subset \mathbb{E}^2$.

9.8 Proposition. *Let X be a complete* CAT(0) *space with basepoint x_0. Let $\xi, \xi' \in \partial X$ and let c, c' be geodesic rays with $c(0) = c'(0) = x_0$, $c(\infty) = \xi$ and $c'(\infty) = \xi'$. Then:*

(1) $\angle(\xi, \xi') = \lim_{t,t' \to \infty} \overline{\angle}_{x_0}(c(t), c'(t')) = \sup\{\overline{\angle}_{x_0}(c(t), c'(t')) \mid t, t' > 0\}$.

(2) *The function $t \mapsto \angle_{c(t)}(\xi, \xi')$ is non-decreasing and*

$$\angle(\xi, \xi') = \lim_{t \to \infty} \angle_{c(t)}(\xi, \xi').$$

(3) *The function*

$$(t, t') \mapsto \left[\pi - \angle_{c(t)}(x_0, c'(t')) - \angle_{c'(t')}(x_0, c(t)\right]$$

is non-decreasing in both t and t', and its limit as $t, t' \to \infty$ is $\angle(\xi, \xi')$.

(4)

$$2\sin(\angle(\xi, \xi')/2) = \lim_{t \to \infty} \frac{1}{t} d(c(t), c'(t)).$$

Proof. We begin with the characterization of $\angle(\xi, \xi')$ given in (1) and explain why this implies (4). Whereas the Alexandrov angle between the rays c and c' is defined to be

$$\angle_{x_0}(\xi, \xi') = \lim_{t,t' \to 0} \overline{\angle}_{\xi_0}(c(t), c'(t')),$$

we now wish to consider the limit as $t, t' \to \infty$,

$$\overline{\angle}_{x_0}(\xi, \xi') := \lim_{t,t' \to \infty} \overline{\angle}_{x_0}(c(t), c'(t)).$$

This limit exists because $(t, t') \mapsto \overline{\angle}_{x_0}(c(t), c'(t'))$ is a non-decreasing function of both t and t' (see (3.1)) and it is bounded above by π. And $2\sin(\overline{\angle}_{x_0}(c(t), c'(t))/2) = d(c(t), c'(t))/t$, so we have

$$2\sin(\overline{\angle}_{x_0}(\xi, \xi')/2) = \lim_{t\to\infty} \frac{1}{t}\, d(c(t), c'(t)).$$

This formula shows that (4) is a consequence of (1). It also shows that $\overline{\angle}_{x_0}(\xi, \xi')$ is independent of the choice of basepoint x_0, for if $\tilde{c}$ and $\tilde{c}'$ are geodesic rays issuing from some other point in X and $\tilde{c}(\infty) = \xi$ and $\tilde{c}'(\infty) = \xi'$, then there is a constant K such that $d(c(t), \tilde{c}(t)) + d(c'(t), \tilde{c}'(t)) < K$ for all $t > 0$, so by the triangle inequality $\lim_{t\to\infty}(d(c(t), c'(t))/t = \lim_{t\to\infty}(d(\tilde{c}(t), \tilde{c}'(t))/t$.

Let $\overline{\angle}(\xi, \xi') := \overline{\angle}_{x_0}(\xi, \xi')$. Note that $\angle(\xi, \xi') \le \overline{\angle}(\xi, \xi')$, because $\angle_{x_0}(\xi, \xi') \le \overline{\angle}_{x_0}(\xi, \xi')$ and $\overline{\angle}_{x_0}(\xi, \xi')$ does not depend on x_0. Thus, in order to prove (1) (and hence (4)) it suffices to show that $\overline{\angle}(\xi, \xi') \le \angle(\xi, \xi')$. The proof of this inequality intertwines naturally with the proofs of (2) and (3).

We use the following notation (see figure 9.2):

$$\alpha_t = \angle_{c(t)}(x_0, c'(t')),\ \ \alpha'_t = \angle_{c'(t')}(x_0, c(t)),\ \ \gamma_t = \angle_{c(t)}(\xi, c'(t'))$$
$$\gamma'_t = \angle_{c'(t')}(\xi', c(t)),\ \ \phi_t = \angle_{c(t)}(\xi, \xi'),\ \ \psi_t = \angle_{c(t)}(\xi', c'(t')).$$

The function $t \mapsto \phi_t$ is non-decreasing because for $s > t$ we have

$$\phi_t + \angle_{c(s)}(x_0, \xi') \le \pi \le \angle_{c(s)}(x_0, \xi') + \phi_s,$$

where the first inequality comes from (9.3) and the second is the triangle inequality for angles together with the fact that the angle at $c(s)$ between the incoming and outgoing germs of the geodesic c is π. This proves the first assertion in (2). The proof that the function described in (3) is non-decreasing is similar.

In the same way as we obtained the displayed inequality above, we get $\gamma_t \le \phi_t + \psi_t$, $\alpha_t + \gamma_t \ge \pi$, $\alpha_{t'} + \gamma_{t'} \ge \pi$ and $\psi_t + \gamma_{t'} \le \pi$, hence

$$\phi_t \ge (\pi - \alpha_t - \alpha'_t).$$

And if we let $\overline{\alpha}_t = \overline{\angle}_{c(t)}(x_0, c'(t'))$ and $\overline{\alpha'}_t = \overline{\angle}_{c'(t')}(x_0, c(t))$, then by the CAT(0) inequality we get

$$(\pi - \alpha_t - \alpha'_t) \ge (\pi - \overline{\alpha}_t - \overline{\alpha}'_t) = \overline{\angle}_{x_0}(c(t), c'(t')).$$

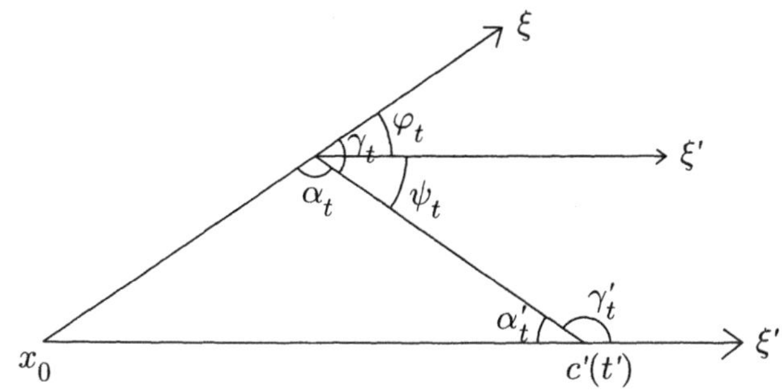

Fig. 9.2 Characterizations of $\angle(\xi, \xi')$

Therefore,

$$\overline{\angle}(\xi, \xi') \geq \angle(\xi, \xi') \geq \phi_t \geq (\pi - \alpha_t - \alpha'_t) \geq \overline{\angle}_{x_0}(c(t), c'(t')).$$

By definition $\lim_{t,t' \to \infty} \overline{\angle}_{x_0}(c(t), c'(t')) = \overline{\angle}(\xi, \xi')$, so passing to the limit as $t, t' \to \infty$, all of the above inequalities become equalities. This proves (1), (2), (3) and (4). $\square$

Proof of (9.7). We first prove (1). Fix $x_0 \in X$ and $\xi \in \partial X$ and let c be the geodesic ray with $c(\infty) = \xi$ that issues from x_0. Consider a basic neighbourhood $U(c, r, \varepsilon)$ of ξ in the cone topology (see (8.7)). Let $\delta = \arcsin(\varepsilon/(2r))$. Given $\xi' \in \partial X$ with $\angle(\xi, \xi') < \delta$, let c' be the geodesic ray with $c'(0) = x_0$ and $c'(\infty) = \xi'$. From part (1) of the above proposition we have $\overline{\angle}_{x_0}(c(r), c'(r)) \leq \delta$ and hence $d(c(r), c'(r)) \leq \varepsilon$. Thus $\angle(\xi, \xi') < \delta$ implies $\xi' \in U(c, r, \varepsilon)$.

The same argument shows that if ξ_n is a Cauchy sequence for the angular metric on ∂X, and c_n is the geodesic ray with $c_n(0) = x_0$ and $c_n(\infty) = \xi_n$, then for every $r > 0$ the sequence $c_n(r)$ is Cauchy and hence convergent in X. Let $c : [0, \infty) \to X$ be the limit ray. Given $\varepsilon > 0$, since ξ_n is Cauchy, we can apply part (1) of the preceding proposition to deduce that for sufficiently large m and n we have $\sup\{\overline{\angle}_{x_0}(c_m(t), c_n(t')) \mid t, t' > 0\} < \varepsilon$. Fixing n and letting $m \to \infty$ we get:

$$\angle(\xi, \xi_n) = \sup\{\overline{\angle}_{x_0}(c(t), c_n(t')) \mid t, t' > 0\} < \varepsilon.$$

Thus $\xi_n \to \xi$. $\square$

9.9 Corollary (Flat Sectors). *Let X be a complete* CAT(0) *space. If for some point $x_0 \in X$ we have $\angle_{x_0}(\xi, \xi') = \angle(\xi, \xi') < \pi$, then the convex hull of the geodesic rays c and c' issuing from x_0 with $c(\infty) = \xi$ and $c'(\infty) = \xi'$ is isometric to a sector in the Euclidean plane $\mathbb{E}^2$ bounded by two rays which meet at an angle $\angle(\xi, \xi')$.*

Proof. The function $t \mapsto \overline{\angle}_{c(t)}(\xi, \xi')$ is a non-decreasing function of t, its limit as $t \to 0$ is $\angle_{x_0}(\xi, \xi')$ and by part (2) of the proposition, its limit as $t \to \infty$ is $\angle(\xi, \xi')$. If $\angle_{x_0}(\xi, \xi') = \angle(\xi, \xi')$ then this function must be constant, so we may apply the

Flat Triangle Lemma (2.9). (The hypothesis $\angle(\xi, \xi') < \pi$ is needed to ensure that the comparison triangles $\overline{\Delta}(x_0, c(t), c'(t))$ are non-degenerate.) □

9.10 Corollary. *Let X be a complete* CAT(0) *space and let c, c' be geodesic rays issuing from x_0 in the direction of $\xi, \xi' \in \partial X$ respectively. For any real numbers $a, a' > 0$ we have*

$$\lim_{t\to\infty} \frac{1}{t} d(c(at), c'(a't)) = (a^2 + a'^2 - 2aa' \cos \angle(\xi, \xi'))^{1/2}.$$

In other words, this limit exists and is equal to the length of the third side of a Euclidean triangle whose other sides have lengths a and a' and meet at an angle $\angle(\xi, \xi')$.

Proof. The Euclidean law of cosines gives

$$d(c(at), c'(a't))^2 = a^2t^2 + a'^2t^2 - 2aa't^2 \cos\left(\overline{\angle}_{\xi_0}(c(at), c'(a't))\right).$$

To obtain the desired formula, we divide by t^2, let $t \to \infty$ and appeal to (9.8(1)). □

9.11 Corollary. *Let X_1 and X_2 be two complete* CAT(0) *spaces. Then $\partial(X_1 \times X_2)$ with the angular metric $\angle$ is isometric to the spherical join $\partial X_1 * \partial X_2$ of $(\partial X_1, \angle)$ and $(\partial X_2, \angle)$. More specifically (in the notation of I.5), given $\xi = (\cos\theta\, \xi_1 + \sin\theta\, \xi_2)$ and $\xi' = (\cos\theta'\, \xi_1' + \sin\theta'\, \xi_2')$ in $\partial(X_1 \times X_2)$, we have*

$$\cos(\angle(\xi, \xi')) = \cos\theta \cos\theta' \cos(\angle(\xi_1, \xi_1')) + \sin\theta \sin\theta' \cos(\angle(\xi_2, \xi_2')).$$

Proof. Choose basepoints $x_1 \in X_1$ and $x_2 \in X_2$ and let c_1, c_2, c_1', c_2' be geodesic rays issuing from these points in the classes $\xi_1, \xi_2, \xi_1', \xi_2'$ respectively. $c(t) = (c_1(t\cos\theta), c_2(t\sin\theta))$ and $c'(t) = (c_1'(t\cos\theta'), c_2'(t\sin\theta'))$ are geodesic rays in $X_1 \times X_2$ with $c(\infty) = \xi$ and $c'(\infty) = \xi'$ and $c(0) = c'(0) = (x_1, x_2)$. The previous corollary implies:

$$\begin{aligned}
&(2 - 2\cos(\angle(\xi, \xi'))\\
&= \lim_{t\to\infty} (d(c(t), c'(t))/t)^2\\
&= \lim_{t\to\infty} (d(c_1(t\cos\theta), c_1'(t\cos\theta'))/t)^2 + \lim_{t\to\infty} (d(c_2(t\sin\theta), c_2'(t\sin\theta'))/t)^2\\
&= \cos^2\theta + \cos^2\theta' - 2\cos\theta\cos\theta' \cos\angle(\xi_1, \xi_1')\\
&\quad + \sin^2\theta + \sin^2\theta' - 2\sin\theta\sin\theta' \cos\angle(\xi_2, \xi_2')\\
&= 2 - 2\cos\theta\cos\theta' \cos\angle(\xi_1, \xi_1') - 2\sin\theta\sin\theta' \cos\angle(\xi_2, \xi_2'),
\end{aligned}$$

as required. □

Later (9.24) we shall prove that the above result admits a converse in the case where geodesics can be extended indefinitely. In other words, if X has the geodesic extension property then one can detect whether or not X is a product by examining the geometry of $(\partial X, \angle)$.

9.12 Examples

(1) If X is a complete CAT(−1) then $\angle(\xi, \xi') = \pi$ for all pairs of distinct points $\xi, \xi' \in \partial X$. To see this, we fix $x_0 \in X$ such that $\angle_{x_0}(\xi, \xi') > 0$ and consider c, c' with $c(0) = c'(0) = x_0$ and $c(\infty) = \xi, c'(\infty) = \xi'$. Let m_t denote the midpoint of $[c(t), c'(t)]$. For a comparison triangle $\Delta(\overline{x}_0, \overline{c(t)}, \overline{c'(t)})$ in $\mathbb{H}^2$ we have

$$\angle_{\overline{x}_0}(\overline{c(t)}, \overline{c'(t)}) \geq \angle_{x_0}(c(t), c'(t)) = \angle_{x_0}(\xi, \xi') > 0.$$

It follows that $d(\overline{x}_0, \overline{m}_t)$ is bounded above by some constant A. Hence,

$$\begin{aligned} 2t = d(x_0, c(t)) + d(x_0, c'(t)) &\leq 2\,d(x_0, m_t) + d(m_t, c(t)) + d(m_t, c'(t)) \\ = 2d(\overline{x}_0, \overline{m}_t) + d(c(t), c'(t)) &\leq 2A + 2t. \end{aligned}$$

Dividing by t and let $t \to \infty$, we get that $\lim_{t\to\infty} d(c(t), c'(t))/t = 2$, and conclude from (9.8(4)) that $\angle(\xi, \xi') = \pi$.

(2) Let $Y = \mathbb{R} \times X$. By (9.11), the angular metric on ∂Y makes it isometric to the suspension of $(X, \angle)$. The case $X = \mathbb{H}^2$ is described in (4). Consider the case where X is a regular metric tree. In this case ∂Y with the cone topology is metrizable as the suspension of the standard Cantor set, and $(\partial Y, \angle)$ is the length space associated to this metric — the metric topology and length-space topology are different in this case.

(3) If $Y = \mathbb{H}^2 \times \mathbb{H}^2$, then $(\partial Y, \angle)$ is the spherical join of two uncountable spaces in each of which all distinct points are a distance π apart.

(4) If $Y = \mathbb{E}^n \times \mathbb{H}^2$, then $(\partial Y, \angle)$ is the geodesic space obtained by taking uncountably many copies of $\mathbb{S}^{n-1}$ and identifying their north poles and south poles. $(\partial Y, \angle)$ would remain unchanged if we were to replace $\mathbb{H}^2$ by the universal cover of any compact space X of strictly negative curvature with the geodesic extension property (other than a circle).

The Boundary $(\partial X, \angle)$ is a CAT(1) Space

The following fundamental property of the angular metrics was proved by Gromov in the case of Hadamard manifolds [BaGS87].

9.13 Theorem. *If X is a complete* CAT(0) *space, then ∂X with the angular metric is a complete* CAT(1) *space.*

First we prove that ∂X is a π-geodesic space. This follows from:

9.14 Proposition. *Let X be a complete* CAT(0) *space. Given $\xi, \eta \in \partial X$ with $\angle(\xi, \eta) < \pi$, there exists $\mu \in \partial X$ such that $\angle(\xi, \mu) = \angle(\eta, \mu) = \angle(\xi, \eta)/2$.*

We follow a proof of B. Kleiner and B. Leeb ([KlL97]); this requires two lemmas.

9.15 Lemma. *Let X be a* CAT(0) *space. Fix $x, y \in X$ with $d(x, y) = 2r$ and let m be the midpoint of the geodesic segment $[x, y]$. If m' is a point such that $d(m', x) \leq r(1+\varepsilon)$ and $d(m', y) \leq r(1+\varepsilon)$, then $d(m, m') \leq r(\varepsilon^2 + 2\varepsilon)^{1/2}$.*

Proof. At least one of the angles $\angle_m(m', x)$ and $\angle_m(m', y)$ is greater than or equal to $\pi/2$, let us say $\angle_m(m', x) \geq \pi/2$. In the comparison triangle $\overline{\Delta}(m, m', x)$, the angle at $\overline{m}$ is also greater than or equal to $\pi/2$ (1.7), hence $r^2(1+\varepsilon)^2 \geq d(m', x)^2 \geq r^2 + d(m, m')^2$. Therefore $d(m, m')^2 \leq r^2(\varepsilon^2 + 2\varepsilon)$. □

9.16 Lemma. *Let X be a complete* CAT(0) *space with basepoint x_0. Let x_n and y_n be sequences of points in X converging to points ξ and η of ∂X in the cone topology. Then,*

$$\liminf_{n\to\infty} \overline{\angle}_{x_0}(x_n, y_n) \geq \angle(\xi, \eta).$$

Proof. By hypothesis the sequences of geodesic segments $[x_0, x_n]$ and $[x_0, y_n]$ converge to the geodesic rays $[x_0, \xi]$ and $[x_0, \eta]$ respectively. Define $x_n' \in [x_0, \xi]$ to be the point furthest from x_0 such that $d(x_n', [x_0, \xi]) \leq 1$. Let x_n'' be the orthogonal projection of x_n' to $[x_0, x_n]$. Let y_n' and y_n'' be defined similarly. Because $x_n \to \xi$ and $y_n \to \eta$, we have $d(x_0, x_n'), d(x_0, y_n') \to \infty$. It follows from 9.8(1) that $\lim_{n\to\infty} \overline{\angle}_{x_0}(x_n', y_n') = \angle(\xi, \eta)$. But $\lim_{n\to\infty} \overline{\angle}_{x_0}(x_n', y_n') = \lim_{n\to\infty} \overline{\angle}_{x_0}(x_n'', y_n'')$ and by the CAT(0) inequality $\overline{\angle}_{x_0}(x_n, y_n) \geq \overline{\angle}_{x_0}(x_n'', y_n'')$. Therefore $\liminf_{n\to\infty}(\xi_n, y_n) \geq \angle(\xi, \eta)$. □

Proof of 9.14. Choose a basepoint $x_0 \in X$ and let $t \mapsto x_t$ and $t \mapsto y_t$ be geodesic rays issuing from x_0 in the classes ξ and η respectively. Let m_t be the midpoint of the geodesic segment $[x_t, y_t]$. We wish to show that the geodesic segments $[x_0, m_t]$ converge to a geodesic ray $[x_0, \mu]$ as t tends to ∞. This will prove the proposition, because $\liminf_t \overline{\angle}_{x_0}(x_t, m_t) \geq \angle(\xi, \mu)$ and $\liminf_t \overline{\angle}_{x_0}(y_t, m_t) \geq \angle(\eta, \mu)$ by (9.16), and $\lim_t \overline{\angle}_{x_0}(x_t, y_t) = \angle(\xi, \eta)$ by (9.8(1)), so letting $t \to \infty$ in $\overline{\angle}_{x_0}(x_t, m_t) = \overline{\angle}_{x_0}(y_t, m_t) \leq (1/2)\overline{\angle}_{x_0}(x_t, y_t)$, we get $\angle(\xi, \mu) = \angle(\eta, \mu) \leq (1/2)\angle(\xi, \eta)$, and the triangle inequality implies that in fact we have equality.

It remains to show that the geodesic segments $[x_0, m_t]$ converge to a geodesic ray as $t \to \infty$. Since X is complete, it is sufficient to prove that for all $R > 0$ and $\varepsilon > 0$ the geodesic segments $[x_0, m_t]$ and $[x_0, m_s]$ intersect the sphere of radius R about x_0 in points a distance at most ε apart whenever t and s are sufficiently large. Note first that $d(x_0, m_t) \to \infty$ as $t \to \infty$, because by (9.8(4)) $\lim_{t\to\infty} d(x_t, m_t)/t = \sin(\angle(\xi, \eta)/2)$, which by hypothesis is less than 1, and by the triangle inequality $t \leq d(x_t, m_t) + d(x_0, m_t)$, so dividing by t and letting $t \to \infty$ we have $\lim_{t\to\infty} d(x_0, m_t)/t \geq 1 - \sin(\angle(\xi, \eta)/2) > 0$.

Given $s > t$ we write $(t/s)\, m_s$ to denote the point on the geodesic segment $[x_0, m_s]$ that is distance $(t/s)\, d(x_0, m_s)$ from x_0. Applying the CAT(0) inequality to the triangle $\Delta(x_0, x_s, m_s)$ we get

$$d(x_t, (t/s)\, m_s) \le \frac{t}{s}\, d(x_s, m_s) = \frac{t}{2s}\, d(x_s, y_s),$$

and from $\Delta(x_0, y_s, m_s)$ we get

$$d(y_t, (t/s)\, m_s) \le \frac{t}{s}\, d(y_s, m_s) = \frac{t}{2s}\, d(x_s, y_s).$$

Therefore

$$d(x_t, y_t) \le d(x_t, (t/s)\, m_s) + d(y_t, (t/s)\, m_s) \le (t/s)\, d(x_s, y_s),$$

for all $s > t$. If we divide all of these inequalities by $d(x_t, y_t)$ and pass to the limit as $t \to \infty$, they all become equalities because $s > t$ and $\lim_{s\to\infty} d(x_s, y_s)/s = \lim_{t\to\infty} d(x_t, y_t)/t = \sin(\angle(\xi, \eta)/2)$. Thus

$$\lim_{s,t\to\infty} \frac{d(x_t, (t/s)\, m_s)}{d(x_t, m_t)} = 1 \text{ and}$$
$$\lim_{s,t\to\infty} \frac{d(y_t, (t/s)\, m_s)}{d(y_t, m_t)} = 1.$$

Using (9.15) we conclude that $\lim_{t,s\to\infty} \frac{d(m_t, (t/s)\, m_s)}{d(x_t, m_t)} = 0$. But we proved in the second paragraph that $\lim_{t\to\infty} d(x_0, m_t)/t > 0$, hence

$$\lim_{s,t\to\infty} \frac{d(m_t, (t/s)\, m_s)}{d(x_0, m_t)} = 0$$

and therefore

$$\lim_{t,s\to\infty} \overline{\angle}_{x_0}(m_t, (t/s)\, m_s) = 0.$$

Given a positive number R and $t > R$ we write ρ_t for the point of $[x_0, m_t]$ a distance R from x_0. The CAT(0) inequality for $\Delta(x_0, m_t, (t/s)\, m_s)$ gives

$$d(\rho_t, \rho_s) \le 2R \sin((1/2)\overline{\angle}_{x_0}(m_t, (t/s)\, m_s),$$

which we have shown converges to 0 as $s, t \to \infty$. Thus the geodesic segments $[x_0, m_t]$ converge to a geodesic ray $[x_0, \mu]$. $\square$

9.17 Remark. There is an easier proof of (9.14) in the case where X is proper: as in the second paragraph of the proof given above one sees that the set of midpoints m_t is unbounded, then using the properness of X one can extract a sequence of points $m_{t(n)}$ from this set so that $[x_0, m_{t(n)}]$ converges to a geodesic ray. The argument of the first paragraph can then be applied.

Proof of Theorem 9.13. We showed in (9.7) that $(\partial X, \angle)$ is a complete metric space, so the existence of midpoints (9.14) implies that it is a π–geodesic space. It remains to show that triangles satisfy the CAT(1) condition.

Consider $\xi_0, \xi_1, \xi_2, \mu \in \partial X$ with $\angle(\xi_0, \xi_1) + \angle(\xi_1, \xi_2) + \angle(\xi_2, \xi_0) < 2\pi$ and $\angle(\xi_0, \mu) = \angle(\xi_1, \mu) = (1/2)\angle(\xi_0, \xi_1)$, let $\Delta(\overline{\xi}_0, \overline{\xi}_1, \overline{\xi}_2) \subset \mathbb{S}^2$ be a comparison triangle and let $\overline{\mu}$ be the midpoint of the segment $[\overline{\xi}_0, \overline{\xi}_1]$. We must show that

$$\angle(\xi_2, \mu) \leq d(\overline{\xi}_2, \overline{\mu}).$$

Let c_0, c_1, c_2, c' be geodesic rays issuing from a fixed point $\xi_0 \in X$ with $c_0(\infty) = \xi_0, c_1(\infty) = \xi_1, c_2(\infty) = \xi_2, c'(\infty) = \mu$, and let $\alpha = \angle(\xi_0, \xi_1)$ and $a = \cos(\alpha/2) > 0$. Let m_t be the midpoint of the segment $[c_0(t), c_1(t)]$. We claim that $\lim_{t\to\infty} d(m_t, c'(at))/t = 0$. For $i = 0, 1$, by (9.8(4)) we have

$$\lim_{t\to\infty} \frac{1}{t}\, d(c_i(t), m_t) = \lim_{t\to\infty} \frac{1}{2t}\, d(c_0(t), c_1(t)) = \sin(\alpha/2),$$

and by (9.10)

$$\lim_{t\to\infty} \frac{1}{t}\, d(c_i(t), c'(at)) = \sin(\alpha/2).$$

Hence $\lim_{t\to\infty} \frac{d(c_i(t), m_t)}{d(c_i(t), c'(at))} = 1$, so by (9.15)

$$\lim_{t\to\infty} d(m_t, c'(at))/t = 0. \tag{1}$$

Consider $\mathbb{S}^2$ as the unit sphere in $\mathbb{R}^3$. Let $\overline{m}$ denote the midpoint of the Euclidean segment $[\overline{\xi}_0, \overline{\xi}_1]$. The spherical distance $d(\overline{\xi}_2, \overline{m})$ is equal to the vertex angle at $0 \in \mathbb{R}^3$ in the Euclidean triangle $\Delta(0, \overline{\xi}_2, \overline{m})$. The law of cosines applied to this triangle yields:

$$\|\overline{\xi}_2 - \overline{m}\|^2 = a^2 + 1 - 2a\cos(d(\overline{\xi}_2, \overline{\mu})). \tag{2}$$

We wish to compare this with the fact that (cf. (9.10))

$$\lim_{t\to\infty} \frac{1}{t^2}\, d(c_2(t), c'(at))^2 = a^2 + 1 - 2a\cos(\angle(\xi_2, \mu)). \tag{3}$$

By the CAT(0) inequality,

$$d(c_2(t), m_t)^2 \leq \frac{1}{2} d(c_2(t), c_0(t))^2 + \frac{1}{2}\, d(c_2(t), c_1(t))^2 - \frac{1}{4}\, d(c_0(t), c_1(t))^2.$$

Combining this with (1), we get:

$$\begin{aligned}
&\lim_{t\to\infty} \frac{1}{t^2}\, d(c_2(t), c'(at))^2 \\
&\leq \lim_{t\to\infty} \frac{1}{2t^2}\, d(c_2(t), c_0(t))^2 + \lim_{t\to\infty} \frac{1}{2t^2}\, d(c_2(t), c_1(t))^2 - \lim_{t\to\infty} \frac{1}{4t^2}\, d(c_0(t), c_1(t))^2 \\
&= \frac{1}{2}\|\overline{\xi}_2 - \overline{\xi}_0\|^2 + \frac{1}{2}\|\overline{\xi}_2 - \overline{\xi}_1\|^2 - \frac{1}{4}\|\overline{\xi}_1 - \overline{\xi}_0\|^2 \\
&= \|\overline{\xi}_2 - \overline{m}\|^2.
\end{aligned}$$

This inequality allows us to compare (2) and (3) and hence deduce that

$$\angle(\xi_2, \mu) \leq d(\overline{\xi}_2, \overline{\mu}).$$

□

The Tits Metric

For many purposes it is convenient to replace the angular metric $\angle$ on the boundary ∂X of a complete CAT(0) space by the corresponding length metric. This metric is named in honour of Jacques Tits.

9.18 Definition. Let X be a CAT(0) space. The *Tits metric* on ∂X is the length metric associated to the angular metric; it is denoted d_T. The length space $(\partial X, d_T)$ is called the *Tits boundary* of X, and is denoted $\partial_T X$.

9.19 Remarks

(1) If ξ and η are points of ∂X which cannot be joined by a path which is rectifiable in the angular metric, then $d_T(\xi, \eta) = \infty$. In particular, if X is a CAT(-1) space, then the Tits distance of any two distinct points in ∂X is infinite (see 9.12(1)).

(2) It follows from Theorem 9.14 that two points of ∂X which are a distance less than π apart are joined by a (unique) geodesic in $(\partial X, \angle)$, so if $\angle(\xi, \eta) < \pi$ then $d_T(\xi, \eta) = \angle(\xi, \eta)$. In particular the identity map from $(\partial X, \angle)$ to $(\partial X, d_T)$ is a local isometry and $(\partial X, d_T)$ is a CAT(1) metric space. If X is proper, then one can say more:

9.20 Theorem. *If X is a complete* CAT(0) *space, then ∂X_T is a complete* CAT(1) *space. Moreover, if X is proper then any two points $\xi_0, \xi_1 \in \partial X$ such that $d_T(\xi_0, \xi_1) < \infty$ are joined by a geodesic segment in $\partial_T X$.*

Proof. As $d_T(\xi_0, \xi_1) < \infty$, there is a sequence of continuous paths $p_n : [0, 1] \to \partial X$, parameterized proportional to arc length, which join ξ_0 to ξ_1 and whose Tits length tends to $d_T(\xi_0, \xi_1)$. We fix N so that for all n sufficiently large $d_T(p_n(k/N), p_n((k+1)/N)) \leq \pi/2$ for $k = 0, \dots, N-1$. As ∂X with the cone topology is compact, we may pass to a subsequence and assume that each of the sequences $p_n(k/N)$ converges in the visual topology, to a point μ_k say. As $d_T(p_n(k/N), p_n((k+1)/N)) = \angle(p_n(k/N, p_n((k+1)/N)$, using the upper semicontinuity of the angular metric (9.2), we have $\angle(\mu_k, \mu_{k+1}) \leq \lim_{n\to\infty} \angle(p_n(k/N), p_n((k+1)/N)$ which is $\lim_{n\to\infty} d_T(p_n(k/N), p_n((k+1)/N)) \leq \pi/2$. Let c_k be a geodesic segment (which exists by 9.14) joining μ_k to μ_{k+1}, for $k = 0, \dots, N-1$, and let c be the path joining ξ_0 to ξ_1 which is the concatenation of $c_0, \dots, c_{N-1}$. Then,

$$d_T(\xi_0, \xi_1) \leq l_{d_T}(c) = \sum_{k=0}^{N-1} l_{d_T}(c_k)$$
$$\leq \sum_{k=0}^{N-1} \angle(\mu_k, \mu_{k+1}) \leq \lim_{n\to\infty} l_{d_T}(p_n) \leq d_T(\xi_0, \xi_1),$$

showing that c is a geodesic segment joining ξ_0 to ξ_1. □

9.21 Proposition. *Let X be a proper* CAT(0) *space and let ξ_0 and ξ_1 be distinct points of ∂X.*

(1) *If $d_T(\xi_0, \xi_1) > \pi$, then there is a geodesic $c : \mathbb{R} \to X$ with $c(\infty) = \xi_0$ and $c(\infty) = \xi_1$.*

(2) *If there is no geodesic $c : \mathbb{R} \to X$ with $c(\infty) = \xi_0$ and $c(-\infty) = \xi_1$, then $d_T(\xi_0, \xi_1) = \angle(\xi_0, \xi_1)$ and there is a geodesic segment in $(\partial X, d_T)$ joining ξ_0 to ξ_1.*

(3) *If $c : \mathbb{R} \to X$ is a geodesic, then $d_T(c(-\infty), c(\infty)) \geq \pi$, with equality if and only if $c(\mathbb{R})$ bounds a flat half-plane.*

(4) *If the diameter of the Tits boundary $\partial_T X$ is π, then every geodesic line in X bounds a flat half-plane.*

We need a lemma.

9.22 Lemma. *Let X be a complete* CAT(0) *space, and let (x_n) and (x_n') be sequences in X converging to distinct points ξ and ξ' of ∂X. Suppose that each of the geodesic segments $[x_n, x_n']$ meets a fixed compact set $K \subset X$. Then, there exists a geodesic $c : \mathbb{R} \to X$ such that $c(\infty) = \xi$ and $c(-\infty) = \xi'$. Moreover, the image of c intersects K.*

Proof. For every $n \in \mathbb{N}$ we fix a point $p_n \in [x_n, x_n'] \cap K$. By passing to a subsequence if necessary, we may assume that the sequence p_n converges to a point $p \in K$. Let c_p and c_p' be the geodesic rays issuing from p which are asymptotic to ξ and ξ' respectively. We claim that the map $c : \mathbb{R} \to X$ defined by $c(t) = c_p(t)$ for $t > 0$ and $c(t) = c_p'(-t)$ for $t < 0$ is a geodesic. Indeed, given $y \in c_p(\mathbb{R}_+)$ and $y' \in c_p'(\mathbb{R}_+)$ there exist sequences of points $y_n \in [p_n, x_n]$ and $y_n' \in [p_n, x_n']$ such that $y_n \to y$ and $y_n' \to y'$ as $n \to \infty$; since $d(y_n, y_n') = d(y_n, p_n) + d(p_n, y_n')$, in the limit we have $d(y, y') = d(y, p) + d(p, y')$. □

Proof of 9.21. (1) follows immediately from (2).

For (2), we consider geodesic rays c_0 and c_1 issuing from the same point and such that $c_0(\infty) = \xi_0$ and $c_1(\infty) = \xi_1$. Let m_t be the midpoint of the segment $[c_0(t), c_1(t)]$. If there is no geodesic joining ξ_0 to ξ_1 then, since X is proper, $d(c_0(0), m_t) \to \infty$ by (9.22). As in 9.17 one sees that there is a point $\mu \in \partial X$ such that $\angle(\xi_0, \mu) = \angle(\mu, \xi_1) = 1/2\,\angle(\xi_0, \xi_1) \leq \pi/2$. As there are geodesic segments in ∂X joining ξ_0 to

μ and μ to ξ_1, we have $d_T(\xi_0, \xi_1) \leq \angle(\xi_0, \mu) + \angle(\mu, \xi_1) = \angle(\xi_0, \xi_1)$. The reverse inequality is obvious, hence $d_T(\xi_0, \xi_1) = \angle(\xi_0, \xi_1)$.

(3) Let $\xi_0 = c(-\infty)$ and $\xi_1 = c(\infty)$. We have $d_T(\xi_0, \xi_1) \geq \angle(\xi_0, \xi_1) = \pi$. Suppose $d_T(\xi_0, \xi_1) = \pi$. According to (9.20), there exists $\eta \in \partial X$ such that $d_T(\xi_0, \eta) = d_T(\xi_1, \eta) = \pi/2$, and hence by (1) $\angle(\xi_0, \eta) = \angle(\xi_1, y) = \pi/2$. Let c' be the geodesic ray issuing from $x = c(0)$ and such that $c'(\infty) = \eta$. As $\angle_x(\xi_i, \eta) \leq \angle(\xi_i, \eta) = \pi/2$ and $\pi = \angle_x(\xi_0, \xi_1) \leq \angle_x(\xi_0, \eta) + \angle_x(\eta, \xi_1) \leq \angle(\xi_o, \eta) + \angle(\eta, \xi_1) = \pi$, we have equality everywhere. In particular, $\angle_x(\xi_i, \eta) = \angle(\xi_i, \eta) = \pi/2$ for $i = 0, 1$, and (9.9) implies that the convex hull of $c'([0, \infty)) \cup c(\mathbb{R})$ is isometric to a flat half-plane bounded by $c(\mathbb{R})$. $\square$

9.23 Examples

(1) Let $X = \{(x, y) \in \mathbb{R}^2 : x \geq 0 \text{ or } y \geq 0\}$ with the length metric induced by the Euclidean metric. Then $(\partial X, d_T)$ is isometric to the interval $[0, 3\pi/2]$.

(2) Let $X = \{(x, y) \in \mathbb{R}^2 : |y| \leq x^2\}$ with the length metric induced by the Euclidean metric. Then $(\partial X, d_T)$ is isometric to the disjoint union of two intervals $[-\pi/2, \pi/2]$. The two half-parabolae $\{(x, y) \mid y = x^2, x \geq 0\}$ and $\{(x, y) \mid -y = x^2, x \geq 0\}$ are the images of two geodesic rays in X that define points of $(\partial X, d_T)$ that are a distance π apart but which cannot be joined by a geodesic in ∂X.

How the Tits Metric Determines Splittings

It follows from (9.11) that if a complete CAT(0) space X splits as a product $X_1 \times X_2$, then its Tits boundary splits as a spherical join $\partial_T X = \partial_T X_1 * \partial_T X_2$. In this section we shall show that, conversely, if X has the geodesic extension property then spherical join decompositions of the Tits boundary $\partial_T X$ give rise to product decompositions of X itself. This theorem is due to V. Schroeder in the case where X is a Hadamard manifold (see [BaGS87], appendix 4).

9.24 Theorem. *If X is a complete* CAT(0) *space in which all geodesic segments can be extended to geodesic lines, and if $\partial_T X$ is isometric to the spherical join of two non-empty spaces A_1 and A_2, then X splits as a product $X_1 \times X_2$, where $\partial X_i = A_i$ for $i = 1, 2$.*

In order to clarify the exposition, we re-express the theorem in a more technical form that indicates the key features of spherical joins.

9.25 Proposition. *Let X be as above and suppose that $A_1, A_2 \subset \partial_T X$ are disjoint, non-empty subspaces such that*

(1) *if $\xi_1 \in A_1$ and $\xi_2 \in A_2$, then $\angle(\xi_1, \xi_2) = \pi/2$, and*

(2) *for every $\xi \in \partial_T X$, there exists $\xi_1 \in A_1$ and $\xi_2 \in A_2$ such that ξ belongs to the geodesic segment $[\xi_1, \xi_2] \subset \partial_T X$.*

Then X splits as a product $X_1 \times X_2$ such that $\partial_T X_i = A_i$ for $i = 1, 2$.

9.26 Remarks. Under the hypotheses of the proposition:

(1) The Tits metric and the angular metric on ∂X coincide.

(2) If $\xi \in A_1$ and $\xi' \in \partial X$ are such that $\angle(\xi, \xi') = \pi$, then $\xi' \in A_1$.

In order to prove (9.25) we shall need the following lemma.

9.27 Lemma. *Let X, A_1 and A_2 be as in the statement of the proposition and consider two geodesic rays $c_1, c_2 : [0, \infty) \to X$ with $c_1(0) = c_2(0) = x_0$. If $c_i(\infty) = \xi_i \in A_i$, for $i = 1, 2$, then $\angle_{x_0}(\xi_1, \xi_2) = \pi/2$ and the convex hull of $c_1([0, \infty) \cup c_2([0, \infty))$ is isometric to a quadrant in the Euclidean plane.*

Proof. By hypothesis, the geodesic ray c_1 is the restriction of a geodesic line $c_1 : \mathbb{R} \to X$. Let $\overline{\xi}_1 = c_1(-\infty)$. As we remarked above, $\overline{\xi}_1 \in A_1$. Hence

$$\pi = \angle_{x_0}(\xi_1, \overline{\xi}_1) \le \angle_{x_0}(\xi_1, \xi_2) + \angle_{x_0}(\overline{\xi}_1, \xi_2) \le \angle(\xi_1, \xi_2) + \angle(\overline{\xi}_1, \xi_2) = \pi,$$

so $\angle_{x_0}(\xi_1, \xi_2) = \angle(\xi_2, \overline{\xi}_1) = \pi/2$, and we can apply Corollary 9.9. □

Proof of Proposition 9.25. We present the proof in four stages. Fix $x_0 \in X$ and let $X_1(x_0)$ be the union of the geodesic rays joining x_0 to the points of A_1.

Claim 1: $X_1(x_0)$ is a closed convex subspace. Any geodesic line which cuts $X_1(x_0)$ in two distinct points is entirely contained in $X_1(x_0')$. And if $x_0' \in X_1(x_0)$ then $X_1(x_0') = X_1(x_0)$.

Our proof relies on the following alternative description of $X_1(x_0)$. Consider a geodesic ray $c : [0, \infty) \to X$ with $c(0) = x_0$ and $c(\infty) \in A_2$, and let b_c be the associated Busemann function. We claim that $X_1(x_0) = \bigcap_c b_c^{-1}(0)$, where c runs over all such geodesic rays. The key observation is that if c_1 is a geodesic ray with $c_1(0) = x_0$ and $c_1(\infty) \in A_1$, then the convex hull of $c([0, \infty))$ and $c_1([0, \infty))$ is a subspace isometric to a Euclidean quadrant (by the lemma) and the intersection of $b_c^{-1}(0)$ with this subspace is $c_1([0, \infty))$ (see 8.24(1)). Since every point of $X_1(x_0)$ lies on some such geodesic ray c_1, we deduce $X_1(x_0) \subset b_c^{-1}(0)$. Conversely, given $x \in X \smallsetminus X_1(x_0)$, we consider a geodesic ray c_x issuing from x_0 and passing through x. By hypothesis, there exist $\xi_1 \in A_1$ and $\xi_2 \in A_2$ such that $c_x(\infty) \in [\xi_1, \xi_2]$. Let c_i be the geodesic rays issuing from x_0 with $c_i(\infty) = \xi_i$. Since x is not in $X_1(x_0)$, the ray c_1 does not pass through x, and so by considering the restriction of b_{c_2} to the Euclidean quadrant yielded by the lemma, we see that $b_{c_2}(x) \neq 0$. Thus $X_1(x_0) = \bigcap_c b_c^{-1}(0)$.

It follows from this equality that for all $x_0' \in X_1(x_0)$ and c with $c(0) = x_0$ and $c(\infty) \in A_2$, the Busemann function of the ray c' issuing from x_0' with $c'(\infty) = c(\infty)$ is the same as b_c. Thus $X_1(x_0) = \bigcap_c b_c^{-1}(0) = X_1(x_0')$.

Given any geodesic ray c with $c(0) = x_0$ and $c(\infty) \in A_2$, we can extend it to a geodesic line with $c(-\infty) \in A_2$, as we noted previously. Let $\overline{c} : [0, \infty) \to X$ be the geodesic ray $t \mapsto c(-t)$. Note that $b_c^{-1}(0) \cap b_{\overline{c}}^{-1}(0) = b_c^{-1}((-\infty, 0]) \cap b_{\overline{c}}^{-1}([0, \infty))$

which is a closed convex set. It follows that $X_1(x_0)$, as the intersection of closed convex sets, is closed and convex.

In the argument given above we described the geodesic ray c_x joining x_0 to an arbitrary $x \in X$ as lying in a subspace isometric to a Euclidean quadrant. The intersection of this subspace with $X_1(x_0)$ is precisely the image of c_1 (one of the axes). It follows that if $x \notin X_1(x_0)$ then c_x meets $X_1(x_0)$ at only one point. This completes the proof of the first claim.

Given $x, y \in X$, define $a_{xy} := d(X_1(x), X_1(y))$, and let $p_{yx} : X_1(x) \to X_1(y)$ denote the restriction of the orthogonal projection $X \to X_1(y)$.

Claim 2: The convex hull of $X_1(x) \cup X_1(y)$ is isometric to $X_1(x) \times [0, a_{xy}]$. And for all $x, y, z \in X$ we have $p_{zx} = p_{zy}p_{yx}$.

Given $x' \in X_1(x)$, we extend the geodesic segment $[x, x']$ to a geodesic line c with $c(0) = x$. It follows from Claim 1 that $c(\mathbb{R}) \subset X_1(x)$ and $c(\infty) \in A_1$. Consider the ray c' issuing from y with $c'(\infty) = c(\infty)$. The function $t \mapsto d(c(t), c'(t))$ is convex and bounded on $[0, \infty)$, hence it is non-increasing and $d(x', X_1(y)) \leq d(x, y)$. Casting $c(t)$ in the rôle of x', it follows that $t \mapsto d(c(t), X_1(y))$ is bounded on the whole of $\mathbb{R}$. But since $X_1(y)$ is convex, this function is convex (2.5), hence constant. Thus $d(x, X_1(y)) = d(x', X_1(y))$.

Because the convex hull of $X_1(x) \cup X_1(y)$ is isometric to $X_1(x) \times [0, a_{xy}]$ by the Sandwich Lemma 2.12(2), the map p_{yx} sends each geodesic line in $X_1(x)$ isometrically onto a geodesic line in $X_1(y)$. And since every point of $X_1(x)$ lies on a geodesic line, the second assertion in Claim 2 is an immediate consequence of Lemma 2.15.

Claim 3: Let $X_2(x_0)$ be the union of the geodesic rays joining x_0 to the points of A_2 and let $p_1 : X \to X_1(x_0)$ be orthogonal projection. We claim that $X_2(x_0) = p_1^{-1}(x_0)$.

Given $x \in X$ distinct from x_0, let $c_x : [0, \infty) \to X$ be a geodesic ray issuing from x_0 and passing through x. By hypothesis, there exist $\xi_i \in A_i$ such that $c_x(\infty) \in [\xi_1, \xi_2]$. Let c_i be the geodesic ray with $c_i(0) = x_0$ and $c_i(\infty) = \xi_i$. According to the lemma, the convex hull of the union of these geodesic rays is a subspace of X isometric to a quadrant in a Euclidean plane. This subspace contains the image of c_x and in this subspace it is obvious that the orthogonal projection of x onto the image of c_1 is x_0 if and only if x lies in the image of c_2; thus if $p(x) = x_0$ then $x \in X_2(x_0)$. Conversely, if $x \in X_2(x_0)$ then in the above construction we may choose c_1 so that it passes through any prescribed point $x' \in X_1(x_0)$, and hence $d(x, x') \geq d(x, x_0)$ (with equality only if $x' = x_0$).

Claim 4: For $i = 1, 2$, let $p_i : X \to X_i(x_0)$ be orthogonal projection. We claim that the map $j = (p_1, p_2) : X \to X_1(x_0) \times X_2(x_0)$ is an isometry, where $X_1(x_0) \times X_2(x_0)$ is endowed with the product metric.

Let $x, y \in X$. It follows from the first assertion in Claim 2 that $d(x, y)^2 = d(X_1(x), X_1(y))^2 + d(x, p_{xy}(y))^2$, and from the second assertion that $d(x, p_{xy}(y)) = d(p_1(x), p_1(y))$. Claim 3 implies that $p_2(x) = X_1(x) \cap X_2(x_0)$ (if we exchange the roles of the subscripts 1 and 2), and we get a similar expression for $p_2(y)$. And Claim 2 (with 1 and 2 exchanged) implies that $d(p_2(x), p_2(y)) = d(X_1(x), X_1(y))$.

Therefore $d(x, y)^2 = d(p_1(x), p_1(y))^2 + d(p_2(x), p_2(y))^2$. It is clear that $\partial X_i(x_0) = A_i$, $i = 1, 2$. □

Visibility Spaces

The notion of visibility for simply connected manifolds of non-positive curvature was introduced by Eberlein and O'Neill [EbON73] as a generalization of strictly negative curvature. This condition can be interpreted in a number of different ways (see [BaGS87, p.54]) most of which can be generalized to CAT(0) spaces.

9.28 Definition. A CAT(0) space is said to be a *visibility space* if for every pair of distinct points ξ and η of the visual boundary ∂X there is a geodesic line $c : \mathbb{R} \to X$ such that $c(\infty) = \xi$ and $c(-\infty) = \eta$.

Note that the geodesic c is in general not unique. For instance, X could be a Euclidean strip $[0, 1] \times \mathbb{R}$, or X could be obtained by gluing such a strip to a hyperbolic half-plane $\mathbb{H}^2_+$ using an isometry of $\{0\} \times \mathbb{R}$ onto the geodesic bounding $\mathbb{H}^2_+$.

9.29 Definition. A CAT(0) space X is said to be *locally visible* if for every $p \in X$ and $\varepsilon > 0$ there exists $R(p, \varepsilon) > 0$ such that if a geodesic segment $[x, y]$ lies entirely outside the ball of radius $R(p, \varepsilon)$ about p then $\angle_p(x, y) < \varepsilon$.

9.30 Definition. A CAT(0) space X is said to be *uniformly visible* if for every $\varepsilon > 0$ there exists $R(\varepsilon) > 0$ such that, given $p \in X$, if a geodesic segment $[x, y]$ lies entirely outside the ball of radius $R(\varepsilon)$ about p then $\angle_p(x, y) < \varepsilon$.

9.31 Remarks

(1) Real hyperbolic space $\mathbb{H}^n$ is uniformly visible and hence so is any CAT(-1) space (because the distance from p to $[x, y]$ does not decrease when one takes a comparison triangle $\overline{\Delta}(x, y, p)$ in $\mathbb{H}^2$ and the comparison angle at p is at least as large as $\angle_p(x, y)$).

(2) There exist complete non-proper CAT(0) spaces which are locally visible but which are not visibility spaces and vice versa. An example of a complete visibility space which is not locally visible can be obtained by taking any complete visibility space, fixing a basepoint and attaching to that point the corner of a Euclidean square of side n for every positive integer n. An example of a complete locally visible space which is not a visibility space is $\{(x, y) \mid 1 > y \geq 1 - (1 + |x|)^{-1}\}$ with the path metric induced from the Euclidean plane. Examples of CAT(0) spaces which are visible but not uniformly visible can be obtained by considering negatively curved 1-connected manifolds whose curvature is not bounded above by a negative constant. The interior of an ideal triangle in the hyperbolic plane provides an example of an incomplete uniformly visible space that is not a visibility space.

(3) In contrast to the remark preceding (9.30), note that by the flat strip theorem, there is a *unique* geodesic line joining any two points at infinity in any proper CAT(-1) space X.

One consequence of this is that if we fix $p \in X$ and $\xi, \xi' \in \partial X$, then there is a well-defined ideal triangle with vertices p, ξ, ξ'. (Exercise: If θ is the angle between the geodesic rays issuing from p with endpoints ξ and ξ', and D is the distance from p to the geodesic line with endpoints ξ and ξ', then $\tan(\theta/4) \leq e^{-D}$.)

9.32 Proposition. *If X is a proper* CAT(0) *space, then the following are equivalent:*

(1) *X is a visibility space;*

(2) *X is locally visible.*

If in addition X is cocompact, then (1) and (2) are equivalent to:

(3) *X is uniformly visible.*

Proof. (2) $\implies$ (1): Let ξ and ξ' be distinct points of ∂X and let x_n and x_n' be sequences of points in X converging to ξ and ξ' respectively. If $p \in X$ is such that $\angle_p(\xi, \xi') > \varepsilon > 0$, then for n big enough we have $\angle_p(x_n, x_n') > \varepsilon$. By (2), there exists $R > 0$ such that $[x_n, x_n']$ meets the compact ball $B(p, R)$. It then follows from Lemma 9.22 that there is a geodesic joining ξ to ξ'.

(1) $\implies$ (2): We argue the contrapositive. If X is not locally visible, then there exists a point $p \in X$, a number $\varepsilon > 0$, and sequences (x_n) and (x_n') such that $d(p, [x_n, x_n'])$ is unbounded but $\angle_p(x_n, x_n') > \varepsilon$. We can assume, by passing to subsequences if necessary, that the sequences x_n and x_n' converge to points ξ and ξ' in ∂X. These points are necessarily distinct because $\angle_p(\xi, \xi') \geq \varepsilon$.

Let p_n be the image of p under projection onto the closed convex set $[x_n, x_n']$. By passing to a subsequence if necessary, we may assume that the sequence p_n converges to a point $\eta \in \partial X$. Reversing the roles of ξ and ξ' if necessary, we assume that $\eta \neq \xi$.

If n is big enough then $p_n \neq x_n$ and hence the angle $\angle_{p_n}(p, x_n)$ is well defined. According to (2.4) this angle is not smaller than $\pi/2$, hence the comparison angle $\overline{\angle}_{p_n}(p, x_n)$ is also no smaller than $\pi/2$, so $\overline{\angle}_p(p_n, x_n) \leq \pi/2$. But then, by the lower semicontinuity of the comparison angle (9.16) we have $\angle(\xi, \eta) \leq \pi/2$. Hence there is no geodesic in X with endpoints ξ and η.

Finally we prove that (2) $\implies$ (3) (the converse is trivial). We argue the contrapositive. Suppose that X is cocompact, covered by the translates of a compact subset K. If X is not uniformly visible then there exists $\varepsilon > 0$ and sequences of points p_n, x_n, y_n in X such that $\angle_{p_n}(x_n, y_n) \geq \varepsilon$ for all n, and $d(p_n, [x_n, y_n]) \to \infty$ as $n \to \infty$. Translating by suitable elements of Isom(X) we may assume that $p_n \in K$ for all n, hence $d(K, [x_n, y_n]) \to \infty$ as $n \to \infty$. Since K and $\overline{X}$ are compact, we may pass to subsequences so that $p_n \to p \in K$ and $x_n \to \xi \in \partial X$, $y_n \to \eta \in \partial X$. By the upper-semicontinuity of angles (9.2), we have $\angle_p(\xi, \eta) \geq \varepsilon$. Hence by the continuity of angles at a fixed p, we have $\angle_p(x_n, y_n) \geq \varepsilon/2$ for all n sufficiently large. Since $d(p, [x_n, y_n]) \geq d(K, [x_n, y_n]) \to \infty$ as $n \to \infty$, this implies that X is not locally visible at p. □

The following theorem was proved by Eberlein [Eb73] in the case of Hadamard manifolds. The version given here was stated by Gromov [Gro87] and a detailed proof was given by Bridson [Bri95]. A generalization to convex metric spaces was proved by Bowditch [Bow95b]. The implications of the stark dichotomy exposed by this theorem will be considered in Part III.

9.33 Theorem. *A proper cocompact* CAT(0) *space X is a visibility space if and only if it does not contain a 2-flat (i.e., a subspace isometric to $\mathbb{E}^2$).*

Proof. It is clear that if X contains a 2-flat then it cannot be a visibility space. On the other hand, if X is not a visibility space then according to (9.21(2)) there exist two distinct points $\xi, \xi' \in \partial X$ such that $\angle(\xi, \xi') \leq \pi/2$. But then, by Proposition 9.5(3), there exist $y \in X$ and $\eta, \eta' \in \partial X$ such that $\angle_y(\eta, \eta') = \angle(\xi, \xi')$. It follows from (9.9) that the convex hull of the geodesic rays issuing from y in the classes η and η' is isometric to a sector in the Euclidean plane; in particular X contains arbitrarily large flat discs. The following general lemma completes the proof of the theorem. $\square$

9.34 Lemma. *Let Y be a separable metric space with basepoint y_0 and let X be a proper cocompact metric space. If for all $n \in \mathbb{N}$ there exists an isometric embedding $\phi_n : B(y_0, n) \hookrightarrow X$ then there exists an isometric embedding $\phi : Y \hookrightarrow X$.*

Proof. Let $y_0, y_1, \dots$ be a countable dense subset of Y. Because X is cocompact, we can replace ϕ_n by a suitable choice of $g_n\phi_n$ with $g_n \in \text{Isom}(X)$, and hence assume that the sequence $(\phi_n(y_0))_n$ is contained in a compact (hence bounded) subset of X. Since X is proper, it follows that for all i the sequence $(\phi_n(y_i))_n$ is contained in a compact subset of X. Hence there exists $x_0 \in X$ and a sequence of integers $(n_0(j))_j$ such that $\phi_{n_0(j)}(y_0) \to x_0$ as $j \to \infty$. And proceeding by recursion on r we can find a sequence of elements $x_n \in X$ and infinite sets of integers $\{n_r(j)\}_j \subset \{n_{r-1}(j)\}_j$ such that $\phi_{n_r(j)}(y_k) \to x_k$ as $j \to \infty$ for all $k \leq r$. The diagonal sequence $m = n_r(r)$ satisfies $\phi_m(y_k) \to x_k$, as $m \to \infty$ for all $k \geq 0$. Since the ϕ_m are isometries, so too is the map $y_k \mapsto x_k$. The set $\{y_k\}_k$ is dense in Y, so $y_k \mapsto x_k$ has a unique extension, which is the desired ϕ. $\square$

We close this section by giving a few other characterizations of proper visibility spaces. In Chapter III.H we shall consider a much more far-reaching reformulation of uniform visibility, Gromov's δ-hyperbolic condition.

9.35 Proposition. *Let X be a proper* CAT(0) *space. The following are equivalent:*

(1) *X is a visibility space.*

(2) *If h, h' are horofunctions centred at different points $\xi, \xi' \in \partial X$, then $h + h'$ assumes its infimum.*

(3) *Let h be a horofunction centred at $\xi \in \partial X$ and let $c : [0, \infty] \to X$ be a geodesic ray with $c(\infty) \neq \xi$. Then $h \circ c(t)$ tends to infinity when t tends to infinity.*

(4) *The intersection of any two horoballs centered at different points of* ∂X *is bounded.*

We first need a lemma.

9.36 Lemma. *Let* $c : \mathbb{R} \to X$ *be a geodesic line in a* CAT(0) *space* X*, and let* b_c *be the Busemann function associated to the geodesic ray* $c|_{[0,\infty]}$*. Let* p *be the projection onto the closed convex subspace* $c(\mathbb{R})$*. Then for all* $x \in X$ *we have*

$$b_c(p(x)) \leq b_c(x).$$

Proof. Let $s = b_c(x)$. For any $t > s$, we have $b_c(c(-t)) = t > s = b_c(c(-s))$, and the projection of $c(-t)$ to the horoball $b_c^{-1}((\infty, s])$ is $c(-s)$. Thus $\angle_{c(-s)}(c(-t), x) \geq \pi/2$, and $d(x, c(-t)) > d(x, c(-s))$. Therefore $p(x) \in c([-s, \infty))$. $\square$

Proof of Proposition 9.35.
(1) $\implies$ (2). Let $c : \mathbb{R} \to X$ be a geodesic such that $c(\infty) = \xi$ and $c(-\infty) = \xi'$, and let b_c (resp. b_{-c}) be the Busemann function associated to the ray $c|_{[0,\infty)}$ (resp. $t \mapsto c(-t)$). These functions are equal up to constants to h and h' respectively, so it is sufficient to show that $b_c + b_{-c}$ assumes its infimum. Note $b_c(c(t)) = -t$ and $b_{-c}(c(t)) = t$, and hence $b_c + b_{-c}$ is identically zero on $c(\mathbb{R})$. Let p be the projection of X to $c(\mathbb{R})$ and $x \in X$. By the lemma, $b_c(x) \geq b_c(p(x))$ and $b_{-c}(x) \geq b_{-c}(p(x)) = -b_c(p(x))$. Thus $b_c(x) + b_{-c}(x) \geq 0$.

(2) $\implies$ (1). Let $x \in X$ be a point such that $h(x) + h'(x)$ is minimum. Let c (resp. c') be the geodesic ray issuing from x such that $c(\infty) = \xi$ (resp. $c'(\infty) = \xi'$). As $h' \in \mathcal{B}$ (cf. 8.22), for $t > 0$, we have

$$h'(c(t)) - h'(x) \leq t,$$

and we have equality if and only if the concatenation of c (reversed) and c' is a geodesic line. As $h(x) - t + h'(c(t)) = h(c(t)) + h'(c(t)) \geq h(x) + h'(x)$, we have

$$h'(c(t)) \geq h'(x) + t,$$

hence the desired equality.

(1) $\implies$ (3). Let $c' : \mathbb{R} \to X$ be a geodesic line with $c'(\infty) = c(\infty) := \xi'$ and $c'(-\infty) = \xi$. We have $h(c'(t)) = h(c'(0)) + t$. If p is the projection on $c'(\mathbb{R})$, by the above lemma we have $h(c(t)) \geq h(p(c(t))$. As $|h(c(t)) - h(c'(t))| \leq d(c(t), c'(t))$, which remains bounded as $t \to \infty$, we have $h(p(c(t)) \to \infty$.

(3) $\implies$ (4). Let h and h' be horofunctions centred at different points ξ and ξ', and let H and H' be horoballs centred at ξ and ξ'. If $H \cap H'$ is not bounded, there is an unbounded sequence (x_n) in $H \cap H'$. As X is proper by hypothesis, we can assume that (x_n) converges to $\eta \in \partial X$. Let $x_0 \in H \cap H'$. As $H \cap H'$ is convex, the geodesic segments $[x_0, x_n]$ converge to a geodesic ray c in $H \cap H'$ with $c(\infty) = \eta$. But the functions $h(c(t))$ and $h'(c(t))$ are both bounded, hence $\eta = \xi = \xi'$, a contradiction.

(4) $\Longrightarrow$ (1). Let ξ and ξ' be distinct points of ∂X and let b_c and $b_{c'}$ be Busemann functions associated to geodesic rays issuing from $x \in X$ and such that $c(\infty) = \xi$ and $c'(\infty) = \xi'$. By hypothesis the intersection C of the closed horoballs $H := b_c^{-1}((-\infty, 0])$ and $H' := b_{c'}^{-1}((-\infty, 0])$ is bounded. For $a \geq 0$, let $H(a) := b_c^{-1}((-\infty, -a])$. If a is bigger than the diameter of C, then $H' \cap H(a) = \emptyset$; indeed for $y \in X$ such that $b_c(y) \leq -a$, we have $a \leq |b_c(y) - b_c(x)| \leq d(x, y)$, hence H' does not intersect $H(a)$.

Let $m := \inf\{a \geq 0 \mid H' \cap H(a) \neq \emptyset\}$. As we assumed X to be proper, $H' \cap H(m) \neq \emptyset$. Given $y \in H' \cap H(m)$, let $\tilde{c}$ and $\tilde{c}'$ be geodesic rays issuing from y and such that $\tilde{c}(\infty) = \xi$ and $\tilde{c}'(\infty) = \xi'$. We have $b_c(\tilde{c}(t)) = -t - m$ and $b_{c'}(\tilde{c}'(t) = -t$. We claim that the union of the images of $\tilde{c}$ and $\tilde{c}'$ is a geodesic line joining ξ to ξ'. It suffices to check that, for $t > 0$, we have $d(\tilde{c}(t), \tilde{c}'(t)) \geq 2t$. Let z be a point on the geodesic segment $[\tilde{c}(t), \tilde{c}'(t)]$ such that $b_c(z) = -m$. Then $b_{c'}(z) \geq 0$, and hence

$$d(\tilde{c}(t), \tilde{c}'(t)) = d(\tilde{c}(t), z) + d(z, \tilde{c}'(t)) \geq |b_c(\tilde{c}(t)) - b_c(z)| + |b_{c'}(z) - b_{c'}(\tilde{c}'(t)| \geq 2t.$$

□

Chapter II.10 Symmetric Spaces

A *symmetric space* is a connected Riemannian manifold M where for each point $p \in M$ there is an isometry σ_p of M such that $\sigma_p(p) = p$ and the differential of σ_p at p is multiplication by -1. Symmetric spaces were introduced by Elie Cartan in 1926 [Car26] and are generally regarded as being among the most fundamental and beautiful objects in mathematics; they play a fundamental role in the theory of semi-simple Lie groups and enjoy many remarkable properties. A comprehensive treatment of symmetric spaces is beyond the scope of this book, but we feel that there is considerable benefit in describing certain key examples from scratch (without assuming any background in differential geometry or the theory of Lie groups), in keeping with the spirit of the book. Simple examples of symmetric spaces include the model spaces M_κ^n that we studied in Part I.

In this chapter we shall be concerned with symmetric spaces that are simply connected and non-positively curved. If such a space has a trivial Euclidean de Rham factor, then it is said to be of *non-compact type*, and in this case the connected component of the identity in the group of isometries of such a space is a semi-simple Lie group G with trivial centre and no compact factors. In fact, there is a precise correspondence between symmetric spaces of non-compact type and such Lie groups: given G one takes a maximal compact subgroup K, forms the quotient $M = G/K$ and endows it with a G-invariant Riemannian metric— this is a symmetric space of non-compact type and the connected component of the identity in Isom(M) is G (see for instance [Eb96, 2.1.1]).

The first class of examples on which we shall focus are the hyperbolic spaces $\mathbb{K}\mathrm{H}^n$, where $\mathbb{K}$ is the real, complex or quaternionic numbers. We shall give a unified treatment of these spaces based on our earlier treatment of the real case (Chapter I.2). We shall define the metric on $\mathbb{K}\mathrm{H}^n$ directly, prove that it is a CAT(-1) space, and then discuss its isometries, Busemann functions and horocyclic coordinates. Besides one exceptional example in dimension 8 (based on the Cayley numbers), the spaces $\mathbb{K}\mathrm{H}^n$ account for all symmetric spaces of rank one, where rank is defined as follows.

The *rank* of a simply connected symmetric space of non-positive curvature is the maximum dimension of flats in M, i.e. subspaces isometric to Euclidean spaces. The results in Chapters 6 and 7 give some indication of the important role that flat subspaces play in CAT(0) spaces; their importance in symmetric spaces is even more pronounced. In particular, there is a clear distinction between the geometry of symmetric spaces of rank one and those of higher rank. Aspects of this distinction will emerge in the course of this chapter. We should also mention one particularly striking aspect of the geometry of symmetric spaces that is not discussed here: if the rank of M is at least 2, then M and lattices[31] in Isom(M) are remarkably rigid (see [Mar90], [Zim84] and (I.8.41)).

We shall exemplify the theory of higher-rank symmetric spaces by focussing on $P(n, \mathbb{R})$, the space of symmetric, positive-definite, real (n, n)-matrices. $P(n, \mathbb{R})$ has a Euclidean de Rham factor of dimension 1 (namely the positive multiples of the identity matrix); the complementary factor $P(n, \mathbb{R})_1$ is the subspace of $P(n, \mathbb{R})$ consisting of matrices of determinant one. $P(n, \mathbb{R})_1$ is irreducible (i.e. cannot be decomposed as a non-trivial product). Following the outline of the treatment given by Mostow [Mos73], we shall prove that $P(n, \mathbb{R})$ is CAT(0) and we shall describe the geometry of its maximal flats (which are of dimension n). We shall also describe the Busemann functions and horospheres in $P(n, \mathbb{R})$. The reason that we chose to explain the example of $P(n, \mathbb{R})$ in some detail is that it plays the following universal role in the theory of symmetric spaces: any simply connected symmetric space of non-positive curvature is, up to rescaling of the irreducible factors in its product decomposition, isometric to a totally geodesic submanifold of $P(n, \mathbb{R})$ for some n (see [Mos73, paragraph 3] and [Eb96, 2.6.5]). Examples of such totally geodesic submanifolds are given in the section entitled "Reductive Subgroups".

In the final section of this chapter we shall discuss the Tits boundary of symmetric spaces (which is interesting only in the case where the rank is ≥ 2). This leads us to a discussion of a remarkable class of polyhedral complexes called *buildings*. These complexes, which were discovered by Jacques Tits, provide important examples of CAT(0) and CAT(1) spaces. We shall not replicate the literature on buildings, but in an appendix to this chapter we shall present their definition and explain how it leads to curvature bounds.

For a concise and elegant introduction to symmetric spaces from the point of view of differential geometry, we recommend the excellent book of Milnor [Mil63]. For the general theory of symmetric spaces, see [Wo67], [Hel78], [Kar65],[Im79], [Eb96]. In this chapter we have been guided by the treatment of Mostow [Mos73]; we also benefitted from the lecture notes of Eberlein [Eb96], as well as conversations with Marc Burger.

Real, Complex and Quaternionic Hyperbolic n-Spaces

In this section we shall construct the complex and quaternionic hyperbolic n-spaces in a manner closely analogous to the way in which we defined real hyperbolic space in (I.2). We shall give as unified a treatment as possible, writing $\mathbb{K}$ to denote either $\mathbb{R}$, $\mathbb{C}$ or the quaternions, and writing $\mathbb{K}H^n$ to denote the $\mathbb{K}$-hyperbolic [32] space of

[31] Discrete subgroups such that $\Gamma\backslash M$ has finite volume. For existence, see [Bo63].

[32] Throughout this book the symbol $\mathbb{H}^n$ is reserved for real hyperbolic n-space, but in this chapter it will be convenient to write $\mathbb{R}H^n$ instead of $\mathbb{H}^n$ so as to give a unified treatment of the real, complex and quaternionic cases. For the same reason, we change the notation of (I.6) by writing $\mathbb{R}^{n,1}$ instead of $\mathbb{E}^{n,1}$ and $\mathbb{R}P^n$ instead of $\mathbb{P}^n$.

dimension n. As in the real case, we shall describe the set of points of $\mathbb{K}\mathrm{H}^n$ in terms of $\mathbb{K}^{n,1}$, the vector space $\mathbb{K}^{n+1}$ equipped with a form of type $(n, 1)$. We shall then define the distance function by means of an explicit formula on this set of points (10.5) and give an explicit description of hyperbolic segments and angles (10.7). Our task, then, will be to show that this 'distance function' is indeed a metric and that our explicit descriptions of hyperbolic segments and angles are precisely the geodesic segments and Alexandrov angles associated to this metric. Proceeding as in (II.2), we shall deduce these facts from an appropriate form of the law of cosines (10.8). This form of the law of cosines also leads to a short proof of the fact that $\mathbb{K}\mathrm{H}^n$ is a CAT(-1) space (10.10).

For an alternative introduction to complex hyperbolic space, see [Ep87], and for the general case of symmetric spaces of rank one see [Mos73, paragraph 19].

Notation. Throughout this section $\mathbb{K}$ will denote either the field $\mathbb{R}$ of real numbers, the field $\mathbb{C}$ of complex numbers, or the non-commutative field of quaternions. (Recall that the quaternions are a 4-dimensional algebra over $\mathbb{R}$ with basis $\{1, i, j, k\}$, where 1 is central, $ij = k, jk = i, ki = j$ and $i^2 = j^2 = k^2 = -1$.)

If $x \in \mathbb{C}$, then we write $\bar{x}$ to denote the complex conjugate of x. Conjugation on $\mathbb{R}$ is trivial. For quaternions, one defines the conjugate of $\lambda = a_0 + a_1 i + a_2 j + a_3 k$ to be $\lambda = a_0 - a_1 i - a_2 j - a_3 k$. The *norm* $|\lambda|$ of $\lambda \in \mathbb{K}$ is the non-negative real number $\sqrt{\lambda\bar{\lambda}}$. The *real part* of x is the real number $\Re\lambda = \frac{\lambda+\bar{\lambda}}{2}$.

In all of the vector spaces which we shall consider, the multiplication by scalars will be on the right.

10.1 Definition of $\mathbb{K}^{n,1}$ and the Form $Q(x, y) = \langle x|y\rangle$. We denote by $\mathbb{K}^{n,1}$ the $\mathbb{K}$-vector space $\mathbb{K}^{n+1}$ endowed with the form $Q(x, y) = \langle x|y\rangle$ of type $(n, 1)$ defined by:

$$\langle x|y\rangle := \sum_{k=1}^{n} \bar{x}_k y_k - \bar{x}_{n+1} y_{n+1},$$

where $x = (x_1, \dots, x_{n+1})$ and $y = (y_1, \dots, y_{n+1})$. If $\langle x|y\rangle = 0$ then x and y are said to be orthogonal. The *orthogonal complement* of $x \in \mathbb{K}^{n,1}$, written $x^\perp$, is $\{u \in \mathbb{K}^{n+1} \mid \langle x|u\rangle = 0\}$.

10.2 Remark. Note that $\langle y|x\rangle = \overline{\langle x|y\rangle}$ and hence $\langle x|x\rangle$ is a real number. If $\langle x|x\rangle < 0$, then the restriction of Q to $x^\perp$ is positive definite, and if $\langle x|x\rangle > 0$ then the restriction of Q to $x^\perp$ is of type $(n-1, 1)$ (see, e.g., [Bour59, paragraph 7]).

10.3 Lemma (The Reverse Schwartz Inequality). *If $\langle x|x\rangle < 0$ and $\langle y|y\rangle < 0$, then*

$$\langle x|y\rangle\langle y|x\rangle \geq \langle x|x\rangle\langle y|y\rangle,$$

with equality if and only if x and y are linearly dependent over $\mathbb{K}$.

Proof. If x and y are linearly dependent then we obviously have equality. The restriction of Q to $x^\perp$ is positive definite and $\langle y|y\rangle < 0$, so $\langle x|y\rangle \neq 0$. Let $\lambda = -\langle x|x\rangle\langle x|y\rangle^{-1}$. Then $x + y\lambda \in x^\perp$, and if x and y are linearly independent then $x + y\lambda \neq 0$, therefore $\langle x + y\lambda | x + y\lambda\rangle = \langle x + y\lambda | y\lambda\rangle > 0$. By expanding this inequality we get

$$0 < -\langle x|x\rangle + \langle x|x\rangle^2 \langle y|y\rangle \langle y|x\rangle^{-1}\langle x|y\rangle^{-1}.$$

After dividing by $\langle x|x\rangle < 0$, this can be rearranged to give the inequality in the statement of the lemma. □

In Part I we described real hyperbolic space as one sheet of the sphere of radius -1 in $\mathbb{R}^{n,1}$. This sheet maps bijectively onto its image under the natural map $p : \mathbb{R}^{n,1} \to \mathbb{R}\mathrm{P}^n$; by endowing the image of the sheet with the metric that makes the restriction of p an isometry one obtains an alternative model for real hyperbolic space. This second (projective) model lends itself readily to the present more general context.

10.4 Notation. Let $\mathbb{K}\mathrm{P}^n$ be the n-dimensional projective space over $\mathbb{K}$, i.e. the quotient of $\mathbb{K}^{n+1} \smallsetminus \{0\}$ by the equivalence relation which identifies $x = (x_1, \dots, x_{n+1})$ with $x\lambda = (x_1\lambda, \dots, x_{n+1}\lambda)$ for all $\lambda \in \mathbb{K}^* = \mathbb{K} \smallsetminus \{0\}$. The class of x is denoted $[x]$ and $(x_1, \dots, x_{n+1})$ are called *homogeneous coordinates* for $[x]$.

10.5 The Hyperbolic n-Space $\mathbb{K}\mathrm{H}^n$ over $\mathbb{K}$. We define $\mathbb{K}\mathrm{H}^n$ to be the set of points $[x] \in \mathbb{K}\mathrm{P}^n$ with $\langle x|x\rangle < 0$. The distance $d([x],[y])$ between two points of $\mathbb{K}\mathrm{H}^n$ is defined by

$$\cosh^2 d([x],[y]) = \frac{\langle x|y\rangle\langle y|x\rangle}{\langle x|x\rangle\langle y|y\rangle}.$$

The preceding lemma shows that the right hand side of this formula is bigger than 1 unless $[x] = [y]$, so the formula makes sense and the distance between each pair of distinct points is non-zero. Thus, in order to show that d is a metric, it only remains to verify the triangle inequality. As in the case $\mathbb{K} = \mathbb{R}$ (see I.2.6), we shall deduce the triangle inequality from an appropriate form of the law of cosines. Following the strategy of (I.2.7), this requires that we first define a primitive notion of hyperbolic segment and hyperbolic angle. This task is complicated in the present setting by the fact that if $\mathbb{K} \neq \mathbb{R}$ then we do not have a canonical way to lift $\mathbb{K}\mathrm{H}^n$ into $\mathbb{K}^{n,1}$, and as a result we cannot describe tangent vectors in as concrete a manner as we did in the real case. We circumvent this difficulty as follows. Given $x \in \mathbb{K}^{n,1}$ with $\langle x|x\rangle < 0$ we regard $x^\perp$ as a model for the tangent space $T_{[x]}\mathbb{K}\mathrm{H}^n$ of $[x] \in \mathbb{K}\mathrm{H}^n$. More precisely:

10.6 Tangent vectors. We identify $x^\perp$ with $T_{[x]}\mathbb{K}\mathrm{H}^n$ using the differential of the natural projection $\mathbb{K}^{n,1} \smallsetminus \{0\} \to \mathbb{K}\mathrm{P}^n$.

If $u \in x^\perp$ is identified with $U \in T_{[x]}\mathbb{K}\mathrm{H}^n$ then we say that u is *the tangent vector* *at x representing U.* (If $\lambda \neq 0$ then $u\lambda$ is the tangent vector at $x\lambda$ representing U, and since $x^{\perp} = (x\lambda)^{\perp}$, one must be careful to specify the choice of x as well as u when describing $U \in T_{[x]}\mathbb{K}\mathrm{H}^n$.)

We note in passing that if $\mathbb{K}$ is the quaternions, $T_{[x]}\mathbb{K}\mathrm{H}^n$ is just a real vector space, whereas if $\mathbb{K} = \mathbb{C}$ then $T_{[x]}\mathbb{K}\mathrm{H}^n$ is naturally a complex vector space. In any case, the symmetric positive-definite $\mathbb{R}$-bilinear form associating to tangent vectors u, v at x the real number $-\Re\langle u|v\rangle/\langle x|x\rangle$ is compatible with the above identifications and therefore defines a scalar product[33] on $T_{[x]}\mathbb{K}\mathrm{H}^n$. In this way $\mathbb{K}\mathrm{H}^n$ is naturally a Riemannian manifold. The interested reader can check that the metric associated to this Riemannian structure is the metric defined in (10.5) (cf. I.6.17).

10.7 Lines, Segments and Angles in $\mathbb{K}\mathrm{H}^n$. Suppose that $\langle x|x\rangle = -1$ and that $u \in x^{\perp}$, with $\langle u|u\rangle = 1$, represents $U \in T_{[x]}\mathbb{K}\mathrm{H}^n$. Then the *hyperbolic geodesic issuing from* $[x]$ *in the direction of* U is defined to be the map $\mathbb{R} \to \mathbb{K}\mathrm{H}^n$ given by $t \mapsto c(t) = [x\cosh t + u\sinh t]$. Given $a > 0$, we write $[c(0), c(a)]$ to denote the image under c of the interval $[0, a]$, and refer to it as a *hyperbolic segment* joining $[x] = c(0)$ to $c(a)$. We call u the *initial vector of* $[c(0), c(a)]$ *at* x.

We claim that, given any two points $[x] \neq [y]$ in $\mathbb{K}\mathrm{H}^n$, there exists a unique hyperbolic geodesic segment joining $[x]$ to $[y]$. To construct this segment one takes $\langle x|x\rangle = -1$ and $\langle y|y\rangle = -1$ then multiplies y by the unique $\lambda \in \mathbb{K}$ with $|\lambda| = 1$ such that $\langle x|y\lambda\rangle$ is real and negative [34]. One then defines $a = \operatorname{arccosh}(-\langle x|y\lambda\rangle)$ and $u = \frac{y\lambda - x\cosh a}{\sinh a}$. The initial segment $[c(0), c(a)]$ of $t \mapsto c(t) = [x\cosh t + u\sinh t]$ is a hyperbolic segment joining $[x]$ to $[y]$. It remains to show that this segment is unique — we leave this as an exercise for the reader.

Given $[x] \in \mathbb{K}\mathrm{H}^n$ and non-zero tangent vectors u, v at x, we define the hyperbolic angle between u and v by

$$\cos \angle_x^H(u, v) := \frac{\Re\langle u|v\rangle}{\sqrt{\langle u|u\rangle}\sqrt{\langle v|v\rangle}}.$$

Since $\angle_x^H(u, v) = \angle_x^H(u\lambda, v\lambda)$ for all $\lambda \in \mathbb{K}$, this formula gives a well-defined notion of angle between vectors $U, V \in T_{[x]}\mathbb{K}\mathrm{H}^n$. The *hyperbolic angle* between two hyperbolic segments issuing from $[x] \in \mathbb{K}\mathrm{H}^n$ is defined to be the angle between their initial vectors. A *hyperbolic triangle* in $\mathbb{K}\mathrm{H}^n$ consists of a choice of three distinct points (its vertices) $A, B, C \in \mathbb{K}\mathrm{H}^n$, and the three hyperbolic segments joining them (its sides). The vertex angle at C is defined to be the hyperbolic angle between the segments $[C, A]$ and $[C, B]$.

10.8 Proposition (The Law of Cosines in $\mathbb{K}\mathrm{H}^n$). *Let Δ be a hyperbolic triangle in $\mathbb{K}\mathrm{H}^n$ with vertices A, B, C, where $C = [x]$ and $\langle x|x\rangle = -1$. Let $a = d(B, C)$, $b = d(C, A)$ and $c = d(A, B)$. Let $u, v \in x^{\perp}$ be the initial vectors at x of the hyperbolic segments $[C, A]$ and $[C, B]$. Then*

[33] If $\mathbb{K} = \mathbb{C}$, there is also a well-defined hermitian form on $T_{[x]}\mathbb{K}\mathrm{H}^n$ given by $-\langle u|v\rangle/\langle x|x\rangle$.

[34] The group of elements (numbers) of norm 1 in $\mathbb{K}$ acts simply transitively (by right multiplication) on each sphere $\{\mu : |\mu| = r > 0\}$ and each such sphere contains a unique real negative number.

$$\cosh c = |\cosh a \ \cosh b - \sinh a \ \sinh b \ \langle u|v\rangle|.$$

Proof. We have $A = [x \cosh a + u \sinh a]$ and $B = [x \cosh b + v \sinh b]$. Hence

$$\begin{aligned}\cosh c &= |\langle x \cosh a + u \sinh a \mid x \cosh b + v \sinh b\rangle| \\ &= |\cosh a \ \cosh b - \sinh a \ \sinh b \ \langle u|v\rangle|,\end{aligned}$$

as required. □

10.9 Corollary (The Triangle Inequality and Geodesics). *For all $A, B, C \in \mathbb{K}\mathrm{H}^n$,*

$$d(A, B) \leq d(A, C) + d(C, B),$$

with equality if and only if C lies on the hyperbolic segment joining A to B. Thus $\mathbb{K}\mathrm{H}^n$ is a uniquely geodesic metric space: the unique geodesic segment joining A to B is the hyperbolic segment $[A, B]$.

Proof. Let $a = d(C, B)$, $b = d(A, C)$, $c = d(A, B)$. We assume that A, B and C are distinct (the case where they are not is trivial). Consider a hyperbolic triangle Δ with vertices A, B, C. With the notations of the preceding proposition we have:

$$\begin{aligned}\cosh c &= |\cosh a \ \cosh b - \sinh a \ \sinh b \ \langle u|v\rangle| \\ &\leq \cosh a \ \cosh b + \sinh a \ \sinh b \ |\langle u|v\rangle| \\ &\leq \cosh a \ \cosh b + \sinh a \ \sinh b = \cosh(a + b),\end{aligned}$$

with equality if and only if $\langle u|v\rangle = -1$. If $\langle u|v\rangle = -1$, then since $\langle u|u\rangle = \langle v|v\rangle = 1$, we have $Q(u+v) = \langle u+v|u+v\rangle = 0$. The restriction of Q to $x^{\perp}$ is positive definite, thus $u = -v$ and C belongs to the geodesic segment $[A, B]$. □

The Curvature of $\mathbb{K}\mathrm{H}^n$

10.10 Theorem. $\mathbb{K}\mathrm{H}^n$ *is a* CAT(-1) *space.*

Proof. We shall prove that $\mathbb{K}\mathrm{H}^n$ is a CAT(-1) space by applying (1.8) with the hyperbolic angle $\angle^H$ in the rôle of A.

Let Δ be as in (10.8) and note that in the light of (10.9) we now know that Δ is a geodesic triangle. Let γ denote the hyperbolic angle at the vertex C in Δ, i.e. $\gamma = \arccos \Re\langle u|v\rangle$.

We consider a geodesic triangle $\Delta(A, B, C)$ in the real hyperbolic plane with $d(B, C) = a$ and $d(A, C) = b$, and with vertex angle γ at C. Let $\overline{c} = d(\overline{A}, \overline{B})$. We must show that $\overline{c} \leq c$. We claim that

$$\begin{aligned}\cosh \overline{c} &= \cosh a \ \cosh b - \sinh a \ \sinh b \ \Re\langle u|v\rangle \\ &\leq |\cosh a \ \cosh b - \sinh a \ \sinh b \ \langle u|v\rangle| = \cosh c.\end{aligned}$$

The equality in the first line is the law of cosines in $\mathbb{H}^2$, the equality in the second line is the law of cosines in $\mathbb{K}\mathrm{H}^n$, and the inequality relating the two lines is a special case of the simple observation that $|\lambda + \mu q| \geq |\lambda + \mu \Re q|$ for all $q \in \mathbb{K}$ and $\lambda, \mu \in \mathbb{R}$ (with equality if and only if $q \in \mathbb{R}$ or $\mu = 0$).

Thus $\overline{c} \leq c$, and we can apply remark (1.8). □

10.11 Exercise. The Alexandrov angle in $\mathbb{K}\mathrm{H}^n$ is equal to the hyperbolic angle (10.7).

We note one further consequence of the law of cosines. This describes the triangles in $\mathbb{K}\mathrm{H}^n$ that are isometric to triangles in the model spaces M^n_κ. Each of these special triangles actually lies in a complete isometrically embedded copy of the appropriate model space (cf. Theorem 10.16).

10.12 Proposition. *Let $\Delta \subset \mathbb{K}\mathrm{H}^n$ be the geodesic triangle considered in (10.8) and suppose that the vertex angle $\angle_C(A, B)$ is not equal to 0 or π. Then:*

(1) *The convex hull of Δ is isometric to the convex hull of its comparison triangle in M^2_{-1} if and only if $\langle u|v\rangle$ is real.*

(2) *The convex hull of Δ is isometric to the convex hull of its comparison triangle in M^2_{-4} if and only if u and v are linearly dependent over $\mathbb{K}$.*

Proof. In the notation of the proof of (10.10), we have $\overline{c} = c$ if and only if $\langle u|v\rangle$ is a real number. If the convex hull of Δ is isometric to that of a geodesic triangle in M^2_{-1} then obviously $\overline{c} = c$. Conversely, if $\overline{c} = c$ then according to (2.10) the convex hull of Δ is isometric to the convex hull of its comparison triangle in M^2_{-1}.

In order to prove the second assertion we compare the given triangle $\Delta(A, B, C)$ in $\mathbb{K}\mathrm{H}^n$ with a triangle $\tilde{\Delta} = (\tilde{A}, \tilde{B}, \tilde{C})$ in the plane M^2_{-4} of constant curvature -4. We choose $d(\tilde{C}, \tilde{A}) = a$, $d(\tilde{C}, \tilde{B}) = b$ and $\angle_{\tilde{C}}(\tilde{A}, \tilde{B}) = \gamma$, where $\gamma = \angle_C(A, B) = \arccos \Re\langle u|v\rangle$. Let $\tilde{c} = d(\tilde{A}, \tilde{B})$.

The law of cosines in M^2_{-4} states that

$$\cosh 2\tilde{c} = \cosh 2a \ \cosh 2b - \sinh 2a \ \sinh 2b \cos \gamma.$$

Squaring this expression and taking account of the identities $\cosh 2d = 2\cosh^2 d - 1$ and $\sinh 2d = 2\sinh d \cosh d$, we get:

$$\cosh^2 \tilde{c} = \cosh^2 a \ \cosh^2 b \ + \sinh^2 a \ \sinh^2 b \ - 2\cosh a \cosh b \sinh a \sinh b \ \cos \gamma.$$

On the other hand, the law of cosines in $\mathbb{K}\mathrm{H}^n$ (10.8) gives

$$\cosh^2 c = \\ \cosh^2 a \ \cosh^2 b \ + |\langle u|v\rangle|^2 \sinh^2 a \ \sinh^2 b \ - 2\cosh a \sinh a \cosh b \sinh b \ \cos \gamma.$$

As $|\langle u|v\rangle| \leq 1$, comparing the above expressions we see that $\tilde{c} \geq c$ with equality if and only if $|\langle u|v\rangle| = 1$.

Since $\langle u|u\rangle = 1$, if $v = u\lambda$ then $\lambda = \langle u|v\rangle$. And since the span of $\{u, v\}$ is contained in $x^\perp$, where the form Q is positive definite, $v = u\lambda$ if and only if

$\langle v - u\lambda \mid v - u\lambda\rangle = 0$. Expanding $\langle v - u\lambda \mid v - u\lambda\rangle$ and setting $\lambda = \langle u|v\rangle$, we get $\langle v|v\rangle + |\langle u|v\rangle|^2\langle u|u\rangle - 2|\langle u|v\rangle|^2$. And since $\langle v|v\rangle = \langle u|u\rangle = 1$, this expression is 0 if and only if $|\langle u|v\rangle| = 1$. Thus $\tilde{c} = c$ if and only if u and v are linearly dependent over $\mathbb{K}$.

In order to complete the proof, as in (1), we would now like to appeal to the analogue 2.10 of the Flat Triangle Lemma. However, since $\mathbb{K}\mathrm{H}^n$ is a CAT(-1) rather than a CAT(-4) space, we cannot appeal to (2.10) directly. Instead, we claim that if u and v are linearly dependent, then Δ is contained in an isometrically embedded copy of M^m_{-4}, where $m = \dim_{\mathbb{R}} \mathbb{K}$. To see this, we identify the one-dimensional $\mathbb{K}$-vector subspace of $T_x\mathbb{K}^{n,1}$ spanned by u with the tangent space at a point $o \in M^m_{-4}$ by means of an isometry $\phi : T_oM^m_{-4} \to \mathbb{K}u$. The conclusion of the preceding paragraph, $\tilde{c} = c$, shows that the map $\phi : M^m_{-4} \to \mathbb{K}\mathrm{H}^n$ sending the point a distance $t \in [0, \infty)$ along the geodesic in M^m_{-4} with initial vector v to the point a distance $t \in [0, \infty)$ along the geodesic in $\mathbb{K}^{n,1}$ with initial vector $\phi(v)$ is an isometry onto its image. $\square$

10.13 Remarks

(1) Consider (10.12) in the case $n = 1$. Given any $[x] \in \mathbb{K}\mathrm{H}^1$, since $x^\perp$ is 1-dimensional, any two vectors $u, v \in x^\perp$ must be dependent. Thus $\mathbb{K}\mathrm{H}^1$ is isometric to M^2_{-4} if $\mathbb{K} = \mathbb{C}$ and to M^4_{-4} if $\mathbb{K}$ is the quaternions.

(2) The argument in the last paragraph of the preceding proof shows that the $\mathbb{K}$-linear map $\mathbb{K}^{1,1} \to \mathbb{K}^{n,1}$ sending $(1, 0)$ to x and $(0, 1)$ to $u \in x^\perp$ induces an isometric embedding $j : \mathbb{K}\mathrm{H}^1 \to \mathbb{K}\mathrm{H}^n$.

(3) The proof of (10.12) shows that the curvature of $\mathbb{K}\mathrm{H}^n$ is bounded below by -4 and above by -1.

The Curvature of Distinguished Subspaces of $\mathbb{K}\mathrm{H}^n$

Continuing the theme of (10.12) and (10.13), our next goal is to characterize those subsets of $\mathbb{K}\mathrm{H}^n$ that are isometric to $\mathbb{H}^k = M^k_{-1}$ and M^k_{-4}. This requires the following definitions.

10.14 Definitions. Regard $\mathbb{K}^{n,1}$ as an $\mathbb{R}$-vector space in the natural way. An $\mathbb{R}$-vector subspace $V \subset \mathbb{K}^{n,1}$ is said to be *totally real* (with respect to Q) if $\langle u|v\rangle \in \mathbb{R}$ for all $u, v \in V$.

Let $p : \mathbb{K}^{n,1} \smallsetminus \{0\} \to \mathbb{K}\mathrm{P}^n$ be the natural projection. If $V \subset \mathbb{K}^{n,1}$ is a totally real subspace of (real) dimension $(k + 1)$, and if there exists $x \in V$ such that $\langle x|x\rangle < 0$, then $p(V \smallsetminus \{0\}) \cap \mathbb{K}\mathrm{H}^n$ is called a *totally real subspace of dimension k in* $\mathbb{K}\mathrm{H}^n$.

If $W \subset \mathbb{K}^{n,1}$ is a $\mathbb{K}$-vector subspace of dimension two that contains a vector v such that $\langle v|v\rangle = -1$, then $p(W \smallsetminus \{0\}) \cap \mathbb{K}\mathrm{H}^n$ is called a $\mathbb{K}$*-affine line*. ($\mathbb{K}$-affine subspaces of dimension k are defined similarly.)

In the course of the next proof we shall need the result of the following simple exercise in linear algebra.

10.15 Exercise. Let $V \subset \mathbb{K}^{n,1}$ be a totally real subspace of dimension $k+1$ and let $x \in V$ be such that $Q(x,x) = -1$. Prove that one can find a real basis $u_1, \dots, u_k, u_{k+1}$ for V such that $u_{k+1} = x$, $\langle u_i | u_i \rangle = 1$ for $i = 1, \dots, k$ and $\langle u_i | u_j \rangle = 0$ for all $i \neq j$. (Hint: Argue that $V = \langle x \rangle \oplus (x^\perp \cap V)$, as a real vector space, and use the fact that the restriction of Q to the second summand is positive definite.)

10.16 Theorem.

(1) *The totally real subspaces of dimension k in $\mathbb{K}\mathrm{H}^n$ are precisely those subsets which are isometric to the real hyperbolic space $\mathbb{R}\mathrm{H}^k = \mathbb{H}^k$.*

(2) *For $\mathbb{K} \neq \mathbb{R}$, the $\mathbb{K}$-affine lines in $\mathbb{K}\mathrm{H}^n$ are precisely those subsets which are isometric to the model space $M^{m_0}_{-4}$ of constant curvature -4, where $m_0 = \dim_{\mathbb{R}}\mathbb{K}$. Moreover, $\mathbb{K}\mathrm{H}^n$ does not contain any subsets isometric to M^m_{-4} for $m > m_0$.*

Proof. In order to prove (1), given a totally real subspace $V \subset \mathbb{K}^{n,1}$ of dimension $k+1$ containing a vector v with $\langle v, v \rangle = -1$, we choose a basis $u_1, \dots, u_{k+1}$ as in exercise (10.15). The $\mathbb{R}$-linear map $\mathbb{R}^{n,1} \to \mathbb{K}^{n,1}$ sending the standard basis of $\mathbb{R}^{n,1}$ to $u_1, \dots, u_{k+1}$ induces an isometry of $\mathbb{R}\mathrm{H}^k$ onto $p(V \smallsetminus \{0\}) \cap \mathbb{K}\mathrm{H}^n$.

For the converse, given an isometric embedding $j : \mathbb{R}\mathrm{H}^k \to \mathbb{K}\mathrm{H}^n$, we choose $k+1$ points $A_0, A_1, \dots, A_k \in \mathbb{R}\mathrm{H}^k$ in general position, write x to denote the element of $\mathbb{K}^{n,1}$ with $Q(x,x) = -1$ and $[x] = j(A_0)$, and then for $i = 1, \dots, n$ we define $u_i \in x^\perp$ to be the initial vector of the geodesic segment $j([A_0, A_i])$. Proposition 10.12 implies that the $\mathbb{R}$-subspace $V \subset \mathbb{K}^{n,1}$ generated by $x, u_1, \dots, u_k$ is totally real (since $\langle x | u_i \rangle = 0$ and $\langle u_i | u_j \rangle \in \mathbb{R}$ for all i, j) and that $p(V \smallsetminus \{0\}) \cap \mathbb{K}\mathrm{H}^n = j(\mathbb{R}H^k)$.

The preimage under $p : \mathbb{K}^{n,1} \to \mathbb{K}\mathrm{P}^n$ of a $\mathbb{K}$-affine line in $\mathbb{K}\mathrm{P}^n$ is a two dimensional $\mathbb{K}$-subspace for which we can choose a basis $\{x, u\}$ with $Q(x,x) = -1$, $Q(u,u) = 1$, and $\langle x | u \rangle = 0$. The $\mathbb{K}$-linear map $\mathbb{K}^{1,1} \to \mathbb{K}^{n,1}$ sending $(1,0)$ to x and $(0,1)$ to u induces an isometric embedding $j : \mathbb{K}\mathrm{H}^1 \to \mathbb{K}\mathrm{H}^n$ whose image is the given affine line. Thus the first assertion in (2) follows from (10.13). The assertion in the second sentence is an immediate consequence of (10.12). □

The Group of Isometries of $\mathbb{K}\mathrm{H}^n$

We describe the isometry group of $\mathbb{K}\mathrm{H}^n$ by means of a sequence of exercises.

Consider the group $GL(n+1, \mathbb{K})$ of invertible $(n+1, n+1)$-matrices with coefficients in $\mathbb{K}$. There is a natural left action of this on $\mathbb{K}^{n,1}$ by $\mathbb{K}$-linear automorphisms: the matrix $A = (a_{i,j})$ sends $x = (x_1, \dots, x_{n+1}) \in \mathbb{K}^{n,1}$ to $A.x = (\sum_{j=1}^{n+1} a_{1,j}x_j, \dots, \sum_{j=1}^{n+1} a_{n+1,j}x_j)$.

10.17 The group $O_{\mathbb{K}}(Q)$. Let $O_{\mathbb{K}}(Q)$ denote[35] the subgroup of $GL(n+1, \mathbb{K})$ that preserves the form Q defining $\mathbb{K}^{n,1}$, i.e. $A \in O_{\mathbb{K}}(Q)$ if and only if $\langle A.x | A.y \rangle = \langle x | y \rangle$ for all $x, y \in \mathbb{K}^{n+1}$.

[35] Classically, $O_{\mathbb{R}}(Q)$ is denoted $O(n,1)$ and $O_{\mathbb{C}}(Q)$ is denoted $U(n,1)$.

Note that the induced action of $O_{\mathbb{K}}(Q)$ on $\mathbb{K}\mathrm{P}^n$ preserves the subset $\mathbb{K}\mathrm{H}^n$ consisting of points $[x]$ with $\langle x|x\rangle < 0$. Note also that the action of $O_{\mathbb{K}}(Q)$ on $\mathbb{K}\mathrm{H}^n$ is by isometries (see 10.5).

The following description of the stabilizer of the point of $\mathbb{K}\mathrm{H}^n$ with homogeneous coordinates $(0, \ldots, 0, 1)$ is simply a matter of linear algebra.

10.18 Exercises

(1) The stabilizer in $O_{\mathbb{K}}(Q)$ of the point p_0 with homogeneous coordinates $(0, \ldots, 0, 1)$ acts transitively on the sphere of unit tangent vectors in $T_{p_0}\mathbb{K}\mathrm{H}^n$.

(2) The stabilizer of p_0 is isomorphic to the group $O(n) \times O(1)$ if $\mathbb{K} = \mathbb{R}$, to $U(n) \times U(1)$ if $\mathbb{K} = \mathbb{C}$, and to $Sp(n) \times Sp(1)$ if $\mathbb{K} = \mathbb{H}$. (Recall that $O(n)$, (resp. $U(n)$, $Sp(n)$) is the subgroup of $GL(n, \mathbb{K})$ leaving invariant the standard scalar product $(x|y) = \sum \overline{x_i} y_i$ on $\mathbb{K}^n$ in the case $\mathbb{K} = \mathbb{R}$ (resp. $\mathbb{C}$, $\mathbb{H}$).)

(Hint: If a matrix fixes p_0 then its last column and last row have zeroes everywhere except their last entry. Let $P_0 = (0, \ldots, 0, 1) \in \mathbb{K}^{n,1}$. In the product decomposition of the stabilizer of p_0, the first factor is the stabilizer of P_0 and the second factor is the pointwise stabilizer of $P_0^{\perp}$.)

(3) In exercise (2), describe the action of the second factor of the stabilizer of p_0 on $\mathbb{K}\mathrm{P}^n$.

With regard to part (1) of the preceding exercise, we remark that (10.12) shows that, in contrast to the real case, if $\mathbb{K} \neq \mathbb{R}$ then the stabilizer of p_0 does not act transitively on the set of pairs of orthogonal vectors in $T_{p_0}\mathbb{K}\mathrm{H}^n$. On the other hand:

10.19 Exercise. $O_{\mathbb{K}}(Q)$ acts transitively on the set of totally real subspaces in $\mathbb{K}\mathrm{H}^n$ of each dimension $k \leq n$. It also acts transitively on the set of $\mathbb{K}$-affine subspaces of each dimension $k \leq n$.

$O_{\mathbb{K}}(Q)$ also contains the one parameter subgroup

$$A(t) = \begin{pmatrix} \cosh t & 0 & \sinh t \\ 0 & I_{n-1} & 0 \\ \sinh t & 0 & \cosh t \end{pmatrix},$$

where I_{n-1} is the identity matrix of size $n - 1$.

10.20 Exercise

(1) Prove that $A(t)$ is a hyperbolic isometry whose axis passes through p_0 and has initial vector $(1, 0, \ldots, 0)$ at x_0. The translation length of $A(t)$ is t.

(2) Deduce (using (1) and 10.18(1)) that the action of $O_{\mathbb{K}}(Q)$ on $\mathbb{K}\mathrm{H}^n$ is transitive. (Compare with 10.28.)

(3) $A \in GL(\mathbb{K}, n + 1)$ belongs to $O_{\mathbb{K}}(Q)$ if and only if the columns $a_1, \ldots, a_n$ of A, viewed as vectors in $\mathbb{K}^{n+1}$, satisfy $\langle a_{n+1}|a_{n+1}\rangle = -1$, $\langle a_i|a_i\rangle = 1$ for $i = 1, \ldots, n$ and $\langle a_i|a_j\rangle = 0$ if $i \neq j$.

(4) Let J be the $(n+1, n+1)$-matrix that differs form I only in that its last diagonal entry is -1. Show that $A \in O_{\mathbb{K}}(Q)$ if and only if ${}^t\bar{A}JA = J$. Deduce that if $A \in O_{\mathbb{K}}(Q)$ then ${}^t\bar{A} \in O_{\mathbb{K}}(Q)$.

The kernel of the action of $O_{\mathbb{K}}(Q)$ on $\mathbb{K}\mathrm{H}^n$ is the subgroup formed by the scalar matrices λI, where $\lambda \in \mathbb{K}$ is central and of norm 1. We write $PO_{\mathbb{K}}(Q)$ to denote the quotient of $O_{\mathbb{K}}(Q)$ by this subgroup [36]. If $\mathbb{K}$ is $\mathbb{R}$ or the quaternions, then this subgroup is simply $\pm I$, while in the case $\mathbb{K} = \mathbb{C}$ it is isomorphic to $U(1)$.

It follows from Exercise 10.18 that the subgroup of $PO_{\mathbb{K}}(Q)$ fixing p_0 is isomorphic to $O(n)$, $U(n)$, or $(Sp(n) \times Sp(1))/\{I, -I\}$, according to whether $\mathbb{K}$ is $\mathbb{R}$, $\mathbb{C}$ or the quaternions.

10.21 Exercise

(1) $PO_{\mathbb{C}}(Q)$ is isomorphic to the quotient of $SU(n, 1)$[37] by the cyclic subgroup generated by $e^{2\pi i/(n+1)}I$.

(2) In addition to $PO_{\mathbb{C}}(Q)$, the group of isometries of $\mathbb{C}\mathrm{H}^n$ also contains the involution given in homogeneous coordinates by $(z_1, \dots, z_{n+1}) \mapsto (\bar{z}_1, \dots, \bar{z}_{n+1})$.

(3) The fixed point set of the involution in (2) is isometric to $\mathbb{R}\mathrm{H}^n$. (It is a totally real subspace of maximal dimension.)

(4) Prove that $PO_{\mathbb{C}}(Q)$ and the involution in (2) generate $\mathrm{Isom}(\mathbb{C}\mathrm{H}^n)$.

(5) Prove that if $\mathbb{K}$ is the quaternions and $n > 1$, then $PO_{\mathbb{K}}(Q) = \mathrm{Isom}(\mathbb{K}\mathrm{H}^n)$.

The Boundary at Infinity and Horospheres in $\mathbb{K}\mathrm{H}^n$

10.22 The Boundary at Infinity. In Chapter 8 we saw that if X is a complete non-positively curved Riemannian manifold, then ∂X is homeomorphic to $\mathbb{S}^{n-1}$, the $(n-1)$-sphere. Thus $\partial\mathbb{K}\mathrm{H}^n$ is a sphere of dimension $n-1$, $2n-1$ or $4n-1$, according to whether $\mathbb{K}$ is $\mathbb{R}$, $\mathbb{C}$ or the quaternions. There is a natural way to realize $\partial\mathbb{K}\mathrm{H}^n$ as a subset of $\mathbb{K}\mathrm{P}^n$: it is the subspace defined by the homogeneous polynomial equation $Q(x, x) = 0$. More precisely, $\mathbb{K}\mathrm{H}^n \subset \mathbb{K}\mathrm{P}^n$ is given by the equation $Q(x, x) < 0$ and the inclusion map $\mathbb{K}\mathrm{H}^n \hookrightarrow \mathbb{K}\mathrm{P}^n$ extends uniquely to a homeomorphism from $\mathbb{K}\mathrm{H}^n \cup \partial\mathbb{K}\mathrm{H}^n$ onto the subspace defined by $Q(x, x) \leq 0$.

It follows from (10.18) that the stabilizer of each point of $\mathbb{K}\mathrm{H}^n$ acts transitively on $\partial\mathbb{K}\mathrm{H}^n$.

10.23 Horospheres in Hyperbolic Spaces. In Chapter 8 we saw that if X is a CAT(0) space, $x \in X$ and $x_n \in X$ is a sequence such that $x_n \to \xi \in \partial X$, then the horosphere centred at ξ that passes through x is the limit of the spheres $S(x_n, r_n)$

[36] Classically, one writes $PO(n, 1)$ (resp. $PU(n, 1)$) when $\mathbb{K} = \mathbb{R}$ (resp. $\mathbb{K} = \mathbb{C}$).

[37] $SU(n, 1) \subset O_{\mathbb{C}}(Q)$ consists of those matrices which preserve Q and have determinant 1.

where $r_n = d(x, x_n)$. In the present setting, spheres about $[x_n] \in \mathbb{K}\mathrm{H}^n$ are the level sets of the function

$$\rho_n([x]) = \frac{\langle x|x_n\rangle\langle x_n|x\rangle}{\langle x|x\rangle}.$$

If $[x_n] \to [y] \in \partial\mathbb{K}\mathrm{H}^n$, then the functions $\rho_n([x])$ converge and the horospheres centred at $[y]$ have the form

$$H_{t,[y]} = \{[x] : t\langle x|x\rangle = \langle x|y\rangle\langle y|x\rangle\},$$

where $t < 0$. An alternative description of these horospheres is given in (10.28) and (10.29).

Other Models for $\mathbb{K}\mathrm{H}^n$

10.24 The Ball Model. We describe an alternative model for $\mathbb{K}\mathrm{H}^n$ corresponding to the model of real hyperbolic space described in (I.6.2). The point set of this model, which we denote $\mathbb{K}\mathrm{B}^n$, is obtained by taking the subset of $\mathbb{K}^n$ that consists of elements $x = (x_1, \dots, x_k)$ such that $\sum_1^n \bar{x}_i x_i < 1$. The distance between points $x, y \in \mathbb{K}\mathrm{B}^n$ is given by the formula

$$\cosh^2 d(x, y) = \frac{(1-(x|y))(1-(y|x))}{(1-(x|x))(1-(y|y))},$$

where $(x|y) = \sum_1^n \bar{x}_i y_i$ is the standard scalar product on $\mathbb{K}^n$.

We leave the reader to check that the map which assigns to $x = (x_1, \dots, x_n) \in \mathbb{K}\mathrm{B}^n$ the point of $\mathbb{K}\mathrm{H}^n$ with homogeneous coordinates $(x_1, \dots, x_n, 1)$ is an isometry and that this map extends to a homeomorphism from the closed unit ball in $\mathbb{K}^n$ to $\mathbb{K}\mathrm{H}^n \cup \partial\mathbb{K}\mathrm{H}^n$.

10.25 The Parabolic Model. The main benefit of the parabolic model is that, like the upper half space model for $\mathbb{H}^n$, it permits a convenient and explicit expression for the horocyclic coordinates associated to a point at infinity (10.29) and the associated subgroup AN which stabilizes the given point (see 10.28). (This gives a way of describing horospheres in $\mathbb{K}\mathrm{H}^n$ that is more amenable to calculation than the description given in (10.23).)

Consider the form Q' on $\mathbb{K}^{n+1}$ given by

$$Q'(x, y) = \langle x|y\rangle' = -\bar{x}_1 y_{n+1} - \bar{x}_{n+1} y_1 + \sum_{i=2}^{n} \bar{x}_i y_i.$$

Q' is equivalent to Q by a linear change of coordinates, namely $x_i \mapsto x_i'$ where $x_1' = (x_1 + x_{n+1})/\sqrt{2}$, $x_{n+1}' = (x_1 - x_{n+1})/\sqrt{2}$, and $x_i = x_i'$ if $i \neq 1, n+1$. The underlying set of the parabolic model for $\mathbb{K}\mathrm{H}^n$ is the subset of $\mathbb{K}\mathrm{P}^n$ defined by $\{[x] \mid Q'(x,x) < 0\}$, and the metric is defined by the formula in (10.5) with Q replaced by Q'.

Recall that if $x \in \mathbb{K}P^n$ has homogeneous coordinates $(x_1, \dots, x_n, 1)$ then its *non-homogeneous* coordinates are, by definition, $(x_1, \dots, x_n)$. These coordinates give the classical identification of $\mathbb{K}^n$ with a subset of $\mathbb{K}P^n$.

For the purposes of calculation it is useful to describe the parabolic model in non-homogeneous coordinates. The subset of $\mathbb{K}P^n$ defined in homogeneous coordinates by $Q'(x, x) < 0$ is precisely the set of points with non-homogeneous coordinates $(x_1, \dots, x_n)$ satisfying the inequality

$$(*) \qquad \sum_{i=2}^{n} \bar{x}_i x_i < x_1 + \bar{x}_1 = 2\Re x_1.$$

Let ∞ denote the point of $\mathbb{K}P^n$ with homogeneous coordinates $(1, 0, \dots, 0)$. The boundary at infinity is

$$\left\{ x \in \mathbb{K}^n \mid \sum_{i=2}^{n} \bar{x}_i x_i = x_1 + \bar{x}_1 \right\} \cup \{\infty\}.$$

∞ is a convenient point to use when giving explicit descriptions of the horocyclic coordinates associated to a point at infinity. (Focussing on ∞ involves no loss of generality because $\text{Isom}(\mathbb{K}H^n)$ acts transitively on $\partial\mathbb{K}H^n$.)

Horocyclic Coordinates and Parabolic Subgroups for $\mathbb{K}H^n$

By definition, $\infty \in \partial\mathbb{K}H^n$ is the endpoint of the geodesic ray given in homogeneous coordinates by $t \mapsto c_0(t) = (e^t, 0, \dots, 0, e^{-t})$. We shall need the following lemma in order to determine the subgroup of $\text{Isom}(\mathbb{K}H^n)$ that fixes all of the horospheres centred at ∞.

10.26 Lemma. *Let X be a* CAT(0) *space and let G be a group acting by isometries on X. Suppose that $h \in G$ leaves invariant a geodesic line $c : \mathbb{R} \to X$ and that $h.c(t) = c(t + a)$ where $a > 0$. Let $x_0 = c(0)$ and let $N \subseteq G$ be the set of elements $g \in G$ such that $h^{-n} g h^n.x_0 \to x_0$ as $n \to \infty$. Then*

(1) *N is a subgroup.*

(2) *N fixes $c(\infty) \in \partial X$ and leaves invariant the Busemann function associated to c.*

Proof. In order to prove (1), given $g, g' \in G$ we write $g_n = h^{-n} g h^n$ and $g'_n = h^{-n} g' h^n$ and calculate:

$$\begin{aligned} d(g_n^{-1} g'_n.x_0, x_0) &= d(g'_n.x_0, g_n.x_0) \\ &\le d(g'_n.x_0, x_0) + d(x_0, g_n.x_0); \end{aligned}$$

the right side converges to 0 as $n \to \infty$.

For (2) we must prove that for any $x \in X$ and $g \in N$ we have $b_c(x) = b_c(g.x)$. Remembering that $x_0 = c(0)$, we have:

$$b_c(g.x) - b_c(x) = \lim_{n\to\infty} d(g.x, h^n.x_0) - d(x, h^n.x_0),$$

and

$$\begin{aligned}|d(g.x, h^n.x_0) - d(x, h^n.x_0)| &= |d(x, g^{-1}h^n.x_0) - d(x, h^n.x_0)| \\ &\leq d(g^{-1}h^n.x_0, h^n.x_0) = d(h^{-n}g^{-1}h^n.x_0, x_0),\end{aligned}$$

which tends to zero as n tends to infinity. □

Let $O_{\mathbb{K}}(Q')$ be the subgroup of $GL(n+1, \mathbb{K})$ that leaves invariant the form Q'. The natural action of $O_{\mathbb{K}}(Q')$ on $\mathbb{K}\mathrm{P}^n$ preserves the parabolic model of $\mathbb{K}\mathrm{H}^n$. Consider the 1-parameter subgroup of $O_{\mathbb{K}}(Q')$ formed by the elements

$$A(t) = \begin{pmatrix} e^t & 0 & 0 \\ 0 & I_{n-1} & 0 \\ 0 & 0 & e^{-t} \end{pmatrix}.$$

This subgroup leaves invariant the geodesic line given in homogeneous coordinates by $c_0(t) = (e^t, 0, \dots, e^{-t})$, and $A(t).c_0(0) = c_0(t)$. (Recall that $c_0(\infty) = \infty$ and note, for future reference, that $c_0(t)$ has non-homogeneous coordinates $(e^{2t}, 0, \dots, 0)$.)

We shall apply the preceding lemma with $A(1)$ in the rôle of h and c_0 in the rôle of c. Let $N \subset O_{\mathbb{K}}(Q')$ be the subgroup formed by the matrices ν such that $\lim_{t\to\infty} A(-t)\nu A(t)$ is the unit matrix. Calculating $A(-t)\nu A(t)$ and noting that the off-diagonal entries must tend to 0 and the diagonal entries must tend to 1, one sees that N consists of matrices of the form

$$\nu = \begin{pmatrix} 1 & M_{12} & M_{13} \\ 0 & I_{n-1} & M_{23} \\ 0 & 0 & 1 \end{pmatrix}.$$

One further calculates (exercise!) that $\langle \nu(x) | \nu(y)\rangle' = \langle x|y\rangle'$ for all x, y with $\langle x|x\rangle' < 0$ and $\langle y'|y'\rangle' < 0$ if and only if M_{12} (which is a $(1, n-1)$-matrix) and M_{23} (which is an $(n-1, 1)$-matrix) satisfy

(♠) $$M_{23} = {}^t\overline{M}_{12} \text{ and } M_{12}M_{23} = M_{13} + \overline{M}_{13},$$

where $\overline{M}$ denotes the standard conjugation on $\mathbb{K}$ and ${}^tM$ is the transpose of M.

Let A denote the 1-parameter subgroup $\{A(t) : t \in \mathbb{R}\}$.

10.27 Lemma.

(1) *$AN := \{A(t)\nu \mid t \in \mathbb{R}, \nu \in AN\}$ is a subgroup of $O_{\mathbb{K}}(Q')$.*

(2) *N is normal in AN.*

(3) *If $\mathbb{K} = \mathbb{R}$ then N is abelian, and if $\mathbb{K}$ is $\mathbb{C}$ or the quaternions then N is nilpotent but is not abelian if $n > 1$.*

Proof. These are simple calculations. □

The following proposition is a special case of a general phenomenon for symmetric spaces (cf. 10.50 and 10.69).

10.28 Proposition.
(1) *The subgroup AN acts simply transitively on $\mathbb{K}\mathrm{H}^n$.*
(2) *The orbits of N are the horospheres centred at the point ∞.*
(3) *The geodesic lines in the parabolic model that are asymptotic to c_0 are precisely the geodesics $t \mapsto \nu.c_0(t)$, where $\nu \in N$ and c_0 is given in homogeneous coordinates by $c_0(t) = (e^t, 0, \dots, 0, e^{-t})$.*

Proof. Recall that ∞ is the endpoint of c_0 in the forwards direction. We have already seen (Lemma 10.26) that $\nu.\infty = \infty$ for all $\nu \in N$ and that the orbits of N are contained in the horospheres centred at ∞. It is clear that A acts simply transitively on $c_0(\mathbb{R})$ and hence on the set of horospheres centred at ∞. Thus, in order to prove (1) and (2) it suffices to show that given an arbitrary point x, with non-homogeneous coordinates $(x_1, \dots, x_n)$ say, there exists a unique $t \in \mathbb{R}$ and a unique $\nu \in N$ such that x is the image under $\nu A(t)$ of the point given in non-homogeneous coordinates by $x_0 = (1, 0, \dots, 0)$.

Writing ν in block form as above, $\nu A(t).x_0 = x$ becomes the system of equations:

$$x_1 = e^{2t} + M_{13}$$
$$X_2 = M_{23}$$

where ${}^tX_2 = (x_2, \dots, x_n)$. Thus X_2 uniquely determines M_{23} and hence M_{12} and $\Re M_{13} = \frac{1}{2}{}^t\overline{X}_2X_2$ (by (♠)). Moreover, ${}^t\overline{X}_2X_2$ is, by definition, equal to the left side of the equation (*) displayed in (10.25), and hence it is less than $2\Re x_1$. There therefore exists a unique $t \in \mathbb{R}$ such that $\Re x_1 = e^{2t} + \Re M_{13}$, and M_{13} is uniquely determined by $x_1 = e^{2t} + M_{13}$. Thus the above system of equations has a unique solution $(t, \nu) \in \mathbb{R} \times N$. □

10.29 Horocyclic Coordinates in the Parabolic Model. Let x be a point in the parabolic model that lies on the horosphere through the point x_t with homogeneous coordinates $(e^t, 0, \dots, e^{-t})$ and let ν be the unique element of N such that $\nu.x_t = x$. Let the submatrices M_{ij} of ν be as above.

The *horocyclic coordinates* of x are, by definition, the entries of the submatrix M_{23} together with the purely non-real part of A_{13}. Thus x is assigned $2n-1$ (resp. $4n-1$) real coordinates when $\mathbb{K}$ is $\mathbb{C}$ (resp. the quaternions). According to (10.28), these coordinates together with t uniquely specify x.

The Symmetric Space $P(n, \mathbb{R})$

We turn now to the study of higher-rank symmetric spaces. For the reasons explained in the introduction, we shall concentrate our attention on $P(n, R)$, the space of positive-definite, symmetric (n, n)-matrices with real coefficients.

Our first task will be to metrize $P(n, R)$; we shall do so by describing a Riemannian metric on it. We shall then describe the natural action of $GL(n, \mathbb{R})$ on $P(n, R)$ and prove that this action is by (Riemannian) isometries. First, though, we pause to fix notation and remind the reader of some basic definitions.

10.30 Notation. We shall write $M(n, \mathbb{R})$ to denote the algebra of (n, n)-matrices with real coefficients. $GL(n, \mathbb{R})$ will denote the group of invertible (n, n)-matrices. The action of matrices on $\mathbb{R}^n$ will be on the left (with elements of $\mathbb{R}^n$ viewed as column vectors). The transpose of a matrix A will be denoted ${}^tA$.

Let $S(n, \mathbb{R})$ be the vector subspace of $M(n, \mathbb{R})$ consisting of symmetric matrices, and let $P(n, \mathbb{R}) \subset S(n, \mathbb{R})$ be the open cone of positive-definite matrices. We remind the reader that a matrix A is symmetric if and only if for all vectors $v, w \in \mathbb{R}^n$ we have $(Av, w) = (v, Aw)$, where $(\ , \ )$ is the usual scalar product on $\mathbb{R}^n$; and A is positive definite if, in addition, $(Av, v) > 0$ for every $v \neq 0$.

Given a symmetric matrix A one can always find an orthonormal basis for $\mathbb{R}^n$ consisting of eigenvectors for A. More precisely, there exists an orthogonal matrix $O \in SO(n)$ of determinant one such that ${}^tOAO$ is diagonal. The entries $\lambda_1, \dots, \lambda_n$ of this diagonal matrix are all positive if and only if A is positive definite (they are the eigenvalues of A).

$P(n, R)$ as a Riemannian Manifold

10.31 The Riemannian Metric on $P(n, \mathbb{R})$. As $P(n, \mathbb{R})$ is an open set of $S(n, \mathbb{R})$, it is naturally a differentiable manifold of dimension $n(n+1)/2$. The tangent space $T_pP(n, \mathbb{R})$ at a point p is naturally isomorphic (via translation) to $S(n, \mathbb{R})$. On $T_pP(n, \mathbb{R})$ we define a scalar product by:

$$(X|Y)_p = \mathrm{Tr}(p^{-1}Xp^{-1}Y),$$

where $X, Y \in T_pP(n, \mathbb{R}) \cong S(n, \mathbb{R})$ and $\mathrm{Tr}(A)$ is the trace of a matrix A. (If $X \in S(n, \mathbb{R})$ has entries x_{ij}, then $(X|X)_I = \|X\|^2 = \sum x_{ij}^2$.)

This formula defines a Riemannian metric on $P(n, \mathbb{R})$.

10.32 The Action of $GL(n, \mathbb{R})$ on $P(n, \mathbb{R})$. The group $GL(n, \mathbb{R})$ acts on $S(n, \mathbb{R})$ and $P(n, \mathbb{R})$ according to the rule

$$g.A := gA\,{}^tg,$$

where $g \in GL(n, \mathbb{R})$ and $A \in S(n, \mathbb{R})$. This action leaves $P(n, \mathbb{R})$ invariant.

10.33 Proposition. *Consider the action of $GL(n, \mathbb{R})$ on $P(n, \mathbb{R})$.*
(1) *The action is transitive.*
(2) *The action is by Riemannian isometries.*
(3) *The stabilizer of $I \in P(n, \mathbb{R})$ is $O(n)$.*
(4) $\{\pm I\} \subset GL(n, \mathbb{R})$ *acts trivially on $P(n, \mathbb{R})$ and $GL(n, \mathbb{R})/\{\pm I\}$ acts effectively.*

Note that in the light of this proposition we may identify $P(n, \mathbb{R})$ with the homogeneous space $GL(n, \mathbb{R})/O(n)$.

Proof. Given $p \in P(n, \mathbb{R})$, there exists $O \in SO(n)$ such that $D = {}^tOpO$ is a diagonal matrix D with positive entries λ_i. Let $p^{1/2} = (O\, D^{1/2}\, {}^tO)$, where $D^{1/2}$ is the diagonal matrix with entries $\sqrt{\lambda_i}$. Then $p = p^{1/2}.I$. This proves (1).

Part (3) is obvious and (4) is an easy exercise. To prove (2) one calculates: given $g \in GL(n, \mathbb{R})$ and $X, Y \in T_pP(n, \mathbb{R})$, the derivative of g at p maps X onto $g.X = gX{}^tg \in T_{g.p}P(n, \mathbb{R})$, so

$$\begin{aligned}(g.X|g.Y)_{g.p} &= \operatorname{Tr}({}^tg^{-1}p^{-1}g^{-1}gX\,{}^tg\,{}^tg^{-1}p^{-1}g^{-1}gY\,{}^tg)\\ &= \operatorname{Tr}(p^{-1}Xp^{-1}Y) = (X|Y)_p.\end{aligned}$$

□

Recall that a *symmetric space* is a connected Riemannian manifold M where for each point $p \in M$ there is a Riemannian isometry σ_p of M such that $\sigma_p(p) = p$ and the differential of σ_p at p is multiplication by -1.

10.34 Proposition. *$P(n, R)$ is a symmetric space.*

Proof. Given $p \in P(n, \mathbb{R})$, the required symmetry σ_p is $q \mapsto pq^{-1}p$. This map is the composition of $\sigma_I : q \mapsto q^{-1}$ and $q \mapsto p.q$. We have already seen that the latter is an isometry. To see that σ_I is an isometry[38], first note that its derivative at q sends $X \in S(n, \mathbb{R}) = T_q(P(n, \mathbb{R}))$ (the initial vector of the curve $t \mapsto q + tX$) to $\overline{X} := q^{-1}Xq^{-1} \in S(n, \mathbb{R}) = T_{q^{-1}}P(n, \mathbb{R})$ (the initial vector of the curve $t \mapsto (q + tX)^{-1}$). Thus we have

$$(\overline{X}|\overline{Y})_{q^{-1}} = \operatorname{Tr}(q\overline{X}q\overline{Y}) = \operatorname{Tr}(Xq^{-1}qq^{-1}Yq^{-1}) = \operatorname{Tr}(q^{-1}Xq^{-1}Y) = (X|Y)_q,$$

and hence σ_I is an isometry. It is obvious that the differential of σ_I at I is multiplication by -1, and hence so is the differential of σ_p at p, because σ_p is the conjugate of σ_I by $q \mapsto p^{1/2}.q$. □

10.35 Remark. In a symmetric space the composition of two symmetries $\sigma_q \circ \sigma_p$ is called a *transvection*; it acts as a translation on any locally geodesic line containing p and q. In the preceding proof we saw that the action of $p \in P(n, \mathbb{R})$, viewed as an

[38] Under the natural identification $P(n, \mathbb{R}) = GL(n, \mathbb{R})/O(n)$, the symmetry σ_I is induced by the map $g \mapsto {}^tg^{-1}$.

element of $GL(n, \mathbb{R})$, is the composition $\sigma_p \circ \sigma_I$, so the action of $P(n, \mathbb{R}) \subset GL(n, \mathbb{R})$ is by transvections.

The Exponential Map exp: $M(n, \mathbb{R}) \to GL(n, \mathbb{R})$

Our next major goal is to prove that $P(n, \mathbb{R})$ is a CAT(0) space. Since we have described the metric in terms of a Riemannian structure, the natural way to prove this is to use (1A.2) from the appendix to Chapter 1. This requires that we understand the divergence properties of geodesics issuing from a point in $P(n, \mathbb{R})$. In order to gain such an understanding we must examine the exponential map.

We have a concrete interpretation of $T_I P(n, \mathbb{R})$ as $S(n, \mathbb{R})$, and the exponential map can be interpreted in a correspondingly concrete manner, namely as the familiar operation of exponentiation for matrices, $\exp : M(n, \mathbb{R}) \to GL(n, R)$, where

$$\exp A := \sum_{k=0}^{\infty} A^n/n!\,.$$

We shall need the following properties of exp.

10.36 Lemma.

(1) $\exp({}^t\!A) = {}^t(\exp A)$.

(2) *If* $AB = BA$, *then* $A(\exp B) = (\exp B)A$ *and* $\exp A \exp B = \exp(A + B)$. *In particular* $\exp(-A) = (\exp A)^{-1}$.

(3) $\exp(gAg^{-1}) = g\exp(A)g^{-1}$ *for all* $g \in GL(n, \mathbb{R})$.

(4) $\frac{d}{dt}(\exp tA) = (\exp tA)A$ *for all* $t \in \mathbb{R}$.

(5) *The map* exp *is differentiable.*

(6) exp *maps* $S(n, \mathbb{R})$ *bijectively onto* $P(n, \mathbb{R})$.

Proof. We shall only prove (6). It is obvious that if A is symmetric then $\exp A$ is symmetric. It is also obvious that every diagonal matrix in $P(n, \mathbb{R})$ is exp of a diagonal matrix, so since every $p \in P(n, \mathbb{R})$ is $O.\Delta$ (action of 10.32) for some $O \in SO(n)$ and some diagonal Δ, the surjectivity of exp follows from (3).

A symmetric matrix A is uniquely determined by its eigenvalues and the associated direct sum decomposition $\bigoplus_{i=1}^k E_i$ of $\mathbb{R}^n$ into the eigenspaces of A. The decomposition associated to $\exp A$ has exactly the same summands, and if the action of A on E_i is multiplication by λ_i then the action of $\exp A$ on E_i is multiplication by $\exp(\lambda_i)$. Thus one can recover the action of A on $\mathbb{R}^n$ from that of $\exp A$, and hence exp is injective on $S(n, \mathbb{R})$. □

10.37 Remark. It follows from the preceding lemma that for every $p \in P(n, \mathbb{R})$ there exists a unique symmetric matrix X such that $\exp X = p$. Let $p^{1/2} = \exp(X/2)$; this is the unique $q \in P(n, \mathbb{R})$ such that $q^2 = p$, and we shall see shortly that it is the midpoint of the unique geodesic joining I to p.

For future reference (10.40), note that $\exp(X/2)$ acts on $P(n, \mathbb{R})$ as the transvection $\sigma_{p^{1/2}} \circ \sigma_I$ which sends p to I.

The following exercises contain some further useful facts concerning the exponential map; some of these facts will be needed in the sequel.

10.38 Exercises

(1) For all $A, B \in M(n, \mathbb{R})$, we have

$$(\exp A)B(\exp -A) = (\exp(\mathrm{ad}_A))(B),$$

where ad_A is the endomorphism of the vector space $M(n, \mathbb{R})$ defined by

$$\mathrm{ad}_A(B) = [A, B] := AB - BA.$$

Hint: Express ad_A as $L_A - R_A$, where L_A (resp. R_A) is the endomorphism of $M(n, \mathbb{R})$ mapping C to AC (resp. CA). Note that $L_A R_A = R_A L_A$, hence

$$(L_A - R_A)^k = \sum_{p+q=k} \frac{k!}{p!q!} L_A^p R_{-A}^q.$$

(2) Verify that the formula $(A|B) = \mathrm{Tr}(A\,{}^tB)$ defines a scalar product on $M(n, \mathbb{R})$. Show that if $X \in S(n, \mathbb{R})$, then ad_X is a self-adjoint operator on $M(n, \mathbb{R})$, i.e. for any $A, B \in M(n, \mathbb{R})$, we have

$$(\mathrm{ad}_X A|B) = (A|\mathrm{ad}_X B).$$

(3) Given $X \in S(n, \mathbb{R})$, consider the endomorphism τ_X of $M(n, \mathbb{R})$ defined by the formula

$$\tau_X(Y) = \frac{d}{dt} \exp(-X/2)\ \exp(X + tY)\ \exp(-X/2)\Big|_{t=0}.$$

Prove that τ_X is self-adjoint.

(Hint: You may first wish to show that

$$\tau_X(Y) = \sum_{k=1}^{\infty} \frac{1}{k!} \sum_{p+q=k-1} \exp(-X/2) X^p Y X^q \ \exp(-X/2)$$

hence

$$(\tau_X(Y)|Z) = \sum_{k=1}^{\infty} \sum_{p+q=k-1} \mathrm{Tr}(\exp(-X/2) X^p Y X^q \ \exp(-X/2)\,{}^tZ).$$

Then use the fact that $\mathrm{Tr}(AB) = \mathrm{Tr}(BA)$.)

(4) Show that $\exp \mathrm{Tr}(A) = \det(\exp A)$.

$P(n, \mathbb{R})$ is a CAT(0) Space

We are now in a position to prove:

10.39 Theorem. *$P(n, \mathbb{R})$ is a proper* CAT(0) *space.*

As we remarked earlier, since we have described the metric on $P(n, \mathbb{R})$ by giving the associated Riemannian structure, the natural way to prove this theorem is to apply (1A.2). In order to do so we must show that the restriction of exp to $S(n, \mathbb{R})$ satisfies the conditions on the map e in (1.A.1). This is the object of the next lemma, for which we require the following notation: given $X \in S(n, \mathbb{R})$, we define an endomorphism $\frac{\sinh \mathrm{ad}_{X/2}}{\mathrm{ad}_{X/2}}$ of $S(n, \mathbb{R})$ by

$$\frac{\sinh \mathrm{ad}_{X/2}}{\mathrm{ad}_{X/2}}(Y) := \sum_{k=0}^{\infty}(-1)^k \frac{\mathrm{ad}_{X/2}^{2k}}{(2k+1)!}(Y).$$

10.40 Lemma.

(1) *Consider the differential τ_X at $X \in S(n, \mathbb{R})$ of the composition of* exp *and the transvection sending* $\exp X$ *to I (see (10.37)). By definition, τ_X is the endomorphism of $S(n, \mathbb{R})$ given by*

$$\tau_X(Y) = \frac{d}{dt}\big(\exp(-X/2)\exp(X+tY)\exp(-X/2)\big)\Big|_{t=0}.$$

We claim that

$$\tau_X = \frac{\sinh \mathrm{ad}_{X/2}}{\mathrm{ad}_{X/2}}.$$

(2) *τ_X does not diminish the norm of any tangent vector. More precisely, for all $Y \in S(n, \mathbb{R})$ we have $|\tau_X(Y)| \geq |Y|$ with equality if and only if $XY = YX$. (Recall that $|Y|^2 = \mathrm{Tr}(Y^2)$.)*

Proof. First we prove (1). Given $Y \in S(n, \mathbb{R})$ we consider the curve $t \mapsto X(t) = X + tY$. By differentiating the identity

$$X(t)\,(\exp X(t)) = (\exp X(t))\,X(t)$$

at $t = 0$, writing $E_0 = \frac{d}{dt}\exp X(t)|_{t=0}$ and noting that $Y = \frac{d}{dt}X(t)|_{t=0}$ and $X = X(0)$, we get:

$$Y\,(\exp X) + X\,E_0 = E_0\,X + (\exp X)\,Y.$$

That is,

$$X\,E_0 - E_0\,X = (\exp X)\,Y - Y\,(\exp X).$$

By definition, $\tau_X(Y) = \exp(-X/2)\,E_0\,\exp(-X/2)$. Thus, multiplying both sides of the above equation on both the left and the right by $\exp(-X/2)$, and noting that $\exp(-X/2)$ commutes with X, we get:

$$X\tau_X(Y) - \tau_X(Y)X = \exp(X/2)\, Y \exp(-X/2) - \exp(-X/2)\, Y \exp(X/2).$$

By definition, $2\sinh(\mathrm{ad}_{X/2}) = \exp(\mathrm{ad}_{X/2}) - \exp(\mathrm{ad}_{-X/2})$. Therefore, applying exercise 10.38(1) to each term on the right-hand side of the above equation we get:

$$X\tau_X(Y) - \tau_X(Y)X = 2\sinh(\mathrm{ad}_{X/2})(Y).$$

Thus

$$(\clubsuit) \qquad \mathrm{ad}_{X/2} \circ \tau_X = \mathrm{ad}_{X/2} \frac{\sinh \mathrm{ad}_{X/2}}{\mathrm{ad}_{X/2}}.$$

Let $F := \tau_X - (\sinh \mathrm{ad}_{X/2})/\mathrm{ad}_{X/2}$. We shall show that $F = 0$ as an endomorphism of $M(n,\mathbb{R})$. Both $\mathrm{ad}_{X/2}$ and τ_X are self-adjoint operators on $M(n,\mathbb{R})$ (exercises 10.38 (3 and 4)), hence F is self-adjoint. Let $N \subset M(n,\mathbb{R})$ be the kernel of $\mathrm{ad}_{X/2}$; it is the vector subspace $\{Y \in M(n,\mathbb{R}) : XY = YX\}$. By ($\clubsuit$) we know that F maps $M(n,\mathbb{R})$ to N. It follows that the restriction of F to $N^\perp$ (the orthogonal complement of N) is 0, because if $x \in N^\perp$ then for all $y \in M(n,\mathbb{R})$ we have $(A(x)|y) = (x|A(y)) = 0$. It only remains to check that F vanishes on N. But this is obvious, because if $XY = YX$ then $\tau_X(Y) = Y$ and $(\sinh \mathrm{ad}_{X/2})/\mathrm{ad}_{X/2}(Y) = Y$.

In order to prove (2) we choose an orthonormal basis $e_1, \dots, e_n$ for $\mathbb{R}^n$ consisting of eigenvectors for X; say $(X/2)e_i = \lambda_i e_i$. Let $E_{ij} \in M(n,\mathbb{R})$ be the endomorphism defined by $E_{ij}e_k = \delta_{jk}e_i$, where $\delta_{jk} = 1$ if $j = k$ and zero otherwise. (The elements E_{ij} form an orthonormal basis of $M(n,\mathbb{R})$.) We have $X/2 = \sum_{i=1}^n \lambda_i E_{ii}$, so since $E_{ij}E_{kl} = \delta_{jk}E_{il}$, it follows that $\mathrm{ad}_{X/2}(E_{ij}) = (\lambda_i - \lambda_j)E_{ij}$, and hence

$$\tau_X(E_{ij}) = \frac{\sinh(\lambda_i - \lambda_j)}{\lambda_i - \lambda_j} E_{ij}.$$

If $Y \in S(n,\mathbb{R})$, then $Y = \sum_{i,j=1}^n y_{ij}E_{ij}$ where $y_{ij} = y_{ji}$, and we have

$$\tau_X(Y) = \sum_{i,j=1}^n y_{ij} \frac{\sinh(\lambda_i - \lambda_j)}{\lambda_i - \lambda_j} E_{ij}.$$

Therefore

$$|\tau_X(Y)|^2 = \mathrm{Tr}(\tau_X(Y)^2) = \sum_{i,j=1}^n y_{ij}^2 \frac{\sinh(\lambda_i - \lambda_j)^2}{(\lambda_i - \lambda_j)^2} \geq \sum_{i,j=1}^n y_{ij}^2 = |Y|^2,$$

because $|(\sinh\lambda)/\lambda| \geq 1$. One gets equality in this expression if and only if $y_{ij} = 0$ whenever $\lambda_i \neq \lambda_j$. And this happens if and only if the action of Y on $\mathbb{R}^n$ leaves the eigenspaces of X invariant, which is the case if and only if $XY = YX$. □

Proof of 10.39. In Lemma 10.36 we saw that $\exp : S(n,\mathbb{R}) \to P(n,\mathbb{R})$ is a diffeomorphism, and the preceding lemma shows that at the point $I \in P(n,\mathbb{R})$, the map exp satisfies the conditions required of e in Lemma 1A.1 (with $\varepsilon > 0$ arbitrary). Since $GL(n,\mathbb{R})$ acts transitively by isometries on $P(n,\mathbb{R})$, it follows from (1A.2) that

$P(n, \mathbb{R})$ is a CAT(0) space. And since exp does not decrease distances, $P(n, \mathbb{R})$ is proper. Thus Theorem 10.39 is proved. □

As a consequence of (10.39) we have:

10.41 Corollary. *Every compact subgroup of* $GL(n, \mathbb{R})$ *is conjugate to a subgroup of* $O(n)$.

Proof. If $K \subset GL(n, \mathbb{R})$ is compact, then every K-orbit in $P(n, \mathbb{R})$ is compact and hence bounded. It follows that K is contained in the stabilizer of the circumcentre p of this orbit (2.7). The action of $GL(n, \mathbb{R})$ is transitive, so there exists $g \in GL(n, \mathbb{R})$ such that $g.I = p$. The stabilizer of I is $O(n)$ (10.33) and therefore $g^{-1}Kg \subset O(n)$. □

We also note two consequences of Lemma 10.40.

10.42 Corollary.
(1) *The geodesic lines* $c : \mathbb{R} \to P(n, \mathbb{R})$ *with* $c(0) = p$ *are the maps* $c(t) = g\,(\exp tX)\,{}^t g$, *where* $X \in S(n, \mathbb{R})$ *with* $\mathrm{Tr}(X^2) = 1$ *and* $p = g\,{}^t g$.
(2) *If* $p = \exp X$ *then* $d(I, p)^2 = \mathrm{Tr}(X^2) = |X|^2$.
(3) *More generally, the restriction of* exp *to any vector subspace of* $S(n, \mathbb{R})$ *consisting of commuting matrices is an isometry.*

10.43 Exercise. Let $A \subset GL(n, \mathbb{R})$ be the group of diagonal matrices with positive entries on the diagonal. The action of A on $P(n, \mathbb{R})$ is by transvections and $I \in \mathrm{Min}(A)$ (see 10.33 and 6.8(1)). Prove that $\mathrm{Min}(A) = A.I = A$ and that $A \subset P(n, \mathbb{R})$ is isometric to $\mathbb{E}^n$ (cf. 6.8(5) and 7.1).

Flats, Regular Geodesics and Weyl Chambers

Flat subspaces play an important role in determining the geometry of symmetric spaces. In this section we shall describe the flat subspaces in $P(n, \mathbb{R})$ (10.45). In contrast to the rank one case, if a CAT(0) symmetric space M has rank ≥ 2, for example $M = P(n, \mathbb{R})$ with $n \geq 2$, then $\mathrm{Isom}(M)$ does not act transitively on the set of geodesics in M (unless $M = \mathbb{E}^n$), however it does act transitively on pairs (F, p), where F is a flat subspace of maximal dimension and $p \in F$. In the case $M = P(n, \mathbb{R})$, one way of seeing that $\mathrm{Isom}(M)$ does not act transitively on the set of geodesics in M is to observe that some geodesics lie in a *unique* maximal flat while others do not (10.45). The set of points $q \in F \setminus \{p\}$ such that F is the unique maximal flat containing $[p, q]$ is open and dense in F; it is the complement of a polyhedral cone of codimension one and its connected components are called Weyl chambers.

The description of the flat subspaces in $P(n, \mathbb{R})$ and their decomposition into Weyl chambers will be important in later sections when we come to describe the geometry of $P(n, \mathbb{R})$ at infinity.

10.44 Definition. A subspace $F \subset P(n, \mathbb{R})$ is called a *flat of dimension k* (more briefly, a *k-flat*) if it is isometric to $\mathbb{E}^k$. (Thus 1-flats are the geodesic lines in $P(n, \mathbb{R})$.) If F is not contained in any flat of bigger dimension, then F is called a *maximal flat*.

In (10.43) we described a particularly nice flat $A \subset P(n, \mathbb{R})$. This is the image under exp of the subspace $\mathfrak{a}_0 \subset S(n, \mathbb{R})$ consisting of diagonal matrices. The following proposition shows that the action of $GL(n, \mathbb{R})$ conjugates every flat in $P(n, \mathbb{R})$ into a subspace of A.

10.45 Proposition.

(1) *Every flat in $P(n, \mathbb{R})$ is contained in a maximal flat, and every maximal flat has dimension n.*

(2) *$GL(n, \mathbb{R})$ acts transitively on the set of pairs (F, p), where $F \subset P(n, \mathbb{R})$ is a maximal flat and $p \in F$.*

(3) *The exponential map* $\exp : S(n, \mathbb{R}) \to P(n, \mathbb{R})$ *induces a bijection from the set of vector subspaces $\{\mathfrak{a} \subset S(n, \mathbb{R}) \mid XY = YX \text{ for all } X, Y \in \mathfrak{a}\}$ to the set of flats* $F = \exp \mathfrak{a}$ *that pass through I.*

(4) *The geodesic line $t \mapsto \exp tX$ through I is contained in a unique maximal flat if and only if the eigenvalues of $X \in S(n, \mathbb{R})$ are all distinct.*

Proof. In any complete Riemannian manifold M of non-positive curvature, if $F \subset M$ contains p and is isometric to $\mathbb{E}^k$, then F is the image under the exponential map of a k-dimensional subspace in T_pM. Thus part (3) of the present proposition is an immediate consequence of 10.40(2).

Let $\mathfrak{a}_0 \subset S(n, \mathbb{R})$ be the subspace of diagonal matrices. If a symmetric matrix commutes with all diagonal matrices then it must be diagonal, so it follows from (3) that $A = \exp \mathfrak{a}_0$ is a maximal flat in $P(n, \mathbb{R})$.

$\mathfrak{a}_0$ has dimension n and $GL(n, \mathbb{R})$ acts transitively on $P(n, \mathbb{R})$, so in order to prove (1) and (2) it suffices (in the light of (3)) to show that if $\mathfrak{a}$ is a vector subspace of $S(n, \mathbb{R})$ such that $XY = YX$ for all $X, Y \in \mathfrak{a}$, then there exists $O \in SO(n)$ such that $O.\mathfrak{a} \subset \mathfrak{a}_0$.

Given $\mathfrak{a}$, we choose a matrix $X \in \mathfrak{a}$ that has the maximum number of distinct eigenvalues, k say. Let $O \in SO(n)$ be such that $OX\,{}^t\!O$ is a diagonal matrix. We must show that $OY\,{}^t\!O$ is diagonal for every $Y \in \mathfrak{a}$. It suffices to show that every eigenvector of X is an eigenvector of Y. (The columns of O are eigenvectors for X.)

Let $\mathbb{R}^n = \oplus_{i=1}^k E_i$ be the decomposition of $\mathbb{R}^n$ into the eigenspaces of X (corresponding to the distinct eigenvalues $\lambda_1 < \cdots < \lambda_k$). If $Y \in \mathfrak{a}$ then Y commutes with X and hence leaves each E_i invariant. If $Y \in \mathfrak{a}$ is close to X, then for each i the eigenvalues of $Y|_{E_i}$ must be close to λ_i. Since Y has at most k eigenvalues and the λ_i are distinct, it follows that $Y|_{E_i}$ (which is diagonalizable) must be a multiple of the identity. This completes the proof of (3).

In (4) it suffices to consider the case where X is diagonal. If two of the diagonal entries of X are the same, then the permutation matrix which interchanges the corresponding basis vectors of $\mathbb{R}^n$ commutes with X. On the other hand, if all of the diagonal entries of X are distinct, then the preceding argument shows that the only symmetric matrices which commute with X are diagonal. Thus A is the only maximal flat containing X if and only if the eigenvalues of X are all distinct. □

The final part of the preceding proposition shows that there are different types of geodesics in $P(n, \mathbb{R})$. The following definition provides a vocabulary for exploring these differences.

10.46 Definition (Singular Geodesics and Weyl Chambers). Let M be a symmetric space. A geodesic line or ray in M is called *regular* if it is contained in a unique maximal flat, and otherwise it is called *singular*.

Let $F \subset M$ be a maximal flat. A *Weyl chamber in F with tip at $p \in F$* is a connected component of the set of points $q \in F \setminus \{p\}$ such that the geodesic line through p and q is regular.

If two geodesic rays c and c' are asymptotic, then c is singular if and only if c' is singular (cf. 10.65). In particular, since singularity is obviously preserved by isometries, the group of isometries of $P(n, \mathbb{R})$ does not act transitively on $\partial P(n, \mathbb{R})$ if $n \geq 2$ (cf. 10.22 and 10.75).

It follows immediately from 10.45(4) that the Weyl chambers in any flat $F \subset P(n, \mathbb{R})$ are open in F. The following proposition gives more precise information.

10.47 Proposition. *Let $F \subset P(n, \mathbb{R})$ be a maximal flat, fix $p \in F$.*

(1) $\{q \in F \setminus \{p\} \mid$ the geodesic line containing $[p, q]$ is regular$\}$ *is the complement in F of $\frac{1}{2}n(n-1)$ flats of dimension $(n-1)$ that pass through p.*
(2) *There are $n!$ Weyl chambers in F with tip at p.*
(3) *$GL(n, \mathbb{R})$ acts transitively on the set of Weyl chambers in $P(n, \mathbb{R})$.*

Proof. In the light of 10.45(2) it suffices to consider the case where $p = I$ and $F = A$ (the space of diagonal matrices with positive entries in the diagonal). Each $q \in A$ can be written uniquely in the form $\text{diag}(e^{t_1}, \dots, e^{t_n}) \in A$; in other words q is the image under exp of the diagonal matrix $\text{diag}(t_1, \dots, t_n) \in A$; we regard the t_i as coordinates for $\mathfrak{a}_0$ (the subspace of $S(n, \mathbb{R})$ comprised of diagonal matrices). According to 10.45(4), the geodesic line through p and $q \neq p$ is regular if and only if all of the t_i are distinct, i.e. if and only if $(t_1, \dots, t_n)$ does not lie in one of the $\frac{1}{2}n(n-1)$ codimension-one subspaces of $\mathfrak{a}_0$ defined by an equation $t_i = t_j$, where $j > i$. The $(n-1)$-dimensional flats in part (1) of the proposition are the images of these subspaces under exp.

The Weyl chambers in A with tip at I are the images under exp of the $n!$ open convex cones defined by $t_{\sigma(1)} > \cdots > t_{\sigma(n)}$, where σ is a permutation of $\{1, \dots, n\}$. The subgroup of $SO(n)$ consisting of monomial[39] matrices acts transitively on these Weyl chambers. Parts (2) and (3) now follow, because $GL(n, \mathbb{R})$ acts transitively on pairs (F, p) and takes Weyl chambers to Weyl chambers (because they are intrinsically defined in terms of the metric on $P(n, \mathbb{R})$). □

[39] matrices with one non-zero entry in each row and column

10.48 Exercise : $P(n, \mathbb{R})$ as a space of ellipsoids in $\mathbb{E}^n$. Given $p \in P(n, \mathbb{R})$ with inverse (p'_{ij}), let $\mathcal{E}_p$ be the ellipsoid in $\mathbb{E}^n$ defined by the equation $\sum p'_{ij}x_i x_j = 1$. Prove that $p \mapsto \mathcal{E}_p$ is a $GL(n, \mathbb{R})$-equivariant map from $P(n, \mathbb{R})$ to the set of (non-degenerate) ellipsoids in $\mathbb{E}^n$ centred at the origin.

This map sends $I \in P(n, \mathbb{R})$ to the sphere $\sum x_i^2 = 1$. Describe the images of the maximal flats through I, the images of the Weyl chambers with tip at I, and the images of the regular geodesics through I.

The Iwasawa Decomposition of $GL(n, \mathbb{R})$

Consider the following subgroups of $GL(n, \mathbb{R})$:

- $K = O(n)$ is the group of orthogonal matrices.
- A is the subgroup of diagonal matrices with positive diagonal entries.
- N is the subgroup of upper-triangular matrices with diagonal entries 1.

Note that AN is the group of upper triangular matrices with positive diagonal entries and that N is a normal subgroup in AN.

The subgroups K, A and N are intimately related to the geometry of the symmetric space $P(n, \mathbb{R})$: we saw in (10.41) that K and its conjugates are the maximal compact subgroups of $GL(n, \mathbb{R})$, and we saw in (10.43) that the orbit of I under the action of A is a maximal flat; N plays a key rôle in describing the horospherical structure of $P(n, \mathbb{R})$ (see (10.50) and compare with the groups A and N considered in (10.28)).

10.49 Proposition (Iwasawa Decomposition). *The map $K \times A \times N \to GL(n, \mathbb{R})$ sending (k, a, n) to kan is a diffeomorphism. More informally,*

$$GL(n, \mathbb{R}) = KAN.$$

Proof. This follows directly from the existence and uniqueness of the Gram-Schmidt orthogonalisation process. If $v_1, \dots, v_n$ are the column vectors of a matrix $g \in GL(n, \mathbb{R})$, there exist unique $\lambda_{ij} \in \mathbb{R}$ for $1 \leq i < j \leq n$ such that the vectors $v'_1 = v_1,\ v'_2 = \lambda_{12}v_1 + v_2, \dots,\ v'_n = \lambda_{1n}v_1 + \cdots + \lambda_{n-1,n}v_{n-1} + v_n$ are mutually orthogonal; if $n' \in N$ is the matrix whose (i, j)-entry is λ_{ij} for $i < j$, then gn' is the matrix whose columns are $v'_1, \dots, v'_n$. Let a be the diagonal matrix whose entries in the diagonal are the norms of the v'_i. Then $gn'a^{-1} = k$ is an orthogonal matrix and we have $g = kan$, where $n = n'^{-1}$. This decomposition is unique and the entries of k, a and n depend differentiably on the entries of g. □

10.50 Corollary. *The map $A \times N \to P(n, \mathbb{R})$ sending (a, n) to $an.I = an\,{}^t\!na$ is a diffeomorphism. In other words the subgroup AN of $GL(n, \mathbb{R})$ acts simply transitively on $P(n, \mathbb{R})$.*

Proof. We know (10.33) that $GL(n, \mathbb{R})$ acts transitively on $P(n, \mathbb{R})$, so given $p \in P(n, \mathbb{R})$ the set $C_p = \{g \in GL(n, \mathbb{R}) | g.p = I\}$ is non-empty. The stabilizer of I

is $K = O(n)$, so C_p is a coset Kan and by the proposition a and n are uniquely determined by p. □

We mention two other closely related and useful decompositions:

10.51 Exercises

(1) Prove that the map $(k, p) \mapsto kp$ from $O(n) \times P(n, \mathbb{R})$ to $GL(n, \mathbb{R})$ is a diffeomorphism (its inverse is the map $g \mapsto (g({}^tgg)^{-1/2}, ({}^tgg)^{1/2})$)

(2) Let $\overline{A}^+$ be the set of diagonal matrices $\operatorname{diag}(\lambda_1, \dots, \lambda_n)$ with $\lambda_1 \geq \dots \geq \lambda_n > 0$. Show that $GL(n) = K\overline{A}^+K$. More precisely, show that every $g \in GL(n, \mathbb{R})$ can be written as $g = kak'$ with $k, k' \in O(n)$ and $a \in \overline{A}^+$ and show that a is uniquely defined by g.

(Hint: To show existence use (1). To show uniqueness, note that the eigenvalues of a are the square roots of the eigenvalues of ${}^tgg$.)

The Irreducible Symmetric Space $P(n, \mathbb{R})_1$

Following the outline of our treatment of hyperbolic spaces, we now turn our attention to the study of convex submanifolds of $P(n, \mathbb{R})$, where the range of such submanifolds is much richer than in the hyperbolic case (see 10.58). The first such submanifold that we consider is $P(n, \mathbb{R})_1$, the subspace consisting of positive definite matrices with determinant one.

A *totally geodesic submanifold* of $P(n, \mathbb{R})$ is, by definition, a differentiable submanifold M of $P(n, \mathbb{R})$ such that any geodesic line in $P(n, \mathbb{R})$ that intersects M in two points is entirely contained in M. Such a submanifold is convex in the sense of (I.1.3) and hence is a CAT(0) space. Moreover, if p lies in such a submanifold then the symmetry σ_p leaves M fixed, and hence M is a symmetric space.

10.52 Lemma.

(1) $P(n, \mathbb{R})_1 \subset P(n, \mathbb{R})$ *is a totally geodesic submanifold.*

(2) $SL(n, \mathbb{R}) \subset GL(n, \mathbb{R})$ *leaves* $P(n, \mathbb{R})_1$ *invariant and acts transitively on it.* $SL(n, \mathbb{R})$ *acts effectively if n is odd and otherwise the kernel of the action is* $\{\pm I\}$.

(3) $P(n, \mathbb{R})_1 = \exp S(n, \mathbb{R})_0$, *where* $S(n, \mathbb{R})_0 \subset S(n, \mathbb{R})$ *is the vector subspace consisting of matrices with trace zero.*

Proof. $P(n, \mathbb{R})_1$ is obviously invariant under the action of $SL(n, \mathbb{R})$, and the proof of 10.33(1) shows that the action of $SL(n, \mathbb{R})$ is transitive; this proves (2). Part (3) is an immediate consequence of 10.36(6) and the (easy) exercise 10.38(4).

It follows from (3) that any geodesic $t \mapsto \exp(tX)$ joining I to a point of $P(n, R)_1$ is entirely contained in $P(n, R)_1$, so by homogeneity (2) we see that $P(n, \mathbb{R})_1$ is totally geodesic. □

Lemma 10.52 illustrates the fact that essentially the whole of our treatment of $P(n, \mathbb{R})$ works equally well if we replace $P(n, \mathbb{R})$ by $P(n, \mathbb{R})_1$, $GL(n, \mathbb{R})$ by $SL(n, \mathbb{R})$, and $S(n, \mathbb{R})$ by $S(n, \mathbb{R})_0$. In particular we have the natural identification

$$P(n, \mathbb{R})_1 = SL(n, \mathbb{R})/SO(n).$$

(Hence $P(2, \mathbb{R})_1 = SL(2, \mathbb{R})/SO(2)$ is isometric to $\mathbb{H}^2$, up to a scaling factor.)

We also have the diffeomorphism

$$SO(n) \times P(n, \mathbb{R})_1 \to SL(n, \mathbb{R})$$

given by $(k, p) \mapsto kp$, and the Iwasawa decomposition

$$SL(n, \mathbb{R}) = KAN$$

where $K = SO(n)$, the subgroup N is the group of upper-triangular matrices with diagonal entries one and A is the group of diagonal matrices with positive diagonal entries and determinant one. And following the proofs of the corresponding results for the action of $GL(n, \mathbb{R})$ on $P(n, R)$, one sees that $SL(n, \mathbb{R})$ acts transitively on the set of Weyl chambers in $P(n, \mathbb{R})_1$ and on the set of pointed flats.

A further important point to note is that $P(n, \mathbb{R})_1$ is one factor in a (metric) product decomposition of $P(n, \mathbb{R})$. To see this, first observe that the centre of $GL(n, \mathbb{R})$ consists of the scalar multiples of the identity and the quotient of this subgroup by $\{\pm I\}$ acts by hyperbolic translations on $P(n, \mathbb{R})$. It follows from (6.15) and (6.16) that $P(n, \mathbb{R})$ has a non-trivial Euclidean de Rham factor, and it is not difficult to see that $P(n, \mathbb{R})_1$ is the complementary factor. More precisely:

10.53 Proposition. *The map* $(s, p) \mapsto e^{s/\sqrt{n}}p$ *gives an isometry*

$$\mathbb{R} \times P(n, \mathbb{R})_1 \cong P(n, \mathbb{R}).$$

Moreover, $P(n, \mathbb{R})_1$ *cannot be expressed as a non-trivial (metric) product.*

We leave the proof as an exercise for the reader. (One way to see that $P(n, \mathbb{R})_1$ is irreducible is to consider how flats intersect in any product and then compare this with the following description of Weyl chambers in $P(n, \mathbb{R})_1$.)

10.54 Flats and Weyl Chambers in $P(n, \mathbb{R})_1$. There is an obvious bijective correspondence between maximal flats in $P(n, \mathbb{R})_1$ and maximal flats in $P(n, \mathbb{R})$: a maximal flat F_1 in $P(n, \mathbb{R})_1$ corresponds (under the decomposition in (10.53)), to the maximal flat $F = \mathbb{R} \times F_1$ in $P(n, \mathbb{R})$. Similarly, there is a correspondence between the Weyl chambers in $P(n, \mathbb{R})_1$ and the Weyl chambers in $P(n, \mathbb{R})$. Following the proof of (10.47) we give an explicit description of the Weyl chambers in $P(n, \mathbb{R})_1$.

Let $\mathfrak{a} \subset S(n, \mathbb{R})$ be the vector space of diagonal matrices with trace zero. Then $A = \exp \mathfrak{a}$ is a maximal flat in $P(n, \mathbb{R})_1$ and $\exp|_{\mathfrak{a}}$ is an isometry. The Weyl chambers in A with tip at I are the images under exp of the $n!$ open convex cones of the form

$\{\text{diag}(t_1, \ldots, t_n) \mid \sum t_i = 0 \text{ and } t_{\sigma(1)} > \cdots > t_{\sigma(n)}\}$, where $\sigma \in \mathcal{S}_n$ (the group of permutations of $\{1, \ldots, n\}$).

The closure of the cone indexed by σ is the intersection of the $(n-1)$ open half-spaces $t_{\sigma(i)} - t_{\sigma(i+1)} < 0$ with the hyperplane $t_1 + \cdots + t_n = 0$. This is a simplicial cone. Its faces of codimension k are obtained by replacing k of the inequalities $t_{\sigma(1)} > \cdots > t_{\sigma(n)}\}$ by equalities. Thus, for example, a codimension one face is defined by the equations $t_1 + \cdots + t_n = t_{\sigma(i+1)} - t_{\sigma(i)} = 0$ and $t_{\sigma(j+1)} - t_{\sigma(j)} \leq 0$ for $j \neq i$.

The Weyl chambers associated to the permutations σ and σ' have a codimension-one face in common if and only if for some $i \in \{1, \ldots, n-1\}$ we have $\sigma'(i) = \sigma(i+1)$, $\sigma'(i+1) = \sigma(i)$ and $\sigma'(j) = \sigma(j)$ for $j \notin \{i, i+1\}$.

The obvious action of the symmetric group $\mathcal{S}_n$ on the coordinates t_i induces a simply transitive action of $\mathcal{S}_n$ on the above set of Weyl chambers. $\mathcal{S}_n$ is called the *Weyl group*; the interested reader can check that it is the quotient of the subgroup of $SO(n)$ that leaves the flat A invariant by the point-wise stabilizer of A.

In order to determine the shape of the Weyl chambers, we calculate the dihedral angles between the codimension one faces of the Weyl chamber corresponding to the trivial permutation. Let F_i be the codimension one face defined by the equation $t_i = t_{i+1}$ and let $e_1, \ldots, e_n$ be the orthonormal basis of the space of diagonal matrices corresponding to the coordinates t_i. The unit normal to F_i is $u_i = \frac{1}{\sqrt{2}}(e_i - e_{i+1})$ and therefore $(u_i|u_j) = -1/2$ if $j = i+1$ and $(u_i|u_j) = 0$ if $j > i+1$. Taking inverse cosines, we see that the dihedral angle between F_i and F_{i+1} is $\pi/3$ and the angle between the faces F_i and F_j is $\pi/2$ if $|i-j| \geq 2$.

10.55 Example. In the notation of (10.54), if $n = 3$ then $\mathfrak{a}$ can be identified with the plane in $\mathbb{R}^3$ defined by the equation $t_1 + t_2 + t_3 = 0$. The Weyl group is $\mathcal{S}_3$, so there are 6 Weyl chambers contained in A with tip at I; they are the images under exp of the sectors in $\mathfrak{a}$ which form the complement of three lines through 0 as indicated in figure (10.1).

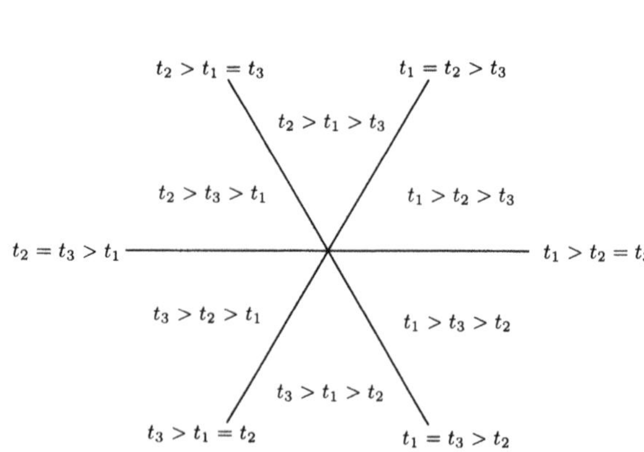

Fig. 10.1 The Weyl chambers in a maximal flat of $P(3, \mathbb{R})_1 = SL(3, \mathbb{R})/SO(3)$

Reductive Subgroups of $GL(n, \mathbb{R})$

In the introduction to this chapter we mentioned that every simply connected symmetric space of non-positive curvature is isometric (after rescaling) to a totally geodesic submanifold of $P(n, \mathbb{R})$. We also alluded to the close connection between the study of such symmetric spaces and the theory of semi-simple Lie groups. In this section we shall present some totally geodesic submanifolds of $P(n, \mathbb{R})$ and describe the connection between such submanifolds and the subgroup structure of $GL(n, \mathbb{R})$ (see 10.58). The subgroups with which we shall be concerned are the following.

10.56 Definition (Reductive and Algebraic Subgroups of $GL(n, \mathbb{R})$)**.** A subgroup G of $GL(n, \mathbb{R})$ is *reductive* if it is closed and it is stable under transposition: if $A \in G$ then ${}^tA \in G$.

A subgroup $G \subset GL(n, \mathbb{R})$ is *algebraic* if there is a finite system of polynomials in the entries of $M(n, \mathbb{R})$ such that G is the intersection of $GL(n, \mathbb{R})$ with the set of common zeros of this system.

Before describing the connection between reductive subgroups and totally geodesic submanifolds of $P(n, \mathbb{R})$, we mention some examples of reductive algebraic subgroups of $GL(n, \mathbb{R})$. Examples (2) and (3) explain why we have restricted attention to real matrices, and (4) will allow us to embed $\mathbb{K}\mathrm{H}^n$ (rescaled) in $P(n, \mathbb{R})$ as a totally geodesic submanifold.

10.57 Examples

(1) $SL(n, \mathbb{R})$, $O(n)$ and $SO(n)$ are all reductive algebraic subgroups of $GL(n, \mathbb{R})$. For instance $O(n)$ is defined by the vanishing of ${}^tMM - I$, the entries of which are quadratic polynomials in the entries of M.

(2) *$GL(n, \mathbb{C})$ as a reductive subgroup of $GL(2n, \mathbb{R})$.* Consider the map from $GL(n, \mathbb{C})$ to $GL(2n, \mathbb{R})$ that sends $A + iB$, where $A, B \in M(n, \mathbb{R})$, to

$$\begin{pmatrix} A & -B \\ B & A \end{pmatrix}.$$

This map is obviously injective and its image is defined by the obvious equalities between entries. Thus $GL(n, \mathbb{C})$ can be realised as a reductive algebraic subgroup of $GL(2n, \mathbb{R})$. Note that transposition in $GL(2n, \mathbb{R})$ induces the conjugate transpose on $GL(n, \mathbb{C})$.

(3) *$GL(n, \mathcal{H})$ as a reductive subgroup of $GL(4n, \mathbb{R})$.* Let $\mathcal{H}$ denote the quaternions. Following the construction of (2), we shall explain how to realize $GL(n, \mathcal{H})$ as a reductive algebraic subgroup of $GL(4n, \mathbb{R})$.

First note that one can embed $\mathcal{H}$ in $M(2, \mathbb{C})$ by sending $q = a_0 + a_1 i + a_2 j + a_3 k$ to the matrix

$$\begin{pmatrix} c & -d \\ \overline{d} & \overline{c} \end{pmatrix} \in M(2, \mathbb{C}),$$

where $c = a_0 + a_1 i$ and $d = a_2 + a_3 i$ (so $q = c + dj$). The map $\mathcal{H} \to M(2, \mathbb{C})$ thus defined is an injective ring homomorphism and the conjugate of q is sent to the conjugate transpose of the image of q. More generally, one can realize $GL(n, \mathcal{H})$ as a subgroup of $GL(2n, \mathbb{C})$ by sending the matrix $C + Dj \in GL(n, \mathcal{H})$, with $C, D \in M(n, \mathbb{C})$, to the matrix

$$\begin{pmatrix} C & -D \\ \overline{D} & \overline{C} \end{pmatrix} \in GL(2n, \mathbb{C}).$$

The conjugate transpose on $GL(2n, \mathbb{C})$ induces the conjugate transpose on $GL(n, \mathcal{H})$.

The composition of the above maps $GL(n, \mathcal{H}) \hookrightarrow GL(2n, \mathbb{C}) \hookrightarrow GL(4n, \mathbb{R})$ embeds $GL(n, \mathcal{H})$ as a reductive algebraic subgroup of $GL(4n, \mathbb{R})$.

(4) Let $\mathbb{K} = \mathbb{R}$, $\mathbb{C}$ or $\mathcal{H}$. We saw in 10.20(4) that $O_{\mathbb{K}}(Q)$, the subgroup of $GL(n+1, \mathbb{R})$ that preserves the form Q on $\mathbb{K}^{n,1}$, is closed under the operation of conjugate transpose. Thus by means of the above embeddings we can realize $O_{\mathbb{K}}(Q)$ as a reductive subgroup of $GL(m, \mathbb{R})$, where $m = (n+1)$, $(2n+2)$ or $(4n+4)$ in the cases $\mathbb{K} = \mathbb{R}$, $\mathbb{C}$ and $\mathcal{H}$ respectively. Moreover, it also follows from 10.20(4) that this reductive subgroup is algebraic.

More generally, instead of Q one can consider a form of type (p, q):

$$Q_{p,q}(x, y) = \sum_{i=1}^{p} \overline{x}_i y_i - \sum_{i=p+1}^{p+q} \overline{x}_i y_i .$$

The subgroup of $GL(n+1, \mathbb{R})$ that preserves $Q_{p,q}$ is closed under the operation of conjugate transpose (it can be described in a manner analogous to 10.20(4)), and its image under the embeddings in (2) and (3) is a reductive algebraic group.

In the course of the following proof we shall need the fact that a closed subgroup G of $GL(n, \mathbb{R})$ is a Lie group, i.e. it is a differentiable submanifold and multiplication and passage to inverses in G are differentiable operations. (This is easy to verify in the above examples.)

Observe that if $G \subseteq GL(n, \mathbb{R})$, then a necessary condition for $G.I \subseteq P(n, \mathbb{R})$ to be totally geodesic is the following (cf. 10.59) :

$(*)$ if $X \in S(n, R)$ and $\exp X \in G$ then $\exp(sX) \in G$, $\forall s \in \mathbb{R}$.

10.58 Theorem. *Let $G \subset GL(n, \mathbb{R})$ be a reductive subgroup satisfying $(*)$. Let $K = G \cap O(n)$ and let $M = G \cap P(n, \mathbb{R})$. Then:*

(1) *M is the G-orbit of I.*

(2) *M is a totally geodesic submanifold of $P(n, \mathbb{R})$; it is diffeomorphic to G/K.*

(3) *M is a* CAT(0) *symmetric space.*

(4) *K is a maximal compact subgroup of G; up to conjugacy in G it is the unique such subgroup.*

(5) *The map $K \times M \to G$ sending (k, m) to km is a diffeomorphism.*

Conversely, if V is a totally geodesic submanifold of $P(n, \mathbb{R})$ and $I \in V$, then $G = \{g \in GL(n, \mathbb{R}) \mid g.V = V\}$ is a reductive subgroup of $GL(n, \mathbb{R})$ and $V = G \cap P(n, \mathbb{R})$.

Proof. If $X \in S(n, \mathbb{R})$ and $p = \exp X$, then $p = g.I$, where $g := \exp(X/2) \in G$. Thus (1) is a consequence of $(*)$.

G is a Lie group and the action of $GL(n, \mathbb{R})$ is differentiable, so $M = G.I$ is a differentiable submanifold of $P(n, \mathbb{R})$. And since the stabilizer of I in $GL(n, \mathbb{R})$ is $O(n)$, M is diffeomorphic to G/K. The image of $t \mapsto \exp(tX)$ is the unique geodesic line containing the segment $[I, p]$ and $(*)$ ensures that this line is entirely contained in M, so by homogeneity M is totally geodesic.

Since M is totally geodesic, if $p \in M$ then the symmetry $\sigma_p : P(n, \mathbb{R}) \to P(n, \mathbb{R})$ leaves M invariant. The existence of the symmetries $\sigma_p|_M$ means that M is a symmetric space.

K is the stabilizer of I in G. It is compact because G is closed in $GL(n, \mathbb{R})$ and $O(n)$ is compact. Thus (4) follows from (2.7), as in (10.41).

We have seen (10.51) that the map $O(n) \times P(n, \mathbb{R})$ sending (k, p) to kp is a diffeomorphism onto $GL(n, \mathbb{R})$, so in order to prove (5) we need only check that if $kp \in G$ then k and p belong to G. We have $p^2 = {}^t(kp)(kp) \in M$, and writing $p^2 = \exp X$ we have $p = \exp(X/2)$. Thus $(*)$ implies $p \in G$, hence $k \in G$ and (5) is proved.

If $V \subset P(n, \mathbb{R})$ is a totally geodesic submanifold then for any $p \in V$ the symmetry σ_p leaves V invariant. Given $p, q \in V$, if m is the midpoint of $[p, q]$ then $\sigma_m(p) = q$. Thus the subgroup G of $GL(n, \mathbb{R})$ leaving V invariant acts transitively on V. Moreover, since $q \mapsto p.q$ is the composition of σ_I and σ_p, we have $V \subseteq G \cap P(n, \mathbb{R})$. And if $p \in G \cap P(n, \mathbb{R})$ then $p.I = p^2 \in V$, so p, which is the midpoint of $[I, p^2]$, is in V. Thus $V = G \cap P(n, \mathbb{R})$.

It remains to check that G is reductive; it is obviously closed so what we must show is that $G = {}^tG$. Given $g \in G$, write $g = pk$ where $k \in O(n)$ and $p \in P(n, \mathbb{R})$. Then $g.I = p.I = p^2 \in V$, hence $p \in V \subseteq G$, therefore $k \in G$ and ${}^tg = k^{-1}p \in G$. □

There are a number of remarks that we should make regarding the above theorem. The first important point to note is that it applies to all reductive *algebraic* subgroups of $GL(n, \mathbb{R})$:

10.59 Lemma. *Let $G \subset GL(n, \mathbb{R})$ be a subgroup and let $X \in S(n, \mathbb{R})$ be such that* $\exp X \in G$. *If G is algebraic, then* $\exp(tX) \in G$ *for all $t \in \mathbb{R}$.*

Proof. It suffices to show that if a polynomial function on $M(n, \mathbb{R})$ vanishes at $\exp(rX)$ for some $X \in S(n, \mathbb{R})$ and all $r \in \mathbb{Z}$, then it vanishes at $\exp(tX)$ for all $t \in \mathbb{R}$.

After conjugation by an element of $O(n)$, we may assume that X is a diagonal matrix $\mathrm{diag}(\lambda_1, \ldots, \lambda_n)$. Thus it suffices to show that if a polynomial

$$P(x_1, \ldots, x_n) = \sum c_{i_1 \ldots i_n} x_1^{i_1} \ldots x_n^{i_n}$$

of degree m in n variables $x_1, \dots, x_n$ is such that $P(e^{r\lambda_1}, \dots, e^{r\lambda_n}) = 0$ for all $r \in \mathbb{Z}$, then $P(e^{t\lambda_1}, \dots, e^{t\lambda_n}) = 0$ for all $t \in \mathbb{R}$.

Let $s_1, \dots, s_k$ be the distinct values taken by the monomials $e^{i_1\lambda_1} \dots e^{i_n\lambda_n}$, on the set of $(i_1, \dots, i_n) \in \mathbb{N}^n$ such that $i_1 + \dots + i_n \leq m$. One can write

$$P(e^{t\lambda_1}, \dots, e^{t\lambda_n}) = \sum_{j=1}^{k} a_j s_j^t,$$

where a_j is the sum of the coefficients $c_{i_1 \dots i_n}$ for which $e^{i_1\lambda_1} \dots e^{i_n\lambda_n} = s_j$.

If we view $\left(\sum_j a_j s_j^r = 0 \mid r = 0, 1, \dots, k-1\right)$ as a system of k equations in the k variables $a_1, \dots, a_k$, then the determinant of the coefficient matrix is $\prod_{j>i}(s_j - s_i)$, the Vandermonde determinant, which is non-zero. By hypothesis $\sum_j a_j s_j^r = 0$ for all $r \in \mathbb{Z}$, hence all the a_j are zero. □

There are several other important remarks to be made concerning (10.58):

10.60 Remarks. In the notation of (10.58):

(1) In general G does not act effectively on M (for example if $M = \{I\}$).

(2) In general an isometry of M cannot be extended to an isometry of $P(n, \mathbb{R})$. For example if $n \geq 2$ and M is a maximal flat in $P(n, \mathbb{R})$, then $\text{Isom}(M)$ acts transitively on the set of geodesics in M whereas G does not (no element of G sends a regular geodesic to a singular one).

(3) If the action of K on the tangent space T_IM is irreducible, then the subspace metric on $M \subset P(n, \mathbb{R})$ is the unique G-invariant Riemannian metric on M up to a scaling factor. To see this, first observe that any G-invariant Riemannian metric is determined by the scalar product given on T_IM, because G acts transitively on M. Fix an isometry from $\mathbb{E}^k$ to T_IM endowed with the metric induced by the Riemannian metric on $P(n, \mathbb{R})$. If there were a non-proportional G-invariant Riemannian metric on M then its unit ball would be an ellipsoid in this Euclidean space different from a sphere, and the subspace spanned by those vectors which had maximal length in the fixed Euclidean metric would span a proper K-invariant subspace of T_IM, contradicting the fact that the action of K is irreducible.

(4) *Embedding* $\mathbb{K}\mathrm{H}^n$ *and Other Symmetric Spaces.* In the light of (4), by applying (10.58) to the reductive algebraic subgroup $G \cong O_{\mathbb{K}}(Q)$ described in 10.57(4) we can realize $\mathbb{K}\mathrm{H}^n$ (rescaled) as a totally geodesic submanifold of $P(m, \mathbb{R})$ where $m = (n+1)\dim_{\mathbb{R}} \mathbb{K}$.

This embedding illustrates a general phenomenon. We mentioned earlier that, after rescaling the metric on its irreducible factors, one can isometrically embed any symmetric space of non-compact type as a totally geodesic submanifold $M \subseteq P(n, \mathbb{R})$ (see for instance [Eb96, 2.6.5]).

(5) The flats in M are the intersection with M of the flats in $P(n, \mathbb{R})$. All of the maximal flats in M have the same dimension (this is called the *rank of* M) and the group G acts transitively on the set of maximal flats.

Semi-Simple Isometries

In the remainder of this chapter we will concentrate on the horospherical structure of $P(n, \mathbb{R})$ and the geometry of $\partial P(n, \mathbb{R})$. However, before turning to this topic we pause to reconcile an ambiguity in our terminology: classically, a matrix $g \in GL(n, \mathbb{R})$ is said to be semi-simple if and only if it is conjugate in $GL(n, \mathbb{C})$ to a diagonal matrix; on the other hand, we have been studying $GL(n, \mathbb{R})$ via its action by isometries on the CAT(0) space $P(n, \mathbb{R})$, and for such isometries we have a different definition of semi-simplicity, *viz.* (6.1). The following proposition shows that these definitions agree. Indeed if $g \in GL(n, \mathbb{R})$ acts as a semi-simple isometry of any totally geodesic submanifold of $P(n, \mathbb{R})$ then it is semi-simple in the classical sense.

10.61 Proposition. *Let M be a totally geodesic submanifold of $P(n, \mathbb{R})$. Suppose that $g \in GL(n, \mathbb{R})$ leaves M invariant and let $\overline{g} \in \operatorname{Isom}(M)$ be the restriction of g. The isometry $\overline{g}$ is semi-simple in the sense of (6.1) if and only if the matrix g is semi-simple.*

Proof. Assume that $\overline{g}$ is semi-simple. If $\overline{g}$ fixes a point $p \in M$, then g is conjugate to an element of $O(n)$, hence it is semi-simple. Otherwise $\overline{g}$ is hyperbolic, which means that there is a geodesic $c : \mathbb{R} \to M$ and $a > 0$ such that $\overline{g}.(c(s)) = c(s+a)$, $\forall s \in \mathbb{R}$. Since M is totally geodesic, c is a geodesic line in $P(n, \mathbb{R})$ and g is hyperbolic. After conjugating g in $GL(n, \mathbb{R})$ we may assume that $c(0) = I$. If we write X to denote the element of $S(n, \mathbb{R})$ such that $\exp(sX) = c(s)$, then $g\,{}^tg = \exp(aX)$ and the equation $g.c(s) = c(s + a)$ becomes $g \exp(sX)\,{}^tg = \exp(sX)\exp(aX)$. Differentiating with respect to s at $s = 0$, we get $gX\,{}^tg = X\exp(aX) = Xg\,{}^tg$, hence $gX = Xg$. Thus $h := \exp(aX/2)$ (which is semi-simple) commutes with g; let $k = gh^{-1} = h^{-1}g$. Both g and h act as translation by a on $c(\mathbb{R})$, therefore k fixes $c(0) = I$, in other words $k \in O(n)$. In particular k is semi-simple, and since it commutes with the semi-simple matrix h, the product $g = hk$ is also semi-simple.

To prove the converse, suppose that $g \in GL(n, \mathbb{R})$ is a semi-simple matrix that preserves a totally geodesic submanifold M. As M is a closed convex subspace, the isometry of $P(n, \mathbb{R})$ defined by g will be semi-simple if and only if its restriction to M is semi-simple (6.2(4)). Thus it is enough to show that g acts semi-simply on $P(n, \mathbb{R})$.

By hypothesis, g is conjugate in $GL(n, \mathbb{C})$ to a diagonal matrix $\operatorname{diag}(\lambda_1, \dots, \lambda_n)$; we may assume that $\lambda_{2i} = \overline{\lambda}_{2i-1}$ if $0 < i \leq k$ and that λ_i is real if $i > 2k$. Standard linear algebra yields elements $k, h \in GL(n, \mathbb{R})$ and a basis $e_1, \dots, e_n$ for $\mathbb{R}^n$ such that $g = kh = hk$, where $h(e_i) = |\lambda_i| e_i$ for all i and k multiplies e_i by $\lambda_i/|\lambda_i|$ if $i > 2k$, and for $i \leq k$ the action of k on the subspace with ordered basis $\{e_{2i-1}, e_{2i}\}$ is rotation by $\arg(\lambda_i/|\lambda_i|)$. Conjugating g by a suitable element of $GL(n, \mathbb{R})$, we may assume that $e_1, \dots, e_n$ is the standard basis, and hence $g = kh = hk$ where $k \in O(n)$ and $h \in P(n, \mathbb{R})$. Now, if $h = I$, then $g \in O(n)$ acts as an elliptic isometry of $P(n, \mathbb{R})$ (fixing I). Otherwise, $h = \exp X$ where $X \in S(n, \mathbb{R})$ is non-zero; in this case h acts by translation on the geodesic $t \mapsto \exp(tX)$ in $P(n, \mathbb{R})$. This geodesic is invariant under the action of k and hence g, therefore g acts on $P(n, \mathbb{R})$ as a hyperbolic isometry. $\square$

While on the subject of reconciling terminology, we should note that in texts on symmetric spaces, [Eb96] for example, the term "hyperbolic isometry" is reserved for transvections. In such texts isometries that are hyperbolic in our sense but which are not transvections are usually called *loxodromic*.

Parabolic Subgroups and Horospherical Decompositions of $P(n, \mathbb{R})$

In this section we shall describe the Busemann functions and horospheres in $P(n, \mathbb{R})$ and calculate the stabilizer $G_\xi \subseteq GL(n, \mathbb{R})$ of each point $\xi \in \partial P(n, \mathbb{R})$. In its broad outline this section follows our earlier discussion of the horospherical structure of $\mathbb{K}\mathrm{H}^n$ (10.26 to 10.29). However, in the present (higher-rank) situation things are less straightforward and more interesting because the parabolic subgroups and horospherical structure associated to a point $c(\infty) \in \partial P(n, \mathbb{R})$ depend very much on the nature of the geodesic c (whether it is singular versus regular for example).

10.62 Definition. Given $\xi \in \partial P(n, \mathbb{R})$, define

$$G_\xi := \{g \in GL(n, \mathbb{R}) \mid g.\xi = \xi\}.$$

If $\xi = c(\infty)$ where c is the geodesic line $c(t) = \exp(tX)$, define

$$N_\xi := \{g \in G_\xi \mid \exp(-tX) g \exp(tX) \to I \text{ as } t \to \infty\}.$$

(We shall show that N_ξ is a normal subgroup of G_ξ.)

G_ξ is called the *parabolic subgroup* associated to ξ, and N_ξ is called the *horospherical subgroup* associated to ξ. It follows from (10.26) that the orbits of N_ξ are contained in the horospheres centred at ξ. These orbits foliate $P(n, \mathbb{R})$ and there is a complementary orthogonal foliation (see 10.69) whose leaves are the totally geodesic subspaces

$$F(c) := \bigcup \{c'(\mathbb{R}) \mid c' : \mathbb{R} \to P(n, \mathbb{R}) \text{ is a geodesic parallel to } c\},$$

where c ranges over the geodesics such that $c(\infty) = \xi$.

We shall give explicit descriptions of G_ξ, N_ξ and $F(c)$. We shall see that the map $N_\xi \times F(c) \to P(n, \mathbb{R})$ sending (v, p) to $v.p = vp\,{}^t v$ is a diffeomorphism, and we shall use this map to describe the Busemann functions of $P(n, \mathbb{R})$.

10.63 Examples. The structure of G_ξ, N_ξ and $F(c)$ depends very much on the nature of $c(\infty) = \xi$. There are two extreme cases: if $c(t) = e^{t/\sqrt{n}} I$ (the direction of the Euclidean de Rham factor (10.53)), then $G_\xi = GL(n, \mathbb{R})$, N_ξ is trivial and $F(c) = P(n, \mathbb{R})$; in contrast, if $t \mapsto c(t)$ is a regular geodesic, then G_ξ is conjugate to the group of upper triangular matrices (10.64), while N_ξ is the subgroup consisting of matrices with ones on the diagonal, and $F(c)$ is the unique maximal flat A containing

$c(\mathbb{R})$. In this second example $P(n, \mathbb{R})$ is foliated by the maximal flats whose boundary at infinity contains $\xi = c(\infty)$ (cf. 10.68).

10.64 Proposition. *Let $\xi \in \partial P(n, \mathbb{R})$ be a point at infinity, and let $c : t \mapsto \exp(tX)$ be the geodesic ray issuing from I with $c(\infty) = \xi$, where $X \in S(n, \mathbb{R})$ and $\|X\| = 1$. Then*

$$G_\xi = \{g \in GL(n, \mathbb{R}) \mid \lim_{t\to\infty} \exp(-tX)\, g\, \exp(tX) \text{ exists}\}.$$

Conjugating by an element of $SO(n)$ we may assume that X is a diagonal matrix $\operatorname{diag}(\lambda_1, \dots, \lambda_n)$ *with $\lambda_1 \geq \cdots \geq \lambda_n$. In this case, if $r_1, \dots, r_k$ are the multiplicities of the eigenvalues of X taken in decreasing order, then $g \in GL(n, \mathbb{R})$ belongs to G_ξ if and only if*

$$g = \begin{pmatrix} A_{11} & A_{12} & \dots & A_{1k} \\ 0 & A_{22} & \dots & A_{2k} \\ \dots & \dots & \dots & \dots \\ 0 & 0 & \dots & A_{kk} \end{pmatrix}$$

where A_{ij} is an (r_i, r_j)-matrix.

Proof. Let $X = \operatorname{diag}(\lambda_1, \dots, \lambda_n)$ where $\lambda_i \geq \lambda_j$ if $i < j$. If the (i, j)-entry of a matrix g is g_{ij}, then the (i, j)-entry of the matrix $\exp(-tX)\, g \exp(tX)$ is $e^{-t(\lambda_i - \lambda_j)} g_{ij}$, which converges as $t \to \infty$ if and only if $\lambda_i \geq \lambda_j$. In particular, the matrices such that $\lim_{t\to\infty} \exp(-tX)\, g \exp(tX)$ exists are precisely those displayed above; it remains to show that these constitute G_ξ.

If $\lim_{t\to\infty} \exp(-tX/2)\, g \exp(tX/2)$ exists, then

$$d\big(\exp(-tX/2)\, g \exp(tX/2).I,\ I\big) = d\big(g \exp(tX/2).I,\ \exp(tX/2).I\big)$$

is bounded for $t > 0$, hence the geodesic ray $t \mapsto g \exp(tX/2).I$ $(t > 0)$ is asymptotic to $c(t) = \exp(tX/2).I = \exp(tX)$; in other words $g.\xi = \xi$.

Conversely, if $g \in G_\xi$, then $d(g \exp(tX/2).I,\ \exp(tX/2).I)$ is bounded for $t > 0$, hence $\exp(-tX/2)\, g \exp(tX/2).I$ remains in a bounded subset. It follows that $\{\exp(-tX/2) g \exp(tX/2) \mid t > 0\} \subset GL(n, \mathbb{R})$ is contained in a compact subset, because the map $GL(n, \mathbb{R}) \to GL(n, \mathbb{R})/O(n) = P(n, \mathbb{R})$ is proper. Hence there exists a sequence $t_n \to \infty$ such that the sequence of matrices $\exp(t_n X)\, g \exp(t_n X)$ converges in $GL(n, \mathbb{R})$ as $n \to \infty$. But, as in the first paragraph, this implies that $g_{i,j} = 0$ if $\lambda_i < \lambda_j$, and hence (by the reverse implication in the first paragraph) $\lim_{t\to\infty} \exp(tX)\, g \exp(tX)$ exists. $\square$

10.65 Remark. For all $\xi \in \partial P(n, \mathbb{R})$ the group G_ξ acts transitively on $P(n, \mathbb{R})$. Indeed if we conjugate G_ξ into the above form, then it contains the group AN of upper triangular matrices with positive entries in the diagonal, and this acts simply transitively on $P(n, \mathbb{R})$ (see 10.49).

Similarly, if ξ is the endpoint of a ray in $P(n, \mathbb{R})_1$, then $G_\xi \cap SL(n, \mathbb{R})$ acts transitively on $P(n, \mathbb{R})_1$.

10.66 Proposition.

(1) *N_ξ is a normal subgroup of G_ξ.*

(2) *Let $X = \mathrm{diag}(\lambda_1, \dots, \lambda_k)$ be as in (10.64), let $c(t) = \exp(tX)$ and let $\xi = c(\infty)$. Then $g \in N_\xi$ if and only if*

$$g = \begin{pmatrix} I_{r_1} & A_{12} & \dots & A_{1k} \\ 0 & I_{r_2} & \dots & A_{2k} \\ \dots & \dots & \dots & \dots \\ 0 & 0 & \dots & I_{r_k} \end{pmatrix}.$$

(3) *N_ξ leaves invariant the Busemann function $b_{c'}$ if $c'(\infty) = \xi$.*

Proof. The inclusion $N_\xi \subset G_\xi$ is immediate from (10.64) or (10.26). It also follows from (10.26) that N_ξ leaves invariant the Busemann functions associated to rays c' with $c'(\infty) = \xi$. Part (2) is an easy calculation, so it only remains to check that N_ξ is normal in G_ξ. If $g \in G_\xi$ and $\nu \in N_\xi$ then $\exp(-tX)\,\nu\,\exp(tX) \to I$ as $t \to \infty$ and $\exp(-tX)\,g\,\exp(tX)$ converges to some element of $GL(n, \mathbb{R})$, therefore $\exp(-tX)\,g\nu g^{-1}\,\exp(tX)$, which is the conjugate of $\exp(-tX)\,\nu\,\exp(tX)$ by $\exp(-tX)\,g\,\exp(tX)$, converges to I. $\square$

Part (3) of the above proposition implies that the orbits of N_ξ are contained in horospheres about ξ. In general N_ξ is not the largest normal subgroup of G_ξ satisfying this property. For example, if $c(t) = \exp(\frac{t}{\sqrt{n}}I)$ then N_ξ is trivial but $SL(n, \mathbb{R})$ preserves the horospheres centred at $c(\infty)$.

Next we characterize $F(c)$, which we defined to be the union of the geodesic lines that are parallel to the given geodesic line $c : \mathbb{R} \to P(n, \mathbb{R})$.

10.67 Proposition. *If $c(t) = \exp(tX)$ is a geodesic ray in $P(n, \mathbb{R})$, then $F(c) \subseteq P(n, \mathbb{R})$ consists of those positive-definite, symmetric matrices which commute with $\exp X$.*

If $X = \mathrm{diag}\,(\lambda_1, \dots, \lambda_k)$ is as in (10.64), then $F(c)$ is the set of matrices $g = (g_{ij}) \in P(n, \mathbb{R})$ such that $g_{ij} = 0$ if $\lambda_i \neq \lambda_j$:

$$F(c) = \begin{pmatrix} P(r_1, \mathbb{R}) & 0 & \dots & 0 \\ 0 & P(r_2, \mathbb{R}) & \dots & 0 \\ 0 & 0 & \dots & 0 \\ 0 & 0 & \dots & P(r_k, \mathbb{R}) \end{pmatrix}.$$

This is a totally geodesic submanifold isometric to the product $\prod_{i=1}^k P(r_i, \mathbb{R})$.

Proof. Let $\xi = c(\infty)$ and let $-\xi = c(-\infty)$. Given $p \in P(n, \mathbb{R})$ we may choose $g \in G_\xi$ such that $p = g.I = g\,{}^t g$ (see 10.65). Since $t \mapsto g.c(t)$ is the unique geodesic line through p such that $g.c(\infty) = \xi$, by the flat strip theorem (2.13), $p \in F(c)$ if and only if $g.c(\infty) = \xi$ and $g.c(-\infty) = -\xi$; in other words, if and only if $g \in G_\xi \cap G_{-\xi}$.

We restrict our attention to the case $c(t) = \exp(tX)$ with $X = \mathrm{diag}(\lambda_1, \dots, \lambda_k)$ as in (10.64). Write $g = (g_{ij})$. By applying (10.64) to $t \mapsto \exp(tX)$ and to $t \mapsto$

$\exp(t(-X))$ we see that $g \in G_\xi \cap G_{-\xi}$ if and only if $g_{ij} = 0$ when $\lambda_i \neq \lambda_j$. This implies that $p = (p_{ij})$ is a symmetric matrix with $p_{ij} = 0$ when $\lambda_i \neq \lambda_j$.

The $p \in P(n,\mathbb{R})$ that commute with $\exp X$ are precisely those with $p_{ij} = 0$ when $\lambda_i \neq \lambda_j$. We have shown that $F(c)$ is contained in this set of matrices. Conversely, any such matrix can be written as $p = \exp Y$, where Y is a symmetric matrix that commutes with X. In this case $d(\exp Y \exp(tX/2).I,\ \exp(tX/2).I) = d(\exp(-tX/2)\exp Y \exp(tX/2).I,\ I) = d(\exp Y.I,\ I)$ is independent of t, hence the geodesic $t \mapsto \exp Y \exp(tX/2).I$, which passes through p, is parallel to $t \mapsto \exp(tX/2).I = c(t)$. □

10.68 Corollary. *If c is a regular geodesic in $P(n,\mathbb{R})$ then $F(c)$ is the unique maximal flat containing $c(\mathbb{R})$.*

Proof. By conjugating we may assume that $c(t) = \exp(tX)$ where all of the eigenvalues of $\exp X$ are positive and distinct. By (10.45), the set of matrices in $P(n,\mathbb{R})$ which commute with $\exp X$ is the maximal flat $A = \exp \mathfrak{a}_0$. □

We know by (2.14) that $F(c) \subset P(n,\mathbb{R})$ is convex and isometric to $F(c)_0 \times \mathbb{R}$, where $F(c)_0$ is the set of points of $F(c)$ whose projection on $c(\mathbb{R})$ is $c(0) = I$. The horospheres centred at $\xi = c(\infty)$ intersect $F(c)$ orthogonally in the parallel translates of $F(c)_0$.

10.69 Proposition (Busemann Functions in $P(n,\mathbb{R})$). *Let $c(t) = \exp(tX)$ be a geodesic line in $P(n,\mathbb{R})$, where $x \in S(n,\mathbb{R})$ and $\|X\| = 1$. Let $\xi = c(\infty)$. Then:*

(1) *The map $N_\xi \times F(c) \to P(n,\mathbb{R})$ sending (ν, p) to $\nu.p = \nu p\, {}^t\nu$ is a diffeomorphism.*

(2) *The Busemann function b_c associated to the geodesic c is given by the formula $b_c(\nu.p) = -\mathrm{Tr}(XY)$, where $p = \exp Y \in F(c)$ and $\nu \in N_\xi$.*

Proof. After conjugation we may assume that X is a diagonal matrix

$$X = \begin{pmatrix} \mu_1 I_{r_1} & 0 & \dots & 0 \\ 0 & \mu_2 I_{r_2} & \dots & 0 \\ 0 & 0 & \dots & 0 \\ 0 & 0 & \dots & \mu_k I_{r_k} \end{pmatrix},$$

with $\mu_1 > \dots > \mu_k$ and $r_1 + \dots + r_k = n$.

To prove that the map $N_\xi \times F(c) \to P(n,R)$ is bijective, we shall argue by induction on k, the number of distinct eigenvalues of X. In the case $k = 1$ this map is trivially a bijection: N_ξ is trivial and $F(c) = P(n,\mathbb{R})$. For the inductive step we write (n,n) matrices $g \in M(n,\mathbb{R})$ in block form $g = \begin{pmatrix} g_{11} & g_{12} \\ g_{21} & g_{22} \end{pmatrix}$, where $g_{22} \in M(r_k,\mathbb{R})$. We want to show that, given $p \in P(n,\mathbb{R})$, the equation $\nu f\, {}^t\nu = p$ has a unique solution $(\nu, f) \in N_\xi \times F(c)$. In the light of (10.66) and (10.67), this equation takes the form:

$$\begin{pmatrix} \nu_{11} & \nu_{12} \\ 0 & I_{r_k} \end{pmatrix} \begin{pmatrix} f_{11} & 0 \\ 0 & f_{22} \end{pmatrix} \begin{pmatrix} {}^t\nu_{11} & 0 \\ {}^t\nu_{12} & I_{r_k} \end{pmatrix} = \begin{pmatrix} p_{11} & p_{12} \\ p_{21} & p_{22} \end{pmatrix},$$

where $p_{21} = {}^t p_{12}$. Thus it is equivalent to the three equations:

$$\begin{aligned} f_{22} &= p_{22} \\ \nu_{12} f_{22} &= p_{12} \\ \nu_{11} f_{11} {}^t\nu_{11} + \nu_{12} f_{22} {}^t\nu_{12} &= p_{11}. \end{aligned}$$

From the first two of these equations we get $f_{22} = p_{22}$ and $\nu_{12} = p_{12} f_{22}^{-1}$, and the third equation becomes

$$\nu_{11} f_{11} {}^t\nu_{11} = p_{11} - p_{12} p_{22}^{-1} p_{21}.$$

If we can show that the symmetric matrix $p_{11} - p_{12} p_{22}^{-1} p_{21}$ is positive-definite then it will follow from our inductive hypothesis that this last equation has a unique solution (ν_{11}, f_{11}) of the required form.

To say that p is positive definite means precisely that ${}^t x p x > 0$ for all non-zero $(n, 1)$-matrices (column vectors) x. Given an $(n - r_k, 1)$-matrix $x_1 \neq 0$, we let $x_2 = -p_{22}^{-1} p_{21} x_1$ and calculate ${}^t x p x$ with ${}^t x = ({}^t x_1, {}^t x_2)$. This yields ${}^t x_1 p_{11} x_1 - {}^t x_1 p_{12} p_{22}^{-1} p_{21} x_1 > 0$, thus $p_{11} - p_{12} p_{22}^{-1} p_{21}$ is positive-definite.

Fix $p \in F(c)$. By 10.66(3) the Busemann function b_c is constant on the orbits of N_ξ, hence $b_c(\nu.p) = b_c(p)$ for all $\nu \in N_\xi$. By (10.67), $p = \exp Y$ where $[X, Y] = 0$, thus using 10.42(2) we get:

$$\begin{aligned} d(p, c(t))^2 &= d(\exp Y, \exp tX)^2 \\ &= d(\exp(-tX/2) \exp Y \exp(-tX/2),\ I)^2 \\ &= d(\exp(Y - tX),\ I)^2 \\ &= \mathrm{Tr}\big((Y - tX)(Y - tX)\big) \\ &= \mathrm{Tr}(YY) - 2t\mathrm{Tr}(XY) + t^2. \qquad (\diamond) \end{aligned}$$

By definition $b_c(p) = \lim_{t\to\infty} d(p, c(t)) - t$. And from the triangle inequality for $\Delta(p, c(0, c(t))$ we get $\lim_{t\to\infty}(d(p, c(t)) + t)/2t = 1$. Therefore

$$b_c(p) = \lim_{t\to\infty} \frac{1}{2t}(d(x, c(t))^2 - t^2),$$

and hence $(\diamond)$ implies that $b_c(p) = -\mathrm{Tr}(XY)$. □

10.70 Exercise. Deduce that if $c_1(\infty) = c_2(\infty) = \xi$ then $F(c_1)$ and $F(c_2)$ are isometric.

The Tits Boundary of $P(n, \mathbb{R})_1$ is a Spherical Building

The purpose of this section is to describe the Tits boundary of $P(n, \mathbb{R})$, as defined in (9.18). Our previous results lead one to expect that $\partial_T P(n, \mathbb{R})$ should enjoy considerable structure: in Chapter 9 we saw that the geometry of flats in a complete CAT(0) space influences the Tits boundary of the space, and in the present chapter we have seen some indication of the extent to which the geometry of $P(n, \mathbb{R})$ is dictated by the nature of its flat subspaces.

10.71 Theorem. *The Tits boundary of $P(n, \mathbb{R})_1$ is (naturally isometric to) a thick spherical building.*

In an appendix to this chapter the reader will find a brief introduction to spherical buildings, a remarkable class of CAT(1) spherical simplicial complexes discovered by Jacques Tits. The defining properties of a spherical building involve the behaviour of its maximal simplices (chambers) and certain distinguished subcomplexes (apartments). In our setting, the role of chambers and apartments will be played by the boundaries at infinity of Weyl chambers and maximal flats (see 10.72).

An Outline of the Proof of Theorem 10.71. First we shall define certain subsets of $\partial P(n, \mathbb{R})_1$ to be apartments and chambers (10.72). We shall then fix our attention on a basic chamber $\overline{A^+}(\infty)$. In the Tits metric $\overline{A^+}(\infty)$ is isometric to a spherical simplex. We metrize $\partial P(n, \mathbb{R})_1$ as a spherical simplicial complex by transporting the metric (and face structure) from $\overline{A^+}(\infty)$ to its translates under the action of $SL(n, \mathbb{R})$ (see 10.75). There are then two key points to check: the intrinsic metric associated to this spherical simplicial structure on $\partial P(n, \mathbb{R})_1$ coincides with the Tits metric (10.78); and the apartments and chambers of $\partial P(n, \mathbb{R})_1$, as defined in (10.72), satisfy the axioms for a (thick) spherical building (10A.1). Both of these points are easy consequences of Proposition (10.77).

10.72 Definition (Apartments and Weyl Chambers at Infinity). A subset of $\partial P(n, \mathbb{R})_1$ is called an *apartment* if it is the boundary at infinity of a maximal flat in $P(n, \mathbb{R})_1$. A subset of $\partial P(n, \mathbb{R})_1$ is called a *Weyl chamber at infinity* if it is the boundary at infinity of the closure of a Weyl chamber in $P(n, \mathbb{R})_1$.

There is not an exact correspondence between Weyl chambers at infinity and Weyl chambers in $P(n, \mathbb{R})_1$. Indeed a Weyl chamber in $P(n, \mathbb{R})_1$ is contained in a unique maximal flat whereas a Weyl chamber at infinity is contained in infinitely many apartments (10.74). On the other hand, the correspondence between apartments and maximal flats is exact: this is an immediate consequence of (10.53) and the following lemma.

10.73 Lemma. *Let A and A' be maximal flats in $P(n, \mathbb{R})$. If $\partial A = \partial A'$ then $A = A'$.*

Proof. Let c be a regular geodesic line in A and let c' be a geodesic line in A' such that $c'(\infty) = c(\infty) = \xi$. Note that $c'(-\infty) = c(-\infty)$, because this is the unique point of $\partial A = \partial A'$ at Tits distance π from ξ. Thus the geodesic lines c and c' are parallel, and hence $F(c') = F(c)$. But $F(c) = A$, by (10.68), so $A' \subset A$. □

10.74 Simplices in $\partial_T P(n, \mathbb{R})_1$. We wish to describe a spherical simplicial structure on $\partial_T P(n, \mathbb{R})_1$ in which the maximal simplices are the Weyl chambers at infinity and the apartments are isometrically embedded.

The apartments of $\partial_T P(n, \mathbb{R})_1$ are isometrically embedded copies of $\mathbb{S}^{n-2}$. Following our discussion in (10.45) and (10.54), we know that every Weyl chamber in $P(n, \mathbb{R})_1$ is a translate by some $g \in SL(n, \mathbb{R})$ of

$$A^+ = \{\exp(sX) \mid s > 0, X = \operatorname{diag}(t_1, \dots, t_n), \sum_{i=0}^{n} t_i = 0 \text{ and } t_1 > \cdots > t_n\}.$$

Therefore every Weyl chamber at infinity is a translate by some $g \in SL(n, \mathbb{R})$ of the closure $\overline{A^+(\infty)}$ of $A^+(\infty) := \{c(\infty) \mid c[0, \infty) \subseteq A^+ \text{ a geodesic ray}\}$, which in the Tits metric is isometric to a spherical simplex of dimension $(n-2)$ (contained in the copy of $\mathbb{S}^{n-2}$ that is the boundary of the maximal flat containing A^+). We described the shape of this simplex in (10.54).

The codimension-one faces of $\overline{A^+(\infty)}$ are obtained by replacing one of the inequalities $t_i > t_{i+1}$ by an equality, and faces of codimension k are obtained by replacing k of these inequalities by equalities.

In order to show that the translates $g.\overline{A^+(\infty)}$ of $\overline{A^+(\infty)}$ by the elements $g \in SL(n, \mathbb{R})$ are the $(n-2)$ simplices of a simplicial subdivision of $\partial_T P(n, \mathbb{R})_1$, we must check that every $\xi \in \partial_T P(n, \mathbb{R})_1$ lies in a translate of $\overline{A^+(\infty)}$ and that $g.\overline{A^+(\infty)} \cap \overline{A^+(\infty)}$ is a (possibly empty) face of $\overline{A^+(\infty)}$.

10.75 Proposition.

(1) *The spherical $(n-2)$-simplex $\overline{A^+(\infty)}$ is a strict fundamental domain for the natural action of $SL(n, \mathbb{R})$ on $\partial P(n, \mathbb{R})_1$ (i.e., $\overline{A^+(\infty)}$ contains exactly one point from each $SL(n, \mathbb{R})$-orbit).*

(2) *For every $g \in SL(n, \mathbb{R})$, the intersection $g.\overline{A^+(\infty)} \cap \overline{A^+(\infty)}$ is a (possibly empty) face of $\overline{A^+(\infty)}$.*

(3) *$SL(n, \mathbb{R})$ acts transitively on pairs $(\partial A, C)$, where ∂A is an apartment and $C \subset \partial A$ is a Weyl chamber at infinity.*

(4) *For every $\xi \in \partial P(n, \mathbb{R})_1$, the stabilizer of ξ in $SL(n, \mathbb{R})$ coincides with the stabilizer of the unique open simplex containing ξ.*

(5) *The orbits of $SL(n, \mathbb{R})$ in $\partial P(n, \mathbb{R})_1$ are the same as the orbits of $SO(n)$.*

Proof. First we prove (5). Given $\xi \in \partial P(n, \mathbb{R})_1$, let $c : [0, \infty) \to P(n, \mathbb{R})_1$ be the geodesic ray with $c(0) = I$ and $c(\infty) = \xi$. Given $g \in SL(n, \mathbb{R})$, let $c' = g \circ c$ and let $\xi' = c'(\infty) = g.\xi$. There is an element $h \in SL(n, \mathbb{R})$ fixing ξ' such that $h.c'(0) = I$ (see 10.65). Therefore hg lies in the stabilizer of I, which is $SO(n)$ (10.33), and $hg.\xi = \xi'$.

Every $\xi \in \partial P(n, \mathbb{R})$ can be represented uniquely as $c(\infty)$ where $c(t) = \exp(tX)$ is a geodesic through I. Now, since X is symmetric, there exists a unique diagonal matrix X' with non-increasing entries along the diagonal such that $X' = gX\,{}^tg$ for some $g \in SO(n)$. It follows that $c'(\infty)$, is the unique point of $\overline{A^+}(\infty)$ in the $SO(n)$-orbit of ξ, where $c'(t) = \exp(tX')$. In the light of (5), this proves (1). Part (4) then follows easily from (10.64). To prove (2), note that if x lies in the given intersection, by (1) we have $g.x = x$ and then by (4) we have that this intersection is a face.

It remains to prove (3). Since $SL(n, \mathbb{R})$ acts transitively on the set of Weyl chambers in $P(n, \mathbb{R})_1$, it acts transitively on the set of Weyl chambers at infinity. Thus it is sufficient to prove (3) for apartments containing $A^+(\infty)$. But in the light of (4), this follows immediately from the fact that for every $\xi \in \partial P(n, \mathbb{R})_1$ the stabilizer of ξ acts transitively on the set of geodesics c with $c(\infty) = \xi$ (10.65) and if ξ is in the interior of $\overline{A^+}(\infty)$ then each such c is contained in a unique maximal flat, therefore G_ξ acts transitively on these flats. □

At this stage we have proved:

10.76 Corollary. *The Weyl chambers at infinity are the maximal simplices in a spherical simplicial complex whose underlying set is $\partial P(n, \mathbb{R})_1$. The natural action of $SL(n, \mathbb{R})$ on $\partial P(n, \mathbb{R})_1$ is by simplicial isometries.*

It remains to check that the intrinsic metric associated to this spherical simplicial structure coincides with the Tits metric. This is the first consequence that we shall draw from the following proposition (whose proof will be completed in 10.80). This proposition also implies that the simplicial structure that we have described satisfies axiom (3) in the definition of a building (10A.1).

10.77 Proposition. *Any two Weyl chambers at infinity are contained in a common apartment.*

Proof. Let $A \subset P(n, \mathbb{R})_1$ be the maximal flat consisting of diagonal matrices. In the light of 10.75(3), it suffices to show that given any Weyl chamber at infinity C there is an element of $SL(n, \mathbb{R})$ that leaves $\overline{A^+}(\infty)$ invariant and maps C into ∂A.

From 10.64 and 10.75(4) we know that the stabilizer of $\overline{A^+}(\infty)$ is the group of upper triangular matrices in $SL(n, \mathbb{R})$, and from 10.54 we know that any other Weyl chamber in ∂A can obtained from $\overline{A^+}(\infty)$ by the action of a permutation matrix. Thus the proposition follows from the fact that any element of $SL(n, \mathbb{R})$ can be written in the form $g_1 s g_2$, where g_1 and g_2 are upper triangular matrices and s is a monomial matrix (this crucial fact will be proved in 10.80). For if $C = g.\overline{A^+}(\infty)$ and $g = g_1 s g_2$, then $g_1^{-1}.C = s.\overline{A^+}(\infty)$, whence both C and $\overline{A^+}(\infty)$ are contained in $g_1.\partial A$. □

10.78 Corollary. *The Tits metric on $\partial P(n, \mathbb{R})_1$ coincides with the intrinsic metric of the spherical simplicial complex obtained by dividing $\partial P(n, \mathbb{R})_1$ into Weyl chambers at infinity.*

Proof. Let d_T denote the Tits metric on $\partial P(n, \mathbb{R})_1$ and let d be the intrinsic metric for the spherical simplicial structure on $\partial P(n, \mathbb{R})_1$. The latter is obtained by taking the infimum of the lengths of piecewise geodesic paths in the spherical simplicial complex. The length of any such path is the same when measured in both the Tits metric and d, because the Tits metric coincides with the simplicial metric on the closure of each Weyl chamber. Thus $d_T(\xi, \xi') \leq d(\xi, \xi')$ for all $\xi, \xi' \in \partial P(n, \mathbb{R})_1$.

On the other hand, in the light of (10.77) we know that ξ, ξ' lie in a common apartment. In the Tits metric this apartment is isometric to $\mathbb{S}^{n-2}$, and any Tits geodesic joining ξ to ξ' obviously has length $d_T(\xi, \xi')$ when measured in the metric d. Thus $d_T(\xi, \xi') \geq d(\xi, \xi')$. □

The Final Step in the Proof of Theorem 10.71

The axioms of a building are given in (10A.1). In the light of the preceding results only axiom (3) requires further argument.

First we consider the case where two apartments $E = \partial A$ and $E' = \partial A'$ contain a chamber C in their intersection. In this case (10.75) implies there is a simplicial isometry $\phi_{E,E'}$ from E' onto E that restricts to the identity on C. We claim that $\phi_{E,E'}$ restricts to the identity on the whole of $E \cap E'$. To see this, note that if $x \in E \cap E'$ then x is a distance less than π from some point y in the open set C; the restriction of $\phi_{E,E'}$ to an initial segment of the unique geodesic $[y, x]$ is the identity, and since $[y, x]$ is the only geodesic of length $d(x, y)$ in E that has this initial segment, $\phi_{E,E'}(x) = x$.

Now suppose that B_1 and B_2 are (possibly empty) simplices in the intersection of two apartments E_1 and E_2. For $i = 1, 2$ we choose chambers $C_i \subset E_i$ such that $B_i \subset C_i$. By (10.77) there is an apartment E' containing both C_1 and C_2. The composition of the maps ϕ_{E,E_1}^{-1} and ϕ_{E,E_2} sends E_2 isometrically onto E_1 and restricts to the identity on $B_1 \cup B_2$. □

$\partial_T P(n, \mathbb{R})$ in the Language of Flags and Frames

The following interpretation of apartments and Weyl chambers at infinity provides a useful language when one wishes to bring the tools of linear algebra to bear on the study of $\partial_T P(n, \mathbb{R})$. For example, the content of (10.80) was crucial in our proof of (10.78).

A *flag* in $\mathbb{R}^n$ is a sequence of subspaces $V_1 \subset V_2 \cdots \subset V_{n-1}$ such that V_i has dimension i. An *unordered frame* in $\mathbb{R}^n$ is a set of n linearly independent subspaces of dimension one, and an *ordered frame* is a sequence of n linearly independent subspaces of dimension one. (Thus every unordered frame gives rise to $n!$ ordered frames.) Associated to each ordered frame one has the flag $V_1 \subset \cdots \subset V_{n-1}$, where V_k is the subspace spanned by the first k elements of the ordered frame.

We write fl_0 to denote the flag associated to the standard (ordered) basis $e_1, \ldots, e_n$ for $\mathbb{R}^n$, and we write fr_0 to denote the associated unordered frame (the set of one-dimensional subspaces $\langle e_i \rangle$). We shall also use the notation:

$\mathcal{A} =$ the set of maximal flats in $P(n, \mathbb{R})_1$ (= the set of apartments in $\partial P(n, \mathbb{R})_1$)

$\mathcal{C} =$ set of Weyl chambers at infinity

$\mathcal{F}r =$ set of unordered frames in $\mathbb{R}^n$

$\mathcal{F}l =$ set of flags in $\mathbb{R}^n$.

The obvious action of $SL(n, \mathbb{R})$ on each of $\mathcal{F}r$ and $\mathcal{F}l$ is transitive, and we saw earlier that the action of $SL(n, \mathbb{R})$ on $\mathcal{A}$ and $\mathcal{C}$ is transitive.

10.79 Proposition. *There are natural $SL(n, \mathbb{R})$-equivariant bijections $\mathcal{A} \to \mathcal{F}r$ and $\mathcal{C} \to \mathcal{F}l$, and an induced bijection from $\{(A, C) \mid A \in \mathcal{A},\ C \subset \partial A\}$ to the set of ordered frames in $\mathbb{R}^n$.*

Proof. The group of upper triangular matrices in $SL(n, \mathbb{R})$ is the stabilizer of both $\overline{A^+}(\infty)$, the basic Weyl chamber at infinity, and the flag fl_0. Thus, since the action of $SL(n, \mathbb{R})$ is transitive on both $\mathcal{C}$ and $\mathcal{F}l$, then $g.\overline{A^+}(\infty) \mapsto g.\mathrm{fl}_0$ is an equivariant bijection from $\mathcal{C}$ to $\mathcal{F}l$.

Let A_0 be the maximal flat in $P(n, \mathbb{R})_1$ consisting of diagonal matrices. The stabilizer in $SL(n, \mathbb{R})$ of A_0 (equivalently ∂A_0) is the subgroup consisting of monomial matrices, i.e. matrices having only one non-zero entry in each row and column. This subgroup is also the stabilizer of the unordered frame fr_0. Thus $g.A \mapsto g.\mathrm{fr}_0$ is an equivariant bijection from $\mathcal{A}$ to $\mathcal{F}r$.

$SL(n, \mathbb{R})$ acts transitively on the set of ordered frames and also on the set of pairs $\{(A, C) \mid A \in \mathcal{A},\ C \subset \partial A\}$. The stabilizer of $(A_0, A^+(\infty))$ is the group of diagonal matrices of determinant one, and this is also the stabilizer of the ordered frame associated to the standard basis of $\mathbb{R}^n$. Hence the required equivariant bijection. □

10.80 Lemma. *Given two flags $V_1 \subset \cdots \subset V_{n-1}$ and $V_1' \subset \cdots \subset V_{n-1}'$, one can find a basis $v_1, \dots, v_n$ for $\mathbb{R}^n$ and a permutation σ of $\{1, \dots, n\}$ such that V_i is the span of $\{v_i \mid j \leq i\}$ for each $i \leq n$ and V_i is the span of $\{v_{\sigma(j)} \mid j \leq i\}$.*

Sketch of proof. In order to prove this lemma one analyzes the standard proof of the Jordan-Hölder Theorem (see [Bro88, p.84]). We shall give the proof for $n = 3$ and indicate how the different cases that arise in the proof are related to the geometry of Weyl chambers as described in 10.55 (and figure 10.1). It follows from the description given there that each apartment $\partial A \subset \partial_T P(n, \mathbb{R})_1$ is an isometrically embedded circle and the Weyl chambers at infinity contained in ∂A are six arcs of length $\pi/3$. The action of the symmetric group $\mathcal{S}_3$ on the set of Weyl chambers in an apartment is described in (10.55).

Let $V_1 \subset V_2$ and $V_1' \subset V_2'$ be the two distinct flags in $\mathbb{R}^n$ and let C and C' be the corresponding Weyl chambers at infinity. We seek a basis $\{v_1, v_2, v_3\}$ of $\mathbb{R}^3$ and a permutation σ such that V_1 (resp. V_1') is spanned by v_1 (resp. $v_{\sigma(1)}$) and V_2 (resp. V_2') is spanned by $\{v_1, v_2\}$ (resp. $\{v_{\sigma(1)}, v_{\sigma(2)}\}$). As we noted above, the present lemma implies that there is an apartment A containing both C and C', and the permutation σ describes the action of $\mathcal{S}_3$ on A (see 10.55). With this geometric picture in mind, one expects there to be essentially three cases to consider: C and C' might be adjacent chambers of A, they might be opposite, or they might be neither.

First case (adjacent chambers): $V_1 = V_1'$ or $V_2 = V_2'$.

If $V_1 = V_1'$, we choose the basis $\{v_1, v_2, v_3\}$ so that $\{v_1, v_2\}$ spans V_2 and $\{v_1, v_3\}$ spans V_2'. The permutation σ is the transposition $(2, 3)$.

If $V_2 = V_2'$, we choose the basis $\{v_1, v_2, v_3\}$ so that v_1 (resp. v_2) spans V_1 (resp. V_1') and $v_3 \notin V_2$. In this case σ is the transposition $(1, 2)$.

Second case (opposite chambers): $V_1 \neq V_1'$, $V_2 \neq V_2'$ and $V_2 \cap V_2' \neq V_1$ or V_1'. This is the generic case.

We choose the basis $\{v_1, v_2, v_3\}$ so that V_1 is spanned by v_1, $V_2 \cap V_2'$ by v_2 and V_1' by v_3. The permutation σ is $(3, 2, 1)$. In this case there is only one apartment containing the given Weyl chambers C and C'.

Third case: $V_2 \cap V_2' = V_1 \neq V_1'$ (the case $V_2 \cap V_2' = V_1' \neq V_1$ is similar).

We choose $\{v_1, v_2, v_3\}$ so that V_1 is spanned by v_1, V_2 by $\{v_1, v_2\}$ and V_1' by v_3. The permutation σ is $(3, 1, 2)$. In this case the Weyl chambers are neither adjacent nor opposite in any common apartment. □

Appendix: Spherical and Euclidean Buildings

Spherical buildings were first introduced by Jacques Tits [Tits74] "as an attempt to give a systematic procedure for the geometric interpretation of the semi-simple Lie groups, in particular the exceptional groups". Euclidean buildings emerged from the study of p-adic Lie groups [IwMa65] and their theory was developed by Bruhat and Tits [BruT72], who established that Euclidean buildings were CAT(0) in the course of proving a version of the Cartan fixed point theorem. (They used the CAT(0) inequality in the guise of 1.9(1c)). Tits later defined abstract buildings as combinatorial objects (chamber systems) whose basic structure can be described in terms of Coxeter groups.

There is an extensive literature approaching the subject of buildings from various perspectives and we shall not attempt to replicate it. Instead, we shall give a rudimentary introduction to spherical and Euclidean buildings with the main objective of proving that the axioms of a building imply upper curvature bounds (Theorem 10A.4). We refer the reader to the books [Bro88], [Ron89] and [Tits74] for a much more comprehensive introduction to buildings. The survey articles of Ronan [Ron92] also contain a lot of information. For a less complete but very readable introduction we suggest [Bro91]. (A useful introduction to Coxeter groups, by Pierre de la Harpe [Har91], can be found in the same volume as [Bro91].) We also recommend to the reader the more metric approach to buildings presented by Mike Davis in [Da98].

The following definition of spherical and Euclidean buildings is not the most usual one, but it fits naturally with the geometric viewpoint of this book. Our definition of a thick building is equivalent to the usual one (see [Bro88, IV, VI] or [Ron89])

10A.1 Definition. A Euclidean (resp. spherical) building of dimension n is a piecewise Euclidean (resp. spherical) simplicial complex X such that:

(1) X is the union of a collection $\mathcal{A}$ of subcomplexes E, called apartments, such that the intrinsic metric d_E on E makes (E, d_E) isometric to the Euclidean space $\mathbb{E}^n$ (resp. the n–sphere $\mathbb{S}^n$) and induces the given Euclidean (resp. spherical) metric on each simplex. The n-simplices of E are called its chambers.

(2) Any two simplices B and B' of X are contained in at least one apartment.

(3) Given two apartments E and E' containing both the simplices B and B', there is a simplicial isometry from (E, d_E) onto $(E', d_{E'})$ which leaves both B and B' pointwise fixed.

The building X is called thick if the following extra condition is satisfied:

(4) *Thickness Condition:* Any $(n-1)$-simplex is a face of at least three n-simplices.

10A.2 Remarks

(1) For the purposes of this definition, the 0-dimensional sphere $\mathbb{S}^0$ is defined to consist of two points a distance π apart.

(2) In condition (3), the simplices B or B' can be empty.

(3) It is usual to say that a building of dimension n has "rank $(n+1)$". This superficially odd convention means that the rank of a symmetric space such as $P(n, \mathbb{R})_1$ is the same as the rank of its Tits boundary. (See also 10A.7.)

10A.3 Examples

(1) Any metric space X such that the distance between any two distinct points is π is a spherical building of dimension 0, where the apartments are the pairs of distinct points of X.

(2) Let X be as in (1). The spherical suspension (as defined in Chapter I.5) of X is a spherical building of dimension 1 whose apartments are circles, namely the suspensions of the apartments of X.

More generally, if X and X' are spherical buildings of dimension n and n', then their spherical join $X * X'$ is a spherical building of dimension $(n + n' + 1)$ whose simplices (resp. apartments) are the spherical join of the simplices (resp. apartments) of X and X'.

(3) If the edges a metric simplicial tree all have length 1 and every vertex of the tree is adjacent to at least two (resp. three) edges, then the tree is a Euclidean building (resp. a thick Euclidean building) of dimension one.

(4) In the final sections of the preceding chapter we showed that, when equipped with the Tits metric, the boundary at infinity of the symmetric space $P(n, \mathbb{R})_1$ is a spherical building of rank $(n - 1)$. In general the Tits boundary of any symmetric space of non-compact type is a spherical building whose rank is the rank of the symmetric space (see [Eb97]).

(5) Let k be a field and consider the simplicial graph X whose vertices are the 1-dimensional and 2-dimensional subspaces of k^3, and which has an edge joining

two vertices if and only if one of the corresponding subspaces is contained in the other. In other words, X is the incidence graph of points and lines in the projective plane over k.

If one metrizes each edge to have length $\pi/3$, then X is a spherical building of dimension one in which the apartments are hexagons (circuits of combinatorial length 6). If k is the field with 2 elements, then X is a graph with 14 vertices and 21 edges, and every vertex is of valence three, corresponding to the fact that every line through the origin lies in exactly three planes and each of these planes contains exactly three lines through the origin. (Exercise: draw this graph.)

More generally, one can consider the poset of proper non-zero vector subspaces in k^n ordered by inclusion. A suitable geometric realization of this set is again a spherical building of rank $(n-1)$, whose apartments are isomorphic to the apartments in $\partial_T P(n, \mathbb{R})_1$; the chambers are in natural correspondence with flags in k^n and the apartments correspond to the unordered frames of k^n (cf. 10.79 and 10.80). There is an obvious action of $SL(n, k)$ on this building.

10A.4 Theorem.

(i) *The intrinsic metric on a Euclidean (resp. spherical) building X is the unique metric inducing the given metric d_E on each apartment E.*

(ii) *With this metric, X is a* CAT(0) *space (resp. a* CAT(1) *space).*

(iii) *If X is thick, then all of the n-simplices in X are isometric to a single geodesic n-simplex in $\mathbb{E}^n$ (resp. $\mathbb{S}^n$). (In this case it follows from (I.7.13) that the intrinsic metric is complete.)*

Proof. The retraction $\rho_{E,C}$. Let C be a chamber contained in an apartment E. If E' is another apartment containing C, then by 10A.1(3) there is a simplicial isometry $\phi_{E,E'}$ of E' onto E fixing C pointwise. This isometry is unique because the image of each geodesic is determined by the image of an initial segment contained in C. This uniqueness forces the restriction of $\phi_{E,E'}$ to $E \cap E'$ to be the identity.

We wish to define a simplicial retraction $\rho = \rho_{E,C}$, which is a simplicial map from X onto E. By 10A.1(2), each point $x \in X$ is contained in an apartment E' that contains C. Define $\rho(x) := \phi_{E,E'}(x)$. This definition is independent of the choice of E' because if E'' is another apartment containing C and x, then $\phi_{E',E''}$ fixes C and x and $\phi_{E,E'} \circ \phi_{E',E''} = \phi_{E,E''}$ because of the uniqueness established above. Note that ρ is a simplicial map and that its restriction to each chamber is an isometry. Moreover its restriction to each apartment containing C is an isometry.

The restriction of the intrinsic metric to E is d_E. Let d denote the intrinsic metric on X (see I.7.4). Because $\rho = \rho_{E,C}$ maps simplices to simplices isometrically, it preserves the length of piecewise geodesic paths (and m-strings). Thus, given $x, y \in E$, since $\rho|_E = \mathrm{id}_E$, when calculating $d(x, y)$ one need only quantify the infimum in the definition of d over paths (strings) contained in E. And this infimum is clearly equal to $d_E(x, y)$. This proves part (i) of the theorem. It also shows that X is a geodesic space (the geodesic joining x to y in E is a geodesic in X). In fact these are the only geodesics.

10A.5 Lemma. *Let E be an apartment in X, let $x, y \in E$, and in the spherical case assume $d(x, y) < \pi$. Then every geodesic in X connecting x to y is contained in E.*

Proof. Suppose that $z \in X$ is such that $d(x, y) = d(x, z) + d(z, y)$. Let z' be the point on the geodesic segment $[x, y]$ joining x to y in E with $d(x, z) = d(x, z')$. We must show that $z = z'$. Let C be a chamber in E containing z' and let $\rho = \rho_{E,C}$. Since ρ does not increase the length of paths, $d(x, \rho(z)) \leq d(x, z)$ and $d(y, \rho(z)) \leq d(y, z)$. And

$$d(x, y) \leq d(x, \rho(z)) + d(\rho(z), y) \leq d(x, z) + d(z, y) = d(x, y).$$

There must be equality everywhere, therefore $\rho(z) = z'$. On the other hand, there is an apartment E' containing both z and C, by 10A.1(2), and by definition $\rho(z) = \phi_{E,E'}(z)$ and $z' = \phi_{E,E'}(z')$. Since $\phi_{E,E'}$ is injective, we have $z = z'$. □

It follows immediately from this lemma that there is a unique geodesic joining each pair of points $x, y \in X$ (assuming $d(x, y) < \pi$ in the spherical case). It also follows that geodesics vary continuously with their endpoints, because if y' is close to y then there is a chamber C containing both y and y', and given x there exists an apartment containing x and C. The desired curvature bounds can thus be deduced from Alexandrov's patchwork (proof of (4.9)) or from (5.4). However, instead of appealing to these earlier results, we prefer to give a direct and instructive proof of the fact that X is CAT(0) (Euclidean case) or CAT(1) (spherical case).

The CAT(0) and CAT(1) Inequalities. Given $x, y, z \in X$ (with $d(x, y)+d(y, z)+d(z, x) < 2\pi$ in the spherical case), let p be a point on the unique geodesic segment $[x, y]$. Let E be an apartment containing $[x, y]$, let $C \subset E$ be a chamber containing p, and let $\rho = \rho_{E,C}$. We have $d(x, \rho(z)) \leq d(x, z)$, $d(y, \rho(z)) \leq d(y, z)$ and $d(p, \rho(z)) = d(p, z)$. Let $\bar{z}$ be a point in E such that $d(x, \bar{z}) = d(x, z)$ and $d(y, \bar{z}) = d(y, z)$. The triangle $\Delta(x, y, \bar{z}) \subset E$ is a Euclidean or spherical comparison triangle for $\Delta(x, y, z)$. Thus it suffices to prove that $d(p, z) \leq d(p, \bar{z})$.

Let $\bar{z}' \in E$ be such that $d(x, \bar{z}') = d(x, \rho(z))$ and $d(y, \bar{z}') = d(y, z)$. As $d(y, \rho(z)) \leq d(y, \bar{z}')$, the vertex angle at x in $\Delta(x, y, \rho(z))$ is less than that in $\Delta(x, y, \bar{z}')$, and hence (by the law of cosines) $d(p, \rho(z)) \leq d(p, \bar{z}')$. Similarly, comparing $\Delta(x, y, \bar{z}')$ with $\Delta(x, y, \bar{z})$, we see that $d(p, \bar{z}') \leq d(p, \bar{z})$, hence $d(p, z) = d(p, \rho(z)) \leq d(p, \bar{z})$. This proves (ii).

The case of a thick building. Let C and C' be distinct chambers in X. They lie in a common apartment and therefore can be connected by a *gallery*, i.e. a sequence of chambers $C = C_0, C_1 \dots, C_n = C'$ such that each consecutive pair C_i, C_{i+1} have an $(n-1)$-dimensional face in common. Thus it suffices to consider the case where C and C' have an $(n-1)$-dimensional face B in common.

By thickness, there exists a chamber C'', distinct from C and C', of which B is a face. Let E and E' be apartments containing $C \cup C'$ and $C \cup C''$ respectively. The restriction to C' of $\rho_{E,C'} \circ \rho_{E',C}$ sends C' isometrically onto C and fixes B. This completes the proof of Theorem 10A.4. □

10A.6 Remark. It follows from (10A.4) that Euclidean buildings are contractible. According to the Solomon-Tits theorem [Sol69], every spherical building of dimension n has the homotopy type of a wedge (one point union) of n-dimensional spheres.

10A.7 Coxeter Complexes. We continue with the notation from the last paragraph of the preceding proof. Let $\mathcal{H}$ be the half-space of E that contains C' and has B in its boundary. It is not difficult to show that the restriction of $\rho_{E,C'} \circ \rho_{E',C}$ to E is simply the reflection of E in the boundary of $\mathcal{H}$ (see [Bro88, IV.7]). It follows that if X is a thick building and E is an apartment, then the subgroup of Isom(E) generated by reflections in codimension-one faces acts transitively on the chambers in E. Further thought shows that any chamber is a strict fundamental domain for this action. This group of reflections, which we denote W, is actually a Coxeter group (as defined in 12.31) with generating system S, where S is the set of reflections in the codimension-one faces of a fixed chamber C (see [Bro88, IV.7]). E is (isomorphic to) the Coxeter complex for (W, S) and $|S|$ is the rank of the building. If W is finite then the building is spherical. If W is infinite then the building is Euclidean.

10A.8 Example. Consider the building $X = \partial_T P(n, \mathbb{R})_1$, which has rank $(n-1)$. In this case each apartment E is tesselated by $(n-2)$-simplices (the closures of the Weyl chambers at infinity), W is the symmetric group on n letters, and the preferred generating set S is the set of reflections in the $(n-1)$ maximal faces of a fixed chamber.

In the case $n = 3$, the apartments are regular hexagons (circles divided into six equal parts — see proof of 10.80), S has two elements and W has presentation $\langle s_1, s_2 \mid s_1^2 = s_2^2 = (s_1 s_2)^3 = 1\rangle$.

10A.9 Remarks

(1) In the definition of a Euclidean building (10A.1), instead of insisting that X be a simplicial complex, we might have assumed instead that the chambers of X were convex Euclidean polyhedra. For instance we might take the cells to be cubes. This leads to a more general notion of a Euclidean building, with the advantage that the product of two such Euclidean buildings is again a Euclidean building (cf. [Da98]).

An examination of the proofs given above reveals that they all remain valid if one adopts this more general definition of a Euclidean building.

(2) One can show that if X is a Euclidean building as defined in 10A.1, the boundary $\partial_T X$ of X with the Tits metric is a spherical building (see [Bro88]), as it is the case when X is a symmetric space of non-compact type. In his Habilitationsschrift [Le97], Bernhard Leeb established characterized irreducible Euclidean building and symmetric spaces in terms of their Tits boundary. He showed that if a proper CAT(0) proper space X has the geodesic extension property (5.7), and the Tits boundary of X is a thick, connected, spherical building, then X is either a symmetric space of non-compact type or a Euclidean building. (X will be a symmetric space if and only if every geodesic segment extends to a *unique line.*)

Chapter II.11 Gluing Constructions

In this chapter we revisit the gluing constructions described in Chapter I.5 and ask under what circumstances one can deduce that the spaces obtained by gluing are of curvature $\leq \kappa$. Our purpose in doing so is to equip the reader with techniques for building interesting new spaces of curvature $\leq \kappa$ out of the basic examples supplied in Chapters 5, 10 and 12.

Gluing CAT(κ) Spaces Along Convex Subspaces

If one glues complete CAT(κ) spaces along complete, convex, isometric subspaces, is the result a CAT(κ) space? We shall show that the answer to this question is yes. In the case of proper metric spaces this result is due to Reshetnyak [Resh60] and the proof is rather short. In the general case one has to overcome the additional problem that the existence of geodesics in the space obtained by gluing is far from obvious (cf. I.5.25(3)).

We maintain the notation established in Chapter I.5; in particular, given two metric spaces X_1 and X_2 and isometries $i_j : A \to A_j$ onto closed subspaces $A_j \subseteq X_j$, we write $X_1 \sqcup_A X_2$ for the quotient of the disjoint union of X_1 and X_2 by the equivalence relation generated by $[i_1(a) \sim i_2(a) \, \forall a \in A]$. (Suppressing mention of the maps i_j in this notation should not cause any confusion.)

11.1 Basic Gluing Theorem. *Let $\kappa \in \mathbb{R}$ be arbitrary. Let X_1 and X_2 be* CAT(κ) *spaces (not necessarily complete) and let A be a complete metric space. Suppose that for $j = 1, 2$, we are given isometries $i_j : A \to A_j$, where $A_j \subseteq X_j$ is assumed to be convex if $\kappa \leq 0$ and is assumed to be D_κ-convex if $\kappa \geq 0$. Then $X_1 \sqcup_A X_2$ is a* CAT(κ) *space.*

Proof.

Step 1. We first prove the theorem assuming that $X = X_1 \sqcup_A X_2$ is a geodesic metric space (which we know to be the case if X_1 and X_2 are proper I.5.24(3)). We shall verify criterion 1.7(4) for all geodesic triangles in X (with perimeter $\leq 2D_\kappa$ if $\kappa > 0$). The only non-trivial case to consider (up to reversing the roles of X_1 and X_2) is that of a geodesic triangle $\Delta([x_1, y], [y, x_2], [x_1, x_2])$ with $x_1, y \in X_1$ and $x_2 \in X_2 \setminus A_2$. For this, one fixes points $z, z' \in A_1$ on the sides $[x_1, x_2]$ and $[y, x_2]$ respectively.

Any such points have the property that $[x_1, z] \subseteq [x_1, x_2]$ and $[y, z'] \subseteq [y, x_2]$ are contained in X_1, and that $[z, x_2] \subseteq [x_1, x_2]$ and $[z', x_2] \subseteq [x_1, x_2]$ are contained in X_2. In order to complete the proof one simply applies the gluing lemma for triangles (4.10), first to $\Delta([x_1, z], [z, z'], [x_1, z'])$ and $\Delta([z, z'], [z', x_2], [z, x_2])$ and then to $\Delta([x_1, x_2], [x_2, z'], [x_1, z'])$ and $\Delta([x_1, y], [y, z'], [x_1, z'])$.

Step 2. We shall now prove that X is a geodesic metric space. As we do not assume that A_1 is proper, according to exercise I.5.25(3) we must use both the fact that X_1 and X_2 are CAT(κ) and the fact that the A_j are convex in order to show that $X = X_1 \sqcup_A X_2$ is a *geodesic* metric space. It suffices to show that for all $x_1 \in X_1$ and $x_2 \in X_2$ (with $d(x_1, x_2) < D_\kappa$ if $\kappa > 0$) there exists $z \in A_1$ such that $d(x_1, z) + d(z, x_2) = d(x_1, x_2)$. We fix a sequence of points $z_n \in A_1$ such that $d(x_1, z_n) + d(z_n, x_2) - d(x_1, x_2) \le 1/n$. Because the numbers $d(x_1, z_n)$ are bounded, we may pass to a subsequence and assume that $d(x_1, z_n) \to \ell_1$ as $n \to \infty$. Thus $d(x_2, z_n) \to \ell_2 := d(x_1, x_2) - \ell_1$ as $n \to \infty$. We claim that (z_n) is a Cauchy sequence; we will then be done because A_1 is complete and $z = \lim_n z_n$ has the desired property.

Before turning to the general case, we give a proof valid for $\kappa \le 0$. Given $\varepsilon > 0$, let $N > 0$ be such that $\max\{|d(x_1, z_n) - \ell_1|, |d(x_2, z_n) - \ell_2|\} < \varepsilon$ for all $n, m > N$. Let p be the midpoint of the geodesic segment $[z_n, z_m] \subseteq A_1$. By the convexity of the metric on X_2 we have:

$$d(x_2, p) \le \max\{d(z_n, x_2), d(z_m, x_2)\} \le l_2 + \varepsilon,$$

hence $d(x_1, p) \ge \ell_1 - \varepsilon$. For a comparison triangle $\overline{\Delta}(x_1, z_n, z_m)$ in $\mathbb{E}^2$, an elementary calculation with the Euclidean scalar product yields:

$$d(\bar{x}_1, \bar{p})^2 = \frac{1}{2} d(\bar{x}_1, \bar{z}_n)^2 + \frac{1}{2} d(\bar{x}_1, \bar{z}_m)^2 - \frac{1}{4} d(\bar{z}_n, \bar{z}_m)^2.$$

We also have $d(x_1, p) \le d(\bar{x}_1, \bar{p})$, by the CAT(0) inequality in X_1. Hence

$$\begin{aligned} d(z_n, z_m)^2 &\le 2\, d(x_1, z_n)^2 + 2\, d(x_1, z_m)^2 - 4\, d(x_1, p)^2 \\ &\le 2\, d(x_1, z_n)^2 + 2\, d(x_1, z_m)^2 - 4(\ell_1 - \varepsilon)^2 \le 16 \ell_1 \varepsilon. \end{aligned}$$

Step 3. We maintain the notation established in Step 2. The following proof that (z_n) is a Cauchy sequence is valid for any $\kappa \in \mathbb{R}$. By reversing the roles of ℓ_1 and ℓ_2 if necessary, we may assume that $\ell_1 \le d(x_1, x_2)/2 < D_\kappa/2$.

We fix a number l with $D_\kappa > l > d(x_1, x_2)$. As in (I.2.25), it follows from (I.2.26) that for every $\varepsilon > 0$ there exists $\delta = \delta(\kappa, l, \varepsilon)$ such that, for all $x, y, w \in M_\kappa^2$, if $d(x, y) < l$ and $d(x, w) + d(w, y) < d(x, y) + \delta$, then $d(w, [x, y]) < \varepsilon$. Shrinking δ, we may assume $d(x_1, x_2) + \delta < l$. We fix an integer N so that for all $n > N$ we have $d(x_1, z_n) < D_\kappa/2$ and $|d(x_1, z_n) - \ell_1| < \varepsilon$, and $d(x_1, z_n) + d(x_2, z_n) < d(x_1, x_2) + \delta$. If $n, m > N$, then $d(z_n, z_m) \le d(x_1, z_n) + d(x_1, z_m) < D_\kappa$, so z_n and z_m are connected by a unique geodesic segment (contained in the convex set A_1) and there are unique geodesic triangles Δ_1 in X_1 and Δ_2 in X_2 with vertices (x_1, z_n, z_m) and (x_2, z_n, z_m) respectively. The perimeter of each of these triangles is bounded by $d(x_1, z_n) + d(x_2, z_n) + d(x_1, z_m) + d(x_2, z_m) < 2l < 2D_\kappa$. We consider comparison triangles $\overline{\Delta}_1 = \Delta(\overline{x}_1, \overline{z}_n, \overline{z}_m)$ and $\overline{\Delta}_2 = \Delta(\overline{x}_2, \overline{z}_n, \overline{z}_m)$ in M^2_κ, joined along the common edge $[\overline{z}_n, \overline{z}_m]$ and arranged so that $\overline{x}_1$ and $\overline{x}_2$ are on opposite sides of the line through z_n and $\overline{z}_m$. (We assume that the triangles $\overline{\Delta}_1$ and $\overline{\Delta}_2$ are non-degenerate; the degenerate case is trivial.)

Consider the unique geodesic segment $[\overline{x}_1, \overline{x}_2]$ in M^2_κ. There are two cases:

(a) $[\overline{x}_1, \overline{x}_2]$ meets $[\overline{z}_n, \overline{z}_m]$ in one point $\overline{y}$, or

(b) $[\overline{x}_1, \overline{x}_2]$ does not meet $[\overline{z}_n, \overline{z}_m]$.

In case (a), if $y \in [z_n, z_m]$ is the point a distance $d(\overline{z}_n, \overline{y})$ from z_n, then by the CAT(κ) inequality,

$$d(x_1, x_2) \leq d(x_1, y) + d(y, x_2) \leq d(\overline{x}_1, \overline{y}) + d(\overline{y}, \overline{x}_2) = d(\overline{x}_1, \overline{x}_2).$$

Hence

$$d(\overline{x}_1, \overline{z}_n) + d(\overline{z}_n, \overline{x}_2) = d(x_1, z_n) + d(z_n, x_2) \leq d(x_1, x_2) + \delta \leq d(\overline{x}_1, \overline{x}_2) + \delta,$$

so by the definition of δ there exists $p_n \in [\overline{x}_1, \overline{x}_2]$ such that $d(\overline{z}_n, p_n) < \varepsilon$. Similarly, there exists $p_m \in [\overline{x}_1, \overline{x}_2]$ such that $d(\overline{z}_m, p_m) < \varepsilon$.

The difference between $d(\overline{x}_1, \overline{z}_n)$ and $d(\overline{x}_1, \overline{z}_m)$ is at most 2ε, so the distance between p_n and p_m is at most 4ε. Hence $d(z_n, z_m) = d(\overline{z}_n, \overline{z}_m) \leq 6\varepsilon$.

In case (b), the sum of the angles of $\overline{\Delta}_1$ and $\overline{\Delta}_2$ at one of the vertices $\overline{z}_n$ or $\overline{z}_m$ is not less than π; suppose this vertex is $\overline{z}_n$. According to I.2.16(1), there exists a triangle in M^2_κ with vertices $(\tilde{x}_1, \tilde{x}_2, \tilde{z}_m)$ such that $d(\tilde{x}_1, \tilde{z}_m) = d(x_1, z_m)$, $d(\tilde{x}_2, \tilde{z}_m) = d(x_2, z_m)$ and $d(\tilde{x}_1, \tilde{x}_2) = d(x_1, z_n) + d(z_n, x_2)$. Let $\tilde{z}_n \in [\tilde{x}_1, \tilde{x}_2]$ be such that $d(\tilde{x}_1, \tilde{z}_n) = d(x_1, z_n)$. From (I.2.16(2)) we have $d(\tilde{z}_n, \tilde{z}_m) \geq d(\overline{z}_n, \overline{z}_m) = d(z_n, z_m)$. And by the definition of δ, the point $\tilde{z}_m$ lies in the ε-neighbourhood of $[\tilde{x}_1, \tilde{x}_2]$. So since $|d(\tilde{x}_1, \tilde{z}_n) - d(\tilde{x}_1, \tilde{z}_m)| < 2\varepsilon$, we have $d(\tilde{z}_n, \tilde{z}_m) < 3\varepsilon$. □

11.2 Remarks

(1) Under the hypotheses of (11.1), X_1 and X_2 are isometrically embedded in $X_1 \sqcup_A X_2$ as D_κ-convex subspaces, and X is complete if and only if X_1 and X_2 are both complete.

(2) (*Successive Gluing*). In the notation of (I.5.26), if each of the metric spaces $X_1, \dots, X_n$ is CAT(κ), and if each of the subspaces A_j along which they are glued is complete and D_κ-convex, then by repeated application of (11.1) we see that the space Y_n obtained by successive gluing is CAT(κ).

We generalize the Basic Gluing Theorem (11.1) to the case of arbitrary families of CAT(κ) spaces.

11.3 Theorem (Gluing Families of CAT(κ) Spaces). *Let $(X_\lambda, d_\lambda)_{\lambda \in \Lambda}$ be a family of CAT(κ) spaces with closed subspaces $A_\lambda \subset X_\lambda$. Let A be a metric space and suppose that for each $\lambda \in \Lambda$ we have an isometry $i_\lambda : A \to A_\lambda$. Let $X = \bigsqcup_A X_\lambda$ be the space obtained by gluing the X_λ along A using the maps i_λ (see I.5.23). If A is a* *complete* CAT(κ) *space (in which case each* A_λ *is* D_κ*-convex and complete), then* X *is a* CAT(κ) *space.*

Proof. We claim that for all $\lambda_1, \lambda_2 \in \Lambda$, the natural inclusion of $X_{\lambda_1} \sqcup_A X_{\lambda_2}$ into X is an isometry. Indeed if $x, y \in X_{\lambda_1} \cup X_{\lambda_2}$ are a distance less than D_κ apart in X then there is a unique geodesic connecting them and this lies in $X_{\lambda_1} \sqcup_A X_{\lambda_2}$. To see this, note that because $A_\lambda \subset X_\lambda$ is D_κ-convex, any geodesic in X_λ of length less than D_κ with endpoints in A_λ lies entirely in A_λ, and thus any chain in X that has endpoints in $X_{\lambda_1} \sqcup_A X_{\lambda_2}$ and has length less than D_κ (in the sense of (I.5.19)) can be shortened by deleting all of its entries that lie in $X_\lambda \setminus A$ for $\lambda \notin \{\lambda_1, \lambda_2\}$.

It follows from this description of geodesics that X is D_κ-geodesic and any geodesic triangle $\Delta \subset X$ with perimeter less than $2D_\kappa$ and vertices $x_i \in X_{\lambda_i}$, $i = 1, 2, 3$ is contained in $(X_{\lambda_1} \sqcup_A X_{\lambda_2}) \sqcup_A X_{\lambda_3}$. Applying (11.1) twice, we see that $(X_{\lambda_1} \sqcup_A X_{\lambda_2}) \sqcup_A X_{\lambda_3}$ is a CAT(κ) space and hence Δ satisfies the CAT(κ) inequality. □

Gluing Using Local Isometries

The preceding constructions were global: we glued CAT(κ) spaces using isometries between complete D_κ-convex subspaces to obtain new CAT(κ) spaces. We now turn to the analogous local situation: we glue spaces of curvature $\leq \kappa$ using local isometries between locally convex and complete subspaces with the expectation that the result will be a metric space of curvature $\leq \kappa$. We also seek more general constructions that will allow us to glue along subspaces which are not locally convex.

At an intuitive level, much of what we expect from local gluing is clear — one can visualize the result of gluing two hyperbolic surfaces along two closed local geodesics of equal length for example. There are, however, a number of technical difficulties associated to showing that the quotient metric has the anticipated properties. These are eased by the following special case of (I.5.27).

11.4 Lemma. *Let* X *be a metric space,* A *a closed subspace of* X *and* $\sigma : A \to A$ *a local isometry such that* σ^2 *is the identity of* A *and* $\sigma(a) \neq a$ *for each* $a \in A$. *Let* $\overline{X}$ *be the quotient of* X *by the equivalence relation* $\sim$ *generated by* $a \sim \sigma(a)$. *Then, the quotient pseudometric on* $\overline{X}$ *is actually a metric, and for every* $x \in X$ *there exists* $\varepsilon(x) > 0$ *such that: if* $x \notin A$ *then* $B(\overline{x}, \varepsilon(x))$ *is isometric to* $B(x, \varepsilon(x))$*; if* $x \in A$ *then the restriction of* σ *to* $\overline{B}(x, 2\varepsilon(x)) \cap A$ *is an isometry and* $B(\overline{x}, \varepsilon(x))$ *is isometric to the ball of radius* $\varepsilon(x)$ *about the image of* x *in the space obtained by gluing* $\overline{B}(x, 2\varepsilon(x))$ *to* $\overline{B}(\sigma(x), 2\varepsilon(x))$ *using this isometry.*

11.5 Lemma. *Let* X, A *and* σ *be as above. If* X *has curvature* $\leq \kappa$ *and* A *is locally convex and complete, then* $\overline{X}$ *also has curvature* $\leq \kappa$.

Proof. Given $x \in A$, by shrinking $\varepsilon = \varepsilon(x)$ if necessary (notation of (11.4)) we may assume that $\overline{B}(x, 2\varepsilon)$ and $\overline{B}(\sigma(x), 2\varepsilon)$ are CAT(κ) and that their intersections with A

are convex and complete. Writing $I = A \cap \overline{B}(x, 2\varepsilon)$, by (11.4) we know that $B(\overline{x}, \varepsilon)$ is isometric to the ball of radius ε about the image of x in $\overline{B}(x, 2\varepsilon) \sqcup_I \overline{B}(\sigma(x), 2\varepsilon)$, where the inclusion $I \to \overline{B}(\sigma(x), 2\varepsilon)$ is the restriction of σ. Thus, by the Basic Gluing Theorem (11.1), $B(x, \varepsilon(x))$ is CAT(κ). □

11.6 Proposition (Simple Local Gluing).

(1) *Let X be a metric space of curvature $\leq \kappa$. Let A_1 and A_2 be two closed, disjoint subspaces of X that are locally convex and complete. If $i : A_1 \to A_2$ is a bijective local isometry, then the quotient of X by the equivalence relation generated by $[\, a_1 \sim i(a_1),\ \forall a_1 \in A_1 \,]$ has curvature $\leq \kappa$.*

(2) *Let X_1 and X_2 be metric spaces of curvature $\leq \kappa$ and let $A_1 \subset X_1$ and $A_2 \subset X_2$ be closed subspaces that are locally convex and complete. If $j : A_1 \to A_2$ is a bijective local isometry, then the quotient of the disjoint union $X = X_1 \coprod X_2$ by the equivalence relation generated by $[\, a_1 \sim j(a_1),\ \forall a_1 \in A_1 \,]$ has curvature $\leq \kappa$.*

Proof. One proves (1) by applying the preceding lemma with $A = A_1 \cup A_2$ and $\sigma|_{A_1} = i$ and $\sigma|_{A_2} = i^{-1}$. And (2) is a special case of (1). □

We highlight three simple but important examples of (11.6), each of which will play a role in Chapter III.Γ.

11.7 Examples

(1) *(Mapping Tori).* Let X be a metric space. The mapping torus of $\phi \in \text{Isom}(X)$ is the quotient of $X \times [0, 1]$ by the equivalence relation generated by $[(x, 0) \sim (\phi(x), 1),\ \forall x \in X]$, it is denoted M_ϕ. If X is non-positively curved, then so is M_ϕ.

(2) *(Doubling along a Subspace).* Given a metric space X and a subspace Y, we write $D(X; Y)$ to denote the metric space obtained by taking the disjoint union of two copies of X and forming the quotient by the equivalence relation that identifies the two copies of Y. The two natural copies of X in $D(X; Y)$ are isometrically embedded and the given identification between them induces a map $D(X; Y) \to X$ which is a left-inverse to both inclusion maps. If X has curvature $\leq \kappa$ and Y is a closed subspace that is locally convex and complete, then $D(X; Y)$ has curvature $\leq \kappa$.

(3) *(Extension over a Subspace).* This construction is a special case of (11.13). Given a metric space X and a subspace Y, we write $X{*_Y}$ to denote the quotient of $X \coprod (Y \times [0, 1])$ by the equivalence relation generated by $[y \sim (y, 0) \sim (y, 1),\ \forall y \in Y]$. The natural inclusion $X \hookrightarrow X{*_Y}$ is an isometry and there is a retraction $X{*_Y} \to X$ induced by the map $(y, t) \mapsto y$ from $Y \times [0, 1]$ to Y. If X has curvature $\leq \kappa$ and Y is a closed subspace that is locally convex and complete, then $X{*_Y}$ has curvature $\leq \kappa$.

11.8 Exercise. For each of the above examples, show that if X is complete then the space obtained by the given construction is complete.

11.9 Remark. One can iterate the construction of (11.6(2)). Let $(X_i, d_i), i = 1, 2 \dots, n$ be a sequence of metric spaces; assume that a bijective local isometry f_2 from a closed subspace A_2 of X_2 onto a closed subspace $f_2(A_2)$ of X_1 is given; let Y_2 be the quotient of the disjoint union of X_1 and X_2 by the equivalence relation generated by $a \sim f_2(a)$ for all $a \in A_2$; proceeding inductively, suppose that for every $i \geq 2$ a bijective local isometry f_i from a closed subspace A_i of X_i onto a closed subspace of Y_{i-1} is given and define Y_i to be the quotient of the disjoint union of X_i and Y_{i-1} by the equivalence relation generated by $a \sim f_i(a)$ for all $a \in A_i$.

If each X_i has curvature $\leq \kappa$ and A_i is locally convex and complete, then Y_n has curvature $\leq \kappa$; according to the description of neighbourhoods in the above proposition, a small ball about each point of Y_n can be obtained by successive gluing of balls in $X_1, \ldots, X_n$.

In (11.4) we required the map σ to be an involution, i.e. we considered the quotient of a metric space X by the equivalence relation associated to a free action of $\mathbb{Z}_2$ by local isometries on a closed subspace $A \subset X$. We restricted our attention to $\mathbb{Z}_2$ actions in order to move as quickly and cleanly as possible to the proof of (11.6). However (I.5.27) applies equally well to the free action of any group Γ by local isometries on a closed subspace $A \subset X$, provided that this action has the property that for each $a \in A$ there exists $\varepsilon > 0$ such that $B(a, \varepsilon) \cap B(\gamma.a, \varepsilon)$ is empty if $\gamma \neq 1$ (where $B(a, \varepsilon)$ is a ball in X, not just A).

In this more general setting, the ball of radius ε about the image of a in the quotient is obtained by gluing the family of balls $B(\gamma.a, \varepsilon)$ using the isometries $B(a, \varepsilon) \cap A \to B(\gamma.a, \varepsilon) \cap A$ obtained by restricting the action of Γ.

Using this description of ε-balls in place of that given in (11.4), and replacing the appeal to (11.1) in the proof of (11.5) with an appeal to (11.3), we obtain:

11.10 Proposition. *Let X be a metric space of curvature $\leq \kappa$, let $A \subseteq X$ be a closed subspace that is locally convex and complete, and suppose that the group Γ acts freely by local isometries on A so that for each $a \in A$ there exists $\varepsilon > 0$ such that $B(a, \varepsilon) \cap B(\gamma.a, \varepsilon)$ is empty for all $\gamma \neq 1$. Then the quotient of X by the equivalence relation generated by $[a \sim \gamma.a, \forall \gamma \in \Gamma\ \forall a \in A]$ has curvature $\leq \kappa$.*

Similar arguments apply to equivalence relations that do not arise from group actions. We mention one other example.

11.11 Proposition (Gluing with Covering Maps). *Let X_1 and X_2 be metric spaces of curvature $\leq \kappa$, and for $i = 1, 2$ let $A_i \subset X_i$ be a closed subspace that is locally convex and complete. Let $p : A_1 \to A_2$ be a local isometry that is a covering map, and suppose that for each $y \in A_2$ there exists $\varepsilon > 0$ such that $B(a, \varepsilon)$ is* CAT(κ) *and $B(a, \varepsilon) \cap B(a', \varepsilon)$ is empty for all distinct points $a, a' \in A_1$ with $p(a) = p(a') = y$.*

Then the quotient of $X_1 \coprod X_2$ by the equivalence relation generated by $[a \sim p(a), \forall a \in A_1]$ has curvature $\leq \kappa$.

Proof. Let X be the quotient of $X_1 \coprod X_2$ by the stated equivalence relation. First suppose that $A_2 = X_2$. If p is a Galois covering, then this proposition reduces to the previous one. The argument in the non-Galois case is essentially the same: one uses p and its local sections to glue the balls $\{B(a, \varepsilon) \mid a \in p^{-1}(y)\}$ along their intersections with A_1 (having shrunk ε to ensure that $B(a, 2\varepsilon) \cap A_1$ is convex and complete for each a). The space obtained by this gluing is CAT(κ), by (11.3), and $B(\overline{y}, \varepsilon/2) \subset X$ is isometric to a ball in this space.

In the general case one first forms $\overline{X}_1 = (X_1 \coprod A_2)/\sim$ and then one constructs X by gluing $\overline{X}_1$ to X_2 along the obvious copies of A_2. The image of A_2 in $\overline{X}_1$ is closed and locally convex and complete, so (11.6) applies. □

11.12 Examples

(1) Let X, A be as in (11.10). If a group Γ acts properly by isometries on X and leaves A invariant, then provided its action on A is free, then the quotient $X/\sim$ described in (11.10) will be a metric space of curvature $\leq \kappa$.

(2) Suppose that X has curvature $\leq \kappa$ and let A be the image of an injective local geodesic $c : \mathbb{S}^1 \to X$. Given any positive integer n, one can apply (11.10) to the action of the cyclic group $\mathbb{Z}_n = \langle \tau \rangle$ on A, where τ acts as $c(\theta) \mapsto c(\theta + 2\pi/n)$.

(3) Let X and A be as in (2) and let Y be a space of curvature $\leq \kappa$ that contains an isometrically embedded line L. Suppose that there exists $\varepsilon > 0$ such that $B(y, \varepsilon)$ is CAT(κ) for every $y \in L$. If $p : L \to A$ is a local isometry, then by (11.11) the quotient of $X \coprod Y$ by the relation generated by $[y \sim p(y),\ \forall y \in L]$ has curvature $\leq \kappa$.

There are many situations in which one wishes to glue along subspaces that are not locally convex, for example totally geodesic submanifolds with self-intersections. When gluing along such subspaces one needs to use *gluing tubes*.

In the following proposition the restriction $\kappa \geq 0$ is necessary because of the need to introduce a product metric on the tube.

11.13 Proposition (Gluing With a Tube). *Let X and A be metric spaces of curvature $\leq \kappa$, where $\kappa \geq 0$. If A is compact and $\varphi, \psi : A \to X$ are local isometries, then the quotient of $X \coprod (A \times [0, 1])$ by the equivalence relation generated by $[\,(a, 0) \sim \varphi(a)$ and $(a, 1) \sim \psi(a),\ \forall a \in A\,]$ has curvature $\leq \kappa$.*

Proof. If a point in the quotient is not in the image of $A \times \{0, 1\}$ then it obviously has a neighbourhood isometric to a neighbourhood of its preimage in X or $A \times [0, 1]$. Consider the image $\overline{y}$ of a point $y \in A \times \{0, 1\}$. By compactness, the preimage in $X \coprod (A \times [0, 1])$ of $\overline{y}$ consists of a finite number of points, and it follows from (I.5.27) that if δ is sufficiently small then $B(\overline{y}, \delta)$ is obtained from the union of the δ-neighbourhoods of these points by successive gluing. Hence (11.2(2)) applies. □

By taking X to be the disjoint union of X_0 and X_1 we obtain:

11.14 Corollary. *Let X_0, X_1 and A be metric spaces of curvature $\leq \kappa$, where $\kappa \geq 0$. If A is compact and $\varphi_i : A \to X_i$ is a local isometry for $i = 0, 1$, then the* *quotient of* $X_0 \coprod (A \times [0,1]) \coprod X_1$ *by the equivalence relation generated by* $[\,(a,0) \sim \varphi_0(a),\ (a,1) \sim \varphi_1(a),\ \forall a \in A\,]$ *has curvature* $\leq \kappa$.

11.15 Examples

(1) Let X_0 and X_1 be metric spaces of curvature $\leq \kappa$ with $\kappa \geq 0$, and for $i = 0, 1$ let $c_i : \mathbb{S}^1 \to X_i$ be closed local geodesics of length ℓ (each c_i may have many self-intersections). Let $S(\ell)$ be a circle of length ℓ and choose arc length parameterizations $\tilde{c}_i : S(\ell) \to X_i$ of the c_i. Then, the quotient of $X_0 \coprod (S(\ell) \times [0,1]) \coprod X_1$ by the relation generated by $[\,(\theta,0) \sim \tilde{c}_0(\theta),\ (\theta,1) \sim \tilde{c}_1(\theta),\ \forall \theta \in S(\ell)\,]$ is a space of curvature $\leq \kappa$. The isometry type of this space depends not only on the curves $c_i(\mathbb{S}^1)$ but also on the parameterizations $\tilde{c}_i$ chosen.

(2) *(Torus Knots).* A torus knot is a smoothly embedded circle which lies on the boundary of an unknotted torus in $\mathbb{R}^3$. The fundamental group of the complement of such a knot has a presentation of the form $\Gamma_{n,m} = \langle x, y \mid x^n = y^m \rangle$, where n and m are positive integers (see [BuZi85], for example). Let X_0 be a circle of length n and let X_1 be a circle of length m. Let c_0 be a local geodesic that wraps m times around X_0 and let c_1 be a circle that wraps n times around X_1. The construction in (1) yields non-positively curved 2-complexes whose fundamental groups are $\Gamma_{n,m}$.

11.16 Exercises

(1) Let $T = \mathbb{S}^1 \times \mathbb{S}^1$, a flat torus. Let c_1 and c_2 be two isometrically embedded circles in T that meet at a single point. Let X be the quotient obtained by gluing two disjoint copies of T by the identity map on $c_1 \cup c_2$. Prove that X is not homeomorphic to any non-positively curved space.

(2) Let F be a closed hyperbolic surface. Let c_1 and c_2 be isometrically embedded circles in F that meet in a finite number of points. Prove that the metric space Y obtained by gluing two disjoint copies of F by the identity map on $c_1 \cup c_2$ (no tube) is not non-positively curved. When is Y homeomorphic to a space with curvature ≤ 0 ? And curvature ≤ -1 ?

(Hint: If F has genus g then one can obtain a hyperbolic structure on it by realizing it as the quotient of a regular $4g$-gon P of vertex angle $\pi/2g$ in $\mathbb{H}^2$ with side pairings (I.5.29(1)); the vertices of the polygon are identified to a single point, v say. If one takes a regular $4g$-gon in $\mathbb{H}^2$ whose edges are shorter than those of P, then the sum of the vertex angles increases, so if one makes the same edge pairings as for P then the metric becomes singular at v, where a concentration of negative curvature manifests itself in the fact that the cone angle (i.e. the length of the link of v) is greater than 2π. Indeed the cone angle tends to $(4g-2)\pi$ as the area of the covering polygon tends to zero. When the cone angle is greater than 4π, there exist sets of four geodesic segments issuing from v in such a way that the Alexandrov angle between any two of them is π; in such a situation, the union of the geodesic segments is locally convex near v.)

In Sections III.Γ.6 and 7 we shall return to the study of constructions involving gluing in the context of determining which groups act properly and cocompactly by isometries on CAT(0) spaces — Proposition 11.13 will be particularly useful in that context. We refer the reader to those sections for explicit examples of spaces obtained by gluing as well as general results of a more group-theoretic nature. We include one such result here because it was needed in our discussion of alternating link complements at the end of Chapter 5, and because it is a precursor to (11.19).

The definition of an amalgamated free product of groups is recalled in (III.Γ.6).

11.17 Proposition. *If each of the groups G_1 and G_2 is the fundamental group of a compact metric space of non-positive curvature, then so too is any amalgamated free product of the form $G_1 *_{\mathbb{Z}} G_2$.*

Proof. Let X_1 and X_2 be compact non-positively curved spaces with $\pi_1 X_i = G_i$. The generators of the subgroups of G_1 and G_2 that are to be amalgamated determine free homotopy classes of loops in X_1 and X_2 respectively, and each of these classes contains a closed local geodesic (I.3.16), which we denote c_i. By rescaling the metric on X_1 we may assume that c_1 and c_2 have the same length and apply the construction of (11.15(1)) to $X = X_1 \cup X_2$, i.e. we glue X_1 and X_2 with a tube whose ends are connected to c_1 and c_2 by arc-length parameterizations. By the Seifert-van Kampen theorem, the fundamental group of the resulting space is (isomorphic to) the given amalgam $G_1 *_{\mathbb{Z}} G_2$. □

Whenever one is presented with a result concerning the group $\mathbb{Z}$ it is natural to ask whether an analogous statement holds for all finitely generated free groups or free abelian groups. In (III.Γ.6) we shall see that (11.17) admits no such extension.

Equivariant Gluing

One can regard (11.17) as a construction for gluing together the actions of G_i on $\widetilde{X}_i$; we were free to describe this construction in terms of the quotients X_i because these actions are free. One would like to have similar results concerning proper actions that are not necessarily free. Actions which are not free are not fully described by the associated quotient spaces alone, so in order to combine such actions one is obliged to work directly with appropriately indexed copies of the spaces on which the groups are acting — that is what we shall do in this section.

Amalgamated free products will be used extensively in Section III.Γ.6. For the purposes of the present section we shall only need their geometric interpretation in the language of the Bass-Serre theory [Ser77]. The Bass-Serre tree T associated to an amalgamated free product $\Gamma = \Gamma_0 *_H \Gamma_1$ (we regard Γ_0, Γ_1 and H as subgroups of Γ in the usual way) is the quotient of the disjoint union $\Gamma \times [0, 1]$ (copies of $[0, 1]$ indexed by Γ) by the equivalence relation generated by

$$(\gamma\gamma_0, 0) \sim (\gamma, 0), \ (\gamma\gamma_1, 1) \sim (\gamma, 1), \ (\gamma h, t) \sim (\gamma, t)$$

for all $\gamma \in \Gamma, \ \gamma_0 \in \Gamma_0, \ \gamma_1 \in \Gamma_1, \ h \in H, \ t \in [0, 1]$. The action of Γ by left translation on the index set of the disjoint union permutes the edges and is compatible with the equivalence relation, therefore it induces an action Γ on T by isometries. The quotient of the tree T by this action is an interval $[0, 1]$; the subgroup $H \subset \Gamma$ is the stabilizer of an edge in T and the stabilizers of the vertices of this edge are the subgroups $\Gamma_0 \subset \Gamma$ and $\Gamma_1 \subset \Gamma$.

11.18 Theorem (Equivariant Gluing). *Let Γ_0, Γ_1 and H be groups acting properly by isometries on complete* CAT(0) *spaces X_0, X_1 and Y respectively. Suppose that for $j = 0, 1$ there exists a monomorphism $\phi_j : H \hookrightarrow \Gamma_j$ and a ϕ_j-equivariant isometric embedding $f_j : Y \to X_j$. Then*

(1) *the amalgamated free product $\Gamma = \Gamma_0 *_H \Gamma_1$ associated to the maps ϕ_j acts properly by isometries on a complete* CAT(0) *space X;*

(2) *if the given actions of Γ_0, Γ_1 and H are cocompact, then the action of Γ on X is cocompact.*

Proof. We wish to define a CAT(0) space X on which Γ will act properly by isometries. We begin with disjoint unions of copies of X_0, X_1 and $[0, 1] \times Y$, each indexed by Γ:

$$(\Gamma \times X_0) \coprod (\Gamma \times [0, 1] \times Y) \coprod (\Gamma \times X_1).$$

Let X be the quotient of the displayed disjoint union by the equivalence relation generated by:

$$(\gamma\gamma_0, x_0) \sim (\gamma, \gamma_0.x_0), \ (\gamma\gamma_1, x_1) \sim (\gamma, \gamma_1.x_1), \ (\gamma h, t, y) \sim (\gamma, t, h.y),$$

$$(\gamma, f_0(y)) \sim (\gamma, 0, y), \ (\gamma, f_1(y)) \sim (\gamma, 1, y)$$

for all $\gamma \in \Gamma, \ \gamma_0 \in \Gamma_0, \ \gamma_1 \in \Gamma_1, \ h \in H, \ x_0 \in X_0, \ x_1 \in X_1, \ t \in [0, 1], y \in Y$.

For $j = 0, 1$, let $\overline{X}_j$ be the quotient of $\Gamma \times X_j$ by the restriction of the above relation, and let $\overline{Y}_j$ be the quotient of $\Gamma \times [0, 1] \times Y$. Note that $\overline{X}_0$ is isometric to a disjoint union of copies of X_0 indexed by Γ/Γ_0: each $\{\gamma\} \times X_0$ contains one element of each equivalence class in $\Gamma \times X_0$; if γ and γ' lie in different cosets then (γ, x_0) and (γ', x_0') are unrelated for all $x_0, x_0' \in X_0$. Similarly, $\overline{X}_1$ is isometric to a disjoint union of copies of X_1 indexed by Γ/Γ_1, and $\overline{Y}$ is isometric to the disjoint union of copies of $[0, 1] \times Y$ indexed by Γ/H.

By definition X is obtained from $\overline{X}_0 \coprod \overline{Y} \coprod \overline{X}_1$ by the following gluing: for each $\gamma H \in \Gamma/H$, one end of the copy of $[0, 1] \times Y$ indexed by γH is glued to the copy of X_0 indexed by $\gamma\Gamma_0$ using a conjugate of the isometry f_0, and the other end is glued to the copy of X_1 indexed by $\gamma\Gamma_1$ using a conjugate of the isometry f_1. It is clear from our earlier results that X is non-positively curved; in order to see that it is a complete CAT(0) space we shall argue that the above gluings can be performed in sequence so that each gluing involves attaching CAT(0) spaces along isometrically embedded subspaces — we shall use the Bass-Serre tree to guide us in choosing the order in which we do these gluings.

The action of Γ by left multiplication on the index sets (i.e. the factors Γ) in the above disjoint union permutes the components of the disjoint union and is compatible with the equivalence relation. Thus we obtain an induced action of Γ by isometries on X. There is a natural Γ-equivariant projection $p : X \to T$, where T is the Bass-Serre tree described prior to the statement of the theorem: the equivalence classes of (γ, x_0), (γ, x_1) and (γ, t, y) are sent by p to the equivalence classes of $(\gamma, 0)$, $(\gamma, 1)$ and (γ, t) respectively, for all $\gamma \in \Gamma$, $x_0 \in X_0$, $x_1 \in X_1$, $t \in [0, 1]$ and $y \in Y$.

For $j = 0, 1$, the components of $\overline{X}_j$ are the preimages in X of the vertices of T that are the equivalence classes of (γ, j), and the edges of T correspond to the components of $\overline{Y}$. It follows that the gluing of $\overline{Y}$ to $\overline{X}_0 \coprod \overline{X}_1$ that yields X can be performed in the following order: begin at the component of $\overline{X}_0$ above a fixed vertex $v \in T$, glue to this the components of $\overline{Y}$ that correspond to the edges of T issuing from v, then glue to the resulting space those components of $\overline{X}_1$ that correspond to the vertices a distance one from $v \in T$; work out radially from v. Repeated application of (11.3) shows that X is a CAT(0) space. (If we were working just with simply connected topological spaces rather than CAT(0) spaces, then essentially the same argument would show that X is simply connected.)

It is clear that the action of Γ on X is proper: the subgroups of Γ leaving the components of $\overline{X}_j \subset X$ invariant are conjugates of Γ_j, these act properly on the components, and if $\gamma \in \Gamma$ does not leave a given component of $\overline{X}_j$ invariant then it moves it to another component, which means each point gets moved a distance at least 2; a similar argument applies to points in $\overline{Y}$.

Finally, the maps $\overline{X}_j \to X_j \to \Gamma_j\backslash X_j$ and $\overline{Y} \to [0, 1] \times Y \to [0, 1] \times (H\backslash Y)$ induce a continuous surjection from $\Gamma_0\backslash X_0 \coprod \left([0, 1] \times (H\backslash Y)\right) \coprod \Gamma_1\backslash X_1$ to $\Gamma\backslash X$, so if the actions of Γ_j and H are cocompact, then the action of Γ on X is cocompact. □

The following result generalizes (11.17).

11.19 Corollary. *If the groups Γ_0 and Γ_1 act properly and cocompactly by isometries on* CAT(0) *spaces X_0 and X_1, and if C is a group that contains a cyclic subgroup of finite index, then any amalgamated free product of the form $\Gamma = \Gamma_0 *_C \Gamma_1$ acts properly and cocompactly by isometries on a* CAT(0) *space.*

Proof. If C is finite, then by (2.8) the image of C in Γ_j fixes a point $p_j \in X_j$ for $j = 0, 1$. Thus we may apply the theorem with Y equal to a single point and $Y \to X_j$ the map with image p_j.

If C contains an infinite cyclic subgroup of finite index, then it contains a normal such subgroup $\langle\tau\rangle$; let τ_j be the image of τ in Γ_j. The action of C on X_j leaves an axis $c_j : \mathbb{R} \to X_j$ of τ_j invariant (6.2). A virtually cyclic group cannot surject to both $\mathbb{Z}$ and $\mathbb{Z}_2 * \mathbb{Z}_2$, so the image of C in $\text{Isom}(X_j)$ either acts on $c_j(\mathbb{R})$ as an infinite dihedral group for $j = 1, 2$, or else it acts on both of these axes as an infinite cyclic group. In each case, the action is determined up to equivariant isometry by the translation length of any element $\gamma \in C$ of infinite order. If we rescale the metric on X_0 so that $|\tau_0| = |\tau_1|$, and if we change the parameterization of c_0 by a suitable translation, then $c_1(t) \mapsto c_2(t)$ is a C-equivariant isometry. Thus we may apply the theorem, taking $Y = \mathbb{R}$ and $f_j(t) = c_j(t)$, where each $\gamma \in C$ acts on Y by $\gamma.t = c_j^{-1}(\gamma.c_j(t))$. □

11.20 Remark. The above argument extends to actions that are not cocompact if one adds the assumption that the images of C in Γ_1 and Γ_0 contain hyperbolic elements.

An argument entirely similar to that of (11.18) yields the following result. (See section III.Γ.6 for basic facts concerning HNN extensions.)

11.21 Proposition. *Let Γ_0 and H be groups acting properly by isometries on* CAT(0) *spaces X_0 and Y respectively. Suppose that for $j = 0, 1$ a monomorphism $\phi_j : H \hookrightarrow \Gamma$ and a ϕ_j-equivariant isometric embedding $f_j : Y \to X_0$ are given. Then the HNN extension $\Gamma = \Gamma_0 *_H$ associated to this data acts properly by isometries on a* CAT(0) *space X.*

Proof. We recall that the HNN extension $\Gamma = \Gamma_0 *_H$ is generated by Γ_0 and an element s such that $\phi_0(h) = s\phi_1(h)s^{-1}$ for all $h \in H$, and that Γ_0 is naturally identified to a subgroup of Γ (i.e. the base group, see III.Γ.6.2). The associated Bass-Serre tree T is the quotient of the disjoint union $\Gamma \times [0, 1]$ (copies of $[0, 1]$ indexed by Γ) by the equivalence relation generated by

$$(\gamma, t) \sim (\gamma\phi_0(h), t), \quad (\gamma, 0) \sim (\gamma\gamma_0, 0) \sim (\gamma s, 1)$$

for all $\gamma \in \Gamma,\ \gamma_0 \in \Gamma_0,\ h \in H,\ t \in [0, 1]$. The action of Γ by left translation on the index set of the disjoint union permutes the edges and is compatible with the equivalence relation, therefore it induces an action of Γ on T by isometries. The quotient of the tree T by this action is a circle; the subgroup $\phi_0(H) \subset \Gamma$ is the stabilizer of an edge in T and Γ_0 is the stabilizer of a vertex.

The space X will be the quotient of the disjoint union

$$(\Gamma \times X_0) \coprod (\Gamma \times [0, 1] \times Y)$$

by the equivalence relation

$$(\gamma\gamma_0, x) \sim (\gamma, \gamma_0.x), \quad (\gamma\phi_0(h), t, y) \sim (\gamma, t, h.y),$$

$$(\gamma, 0, y) \sim (\gamma, f_0(y)) \sim (\gamma s, f_1(y))$$

for all $\gamma \in \Gamma,\ , \gamma_0 \in \Gamma_0,\ h \in H,\ t \in [0, 1]$.

The rest of the proof is entirely analogous to the proof of (11.18). □

As in (11.19) we deduce:

11.22 Corollary. *Let Γ_0 be a group that acts properly by isometries on a* CAT(0) *space, let C and C' be subgroups of Γ_0 that contain cyclic subgroups of finite index and let $\phi : C \to C'$ by an isomorphism. If C has an element γ of infinite order then suppose that γ and $\phi(\gamma)$ are hyperbolic isometries with the same translation length. Then the HNN extension $\Gamma = \Gamma_0 *_\phi$ acts properly by isometries on a* CAT(0) *space, and if the action of Γ_0 is cocompact then so is the action of Γ.*

11.23 Remark. The hypothesis of properness was only used in (11.18) to deduce that the action of Γ was proper. The theorem remains valid if one replaces properness by semi-simplicity or faithfulness (as both an hypothesis and a conclusion).

Gluing Along Subspaces That Are Not Locally Convex

Our next set of examples illustrates the fact that there are circumstances in which one can obtain a CAT(0) space by gluing spaces (which need not be CAT(0)) along subspaces which are not locally convex.

We showed in (I.6.15) that, as Riemannian manifolds, spheres of radius r in $\mathbb{H}^n$ are isometric to spheres of radius $\sinh r$ in $\mathbb{E}^n$. Let $D \subset \mathbb{E}^n$ be a closed ball of radius $\sinh r$ and let $D' \subset \mathbb{H}^n$ be an open ball enclosed by a sphere S of radius r. Endow $Y = \mathbb{H}^n \smallsetminus D'$ with the induced path metric and let $i : \partial D \to S$ be a Riemannian isometry. If X is the quotient of $Y \coprod D$ by the equivalence relation generated by $[x \sim i(x),\ \forall x \in \partial D]$, we say "$X$ is obtained from $\mathbb{H}^n$ by replacing a hyperbolic ball with a Euclidean ball".

In the course of the next proof we shall need the following observation.

11.24 Remark. Let (X, d) be a metric space and let $(Z, \overline{d})$ be a subspace endowed with the induced length metric. If c and c' are geodesics in X that have a common origin, and if the images of c and c' are contained in Z, then $\angle(c, c')$ as measured in $(Z, \overline{d})$ is no less than $\angle(c, c')$ as measured in (X, d), because $\overline{d} \geq d$. In particular, if a geodesic triangle $\Delta \subset X$ is contained in Z and satisfies the CAT(0) angle condition (1.7(4)) with respect to $\overline{d}$, then it also satisfies condition (1.7(4)) with respect to d.

11.25 Proposition. *If X is obtained from $\mathbb{H}^n$ by replacing a hyperbolic ball with a Euclidean ball in the manner described above, then X is a* CAT(0) *space.*

Proof. We begin by describing the geodesics in X. We maintain the notation established in the discussion preceding (11.24).

S is convex in Y, by 2.6(2), and the map $i : \partial D \to S$ preserves the length of curves. Since the restriction to ∂D of the metric on D is Lipschitz equivalent to the induced path metric on ∂D, it follows that X is a proper length space (homeomorphic to $\mathbb{R}^n$) and hence a geodesic space (see I.5.20 and I.5.22(4)).

For each pair of distinct points $p, q \in \partial D$, we have $d_{\mathbb{E}}(p, q) < d_Y(i(p), i(q))$. This, together with the fact that S is convex in Y (and the definition of quotient metrics in terms of chains, I.5.19), yields the following description of geodesics in X: if $x, x' \in D \subset X$, the Euclidean segment $[x, x']$ is the unique geodesic joining x to x' in X; if $x \in D \subset X$ and $y \in Y \subset X$, each geodesic joining x to y in X is the concatenation of a Euclidean segment $[x, x'] \subset D$ and a hyperbolic segment $[x', y] \subset Y$ that meets S only at x'; and any geodesic in X joining a pair of points $y, y' \in Y$ is either a hyperbolic segment in Y or else the concatenation of three segments, two hyperbolic ones $[y, x]$, $[x', y'] \subset Y$ separated by a Euclidean one $[x, x'] \subset D$.

We shall verify that all geodesic triangles $\Delta \subset X$ satisfy the CAT(0) condition as phrased in (1.7(4)). If Δ is Euclidean, i.e. is contained in $D \subset X$, then Δ obviously satisfies the CAT(0) condition, and likewise (in the light of 1.12) if Δ is hyperbolic, i.e. $\Delta \subset Y \setminus S$. (If $n \geq 3$ then $\Delta \subset Y$ does not imply that the convex hull of Δ is contained in Y, but this does not effect the calculation of vertex angles.)

We define a *dart* to be a geodesic triangle in X that has one edge in D and two edges in Y (figure 11.1). Assume for a moment that we know that darts satisfy the CAT(0) inequality. Figure 11.2 (where the cases are to be taken in order) indicates how arbitrary geodesic triangles in X can be decomposed into smaller triangles that are already known to satisfy the CAT(0) inequality. By applying the gluing lemma for triangles (10.4) one sees that the given triangle satisfies the CAT(0) inequality.

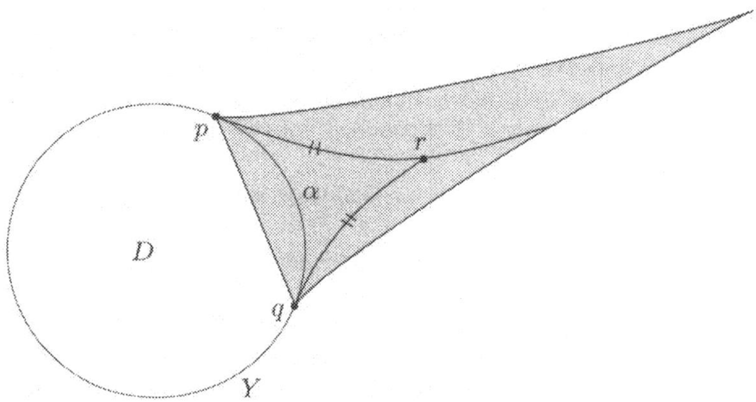

Fig. 11.1 Darts

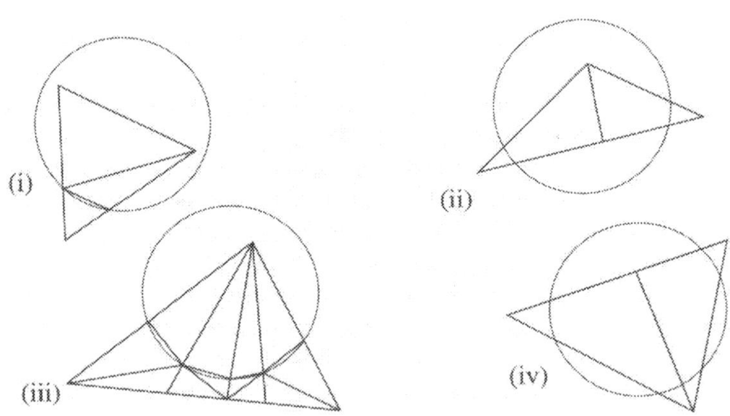

Fig. 11.2 Decomposing arbitrary triangles for (11.25)

It remains to prove that every dart Δ satisfies the CAT(0) inequality. Let H be the intersection of Y with the hyperbolic 2-plane in $\mathbb{H}^n$ spanned by the vertices of Δ and let $X_0 \subset X$ be the union of H and the 2-disc in D spanned by the boundary of H. With the induced length metric, X_0 is isometric to the space obtained from $\mathbb{H}^2$ by replacing a hyperbolic disc by a Euclidean disc. The inclusion $X_0 \hookrightarrow X$ does not increase distances and its restriction to each side of Δ is an isometry, hence it is sufficient to show that Δ verifies the CAT(0) inequality in X_0 (with the induced length metric). By making two applications of the gluing lemma for triangles (in X_0), we can reduce to the case of darts in which the two sides that are hyperbolic segments meet the circle tangentially (see figure 11.1) — suppose that $\Delta = \Delta(p, q, r)$ has this property.

The first inequality in Exercise I.6.19(4) shows that the vertex angles at $\overline{p}$ and $\overline{q}$ in any (isosceles) comparison triangle $\overline{\Delta}(p, q, r) \subset \mathbb{E}^2$ are greater than those at $p, q \in \Delta \subset X_0$. And the second inequality in I.6.19(4) shows that the angle at $\tilde{r}$ in a comparison triangle $\tilde{\Delta}(p, q, r) \subset \mathbb{H}^2$ is less than the vertex angle at $r \in \Delta$. By (1.12), the angle at $\tilde{r}$ in $\tilde{\Delta}(p, q, r) \subset \mathbb{H}^2$ is less than the angle at $\overline{r}$ in $\overline{\Delta}(p, q, r) \subset \mathbb{E}^2$, so we are done. □

11.26 Exercises

(1) To obtain an alternative proof of (11.25), one can replace the above considerations of darts by the following argument for the case $n = 2$.

Show that if $n = 2$, then X is a Gromov-Hausdorff limit of the CAT(0) spaces X_n obtained from $\mathbb{H}^2$ by deleting a regular $2n$-sided polygon inscribed in S and replacing it with a regular Euclidean polygon with the same side lengths. (To see that X_n is CAT(0), assemble it from $2n$ sectors by a sequence of gluings.)

(2) Let $c : \mathbb{S}^1 \to \mathbb{E}^2$ be a simple closed curve of length 2π that is not a circle. Show that the space obtained by replacing the interior of the disc bounded by c with a round Euclidean disc of radius 1 is *not* a CAT(0) space.

(Hint: Enclose c in a large square. If the described surgery on the plane gave a metric of non-positive curvature, we could get a metric of non-positive curvature on the torus by identifying opposite sides of the square. The Flat Torus Theorem would then imply that the surgered plane is isometric to the Euclidean plane. To see that this cannot be the case, recall that a simple closed curve of length 2π in the Euclidean plane encloses an area at most π, with equality if and only if the curve is a round circle.)

(3) The space obtained by removing a round disc from the Euclidean plane and replacing it with a hyperbolic disc of the same circumference is not non-positively curved.

Truncated Hyperbolic Space

In this section we shall describe the geometry of truncated hyperbolic spaces, i.e. geodesic spaces obtained from real hyperbolic space by deleting a disjoint collection of open horoballs and endowing the resulting subspace with the induced length metric. The following result is a special case of the Alexander–Berg–Bishop characterization of curvature for manifolds with boundary [ABB93b]. We present this special case in detail because it provides a concrete example derived from a familiar context, and also because it admits an elementary and instructive proof. The construction does not involve gluing, but it is natural to include it at this point because the arguments involved are very similar to those employed in the preceding section.

11.27 Theorem. *Let $X \subseteq \mathbb{H}^n$ be a subspace obtained by deleting a family of disjoint open horoballs. When endowed with the induced length metric, X is a complete* CAT(0) *space.*

This result does not extend to more general rank 1 symmetric spaces, as one can see by looking at the stabilizers of horospheres (11.35). Indeed, because of the nilpotent subgroups corresponding to such stabilizers, non-uniform lattices in rank 1 Lie groups other than $SO(n, 1)$ cannot act properly by semi-simple isometries on any CAT(0) space (7.16). In contrast, from (11.27) we get:

11.28 Corollary. *Every lattice $\Gamma \subset SO(n, 1)$ acts properly and cocompactly by isometries on a* CAT(0) *space X.*

Proof. If Γ is a cocompact lattice, take $X = \mathbb{H}^n$. If Γ is not cocompact, then one removes a Γ-equivariant set of disjoint open horoballs about the parabolic fixed points of Γ; the action of Γ on the complement X is cocompact and by the above theorem X is CAT(0) (cf. page 266 of [Ep+92]). □

Throughout this paragraph we shall work with the upper half space model for $\mathbb{H}^n$. Thus $\mathbb{H}^n$ is regarded as the submanifold of $\mathbb{R}^n$ consisting of points $(x_1, \dots, x_n)$ with $x_n > 0$, and this manifold is endowed with the Riemannian metric $\frac{ds}{x_n}$, where ds is the Euclidean Riemannian metric on $\mathbb{R}^{n+1}$.

11.29 Horoballs in $\mathbb{H}^n$. An open (resp. closed) *horoball* in $\mathbb{H}^n$ is a translate of $B_0 = \{(x_1, \dots, x_n) \mid x_n > 1\}$ (resp. $x_n \geq 1$) by an element of $\text{Isom}(\mathbb{H}^n)$. A *horosphere* is any translate of the set $H_0 = \{(x_1, \dots, x_n) \mid x_n = 1\}$. The closed horoballs in $\mathbb{H}^n$ are the subspaces $x_n >$ const and the Euclidean balls tangent to $\partial\mathbb{H}^n$. In the Poincaré ball model, the horoballs are Euclidean balls tangent to the boundary sphere.

11.30 Exercises

(1) Prove that every disjoint collection of closed horoballs in $\mathbb{H}^n$ is locally finite, i.e., only finitely many of the horoballs meet any compact subset of $\mathbb{H}^n$.

(Hint: In the Poincaré ball model, if a horoball intersects the ball of Euclidean radius r about the centre of the model, then the horoball has Euclidean radius at least $(1 - r)/2$.)

(2) Let X be as in the statement of Theorem 11.27. Prove that X, equipped with the induced path metric, is simply connected.

In the light of the Cartan-Hadamard Theorem (4.1) and the preceding exercise, in order to prove Theorem 11.27, it suffices to show that the truncated space X is non-positively curved. Since this is a local problem, there is no loss of generality in restricting our attention to the case where X is obtained from $\mathbb{H}^n$ by deleting the single horoball $B_0 = \{(x_1, \dots, x_n) \mid x_n > 1\}$. Let $X_0 = \mathbb{H}^n \smallsetminus B_0$, endowed with the induced path metric from $\mathbb{H}^n$, and let $H_0 = \{(x_1, \dots, x_n) \mid x_n = 1\}$.

11.31 Lemma. *X_0 is uniquely geodesic.*

Proof. Fix $x, y \in X_0$. We extend the natural action of $O(n-1)$ on $\mathbb{R}^{n-1}$ to $\mathbb{R}^n$ so that the action is trivial on the last coordinate. This action preserves the Riemannian metric $\frac{ds}{x_n}$ and hence restricts to an action by isometries on $\mathbb{H}^n$ and X_0. Transporting x and y by a suitable element of $O(n-1)$, we may assume that they lie in $P = \{0\} \times \mathbb{R}^2$.

Because the metric on $\mathbb{H}^n$ arises from the Riemannian metric $\frac{ds}{x_n}$, it is clear that the map $\mathbb{H}^n \to \mathbb{H}^n \cap P$ given by $(x_1, \dots, x_n) \mapsto (0, \dots, 0, x_{n-1}, x_n)$ strictly decreases the length of any path joining x to Y unless the path is entirely contained in P. Thus $P \cap X_0$ is a (strictly) convex subset of X_0.

We have reduced to showing that there is a unique geodesic connecting x to y in $P \cap X_0$. But $P \cap X_0$ is isometric to $\mathbb{H}^2$ minus an open horoball (with the induced path metric), and this is a CAT(-1) space (1.16(5)). $\square$

11.32 Lemma. *The bounding horosphere H_0 is a convex subspace of X_0, and with the induced path metric it is isometric to $\mathbb{E}^{n-1}$.*

Proof. The convexity of H_0 is a special case of observation 2.6(2) — the point is that the projection $X_0 \to H_0$ given by $\pi : (x_1, \dots, x_n) \mapsto (x_1, \dots, 1)$ decreases the length of any path not contained in H_0. The length of paths in H_0 is measured using the Euclidean Riemannian metric ds (because $x_n = 1$ on H_0), so H_0 is isometric to $\mathbb{E}^{n-1}$. $\square$

As in (11.31), one can reduce the proof of the next lemma to the case $n = 2$, where the result is clear.

11.33 Lemma. *If the hyperbolic geodesic connecting two points $x, y \in X_0$ is not contained in X_0, then there exist points $x', y' \in H_0$ such that the unique geodesic connecting x to y in X_0 is the concatenation of the hyperbolic geodesic joining x to x', the (Euclidean) geodesic joining x' to y' in H_0, and the hyperbolic geodesic joining y' to y. Moreover the hyperbolic geodesics $[x', x]$ and $[y', y]$ meet H_0 tangentially.*

Proof of Theorem 11.27. We must show that for every geodesic triangle $\Delta = \Delta(p, q, r)$ in X_0 the angle at each vertex is no larger than the angle at the corresponding vertex of a comparison triangle in $\mathbb{E}^2$. We may assume that p, q and r are distinct. The proof divides into cases according to the position of the vertices relative to the horosphere H_0.

Case 0. In the case where Δ is contained in H_0 there is nothing to prove, because H_0 is isometric to $\mathbb{E}^{n-1}$. The case $\Delta \subset X_0 \smallsetminus H_0$ is also obvious (in the light of (1.12)).

Case 1. Suppose that just one vertex, q say, of the triangle $\Delta(p, q, r)$ lies on H_0, and suppose that the sides of the triangle incident at this vertex are hyperbolic geodesics. (This happens precisely when p and r lie on or below the unique hyperplane in $\mathbb{H}^n$ that is tangent to H_0 at q.) In this case the convex hull in X of the points p, q, r is equal to their convex hull in $\mathbb{H}^n$, which is CAT(0).

Case 2. Now assume that Δ has one vertex in $X_0 \smallsetminus H_0$ and two vertices on H_0. Suppose also that each of the geodesic segments $[p, q]$ and $[p, r]$ is hyperbolic, i.e. intersects H_0 only at q and r respectively. Let Y be the intersection of X_0 with the unique 2-plane (copy of $\mathbb{H}^2$) in $\mathbb{H}^n$ that contains p, q and r. If the boundary of this 2-plane contains the centre of the horosphere X_0, then we are in the setting of 1.16(6) and otherwise $Y \cap H_0$ is a circle — assume that we are in the latter case. Let $D \subset H_0$ be the Euclidean disc bounded by the circle $Y \cap H_0$, and let $Z = Y \cup D$. Lemma 11.33 shows that Δ is contained in Z, and (11.25) shows that Z, equipped with the induced path metric from X, is a CAT(0) space. Thus the vertex angles of Δ are no greater than the corresponding comparison angles (cf. 11.24).

We list the remaining cases and leave the reader the exercise of subdividing the given triangles (using Lemma 11.33) so that by using the gluing lemma one can reduce to earlier cases (cf. figure 11.2).

Case 3. Assume that $p \in X_0 \smallsetminus H_0$ and $q, r \in H_0$, and that at least one of the geodesic segments $[p, q]$ and $[p, r]$ meets H_0 in more than one point.

Case 4. Suppose that $q \in H_0$ and $p, r \in X \smallsetminus H_0$ but we are not in Case 2. Consider the cases $[p, r] \cap H_0 \neq \emptyset$ and $[p, r] \cap H_0 = \emptyset$ separately.

Case 5. $p, q, r \in X_0 \smallsetminus H_0$ but $\Delta(p, q, r) \cap H_0 \neq \emptyset$. □

11.34 Corollary (Geodesics in Truncated Hyperbolic Spaces). *Let $X \subseteq \mathbb{H}^n$ be as in Theorem 11.27. A path $c : [a, b] \to X$ parameterized by arc length is a geodesic in X if and only if it can be expressed as a concatenation of non-trivial paths $c_1, \dots, c_n$ parameterized by arc length, such that:*

(1) *each of the paths c_i is either a hyperbolic geodesic or else its image is contained in one of the horospheres bounding X and in that horosphere it is a Euclidean geodesic;*

(2) *if c_i is a hyperbolic geodesic then the image of c_{i+1} is contained in a horosphere, and vice versa;*

(3) *when viewed as a map $[a, b] \to \mathbb{H}^n$, the path c is C^1.*

Proof. Since X is a CAT(0) space, it suffices to check that this description characterizes local geodesics in X. And for this local problem it is enough to consider the case $X = X_0$, where the desired result is a restatement of Lemma 11.33. □

11.35 Remark. If one removes an open horoball from a rank 1 symmetric space other than $\mathbb{H}^n$ then in the induced path metric on the complement the corresponding horosphere will be convex (2.6(2)). But this horosphere admits a proper cocompact action by a nilpotent group that is not virtually abelian (cf. 10.28) and therefore the horosphere is not a CAT(0) space (7.16). Hence the symmetric space minus an open horoball is not a CAT(0) space.

11.36 Remark (The Geometry of Cusps). If $\Gamma \subset SO(n, 1)$ is a torsion-free non-uniform lattice, then $M = \Gamma\backslash\mathbb{H}^n$ is a complete manifold of finite volume. The space $X \subset \mathbb{H}^n$ described in (11.18) projects to a subspace $N \subset M$ that is a compact manifold with boundary. The boundary components of N are in 1-1 correspondence with the Γ-orbits of the horospheres in $\mathbb{H}^n$ that bound X: each boundary component C of N is the quotient of such a horosphere by the subgroup of Γ that stabilizes it; C is a closed manifold on which the induced path metric is flat (i.e. locally Euclidean), and $C \hookrightarrow N$ is a local isometry (where N is equipped with the induced path-metric from M). The complement of N in M consists of a finite number of cusps, one for each boundary component of N. If C is the quotient of a horosphere S by $\Gamma_0 \subset \Gamma$, then the cusp corresponding to C is the quotient by Γ_0 of the horoball bounded by S; this cusp is homeomorphic to $C \times [0, \infty)$.

By replacing each of the horoballs in the definition of X by a smaller concentric horoball in a Γ-equivariant manner, one obtains a Γ-invariant subspace $X' \subset \mathbb{H}^n$ that contains X and hence a subspace $N' = \Gamma\backslash X' \subset M$ containing N. The components of the complement of N in N' correspond to the components of ∂N; the component corresponding to $C \subset \partial N$ is homeomorphic to $C \times (0, 1]$. If $C' \subset \partial N'$ corresponds to C, then the path metric on C' is a scalar multiple of the path metric on C. Thus, replacing N by a suitable choice of N', one can decrease the volume of each boundary component of N arbitrarily by extending down the corresponding cusp of M.

As one extends N down the cusps of M in the above manner, the shape[40] of each boundary component C remains constant. In order to vary the shape of the cusps while retaining non-positive curvature, one must consider more general metrics on N. The appropriate techniques are described in [Sch89][41]: one regards each cusp as a warped product $C \times [0, \infty)$ with the Riemannian metric $ds^2 = e^{-2t}\, ds_0^2 + dt^2$, where ds_0 is the given Riemannian metric on C. One can vary this metric on $C \times [1, \infty)$ so that negative curvature is retained and for suitable $T_0 > 1$ the metric on $C \times [T_0, \infty)$ becomes $ds^2 = e^{-2t}\, ds_1^2 + dt^2$, where ds_1 is a chosen flat metric on C. A local version of (11.27) shows that the length metric on the manifold with boundary obtained by deleting $C \times (T, \infty)$ from M is non-positively curved if $T > T_0$.

[40] The shape of C is the isometry type of the flat manifold of volume one that is obtained from C by multiplying the metric with a constant.

[41] Schroeder informs us that a detailed account of the construction required in the present context was written by Buyalo as an appendix to the Russian edition of [Wo67].

11.37 Exercises

(1) Let Γ be as in (11.36) and suppose that the boundary components of $N = \Gamma\backslash X$ are all tori. Show that if a group G acts properly and cocompactly by isometries on CAT(0) space, then so too does any amalgamated free product of the form $\Gamma *_A G$, where A is abelian.

(Hint: If A is not cyclic, then since X is locally CAT(-1) away from its bounding horospheres, the A-invariant flats F yielded by (7.1) must be contained in one of these horospheres. Adjust the size and shape of $A\backslash F$ using the last paragraph of (11.36) and apply (11.18).)

(2) The purpose of this exercise is to extend (11.28) to all *geometrically finite* subgroups $\Gamma \subset \mathrm{SO}(n, 1)$. Let $\Gamma \subset \mathrm{SO}(n, 1)$ be an infinite subgroup, let $\Lambda(\Gamma)$ be its limit set (the set of points in $\partial\mathbb{H}^n$ that lie in the closure of an orbit of Γ) and let $C(\Gamma)$ be the convex hull in $\mathbb{H}^n$ of the union of geodesic lines with both endpoints in $\Lambda(\Gamma)$. If Γ is geometrically finite then $C(\Gamma)$ is non-empty and there is a Γ-equivariant collection of disjoint open horoballs, with union U, such that the action of Γ on $X = C(\Gamma) \cap (\mathbb{H}^n \smallsetminus U)$ is proper and cocompact.

Prove that when endowed with the induced path metric, X is a CAT(0) space.

(We refer the reader to the papers of Bowditch for the general theory of geometrically finite groups, and to Chapter 11 of [Ep+92] for results related to this exercise.)

Chapter II.12 Simple Complexes of Groups

In this chapter we describe a construction that allows one to build many interesting examples of group actions on complexes (12.18). This construction originates from the observation that if an action of a group G by isometries on a complex X has a strict fundamental domain[42] Y, then one can recover X and the action of G directly from Y and the pattern of its isotropy subgroups. (The isotropy subgroups are organised into a *simple complex of groups* (12.11).)

With this observation in mind, one might hope to be able to start with data resembling the quotient of a group action with strict fundamental domain (i.e. a simple complex of groups) and then construct a group action giving rise to the specified data. Complexes of groups for which this can be done are called *strictly developable*. A simple complex of groups consists of a simplicial complex with groups G_σ associated to the individual simplices σ, and whenever one cell τ is contained in another σ a monomorphism $\psi_{\tau\sigma} : G_\sigma \to G_\tau$ is given. If this data arose from the action of a group G then the inclusion maps $\varphi_\sigma : G_\sigma \to G$ would be a family of monomorphisms that were compatible in the sense that $\varphi_\tau \psi_{\tau\sigma} = \varphi_\sigma$ whenever $\tau \subset \sigma$. The main result of this section is that the existence of such a family of monomorphisms is not only a necessary condition for strict developability, it is also sufficient. More precisely, given any group G, the Basic Construction described in (12.18) associates to each compatible family of monomorphisms $\varphi_\sigma : G_\sigma \to G$ an action of G on a simplicial complex such that the (strict) fundamental domain and isotropy groups of this action are precisely the simple complex of groups with which we began.

The question of whether the local groups of a simple complex of groups inject into some group in a compatible way is a property that appears to require knowledge of the whole complex. Remarkably though, in keeping with the local-to-global theme exemplified by the Cartan-Hadamard theorem, there is a local criterion for developability: if one has a metric on the underlying complex, and this metric has the property that when one constructs a model for the star of each vertex in a putative development, all of these local models are non-positively curved, then the complex of groups is indeed developable (12.28). This will proved in Part III, Chapter $\mathcal{G}$.

The construction outlined above is due essentially to Tits [Tits75 and Tits86b], and has been used extensively by Davis and others in connection with constructions that involve Coxeter groups and buildings (see in particular [Da83]). We shall give

[42] i.e. a subcomplex that meets each orbit in exactly one point.

a variety of concrete examples in an attempt to illustrate the utility of some of the ideas involved in these and other applications. If the reader wishes to move quickly through the main points of this chapter, then he should direct his attention first to the basic definitions in 12.11-15, the main construction in 12.18 and the list of its properties in 12.20.

As well as providing us with a basic tool for constructing examples of non-positively curved spaces, this chapter also serves as an introduction to Chapter III.$\mathcal{C}$, where the same issues are addressed in a more serious fashion: a more sophisticated notion of complexes of groups is introduced and used as a tool for (re)constructing group actions that may not have a strict fundamental domain. (Requiring a group action to have a strict fundamental domain is a rather restrictive condition.)

Stratified Spaces

Most of the examples that we shall consider will involve groups acting on simplicial or polyhedral complexes, but it is both useful and convenient to work in the more general setting of stratified spaces. A useful example to bear in mind is that of a metric polyhedral complex K in which the characteristic map of each closed cell is injective: one can stratify K by defining the closed cells of K to be the strata, or alternatively one can take the dual stratification of K (see (12.2(1)).

12.1 Definition (Stratified Sets and Spaces). A *stratified set* $(X, \{X^\sigma\}_{\sigma\in\mathcal{P}})$ consists of a set X and a collection of subsets X^σ called *strata*, indexed by a set $\mathcal{P}$, such that:

(1) X is a union of strata,
(2) if $X^\sigma = X^\tau$ then $\sigma = \tau$,
(3) if an intersection $X^\sigma \cap X^\tau$ of two strata is non-empty, then it is a union of strata,
(4) for each $x \in X$ there is a unique $\sigma(x) \in \mathcal{P}$ such that the intersection of the strata containing x is $X^{\sigma(x)}$.

The inclusion of strata gives a partial ordering on the set $\mathcal{P}$, namely $\tau \leq \sigma$ if and only if $X^\tau \subseteq X^\sigma$. We shall often refer to $(X, \{X^\sigma\}_{\sigma\in\mathcal{P}})$ as "a stratified space X with strata indexed by the poset $\mathcal{P}$" or (more casually) "a stratified space over $\mathcal{P}$".

Stratified topological spaces: Suppose that each stratum X^σ is a topological space, that $X^\sigma \cap X^\tau$ is a closed subset of both X^σ and X^τ for each $\sigma, \tau \in \mathcal{P}$, and that the topologies which X^σ and X^τ induce on $X^\sigma \cap X^\tau$ are the same. We can then define a topology on X that is characterized by the property that a subset of X is closed if and only if its intersection with each stratum X^σ is a closed subspace of X^σ. (This is called the weak topology associated to $\{X^\sigma\}_{\sigma\in\mathcal{P}}$, [Spa66, p.5].)

If X has the additional property that each stratum contains only a finite number of strata (which ensures that any union of strata is a closed set), then X is called a *stratified topological space*.

Stratified simplicial complexes: If a stratified set X is the geometric realization of an abstract simplicial complex and each stratum X^σ is a simplicial subcomplex, then X is called a *stratified simplicial complex*. If we endow each of the strata with its weak topology (I.7A.5), then the topology on X arising from the stratification will coincide with the (usual) weak topology on X. We shall implicitly assume that all simplicial complexes considered in this section are endowed with this topology, but it is also worth noting that all of the statements that we shall make remain valid if one replaces the weak topology on X by the metric topology of (I.7A.5).

Stratified M_κ-polyhedral complexes: If a stratified set X is an M_κ-polyhedral complex with Shapes(X) finite, and if each stratum X^σ is a subcomplex (i.e. a union of closed faces of cells), then X is called a *stratified M_κ-polyhedral complex*. As in the simplicial case, one has to decide whether to use the weak topology or metric topology in this case, but all statements that we shall make will be valid for either choice of topology. (Note that in the simplicial and polyhedral cases one does not need to assume that each stratum contains only finitely many strata in order to ensure that arbitrary unions of strata are closed.)

Throughout this section, the term *stratified space* will be used to mean one of the three special types of stratified sets defined above. Morphisms of stratified spaces are defined in (12.5).

st(x): Let X be a stratified space and let $x \in X$. The complement of the union of strata which do not contain x is an open set; this set is denoted st(x), and is called the *open star of* x. (In the case where X is simplicial or polyhedral, this terminology agrees with that of (I.7.3) if one takes as strata the closed cells of X.) The closure of st(x) will be denoted St(x).

Note that $\mathrm{st}(x) = \{y \in X \mid \forall\sigma,\ y \in X^\sigma \implies x \in X^\sigma\}$.

12.2 Examples

(1) Let K be a simplicial complex or an M_κ–polyhedral complex K with Shapes(K) finite. Let $\mathcal{P}$ be the poset of the cells of K ordered by inclusion. As we noted above, K has a natural stratification indexed by $\mathcal{P}$, namely one takes as strata the cells of K. Dual to this, one has a stratification indexed by $\mathcal{P}^{\mathrm{op}}$ (the set $\mathcal{P}$ with the order reversed): the stratum indexed by the cell σ of K is the union of those simplices in the barycentric subdivision of K whose vertices are barycentres of cells containing σ (figure 12.1).

(2) One can regard a 2-dimensional manifold X with a finite set of disjoint closed curves as a stratified space whose strata are the curves and the closures of the connected components of the complement of their union (figure 12.2 and 12.10).

(3) Let M be a compact manifold with boundary and let L be a triangulation of its boundary as in 5.23. The stratification associated to this situation in 5.23 is indexed by a poset $\mathcal{Q}$. The stratum $M^\emptyset$ is equal to M. The stratum M^σ (denoted L^σ in 5.23) is the union of simplices in the barycentric subdivision L' of L; a simplex of L' is in M^σ if all of its vertices are the barycentres of simplices in L that contain σ.

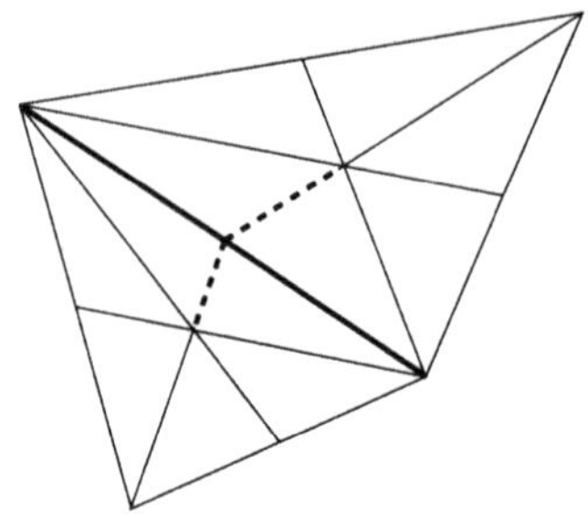

Fig. 12.1 A stratum and its dual in the union of two triangles

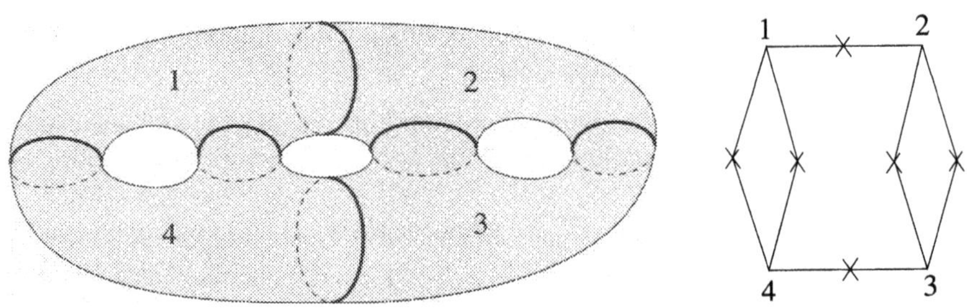

Fig. 12.2 A stratified surface and the geometric realization of the associated poset

12.3 Affine Realization of Posets

Associated to any poset $\mathcal{P}$ one has a simplicial complex[43] whose set of vertices is $\mathcal{P}$ and whose k-simplices are the strictly increasing sequences $\sigma_0 < \cdots < \sigma_k$ of elements of $\mathcal{P}$. The affine realization of this simplicial complex will be called the *affine realization*[44] *of the poset* $\mathcal{P}$, denoted $|\mathcal{P}|$. It has a natural stratification indexed by $\mathcal{P}$: the stratum $|\mathcal{P}|^\sigma$ indexed by $\sigma \in \mathcal{P}$ is the union of the k-simplices $\sigma_0 < \cdots < \sigma_k$ with $\sigma_k \leq \sigma$. The closed star $\mathrm{St}(\sigma)$ is the union of the simplices with σ as a vertex, and the open star $\mathrm{st}(\sigma)$ is the union of the interiors of these simplices.

For each $\sigma \in \mathcal{P}$ we consider the following sub-posets:

$$\begin{aligned} \mathcal{P}(\sigma) &:= \{\tau \in \mathcal{P} \mid \tau \leq \sigma \text{ or } \tau > \sigma\}, \\ \mathcal{P}^\sigma &:= \{\tau \in \mathcal{P} \mid \tau \leq \sigma\}, \text{ and} \\ Lk_\sigma(\mathcal{P}) &:= \{\rho \in \mathcal{P} \mid \rho > \sigma\}. \end{aligned}$$

$Lk_\sigma(\mathcal{P})$ (often abbreviated to Lk_σ) is called the *upper link* of σ. The affine realization of $\mathcal{P}^\sigma$ is the stratum $|\mathcal{P}|^\sigma$, while the affine realization of $\mathcal{P}(\sigma)$ is $\mathrm{St}(\sigma)$. The affine realization of Lk_σ is related to the others by means of the join construction, which is defined as follows.

[43] See the appendix to Chapter I.7 for basic notions concerning simplicial complexes.
[44] Sometimes we shall use "geometric realization" instead of "affine realization".

Given posets $\mathcal{P}$ and $\mathcal{P}'$, one can extend the given partial orderings to the disjoint union $\mathcal{P} \coprod \mathcal{P}'$ by decreeing that $\sigma < \sigma'$ for all $\sigma \in \mathcal{P}, \sigma' \in \mathcal{P}'$; the resulting poset, denoted $\mathcal{P} * \mathcal{P}'$, is called the *join* of $\mathcal{P}$ and $\mathcal{P}'$. The affine realization of $\mathcal{P} * \mathcal{P}'$ is the join of the simplicial complexes $|\mathcal{P}|$ and $|\mathcal{P}'|$ (cf. I.7A.2). For all $\sigma \in \mathcal{P}$, we have $\mathcal{P}(\sigma) = \mathcal{P}^\sigma * Lk_\sigma$, and hence $St(\sigma)$ is the simplicial join of $|\mathcal{P}|^\sigma$ and $|Lk_\sigma|$.

A map between posets $f : \mathcal{P} \to \mathcal{Q}$ is called a *morphism* if it is order preserving. Such a morphism is called *non-degenerate* if for each $\sigma \in \mathcal{P}$ the restriction of f to $\mathcal{P}^\sigma$ is a bijection onto $\mathcal{Q}^{f(\sigma)}$ (this condition ensures that the affine realization $|f| : |\mathcal{P}| \to |\mathcal{Q}|$, which was defined in (I.7A.3), sends each stratum $|P^\sigma|$ homeomorphically onto the stratum $|\mathcal{Q}|^{f(\sigma)}$).

Note that, as simplicial complexes, $|\mathcal{P}| = |\mathcal{P}^{\rm op}|$, but the natural stratifications on these complexes are different. We write $|\mathcal{P}|_\sigma$ to denote the stratum in $|\mathcal{P}^{\rm op}|$ indexed by σ.

12.4 Examples of Poset Realizations

(1) Let K be an M_κ–polyhedral complex for which the inclusion maps of the cells are injective (e.g. a simplicial or cubical complex), and let $\mathcal{P}$ be the poset of the cells of K ordered by inclusion. Then $|\mathcal{P}|$ is simplicially isomorphic to the barycentric subdivision of K; the stratum $|\mathcal{P}|^\sigma$ is the cell $\sigma \subset K$, while the stratum $|\mathcal{P}|_\sigma = |\mathcal{P}^{op}|^\sigma$ is the dual stratum described in (12.2(1)).

(2) Let n be a positive integer. Let S be a finite set with n elements, let $\mathcal{S}$ be the poset of its subsets ordered by reverse-inclusion: for $\sigma, \tau \in \mathcal{S}$, if $\sigma \subseteq \tau$ then $\tau \leq \sigma$. Let $\mathcal{T}$ be the subposet consisting of the proper[45] subsets of S. The affine realization of $\mathcal{T}$ is the barycentric subdivision of an $(n-1)$-simplex (see figure 12.3).

The affine realization $|\mathcal{S}|$ is a simplicial subdivision of the n-cube spanned by an orthonormal basis $\{e_s\}_{s\in S}$ in $\mathbb{R}^n$; the subset $\sigma = \{s_1, \dots, s_k\}$ corresponds to the vertex $\sum_{i=1}^k e_{s_i}$. This subdivision is isomorphic to the simplicial cone over the barycentric subdivision $|\mathcal{T}|$ of the $(n-1)$-simplex; the cone vertex corresponds to subset $S \subseteq S$. For $\sigma = \{s_1, \dots, s_k\} \subseteq S$, the stratum $|\mathcal{S}|^\sigma$ is the $(n-k)$-dimensional sub-cube consisting of those points $x = \sum_{s\in S} x_s e_s$ of the cube with $x_{s_1} = \dots = x_{s_k} = 1$. The dual stratum $|\mathcal{S}|_\sigma$ is the k-dimensional sub-cube spanned by the basis vectors $e_{s_1}, \dots, e_{s_k}$.

(3) Consider the poset of subsets of the set of vertices of a simplicial complex L, ordered by reverse-inclusion. Let $\mathcal{Q}$ be the subposet consisting of the empty subset and those subsets which span a simplex of L. The affine realization of $\mathcal{Q}$ is the simplicial cone over the barycentric subdivision L' of L. As in the previous example, one can identify the cone over the barycentric subdivision of each $(k-1)$-simplex of L with a certain simplicial subdivision of a k-cube, and thus $|\mathcal{Q}| = CL'$ is isomorphic to the cubical complex F considered in 5.23. The stratum $|\mathcal{Q}|^\sigma$ is the subcomplex denoted F^σ in example 5.23 (if $\sigma = \emptyset$, then $|\mathcal{Q}|^\emptyset = |\mathcal{Q}| = F$).

[45] By convention, the empty subset is proper; S itself is not.

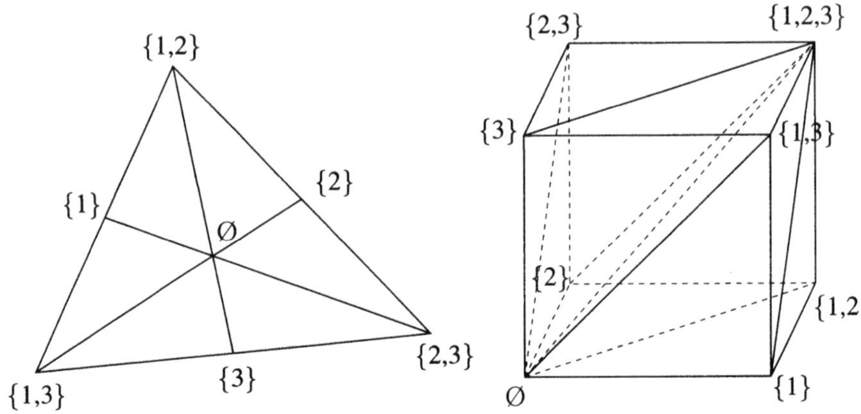

Fig. 12.3 The affine realization of the poset of proper subsets of $\{1, 2, 3\}$, and of the subposet of all subsets.

Group Actions with a Strict Fundamental Domain

The main objects that we wish to study with the techniques developed in this chapter are group actions on polyhedral complexes where the action respects the cell structure. If a subcomplex Y contains exactly one point of each orbit for such an action, then Y is called a *strict fundamental domain*. In this section we generalize the notion of cellular action to *strata preserving action*, and define strict fundamental domains for such actions.

12.5 Strata Preserving Maps and Actions. Given two stratified sets $(X, \{X^\sigma\}_{\sigma\in\mathcal{P}})$ and $(Y, \{Y^\tau\}_{\tau\in\mathcal{Q}})$, a map $f : X \to Y$ is called *strata preserving* if it maps each stratum X^σ of X bijectively onto some stratum of Y. The map $f : \mathcal{P} \to \mathcal{Q}$ obtained by defining $f(\sigma)$ to be the index of the stratum $f(X^\sigma)$ is a morphism of posets, as defined in (12.3).

If X and Y are stratified spaces of the same type, then f is called a *strata preserving morphism* if its restriction to each stratum preserves the specified structure (i.e. in the topological case the restriction must be a homeomorphism, in the simplicial case it must be a simplicial isomorphism, and in the polyhedral case it must be a polyhedral isometry). If there is a strata preserving morphism inverse to f, then f is called an isomorphism of stratified spaces.

The action of a group G on a stratified space $(X, \{X^\sigma\}_{\sigma\in\mathcal{P}})$ is said to be *strata preserving* if for each $g \in G$ the map $x \mapsto g.x$ is a strata preserving morphism.

12.6 Exercise. Let G be a finite subgroup of $O(3)$. For a subgroup $H \subseteq G$, let $\mathbb{R}^3_H$ (resp. $\mathbb{S}^2_H$) be the set of points x in $\mathbb{R}^3$ (resp. $\mathbb{S}^2$) such that the isotropy subgroup of x is H. Show that the closures of the connected components of $\mathbb{R}^3_H$ (resp. $\mathbb{S}^2_H$), $H \subseteq G$, form a stratification of $\mathbb{R}^3$ (resp. $\mathbb{S}^2$), and that the action of G is strata preserving.

12.7 Definition (Strict Fundamental Domain). Let G be a group acting by strata preserving morphisms on a stratified set $(X, \{X^\sigma\}_{\sigma\in\mathcal{P}})$. A subset $Y \subseteq X$ is called a *strict fundamental domain* for the action if it contains exactly one point from each orbit and if for each $\sigma \in \mathcal{P}$, there is a unique $p(\sigma) \in \mathcal{P}$ such that $g.X^\sigma = X^{p(\sigma)} \subseteq Y$ for some $g \in G$. (Note that Y is a union of strata and hence is closed.)

The set of indices $\mathcal{Q} \subseteq \mathcal{P}$ for the strata contained in Y is a strict fundamental domain for the induced action of G on $\mathcal{P}$ in the following sense: $\mathcal{Q}$ is a subposet of $\mathcal{P}$ which intersects each orbit in exactly one element, and $[\sigma \in \mathcal{Q},\ \tau < \sigma]$ implies $\tau \in \mathcal{Q}$. Conversely, given a strata preserving action of G on X, if the induced action on $\mathcal{P}$ has a strict fundamental domain $\mathcal{Q}$ in this sense, then the action of G on X has a strict fundamental domain, namely $Y = \bigcup_{\sigma\in\mathcal{Q}} X^\sigma$.

12.8 Lemma. *Consider a strata-preserving action of a group G on a stratified space $(X, \{X^\sigma\}_{\sigma\in\mathcal{P}})$ with strict fundamental domain $Y = \bigcup_{\sigma\in\mathcal{Q}} X^\sigma$. Let $x \mapsto p(x)$ (resp. $\sigma \mapsto p(\sigma)$) be the map that associates to each $x \in X$ (resp. $\sigma \in \mathcal{P}$) the unique point of Y (resp. $\mathcal{Q}$) in its G-orbit. Then:*

(1) *$x \mapsto p(x)$ is a strata preserving morphism $X \to Y$.*

(2) *$\sigma \mapsto p(\sigma)$ is a non-degenerate morphism of posets $\mathcal{P} \to \mathcal{Q}$.*

(3) *For each $\sigma \in \mathcal{P}$, the following subgroups of G are equal: $G_\sigma = \{g \in G \mid g.X^\sigma = X^\sigma\}$ and $\{g \in G \mid g.x = x,\ \forall x \in X^\sigma\}$.*

(4) *If $X^\tau \subseteq X^\sigma$ then $G_\sigma \subseteq G_\tau$.*

(5) *The isotropy subgroup (stabilizer) of each $x \in X$ is $\{g \in G \mid g.st(x)\cap st(x) \neq \emptyset\}$.*

Proof. (1) and (2) are immediate from the definitions. If $g \in G_\sigma$ and $x \in X^\sigma$, then $p(x) = p(g.x)$, since the image of p contains only one point from each orbit. But $x \mapsto p(x)$ is a bijection from X^σ to $X^{p(\sigma)}$ and both x and $g.x$ are in X^σ, therefore $x = g.x$. This proves (3), from which (4) follows immediately.

For (5), recall that $st(x) = \{z \mid \forall\sigma,\ z \in X^\sigma \implies x \in X^\sigma\}$. We can assume that $x \in Y$. Suppose that $g.z = z'$ for some $z, z' \in st(x)$, and that $z \in X^\sigma$. Then $x \in X^\sigma \cap g.X^\sigma$. Let $h \in G$ be such that $h.X^\sigma \subseteq Y$. We have $h.x = x$. As hg^{-1} maps $g.X^\sigma$ to Y, we also have $hg^{-1}.x = x$, thus $g.x = x$. □

12.9 Examples

(1) Consider the tesselation of the Euclidean plane by equilateral triangles (see figure 12.4). We regard $\mathbb{E}^2$ as a stratified space whose strata are the simplices of the barycentric subdivision of this tesselation. Let G be the subgroup of $\text{Isom}(\mathbb{E}^2)$ that preserves the tesselation. The action of G is strata preserving and each 2-simplex is a strict fundamental domain for the action.

Let G_0 be the orientation-preserving subgroup of index two in G and let G_1 be the subgroup of index six in G generated by the reflections in the lines of the tesselation.

Any triangle of the tesselation serves as a strict fundamental domain for the action of G_1, but there does not exist a strict fundamental domain for the action of G_0.

(2) Consider the construction due to Davis that we described in 5.23 and the associated stratifications described in 12.(3) and 12.2(3). In this setting the action of

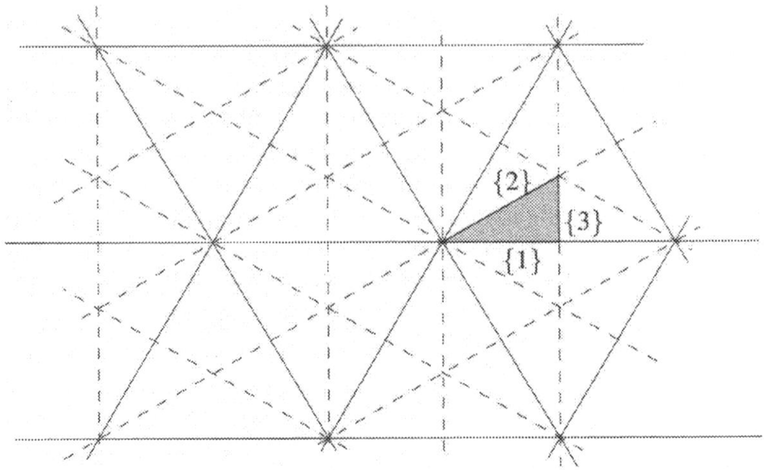

Fig. 12.4 A strict fundamental domain for the action of G in 12.9(1).

G on K (resp. N) is strata preserving and has as a strict fundamental domain F (resp. M).

(3) Let $\Sigma(n)$ be the group of permutations of the set $\{1, \dots, n\}$. Let $\Sigma(n)$ act on $\mathbb{R}^n$ by permuting the coordinates. The subset of $\mathbb{R}^n$ consisting of points $x = (x_1, \dots, x_n)$ with $x_1 \geq \dots \geq x_n$ is a strict fundamental domain for this action: it is a convex cone C, the intersection of the $(n-1)$ half-spaces defined by the equations $x_i \geq x_{i+1}$. Let H_i be the hyperplane defined by $x_i = x_{i+1}$, where $i = 1, \dots, n-1$. The boundary of C is the union of the faces $C \cap H_i$. Let $s_i \in \Sigma(n)$ be the transposition exchanging i and $i+1$: it acts on $\mathbb{R}^n$ as the reflection fixing the hyperplane H_i. The angle between two consecutive hyperplanes H_i and H_{i+1} is $\pi/3$, and if $|i-j| > 1$ then the angle between H_i and H_j is $\pi/2$. Let S^{n-2} be the intersection of the unit sphere in $\mathbb{R}^n$ with the hyperplane $\sum_{i=1}^n x_i = 0$. The intersection of the cone C with S^{n-2} is a spherical $(n-2)$-simplex Δ, with dihedral angles $\pi/3$ and $\pi/2$, which is a strict fundamental domain for the action of $\Sigma(n)$ restricted to S^{n-2}. The translates of Δ by the action of $\Sigma(n)$ are the $(n-2)$-simplices of a triangulation T of S^{n-2} and the action of $\Sigma(n)$ is strata preserving with respect to the stratification whose strata are the simplices of this triangulation.

(4) Consider the Tits boundary $\partial_T P(n, \mathbb{R})_1$ of the symmetric space $P(n, 1)$. It has a natural stratification by the Weyl chambers at infinity and their faces (see 10.75). The natural action of $\mathrm{SL}(n, \mathbb{R})$ is strata preserving. Any Weyl chamber at infinity serves as a strict fundamental domain for this action, for instance the set of points at infinity of the subset consisting of diagonal matrices $\mathrm{diag}(e^{t_1}, \dots, e^{t_n})$ with $t_1 \geq \dots \geq t_n$ and $\sum t_i = 0$. This fundamental domain is isometric to the spherical $(n-2)$-simplex Δ described above.

Notice that it follows from part (4) of the following lemma that if a group action on a connected stratified space is proper and has a strict fundamental domain, then the group is generated by elements of finite order. (In particular, a torsion-free group never admits such an action.)

12.10 Lemma. *Consider a strata-preserving action of a group G on a stratified space $(X, \{X^\sigma\}_{\sigma\in\mathcal{P}})$ with strict fundamental domain $Y = \bigcup_{\mathcal{Q}} X^\sigma$. Let $G_0 \subset G$ be the subgroup generated by the isotropy groups G_σ, $\sigma \in \mathcal{Q}$, and let $X_0 = G_0.Y$.*

(1) *If $g.Y \cap Y \neq \emptyset$, then $g \in G_\sigma$ for some $\sigma \in \mathcal{Q}$.*

(2) *If $g.X_0 \cap X_0 \neq \emptyset$, then $g \in G_0$, hence $g.X_0 = X_0$.*

(3) *X_0 is open and closed in X.*

(4) *X is connected if and only if Y is connected and $G = G_0$.*

Proof. Part (1) follows immediately from the fact that each orbit intersects Y in only one point.

(2) If $x \in g.X_0 \cap X_0$, then there are points $y, y' \in Y$ and elements $g_0, g_0' \in G_0$ such that $x = g_0.y = gg_0'.y'$. As in (1), we then have $y = y'$ and hence $g_0^{-1}gg_0' \in G_{\sigma(y)}$.

(3) X_0 is a union of strata, hence it is closed. Given $x \in X^\sigma$ not in X_0, choose $g \in G$ so that $g.X^\sigma \subseteq Y$. If $X^\sigma \cap X_0$ were not empty, then $g.X_0 \cap X_0$ would not be empty, so by (2) we would have $g.X_0 = X_0$, which is nonsense because $x \in g.X_0$. It follows that $X \smallsetminus X_0$ is a union of strata, hence it is closed.

(4) Suppose that Y is connected. Each $g \in G_0$ is a product of elements $g_1, \dots, g_k$ from the isotropy subgroups of the points of Y. Let $\gamma_n = g_1 \dots g_n$. Each pair of consecutive sets in the sequence $\gamma_k.Y, \dots, \gamma_1.Y, \gamma_0.Y = Y$ have a non-empty intersection, hence their union is connected. Therefore $g.Y$ is in the same connected component of X as Y is, and so $X_0 = G_0.Y$ is connected. If $G_0 = G$ then X is also connected.

Conversely, if X is connected then $X = X_0$, by (3), and hence $G = G_0$, by (2). Moreover, since Y is a continuous image of X (12.8(1)), it too is connected. □

Simple Complexes of Groups: Definitions and Examples

In the introduction we indicated how simple complexes of groups were to be used to reconstruct group actions on stratified spaces from a knowledge of the isotropy subgroups on a strict fundamental domain. A more sophisticated notion of complex of groups, adapted to describe group actions that do not have strict fundamental domains, will be described in Part III, Chapter $\mathcal{C}$.

12.11 Definition (Simple Complex of Groups). A *simple complex of groups*[46] $G(\mathcal{Q}) = (G_\sigma, \psi_{\tau\sigma})$ over a poset $\mathcal{Q}$ consists of the following data:

[46] Equivalent definition: A poset can be regarded as a small category (III.$\mathcal{C}$) whose set of objects is $\mathcal{Q}$ and which has an arrow $\sigma \to \tau$ whenever $\tau \leq \sigma$. A simple complex of groups is a functor from $\mathcal{Q}$ to the category of groups and monomorphisms. In this definition we assume that the homomorphisms $\psi_{\tau\sigma}$ are injective to ensure that simple complexes of groups are always locally developable (12.24). In all respects other than local developability, the theory would go through without any essential changes if we did not assume $\psi_{\tau\sigma}$ to be injective.

(1) for each $\sigma \in \mathcal{Q}$, a group G_σ, called the *local group* at σ;
(2) for each $\tau < \sigma$, an injective homomorphism $\psi_{\tau\sigma} : G_\sigma \to G_\tau$ such that if $\tau < \sigma < \rho$, then

$$\psi_{\tau\rho} = \psi_{\tau\sigma}\psi_{\sigma\rho}.$$

Two simple complexes of groups $(G_\sigma, \psi_{\tau\sigma})$ and $(G'_\sigma, \psi'_{\tau\sigma})$ over a given poset $\mathcal{Q}$ are said to be *simply isomorphic* if there is a family of isomorphisms $\varphi_\sigma : G_\sigma \to G'_\sigma$ such that $\psi'_{\tau\sigma}\varphi_\sigma = \varphi_\tau\psi_{\tau\sigma}$ for all $\tau < \sigma$.

Let G be a group. A *simple morphism* $\varphi = (\varphi_\sigma)$ from $G(\mathcal{Q})$ to G, written $\varphi : G(\mathcal{Q}) \to G$, is a map that associates to each $\sigma \in \mathcal{Q}$ a homomorphism $\varphi_\sigma : G_\sigma \to G$ such that if $\tau < \sigma$, then $\varphi_\sigma = \varphi_\tau\psi_{\tau\sigma}$. We say that φ is *injective on the local groups* if φ_σ is injective for each $\sigma \in \mathcal{Q}$.

12.12 Definition. Associated to any simple complex of groups $G(\mathcal{Q})$ over $\mathcal{Q}$, one has the group

$$\widehat{G(\mathcal{Q})} := \varinjlim_{\sigma \in \mathcal{Q}} G_\sigma,$$

which is the direct limit [47] of the system of groups and monomorphisms $(G_\sigma, \psi_{\tau\sigma})$. The group $\widehat{G(\mathcal{Q})}$ is characterized up to canonical isomorphism by the universal property given below. It is obtained by taking the free product of the groups G_σ and making the identifications $\psi_{\tau\sigma}(h) = h,\ \forall h \in G_\sigma,\ \forall \tau < \sigma$. Thus, given presentations $G_\sigma = \langle \mathcal{A}_\sigma \mid \mathcal{R}_\sigma \rangle$, one can present $\widehat{G(\mathcal{Q})}$ as $\langle \coprod_\sigma \mathcal{A}_\sigma \mid \mathcal{R}_\sigma,\ \psi_{\tau\sigma}(a) = a,\ \forall a \in \mathcal{A}_\sigma,\ \forall \tau < \sigma \rangle$ (where $\psi_{\tau\sigma}(a)$ is given as a word in the generators $\mathcal{A}_\tau$).

The natural homomorphisms $\iota_\sigma : G_\sigma \to \widehat{G(\mathcal{Q})}$ give a canonical simple morphism $\iota : G(\mathcal{Q}) \to \widehat{G(\mathcal{Q})}$, where $\iota = (\iota_\sigma)$. In general ι_σ is not injective (see for example 12.17(5)).

12.13 The Universal Property. *Given any group G and any simple morphism $\varphi : G(\mathcal{Q}) \to G$, there is a unique homomorphism $\hat{\varphi} : \widehat{G(\mathcal{Q})} \to G$ such that $\hat{\varphi} \circ \iota = \varphi$ (i.e. $\hat{\varphi} \circ \iota_\sigma = \varphi_\sigma$ for every $\sigma \in \mathcal{Q}$).*

(Conversely, every homomorphism $\hat{\varphi} : \widehat{G(\mathcal{Q})} \to G$ induces a simple morphism $\varphi : G(\mathcal{Q}) \to G$, given by $\varphi_\sigma := \hat{\varphi} \circ \iota_\sigma$.)

12.14 Remark. There exists a simple morphism φ from $G(\mathcal{Q})$ to some group G that is injective on the local groups if and only if ι is injective on the local groups.

12.15 Strict Developability. Consider a strata preserving action of a group G on a stratified space $(X, \{X^\sigma\}_{\sigma \in \mathcal{P}})$ with a strict fundamental domain Y. Let $\mathcal{Q} = \{\sigma \in \mathcal{P} \mid X^\sigma \subseteq Y\}$. To this action and choice of Y, one associates the simple complex of groups $G(\mathcal{Q}) = (G_\sigma, \psi_{\tau\sigma})$ over $\mathcal{Q}$, where G_σ is the isotropy subgroup of the stratum $X^\sigma \subseteq Y$ and $\psi_{\tau\sigma} : G_\sigma \to G_\tau$ is the natural inclusion associated to $X^\tau \subseteq X^\sigma$. The inclusions $\varphi_\sigma : G_\sigma \to G$ define a simple morphism $\varphi : G(\mathcal{Q}) \to G$ that is injective on the local groups.

[47] Sometimes called the amalgamated sum or amalgam.

A simple complex of groups $G(\mathcal{Q})$ is called *strictly developable* if it arises from an action with a strict fundamental domain in this way. A necessary condition for the strict developability is that there should exist a simple morphism $\widehat{G(\mathcal{Q})} \to G$ that is injective on the local groups. In (12.18) we shall show that this condition is also sufficient.

12.16 Remark

The notion of a simple complex of groups is a particular case of the more general notion of complex of groups explained in III.$\mathcal{C}$. In the more general framework, one considers morphisms $G(\mathcal{Q}) \to G$ that are not simple and there is a correspondingly weaker notion of developability (cf. III.$\mathcal{C}$, 2.15-16). If the affine realization of $\mathcal{Q}$ is not simply connected, then it can happen that a simple complex of groups over $\mathcal{Q}$ is developable but not strictly developable (this is the case in 12.17(5ii)).

12.17 Examples

(1) **Triangles of Groups.** Let $\mathcal{T}$ be the poset of proper subsets of the set $S_3 = \{1, 2, 3\}$, ordered by reverse-inclusion. A simple complex of groups $G(\mathcal{T})$ over $\mathcal{T}$ is called a *triangle of groups*. (A k-simplex of groups is defined similarly.) Thus a triangle of groups is a commutative diagram of group monomorphisms as in figure 12.5. Such diagrams have been studied by Gersten and Stallings [St91] and others. Tits considered more generally simplices of groups, i.e. complexes of groups over the poset of faces of a simplex. In general triangles of groups are not (strictly) developable (e.g. example(6)).

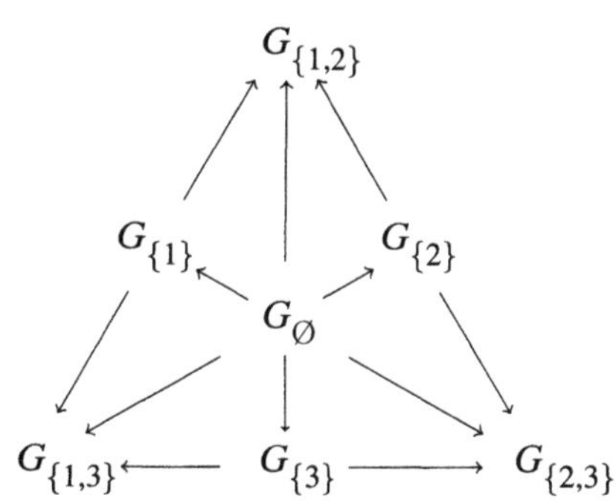

Fig. 12.5 A triangle of groups

(2) A 1-simplex of groups is a diagram $G_1 \leftarrow H \to G_2$ of monomorphisms of groups. The direct limit of this diagram is called the amalgamated free product of G_1 and G_2 over H and is often denoted $G_1 *_H G_2$ (see III.Γ.6.1). The natural homomorphisms of G_1 and G_2 in $G_1 *_H G_2$ are injective and $G_1 *_H G_2$ acts on a simplicial tree [Ser77], called the associated Bass-Serre tree, with strict fundamental domain a 1-simplex (see 11.18). The isotropy subgroups of the vertices and the edge

of this simplex are respectively G_1, G_2 and H (see 12.20(4)). More generally, any simple complex of groups whose underlying complex is a tree is strictly developable.

(3) Consider the action of G on $\mathbb{E}^2$ described in 12.9(1). Let s_1, s_2, s_3 be the reflections fixing the sides of a fundamental triangle, where the labelling $\{1\}, \{2\}, \{3\}$ is chosen as in figure (12.5); the vertex angles are $\pi/6,\ \pi/3,\ \pi/2$. The triangle of groups $G(\mathcal{T})$ associated to this action has the following local groups: $G_{\{i\}}$ is the cyclic group of order two generated by the reflections s_i, for $i = 1, 2, 3$; the local groups at the vertices $G_{\{1,2\}}, G_{\{2,3\}}, G_{\{1,3\}}$ are dihedral [48] with orders 12, 6 and 4 respectively; and the group $G_\emptyset$ in the middle of the triangle is trivial (see figure 12.6).

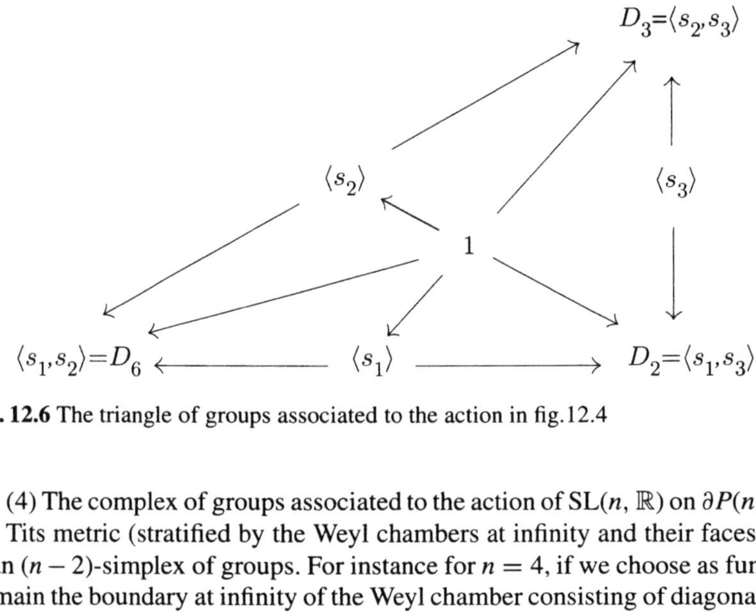

Fig. 12.6 The triangle of groups associated to the action in fig.12.4

(4) The complex of groups associated to the action of $\mathrm{SL}(n,\mathbb{R})$ on $\partial P(n,\mathbb{R})_1$ with the Tits metric (stratified by the Weyl chambers at infinity and their faces,cf.10.75) is an $(n-2)$-simplex of groups. For instance for $n = 4$, if we choose as fundamental domain the boundary at infinity of the Weyl chamber consisting of diagonal matrices of the form $\mathrm{diag}(e^{t_1}, \dots, e^{t_4})$ with $t_1 \geq \dots \geq t_4$, we get a triangle of subgroups of $\mathrm{SL}(4,\mathbb{R})$ as indicated in figure 12.7.

(5) **A Non-Developable Complex of Groups.** Let $\mathcal{Q}$ be the poset consisting of 5 elements $\{\rho, \sigma_1, \sigma_2, \tau_1, \tau_2\}$, with the ordering $\tau_i < \sigma_j < \rho$ for $i, j = 1, 2$. Consider the following complex of groups over $\mathcal{Q}$: let G_ρ be the trivial group, let $G_{\tau_2} \cong G_{\sigma_1} \cong G_{\sigma_2} \cong \mathbb{Z}_2$, and let G_{τ_1} be a group H containing two distinct elements of order two, t_1 and t_2 say; for $i = 1, 2$, the homomorphism $\psi_{\tau_1\sigma_i} : G_{\sigma_i} \to G_{\tau_1}$ sends the generator of G_{σ_i} to t_i. This complex of groups is not strictly developable, because the kernel of the canonical homomorphism $\iota_{\tau_1} : G_{\tau_1} \to \widehat{G(\mathcal{Q})}$ contains $t_1t_2^{-1}$; indeed $\widehat{G(\mathcal{Q})}$ is the quotient of H by the normal subgroup generated by $t_1t_2^{-1}$. In particular, if we take H to be a simple group such as A_5, then $\widehat{G(\mathcal{Q})}$ will be the trivial group. (See figure 12.8.)

[48] Our convention will be to write D_n to denote the dihedral group of order $2n$.

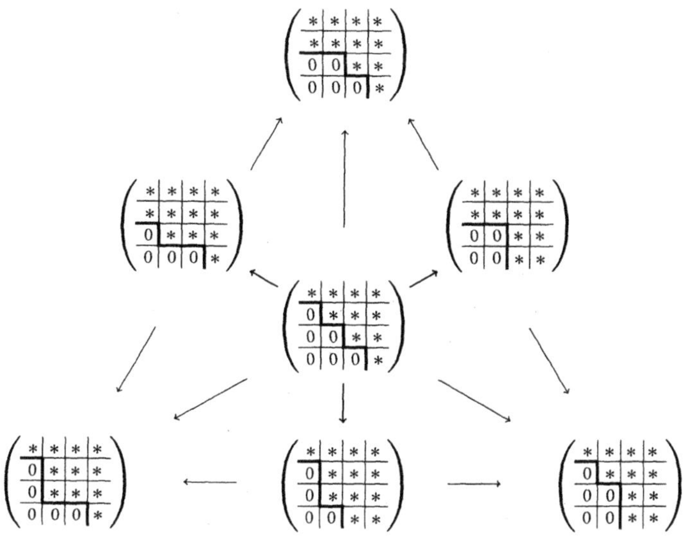

Fig. 12.7 The triangle of groups associated to the action of SL(4, $\mathbb{R}$) on $\partial P(4, \mathbb{R})_1$

Let $\mathcal{Q}'$ be the subposet obtained from $\mathcal{Q}$ by removing ρ. The simple complex of groups $G(\mathcal{Q}')$ obtained by restricting $G(\mathcal{Q})$ to $\mathcal{Q}'$ is developable in the sense of (III.$\mathcal{C}$.2.11), but it is not strictly developable. Indeed, although $G(\mathcal{Q}')$ does not arise from an action with strict fundamental domain, it is the complex of groups associated to the simplicial action of a group $H*_{\mathbb{Z}_2}$ on a tree such that the quotient is the affine realization of $\mathcal{Q}'$ and the isotropy subgroups correspond to the local groups in $\mathcal{Q}'$. (There is some ambiguity concerning the choice of maps $\psi_{\tau_i\sigma_j}$.)

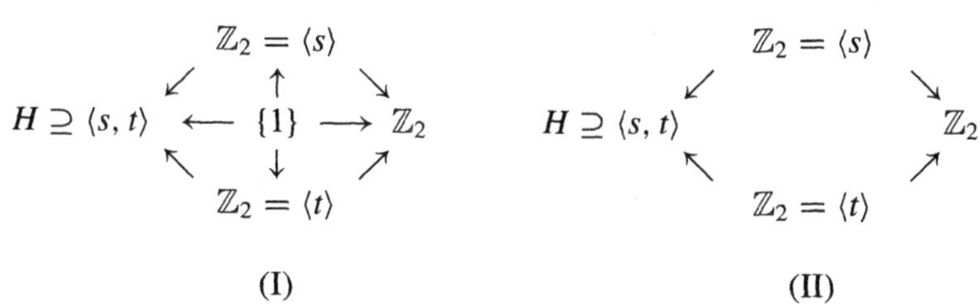

Fig. 12.8 (I) A non-developable complex of groups with $\widehat{G(\mathcal{Q})} = \mathbb{Z}_2$. (II) A graph of groups which is not strictly developable

(6) ***n*-Gons of Groups.** Let $n \geq 3$ be an integer. An *n-gon of groups* is a commutative diagram of groups and monomorphisms of the type shown in Figure 12.9. More precisely, it is a complex of groups over $\mathcal{Q}_n$, where $\mathcal{Q}_n$ is the poset of faces of a regular n-gon P in $\mathbb{E}^2$ ordered by inclusion.

In order to give examples, we introduce the notation $\mathcal{Q}_n = \{\rho, \sigma_i, \tau_i \mid i \in \mathbb{Z} \bmod n\}$, where the σ_i and τ_i correspond to the sides and vertices of P and ρ is P

itself; the indices are chosen so that $\tau_i < \sigma_i < \rho$ and $\tau_i < \sigma_{i+1} < \rho$. (See figure 12.9.)

An interesting class of examples is obtained by taking $G_\rho = \{1\}$, $G_{\sigma_i} = \langle x_i \rangle \cong \mathbb{Z}$ and $G_{\tau_i} = \langle x_i, x_{i+1} \mid x_i x_{i+1} x_i^{-1} = x_{i+1}^2 \rangle$, and taking homomorphisms $\psi_{\tau_i \sigma_i}$ and $\psi_{\tau_i \sigma_{i+1}}$ that send the generators $x_i \in G_{\sigma_i}$ and $x_{i+1} \in G_{\sigma_{i+1}}$ to the elements of G_{τ_i} that have the same name[49]. Consider $\widehat{G(\mathcal{Q}_n)} = \langle x_1, \dots, x_n \mid x_i x_{i+1} x_i^{-1} = x_{i+1}^2,\ i \in \mathbb{Z} \mod n \rangle$.

$\widehat{G(\mathcal{Q}_3)}$ is the trivial group (this is not obvious!). If $n \geq 4$ then the local groups G_{τ_i} inject into $\widehat{G(\mathcal{Q}_n)}$, and hence $G(\mathcal{Q}_n)$ is developable. The case $n = 4$ was studied in detail by Higman, who showed that $\widehat{G(\mathcal{Q}_4)}$ has no proper subgroups of finite index (see [Ser77, 1.4]). Because $\widehat{G(\mathcal{Q}_4)}$ is infinite, one can construct finitely generated infinite simple groups by taking the quotient of $\widehat{G(\mathcal{Q}_4)}$ by a maximal normal subgroup.

$$\begin{array}{ccccccccc}
 & & & & \langle x_2 \rangle & & & & \\
 & & & & \| & & & & \\
\langle x_2, x_3 \rangle & = & \mathbf{G}_{\tau_2} & \longleftarrow & \mathbf{G}_{\sigma_2} & \longrightarrow & \mathbf{G}_{\tau_1} & = & \langle x_1, x_2 \rangle \\
 & & \uparrow & \nwarrow & \uparrow & \nearrow & \uparrow & & \\
\langle x_3 \rangle & = & \mathbf{G}_{\sigma_3} & \longleftarrow & \mathbf{G}_\rho = \{1\} & \longrightarrow & \mathbf{G}_{\sigma_1} & = & \langle x_1 \rangle \\
 & & \downarrow & \swarrow & \downarrow & \searrow & \downarrow & & \\
\langle x_3, x_4 \rangle & = & \mathbf{G}_{\tau_3} & \longleftarrow & \mathbf{G}_{\sigma_4} & \longrightarrow & \mathbf{G}_{\tau_4} & = & \langle x_4, x_1 \rangle \\
 & & & & \| & & & & \\
 & & & & \langle x_4 \rangle & & & &
\end{array}$$

Fig. 12.9 A 4-gon (square) of groups

(7) **Extending an Action with Strict Fundamental Domain.** Let $G(\mathcal{Q})$ be the simple complex of groups associated to a strata preserving action of a group G on a stratified space X with strict fundamental domain Y. Let $\varphi : G(\mathcal{Q}) \to G$ be the associated simple morphism. Assume that X and its strata are arcwise connected and that Y is simply connected. Let $p : \widetilde{X} \to X$ be a universal covering of X. Let $\widetilde{G}$ be the set pairs (h, g), where $g \in G$ and $h : \widetilde{X} \to \widetilde{X}$ is a homeomorphism such that $p(h.\tilde{x}) = g.p(\tilde{x})$ for all $\tilde{x} \in \widetilde{X}$. The operation $(g_1, h_1).(g_2, h_2) = (g_1 g_2, h_1 h_2)$ defines a group structure on $\widetilde{G}$. This group acts on $\widetilde{X}$ by $(g, h).\tilde{x} = h(\tilde{x})$.

Standard covering space theory (see e.g. [Mass91]) tells us that if $(g, h), (g, h') \in \widetilde{G}$ then $h^{-1}h'$ acts as a covering translation, and the map $\pi : \widetilde{G} \to G$ that sends (h, g) to $g \in G$ is a surjective homomorphism whose kernel is naturally isomorphic to the fundamental group of X:

$$1 \to \pi_1(X) \to \widetilde{G} \overset{\pi}{\to} G \to 1.$$

[49] The group $\langle y, x \mid yxy^{-1} = x^2 \rangle$ has several interesting manifestations, for example as the subgroup of $\mathrm{SL}(2, \mathbb{Z}[\frac{1}{2}])$ generated by $y = \begin{pmatrix} 2 & 0 \\ 0 & 1 \end{pmatrix}$ and $x = \begin{pmatrix} 1 & 1 \\ 0 & 1 \end{pmatrix}$, which acts as affine transformations on $\mathbb{R}$ by $x(t) = t + 1$ and $y(t) = 2t$.

If we use π to define an action of $\widetilde{G}$ on X, then p becomes a $\widetilde{G}$-equivariant map. As Y is simply connected, there is a continuous map $s : Y \to \widetilde{X}$ such that $p \circ s$ is the identity map on Y.

Let $(X, \{X^\sigma\})$ be the given stratification on X. We stratify $\widetilde{X}$ by choosing as strata the connected components of the inverse images of the strata of X. Note that the action of $\widetilde{G}$ is obviously strata-preserving and $s(Y)$ is a strict fundamental domain for this action. Moreover, each stratum of X is contained in some translate of $s(Y)$, which is assumed to be simply connected, and hence p maps each stratum of $\widetilde{X}$ homeomorphically onto a stratum of X.

For each $\sigma \in \mathcal{Q}$, consider the isotropy subgroup $\widetilde{G}_\sigma$ of $s(Y^\sigma)$. The restriction of π to $\widetilde{G}_\sigma$ is an isomorphism onto G_σ. These isomorphisms on local groups give a simple isomorphism (in the sense of (12.11)) from $G(\mathcal{Q})$ to the simple complex of groups associated to the action of $\widetilde{G}$ on $\widetilde{X}$ with strict fundamental domain $s(Y)$. And the monomorphisms $\tilde{\varphi}_\sigma : G_\sigma \to \widetilde{G}$ obtained by composing the isomorphisms $G_\sigma \cong \widetilde{G}_\sigma$ with the natural inclusions $\widetilde{G}_\sigma \hookrightarrow \widetilde{G}$ define a simple morphism $\tilde{\varphi} : G(X) \to \widetilde{G}$ that is injective on the local groups, and $\varphi = \pi\tilde{\varphi}$.

The Basic Construction

We shall now describe the Basic Construction which we talked about in the introduction. This construction allows one to reconstruct a group action with strict fundamental domain from the associated simple complex of groups. It also shows that a simple complex of groups $G(\mathcal{Q})$ is strictly developable if and only if the natural simple morphism $\iota : G(\mathcal{Q}) \to \widehat{G(\mathcal{Q})}$ is injective on the local groups.

Following the Basic Construction we shall give a number of explicit examples to try to give a sense of how one uses it, then we shall gather some of its basic properties. Further examples will be given in subsequent sections.

12.18 Theorem (The Basic Construction). *Let $(Y, \{Y^\sigma\}_{\sigma\in\mathcal{Q}})$ be a stratified space indexed by the poset $\mathcal{Q}$. (The three types of stratified spaces were described in (12.1).) Let $G(\mathcal{Q}) = (G_\sigma, \psi_{\tau\sigma})$ be a simple complex of groups over $\mathcal{Q}$. Let G be a group and let $\varphi : G(\mathcal{Q}) \to G$ be a simple morphism that is injective on the local groups. Then:*

(1) *Canonically associated to φ there is a poset $D(\mathcal{Q}, \varphi)$, called the development of $\mathcal{Q}$ with respect to φ, and a stratified space $D(Y, \varphi)$ over $D(\mathcal{Q}, \varphi)$ called the development of Y with respect to φ.*

(2) *Y is contained in $D(Y, \varphi)$ and there is a strata preserving action of G on $D(Y, \varphi)$ with strict fundamental domain Y.*

(3) *The simple complex of groups associated to this action is canonically isomorphic to $G(\mathcal{Q})$, and the simple morphism $G(\mathcal{Q}) \to G$ associated to the action is φ. Therefore $G(\mathcal{Q})$ is developable.*

(4) *If $Y = |\mathcal{Q}|$ then $D(Y, \varphi) = |D(\mathcal{Q}, \varphi)|$.*

Proof. As the simple morphism $\varphi : G(\mathcal{Q}) \to G$ is injective on the local groups, we can identify each G_σ with its image $\varphi_\sigma(G_\sigma)$ in G; if $\tau < \sigma$, then the subgroup G_σ is contained in G_τ.

Define $\mathcal{P} = D(\mathcal{Q}, \varphi) := \coprod_{\mathcal{Q}} G/G_\sigma$, the set of pairs (gG_σ, σ) with $\sigma \in \mathcal{Q}$ and $gG_\sigma \in G/G_\sigma$, a coset of $G_\sigma \subseteq G$. We define a partial ordering on $\mathcal{P}$ by: $(gG_\tau, \tau) < (g'G_\sigma, \sigma)$ if and only if $\tau < \sigma$ and $g^{-1}g' \in G_\tau$. We have a natural action of G on $D(\mathcal{Q}, \varphi)$ defined, for $g' \in G$, by $g'.(gG_\sigma, \sigma) = (g'gG_\sigma, \sigma)$. The map $\sigma \mapsto (G_\sigma, \sigma)$ identifies $\mathcal{Q}$ with a subposet of $D(\mathcal{Q}, \varphi)$ which is a strict fundamental domain for this action.

Let $D(Y, \varphi)$ be the set which is the quotient of $G \times Y$ by the equivalence relation: $[(g, y) \sim (g', y') \iff y = y'$ and $g^{-1}g' \in G_{\sigma(y)}]$, where $X^{\sigma(y)}$ is the smallest stratum containing y. We write $[g, y]$ to denote the equivalence class of (g, y). The set $G \times Y$ is a disjoint union $\coprod_G Y$ and as such inherits the (topological, simplicial or polyhedral) character of Y. We endow $D(Y, \varphi)$ with the quotient structure. $D(Y, \varphi)$ is then a stratified space over $\mathcal{P}$ where the stratum indexed by (gG_σ, σ) is $\bigcup_{y \in Y^\sigma} [g, y]$. The group G acts by strata preserving automorphisms according to the rule $g'.[g, y] = [g'g, y]$. And if we identify Y with the image of $\{1\} \times Y$ in $D(Y, \varphi)$ (where 1 is the unit element of G), then Y is a strict fundamental domain for the action, and (modulo the natural identifications of the G_σ with subgroups of G, and of Y with the fundamental domain) the associated complex of groups is $G(\mathcal{Q})$.

To prove (4), we note that the k-simplices in $|D(\mathcal{Q}, \varphi)|$ that project to the k-simplex of $|\mathcal{Q}|$ with vertices $\sigma_0 < \cdots < \sigma_k$ have vertices $(gG_{\sigma_0}, \sigma_0) < \cdots < (gG_{\sigma_k}, \sigma_k)$, where $g \in G$ is well-defined modulo G_{σ_k}. In particular these simplices are in bijection with the elements of G/G_{σ_k}. □

12.19 Examples of the Basic Construction

(1) Let $\mathcal{Q} = \{\sigma, \tau\}$, with $\tau < \sigma$. Let Y be the 2-dimensional disc with two strata, the disc itself Y^σ and its boundary Y^τ. Let $G(\mathcal{Q})$ be the complex of groups over $\mathcal{Q}$ with $G_\sigma = \{1\}$ and $G_\tau = \mathbb{Z}_2$. There is only one simple morphism $\varphi : G(X) \to \mathbb{Z}_2$ that is injective on the local groups. The stratified space $D(Y, \varphi)$ is the 2-sphere with three strata, namely the two closed hemispheres and the equator. The generator of $\mathbb{Z}_2$ fixes the equator and exchanges the hemispheres. $\widehat{G(\mathcal{Q})} \cong \mathbb{Z}_2$.

(2) Let $\mathcal{Q} = \{\sigma, \tau_0, \tau_1\}$, with $\tau_0 < \sigma$ and $\tau_1 < \sigma$. Let Y be the segment $[0, 1]$ with three strata $Y^\sigma = [0, 1]$, $Y^{\tau_0} = \{0\}$, and $Y^{\tau_1} = \{1\}$. Consider the complex of groups $G(\mathcal{Q})$ with $G_\sigma = \{1\}$, $G_{\tau_0} \cong \mathbb{Z}_3$, and $G_{\tau_1} \cong \mathbb{Z}_2$. Let $\varphi : G(\tau) \to G$ be a simple morphism that is injective on the local groups. Then $D(Y, \varphi)$ is the barycentric subdivision of a graph with vertices of valence 2 and 3; the vertices of valence 3 are indexed by pairs $(\gamma G_{\tau_0}, \tau_0)$ (notation as in the proof of (12.18)); the vertices of valence 2 are indexed by pairs $(\gamma G_{\tau_1}, \tau_1)$; the edges are indexed by pairs $(\gamma G_\sigma, \sigma)$.

For instance, suppose that $G \cong \mathbb{Z}_6$ is generated by γ and define φ by sending a fixed generator of $\mathbb{Z}_3$ (resp. $\mathbb{Z}_2$) to γ^2 (resp. γ^3). Then $D(Y, \varphi)$ is the barycentric subdivision of the graph obtained by taking the simplicial join of the discrete sets $\{\tau_{0,0}, \tau_{0,1}\}$ and $\{\tau_{1,0}, \tau_{1,1}, \tau_{1,2}\}$. The generator γ acts on $D(Y, \varphi)$ by exchanging the vertices $\tau_{0,0}$ and $\tau_{0,1}$ and cyclically permuting $\{\tau_{1,0}, \tau_{1,1}, \tau_{1,2}\}$.

As another example, let $G = \langle s, r \mid s^2 = 1, r^3 = 1\rangle \cong D_3$ and define $\varphi : G(\mathcal{Q}) \to G$ by sending a generator of $\mathbb{Z}_3$ to r and the generator of $\mathbb{Z}_2$ to s. Then $D(Y, \varphi)$ is the same complex as above, but the action is different: r fixes the vertices $\tau_{0,0}$ and $\tau_{0,1}$ and cyclically permutes the others, while s exchanges $\tau_{0,0}$ and $\tau_{0,1}$, fixes one of the $\tau_{1,i}$, and exchanges the other two.

The group $\widehat{G(\mathcal{Q})}$ is the free product $\mathbb{Z}_2 * \mathbb{Z}_3$. The space $D(Y,)$ is the barycentric subdivision of a regular tree (cf. 12.20 (4)) with two kinds of vertices, those of valence 2 adjacent to those of valence 3.

(3) Let $\mathcal{Q} = \{\rho, \sigma_1, \sigma_2, \sigma_3\}$, with $\sigma_i < \rho$ for $i = 1, 2, 3$. Let Y be a pair of pants, i.e. the complement in a 2-sphere of the interior of three disks whose closures are disjoint. We stratify Y over $\mathcal{Q}$, with $Y^\rho = Y$ and each Y^{σ_i} a boundary circle. Let $G(\mathcal{Q})$ be the simple complex of groups with $G_\rho = \{1\}$ and $G_{\sigma_i} \cong \mathbb{Z}_2$. In this case $\widehat{G(\mathcal{Q})}$ is a free product $\mathbb{Z}_2 * \mathbb{Z}_2 * \mathbb{Z}_2$, and $D(Y,)$ is an infinite surface stratified by pairs of pants and their boundary curves.

Let $G = \mathbb{Z}_2 \times \mathbb{Z}_2$, with generators a and b. Let $\varphi : G(\mathcal{Q}) \to \mathbb{Z}_2 \times \mathbb{Z}_2$ be the simple morphism that sends the generators of G_{σ_1}, G_{σ_2}, and G_{σ_3} to a, b, and ab respectively. The geometric realization of $D(\mathcal{Q}, \varphi)$ is the barycentric subdivision of the 1-skeleton of a regular tetrahedron. The non-trivial elements of G act as rotations around axes joining the midpoints of opposite edges. $D(Y, \varphi)$ is homeomorphic to the boundary of a tubular neighbourhood of the 1-skeleton of the regular tetrahedron in $\mathbb{R}^3$. Thus $D(Y, \varphi)$ is a closed surface of genus 3, stratified as the union of four pairs of pants (see figure 12.10).

Let $\varphi' : G(\mathcal{Q}) \to G$ be the simple morphism that sends the generators of G_{σ_1} and G_{σ_2} to a, and sends the generator of G_{σ_3} to b. Then $Y_{\varphi'}$ is also a surface of genus 3, but the decomposition in pairs of pants is different — it is shown in figure 12.2. The affine realizations of $D(\mathcal{Q}, \varphi)$ and $D(\mathcal{Q}, \varphi')$ are not homeomorphic.

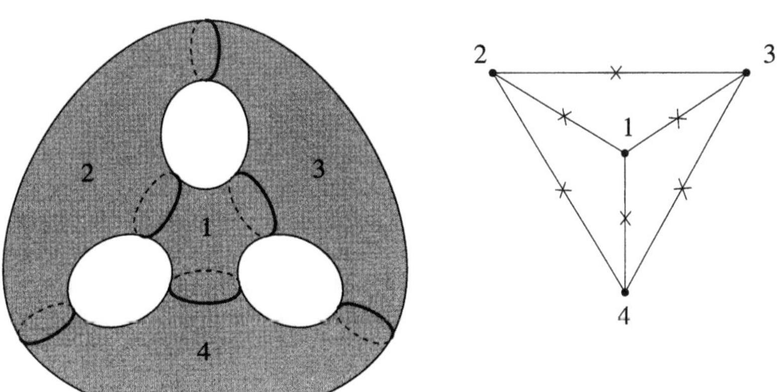

Fig. 12.10 The surface $D(Y, \varphi)$ in 12.19(3) and the affine realization of $D(\mathcal{Q}, \varphi)$

(4) The examples of Mike Davis described in 5.22 and 5.23 are developments $D(F, \varphi)$ and $D(M, \varphi)$ of the spaces F and M with the stratifications described in 12.4(3) and 12.2(3) respectively. To see this, take $\mathcal{Q}$ to be the poset associated to these stratifications, let $G(\mathcal{Q})$ be the complex of groups that associates to $\sigma = (s_1, \dots, s_k) \in \mathcal{Q}$ the group $G_\sigma = \prod_{i=1}^k G_{s_i}$ defined in 5.23, and let $\varphi : G(\mathcal{Q}) \to G$ be the morphism defined by the inclusions φ_σ of the groups G_σ into the group $G = \prod_{s \in S} G_s$.

(5) **Constructing Tesselations of $\mathbb{H}^2$.** Let P be a k-gon in the hyperbolic plane with vertices $\tau_1, \dots, \tau_\kappa$ (in order). Assume that the angle at the vertex τ_i is of the form π/p_i, where p_i is an integer ≥ 2 (such polygons exist if $k > 4$ and in general if $\sum_i 1/p_i < k - 2$). Let σ_i be the side of P joining the vertex τ_{i-1} to the vertex τ_i (the indices are taken modulo k), and let s_i be the reflection of $\mathbb{H}^2$ fixing σ_i. Let G be the subgroup of $\text{Isom}(\mathbb{H}^2)$ generated by $s_1, \dots, s_k$. We claim that *G acts on $\mathbb{H}^2$ with strict fundamental domain P* (and thus the images of P by the elements of G form a tesselation of $\mathbb{H}^2$ by polygons congruent to P). This is a particular case of a classical theorem of Poincaré.

To prove our assertion, we consider a complex of groups $G(\mathcal{Q}_k)$ over the poset $\mathcal{Q}_k = \{\rho, \sigma_i, \tau_i\}$ of faces of P where ρ corresponds to P. Define $G_\rho = 1$, G_{σ_i} to be the subgroup of G generated by s_i, and G_{τ_i} the subgroup generated by s_{i-1} and s_i. Let $\varphi : G(\mathcal{Q}) \to G$ be the morphism defined by the inclusions into G. The development $D(P, \varphi)$ is a connected hyperbolic polyhedron. There is a natural G-equivariant projection $f : D(P, \varphi) \to \mathbb{H}^2$ mapping $[g, y]$ to $g.y$ and we claim that this projection is a local isometry. This is trivial except at the vertices $\tau_i = [1, \tau_i]$ of $P \subset D(P, \varphi)$, where it follows from our hypothesis that the angle of P at τ_i is of the form π/p_i. Indeed this hypothesis implies that G_{τ_i} is the dihedral group of order $2k$ which acts on $\mathbb{H}^2$ with a strict fundamental domain which is the sector bounded by the rays issuing from τ_i and containing the sides σ_{i-1} and σ_i. As $D(P, \varphi)$ is complete, the map f is a covering of $\mathbb{H}^2$, hence an isometry (cf. I.3.25), which means that we can identify $D(P, \varphi)$ with $\mathbb{H}^2$.

The same construction using the morphism $\iota : G(\mathcal{Q}_k) \to \widehat{G(\mathcal{Q}_k)}$ would lead to the same conclusion, and therefore G is isomorphic to the group $\widehat{G(\mathcal{Q}_k)}$ obtained by amalgamating the groups G_{τ_i} and G_{σ_i} as in (12.12) (see also 12.22).

The same argument works when the hyperbolic plane is replaced by $\mathbb{S}^2$ or $\mathbb{E}^2$. Indeed we could replace P by any convex polyhedral n-cell P in M_κ^n such that the dihedral angle between each pair of $(n-1)$-dimensional faces is of the form π/m where m is an integer ≥ 2. The reflections fixing the $(n-1)$-faces of P generate a subgroup G of $\text{Isom}(M_\kappa^n)$ and the images of P by the elements of G form a tesselation of M_κ^n. The argument is similar to the preceding one and uses induction on the dimension n (for details, see de la Harpe [Har91, p.221-238]).

12.20 Proposition (Properties of the Basic Construction). *We maintain the notation established in (12.18). Thus we have a simple complex of groups $G(\mathcal{Q})$, a stratified space $(Y, \{Y^\sigma\}_{\sigma \in \mathcal{Q}})$, and we consider simple morphisms $\varphi : G(\mathcal{Q}) \to G$ that are injective on the local groups. (G is an arbitrary group.)*

(1) (Uniqueness) *Let X be a stratified space over* $\mathcal{Q}$*. If G acts by strata preserving morphisms on X with strict fundamental domain Y, and the associated simple complex of groups is simply isomorphic to* $G(\mathcal{Q})$*, then the identity map on Y extends uniquely to a strata preserving G-equivariant isomorphism* $D(Y,\varphi) \to X$.

(2) *Let* $G_0 \subset G$ *be the group generated by the subgroups* $\varphi_\sigma(G_\sigma)$*. If Y is connected, then the set of connected components of* $D(Y,\varphi)$ *is in bijection with the set of cosets* G/G_0*. In particular* $D(Y,\varphi)$ *is connected if and only if* $G_0 = G$.

(3) (Functoriality) *Let* $\pi : G \to G'$ *be a surjective homomorphism of groups. Let* $\varphi' : G(\mathcal{Q}) \to G'$ *be the simple morphism defined by* $\varphi'_\sigma := \pi\varphi_\sigma$*. If* φ' *is injective on the local groups, then the identity map on Y extends uniquely to a strata preserving G-equivariant map* $p : D(Y,\varphi) \to D(Y,\varphi')$*. The map p is a covering projection, and the kernel of* π *acts freely and transitively on the fibres of p.*

(4) *Suppose that Y is connected and simply connected and that each stratum of Y is arcwise connected. Then the development* $D(Y,\iota)$ *of Y with respect to the canonical morphism* $\iota : G(\mathcal{Q}) \to \widehat{G(\mathcal{Q})}$ *is simply connected. If the canonical homomorphism* $\hat{\varphi} : \widehat{G(\mathcal{Q})} \to G$ *associated to* φ *is surjective, then* $\pi_1(D(Y,\varphi)) \cong \ker\hat{\varphi}$*. (In the topological case one has to assume that* $D(Y,\varphi)$ *is locally simply connected.)*

Proof. We use the notation established in the proof of (12.18).

(1) The map $[g,y] \to g.y$ is a stratum preserving G-equivariant isomorphism from $D(Y,\varphi)$ to X.

(2) follows from (12.10).

(3) Let $p : D(Y,\varphi) \to D(Y,\varphi')$ be the map sending $x = [g,y]$ to $x' = [\pi(g),y]$; it is G-equivariant and maps the stratum indexed by $(g\varphi_\sigma(G_\sigma),\sigma)$ onto the stratum indexed by $(\pi(g)\varphi'_\sigma(G_\sigma),\sigma)$. The inverse image of x' is the set of points of the form $k.x$, where $k \in \ker\pi$. Let $Y^{\sigma(y)}$ be the smallest stratum of Y containing y and let $G_y = \varphi_{\sigma(y)}(G_{\sigma(y)})$. If $st(y)$ is the open star of y in Y, the open star $st(x)$ of x in $D(Y,\varphi)$ is $gG_y.st(y)$ (see 12.8(5)). Similarly $star(x') = \pi(g)G'_y.st(y)$, where $G'_y = \varphi'_{\sigma(y)}(G_{\sigma(y)})$. Thus p maps $st(x)$ homeomorphically onto $st(x')$, and $p^{-1}(st(x'))$ is the disjoint union of the open sets $k.st(x)$, where $k \in \ker\pi$. Therefore p is a Galois covering with Galois group $\ker\pi$.

(4) We use the construction of 12.17(7) applied to the action of G on $X = D(Y,\varphi)$. Thus we get a group $\tilde{G}$ which is an extension of G by the fundamental group of $D(Y,\varphi)$, a stratification of the universal covering $\tilde{D}(Y,\varphi)$ of $D(Y,\varphi)$, and a strata preserving action of $\tilde{G}$ on $\tilde{D}(Y,\varphi)$ with a strict fundamental domain one can identify to Y so that the associated complex of groups is canonically isomorphic to $G(\mathcal{Q})$. Let $\tilde{\varphi}$: $G(\mathcal{Q}) \to \tilde{G}$ be the associated morphism and let $\hat{\tilde{\varphi}} : \widehat{G(\mathcal{Q})} \to \tilde{G}$ be the homomorphism associated to $\tilde{\varphi}$. By (1), there is a $\tilde{G}$-equivariant isomorphism $\tilde{D}(Y,\varphi) \to D(Y,\tilde{\varphi})$, and by (3), a $\widehat{G(\mathcal{Q})}$-equivariant covering map $D(Y,\iota) \to D(Y,\tilde{\varphi})$; as $D(Y,\tilde{\varphi})$ is simply connected, this is an isomorphism. Hence $D(Y,\iota)$ is simply connected and

$\widehat{G(\mathcal{Q})} \to \tilde{G}$ is an isomorphism. The last claim follows from (3) applied to the homomorphism $\hat{\tilde{\varphi}}$. □

12.21 Corollary. *Let $G(\mathcal{Q})$ be a simple complex of groups over a poset $\mathcal{Q}$. Let $Y = |\mathcal{Q}|$. If the canonical simple morphism $\iota : G(\mathcal{Q}) \to \widehat{G(\mathcal{Q})}$ is injective on the local groups and if Y is simply connected, then the space $D(Y, \iota)$ given by the Basic Construction is also simply connected. (And by 12.18(5) it is the realization of the poset $D(\mathcal{Q}, \varphi)$.)*

The proposition can sometimes be used to find presentations of groups.

12.22 Corollary. *Let G be a group acting on a simply connected stratified space X with a strict fundamental domain Y. Then G is the direct limit (amalgam) of the isotropy subgroups of the various strata of Y.*

Proof. Let $G(\mathcal{Q})$ be the complex of groups associated to this action, let $\varphi : G(\mathcal{Q}) \to G$ be the corresponding morphism and let $\hat{\varphi} : \widehat{G(\mathcal{Q})} \to G$ be the associated homomorphism. According to (1), the space X is G-equivariantly isomorphic to $D(Y, \varphi)$, by (2) $\hat{\varphi}$ is surjective, and by (3) the $\hat{\varphi}$-equivariant map $D(Y, \iota) \to D(Y, \varphi)$ is a covering. As $D(Y, \varphi)$ is simply connected, this last map is an isomorphism, and so too is $\hat{\varphi}$. □

12.23 Examples

(1) We revisit the group action and triangle of groups described in 12.17(3), maintaining the same notation. Thus we have a group G generated by reflections s_1, s_2, s_3 acting on $\mathbb{E}^2$. As $\mathbb{E}^2$ is simply connected, we can appeal to 12.22 (or to 12.19(5)) to see that the group G is isomorphic to $\widehat{G(\mathcal{Q})}$, which has presentation

$$\langle s_1, s_2, s_3 \mid s_1^2 = s_2^2 = s_3^2 = (s_1 s_2)^6 = (s_2 s_3)^3 = (s_1 s_3)^2 = 1 \rangle.$$

Let $\varphi : G(\tau) \to D_6$ be the simple morphism mapping s_1, s_2 to the corresponding generators of the dihedral group D_6 and s_3 to $s_2(s_1 s_2)^2$. If Y is the triangle chosen as strict fundamental domain for the action, then $D(Y, \varphi)$ is the 2-torus obtained by identifying opposite sides of the parallelogram formed by two adjacent triangles of the tesselation. The group D_6 acts on this 2-torus with a strict fundamental domain isometric to Y. In general, given any surjective homomorphism $\hat{\varphi} : \widehat{G(\mathcal{Q})} = G \to H$ that is injective on the local groups, if $\ker \hat{\varphi}$ is non trivial, then $D(Y, \varphi)$ is isometric to a 2-torus obtained by multiplying the distance function on the one described above by a positive integer.

(2) We can apply corollary 12.22 to check that the group of permutations $\Sigma(n)$ (in the notations of example 12.9(3)) has the following presentation for $n \geq 4$: it is generated by the elements $s_1, \dots, s_{n-1}$, subject to the relations $s_i^2 = 1$ for $i \leq n-1$, $(s_i s_{i+1})^3 = 1$ for $i < n-1$ and $(s_i s_j)^2 = 1$ for $j \neq i \pm 1$.

Let T be the triangulation of S^{n-2} described in 12.9(3), and let X be the 2-dimensional subcomplex of the barycentric subdivision formed by the union of the simplices whose vertices are barycentres of those simplices of T that have dimension

$n-2, n-3$ or $n-4$. This complex X is the union of the cells of dimension ≤ 2 in the dual subdivision of T, hence it is simply connected. The action of $\Sigma(n)$ on S^{n-2} preserves X and has a strict fundamental domain $X \cap \Delta$. The isotropy subgroups of faces of dimension $n-3$ (resp. $n-4$) of Δ are the cyclic subgroups generated by the s_i (resp. the dihedral groups generated by pairs (s_i, s_j), $i \neq j$). By applying 12.22 to the action of $\Sigma(n)$ on X we get the desired presentation.

Local Development and Curvature

The results of the previous section show that one can construct interesting group actions from simple complexes of groups if there exist simple morphisms that are injective on the local groups. However, we have not yet developed an effective criterion that is sufficient to show that the local groups inject; that is the purpose of this section. We shall explain how one can use the local structure of a simple complex of groups to understand what the closed star of a point in the development would look like if it were to exist – this is called local development. It is a remarkable fact (akin to the Cartan-Hadamard Theorem) that if at every point in the complex the putative star is non-positively curved (and the local groups act by isometries), and if $|\mathcal{Q}|$ is simply connected, then the complex is strictly developable. This will be proved in Part III.

12.24 Construction of the Local Development. We shall describe an explicit model for the closed star $St(\tilde{\sigma})$ of the vertex $\tilde{\sigma} = [1, \sigma]$ in the affine realization of the posets $D(\mathcal{Q}, \varphi)$ constructed in the Basic Construction (12.18). This description is independent of φ. The local group G_σ acts naturally on $St(\tilde{\sigma})$ with a strict fundamental domain naturally isomorphic to $St(\sigma)$.

We define the complex $St(\tilde{\sigma})$ to be the affine realization of the subposet of $D(\mathcal{Q}, \varphi)$ whose elements are pairs that are either of the form $(g\psi_{\sigma\rho}(G_\rho), \rho)$, where $\rho > \sigma$ and $g \in G_\sigma$, or else are of the form (G_σ, τ), where $\tau \leq \sigma$ (in such pairs, the first member is considered as a coset in G_σ). The partial ordering is defined by:

$$\text{for } \sigma < \rho, \rho' : \ (g\psi_{\sigma\rho}(G_\rho), \rho) < (g'\psi_{\sigma\rho'}(G_{\rho'}), \rho') \text{ if } \rho < \rho' \text{ and } g^{-1}g' \in \psi_{\sigma\rho}(G_\rho),$$

$$\text{for } \tau, \tau' \leq \sigma : \ (G_\sigma, \tau) < (G_\sigma, \tau') \text{ if } \tau < \tau',$$

$$\text{for } \tau \leq \sigma < \rho : \ (G_\sigma, \tau) < (g\psi_{\sigma\rho}(G_\rho), \rho).$$

Using the notations of 12.3, this poset can also be described as follows.

Let $G(Lk_\sigma)$ be the restriction of the simple complex of groups $G(\mathcal{Q})$ to the subposet $Lk_\sigma(\mathcal{Q}) = \{\rho \in \mathcal{Q} : \rho > \sigma\}$. One has a canonical simple morphism $\psi_\sigma : G(Lk_\sigma) \to G_\sigma$ given by the injective homomorphisms $(\psi_\sigma)_\rho := \psi_{\sigma\rho} : G_\rho \to G_\sigma$. The upper link of $\tilde{\sigma} = (G_\sigma, \sigma)$ in the poset $D(\mathcal{Q}, \varphi)$ can be defined as

$$Lk_{\tilde{\sigma}} := D(Lk_\sigma, \psi_\sigma),$$

and $St(\tilde{\sigma})$ is the affine realization of the poset

$$\mathcal{Q}^{\sigma} * D(Lk_{\sigma}, \psi_{\sigma}).$$

The map that associates to each $\rho > \sigma$ the pair $\tilde{\rho} = (\psi_{\sigma\rho}(G_{\rho}), \rho)$, and to each $\tau \leq \sigma$ the pair $\tilde{\tau} = (\Gamma_{\sigma}, \tau)$, gives an identification of $St(\sigma)$ with a subspace of $St(\tilde{\sigma})$. There is also a natural projection $p_{\sigma} : St(\tilde{\sigma}) \to St(\sigma)$, induced by the poset map sending each pair to its second coordinate.

The group G_{σ} acts in an obvious way on $St(\tilde{\sigma})$. Moreover it is clear that the preceding construction does not involve the homomorphism φ. Thus we have the following local developability property in complete generality.

12.25 Proposition. *Let $G(\mathcal{Q}) = (G_{\sigma}, \psi_{\tau\sigma})$ be a complex of groups over a poset $\mathcal{Q}$. Canonically associated to each $\sigma \in \mathcal{Q}$ there is a poset on whose affine realization $St(\tilde{\sigma})$ the group G_{σ} acts with strict fundamental domain $St(\sigma)$. In this action, the isotropy subgroup of the stratum indexed by each $\tau \leq \sigma$ is G_{σ}, and the isotropy subgroup of the stratum indexed by $\rho > \sigma$ is $\psi_{\sigma\rho}(G_{\rho})$.*

$St(\tilde{\sigma})$ together with the action of G_{σ} is called the *local development* of $G(\mathcal{Q})$ at the vertex σ.

Assume that the affine realization $|\mathcal{Q}|$ of $\mathcal{Q}$ is endowed with the structure of an M_{κ}-simplicial complex with finite shapes. Then for each $\sigma \in \mathcal{Q}$, the simplicial complex $St(\tilde{\sigma})$ is also an M_{κ}-simplicial complex, and the projection p_{σ} is an isometry on each simplex.

12.26 Definition (Simple Complex of Groups of Curvature $\leq \kappa$). Let $\kappa \in \mathbb{R}$. A complex of groups $G(\mathcal{Q})$ over an M_{κ}-simplicial complex $|\mathcal{Q}|$ is said to be of curvature $\leq \kappa$ if the induced metric on the open star $st(\tilde{\sigma})$ in the local development is of curvature $\leq \kappa$ for each $\sigma \in \mathcal{Q}$.

We shall be most interested in the case $\kappa = 0$ where, of course, we say "non-positively curved" instead of curvature ≤ 0.

12.27 Remarks

(1) If $|\mathcal{Q}|$ is one dimensional, and the length of each of its edges is one, say, then any complex of groups over $\mathcal{Q}$ is of curvature $\leq \kappa$ for every $\kappa \in \mathbb{R}$.

(2) Suppose that $\mathcal{Q}$ is the poset of cells of an M_{κ}-polyhedral complex K with Shapes(K) finite and suppose that there are no identifications on the boundaries of cells in K. Then the affine realization $|\mathcal{Q}|$ of $\mathcal{Q}$ is isomorphic to the barycentric subdivision K' of K, and as such inherits the structure of an M_{κ}-simplicial complex. And for each vertex τ of K, the geometric link $\text{Lk}(\tau, K)$ inherits a spherical metric: it is a (spherical) geometric realization of the subposet $Lk_{\tau} \subset \mathcal{Q}$.

To check that a simple complex of groups $G(\mathcal{Q})$ over $\mathcal{Q}$ is of curvature $\leq \kappa$, it is sufficient to check that for each vertex $\tau \in K$, the affine realization of $Lk_{\tilde{\tau}} = D(Lk_{\tau}, \psi_{\tau})$, with its natural structure as a spherical complex, is CAT(1) (see I.7.56 and 5.2).

The following theorem, first proved by Gersten-Stallings [St91] in the particular case of triangles of groups (see also Gromov [Gro87, p.127-8], Haefliger [Hae90 and 91], Corson [Cors92] and Spieler [Spi92]), will be proved later in a more general framework (III.$\mathcal{G}$).

12.28 Theorem. *Let $\mathcal{Q}$ be a poset whose affine realization $|\mathcal{Q}|$ is simply connected, and suppose that $|\mathcal{Q}|$ is endowed with an M_κ-simplicial structure with finite shapes, where $\kappa \leq 0$. Let $G(\mathcal{Q})$ be a simple complex of groups over $\mathcal{Q}$ which is of curvature $\leq \kappa$. Then $G(\mathcal{Q})$ is strictly developable.*

If $\mathcal{Q}$ is finite and each local group G_σ is finite, this implies that $\widehat{G(\mathcal{Q})}$ acts cocompactly on the CAT(κ) space $|D(\mathcal{Q}, \iota)|$ and that any finite subgroup of $\widehat{G(\mathcal{Q})}$ is conjugate to a subgroup of one of the local groups G_σ.

12.29 The Local Development at the Corner of an n-Gon of Groups. We wish to describe the local development at the vertices of polygons of groups. If one focuses on the corner of a polygon of groups, one has the vertex group, the two incoming edge groups, and the local group of the 2-cell. With this picture in mind, we consider simple complexes of groups over the poset $\mathcal{Q}_0 = \{\rho, \sigma_1, \sigma_2, \tau\}$ with $\rho < \sigma_1,\ \rho < \sigma_2$ and $\sigma_1 < \tau,\ \sigma_2 < \tau$. This is equivalent to considering a group G (playing the role of G_τ), with two subgroups G_1 and G_2 (corresponding to σ_1 and σ_2) and a subgroup $H \subseteq G_1 \cap G_2$ (namely G_ρ).

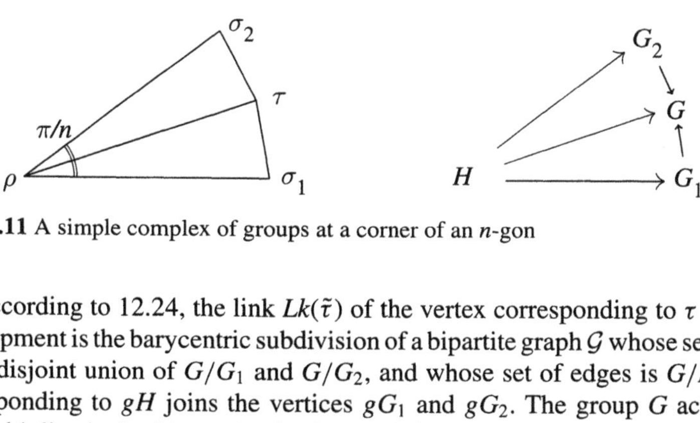

Fig. 12.11 A simple complex of groups at a corner of an n-gon

According to 12.24, the link $Lk(\tilde{\tau})$ of the vertex corresponding to τ in the local development is the barycentric subdivision of a bipartite graph $\mathcal{G}$ whose set of vertices is the disjoint union of G/G_1 and G/G_2, and whose set of edges is G/H; the edge corresponding to gH joins the vertices gG_1 and gG_2. The group G acts on $\mathcal{G}$ (by left multiplication) with a strict fundamental domain which is an edge. This graph is connected if and only if G is generated by $G_1 \cup G_2$. The girth $girth(\mathcal{G})$ of this graph is by definition the minimal number of edges contained in a simple closed curve in $\mathcal{G}$ (in the case of a bipartite graph, the girth is even or infinity). For instance the girth is 2 if and only if $H \neq G_1 \cap G_2$, and it is infinity if and only if the subgroup of G generated by $G_1 \cup G_2$ is the amalgamated product $G_1 *_H G_2$. If $G_1 = \langle s_1 \rangle$ and $G_2 = \langle s_2 \rangle$ have order 2 and G is the dihedral group D_n of order $2n$ generated by $\{s_1, s_2\}$, then the girth is $2n$. If $G = G_1 \times G_2$, the girth of $\mathcal{G}$ is 4.

The affine realization of the poset $\mathcal{Q}_0$ is the union of two triangles glued along the side joining ρ to τ. If we metrize $|\mathcal{Q}_0|$ so that it is the union of two triangles in M_κ^2 such that the total angle at the vertex τ is α, then the development at τ is of curvature $\leq \kappa$ if and only if $2\pi \geq \alpha\ girth(\mathcal{G})$. (Stallings [St91] calls $2\pi/girth(\mathcal{G})$ the (group theoretic) angle of the n-gon of groups at the vertex τ.)

For instance, if P is a regular n-gon in $\mathbb{E}^2$, the specific n-gon of groups considered in 12.17(6) is non-positively curved if $n \geq 4$, because the girth of the link of the local development at each vertex τ_i is 4 (a circuit of length four is formed by the edges labelled $1, x_i, x_i x_{i+1}, x_i x_{i+1} x_i^{-1} = x_{i+1}^2$) and the angle at a vertex is $\geq \pi/2$. If we retain the same underlying simple complex of groups but now regard P as a regular n-gon in the hyperbolic plane with vertex angles greater than $\pi/2$ (this is possible if $n \geq 5$), then this complex of groups is of curvature ≤ -1.

More generally, we might consider a complex of groups over the poset $\mathcal{Q}_n$ of faces of a regular Euclidean n-gon P such that, in the local groups at the vertices of P, the intersection of the two subgroups coming from the adjacent edges is equal to the subgroup coming from the barycentre of P. Any such n-gon of groups is non-positively curved if $n \geq 4$, hence is developable, in the light of (12.28).

We now turn to more explicit examples of non-positively curved spaces with cocompact group actions.

12.30 Examples

(1) *Hyperbolic Polyhedra.* Let P be a right-angled regular p-gon in $\mathbb{H}^2$ (such a polygon exists if $p > 4$). We consider the following complex of groups $G(\mathcal{Q})$ over the poset $\mathcal{Q}$ of faces of P: the group associated to P is trivial, the group associated to each edge e_i of P is a cyclic group $\mathbb{Z}_q$ of order q (generated by an element s_i, say) and at the vertex of P adjacent to the edges e_i and e_{i+1}, $i \in \mathbb{Z} \bmod p$, we have the product $\mathbb{Z}_q \times \mathbb{Z}_q$ generated by s_i and s_{i+1} (the labelling of the generators defines the inclusions of the edges groups). This complex of groups is developable, because there is an obvious simple morphism of $G(\mathcal{Q})$ to the product $\prod_1^p \mathbb{Z}_q$ mapping s_i to a generator of the ith factor. The group $\Gamma_{p,q} := \widehat{G(\mathcal{Q})}$ is generated by the elements s_i subject to the relations $s_i^q = 1$ and $s_i s_{i+1} = s_{i+1} s_i$ for $i \in \mathbb{Z} \bmod p$. The link of the local development at each vertex is the complete bipartite graph obtained by taking the join of two copies of $\mathbb{Z}_q$; its girth is 4, therefore the M_{-1}-complex $I_{p,q} := D(P, \iota)$ on which $\Gamma_{p,q}$ acts properly and cocompactly is CAT(-1). Marc Bourdon [Bou97a] has determined the conformal dimension (in the sense of Pansu [Pan90]) of the boundary at infinity of $I_{p,q}$. This boundary is homeomorphic to a Menger curve [Ben92].

(2) *Graph Products of Groups.* Let L be a simplicial graph with vertex set S and let $(G_s)_{s\in S}$ be a family of groups indexed by the set of vertices of L. The *graph product* G_L of the family of groups G_s is the abstract group generated by the elements of the groups G_s subject to the relations in G_s and the relations $g_s g_t = g_t g_s$ if $g_s \in G_s$, $g_t \in G_t$ and s and t are joined by an edge in L. Let G be the direct product $\prod_{s\in S} G_s$ of the groups G_s. There is a unique homomorphism $\varphi : G_L \to G$ such that the natural homomorphism of G_s into G_L composed with φ then the projection onto the factor G_s is the identity.

Following [Da98] and [Mei96], we construct a CAT(0) piecewise Euclidean cubical complex X on which G_L acts by isometries (preserving the cell structure) with a strict fundamental domain.

Let K be the flag simplicial complex whose 1-skeleton is the graph L. let $\mathcal{Q}$ be the poset of those subsets of S (the vertex set of K) which span a simplex of K together with the empty set, ordered by reverse-inclusion, $\sigma \leq \tau$ if $\tau \subseteq \sigma$. To each $\sigma = \{s_1, \dots, s_k\} \in \mathcal{Q}$, one associates the product $G_\sigma = \prod_{i=1}^k G_{s_i}$ (define $G_\emptyset = \{1\}$), and for $\tau \subset \sigma$ we denote by $\psi_{\sigma\tau}$ the natural inclusion $G_\tau \hookrightarrow G_\sigma$. This defines on $\mathcal{Q}$ a simple complex of groups $G(\mathcal{Q})$. It is strictly developable thanks to the existence of the homomorphism φ, and $\widehat{G(\mathcal{Q})}$ is the graph product of the groups G_s. The affine realization of the poset $\mathcal{Q}$ is the simplicial cone over the barycentric subdivision of K (see 12.4(3)) which, in a natural way, can be considered as a cubical complex; let F denote this cubical complex with its natural piecewise Euclidean metric.

Let X be the cubical complex $D(F, \iota)$, where $\iota : G(\mathcal{Q}) \to G_L$ is the canonical simple morphism. As F is simply connected, X is also simply connected. One can check that the link of each vertex of X is a flag complex, hence X is CAT(0) by Gromov's criterion (5.20) and the Cartan-Hadamard theorem.

(3) We should mention that Tits has constructed triangles of finite groups ([Tits86b], [Ron89, p.48-49]) whose simply connected developments provide remarkable examples of Euclidean buildings which have the same type and are locally isomorphic, but enjoy quite different properties.

We should also mention the striking work of Floyd and Parry [FP97]. They give explicit calculations of the (rational) growth functions of the direct limits of certain non-positively curved triangles of finite groups. They also exhibit examples of pairs of such triangles such that all of the local groups are the same in each case, but the direct limit of one triangle contains $\mathbb{Z}^2$ while the other does not (it is δ-hyperbolic in the sense of (III.Γ.2.1)). Remarkably, these two groups have the same growth function.

Constructions Using Coxeter Groups

12.31 Definition. A Coxeter group is a group W with a specified finite generating set S and defining relations of the form $(st)^{m_{st}}$, one for each pair $s, t \in S$, where $m_{ss} = 1$ and if $s \neq t$ then m_{st} is an integer ≥ 2 or $m_{st} = \infty$ (which this means no relation between s and t is given).

The pair (W, S) is called a Coxeter system.

The matrix $M = (m_{st})_{s,t \in S}$ describing the defining relations can be reconstructed from a weighted graph, called the *Coxeter graph*. The nodes (vertices) of this graph are labelled by the elements of S. The vertices labelled s and t are joined by an unweighted edge if $m_{st} = 3$ and by an edge weighted (i.e. labelled) by m_{st} if $m_{st} > 3$ or $m_{st} = \infty$. (If s and t are not joined by an edge, this means that $m_{st} = 2$, i.e. the generators s and t commute.)

Coxeter Graphs Coxeter Groups

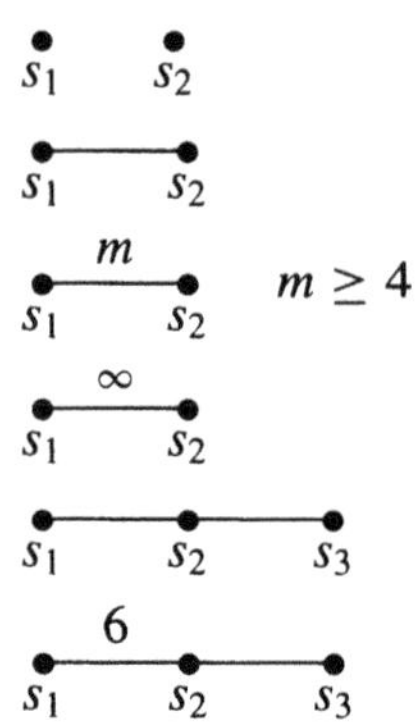

$D_2 = \mathbb{Z}_2 \times \mathbb{Z}_2$

$D_3 = S(3)$

D_m

$D_\infty = \mathbb{Z}_2 \star \mathbb{Z}_2$

$S(4)$

Group of isometries of $\mathbb{E}^2$ preserving a tessellation by equilateral triangles (see 12.17(3))

Fig. 12.12 Some Coxeter graphs and the corresponding Coxeter groups.

12.32 Basic Property of Coxeter Systems. *Let (W, S) be a Coxeter system. Given a subset T of S, let W_T be the Coxeter group with generating set T and relations $(tt')^{m_{tt'}}$ for $t, t' \in T$ (where the exponents $m_{tt'}$ are as in the defining relations for (W, S)). Then the natural homomorphism $W_T \to W$ induced by the inclusion $T \hookrightarrow S$ is injective.*

For a proof, see Bourbaki [Bou81].

Note that in particular this result says that each element s of S generates a cyclic subgroup of order two in W, that two distinct elements $s, s' \in S$ with $m_{ss'} = \infty$ generate an infinite dihedral group, and if $m_{ss'}$ is the integer $m \geq 2$, then s and s' generate a dihedral group of order $2m$ (denoted D_m). The subgroups of order 2 generated by different elements of S are all distinct.

12.33 Complexes of Groups Associated to Coxeter Systems. In our language, 12.32 can be reformulated as follows. Let $\mathcal{S}$ be the poset of subsets of S ordered by reverse-inclusion, that is $T_1 \leq T_2$ if $T_2 \subseteq T_1$. We define a simple complex of groups $W(\mathcal{S})$ over the poset $\mathcal{S}$: to $T \subseteq S$ we associate the group W_T; if $T_2 \subseteq T_1 \subseteq S$, then $\psi_{T_1T_2}$ is the inclusion $W_{T_2} \hookrightarrow W_{T_1}$ induced by $T_1 \hookrightarrow T_2$. For $T = \emptyset$ the associated group $W_\emptyset$ is the trivial group, and $W_S = W$. Note that W is the direct limit of the groups W_T, so we have a canonical morphism $\iota : W(\mathcal{S}) \to W$ (see 12.12).

The affine realization of $\mathcal{S}$ is a simplicial subdivision of an n-cube, where n is the cardinality of S (if T has cardinality k, the stratum indexed by T is an $(n-k)$-cube, (see 12.4(2))). Let $\mathcal{T}$ be a subposet of $\mathcal{S}$ and let $W(\mathcal{T})$ be the simple complex of groups obtained from $W(\mathcal{S})$ by restriction to $\mathcal{T}$. The restriction to $W(\mathcal{T})$ of the morphism ι (which is injective on the local groups) will also be denoted $\iota : W(\mathcal{T}) \to W$. Let $|\mathcal{T}|$ be the affine realization of $\mathcal{T}$ with its natural stratification and let $\Sigma(W, \mathcal{T})$ be the stratified space $D(|\mathcal{T}|, \iota)$ given by the basic construction (12.18). The group W acts on $\Sigma(W, \mathcal{T})$ with strict fundamental domain $|\mathcal{T}|$. We consider various special cases of this construction.

12.34 Examples

(1) If $\mathcal{T}$ is the poset of proper subsets of S, then $\Sigma(W, \mathcal{T})$ is sometimes called the Coxeter complex of (W, S) (cf. [Bro88]). If S has cardinality n, this complex has dimension $n-1$. If W is finite, then $\Sigma(W, \mathcal{T})$ is an $(n-1)$-sphere tesselated by the translates of the simplex $|\mathcal{T}|$.

(2) M. Davis [Da98] introduced a different (inequivalent) notion of Coxeter complex. He did so by considering the subposet $\mathcal{S}^f$ of $\mathcal{S}$ consisting of those subsets $T \subseteq S$ such that W_T is a finite group. The geometric realization $|\mathcal{S}^f|$ is a cubical complex, and the stratified space $\Sigma(W, \mathcal{S}^f)$ given by the Basic Construction is a simply connected cubical complex on which W acts properly. Indeed the geometrical realization of $\mathcal{S}^f$ is always contractible, because it is a cone (the cone point corresponds to the empty subset).

To obtain an explicit example, we consider a polygon P in the hyperbolic plane $\mathbb{H}^2$ such that the angle between each consecutive pair of sides s and s' is of the form $\pi/m_{ss'}$, where $m_{ss'} \geq 2$ is an integer. Let S be the set of sides of the polygon and define $m_{ss} = 1$ and $m_{ss'} = \infty$ if the sides are distinct and not adjacent. Let W be the Coxeter group determined by the Coxeter matrix $M = (m_{ss'})$. In this case $|\mathcal{S}^f|$ is isomorphic to the barycentric subdivision of P and $\Sigma(W, \mathcal{S}^f)$ is isomorphic to the barycentric subdivision of a tesselation of $\mathbb{H}^2$ by polygons congruent to P (see 12.19(5)).

More generally, if P has some vertices at infinity (and we define $m_{ss'} = \infty$ for the corresponding consecutive sides), then $\mathcal{S}^f$ is the poset associated to the stratification of P in which the strata are P, the sides of P, and the vertices that are not at infinity. In this case $|\mathcal{S}^f|$ is a subcomplex of the barycentric subdivision of P. The stratified space $D(P, \iota)$ given by the basic construction is isometric to a tesselation of $\mathbb{H}^2$ by polygons congruent to P, while $\Sigma(W, \mathcal{S}^f)$ is isomorphic to a proper subspace of $\mathbb{H}^2$. For instance, if all the vertices of P are at infinity, then $\Sigma(W, \mathcal{S}^f)$ is the barycentric subdivision of a regular tree; the number of edges issuing from a vertex is equal to the number of sides of P.

If (W, S) is a right angled Coxeter system (meaning that $m_{st} = 2$ or ∞ for all $s \neq t$), then $\Sigma(W, \mathcal{S}^f)$ is the universal covering of the complex constructed in 5.22.

In his thesis, Gabor Moussong [Mou88] proved that, for any Coxeter system (W, S), one can metrize the cells of $|\mathcal{S}^f|$ as Euclidean polyhedra so that in the induced piecewise Euclidean polyhedral structure, $\Sigma(W, \mathcal{S}^f)$ is CAT(0).

(3) Let L be a finite simplicial graph with vertex set S and suppose that each edge is assigned an integer $m_{ss'} \geq 2$ (where s and s' are the endpoints of the edge). There is a natural way to associate to such a labelled graph a two-dimensional polyhedron K and a group W acting simply transitively on the set of vertices of K, so that the link of each vertex of K is isomorphic to L; in this isomorphism the edge of L joining s to s' represents the corner of a polygonal face with $2m_{s,s'}$ sides that is incident to the vertex (see Benakli [Ben91a] and Haglund [Hag91]). To construct K and W, consider $M = (m_{ss'})$ to be a Coxeter matrix with $m_{ss'} = \infty$ if s and s' are not joined by an edge in L, and let W be the associated Coxeter group. Let $\mathcal{T} \subseteq \mathcal{S}^f$ be those subsets with cardinality ≤ 2. The geometric realization of $|\mathcal{T}|$ is the cone over the barycentric subdivision of L (see fig.12.13, where L is the union of two edges having a vertex

in common). The first barycentric subdivision of K will be the complex $\Sigma(W, \mathcal{T})$. A strict fundamental domain for the action of W on $\Sigma(W, \mathcal{T})$ is the star of the vertex corresponding to the empty set.

(4) **Haglund's Construction of Gromov Polyhedra.** Recall that a simplicial graph is said to be complete if every pair of vertices is joined by an edge. In this paragraph we descibe how to construct simply connected polyhedra in dimension two where the faces are k-gons and the link of each vertex is a complete graph with l vertices. The existence of such polyhedra was asserted by Gromov [Gro83]. In fact, as shown by Haglund [Hag91] and independently by Ballmann-Brin [BaBr94], there is an uncountable number of isomorphisms classes of such polyhedra when $k \geq 6$ and $l \geq 4$. When $l = 3$, such a polyhedron is a simply connected surface tesselated by k-gons: if $k \leq 5$ this is a sphere, if $k = 6$ the Euclidean plane, and if $k > 6$ the hyperbolic plane.

The construction that we shall describe gives a polyhedron $X_{k,l}$ with an action of a group W which is transitive on triples (vertex, edge, face), where, of course, the vertex is in the edge, and the edge is in the face. A strict fundamental domain for this action will be a 2-simplex T of the barycentric subdivision of $X_{k,l}$.

We start from the Coxeter system (W, S) with $S = \{s_0, s_1, \dots, s_{l-1}\}$ and $m_{s_0 s_1} = k$, $m_{s_i s_{i+1}} = 3$ for $i = 1, \dots, l-2$. The other non-diagonal entries are all equal to 2. In other words, the Coxeter graph is:

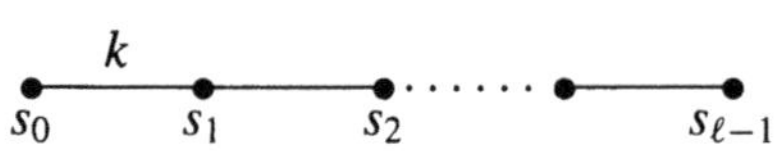

Let $\mathcal{S}$ be the poset of subsets of S ordered by reverse-inclusion. Let $\mathcal{T}$ be the subposet of $\mathcal{S}$ consisting of the proper subsets of S containing $\rho = \{s_3, \dots, s_{l-1}\}$. The geometric realization of $\mathcal{T}$ is the barycentric subdivision of a triangle T whose three vertices τ_0, τ_1 and τ_2 correspond to the subsets of S of cardinality $(l-1)$ not containing respectively s_0, s_1 and s_2. The barycentre ρ of T corresponds to the intersection of these subsets and the barycentre σ_i of the side joining τ_j, τ_k corresponds to the intersection of the subsets τ_j and τ_k (where $\{i, j, k\} = \{0, 1, 2\}$). (See figure 12.14 where each group is represented by the corresponding Coxeter graph).

The second barycentric subdivision of the polyhedron $X_{k,l}$ will be $\Sigma(W, \mathcal{T})$, on which we have the usual action of W with T as strict fundamental domain. The vertices τ_0, τ_1, τ_2 of T are, respectively, a vertex of $X_{k,l}$, a barycentre of a side of $X_{k,l}$, and a barycentre of a face of $X_{k,l}$. These data define $X_{k,l}$ uniquely.

Let us check that the faces of $X_{k,l}$ are k-gons. The isotropy subgroup of τ_2 is $W_{\tau_2} = W_{\{s_0, s_1, s_3, \dots, s_{l-1}\}}$, which is the direct product of a dihedral group of order $2k$ (generated by s_0 and s_1) and the subgroup generated by the s_i with $i > 2$. This second factor acts trivially on T and the orbit of T under the dihedral group is a union of $2k$ triangles forming the barycentric subdivision of a k-gon P.

To check that the link in $X_{k,l}$ of the vertex τ_0 is the complete graph on l vertices, we note that the isotropy subgroup of τ_0 is $W_{\tau_0} = W_{\{s_1, \dots, s_{l-1}\}}$, which can be represented as the group of permutations of the set $\{1, \dots, l\}$, where s_i acts as the transposition $(i, i+1)$ (see 12.9(3) and 12.23(2)). This action extends to a simplicial action on the

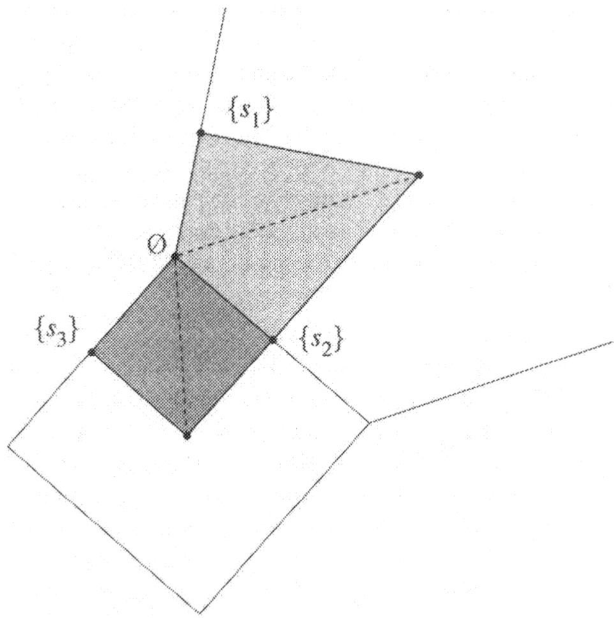

Fig. 12.13 The fundamental domain $|\mathcal{T}|$ from 12.34(3) in the case $S = \{s_1, s_2, s_3\}$, $m_{s_1 s_2} = 3$, $m_{s_2,s_3} = 2$, $m_{s_1 s_3} = \infty$.

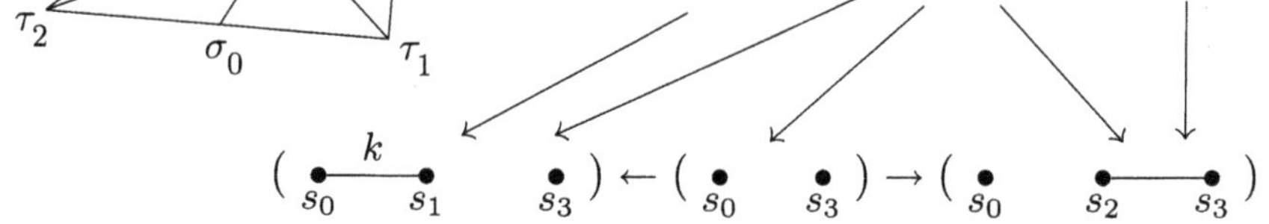

Fig. 12.14 The triangle of groups $W(\mathcal{T})$ for $l = 4$.

complete graph $K(l)$ with vertices $1, \ldots, l$. A fundamental domain for this action is the edge of the barycentric subdivision $K(l)'$ joining the vertex 1 to the barycentre of the simplex joining 1 to 2. The natural action of W_{τ_0} on the cone over the second barycentric subdivision of $K(l)$ has the cone over this edge as strict fundamental domain, and the associated complex of groups is just the restriction of $W(\tau)$ to the star of the vertex τ_0 in T. We leave to the reader to check that the stars of the other vertices of the fundamental domain T are isomorphic to the stars of the vertices in the second barycentric subdivision of any polyhedron satisfying the required local conditions.

It is natural to identify T with a Euclidean triangle whose angles at the vertices τ_0, τ_1, τ_2 are, respectively, $\frac{k-2}{2k}\pi$, $\frac{1}{2}\pi$, $\frac{1}{k}\pi$. In this way the 2-cells of $X_{k,l}$ become regular Euclidean k-gons. If $k \geq 6$, then the Euclidean polyhedron $X_{k,l}$ is CAT(0). If $k > 6$, then we can metrize the 2-cells of K as regular hyperbolic k-gons with vertex angles $\pi/3$. The resulting piecewise hyperbolic structure on $X_{k,l}$ is CAT(-1).

Nadia Benakli [Ben92] has shown that, if k is even and at least 8, the boundary at infinity of $X_{k,l}$ is homeomorphic to the Sierpinski curve if $l = 4$, and to the Menger curve if $l > 4$.

(5) For $l = 4$, Haglund has constructed a twisted variant $M_{k,l}$ of the above complex $X_{k,l}$. This arises from a triangle of groups obtained from the one considered above by replacing the local group at the vertex τ_2 by the dihedral group D_k generated by the elements s_0 and s_1, and then mapping s_3 to the central element $(s_0s_1)^k \in D_k$.

The local developments for this triangle of groups are the same as in the previous one, hence if $k \geq 6$ and we metrize T as a Euclidean triangle as above, the triangle of groups is non-positively curved, therefore strictly developable by 12.28. Its simply connected development is a complex $M_{k,4}$ satisfying the same local properties as $X_{k,l}$. The direct limit of the triangle of groups is the group G generated by the elements $s_0, \dots, s_3$ subject to the relations $(s_i)^2 = (s_1s_2)^3 = (s_2s_3)^3 = (s_0s_1)^{2k} = (s_1s_3)^2 = (s_0s_2)^2 = (s_0s_1)^k s_3^1 = 1$. In fact Haglund showed directly that this triangle of groups is strictly developable for $k \geq 4$. For $k = 3$ it is not developable.

To see that the complexes $X_{k,l}$ and $M_{k,l}$ are not isomorphic, let us fix a polygonal 2-cell P in one of these complexes. Let $e_1, \dots, e_k$ be the sides of P in order. Let P_1 be a 2-cell adjacent to P along the edge e_1 (there are two such cells). Let P_2 be the 2-cell adjacent to P_1 and to P along e_2. By induction we define P_i to be the 2-cell adjacent to P_{i-1} and to P along e_i. We claim that the polygon P_{k+1} adjacent to P_k and to P along e_1 is equal to P_1 in $X_{k,l}$ and is different from P_1 in $M_{k,l}$.

To see this, we fix our attention on the 2-cell P which contains the fundamental domain T. It is a k-gon which is the union of the $2k$ triangles $\bigcup_{i=0}^{k-1} r^i.(T \cup s_0.T)$, where $r = s_1s_0$ acts as a rotation of angle $2\pi/k$ on P. Let e_1 be the side of P containing the vertices τ_0 and τ_1 of T. The k sides of P are (in order) $e_1, e_2 = r.e_1, \dots, e_k = r^{k-1}.e_1$. The element s_1 fixes τ_0 and exchanges e_1 and e_2. The element s_3 fixes e_1 and e_2 and permutes the two other edges issuing from τ_0. The element s_2 fixes e_1 and maps e_2 to an edge $s_2.e_2 \neq e_1$, hence $P_1 := s_2.P$ is adjacent to P along e_1. As s_0 preserves P, reverses the orientation of e_1, and commutes with s_2, we have $s_0.P_1 = P_1$ and hence the two other edges of P_1 adjacent to e_1 are $s_2.e_2$ and $(s_0s_2).e_2$. The face $P_2 := (s_1s_0).P_1 = r.P_1$ is adjacent to P along e_2 and to P_1 along the side $(rs_0s_2).e_2 = s_2.e_2$. It follows that $P_i := r^{i-1}.P_1$ is adjacent to P along e_i and to P_{i-1} for $i \leq k$. The face $P_{k+1} = r^k P_1$ is adjacent to P along e_1 and to P_k. In the complex $X_{k,4}$ this face is equal to P_1, while in $M_{k,4}$ it is equal to $(s_3s_2).P = s_3.P_2 \neq P_1$.

Part III. Aspects of the Geometry of Group Actions

The unifying theme in this part of the book is the analysis of group actions, particularly actions by isometries on CAT(0) spaces. The material presented here is of a more specialized nature than that in Parts I and II and the reader may find the style a little less pedestrian.

Part III is divided into four lengthy chapters, the contents of which were explained briefly in the Introduction. Each chapter begins with an overview. Chapters III.$\mathcal{C}$ and III.$\mathcal{G}$ are independent of Chapters III.H and III.Γ. The main arguments in the chapter on groupoids, III.$\mathcal{G}$, rely only on material from Part II, but the reader who digests the chapter on complexes of groups first, III.$\mathcal{C}$, will have access to a greater range of examples in III.$\mathcal{G}$ and may find the ideas more transparent. Likewise, an understanding of the main points in the chapter on hyperbolic metric spaces, III.H, is required in order to appreciate certain sections in the chapter on non-positive curvature in group theory, III.Γ.

Chapter III.H δ-Hyperbolic Spaces and Area

In Part II we explored the geometry of spaces whose curvature is bounded above in a strict, local, sense by means of the CAT(κ) inequality. In the non-positively curved case, the Cartan-Hadamard Theorem (II.4.1) allowed us to use this local information to make deductions about the global geometry of the universal coverings of the spaces under consideration. In this way we were able to generalize classical results concerning the global geometry of complete, simply connected manifolds of negative and non-positive curvature.

M. Gromov's theory of *δ-hyperbolic spaces*, as set out in [Gro87], is based on a completely different method of generalization. Ignoring the local structure, Gromov identified a robust condition that encapsulates many of the global features of the geometry of complete, simply connected manifolds of negative curvature. He then showed that the geodesic spaces which satisfy this condition (δ-hyperbolic spaces) display many of the elegant features that one associates with the large-scale geometry of such manifolds. Moreover, the robustness of this condition makes it an invariant of quasi-isometry among geodesic spaces.

In this chapter we shall present the foundations of the theory of δ-hyperbolic spaces, and in sections 2 and 3 of the next chapter we shall study the class of finitely generated groups whose Cayley graphs are δ-hyperbolic.

δ-Hyperbolic spaces form a natural context in which to explore the dichotomy in the large-scale geometry of CAT(0) spaces exposed by the Flat Plane Theorem (II.9.33): a proper cocompact CAT(0) space is hyperbolic if and only if it does not contain an isometrically embedded copy of the Euclidean plane (Theorem 1.5). An important aspect of this dichotomy concerns the behaviour of quasi-geodesics. In δ-hyperbolic spaces the large-scale geometry of quasi-geodesics mimics that of geodesics rather closely (Theorem 1.7), whereas this is not the case in spaces that contain a flat plane (I.8.23). The stability properties of quasi-geodesics and local geodesics in hyperbolic spaces play an important role throughout this chapter. In particular this is true in section 3, where we describe the Gromov boundary at infinity for hyperbolic spaces. If the space is CAT(0), the Gromov boundary coincides with the visual boundary introduced in (II.8), and our construction of the Gromov boundary is modelled on our earlier treatment of this special case.

In the second section of this chapter we discuss a coarse notion of filling-area for loops in geodesic spaces. Our purpose in doing so is to characterize hyperbolicity in terms of isoperimetric inequalities (2.7 and 2.9).

1. Hyperbolic Metric Spaces

The Slim Triangles Condition

1.1 Definition (Slim Triangles). Let $\delta > 0$. A geodesic triangle in a metric space is said to be δ-*slim* if each of its sides is contained in the δ-neighbourhood of the union of the other two sides. A geodesic space X is said to be δ-hyperbolic if every triangle in X is δ-slim. (If X is δ-hyperbolic for some $\delta > 0$, one often says simply that X is hyperbolic.)

There are a number of equivalent ways of formulating the hyperbolicity condition (see (1.17) and (1.22)). Gromov [Gro87] attributes the above (slim triangles) formulation to Rips. A geodesic space is 0-hyperbolic if and only if it is an $\mathbb{R}$-tree, and it is often useful to think of δ-hyperbolic spaces as thickened versions of metric trees.

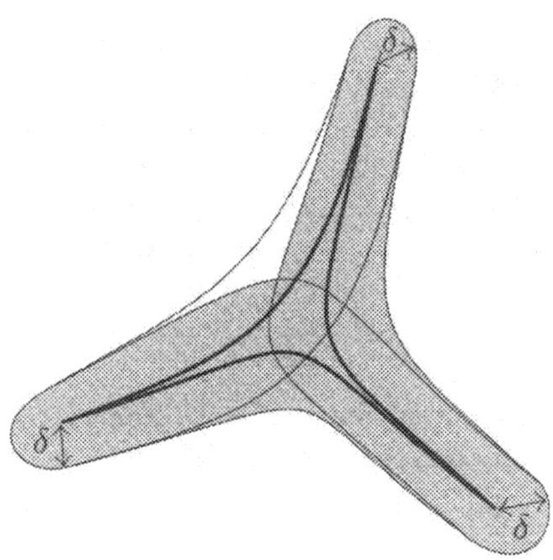

Fig. H.1 A δ-slim triangle

We shall deal with several aspects of the theory of δ-hyperbolic spaces in this chapter, and in the next chapter we shall study the class of groups whose Cayley graphs are δ-hyperbolic, but our account of this active field is by no means comprehensive. It also owes much to the earlier works [CDP90], [GhH90] and [Sho91].

We begin by relating hyperbolicity to the ideas of negative curvature considered in earlier sections.

1.2 Proposition. *If* $\kappa < 0$ *then every* $\mathrm{CAT}(\kappa)$ space is δ-hyperbolic, where δ depends only on κ.

Proof. This follows immediately from the $\mathrm{CAT}(\kappa)$ inequality and the fact that the hyperbolic plane is a δ-hyperbolic space for a suitable δ. To see that $\mathbb{H}^2$ is hyperbolic, note that since the area of geodesic triangles in $\mathbb{H}^2$ is bounded by π, there is a bound on the radius of semicircles that can be inscribed in a geodesic triangle. $\square$

1.3 Exercise. $\mathbb{H}^2$ is hyperbolic — find the best δ.

1.4 Proposition. *Let X be a proper* CAT(0) *space. X is hyperbolic if and only if it is uniformly visible (see II.9.30).*

Proof. Suppose that X is δ-hyperbolic. Consider a geodesic triangle in X with vertices p, x, y, and suppose that the distance from p to $[x, y]$ is greater than $\rho \geq 3\delta$. Because X is δ-hyperbolic, we can find z on $[x, y]$, u on $[p, x]$, and v on $[p, y]$ such that $d(z, u), d(z, v) \leq \delta$. Thus $d(u, v) \leq 2\delta$ and $d(p, u), d(p, v) \geq \rho - \delta$. Applying the CAT(0) inequality for comparison angles (II.1.7(4)) to $\Delta(p, u, v)$, we see that $d(u, v) \geq (\rho - \delta)\sin(\angle_p(x, y)/2)$. Thus

$$\angle_p(x, y) \leq 2 \arcsin[2\delta/(\rho - \delta)],$$

which is independent of p and tends to 0 as $\rho \to \infty$. Therefore X is uniformly visible.

Conversely, if X is uniformly visible then X is δ-hyperbolic with $\delta = R(\pi/2)$ (where R is as in definition (II.9.30)). To see this, consider a geodesic triangle with vertices x, y, z and a point p on the side $[x, y]$. The angles $\angle_p(x, z)$ and $\angle_p(y, z)$ must sum to at least π, so one of these angles, $\angle_p(x, z)$ say, must be at least $\pi/2$. But then, by the definition of uniform visibility, $d(p, [x, z]) \leq R(\pi/2)$. $\square$

As a corollary of (II.9.32 and 33) and (1.4), we get the following theorem.

1.5 Flat Plane Theorem. *A proper cocompact* CAT(0) *space X is hyperbolic if and only if it does not contain a subspace isometric to $\mathbb{E}^2$.*

Quasi-Geodesics in Hyperbolic Spaces

One of the most important facets of the dichotomy highlighted by the Flat Plane Theorem concerns the lengths of paths which are not close to geodesics. For example, the following result is blatantly false in any space that contains an isometrically embedded plane.

1.6 Proposition. *Let X be a δ-hyperbolic geodesic space. Let c be a continuous rectifiable path in X. If $[p, q]$ is a geodesic segment connecting the endpoints of c, then for every $x \in [p, q]$*

$$d(x, \text{im}(c)) \leq \delta \,|\log_2 l(c)| + 1.$$

(Here, as usual, $l(c)$ denotes the length of c.)

Proof. If $l(c) \leq 1$, the result is trivial. Suppose that $l(c) > 1$. Without loss of generality we may assume that c is a map $[0, 1] \to X$ that parameterizes its image proportional to arc length. Thus $p = c(0)$ and $q = c(1)$. Let N denote the positive integer such that $l(c)/2^{N+1} < 1 \leq l(c)/2^N$.

Let $\Delta_1 = \Delta([c(0), c(1/2)], [c(1/2), c(1)], [c(0), c(1)])$ be a geodesic triangle in X containing the given geodesic $[c(0), c(1)]$. Given $x \in [c(0), c(1)]$, we choose $y_1 \in [c(0), c(1/2)] \cup [c(1/2), c(1)]$ with $d(x, y_1) \leq \delta$. If $y_1 \in [c(0), c(1/2)]$ then we consider a geodesic triangle $\Delta([c(0), c(1/2)], [c(1/4), c(1/2)], [c(0), c(1/4)])$, which has the edge $[c(0), c(1/2)]$ in common with Δ_1 and call this triangle Δ_2. If on the other hand $y_1 \in [c(1/2), c(1)]$, then we consider $\Delta([c(1/2), c(3/4)], [c(3/4), c(1)], [c(1/2), c(1)])$ and call this triangle Δ_2. In either case we can choose $y_2 \in \Delta_2 \smallsetminus \Delta_1$ such that $d(y_1, y_2) \leq \delta$.

We proceed inductively: at the $(n + 1)$-st stage we consider a geodesic triangle Δ_{n+1} which has in common with Δ_n the side $[c(t_n), c(t'_n)]$ containing y_n, and which has as its third vertex $c(t_{n+1})$, where $t_{n+1} = (t_n + t'_n)/2$. We choose $y_{n+1} \in \Delta_{n+1} \smallsetminus [c(t_n), c(t'_n)]$ with $d(y_n, y_{n+1}) \leq \delta$.

At the N-th stage of this construction we obtain a point y_N which is a distance at most δN from x, and which lies on an interval of length $l(C)/2^N$ with endpoints in the image of c. If we define y to be the closest endpoint of this interval, then since $l(C)/2^{N+1} < 1$ and $2^N \leq l(c)$ we have $d(x, y) \leq \delta \mid \log_2 l(c)| + 1$. $\square$

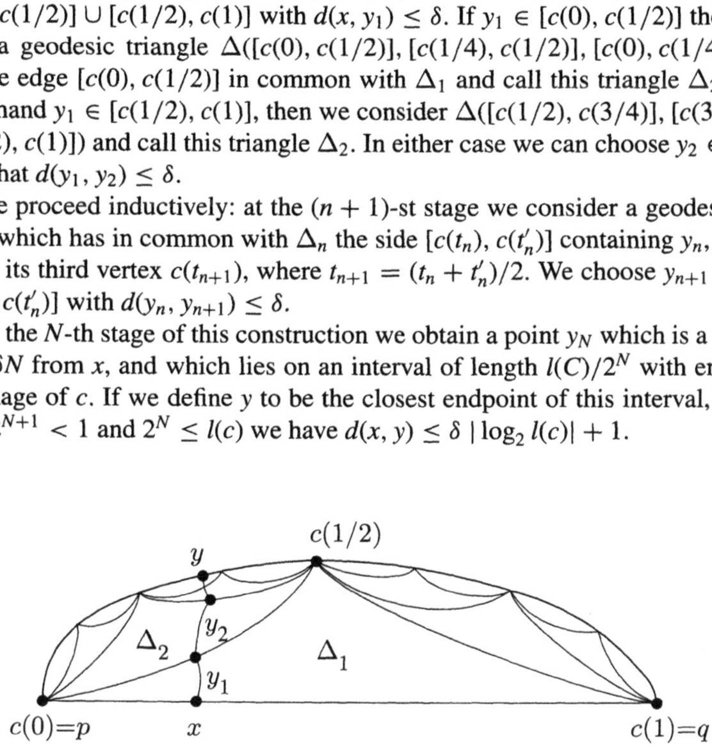

Fig. H.2 Geodesics stay close to short curves (1.6)

We shall develop the theme of the preceding proposition by studying quasi-geodesics in hyperbolic spaces. Quasi-geodesics were defined in (I.8.22). In arbitrary CAT(0) spaces quasi-geodesics can be fairly wild; in particular the image of a quasi-geodesic need not be Hausdorff close to any geodesic (cf. I.8.23). In contrast, the large-scale behaviour of quasi-geodesics in δ-hyperbolic spaces mimics that of geodesics rather closely:

1.7 Theorem (Stability of Quasi-Geodesics). *For all $\delta > 0, \lambda \geq 1, \varepsilon \geq 0$ there exists a constant $R = R(\delta, \lambda, \varepsilon)$ with the following property:*

If X is a δ-hyperbolic geodesic space, c is a (λ, ε)-quasi-geodesic in X and $[p, q]$ is a geodesic segment joining the endpoints of c, then the Hausdorff distance between $[p, q]$ and the image of c is less than R.

We defer the proof of this theorem for a moment while we consider some of its consequences.

By definition, a (λ, ε)-quasi-geodesic triangle in a metric space X consists of three (λ, ε)-quasi-geodesics (its sides) $p_i : [0, T_i] \to X$, $i = 1, 2, 3$ with $p_i(T_i) = p_{i+1}(0)$ (indices mod 3). Such a triangle is said to be *k–slim* (where $k \geq 0$) if for each $i \in \{1, 2, 3\}$ every point $x \in \text{im}(p_i)$ lies in the k-neighbourhood of $\text{im}(p_{i-1}) \cup \text{im}(p_{i+1})$ (indices mod 3).

As an immediate consequence of (1.7) we have:

1.8 Corollary. *A geodesic metric space X is hyperbolic if and only if, for every $\lambda \geq 1$ and every $\varepsilon \geq 0$, there exists a constant M such that every (λ, ε)-quasi-geodesic triangle in X is M–slim. (If X is δ-hyperbolic, then M depends only on δ, λ and ε.)*

The following result should be regarded as the analogue in coarse geometry of the fact that a convex subspace of a CAT(−1) space is a CAT(−1) space. It shows in particular that hyperbolicity is an invariant of quasi-isometry.

1.9 Theorem. *Let X and X' be geodesic metric spaces and let $f : X' \to X$ be a quasi-isometric embedding. If X is hyperbolic then X' is hyperbolic. (If X is δ-hyperbolic and $f : X' \to X$ is a (λ, ε)-quasi-isometric embedding, then X' is δ'-hyperbolic, where δ' is a function of δ, λ and ε.)*

Proof. Let Δ be a geodesic triangle in X' with sides p_1, p_2, p_3. According to (1.8), there is a constant $M = M(\delta, \lambda, \varepsilon)$ such that the (λ, ε)-quasi-geodesic triangle in X with sides $f \circ p_1, f \circ p_2, f \circ p_3$ is M–slim, i.e. for all $x \in \text{im}(p_1)$ there exists $y \in \text{im}(p_2) \cup \text{im}(p_3)$ such that $d(f(x), f(y)) \leq M$. Since f is a (λ, ε)-quasi-isometric embedding,

$$d(x, y) \leq \lambda\, d(f(x), f(y)) + \varepsilon\lambda \leq \lambda M + \lambda\varepsilon.$$

Repeating this argument with p_2 and p_3 in place of p_1, we see that Δ is δ'–slim, where $\delta' = \lambda M + \lambda\varepsilon$. □

Before proving (1.7) we mention one other consequence of it. The spiral described in (I.8.23) shows that in general quasi-geodesic rays in CAT(0) spaces do not tend to a definite point at infinity, whereas (1.7) implies that in δ-hyperbolic spaces they do. (This observation will play an important role in section 3.)

1.10 Proposition. *Let X_1 and X_2 be complete* CAT(0) *spaces that are δ-hyperbolic. Consider $\overline{X}_i = X_i \cup \partial X_i$ with the cone topology (II.8.6).*

(1) *If $c : [0, \infty) \to X_1$ is a quasi-geodesic, then there exists a point of ∂X_1, denoted $c(\infty)$, such that $c(t) \to c(\infty)$ as $t \to \infty$.*

(2) *If $f : X_1 \to X_2$ is a quasi-isometry, then the map $f_\partial : \partial X_1 \to \partial X_2$, which sends the equivalence class of each geodesic ray $c : [0, \infty) \to X_1$ to the endpoint $(f \circ c)(\infty)$ of the quasi-geodesic ray $f \circ c$, is a homeomorphism.*

Proof. Exercise. (Stronger results will be proved in section 3 of this chapter.) □

Remark. If a finitely generated group Γ acts properly and cocompactly by isometries on geodesic spaces X_1 and X_2, then Γ, X_1 and X_2 are all quasi-isometric (I.8.19). In particular, if X_1 and X_2 are CAT(0) spaces that are hyperbolic, then ∂X_1 and ∂X_2 will be homeomorphic, by (1.10). Chris Croke and Bruce Kleiner have recently shown[50] that in the non-hyperbolic case ∂X_1 need not be homeomorphic to ∂X_2.

We now turn to the deferred proof of Theorem 1.7. The following technical result allows one to circumvent the difficulties posed by some of the more pathological traits of quasi-geodesics.

1.11 Lemma (Taming Quasi-Geodesics). *Let X be a geodesic space. Given any (λ, ε) quasi-geodesic $c : [a, b] \to X$, one can find a continuous (λ, ε') quasi-geodesic $c' : [a, b] \to X$ such that:*

(1) $c'(a) = c(a)$ *and* $c'(b) = c(b)$;

(2) $\varepsilon' = 2(\lambda + \varepsilon)$;

(3) $l(c'|_{[t,t']}) \leq k_1\, d(c'(t), c'(t')) + k_2$, *for all* $t, t' \in [a, b]$, *where* $k_1 = \lambda(\lambda + \varepsilon)$ *and* $k_2 = (\lambda\varepsilon' + 3)(\lambda + \varepsilon)$;

(4) *the Hausdorff distance between the images of c and c' is less than* $(\lambda + \varepsilon)$.

Proof. Define c' to agree with c on $\Sigma = \{a, b\} \cup (\mathbb{Z} \cap (a, b))$. Then choose geodesic segments joining the images of successive points in Σ and define c' by concatenating linear reparameterizations of these geodesic segments. Note that the length of each of the geodesic segments is at most $(\lambda + \varepsilon)$. Every point of $\operatorname{im}(c) \cup \operatorname{im}(c')$ lies in the $(\lambda + \varepsilon)/2$ neighbourhood of $c(\Sigma)$, thus (4) holds.

Let $[t]$ denote the point of Σ closest to $t \in [a, b]$. Because c is a (λ, ε)-quasi-geodesic and $c([t]) = c'([t])$ for all $t \in [a, b]$,

$$\begin{aligned} d(c'(t), c'(t')) &\leq d(c'([t]), c'([t'])) + (\lambda + \varepsilon) \\ &\leq \lambda\big|[t] - [t']\big| + \varepsilon + (\lambda + \varepsilon) \\ &\leq \lambda(|t - t'| + 1) + (\lambda + 2\varepsilon); \end{aligned}$$

and

$$\begin{aligned} \frac{1}{\lambda}|t - t'| - 2(\lambda + \varepsilon) &\leq \frac{1}{\lambda}\big(|t - t'| - 1\big) - (\lambda + 2\varepsilon) \\ &\leq \frac{1}{\lambda}\big|[t] - [t']\big| - (\lambda + 2\varepsilon) \\ &\leq d(c'([t]), c'([t'])) - (\lambda + \varepsilon) \\ &\leq d(c'(t), c'(t')). \end{aligned}$$

This proves that c' is a (λ, ε') quasi-geodesic with ε' as in (2).

[50] "Boundaries of spaces with non-positive curvature", to appear in *Topology*.

For all integers $n, m \in [a, b]$,

$$l(c'|_{[n,m]}) = \sum_{i=n}^{m-1} d(c(i), c(i+1)) \leq (\lambda + \varepsilon)|m - n|,$$

and similarly $l(c'|_{[a,m]}) \leq (\lambda + \varepsilon)(m - a + 1)$ and $l(c'|_{[n,b]}) \leq (\lambda + \varepsilon)(b - n + 1)$. Thus for all $t, t' \in [a, b]$ we have:

$$l(c'|_{[t,t']}) \leq (\lambda + \varepsilon)(|\,[t] - [t']\,| + 2),$$

and

$$d(c'(t), c'(t')) \geq \frac{1}{\lambda}|t - t'| - \varepsilon' \geq \frac{1}{\lambda}\big(|[t] - [t']| - 1\big) - \varepsilon'.$$

By combining these inequalities and noting that $\varepsilon \leq \varepsilon'$ we obtain (3). □

Remark. A slight modification of the preceding argument allows one to extend the result to length spaces.

Proof of Theorem 1.7. First one tames c, in other words one replaces it by c' as in the preceding lemma. We write $\mathrm{im}(c')$ for the image of c' and $[p, q]$ for a choice of geodesic segment joining its endpoints. Let $D = \sup\{d(x, \mathrm{im}(c')) \mid x \in [p, q]\}$ and let x_0 be a point on $[p, q]$ at which this supremum is attained. The open ball of radius D with centre x_0 does not meet $\mathrm{im}(c')$. We shall use (1.6) to bound D.

Let y be the point of $[p, x_0] \subset [p, q]$ that is a distance $2D$ from x_0 (if $d(x_0, p) < 2D$ then let $y = p$). Choose $z \in [x_0, q]$ similarly. We fix $y', z' \in \mathrm{im}(c')$ with $d(y, y') \leq D$ and $d(z, z') \leq D$ and choose geodesic segments $[y, y']$ and $[z', z]$. (See figure H.3.) Consider the path γ from y to z that traverses $[y, y']$ then follows c' from y' to z', then traverses $[z', z]$. This path lies outside $B(x_0, D)$.

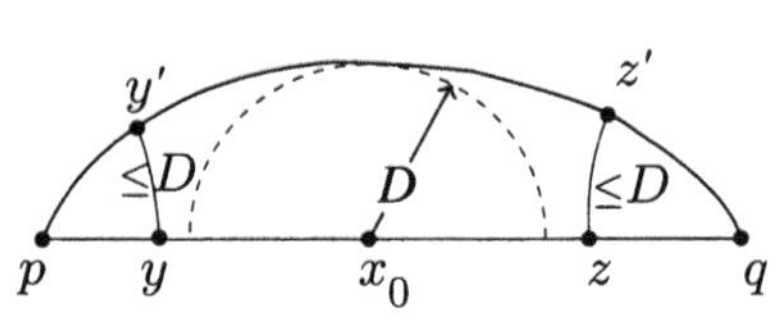

Fig. H.3 Quasi-geodesics are close to geodesics

Since $d(y', z') \leq d(y', y) + d(y, z) + d(z, z') \leq 6D$, from 1.11(3) we have $l(\gamma) \leq 6Dk_1 + k_2 + 2D$. And from (1.6), as $d(x_0, \mathrm{im}(\gamma)) = D$ we have $D - 1 \leq \delta \mid \log_2 l(\gamma)|$. Thus

$$D - 1 \leq \delta\ \log_2(6Dk_1 + k_2 + 2D),$$

whence an upper bound on D depending only on λ, ε and δ. Let D_0 be such a bound.

We claim that $\mathrm{im}(c')$ is contained in the R'-neighbourhood of $[p, q]$, where $R' = D_0(1 + k_1) + k_2/2$. Consider a maximal sub-interval $[a', b'] \subset [a, b]$ such that $c'([a', b'])$ lies outside the D_0-neighbourhood $V_{D_0}[p, q]$. Every point of $[p, q]$ lies in

$V_{D_0}(\text{im}(c'))$, so by connectedness there exist $w \in [p, q]$, $t \in [a, a']$ and $t' \in [b', b]$ such that $d(w, c'(t)) \leq D_0$ and $d(w, c'(t')) \leq D_0$. In particular $d(c'(t), c'(t')) \leq 2D_0$, so $l(c'|_{[t,t']}) \leq 2k_1 D_0 + k_2$, by (1.11(3)). Hence $\text{im}(c')$ is contained in the R' neighbourhood of $[p, q]$. From this and (1.11(4)) it follows that $R = R' + \lambda + \varepsilon$ satisfies the statement of the theorem. □

k-Local Geodesics

The following result provides a useful companion to Theorem 1.7. It gives a local criterion for recognizing quasi-geodesics.

1.12 Definition. Let X be a metric space and fix $k > 0$. A path $c : [a, b] \to X$ is said to be a *k-local geodesic* if $d(c(t), c(t')) = |t - t'|$ for all $t, t' \in [a, b]$ with $|t - t'| \leq k$.

1.13 Theorem (k-Local Geodesics are Quasi-Geodesics). *Let X be a δ-hyperbolic geodesic space and let $c : [a, b] \to X$ be a k-local geodesic, where $k > 8\delta$. Then:*

(1) $\text{im}(c)$ *is contained in the 2δ-neighbourhood of any geodesic segment* $[c(a), c(b)]$ *connecting its endpoints,*

(2) $[c(a), c(b)]$ *is contained in the 3δ-neighbourhood of* $\text{im}(c)$, *and*

(3) *c is a (λ, ε)-quasi-geodesic, where $\varepsilon = 2\delta$ and $\lambda = (k + 4\delta)/(k - 4\delta)$.*

Proof. First we prove (1). Let $x = c(t)$ be a point of $\text{im}(c)$ that is at maximal distance from $[c(a), c(b)]$. First suppose that both $(t - a)$ and $(b - t)$ are greater than 4δ. Then we may suppose that there is a subarc of c centred at x of length strictly greater than 8δ but less than k. Let y and z be the endpoints of this arc and let y' and z' be points on $[c(a), c(b)]$ that are closest to y and z respectively. Consider a geodesic quadrilateral with vertices y, z, y', z' such that the sides $[y, z]$ and $[y', z']$ are the obvious subarcs of c and $[c(a), c(b)]$. By dividing this quadrilateral with a diagonal and applying the δ-hyperbolic criterion to each of the resulting triangles, we see that there exists w on one of the sides other than $[y, z]$ such that $d(w, x) \leq 2\delta$. If $w \in [y, y']$ then there would be a path through w joining x to y' that was shorter than $d(y, y')$:

$$\begin{aligned} d(x, y') - d(y, y') &\leq [d(x, w) + d(w, y')] - [d(y, w) + d(w, y')] \\ &= d(x, w) - d(y, w) \\ &\leq d(x, w) - [d(y, x) - d(x, w)] \\ &= 2\, d(x, w) - d(x, y) \\ &< 4\delta - 4\delta = 0, \end{aligned}$$

and this contradicts the choice of x. If $w \in [z, z']$ then we obtain a similar contradiction. Thus $w \in [y', z']$ and hence the distance from x to $[c(a), c(b)]$ is at most 2δ. A similar argument applies in the cases where at least one of $(t - a)$ or $(b - t)$ is less than 4δ. (In fact, in those cases x lies in the δ-neighbourhood of $[c(a), c(b)]$.)

We now prove (2). Suppose that $p \in [c(a), c(b)]$. Every point of $\mathrm{im}(c)$ lies in one of the two open sets that are the 2δ-neighbourhoods of $[c(a), p]$ and $[p, c(b)]$. Since $\mathrm{im}(c)$ is connected, some $x \in \mathrm{im}(c)$ lies in both; choose $q \in [c(a), p]$, $r \in [p, c(b)]$ such that $d(x, q) \leq 2\delta$ and $d(x, r) \leq 2\delta$. By applying the δ-hyperbolic condition to a geodesic triangle with vertices x, q, r, we see that $p \in [q, r]$ is in the δ-neighbourhood of $[x, q] \cup [x, r]$ and hence $d(p, x) \leq 3\delta$.

For (3), note first that $d(c(t), c(t')) \leq |t - t'|$ for all $t, t' \in [a, b]$. In order to bound $d(c(t), c(t'))$ below by a linear function of $|t - t'|$, we shall argue that if one divides c into subpaths of length $k' = k/2 + 2\delta$ and projects the endpoints of these subarcs onto $[c(a), c(b)]$, then the points of projection form a monotone sequence. To this end, we consider a subarc of c with length $2k'$. Let x and y be the endpoints of this arc and let m be the midpoint of the arc; let x', y' and m' be points of $[c(a), c(b)]$ that are a distance at most 2δ from x and y and m respectively. We must show that m' lies between x' and y'.

Let x_0 (resp. y_0) be the point on the image of c that is a distance 2δ from x (resp. y) in the direction of m. By δ-hyperbolicity, any geodesic triangle $\Delta(x, x', x_0)$ is contained in the 3δ-neighbourhood of x and therefore (since $d(x, m) = k' > 6\delta$) any such triangle lies outside the 3δ-neighbourhood of m. Similarly, there exists a geodesic triangle $\Delta(y, y', y_0)$ outside the 3δ-neighbourhood of m. By applying the δ-hyperbolic condition to the (obvious) geodesic quadrilateral (x', x_0, y_0, y') (divided into two triangles by a diagonal), we deduce that m lies within a distance 2δ of some point $m'' \in [x', y'] \subset [c(a), c(b)]$. Any point between m' and m'' is a distance at most 3δ from m (by the hyperbolicity of $\Delta(m, m', m'')$). In particular, neither x' nor y' lie between m' and m'', which means that $m' \in [x', y']$, as we wished to prove.

We express c as a concatenation of $M \leq (b - a)/k'$ geodesics of length k' with a smaller piece, of length η say, at the end. By the preceding argument, the projections (choice of closest points) of the endpoints of these geodesics on $[c(a), c(b)]$ form a monotone sequence. By (1), the distance between successive projections is at least $k' - 4\delta$. And by the triangle inequality, the distance from the last projection point to $c(b)$ is at least $\eta - 2\delta$.

Thus we have

$$b - a = Mk' + \eta,$$

and

$$d(c(a), c(b)) \geq M(k' - 4\delta) + \eta - 2\delta = (b - a) - 4\delta M - 2\delta.$$

Since $M \leq (b - a)/k'$, we deduce that

$$d(c(a), c(b)) \geq \frac{k' - 4\delta}{k'}(b - a) - 2\delta.$$

This together with the (trivial) remark that a subpath of a k-local geodesic is again a k-local geodesic proves (3). □

The above theorem shows that if $k > 8\delta$ then there are no non-trivial closed k-local geodesics in a δ-hyperbolic space:

1.14 Corollary. *If X is a δ-hyperbolic geodesic metric space and $c : [a, b] \to X$ is a k-local geodesic for some $k > 8\delta$, then either c is constant (i.e., $a = b$) or else $c(a) \neq c(b)$.*

Proof. If $c(a) = c(b)$ then by the above theorem $\text{im}(c)$ is contained in the ball of radius 2δ about $c(a)$. Since c is an 8δ-local geodesic, this implies that c is constant. □

Remark. By combining (1.7) and (1.13) in the obvious way one gets a criterion for seeing that paths which are locally quasi-geodesic (with suitable parameters) are actually quasi-geodesics.

Although quasi-geodesics stay Hausdorff close to geodesics in δ-hyperbolic spaces, trivial examples show that they need not be uniformly close. On the other hand, in any metric space, if geodesics with a common origin are Hausdorff close then they must be uniformly close. In hyperbolic spaces one can say more:

1.15 Lemma. *Let X be a geodesic space that is δ-hyperbolic and let $c, c' : [0, T] \to X$ be geodesics with $c(0) = c'(0)$. If $d(c(t_0), \text{im}(c')) \leq K$, for some $K > 0$ and $t_0 \in [0, T]$, then $d(c(t), c'(t)) \leq 2\delta$ for all $t \leq t_0 - K - \delta$.*

Let $c_1 : [0, T_1] \to X$ and $c_2 : [0, T_2] \to X$ be geodesics with $c_1(0) = c_2(0)$. Let $T = \max\{T_1, T_2\}$ and extend the shorter geodesic to $[0, T]$ by the constant map. If $k = d(c_1(T), c_2(T))$, then $d(c_1(t), c_2(t)) \leq 2(k + 2\delta)$ for all $t \in [0, T]$.

Proof. To prove the first assertion, we choose a geodesic c_0 joining $c(t_0)$ to a closest point $c'(t_1)$ on the image of c'. By the triangle inequality $|t_0 - t_1| \leq K$. Note that $c(t)$ is not δ-close to any point on c_0 if $t < t_0 - K - \delta$. It follows from the δ-slimness of the triangle with sides $c_0, c([0, t_0]), c'([0, t_1])$ that $d(c(t), c'(t')) \leq \delta$ for some t'. By the triangle inequality $|t - t'| \leq \delta$. Therefore $d(c(t), c'(t)) \leq 2\delta$.

To prove the second assertion we consider a geodesic triangle two of whose sides are c_1 and c_2. If $c_1(t)$ is δ-close to a point $c_2(t')$, then as above $|t - t'| \leq \delta$ and hence $d(c_1(t), c_2(t)) \leq 2\delta$. If $c_1(t)$ is δ-close to a point on the third side of the triangle, then it is $(k + \delta)$-close to the endpoint of c_2 and, as in the first case, this implies that $d(c_1(t), c_2(t)) \leq 2(k + \delta)$. □

Reformulations of the Hyperbolicity Condition

In this section we shall describe some alternative ways of phrasing Gromov's hyperbolicity condition for geodesic spaces. Each of the reformulations that we shall discuss is mentioned in Gromov's original article [Gro87].

Thin Triangles

It is often useful to think of δ-hyperbolic spaces as being fattened versions of trees[51]. With this in mind, we wish to compare triangles in arbitrary metric spaces to triangles in metric trees.

Given any three positive numbers a, b, c, we can consider the metric tree $T(a, b, c)$ that has three vertices of valence one, one vertex of valence three, and edges of length a, b and c. Such a tree is called a *tripod*. For convenience, we extend the definition of tripod in the obvious way to cover the cases where a, b and c are allowed to be zero. Thus a tripod is a metric simplicial tree with at most three edges and at most one vertex of valence greater than one.

Given any three points x, y, z in a metric space, the triangle equality tells us that there exist unique non-negative numbers a, b, c such that $d(x, y) = a + b$, $d(x, z) = a + c$ and $d(y, z) = b + c$; in the notation of (1.19), $a = (y \cdot z)_x$, $b = (x \cdot z)_y$ and $c = (x \cdot y)_z$. There is an isometry from $\{x, y, z\}$ to a subset of the vertices of $T(a, b, c)$ (the vertices of valence one in the non-degenerate case); we label these vertices v_x, v_y, v_z in the obvious way.

Given a geodesic triangle, $\Delta = \Delta(x, y, z)$, we define $T_\Delta := T(a, b, c)$, and write o_Δ to denote the central vertex of T_Δ (i.e., the point a distance a from v_x). The above map $\{x, y, z\} \to \{v_x, v_y, v_z\}$ extends uniquely to a map $\chi_\Delta : \Delta \to T_\Delta$ whose restriction to each side of Δ is an isometry.

A natural measure of thinness for Δ is the diameter of the fibres of χ_Δ. The only fibre that may have more than two points is $\chi_\Delta^{-1}(o_\Delta)$, which contains one point on each of the sides of Δ. The points of $\chi_\Delta^{-1}(o_\Delta)$ are called the *internal points of* Δ. Note that these internal points are the images in Δ of the points at which the comparison triangle $\overline{\Delta} \subset \mathbb{E}^2$ meets its inscribed circle (figure H.4).

1.16 Definition (δ-Thinness and Insize). Let Δ be a geodesic triangle in a metric space X and consider the map $\chi_\Delta : \Delta \to T_\Delta$ defined above. Let $\delta \geq 0$.

Δ is said to be *δ-thin* if $p, q \in \chi_\Delta^{-1}(t)$ implies $d(p, q) \leq \delta$, for all $t \in T_\Delta$.

The diameter of $\chi_\Delta^{-1}(o_\Delta)$ is denoted *insize* Δ.

1.17 Proposition. *Let X be a geodesic space. The following conditions are equivalent.*

(1) *There exists $\delta_0 > 0$ such that every geodesic triangle in X is δ_0-slim (definition 1.1).*

(2) *There exists $\delta_1 > 0$ such that every geodesic triangle in X is δ_1-thin (definition 1.16).*

(3) *There exists $\delta_2 > 0$ such that insize $\Delta \leq \delta_2$ for every geodesic triangle Δ in X (definition 1.16).*

[51] See [Bow91], [CDP91], [Gro87] and [GhH90] for precise results regarding approximation by trees.

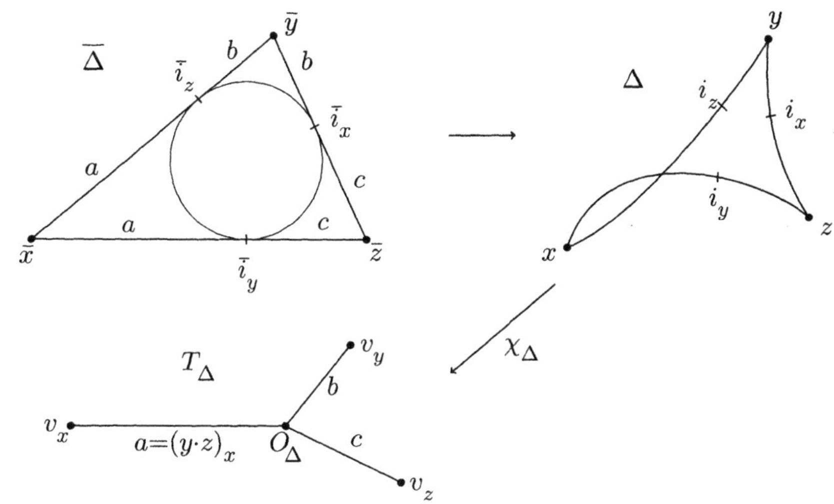

Fig. H.4 The internal points of Δ and the tripod T_Δ

Proof. It is obvious that (2) implies (1). To prove that (1) implies (3), we assume that every triangle in X is δ_0-slim and consider a geodesic triangle $\Delta(x, y, z)$ with internal points i_x, i_y, i_z (where $i_x \in [y, z]$, etc.). By hypothesis, i_x lies in the δ_0-neighbourhood of either $[y, x]$ or $[z, x]$; say $d(x, p) \leq \delta_0$ where $p \in [y, x]$. By the triangle inequality, $|d(y, p) - d(y, i_x)| \leq \delta_0$. Since $d(y, i_x) = d(y, i_z)$, it follows that $d(i_z, p) \leq \delta_0$ and hence $d(i_x, i_z) \leq 2\delta_0$. Similarly, $d(i_y, \{i_x, i_z\}) \leq 2\delta_0$ and $d(i_z, \{i_x, i_y\}) \leq 2\delta_0$. Hence *insize* $\Delta \leq 4\delta_0$.

Now assume that (3) holds and let $\Delta(x, y, z)$ be a geodesic triangle in X with internal points i_x, i_y, i_z. If $p \in [y, z]$ is such that $d(y, p) < d(y, i_x)$, then the fibre of χ_Δ containing p is $\{p, q\}$, where $q \in [y, x]$ is the point with $d(y, p) = d(y, q)$. We shall bound $d(p, q)$ by constructing a geodesic triangle for which p and q are internal points.

Let $c : [0, 1] \to X$ be a monotone parameterization of $[y, z]$ and for each $t \in [0, 1]$ consider a geodesic triangle $\Delta_t = \Delta(x, y, c(t))$ two of whose sides are $[y, x] \subset \Delta$ and $c([0, t])$. The internal point of Δ_t on the side $c([0, t])$ varies continuously (though perhaps not monotonically) as a function of t. At $t = 0$ this point is y and at $t = 1$ it is i_x, so for some $t_0 \in [0, 1]$ it is p. And since $d(y, p) = d(y, q)$, it follows that q is also an internal point of Δ_{t_0}. □

1.18 Exercise. Prove that the above conditions are equivalent to: there exists $\delta_3 > 0$ such that for every geodesic triangle $\Delta(x, y, z)$ in X,

$$\inf\{\operatorname{diam}(x', y', z') \mid x' \in [y, z],\ y' \in [x, z],\ z' \in [x, y]\} \leq \delta_3.$$

The Gromov Product and a 4-Point Condition

The Gromov product on a metric space (sometimes called the overlap function) measures the lengths of the edges of the tripods described in the paragraph on thin triangles.

1.19 Definition. Let X be a metric space and let $x \in X$. The Gromov product of $y, z \in X$ with respect to x is defined to be

$$(y \cdot z)_x = \frac{1}{2}\{d(y, x) + d(z, x) - d(y, z)\}.$$

Equivalently, $(y \cdot z)_x$ is the distance from the comparison point for x to the adjacent internal points in a comparison triangle $\overline{\Delta}(x, y, z) \subset \mathbb{E}^2$. (See figure H.4.)

Note that if $\Delta = \Delta(x, y, z)$ is a geodesic triangle in any metric space X, then $d(x, [y, z]) \leq (y \cdot z)_x + \textit{insize}\ \Delta$. And if Δ is δ-thin, then $|d(x, [y, z]) - (y \cdot z)_x| \leq \delta$.

In the following definition X is not required to be a geodesic space.

1.20 Definition. Let $\delta \geq 0$. A metric space X is said to be (δ)-hyperbolic if

$$(x \cdot y)_w \geq \min\{(x \cdot z)_w, (y \cdot z)_w\} - \delta$$

for all $w, x, y, z \in X$.

1.21 Remark. It is not difficult to show that if there exists *some* $w \in X$ such that the above inequality holds for all $x, y, z \in X$, then X is (2δ)-hyperbolic, i.e. the inequality holds for *all* x, y, z, w if one replaces δ by 2δ. See [Gro87, 1.1B].

Henceforth, when we say that a metric space is hyperbolic we shall mean that it is (δ)-hyperbolic for some $\delta > 0$. Proposition 1.22 shows that this usage agrees with our previous convention (1.1) in the case where X is a geodesic space.

There is a disturbing asymmetry about the respective roles of w, x, y and z in the above definition. To understand this, we first unravel the definition of the Gromov product, rewriting the above inequality as a 4-point condition:

$$(\mathrm{Q}(\delta)) \qquad d(x, w) + d(y, z) \leq \max\{d(x, y) + d(z, w), d(x, z) + d(y, w)\} + 2\delta$$

for all $w, x, y, z \in X$.

The geometry behind this equality becomes apparent if we think of w, x, y, z as the vertices of a tetrahedron; $d(x, y) + d(z, w)$, $d(x, z) + d(y, w)$ and $d(x, w) + d(y, z)$ correspond to the sums of the lengths of the three opposite pairs of edges. With this picture in mind, we call these three sums *the pair sizes* of $\{w, x, y, z\}$. The inequality $\mathrm{Q}(\delta)$ states that if we list the pair sizes in increasing order, say $S \leq M \leq L$, then $L - M \leq 2\delta$.

Suppose $S = d(x, z) + d(y, w)$, $M = d(x, y) + d(z, w)$ and $L = d(x, w) + d(y, z)$. In terms of comparison triangles, the inequality $S \leq M$ means that by choosing adjoining comparison triangles $\overline{\Delta}(x, w, y)$ and $\overline{\Delta}(x, w, z)$ in $\mathbb{E}^2$, we obtain the configuration shown in (H.5), with $l \geq 0$. (For convenience, we have omitted the overbars in labelling the comparison points.) We shall examine the inequality $L - M \leq 2\delta$ in terms of this comparison figure.

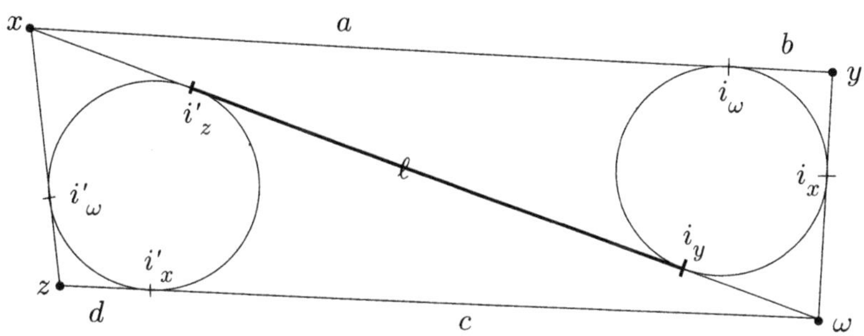

Fig. H.5 The case $d(x, y) + d(z, w) \geq d(x, z) + d(y, w)$

1.22 Proposition. *Let X be a geodesic space. X is hyperbolic in the sense of (1.17) if and only if there is a constant $\delta > 0$ such that X is (δ)-hyperbolic in the sense of (1.20).*

Proof. First we shall show that if *insize* $\Delta \leq \delta$ for all geodesic triangles Δ in X, then X is (δ)-hyperbolic.

Given $w, x, y, z \in X$, we may assume without loss of generality that $S := d(x, z) + d(y, w) \leq M := d(x, y) + d(z, w) \leq L := d(x, w) + d(y, z)$. What we must show is that $L \leq M + 2\delta$. Let $\Delta = (x, w, y)$ and $\Delta' = \Delta(x, w, z)$ be geodesic triangles, and denote their interior points by (i_x, i_w, i_y) and (i'_x, i'_w, i'_z) respectively (see figure H.5). By considering the path from y to z that proceeds via i_x, i_y, i'_z and i'_w, we get $d(y, z) \leq d(y, i_x) + d(i_x, i_y) + d(i_y, i'_z) + d(i'_z, i'_w) + d(i'_w, z)$, which, in the notation of the diagram, is $(b + d(i_x, i_y) + l + d(i'_z, i'_w) + d)$. Also $d(x, w) = a + c - l$. Thus $L \leq (a + b + c + d) + 2\delta = M + 2\delta$.

Now we shall assume that X is (δ)-hyperbolic and deduce that *insize* $\Delta \leq 6\delta$ for all geodesic triangles $\Delta = \Delta(x, y, z)$ in X. We first focus our attention on the internal point $i_x \in [y, z]$ of Δ and apply the condition $Q(\delta)$ to the four points $\{x, y, z, i_x\}$. Note that $d(x, i_x) + d(y, z)$ is the largest of the three pair sizes for $\{x, y, z, i_x\}$. Indeed $2(d(x, i_x) + d(y, z)) = [d(x, i_x) + d(i_x, z)] + [d(x, i_x) + d(i_x, y)] + d(y, z)$ is greater than the perimeter P of Δ, whereas the other pair sizes are $(d(z, i_x) + d(x, y)) = (d(y, i_x) + d(x, z)) = P/2$. Therefore $(d(x, i_x) + d(y, z)) \leq P/2 + 2\delta$. Since $d(y, z) + d(x, i_z) = P/2$, we deduce that $|d(x, i_z) - d(x, i_x)| \leq 2\delta$. Similarly, $|d(z, i_x) - d(z, i_z)| \leq 2\delta$.

Now consider the four points $\{x, z, i_x, i_z\}$ (see figure H.6). In this case the three pair sizes are $d(x, z) = d(x, i_z) + d(z, i_x)$, $d(i_x, i_z) + d(x, z)$, and $d(x, i_x) + d(z, i_z)$. By the inequalities above, the third of these terms is $\leq d(x, i_z) + d(z, i_x) + 4\delta = d(x, z) + 4\delta$. Thus we see that the last two of the three pair sizes listed are the largest. And applying $Q(\delta)$ we deduce that $d(i_x, i_z) \leq 6\delta$. Similarly, $d(i_x, i_y) \leq 6\delta$ and $d(i_y, i_z) \leq 6\delta$. $\square$

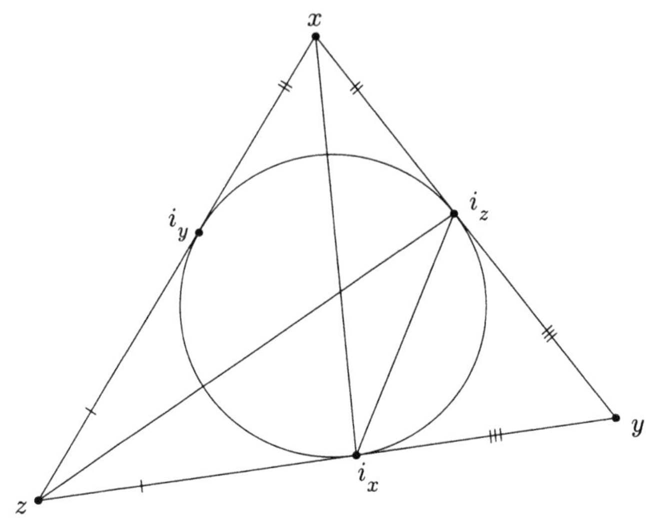

Fig. H.6 Proving that (δ)-hyperbolic implies δ'-hyperbolic

1.23 Exercise. Show that there exist quasi-geodesic rays in the Euclidean plane whose images, equipped with the induced metric from $\mathbb{E}^2$, are not (δ)-hyperbolic for any $\delta > 0$. (The spiral described in (I.8.23) provides one such example; the ray given in rectangular coordinates by $t \mapsto (\sqrt{t}, t)$ is another.) Deduce that hyperbolicity is not an invariant of quasi-isometry among arbitrary metric spaces (cf. 1.9).

Divergence of Geodesics

A key difference between Euclidean and hyperbolic geometry is that in the Euclidean plane if two distinct geodesic rays c_1 and c_2 have a common origin p then for all $t > 0$ one can connect $c_1(t)$ to $c_2(t)$ by a path outside $B(p, t)$ that has length at most πt, whereas in the hyperbolic plane the length of the shortest such path increases as an exponential function of t. In order to discuss such divergence properties in greater generality we need the following definition.

1.24 Definition. Let X be a metric space. A map $e : \mathbb{N} \to \mathbb{R}$ is said to be a *divergence function* for X if the following condition holds for all $R, r \in \mathbb{N}$, all $x \in X$ and all geodesics $c_1 : [0, a_1] \to X$ and $c_2 : [0, a_2] \to X$ with $c_1(0) = c_2(0) = x$.

If $R + r \leq \min\{a_1, a_2\}$ and $d(c_1(R), c_2(R)) > e(0)$, then any path connecting $c_1(R + r)$ to $c_2(R + r)$ outside the ball $B(x, R + r)$ must have length at least $e(r)$.

1.25 Proposition. *Let X be a geodesic space. If X is hyperbolic, then X has an exponential divergence function.*

Proof. Suppose that triangles in X are δ-thin. We claim that $e(n) = \max\{3\delta, 2^{\frac{n-1}{\delta}}\}$ is a divergence function for X. Let c_1 and c_2 be as in the preceding definition and suppose that $R, r \in \mathbb{N}$ are such that $d(c_1(R), c_2(R)) > 3\delta$ and $R + r \leq \min\{a_1, a_2\}$. Let $y_i = c_i(R + r)$ for $i = 1, 2$, and consider a geodesic triangle Δ with sides $c_1([0, R + r])$, $c_2([0, R + r])$ and $[y_1, y_2]$. Note that since $d(c_1(R), c_2(R)) > 3\delta$, the internal points of Δ on the sides $c_i([0, R + r])$ are a distance less than $R - \delta$ from $x = c_i(0)$, and hence the internal point on $[y_1, y_2]$ (which is actually the midpoint m) is a distance less than R from x. Thus $B(m, r) \subset B(x, R + r)$, and hence any path joining y_1 to y_2 in the complement of $B(x, R + r)$ lies outside $B(m, r)$. According to (1.6), such a path has length at least $2^{\frac{r-1}{\delta}}$. □

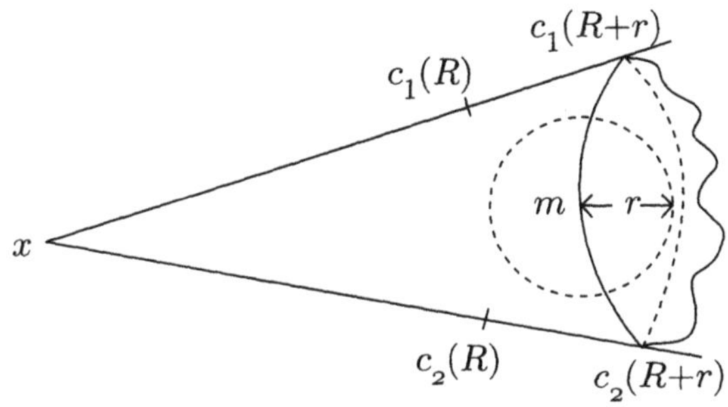

Fig. H.7 Exponential divergence of geodesics

There is a (strong) converse to the preceding proposition:

1.26 Proposition. *Let X be a geodesic space. If X has a divergence function e such that* $\liminf_{n\to\infty} e(n)/n = \infty$, *then X is δ-hyperbolic for some $\delta > 0$.*

1.27 Exercise (The Proof of (1.26)). We sketch the proof, leaving the details as an exercise for the reader (cf. [Sho91, p.37]). Consider a geodesic triangle $\Delta = \Delta(x, y, z)$. Let $c_i^x : [0, a_i] \to X$, $i = 1, 2$ be the sides of Δ issuing from x and define T_x to be $\sup\{t_0 \mid d(c_1^x(t), c_2^x(t)) < e(0)\ \forall t \leq t_0\}$. Define T_y and T_z similarly. Show that if $T_x + T_y \geq d(x, y)$ then Δ is δ-slim, where δ depends only on e. Now assume that if we delete from Δ the images of $c_i^w([0, T_w])$, $i = 1, 2$, $w \in \{x, y, z\}$, then we are left with three non-empty segments. Without loss of generality we may suppose that the segment contained in $[x, y]$ is the longest of the three and that the one contained in $[y, z]$ is the second longest; let p be the midpoint of the former and let q be the point of $[y, z]$ that is a distance $L := d(y, p)$ from y. Show that there is a path of length $\leq 2L + 6e(0)$ from p to q outside the open ball of radius L about y and use the fact that $\liminf_{n\to\infty} e(n)/n = \infty$ to bound L. Deduce from this bound that triangles in X are uniformly thin.

2. Area and Isoperimetric Inequalities

In this section we shall describe notions of area that are useful in the context of geodesic metric spaces and consider the relationship between curvature and *isoperimetric inequalities*, which relate the length of closed curves to the infimal area of the discs which they bound. It is well-known that every closed loop of length ℓ in the Euclidean plane bounds a disc whose area is less than $\ell^2/4\pi$, and this bound is optimal. Thus one has a quadratic isoperimetric inequality for loops in Euclidean space. In contrast, loops in real hyperbolic space satisfy a linear isoperimetric inequality: there is a constant C such that every closed loop of length ℓ in $\mathbb{H}^n$ bounds a disc whose area is less than or equal to $C\ell$. The main goal of this section is to prove that (with a suitable notion of area) a geodesic space X is δ-hyperbolic if and only if loops in X satisfy a linear isoperimetric inequality. We shall also see that there is a quadratic isoperimetric inequality for loops in arbitrary CAT(0) spaces (2.4), and we shall see that the sharp distinction between linear and quadratic isoperimetric inequalities is very general (2.13).

A Coarse Notion of Area

In order to begin a serious discussion of isoperimetric inequalities we must define a notion of area that is sufficiently robust to withstand the lack of regularity in arbitrary geodesic spaces.

2.1 Definitions (ε-Filling, Area_ε and Isoperimetric Inequalities). Let D^2 denote the unit disc in the Euclidean plane (so $\partial D^2 = \mathbb{S}^1$). A *triangulation of* D^2 is a homeomorphism P from D^2 to a combinatorial 2-complex in which every 2-cell is a 3-gon (i.e. its attaching map has combinatorial length three — see I.8A for definitions). We endow D^2 with the induced cell structure and refer to the preimages under P of 0-cells, 1-cells and 2-cells as, respectively, the vertices, edges and faces of P.

Let X be a metric space. Let $c : \mathbb{S}^1 \to X$ be a rectifiable loop in X. An *ε-filling* (P, Φ) of c consists of a triangulation P of D^2 and a (not necessarily continuous) map $\Phi : D^2 \to X$ such that $\Phi|_{\mathbb{S}^1} = c$ and the image under Φ of each face of P is a set of diameter at most ε. We write $|\Phi|$ to denote the number of faces of P and refer to this as the *area of the filling*. One says that *c spans* Φ. The *ε-area* of c is defined to be:

$$\text{Area}_\varepsilon(c) := \min\{|\Phi| \mid \Phi \text{ an } \varepsilon\text{- filling of } c\}.$$

(If there is no ε-filling then $\text{Area}_\varepsilon(c) := \infty$.) An ε-filling Φ of c is called a *least ε-area filling* if $|\Phi| = \text{Area}_\varepsilon(c)$.

A function $f : [0, \infty) \to [0, \infty)$ is called a *coarse isoperimetric bound* for X if there exists $\varepsilon > 0$ such that every rectifiable loop c in X has an ε-filling and $\text{Area}_\varepsilon(c) \leq f(l(c))$.

If f is linear (resp. quadratic, polynomial, exponential, etc.) then we say that X satisfies a (coarse) linear (resp. quadratic, polynomial, exponential etc.) *isoperimetric inequality*.

Recall that, given two functions $f, g : [0, \infty) \to [0, \infty)$, one writes $f \preceq g$ if there exists a constant $K > 0$ such that $f(x) \leq K\, g(Kx + K) + Kx + K$ for all $x \in [0, \infty)$. One writes $f \simeq g$ if, in addition, $g \preceq f$.

2.2 Proposition. *Let X' and X be quasi-isometric length spaces. If there exists $\varepsilon > 0$ such that every loop in X has an ε-filling, and if* $\text{Area}_\varepsilon(c) \preceq f(l(c))$ *for every rectifiable loop c in X, then X' admits a coarse isoperimetric bound $f' \preceq f$.*

Proof. Exercise: Follow the proof of (I.8.24) to construct fillings, and count the number of faces. □

2.3 Remarks

(1) The triangulations P in definition 2.1 are not required to be simplicial, for instance the attaching map of a 2-cell need not be injective, and a pair of closed 2-cells may have two faces in common.

The number of vertices, edges and faces in any triangulation of a disc satisfy the obvious relations $V \leq 2E$, $E \leq 3F$, $F \leq 2E/3$ and hence, by Euler characteristic, $V \geq 1 + 2E/3$. Thus if one were to define area by counting the minimal number of edges or vertices in a triangulation, instead of faces, the resulting notions would be $\simeq$ equivalent to the one that we have adopted.

(2) We shall use the following observation implicitly on a number of occasions. Since ε-fillings need not be continuous, if one has specified a filling map Φ on the vertices of a triangulation of the disc, then one can extend Φ across edges and triangles in the interior of the disc by simply sending them to the image of a vertex in their boundary. Thus, given a loop, if one wishes to exhibit the existence of an ε-filling of a given area, then one need only specify a triangulation of the disc and specify a map on the vertices of the triangulation, ensuring that the vertex set of each triangle in the interior of the disc is sent to a set of diameter at most ε, and ensuring that if a triangle has sides in the boundary circle then the image of these sides together with the set of vertices has diameter at most ε.

(3) If $\varepsilon' > \varepsilon$ then obviously $\text{Area}_{\varepsilon'}(c) \leq \text{Area}_\varepsilon(c)$ for all rectifiable loops c in a given space X. Thus as one increases ε, the function $f_\varepsilon^X(l) = \sup\{\text{Area}_\varepsilon(c) \mid l(c) \leq l\}$ (which need not be finite) may decrease. For example, consider the closed unit disc $D \subset \mathbb{E}^2$ and the loops $c_n : \mathbb{S}^1 \to X$ given in complex coordinates by $c(z) = z^n$. If $\varepsilon > 0$ is small, then for a suitable constant $k > 0$ we get $\text{Area}_\varepsilon(c_n) \geq kn$ for all $n > 0$, whereas if $\varepsilon \geq 2$ then $\text{Area}_\varepsilon(c) = 1$ for all loops in D.

(4) Instead of insisting that ε-fillings should be given by triangulations of the disc, one could allow decompositions of the disc into combinatorial complexes for which there is an integer $k \geq 3$ such that the boundary of each 2-cell has at most k faces. This leads to an $\simeq$ equivalent notion of area, for if we write $\text{Area}_\varepsilon^{(k)}$ for the resulting notion of area, then an obvious subdivision shows that $\text{Area}_\varepsilon(c) \leq (k-2)\text{Area}_\varepsilon^{(k)}(c)$ for all rectifiable loops, and trivially $\text{Area}_\varepsilon(c) \geq \text{Area}_\varepsilon^{(k)}(c)$.

(5) Let X be a metric graph in which all edges have length one. Given a loop $c : \mathbb{S}^1 \to X$, proceeding along c from a vertex v_1 in the image we record the vertices

that c visits $v_1, v_2 \dots, v_n = v_1$, where we record v_i only if it is distinct from v_{i-1} (in other words we do not record the occasions on which c returns to a vertex without visiting any other vertices). Say $v_i = c(t_i)$.

Let c' be the combinatorial loop in X obtained by concatenating the edges $[v_{i-1}, v_i]$. If $\varepsilon \geq 2$, then in order to construct an ε-filling of $c : \mathbb{S}^1 \to X$ we can inscribe an n-gon in $\mathbb{S}^1$ with vertices t_i, map this n-gon to X by c' in the obvious way, fill the n-gon with an ε-filling of c', and then extend the given triangulation of the n-gon to a triangulation of the disc by introducing extra edges and vertices (as necessary) in the sectors bounded by the arcs and chords $[t_{i-1}, t_i]$: the image of the boundary of each of these sectors has diameter less than 2, so it suffices (see (2)) to introduce a new vertex on the arc joining t_i to t_{i+1} and to introduce an edge joining this vertex to each of the vertices of the filling of c' that lie on $[t_{i-1}, t_i]$.

This construction bounds $\text{Area}_\varepsilon(c)$ by a linear function of $\text{Area}_\varepsilon(c')$ and $l(c)$. Thus for any metric graph with unit edge lengths, if $\varepsilon \geq 2$ then the function f_ε^X described in (3) is $\simeq$ equivalent to the function $\sup\{\text{Area}_\varepsilon(c') \mid l(c') \leq l,\ c' \text{ an edge-loop}\}$.

2.4 Example. Every CAT(0) space satisfies a quadratic isoperimetric inequality. Indeed if X is a CAT(0) space, for every $\varepsilon > 0$ and every rectifiable loop $c : \mathbb{S}^1 \to X$ one has $\text{Area}_\varepsilon(c) \leq 2(4l(c)/\varepsilon + 1)^2$. To see this, divide $\mathbb{S}^1$ into L arcs of equal length, where L is the least integer such that $4l(c) < \varepsilon L$; let $\theta_1, \dots, \theta_{L+1} = \theta_1$ be the endpoints of these arcs. To define the 1-skeleton of an ε-filling of c, for $i = 2, \dots, L$ one sends each of the Euclidean segments $\sigma_i = [\theta_1, \theta_i]$ to the geodesic segment $[c(\theta_1), c(\theta_i)] \subset X$, then one introduces equally spaced vertices $\theta_1 = v_{i,0}, v_{i,1}, \dots, v_{i,L} = \theta_i$ along each σ_i, for $i < L$ one connects each $v_{i,j}$ to both $v_{i+1,j}$ and $v_{i+1,j+1}$ by line segments, and one connects each $v_{2,j}$ and $v_{L,j}$ to θ_1. The convexity of the metric on X ensures that this is an ε-filling, and there are less than $2L^2$ triangular faces in the filling.

2.5 Exercises

(1) An ε-filling (P, Φ) of a loop c in a metric space X is said to be piecewise geodesic if the restriction of Φ to each edge in the interior of P is a linear parameterization of a geodesic segment in X. Suppose that X is a geodesic space and that c is rectifiable. If c admits an ε-filling, then it admits a piecewise geodesic $3\varepsilon/2$-filling Φ with $|\Phi| = \text{Area}_\varepsilon(c)$.

(2) Let X be a geodesic space. Given $\varepsilon > \varepsilon' > 0$, if there exists a constant $K(\varepsilon, \varepsilon') > 0$ such that $\text{Area}_{\varepsilon'}(\gamma) \leq K$ for all rectifiable loops γ with $l(\gamma) \leq 3\varepsilon$, then there is a constant K' such that $\text{Area}_{\varepsilon'}(c) \leq K' \,\text{Area}_\varepsilon(c)$ for all rectifiable loops c in X.

(Hint: Given c, consider a piecewise geodesic filling as in (1) and choose a least-area ε'-filling of the boundary of each face. The union of these ε'-fillings gives a combinatorial structure on the disc in which each 2-cell has (comfortably) less than $6K$ edges in its boundary — triangulate each of these 2-cells without introducing new vertices.)

(3) Let X be a CAT(κ) space. Prove that for all $\varepsilon' < \varepsilon < 2\pi/\sqrt{\kappa}$ there exists a constant $k = k(\varepsilon, \varepsilon', \kappa)$ such that $\text{Area}_{\varepsilon'}(c) \leq k \,\text{Area}_\varepsilon(c)$ for all rectifiable curves c in X. (See (2) and (2.4). Compare with (2.17).)

(4) Let X be a geodesic space. Suppose that there exists $\varepsilon_0 > 0$ such that $\text{Area}_{\varepsilon_0}(c)$ is finite for all rectifiable loops in X, and for all $\varepsilon > \varepsilon_0$ there is a constant $K(\varepsilon, \varepsilon_0)$ satisfying (2). Use (2.2) to show that if X' is a metric space quasi-isometric to X, then for all sufficiently large ε' and ε'' the functions $f_{\varepsilon'}^{X'}$ and $f_{\varepsilon''}^{X'}$ (as defined in 2.3(2)) take finite values and are $\simeq$ equivalent.

(5) Let X be the Cayley graph of a finitely presented group Γ. Prove that if $N \in \mathbb{N}$ is sufficiently large then $f_N^X(l)$, as defined in (2.3(3)), is $\simeq$ equivalent to the Dehn function of Γ, as defined in (I.8A).

The Linear Isoperimetric Inequality and Hyperbolicity

The purpose of this section is to characterize δ-hyperbolic geodesic spaces as those which satisfy a linear isoperimetric inequality (in the coarse sense defined in (2.1)). In the light of (2.2) and the quasi-isometry invariance of hyperbolicity (1.9), there is no loss of generality in assuming that the spaces under consideration are metric graphs with integer edge lengths. For hyperbolic graphs there is an efficient way of filling edge-loops that is essentially due to Max Dehn[52] [Dehn12a]; the proof of (2.7) is a straightforward implementation of this method.

2.6 Lemma. *Let X be a metric graph whose edges all have integer lengths, and suppose that X is δ-hyperbolic where $\delta > 0$ is an integer. Given any non-trivial locally-injective (hence rectifiable) loop $c : [0, 1] \to X$ beginning at a vertex, one can find $s, t \in [0, 1]$ such that $c(s)$ and $c(t)$ are vertices of X, $d(c(s), c(t)) \leq l(c|_{[s,t]})-1$ and $d(c(s), c(t)) + l(c|_{[s,t]}) \leq 16\delta$.*

Proof. Note that the difference in length between any two locally-injective paths with common endpoints at vertices of X is an integer. According to (1.13), X contains no closed loops which are k-local geodesics for $k = 8\delta + \frac{1}{2}$. Choose a non-geodesic subarc $c|_{[s_0,t_0]}$ of c that has length less than $8\delta + \frac{1}{2}$ and choose a geodesic connecting $c(s_0)$ to $c(t_0)$. Define $c(s)$ and $c(t)$ to be the first and last vertices of X through which this geodesic passes. (If $c(s_0)$ is a vertex then $s = s_0$, if not then the edge containing $c(s_0)$ is the image under c of an arc one of whose endpoints is s. Likewise for t.) □

2.7 Proposition. *Let X be a geodesic space. If X is δ-hyperbolic, then it satisfies a linear isoperimetric inequality.*

Proof. According to (I.8.44), X is quasi-isometric to a metric graph X' with unit edge lengths, and according to (1.9) this graph is δ'-hyperbolic for some $\delta' > 0$. If we

[52] In his foundational paper [Dehn12a], Dehn applied this method to graphs which arise as the 1-skeleton of a periodic tesselation of the hyperbolic plane, and used it to solve the word problem for surface groups (see Γ.2.4).

can show that X' satisfies a linear isoperimetric inequality, then it will follow from (2.2) (where the roles of X and X' are reversed) that X satisfies a linear isoperimetric inequality.

Assume that X is a metric graph with unit edge lengths. Assume that X is δ-hyperbolic where $\delta \geq 1$ is an integer. In what follows it is convenient to use the term *edge-loop* in a metric graph to describe loops which are the concatenation of a finite number of paths each of which is either a constant speed parameterization of an edge or a constant map at a vertex. As usual, we write $l(c)$ for the length of such a loop c. We write $l_0(c)$ to denote the number of maximal non-trivial arcs where $c : \mathbb{S}^1 \to X$ is constant. Note that $l_0(c) \leq l(c)+1$. A *standard* ε-filling of an edge-loop is an ε-filling given by a triangulation of the disc such that all of the vertices on the boundary circle are points of concatenation of the given edge-loop, and each edge of the triangulation is either mapped to a concatenation of edges in X or else is sent to a vertex of X by a constant map. We shall prove by induction on $l(c) + l(c_0)$ that every edge-loop in X admits a standard 16δ-filling of area $(8\delta + 2)(l_0(c) + l(c))$. In the light of 2.3(5), it will follow that X satisfies a linear isoperimetric inequality.

The first few steps of the induction are trivial. For the inductive step, given an edge-loop $c : \mathbb{S}^1 \to X$ with $l(c) \geq 2$, we consider how to reduce $l_0(c) + l(c)$. If $l_0(c) = 0$, then c is either locally injective or else it contains a subpath which backtracks (i.e. traverses an edge and then immediately returns along that edge). To begin our construction of a filling for c, in the latter case we connect the endpoints s, t of a backtracking subpath by a Euclidean segment in the disc and send this segment to X by a constant map; in the former case we choose s and t as in Lemma 2.6 and map the segment connecting s to t to a constant speed parameterization of a geodesic segment joining $c(s)$ to $c(t)$ in X. If $l_0(c) \geq 1$ then we focus on a subpath $c|_{[s,t]}$ which is the concatenation of a maximal subpath of length zero (defined on a non-trivial arc) and a subpath of length one. We again connect s to t by a Euclidean segment in the disc and map this segment to X as a constant speed parameterization of a geodesic segment $[c(s), c(t)]$.

In each of these three situations, we have begun to fill c by dividing the disc into two sectors, one (the "small" sector) whose boundary maps to an edge-loop (containing $c([s, t])$) of length at most 16δ, and one (the "big" sector) whose boundary map is an edge-loop c' with $l_0(c')+l(c') < l_0(c)+l(c)$. By induction, we may fill this big sector with a standard 16δ-filling (P, Φ) of c' that has at most $(8\delta+2)(l_0(c')+l(c'))$ faces. Subdividing and adding two extra faces if necessary, we may assume that s and t are vertices of P. The restriction of Φ to the Euclidean segment $[s, t]$ is a concatenation of at most 8δ edges and hence its interior contains fewer than 8δ vertices from the triangulation of the filling. To complete the desired standard 16δ-filling of c, we introduce edges connecting this set of vertices to a vertex introduced on $\mathbb{S}^1$ between s and t. □

2.8 Remarks

(1) By unraveling the various components of the above proof, one can show that there exist universal constants $\varepsilon, \mu > 0$ such that for every $\delta > 0$, every δ-hyperbolic geodesic space X and every rectifiable curve c in X, one has $\mathrm{Area}_{\varepsilon(\delta+1)}(c) \leq \mu(l(c)+1)$.

The following result provides a converse to (2.7) and thereby completes the characterization of hyperbolic spaces as those which satisfy a linear isoperimetric inequality. Our proof is based on that of Short [Sho91]. (See also [Ol91] and [Lys90].)

2.9 Theorem (Linear Isoperimetric Inequality Implies δ-Hyperbolic). *Let X be a geodesic metric space. If there exist constants $K, N > 0$ such that* $\text{Area}_N(c) \leq Kl(c) + K$ *for every piecewise geodesic loop c in X, then X is δ-hyperbolic, where δ depends only on K and N.*

Proof. Again, in the light of (I.8.44), (1.9) and (2.2) it suffices to consider the case where X is a metric graph with unit edge lengths. By increasing K and N if necessary, we may assume that $\text{Area}_N(c) \leq Kl(c)$ for every edge-loop c in X, and we may assume that K and N are integers.

In order to show that there exists $\delta > 0$ such that X is δ-hyperbolic, we must bound the size of integers $n > 0$ for which there is a geodesic triangle in X which is not $(n+1)$ slim. To this end, we fix n and suppose that there is a geodesic triangle $\Delta = \Delta(p, q, r)$ in X and a point $a \in [p, q]$ so that the distance from a to the union of the other two sides is greater than $n + 1$. We may assume that the sides of this triangle are edge-paths. We may also assume that the length of the perimeter of Δ is minimal. We replace a by an adjacent vertex v, which is a distance more than n from $[p, r] \cup [r, q]$.

Let $k = 3KN^2$ and let $m = 3KN$. We suppose that $n > 6k$ in order to avoid degeneracies in the following argument.

Reversing the roles of p and q if necessary, there are only two cases to consider: either $[p, v]$ is disjoint from the $4k$ neighbourhood of $[r, q]$ and $[v, q]$ is disjoint from the $4k$ neighbourhood of $[p, r]$, or else there exists $w \in [v, q]$ and $w' \in [p, r]$ with $d(w, w') = 4k$.

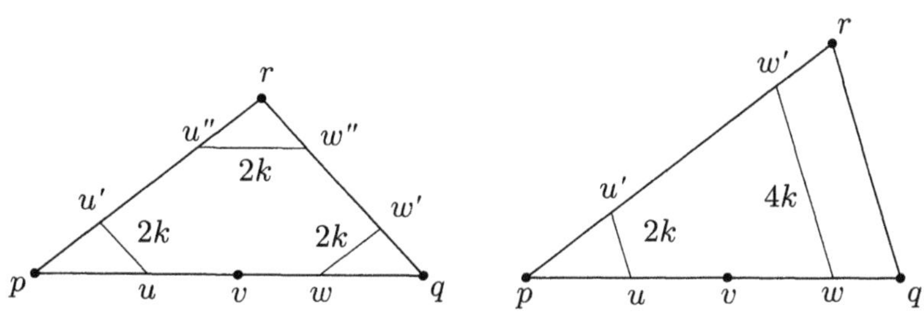

Fig. H.8 (i) The hexagonal case; (ii) The quadrilateral case

In the first case (Case (*i*) of figure H.8) we consider the minimal subarc $[u, w] \subset [p, q]$ that contains v and has endpoints a distance exactly k from the union of the other sides of Δ; let $u' \in [p, r]$ and $w' \in [r, q]$ be the points closest to r with $d(u, u') = 2k$ and $d(w, w') = 2k$. Let $[u', u''] \subset [u', r]$ and $[w', w''] \subset [w', r]$ be the maximal subarcs for which the k neighbourhoods of the interiors of these

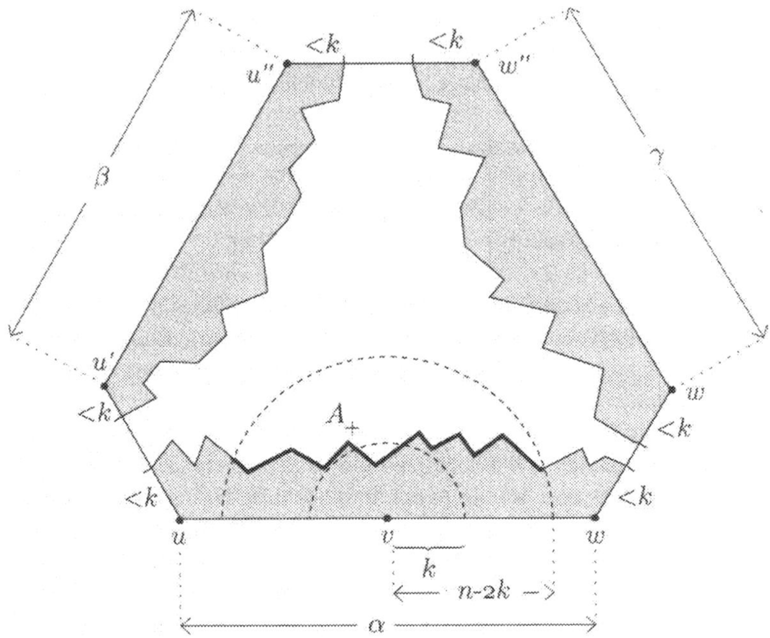

Fig. H.9 Obtaining a lower bound on area

arcs are disjoint. Hence $d(u'', w'') = 2k$. (Because $d(u, w'), d(w, u') > 4k$ and $d(u, u'), d(w, w') = 2k$, the points u', u'', w', w'' are all distinct.)

In the second case (Case (*ii*) of figure H.8) we consider maximal subarcs $[u, w] \subset [p, q]$ and $[u', w'] \subset [p, r]$ for which the k neighbourhoods of the interiors of these arcs are disjoint, $d(u, u') = 2k$, $d(w, w') = 4k$ and $v \in [u, w]$.

We deal with Case (*i*) in detail. Case (*ii*) is only slightly different.

We consider a geodesic hexagon $\mathcal{H}$ with vertices u, u', u'', w'', w', w; the sides $[u, w]$, $[u', u'']$ and $[w'', w']$ are subarcs of the sides of Δ, and the other sides have length $2k$. Let (P, Φ) be a least-area piecewise geodesic N-filling of $\mathcal{H}$ (see 2.5(1)). It is convenient to subdivide the edges of P by adding vertices along the edges so as to ensure that the restriction of Φ to each open edge is either a constant map or a homeomorphism onto an open edge of X; since each edge in the original triangulation was mapping to a path of length at most N, this subdivision can be performed in such a way that the boundary of each 2-cell in the resulting combinatorial structure on the disc contains at most $3N$ edges. We shall continue to refer to the disc with this combinatorial structure as P. It is also convenient to identify points on $\mathcal{H}$ with their preimages under Φ.

We wish to estimate the number of faces in P. By construction, the k neighbourhoods of the segments $[u, w]$, $[u', w']$ and $[u'', w'']$ are disjoint (in X and hence P). We begin by estimating the number of faces in each of these neighbourhoods. Let $\alpha = d(u, w)$, let $\beta = d(u', u'')$ and let $\gamma = d(w', w'')$.

First we consider the union of those 2-cells in P which intersect $[u, w]$ (i.e., the star neighbourhood $[u, w]$); we take the minimal combinatorial subdisc $D_1 \subset P$ containing this union. We then iterate this process: take the star neighbourhood of D_1, 'fill in the holes' to form the minimal subdisc D_2 containing this star neighbourhood; repeat m times. The boundary of D_m lies in the $k = mN$ neighbourhood of $[u, w]$. (Recall that $m = 3KN$ and $k = 3KN^2$.)

We shall estimate how many 2-cells are added at each stage of the above process. Because $d(u, w) = \alpha$ and the perimeter of each face of P has combinatorial length at most $3N$, there are at least $\alpha/(3N)$ faces in D_1. For $i = 1, \ldots, m - 1$ there is a (unique) injective edge-path connecting u to w in ∂D_i with no edge in $[u, w]$. This edge path has combinatorial length at least $\alpha = d(u, w)$. Because $m = k/N$, if any of the edges of the path lie in $\mathcal{H} = \partial P$ then they lie on $[u, u']$ or $[w, w']$, and hence there are at most $2k$ of them. So since the boundary of each face in P contains at most $3N$ edges, there are at least $(\alpha - 2k)/(3N)$ faces in $P \smallsetminus D_i$ that abut this path, and hence at least this many faces in $D_{i+1} \smallsetminus D_i$. Summing, we get a lower bound of

$$\frac{m}{3N}(\alpha - 2k)$$

on the number of faces in the k-neighbourhood of $[u, w]$ in P. Since $m = KN$, this simplifies to $K(\alpha - 2k)$. Similarly, we get lower bounds of $K(\beta - 2k)$ and $K(\gamma - 2k)$ on the number of faces in the k-neighbourhoods of $[u', u'']$ and $[w', w'']$ respectively. By construction of $\mathcal{H}$, these neighbourhoods are disjoint, so we may add these estimates to obtain a lower bound on $\mathrm{Area}_N(\mathcal{H})$:

$$\mathrm{Area}_N(\mathcal{H}) \geq K(\alpha + \beta + \gamma) - 6kK.$$

To improve this lower bound we note that $\partial D_m \smallsetminus \mathcal{H}$ intersects the k-neighbourhood of v but $d(v, \mathcal{H} - [u, w]) \geq n - 2k$. It follows that there is an arc A_+ of combinatorial length at least $n - 3k$ in $\partial D_m - \mathcal{H}$ and hence at least $(n - 3k)/(3N)$ faces of P that abut this arc but which do not lie in the neighbourhoods of $[u, w]$, $[u', u'']$, $[w'', w']$ considered above.

Thus we obtain a lower bound

$$\mathrm{Area}_N(\mathcal{H}) \geq K(\alpha + \beta + \gamma) - 6kK + \frac{n - 3k}{3N}.$$

But $\mathcal{H}$ has length $l(\mathcal{H}) = \alpha + \beta + \gamma + 6k$, and according to the linear isoperimetric inequality for X,

$$\mathrm{Area}_N(\mathcal{H}) \leq K\, l(\mathcal{H}).$$

Thus,

$$\frac{n - 3k}{3N} \leq 12kK.$$

K and N are constant and $k = 3KN^2$, so this inequality is clearly nonsense if n is large. Thus we have bounded n and therefore can deduce that X is δ-hyperbolic, where δ depends only on K and N. □

Subquadratic Implies Linear

It follows easily from the Flat Plane Theorem and the characterization of hyperbolic spaces by the linearity of their isoperimetric functions (2.7 and 2.9) that the optimal isoperimetric inequality for any cocompact CAT(0) space is either linear or quadratic. The following result shows that this dichotomy holds in much greater generality. This insight is due to Gromov [Gro87], and was clarified by Bowditch [Bow95a], Ol'shanskii and others (see [Ol91], [Paps95a] and references therein). Following Bowditch, we consider notions of area that satisfy the following axiomatic scheme:

2.10 Definition. Let c_1, c_2, c_3 be rectifiable curves which have a common initial point and a common terminal point. Let γ_i denote the concatenation $c_i\overline{c_{i+1}}$, where the overline denotes reversed orientation and indices are taken mod 3. Then, $\{\gamma_1, \gamma_2, \gamma_3\}$ is said to be a *theta curve* spanned by c_1, c_2, c_3. Under the same circumstances, we say that γ_1 and γ_2 are obtained by *surgering* $\overline{\gamma}_3$ *along* c_2.

Let X be a geodesic space and let Ω be a set of rectifiable loops in X. We consider *area functionals* $A : \Omega \to \mathbb{R}_+$ such that:

(A1) (Triangle inequality) If $\gamma_1, \gamma_2, \gamma_3 \in \Omega$ form a theta-curve, then $A(\gamma_1) \leq A(\gamma_2) + A(\gamma_3)$.

(A2) (Quadrangle inequality) There is a constant $K > 0$ such that if $\gamma \in \Omega$ is the concatenation of four paths, $\gamma = c_1c_2c_3c_4$, then $A(\gamma) \geq K\,d_1d_2$ where $d_1 = d(\text{im}(c_1), \text{im}(c_3))$ and $d_2 = d(\text{im}(c_2), \text{im}(c_4))$.

The usual notion of infimal area for spanning discs of piecewise smooth curves in complete Riemannian manifolds satisfies these axioms. A counting argument closely analogous to the proof of (2.9) shows that for the set of edge-loops in a metric graph with integer edge lengths, the notion of area defined in (2.1) satisfies axiom (A2). It follows that the notion of area that one gets by counting vertices instead of faces also satisfies this axiom (see 2.3(1)), and this latter notion also satisfies (A1), trivially. Thus, in the light of (2.9), the following theorem implies in particular that if a metric graph with integer edge lengths satisfies a subquadratic isoperimetric inequality, then the graph is δ-hyperbolic.

A function $f : [0, \infty) \to [0, \infty)$ is said to be *subquadratic* if it is $o(x^2)$, that is $\lim_{x\to\infty} f(x)/x^2 = 0$. A class of loops Ω is said to satisfy a subquadratic isoperimetric inequality with respect to an area functional A if

$$f_{A,\Omega}(x) = \sup\{A(\gamma) \mid \gamma \in \Omega,\, l(\gamma) \leq x\}$$

is a subquadratic function.

2.11 Theorem (Subquadratic Implies Linear). *Let X be a metric space, let Ω be a class of rectifiable loops in X that is closed under the operation of surgery along geodesic arcs, and suppose that $A : \Omega \to \mathbb{R}_+$ satisfies axioms (A1) and (A2). If Ω*

satisfies a subquadratic isoperimetric inequality with respect to A then it satisfies a linear isoperimetric inequality.

Again following Bowditch, we present the proof in the form of two technical lemmas. There is no loss of generality in assuming that $K = 1$ in (A2), and we shall do so in what follows. Let $f(x) = f_{A,\Omega}(x)$ be as above.

2.12 Lemma. *For every $x \in [0, \infty)$ there exist $p, q \in [0, \infty)$ such that:*

$$\begin{aligned} f(x) &\leq f(p) + f(q), \\ p, q &\leq \frac{3}{4}x + 3\sqrt{f(x)}, \\ p + q &\leq x + 6\sqrt{f(x)}. \end{aligned}$$

Proof. In order to clarify the idea of the proof, we only consider the case where the supremum in the definition of $f(x)$ is attained, by γ say. (The general case follows by an obvious approximation argument.) Let $\Sigma \subset S^1 \times S^1$ be the set of points (t, u) such that the restriction of γ to each of the two subarcs with endpoints $\{t, u\}$ has length is at least $x/4$. Let $L = \min\{d(\gamma(t), \gamma(u)) \mid (t, u) \in \Sigma\}$ and $(t_0, u_0) \in \Sigma$ be such that $d(\gamma(t_0), \gamma(u_0)) = L$. Let γ_+ and γ_- be the paths obtained by restricting γ to the connected components of $S^1 - \{t, u\}$; these are chosen so that $l(\gamma_-) \leq l(\gamma_+) \leq 3x/4$ and oriented so that each has initial point $\gamma(t_0)$ and terminal point $\gamma(u_0)$.

Consider the theta-curve spanned by γ_+, γ_- and a choice of geodesic segment $[\gamma(t_0), \gamma(u_0)]$; let β_+ and β_- be the loops obtained by surgering γ along $[\gamma(t_0), \gamma(u_0)]$. Let $a, b \in [\gamma(t_0), \gamma(u_0)]$ be such that $d(\gamma(t_0), a) = d(a, b) = d(b, \gamma(u_0)$. Let $\alpha_1, \delta, \alpha_2$ be geodesics whose images are the subarcs $[\gamma(t_0), a]$, $[a, b]$, $[\gamma(t_0), \gamma(u_0)]$ of $[\gamma(t_0), \gamma(u_0)]$, respectively.

By construction $d(\text{im}\alpha_1, \text{im}\alpha_2) = L/3$. We claim that $d(\text{im}\delta, \text{im}\gamma_+) \geq L/3$. Once we have proved this claim, by Axiom A2 we will have $A(\beta_+) \geq (L/3)^2$, whereas by Axiom A1:

$$f(x) = A(\gamma) \leq A(\beta_+) + A(\beta_-) \leq f(l(\beta_+)) + f(l(\beta_-)).$$

It therefore suffices to let $p = l(\beta_+)$ and $q = l(\beta_-)$.

It remains to prove the claim. Given $z = \gamma(v) \in \text{im}\gamma_+$ and $z' \in \text{im}\delta$, we divide γ_+ into subarcs σ_1 and σ_2 with endpoint z. Without loss of generality we may suppose that $l(\sigma_1) \leq l(\sigma_2)$. Note that $l(\sigma_1) + l(\sigma_2) + l(\gamma_-) = l(\gamma) = x$ while $l(\gamma_-) \leq l(\sigma_1) + l(\sigma_2)$. Since $l(\sigma_1) \leq l(\sigma_2)$ and $l(\gamma_-) \leq x/2$, we see that $(v, u_0) \in \Sigma$. Thus $d(z, \gamma(u_0)) \geq L = d(\gamma(t_0), \gamma(u_0))$. Since $z' \in [a, \gamma(u_0)]$ it follows that $d(z, z') \geq d(z, \gamma(u_0)) - d(\gamma(u_0), a) \geq L - 2L/3$. □

2.13 Lemma. *Let $g : [0, \infty) \to [0, \infty)$ be an increasing function and suppose that there exist constants $k > 0$ and $\lambda \in (0, 1)$ such that for every $x \in [0, \infty)$ one can find $p, q \in [0, \infty)$ with*

$$\begin{aligned} g(x) &\le g(p) + g(q), \\ p, q &\le \lambda x + k\sqrt{g(x)}, \\ p + q &\le x + k\sqrt{g(x)}. \end{aligned}$$

If $g(x) = o(x^2)$ *then* $g(x) = O(x)$.

Proof. By replacing g with a constant multiple we may assume that $k = 1$. Let $\mu = (1+\lambda)/2$ and fix $x_0 \in [0, \infty)$ such that if $x > x_0$ then $g(x) \le (1-\mu)^2 x^2$. Note that if $x > x_0$ then $p, q \le \lambda x + (1-\mu)x = \mu x$. Increasing x_0 if necessary, we may assume that it is bigger than 1.

Let $h(x) = g(x)/x$. If $x > x_0$ then $xh(x) \le ph(p) + qh(q)$, so

$$h(x) \le \frac{p}{x}\, h(p) \;+\; \frac{q}{x}\, h(q).$$

Without loss of generality we may assume that $h(q) \le h(p)$, so

$$h(x) \le \left(\frac{p+q}{x}\right) h(p) \le \left(\frac{x + \sqrt{g(x)}}{x}\right) h(p) \le \left(1 + \sqrt{\frac{h(x)}{x}}\right) h(p). \tag{13.1}$$

Thus if $x > x_0$, then there exists $p \in [1, \mu x]$ with $h(x) \le (1 + \sqrt{h(x)/x})\, h(p)$. We want to show that it $h(x) = o(x)$ ('little oh'), then h is bounded.

For this we fix $\varepsilon > 0$ and $x_1 > 0$ so that $h(x) \le \varepsilon^2 x$ for all $x > x_1$. Let $B = \max\{h(x) \mid 1 \le x \le x_1\}$. We have, for $x > x_1$, that $h(x) \le (1+\varepsilon)\, h(p)$ with $p \le \mu x$. By using this inequality $n \le 1 + \log(x/x_1)/\log(1/\mu)$ times, we obtain:

$$h(x) \le B(1+\varepsilon)^n \le B(1+\varepsilon)(x/x_1)^r,$$

where $r = \log(1+\varepsilon)/\log(1/\mu)$. We choose ε small enough to ensure that $r < 1$.

Fix $s > 0$ with $r < 1 - 2s$. Then, $h(x) = o(x^{1-2s})$, and so we may choose x_2 with $h(x) \le x^{1-2s}$ for all $x > x_2 \ge x_1$. But then, by equation (13.1):

$$h(x) \le (1 + x^{-s})\, h(p).$$

Again we iterate our estimate, k times,

$$h(x) \le C(1 + x^{-s})(1 + \mu^{-s}x^{-s}) \cdots (1 + \mu^{-ks}x^{-ks}),$$

where $C = \max\{h(y) \mid 1 \le y \le x_2\}$ and k is the greatest integer such that $\mu^k x > x_2$. Therefore, since $\log(1+y) \le y$ for all $y > 0$,

$$\begin{aligned} \log h(x) &\le \log C + x^{-s}\big(1 + \mu^{-s} + \cdots + \mu^{-ks}\big) \\ &= \log C + \frac{(\mu^k x)^{-s} - \mu^s x^{-s}}{1 - \mu^s} \\ &\le \log C + \frac{x_2^{-s}}{1 - \mu^s}. \end{aligned}$$

Thus h is bounded. □

More Refined Notions of Area

The notions of area considered above are rather crude but are well suited to our purposes since they are technically simple and relatively stable under quasi-isometry — they describe something of the large-scale geometry of spaces, which is the main theme of our book. However, there is a striking body of work by the Russian school which deals with area in a more sophisticated and local sense. In this paragraph we describe some of the main concepts and results from this work, without giving proofs. For a more complete introduction, see the survey article of Berestovskii and Nikolaev [BerN93].

The following definition of area was given by Nikolaev [Ni79] and is adapted from an earlier definition of Alexandrov [Ale57]. The basic idea is to define the area of a surface to be the limiting area of approximating polyhedral surfaces built out of Euclidean triangles. Thus, in particular, this coincides with Lebesgue area for surfaces in Euclidean space.

2.14 Definitions. Let X be a geodesic space and let D^2 denote the unit disc in the Euclidean plane. By definition, a *parameterized surface* in X is a continuous map $f : D^2 \to X$, and a *non-parameterized* surface is the image F of such a map.

Let f be a parameterized surface. Intuitively speaking, *a complex for the surface* f is a polyhedral approximation to f. More precisely, given a triangulation of D^2 with vertices $A_1, \dots, A_n$, we choose points $\tilde{A}_i \in X$ such that $\tilde{A}_i = \tilde{A}_j$ if and only if $f(A_i) = f(A_j)$; if the vertices A_i, A_j are adjacent, we connect $\tilde{A}_i$ to $\tilde{A}_j$ by a geodesic in X; thus we obtain a set of triangles in X. This set of triangles is called *a complex for the surface* f and the $\tilde{A}_i$ are called the vertices of the complex. Let $\Psi(f)$ denote the set of sequences of complexes (Ψ_m) which have the property that as $m \to \infty$ the maximum of the distances $d(\tilde{A}_i, f(A_i))$ tends to zero and the maximum of the lengths of the sides of the triangles in Ψ_m tends to zero.

The area of a complex Ψ, denoted $\sigma(\Psi)$, is obtained by replacing the triangles in Ψ with Euclidean comparison triangles and summing the area of the latter. For a parameterized surface f one defines:

$$\text{Area}(f) := \inf\{\liminf \sigma(\Psi_m) \mid (\Psi_m) \in \Psi(f)\}.$$

And for a non-parameterized surface F one defines:

$$\text{Area}(F) := \inf\{\text{Area}(f) \mid f(D^2) = F\}.$$

(Of course, these areas may be infinite.)

One has the following basic properties:

2.15 Lemma. *Let X and Y be geodesic spaces.*

(1) (Semicontinuity) *If $f_n : D^2 \to X$ is a sequence of parameterized surfaces in X and $f_n \to f$ uniformly, then* $\text{Area}(f) \leq \liminf \text{Area}(f_n)$.

(2) (Kolmogorov's Principle) *If $p : X \to Y$ does not increase distances then for every parameterized surface in X one has:*

$$\text{Area}(p \circ f) \leq \text{Area}(f).$$

Let X be a CAT(κ) space and let $c : S^1 \to X$ be a rectifiable loop in X (of length $< 2\pi/\sqrt{\kappa}$ if $\kappa > 0$). A *ruled surface bounded by* c is a parameterized surface $f : D^2 \to X$ obtained by choosing a basepoint $o \in S^1$ and defining f so that for each $\theta \in S^1$ the restriction of f to the Euclidean segment $[o, \theta]$ is a linear parameterization of the unique geodesic segment joining $c(o)$ to $c(\theta)$ in X.

The following result, which is due to Alexandrov [Ale57], follows from the Flat Triangle Lemma (II.2.9) and Kolmogorov's principle.

2.16 Proposition. *The area of any ruled surface bounded by a triangle Δ (of perimeter less then $2\pi/\sqrt{\kappa}$) in a* CAT(κ) *space is no greater than the area of the comparison triangle $\overline{\Delta}$ in M_κ^2, and is equal to it if and only if Δ and Δ' are isometric.*

From this Alexandrov deduces:

2.17 Theorem. *Let X be a* CAT(κ) *space and let c be a rectifiable loop in X. (If $\kappa > 0$ assume that $l(c) < 2\pi/\sqrt{\kappa}$.) Then, the area of any ruled surface bounded by c is no greater than the area of a disc in M_κ^2 whose boundary is a circle of length $l(c)$. Moreover, equality holds only if the disc and the ruled surface are isometric.*

Notice that by reference to the classical cases $\mathbb{E}^2$ and $\mathbb{H}^2$, this result gives a quadratic isoperimetric inequality for loops in CAT(0) spaces and a linear isoperimetric inequality in CAT(-1) spaces (with respect to the notion of area defined in (2.14)). In the same vein, the following theorem of Reshetnyak [Resh68] can also be viewed as a strong isoperimetric result (via Kolmogorov's principle).

A convex domain $V \subset M_\kappa^2$ is said to *majorize* a rectifiable loop c in a metric space X if there is a non-expanding map $V \to X$ which restriction to ∂V is an arc length parameterization of c.

2.18 Theorem. *For any rectifiable loop c in a* CAT(κ) *space X (with $l(c) < 2\pi/\sqrt{\kappa}$ if $\kappa > 0$) there exists a convex domain $V \subset M_\kappa^2$ which majorizes c.*

2.19 Remark (Plateau's Problem). Plateau's problem asks about the existence of minimal-area fillings for rectifiable loops in a given space (see [Alm66]). Nikolaev [Ni79] solved Plateau's problem in the context of CAT(κ) spaces.

3. The Gromov Boundary of a δ-Hyperbolic Space

In this section we describe the Gromov boundary ∂X of a δ-hyperbolic space X. If X is a proper geodesic space, then there is a natural topology on $X \cup \partial X$ making it a compact metrizable space and there is a natural family of "visual" metrics on ∂X. The topological space ∂X is an invariant of quasi-isometry among geodesic spaces, as is the quasi-conformal structure associated to its visual metrics. If X is a proper CAT(-1) space, or more generally a CAT(0) visibility space, the Gromov boundary is the same as the visual boundary and the topology on $X \cup \partial X$ is the cone topology (II.8.6).

This section is organised as follows. First we shall consider the case where X is a geodesic space, describing ∂X in terms of equivalence classes of geodesics rays as we did for CAT(0) spaces in Chapter II.8. We shall then interpret ∂X in terms of sequences of points that converge at infinity. As well as providing a definition of ∂X in the case where X is not geodesic, the language of sequences provides a vocabulary with which to discuss the extension of the Gromov product from X to $X \cup \partial X$. Taking the case $X = \mathbb{H}^n$ as motivation, we shall explain how this extended product can be used to define metrics on ∂X.

The Boundary ∂X as a Set of Rays

Recall that two geodesic rays $c, c' : [0, \infty) \to X$ in a metric space X are said to be asymptotic if $\sup_t d(c(t), c'(t))$ is finite; this condition is equivalent to saying that the Hausdorff distance between the images of c and c' is finite. We define *quasi-geodesic rays* to be *asymptotic* if the Hausdorff distance between their images is finite. Being asymptotic is an equivalence relation on quasi-geodesic rays. We write ∂X to denote the set of equivalence classes of geodesic rays in X and we write $\partial_q X$ to denote the set of equivalence classes of quasi-geodesic rays. In each case we write $c(\infty)$ to denote the equivalence class of c.

3.1 Lemma. *If X is a proper geodesic space that is δ-hyperbolic, then the natural map from ∂X to $\partial_q X$ is a bijection.*

For each $p \in X$ and $\xi \in \partial X$ there exists a geodesic ray $c : [0, \infty) \to X$ with $c(0) = p$ and $c(\infty) = \xi$.

Proof. The natural map $\partial X \to \partial_q X$ is obviously injective. To prove the remaining assertions, given $p \in X$ and a quasi-geodesic ray $c : [0, \infty) \to X$, let c_n be a geodesic with $c_n(0) = p$ that joins p to $c(n)$. Since X is proper, a subsequence of the c_n converges to a geodesic ray $c_\infty : [0, \infty) \to X$ (by the Arzelà-Ascoli Theorem (I.3.10)). Theorem 1.7 provides a constant k such that the Hausdorff distance between $c([0, n])$ and the image of c_n is less than k; thus we obtain a bound on the Hausdorff distance between c and c_∞. $\square$

3.2 Lemma (Visibility of ∂X). *If the metric space X is proper, geodesic and δ-hyperbolic, then for each pair of distinct points $\xi_1, \xi_2 \in \partial X$ there exists a geodesic line $c : \mathbb{R} \to X$ with $c(\infty) = \xi_1$ and $c(-\infty) = \xi_2$.*

Proof. Fix $p \in X$ and choose geodesic rays $c_1, c_2 : [0, \infty) \to X$ issuing from p with $c_1(\infty) = \xi_1$ and $c_2(\infty) = \xi_2$. Let T be such that the distance from $c_1(T)$ to the image of c_2 is greater than δ. For each $n > T$ we choose a geodesic segment $[c_1(n), c_2(n)]$ and consider the geodesic triangle with sides $c_1([0, n])$, $c_2([0, n])$ and $[c_1(n), c_2(n)]$. Since this triangle is δ-slim, $[c_1(n), c_2(n)]$ must intersect the closed (hence compact) ball of radius δ about $c_1(T)$, at a point p_n say. By the Arzelà-Ascoli Theorem, as $n \to \infty$ a subsequence of the geodesics $[p_n, c_2(n)] \subset [c_1(n), c_2(n)]$ will converge. By passing to a further subsequence we may assume that the sequence $[c_1(n), c_2(n)]$ converges. The limit is a geodesic line which we call c.

Since each $[c_1(n), c_2(n)]$ is contained in the δ-neighbourhood of the union of the images of c_1 and c_2, the image of c is also contained in this neighbourhood. Thus the endpoints of c are ξ_1 and ξ_2. □

3.3 Lemma (Asymptotic Rays are Uniformly Close). *Let X be a proper δ-hyperbolic space and let $c_1, c_2 : [0, \infty) \to X$ be geodesic rays with $c_1(\infty) = c_2(\infty)$.*

(1) *If $c_1(0) = c_2(0)$ then $d(c_1(t), c_2(t)) \leq 2\delta$ for all $t > 0$.*

(2) *In general, there exist $T_1, T_2 > 0$ such that $d(c_1(T_1 + t), c_2(T_2 + t)) \leq 5\delta$ for all $t \geq 0$.*

Proof. (1) follows immediately from (1.15).

In order to prove (2), we apply Arzelà-Ascoli to obtain a subsequence of the geodesics $c_n = [c_1(0), c_2(n)]$ that converges to a geodesic ray c_1' with $c_1'(0) = c_1(0)$. Since the triangles $\Delta(c_1(0), c_2(0), c_2(n))$ are δ-slim, all but a uniformly bounded initial segment of each c_n is contained in the δ-neighbourhood of the image of c_2, and hence a terminal segment of c_1' is also contained in this neighbourhood. In other words, there exist $T_1, T_2 > 0$ with $d(c_2(T_2), c_1'(T_1)) \leq \delta$ such that for all $t \geq 0$ one can find t' with $d(c_2(T_2 + t'), c_1'(T_1 + t)) \leq \delta$. By the triangle inequality, t and t' differ by at most 2δ. Thus for all $t \geq 0$ we have $d(c_2(T_2 + t), c_1'(T_1 + t)) \leq 3\delta$. And from (1) we know that $d(c_1(T_1 + t), c_1'(T_1 + t)) \leq 2\delta$ for all $t \geq 0$. □

3.4 Remark (Busemann Functions and Horospheres). Let X be a proper geodesic space that is δ-hyperbolic and let $c : [0, \infty) \to X$ be a geodesic ray. The Busemann function of c is $b_c(x) := \limsup_{t\to\infty} d(x, c(t)) - t$. (The triangle inequality ensures that $|b_c(x)| \leq d(c(0), x)$.) It follows from the preceding lemma that if c' is a geodesic ray with $c(\infty) = c'(\infty)$, then $|b_c(x) - b_{c'}(x)|$ is a bounded function of $x \in X$. On the other hand, if $c(\infty) \neq c'(\infty)$ then $|b_c(x) - b_{c'}(x)|$ is obviously not a bounded function. This observation allows one to construct ∂X as a space of equivalence classes of Busemann functions in analogy with (II.8.16) — see section 7.5 of [Gro87].

If we choose a basepoint $p \in X$ and alter each b_c be an additive constant so that $b_c(p) = 0$, then the (modified) Busemann functions associated to asymptotic rays differ by at most a fixed multiple of δ. (To remove this mild ambiguity one might take the supremum or infimum over all rays c with $c(\infty) = \xi$ to define "the" Busemann function of ξ.) Horospheres and horoballs about ξ (the level sets and sub-level sets of Busemann functions) are well-defined in the same sense. See [CDP90], Chapter 8 in [GhH90] and [Gro87] for applications.

The Topology on $X \cup \partial X$

The description that we are about to give of the topology on $\overline{X} = X \cup \partial X$ is closely analogous to (II.8.6). As was the case there, it is convenient to consider generalized rays.

Notation. A *generalized ray* is a geodesic $c : I \to X$, where either $I = [0, R]$ for some $R \geq 0$ or else $I = [0, \infty)$. In the case $I = [0, R]$ it is convenient to define $c(t) = c(R)$ for $t \in [R, \infty]$. Thus $\overline{X} := X \cup \partial X$ is the set $\{c(\infty) \mid c \text{ a generalized ray}\}$.

3.5 Definition (The Topology on $\overline{X} = X \cup \partial X$). Let X be a proper geodesic space that is δ-hyperbolic. Fix a basepoint $p \in X$. We define convergence in $\overline{X}$ by: $x_n \to x$ as $n \to \infty$ if and only if there exist generalized rays c_n with $c_n(0) = p$ and $c_n(\infty) = x_n$ such that every subsequence of (c_n) contains a subsequence that converges (uniformly on compact subsets) to a generalized ray c with $c(\infty) = x$. This defines a topology on $\overline{X}$: the closed subsets $B \subset \overline{X}$ are those which satisfy the condition $[x_n \in B,\ \forall n > 0 \text{ and } x_n \to x] \implies x \in B$.

3.6 Lemma (Neighbourhoods at Infinity). *Let X and $p \in X$ be as above. Let $k > 2\delta$. Let $c_0 : [0, \infty) \to X$ be a geodesic ray with $c_0(0) = p$ and for each positive integer n let $V_n(c_0)$ be the set of generalized rays c such that $c(0) = p$ and $d(c(n), c_0(n)) < k$. Then $\{V_n(c_0) \mid n \in \mathbb{N}\}$ is a fundamental system of (not necessarily open) neighbourhoods of $c(\infty)$ in $\overline{X}$.*

Proof. Let c' be a ray in X with $c'(0) = p$. It follows from (3.3) that $c'(\infty) = c_0(\infty)$ if and only if $c'(n) \in V_n(c_0)$ for all $n > 0$. And if c_i is a sequence of generalized rays in X with $c_i(0) = p$ and $c_i \notin V_n(c_0)$, then by the Arzelà-Ascoli theorem there is a subsequence $c_{i(j)}$ that converges to some $c \notin V_n(c_0)$, hence c_i does not converge to c_0 in $\overline{X}$. Thus $c_i \to c_0$ in $\overline{X}$ if and only if for every $n > 0$ there exists $N_n > 0$ such that $c_i \in V_n(c_0)$ for all $i > N_n$. □

3.7 Proposition. *Let X be a proper geodesic space that is δ-hyperbolic.*

(1) *The topology on $\overline{X} = X \cup \partial X$ described in (3.5) is independent of the choice of basepoint, and*

(2) *if X is a* CAT(0) *space this is the cone topology (II.8.6).*

(3) *$X \hookrightarrow \overline{X}$ is a homeomorphism onto its image and $\partial X \subset \overline{X}$ is closed.*
(4) *$\overline{X}$ is compact.*

Proof. (1) follows easily from Lemmas 3.6 and 3.3(2).

Suppose that X is CAT(0). In (II.8.6) we described a fundamental system of neighbourhoods $U(c_0, r, \varepsilon)$ for $c_0(\infty) \in \partial X$. Fix such a neighbourhood. Also fix $k > 2\delta$. If $N > n > r$ and $c : [0, N] \to X$ lies in $V_n(c_0)$ (the neighbourhood of $c_0(\infty)$ defined in (3.6)), then $d(c(r), c_0(r)) \leq r/n\, d(c(n), c_0(n))$, because of the convexity of the metric on X. It follows that if n is sufficiently large, then $V_n(c_0) \subset U(c_0, r, \varepsilon)$. Conversely, if $\varepsilon < k$ then $U(c_0, r, \varepsilon) \subset V_r(c_0)$. This proves (2).

(3) Let $x \in X$ and let $c_n : [0, T_n] \to X$ be a sequence of geodesics issuing from a fixed point $p \in X$. By the Arzelà-Ascoli theorem, if $c_n(T_n) \to x$ as $n \to \infty$ then every subsequence of (c_n) contains a subsequence that converges to a geodesic joining p to x. Conversely, if $(c_n(T_n))$ does not converge to x as $n \to \infty$, then some subsequence of (c_n) does not contain a subsequence that converges to a generalized ray with terminal point x. Thus $X \hookrightarrow \overline{X}$ is a homeomorphism onto its image. X is obviously open in $\overline{X}$.

(4) The balls $B(x, r)$, with $r > 0$ rational, form a fundamental system of neighbourhoods about $x \in X \subset \overline{X}$. This observation, together with the preceding lemma, shows that the topology on $\overline{X}$ satisfies the first axiom of countability. Thus it suffices to prove that $\overline{X}$ is sequentially compact, and this is obvious by Arzelà-Ascoli. □

3.8 Exercise. Let X be a geodesic space that is proper and hyperbolic. Prove that the natural map $\partial X \to \text{Ends}(X)$ is continuous and that the fibres of this map are the connected components of ∂X.

3.9 Theorem. *Let X and X' be proper δ-hyperbolic geodesic spaces. If $f : X \to X'$ is a quasi-isometric embedding, then $c(\infty) \mapsto (f \circ c)(\infty)$ defines a topological embedding $f_\partial : \partial X \to \partial X'$. If f is a quasi-isometry, then f_∂ is a homeomorphism.*

Proof. Suppose that $f : X \to X'$ is a (λ, ε)-quasi-isometric embedding, fix $p \in X$ and let $p' = f(p)$. If $c_1, c_2 : [0, \infty) \to X$ are geodesic rays with $c_i(0) = p$, then $f \circ c_1$ and $f \circ c_2$ are (λ, ε)-quasi-geodesics with $f \circ c_i(0) = p'$. These quasi-geodesic rays are asymptotic (i.e. their images are within finite Hausdorff distance of each other) if and only if c_1 and c_2 are asymptotic. Thus, in the light of (3.1), f_∂ is well-defined and injective.

Fix $k > 2\delta$. In order to see that f_∂ is continuous, for $i = 1, 2$ we choose a geodesic ray c_i' with $c_i'(0) = p'$ and $c_i'(\infty) = f \circ c_i(\infty)$. The Hausdorff distance between c_i' and $f \circ c_i$ is bounded by a constant K depending only on δ, λ and ε (see 1.8). If $c_1(n)$ lies in the k-neighbourhood of the image of c_2, then $f \circ c_1(n)$ lies in the $(\lambda k + \varepsilon)$-neighbourhood of the image of $f \circ c_2$ and hence $c_1'(t_0)$ lies in the $(2K + \lambda k + \varepsilon)$-neighbourhood of the image of c_2' for some $t_0 > K - \varepsilon + n/\lambda$. Let n' denote the least integer greater than $n - (K - \varepsilon + n/\lambda + \delta)$. As in (1.15) we have $d(c_1'(t), c_2'(t)) \leq 2\delta$ for all $t < n'$. Thus, in the notation of Lemma 3.6, $c_1 \in V_n(c_2)$

implies $c_1' \in V_{n'}(c_2')$, and hence f_∂ is continuous. A similar argument shows that $f_\partial c(\infty) \mapsto c(\infty)$ is also continuous.

If f is a quasi-isometry then it has a quasi-inverse $f' : X' \to X$ (see I.8.16), and since $d(f'f(x), x)$ is bounded, $(f'f)_\partial = f'_\partial f_\partial$ is the identity on ∂X. □

3.10 Corollary. *The real hyperbolic spaces $\mathbb{H}^n$ and $\mathbb{H}^m$ are quasi-isometric if and only if $m = n$.*

Proof. $\partial\mathbb{H}^n$ is an $(n-1)$-dimensional sphere. □

∂X as Classes of Sequences

Let X be a CAT(-1) space with basepoint p and consider a sequence (x_n) in X. In order for there to exist a point ξ in the visual boundary of X such that $x_n \to \xi$, it is necessary and sufficient that the distance from p to the geodesic $[x_i, x_j]$ should tend to infinity as $i, j \to \infty$ (see II.9.30). And as we indicated after (1.19), this is equivalent to saying that $(x_i \cdot x_j)_p \to \infty$ as $i, j \to \infty$. Using the description of the basic neighbourhoods of points at infinity given in (3.6) it is easy to generalize this observation:

3.11 Exercise. Let X be a geodesic space that is proper and δ-hyperbolic. Fix $p \in X$. Show that a sequence (x_n) in X converges to a point of $\partial X \subset \overline{X}$ if and only if $(x_i \cdot x_j)_p \to \infty$ as $i, j \to \infty$.

Motivated by this exercise, we define:

3.12 Definition. Let X be an arbitrary (δ)-hyperbolic space. A sequence (x_n) in X *converges at infinity* if $(x_i \cdot x_j)_p \to \infty$ as $i, j \to \infty$. Two such sequences (x_n) and (y_n) are said to be *equivalent* if $(x_i \cdot y_j)_p \to \infty$ as $i, j \to \infty$. The equivalence class of (x_n) is denoted $\lim x_n$ and the set of equivalence classes is denoted $\partial_s X$. (These definitions are independent of the choice of basepoint p.)

Remark. The "equivalence" of sequences defined above is obviously a reflexive and symmetric relation, but for arbitrary metric spaces it is not transitive (consider $\mathbb{E}^2$ for example). However, it follows immediately from (1.20) that in hyperbolic spaces this is an equivalence relation.

3.13 Lemma. *If X is a proper geodesic space that is δ-hyperbolic, then there is a natural bijection $\partial_s X \to \partial X$.*

Proof. It follows from exercise 3.11 that $(x_n) \mapsto \lim_n x_n$ induces a well-defined map $\partial_s X \to \partial X$ that is injective. To see that this map is surjective, note that given any geodesic ray c in X, the point $\lim c(n) \in \partial_s X$ maps to $c(\infty)$. □

3.14 Example. In general $\partial_s X$ is not a quasi-isometry invariant among (non-geodesic) hyperbolic spaces. For example, consider the spiral $c : [0, \infty) \to \mathbb{E}^2$ given in polar coordinates by $c(t) = (t, \log(1+t))$. In (1.23) we noted that the image of c, with the induced metric from $\mathbb{E}^2$, is not δ-hyperbolic. However, the intersection of im(c) with any line through the origin in $\mathbb{E}^2$ is hyperbolic; indeed, since it is a subset of a line, it is (0)-hyperbolic. Let S denote such an intersection.

We claim that although $c^{-1}(S)$ and S are quasi-isometric, their boundaries are different. Indeed, since the spiral c crosses any half-line infinitely often, $\partial_s S$ has two elements, whereas $c^{-1}(S) \subset [0, \infty)$ has only one element.

Metrizing ∂X

In order to discuss metrics on the boundary of a hyperbolic space X, we need to extend the Gromov product to ∂X. For this it is convenient to consider the boundary in the guise of $\partial_s X$, and the symbol ∂X shall have that meaning throughout this section. If X is a CAT(-1) space, the product on the boundary can be defined by the simple formula $(\lim x_i \cdot \lim y_j)_p := \lim_{i,j}(x_i \cdot y_j)_p$ (see 3.18). But for hyperbolic spaces in general this limit may not exist (3.16).

3.15 Definition. Let X be a (δ)-hyperbolic space with basepoint p. We extend the Gromov product to $\overline{X} = X \cup \partial X$ by:

$$(x \cdot y)_p := \sup \liminf_{i,j\to\infty} (x_i \cdot y_j)_p,$$

where the supremum is taken over all sequences (x_i) and (y_j) in X such that $x = \lim x_i$ and $y = \lim y_j$.

The following example illustrates the need to take lim inf and sup rather than just lim in the above definition.

3.16 Example. Let X be the Cayley graph of $\mathbb{Z} \times \mathbb{Z}_2 = \langle a, b \mid [a, b] = b^2 = 1\rangle$. Note that X is hyperbolic and that ∂X consists of two points (which we call a_- and a_+). Consider the following sequences: $x_n := ba^{-n}$, $y_n = a^n$, $z_n = ba^n$. Define w_n to be equal y_n if n is even and z_n if n is odd. As $n \to \infty$, the sequences y_n, z_n and w_n converge to a_+ while x_n converges to a_-. For all positive integers i, j we have $(x_i \cdot y_j)_1 = 0$ and $(x_i \cdot z_j)_1 = 1$. Thus $\liminf_{i,j\to\infty}(x_i \cdot y_j)_1$ is not equal to $\liminf_{i,j\to\infty}(x_i \cdot z_j)_1$ and in particular $\lim_{i,j\to\infty}(x_i \cdot w_j)_1$ does not exist.

3.17 Remarks. Let X be a (δ)-hyperbolic space and fix $p \in X$. Let $\overline{X} = X \cup \partial X$.

(1) $(\cdot)_p$ is continuous on X (not $\overline{X}$).

(2) $(x \cdot y)_p = \infty$ if and only if $x = y \in \partial X$.

(3) For all $x, y \in \overline{X}$ there exist sequences (x_n) and (y_n) in X such that $x = \lim x_n$, $y = \lim y_n$ and $(x \cdot y)_p = \lim_n (x_n \cdot y_n)_p$. If $x \in X$ then one can take (x_n) to be

the constant sequence, and in the cases where x or y belongs to ∂X, one can construct (x_n) and (y_n) by a diagonal sequence argument.

(4) For all $x, y, z \in \overline{X}$ we have $(x \cdot y)_p \geq \min\{(x \cdot z)_p, (z \cdot y)_p\} - 2\delta$. To see this, choose sequences in X with $x = \lim x_n$, $y = \lim y_n$ and $z = \lim z_n = \lim z'_n$, such that $\lim_n (x_n \cdot z_n)_p = (x \cdot z)_p$ and $\lim_n (z'_n \cdot y_n)_p = (z \cdot y)_p$, then take $\liminf_{i,j}$ through $(x_i \cdot y_j)_p \geq \min\{(x_i \cdot z_i)_p, (z_i \cdot z'_j)_p, (z'_j \cdot y_j)_p\} - 2\delta$, noting that $\liminf_{i,j}(z_i \cdot z'_j)_p = \infty$.

(5) For all $x, y \in \partial X$ and all sequences (x'_i) and (y'_j) in X with $x = \lim x'_i$ and $y = \lim y'_j$,

$$(x \cdot y)_p - 2\delta \leq \liminf_{i,j} (x'_i \cdot y'_j)_p \leq (x \cdot y)_p.$$

The second inequality is true by definition, and the first is obtained by taking $\liminf_{i,j}$ in $(x'_i \cdot y'_j)_p \geq \min\{(x'_i \cdot x_i)_p, (x_i \cdot y_i)_p, (y_i \cdot y'_j)_p\} - 2\delta$, where (x_i) and (y_i) are as in remark (3). (Note that $(x'_i \cdot x_i)_p$ and $(y_i \cdot y'_j)_p$ tend to infinity as $i, j \to \infty$.)

(6) If X is proper and geodesic, and if (ξ_n) is a sequence in ∂X (identified with $\partial_s X$ as in (3.13)), then $\xi_n \to \xi$ in the sense of (3.5) if and only if $(\xi_n \cdot \xi) \to \infty$ as $n \to \infty$.

To see this, choose geodesics c_n issuing from p with $c_n(\infty) = \xi_n$ and define $x_n = c_n(n)$. Note that $\xi_n \to \xi$ as $n \to \infty$ if and only if $x_n \to \xi$, which is equivalent to $(x_n \cdot \xi) \to \infty$. As $n \to \infty$, $(\xi_n \cdot x_n)_p \to \infty$, and from (4) we have $(\xi_n \cdot \xi) \geq \min\{(\xi_n \cdot x_n)_p, (x_n \cdot \xi)_p\} - 2\delta$ and $(x_n \cdot \xi) \geq \min\{(\xi_n \cdot x_n)_p, (\xi_n \cdot \xi)_p\} - 2\delta$.

3.18 Exercises

(1) Let X be a (δ)-hyperbolic space and fix $p \in X$. Show that there exists a constant η such that if (x_i) and (y_j) are sequences in X that converge in $\overline{X}$, then $\eta > |(x_i \cdot y_j)_p - (x_{i'} \cdot y_{j'})_p|$ for all i, i', j, j' sufficiently large.

(2) Show that if X is a CAT(-1) space, $p \in X$, and (x_i), (y_j) are sequences in X converging to ξ, ξ', then $(\xi \cdot \xi')_p = \lim_i (x_i \cdot y_i)_p$. Show further that in the case $X = \mathbb{H}^n$ this limit is the distance from p to two of the internal points of the ideal triangle $\Delta(p, \xi, \xi')$, i.e. the points at which the inscribed circle meets the rays $[p, \xi]$ and $[p, \xi']$. (The third internal point of $\Delta(p, \xi, \xi')$ is the foot of the perpendicular from p to $[\xi, \xi']$.)

(3) Fix $\delta > 0$. Show that there is a constant k such that if X is a proper geodesic space that is δ-hyperbolic, then $|d(p, \text{im}(c)) - (\xi \cdot \xi')_p| < k$ for all $p \in X$, all $\xi, \xi' \in \partial X$ and every geodesic line c in X with $c(-\infty) = \xi$ and $c(\infty) = \xi'$.

(4) *(Uniform Structure on $\overline{X}$).* The purpose of this exercise is to indicate that, for rather general reasons, if X is a proper δ-hyperbolic geodesic space, then $\overline{X}$ is a compact metrizable space (cf. chapter 7 of [GhH90]). To see this, recall that if X is a set then a family $\mathcal{B}$ of symmetric subsets $B \subseteq X \times X$ form a base ("are the basic entourages") for a uniform structure on X if each $B \in \mathcal{B}$ contains the diagonal $\Delta \subset X \times X$, for each $B \in \mathcal{B}$ there exists $E \in \mathcal{B}$ such that $(x, y), (y, z) \in E$ implies $(x, z) \in B$, and for all $B_1, B_2 \in \mathcal{B}$ there exists $B_3 \in \mathcal{B}$ such that $B_3 \subseteq B_1 \cap B_2$. The uniform structure is said to be separated if $\bigcap_{\mathcal{B}} B = \Delta$. A separated uniform structure with a countable base gives rise to a metric by a well-known construction (see sections II.1 and IX.4 of [Bour53] or [Jam89]).

Show that if X is a proper δ-hyperbolic geodesic space and $p \in X$, then the sets $B_r = \{(\xi, \xi') \mid (\xi \cdot \xi')_p \geq r\}$, with $r > 0$ rational, form a base for a separated uniform structure on ∂X whose associated topology is that defined in (3.5). Describe a base for a separated uniform structure on $\overline{X}$.

Visual Metrics on ∂X

In this subsection we shall construct explicit metrics on the boundary of proper δ-hyperbolic spaces X. In order to motivate the construction we first consider the example $X = \mathbb{H}^n$.

3.19 Constructing Visual Metrics on $\mathbb{H}^n$. If we fix $p \in \mathbb{H}^n$, then $(\xi, \xi') \mapsto \angle_p(\xi, \xi')$ defines a metric on $\partial\mathbb{H}^n$ making it isometric to $\mathbb{S}^{n-1}$; since $\angle_p$ describes the geometry of $\partial\mathbb{H}^n$ as seen from p, it is called a *visual metric* on $\partial\mathbb{H}^n$. We wish to interpret $\angle_p$ in terms of the Gromov product and the distances between points and geodesics in $\mathbb{H}^n$.

To this end, we apply the basic construction of (I.1.24) to the function ρ on $\partial\mathbb{H}^n \times \partial\mathbb{H}^n$ defined by

$$\rho(\xi, \xi') = e^{-D_p(\xi, \xi')},$$

where $D_p(\xi, \xi')$ is the distance from p to the geodesic line in $\mathbb{H}^n$ with endpoints ξ and ξ'. Associated to ρ we have the (pseudo)metric

$$d_\rho(\xi, \xi') := \inf \sum_{i=0}^{n-1} \rho(\xi_i, \xi_{i+1}),$$

where the infimum is taken over all chains $(\xi = \xi_0, \dots, \xi_n = \xi')$, no bound on n.

According to (I.6.19(3)), $\tan(\angle(\xi, \xi')/4) = e^{-D_p(\xi, \xi')}$. If $a, b \in [0, \pi/4]$ then $\tan(a + b) \geq \tan a + \tan b$. And $\lim_{t \to 0} \frac{1}{t} \tan t = 1$. It follows from (I.3.6) that d_ρ is the length metric associated to $d_p(\xi, \xi') := \frac{1}{4}\angle_p(\xi, \xi')$. Since d_p is itself a length metric, we have $d_\rho = d_p$. Thus the visual metric $\angle_p$ on $\partial\mathbb{H}^n$ can be constructed from the function D_p. Moreover, since $\tan\theta \in [\theta, 2\theta]$ if $\theta \in [0, \pi/4]$, and $\tan \frac{1}{4}\angle_p(\xi, \xi') = e^{-D_p(\xi, \xi')}$, we have

$$\frac{1}{2} e^{-D_p(\xi, \xi')} \leq \angle_p(\xi, \xi') \leq 4 e^{-D_p(\xi, \xi')}$$

for all $\xi, \xi' \in \partial\mathbb{H}^n$.

According to exercise (3.18(3)), there is a universal bound on $|D_p(\xi, \xi') - (\xi \cdot \xi')_p|$. Hence there exist constants k_1, k_2 independent of p such that

$$k_1 e^{-(\xi \cdot \xi')_p} \leq \angle_p(\xi, \xi') \leq k_2 e^{-(\xi \cdot \xi')_p}$$

for all $\xi, \xi' \in \partial\mathbb{H}^n$.

3.20 Definition. Let X be a hyperbolic space with basepoint p. A metric d on ∂X is called a *visual metric* with parameter a if there exist constants $k_1, k_2 > 0$ such that

$$k_1 a^{-(\xi \cdot \xi')_p} \leq d(\xi, \xi') \leq k_2 a^{-(\xi \cdot \xi')_p}$$

for all $\xi, \xi' \in X$.

In [Bou96] Marc Bourdon shows that if X is a proper CAT($-b^2$) space, then for each number $a \in (1, e^b]$ and each $p \in X$, the formula $d_p(\xi, \xi') := a^{-(\xi \cdot \xi')_p}$ defines a metric on ∂X. However one cannot construct visual metrics on the boundary of arbitrary hyperbolic spaces in such a direct manner. To circumvent this difficulty we mimic the construction of (3.19).

The following discussion follows section 7.3 of [GhH90].

Constructing Visual Metrics. Let X be a hyperbolic space with basepoint p. Let $\varepsilon > 0$ and consider the following measure of separation for points in ∂X:

$$\rho_\varepsilon(\xi, \xi') := e^{-\varepsilon(\xi \cdot \xi')_p}.$$

It is clear that $\rho_\varepsilon(\xi, \xi') = \rho_\varepsilon(\xi', \xi)$. From remark 3.17(2) we see that $\rho_\varepsilon(\xi, \xi') = 0$ if and only if $\xi = \xi'$, and from remark 3.17(4) we see that

$$(\#) \qquad \rho_\varepsilon(\xi, \xi') \leq (1 + \varepsilon') \max\{\rho_\varepsilon(\xi, \xi''), \rho_\varepsilon(\xi'', \xi')\}$$

for all $\xi, \xi', \xi'' \in \partial X$, where $\varepsilon' = e^{2\delta\varepsilon} - 1$.

It may be that ρ_ε does not satisfy the triangle inequality, but we can apply the general construction of (I.1.24) with $\rho = \rho_\varepsilon$ to obtain a pseudo-metric; since we wish to retain ε in the notation, we write d_ε rather than d_ρ for this pseudo-metric. Thus on ∂X we consider

$$d_\varepsilon(\xi, \xi') = \inf \sum_{i=1}^{n} \rho_\varepsilon(\xi_{i-1}, \xi_i),$$

where the infimum is taken over all chains ($\xi = \xi_0, \dots, \xi_n = \xi'$), no bound on n.

3.21 Proposition. *Let X be a (δ)-hyperbolic space. Let $\varepsilon > 0$ and let $\varepsilon' = e^{2\delta\varepsilon} - 1$. If $\varepsilon' \leq \sqrt{2} - 1$, then d_ε is a visual metric on ∂X, indeed*

$$(1 - 2\varepsilon')\rho_\varepsilon(\xi, \xi') \leq d_\varepsilon(\xi, \xi') \leq \rho_\varepsilon(\xi, \xi')$$

for all $\xi, \xi' \in \partial X$.

If X is a proper geodesic space, then the topology which d_ε induces on ∂X is the same as the topology described in (3.5).

Proof. The inequality $d_\varepsilon \leq \rho_\varepsilon$ is obvious. The proof of the other inequality between ρ_ε and d_ε is based on the standard technique for constructing a metric from a uniform structure. Following [Bour53] (IX.1.4) and [GhH90], we proceed by induction on n, the size of chains ($\xi_0, \dots, \xi_n$), to prove that

$$(1 - 2\varepsilon')\rho_\varepsilon(\xi_0, \xi_n) \leq \sum_{i=1}^{n} \rho_\varepsilon(\xi_{i-1}, \xi_i).$$

For brevity we write $S(p) = \sum_{i=1}^{p} \rho_\varepsilon(\xi_{i-1}, \xi_i)$. The inequality is obvious if $n = 1$ or $S(n) \geq 1 - 2\varepsilon'$, so we suppose that $n \geq 2$ and $S(n) \leq 1 - 2\varepsilon'$. Let p be the greatest integer such that $S(p) \leq S(n)/2$. (So $S(n) - S(p+1) \leq S(n)/2$.) By our inductive hypothesis, both $\rho_\varepsilon(\xi_0, \xi_p)$ and $\rho_\varepsilon(\xi_{p+1}, \xi_n)$ are no greater than $S(n)/(2 - 4\varepsilon')$. Also $\rho_\varepsilon(\xi_p, \xi_{p+1}) \leq S(n)$. And applying inequality (#) twice we have

$$\rho_\varepsilon(\xi_0, \xi_n) \leq (1 + \varepsilon')^2 \max\{\rho_\varepsilon(\xi_0, \xi_p), \rho_\varepsilon(\xi_p, \xi_{p+1}), \rho_\varepsilon(\xi_{p+1}, \xi_n)\}.$$

Hence $(1 - 2\varepsilon')\rho_\varepsilon(\xi_0, \xi_n) \leq (1 + \varepsilon')^2 S(n) \max\{1 - 2\varepsilon', 1/2\}$. In order to complete the induction, it only remains to note that $(1 + \varepsilon')^2(1 - 2\varepsilon') \leq 1$ for all $\varepsilon' \geq 0$ and $(1 + \varepsilon')^2 \leq 2$ for all positive $\varepsilon' \leq \sqrt{2} - 1$.

The topology associated to d_ε is the same as the topology defined in (3.5), because by definition $\rho_\varepsilon(\xi, \xi_i) \to 0$ as $i \to \infty$ if and only if $(\xi_i \cdot \xi)_p \to \infty$, and by (3.17(6)) this is equivalent to convergence in the sense of (3.5). □

Finer Structure

We shall briefly describe, without giving proofs, the natural quasi-conformal structure on the boundary of a (δ)-hyperbolic space.

Let (X, d) be a metric space and let $k \geq 1$. A *k-ring of d in X* is a pair of concentric balls $(B(x, r), B(x, kr))$. Following Pansu, we say that two (pseudo)metrics d_1, d_2 on X are *quasi-conformally equivalent* if there exist functions $f_1, f_2 : [1, \infty) \to [1, \infty)$ such that for every k-ring (B_1, B_2) of d_1 (resp. d_2) there is an $f_1(k)$-ring of d_2 (resp. an $f_2(k)$-ring of d_1) (B_1', B_2') such that

$$B_1' \subseteq B_1 \subseteq B_2 \subseteq B_2'.$$

A *quasi-conformal structure* on X is a class of metrics which are equivalent in this sense, and a map between metric spaces $\phi : (X, d) \to (X', d')$ is called *quasi-conformal* if d is equivalent to $d_\phi(x, y) := d'(\phi(x), \phi(y))$.

Let X be a (δ)-hyperbolic space. Note that all visual metrics on ∂X are equivalent. Thus X has a *canonical quasi-conformal structure*, namely that associated to the metrics d_ε described in Proposition 3.21. The following is a generalization of a classical result of Margulis (see [GrPan91] section 3.10). It was first stated by Gromov and a detailed proof was given by Bourdon [Bou96b].

3.22 Theorem. *Let X and Y be proper geodesic spaces that are δ-hyperbolic and equip ∂X and ∂Y with their canonical quasi-conformal structures. If $f : Y \to X$ is a quasi-isometric embedding, then the associated topological embedding $\partial Y \to \partial X$ (see 1.9) is a quasi-conformal map.*

3.23 Corollary. *The canonical quasi-conformal structure on ∂X is an invariant of quasi-isometry among hyperbolic spaces that are proper and geodesic.*

3.24 Remark. In [Pau96] Paulin proves that under certain additional hypotheses (which are satisfied if X is the Cayley graph of a finitely generated group for example), the quasi-conformal structure on ∂X uniquely determines the space X up to quasi-isometry.

There are many interesting aspects of the boundary of hyperbolic spaces that we have not discussed at all. In particular we have not said anything about the Hausdorff and conformal dimensions of the boundary, nor the action of Isom(X) on ∂X and the structure of limit sets, nor the construction of measures at infinity and their relationship to rigidity results. We refer the reader to [CDP90] and [GhH90] for basic facts in this direction and [Gro87], [Gro93], [Pan89,90], [Bou95] and [BuM96] for further reading.

Bowditch has given a topological characterization of groups whose Cayley graphs X are δ-hyperbolic by examining the nature of the action of the group on ∂X, and he and Swarup have shown that the structure of local cutpoints in ∂X tell one a great deal about the graph of groups decompositions of the group (see [Bow98a], [Bow98b], [Sw96] and references therein). Also in the case where X is a Cayley graph, Bestvina and Mess [BesM91] have explained how the dimension of ∂X is related to the virtual cohomological dimension of the group. Various other notions of dimension for ∂X are discussed at length in [Gro93].

Chapter III.Γ Non-Positive Curvature and Group Theory

We have already seen that one can say a good deal about the structure of groups which act properly by isometries on CAT(0) spaces, particularly if the action is cocompact. One of the main goals of this chapter is to add further properties to the list of things that we know about such groups. In particular, in Section 1, we shall show that if a group acts properly and cocompactly by isometries on a CAT(0) space, then the group has a solvable word problem and a solvable conjugacy problem. Decision problems also form the focus of much of Section 5, the main purpose of which is to demonstrate that the class of groups which act properly by semi-simple isometries on complete CAT(0) spaces is much larger and more diverse than the class of groups which act properly and cocompactly by isometries on CAT(0) spaces; this diversity can already be seen among the finitely presented subgroups of groups that act cocompactly.

The natural framework in which to address the issues that shall concern us in this chapter is that of geometric and combinatorial group theory. We shall be concerned mostly with the geometric side of the subject, regarding finitely generated groups as geometric objects, as in Chapter I.8. Within this framework, we shall examine the extent to which the geometry of CAT(0) spaces is reflected in the large-scale geometry of the groups which act properly and cocompactly on them by isometries, and the extent to which the basic properties of such groups can be deduced directly from features of the large-scale geometry of their Cayley graphs. Once the salient features have been identified, results concerning such groups of isometries can be extended to larger classes of groups whose Cayley graphs share these features.

This approach leads us to consider what it should mean for a finitely generated group to be negatively curved, or non-positively curved, on the large scale. We saw in the previous chapter that Gromov's notion of δ-hyperbolicity captures the essence of negative curvature in a manner that is invariant under quasi-isometry. In Section 2 of the present chapter we shall discuss the algorithmic properties of groups whose Cayley graphs are δ-hyperbolic (i.e. hyperbolic groups) and in Section 3 we shall present further properties of these groups. Our presentation is phrased in such a way as to emphasize the close parallels with results concerning groups that act properly and cocompactly by isometries on CAT(-1) spaces. The study of hyperbolic groups is an active area of current research and our treatment if far from exhaustive.

In Section 4 we present the basic theory of semihyperbolic groups, following [AloB95]. The theory of finitely generated groups which are, in a suitable sense, non-positively curved is less developed than in the negatively curved (hyperbolic) case. Nevertheless, we shall see that most of the results concerning groups which act properly and cocompactly by isometries on CAT(0) spaces can be recovered in the setting of semihyperbolic groups.

In Section 6 of this chapter we shall examine the circumstances in which the gluing constructions from (II.11) can be used to amalgamate group actions on CAT(0) spaces. In Section 7 we discuss the (non)existence of finite-sheeted covering spaces of compact non-positively curved spaces.

1. Isometries of CAT(0) Spaces

Let us begin by compiling a list of what we already know about groups which act by isometries on CAT(0) spaces.

A Summary of What We Already Know

1.1 Theorem. *If a group Γ acts properly and cocompactly by isometries on a* CAT(0) *space X, then:*

(1) *Γ is finitely presented.*

(2) *Γ has only finitely many conjugacy classes of finite subgroups.*

(3) *Every solvable subgroup of Γ is virtually abelian.*

(4) *Every abelian subgroup of Γ is finitely generated.*

(5) *If Γ is torsion-free, then it is the fundamental group of a compact cell complex whose universal cover is contractible.*

If H is a finitely generated group that acts properly (but not necessarily cocompactly) by semi-simple isometries on X, then:

(i) *Every polycyclic subgroup of H is virtually abelian.*

(ii) *Every finitely generated abelian subgroup of H is quasi-isometrically embedded (with respect to any choice of word metrics).*

(iii) *H does not contain subgroups of the form $\langle a, t \mid t^{-1}a^p t = a^q\rangle$ with $|p| \neq |q|$.*

(iv) *If $A \cong \mathbb{Z}^n$ is central in H then there exists a subgroup of finite index in H that contains A as a direct factor.*

(v) *If X is δ-hyperbolic then every element of infinite order $\gamma \in H$ has finite index in its centralizer.*

The class of groups which act properly and cocompactly by isometries on CAT(0) *spaces is closed under the following operations:*

(a) *direct products,*

(b) *free products with amalgamation and HNN extensions along finite subgroups,*

(c) *free products with amalgamation along virtually cyclic subgroups.*

Proof. (1) is a special case of (I.8.11) and (2) was proved in (II.6.11). Parts (3), (4) and (i) were proved in Chapter II.7, and (5) was proved in (II.5.13).

Part (ii) is a consequence of the Flat Torus Theorem (II.7.1), as we shall now explain. By passing to a subgroup of finite index we can reduce to the case $\mathbb{Z}^n \cong A \hookrightarrow H$. We choose a minimal generating set for A and extend this to a set of generators for H (thus ensuring that $d_H(a, a') \leq d_A(a, a')$ for all $a, a' \in A$). By the Flat Torus Theorem, Min(A) contains an isometrically embedded subspace $F \cong \mathbb{E}^n$ on which A acts as a lattice of translations. Fix a basepoint $x_0 \in F$. By applying the Švarc-Milnor Lemma (I.8.19) to the action of A on F, we obtain a constant $\lambda > 0$ such that $d_X(a.x_0, a'.x_0) \geq \lambda d_A(a, a')\varepsilon$ for all $a, a' \in A$ (in fact one can take $\varepsilon = 0$). And (I.8.20) yields a constant $\mu > 0$ such that $d_H(a, a') \geq \mu\, d_X(a.x_0, a'.x_0)$. Thus $d_H(a, a') \geq \mu\lambda\, d_A(a, a') - \mu\varepsilon \geq \mu\lambda d_H(a, a') - \mu\varepsilon$.

To prove (iii), first note that $\langle a, t \mid t^{-1}a^p t = a^q \rangle$ is an HNN extension of $\mathbb{Z}$, and hence it is torsion free (6.3). In particular a has infinite order. The action of H on X is proper and semisimple, so if such a subgroup were to exist then a would be a hyperbolic isometry. By (II.6.2(2)) we would have $|a^p| = |a^q|$, and (II.6.8(1)) would then imply that $|p| = |q|$, contrary to hypothesis.

Part (iv) is (II.6.12). Statement (v) is a consequence of (iv) and (II.6.8(5)): from (iv) we know that $C_H(\gamma)$ has a subgroup of finite index of the form $K \times \langle \gamma \rangle$; according to (II.6.8(5)), this acts on Min(γ), which splits isometrically as $Y \times \mathbb{R}$; because the action is proper, if K were infinite then Y would have to be unbounded; but X is δ-hyperbolic, so there is a bound on the width of flat strips in X.

Items (b) and (c) were proved in Chapter II.11. To prove (a) one simply notes that, given a proper cocompact action of a group Γ_i by isometries on a space X_i, for $i = 1, 2$, the induced action $(\gamma_1, \gamma_2).(x_1, x_2) := (\gamma_1.x_1, \gamma_2.x_2)$ of $\Gamma_1 \times \Gamma_2$ on $X_1 \times X_2$ is also proper and cocompact. □

Decision Problems for Groups of Isometries

1.2 Dehn's Formulation of the Basic Decision Problems

Combinatorial Group Theory is the study of groups given by generators and defining relations. This method of describing groups emerged at the end of the nineteenth century. Much of the subject revolves around the three basic decision problems that were first articulated by Max Dehn in 1912. Dehn was working on the basic problems of recognition and classification for low-dimensional manifolds (see [Dehn87]). In that setting, the key invariant for many purposes is the fundamental group of the space at hand. When one is presented with the space in a concrete way, the fundamental group often emerges in the form of a presentation. In the course of his attempts to recover knowledge about fundamental groups (and hence manifolds) from such presentations, Dehn came to realise that the problems he was wrestling with were manifestations of fundamental problems in the theory of groups, which he formulated as follows (see [Dehn12b]).

"The general discontinuous group is given by n generators and m relations between them, as defined by Dyck (Math. Ann., 20 and 22). The results of those works, however, relate essentially to finite groups. The general theory of groups defined in this way at present appears very undeveloped in the infinite case. Here there are above all three fundamental problems whose solution is very difficult and which will not be possible without a penetrating study of the subject.

1. The Identity [Word] Problem: *An element of the group is given as a product of generators. One is required to give a method whereby it may be decided in a finite number of steps whether this element is the identity or not.*

2. The Transformation [Conjugacy] Problem: *Any two elements S and T of the group are given. A method is sought for deciding the question whether S and T can be transformed into each other, i.e. whether there is an element U of the group satisfying the relation*

$$S = UTU^{-1}.$$

3. The Isomorphism Problem: *Given two groups, one is to decide whether they are isomorphic or not (and further, whether a given correspondence between the generators of one group and elements of the other is an isomorphism or not).*

These three problems have very different degrees of difficulty. [...] One is already led to them by necessity with work in topology. Each knotted space curve, in order to be completely understood, demands the solution of the three above problems in a special case."

1.3 Remark. Given the nature of our study, we should augment Dehn's last remark with the observation that he was led to these problems while studying surfaces and knots, and all surfaces of positive genus and all knot-complements (viewed as compact manifolds with boundary) support metrics of non-positive curvature.

We should also remark that the word and conjugacy problems are closely related to filling problems in Riemannian geometry (cf. [Gro93]).

Terminology. The explicit nature of the decision problems that we shall discuss here is such that we shall not need to delve deeply into the question of what it means for an algorithm or "decision process" to exist. (See [Mill71] for background.) For completeness, though, we should mention that a group Γ with a finite generating set $\mathcal{A}$ is said to have a solvable word problem if and only if the set of words w such that $w = 1$ in Γ, and the set of words w such that $w \neq 1$ in Γ, are both recursively enumerable subsets of the free monoid on $\mathcal{A} \cup \mathcal{A}^{-1}$.

A finitely presented group has a solvable word problem if and only if its Dehn function, as defined in (I.8A), is a computable function.

The main result of this section is:

1.4 Theorem. *If a group Γ acts properly and cocompactly by isometries on a* CAT(0) *space, then its word and conjugacy problems are solvable.*

We shall treat the word and conjugacy problems separately, obtaining estimates on complexity in each case. In the next section we shall see that the efficiency of the solutions can be sharpened considerably if one assumes that the CAT(0) space on which Γ acts does not contain a flat plane. In Section 5 we shall see that the above theorem does not remain valid if one replaces the hypothesis that the action of Γ is cocompact by the hypothesis that it is semi-simple.

The Word Problem

Let Γ be a group acting properly and cocompactly by isometries on a CAT(0) space X. Fix $x_0 \in X$ and let $D > 0$ be such that X is the union of the balls $\gamma.B(x_0, D/3)$, $\gamma \in \Gamma$.

1.5 Lemma. *If Γ and D are as above, then $\mathcal{A} = \{a \in \Gamma \mid d(a.x_0, x_0) \leq D + 1\}$ generates Γ. And given $\gamma \in \Gamma$, if $d(x_0, \gamma.x_0) \leq 2D + 1$ then $\gamma = a_1 a_2 a_3 a_4$ for some $a_i \in \mathcal{A}$.*

Proof. To say that $\mathcal{A}$ generates is a weak form of (I.8.10). We leave the proof of the assertion concerning elements with $d(x_0, \gamma.x_0) \leq 2D + 1$ as an exercise for the reader. □

Given a set $\mathcal{A}$, to check if two words represent the same element of the free group $F(\mathcal{A})$ one simply looks at the reduced words obtained by deleting all adjacent pairs of letters aa^{-1}; the original words represent the same element of $F(\mathcal{A})$ if and only if the corresponding reduced words are identical. It follows that one can easily decide if an arbitrary word belongs to a given finite subset of $F(\mathcal{A})$. We shall solve the word problem in Γ by showing that in order to decide if a word represents $1 \in \Gamma$, one need only check if the word belongs to a certain finite subset of the free group on the generators of Γ.

We shall always write $|w|$ to denote the number of letters in a word w.

1.6 Proposition. *Let Γ and $\mathcal{A}$ be as in the preceding lemma, and let $\mathcal{R} \subset F(\mathcal{A})$ be the set of reduced words of length at most ten that represent the identity in Γ. A word w in the letters $\mathcal{A}^{\pm 1}$ represents the identity in Γ if and only if in the free group $F(\mathcal{A})$ there is an equality*

$$w = \prod_{i=1}^{N} x_i r_i x_i^{-1},$$

where $N \leq (D + 1)|w|^2$, each $r_i \in \mathcal{R}$, and $|x_i| \leq (D + 1)|w|$. In particular Γ has a solvable word problem.

Proof. To each $\gamma \in \Gamma$ we associate a word σ_γ in the generators $\mathcal{A}$ as follows. Let c_γ be the unique geodesic joining $x_0 = c_\gamma(0)$ to $\gamma.x_0$ in X. For each positive integer $i < d(x_0, \gamma.x_0)$ let $\sigma_\gamma(i) \in \Gamma$ be such that $d(c_\gamma(i), \sigma_\gamma(i).x_0) \leq D/3$, define $\sigma_\gamma(0) = 1$ and $\sigma_\gamma(i) = \gamma$ for $i \geq d(x_0, \gamma.x_0)$. Note that $a_i := \sigma_\gamma(i-1)^{-1}\sigma_\gamma(i) \in \mathcal{A}$. Define σ_γ

to be $a_1 \dots a_n$, where n is the least integer greater than $d(x_0, \gamma.x_0)$. Note that $\gamma = \sigma_\gamma$ in Γ. We choose σ_1 to be the empty word.

We fix $\gamma \in \Gamma$ and $b \in \mathcal{A}$ and compare σ_γ with $\sigma_{\gamma'}$ where $\gamma' = \gamma b$. By appending letters a_i that represent the identity if necessary, we may write $\sigma_\gamma = a_1 \dots a_n$ and $\sigma_{\gamma'} = a'_1 \dots a'_n$, where $n = n(\gamma, \gamma') = \max\{|\sigma_\gamma|, |\sigma_{\gamma'}|\}$. Now $d(\gamma.x_0, \gamma'.x_0) \le D+1$ (because $b \in \mathcal{A}$), so from the convexity of the metric on X we have $d(c_\gamma(i), c_{\gamma'}(i)) < D+1$ for all i. Hence $d(\sigma_\gamma(i).x_0, \sigma_{\gamma'}(i).x_0) < 2D+1$. For each i we choose a word $\alpha(i)$ of length at most 4 that is equal to $\sigma_\gamma(i)^{-1}\sigma_{\gamma'}(i)$. Note that $\alpha(i)^{-1}a_{i+1}\alpha(i+1){a'_{i+1}}^{-1} \in \mathcal{R}$. We choose $\alpha(n)$ to be b (the difference between γ and γ') and $\alpha(0)$ to be the empty word.

If we write $p_\gamma(i)$ for the word $a_1 \dots a_i$ (the i-th prefix of σ_γ), and similarly for γ', and we define $p_{\gamma'}(0)$ to be the empty word, then we have the following equality in the free group on $\mathcal{A}$ (see figure Γ.1):

$$\sigma_\gamma b \sigma_{\gamma'}^{-1} = \prod_{i=0}^{n(\gamma,\gamma')-1} p_{\gamma'}(i)\left[\alpha(i)^{-1}a_{i+1}\alpha(i+1){a'_{i+1}}^{-1}\right]p_{\gamma'}(i)^{-1}. \tag{1.6.1}$$

Each of the words in square brackets belongs to $\mathcal{R}$.

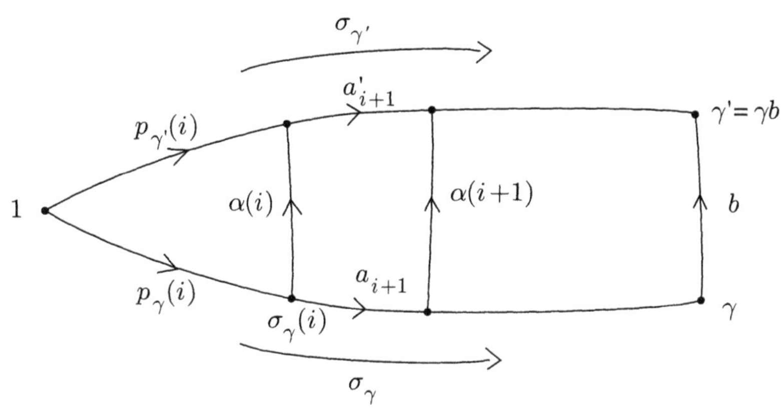

Fig. Γ.1 The Equality in Equation 1.6.1

Finally, we consider the given word $w = b_1 \dots b_m$ that represents the identity in Γ, where each $b_i \in \mathcal{A}^{\pm 1}$ (so $m = |w|$). Let $\gamma_0 = 1$ and let $\gamma_j \in \Gamma$ be the element represented by $b_1 \dots b_j$. Note that $\gamma_m = 1$ and $\sigma_{\gamma_m} = \sigma_{\gamma_0}$ is the empty word. Trivially, in $F(\mathcal{A})$ we have the equality

$$w = \prod_{j=1}^{m} \sigma_{\gamma_{j-1}} b_j \sigma_{\gamma_j}^{-1}. \tag{1.6.2}$$

By replacing each factor on the right hand side of this equality with the right hand side of the previous equation (with $\gamma = \gamma_{j-1}$, $b = b_j$, and $\gamma' = \gamma_j$), we obtain the desired expression for w as a product of conjugates of defining relations. We claim that this product has the desired number of factors.

To see this, consider the image in X of Γ under the map $\gamma \mapsto \gamma.x_0$. Under this map the sequence of elements γ_j form a sequence of m points beginning and ending at x_0. Since each successive pair of elements differ by right multiplication by a single element of $\mathcal{A}$, their images in X are a distance at most $D + 1$ apart. Hence $d(\gamma_j.x_0, x_0) \leq (D + 1)|w|/2$ for $j = 1, \ldots, m$. It follows that each of the integers $n(\gamma_j, \gamma_{j+1})$ in equation (1.6.1) is no greater than $1 + (D + 1)|w|/2$. The final equality for w is a product of $m = |w|$ terms, the jth of which (coming from (1.6.1)) consists of $n(\gamma_j, \gamma_{j+1})$ conjugates of relators. Thus we have less than $(D + 1)|w|^2$ factors in total.

The conjugating elements in our final equality are the prefixes $p_{\gamma_j}(i)$ coming from (1.6.1), and the lengths of these are bounded by the integers $n(\gamma_j, \gamma_{j+1})$. □

1.7 Remarks

(1) The equality displayed in the statement of the proposition shows that w is in the normal closure of $\mathcal{R} \subset F(\mathcal{A})$ and hence $\langle \mathcal{A} \mid \mathcal{R} \rangle$ is a finite presentation for Γ.

(2) Proposition 1.6 states that the Dehn function for the presentation $\mathcal{R} \subset F(\mathcal{A})$, as defined in (I.8A), is bounded above by a quadratic function. It follows that the Dehn function for any finite presentation of Γ is bounded above by a quadratic function (see [Alo90], H.2.5(5)).

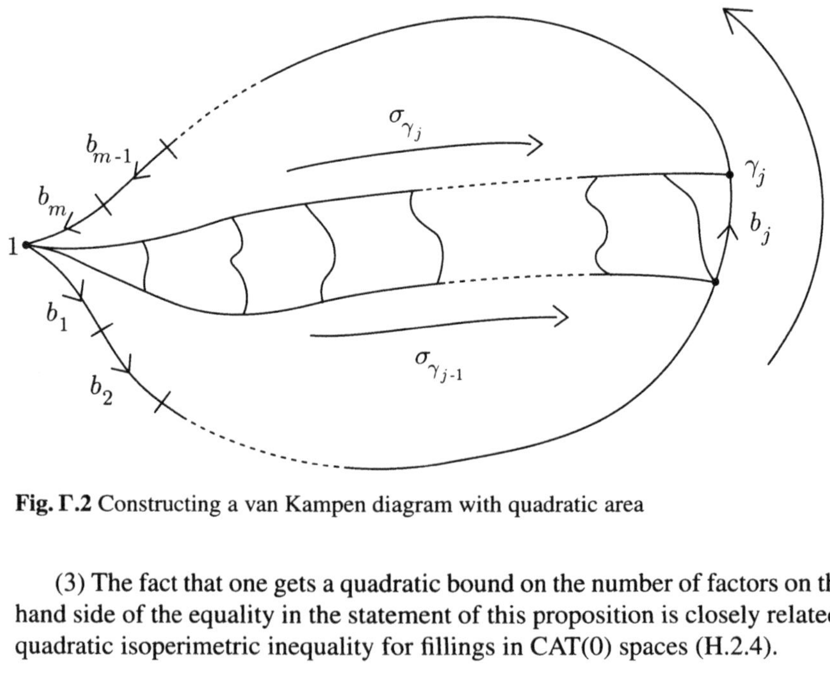

Fig. Γ.2 Constructing a van Kampen diagram with quadratic area

(3) The fact that one gets a quadratic bound on the number of factors on the right hand side of the equality in the statement of this proposition is closely related to the quadratic isoperimetric inequality for fillings in CAT(0) spaces (H.2.4).

Figure (Γ.2) illustrates the scheme by which relations were applied in the proof. By removing faces whose boundaries map to the trivial path in $\mathcal{C}_{\mathcal{A}}(\Gamma)$ one obtains a van Kampen diagram for w.

The Conjugacy Problem

The way in which we shall solve the conjugacy problem for groups that act properly and cocompactly by isometries on CAT(0) spaces is motivated by the following geometric observation.

1.8 Proposition. *Let Y be a compact non-positively curved space. If two closed rectifiable loops $c_0, c_1 : \mathbb{S}^1 \to Y$ are freely homotopic, then there is a homotopy $\mathbb{S}^1 \times [0,1] \to Y$ from c_0 to c_1 through loops c_t such that $l(c_t) \leq \max\{l(c_0), l(c_1)\}$ (where l denotes length).*

Proof. Parameterize both c_0 and c_1 proportional to arc length. By hypothesis, there is a homotopy $H : \mathbb{S}^1 \times [0,1] \to Y$ from c_0 to c_1. For each $\theta \in \mathbb{S}^1$ we replace the path $t \mapsto H(\theta, t)$ by the unique constant-speed local geodesic $p_\theta : [0,1] \to Y$ in the same homotopy class (rel. endpoints). Let $c_t(\theta) = p_\theta(t)$.

We fix an arc $[\theta, \theta']$ in $\mathbb{S}^1$ and in $\widetilde{Y}$, the universal covering of Y, we consider the geodesic rectangle obtained by lifting the concatenation of the four paths $c_0|_{[\theta,\theta']}$, $p_{\theta'}$, $\overline{c}_1|_{[\theta,\theta']}$ and $\overline{p}_\theta$, where the overline denotes reversed orientation. By the convexity of the metric on $\widetilde{Y}$, for sufficiently small $|\theta - \theta'|$ we have $d(c_t(\theta), c_t(\theta')) = d(p_\theta(t), p_{\theta'}(t)) \leq (1-t)\, d(c_0(\theta), c_0(\theta')) + t\, d(c_1(\theta), c_1(\theta'))$ (see II.2.2 and II.4.1). Hence $l(c_t) \leq (1-t)\, l(c_0) + t\, l(c_1)$. □

Free homotopy classes of loops in a connected space correspond to conjugacy classes of elements in the fundamental group of the space. With this correspondence in mind, we define an algebraic property motivated by the above proposition. In this definition, the elements $w_i^{-1} u w_i \in \Gamma$ play the role of the intermediate loops c_t in (1.8).

1.9 Definition (q.m.c.). A group Γ with finite generating set $\mathcal{A}$ is said to have the *quasi-monotone conjugacy property (q.m.c.)* if there is a constant $K > 0$ such that whenever two words $u, v \in F(\mathcal{A})$ are conjugate in Γ, one can find a word $w = a_1 \dots a_n$ with $a_i \in \mathcal{A}^{\pm 1}$, such that $w^{-1} u w = v$ and $d(1, w_i^{-1} u w_i) \leq K \max\{|u|, |v|\}$ for $i = 1, \dots, n$, where $w_i = a_1 \dots a_i$.

In this definition, the existence of K does not depend on the choice of generating set $\mathcal{A}$ but its value does.

1.10 Lemma. *If a group Γ acts properly and cocompactly by isometries on a* CAT(0) *space X, then Γ has the q.m.c. property.*

Proof. Fix $x_0 \in X$. We choose generators $\mathcal{A}$ for Γ as in (1.5) and represent elements $\gamma \in \Gamma$ by words σ_γ as in the proof of (1.6). We claim that there is a constant $K > 0$, depending only on the parameters of the quasi-isometry $\gamma \mapsto \gamma.x_0$, such that if two words u and v are such that $\gamma u \gamma^{-1} = v$ in Γ, then $w := \sigma_\gamma$ satisfies the requirements of (1.9).

To see this one uses the convexity of the metric on X and compares the geodesic quadrilateral in X with sides $[x_0, \gamma.x_0]$, $[x_0, u.x_0]$, $\gamma.[x_0.u.x_0]$ and $[u.x_0, \gamma u.x_0]$ to a quadrilateral Q in the Cayley graph $\mathcal{C}_\mathcal{A}(\Gamma)$. The vertices of Q are $\{1, \gamma, u, \gamma u = v\gamma\}$, two of its sides are labelled σ_γ and the other sides labelled u and v. We leave the (instructive) verification of the details to the reader (cf. 4.9(3)). □

1.11 An Algorithm to Determine Conjugacy. Let Γ be a group with finite generating set $\mathcal{A}$. Suppose that Γ has a solvable word problem and also has the q.m.c. property. Let the constant K be as in (1.9).

For each positive integer n, we consider the set $B(n)$ of words in $F(\mathcal{A})$ that have length at most n. Because Γ has a solvable word problem, given a pair of words $v_1, v_2 \in B(n)$ one can decide if there exists $a \in \mathcal{A}^{\pm 1}$ such that $a^{-1}v_1a = v_2$ in Γ; if such an a exists we write $v_1 \sim v_2$.

Consider the algorithmically constructed finite graph $\mathcal{G}(n)$ with vertex set $B(n)$ that has an edge joining v_1 to v_2 if and only if $v_1 \sim v_2$. The q.m.c. property says precisely that two words u and v are conjugate in Γ if and only if u and v lie in the same path connected component of $\mathcal{G}(n)$, where $n = K\max\{|u|, |v|\}$. Thus we may decide if u and v represent conjugate elements of Γ.

Note that $\mathcal{G}(n)$ has less than $(2|\mathcal{A}|)^n$ vertices, and hence any injective edge-path in $\mathcal{G}(n)$ has length less than $(2|\mathcal{A}|)^n$. Any path of length l joining u to v in $\mathcal{G}(n)$ determines a word of length l which conjugates u to v in Γ, and therefore u is conjugate to v in Γ if and only if there is a word w of length $\le \mu^{\max\{|u|,|v|\}}$ such that $w^{-1}uw = v$ in Γ, where $\mu = (2|\mathcal{A}|)^K$.

1.12 Theorem. *If Γ acts properly and cocompactly by isometries on a* CAT(0) *space X, then for any choice of finite generating set $\mathcal{A}$ there exists a constant $\mu > 0$ such that words $u, v \in F(\mathcal{A})$ represent conjugate elements of Γ if and only if there is a word w of length $\le \mu^{\max\{|u|,|v|\}}$ such that $w^{-1}uw = v$ in Γ. In particular, Γ has a solvable conjugacy problem.*

Proof. Γ has a solvable word problem (1.6) and the q.m.c. property (1.10), so algorithm (1.11) applies. □

Conjugacy of Torsion Elements

It is unknown whether, in general, one can significantly improve the exponential bound on the length of the conjugating element in (1.12). However one can do better for elements of finite order.

1.13 Proposition. *Let Γ be a group which acts properly and cocompactly by isometries on a* CAT(0) *space X. If $\mathcal{A}$ is a finite generating set for Γ, there exists a finite subset Σ of the free group $F(\mathcal{A})$ and a constant K such that a word $u \in F(\mathcal{A})$ represents an element of finite order in Γ if and only if there exists $\sigma \in \Sigma$ and $w \in F(\mathcal{A})$, with $|w| \leq K\,|u|$, such that $w^{-1}uw = \sigma$ in Γ.*

Proof. Choose a ball $B = \overline{B}(x_0, D)$ big enough so that its translates by Γ cover X. By the Švarc-Milnor Lemma (I.8.19), there exist positive constants λ and ε such that

$$\frac{1}{\lambda}d(x_0, \gamma.x_0) - \varepsilon \leq d(1, \gamma) \leq \lambda\, d(x_0, \gamma.x_0) + \varepsilon,$$

for all $\gamma \in \Gamma$, where Γ is equipped with the word metric associated to $\mathcal{A}$.

Let Σ be a subset of $F(\mathcal{A})$ which maps bijectively under $F(\mathcal{A}) \to \Gamma$ to the union of the stabilizers of the points of B. This set is finite because the action is proper.

Suppose that $\gamma \in \Gamma$ is an element of finite order, represented by the word $u \in F(\mathcal{A})$. We know that γ fixes a point of X (II.2.8); choose a fixed point x_1 closest to x_0 and let $g \in \Gamma$ be such that $g^{-1}.x_1 \in B$. Note that $g^{-1}\gamma g$ lies in the image of Σ. It suffices to bound $d(1, g)$ as a linear function of $|u|$.

Writing $d(x_0, x_1) = \rho$, we have:

$$d(g.x_0,\ x_0) \leq d(g.x_0,\ x_1) + d(x_1, x_0) \leq D + \rho,$$

and hence $d(1, g) \leq \lambda(\rho + D) + \varepsilon$. Thus it only remains to bound ρ by a linear function of $|u| \geq d(1, \gamma)$.

Claim. There exist positive constants α and β such that for all torsion elements $\gamma \in \Gamma$, if $d(x_0, Fix(\gamma)) = \rho \geq 1$, then $d(1, \gamma) \geq \alpha\rho - \beta$.

We write $V_1(\gamma)$ to denote the set of points a distance exactly 1 from $Fix(\gamma)$. Consider the (finite) set $S \subset \Sigma$ consisting of elements σ such that $V_1'(\sigma) := V_1(\sigma) \cap \overline{B}(x_0, D + 1)$ is non-empty. Let $L_\sigma = \inf\{d(x, \sigma.x) \mid x \in V_1'(\sigma)\} > 0$, and let $L = \min\{L_\sigma \mid \sigma \in S\}$.

Given a torsion element $\gamma \in \Gamma$ and a point $x \in V_1(\gamma)$, let $y \in Fix(\gamma)$ be the point closest to x, and fix $h \in \Gamma$ such that $h.y \in B$. Let $\sigma = h\gamma h^{-1}$; so $h.x \in V_1'(\sigma)$. We have $d(x, \gamma.x) = d(h.x, h\gamma h^{-1}.(h.x)) = d(h.x, \sigma.(h.x))$. Thus $d(x, \gamma.x) \geq L$.

Now let $z \in X$ be any point such that $\rho = d(z, Fix(\gamma)) \geq 1$, let y be the closest point of $Fix(\gamma)$ and let x be the point a distance 1 from y on the geodesic $[y, z]$. The CAT(0) inequality for the triangle $\Delta(y, z, \gamma.z)$ gives $d(z, \gamma.z) \geq \rho\, d(x, \gamma.x) \geq \rho L$.

To complete the proof of the claim, we cast x_0 in the role of z, recall that $\gamma \mapsto \gamma.x_0$ is a (λ, ε) quasi-isometry, and let $\alpha = L/\lambda$ and $\beta = \varepsilon$. □

1.14 Corollary. *Let Γ be a group with finite generating set $\mathcal{A}$. Suppose that Γ acts properly and cocompactly by isometries on a* CAT(0) *space. Then:*

(1) *There is an algorithm to decide which words $u \in F(\mathcal{A})$ represent elements of finite order in Γ.*

(2) *There is a constant k such that in order to decide if two words u and v in the generators represent torsion elements of Γ that are conjugate, it suffices to check if $w^{-1}uw = v$ in Γ for some word w with $|w| \leq k \max\{|u|, |v|\}$.*

2. Hyperbolic Groups and Their Algorithmic Properties

We now set about the second main task of this chapter: we want to use the tools of geometric group theory to study the structure of groups which act properly and cocompactly by isometries on CAT(0) spaces and, more generally, groups which in a strict metric sense resemble such groups of isometries. We begin with the δ-hyperbolic case. In this section we shall use the results of the previous chapter (in particular results concerning quasi-geodesics) to examine the algorithmic structure of groups whose Cayley graphs are δ-hyperbolic (i.e. hyperbolic groups). We shall describe solutions to the word and conjugacy problems for such groups — these are considerably more efficient than the solutions described in the previous section — and we shall explain a result of Jim Cannon which shows that geodesics in the Cayley graphs of hyperbolic groups can be described in a remarkably simple algorithmic manner (2.18).

Hyperbolic Groups

We saw in (H.1.10) that being hyperbolic is an invariant of quasi-isometry among geodesic spaces. Hence the following definition does not depend on the choice of generators (although the value of δ does).

2.1 Definition of a Hyperbolic Group. A finitely generated group is *hyperbolic* (in the sense of Gromov) if its Cayley graph is a δ-hyperbolic metric space for some $\delta > 0$.

2.2 Proposition. *Every hyperbolic group is finitely presented.*

Proof. This is a weak form of (2.6). We give an alternative (more direct) proof, the scheme of which is portrayed in figures (Γ.1) and (Γ.2).

Fix a finite generating set $\mathcal{A}$ with respect to which Γ is δ-hyperbolic. Given a word $w \in F(\mathcal{A})$ which represents $1 \in \Gamma$, consider the loop in the Cayley graph $\mathcal{C}_{\mathcal{A}}(\Gamma)$ that begins at 1 and is labelled w. Let $w(i)$ be the ith vertex which this loop visits (i.e. the image in Γ of the ith prefix of w). Let σ_i be a geodesic from 1 to $w(i)$. Suppose that $w(i+1) = w(i)b$. Note that $d(\sigma_i(t), \sigma_{i+1}(t)) \leq 2(\delta + 1)$ for all integers

$i \leq |w|$ and all $t > 0$ for which $\sigma_i(t)$ and $\sigma_{i+1}(t)$ are defined (H.1.15). As in equation (1.6.1), one can express $\sigma_i b \sigma_{i+1}^{-1}$ in $F(\mathcal{A})$ as a product of conjugates of relations of length at most $(4\delta + 6)$. And from equation (1.6.2) it follows that w is also equal in $F(\mathcal{A})$ to a product of conjugates of relations of length at most $(4\delta + 6)$. □

Since much of the essence of a CAT(−1) space is encoded in the geometry of its geodesics, and since we know from the previous chapter that the geometry of quasi-geodesics in hyperbolic spaces mimics the geometry of geodesics rather closely, a great deal of the geometry of CAT(−1) spaces ought to be transmitted to the (quasi-isometric) groups which act properly and cocompactly on them by isometries. With this in mind, when attempting to understand hyperbolic groups in general, it is often useful to regard them as coarse versions of CAT(−1) spaces. This viewpoint will be the dominant one in Section 3.

An alternative perspective, which can be useful in motivating proofs, e.g. (2.8), is to regard hyperbolic groups as fattened-up versions of free groups. After all, a geodesic space is 0-hyperbolic if and only if it is a metric tree, and if the Cayley graph of a group is a tree then the group is free.

2.3 Remarks

(1) In a certain statistical sense (see [Gro87], [Cham93] and [Ol92]), almost all finitely presented groups are hyperbolic.

(2) At the time of writing, it is unknown whether every hyperbolic group Γ acts properly and cocompactly by isometries on some CAT(0) (or even CAT(−1)) space. In the torsion-free case, this would mean that Γ would be the fundamental group of a compact non-positively curved space. In general Γ will not be the fundamental group of a compact non-positively curved n-manifold with empty or locally convex boundary, because such a fundamental group has cohomological dimension[53] n (if the boundary is empty) or $n - 1$, and there are hyperbolic groups of cohomological dimension two, for example, that are not the fundamental group of any manifold of dimension two or three — the construction of (II.5.45) gives many such groups.

Dehn's Algorithm

Dehn's algorithm is perhaps the most direct approach that one can hope for whereby the information in a finite presentation is used directly to solve the word problem in the group presented. It is the algorithm that Dehn used to solve the word problem in Fuchsian groups [Dehn12a].

2.4 Dehn's Algorithm for Solving the Word Problem. Given a finite set of generators $\mathcal{A}$ for a group Γ, one would have a particularly efficient algorithm for solving the word problem if one could construct a finite list of words $u_1, v_1, u_2, v_2, \ldots, u_n, v_n$,

[53] We refer to [Bro82] for basic facts concerning the cohomology of groups.

with $u_i = v_i$ in Γ, whose lengths satisfy $|v_i| < |u_i|$ and which have the property that if a word w represents the identity in Γ then at least one of the u_i is a subword of w.

If such a list of words exists, then given an arbitrary word w one looks for a subword of the form u_i; if there is no such subword, one stops and declares that w does not represent the identity; if u_i occurs as a subword then one replaces it with v_i and repeats the search for subwords of the (shorter) word obtained from w — this new word represents the same element of the group as w. After at most $|w|$ steps one will have either reduced to the empty word (in which case $w = 1$ in Γ) or else verified that w does not represent the identity.

2.5 Definition. A finite presentation $\langle \mathcal{A} \mid \mathcal{R} \rangle$ of a group Γ is called a *Dehn presentation* if $\mathcal{R} = \{u_1 v_1^{-1}, \dots, u_n v_n^{-1}\}$, where the words $u_1, v_1, \dots, u_n, v_n$ satisfy the conditions of Dehn's algorithm.

2.6 Theorem. *A group is hyperbolic if and only if it admits a (finite) Dehn presentation.*

As we noted above, Dehn proved that Fuchsian groups admit Dehn presentations. Jim Cannon extended Dehn's theorem to include the fundamental groups of all closed negatively curved manifolds [Ca84]. Theorem 2.6 is due to Mikhael Gromov [Gro87]. (Cannon gave a proof in [Ca91] and alternative proofs can be found elsewhere, e.g. [Sho91].)

Proof. Suppose that $\langle \mathcal{A} \mid \mathcal{R} \rangle$ is a Dehn presentation for Γ and let ρ be the length of the longest word in $\mathcal{R}$. Consider an edge-loop c of length n in the Cayley graph $\mathcal{C}_{\mathcal{A}}(\Gamma)$. This is labelled by a word w in the generators and their inverses, and this word represents the identity in Γ. The Dehn algorithm of the presentation yields a subpath (corresponding to a subword u_i of w) which is not geodesic — there is a shorter path with the same endpoints that is labelled v_i.

Let c' be the edge-loop obtained by replacing the subpath of c labelled u_i by the path labelled v_i. Given a standard ρ-filling $D^2 \to \mathcal{C}_{\mathcal{A}}(\Gamma)$ of c' (terminology of H.2.7), one can obtain a standard ρ-filling of c by adding an extra polyhedral face to the filling in the obvious way (figure Γ.3) and adding extra edges to divide this face into $\leq \rho$ triangles. By induction on $n = l(c)$, we may suppose that there is a standard ρ-filling whose ρ-area is $\leq \rho l(c')$. Hence $\text{Area}_\rho(c) \leq \rho l(c)$. This implies that $\mathcal{C}_{\mathcal{A}}(\Gamma)$ is hyperbolic (H.2.9).

Conversely, suppose that the Cayley graph of Γ with respect to some finite generating set $\mathcal{A}$ is δ-hyperbolic. Fix an integer $k > 8\delta$. Every edge-loop in $\mathcal{C}_{\mathcal{A}}(\Gamma)$ contains a subpath p of length at most k that has its endpoints at vertices and is not a geodesic (see H.2.6). If a word in the generators $\mathcal{A}^{\pm 1}$ represents $1 \in \Gamma$ then it is the label on an edge-loop in $\mathcal{C}_{\mathcal{A}}(\Gamma)$, and the non-geodesic subpath p described in the preceding sentence is labelled by a subword u that is equal in Γ to a shorter word v (the label on a geodesic in $\mathcal{C}_{\mathcal{A}}(\Gamma)$ with the same endpoints as p). Thus we obtain a Dehn presentation $\langle \mathcal{A} \mid \mathcal{R} \rangle$ for Γ by defining $\mathcal{R}$ to be the set of words $u_i v_i^{-1}$, where

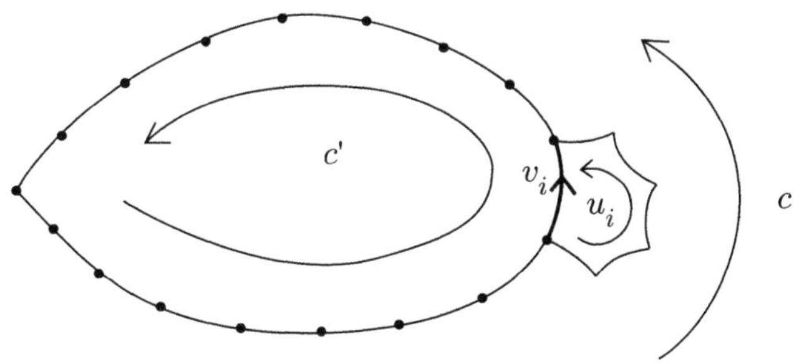

Fig. Γ.3 Applying a relation in Dehn's algorithm

u_i varies over all words of length at most k in the generators and their inverses and v_i is a word of minimal length that is equal to u_i in Γ. □

2.7 Remarks

(1) As a consequence of the above theorem, we see that if a group admits a Dehn algorithm with respect to one set of generators then it admits such an algorithm with respect to every set of generators. (This is not obvious *a priori*.)

(2) It follows from the above theorem and (H.2.7,9 and 11) that a group admits a finite Dehn presentation if and only if its Dehn function (as defined in I.8A) is linear (equivalently, sub-quadratic).

(3) The reader might like to prove directly that $\mathbb{Z}^2$ does not have a (potentially obscure) Dehn presentation.

The Conjugacy Problem

2.8 Theorem. *Every hyperbolic group has a solvable conjugacy problem.*

Hyperbolic groups have the q.m.c. property (1.9), hence they have a solvable conjugacy problem. Indeed it is easy to check that if two words u and v in the generators of a δ-hyperbolic group Γ are such that $\gamma u \gamma^{-1} = v$ in Γ, and if w is a geodesic word representing γ, then w satisfies the requirements of (1.9) (with respect to a constant K that depends only on δ).

The bound on the length of the minimal conjugating word that one obtains from the algorithm based on the q.m.c. property is an exponential function of $|u|$ and $|v|$ (see (1.12)). We shall describe a more efficient algorithm that gives a linear bound. A still more efficient algorithm was recently discovered by David Epstein. In order to motivate our algorithm, we reflect on our earlier remark that it is often worthwhile to regard hyperbolic groups as fattened-up free groups.

In a free group one solves the conjugacy problem in the following manner. By definition, a word $w = a_0 \dots a_n$ in the free group on $\mathcal{A}$ is *cyclically reduced* if

$a_i^{-1} \neq a_{i+1}$ for $1 = 0, \dots, n-1$ and $a_0^{-1} \neq a_n$. Given a word in the letters $\mathcal{A}$ and their inverses, one can cyclically reduce it by repeatedly performing the following operations: delete any subword of the form $a_i^{-1}a_i$, and delete the first and last letters of the word if $u = au'a^{-1}$. The cyclically reduced word that one obtains in this way is unique up to cyclic permutation of its letters. (For example, $bab^{-1}babb^{-1}$ can be reduced to baa or aab.) Given two words u and v, in order to check if they define conjugate elements of $F(\mathcal{A})$, one cyclically reduces both words and then looks to see if one of the resulting words is a cyclic permutation of the other.

Motivated by the example of free groups, given a group Γ with a fixed finite generating set, we say that a word w in the generators and their inverses is *fully reduced* if w and all of its cyclic permutations label geodesics[54] will henceforth in the Cayley graph of Γ.

2.9 Lemma. *Let Γ be a group that is δ-hyperbolic with respect to the finite generating set $\mathcal{A}$. If two fully reduced words $u, v \in F(\mathcal{A})$ represent conjugate elements of Γ, then*

(1) $\max\{|u|, |v|\} \leq 8\delta + 1$, *or else*

(2) *there exist cyclic permutations u' and v' of u and v and a word $w \in F(\mathcal{A})$ of length at most $2\delta + 1$ such that $wu'w^{-1} = v'$ in Γ.*

Proof. Let w be a geodesic word in $F(\mathcal{A})$ such that $wuw^{-1} = v$. Consider a geodesic quadrilateral Q in the Cayley graph of Γ whose sides (read in order from a vertex) trace out edge paths labelled w, u, w^{-1} and v^{-1} in that order. We shall refer to the sides of Q that are labelled $w^{\pm 1}$ as the vertical sides and we shall refer to the other sides as the top and bottom.

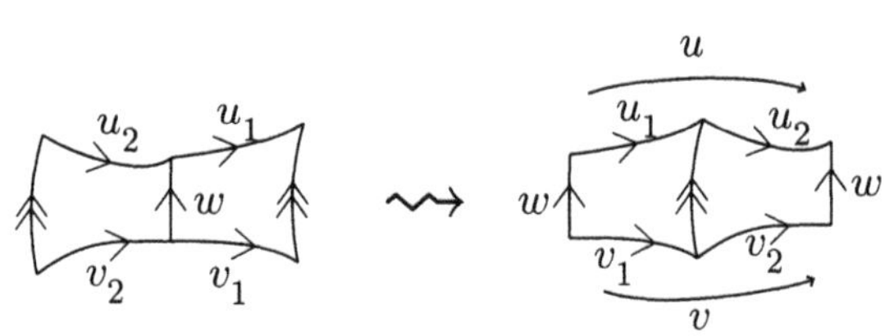

Fig. Γ.4 Arranging the conjugacy diagram

By replacing u and v with suitable cyclic permutations if necessary (see fig. Γ.4), we may assume that each vertex on the top side of Q is a distance at least $|w|$ from each vertex on the bottom. Consider the midpoint p of the top side. This is a distance at most 2δ from a point on one of the other three sides. If it were within 2δ of a point

[54] Such words are often called *geodesic words*. Likewise, it is common to speak of a word as being a k-local geodesic, meaning that the path which it labels in the Cayley graph is a k-local geodesic.

p' on the bottom, then the vertices closest to p and p' would be a distance at most $2\delta + 1$ apart; if $|w| > 2\delta + 1$, then this cannot happen.

Suppose now that p is within 2δ of a point q on one of the vertical sides. Let x and y be, respectively, the top and bottom vertices of the side containing q. From the inequalities $|w| - (1/2) \leq d(p, y) \leq 2\delta + d(q, y)$ and $d(q, y) = |w| - d(x, q)$, we get $d(x, q) \leq 2\delta + (1/2)$, and therefore $d(p, x) \leq d(p, q) + d(x, q) \leq 4\delta + (1/2)$. And $d(p, x) = |u|/2$.

An entirely similar argument shows that if $|w| > 2\delta + 1$, then the bottom side of Q also has length at most $8\delta + 1$. □

If applied naively, the preceding lemma would yield a solution to the conjugacy problem involving an unacceptably large number of verifications to see if the words considered were fully reduced. To circumvent this difficulty, instead of using (2.9) directly we use (2.11), whose proof is essentially the same as that of (2.9) once one has observed the following consequence of (H.1.7) and (H.1.13).

2.10 Lemma. *Let X be a δ-hyperbolic geodesic space. There is a constant C, depending only on δ, such that if the sides of a quadrilateral $Q \subset X$ are all $(8\delta + 1)$-local geodesics, then every side of Q lies in the C-neighbourhood of the union of the other three sides.*

2.11 Lemma. *Let Γ be a group that is δ-hyperbolic with respect to the finite generating set $\mathcal{A}$. There is a positive constant K, depending only on δ, such that if $u, v \in F(\mathcal{A})$ represent conjugate elements of Γ, and if u, v and all of their cyclic permutations are $(8\delta + 1)$-local geodesics, then*

(1) $\max\{|u|, |v|\} \leq K$, *or else*

(2) *there exists a word $w \in F(\mathcal{A})$ of length at most K such that $w^{-1}u'w = v'$ in Γ, where u' and v' are cyclic permutations of u and v.*

2.12 An Algorithm to Determine Conjugacy in Hyperbolic Groups. Let Γ be a group that is δ-hyperbolic with respect to the finite generating set $\mathcal{A}$. Given two words u and v over the alphabet $\mathcal{A}^{\pm 1}$, one looks in u, v and their cyclic permutations to find subwords of length at most $8\delta + 1$ that are not geodesic. If one finds such a subword, one replaces it with a geodesic word representing the same group element. One continues in this manner until u and v have been replaced by (conjugate) words u' and v' all of whose cyclic permutations are $(8\delta + 1)$-local geodesics. (Working with cyclic words, this requires the application of less than $|u| + |v|$ relations from a Dehn presentation of Γ.) Lemma 2.11 provides a *finite* set of words Σ such that u is conjugate to v in Γ if and only if $w^{-1}u'w = v'$ in Γ for some $w \in \Sigma$. (One can take Σ to be the set of words of length at most K together with a choice of one conjugating element for each pair of conjugate elements u_0, v_0 with $\max\{|u_0|, |v_0|\} \leq K$.) Using Dehn's algorithm, one can decide whether any of the putative relations $w^{-1}u'w = v'$ is valid in Γ. □

2.13 Remark (Annular Diagrams). Let $\mathcal{P}$ be a finite presentation of a group Γ, and let K be the corresponding 2-complex (as described in I.8A). Showing that a word w in the generators represents $1 \in \Gamma$ is equivalent to showing that the edge-loop in $K^{(1)}$ labelled w is homotopic to a constant map. We explained in (I.8A) how one can construct a planar van Kampen diagram that portrays this homotopy, and the number of 2-cells in a minimal such diagram is the minimal number of defining relations that one must apply to show that $w = 1$ in Γ.

Showing that two words u and v in the generators represent conjugate elements of Γ is equivalent to showing that the corresponding edge-loops in $K^{(1)}$ are freely homotopic, in other words there is a continuous map from an annulus into K that sends the boundary curves to these edge-loops. As in the case of a disc, one can use cellular approximation and combinatorial arguments to deduce the existence of *annular van Kampen diagrams*.

Such a diagram is a finite, planar, combinatorial 2-complex, homotopy equivalent to an annulus, that portrays the homotopy between the curves representing u and v. The 1-cells of the complex are labelled by generators, the word labelling the boundary cycle of each 2-cell is one of the defining relations, and the words labelling the two boundary cycles of the diagram (suitably oriented) are u and v. The number of 2-cells in a minimal such diagram is the minimal number N of factors among all free equalities of the form

$$u^{-1}(x_0 v x_0^{-1}) = \prod_{i=1}^{N} x_i r_i x_i^{-1},$$

where the x_i are any words and each r_i is one of the defining relations (or its inverse).

The algorithm for solving the conjugacy problem in hyperbolic groups that is described above yields the following analogue of the linear isoperimetric inequality for the word problem. We leave the proof as an exercise for the interested reader.

2.14 Proposition. *If Γ is a hyperbolic group with finite presentation $\mathcal{P} = \langle \mathcal{A} \mid \mathcal{R} \rangle$, then there exists a constant M such that if two words u, v in the letters $\mathcal{A}^{\pm 1}$ define conjugate elements of Γ, then one can construct an annular van Kampen diagram over $\mathcal{P}$, with boundary labels u and v, that has at most $M \max\{|u|, |v|\}$ 2-cells.*

2.15 The Isomorphism Problem. Zlil Sela [Sel95] has shown that there is an algorithm that decides isomorphism among torsion-free hyperbolic groups. More precisely, there exists an algorithm which takes as input two finite group presentations of torsion-free hyperbolic groups, and which (after a finite amount of time) will stop and answer *yes* or *no* according to whether or not the groups being presented are isomorphic. One requires no knowledge of δ: the simple fact that the groups being presented are hyperbolic and torsion-free is enough to ensure that the algorithm will terminate.

The proof of this result is beyond the scope of the techniques that we have described.

Cone Types and Growth

The purpose of this subsection is to show that there is a simple algorithmic procedure for recognizing geodesics in hyperbolic groups. More precisely, given a finite generating set $\mathcal{A}$ for a hyperbolic group Γ, using the notion of *cone type* one can construct a finite state automaton which accepts precisely those words in the free monoid on $\mathcal{A}^{\pm 1}$ that label geodesics in the Cayley graph $\mathcal{C}_{\mathcal{A}}(\Gamma)$. It follows that $\Sigma_{n=0}^{\infty}\beta_{\mathcal{A}}(n)t^n$ is a rational function of t, where $\beta_{\mathcal{A}}(n)$ is the number of vertices in the ball of radius n about $1 \in \mathcal{C}_{\mathcal{A}}(\Gamma)$ (see 2.21). This important discovery is due to Jim Cannon [Ca84]. In [Gro87] Gromov discusses related matters under the heading "Markov properties" (see also chapter 9 of [GhH90]).

Cannon's insights concerning the algorithmic structure of groups of isometries of real hyperbolic space were the starting point for the theory of automatic groups developed in the book by Epstein et al. [Ep+92].

2.16 Definition (Cone Types)**.** Let Γ be a group with finite generating set $\mathcal{A}$ and corresponding word metric d. The *cone type* of an element $\gamma \in \Gamma$ is the set of words $v \in F(\mathcal{A})$ such that $d(1, \gamma v) = d(1, \gamma) + |v|$. In other words, if γ is represented by a geodesic word u, then the cone type of γ is the set of words v such that uv is also a geodesic.

2.17 Example. A free group of rank m has $(2m + 1)$ cone types with respect to any set of free generators, and so does a free abelian group of rank m.

2.18 Theorem. *If a group Γ is hyperbolic, then it has only finitely many cone types (with respect to any finite generating set).*

Proof. We follow the proof of Cannon [Ca84] (see also [Ep+92, p.70]). The idea of the proof is to show that in order to determine the cone type of $\gamma \in \Gamma$, we only need to know which vertices near γ in the Cayley graph are closer to $1 \in \Gamma$ than γ is. Thus, given $k > 0$, we consider the *k-tail* of $\gamma \in \Gamma$, which is the set of elements $h \in \Gamma$ such that $d(1, \gamma h) < d(1, \gamma)$ and $d(1, h) \leq k$.

Let $\mathcal{C}$ be the Cayley graph of Γ with respect to a fixed finite generating set $\mathcal{A}$; suppose that it is δ-hyperbolic. Let $k = 2\delta + 3$. We claim that the k-tail of each element $\gamma \in \Gamma$ determines which words v belong to the cone type of γ. We shall prove this by induction on the length of v. (The first few steps of the induction are trivial.) We fix an arbitrary $\gamma' \in \Gamma$ that has the same k-tail as γ and choose a geodesic word u representing γ'. In order to complete the induction we must show that if a word v is in the cone type of both γ and γ', and if $a \in \mathcal{A}$ is such that va is in the cone type of γ, then va is also in the cone type of γ'.

If va were not in the cone type of γ', then there would exist a geodesic word $w \in F(\mathcal{A})$ of length less than $d(1, \gamma') + |v| + 1$ such that $w = \gamma' va$. We write w as $w_1 w_2$, where $|w_1| = d(1, \gamma') - 1$ and $|w_2| \leq |v| + 1$.

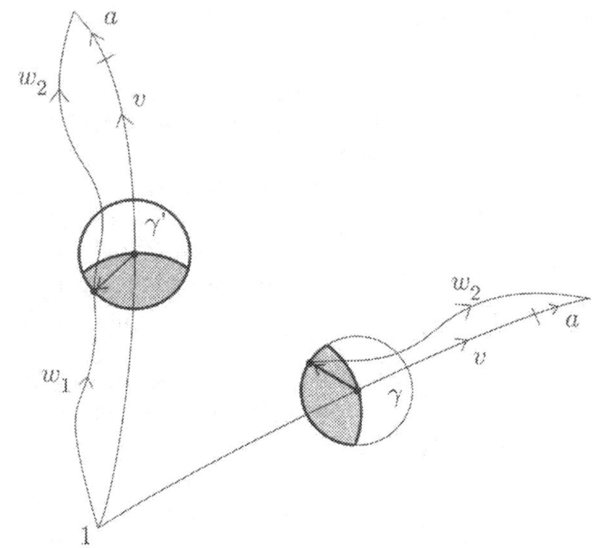

Fig. Γ.5 Cone types are determined by k-tails

The edge-paths determined by w and uv are geodesics in $\mathcal{C}$ that issue from the identity and end a distance one apart. As such, these paths stay uniformly $2(\delta + 1)$-close (see (H.1.15)). In particular, since $|w_1| = d(1, \gamma') - 1$ we have $d(w_1, \gamma') \leq 2\delta+3$. Thus $\gamma'^{-1}w_1$ lies in the k-tail of γ'. But then, by hypothesis, $\gamma'^{-1}w_1$ must lie in the k-tail of γ, which leads to the following contradiction: concatenate any geodesic from 1 to $\gamma\gamma'^{-1}w_1$ with the edge-path from $\gamma\gamma'^{-1}w_1$ that is labelled w_2; the result is a path from 1 to γva that has length strictly less than $d(1, \gamma)+|w_2| \leq d(1, \gamma)+|v|+1$, contradicting the assumption that va is in the cone type of γ. □

2.19 The Geodesic Automaton. Let Γ be a hyperbolic group with finite generating set $\mathcal{A}$ and consider the finite graph whose vertices are the cone types of Γ and which has a directed edge connecting the cone type of $\gamma \in \Gamma$ to the cone type of γa if and only if $a \in \mathcal{A} \cup \mathcal{A}^{-1}$ and a belongs to the cone type of γ.

Consider the set of edge-paths in this graph that begin at the cone type of the identity and follow only positively directed edges; such paths have a natural labelling by words in the letters $\mathcal{A}^{\pm 1}$. It follows immediately from the definition of cone type that the set of words that occur as such labels is precisely the set of words that label the geodesic edge-paths in $\mathcal{C}_{\mathcal{A}}(\Gamma)$.

The graph described above is an example of a finite state automaton. A *finite state automaton* over an alphabet $\mathcal{B}$ is a finite graph whose edges are directed and labelled by elements of $\mathcal{B}$; the vertices of the graph are divided into two sets "accept" and "reject", and there is a distinguished vertex s_0, called the initial vertex. The set of words which occur as labels on the directed edge paths that begin at s_0 and end at an accept vertex is called the accepted language of the automaton. In (2.19) all of the vertices are accept vertices and the initial state is the cone type of $1 \in \Gamma$.

A subset of the free monoid on $\mathcal{B}$ is called a *regular language* if it is the accepted language of some finite state automaton over $\mathcal{B}$. We refer the reader to [Ep+92] for a detailed account of the role which regular languages play in geometry and group theory.

2.20 Corollary. *Let Γ be a hyperbolic group with finite generating set $\mathcal{A}$.*

(1) *The set of words in the free monoid $(\mathcal{A} \cup \mathcal{A}^{-1})^*$ that label geodesics in the Cayley graph $\mathcal{C}_{\mathcal{A}}(\Gamma)$ is a regular language.*

(2) *Γ is biautomatic.*

Proof. A group with finite generating set $\mathcal{A}$ is biautomatic if and only if there is a regular language $\mathcal{L}$ in the free monoid $(\mathcal{A} \cup \mathcal{A}^{-1})^*$ that satisfies the following two properties. First, every $\gamma \in \Gamma$ must be represented by at least one member of the language, i.e. the restriction to $\mathcal{L}$ of the natural map $\mu : (\mathcal{A} \cup \mathcal{A}^{-1})^* \to \Gamma$ must be surjective. Secondly, one requires the language to satisfy the "fellow-traveler property": there must exist a constant $K > 0$ such that if $u, v \in \mathcal{L}$ are such that $d(a.\mu(u), \mu(v)) \leq 1$, where $a \in \mathcal{A} \cup \{1\}$, then the edge-path in $\mathcal{C}_{\mathcal{A}}(\Gamma)$ that begins at the identity and is labelled v must remain uniformly K-close to the path that begins at the vertex a and is labelled u.

We have just seen (2.19) that the language of geodesics in any hyperbolic group is a regular language, and for the fellow-traveler property it suffices to take $K = 2(2\delta + 1)$ (cf. H.1.15). $\square$

For an alternative proof of (2.20), due to Bill Thurston, see [BGSS91].

In Chapter I.8 we discussed the growth function of a group Γ with respect to a finite generating set: $\beta_{\mathcal{A}}(n)$ is the number of elements $\gamma \in \Gamma$ such that $d(1, \gamma) \leq n$, where d is the word metric associated to $\mathcal{A}$. We now also wish to consider the function $\sigma_{\mathcal{A}}(n) := \{\gamma \in \Gamma \mid d(1, \gamma) = n\}$. It is convenient to encode the sequences $(\beta_{\mathcal{A}}(n))$ and $(\sigma_{\mathcal{A}}(n))$ as the coefficients of formal power series:

$$f_{\mathcal{A}}(t) = \sum_{n=0}^{\infty} \beta_{\mathcal{A}}(n) t^n \quad \text{and} \quad \zeta_{\mathcal{A}}(t) = \sum_{n=0}^{\infty} \sigma_{\mathcal{A}}(n) t^n.$$

Note that, as formal power series, $f_{\mathcal{A}}(t) = \zeta_{\mathcal{A}}(t)/(1-t)$.

Γ is said to have *rational growth* with respect to $\mathcal{A}$ if $\zeta_{\mathcal{A}}(t)$ is a rational function, i.e. it is the power series expansion of $p(t)/q(t)$, where $p(t)$ and $q(t)$ are polynomials with integer coefficients.

2.21 Theorem. *If Γ is hyperbolic then the growth of Γ with respect to any finite generating set is rational.*

Proof. Let $\mathcal{A}$ be a finite generating set for Γ. We choose a linear ordering on the alphabet $\mathcal{A} \cup \mathcal{A}^{-1}$ and impose the associated lex-least ordering $\prec$ on finite words over $\mathcal{A} \cup \mathcal{A}^{-1}$, that is $w \prec v$ if and only if $|w| < |v|$ or $|w| = |v|$ and w precedes v in the dictionary.

Let $\mathcal{L}$ be the set of words in the free monoid on $\mathcal{A} \cup \mathcal{A}^{-1}$ that occur as labels on geodesic edge-paths in $\mathcal{C}_{\mathcal{A}}(\Gamma)$ (i.e. geodesic words). We have seen that $\mathcal{L}$ is a regular language. General considerations concerning regular languages imply that $\mathcal{L}' = \{w \in \mathcal{L} \mid \forall v \in \mathcal{L},\ w =_\Gamma v \text{ implies } w \preceq v\}$ is also a regular language (see page 57 of [Ep+92]). Note that the natural map $(\mathcal{A} \cup \mathcal{A}^{-1})^* \to \Gamma$ restricts to a bijection from the set of words of length n in $\mathcal{L}'$ to the set of elements a distance n from the identity in Γ.

Since $\mathcal{L}'$ is regular, it is the accepted language of a connected finite state automaton over $\mathcal{A} \cup \mathcal{A}^{-1}$. Let v_{i_0} be the initial vertex of the automaton and note that since every prefix of a word in $\mathcal{L}'$ is again in $\mathcal{L}'$, every vertex of the automaton is an accept vertex. We consider the transition matrix M of this automaton. The rows and columns of M are indexed by the vertices v_i of the automaton and the (i, j)-entry, which we denote $M(i, j)$, is the number of directed edges from v_i to v_j.

The number of words of length n in $\mathcal{L}'$ is the number of distinct edge-paths of length n that begin at the initial vertex v_{i_0}, and this is the sum of the entries in the i_0th row of M^n. Thus

$$\zeta_{\mathcal{A}}(t) = \sum_{n=0}^{\infty} \sum_j M^n(i_0, j) t^n.$$

Let $c_0 + c_1 t + \cdots + c_k t^k$ be the minimum polynomial of M over $\mathbb{Z}$ and let $q(t) = c_k + c_{k-1} t + \cdots + c_0 t^k$. For all integers $n \geq 0$, we have $c_0 M^n + c_1 M^{n+1} + \cdots + c_k M^{n+k} = 0$. It follows that $q(t)\zeta_{\mathcal{A}}(t)$ is a polynomial of degree at most $k - 1$, because for $m \geq k$ the coefficient of t^m is

$$\sum_j c_0 M^{m-k}(i_0, j) + \cdots + c_k M^m(i_0, j),$$

which equals zero. □

We close this section by noting a further consequence of (2.18).

2.22 Proposition. *If a hyperbolic group is infinite then it contains an element of infinite order. (More generally, this is true of any group Γ with finitely many cone types).*

Proof. Because Γ is infinite, there is a geodesic edge-path in its Cayley graph that begins at the identity and has length greater than the number of cone types. Let w be the word labelling such a geodesic and decompose w as $u_1 u_2 u_3$, where u_2 is the label on a path that connects two vertices of the Cayley graph that have the same cone type. By definition, $u_2 u_3$ lies in the cone type of the first vertex, which is the same as the cone type of the second vertex. Hence $u_2^2 u_3$ lies in the cone type of the first vertex.

Iterating this argument we see that $u_1 u_2^n u_3$ is a geodesic word for every positive integer n. Since a subword of a geodesic word is geodesic, it follows that u_2^n is a geodesic word for every $n > 0$. In particular u_2^n does not represent the identity, and hence the image of u_2 in Γ is an element of infinite order. □

3. Further Properties of Hyperbolic Groups

In this section we shall present more of the basic properties of hyperbolic groups. We shall see that our earlier results concerning isometries of CAT(-1) spaces, including parts (1) to (5) and (i) to (v) of Theorem 1.1, all remain valid in the context of hyperbolic groups. Specifically, we shall prove that if a group Γ is hyperbolic then it contains only finitely many conjugacy classes of finite subgroups, the centralizer of every element of infinite order contains a cyclic subgroup of finite index, and if Γ is torsion-free then it is the fundamental group of a finite cell complex whose universal cover is contractible. We shall also describe an algebraic notion of translation number and show that the translation numbers of elements of infinite order in hyperbolic groups are positive and form a discrete subset of $\mathbb{R}$ (cf. II.6.10(3)); in fact they are rational numbers with bounded denominators (3.17).

The results in this section all originate from [Gro87].

Let's begin by noting the direct connection between hyperbolic groups and isometries of CAT(0) spaces. This comes from the Flat Plane Theorem (H.1.5) and the quasi-isometry invariance of hyperbolicity (H.1.10).

3.1 Theorem. *If a group Γ acts properly and cocompactly by isometries on a* CAT(0) *space X, then Γ is hyperbolic if and only if X does not contain an isometrically embedded copy of the Euclidean plane.*

Finite Subgroups

The proof of the following theorem illustrates how arguments concerning groups of isometries acting on CAT(κ) spaces, $\kappa < 0$, can be transported into the world of hyperbolic and related groups. The key to such adaptations is that one must find an appropriate way to "quasify" the key role that negative curvature is playing in the classical setting; one then attempts to encapsulate a robust form of the salient feature of curvature in the more relaxed world of hyperbolic spaces [55]. We apply this general philosophy to the study of finite subgroups. The following result should be compared with (II.2.8).

3.2 Theorem. *If a finitely generated group Γ is hyperbolic, then it contains only finitely many conjugacy classes of finite subgroups.*

As in (II.2.8), we shall deduce this result from the existence of an appropriate notion of centre for bounded sets.

[55] The first example of this is the very definition of a δ-hyperbolic space: one observes that many of the global implications of the CAT(-1) inequality stem from the slim triangles condition.

3.3 Lemma (Quasi-Centres). *Let X be a δ-hyperbolic geodesic space. Let $Y \subset X$ be a non-empty bounded subspace. Let $r_Y = \inf\{\rho > 0 \mid Y \subseteq B(x,\rho),\ \text{some } x \in X\}$. For all $\varepsilon > 0$, the set $C_\varepsilon(Y) = \{x \in X \mid Y \subseteq B(x, r_Y + \varepsilon)\}$ has diameter less than $(4\delta + 2\varepsilon)$.*

Proof. Given $x, x' \in C_\varepsilon(Y)$, let m be the midpoint of a geodesic segment $[x, x']$. For each $y \in Y$ we consider a geodesic triangle with vertices x, x', y and with $[x, x']$ as one of the sides. Because X is δ-hyperbolic, m is within a distance δ of some $p \in [x, y] \cup [x', y]$; suppose that $p \in [x, y]$. Then, since $d(x, m) = d(x, x')/2$ and $d(p, x) \geq d(x, m) - \delta$, we have $d(y, p) = d(y, x) - d(p, x) \leq d(y, x) + \delta - d(x, x')/2$, and since $d(x, y) < r_Y + \varepsilon$,

$$d(y, m) \leq d(y, p) + d(p, m) < r_Y + \varepsilon + 2\delta - \frac{1}{2}d(x, x').$$

But $d(y, m) \geq r_Y$ for some $y \in Y$, hence $\varepsilon + 2\delta - \frac{1}{2}d(x, x') > 0$. □

Proof of Theorem 3.2. Let Γ be a group whose Cayley graph $\mathcal{C}$ with respect to some finite generating set is δ-hyperbolic. Let $H \subset \Gamma$ be a finite subgroup, and let $C_1(H) \subseteq \mathcal{C}$ be as in the lemma. $C_1(H)$ contains at least one vertex and the action of H leaves $C_1(H)$, and hence its vertex set, (set-wise) invariant. If x is one of the vertices of $C_1(H)$, then $x^{-1}Hx$ leaves $x^{-1}C_1(H)$ invariant. Since $x^{-1}C_1(H)$, which is a set of diameter less than $(4\delta + 2)$, contains the identity 1, it also contains $x^{-1}Hx = (x^{-1}Hx).1$. Thus every finite subgroup of Γ is conjugate to a subset of the ball of radius $(4\delta + 2)$ about the identity. □

The proof given above does not appear in the literature but the idea behind it was known to a number of researchers in the field, in particular Brian Bowditch, Noel Brady and Ilya Kapovich. Alternative proofs have been given by Ol'shanskii and by Bogopolskii and Gerasimov [BoG95].

Quasiconvexity and Centralizers

Recall that a subspace C of a geodesic space X is said to be *convex* if for all $x, y \in C$ each geodesic joining x to y is contained in C. Following Gromov [Gro87], we quasify this notion.

3.4 Definition. A subspace C of a geodesic metric space X is said to be *quasiconvex* if there exists a constant $k > 0$ such that for all $x, y \in C$ each geodesic joining x to y is contained in the k-neighbourhood of C.

3.5 Lemma. *Let G be a group with finite generating set $\mathcal{A}$ and let $H \subset \Gamma$ be a subgroup. If H is quasiconvex in the Cayley graph $\mathcal{C}_{\mathcal{A}}(\Gamma)$, then it is finitely generated and $H \hookrightarrow \Gamma$ is a quasi-isometric embedding (with respect to any choice of word metrics).*

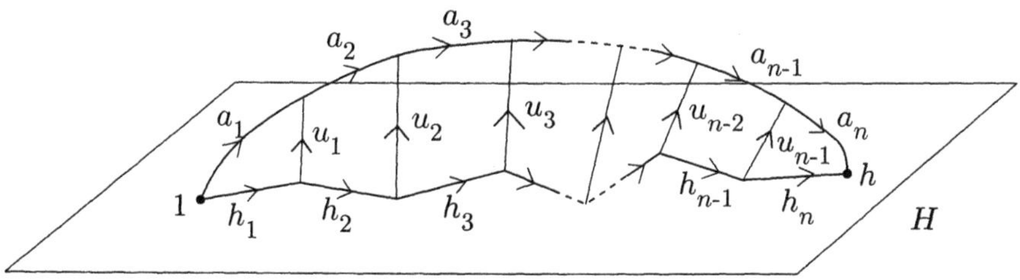

Fig. Γ.6 A Quasiconvex Subgroup

Proof. Let k be as in (3.4). Given $h \in H$, we choose a geodesic from 1 to h in $\mathcal{C}_{\mathcal{A}}(\Gamma)$; suppose that it is labelled $a_1 \dots a_n$. For $i = 1, \dots, n$ we choose a word u_i of length at most k so that $h_i := u_{i-1}a_iu_i^{-1} \in H$ (where u_0 and u_n are defined to be the empty word). (See figure Γ.6.)

We have $h = h_1 \dots h_n$. It follows that H is generated by the (finite) set of elements $h_i \in H$ that lie in the ball of radius $(2k + 1)$ about $1 \in \Gamma$. It also follows that the distance from 1 to h in the word metric associated to this generating set is at most $n = d_{\mathcal{A}}(1, h)$, and hence $H \hookrightarrow \Gamma$ is a quasi-isometric embedding. □

3.6 Corollary. *Let Γ be a hyperbolic group and let $H \subset \Gamma$ be a finitely generated subgroup.*

(1) *If H is quasiconvex with respect to one finite generating set, then it is quasiconvex with respect to all finite generating sets. (Thus we may unambiguously say that H is a quasiconvex subgroup of Γ.)*

(2) *$H \subset \Gamma$ is quasiconvex if and only if it is quasi-isometrically embedded.*

Proof. One direction of (2) is proved in the lemma. For the converse, we fix finite generating sets $\mathcal{A}$ for Γ and $\mathcal{B}$ for H and suppose that H is quasi-isometrically embedded in Γ, which implies that there is a quasi-isometric embedding $\phi : \mathcal{C}_{\mathcal{B}}(H) \to \mathcal{C}_{\mathcal{A}}(\Gamma)$. Given two points $h, h' \in H$, we join them by a geodesic c in $\mathcal{C}_{\mathcal{B}}(H)$ and consider the quasi-geodesic $\phi \circ c$ joining h to h' in $\mathcal{C}_{\mathcal{A}}(\Gamma)$. According to (H.1.7), any geodesic joining h to h' in $\mathcal{C}_{\mathcal{A}}(\Gamma)$ is k-close to this quasi-geodesic, where k depends only on the hyperbolicity constant of Γ and the parameters of the quasi-isometry ϕ. Thus H is quasiconvex in $\mathcal{C}_{\mathcal{A}}(\Gamma)$, and (2) is proved.

The statement "$H \hookrightarrow \Gamma$ is a quasi-isometric embedding" does not depend on a choice of generating sets, so (1) follows from (2). □

By combining (H.1.9) and (3.6) we get:

3.7 Proposition. *The quasiconvex (equivalently, quasi-isometrically embedded) subgroups of hyperbolic groups are hyperbolic.*

3.8 Remarks

(1) The converse of (3.7) is not true: there exist pairs of hyperbolic groups $H \subset \Gamma$ such that H is not quasi-isometrically embedded in Γ (examples are given in (6.21)).

(2) Proposition 3.7 shows that quasi-isometrically embedded subgroups of hyperbolic groups are finitely presented; the same is not true of semihyperbolic groups, for example the direct product of two finitely generated free groups (see 5.12(3)).

(3) The construction described in (II.5.45) yields examples of hyperbolic groups with finitely generated subgroups that are not finitely presented and hence not hyperbolic. It is more difficult to construct finitely presented subgroups that are not hyperbolic, the only know examples are due to Noel Brady[56].

The following results are due to Gromov [Gro87]. The ideas underlying them have been used by other authors to obtain similar results in wider contexts (e.g., [Ep+92], [GeS91], [AloB95]). In the next section we shall provide proofs in a wider context — see (4.13) and (4.14).

3.9 Proposition. *Let Γ be a hyperbolic group.*

(1) *The centralizer $C(\gamma)$ of every $\gamma \in \Gamma$ is a quasiconvex subgroup.*

(2) *If the subgroups $H_1, H_2 \subset \Gamma$ are quasiconvex then so is $H_1 \cap H_2$.*

3.10 Corollary. *Suppose that Γ is a hyperbolic group and that $\gamma \in \Gamma$ has infinite order.*

(1) *The map $\mathbb{Z} \to \Gamma$ given by $n \mapsto \gamma^n$ is a quasi-geodesic.*

(2) *$\langle\gamma\rangle$ has finite index in $C(\gamma)$. In particular, Γ does not contain $\mathbb{Z}^2$.*

Proof of Corollary. $C(\gamma)$ is quasiconvex, hence finitely generated and hyperbolic. By intersecting the centralizers of a finite generating set for $C(\gamma)$, we see that the centre $Z(C(\gamma))$ is also quasiconvex, hence finitely generated and hyperbolic. It is easy to see that a finitely generated abelian group is hyperbolic if and only if it contains a cyclic subgroup of finite index. Hence $Z(C(\gamma))$ contains $\langle\gamma\rangle$ as a subgroup of finite index. Moreover, since $Z(C(\gamma)) \subset C(\gamma)$ and $C(\gamma) \subset \Gamma$ are quasiconvex, the maps $\langle\gamma\rangle \hookrightarrow Z(C(\gamma)) \hookrightarrow C(\gamma)$ and $C(\gamma) \hookrightarrow \Gamma$ are quasi-isometric embeddings, by (3.5), and hence so is $\langle\gamma\rangle \hookrightarrow \Gamma$. This proves (1).

Fix a finite generating set $\mathcal{A}$ with respect to which Γ is δ-hyperbolic and suppose that $\gamma \in \Gamma$ has infinite order. It follows from (1) that if γ^p is conjugate to γ^q then $|p| = |q|$. For if $t^{-1}\gamma^p t = \gamma^q$, then $t^{-m}\gamma^{p^m}t^m = \gamma^{q^m}$ for all integers $m \geq 1$, which means $d(1, \gamma^{q^m}) \leq 2m\, d(1, t) + |p|^m\, d(1, \gamma)$, and if $|p|$ were less than $|q|$ this would contradict the fact that $n \mapsto \gamma^n$ is a quasi-geodesic.

Since the positive powers of γ define distinct conjugacy classes, by replacing γ with a suitable power if necessary, we may assume that γ is not conjugate to any element a distance 4δ or less from the identity. We claim that if $g \in \Gamma$ commutes with γ then g lies within a distance $K := 2d(1, \gamma) + 4\delta$ of $\langle\gamma\rangle$. Suppose that this were not the case and suppose that $d(g, \langle\gamma\rangle) = d(g, \gamma^r)$. Replacing g by $\gamma^{-r}g$, we

[56] "Branched coverings of cubical complexes and subgroups of hyperbolic groups", J. London Math. Soc, to appear.

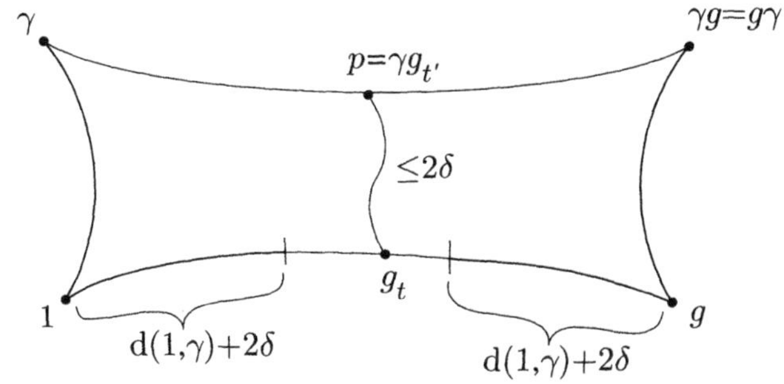

Fig. Γ.7 Centralizers are virtually cyclic

may assume that $d(g, \langle\gamma\rangle) = d(g, 1) > K$. Consider a geodesic quadrilateral Q in $\mathcal{C}_{\mathcal{A}}(\Gamma)$ with sides $\{[1, \gamma], [1, g], g.[1, \gamma], \gamma.[1, g]\}$. Let g_t denote the point a distance t from 1 on $[1, g]$. (See figure Γ.7.)

Because $\mathcal{C}_{\mathcal{A}}(\Gamma)$ is δ-hyperbolic, g_t lies a distance at most 2δ from some point p on one of the remaining three sides of Q. If we choose t so that $d(1, g) - d(1, \gamma) - 2\delta > t > d(1, \gamma) + 2\delta$, then p must belong to $[\gamma, \gamma g]$, say $p = \gamma g_{t'}$. Since $t' = d(\gamma g_{t'}, \langle\gamma\rangle)$ and $t = d(g_t, \langle\gamma\rangle)$, we have $|t - t'| \leq 2\delta$ and hence $d(g_t, \gamma g_t) \leq 4\delta$. But this means $d(1, g_t^{-1}\gamma g_t) \leq 4\delta$, contrary to our hypothesis on γ. This contradiction proves (2). □

As a further illustration of the general approach of quasification, we adapt the existence of projection maps in CAT(0) spaces (II.2.4).

3.11 Proposition (Quasi-Projection). *If X is a δ-hyperbolic geodesic space and $Y \subseteq X$ is a quasiconvex subspace, then there exists a constant $K > 0$ with the following property: given $\varepsilon \geq 0$, if $\pi : X \to Y$ assigns to each $x \in X$ a point $\pi(x) \in Y$ such that $d(x, \pi(x)) \leq d(x, Y) + \varepsilon$, then $d(\pi(x), \pi(x')) \leq d(x, x') + K + 2\varepsilon$ for all $x, x' \in X$.*

Proof. Let $k > 0$ be as in definition (3.4). We shall show that it suffices to take $K = 8\delta + 2k$.

Given $x, x' \in X$, we choose geodesic segments joining each pair of the points $\{x, x', \pi(x), \pi(x')\}$ and consider the geodesic quadrilateral formed by the triangles $\Delta_1 = \Delta([x, \pi(x)], [\pi(x), x'], [x, x'])$ and $\Delta_2 = \Delta([\pi(x), \pi(x')], [\pi(x), x'], [x', \pi(x')])$. There is a point $p \in [\pi(x), x']$ which is within δ of both $[\pi(x), \pi(x')]$ and $[x', \pi(x')]$ (the δ-hyperbolic condition for Δ_2), and p is also within δ of a point on $[x, \pi(x)] \cup [x, x']$ (the δ-hyperbolic condition for Δ_1). Each point on $[\pi(x), \pi(x')]$ is within k of Y. Hence the point $q' \in [x', \pi(x')])$ closest to p is within $2\delta + k$ of Y. Therefore, since $d(x', \pi(x')) = d(x', q') + d(q', \pi(x'))$ and $d(x', \pi(x')) \leq d(x', Y) + \varepsilon$, we have $d(q', \pi(x')) \leq 2\delta + k + \varepsilon$. For future reference we also note that

$$(\#) \qquad d(p, \pi(x')) \leq 3\delta + k + \varepsilon.$$

If p is within δ of $q \in [x, \pi(x)]$, then, as above, $d(q, \pi(x)) \leq 2\delta + k + \varepsilon$. Hence

$$\begin{aligned} d(\pi(x), \pi(x')) &\leq d(\pi(x), q) + d(q, p) + d(p, q') + d(q', \pi(x')) \\ &\leq (2\delta + k + \varepsilon) + 2\delta + (2\delta + k + \varepsilon) \\ &< K + 2\varepsilon. \end{aligned}$$

So we are done if p is a distance at most δ from $[x, \pi(x)]$. If it is not, then instead of considering Δ_1 and Δ_2 we consider $\Delta_1' = \Delta([x, \pi(x)], [\pi(x), \pi(x')], [x, \pi(x')])$ and $\Delta_2' = \Delta([x, x'], [x, \pi(x')], [x', \pi(x')])$. Let p' be a point of $[x, \pi(x')]$ that is a distance at most δ from both $[x, \pi(x)]$ and $[\pi(x), \pi(x')]$. As above, if $d(p', [x', \pi(x')])$ is at most δ, then we are done.

It remains to consider the case where $p \in [x', \pi(x)]$ is within δ of a point $r \in [x, x']$, and $p' \in [x, \pi(x')]$ is within δ of some $r' \in [x, x']$. Since r and r' both lie on $[x, x']$, we have $d(r, r') \leq d(x, x')$. Thus, using (#) twice, we have:

$$\begin{aligned} d(\pi(x), \pi(x')) &\leq d(\pi(x), p') + d(p', r') + d(r', r) + d(p, r) + d(p, \pi(x')) \\ &\leq 2\big(3\delta + k + \varepsilon\big) + 2\delta + d(x, x'). \end{aligned}$$

□

In the special case $Y = \mathbb{Z}$, the following corollary answers a question raised by Alonso et al. [Alo+98] in the course of their work on higher order Dehn functions.

3.12 Corollary. *Let H and Γ be hyperbolic groups. The image of every quasi-isometric embedding $f : H \to \Gamma$ is a quasi-retract, i.e. there exists a map $\rho : \Gamma \to H$ such that, given any choice of word metrics d_Γ and d_H, there is a constant M such that $d_H(\rho f(y), y) \leq M$ for all $y \in H$ and $d_H(\rho(x), \rho(x')) \leq M\, d_\Gamma(x, x')$ for all $x, x' \in \Gamma$.*

Proof. The image of f is quasiconvex in Γ. Let $\pi : \Gamma \to f(H)$ be a choice of closest point and define ρ to be $f' \circ \pi$, where f' is a quasi-inverse for f in the sense of (I.8.16). □

Translation Lengths

In Chapter II.6 we saw that one can deduce a good deal about the structure of a group Γ acting by isometries on a CAT(0) space X by looking at the translation numbers $|\gamma| = \inf\{d(x, \gamma.x) \mid x \in X\}$. We also noted (II.6.6) that $|\gamma| = \lim_{n\to\infty} \frac{1}{n} d(x, \gamma^n.x)$ and that if Γ acts properly and cocompactly then the set of numbers $\{|\gamma| : \gamma \in \Gamma\}$ is discrete (II.6.10(3)).

3.13 Definition. Let Γ be a group with finite generating set $\mathcal{A}$ and associated word metric d. The *algebraic translation number* of $\gamma \in \Gamma$ is defined to be

$$\tau_{\Gamma,\mathcal{A}}(\gamma) := \lim_{n\to\infty} \frac{1}{n} d(1, \gamma^n).$$

(The function $n \mapsto d(1,\gamma^n)$ is sub-additive and hence $\lim_{x\to\infty}\frac{1}{x}f(x)$ exists — see II.6.6.) Normally we shall write τ_Γ or τ instead of $\tau_{\Gamma,\mathcal{A}}$, thus suppressing the dependence on $\mathcal{A}$.

3.14 Remarks

(1) $\tau(\gamma)$ depends only on the conjugacy class of γ.

(2) $\tau(\gamma^m) = |m|\,\tau(\gamma)$ for every $m \in \mathbb{Z}$.

(3) If a finitely generated subgroup $H \subset \Gamma$ is quasi-isometrically embedded, then for any choice of word metrics there is a constant K such that $\frac{1}{K}\tau_\Gamma(h) \le \tau_H(h) \le K\tau_\Gamma(h)$ for all $h \in H$.

3.15 Proposition. *If Γ is a hyperbolic group with finite generating set $\mathcal{A}$, then for every $R > 0$ there exist only finitely many conjugacy classes $[\gamma]$ such that $\tau_{\Gamma,\mathcal{A}}(\gamma) < R$.*

Proof. Suppose that $\mathcal{C}_\mathcal{A}(\Gamma)$ is δ-hyperbolic. Let u be the shortest word among all those which represent elements of the conjugacy class $[\gamma]$. If u has length at least $(8\delta+1)$, then for every integer $n > 0$ the edge-paths in $\mathcal{C}_\mathcal{A}(\Gamma)$ labelled u^n are $(8\delta+1)$-local geodesics. It follows from (H.1.13) that these edge-paths are (λ,ε)-quasi-geodesics, where λ and ε depend only on δ. Therefore $\tau(u) = \tau(\gamma) \ge |u|/\lambda$. □

3.16 Proposition. *Let Γ be a hyperbolic group. If a subgroup $H \subset \Gamma$ is infinite and quasiconvex, then it has finite index in its normalizer.*

Proof. Fix finite generating sets for H and Γ, and let d be the associated word metric on Γ. We know from (2.22) that H contains an element of infinite order, α say, and we know from (3.10) that $\langle\alpha\rangle$ has finite index in the centralizer $C_\Gamma(\alpha)$, i.e. there exists a constant $k > 0$ such that if $g \in C_\Gamma(\alpha)$ then $d(g,\langle\alpha\rangle) \le k$.

Since H is quasiconvex, there is also a constant $K > 0$ such that if $\gamma^{-1}\alpha\gamma$ is in H, then $\tau_H(\gamma^{-1}\alpha\gamma) \le K\,\tau_\Gamma(\gamma^{-1}\alpha\gamma) = K\,\tau_\Gamma(\alpha)$. By applying the preceding proposition to H, we see $\gamma^{-1}\alpha\gamma$ must belong to one of finitely many conjugacy classes in H; we choose representatives $c_1^{-1}\alpha c_1, \dots, c_n^{-1}\alpha c_n$ for these classes, where $c_i \in \Gamma$ (only one of the c_i is in H).

Given an element $\gamma_0 \in \Gamma$ in the normalizer of H, there exists c_i such that $\gamma_0^{-1}\alpha\gamma_0 = h^{-1}c_i^{-1}\alpha c_i h$ for some $h \in H$, which implies that $\gamma_0 h^{-1} c_i^{-1}$ belongs to $C_\Gamma(\alpha)$. Thus $d(\gamma_0 h^{-1}c_i^{-1},\langle\alpha\rangle) \le k$, and writing $\gamma_0 h^{-1} = h'\gamma_0$ we have:

$$\begin{aligned} d(\gamma_0, H) &= d(h'\gamma_0, H) \\ &\le d(h'\gamma_0, h'\gamma_0 c_i^{-1}, H) + d(h'\gamma_0 c_i^{-1}) \\ &= d(1, c_i) + d(\gamma_0 h^{-1} c_i^{-1}, H) \\ &\le \max_j d(1, c_j) + k. \end{aligned}$$

Thus $d(\gamma_0, H)$ is uniformly bounded, i.e. H has finite index in its normalizer. □

The following remarkable theorem is due to Gromov [Gro87].

3.17 Theorem (Translation Lengths are Discrete). *If Γ is δ-hyperbolic with respect to a fixed generating set $\mathcal{A}$, then $\{\tau(\gamma) \mid \gamma \in \Gamma\}$ is a discrete set of rational numbers. Indeed there is a positive integer N such that $N\tau(\gamma) \in \mathbb{N}$ for every $\gamma \in \Gamma$.*

Proof. We follow a proof of Thomas Delzant [Del96]. Given γ, let u be the shortest word among those which represent elements in the conjugacy class of γ. Ignoring finitely many conjugacy classes, we may assume that u has length at least $(8\delta + 1)$. Consider the bi-infinite path p_u in the Cayley graph $\mathcal{C}_{\mathcal{A}}(\Gamma)$ that begins at the identity and is labelled by the powers of u. This is an $(8\delta + 1)$-local geodesic, and hence it is a quasi-geodesic. Let $U_-, U_+ \in \partial\Gamma = \partial\mathcal{C}_{\mathcal{A}}(\Gamma)$ be the endpoints of this quasi-geodesic. (H.3.2) guarantees the existence of a geodesic line with endpoints U_- and U_+, and (H.1.7) implies that any such geodesic is contained in the R-neighbourhood of p_u, where R depends only on δ.

We fix a linear ordering on the generators $\mathcal{A}^{\pm 1}$ and consider the induced lex-least ordering $\prec$ on the finite geodesic edge-paths in $\mathcal{C}_{\mathcal{A}}(\Gamma)$: one path is less than another if it is shorter or else has the same length but its label comes before that of the other path in the dictionary.

For each positive integer m, let I_m be the geodesic that is lex-least among those geodesic edge-paths that are contained in the K-neighbourhood of p_u and have endpoints within a distance K of u^{-m} and u^m respectively. For each $n < m$, let $I_{m,n}$ be the minimal sub-segment of I_m that has endpoints within a distance K of u^{-n} and u^n respectively. For fixed n, there are only finitely many possibilities for $I_{m,n}$, so by a diagonal sequence argument we may extract a subsequence of the I_m such that $I_{m,n}$ remains constant for all n as $m \to \infty$. The union of the geodesic segments in this subsequence $\bigcup_n I_{m,n}$ gives a geodesic line joining U_- to U_+. This (oriented) line has the property that all of its subsegments are minimal in the lexicographical ordering $\prec$ on segments in the Cayley graph that have the given length. (Delzant calls such geodesics "special".)

Through each point of $\mathcal{C}_{\mathcal{A}}(\Gamma)$ there is at most one special geodesic with endpoints U_- and U_+, because a segment of such a geodesic cannot have two extensions (both forwards or both backwards) of a given length that are both lexicographically least. Since any geodesic joining U_- to U_+ is contained in the R-neighbourhood of p_u, it follows that there are at most V special geodesics joining U_- to U_+, where V is the cardinality of a ball of radius R in Γ. The action of u by left multiplication permutes this finite set. Therefore some power u^r, where r divides $V!$, fixes a special geodesic and acts by translations on it; let $a \in \mathbb{N}$ be the translation distance of the action of u^r on this geodesic. If x is a vertex on this line, then $d(u^{rn}.x, x) = na$ for all $n > 0$ and hence $d((x^{-1}u^{rn}x), 1) = na$. Dividing by n and taking the limit, we get $r\,\tau(\gamma) = r\,\tau(u) = \tau(u^r) = a \in \mathbb{N}$. Thus we may take the integer N in the statement of the theorem to be $V!$. □

Free Subgroups

The following result is due to Gromov [Gro87, 5.3B]. The first detailed proof was given by Delzant [Del91]. One might think of this result as a strong form of the assertion that two generic elements in a hyperbolic group generate a free subgroup.

3.18 Theorem. *If Γ is a torsion-free hyperbolic group and H is a two-generator group that is not free, then up to conjugacy there are only finitely many embeddings $H \hookrightarrow \Gamma$.*

We recall the "ping-pong" construction of Fricke and Klein [FrKl12] (cf. [Har83]).

3.19 Lemma (Ping-Pong Lemma). *Let $h_1, \dots, h_r$ be bijections of a set Ω and suppose that there exist non-empty disjoint subsets $A_{1,1}, A_{1,-1}, \dots, A_{r,1}, A_{r,-1} \subset \Omega$ such that $h_i^{\varepsilon}(\Omega \smallsetminus A_{i,\varepsilon}) \subset A_{i,-\varepsilon}$ for $\varepsilon = \pm 1$, $i = 1, \dots, r$. Then $h_1, \dots, h_r$ generate a free subgroup of rank r in* Perm(Ω).

Proof. Exercise. □

3.20 Proposition. *If Γ is hyperbolic, then for every finite set of elements $h_1, \dots, h_r \in \Gamma$ there exists an integer $n > 0$ such that $\{h_1^n, \dots, h_r^n\}$ generates a free subgroup of rank at most r in Γ.*

Proof. Let $\mathcal{C}$ be the Cayley graph of Γ with respect to a fixed finite generating set and suppose that $\mathcal{C}$ is δ-hyperbolic. By replacing each h_i with a sufficiently high power and throwing away those which become trivial, we may assume that each of the h_i has infinite order. Define $t_i = \tau(h_i)$.

As in the proof of (3.17), by raising the h_i to a further power we may assume that there is a geodesic line $c_i : \mathbb{R} \to \mathcal{C}$ which is invariant under the action of h_i; the action is $h_i.c_i(s) = c_i(s + t_i)$. Define $h_i^{\pm\infty} = c_i(\pm\infty)$.

We claim that if $c_i(\infty) = c_j(\infty)$, then some powers of h_i and h_j generate a cyclic subgroup of Γ (and hence $c_i(-\infty) = c_j(-\infty)$). To see this, first note that according to (H.3.3), we may parameterize the lines c_i and c_j so that $d(c_i(s), c_j(s)) \le 5\delta$ for all $s > 0$. For each integer $r > 0$ we shall estimate $d(h_i^{-r}h_jh_i^r.c_i(0), c_j(t_j))$. For this purpose it is convenient to use the notation $a \sim_\varepsilon b$ to mean $|a - b| \le \varepsilon$; thus $d(\gamma, c_i(s)) \sim_{5\delta} d(\gamma, c_j(s))$ for all $s > 0$ and all $\gamma \in \Gamma$.

$$\begin{aligned} d(h_i^{-r}h_jh_i^r.c_i(0), c_j(t_j)) &= d(h_j.c_i(rt_i), h_i^r.c_j(t_j)) \\ &\sim_{5\delta} d(h_j.c_i(rt_i), h_i^r.c_i(t_j)) \\ &\sim_{5\delta} d(h_j.c_j(rt_i), h_i^r.c_i(t_j)) \\ &= d(c_j(t_j + rt_i), c_i(t_j + rt_i)) \le 5\delta\,. \end{aligned}$$

Thus $d(h_i^{-r}h_jh_i^r.c_i(0), c_j(t_j)) \le 15\delta$ for all $r > 0$, which means that $h_i^{-r}h_jh_i^r.c_i(0) = h_i^{-s}h_jh_i^s.c_i(0)$ for some integers r, s with $0 < r - s < v$, where v is the number of elements in a ball of radius 15δ in Γ. Since the action of Γ on $\mathcal{C}$ is free, it follows that h_i^{s-r} lies in the centralizer h_j, which according to (3.10) is virtually cyclic. This proves the claim.

If h_i and h_j have powers that generate a cyclic subgroup, then we may replace them with a generator of that cyclic subgroup. Thus it only remains to consider the case where the $2r$ points $h_1^{\pm\infty}, \dots, h_r^{\pm\infty} \in \partial\Gamma$ are distinct.

For each $x \in \Gamma$, let x_i be a point on the image of c_i that is closest to x. We claim that if $R > 0$ is sufficiently large then the following $2r$ sets are disjoint: $A_{i,-1} = \{x \mid x_i \in c_i(-\infty, -R]\}$ and $A_{i,1} = \{x \mid x_i \in c_i[R, \infty)\}$, where $i = 1, \dots, r$. It is clear that by replacing the h_i with sufficiently high powers we can ensure that their action on the above sets satisfies the hypotheses of the ping-pong lemma, thus it only remains to prove that the $A_{i,\varepsilon}$ really are disjoint.

To see this, we fix a constant $\rho > 0$ that is sufficiently large to ensure that the closed ball B of radius ρ about $1 \in \Gamma$ contains $c_i(0)$ for $i = 1, \dots, r$, and every geodesic segment of the form $[x_i, x_j]$ intersects B. Given $x \in \Gamma$, since $\Delta(x, x_i, x_j)$ is δ-slim, there is a point p on $[x, x_i]$ or $[x, x_j]$ that is within a distance δ of $[x_i, x_j] \cap B$. Suppose that $p \in [x, x_i]$. Since $d(c_i(0), p) \geq d(x_i, p)$ (by definition of x_i), and $d(c_i(0), p) \leq d(c_i(0), 1) + d(1, p) \leq 2\rho + \delta$, we have $d(c_i(0), x_i) \leq d(c_i(0), p) + d(p, x_i) \leq 2\, d(c_i(0), p) \leq 4\rho + 2\delta$.

It follows that if $R > 4\rho + 2\delta$, then each $x \in \Gamma$ can belong to at most one of the sets $A_{i,\varepsilon}$ defined above. □

The Rips Complex

We continue the work of deciding which of the properties listed in (1.1) are enjoyed by all hyperbolic groups. This paragraph concerns (1.1(5)). We describe a construction due to E. Rips which shows that every torsion-free hyperbolic group has a finite Eilenberg-MacLane space. It also shows that every hyperbolic group has an Eilenberg-MacLane space with finitely many cells in each dimension (see 3.26).

3.21 Theorem. *Every hyperbolic group Γ acts on a simplicial complex P such that:*

(1) *P is finite-dimensional, contractible and locally finite;*

(2) *Γ acts simplicially, with compact quotient and finite stabilizers;*

(3) *Γ acts freely and transitively on the vertex set of P.*

In particular, if Γ is torsion-free then it has a finite Eilenberg-MacLane space $K(\Gamma, 1)$, namely $\Gamma \backslash P$.

Remark. The complex described in this theorem has no particular local geometric features; in particular there is no reason to expect it to support a CAT(0) metric.

3.22 Definition. Let X be a metric space and let $R > 0$. The *Rips complex* $P_R(X)$ is the geometric realization of the simplicial complex with vertex set X whose n-simplices are the $(n+1)$-element subsets $\{x_0, \dots, x_n\} \subset X$ of diameter at most R.

One usually endows $P_R(X)$ with the weak topology, which is the same as the metric topology (I.7) if each vertex lies in only finitely many cells, i.e. $B(x, R)$ is finite for every $x \in X$. (In any case, the weak topology and the metric topology define the same homotopy type [Dow52]). Note that $P_R(X)$ is finite dimensional only if there is a bound on $|B(x, R)|$ as x varies over X.

Let $r > 0$. Recall that a subset X of a metric space Y is said to be *r-dense* if for every $y \in Y$ there exists $x \in X$ with $d(x, y) \leq r$.

3.23 Proposition. *Let Y be a geodesic space and let X be an r-dense subset. If Y is δ-hyperbolic then $P_R(X)$ is contractible whenever $R \geq 4\delta + 6r$.*

Proof. The complex $P_R(X)$ is contractible if and only if all of its homotopy groups are trivial (see [Spa66]). The image of any continuous map of a sphere into $P_R(X)$ (with the weak topology) lies in a finite subcomplex (because it is compact), so it suffices to prove that any finite subcomplex $L \subset P_R(X)$ can be contracted to a point in $P_R(X)$. We fix a basepoint $x_0 \in X$.

Case 1: If the distance in X between x_0 and each vertex of L is at most $R/2$ then L is contained in a face of a simplex of $P_R(X)$, and hence it is contractible.

Case 2: Suppose that there is a vertex $v \in L$ such that $d(x_0, v) > R/2$, and choose v so that $d(x_0, v)$ is maximal. (Beware: the metric d is on X, not L.) The idea of the proof is to homotop L by pushing v towards x_0 while leaving the remaining vertices of L alone. If we are able to do this, then by repeating a finite number of times we can homotop L to a complex covered by Case 1.

We choose a point y on a geodesic $[x_0, v] \subset Y$ with $d(v, y) = R/2$, and then choose $v' \in X$ with $d(y, v') \leq r$. Let $\rho = d(v, v')$ and note that $\rho \in [2\delta+2r, (R/2)+r]$. We claim that if u is a vertex of L and $d(u, v) \leq R$ then $d(u, v') \leq R$.

Consider a geodesic triangle $\Delta(x_0, u, v) \subset Y$ with $y \in [x_0, v]$. This is δ-slim, so either $d(y, u') \leq \delta$ for some $u' \in [x_0, u]$ or else $d(y, w) \leq \delta$ for some $w \in [u, v]$. (Readers will find the ensuing inequalities easier to follow if they draw a picture of each situation and label the lengths of arcs.)

In the first case, by hypothesis,

$$d(x_0, v') + d(v', v) \geq d(x_0, v) \geq d(x_0, u) = d(x_0, u') + d(u', u),$$

and $d(x_0, v') \leq d(x_0, u') + d(u', v') \leq d(x_0, u') + (\delta + r)$. Thus $\rho = d(v, v') \geq d(u', u) - (\delta+r)$, and hence $d(u, v') \leq d(u, u') + d(u', v') \leq (\rho+\delta+r)+(\delta+r) \leq R$.

In the second case, $d(v', w) \leq r + \delta$ and we have: $\rho = d(v, v') \leq d(v, w) + d(w, v')$, so $d(v, w) \geq \rho - (\delta+r)$. Whence $d(u, w) = d(u, v) - d(v, w) \leq R - \rho + \delta + r$, and $d(u, v') \leq d(u, w) + d(w, v') \leq R - \rho + 2(\delta + r) \leq R$.

Thus we have shown that if a vertex u of L is in the star of v then it is also in the star of v'. Moreover, v' is also in the link of v. Let L' be the complex obtained from L by replacing each simplex of the form $\{v, x_1, \dots, x_r\}$ with $\{v', x_1, \dots, x_r\}$. The obvious (affine) homotopies moving each $\{v, x_1, \dots, x_r\}$ to $\{v', x_1, \dots, x_r\}$ in the simplex $\{v, v', x_1, \dots, x_r\}$, together with the identity map on $L \smallsetminus \text{st}(v)$, give a homotopy from $L \hookrightarrow P_R(X)$ to $L' \hookrightarrow P_R(X)$, and (as we noted above) by iterating this operation we can contract L to x_0. □

Proof of Theorem (3.21). The inclusion of Γ as the vertex set of its Cayley graph is 1-dense, so we can apply the preceding proposition to deduce that $P_R(\Gamma)$ is contractible when R is large enough. Because there is a bound on the number of points in any ball of a given radius in Γ, the complex $P_R(\Gamma)$ is finite dimensional and locally finite.

One extends the action of Γ on itself by left multiplication to a simplicial action on $P_R(\Gamma)$ using the affine structure (I.7A) in the obvious way. The action of Γ on $P_R(\Gamma)$, which is obviously free and transitive on vertices, is simplicial in the sense that it takes simplices to simplices setwise, but if the group has torsion then certain simplices will be sent to themselves without being fixed pointwise (and as a result the quotient space will not be a simplicial complex).

Consider the stabilizer of a simplex $\sigma = \{x_1, \dots, x_n\}$. Because the action of Γ is free on vertices, the representation of Stab(σ) into the symmetric group on the vertices is faithful, and hence $|\text{Stab}(\sigma)| \leq n!$. □

3.24 Remarks

(1) The dimension of the complex $P_R(\Gamma)$ is one less than the cardinality of the largest set in Γ of diameter R. This is bounded (crudely) by the number of words of length R in the generators and their inverses, which is less than $(2|\mathcal{A}|)^R$.

(2) Let Γ be a group equipped with the word metric associated to the finite generating set $\mathcal{A}$, and let $\mathcal{B}$ be the set of non-trivial elements in the ball of radius R about the identity. The 1-skeleton of the Rips complex $P_R(\Gamma)$ is the Cayley graph of Γ with respect to $\mathcal{B}$.

3.25 The Finiteness Conditions FP_n, FL_n and F_n. We remind the reader that an Eilenberg-MacLane complex $K(\Gamma, 1)$ for a group Γ is a CW complex with fundamental group Γ and contractible universal cover. Such a space always exists and its homotopy type depends only on Γ.

If there exists a $K(\Gamma, 1)$ with a finite n-skeleton, then one says that Γ is *of type* F_n. Being of type F_0 is an empty condition, being of type F_1 is equivalent to being finitely generated, and being of type F_2 is equivalent to being finitely presented.

A group Γ is said to be of type FP_n (resp. FL_n) if $\mathbb{Z}$, regarded as a trivial module over the group ring $\mathbb{Z}\Gamma$, admits a projective (resp. free) resolution in which the first $(n+1)$ resolving modules are finitely generated. Γ is said to be of type FP_∞ if it is FP_n for every n. A finitely presented group Γ is of type FP_∞ if and only if it has a $K(\Gamma, 1)$ with finitely many cells in each dimension (see [Bro82] or [Bi76b]).

The cohomology of Γ with coefficients in a ring R can be defined as $H^*(\Gamma, R) := H^*(K(\Gamma, 1), R)$.

3.26 Corollary. *Let Γ be a hyperbolic group.*

(1) *If Γ is torsion-free then it has a finite $K(\Gamma, 1)$.*

(2) *In general, Γ is of type FP_∞, and*

(3) $H^n(\Gamma, \mathbb{Q}) = 0$ *for all n sufficiently large.*

Proof. (1) follows immediately from Theorem 3.21. Parts (2) and (3) are standard spectral sequence arguments (see [Bro87], for example), we omit the details. □

3.27 Remark. We mentioned earlier that if a group Γ acts properly and cocompactly by isometries on a CAT(0) space X then X is Γ-equivariantly homotopy equivalent to a finite-dimensional cell complex on which Γ acts cellularly and cocompactly with finite stabilizers (cf. II.5.13). As in the hyperbolic case, this implies that Γ is finitely presented and of type FP_∞, and $H^*(\Gamma, \mathbb{Q})$ is finite dimensional.

4. Semihyperbolic Groups

Let Γ be a finitely generated group that acts properly and cocompactly by isometries on a CAT(0) space X. By fixing a basepoint $x_0 \in X$ we obtain a quasi-isometry $\gamma \mapsto \gamma.x_0$ from Γ to X. One of the main goals of this chapter is to understand how the geometry that is transmitted from X to Γ by such quasi-isometries is reflected in the structure of Γ. In the case where X is a visibility space, we have seen that a great deal of the geometry of X is transmitted to Γ through the slim triangles condition. In this section we consider the case where X is an arbitrary CAT(0) space. We shall describe a weak convexity condition that Γ inherits from the convexity of the metric on X. The groups that satisfy this condition are called *semihyperbolic groups*. Following the treatment of [AloB95], we shall see that most of our previous results concerning isometries of CAT(0) spaces can be extended to semihyperbolic groups.

If Γ acts properly and cocompactly by isometries on a CAT(0) space X, then there is a natural correspondence between the sets of quasi-geodesics in X and Γ. However we have seen that in general the set of all quasi-geodesics in a CAT(0) space X is not a very manageable object (cf. I.8.23). With this is mind, given a quasi-isometry $X \to \Gamma$ associated to an action of Γ on X, we restrict our attention to those quasi-geodesics in Γ that are the images of geodesics in X. In a coarse sense, the convexity properties of the metric on X will be reflected in this set of quasi-geodesics. In order to describe the resulting weak convexity properties, we need the following definitions.

Definition of a Semihyperbolic Group

4.1 Definitions. Let X be a metric space and let $\mathcal{P}(X)$ be the set of eventually constant maps $p : \mathbb{N} \to X$, thought of as finite discrete paths in X. Let T_p denote the integer at which p becomes constant, i.e. the greatest integer such that $p(T_p) \neq p(T_p - 1)$. A *(discrete) bicombing* of X consists of a choice of a path $s_{(x,y)} \in \mathcal{P}(X)$ joining each pair of points $x, y \in X$; in other words a bicombing is a map $s : X \times X \to \mathcal{P}(X)$ such that $e \circ s$ is the identity on $X \times X$, where $e : \mathcal{P}(X) \to X \times X$ is the endpoints map $e(p) = (p(0), p(T_p))$. The path $s_{(x,y)}$ is called the *combing line from x to y*.

A bicombing s is said to be *quasi-geodesic* if there exist constants λ and ε such that, for all $x, y \in X$, the restriction of $s_{(x,y)}$ to $[0, T_{s_{(x,y)}}]$ is a (λ, ε)-quasi-geodesic.

A bicombing s is said to be *bounded* if there exist constants $k_1 \geq 1, k_2 \geq 0$ such that for all $x, y, x', y' \in X$ and all $t \in \mathbb{N}$

$$d(s_{(x,y)}(t), s_{(x',y')}(t)) \leq k_1 \max\,\{d(x, x'), d(y, y')\} + k_2. \tag{4.1.1}$$

If s is a bounded bicombing by (λ, ε)-quasi-geodesics and the constants k_1 and k_2 are as above, then we say that *s has parameters* $[\lambda, \varepsilon, k_1, k_2]$.

Let Γ be a finitely generated group. We use the term *metric* Γ*–space* to mean a metric space X together with an action of Γ on X by isometries. X is said to be Γ-*semihyperbolic* if it admits a bounded quasi-geodesic bicombing which is equivariant with respect to the action of Γ, that is $\gamma.s_{(x,y)}(t) = s_{(\gamma x, \gamma y)}(t)$ for all $x, y \in X$ and $t \in \mathbb{N}$.

Fix a word metric on Γ. The action of Γ on itself by left multiplication makes it a metric Γ–space. Γ is said to be *semihyperbolic* if, when viewed in this way, it is a Γ-semihyperbolic metric space, with bicombing s, say. Let $[\lambda, \varepsilon, k_1, k_2]$ be the parameters of s. Because s is equivariant, it is entirely determined by the map $\sigma : \Gamma \to \mathcal{P}(\Gamma)$ that sends γ to $s_{(1,\gamma)}$. One calls σ a *semihyperbolic structure* for Γ with parameters $[\lambda, \varepsilon, k_1, k_2]$.

4.2 Lemma. *The definition of a semihyperbolic group does not depend on the choice of generators (word metric).*

Proof. Let Γ be a finitely generated group. Any two word metrics d, d' on Γ are Lipschitz equivalent, i.e. there exists a constant $K \geq 1$ such that $\frac{1}{K} d(x, y) \leq d'(x, y) \leq K\, d(x, y)$ for all $x, y \in \Gamma$. Therefore any semihyperbolic structure σ for (Γ, d), with parameters $[\lambda, \varepsilon, k_1, k_2]$, is a semihyperbolic structure for (Γ, d') with parameters $[K\lambda, K\varepsilon, K^2 k_1, K k_2]$. □

4.3 Example. If Γ acts by isometries on a CAT(0) space X, then X is Γ-semihyperbolic: for all $x, y \in X$ there is a unique geodesic $c_{x,y} : [0, d(x, y)] \to X$ joining x to y, and the desired Γ-equivariant bicombing $s : X \times X \to \mathcal{P}(X)$ is $s_{(x,y)}(t) := c_{x,y}(t)$ for all $t \in \mathbb{N} \cap [0, d(x, y)]$ and $s_{(x,y)}(t) = y$ for all $t \geq d(x, y)$.

4.4 Exercise. Let Γ be a group with finite generating set $\mathcal{A}$. There is natural map from the free monoid $(\mathcal{A}^{\pm 1})^*$ to $\mathcal{P}(\Gamma)$ that sends each word w to the path $t \mapsto w(t)$, where $w(t)$ is the image in Γ of the prefix of length t in w.

Show that if Γ admits a semihyperbolic structure $\sigma : \Gamma \to \mathcal{P}(\Gamma)$, then it admits a semihyperbolic structure $\sigma' : \Gamma \to \mathcal{P}(\Gamma)$ (close to σ) whose image lies in $(\mathcal{A}^{\pm 1})^* \subset \mathcal{P}(\Gamma)$. (cf. [AloB95] Section 10.)

This exercise shows that one can express the definition of a semihyperbolic group in purely algebraic terms:

4.5 Proposition. *Let Γ be a group with finite generating set $\mathcal{A}$. Then Γ is semihyperbolic if and only if there exist positive constants λ, ε, k and a choice of words $\{w_\gamma \mid \gamma \in \Gamma\} \subset (\mathcal{A}^{\pm 1})^*$, such that $w_\gamma = \gamma$ in Γ, the discrete paths $t \mapsto w_\gamma(t)$ determined by the words w_γ are (λ, ε)-quasi-geodesics, and*

$$d(w_\gamma(t), a.w_{a^{-1}\gamma a'}(t)) \leq k,$$

for all $a, a' \in \mathcal{A}^{\pm 1} \cup \{1\}$ and all $t \in \mathbb{N}$.

Proof. The displayed condition is illustrated in figure (Γ.8). If there exist words w_γ as described then the triangle inequality in Γ implies that the paths $s_{(x,y)}(t) := x.w_{x^{-1}y}(t)$ satisfy condition (4.1.1) with $k = k_1$ and $k_2 = 0$. Thus $\gamma \mapsto w_\gamma$ is a semihyperbolic structure for Γ. The converse is the content of the preceding exercise. □

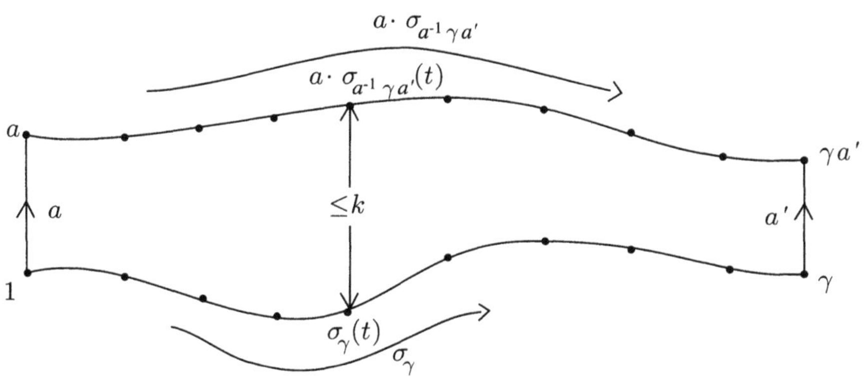

Fig. Γ.8 A semihyperbolic structure for Γ

4.6 Proposition. *Every hyperbolic group is semihyperbolic.*

Proof. Let Γ be a hyperbolic group with finite generating set $\mathcal{A}$. For each $\gamma \in \Gamma$, choose a geodesic word w_γ that represents it. In any δ-hyperbolic geodesic space, geodesics that begin and end a distance at most one apart are uniformly $(2\delta + 2)$-close. By applying this observation to the Cayley graph of Γ, we deduce $\gamma \mapsto w_\gamma$ is a semihyperbolic structure for Γ. □

Basic Properties of Semihyperbolic Groups

Since the definition of semihyperbolicity is given in terms of Γ–spaces and not just metric spaces, the natural morphisms under which to expect this property to be preserved are not quasi-isometries but rather Γ-equivariant quasi-isometries. Thus the following result provides an analogue of the quasi-isometric invariance of hyperbolicity (H.1.9). In contrast to the hyperbolic case, the following theorem does not extend to the case of quasi-isometric embeddings (cf. 5.12(3)).

4.7 Theorem (Invariance Under Equivariant Quasi-Isometries). *Let X_1 and X_2 be metric Γ-spaces and suppose that there exists a Γ-equivariant quasi-isometry $f : X_1 \to X_2$. If the action of Γ on X_1 is free and X_2 is Γ-semihyperbolic, then X_1 is also Γ-semihyperbolic. Conversely, if Γ acts freely on X_2 and X_1 is Γ-semihyperbolic, then X_2 is Γ-semihyperbolic.*

The proof of this result is straightforward but involves some quite lengthy verifications so we omit it — see [AloB95]. Our main interest lies with the following corollaries.

4.8 Corollary.

(1) *If Γ acts properly and cocompactly by isometries on a* CAT(0) *space X, then Γ is semihyperbolic.*

(2) *If G is semihyperbolic and $H \subset G$ is a subgroup of finite index, then H is semihyperbolic.*

(3) *Let $1 \to F \to \Gamma \overset{\pi}{\to} Q \to 1$ be a short exact sequence. If F is finite and Q is semihyperbolic, then Γ is semihyperbolic.*

Proof. (1) Fix $x_0 \in X$. In the light of (4.3), we may apply the theorem with $X_1 = \Gamma$, $X_2 = X$ and $f(\gamma) = \gamma.x_0$.

(2) If we fix a word metric on G, then left multiplication by H is an action by isometries. If G is G-semihyperbolic then it is *a fortiori* H-semihyperbolic. The action of H is free and $H \hookrightarrow G$ is an H-equivariant quasi-isometry, so we may apply the theorem with $H = \Gamma$, $X_1 = H$ and $X_2 = G$.

(3) π is a Γ-equivariant quasi-isometry, where Γ acts on Q by $\gamma.q = \pi(\gamma)q$. Thus we may apply the theorem with $X_1 = \Gamma$ and $X_2 = Q$. □

Finite Presentability and the Word and Conjugacy Problems

4.9 Theorem. *Let Γ be a group with finite generating set $\mathcal{A}$. If Γ is semihyperbolic then there exist positive constants K_1 and K_2 with the following properties.*

(1) *If $\mathcal{R}$ is the set of words in $F(\mathcal{A})$ that have length at most K_1 and represent $1 \in \Gamma$, then $\langle \mathcal{A} \mid \mathcal{R} \rangle$ is a presentation of Γ.*

(2) *A word w in the letters $\mathcal{A}^{\pm 1}$ represents the identity in Γ if and only if in the free group $F(\mathcal{A})$ there is an equality*

$$w = \prod_{i=1}^{N} x_i r_i x_i^{-1},$$

where $N \le K_2|w|^2$, each $r_i \in \mathcal{R}$, and $|x_i| \le K_2|w|$. In particular Γ has a solvable word problem.

(3) *Γ has the quasi-monotone conjugacy property (1.9). In particular Γ has a solvable conjugacy problem (1.11); indeed there exists a constant $\mu > 0$ such*

that words $u, v \in F(\mathcal{A})$ represent conjugate elements of Γ if and only if there is a word $w \in F(\mathcal{A})$ of length $\leq \mu^{\max\{|u|,|v|\}}$ with $w^{-1}uw = v$ in Γ.

Proof. We arranged the proofs in the special case of groups which act properly and cocompactly on CAT(0) spaces precisely so that they would generalize immediately to the present setting. Thus the proof of (1.6) remains valid with an arbitrary semihyperbolic structure $\gamma \mapsto \sigma_\gamma$ in place of the explicit one constructed in that proof: the distance between $\sigma_\gamma(t)$ and $\sigma_{\gamma a}(t)$ is now bounded by the parameters of the semihyperbolic structure rather than in terms of D (notation of (1.6)), but modulo a change of constant the proof is exactly the same. This proves (1) and (2).

The fact that Γ has the q.m.c. property follows immediately from the definition of semihyperbolicity: if σ is a semihyperbolic structure for Γ with parameters $[\lambda, \varepsilon, k_1, k_2]$ and u and v are words in the generators such that $\gamma^{-1}u\gamma = v$ for some $\gamma \in \Gamma$, then $w = \sigma_\gamma$ satisfies the requirements of (1.9) with constant $K = k_1 + k_2$. □

Subgroups of Semihyperbolic Groups

In this section we generalize the subgroup results in (1.1) to the class of semihyperbolic groups. In particular we shall prove the following algebraic analogue of the Flat Torus Theorem (II.7.1), and we shall also prove an analogue (4.21) of the Splitting Theorem (II.6.21).

4.10 Algebraic Flat Torus Theorem. *If Γ is semihyperbolic and A is a finitely generated abelian group, then every monomorphism $\phi : A \hookrightarrow \Gamma$ is a quasi-isometric embedding (with respect to any choice of word metrics).*

We defer the proof to (4.16).

When is a subgroup of a semihyperbolic group itself semihyperbolic? Motivated by the hyperbolic case (3.7), one might ask if being quasi-isometrically embedded or quasi-geodesic is a sufficient condition. However, being quasi-isometrically embedded does not force a subgroup to be semihyperbolic; it does not even force it to be finitely presented (5.12(3)). However, when suitably adapted to the semihyperbolic structure, the notion of *quasiconvexity* does provide a useful criterion for showing that subgroups of semihyperbolic groups are semihyperbolic. This was recognised by Gersten and Short, who were working in the context of biautomatic groups [GeS91].

4.11 Definition. Let X be a metric space with a bicombing s. A subset $C \subset X$ is said to be quasiconvex with respect to s if there is a constant $k > 0$ such that $d(s_{(x,y)}(t), C) \leq k$ for all $x, y \in C$ and $t \in \mathbb{N}$.

If Γ is a group with semihyperbolic structure σ, then a subgroup $H \subseteq \Gamma$ is said to be *σ-quasiconvex* if it is quasiconvex with respect to $s_{(x,y)}(t) := x.\sigma_{x^{-1}y}(t)$. This is equivalent to requiring that there exist a constant $k > 0$ such that $d(\sigma_h(t), H) \leq k$ for all $h \in H$ and $t \in \mathbb{N}$.

As in the hyperbolic case (3.7) we have:

4.12 Proposition. *Let Γ be a group with a semihyperbolic structure σ. If a subgroup $H \subset \Gamma$ is σ-quasiconvex, then it is finitely generated, quasi-isometrically embedded, and semihyperbolic.*

Proof. Suppose that with respect to a fixed word metric $d_{\mathcal{A}}$, each of the paths σ_γ, $\gamma \in \Gamma$, is a (λ, ε)-quasi-geodesic. Let k be the constant of quasiconvexity for H (as in 4.11) and let $\mathcal{B}$ be the intersection of H with the closed ball of radius $(2k + \lambda + \varepsilon)$ centred at $1 \in \Gamma$.

The idea of the proof is as shown in figure (Γ.6). Given $h \in H$ with combing line $\sigma_h : [0, T_h] \to \Gamma$, for each integer $t \leq T_h$ we choose a point $p_t \in H$ closest to $\sigma_h(t)$. Let $b_i = p_{i-1}^{-1} p_i$. Note that $b_i \in H$ and $d_{\mathcal{A}}(1, b_i) = d_{\mathcal{A}}(p_i, p_{i-1}) \leq d_{\mathcal{A}}(p_i, \sigma_h(i)) + d_{\mathcal{A}}(\sigma_h(i), \sigma_h(i-1)) + d_{\mathcal{A}}(\sigma_h(i-1), p_{i-1}) \leq k + (\lambda + \varepsilon) + k$, therefore $b_i \in \mathcal{B}$.

The word $\sigma_h^H = b_1 \dots b_{T_h}$ is equal to h in H, therefore $\mathcal{B}$ is a finite generating set for H. Moreover, $d_{\mathcal{B}}(1, h) \leq T_h \leq \lambda d_{\mathcal{A}}(1, h) + \varepsilon$, and hence H is quasi-isometrically embedded in Γ. Finally, since $d_{\mathcal{A}}(\sigma_h^H(t), \sigma_h(t)) \leq k$ for all $h \in H$ and $t \in \mathbb{N}$, we see that σ^H is a semihyperbolic structure for H with parameters that depend only on k and the parameters of σ. □

The following result is due to Hamish Short[57].

4.13 Proposition. *Let Γ be a group with a semihyperbolic structure σ. The intersection of any two σ-quasiconvex subgroups of Γ is σ-quasiconvex (and hence semihyperbolic).*

Proof. See figure (Γ.9). Suppose that $H_1, H_2 \subset \Gamma$ are σ-quasiconvex. If $\mathcal{A}_0$ is a finite generating set for Γ, then for some $\lambda \geq 1$ and $\varepsilon \geq 0$ each of the combing lines $\sigma_\gamma(t)$ is a (λ, ε)-quasi-geodesic in the word metric associated to $\mathcal{A}_0$. Let $\mathcal{A} \subset \Gamma$ be the set of elements with $d_{\mathcal{A}_0}(1, a) \leq \lambda + \varepsilon$. Note that $\mathcal{A}$ is finite and generates Γ; we shall work with the associated word metric d. Note that $d(\sigma_\gamma(t), \sigma_\gamma(t+1)) \leq 1$ for all $\gamma \in \Gamma$ and $t \in \mathbb{N}$. Let k be a constant of quasiconvexity (as in 4.11) for both H_1 and H_2.

Fix $h \in H_1 \cap H_2$ and consider the combing line $\sigma_h : [0, T_h] \to \Gamma$. For each integer $t_0 \leq T_h$ and $j = 1, 2$ we choose a point $h_j(t_0) \in H_i$ such that $d(\sigma_h(t_0), h_j(t_0)) \leq k$. Note that $d(h_1(t_0), h_2(t_0)) \leq 2k$.

We claim that $d(\sigma_h(t_0), H_1 \cap H_2) \leq V^2$, where V is the number of points in the ball of radius k about $1 \in \Gamma$. To see why, we write $\gamma = \sigma_h(t_0)$ and consider those words w over the alphabet $\mathcal{A}^{\pm 1}$ which have the property that $\gamma w \in H_1 \cap H_2$ and for every prefix u of w we have $d(\gamma u, H_j) \leq k$ for $j = 1, 2$. Let Ω denote the set of

[57] "Groups and combings", preprint, ENS Lyon, 1990

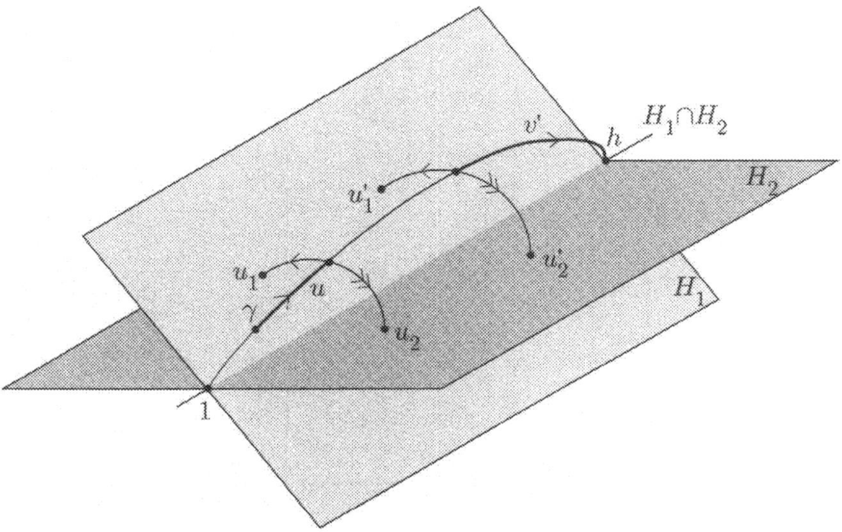

Fig. Γ.9 The intersection of σ-quasiconvex subgroups

such words. Ω is non-empty; indeed it contains the label on any geodesic edge-path connecting the vertices of $\sigma_h([t_0, T]) \subset \mathcal{C}_{\mathcal{A}}(\Gamma)$ (in order). Let $w_0 \in \Omega$ be a word of minimal length. We will be done if we can show that $|w_0| \leq V^2$.

For each prefix u of w_0 and $j = 1, 2$ we choose a point $u_j \in H_j$ closest to γu. Note there are at most V^2 possibilities for $(u^{-1}u_1, u^{-1}u_2) \in \Gamma \times \Gamma$. If $|w_0| > V$ then there would exist prefixes u, u' of w_0 with $|u| < |u'|$ and $(u^{-1}u_1, u^{-1}u_2) = (u'^{-1}u_1', u'^{-1}u_2')$. If this were the case, then for all $g \in \Gamma$ and $i = 1, 2$ we would have $d(ug, H_i) = d(g, u^{-1}u_iH_i) = d(g, u'^{-1}u_i'H_i) = d(u'g, H_i)$. Thus if $w_0 = uv = u'v'$, then the word $w = uv'$ would be in Ω. But $|w| < |w_0|$, contradicting the minimality of $|w_0|$. □

We shall now present a series of results that reveal an intimate connection between quasiconvexity and the structure of centralizers in semihyperbolic groups.

4.14 Proposition. *Let Γ be a finitely generated group and let σ be a semihyperbolic structure for Γ. The centralizer of every element in Γ is σ-quasiconvex.*

Proof. Let σ be a semihyperbolic structure with parameters $[\lambda, \varepsilon, k_1, k_2]$. Let the constant μ be as in (4.9(3)).

If $\gamma \in C_\Gamma(g)$ then the combing lines σ_γ and $g.\sigma_\gamma$ begin and end a distance $d(1, g)$ apart, so $d(\sigma_\gamma(t), g.\sigma_\gamma(t)) \leq k_1\, d(1, g) + k_2$ for all $t \in \mathbb{N}$. In particular, if we fix t_0 and let $h = \sigma_\gamma(t_0)^{-1} g \sigma_\gamma(t_0)$ then $d(1, h) \leq k_1\, d(1, g) + k_2$. The elements g and h are obviously conjugate, so according to (4.9(3)) there is an element $x \in \Gamma$ with $hx = xg$ and

$$d(1, x) \leq \mu^{\max\{d(1,g), d(1,h)\}} \leq \mu^{k_1\, d(1,g) + k_2}.$$

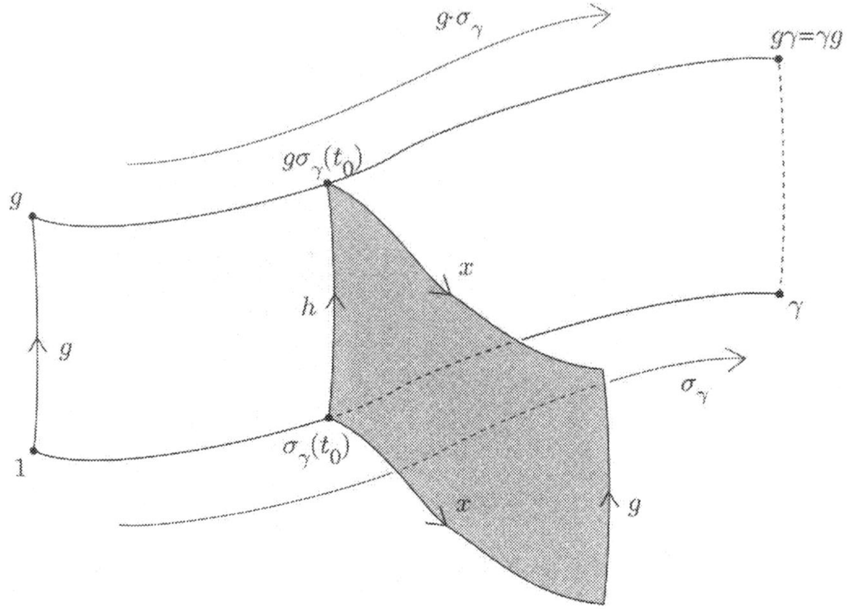

Fig. Γ.10 Centralizers are σ-quasiconvex

We will be done if we can show that $\sigma_\gamma(t_0)x \in C_\Gamma(g)$.

$$\begin{aligned}\sigma_\gamma(t_0)xg &= \sigma_\gamma(t_0)hx \\ &= \sigma_\gamma(t_0)\,(\sigma_\gamma(t_0)^{-1}g\sigma_\gamma(t_0))\,x \\ &= g\,\sigma_\gamma(t_0)\,x.\end{aligned}$$

The idea of the proof is illustrated in figure (Γ.10). □

4.15 Proposition. *Let Γ be a group with semihyperbolic structure σ.*

(1) *If $S \subset \Gamma$ is a finite subset, then $C_\Gamma(S) := \bigcap_{x\in S} C_\Gamma(x)$ is σ-quasiconvex; in particular it is finitely generated and semihyperbolic.*

(2) *If $H \subset \Gamma$ is a finitely generated subgroup, then $C_\Gamma(H)$ is σ-quasiconvex.*

(3) *The centre $Z(\Gamma)$ of Γ is a finitely generated abelian group, and $Z(\Gamma) \subset \Gamma$ is σ-quasiconvex.*

Proof. Part (1) follows immediately from (4.12), (4.13) and (4.14). To prove (2), apply (1) to a finite generating set S for H, and to prove (3) take $H = \Gamma$. □

4.16 The Proof of Theorem (4.10). We must show that if Γ is semihyperbolic and A is a finitely generated abelian group, then every monomorphism $\phi : A \hookrightarrow \Gamma$ is a quasi-isometric embedding. The map ϕ can be factored as follows.

$$A \hookrightarrow Z(C_\Gamma(\phi(A))) \hookrightarrow C_\Gamma(\phi(A)) \hookrightarrow \Gamma.$$

In the light of (4.12) and (4.15) we know that the last two of these maps are quasi-isometric embeddings, so since the composition of quasi-isometric embeddings is a quasi-isometric embedding, it suffices to show that $A \hookrightarrow Z(C_\Gamma(\phi(A)))$ is a quasi-isometric embedding. But $Z(C_\Gamma(\phi(A)))$ is a finitely generated abelian group, and it is easy to see that any monomorphism from one finitely generated abelian group to another is a quasi-isometric embedding. □

In Chapter II.7 we used the Flat Torus Theorem to see that a polycyclic group cannot act properly by semisimple isometries on a CAT(0) space unless it is virtually abelian (II.7.16). One can use the Algebraic Flat Torus Theorem in much the same way.

4.17 Proposition. *A polycyclic group P is a subgroup of a semihyperbolic group if and only if P is virtually abelian.*

Proof. If P is virtually abelian then it acts properly and cocompactly by isometries on a Euclidean space of some dimension (II.7.3), and hence it is semihyperbolic.

To show that if P is not virtually abelian then it is not a subgroup of a semihyperbolic group, we use an idea due to Gersten and Short [GeS91]. As in (II.7.16), induction on the Hirsch length reduces the problem to that of showing that if $\phi \in \mathrm{GL}(n, \mathbb{Z})$ has infinite order then $H = \mathbb{Z}^n \rtimes_\phi \langle t \rangle$ cannot be a subgroup of any semihyperbolic group.

Consider the ℓ_1-norm on $\mathrm{GL}(n, \mathbb{Z}) \subset \mathbb{R}^{n^2}$. Since ϕ has infinite order, the function $k \mapsto \|\phi^k\|$ grows at least linearly, and hence for some basis element $a \in \mathbb{Z}^n$ the function $k \mapsto d_{\mathbb{Z}^n}(\phi^k(a^k), 1)$ grows at least quadratically. But in H we have $t^{-k}a^k t^k = \phi^k(a^k)$, showing that $k \mapsto d_H(\phi^k(a^k), 1)$ grows linearly. Thus $\mathbb{Z}^n$ is not quasi-isometrically embedded in H (nor any group containing it, *a fortiori*), and hence H cannot be a subgroup of any semihyperbolic group. □

In (1.1) we showed that if $|p| \neq |q|$ then groups of the form $\langle x, t \mid t^{-1}x^p t = x^q \rangle$ cannot act properly by semisimple isometries on any CAT(0) space. We did so by means of a simple calculation involving translation lengths. This is one of a number of results that one can recover in the setting of semihyperbolic groups using algebraic translation lengths (3.13) in place of geometric ones.

4.18 Lemma. *If Γ is semihyperbolic and $\gamma \in \Gamma$ has infinite order then $\tau(\gamma) > 0$.*

Proof. The inclusion into Γ of the infinite cyclic subgroup generated by γ is a quasi-isometric embedding (4.10). □

4.19 Corollary. *If $|p| \neq |q|$ then $\langle x, t \mid t^{-1}x^p t = x^q \rangle$ cannot be a subgroup of a semihyperbolic group.*

Proof. In any finitely generated group Γ, the number $\tau(\gamma)$ depends only on the conjugacy class of $\gamma \in \Gamma$ and $\tau(\gamma^n) = |n|\,\tau(\gamma)$ for all $n \in \mathbb{Z}$. Thus if γ^p is conjugate to γ^q in Γ, then

$$|p|\,\tau(\gamma) = \tau(\gamma^p) = \tau(\gamma^q) = |q|\,\tau(\gamma).$$

Thus $|p| \neq |q|$ implies $\tau(\gamma) = 0$. Since x has infinite order in $B = \langle x, t \mid t^{-1}x^p t = x^q\rangle$ (as we explained in the proof of (1.1)), it follows from (4.18) that $B \subset \Gamma$ implies Γ is not semihyperbolic. □

Direct Products

In analogy with (1.1(a)) we have:

4.20 Exercise. The direct product of any two semihyperbolic groups is semihyperbolic.

And in analogy with the Splitting Theorem (II.6.21) we have:

4.21 Theorem (Algebraic Splitting). *If $\Gamma = \Gamma_1 \times \Gamma_2$ is semihyperbolic and the centre of Γ_2 is finite, then Γ_1 is semihyperbolic.*

Proof. Since Γ is finitely generated, so too are Γ_1 and Γ_2. Let $\mathcal{A}$ be a finite generating set for Γ_2. Let $S = \{(1, a) \mid a \in \mathcal{A}\} \subset \Gamma$ and note that $C_\Gamma(S) = \Gamma_1 \times Z(\Gamma_2)$. From (4.15) we know that $C_\Gamma(S)$ is semihyperbolic, and since Γ_1 has finite index in $C_\Gamma(S)$, it too is semihyperbolic. □

4.22 Remarks. At this stage we have established that most but not all of the properties listed in (1.1) remain valid in the context of semihyperbolic groups. We conclude by saying what is known about the remaining points.

The class of semihyperbolic groups is closed under free products with amalgamation and HNN extensions along finite subgroups (see [AloB95]). It is also known that every semihyperbolic group Γ is of type FP_∞, but it is unknown whether Γ has a finite $K(\Gamma, 1)$ if it is torsion-free. It is unknown if semihyperbolic groups can contain solvable subgroups that are not virtually abelian or abelian subgroups that are not finitely generated.

In contrast to (1.1(iv)), if Γ is semihyperbolic and $\gamma \in \Gamma$ has infinite order, then in general $\langle\gamma\rangle$ will not be a direct factor of a subgroup of finite index in $C_\Gamma(\gamma)$. Indeed, following a suggestion of Thurston and Gromov, Neumann and Reeves [NeR97] showed that if G is a hyperbolic group and A is a finitely generated abelian group, then every central extension Γ of the form $1 \to A \to \Gamma \to G \to 1$ is semihyperbolic (indeed biautomatic) — in general such central extensions Γ do not have subgroups of finite index that are non-trivial direct products.

In the above construction, Γ is quasi-isometric to $G \times A$ (although in general they are not commensurable, cf. I.8.21). Using this fact one can show that there exist semihyperbolic groups Γ_1 and Γ_2 such that Γ_1 and Γ_2 are quasi-isometric and Γ_1 is the fundamental group of a closed non-positively curved manifold, yet Γ_2 cannot act properly by semi-simple isometries on any CAT(0) space. For example, if Σ is a closed hyperbolic surface, then one can take Γ_1 to be $\pi_1\Sigma \times \mathbb{Z}$ and Γ_2 to be the fundamental group of the unit tangent bundle to Σ. (This example is explained in detail in various places, including [Ep+92] and Section 8 of [AloB95].)

5. Subgroups of Cocompact Groups of Isometries

We have said a great deal about groups which act properly and compactly by isometries on CAT(0) spaces. In the case of some results it was sufficient to assume that the groups concerned were acting properly and semisimply (see (1.1)), in other cases we definitely used the fact that the groups concerned were acting cocompactly. The main purpose of this section is to show that in the latter case the results concerned do not remain valid if one assumes only that the group is acting properly by semisimple isometries. Specifically, we shall prove that the finiteness properties that we established for cocompact groups of isometries are not inherited by finitely presented subgroups, and we shall show that the solutions to the word and conjugacy problem established in Section 1 are not inherited by finitely presented subgroups. In the course of our discussion of decision problems we shall also prove that there are compact non-positively curved spaces X for which there is no algorithm to decide isomorphism among the finitely presented subgroups of $\pi_1 X$.

The underlying theme throughout this section is that subgroups of groups which act properly and cocompactly by isometries on CAT(0) spaces can be rather complicated.

Our discussion of finiteness properties is motivated by examples of John Stallings [St63] and Robert Bieri [Bi76b]. Our discussion of decision problems is based on that in the papers of Baumslag, Bridson, Miller and Short ([BBMS98] and subsequent preprints).

Finiteness Properties

The most common and useful finiteness properties of groups are finite generation and finite presentability. In (3.25) we described three conditions that measure higher-dimensional finiteness, F_n, FP_n and FL_n. (See [Bro82] or [Bi76b] for a detailed introduction to these conditions.)

If we write $H_n(\Gamma)$ to denote the nth homology group of Γ with coefficients in the trivial Γ-module $\mathbb{Z}$, then

$$F_n \implies FL_n \implies FP_n \implies H_n(\Gamma) \text{ f.g.}$$

Bestvina and Brady [BesB97] recently proved that there exist finitely generated groups which are of type FP_2 but not of type F_2.

We saw in (II.5.13) that if Γ is the fundamental group of a compact non-positively curved space, then Γ is of type F_n for every $n \in \mathbb{N}$. Our purpose in this subsection is to show that this property is not inherited by arbitrary finitely presented subgroups of Γ. In the process of doing so we shall construct explicit examples of finitely presented subgroups that cannot be made to act properly and *cocompactly* by isometries on any CAT(0) space (cf. 5.10).

Any direct product of finitely generated free groups is the fundamental group of a compact non-positively curved space, namely the cube complex obtained by taking the Cartesian product of graphs which have one vertex and the appropriate number of loops. Such direct products have a surprising variety of finitely presented subgroups, as the following theorem illustrates. We shall use this theorem as a convenient focus for our discussion of finiteness properties, but we should offset this by emphasizing that results of this type apply to many groups other than direct products of free groups (see [BesB97], [MeiMV97], [Bi98]).

Given a group G, a subgroup $H \subset G$ and a positive integer m, we write $\Delta_m(G; H)$ to denote the free product of m copies of G amalgamated along H (see 6.5) and refer to the m distinguished copies of G as the *factors* of $\Delta_m(G; H)$. There is an implicit isomorphism from each factor to G and these isomorphisms agree on H. It is sometimes convenient to regard $\Delta_m(G; H)$ as a subgroup of $\Delta_n(G; H)$ if $m < n$, namely the subgroup generated by the first m factors. Note that there is then a natural retraction $\Delta_n(G; H) \to \Delta_m(G; H)$ — this retraction is the identity on the first m factors and sends each of the remaining factors to the first factor by means of the implicit isomorphisms with G.

5.1 Theorem. *L be a free group of rank $m \geq 2$ and $L^{(n)}$ denote the direct product of n copies of L. Let $\phi : L \twoheadrightarrow \mathbb{Z}$ be an epimorphism and let K_n be the kernel of the induced homomorphism $L^{(n)} \twoheadrightarrow \mathbb{Z}$. For every $n \geq 1$,*

(1) $K_{n+1} \cong \Delta_m(L^{(n)}; K_n)$, *and*

(2) *K_n is of type F_{n-1} but $H_n(K_n)$ is not finitely generated.*

5.2 Exercises. With the notation of (5.1):

(1) Show that up to isomorphism K_n is independent of the epimorphism ϕ.

(2) Let $\Psi : L^{(n)} \to \mathbb{Z}$ be a homomorphism. Show that if $\ker \Psi$ is of type F_r but not type F_{r+1}, then the restriction of Ψ to exactly $(n - r - 1)$ of the direct factors of $L^{(n)}$ is trivial. (Hint: If H has finite index in G, then G is of type F_r if and only if H is of type F_r.)

We shall present the proof of (5.1) as a series of lemmas. The following simple lemma indicates how the lack of finiteness described in 5.1(2) begins.

5.3 Lemma. *Let L be a finitely generated free group and let $N \subset L$ be a non-trivial normal subgroup. If L/N is infinite then $H_1(N)$ is not finitely generated.*

Proof. L is the fundamental group of a finite graph X with a single vertex. Consider the covering space $\widehat{X} = \widetilde{X}/N$ corresponding to N. Because L/N acts freely and transitively on the vertices of $\widehat{X}$, if L/N is infinite then $\widehat{X}$ contains infinitely many disjoint loops. Thus the lemma follows from the fact that the fundamental group of a connected graph is freely generated by the set of edges in the complement of any maximal tree. □

5.4 Example (The Groups of Bieri and Stallings).

Let L be a free group of rank two with basis $\{a_1, a_2\}$. Let $L^{(n)}$ denote the direct product of n copies of L, and let $h_n : L^{(n)} \to \langle t \rangle \cong \mathbb{Z}$ be the homomorphism that sends each of the generators $(1, \dots, a_i, \dots, 1)$ to t. Let $\mathrm{SB}_n = \ker h_n$; in the notation of (5.1), this is K_n in the case $m = 2$.

Lemma 5.3 says that SB_1 is not finitely generated. If $n \geq 2$ then SB_n is finitely generated, for example SB_2 is generated by

$$\{(a_1a_2^{-1}, 1), (1, a_1a_2^{-1}), (a_1, a_1^{-1}), (a_2, a_2^{-1})\}.$$

Theorem 5.1 says that SB_2 is not finitely presentable and that SB_3 is finitely presentable but $H_3(\mathrm{SB}_3)$ is not finitely generated. SB_3 was the first example of a group with these properties; it was discovered by John Stallings [St63]. Robert Bieri [Bi76b] recast Stallings's example in the terms we have used and proved that SB_n is of type F_{n-1} but not type F_n; this was the first such sequence of groups to be discovered. There is now a substantial theory of finiteness properties of groups — see [Bi98] for a recent survey and references.

The first of the following lemmas provides a tool for showing that groups are *not* of type F_n, the second provides a tool for showing that groups *are* of type F_n.

5.5 Lemma. *Let $\Gamma = A *_C B$. If A and B are of type F_n but $H_{n-1}(C)$ is not finitely generated, then $H_n(\Gamma)$ is not finitely generated.*

Proof. Consider the Mayer-Vietoris sequence for $\Gamma = A *_C B$:

$$\cdots \longrightarrow H_n(A) \oplus H_n(B) \longrightarrow H_n(\Gamma) \longrightarrow H_{n-1}(C) \longrightarrow H_{n-1}(A) \oplus H_{n-1}(B) \longrightarrow \dots$$

□

5.6 Corollary. *Let A and C be finitely generated groups, let $m \geq 2$ be an integer, and let $D_m = \Delta_m(A; C)$. If A is of type F_n but $H_{n-1}(C)$ is not finitely generated, then $H_n(D_m)$ is not finitely generated.*

Proof. As we noted before stating (5.1), there is a retraction of D_m onto $D_2 = A *_C A$. Thus $H_n(D_2)$ is a direct summand of $H_n(D_m)$, and we can apply the lemma. □

5.7 Lemma. *If A is of type F_n and C is of type F_{n-1}, then $\Delta_m(A; C)$ is of type F_n.*

Proof. Let X_A be an Eilenberg-MacLane complex for A that has a finite n-skeleton. Let X_C be an Eilenberg-MacLane space for C that has a finite $(n-1)$-skeleton. Let $f : X_C \to X_A$ be a cellular map that induces the inclusion $C \hookrightarrow A$. One obtains an Eilenberg-MacLane complex X for $\Delta_m(A; C)$ by taking $(m-1)$ disjoint copies of the double mapping cylinder for f and joining them at one end:

$$\frac{\coprod_{i=1}^{m-1} \left((X_A, 0, i) \cup \left(X_C \times [0, 1], i\right) \cup (X_A, 1, i)\right)}{(x, 0, 1) \sim (x, 0, i),\ \ (y, 0, i) \sim (f(y), 0, i) \text{ and } (y, 1, i) \sim (f(y), 1, i)},$$

where $x \in X_A$, $y \in X_C$ and $i = 1, \dots, m-1$.

The open n-cells in X are of two kinds: there are the open n-cells in the images of X_A, and there are the products $e \times (0, 1)$ where e is an open $(n-1)$-cell in X_C. Thus X has only finitely many n-cells. □

We need two more lemmas.

5.8 Lemma. *If $N \subset G$ is normal then there is an embedding*

$$\Phi : \Delta_m(G; N) \hookrightarrow \left(*_{i=1}^m G/N\right) \times G.$$

Proof. Let $\phi : \Delta_m(G; N) \to *_{i=1}^m G/N$ be the quotient by N, and let $\psi : \Delta_m(G; N) \to G$ be the natural retraction onto the first factor (copy of G). $\ker \phi = N$ and the restriction of ψ to N is injective. Thus $\Phi := (\phi, \psi)$ is injective. □

In the following lemma it is convenient to write the infinite cyclic group $\langle\tau\rangle$ additively.

5.9 Lemma. *Let G be a group. Let $h : G \to \langle\tau\rangle$ be an epimorphism to an infinite cyclic group and let $N = \ker h$. Choose $a \in G$ with $h(a) = \tau$ and let $\overline{a}$ denote the image of a in G/N. Let $\hat{h} : \left(*_{i=1}^m G/N\right) \times G \to \langle\tau\rangle$ be the map that is given on each of the free factors of $*_{i=1}^m G/N$ by $\hat{h}(\overline{a}) = -\tau$ and is such that $\hat{h}|_G = h$. Let Φ be as in (5.8). Then* $\ker \hat{h} = \operatorname{im} \Phi$.

Proof. $\Delta_m(G; N)$ is generated by N and $a_1, \dots, a_m$, where $\Phi(a_i) = (\overline{a}_i, a)$ and $\overline{a}_i$ generates the i-th free factor of $*_{i=1}^m G/N$. For each $g \in N$ we have $\Phi(g) = (1, g)$, and hence $\hat{h}\Phi(g) = h(g) = 0$. For each a_i we have: $\hat{h}\Phi(a) = \hat{h}(\overline{a}_i, a) = -\tau + \tau$. Thus $\operatorname{im} \Phi \subset \ker \hat{h}$.

Now suppose $(u, v) \in \ker \hat{h}$. Write u as a word $\overline{a}_{i(1)}^{s_1} \dots \overline{a}_{i(r)}^{s_r}$ and let

$$\gamma = (a_{i(1)}^{s_1} \dots a_{i(r)}^{s_r})\, a_1^{-h(v)} v.$$

Obviously $a_1^{-h(v)} v$ lies in $N = \ker h$, so the first coordinate of $\Phi(\gamma)$ is u. On the other hand, because $-\hat{h}(u) = \hat{h}(v) = h(v)$, we have $\sum s_i = h(v)$, and therefore the second coordinate of $\Phi(\gamma)$ is v. □

Proof of Theorem 5.1

In the notation of (5.1): Lemma 5.8, with $G = L^{(n)}$ gives an embedding $\Delta_m(L^{(n)}; K_n) \hookrightarrow \left(*_{i=1}^m \mathbb{Z}\right) \times L^{(n)} \cong L^{(n+1)}$, and Lemma 5.9 shows that (modulo the natural identifications) the image of this embedding is K_{n+1}. This proves 5.1(1).

We shall prove 5.1(2) by induction on n, using 5.1(1). In the induction we shall use the fact that $L^{(n)}$ is of type F_r for every $r \in \mathbb{N}$. (It is the fundamental group of the compact non-positively curved cube complex obtained by taking the product of n metric graphs each of which has one vertex and m edges.)

The case $n = 1$ is covered by Lemma 5.3. Thus we may suppose that $n \geq 2$ and that $K_n \subset L^{(n)}$ is of type F_{n-1} but that $H_n(K_n)$ is not finitely generated. According to (5.6), the $(n + 1)$st integral homology of $\Delta_m(L^{(n)}; K_n)$ is not finitely generated. And according to (5.7), $\Delta_m(L^n; K_n)$ is of type F_n. This completes the induction. □

The first part of Theorem 5.1 enables one to write down a simple presentation of K_n. We give this description in the case $m = 2$, in which case $K_n \cong \mathrm{SB}_n$, the n-th Stallings-Bieri group.

5.10 Proposition (A Presentation for SB_n). *If $n \geq 2$ then SB_{n+1} is generated by $3n$ elements $x_1, \ldots, x_n, y_1, y_1', \ldots, y_n, y_n'$ subject to the relations*

$$y_1^{-1}y_j = y_1'^{-1}y_j', \quad [x_i, x_j] = [x_i, y_j] = [x_i, y_j'] = [y_i, y_j] = [y_i', y_j'] = 1$$

for $1 \leq i < j \leq n$.

Proof. Let L_2 be a free group with basis $\{x, y\}$ and let $x_i, y_i, i = 1, \ldots, n$ be the obvious generators for $L_2^{(n)}$ (that is, $x_i = (1, \ldots, x, \ldots, 1)$ etc.). The group SB_n is the kernel of the map $\Psi : L_2^{(n)} \twoheadrightarrow \mathbb{Z}$ that sends each x_i to the identity and each y_i to a fixed generator of $\mathbb{Z}$.

The presentation displayed above describes $\Delta_2(L_2^{(n)}; H)$, where H is the subgroup generated by $\{x_1, \ldots, x_n, y_1^{-1}y_2, \ldots, y_1^{-1}y_n\}$. According to 5.1(1), in order to prove the present proposition, it suffices to show that $H = \ker \Psi$. It is clear that $H \subset \ker \Psi$. Conversely, suppose that $\gamma = \gamma_1 \ldots \gamma_n \in \ker \Psi$, where γ_i is a word in the generators x_i and y_i.

Each γ_i is equal in the free group on $\{x_i, y_i\}$ to a word of the form

$$\left(\prod_{k=1}^{N_i} y_i^{p_{i,k}} x_i^{r_{i,k}} y_i^{-p_{i,k}} \right) y_i^{q_i}.$$

For example, $y_i^2 x_i y_i x_i^2 = (y_i^2 x_i y_i^{-2})(y_i^3 x_i^2 y_i^{-3}) y_i^3$.

If $i \geq 2$, then $(y_1^{-1}y_i)^p x_i^r (y_1^{-1}y_i)^{-p} = y_i^p x_i^r y_i^{-p}$, and we also have $(y_2^{-1}y_1)^p x_1^r (y_2^{-1}y_1)^{-p} = y_1^p x_1^r y_1^{-p}$. Hence all elements of the form $y_i^{p_{i,k}} x_i^{r_{i,k}} y_i^{-p_{i,k}}$ are contained in H. Thus in order to show that $\ker \Psi \subset H$, we need only show that H contains all elements of the form $\gamma = y_1^{q_1} \ldots y_n^{q_n}$ with $\sum q_i = 0$. But in this case, in $L_2^{(n)}$ we can write

$$\gamma = \prod_{i=2}^{n} (y_1^{-1} y_i)^{q_i},$$

and since each $y_1^{-1}y_i$ is in H, we are done. □

5.11 Remarks

(1) The paper of Bestvina and Brady that we mentioned earlier [BesB97] clarified the relationship between various finiteness properties of groups. Their main tool for doing so was a Morse theory that they developed for CAT(0) cube complexes. If Γ is the fundamental group of a compact non-positively curved cube complex X and $h : \Gamma \to \mathbb{Z}$ is an epimorphism for which there is an h-equivariant Morse function[58] $\mu : \widetilde{X} \to \mathbb{R}$, then Bestvina and Brady are able to relate the finiteness properties of $\ker h$ to the connectedness properties of the sub-level sets of μ.

The maps $h_n : L^{(n)} \to \mathbb{Z}$ considered in (5.4) were the prototypes for this approach; the complex X is a product of graphs each with one vertex and two edges; $L^{(n)}$ can be identified with the vertex set of $\widetilde{X}$ and if one identifies $L^{(n)}$ with the zero-skeleton of the universal covering $\widetilde{X}$, then h_n extends uniquely to a Morse function $\widetilde{X} \to \mathbb{R}$.

(2) For $n \geq 2$, the Dehn functions of the groups K_n described in Theorem 5.1 are quadratic [Bri99b]. Thus the existence of a quadratic isoperimetric inequality for a group does not constrain the higher finiteness properties of a group, in contrast to the subquadratic (i.e. hyperbolic) case (3.21).

5.12 Exercises

(1) Let L and K_2 be as in (5.1). We have shown that K_2 is not finitely presentable. There is a theorem of Baumslag and Roseblade [BR84] which states that a subgroup of $L^{(2)}$ is finitely presentable if and only of it contains a subgroup of finite index that is either free or a direct product of free groups. By examining the structure of centralizers, verify directly that K_2 does not contain such a subgroup of finite index.

(2) (The HNN analogue of Lemma 5.8.) Let G be a group and let $A \subset G$ be a normal subgroup. Show that the HNN extension $(G, t \mid t^{-1}at = a, \ \forall a \in A)$ can be embedded in the direct product of G and $(G/A) * \mathbb{Z}$.

(3) In the notation of (5.1): Show that K_n is an isometrically embedded subgroup of $L^{(n)}$ if $n \geq 2$. (Note that in the case $n = 2$ this shows that an isometrically embedded subgroup of a semihyperbolic group need not be finitely presented — cf. 3.7.)

(Hint: Follow the last part of the proof of (5.10).)

[58] i.e. a map $\mu : \widetilde{X} \to \mathbb{R}$ that is affine on cells, constant only on 0-cells, and has the property that the image of the 0-skeleton is discrete

The Word, Conjugacy and Membership Problems

We now turn our attention to decision problems for finitely presented subgroups of groups which act properly and cocompactly by isometries on spaces of non-positive curvature. We shall say something about each of the basic decision problems that lie at the heart of combinatorial group theory: the word problem, the conjugacy problem, and the isomorphism problem. We shall also say something about the generalized word problem. At the end of the section we shall explain how these group theoretic results impinge on the question of whether or not there exists an algorithm to determine homeomorphism among compact non-positively curved manifolds.

The Word Problem

If a group has a solvable word problem then so do all of its finitely generated subgroups: any word in the generators of the subgroup can be rewritten in terms of the generators of the ambient group, and by applying the decision process in the ambient group one can decide if the given word represents the identity.

The complexity of the word problem is a more complicated matter. The Dehn function of a finitely presented group, as defined in (I.8A), estimates the number of relations that one must apply in order to decide if a given word in the generators and their inverse represents the identity in the group. If Γ acts properly and cocompactly by isometries on a CAT(0) space, then the Dehn function for any finite presentation of Γ is bounded above by a quadratic function (see 1.7). Since there is no obvious way to bound the Dehn function of a finitely presented subgroup in terms of the Dehn function of the ambient group, given a group that acts properly and cocompactly by isometries on a CAT(0) space one might be able to use Dehn functions to identify finitely presented subgroups that cannot themselves act properly and cocompactly by isometries on any CAT(0) space. Here is an example of this phenomenon.

5.13 Theorem. *There exist closed non-positively curved 5-dimensional manifolds and finitely presented subgroups $H \subset \pi_1 M$ such that the Dehn function of H is exponential.*

Proof. Let N be a closed hyperbolic 3-manifold that fibres over the circle with compact fibre Σ (see [Ot96] for examples). Let $S = \pi_1\Sigma$. The Dehn function of the double $\Delta_2(\pi_1 N; S)$ is exponential (see 6.22). Lemma 5.8 gives an embedding of $\Delta_2(\pi_1 N; S)$ into the direct product $\pi_1 N \times F$, where F is a free group of rank 2. This direct product is a subgroup of any closed non-positively curved 5-manifold of the form $M = N \times Y$ where Y is a closed surface of genus at least two. $\square$

5.14 Remark. Examples such as (5.13) show that the difficulty of the word problem in a group G is not completely described by the Dehn function: the Dehn function measures the complexity of the challenge that one faces when trying to solve the word problem using only the information apparent in a given presentation, but extrinsic information, such as the existence of a nice embedding, might facilitate a more efficient solution.

The Membership Problem

The *generalized word problem* (often called the Magnus problem or *membership problem*) asks about the existence of an algorithm to decide whether words in the generators of a group G define elements of a fixed subgroup H. If H and G are finitely generated and G has a solvable word problem, then the generalized word problem for $H \subset G$ is solvable if and only if the distortion of H in G, as defined in (6.17), is not bounded above by any recursive function.

5.15 Proposition. *There exist compact negatively curved 2-complexes K and finitely generated subgroups $N \subset \pi_1 K$ for which there is no algorithm to decide membership of N (equivalently, the distortion of N in $\pi_1 K$ is non-recursive).*

Proof. There exist finitely presented groups G with unsolvable word problem (see [Rot95]). Let G be such a group and let $\langle x_1, \dots, x_n \mid r_1, \dots, r_m \rangle$ be a finite presentation of G. The generalized Rips construction (II.5.45) associates to this presentation a short exact sequence

$$1 \to N \to \pi_1 K \to G \to 1.$$

K is a compact, negatively curved, piecewise-hyperbolic 2-complex whose fundamental group $\pi_1 K$ has a finite presentation

$$\langle x_1, \dots, x_n, a_1, \dots, a_M \mid r_k = v_k,\ x_j^{\varepsilon} a_i x_j^{-\varepsilon} = u_{i,j,\varepsilon},$$
$$1 \le i \le M,\ 1 \le j \le n,\ 1 \le k \le m,\ \varepsilon = \pm 1 \rangle,$$

where v_k and $u_{i,j,\varepsilon}$ are words in the generators $\underline{a} = \{a_1, \dots, a_M\}$ and N is the subgroup generated by $\underline{a}$.

The key thing to observe is that a word w in the letters $\underline{x} = \{x_1, \dots, x_n\}$ and their inverses represents an element of $N \subset \Gamma$ if and only if $w = 1$ in G. We chose G specifically so that there is no algorithm to decide if $w = 1$ in G, and therefore there is no algorithm to decide whether words in the letters $\underline{x}$ represent elements of $N \subset \Gamma$. □

The groups N considered in (5.15) are never finitely presented (II.5.47). In order to construct finitely presented examples of a similar nature we shall use the following theorem of Baumslag, Bridson, Miller and Short [59]. The name of this theorem comes from the fact that the groups appearing in the short exact sequence are assumed to be of types F_1, F_2 and F_3 respectively. (Type F_n was defined in (3.25).)

---

[59] *"Fibre products, non-positive curvature, and decision problems"*, Preprint 1998.

5.16 The 1-2-3 Theorem. *Suppose that* $1 \to N \to \Gamma \xrightarrow{p} G \to 1$ *is exact, and consider the fibre product*

$$P := \{(\gamma_1, \gamma_2) \mid p(\gamma_1) = p(\gamma_2)\} \subset \Gamma \times \Gamma.$$

Suppose that N *is finitely generated, that* Γ *is finitely presented and that* G *of type* F_3*. Then* P *is finitely presented.*

It is easy to show that if N has finite generating set $\underline{a}$ and Γ is generated by $(\underline{a} \cup \underline{x})$, then P is generated by the elements $a_i^L := (a_i, 1)$, $a_i^R := (1, a_i)$, and (x_j, x_j), where $a_i \in \underline{a}$ and $x_j \in \underline{x}$. The key to proving the theorem lies in understanding all of the relations that hold in $N \subset \Gamma$. The proof is long, technical, and purely algebraic; since it would add little to the present discussion, we omit it.

5.17 Proposition. *There exist compact non-positively curved 4-dimensional complexes* X *and finitely presented subgroups* $P \subset \pi_1 X$ *for which there is no algorithm to decide membership of* P *(i.e., the distortion of* P *in* $\pi_1 X$ *is non-recursive).*

Proof. Collins and Miller [CoMi98] have constructed examples of finitely presented groups of type F_3 that are torsion-free and have an unsolvable word problem. By applying the construction of (II.5.45) to a finite presentation of such a group G we obtain a short exact sequence

$$1 \to N \to \pi_1 K \xrightarrow{p} G \to 1,$$

where K and N are as in the proof of (5.15). Let $X = K \times K$.

Take generators x_j and a_i for $\pi_1 K$ as in (5.15) and consider the associated generators $(x_j, 1), (1, x_j), (a_i, 1), (1, a_i)$ for $\pi_1 X$. Theorem 5.16 tells us that the subgroup $P = \{(\gamma_1, \gamma_2) \mid p(\gamma_1) = p(\gamma_2)\} \subset \pi_1 X$ is finitely presented. Given a word w in the generators $(x_j, 1)$, we ask if this word defines an element of P. The answer is "yes" if and only if the word obtained from w by replacing each $(x_j, 1)$ by x_j represents the identity in G. Thus membership of P is undecidable. □

The Conjugacy Problem

In general, the conjugacy problem is less robust than the word problem. For example, there exist finitely presented groups which have a solvable conjugacy problem but which contain subgroups of index two that have an unsolvable conjugacy problem [CoMi77]. In (1.12) we showed that if a group Γ acts properly and cocompactly by isometries on a CAT(0) space then it has a solvable conjugacy problem. Thus we might use the (un)solvability of the conjugacy problem as an invariant for identifying finitely presented subgroups of Γ that cannot act properly and cocompactly on any CAT(0) space.

Notation. We shall continue our earlier practice of writing $C_\Gamma(g)$ for the centralizer of the element g in the group Γ.

5.18 Lemma. *Let $H \subset Q \subset \Gamma$ be finitely generated groups. Suppose that H is normal in Γ and that there exists $h_0 \in H$ such that $C_\Gamma(h_0) \subset Q$.*

If there is no algorithm to decide membership of Q, then Q has an unsolvable conjugacy problem.

Proof. We fix finite generating sets $\mathcal{B}$ for Γ and $\mathcal{A}$ for H, and for each $a \in \mathcal{A}$, $b \in \mathcal{B}$ and $\varepsilon = \pm 1$ we choose a word $u_{a,b,\varepsilon}$ in the generators $\mathcal{A}^{\pm 1}$ so that $b^\varepsilon a b^{-\varepsilon} = u_{a,b,\varepsilon}$ in H. Fix a word in the generators $\mathcal{A}$ that equals h_0 in Γ.

Given an arbitrary word w in the generators $\mathcal{B}^{\pm 1}$, we can use the relations $b^\varepsilon a b^{-\varepsilon} = u_{a,b,\epsilon}$ to write $w h_0 w^{-1}$ as a word w' in the generators $\mathcal{A}^{\pm 1}$. (The length of w' is bounded by an exponential function of the length of w and the process of passing from w to w' is entirely algorithmic.)

Now ask if w' is conjugate to h_0 in Q. Well, if there exists $q \in Q$ such that $q^{-1} h_0 q = w'$, then $qw \in C_\Gamma(h_0) \subset Q$, whence $w \in Q$. Thus w' is conjugate to h_0 in Q if and only if $w \in Q$. And there is no algorithm to decide membership of Q. $\square$

5.19 Theorem. *There exists a negatively curved 2-dimensional complex K and a finitely presented subgroup $P \subset \pi_1(K \times K)$ such that P has an unsolvable conjugacy problem.*

Proof. Let K, N and P be as in the proof of (5.17). Fix one of the generators a of N (each generator is non-trivial, II.5.46). Let $h_0 = (a, a) \in \pi_1 K \times \pi_1 K$. Because $\pi_1 K$ is torsion-free (II.4.13) and hyperbolic, the centralizer of a in $\pi_1 K$ is cyclic (3.10). Moreover, a is not a proper power, because the closed geodesic in K representing a is the shortest homotopically non-trivial loop in K (II.5.46). Thus the centralizer of a in $\pi_1 K$ is $\langle a \rangle$ and the centralizer of h_0 in $\pi_1(K \times K)$ is $\{(a_i^n, a_i^m) \mid n, m \in \mathbb{Z}\}$, which is contained in $N \times N \subset P$. Therefore we may apply Lemma 5.18 with $H = N \times N$ and $Q = P$. $\square$

5.20 Remarks

(1) There exists a negatively curved 2-dimensional complex K and a finitely generated normal subgroup $N \subset \pi_1 K$ such that N has an unsolvable conjugacy problem. To see this, let K and N be as in the proof of (5.15) and apply Lemma 5.18 with $N = H = Q$ and one of the generators $a \in N$ in the role of h_0.

(2) At the time of writing, there are only a few known examples of non-hyperbolic finitely presented subgroups in hyperbolic groups. The known examples all have a solvable conjugacy problem.

Isomorphism Problems

The main result of this subsection is:

5.21 Theorem. *There exists a compact, non-positively curved 4-dimensional polyhedral complex X and a recursive sequence of finitely presented groups P_n ($n \in \mathbb{N}$) with explicit monomorphisms $P_n \to \pi_1 X$, such that there is no algorithm to determine if P_n is (abstractly) isomorphic to P_1.*

This theorem is due to Baumslag, Bridson, Miller and Short. The original proof was algebraic. We shall give a shorter geometric proof based on the Flat Torus Theorem (II.7.1).

Whenever one is trying to prove that the isomorphism problem is undecidable in a certain class of groups, one is invariably faced with the difficulty of ruling out "accidental" isomorphisms – one needs invariants that allow one to deduce that if some obvious map is not an isomorphism then the groups in question are not isomorphic. The invariants that we shall use for this purpose are the centralizers of elements; our previous results concerning centralizers in groups that act by semi-simple isometries on CAT(0) spaces will be useful in this regard.

In the following lemma we can ignore the obvious basepoint difficulties associated to statements about π_1 because we are only interested in specifying elements and subgroups up to conjugacy.

5.22 Lemma. *Let H be a group acting freely and properly by semisimple isometries on a complete* CAT(-1) *space X. Let $c_0, c_1 : S \to H\backslash X$ be isometrically embedded circles (of the same length) and let Y be the space obtained by gluing the ends of a cylinder $C = S \times [0, 1]$ to $H\backslash X$ along c_0 and c_1 respectively. Let $\overline{C}$ be the image of C in Y.*

(1) *If the images of c_0 and c_1 are not the same, then $\pi_1 Y$ does not contain a subgroup isomorphic to $\mathbb{Z}^2$.*

(2) *If the images of c_0 and c_1 are the same, then $\pi_1 Y$ contains subgroups isomorphic to $\mathbb{Z}^2$, and any such subgroup is conjugate to a subgroup of $\pi_1 \overline{C} \subset \pi_1 Y$.*

Proof. Note that the local gluing theorem (II.11.6) implies that Y is non-positively curved in the quotient path metric, and $H\backslash X \hookrightarrow Y$ is a local isometry.

(2) If the images of c_0 and c_1 coincide, then $\overline{C}$ is an isometrically embedded flat torus or Klein bottle, and by (II.4.13) $\pi_1 \overline{C} \subset \pi_1 Y$ is isomorphic to the fundamental group of the surface. Moreover, since $H\backslash X$ is locally CAT(-1), the image of any local isometry from a torus to Y must be contained in $\overline{C}$. The Flat Torus Theorem tells us that every monomorphism $\phi : \mathbb{Z}^2 \to \pi_1 Y$ is represented by such a map from a torus, and therefore the image of ϕ is conjugate to a subgroup of $\pi_1 \overline{C} \subset \pi_1 Y$.

(1) Suppose that the images of c_0 and c_1 do not coincide, and consider a point p that is in the image of c_0 but not in the image of c_1. In Y, this point p has a neighbourhood that is obtained by gluing a Euclidean half-disc (a neighbourhood of a boundary

point in C) to a CAT(-1) space $B(p, \varepsilon) \subset H\backslash X$. In particular, given any Euclidean disc $B(0, r) \subset \mathbb{E}^2$, there does not exist an isometric embedding $f : B(0, r) \to Y$ with $f(0) = p$. Thus if $\tilde{c}_0$ is a geodesic line in $\widetilde{Y}$ that covers c_0, then $\tilde{c}_0$ cannot be contained in any flat plane. Hence the strip $\mathbb{R} \times [0, 1]$ in the pre-image of $\overline{C}$ that is attached to $\tilde{c}_0$ is not contained in any flat plane. Since the complement of the preimage of $\overline{C}$ is locally CAT(-1), we have proved that there are no flat planes in $\widetilde{Y}$, and therefore, by the Flat Torus Theorem, no subgroup of $\pi_1 Y$ is isomorphic to $\mathbb{Z}^2$. □

5.23 Corollary. *Let N and K be as in the modified Rips construction (II.5.45) and fix one of the generators $a \in \underline{a}$ described there.*

(1) *If a subgroup of the HNN-extension*

$$\left(\pi_1 K, \tau \mid \tau^{-1} a \tau = a\right)$$

is isomorphic to $\mathbb{Z}^2$, then it is conjugate to a subgroup of $\langle a, \tau \rangle$.

(2) *If $\gamma \in \pi_1 K$, then the HNN-extension*

$$N_\gamma := \left(N, t \mid t^{-1} a t = \gamma^{-1} a \gamma\right)$$

contains a copy of $\mathbb{Z}^2$ if and only if $\gamma \in N$.

Proof. We pointed out in (II.5.46) that the free homotopy class of loops representing the conjugacy class of $a \in \pi_1 K$ is represented by a unique isometrically embedded circle, $c : S \to K$ say. By the Seifert-van Kampen Theorem, the HNN extension displayed in (1) is the fundamental group of the space obtained from K by attaching both ends of a cylinder $S \times [0, 1]$ along c. Thus assertion (1) is an immediate consequence of 5.22(2).

(2) In $N\backslash\widetilde{K}$, the conjugacy classes in N of a and $\gamma^{-1} a \gamma \in N$ are each represented by a unique isometrically embedded circle — call these circles c_0 and c_1, respectively. These circles have the same image if and only if a and $\gamma^{-1} a \gamma$ are conjugate in N.

$\widetilde{K}$ is a CAT(-1) space, so we can apply the lemma with $X = \widetilde{K}$ and $H = N$. By the Seifert-van Kampen Theorem, the space Y obtained by attaching the ends of a cylinder to c_0 and c_1 is N_γ. According to (5.22), this group contains $\mathbb{Z}^2$ if and only if a is conjugate to $\gamma^{-1} a \gamma$ in N.

If there exists $n \in N$ such that $n^{-1} a n = \gamma^{-1} a \gamma$, then $n \gamma^{-1} \in C_\Gamma(a)$. But, as we noted in the proof of (5.19), $C_\Gamma(a) = \langle a \rangle \subset N$. Thus a is conjugate to $\gamma^{-1} a \gamma$ in N if and only if $\gamma \in N$. □

We need one more lemma in order to prove Theorem 5.21.

5.24 Lemma. *Let H be a group. For all $a, h \in H$, the following HNN extensions are isomorphic:*

$$(H, t \mid t^{-1} a t = h^{-1} a h) \cong (H, \tau \mid \tau^{-1} a \tau = a).$$

Proof. The desired isomorphism sends H to itself by the identity and sends t to τh. The inverse sends τ to $t h^{-1}$. □

The Proof of Theorem 5.21. Let N, K and P be as in (5.17) and let $\Gamma = \pi_1 K$. We shall work with the presentation of $\Gamma = \pi_1 K$ described in the proof of (5.15); in particular Γ has generators $(\underline{a} \cup \underline{x})$. We write $\mathcal{B}$ to denote the set of generators for P described immediately following (5.16), namely $a_i^L = (a_i, 1)$, $a_i^R = (1, a_i)$, with $a_i \in \underline{a}$, and (x, x), $x \in \underline{x}$. Let $\langle \mathcal{B} \mid \mathcal{R} \rangle$ be a finite presentation for P.

Given a reduced word in the letters $\underline{x}^{\pm 1}$, we can use the relations $x_j^{\varepsilon} a_i x_j^{-\varepsilon} = u_{i,j,\varepsilon}$ in our presentation of Γ to rewrite $w^{-1} a_1 w$ as a word W in the letters $\underline{a}^{\pm 1}$ that represents the same element of $N \subset \Gamma$. We write W^L for the word in the generators of P obtained from W by replacing each a_i by $a_i^L = (a_i, 1)$.

The length of W is bounded by an exponential function of the length of w and the process of passing from w to W is entirely algorithmic. Thus we obtain a recursive family of group presentations indexed by the set of reduced words w in the letters $\underline{x}^{\pm 1}$:

$$\mathcal{P}_w = \langle \mathcal{B}, t_w \mid \mathcal{R},\ t_w^{-1} a_1^L t_w = W^L,\ t_w^{-1} a_i^R t_w = a_i^R\ \forall a_i \in \underline{a} \rangle.$$

Note that $\mathcal{P}_w$ is a presentation for the HNN extension of P by a single stable letter t_w that commutes with $P \cap (\{1\} \times \Gamma)$ and conjugates a_1^L to W^L. Let $\tilde{P}_w$ denote this HNN extension.

We also consider the following HNN extensions of Γ:

$$\Gamma_w = (\Gamma, \bar{t}_w \mid \bar{t}_w^{-1} a_1 \bar{t}_w = w^{-1} a_1 w = W),$$

and regard $\Gamma_w \times \Gamma$ as an HNN extension of $\Gamma \times \Gamma$ with stable letter $t_w := (\bar{t}_w, 1)$. Let P_w be the subgroup of $\Gamma_w \times \Gamma$ generated by P and t_w.

Each element of $P \subset \Gamma \times \Gamma$ is of the form $p = (\gamma, \gamma\alpha)$ where $\alpha \in N$. Thus $p \in \langle W^L, \underline{a}^R \rangle$ if and only if $p \in \langle W^L \rangle \times \Gamma$. Similarly, $p \in \langle a_1^L, \underline{a}^R \rangle$ if and only if $p \in \langle a_1 \rangle \times \Gamma$. It follows that if an element of the HNN extension $\tilde{P}_w$ is in reduced form (in the sense of (6.4)), then its image in Γ_w is also in reduced form. Hence, for each word w, the natural map $\tilde{P}_w \to P_w \subset \Gamma_w \times \Gamma$ is injective.

By combining these natural maps with the explicit isomorphisms $\Gamma_w \to \Gamma$ described in Lemma 5.24, we get explicit embeddings into $\Gamma_1 \times \Gamma$ of the groups given by the presentations $\mathcal{P}_w$:

$$\tilde{P}_w \xrightarrow{\cong} P_w \hookrightarrow \Gamma_w \times \Gamma \xrightarrow{\cong} \Gamma_1 \times \Gamma.$$

Γ_1 is the fundamental group of the compact non-positively curved complex K_1 obtained by attaching both ends of a cylinder along the closed local geodesic in K that represents a_1. We claim that a space X as required in the statement of Theorem 5.21 can be obtained by setting $X = K_1 \times K$, in which case $\pi_1 X = \Gamma_1 \times \Gamma$. We will be done if we can show that there does not exist an algorithm to decide if P_w is isomorphic to P_1.

There is no algorithm to decide if a word w in the generators $\underline{x}^{\pm 1}$ represents an element of $N \subset \Gamma$ (see 5.15), and therefore the following lemma completes the proof of Theorem 5.21.

5.25 Lemma. *P_w is isomorphic to P_1 if and only if P_w contains a subgroup isomorphic to $\mathbb{Z}^3$, and it contains such a subgroup if and only if w represents an element of $N \subset \Gamma$.*

Proof. Since Γ is torsion-free and hyperbolic, any homomorphism $\mathbb{Z}^3 \to \Gamma$ must have cyclic image. Applying this observation to the projection of $P_w \subset \Gamma_w \times \Gamma$ onto the second factor, we see that P_w contains a subgroup isomorphic to $\mathbb{Z}^3$ if and only if $I_w := P_w \cap (\Gamma_w \times \{1\})$ contains $\mathbb{Z}^2$. Note that since the action of $(x, x) \in P$ by conjugation on $\Gamma_w \times \{1\}$ is the same as the action of $(x, 1)$, in fact I_w is normal in $\Gamma_w \times \{1\}$; the quotient is G, in the notation of (5.17), which is torsion-free.

Applying (5.23(1)) to Γ_w (via the isomorphism $\Gamma_w \cong \Gamma_1$ in (5.24)), we see that I_w contains $\mathbb{Z}^2$ if and only if P_w contains a conjugate of a subgroup of finite index in $\langle a_1^L, t_w(w^L)^{-1}\rangle$, where $w^L = (w, 1) \in \Gamma_w \times \Gamma$. Since I_w is normal and contains a_1^L, it contains such a conjugate only if it contains a power of $t_w(w^L)^{-1}$. Since G is torsion-free, this happens only if $t_w(w^L)^{-1} \in P_w$. And since $t_w \in P_w$, this is equivalent to saying that $w^L = (w, 1)$ lies in P, in other words $(w, 1) \in N \times \{1\} = P \cap (\Gamma \times \{1\})$. □

5.26 Remarks

(1) One can subdivide the complex constructed in (II.5.45) and remetrize it as a cubical complex. If one does so, then one gets an induced cubical structure of non-positive curvature on the complex X constructed in the proof of (5.21).

(2) In the notation of the proof of (5.21), the group N_w of (5.23(2)) is given by the relative presentation $(N, t_w \mid t_w^{-1} a t_w = W)$. As above, for each word w in the generators $\underline{x}$ there is a natural monomorphism $N_w \hookrightarrow \pi_1 K_1$, and there is no algorithm to decide if N_w is isomorphic to N_1. Indeed $N_w \cong N_1$ if and only if N_w contains $\mathbb{Z}^2$, and by (5.23(2)) we know that this happens if and only if $w \in N$, and membership of N is algorithmically undecidable.

Distinguishing Among Non-Positively Curved Manifolds

Closed 2-manifolds were classified in the nineteenth century — they are determined up to homeomorphism by orientability and Euler characteristic. If Thurston's Geometrization Conjecture [Thu82] is true then the homeomorphism problem for compact 3-manifolds is also solvable. In other words, there is an algorithm which takes as input pairs of compact 3-manifolds and answers YES or NO, after a finite amount of time, according to whether or not the manifolds are homeomorphic. An easy consequence of this is that there is an algorithm that will produce a list of compact 3-manifolds such that every compact 3-manifold is homeomorphic to exactly one member of the list. (We are implicitly assuming that the manifolds under consideration are described as finite objects. For the sake of argument, since all 3-manifolds are triangulable, let us suppose that they are specified as finite simplicial complexes.)

Markov [Mark58] showed that in dimensions $n \geq 4$ there can be no algorithm to decide homeomorphism among closed (smooth, PL or topological) manifolds, because the existence of such an algorithm would contradict the fact that the isomorphism problem for finitely presented groups is unsolvable[60]. However, if one restricts one's attention to closed manifolds of negative curvature then the homeomorphism problem is solvable in dimensions $n \geq 5$. To see this one needs the following deep topological rigidity theorem of Farrell and Jones [FJ93].

5.27 Theorem. *Let $n \geq 5$ and let M and N be closed non-positively curved manifolds. If $\pi_1 M \cong \pi_1 N$ then M and N are homeomorphic.*

Farrell and Jones worked in the smooth category. For an extension to the polyhedral case see [Hu(B)93]. Fundamental groups of closed negatively curved manifolds are hyperbolic and torsion-free so, as Sela points out in [Sel95], by combining the solution to the isomorphism problem for torsion-free hyperbolic groups [Sel95] with (5.27) one gets:

5.28 Theorem. *Let $n \geq 5$ be an integer. There exists an algorithm which takes as input two closed n-manifolds that support metrics of negative curvature, and which (after a finite amount of time) will stop and answers YES or NO according to whether or not the manifolds are homeomorphic.*

There is a technical problem here with how the manifolds are given. They must be given by a finite amount of information (from which one can read off a presentation of the fundamental group). Cautious readers should interpret (5.28) as a statement regarding the homeomorphism problem for any recursive class of negatively curved manifolds.

5.29 Remarks

(1) Let $n \in \mathbb{N}$. At the time of writing it is unknown whether or not there exists an algorithm which takes as input two compact non-positively curved n-manifolds, and (after a finite amount of time) stops and answers YES or NO according to whether or not the manifolds are homeomorphic.

(2) By a process of relative hyperbolization, one can embed any compact non-positively curved piecewise Euclidean complex isometrically into a closed manifold of higher dimension that has a piecewise Euclidean metric of non-positive curvature (see [Gro87], [Hu(B)93]). Thus, in the light of (5.26(1)), the space X in Theorem 5.21 can be taken to be a closed manifold with a piecewise Euclidean metric. The connected covering spaces of this manifold are Eilenberg-MacLane complexes and hence the homotopy type of each is determined by its fundamental group. Thus (5.21) implies that there exists a closed non-positively curved manifold M and a sequence of covering spaces $\widehat{M}_i \to M$, $i \in \mathbb{N}$, with each $\pi_1 \widehat{M}_i$ finitely presented, such that there does not exist an algorithm to determine whether or not $\widehat{M}_i$ is homotopy equivalent

[60] For each $n \geq 4$, there is an algorithm that associates to any finite presentation $\mathcal{P}$ a closed n-manifold with fundamental group $|\mathcal{P}|$. Markov shows that if one could decide homeomorphism among the resulting manifolds then one could decide isomorphism among the groups being presented.

to $\widehat{M}_0$. (Again, one has to be careful about how the sequence $\widehat{M}_i$ is given in this theorem.)

In this section we have seen that the finitely presented subgroups of the fundamental groups of compact non-positively curved spaces form a much more diverse class than the fundamental groups themselves. We have essentially shown that if a property of cocompact groups is not *obviously* inherited by all finitely presented subgroups, then in general it will not be inherited. However we should offset this statement by pointing out that in general it is difficult to distinguish those subgroups which can act cocompactly by isometries on CAT(0) spaces from those that cannot. For instance:

5.30 Theorem. *Let M be a closed hyperbolic 3-manifold that fibres over the circle. There exist subgroups $G \subset \pi_1(M \times M \times M)$ such that*

(1) *G has a compact $K(G, 1)$,*

(2) *G has a quadratic Dehn function,*

(3) *G has a solvable conjugacy problem,*

(4) *the centralizer C of every element of G has a compact $K(C, 1)$, but*

(5) *G is not semihyperbolic.*

The point here is that G acts properly by semi-simple isometries on a CAT(0) space and satisfies most of the properties associated with groups that act cocompactly, and yet G cannot be made to act properly and cocompactly by isometries on any CAT(0) space.

We should also mention that, in contrast to the general theme of this section, there are certain classes of non-positively curved spaces for which it is true that if X is in the class then any finitely presented subgroup of $\pi_1 X$ is also the fundamental group of a space in the class. We saw in (II.5.27) that 2-dimensional M_κ-complexes have this property. Compact 3-dimensional manifolds [61] form another such class — this is explained in [Bri98b].

6. Amalgamating Groups of Isometries

A basic way of constructing interesting new groups is to combine known examples using amalgamated free products and HNN extensions. The Seifert-van Kampen theorem tells us that these processes appear naturally in geometry and topology: they describe exactly what happens when one starts gluing spaces along π_1-injective subspaces.

In this section we shall use our earlier results concerning gluing and isometries of CAT(0) spaces to address the following:

[61] The manifolds may have boundary, and the boundary need not be convex.

6.1 Question. *If the groups Γ_1 and Γ_2 act properly and cocompactly by isometries on* CAT(0) *spaces, then under what circumstances can one deduce that various amalgamated free products of Γ_1 and Γ_2 also act properly and cocompactly by isometries on a* CAT(0) *space? Similarly, if Γ acts properly and cocompactly by isometries then under what circumstances can one deduce that a given HNN extension of Γ acts in the same way?*

The analogous questions for groups which act semisimply will also be considered, but only in passing.

We shall give criteria for showing that certain amalgamated free products and HNN extensions do act nicely on CAT(0) spaces, and we shall also present some sobering counterexamples to illustrate what can go wrong.

For results concerning the question of when an amalgamated free product of δ-hyperbolic groups is δ-hyperbolic, see [BesF92].

Amalgamated Free Products and HNN Extensions

Amalgamated free products and HNN extensions have appeared a number of times at earlier points in this book, but since they are central to the present discussion we take this opportunity to recall their definitions.

6.2 Definitions. Let H be a group and let $(\Gamma_\lambda : \lambda \in \Lambda)$ be a family of groups. Associated to any family of monomorphisms $(\phi_\lambda : H \to \Gamma_\lambda : \lambda \in \Lambda)$ one has an *amalgamated free product*, which is the quotient of the free product $*_{\lambda\in\Lambda}\Gamma_\lambda$ by the normal subgroup generated by the conjugates of the elements $\{\phi_\lambda(h)\phi_{\lambda'}(h)^{-1} \mid h \in H,\ \lambda, \lambda' \in \Lambda\}$. The natural map from each Γ_λ to the amalgamated free product is injective, and we identify Γ_λ with its image in order to realise it as a subgroup of the amalgamated free product.

In the case $\Lambda = \{1, 2\}$, it is usual to write $\Gamma_1 *_H \Gamma_2$ to denote an amalgamated free product (suppressing mention of the given maps $H \hookrightarrow \Gamma_i$), and to refer to $\Gamma_1 *_H \Gamma_2$ as "an amalgamated free product of Γ_1 and Γ_2 along H."

Let Γ be a group and let $\phi : A_1 \to A_2$ be an isomorphism between subgroups of Γ. Associated to this data one has an HNN *extension*[62] of G: this is the quotient of $\Gamma * \langle t\rangle$ by the smallest normal subgroup containing $\{a^{-1}t\phi(a)t^{-1} \mid a \in A_1\}$. This quotient may be described by the relative presentation

$$\Gamma *_\phi = (\Gamma, t \mid t^{-1}at = \phi(a),\ \forall a \in A_1),$$

or by a phrase such as *"an HNN extension of Γ by a stable letter t conjugating A_1 to A_2"*.

[62] The initials HNN are in honour of Graham Higman, Bernhard Neumann and Hanna Neumann, who first studied these extensions [HNN49]. See [ChaM82] for historical details.

The group Γ is called the *base* of the extension. The natural map $\Gamma \to \Gamma *_\phi$ is injective, and thus we regard Γ as a subgroup of $\Gamma *_\phi$.

Following common practice, we shall normally use the more casual notation $\Gamma *_A$ to describe $\Gamma *_\phi$, where A is an abstract group isomorphic to A_1 and A_2. And we shall refer to $\Gamma *_A$ as "an HNN extension of Γ over A." In the special case where $A \subset \Gamma$ is a specific subgroup and ϕ is the identity map, one has the *trivial* HNN *extension of* Γ *over* A, which we denote

$$\Gamma \dot{*}_A = (\Gamma, t \mid t^{-1}at = a, \ \forall a \in A).$$

6.3 The Bass-Serre Tree. Free products with amalgamation and HNN extensions are the basic building blocks of graphs of groups[63] in the sense of Bass and Serre [Ser77]. (The fundamental group of any graph of groups can be described by taking iterated amalgamated free products and HNN extensions.)

An amalgamated free product of the form $\Gamma_1 *_H \Gamma_2$ is the fundamental group of a 1-simplex of groups, where Γ_1 and Γ_2 are the local groups at the vertices of the simplex ("the vertex groups") and H is the local group at the barycentre of the 1-cell ("the edge group"). An HNN extension $\Gamma *_A$ is the fundamental group of a graph of groups with a single vertex group Γ and a single edge group A. The homomorphisms between the local groups are those implicit in the notations $\Gamma_1 *_H \Gamma_2$ and $\Gamma *_A$.

All graphs of groups are developable in the sense of Chapter $\mathcal{C}$ (see $\mathcal{C}$.2.17). The universal development gives an action of the fundamental group of the graph of groups on a tree called the *Bass-Serre tree*. The tree for $\Gamma_1 *_H \Gamma_2$ was described explicitly prior to (II.11.18), and the tree for $\Gamma *_A$ was described explicitly in the proof of (II.11.21). The vertex stabilizers for these actions are the conjugates of the given vertex groups. Thus, as a special case of (II.2.8), we see that every finite subgroup of $\Gamma_1 *_H \Gamma_2$ is conjugate to a subgroup of Γ_1 or Γ_2, and every finite subgroup of $\Gamma *_A$ is conjugate to a subgroup of the base group Γ.

We shall need the following well-known fact (see [Ser77], 5.2, Theorem 11).

6.4 Lemma. *Let* $G = \Gamma_1 *_H \Gamma_2$. *Let* $x_0, \dots, x_n \in \Gamma_1$ *and* $y_0, \dots, y_n \in \Gamma_2$ *be elements such that* $x_i \notin H$ *if* $i > 0$ *and* $y_i \notin H$ *if* $i < n$. *Then* $x_0 y_0 \dots x_n y_n \neq 1$ *in* G. *When an element of* G *is expressed as such a product, it is said to be in reduced form. Every element of* $G \smallsetminus \{1\}$ *can be written in reduced form.*

In $G *_\phi = (\Gamma, t \mid t^{-1}at = \phi(a), \forall a \in A)$, *consider a product* $z_0 t^{m_1} z_1 \dots z_m t^{m_n} z_n$, *where the* m_i *are non-zero integers. Suppose* $z_i \notin A$ *if* $m_i < 0$ *and* $z_i \notin \phi(A)$ *if* $m_i > 0$. *Then* $z_0 t^{m_1} z_1 \dots z_m t^{m_n} z_n \neq 1$ *in* $G *_\phi$. *When an element of* $G *_\phi$ *is expressed as such a product, it is said to be in reduced form. Every element of* $G *_\phi$ *can be written in reduced form.*

The above fact concerning HNN extensions is often called Britton's Lemma.

---

[63] 1-complexes of groups in the language of Chapter $\mathcal{C}$.

6.5 Doubling Along Subgroups and Extending Over Them. In (II.11.7) we discussed the construction of *doubling* a space X along a closed subspace Y. As a set, the double of X along Y is the disjoint union of two copies of X (denoted X and $\overline{X}$) modulo the equivalence relation generated by $[y \sim \overline{y},\ \forall y \in Y]$. When endowed with the quotient topology, this space is denoted $D(X; Y)$. More generally, one can take m copies of X and identify them along Y; the resulting space is denoted $D_m(X; Y)$. We also considered the analogous construction $X*_Y$, which is the disjoint union of X and $Y \times [0, 1]$ modulo the equivalence relation generated by $[y \sim (y, 0) \sim (y, 1)\, \forall y \in Y]$.

There are analogous constructions for groups. As in (5.1), given a group Γ and a subgroup H, one can consider the amalgamated free product of m copies of Γ along H, denoted $\Delta_m(\Gamma; H)$. In the terminology of (6.2), this is an amalgamated free product whose index set Λ has m elements, each Γ_λ is Γ, and each of the maps ϕ_λ is the inclusion $H \hookrightarrow \Gamma$. One can also consider the trivial HNN extension $\Gamma*_H$.

If X is a path-connected topological space, Y is a path-connected subspace that has an open neighbourhood which retracts onto it, and $Y \hookrightarrow X$ induces an injection of fundamental groups, then by the Seifert-van Kampen Theorem the fundamental group of $D_m(X; Y)$ is $\Delta_m(\pi_1 X; \pi_1 Y)$ and $\pi_1(X*_Y) = (\pi_1 X)*_{\pi_1 Y}$.

6.6 Proposition. *Let Γ be the fundamental group of a compact non-positively curved space X, let Z be a compact geodesic space and let $f : Z \to X$ be a local isometry. Let $H = f_*(\pi_1 Z) \subset \Gamma$. Then $\Gamma*_H$ and $\Delta_m(\Gamma; H)$ are the fundamental groups of compact non-positively curved spaces.*

Proof. These are special cases of our earlier results concerning equivariant gluing, (11.18) and (11.21), but for clarity of exposition we repeat the details in the present simpler setting.

(II.4.14) and (II.11.6) show that the quotient Q of $X \coprod (Z \times [0, 1])$ by the equivalence relation generated by $f(z) \sim (z, 0)$ is non-positively curved. Let Y denote the image of $Z \times \{1\}$ in Q and note that $D_m(Q; Y)$ and $Q*_Y$ are non-positively curved (II.11.7). The obvious homotopy equivalence between Q and X gives an isomorphism $\pi_1 Q \cong \Gamma$ identifying $\pi_1 Y \subset \pi_1 Q$ with H. We know from (II.4.13) that $\pi_1 Y \to \pi_1 Q$ is an injection, so the general remarks immediately preceding the proposition apply. □

6.7 Exercises

(1) Show that if the trivial HNN extension $\Gamma*_A$ acts properly and cocompactly by isometries on a CAT(0) space, then A is quasi-isometrically embedded in Γ. (Hint: Argue, perhaps using (6.4), that the centralizer of the stable letter t is $A \times \langle t \rangle$. Apply (4.14).)

(2) If H is normal in Γ and $\Gamma/H \cong \mathbb{Z}$, then $\Delta_2(\Gamma; H)$ is isomorphic to the trivial HNN extension $\Gamma*_H$.

We shall show in (6.16) that the converse to (6.7(1)) fails: there exist groups Γ that act properly and cocompactly by isometries on a CAT(0) space and quasi-isometrically embedded subgroups $A \subset \Gamma$ such that $\Gamma*_A$ does not act properly and cocompactly by isometries on any CAT(0) space.

Amalgamating along Abelian Subgroups

In Chapter II.11 we proved that if two groups Γ_1 and Γ_2 act properly and cocompactly by isometries on CAT(0) spaces, then so too does any amalgamated free product of the form $\Gamma_1 *_{\mathbb{Z}} \Gamma_2$. In this subsection we examine the extent to which this result can be generalized to the case of amalgamations along other abelian groups. In the next subsection we shall consider amalgamations along finitely generated free groups.

6.8 Proposition. *If* Γ *acts properly and cocompactly by isometries on a* CAT(0) *space, then so too does every amalgamated free product of the form*

$$\Delta_m(\Gamma; A)$$

and every HNN extension of the form

$$(\Gamma, t \mid t^{-1}at = a,\ a \in A),$$

where A *is virtually abelian.*

Proof. We proved in (II.7.2) that if Γ acts properly and cocompactly by isometries on a CAT(0) space, and $A \subset \Gamma$ is a virtually abelian subgroup of Γ, then there is a proper cocompact action of A on a Euclidean space $\mathbb{E}^m$ and an A-equivariant isometry $\mathbb{E}^m \to X$. Thus we may apply the equivariant gluing techniques described in (II.11.18) and (II.11.21). □

One would like to generalize the preceding proposition to cover, for example, the case of arbitrary amalgamations of the form $\Gamma_1 *_{\mathbb{Z}^n} \Gamma_2$, where the Γ_i are the fundamental groups of compact non-positively curved spaces. One can do so in certain special cases — for example (II.11.37) where the groups Γ_i are non-uniform lattices in SO$(n, 1)$. However, if $n \geq 2$ then in general one cannot make $\Gamma_1 *_{\mathbb{Z}^n} \Gamma_2$ the fundamental group of a compact non-positively curved space. Let us consider why.

In the case $n = 1$, one can take compact non-positively curved spaces X_1 and X_2 with $\Gamma_i = \pi_1 X_i$ and then scale the metrics on them so that the closed geodesics in the free homotopy class of the generators of the subgroups being amalgamated have the same length; one then attaches the ends of a tube to these closed geodesics. Consider what happens when we try to imitate this construction with subgroups $A_i \cong \mathbb{Z}^n$, where $n \geq 2$: the Flat Torus Theorem provides us with local isometries $\mathbb{E}^n/A_i \to X_i$ that realise the inclusions $A_i \hookrightarrow \Gamma_i$ (these maps play the role of the closed geodesics in the case $n = 1$), but in general one cannot make the flat tori $\mathbb{E}^n/A_1$ and $\mathbb{E}^n/A_2$ isometric simply by scaling the metric on each by a linear factor — it might be that they are not conformally equivalent.

In some cases one can vary the metric on the spaces that are to be glued and thus realise the desired amalgamated free product as the fundamental group of a compact non-positively curved space (cf. II.11.36). But in other cases the shape of the torus associated to a particular abelian subgroup is an invariant of the group rather than simply the particular non-positively curved space at hand. We illustrate this point with some specific examples.

6.9 The Groups $T(n)$. Let

$$T(n) = \langle a, b, t_a, t_b \mid [a, b] = 1,\ t_a^{-1} a t_a = (ab)^n,\ t_b^{-1} b t_b = (ab)^n \rangle.$$

As Dani Wise pointed out in [Wi96b], $T(n)$ is the fundamental group of the non-positively curved 2-complex $X(n)$ that one constructs as follows: take the (skew) torus Σ_n formed by identifying opposite sides of a rhombus whose sides have length n and whose small diagonal has length 1; the loops formed by the images of the sides of the rhombus are labelled a and b respectively; we attach to Σ_n two tubes $S \times [0, 1]$, where S is a circle of length n; one end of the first tube is attached to the loop labelled a and one end of the second tube is attached to the loop labelled b; in each case the other end of the tube wraps around the image of the small diagonal n times.

There is some flexibility in how one attaches the tubes in this example, but because a is conjugate to b and to $(ab)^n$, the shape of the torus $\Sigma_n \subset X(n)$ supporting the abelian subgroup $A_n = \langle a, b \rangle$ is entirely determined by the algebra of the group. More precisely, given any proper action of $T(n)$ by semisimple isometries on a CAT(0) space one considers the induced action of A_n on each of the flat planes E yielded by the Flat Torus Theorem, modulo a constant scaling factor, the quotient metric on $A_n \backslash E$ must make it isometric to the torus Σ_n described above, because the translation lengths of a, b and $(ab)^n$ have to be the same.

If $n \neq m$, then the tori Σ_n and Σ_m cannot be made isometric simply by scaling the metric on each. Thus we have:

6.10 Proposition. *In the notation of (6.9), if $n \neq m$ and if Γ is an amalgamated free product of $T(n)$ and $T(m)$ obtained by identifying A_n with A_m, then Γ does not act properly by semisimple isometries on any* CAT(0) *space.*

The interested reader should find little difficulty in producing many variations on this result, we note one other:

6.11 Example. Let $\Gamma = \langle a, b, t \mid [a, b] = 1,\ t^{-1} a t = b^2 \rangle$. This is the fundamental group of the non-positively curved 2-complex obtained as follows: one takes the torus formed by identifying opposite sides of a rectangle whose sides c_1 and c_2 have length 1 and 2, respectively; to this one attaches a cylinder $S \times [0, 1]$, where S is a circle of length 2; the end $S \times \{0\}$ wraps twice around the image of c_1 and the end $S \times \{1\}$ is identified with the image of c_2.

We take two copies of Γ and consider the amalgamated free product $G = \Gamma *_{\mathbb{Z}^2} \overline{\Gamma}$ obtained by making the identifications $a = \overline{b}$ and $b = \overline{a}$. Writing $\sim$ to denote conjugacy in G, we have $a \sim b^2 = \overline{a}^2 \sim \overline{b}^4 = a^4$. We claim that G cannot act properly by semisimple isometries on any CAT(0) space. Indeed, given any action of G on a CAT(0) space, the translation lengths (II.6.3) of a and b satisfy:

$2|b| = |b^2| = |a| = |a^4| = 4|a|$, thus $|a| = |b| = 0$. Thus if the action is semisimple, a and b must be elliptic, and since they have infinite order in G, the action will not be proper.

The same calculation with algebraic translation numbers in place of geometric ones shows that if Q is semihyperbolic, then the image of $\langle a, b\rangle$ under any homomorphism $G \to Q$ must be finite (see 4.18).

For our last example of a bad amalgamation along abelian subgroups we take two n-torus bundles over the circle with finite holonomy and equip them with flat metrics, then we ask when they can be glued along their fibres so as to produce a compact non-positively curved space.

6.12 Proposition. *Let $\phi_i \in \mathrm{GL}(n, \mathbb{Z})$, $i = 1, 2$ be elements of finite order and consider the corresponding semidirect products $G_i = \mathbb{Z}^n \rtimes_{\phi_i} \mathbb{Z}$. The amalgamated free product $\Gamma(\phi_1, \phi_2) := G_1 *_{\mathbb{Z}^n} G_2$ is the fundamental group of a compact non-positively curved space if and only if the subgroup $\langle\phi_1, \phi_2\rangle \subset \mathrm{GL}(n, \mathbb{Z})$ is finite. Moreover, if $\langle\phi_1, \phi_2\rangle$ is infinite then $\Gamma(\phi_1, \phi_2)$ does not act properly by semi-simple isometries on any* CAT(0) *space, it is not semihyperbolic, and its Dehn function grows at least cubically.*

Proof. The quotient of $\Gamma(\phi_1, \phi_2)$ by the normal subgroup $\mathbb{Z}^n$ is free of rank two, so $\Gamma(\phi_1, \phi_2) = \mathbb{Z}^n \rtimes_\Phi F_2$, where the image of $\Phi : F_2 \to \mathrm{GL}(n, \mathbb{Z})$ is $H := \langle\phi_1, \phi_2\rangle$. If H is finite then we can choose an H-equivariant flat metric on the n-torus Σ. Regard each ϕ_i as an isometry of Σ. For $i = 1, 2$, the mapping torus $M_i := \Sigma \times [0, 1]/[(x, 0) \sim (\phi_i(x), 1)]$ is a compact non-positively curved space with fundamental group G_i, $i = 1, 2$. By gluing M_1 to M_2 along $\Sigma \times \{0\}$ we obtain a non-positively curved space with fundamental group $\Gamma(\phi_1, \phi_2)$.

If H is not finite, then the normal subgroup $\mathbb{Z}^n$ is not virtually a direct factor of its normalizer in $\Gamma(\phi_1, \phi_2)$, and therefore $\Gamma(\phi_1, \phi_2)$ does not act properly by semi-simple isometries on any CAT(0) space (II.7.17). To see that $\Gamma(\phi_1, \phi_2)$ is not semihyperbolic in this case, one can argue that $H = \langle\phi_1, \phi_2\rangle$ contains an element of infinite order and appeal to (4.17). Alternatively[64], since we know that the Dehn function of a semihyperbolic group is linear or quadratic (4.9(2)), we can appeal to the Main Theorem of [Bri95b], which implies that if $\langle\phi_1, \phi_2\rangle$ is infinite, then the Dehn function of $\Gamma(\phi_1, \phi_2)$ will be either exponential or polynomial of degree d, where $3 \le d \le n + 1$ (it will be exponential if and only if H contains an element with an eigenvalue of absolute value bigger 1). □

6.13 Example. SL(2, $\mathbb{Z}$) is generated by the following two matrices of finite order

$$\phi_1 = \begin{pmatrix} 0 & 1 \\ -1 & 0 \end{pmatrix} \text{ and } \phi_2 = \begin{pmatrix} 0 & 1 \\ -1 & 1 \end{pmatrix}.$$

[64] This argument was discovered independently by Natasa Macura and Christian Hidber [Hid97].

Since SL(2, $\mathbb{Z}$) contains matrices that have eigenvalues of absolute value bigger than 1, the group $\Gamma(\phi_1, \phi_2)$, which is the fundamental group of a space obtained by gluing two flat 3-manifolds along tori, has an exponential Dehn function and does not act properly by semi-simple isometries on any CAT(0) space.

Amalgamating Along Free Subgroups

Let F be a finitely generated free group. In order to realise $\Gamma_1 *_F \Gamma_2$ as the fundamental group of a compact non-positively curved space, one should look for compact non-positively curved spaces X_i with $\Gamma_i = \pi_1 X_i$ where the inclusions $F \hookrightarrow \Gamma_i$ can be realized by local isometries $Y \to X_i$, where Y is a metric graph with fundamental group F. A simple instance of this is:

6.14 Lemma. *Let $B = B(l_1, \dots, l_n)$ be a metric graph that has one vertex and n edges with lengths $l_1, \dots, l_n$. Let $\phi_1 : B \to X_1$ and $\phi_2 : B \to X_2$ be local isometries to compact non-positively curved spaces. The double mapping cylinder*

$$\frac{X_1 \cup (B \times [0,1]) \cup X_2}{(b,0) \sim \phi_1(b),\ (b,1) \sim \phi_2(b)}$$

is non-positively curved (when endowed with the quotient path metric).

Such embeddings $B \to X_i$ arise naturally in the case of spaces built by attaching gluing tubes as in (II.11.13).

Let X and $Y_1, \dots, Y_n$ be compact non-positively curved spaces with basepoints p and $q_1, \dots, q_n$ respectively, and let $\phi^{0,i}, \phi^{1,i} : Y_i \to X$ be local isometries with $\phi^{0,i}(q_i) = \phi^{1,i}(q_i) = p$ for all i. (II.11.13) tells us that

$$Q = \frac{X \cup (Y_1 \times [0,1]) \cup \dots \cup (Y_n \times [0,1])}{(y_i, 0) \sim \phi^{0,i}(y_i),\ (y_i, 1) \sim \phi^{1,i}(y_i)\ \forall y_i \in Y_i,\ i = 1, \dots, n}$$

is non-positively curved when endowed with the quotient path metric. Associated to this description of Q we have the relative presentation

$$\pi_1(Q,p) = \big(\pi_1(X,p), \pi_1(Y_i,q_i), s_i, i = 1, \dots, n \mid s_i^{-1}\phi_*^{0,i}(g_i)s_i = \phi_*^{1,i}(g_i),\ \forall g_i \in \pi_1(Y_i,q_i)\big).$$

The homotopy class s_i contains the loop based at $p \in Q$ which is the image of the path $t \mapsto (q_i, t)$ in $Y_i \times [0,1]$; let σ_i denote this loop.

6.15 Lemma. *Let Q be as in the preceding paragraph. Let B be the metric simplicial graph that has one vertex v and n edges of length 1. Let $\psi : B \to X$ be the map that sends v to $p \in Q$ and sends the i-th edge of B isometrically onto the loop σ_i. Then ψ is an isometry onto its image.*

Proof. By construction ψ is length-preserving and injective. By composing the projection $Y_i \times [0, 1] \to [0, 1]$ with the arc length parameterization of the i-th edge in B one obtains maps $Y_i \to B$. These maps, together with the constant map $X \to \{v\}$ induce a map $f : Q \to B$ that does not increase lengths and which is such that $f\psi = \mathrm{id}_B$. Thus ψ is an isometry. □

An application of this lemma is given in (7.9).

Bad Amalgamations Along Free Subgroups

Recall that a subgroup $H \subset \Gamma$ is said to be a *retract* if there is a homomorphism $\Gamma \to H$ that restricts to the identity on H. Given any finite generating set for H, one can extend it to a finite generating set for Γ by adding generators for the kernel of the retraction $\Gamma \to H$. With respect to the resulting word metrics, $H \hookrightarrow \Gamma$ is an isometry.

Part (2) of the following proposition illustrates the point that we made following (6.7): in order for $G{*}_A$ to be semihyperbolic it is necessary for A to be quasi-isometrically embedded in G, but this condition is not sufficient.

6.16 Proposition. *Let F_1 and F_2 be free groups of rank two. Choose generators $a_1, b_1 \in F_1$ and $a_2, b_2 \in F_2$ and let L denote the subgroup of $F_1 \times F_2$ generated by $\alpha = a_1a_2$ and $\beta = a_1b_2$. (By projecting into F_2 one sees that L is free.)*

(1) *$L \subset F_1 \times F_2$ is a retract. (In particular, for a suitable choice of word metrics $L \hookrightarrow F_1 \times F_2$ is an isometry.)*

(2) *The trivial HNN extension $E = \big(F_1 \times F_2,\ t \mid t^{-1}lt = l\ \forall l \in L\big)$ is not semihyperbolic.*

(3) *The double $\Delta = \Delta_2(F_1 \times F_2; L)$ is not semihyperbolic.*

(In fact both E and Δ have cubic Dehn functions.)

Proof. One obtains a retraction from $F_1 \times F_2$ to L by sending F_1 to the identity, a_2 to α and b_2 to β.

In order to see that E is not semihyperbolic, we look at centralizers (4.15). We claim that the intersection of the centralizers of t, a_1 and b_1 is $L \cap F_2$, which is the kernel of the map $L \to \mathbb{Z}$ that sends each of α and β to a fixed generator. This kernel is not finitely generated (see 5.3), so (4.15) tells us that E is not semihyperbolic.

To see that the intersection of centralizers is what we claim, note first that by (6.4) the centralizer of t is L, then note that the centralizers of a_1 and b_1 in E are the same as their centralizers in $F_1 \times F_2$, and that the centralizer of a_1 in $F_1 \times F_2$ is $\langle a_1 \rangle \times F_2$ while that of b_1 is $\langle b_1 \rangle \times F_2$. The second assertion is a consequence of (6.4), but can also be seen by looking at the action of E on the Bass-Serre tree: each of a and b is an elliptic element with a single fixed point, the stabilizer of which is $F_1 \times F_2$; and the centralizer of any elliptic isometry of a CAT(0) space must preserve its fixed point set.

A similar argument shows that in Δ the intersection of the centralizers of the two copies of a_1 and the two copies of b_1 is again $L \cap F_2$.

We shall sketch a proof of the parenthetical assertion concerning the Dehn functions of E and Δ. The required cubic upper bound is proved in (6.20). It is enough to establish the lower bound for E, because Δ retracts onto a copy of E and the Dehn function of a group gives an upper bound on the Dehn function of any retract of that group [Alo90]. (To get a retraction from Δ onto a subgroup isomorphic to E, leave one of the natural copies of $F_1 \times F_2$ in Δ alone and in the other send the generator b_1 to 1 and the other generators to themselves.)

To obtain a cubic lower bound on the Dehn function of E we work with the following presentation:

$$\langle a_1, a_2, b_1, b_2, t \mid [a_1, a_2], [a_1, b_2], [b_1, a_2], [b_1, b_2], [t, a_1a_2], [t, a_1b_2]\rangle.$$

Let $u_n = (a_2^n b_2^{-n})$, let $v_n = (tb_1)^n$ and let $w_n = v_n u_n v_n^{-1} u_n^{-1}$. Note that $w_n = 1$ in E. We claim that the area of any van Kampen diagram for w_n is bounded below by a cubic function of n.

There are two important observations to be made. First, because no subword of w_n equals 1 in E, any van Kampen diagram for w_n must be a topological disc. Second, since the only relations involving t are of the form $[t, x]$, if one enters a 2-cell in a van Kampen diagram for w_n by crossing an edge labelled t, then there is a unique other edge labelled t which you can cross in order to exit the 2-cell; thus one enters a second 2-cell, from which there is a unique way to exit crossing an edge labelled t, and so on. Continuing in this way, one obtains a chain of 2-cells crossing the diagram (see figure Γ.11). The chains of 2-cells that one obtains in this manner are called *t-corridors*. There is one such t-corridor incident at each edge labelled t in the boundary of the diagram. See [BriG96] for a careful treatment of t-corridors.

In exactly the same way, one obtains a b_1-corridor incident at each edge labelled b_1 in the boundary of the diagram. The orientations on the ends of the corridors force them to cross the diagram in the manner shown in figure Γ.11.

The key parts of the diagram on which to focus are the gaps between corridors. The top and bottom of each t-corridor is labelled by a word in the free group on (a_1b_2) and (a_1a_2) that freely reduces to $(a_1a_2)^n(a_1b_2)^{-n}$. The side of each b_1-corridor is labelled by a word in the free group on a_2 and b_2 that freely reduces to $a_2^n b_2^{-n}$. The subdiagram between each pair of adjacent corridors is as drawn in figure Γ.12. It may be regarded as a diagram over the natural presentation of $\langle a_2, b_2\rangle \times \langle a_1\rangle \cong F_2 \times \mathbb{Z}$, and as such it has area $\simeq n^2$. Hence the area of the whole diagram in figure Γ.11 is $\succeq n^3$. $\square$

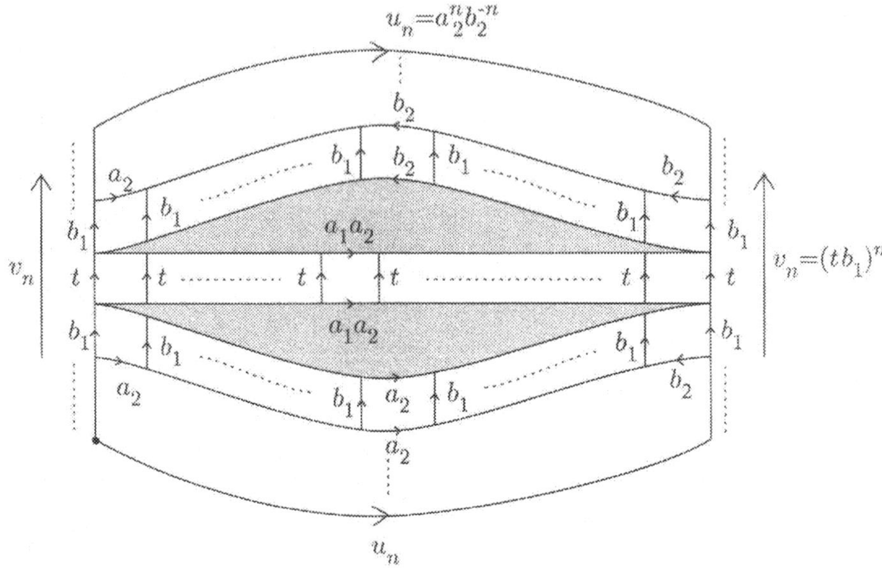

Fig. Γ.11 A van Kampen diagram for $u_n v_n u_n^{-1} v_n^{-1}$.

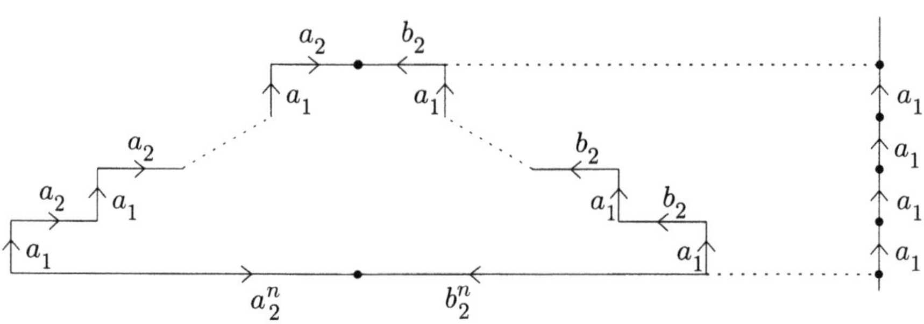

Fig. Γ.12 The subdiagram showing that $a_2^n b_2^{-n} = (a_1 a_2)^n (a_1 b_2)^{-n}$.

Subgroup Distortion and the Dehn Functions of Doubles

The following notion of distortion has been lurking in the background of several earlier proofs (for example the preceding proposition and (5.15)).

6.17 Definition (Subgroup Distortion). Let $H \subset \Gamma$ be a pair of finitely generated groups, and let d_Γ and d_H be the word metrics associated to a choice of finite generating set for each. The *distortion* of H in Γ is the function

$$\delta_H^\Gamma(n) = \max\{d_H(1, h) \mid h \in H \text{ with } d_\Gamma(1, h) \leq n\}.$$

One checks easily that, up to Lipschitz equivalence, this function is independent of the choice of word metrics d_Γ and d_H.

6.18 Remarks

(1) $H \hookrightarrow \Gamma$ is a quasi-isometric embedding if and only if there exists a constant K such that $\delta_H^\Gamma(n) \leq Kn$.

(2) The definition of distortion that we have adopted differs from that of Gromov [Gro93] in that we have omitted a normalizing factor of $1/n$ on the right hand side. Authors vary in this convention.

6.19 Exercises

(1) Suppose that H and Γ are finitely generated and that $H \subset \Gamma$ is normal. Prove that there is a constant $k > 0$ (depending on the choice of word metrics) such that distortion of H in Γ is bounded above by $n \mapsto k^n$. (Hint: Take k to be the maximum of the distances $d_H(1, b^{-1}ab)$ where b runs over a set of generators $\mathcal{B} = \mathcal{B}^{-1}$ for Γ and a runs over the generators of H.)

(2) Let $f : N \to M$ be a bi-Lipschitz map from one closed Riemannian manifold into another. Suppose that f is π_1-injective and consider a lifting $\tilde{f} : \tilde{N} \to \tilde{M}$ to the universal cover. Consider the relationship between the length metric $\tilde{d}_{\tilde{N}}$ on $\tilde{N}$ given by its Riemannian structure and the metric $d(x, y) := d_{\tilde{M}}(\tilde{f}(x), \tilde{f}(y))$. After taking note of (I.8.19), bound $\tilde{d}_{\tilde{N}}(x, y)$ in terms of $d(x, y)$ and the distortion of $\pi_1 N \subset \pi_1 M$.

We have introduced subgroup distortion at this time so that we can formulate the following theorem. We recall that $\Delta_m(\Gamma; A)$ is the amalgamated free product of m copies of Γ along A. As we noted prior to (5.1), if $m \geq 3$ then there is a canonical retraction of $\Delta_m(\Gamma; A)$ onto $\Delta_2(\Gamma; A)$, sending the first two copies of Γ to themselves and sending the other $(m - 2)$ copies of Γ to the first copy by the identification implicit in the notation. Similarly, by identifying the m natural copies of Γ, one obtains a retraction from $\Delta_m(\Gamma; A)$ onto (the first copy of) Γ.

We recall that modulo the equivalence relation $\simeq$ described in (I.8A.4), the Dehn function of a finitely presented group does not depend on the choice of finite presentation. And if H is a retract of a finitely presented group Γ, then the Dehn function of H is $\preceq$-bounded above by that of Γ in the sense of (I.8A.4) — see [Alo90].

6.20 Theorem. *Let $m \geq 2$ be an integer. Let Γ be a finitely presented group with Dehn function f_Γ and let $H \subset \Gamma$ by a finitely presented subgroup. Let $\delta(n)$ be the distortion of H in Γ with respect to some choice of word metrics. The Dehn function f_{Δ_m} of $\Delta_m = \Delta_m(\Gamma; H)$ satisfies*

$$\max\{f_\Gamma(n),\ \delta(n)\} \preceq f_{\Delta_m}(n) \preceq n f_\Gamma(\delta(n)),$$

and the Dehn function of the trivial HNN extension $\hat{\Gamma} = \Gamma *_H$ satisfies

$$\max\{f_\Gamma(n),\ n\,\delta(n)\} \preceq f_{\hat{\Gamma}}(n) \preceq n f_\Gamma(\delta(n)).$$

Proof. First we consider the lower bound for Δ_m. Because Δ_m retracts onto Γ, we have $f_\Gamma(n) \preceq f_{\Delta_m}(n)$, so it only remains to show that $\delta(n) \preceq f_{\Delta_m}(n)$. Because Δ_m retracts onto Δ_2, we may assume that $m = 2$.

We fix a finite presentation $\langle \mathcal{A} \mid \mathcal{R}\rangle$ for Γ, where $\mathcal{A}$ includes a generating set $\mathcal{B}$ for H. Let d_H be the word metric on H associated to $\mathcal{B}$. We shall work with the following presentation of Δ_2:

$$\langle \mathcal{A}, \mathcal{A}' \mid \mathcal{R}, \mathcal{R}', \ h^{-1}h' = 1 \ \forall h \in \mathcal{B}\rangle.$$

We choose a sequence of geodesic words w_n in the generators $\mathcal{A}$ so that $w_n \in H$, $|w_n| \leq n$ and $d_H(1, w_n) = \delta(n)$. Let w_n' be the same word in the alphabet $\mathcal{A}'$ and consider a least-area van Kampen diagram D_n for the word $w_n^{-1}w_n'$ (which represents the identity in Δ_2).

Across the interior of each 2-cell (digon) corresponding to relators of the form $h^{-1}h'$ we draw an arc connecting the vertices. These arcs form a graph $\mathcal{G} \subset D_n$, whose edges we label with letters $h \in \mathcal{B}$ in the obvious way. The basepoint of the diagram D_n and the endpoint of the arc of ∂D_n labelled w_n are two vertices of $\mathcal{G}$. We claim that these two vertices lie in the same connected component of $\mathcal{G}$. If this claim were false, then there would be an arc through the interior of D_n that joined a point on the arc of ∂D_n labelled w_n to a point on the arc labelled w_n' and did not pass through any digon labelled $h^{-1}h'$. But this is clearly impossible, because each edge in the portion of the boundary labelled w_n lies in the closure of a 2-cell labelled by a relator from $\mathcal{R}$ and each edge in the portion of the boundary labelled w_n' lies in the closure of a 2-cell labelled by a relator from $\mathcal{R}'$, and the only relators involving both primed and unprimed letters are those of the form $h = h'$.

Thus we can connect the basepoint of D_n to the endpoint of the arc of ∂D_n labelled w_n by a path in $\mathcal{G}$; we choose a shortest such path σ. The edges of σ are labelled by generators $h_{i_1}, \dots, h_{i_m} \in \mathcal{B}^{\pm 1}$ such that

$$h_{i_1} \dots h_{i_m} = w_n$$

in Γ. Thus $m \geq \delta(n)$, and since the 2-cells (digons) of D_n corresponding to the edges of σ are all distinct, we deduce that D_n has at least $\delta(n)$ 2-cells. The integer n is arbitrary and ∂D_n has length $\leq 2n$, thus the Dehn function for the given presentation of $\Delta_2(\Gamma; H)$ satisfies $f_{\Delta_2}(2n) \geq \delta(n)$ for all n.

We now consider the case of HNN extensions. Again, since Γ is obviously a retract of $\hat{\Gamma}$, the inequality $f_\Gamma(n) \preceq f_{\hat{\Gamma}}(n)$ is clear.

$\Gamma{*}_H$ has finite presentation $\langle \mathcal{A}, t \mid \mathcal{R}, [t, h] = 1 \ \forall h \in \mathcal{B}\rangle$. For each positive integer n consider a least-area van Kampen diagram D_n' for $t^n w_n t^{-n} w_n^{-1}$, where the words w_n are chosen as above. One obtains a lower bound of $n\delta(n)$ on these diagrams by getting lower bounds on the length of t-corridors as in (6.16) — we omit the details.

We now turn our attention to the upper bounds. We give the details only in the case of Δ_2. The argument for Δ_m is entirely similar (but notationally a bit more complicated). The HNN case is also entirely similar, except that one uses the reduced form for elements in HNN extensions in place of the reduced form for amalgamated free products (see (6.4)).

We continue to work with the presentation $\langle \mathcal{A}, \mathcal{A}' \mid \mathcal{R}, \mathcal{R}', h^{-1}h' = 1 \ \forall h \in \mathcal{B}\rangle$ of Δ_2. Let $\delta(n)$ be the distortion of $H \subset \Gamma$ with respect to the word metrics associated to $\mathcal{B}$ and $\mathcal{A}$.

Any word in the given generators is equal (purely as a word) to a product of the form $w = u_1 v_1 \dots u_l v_l$, where each u_i is a word in the generators $\mathcal{A}$ and each v_i in the generators $\mathcal{A}'$, and all but possibly u_1 and v_l are non-empty. The integer l is called the *alternating length* of w. *The idea of the proof is to reduce the alternating length of words representing the identity by applying a controlled number of relations.*

Let $n = |w|$ and suppose that $w = 1$ in Δ_2. Lemma 6.4 implies that one of the u_i or v_i represents an element of H. Suppose that it is u_i and let U_i be the shortest word in the generators $\mathcal{B}$ that represents the same element of H. By definition, $|U_i| \leq \delta(|u_i|)$. Let $m = |U_i| + |u_i|$. Let U_i' be the word obtained by replacing the letters of U_i by the same letters primed.

By the definition of the Dehn function, in the free group $F(\mathcal{A})$ there is a product P of at most $f_\Gamma(m)$ conjugates of relators $r \in \mathcal{R}^{\pm 1}$ such that $u_i U_i^{-1} = P$. And there is an obvious way to write $U_i U_i'^{-1}$ as the product Q of $|U_i| \leq m$ conjugates of relators $b^{-1}b'$. Thus PQP'^{-1} is a product of at most $2m + 2f_\Gamma(m)$ conjugates of relations.

In the free group on $\mathcal{A} \cup \mathcal{A}'$ we have: $u_i = (u_i U_i^{-1})(U_i U_i'^{-1})(U_i' u_i'^{-1}) u_i' = PQP'^{-1}u_i'$. Thus

$$w = u_1 v_1 \dots v_{i-1} PQP'^{-1} u_i' v_i \dots v_l.$$

We move the subword PQP'^{-1} by conjugating it with the suffix s of w that follows it. As a result we obtain the following equality in the free group

$$w = u_1 v_1 \dots [v_{i-1} u_i' v_i] \dots v_l (s^{-1} PQP'^{-1} s).$$

The subword of the right hand side that precedes $(s^{-1}PQP'^{-1}s)$ has smaller alternating length than w because the term in square brackets is a word over $\mathcal{A}'$. Repeating this argument some number $N \leq n = |w|$ times, we obtain $w = W\Pi$, where W is a word of length at most n in either the generators $\mathcal{A}$ or the generators $\mathcal{A}'$, and where Π is the product of at most $2N(m + f_\Gamma(m))$ conjugates of relations obtained by gathering the products PQP'^{-1} at the right after each stage of the argument.

Finally, by the definition of the Dehn function of $\langle \mathcal{A} \mid \mathcal{R} \rangle$, we may write W as a product of at most $f_\Gamma(n)$ conjugates of relations from $\mathcal{R}$ or $\mathcal{R}'$. Thus w can be expressed in the free group on $\mathcal{A} \cup \mathcal{A}'$ as a product of at most $f_\Gamma(n) + 2N(m + f_\Gamma(m))$ conjugates of relators. By definition, f_Γ is non-decreasing, $n \preceq \delta(n)$, $N \leq n$, and $m = |u_i| + |U_i| \leq n + \delta(n)$. Thus the theorem is proved. □

In the light of 6.19(1), the following proposition shows that if Γ is hyperbolic then the distortion in Γ of any infinite, finitely generated, subgroup of infinite index is actually $\simeq k^n$, where $k > 1$. (Recall that $k^n \simeq 2^n$ for any constant $k > 1$.)

6.21 Proposition. *Let Γ be a hyperbolic group and let $N \subset \Gamma$ be a finitely generated normal subgroup. Suppose that N and Γ/N are both infinite. Then there exists a constant $k > 1$ such that $\delta_N^\Gamma(n) \geq k^n$. (The value of k depends on the choice of word metric, but its existence does not.)*

Proof. Let $\mathcal{A}$ be a generating set for N and extend this to a generating set $\mathcal{A} \cup \mathcal{B}$ for Γ. We also regard $\mathcal{B}$ as a generating set for Γ/N. Let $\mathcal{C}$ be the Cayley graph of Γ with respect to the generators $\mathcal{A} \cup \mathcal{B}$ and let $\overline{\mathcal{C}}$ be the Cayley graph of Γ/N with respect to the generators $\mathcal{B}$. There is a unique way to extend the quotient map $\Gamma \to \Gamma/N$ to a map $\pi : \mathcal{C} \to \overline{\mathcal{C}}$ that sends directed edges labelled by each $b \in \mathcal{B}$ isometrically onto edges labelled b and collapses edges labelled $a \in \mathcal{A}$. Note that $d(\pi(x), \pi(y)) \leq d(x, y)$ for all $x, y \in \mathcal{C}$.

Assume $\alpha \in \mathcal{A} \subset N$ is an element of infinite order. $n \mapsto \alpha^n$ is a quasi-geodesic (3.10), hence there exists an integer $\lambda > 0$ such that $d(1, \alpha^{\lambda r}) \geq 4|r|$ for all $r \in \mathbb{Z}$. Moreover, since $\langle \alpha \rangle$ is quasiconvex (3.6), there exists a constant $K > 0$ such that for every $r \in \mathbb{Z}$ and $\gamma \in \Gamma$, any geodesic in $\mathcal{C}$ joining γ to $\gamma\alpha^r$ must lie in the K-neighbourhood of the edge-path labelled α^r that begins at the vertex γ.

For each integer $n > K$ we choose a geodesic edge path of length n beginning at $1 \in \overline{\mathcal{C}}$. Let u_n be the word in the letters $\mathcal{B}^{\pm 1}$ that labels this path. Let w_n be a word of minimal length in the generators $\mathcal{A}$ that is equal to $u_n \alpha^{\lambda n} u_n^{-1}$ in Γ. Our goal is to show that the length of w_n is bounded below by an exponential function of n. Since $u_n \alpha^{\lambda n} u_n^{-1}$ is a word of length $n(2 + \lambda)$, this will prove the proposition.

We fix a geodesic segment in $\mathcal{C}$ joining $u_n \in \Gamma$ to $u_n \alpha^{\lambda n} \in \Gamma$; let m be the midpoint of this segment. In the light of (H.1.16), we will be done if we can show that the edge-path in $\mathcal{C}$ that begins at $u_n \in \Gamma$ and is labelled $u_n^{-1} w_n u_n$ lies outside the ball of radius $(n - K)$ about m (see figure Γ.13). For the subpaths labelled u_n and u_n^{-1} this is clear, because they have length n and they have an endpoint a distance at least $2n$ from m. And the subpath labelled w_n is sent by π to $1 \in \overline{\mathcal{C}}$, whereas the geodesic containing m is send to a point a distance at most K from $u_n \in \overline{\mathcal{C}}$. Since π does not increase distances, the arc labelled w_n must lie entirely outside the ball of radius $(n - K)$ about m. □

Hyperbolic Surface Bundles. There are many hyperbolic 3-manifolds which fibre over the circle with compact fibre (see [Ot96]), indeed Bill Thurston conjectured that every closed hyperbolic 3-manifold has a finite-sheeted covering that fibres in this way [Thu82]. If M^3 admits such a fibration then from the long exact sequence of the fibration we get a short exact sequence $1 \to S \to \pi_1 M^3 \to \mathbb{Z} \to 1$, where S is the fundamental group of the surface fibre. The previous proposition shows that the distortion of S in $\pi_1 M^3$ is exponential, so as a consequence of (6.20) we have:

6.22 Corollary. *Let M^3 be a hyperbolic 3-manifold that fibres over the circle with compact fibre Σ and let $S = \pi_1 \Sigma$. The Dehn function of $\Delta_2(\pi_1 M^3; S)$ is $\simeq 2^n$.*

Similar arguments apply to hyperbolic knot complements that fibre over the circle, and to free-by-cyclic hyperbolic groups, examples of which are given by [BesF92].

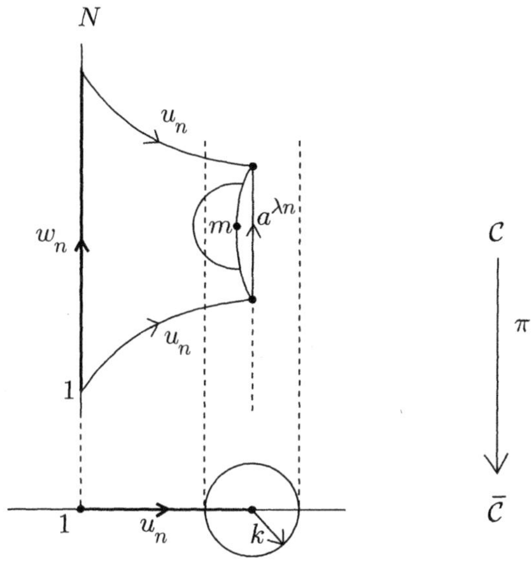

Fig. Γ.13 Normal subgroups are exponentially distorted

7. Finite-Sheeted Coverings and Residual Finiteness

In this section we shall explain how to construct compact non-positively curved spaces that have no non-trivial, connected, finite-sheeted coverings. We shall also construct groups which act properly and cocompactly by isometries on CAT(0) spaces but contain no torsion-free subgroups of finite index.

Residual Finiteness

A connected, locally simply-connected space X has no non-trivial, connected, finite-sheeted coverings if and only if $\pi_1 X$ has no non-trivial finite quotients[65]. Thus as a first step towards trying to build spaces with no connected, non-trivial, finite-sheeted coverings, we look for spaces whose fundamental groups do not satisfy the following condition.

7.1 Definition (Residual Finiteness). A group G is *residually finite* if for every $g \in G \smallsetminus \{1\}$ there is a finite group Q and an epimorphism $\phi : G \to Q$ such that $\phi(g) \neq 1$. Equivalently, the intersection of all subgroups of finite index in G is $\{1\}$.

Let X be a connected complex and suppose that $\pi_1 X$ is residually finite. The topological content of residual finiteness is that given any homotopically non-trivial

[65] If G is a group and $H \subset G$ is a subgroup of finite index, then there is a subgroup of finite index $N \subset H$ that is normal in G.

loop in X, there is a connected finite-sheeted covering of X to which this loop does not lift. And the reformulation of residual finiteness given in the second sentence of the above definition implies that if $\pi_1 X$ is residually finite then the universal covering of X is the inverse limit of a sequence of finite-sheeted coverings. (If X is compact and $\pi_1 X$ is infinite, then this can be useful because the finite-sheeted coverings are compact whereas the universal cover is not.)

The fundamental groups of the most classical examples of non-positively curved spaces, quotients of symmetric spaces of non-compact type, are residually finite. In the light of this, it may seem natural to ask whether the fundamental group of every compact non-positively curved space is residually finite. However, in the classical setting residual finiteness appears as an artifact of linearity rather than considerations of curvature *per se*: the fundamental groups in question can be realised as groups of matrices, and a theorem of Mal'cev [Mal40] shows that every finitely generated linear group is residually finite.

Dani Wise produced the first examples of compact non-positively curved spaces whose fundamental groups are not residually finite [Wi96a]. Some of the groups he constructed have no non-trivial finite quotients. Subsequently, Burger and Mozes [BuM97] constructed compact non-positively curved 2-complexes whose fundamental groups are simple. The situation for negatively curved spaces is less clear: it is known that hyperbolic groups are never simple [Gro87], [Ol95], but it is unknown whether they are always residually finite.

In the following statement the term *non-trivial* is used to mean that a covering $\widehat{Z} \to Z$ is connected and $\widehat{Z} \to Z$ is not a homeomorphism.

7.2 Embedding Theorem. *Every compact, connected, non-positively curved space X admits an isometric embedding into a compact, connected, non-positively curved space $\overline{X}$ such that $\overline{X}$ has no non-trivial finite-sheeted coverings. If X is a polyhedral complex of dimension $n \geq 2$, then one can arrange for $\overline{X}$ to be a complex of the same dimension.*

We shall give a self-contained proof of this theorem, following [Bri98a]. The main work goes into the construction of a compact non-positively curved 2-complex K whose fundamental group has no finite quotients (7.9). With K in hand, the proof of (7.2) becomes straightforward.

Proof. Choose a finite set of generators $\gamma_1, \dots, \gamma_N$ for $\pi_1 X$, where no $\gamma_i = 1$, and let $c_1, \dots, c_N$ be closed local geodesics in X representing the conjugacy classes of these elements. Lemma 7.9 (alternatively [Wi96a] or [BuM97]) gives a compact non-positively curved 2-complex K whose fundamental group has no finite quotients; fix a closed local geodesic c_0 in K. Take N copies of K and scale the metric on the i-th copy so that the length of c_0 in the scaled metric is equal to the length $l(c_i)$ of c_i, then glue the N copies of K to X using cylinders $S_i \times [0, L]$ where S_i is a circle of length $l(c_i)$; the ends of $S_i \times [0, L]$ are attached by arc length parameterizations of c_0 and c_i respectively (cf. II.11.17). Call the resulting space $\overline{X}$.

(II.11.13) assures us that $\overline{X}$ is non-positively curved. Moreover, if the length L of the gluing tubes is sufficiently large, then the natural embedding $X \hookrightarrow \overline{X}$ will be an isometry. The Seifert-van Kampen Theorem describes the fundamental group of $\overline{X}$: it is an iterated amalgamated free product; one first amalgamates $\pi_1 X$ and $\pi_1 K$ by identifying $\langle \gamma_1 \rangle \subset \pi_1 X$ with an infinite cyclic subgroup $C \subset \pi_1 K$ represented by the closed local geodesic c_0; the result of this first amalgamation is then amalgamated with a further copy of $\pi_1 K$, identifying $\langle \gamma_2 \rangle \subset \pi_1 X$ with the copy of C in this second $\pi_1 K$, and so on. The key point to note is that $\pi_1 \overline{X}$ is generated by the N (obvious) copies of $\pi_1 K$ that it contains, and since each of these copies of $\pi_1 K$ must have trivial image under any homomorphism from $\pi_1 \overline{X}$ to a finite group, $\pi_1 \overline{X}$ has no non-trivial finite quotients. □

By using equivariant gluing (II.11.19) instead of local gluing (II.11.13), one can extend the above result as follows:

7.3 Theorem. *If a group Γ acts properly and cocompactly by isometries on a* CAT(0) *space Y then one can embed Γ in a group $\widehat{\Gamma}$ that acts properly and cocompactly by isometries on a* CAT(0) *space $\overline{Y}$ and has no proper subgroups of finite index. If Y is a polyhedral complex of dimension $n \geq 2$ then so is $\overline{Y}$.*

If the group Γ in (7.3) is not torsion-free, then $\widehat{\Gamma}$ will be a semihyperbolic group that does not contain a torsion-free subgroup of finite index, and the complex of groups associated to the action of $\widehat{\Gamma}$ on $\overline{Y}$ will be a finite, non-positively curved, complex of finite groups that is not covered (in the sense of Section $\mathcal{C}$.5) by any compact polyhedron (i.e. a complex of groups whose local groups are trivial). An explicit example of such a group is given in (7.10).

The Hopf Property

An effective way of showing that certain finitely generated groups are not residually finite is to exhibit an isomorphism between the given group and one of its proper quotients.

7.4 Definition (Hopfian). A group H is said to be *Hopfian* if every epimorphism $H \twoheadrightarrow H$ is an isomorphism. In other words, if $N \subset H$ is normal and $H/N \cong H$ then $N = \{1\}$.

7.5 Proposition. *If a finitely generated group is residually finite then it is Hopfian.*

Proof. Let G be a finitely generated group and suppose that there is an epimorphism $\phi : G \to G$ with non-trivial kernel. We fix $g_0 \in \ker \phi \smallsetminus \{1\}$ and for every $n > 0$ we choose $g_n \in G$ such that $\phi^n(g_n) = g_0$.

If there were a finite group Q and a homomorphism $p : G \to Q$ such that $p(g_0) \neq 1$, then all of the maps $\phi_n := p\phi^n$ would be distinct, because $\phi_n(g_n) \neq 1$ whereas $\phi_m(g_n) = 1$ if $m > n$. But there are only finitely many homomorphisms from any finitely generated group to any finite group (because the image of the generators determines the map). □

The Hopf property is named in honour of H. Hopf. It arose in connection with his investigations into the problem of deciding when degree one maps between closed manifolds give rise to homotopy equivalences. Zlil Sela [Sel99] has shown that if a torsion-free hyperbolic group does not decompose as a free-product, then it is Hopfian. In particular this shows that the fundamental groups of closed negatively curved manifolds are Hopfian.

7.6 Examples. The following group was discovered by Baumslag and Solitar [BSo62]:

$$\mathrm{BS}(2,3) = \langle a, t \mid t^{-1}a^2t = a^3 \rangle.$$

The map $a \mapsto a^2$, $t \mapsto t$ is onto; a is in the image because $a = a^3a^{-2} = (t^{-1}a^2t)a^{-2}$. However this map is not an isomorphism: $[a, t^{-1}at]$ is a non-trivial element of the kernel. Meier [Me82] noticed that the salient features of this example are present in many other HNN extensions of abelian groups. Some of these groups were later studied by Wise [Wi96b], among them

$$T(n) = \langle a, b, t_a, t_b \mid [a,b] = 1,\ t_a^{-1}at_a = (ab)^n,\ t_b^{-1}bt_b = (ab)^n \rangle,$$

which is the fundamental group of the compact non-positively curved 2-complex $X(n)$ described in (6.9). If $n \geq 2$ then certain commutators, for example $g_0 = [t_a(ab)t_a^{-1}, b]$, lie in the kernel of the epimorphism $T(n) \twoheadrightarrow T(n)$ given by $a \mapsto a^n$, $b \mapsto b^n$, $t_a \mapsto t_a$, $t_b \mapsto t_b$. Britton's Lemma (6.4) implies that $g_0 \neq 1$ in $T(n)$. The proof of (7.5) shows that g_0 has trivial image in every finite quotient of $T(n)$.

Groups Without Finite Quotients

In this subsection we describe a technique for promoting the absence of residual finiteness to the absence of subgroups of finite index.

7.7 Proposition. *Let $\mathcal{G}$ be a class of groups that is closed under the formation of HNN extensions and amalgamated free products along finitely generated free groups. If $G \in \mathcal{G}$ is finitely generated, then it can be embedded in a finitely generated group $\widehat{G} \in \mathcal{G}$ that has no proper subgroups of finite index.*

The following proof is not the most direct possible. It is chosen specifically so that each step can be replicated in the context of compact non-positively curved 2-complexes (Lemma 7.9). In [Bri98a] this construction is used to prove other controlled embedding theorems. A similar construction was also used in [Wi96a].

Proof. We may assume that G contains an element of infinite order $g_0 \in G$ whose image in every finite quotient of G is trivial, for if it is does not then we can replace it with $G * \mathrm{BS}(2,3) \in \mathcal{G}$ by forming $G *_{\{1\}} = G * \langle a \rangle$ and then taking an HNN extension by a stable letter t conjugating a^2 to a^3. We may also assume that G is generated by elements of infinite order, because if $\{b_1, \dots, b_n\}$ generates G then $\{a, t, tb_1, \dots, tb_n\}$ generates $G * \mathrm{BS}(2,3)$, and the natural retraction $G * \mathrm{BS}(2,3) \twoheadrightarrow \mathrm{BS}(2,3)$ sends tb_i to t, which has infinite order in $\mathrm{BS}(2,3)$.

Let $\mathcal{A} = \{a_1, \dots, a_n\}$ be a generating set for G where each a_i has infinite order.

Step 1: We take an HNN extension of G with n stable letters:

$$E_1 = \langle G, s_1, \dots, s_n \mid s_i^{-1} a_i s_i = g_0^{p_i},\ i = 1, \dots, n \rangle,$$

where the p_i are any non-zero integers. Now, since each a_i is conjugate to a power of g_0 in E_1, the only generators of E_1 that can survive in any finite quotient are the s_i. However, since there is an obvious retraction of E_1 onto the free subgroup generated by the s_i, the group E_1 still has plenty of finite quotients.

Step 2: We repeat the extension process, this time introducing stable letters τ_i to make the generators s_i conjugate to g_0:

$$E_2 = \langle E_1, \tau_1, \dots, \tau_n \mid \tau_i^{-1} s_i \tau_i = g_0,\ i = 1, \dots, n \rangle.$$

Step 3: We add a single stable letter σ that conjugates the free subgroup of E_2 generated by the s_i to the free subgroup of E_2 generated by the τ_i:

$$E_3 = \langle E_2, \sigma \mid \sigma^{-1} s_i \sigma = \tau_i,\ i = 1, \dots, n \rangle.$$

At this stage we have a group in which all of the generators except σ are conjugate to g_0. In particular, every finite quotient of E_3 is cyclic.

Step 4: Because no power of a_1 lies in either of the subgroups of E_2 generated by the s_i or the τ_i, the normal form theorem for HNN extensions (6.4) implies that $\{a_1, \sigma\}$ freely generates a free subgroup of E_3.

We define $\widehat{G}$ to be an amalgamated free product of two copies of E_3,

$$\widehat{G} = E_3 *_F \overline{E_3},$$

where $F = F(x, y)$ is a free group of rank two; the inclusion of F into E_3 is $x \mapsto a_1$ and $y \mapsto \sigma$, and the inclusion into $\overline{E}_3$ is $x \mapsto \overline{\sigma}$ and $y \mapsto \overline{a}_1$. All of the generators of $\widehat{G}$ are conjugate to a power of either g_0 or $\overline{g_0}$, and therefore cannot survive in any finite quotient. In other words, $\widehat{G}$ has no finite quotients. $\square$

Complexes With No Non-Trivial Finite-Sheeted Coverings

In (7.2) we reduced the Embedding Theorem to the claim that there exist compact non-positively curved 2-complexes whose fundamental groups have no non-trivial finite quotients. We shall construct such a complex by following the scheme of the proof of (7.7), but first we must condition our complexes in the following manner.

Recall that a closed local geodesic in a metric space is a local isometry from a circle to the space.

7.8 Lemma. *Given a compact, connected, non-positively curved 2-complex X and a non-trivial element $g_1 \in \pi_1(X, x_0)$, one can isometrically embed X in a compact, connected, non-positively curved 2-complex K with basepoint $x_0 \in K$ such that:*

(1) *$\pi_1(K, x_0)$ is generated by a finite set of elements each of which is represented by a closed local geodesic that passes through x_0 and has integer length;*

(2) *a power of g_1 is represented by a closed local geodesic of length 1 that passes through x_0.*

Proof. Fix a point x_0 on a closed local geodesic that represents the conjugacy class of g_1. Suppose that $\pi_1(X, x_0)$ is generated by the non-trivial elements $\{b_1, \dots, b_n\}$, let β_i be the shortest loop based at x_0 in the homotopy class b_i, and let l_i be the length of β_i. Let l_0 be the length of the closed local geodesic representing g_1. Replacing g_1 by an appropriate power, we may assume that $l_0 > l_i$ for $i = 1, \dots, n$.

Consider the following metric graph Λ: there are $(n+1)$ vertices $\{v_0, \dots, v_n\}$ and $2n$ edges $\{e_1, \varepsilon_1, \dots, e_n, \varepsilon_n\}$; the edge e_i connects v_0 to v_i and has length $(l_0 - l_i)/2$; the edge ε_i is a loop of length l_0 based at v_i. We obtain the desired complex K by gluing Λ to X, identifying v_0 with x_0, and then scaling the metric by a factor of l_0 so that the closed local geodesic representing $g_1 \in \pi_1(K, x_0)$ has length 1. See figure Γ.14.

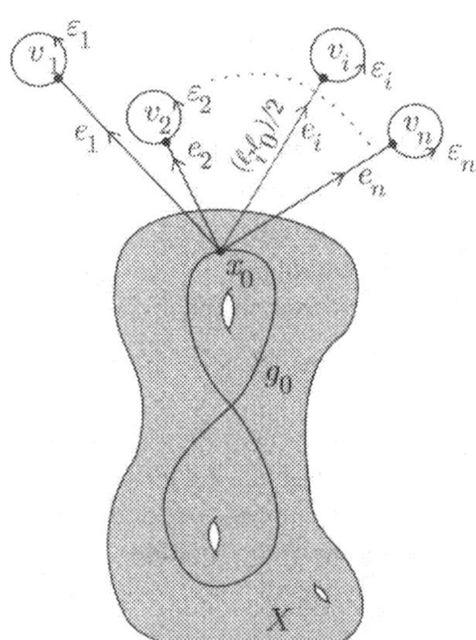

Fig. Γ.14 Adding the lollipop graph Λ to X

Let $\gamma_i \in \pi_1(K, x_0)$ be the element given by the geodesic c_i that traverses e_i, crosses ε_i, and then returns along e_i, that is $c_i = e_i \varepsilon_i \overline{e}_i$ where the overline denotes reversed orientation. Note that $\pi_1(K, x_0)$ is the free product of $\pi_1(X, x_0)$ and the free group generated by $\{\gamma_1, \ldots, \gamma_n\}$. As generating set for $\pi_1(K, x_0)$ we choose $\{b_i\gamma_i, b_i\gamma_i^2 \mid i = 1, \ldots, n\}$.

The concatenation of any non-trivial locally geodesic loop in X, based at x_0, and any non-trivial locally geodesic loop in Λ based at v_0 is a closed local geodesic in K. Thus $\beta_i c_i$ and $\beta_i e_i \varepsilon_i^2 \overline{e}_i$ are closed local geodesics in K; the former has length 2 and the latter has length 3; the former represents $b_i\gamma_i$ and the latter represents $b_i\gamma_i^2$. □

The following lemma completes the proof of The Embedding Theorem (7.2).

7.9 Lemma. *There exists a compact, connected, non-positively curved, 2-complex whose fundamental group has no finite quotients.*

Proof. Let X be a compact, connected, non-positively curved, 2-complex whose fundamental group is not residually finite (for example one of the complexes $X(n)$ in (7.6)). Let $g_1 \in \pi_1 X$ be an element whose image in every finite quotient is trivial. Let K be the 2-complex obtained by applying the construction of the preceding lemma, and let g_0 be the power of g_1 that appears in item (2) of the statement of that lemma.

We shall mimic the proof of Proposition 7.7 with $\pi_1(K, x_0)$ in the role of G. At each stage we shall state what the fundamental group of the complex being constructed is; in each case this is a simple application of the Seifert-van Kampen theorem.

Let c_0 be the closed local geodesic of length 1 representing g_0. Let $\{a_1, \ldots, a_n\}$ be the generators given by 7.7(1), let α_i be the closed local geodesic through x_0 that represents a_i, and suppose that α_i has length λ_i. For each i, we glue to K a cylinder $S_{\lambda_i} \times [0, 1]$, where S_{λ_i} is a circle of length λ_i with basepoint v_i; one end of the cylinder is attached to α_i and the other end wraps λ_i-times around c_0, and $v_i \times \{0, 1\}$ is attached to x_0. Let K_1 be the resulting complex. In the notation of (7.7) we have $\pi_1(K_1, x_0) = E_1$. Moreover K_1 is non-positively curved (II.11.6).

The images in K_1 of the paths $v_i \times [0, 1]$ give an isometric embedding into K_1 of the metric graph Y that has one vertex and n edges of length 1; call the corresponding free subgroup $F_1 \subset E_1$ (it is the subgroup generated by the s_i in (7.7)).

Step 2 of (7.7) is achieved by attaching n cylinders of unit circumference and length to K_1, the ends of the i-th cylinder being identified to c_0 and to the image of $v_i \times [0, 1]$. The resulting complex K_2 has $\pi_1(K_2, x_0) = E_2$. As in the previous step, the free subgroup $F_2 \subset E_2$ generated by the basic loops that run along the new cylinders is the π_1-image of an isometric embedding $Y \to K_2$ (it is the subgroup generated by the τ_i in (7.7)).

To achieve Step 3 of (7.7), we now glue $Y \times [0, L]$ to K_2 by attaching the ends according to the isometric embeddings that realise the embeddings $F_1, F_2 \subset \pi_1(K_2, x_0)$. This gives us a compact non-positively curved complex K_3 with fundamental group E_3 (in the notation of (7.7)). Let v be the vertex of Y, observe that $v \times \{0, L\}$ is attached to $x_0 \in K_3$, and let $\sigma \in \pi_1(K_3, x_0)$ be the homotopy class of the loop $[0, L] \to K_3$ given by $t \mapsto (v, t)$.

We left open the choice of L, the length of the mapping cylinder in Step 3, we now specify that it should be λ_1, the length of the geodesic representing the generator a_1. An important point to observe is that the angle at x_0 between the image of $v \times [0, L]$

and any path in $K_1 \subset K_3$ is π. Thus the free subgroup $\langle a_1, \sigma \rangle$ is the π_1-image in $\pi_1(K_3, x_0)$ of an isometry from the metric graph Z with one vertex (sent to x_0) and two edges of length $L = \lambda_1$ (cf. 6.15). In fact, we have two such isometries $Z \to K_3$, corresponding to the free choice of which edge of Z to send to the image of $v \times [0, L]$. We use these two maps to realise Step 4 of the construction on (7.7), appealing to (6.14) to see that the resulting 2-complex is non-positively curved. □

7.10 A Semihyperbolic Group That Is Not Virtually Torsion-Free. We close this section with an explicit example to illustrate the remark that we made following Theorem 7.3. This example is essentially contained in Wise's thesis [Wis96a].

In the hyperbolic plane we consider a regular geodesic quadrilateral Q with vertex angles $\pi/4$. Let α and β be hyperbolic isometries that identify the opposite sides of Q. Then Q is a fundamental domain for the action of $\Gamma = \langle \alpha, \beta \rangle$. The commutator $[\alpha, \beta]$, acts as a rotation through an angle π at one vertex of Q, and away from the orbit of this vertex the action of Γ is free. Thus the quotient orbifold $V = \Gamma \backslash \mathbb{H}^2$ is a torus with one singular point, and at that singular point the local group is $\mathbb{Z}_2$.

Let $X(n)$ and $T(n)$ be as in (7.6), let g_0 be a non-trivial element in the kernel of a self-surjection $T(n) \twoheadrightarrow T(n)$ and consider the amalgamated free product $G = T(n) *_{\mathbb{Z}} \Gamma$ in which g_0 is identified with β. Note that since g_0 has trivial image in every finite quotient of $T(n)$, the commutator $[\alpha, \beta] = [\alpha, g_0]$, which has order two, has trivial image in every finite quotient of G.

The equivariant gluing described in (II.11.19) yields a proper cocompact action of G by isometries on a CAT(0) space. The quotient space Y can be described as follows: scale the metric on $X(n)$ so that the closed local geodesic c representing g_0 has length $l = |\alpha| = |\beta|$, then glue $X(n)$ to V using a tube $S_l \times [0, 1]$ one end of which is glued to c and the other end of which is glued to the image in V of the axis of β. When viewed as a complex of groups, Y has only one non-trivial local group, namely the $\mathbb{Z}_2$ at the singular point of V.

In the case $n = 2$, if we choose g_0 as in (7.6), then G can be presented as follows:

$$\langle a, b, s, t, \alpha, \beta \mid \beta = [s(ab)s^{-1}, b],\ [a, b] = [\alpha, \beta]^2 = 1,\ tbt^{-1} = sas^{-1} = (ab)^2 \rangle.$$

Chapter III.$\mathcal{C}$ Complexes of Groups

Roughly speaking, *complexes of groups* were introduced to describe actions of groups on simply connected simplicial complexes in terms of suitable local data on the quotient. They are natural generalizations of the concept of graphs of groups due to Bass and Serre. A technical problem arises from the fact the quotient of a simplicial complex by a simplicial group action will not be a simplicial complex in general. Indeed, even if the set-wise stabilizer of each simplex is equal to its point-wise stabilizer, faces of a given simplex might get identified. (For example, regard the real line as a one-dimensional complex with vertices at the integers and consider the action of $\mathbb{Z}$ by translations.) Because of this problem, it is more natural to work with polyhedral complexes.

To describe a polyhedral complex K combinatorially, it is natural to take its first barycentric subdivision K' and view it as the geometric realization of the *nerve of a category* whose objects are the barycentres of cells of K and whose non-trivial morphisms correspond to the edges of K'. Such a category is a particular case of what we call a *small category without loops* (scwol). The geometric realization of a category without loops is like a simplicial complex, but the intersection of two distinct simplices may contain more than a common face. Small categories without loops serve as combinatorial descriptions of polyhedral complexes.

In the first section of this chapter we introduce scwols and their *geometric realizations*, and we define what it means for a group to act on a scwol. The reader in a hurry may wish to read only the definitions (1.1), (1.5) and (1.11) and then proceed to the next section.

In the second section we give the basic definitions of a *complex of groups* $G(\mathcal{Y})$ *over a scwol* $\mathcal{Y}$ and of morphisms of complexes of groups. Associated to each action (in a suitable sense) of a group G on a scwol $\mathcal{X}$ there is a complex of groups $G(\mathcal{Y})$ over the quotient scwol $\mathcal{Y} = G\backslash\mathcal{X}$. The construction of $G(\mathcal{Y})$ depends on some choices but it is unique up to isomorphism. A complex of groups that arises in this way is called *developable*. In contrast to the case of graphs of groups, in general a complex of groups $G(\mathcal{Y})$ does not arise from an action of a group when the dimension of $\mathcal{Y}$ is bigger than one. We give a quick proof of the developability of complex of groups over scwols of dimension one (a result due to Bass and Serre [Ser77]).

The fundamental group $\tilde{G}$ of a complex of groups $G(\mathcal{Y})$ is defined in the third section and a presentation of this group is given. In the case of a developable complex of groups $G(\mathcal{Y})$, it is possible to construct a simply connected category without loops

$\tilde{\mathcal{X}}$ and an action of the fundamental group $\tilde{G}$ on $\tilde{\mathcal{X}}$ so that $G(\mathcal{Y})$ is the complex of groups associated to this action.

Every complex of groups is locally developable; this is explained in the fourth section where the local development is constructed explicitly; thus non-developability is a global phenomenon. In the next chapter we shall prove that if a complex of groups carries a metric of non-positive curvature, then it is always developable.

The theory of coverings of complexes of groups is explained in the last section. It should be compared to the theory of coverings of graphs of groups in the sense of Bass [Bass93].

It turns out that many concepts such as morphisms of complexes of groups, coverings of complexes of groups, and so on, become more natural if we interpret them in the framework of category theory. For this reason we recall in an appendix the basic notions of small category, the fundamental group of such a category and coverings. We hope that this will help the reader to realize that many complicated formulas displayed elsewhere in this chapter are simply explicit descriptions of elementary concepts.

Our treatment of complexes of groups is very similar to the combinatorial approach of Bass-Serre to graphs of groups. A special case of complexes of groups, triangles of groups, was first studied by Gersten and Stallings [Sta91]. They proved the developability theorem for non-positively curved triangles of groups. The general case of complexes of groups was considered by Haefliger [Hae91] and (in dimension 2) independently by J.Corson [Cors92].

There is one important aspect of the theory of complexes of groups that we have neglected here, namely the construction of their geometric realizations and their relationship to complexes of spaces. This interpretation is useful for topological and homological considerations. For this aspect of the theory we refer the reader to [Hae91,92], [Cor92] and, in the case of graphs of groups, to [ScoW79].

1. Small Categories Without Loops (scwols)

Small categories without loops (abbreviated *scwol*) are algebraic objects which can serve as combinatorial substitutes for polyhedral complexes. Canonically associated to each polyhedral complex K there is a scwol $\mathcal{X}$ whose set of vertices is in bijection with the set of cells of K. The geometric realization of $\mathcal{X}$ is a cell complex isomorphic to the first barycentric subdivision of K. In this section we first give the basic definitions of scwols, their geometric realizations, and morphisms between them. The notions of fundamental group and covering for scwols are then introduced; these correspond to the usual topological notions for their geometric realizations. We also define what it means for a group G to act on a scwol $\mathcal{X}$; when $\mathcal{X}$ is associated to a polyhedral complex K, actions of G on $\mathcal{X}$ correspond to actions of G on K by homeomorphisms that preserve the affine structure of the cells and are such that if an element of G maps a cell to itself, then it fixes this cell pointwise. At the end we analyse the local structure of scwols (this will be needed in sections 4 and 5).

Scwols and Their Geometric Realizations

1.1 Definitions. A *small category without loops* $\mathcal{X}$ (briefly a *scwol*)[66] is a set $\mathcal{X}$ which is the disjoint union of a set $V(\mathcal{X})$, called the vertex set of $\mathcal{X}$ (the elements of which will be denoted by Greek letters $\tau, \sigma, \rho, \dots$), and a set $E(\mathcal{X})$, called the set of edges of $\mathcal{X}$ (the elements of which will be denoted by Latin letters $a, b, c, \dots$). Two maps are given

$$i : E(\mathcal{X}) \to V(\mathcal{X}) \text{ and } t : E(\mathcal{X}) \to V(\mathcal{X});$$

$i(a)$ is called the initial vertex of $a \in E(\mathcal{X})$ and $t(a)$ is called the terminal vertex of a.

Let $E^{(2)}(\mathcal{X})$ denote the set of pairs $(a, b) \in E(\mathcal{X}) \times E(\mathcal{X})$ such that $i(a) = t(b)$. A third map

$$E^{(2)}(\mathcal{X}) \to E(\mathcal{X})$$

is given that associates to each pair (a, b) an edge ab called their *composition*[67]. These maps are required to satisfy the following conditions:

(1) for all $(a, b) \in E^{(2)}(\mathcal{X})$, we have $i(ab) = i(b)$ and $t(ab) = t(a)$;

(2) *associativity:* for all $a, b, c \in E(\mathcal{X})$, if $i(a) = t(b)$ and $i(b) = t(c)$, then $(ab)c = a(bc)$ (thus the composition may be denoted abc);

(3) *no loops condition:* for each $a \in E(\mathcal{X})$, we have $i(a) \neq t(a)$.

If we exchange the roles of i and t, we get the *opposite scwol.*

Consider the equivalence relation on $V(\mathcal{X})$ generated by $\tau \sim \sigma$ if there is an edge a such that $i(a) = \sigma$ and $t(a) = \tau$. The scwol $\mathcal{X}$ is *connected* if there is only one equivalence class.

A *subscwol* $\mathcal{X}'$ of a scwol $\mathcal{X}$ is given by subsets $V(\mathcal{X}') \subseteq V(\mathcal{X})$ and $E(\mathcal{X}') \subseteq E(\mathcal{X})$ such that if $a \in E(\mathcal{X}')$, then $i(a), t(a) \in V(\mathcal{X}')$, and if $a, b \in E(\mathcal{X}')$ are such that $i(a) = t(b)$, then $ab \in E(\mathcal{X}')$. Every scwol $\mathcal{X}$ is the disjoint union of connected subscwols, called its connected components.

We can consider $\mathcal{X}$ as the set of arrows of a small category [68] with set of objects $V(\mathcal{X})$: each vertex σ is identified to a unit element 1_σ and we define $i(1_\sigma) = t(1_\sigma) = \sigma$; each edge corresponds to an element of $\mathcal{X}$ which is not a unit. When we view $\mathcal{X}$ in this way, unspecified elements of $\mathcal{X}$ will be denoted by Greek letters $\alpha, \beta, \dots$; an element $\alpha \in \mathcal{X}$ is either an edge a or the unit 1_σ identified to an object σ. A subscwol corresponds to a subcategory.

1.2 Examples

(1) To any poset $\mathcal{Q}$ we can associate a scwol as follows. The set of vertices is $\mathcal{Q}$ and the edges are pairs $(\tau, \sigma) \in \mathcal{Q} \times \mathcal{Q}$ such that $\tau < \sigma$; the initial (resp. terminal)

[66] In [Hae92] a scwol (or rather its geometric realization) was called an *ordered simplicial cell complex.*

[67] If $(a, b) \in E^{(2)}(\mathcal{X})$, one says that a and b are *composable.* One thinks of ab as "a following b".

[68] See the appendix for basic definitions concerning categories.

vertex of the edge (τ, σ) is σ (resp. τ); the composition $(\tau, \sigma)(\sigma, \rho)$ is defined to be (τ, ρ).

(2) *Product of scwols.* Given two scwols $\mathcal{X}$ and $\mathcal{X}'$, their product $\mathcal{X} \times \mathcal{X}'$, is the scwol defined as follows: $V(\mathcal{X} \times \mathcal{X}') := V(\mathcal{X}) \times V(\mathcal{X}')$,

$$E(\mathcal{X} \times \mathcal{X}') := (E(\mathcal{X}) \times V(\mathcal{X}')) \coprod (E(\mathcal{X}) \times E(\mathcal{X}')) \coprod (V(\mathcal{X}) \times E(\mathcal{X}')).$$

The maps $i, t : \mathcal{X} \times \mathcal{X}' \to V(\mathcal{X}) \times V(\mathcal{X}')$ are defined by $i((\alpha, \alpha')) = (i(\alpha), i(\alpha'))$ and $t((\alpha, \alpha')) = (t(\alpha), t(\alpha'))$, and the composition $(\alpha, \alpha')(\beta, \beta')$, whenever defined, is equal to $(\alpha\beta, \alpha'\beta')$. This is a particular case of the product of two small categories (see A.2(2)).

Notations. For each integer $k > 0$, let $E^{(k)}(\mathcal{X})$ be the set of sequences $(a_1, \dots, a_k)$ of composable edges — $(a_i, a_{i+1}) \in E^{(2)}(\mathcal{X})$ for $i = 1, \dots, k-1$. By convention $E^{(0)}(\mathcal{X}) = V(\mathcal{X})$. The *dimension k* of $\mathcal{X}$ is the supremum of the integers $k \geq 0$ such that $E^{(k)}(\mathcal{X})$ is non empty.

1.3 The Geometric Realization. For an integer $k \geq 0$, let Δ^k be the standard k-simplex, i.e. the set of points $(t_0, \dots, t_k) \in \mathbb{R}^{k+1}$ such that $t_i \geq 0$ and $\sum_{i=0}^{k} t_i = 1$. Recall that the t_i are the barycentric coordinates in Δ^k.

We shall define the geometric realization $|\mathcal{X}|$ of $\mathcal{X}$ as a polyhedral complex whose cells of dimension k are standard k-simplices indexed by the elements of $E^{(k)}(\mathcal{X})$. For $k > 1$ and $i = 0, \dots, k$, let $\partial_i : E^{(k)}(\mathcal{X}) \to E^{(k-1)}(\mathcal{X})$ be the maps defined by

$$\begin{aligned} \partial_0(a_1, \dots, a_k) &= (a_2, \dots, a_k) \\ \partial_i(a_1, \dots, a_k) &= (a_1, \dots, a_i a_{i+1}, \dots, a_k) \quad 1 \leq i < k \\ \partial_k(a_1, \dots, a_k) &= (a_1, \dots, a_{k-1}). \end{aligned}$$

For $k = 1$, we define $\partial_0(a_1) = i(a_1)$ and $\partial_1(a_1) = t(a_1)$. The following relations follow from the associativity axiom 1.1(2):

$$\partial_i \partial_j = \partial_{j-1} \partial_i, \quad i < j. \tag{1.3-1}$$

We also define maps

$$d_i : \Delta^{k-1} \to \Delta^k$$

for $i = 0, \dots, k$.

By definition d_i sends $(t_0, \dots, t_{k-1}) \in \Delta^{k-1}$ to $(t_0, \dots, t_{i-1}, 0, t_i, \dots, t_{k-1}) \in \Delta^k$. Thus $d_j(\Delta^{k-1})$ is the face of Δ^k consisting of points $(t_0, \dots, t_k)$ such that $t_j = 0$. More generally, the face of codimension r defined by $t_{i_1} = \dots = t_{i_r} = 0$, $i_1 > \dots > i_r$, is the subset $d_{i_1} \dots d_{i_r}(\Delta^{k-r})$.

We have the obvious relations:

$$d_i d_j = d_{j+1} d_i, \quad i \leq j. \tag{1.3-2}$$

The Construction of the Geometric Realization

On the disjoint union

$$\bigcup_{k,A} \Delta^k \times \{A\},$$

where $k \geq 0$ and $A \in E^{(k)}(\mathcal{X})$, we consider the equivalence relation generated by

$$(d_i(x), A) \sim (x, \partial_i(A)),$$

where $(x, A) \in \Delta^{k-1} \times E^{(k)}(\mathcal{X})$. We define $|\mathcal{X}|$, the *geometric realization of* $\mathcal{X}$, to be the quotient by this equivalence relation; it is a piecewise Euclidean complex whose k-cells are isometric to Δ^k. The simplex $\Delta^k \times \{A\}$ maps injectively to the quotient (see exercise below) and its image will be called the *k-simplex of* $|\mathcal{X}|$ *labelled by A, or simply "the k-simplex A"*. Thus the letters denoting elements of $V(\mathcal{X})$ and $E(\mathcal{X})$ will also be used to denote the corresponding vertices (0-simplices) and 1-simplices of $|\mathcal{X}|$. If $A = (a_1, \dots, a_k)$, we call $i(a_k)$ the initial vertex of A.

1.3.3 Remarks

(1) Note that in general the intersection of two simplices in $|\mathcal{X}|$ is not a common face, rather it is a union of faces.

(2) In general we consider $|\mathcal{X}|$ as a piecewise Euclidean complex with the topology associated to its intrinsic metric. Later we shall identify each simplex $\Delta_k \times \{A\}$ to a simplex in some M^n_κ in such a way that the inclusions $(x, \partial_i A) \mapsto (d_i(x), A)$ are isometries, thus metrizing $|\mathcal{X}|$ as a M_κ-polyhedral complex. If the set of isometry types of simplices is finite, then the topology associated to this new metric is the same as the topology defined above.

(3) The geometric realization of the scwol associated to a poset $\mathcal{Q}$ is the same as the geometric (or affine) realization of $\mathcal{Q}$ as defined in (II.12.3). The geometric realization of the product of two scwols is homeomorphic to the product of their geometric realizations.

(4) The dimension of $\mathcal{X}$ is equal to the dimension of its geometric realization. As observed above, the 1-simplices of $|\mathcal{X}|$ correspond bijectively to the edges of $\mathcal{X}$; they inherit an orientation from this correspondence. If one follows an edge path consisting of 1-simplices in the direction of the given orientations, one never gets a closed circuit. The set of vertices of a k-simplex is ordered.

1.3.4 Exercise. We use the notations of 1.3. Prove that the map λ_A associating to $(x, A) \in \Delta^k \times E^{(k)}$ its equivalence class in $|\mathcal{X}|$ is injective on $\Delta^k \times \{A\}$.

(Hint: Suppose that $x \in \Delta^k$ is contained in the interior of a face of codimension r defined by $t_{i_1} = \cdots = t_{i_r} = 0$, $i_1 > \cdots > i_r$. Let $\bar{x}$ be the unique point of Δ^{k-r} such that $d_{i_1} \dots d_{i_r}(\bar{x}) = x$ and let $\bar{A} = \partial_{i_1} \dots \partial_{i_r}(A)$. Using the relations 1.3-1 and 1.3-2, show that $(\bar{x}, \bar{A})$ is the unique representative of (x, A) in $\Delta^{k-r} \times E^{(k-r)}(\mathcal{X})$. This implies that the restriction of λ_A to $\{(x, A) \mid x \in \text{interior of } \Delta^k\}$ is injective. The injectivity of λ_A follows from 1.1(3).)

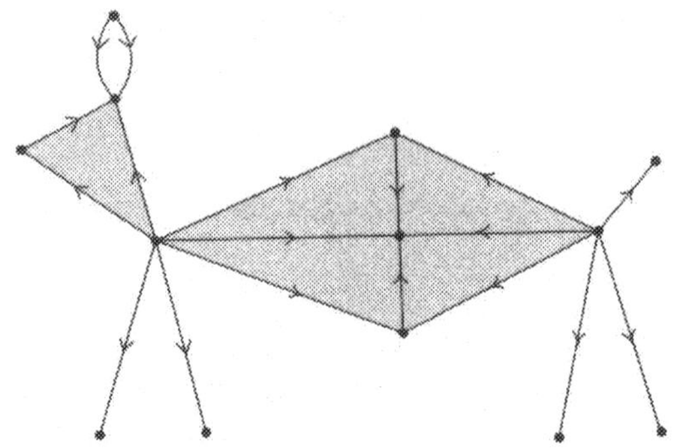

Fig. $\mathcal{C}$.1 The geometric realization of a scwol of dimension 2

1.4 Examples

(1) If $\mathcal{X}$ is connected and 1-dimensional, its geometric realization is a connected graph with two kinds of vertices: *sources* are initial vertices of edges and *sinks* are terminal vertices of edges. This graph is bipartite and oriented. There are no edge loops, but it is possible to have several edges connecting a given source to a given sink.

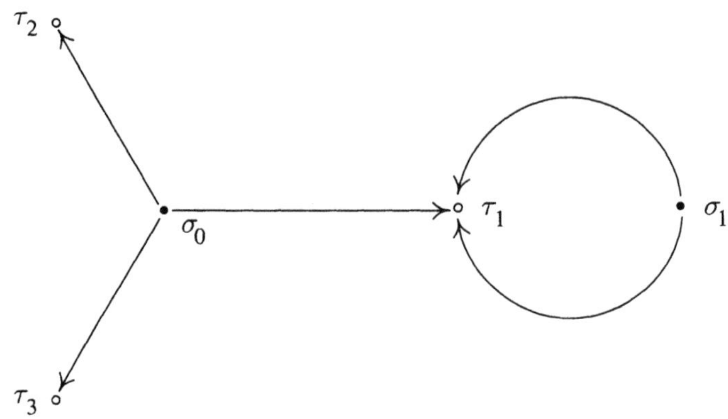

Fig. $\mathcal{C}$.2 A scwol of dimension one

(2) Naturally associated to any M_κ-polyhedral complex K there is scwol $\mathcal{X}$ whose geometric realization is the first barycentric subdivision of K. The set of vertices $V(\mathcal{X})$ of $\mathcal{X}$ is the set of cells of K (equivalently the set of barycentres of the cells of K). The edges of $\mathcal{X}$ are the 1-simplices of the barycentric subdivision K' of K: each 1-simplex a of K' corresponds to a pair of cells $T \subset S$; we define $i(a)$ to be the barycentre of S and $t(a)$ to be the barycentre of T.

A connected, one-dimensional scwol $\mathcal{X}$ is associated to a 1-dimensional polyhedral complex (i.e. a graph) if and only if each vertex of $\mathcal{X}$ which is a source is the initial vertex of exactly two edges.

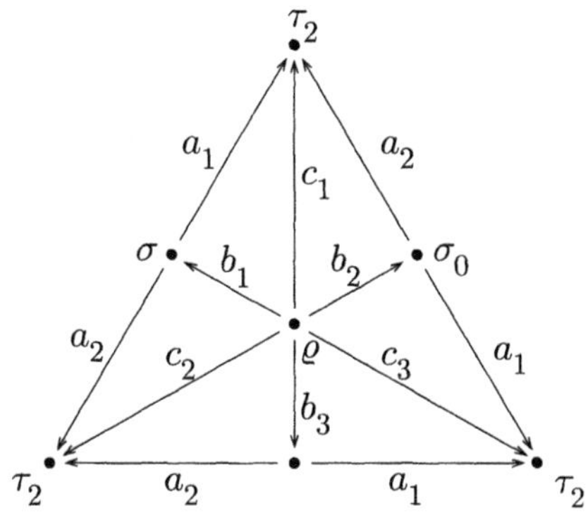

$V(\mathcal{Y}) = \{\rho, \sigma, \tau\}$
$E(\mathcal{Y}) = \{a_1, a_2, b_1, b_2, b_3, c_1, c_2, c_3\}$

(elements with the same label should be identified)

Fig. $\mathcal{C}$.3 The scwol associated to the dunce hat (cf. I.7.41(2))

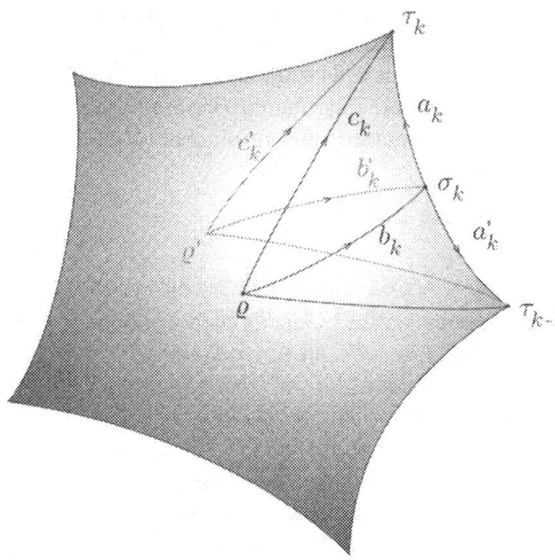

Fig. $\mathcal{C}$.4 The scwol associated to two congruent n-gons glued along their boundary

(3) Figure $\mathcal{C}$.4 shows the scwol $\mathcal{X}$ associated to the polyhedral complex obtained by gluing two congruent n-gons in M_κ^2 along their boundary. There are n vertices τ_k corresponding to the vertices of the polygons, n vertices σ_k corresponding to the barycentres of the sides and two vertices ρ and ρ' corresponding to the barycentres of the two polygons. There are $6n$ edges, $a_k, a_k', b_k, b_k', c_k, c_k'$ with $k = 1, \dots, n$, and $i(b_k) = i(c_k) = \rho$, $i(b_k') = i(c_k') = \rho'$, $t(b_k) = t(b_k') = \sigma_k$, $t(c_k) = t(c_k') = \tau_k$, $i(a_k) = i(a_k') = \sigma_k$ and $t(a_k) = \tau_k$, $t(a_k') = \tau_{k-1}$ mod n. We have $c_k = a_k b_k = a_{k+1}' b_{k+1}$.

1.5 Morphisms of Scwols. Let $\mathcal{X}$ and $\mathcal{Y}$ be two scwols. A *non-degenerate morphism* $f : \mathcal{X} \to \mathcal{Y}$ is a map that sends $V(\mathcal{X})$ to $V(\mathcal{Y})$, sends $E(\mathcal{X})$ to $E(\mathcal{Y})$ and is such that

(1) for each $a \in E(\mathcal{X})$, we have $i(f(a)) = f(i(a))$ and $t(f(a)) = f(t(a))$,

(2) for each $(a, b) \in E^{(2)}(\mathcal{X})$ we have $f(ab) = f(a)f(b)$, and

(3) for each vertex $\sigma \in V(\mathcal{X})$, the restriction of f to the set of edges with initial vertex σ is a bijection onto the set of edges of $\mathcal{Y}$ with initial vertex $f(\sigma)$.

More generally, a *morphism* $f : \mathcal{X} \to \mathcal{Y}$ is a functor from the category $\mathcal{X}$ to the category $\mathcal{Y}$ (see A.1). An *automorphism* of $\mathcal{X}$ is a morphism from $\mathcal{X}$ to $\mathcal{X}$ that has an inverse, i.e. a functor $f^{-1} : \mathcal{X} \to \mathcal{X}$ such that $f^{-1}f = ff^{-1}$ is the identity of $\mathcal{X}$.

A morphism $f : \mathcal{X} \to \mathcal{Y}$ induces a map $|f| : |\mathcal{X}| \to |\mathcal{Y}|$ of the geometric realizations: $|f|$ maps each vertex σ of $|\mathcal{X}|$ to the vertex $f(\sigma)$ of $|\mathcal{Y}|$ and its restriction to each simplex of $|\mathcal{X}|$ is affine. In the particular case where f maps $E(\mathcal{X})$ to $E(\mathcal{Y})$ (for instance when f is non-degenerate), $|f|$ maps a point of the k-simplex of $|\mathcal{X}|$ labelled $(a_1, \dots, a_k)$ to the point in the simplex of $|\mathcal{Y}|$ labelled $(f(a_1), \dots, f(a_k))$ that has the same barycentric coordinates. If f is an isomorphism, then $|f|$ is an isometry.

If $\mathcal{X}$ and $\mathcal{Y}$ are associated to polyhedral complexes K and L, then a morphism $f : \mathcal{X} \to \mathcal{Y}$ is non-degenerate if and only the restriction of $|f|$ to the interior of each cell of K is a homeomorphism onto the interior of a cell of L.

The Fundamental Group and Coverings

1.6 Edge Paths. Let $\mathcal{X}$ be a scwol. Let $E^{\pm}(\mathcal{X})$ be the set of symbols a^+ and a^-, where $a \in E(\mathcal{X})$. The elements e of $E^{\pm}(\mathcal{X})$ can be considered as *oriented edges* of $\mathcal{X}$. For $e = a^+$, we define $i(e) = t(a)$, $t(e) = i(a)$ and $e^{-1} = a^-$; for $e = a^-$, we define $i(e) = i(a)$, $t(e) = t(a)$ and $e^{-1} = a^+$. An *edge path* in $\mathcal{X}$ (or a path in $\mathcal{X}$) joining the vertex σ to the vertex τ is a sequence $c = (e_1, \dots, e_k)$ of elements of $E^{\pm}(\mathcal{X})$ such that $i(e_1) = \sigma$, $t(e_i) = i(e_{i+1})$ for $1 \le i \le k-1$, and $t(e_k) = \tau$. We call σ (resp. τ) the initial vertex $i(c)$ of c (resp. the terminal vertex $t(c)$ of c). If $\sigma = \tau$, we allow k to be 0, giving the *constant path at* σ. Note that $\mathcal{X}$ is connected if and only if for any $\sigma, \tau \in V(\mathcal{X})$, there is a path joining σ to τ. If $c' = (e'_1, \dots, e'_{k'})$ is a path in $\mathcal{X}$ joining $\sigma' = \tau$ to τ', then one can compose c and c' to obtain the path $c * c' = (e_1, \dots, e_k, e'_1, \dots, e'_k)$ joining σ to τ', called the *concatenation* of c and c'. Note that this concatenation operation is associative. The *inverse* of the path c is defined to be the path $c^{-1} = (e'_1, \dots, e'_k)$ joining τ to σ, where $e'_j = e^{-1}_{k-j+1}$.

If $i(c) = t(c) = \sigma$, then c is called a *loop* at σ.

1.7 Homotopy of Edge Paths. Let $c = (e_1, \dots, e_k)$ be an edge path in $\mathcal{X}$ joining σ to τ. Consider the following two operations on c:

(i) Assume that for some $j < k$, we have $e_j = e^{-1}_{j+1}$. Then we get a new path c' by deleting from c the subsequence e_j, e_{j+1}.

(ii) Assume that for some $j < k$, we have $e_j = a^+$, $e_{j+1} = b^+$ (resp. $e_j = b^-$, $e_{j+1} = a^-$), and therefore the composition ab is defined; we get a new path c' in $\mathcal{X}$ by replacing the subsequence e_j, e_{j+1} of c by $(ab)^+$ (resp. $(ab)^-$).

Under the circumstances of (i) or (ii), we say that c and c' are obtained from each other by an elementary homotopy (thus implicitly allowing the inverses of operations (i) and (ii)). Two paths joining σ to τ are said to be *homotopic* if one can pass from the first to the second by a sequence of elementary homotopies (cf. I.8A and A.11). The set of homotopy classes of edge paths in $\mathcal{X}$ joining σ to τ is denoted $\pi_1(\mathcal{X}, \sigma, \tau)$. The set $\pi_1(\mathcal{X}, \sigma, \sigma)$ will also be denoted $\pi_1(\mathcal{X}, \sigma)$.

Note that the path $c * c^{-1}$ is homotopic to the constant path at σ and that $c^{-1} * c$ is homotopic to the constant path at τ. The homotopy class of the concatenation $c * c'$ of two paths depends only on the homotopy classes of c and c'.

1.8 Definition of the Fundamental Group. Let $\mathcal{X}$ be a connected scwol and let $\sigma \in V(\mathcal{X})$. The set $\pi_1(\mathcal{X}, \sigma_0)$ of homotopy classes of edge loops at σ_0, with the law of composition induced by concatenation, is a group called the *fundamental group* of $\mathcal{X}$ at σ_0.

If c is an edge path in $\mathcal{X}$ joining σ to τ, then the map that associates to each loop l at σ the loop $c^{-1} * l * c$ at τ induces an isomorphism of $\pi_1(\mathcal{X}, \sigma)$ onto $\pi_1(\mathcal{X}, \tau)$. A scwol $\mathcal{X}$ is said to be *simply connected* if it is connected and $\pi_1(\mathcal{X}, \sigma)$ is the trivial group for some (hence all) $\sigma \in V(\mathcal{X})$.

It is a classical fact that there is a natural isomorphism from $\pi_1(\mathcal{X}, \sigma_0)$ to $\pi_1(|\mathcal{X}|, \sigma_0)$ [Mass91].

1.9 Definition of a Covering. Let $\mathcal{X}$ be a connected scwol and let $\mathcal{X}'$ be a (non-empty) scwol. A morphism $p : \mathcal{X}' \to \mathcal{X}$ is called a *covering* of $\mathcal{X}$ if, for every $\sigma' \in V(\mathcal{X}')$, p sends the set $\{a' \in E(\mathcal{X}') \mid t(a') = \sigma'\}$ bijectively onto $\{a \in E(\mathcal{X}) \mid t(a) = p(\sigma')\}$ and sends the set $\{a' \in E(\mathcal{X}') \mid i(a') = \sigma'\}$ bijectively onto $\{a \in E(\mathcal{X}) \mid i(a) = p(\sigma')\}$. If $p : \mathcal{X}' \to \mathcal{X}$ is a covering then one says that $\mathcal{X}'$ covers $\mathcal{X}$.

Remarks

(1) Note that the geometric realization $|p| : |\mathcal{X}'| \to |\mathcal{X}|$ of p is a topological covering. If $\mathcal{X}'$ covers $\mathcal{X}$ and $\mathcal{X}$ is a scwol associated to a poset, then $\mathcal{X}'$ is also associated to a poset. The converse is not true in general.

(2) Let $p : \mathcal{X}' \to X$ be a covering. An edge path $c' = (e'_1, \dots, e'_k)$ in $\mathcal{X}'$ with $i(c) = \sigma'$ is called a *lifting* at σ' of the path $c = p(c') = (e_1, \dots, e_k)$ if $e_i = p(e'_i)$ (where, by definition, $p(a'^{\pm}) = (p(a'))^{\pm}$ if $a' \in E(\mathcal{X}')$). It follows immediately from the definition of a covering that if $p(\sigma') = \sigma$, then any edge path in $\mathcal{X}$ issuing from σ can be lifted uniquely to an edge path in $\mathcal{X}'$ at σ'. In particular, as $\mathcal{X}$ is assumed to be connected, p is surjective. Moreover if two edge paths c'_1, c'_2 in $\mathcal{X}'$ issuing from σ' are mapped by p to homotopic edge paths, then c'_1 and c'_2 are also homotopic. Therefore the map associating to the homotopy class of an edge loop c' at σ' the homotopy class of its image under p induces an injective homomorphism $\pi_1(\mathcal{X}', \sigma') \to \pi_1(\mathcal{X}, \sigma)$.

(3) If $\mathcal{X}$ is not connected, it is reasonable to define a morphism $p : \mathcal{X}' \to \mathcal{X}$ to be a covering if p is *surjective* and if, for each connected component $\mathcal{X}_0$ of $\mathcal{X}$, the restriction of p to $p^{-1}(\mathcal{X}_0)$ is a covering of $\mathcal{X}_0$ as defined above.

1.10 Construction of a Simply Connected Covering. Let $\mathcal{X}$ be a connected scwol and let σ_0 be a base vertex. We construct a simply connected scwol $\tilde{\mathcal{X}}$ and a covering map $p : \tilde{\mathcal{X}} \to \mathcal{X}$ as follows. The set of vertices $V(\tilde{\mathcal{X}})$ of $\tilde{\mathcal{X}}$ is the set of pairs $(\sigma, [c])$, where $\sigma \in V(\mathcal{X})$ and $[c]$ is the homotopy class of an edge path c in $\mathcal{X}$ that joins σ_0 to σ. The set $E(\tilde{\mathcal{X}})$ consists of pairs $(a, [c])$ where $a \in E(\mathcal{X})$ and $[c]$ is the homotopy class of an edge path c joining σ_0 to $i(a)$. The initial vertex of the edge $(a, [c])$ is $(i(a), [c])$ and its terminal vertex is $(t(a), [c * a^-])$; the image under p of $(\alpha, [c])$ is α. The composition $(a, [c])(a', [c'])$ is defined if $[c] = [c' * a'^-]$, in which case it is $(aa', [c'])$. It is clear that p is a morphism which is a covering. For instance, if $[c]$ and a are such that $t(c) = t(a)$, then $(a, [c * a^+])$ is the unique edge of $\tilde{\mathcal{X}}$ with terminal vertex $(t(a), [c])$ that is sent by p to a. It is clear that $\tilde{\mathcal{X}}$ is connected. To see that $\tilde{\mathcal{X}}$ is simply connected, consider $\tilde{\sigma}_0 = (\sigma_0, [c_0]) \in V(\tilde{X})$, where $[c_0]$ is the homotopy class of the trivial loop at σ_0; the lifting at $\tilde{\sigma}_0$ of an edge loop c in $\mathcal{X}$ based at σ_0 will be an edge loop if and only if the homotopy class of $[c]$ is trivial.

Group Actions on Scwols

1.11 Definition. An action of a group G on a scwol $\mathcal{X}$ is a homomorphism from G to the group of automorphisms of $\mathcal{X}$ such that the following two conditions are satisfied:

(1) for all $a \in E(\mathcal{X})$ and $g \in G$, we have $g.i(a) \neq t(a)$;

(2) for all $g \in G$ and $a \in E(\mathcal{X})$, if $g.i(a) = i(a)$ then $g.a = a$.

(Here $g.\alpha$ denotes the image of $\alpha \in \mathcal{X}$ under the automorphism of $\mathcal{X}$ associated to $g \in G$; thus condition (2) means that the stabilizer of $i(a)$ is contained in the stabilizer of a.)

1.12 Remarks

(a) Condition (1) is automatically satisfied if $\mathcal{X}$ is finite dimensional. It is not satisfied, for example, if $\mathcal{X}$ is the category associated to the poset $\mathbb{Z}$ with its natural ordering and G is the infinite cyclic group whose generator maps n to $n + 1$.

(b) *The induced action on* $|\mathcal{X}|$. The action of G on $\mathcal{X}$ induces an action of G by isometries of the geometric realization $|\mathcal{X}|$ in the obvious way; the isometry induced by $g \in G$ is denoted $|g|$. Geometrically, condition (2) in the above definition means that if $|g|$ fixes a vertex σ, then it fixes (pointwise) the union of the simplices corresponding to composable sequences $(a_1, \dots, a_k)$ with $i(a_k) = \sigma$. In particular, if $\mathcal{X}$ is the scwol associated to a polyhedral complex K and $|g|$ leaves a cell of K invariant, then its restriction to that cell is the identity. (When K is a graph, an action satisfying condition (2) is called an action without inversions.) This condition is always satisfied for the action induced on the barycentric subdivision of K by a polyhedral action on K.

1.13 The Quotient by an Action. Condition (1) ensures that the quotient $\mathcal{Y} = G\backslash\mathcal{X}$ of $\mathcal{X}$ by the action of G is naturally a scwol $\mathcal{Y}$ with $V(\mathcal{Y}) = G\backslash V(\mathcal{X})$ and $E(\mathcal{Y}) = G\backslash E(\mathcal{X})$. The initial (resp. terminal) vertex of the G-orbit of an edge a is the G-orbit of $i(a)$ (resp. $t(a)$), and if the composition ab is defined, then the orbit of ab is the composition of the orbit of a with the orbit of b. Let $p : \mathcal{X} \to \mathcal{Y}$ denote the natural projection. Condition (2) implies that p is a non-degenerate morphism. Note that if $\mathcal{Y}$ is connected, then $p : \mathcal{X} \to \mathcal{Y}$ is a covering provided the action of G on $V(\mathcal{X})$ is free.

A *Galois covering of a scwol $\mathcal{Y}$ with Galois group G* is a covering $p : \mathcal{X} \to \mathcal{Y}$ together with a free action of the group G on $\mathcal{X}$ such that p induces an isomorphism from $G\backslash\mathcal{X}$ to $\mathcal{Y}$. In particular G acts simply transitively on the fibres of p.

1.14 Examples

(0) Let $\mathcal{X}$ be a connected scwol with a base vertex σ_0 and let $p : \tilde{\mathcal{X}} \to \mathcal{X}$ be a covering such that $\tilde{\mathcal{X}}$ is simply connected. Then p can be considered as a Galois covering with Galois group $\pi_1(\mathcal{X}, \sigma_0)$.

To see this, let $\tilde{\sigma}_0 \in V(\tilde{\mathcal{X}})$ be such that $p(\tilde{\sigma}_0) = \sigma_0$. For each $\tilde{\sigma} \in V(\tilde{\mathcal{X}})$, there is an edge path $\tilde{c}$ joining $\tilde{\sigma}_0$ to $\tilde{\sigma}$; this edge path is unique up to homotopy. Given $[c] \in \pi_1(\mathcal{X}, \sigma_0)$, by definition the image $[c].\tilde{\sigma}$ of $\tilde{\sigma}$ under the action of $[c]$ is the terminal vertex of the lift at $\tilde{\sigma}_0$ of the edge path $c * p(\tilde{c})$. For each $\tilde{a} \in E(\tilde{\mathcal{X}})$ with $p(\tilde{a}) = a$ and $i(\tilde{a}) = \tilde{\sigma}$, we define $[c].\tilde{a}$ to be the unique edge that projects by p to a and has initial vertex $[c].\tilde{\sigma}$. This completes the definition of the action of $\pi_1(\mathcal{X}, \sigma_0)$ on $\tilde{\mathcal{X}}$. The action is simply transitive on each fibre of p.

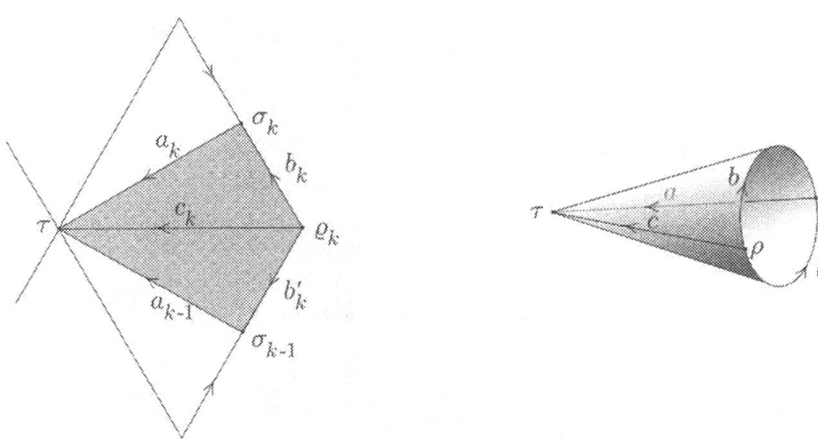

Fig. $\mathcal{C}$.5 The quotient of an n-gon by a rotation

(1) Let P be a regular n-gon in $\mathbb{E}^2$ centred at 0 (see figure $\mathcal{C}$.5) with vertices $\rho_1, \ldots, \rho_n$ (in cyclic order). Let σ_k be the midpoint of the side joining ρ_k to ρ_{k+1} (the indices are taken modulo n). Let $\mathcal{X}$ be the scwol whose set of vertices $V(\mathcal{X})$ is the

union of the origin $\tau = 0$, the n points σ_k and the n points ρ_k, where $1 \leq k \leq n$. The set of edges $E(\mathcal{X})$ contains $4n$ elements: n edges a_k with $i(a_k) = \sigma_k$ and $t(a_k) = \tau$; n edges b_k (resp. b'_k) with $i(b_k) = \rho_k$ and $t(b_k) = \sigma_k$ (resp. $i(b'_k) = \rho_k$ and $t(b'_k) = \sigma_{k-1}$); and n edges c_k with $i(c_k) = \rho_k$ and $t(c_k) = \tau$. Moreover, $c_k = a_k b_k = a_{k-1} b'_k$. The geometric realization of $\mathcal{X}$ is isomorphic (as a simplicial complex) to the barycentric subdivision of P.

Let G be the cyclic group of order n. The action of G by rotations of $\mathbb{E}^2$ fixing the origin induces an action on $\mathcal{X}$; the quotient is the scwol $\mathcal{Y}$ with 3 vertices τ, σ, ρ and 4 edges a, b, b', c, where $\tau = t(a) = t(c)$, $\sigma = i(a) = t(b) = t(b')$ and $\rho = i(b) = i(b')$. The geometric realization of $\mathcal{Y}$ is a union of two triangles whose intersection is the union of two sides. Although the category $\mathcal{X}$ is associated to a poset, $\mathcal{Y}$ is not associated to a poset.

Note that if one were to replace $\mathcal{X}$ by the opposite category (interchanging i and t), then condition (2) in the definition of an action would not be satisfied.

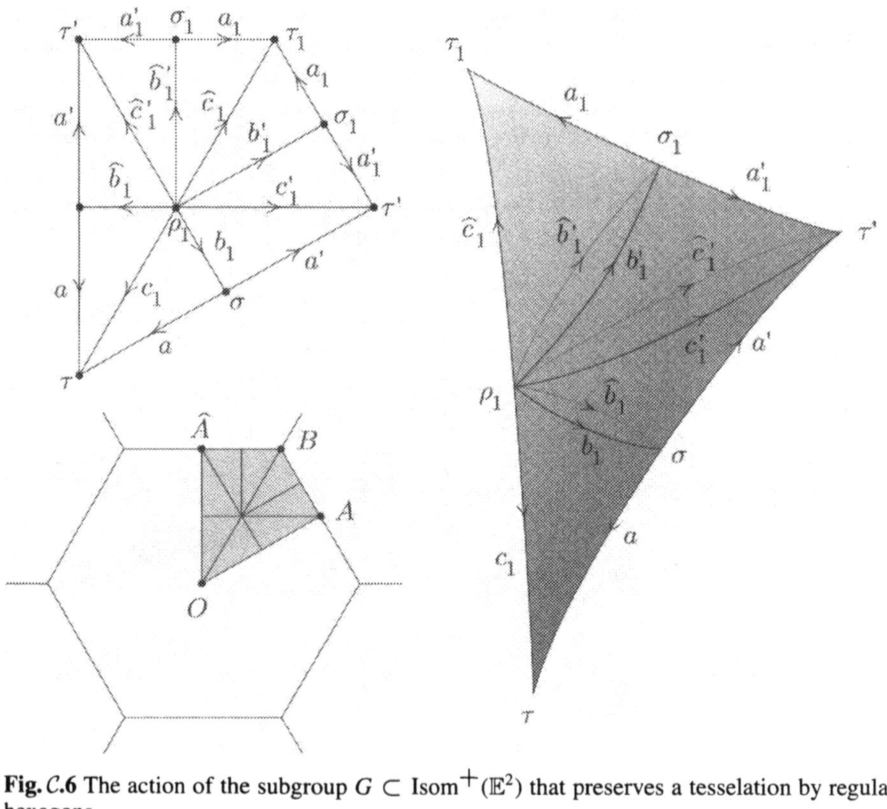

Fig. $\mathcal{C}$.6 The action of the subgroup $G \subset \mathrm{Isom}^+(\mathbb{E}^2)$ that preserves a tesselation by regular hexagons

(2) Suppose that we are given a tesselation of M^2_κ by regular n-gons with vertex angles $2\pi/3$ and let G be the group of orientation preserving isometries of M^2_κ that

preserve this tesselation. We consider an associated polyhedral subdivision K of M^2_κ: the 2-cells of K are the quadrilaterals formed by pairs of triangles from the barycentric subdivision of the tesselation that meet along an edge joining the barycentre of an n-gon of the tesselation to a vertex of that n-gon (figure $\mathcal{C}.6$ shows the case of the tesselation of $\mathbb{E}^2$ by regular hexagons).

The group G preserves the cell structure of K and if an element of G leaves a cell invariant, then it fixes it pointwise. It follows that G acts on the scwol $\mathcal{X}$ associated to K (in the sense of 1.11). The quotient $\mathcal{Y}$ of $\mathcal{X}$ by this action is the scwol associated to the polyhedral complex L obtained by identifying two pairs of adjacent sides in a quadrilateral 2-cell $OAB\hat{A}$ in figure $\mathcal{C}.6$.

$\mathcal{Y}$ has 6 vertices, three of them, τ, τ', τ_1, correspond to the three vertices of L, two, σ and σ_1, correspond to the 1-cells of L, and one, ρ, corresponds to the 2-cell of L. And $\mathcal{Y}$ has 12 edges: eight of them $b_1, \hat{b}_1, b'_1, \hat{b}'_1, c_1, \hat{c}_1, c'_1, \hat{c}'_1$ have initial vertex ρ, and $t(b_1) = t(\hat{b}_1) = \sigma$, $t(b'_1) = t(\hat{b}'_1) = \sigma_1$, $t(c_1) = \tau$, $t(\hat{c}_1) = \tau_1$, $t(c'_1) = t(\hat{c}'_1) = \tau'$.

1.15 Lifting Actions to a Simply Connected Covering. Let $\mathcal{X}$ be a scwol with base vertex σ_0 and let $p : \tilde{\mathcal{X}} \to \mathcal{X}$ be the simply connected covering constructed in 1.10. Given a group G acting on $\mathcal{X}$, we consider a group $\tilde{G}$ whose elements are the pairs $([c], g)$, where $g \in G$ and $[c]$ is the homotopy class of an edge path in $\mathcal{X}$ joining σ_0 to $g.\sigma_0$; the composition of elements in $\tilde{G}$ is defined by $([c], g)([c'], g) = ([c * g.c'], gg')$. The group $\tilde{G}$ acts on $\tilde{\mathcal{X}}$ as follows: given $(\alpha', [c']) \in \tilde{\mathcal{X}}$, the action of $([c], g) \in \tilde{G}$ is $([c], g).(\alpha', [c']) = (g.\alpha', [c * g.c'])$. Let $\Phi : \tilde{G} \to G$ be the homomorphism mapping $(g, [c])$ to g. We have an obvious exact sequence:

$$1 \to \pi_1(\mathcal{X}, \sigma_0) \to \tilde{G} \to G \to 1.$$

The projection $p : \tilde{\mathcal{X}} \to \mathcal{X}$ is Φ-equivariant and induces an isomorphism of $\tilde{G}\backslash\tilde{\mathcal{X}}$ to $G\backslash\mathcal{X}$. The subgroup $\pi_1(\mathcal{X}, \sigma_0)$ of $\tilde{G}$ acts freely on $\tilde{\mathcal{X}}$ with quotient $\mathcal{X}$.

The Local Structure of Scwols

1.16 Definition of the Join. Let $\mathcal{X}$ and $\mathcal{X}'$ be two scwols. The *join*[69] $\mathcal{X} * \mathcal{X}'$ of $\mathcal{X}$ and $\mathcal{X}'$ is the following scwol, which contains the disjoint union of $\mathcal{X}$ and $\mathcal{X}'$ as a subcategory:

$$V(\mathcal{X} * \mathcal{X}') := V(\mathcal{X}) \coprod V(\mathcal{X}'), \quad E(\mathcal{X} * \mathcal{X}') := E(\mathcal{X}) \coprod (V(\mathcal{X}) \times V(\mathcal{X}')) \coprod E(\mathcal{X}')$$

(the edge corresponding to $(\sigma, \sigma') \in V(\mathcal{X}) \times V(\mathcal{X}')$ is denoted $\sigma * \sigma'$); the maps $i, t : E(\mathcal{X} * \mathcal{X}') \to V(\mathcal{X} * \mathcal{X}')$ restrict to $E(\mathcal{X})$ and $E(\mathcal{X}')$ in the obvious way and $i(\sigma * \sigma') := \sigma'$ and $t(\sigma * \sigma') := \sigma$. The composition of the edge $a \in E(\mathcal{X})$ with the edge $i(a) * \sigma' \in V(\mathcal{X}) \times V(\mathcal{X}')$ is given by $a(i(a) * \sigma') := t(a) * \sigma'$ and the composition of the edge $\sigma * t(a') \in V(\mathcal{X}) \times V(\mathcal{X}')$ with the edge $a' \in E(\mathcal{X}')$ is defined by $(\sigma * t(a'))a' := \sigma * i(a')$.

[69] In general $\mathcal{X} * \mathcal{X}'$ is not isomorphic to $\mathcal{X}' * \mathcal{X}$

Given two morphisms of scwols $f : \mathcal{X} \to \mathcal{Y}$ and $f' : \mathcal{X}' \to \mathcal{Y}'$, their join is the unique morphism $f * f' : \mathcal{X} * \mathcal{X}' \to \mathcal{Y} * \mathcal{Y}'$ which extends f and f' and which maps the edge $\sigma * \sigma'$ to the edge $f(\sigma) * f'(\sigma')$.

The geometric realization of $\mathcal{X} * \mathcal{X}'$ is affinely isomorphic to the simplicial join of the geometric realizations of $\mathcal{X}$ and $\mathcal{X}'$.

1.17 Definitions of $\mathrm{Lk}_\sigma(\mathcal{X}),\ \mathrm{Lk}^\sigma(\mathcal{X}),\ \mathcal{X}_\sigma,\ \mathcal{X}^\sigma,\ \mathcal{X}(\sigma)$.

For $\sigma \in V(\mathcal{X})$, we define a scwol $\mathrm{Lk}_\sigma(\mathcal{X}) = \mathrm{Lk}_\sigma$, called the *upper link* of σ, by setting

$$V(\mathrm{Lk}_\sigma(\mathcal{X})) = \{a \in E(\mathcal{X}) : \ t(a) = \sigma\}$$
$$E(\mathrm{Lk}_\sigma(\mathcal{X})) = \{(a, b) \in E^{(2)}(\mathcal{X}) : \ t(a) = \sigma\},$$

defining maps $i, t : E(\mathrm{Lk}_\sigma(\mathcal{X})) \to V(\mathrm{Lk}_\sigma(\mathcal{X}))$ by $i((a, b)) = ab,\ t((a, b)) = a$, and defining composition when $ab = a'$ by

$$(a, b)(a', b') = (a, bb').$$

The natural morphism $\mathrm{Lk}_\sigma \to \mathcal{X}$ mapping $(a, b) \in E(\mathrm{Lk}_\sigma)$ to $b \in E(\mathcal{X})$ is in general not injective on the set of vertices.

The *lower link* $\mathrm{Lk}^\sigma(\mathcal{X})$ is a scwol defined similarly:

$$V(\mathrm{Lk}^\sigma(\mathcal{X})) = \{a \in E(\mathcal{X}) : \ i(a) = \sigma\}$$
$$E(\mathrm{Lk}^\sigma(\mathcal{X})) = \{(a, b) \in E^{(2)}(\mathcal{X}) : \ i(b) = \sigma\},$$

with $i((a, b)) = b,\ t((a, b)) = ab$ and composition is defined by

$$(a, b)(a', b') = (aa', b').$$

There is a natural morphism $\mathrm{Lk}^\sigma \to \mathcal{X}$ mapping $(a, b) \in E(\mathrm{Lk}^\sigma)$ to $a \in E(\mathcal{X})$.

Regarding σ as a scwol with a single element σ, we define

$$\mathcal{X}_\sigma = \sigma * \mathrm{Lk}_\sigma(\mathcal{X}), \quad \mathcal{X}^\sigma = \mathrm{Lk}^\sigma(\mathcal{X}) * \sigma$$

$$\mathcal{X}(\sigma) = \mathrm{Lk}^\sigma(\mathcal{X}) * \sigma * \mathrm{Lk}_\sigma(\mathcal{X}) = \mathcal{X}^\sigma * \mathrm{Lk}_\sigma(\mathcal{X}) = \mathrm{Lk}^\sigma(\mathcal{X}) * \mathcal{X}_\sigma.$$

A non-degenerate morphism $f : \mathcal{X} \to \mathcal{Y}$ induces, for each $\sigma \in V(\mathcal{X})$, an isomorphism from $\mathcal{X}^\sigma$ onto $\mathcal{Y}^{f(\sigma)}$. Note that if $\mathcal{Y}$ is connected and $\mathcal{X}$ non-empty, then f is a covering if and only if f induces an isomorphism from $\mathcal{X}(\sigma)$ to $\mathcal{Y}(f(\sigma))$ for all $\sigma \in V(\mathcal{X})$.

Note that if we replace $\mathcal{X}$ by the opposite scwol, then the roles of $\mathcal{X}_\sigma$ and $\mathcal{X}^\sigma$ are exchanged.

We have a natural morphism

$$\mathcal{X}_\sigma = \sigma * \mathrm{Lk}_\sigma \to \mathcal{X},$$

mapping σ to σ, whose restriction to Lk_σ is as above and which maps the edge $\sigma * a$ to a. Similarly we have a morphism

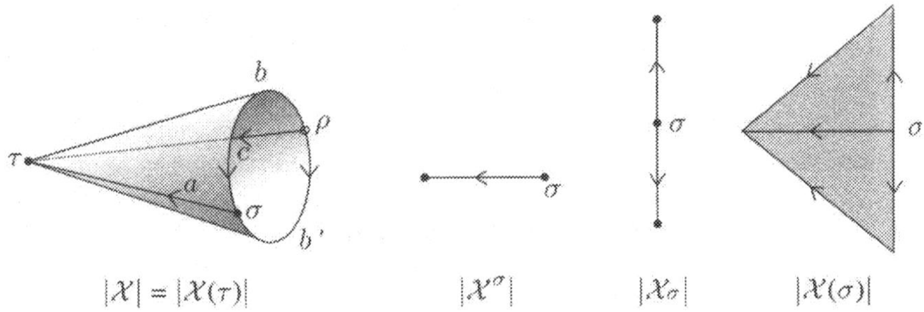

Fig. $\mathcal{C}$.**7** Examples

$$\mathcal{X}^\sigma \to \mathcal{X}.$$

We also have a morphism

$$\mathcal{X}(\sigma) \to \mathcal{X}$$

that extends the morphisms defined above and maps the edge $a*b \in E(\mathrm{Lk}^\sigma * \mathrm{Lk}_\sigma)$ to the edge $ab \in E(\mathcal{X})$, where $i(a) = t(b) = \sigma$. This morphism is injective if and only if any two distinct edges with initial vertex σ (resp. terminal vertex σ) have distinct terminal vertices (resp. initial vertices).

1.18 Examples

(1) If $\mathcal{X}$ is associated to a poset $\mathcal{P}$ (as in 1.2(1)), then $\mathrm{Lk}_\sigma(\mathcal{X})$ (resp. $\mathcal{X}^\sigma$ or $\mathcal{X}(\sigma)$) is the category associated to the subposet $\mathrm{Lk}_\sigma(\mathcal{P})$ (resp. $\mathcal{P}^\sigma$ or $\mathcal{P}(\sigma)$) as defined in II.12.3.

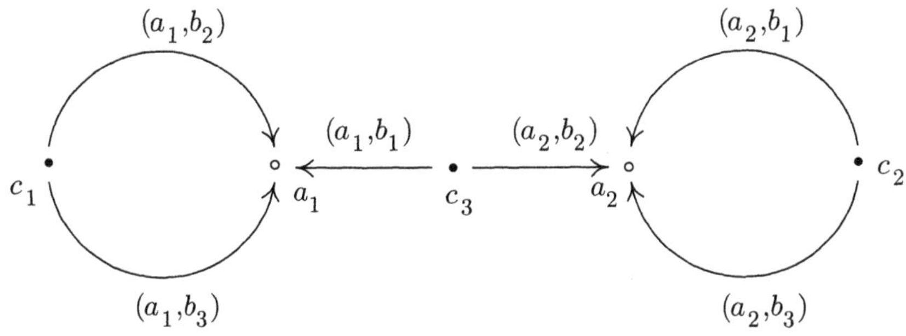

Fig. $\mathcal{C}$.**8** $\mathrm{Lk}_\tau(\mathcal{X})$ in the scwol associated to the dunce hat (see figure $\mathcal{C}$.3)

(2) Let $\mathcal{X}$ be the scwol associated to an M_κ-polyhedral complex K (see example 1.4(2)). Let $\sigma \in V(X)$ be the barycentre of a cell of K which is the image under p_λ of a convex cell C_λ (in the notations of I.7.37). Note that the geometric realization of $\mathcal{X}^\sigma$ is the barycentric subdivision of C_λ. Therefore, the morphism $\mathcal{X}^\sigma \to \mathcal{X}$ is injective if and only if the map p_λ is injective.

If σ is a 0-cell of K, then there is a natural bijection from the geometric realization of Lk_σ to the geometric link of σ in the barycentric subdivision of K as defined in I.7.38. (See figure $\mathcal{C}$.8.)

(3) Let $\mathcal{I}$ be the scwol whose geometric realization is the barycentric subdivision of the unit interval. Thus $\mathcal{I}$ has three vertices τ_0, τ_1, σ and two edges a_0, a_1 with $t(a_0) = \tau_0,\ t(a_1) = \tau_1$, and $i(a_0) = i(a_1) = \sigma$. The geometric realization of the n-fold product $\mathcal{I}^n$ is the barycentric subdivision of the n-cube.

A scwol $\mathcal{X}$ is associated to a cubed complex if and only if, for each $\sigma \in V(\mathcal{X})$, the scwol $\mathcal{X}^\sigma$ is isomorphic to $\mathcal{I}^k$ for some k.

1.19 Exercise. Show that the geometric realization of the morphism $\mathcal{X}(\sigma) \to \mathcal{X}$ induces an affine isomorphism of the open star of $\sigma \in |\mathcal{X}(\sigma)|$ onto the open star of $\sigma \in |\mathcal{X}|$.

2. Complexes of Groups

Let G be a group acting on an M_κ-polyhedral complex K by isometries that preserve the cell structure. Assume that if an element of G leaves a cell of K invariant, then it fixes the cell pointwise. The quotient of K by this action is again an M_κ-polyhedral complex, which we denote L. For each cell σ of L, we select a cell $\overline{\sigma}$ of K in the orbit σ. Let G_σ be the isotropy subgroup of $\overline{\sigma}$. If τ is a face of σ, the isotropy subgroup of the corresponding face of $\overline{\sigma}$ is a subgroup of G_σ, therefore the isotropy subgroup G_τ of the chosen representative $\overline{\tau}$ of τ is conjugate to a subgroup of G_σ. The family of abstract groups (G_σ) together with the injective homomorphisms $G_\sigma \to G_\tau$ and some additional local data define on L what we call a *complex of groups $G(L)$ over L*. The G_σ are called the *local groups* of $G(L)$. When L is one dimensional, $G(L)$ is a graph of groups. Remarkably, if K is simply connected then one can reconstruct from $G(L)$ the complex K, the group G, and its action on K. On the other hand, when L is of dimension greater than one, it is in general not the case that an abstract complex of groups $G(L)$ over L is *developable*, i.e. that it arises from an action of a group on a polyhedral complex K.

In order to describe complexes of groups over polyhedral complexes combinatorially, it is convenient to pass to the associated scwol and to develop the theory in this framework. We can then come back to geometry using geometric realizations of scwols.

The essential points of this section are the definition (2.1) of a complex of groups $G(\mathcal{Y})$ over a scwol $\mathcal{Y}$, the definition (2.4) of morphisms of complexes of groups, and the construction in (2.9) of the complex of groups $G(\mathcal{Y})$ associated to an action of a group G on a scwol $\mathcal{X}$ (where $\mathcal{Y}$ is the quotient of $\mathcal{X}$ by G), and finally the notion of *developability*. If $G(\mathcal{Y})$ is a complex of groups associated to an action of a group G on a scwol $\mathcal{X}$, then there is a morphism $\phi : G(\mathcal{Y}) \to G$ which is injective on the local groups. Conversely, one can associate to any such morphism an action of G on a scwol $D(\mathcal{Y}, \phi)$ called the *development of* $\mathcal{Y}$ *with respect to* ϕ. (This construction, described in (2.13), generalizes the Basic Construction in II.12.18.)

Basic Definitions

The next definition will be justified by the considerations in (2.9).

2.1 Definition of a Complex of Groups

Let $\mathcal{Y}$ be a scwol. A complex of groups $G(\mathcal{Y}) = (G_\sigma, \psi_a, g_{a,b})$ over $\mathcal{Y}$ is given by the following data:

(1) for each $\sigma \in V(\mathcal{Y})$, a group G_σ called the *local group at* σ,

(2) for each $a \in E(\mathcal{Y})$, an injective[70] homomorphism $\psi_a : G_{i(a)} \to G_{t(a)}$,

(3) for each pair of composable edges $(a, b) \in E^{(2)}(\mathcal{Y})$, a twisting element $g_{a,b} \in G_{t(a)}$,

with the following compatibility conditions:

(i) $$\mathrm{Ad}(g_{a,b})\psi_{ab} = \psi_a\psi_b,$$

where $\mathrm{Ad}(g_{a,b})$ is the conjugation by $g_{a,b}$ in $G_{t(a)}$;
and for each triple $(a, b, c) \in E^{(3)}(\mathcal{Y})$ of composable elements we have the *cocycle condition*

(ii) $$\psi_a(g_{b,c})g_{a,bc} = g_{a,b}g_{ab,c}.$$

A *simple* complex of groups over $\mathcal{Y}$ is a complex of groups over $\mathcal{Y}$ such that all the twisting elements $g_{a,b}$ are trivial. A complex of groups over a poset is by definition a complex of groups over the associated scwol. A simple complex of groups over a poset as defined in (II.12.11) is a simple complex of groups over the scwol associated to this poset.

Note that condition (i) (resp. condition (ii)) is empty if $\mathcal{Y}$ is of dimension 1 (resp. ≤ 2) because in that case $E^{(2)}(\mathcal{Y})$ (resp. $E^{(3)}(\mathcal{Y})$) is empty.

Suppose that we are given a map that associates to each edge $a \in E(\mathcal{Y})$ an element $g_a \in G_{t(a)}$. The complex of groups $G'(\mathcal{Y})$ over $\mathcal{Y}$ given by $G'_\sigma = G_\sigma$, $\psi'_a = \mathrm{Ad}(g_a)\psi_a$ and $g'_{a,b} = g_a\psi_a(g_b)g_{a,b}g_{ab}^{-1}$ is said to be *deduced* from $G(\mathcal{Y})$ by the *coboundary* of (g_a). Note that $G(\mathcal{Y})$ is deduced from $G'(\mathcal{Y})$ by the coboundary of (g_a^{-1}).

2.2 Notation. We wish to use the description of a scwol as (the arrows of) a category, identifying vertices σ to units 1_σ. We define $\psi_\alpha : G_{i(\alpha)} \to G_{t(\alpha)}$ as before if $\alpha \in E(\mathcal{Y})$ and as the identity of G_σ if $\alpha = 1_\sigma$. For composable morphisms $\alpha, \beta \in \mathcal{Y}$, we define $g_{\alpha,\beta}$ as above if $\alpha, \beta \in E(\mathcal{Y})$ and as the unit element of $G_{t(\alpha)}$ otherwise. Conditions (i)

[70] One could develop the theory of complexes of groups without this injectivity assumption, the only serious modification would be that local developability (in the sense of section 4) would then be equivalent to the injectivity of the ψ_a.

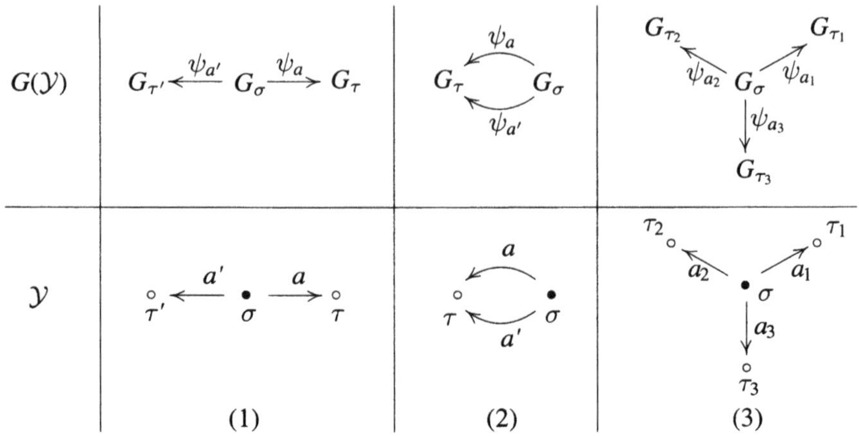

Fig. $\mathcal{C}$**.9** Complexes of groups over scwols of dimension one

and (ii) of 2.1 are still satisfied if we replace a, b, c in those formulae by composable elements $\alpha, \beta, \gamma \in \mathcal{X}$.

2.3 Remark. Every complex of groups over the poset associated to the faces of an n-simplex can be deduced from a simple complex of groups over this poset by a suitable coboundary.

More generally, let $\mathcal{Y}$ be a scwol which has a vertex σ_0 such that for each vertex $\sigma \neq \sigma_0$, there is a unique edge a_σ with $i(a_\sigma) = \sigma_0$, $t(a_\sigma) = \sigma$. Then any complex of groups $G(\mathcal{Y})$ over $\mathcal{Y}$ can be deduced from a simple complex of groups over $\mathcal{Y}$ by a coboundary. Indeed it is straightforward to check that the complex of groups deduced from $G(\mathcal{Y})$ by the coboundary of $(g^{-1}_{a,a_{i(a)}})$ is simple.

2.4 Morphisms of Complexes of Groups. Let $G(\mathcal{Y}') = (G_{\sigma'}, \psi_{a'}, g_{a',b'})$ be a complex of groups over a scwol $\mathcal{Y}'$. Let $f : \mathcal{Y} \to \mathcal{Y}'$ be a (possibly degenerate) morphism of scwols. A *morphism* $\phi = (\phi_\sigma, \phi(a))$ from $G(\mathcal{Y})$ to $G(\mathcal{Y}')$ over f consists of:

(1) a homomorphism $\phi_\sigma : G_\sigma \to G_{f(\sigma)}$ of groups, for each $\sigma \in V(\mathcal{Y})$, and

(2) an element $\phi(a) \in G_{t(f(a))}$ for each $a \in E(\mathcal{Y})$, such that

(i) $$\mathrm{Ad}(\phi(a))\psi_{f(a)}\phi_{i(a)} = \phi_{t(a)}\psi_a,$$

and for all $(a, b) \in E^{(2)}(\mathcal{Y})$,

(ii) $$\phi_{t(a)}(g_{a,b})\phi(ab) = \phi(a)\psi_{f(a)}(\phi(b))g_{f(a),f(b)}.$$

If f is an isomorphism of scwols and ϕ_σ is an isomorphism for every $\sigma \in V(\mathcal{Y})$, then ϕ is called an *isomorphism*.

With the point of view of (2.2) (which we shall often take implicitly in what follows), given $\alpha \in \mathcal{Y}$, we define $\phi(\alpha)$ as above if $\alpha \in E(\mathcal{Y})$ and as the identity

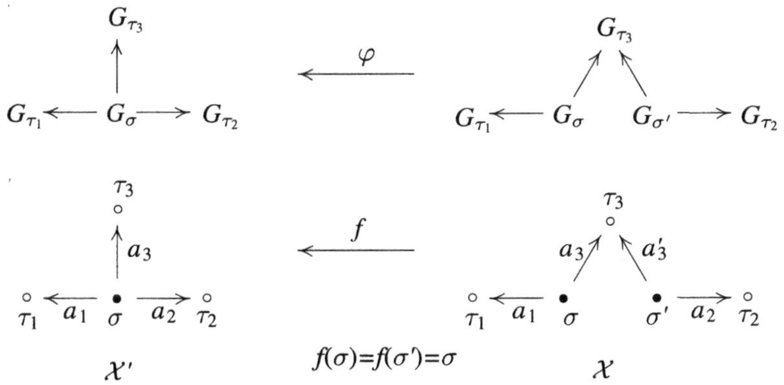

Fig. C.10 A morphism of one-dimensional complexes of groups (homomorphisms between local groups with the same label are isomorphisms)

element of $G_{f(\sigma)}$ if $\alpha = 1_\sigma$. Conditions (i) and (ii) are still satisfied if a, b are replaced by composable elements $\alpha, \beta \in \mathcal{Y}$.

Homotopy. Let ϕ and ϕ' be two morphisms from $G(\mathcal{Y})$ to $G(\mathcal{Y}')$ over f given respectively by $(\phi_\sigma, \phi(a))$ and $(\phi'_\sigma, \phi'(a))$. A *homotopy* from ϕ to ϕ' is given by a family of elements $s_\sigma \in G_{f(\sigma)}$ indexed by $\sigma \in V(\mathcal{Y})$ such that

$$\phi'_\sigma = \mathrm{Ad}(k_\sigma)\phi_\sigma \text{ and } \phi'(a) = k_{t(a)}\phi(a)\psi_{f(a)}(k_{i(a)}^{-1}) \ \forall \sigma \in V(\mathcal{Y}),\ a \in E(\mathcal{Y}).$$

If such a homotopy exists, we say that ϕ and ϕ' are *homotopic*.

Composition. If $\phi = (\phi_\sigma, \phi(a)) : G(\mathcal{Y}) \to G(\mathcal{Y}')$ and $\phi' = (\phi_{\sigma'}, \phi'(a')) : (G(\mathcal{Y}') \to G(\mathcal{Y}'')$ are morphisms of complexes of groups over morphisms $f : \mathcal{Y} \to \mathcal{Y}'$ and $f' : \mathcal{Y}' \to \mathcal{Y}''$, then the composition $\phi' \circ \phi : G(\mathcal{Y}) \to G(\mathcal{Y}'')$ is the morphism over $f'f$ defined by the homomorphisms $(\phi' \circ \phi)_\sigma = \phi'_{f(\sigma)} \circ \phi_\sigma$ and the elements $(\phi' \circ \phi)(a) = \phi'_{f(t(a))}(\phi(a))\phi'(f(a))$.

Let us repeat the definition of morphism in the important special case where $\mathcal{Y}'$ has only one vertex, i.e. $G(\mathcal{Y}')$ is simply a group G.

2.5 Definition. A morphism $\phi = (\phi_\sigma, \phi(a))$ from a complex of groups $G(\mathcal{Y})$ to a group G consists of a homomorphism $\phi_\sigma : G_\sigma \to G$ for each $\sigma \in V(\mathcal{Y})$ and an element $\phi(a) \in G$ for each $a \in E(\mathcal{Y})$ such that

$$\phi_{t(a)}\psi_a = \mathrm{Ad}(\phi(a))\phi_{i(a)} \text{ and } \phi_{t(a)}(g_{a,b})\phi(ab) = \phi(a)\phi(b).$$

We say that ϕ is *injective on the local groups* if each homomorphism ϕ_σ is injective.

2.6 Remark. Let $G(\mathcal{Y})$ be a simple complex of groups. A morphism $\phi : G(\mathcal{Y}) \to G$ is called *simple* if $\phi(a) = 1$ for all $a \in E(\mathcal{Y})$. We shall see later (3.10(3)) that if $\mathcal{Y}$ is simply connected, then every morphism $\phi : G(\mathcal{Y}) \to G$ is homotopic to a simple one.

2.7 Induced Complex of Groups. If $f : \mathcal{Y} \to \mathcal{Y}'$ is a morphism of scwols and if $G(\mathcal{Y}') = (G_{\sigma'}, \psi_{a'}, g_{a',b'})$ is a complex of groups over $\mathcal{Y}'$, then the *induced complex of groups* $f^*(G(\mathcal{Y}'))$ is the complex of groups $(G_\sigma, \psi_a, g_{a,b})$ over $\mathcal{Y}$, where

$$G_\sigma = G_{f(\sigma)}, \ \psi_a = \psi_{f(a)}, \ g_{a,b} = g_{f(a),f(b)}.$$

(Note that there is an obvious morphism $\phi : f^*G(\mathcal{Y}') \to G(\mathcal{Y}')$ over f, where ϕ_σ is the identity and $\phi(a)$ is the unit element of $G_{t(f(a))}$.)

If $\mathcal{Y}$ is a subscwol of $\mathcal{Y}'$ and f is the inclusion, then $f^*G(\mathcal{Y}')$ is called the *restriction of* $G(\mathcal{Y}')$ *to* $\mathcal{Y}$.

2.8 The Category $CG(\mathcal{Y})$ Associated to a Complex of Groups $G(\mathcal{Y})$

The definitions of complexes of groups and morphisms between them appear more natural if we interpret them in the framework of category theory (see A.1).

We associate to the complex of groups $G(\mathcal{Y})$ the small category $CG(\mathcal{Y})$ whose set of objects is $V(\mathcal{Y})$ and whose set of elements (arrows) are the pairs (g, α), where $\alpha \in \mathcal{Y}$ and $g \in G_{t(\alpha)}$. We define maps $i, t : CG(\mathcal{Y}) \to V(\mathcal{Y})$ by $i((g, \alpha)) = i(\alpha)$ and $t((g, \alpha)) = t(\alpha)$. The composition $(g, \alpha)(h, \beta)$ is defined if $i(\alpha) = t(\beta)$ and then it is equal to

$$(g, \alpha)(h, \beta) = (g\psi_\alpha(h)g_{\alpha,\beta}, \alpha\beta).$$

Conditions (i) and (ii) in (2.1), when taken together, are equivalent to the associativity of this law of composition. The map $(g, \alpha) \mapsto \alpha$ is a functor $CG(\mathcal{Y}) \to \mathcal{Y}$.

A morphism $\phi : G(\mathcal{Y}) \to G(\mathcal{Y}')$ of complexes of groups over $f : \mathcal{Y} \to \mathcal{Y}'$ gives a functor $CG(\mathcal{Y}) \to CG(\mathcal{Y}')$ of the associated categories, namely the functor (also denoted ϕ) mapping $(g, \alpha) \in CG(\mathcal{Y})$ to $(\phi_{t(\alpha)}(g)\phi(\alpha), f(\alpha)) \in CG(\mathcal{Y}')$. Conversely, any functor $\phi : CG(\mathcal{Y}) \to CG(\mathcal{Y}')$ projects to a functor $f : \mathcal{Y} \to \mathcal{Y}'$ and defines a morphism (also denoted ϕ) from $G(\mathcal{Y})$ to $G(\mathcal{Y}')$ over f — the homomorphisms ϕ_σ and the elements $\phi(a)$ are determined by the relations: $\phi(g, 1_\sigma) = (\phi_\sigma(g), 1_{f(\sigma)})$ and $\phi(1_{t(a)}, a) = (\phi(a), f(a))$.

If $\phi' : G(\mathcal{Y}') \to G(\mathcal{Y}'')$ is a morphism over $f' : \mathcal{Y}' \to \mathcal{Y}''$, then the composition $\phi' \circ \phi$ is the morphism $G(\mathcal{Y}) \to G(\mathcal{Y}'')$ over $f' \circ f : \mathcal{Y} \to \mathcal{Y}''$ corresponding to the composition of the associated functors.

Developability

2.9 The Complex of Groups Associated to an Action

(1) Definition. Let G be a group acting on a scwol $\mathcal{X}$ (in the sense of definition 1.11). Let $\mathcal{Y} = G\backslash\mathcal{X}$ be the quotient scwol and let $p : \mathcal{X} \to \mathcal{Y}$ be the natural projection.

For each vertex $\sigma \in V(\mathcal{Y})$ choose a vertex $\overline{\sigma} \in V(\mathcal{X})$ such that $p(\overline{\sigma}) = \sigma$. For each edge $a \in E(\mathcal{X})$ with $i(a) = \sigma$, condition (2) of (1.11) implies that there is a unique edge $\overline{a} \in \mathcal{X}$ such that $p(\overline{a}) = a$ and $i(\overline{a}) = \overline{\sigma}$. If $\tau = t(a)$, then in general $\overline{\tau} \neq t(\overline{a})$; choose $h_a \in G$ such that $h_a.t(\overline{a}) = \overline{\tau}$. For $\sigma \in V(\mathcal{Y})$, let G_σ be the isotropy

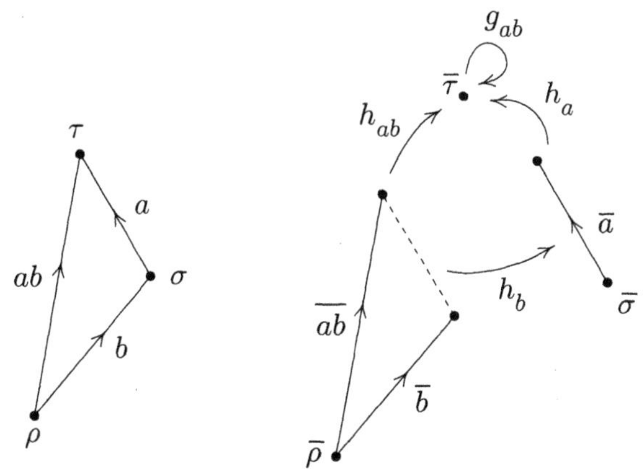

Fig. $\mathcal{C}$.11 Associating a complex of groups to an action

subgroup of $\overline{\sigma}$, and for each $a \in E(\mathcal{Y})$, let $\psi_a : G_{i(a)} \to G_{t(a)}$ be the homomorphism defined by

$$\psi_a(g) = h_a g h_a^{-1}.$$

Condition (2) of (1.11) implies that the image of ψ_a lies in $G_{t(a)}$. For composable edges $(a, b) \in E^{(2)}(\mathcal{Y})$, define

$$g_{a,b} = h_a h_b h_{ab}^{-1},$$

an element of $G_{t(a)}$ (see figure $\mathcal{C}$.11). The *complex of groups over $\mathcal{Y}$ associated to the action of G on $\mathcal{X}$* (and the choices above) is

$$G(\mathcal{Y}) = (G_\sigma, \psi_a, g_{a,b}).$$

It is easy to check that the two conditions (i) and (ii) of (2.1) are satisfied.

There is a natural morphism associated to the action:

$$\phi : G(\mathcal{Y}) \to G,$$

$\phi = (\phi_\sigma, \phi(a))$, where $\phi_\sigma : G_\sigma \to G$ is the natural inclusion and $\phi(a) = h_a$. This morphism is injective on the local groups. (In (2.13) we shall see that all morphisms with this local injectivity property arise from the construction described above.)

(2) Different Choices. A different choice of elements h_a in the above construction would lead to a complex of groups deduced from $G(\mathcal{Y})$ by a coboundary. A different choice of representatives $\overline{\sigma}'$ in the G-orbits of σ and associated choices h'_a would lead to a complex of groups $G'(\mathcal{Y}) = (G'_\sigma, \psi'_a, g'_{a,b})$ over $\mathcal{Y}$ that is isomorphic to $G(\mathcal{Y})$. An isomorphism $\lambda = (\lambda_\sigma, \lambda(a))$ from $G(\mathcal{Y})$ to $G'(\mathcal{Y})$ is obtained by choosing elements $k_\sigma \in G$ such that $k_\sigma.\overline{\sigma} = \overline{\sigma}'$ and defining $\lambda_\sigma = \mathrm{Ad}(k_\sigma)|_{G_\sigma}$ and $\lambda(a) = k_{t(a)} h_a k_{i(a)}^{-1} {h'_a}^{-1}$. (Different choices of the elements k_σ give homotopic isomorphisms.) Note that the

k_σ determine a homotopy from ϕ to $\phi'\lambda$, where $\phi' : G'(\mathcal{Y}) \to G$ is the natural morphism described above.

Note also that if one can find a strict fundamental domain for the action of G, i.e. if one can choose the liftings such that for each $a \in E(\mathcal{Y})$ with $\tau = t(a)$ we have $t(\overline{a}) = \overline{\tau}$, then by choosing the h_a to be trivial, we get a simple complex of groups.

(3) When $\mathcal{Y}$ is connected, one way of choosing the liftings $\overline{\sigma}$ is to first construct a maximal tree T in the 1-skeleton $|\mathcal{Y}|^{(1)}$ of the geometric realization of $\mathcal{Y}$ and to lift it to a tree $\overline{T}$ in $|\mathcal{X}|^{(1)}$ so that $p(\overline{T}) = T$ (this is possible because p is non-degenerate). Then choose $\overline{\sigma}$ to be a vertex of $\overline{T}$ and h_a to be the identity if $i(a) = \sigma$ and $t(\overline{a})$ is a vertex of $\overline{T}$ (where as above $\overline{a}$ is the unique edge in $\mathcal{X}$ projecting to a with $i(\overline{a}) = \overline{\sigma}$). If $\mathcal{X}_0$ is the connected component of $\mathcal{X}$ containing $\overline{T}$ and if G_0 is the subgroup of G leaving $\mathcal{X}_0$ invariant, then the inclusion $\mathcal{X}_0 \to \mathcal{X}$ induces a bijection $G\backslash\mathcal{X}_0 \to G\backslash\mathcal{X} = \mathcal{Y}$ and the complex of groups associated to the action of G_0 on $\mathcal{X}_0$ is canonically isomorphic to $G(\mathcal{Y})$.

(4) Equivariant Actions. Let G' be a group acting on a scwol $\mathcal{X}'$ and let p' be the natural projection from $\mathcal{X}'$ to $\mathcal{Y}' := G'\backslash\mathcal{X}'$. Let $G(\mathcal{Y}') = (G_{\sigma'}, \psi_{a'}, g_{a',b'})$ be the complex of groups associated to this action with respect to choices $\overline{\sigma'} \in V(\mathcal{X}')$ and $h_{a'} \in G'$ as above. Let $\phi' : G(\mathcal{Y}') \to G'$ be the associated morphism. Suppose that we have a morphism $L : \mathcal{X} \to \mathcal{X}'$ that is equivariant with respect to a homomorphism $\Lambda : G \to G'$ (i.e. $L(g.\alpha) = \Lambda(g).L(\alpha)$ for all $\alpha \in \mathcal{X}$ and $g \in G$). This induces a morphism $l : \mathcal{Y} \to \mathcal{Y}'$, for which there is a commutative diagram

$$\begin{array}{ccc} \mathcal{X} & \xrightarrow{L} & \mathcal{X}' \\ {\scriptstyle p}\downarrow & & {\scriptstyle p'}\downarrow \\ \mathcal{Y} & \xrightarrow{l} & \mathcal{Y}'. \end{array}$$

For each $\sigma \in V(\mathcal{Y})$ we choose an element $k_\sigma \in G'$ such that $k_\sigma.L(\overline{\sigma}) = \overline{l(\sigma)}$. For each $\sigma \in V(\mathcal{Y})$ let $\lambda_\sigma : G_\sigma \to G_{l(\sigma)}$ be the homomorphism $g \mapsto k_\sigma\Lambda(g)k_\sigma{}^{-1}$, and for each $a \in E(\mathcal{Y})$ let $\lambda(a) = k_{t(a)}\Lambda(h_a)k_{i(a)}^{-1}h_{l(a)}^{-1} \in G'_{t(l(a))}$.

It is straightforward to check that $\lambda = (\lambda_\sigma, \lambda(a))$ defines a morphism $G(\mathcal{Y}) \to G(\mathcal{Y}')$ over $l : \mathcal{Y} \to \mathcal{Y}'$ and that we get a diagram that commutes up to homotopy

$$\begin{array}{ccc} G(\mathcal{Y}) & \xrightarrow{\lambda} & G(\mathcal{Y}') \\ {\scriptstyle \phi}\downarrow & & {\scriptstyle \phi'}\downarrow \\ G & \xrightarrow{\Lambda} & G' \end{array}$$

The family (k_σ) gives the homotopy from $\Lambda\phi$ to $\phi'\lambda$.

2.10 Examples

(1) We describe the complex of groups associated to the action considered in example 1.14 (1) (maintaining the notation established there). As representatives for

$0 := \tau, \sigma, \rho$, choose respectively O, σ_1, ρ_1. Then $\overline{b} = b_1$, $\overline{b}' = b_1'$, and $h_{b_1'}$ is the rotation R about O that sends $\sigma_0 = \sigma_n$ to σ_1. All the local groups are trivial except $G_\tau = \langle R \rangle$, and the only non-trivial twisting element is $g_{a,b'} = R$.

(2) Let $G(\mathcal{Y})$ be the complex of groups associated to an action of a group G on a connected scwol $\mathcal{X}$ with choices $\overline{\sigma}$ and h_a as in (2.9(1)). For convenience, as in (2.9(3)), we assume that the vertices $\overline{\sigma}$ belong to a tree $\overline{T}$ that projects to a maximal tree T in the 1-skeleton of the geometric realization of $\mathcal{Y} = G\backslash\mathcal{X}$. We pass to the universal covering $p : \tilde{\mathcal{X}} \to \mathcal{X}$ of $\mathcal{X}$, and as in (1.15) we construct an action of a group $\tilde{G}$ on $\tilde{\mathcal{X}}$ and a homomorphism $\Phi : \tilde{G} \to G$ so that p is Φ-equivariant and the kernel of Φ acts freely on $\tilde{\mathcal{X}}$ with quotient $\mathcal{X}$. There is an isomorphism $\tilde{G}\backslash\tilde{\mathcal{X}} \to G\backslash\mathcal{X} = \mathcal{Y}$ induced by p, and we identify $\tilde{G}\backslash\tilde{\mathcal{X}}$ with $G\backslash\mathcal{X}$ by means of this map. To construct a complex of groups associated to this action of $\tilde{G}$ on $\tilde{\mathcal{X}}$, we can lift $\overline{T}$ to a tree $\tilde{T}$ in the 1-skeleton of the geometric realization of $\tilde{\mathcal{X}}$ and for each $\sigma \in V(\mathcal{Y})$ define the vertex $\tilde{\sigma} \in V(\tilde{\mathcal{X}})$ by: $p(\tilde{\sigma}) = \overline{\sigma}$ and $\tilde{\sigma} \in \tilde{T}$. For each $a \in E(\mathcal{Y})$ with $i(a) = \sigma$, let $\tilde{a}$ be the edge in $\tilde{\mathcal{X}}$ with $i(\tilde{a}) = \tilde{\sigma}$ and $p(\tilde{a}) = \overline{a}$. Then choose an element $\tilde{h}_a \in \tilde{G}$ such that $\Phi(\tilde{h}_a) = h_a$ and $\tilde{h}_a.t(\tilde{a}) = \tilde{\tau}$ where $\tau = t(a)$. Note that Φ maps the isotropy subgroup $\tilde{G}_{\tilde{\sigma}}$ isomorphically onto the isotropy subgroup $G_{\overline{\sigma}} = G_\sigma$.

If we identify these subgroups by this isomorphism, then the complex of groups over $\mathcal{Y}$ associated to this action of $\tilde{G}$ on $\tilde{\mathcal{X}}$, with respect to the choices that we have made, is exactly $G(\mathcal{Y})$.

(3) We describe the complex of groups associated to the action considered in example 1.14(2). Let C be a 2-cell of K which is the union of two adjacent triangles in the barycentric subdivision of the tesselation with vertices $O, A, B, \hat{A}$, where O is the barycentre of an n-gon of the tesselation and B a vertex of this n-gon (see figure $\mathcal{C}$.6). Let s (resp. r^{-1}) be the element of G fixing O (resp. B) and mapping A to $\hat{A}$, and let t be the half turn fixing $\hat{A}$. As representatives for the vertices of $\mathcal{Y}$ we choose the barycentres of those faces of C that are contained in the triangle $O\hat{A}B$. Then the non-trivial local groups are $G_\tau = \mathbb{Z}_n$, $G_{\tau_1} = \mathbb{Z}_3$, $G_{\tau'} = \mathbb{Z}_2$, generated respectively by s, r, t. The non-trivial twisting elements are $g_{a,b_1} = s$, $g_{a_1',b_1'} = t$, $g_{a_1,b_1'} = r^{-1}$. (See figure $\mathcal{C}$.12.)

2.11 Developability. A complex of groups $G(\mathcal{Y})$ is called *developable* if it is isomorphic to a complex of groups that is associated, in the sense of 2.9(1), to an action of a group G on a scwol $\mathcal{X}$ with $\mathcal{Y} = G\backslash\mathcal{X}$.

In general, complexes of groups are not developable if the dimension of $\mathcal{Y}$ is greater than one. Conditions for developability will be discussed in (2.15) and Chapter III.$\mathcal{G}$.

2.12 Remark. If a complex of groups $G(\mathcal{Y})$ over a connected scwol $\mathcal{Y}$ is developable, then it is isomorphic to a complex of groups associated to an action of a group G on a *simply connected* scwol.

Indeed, given an action whose associated complex of groups is isomorphic to $G(\mathcal{Y})$, we can first restrict this action to a connected component as indicated in

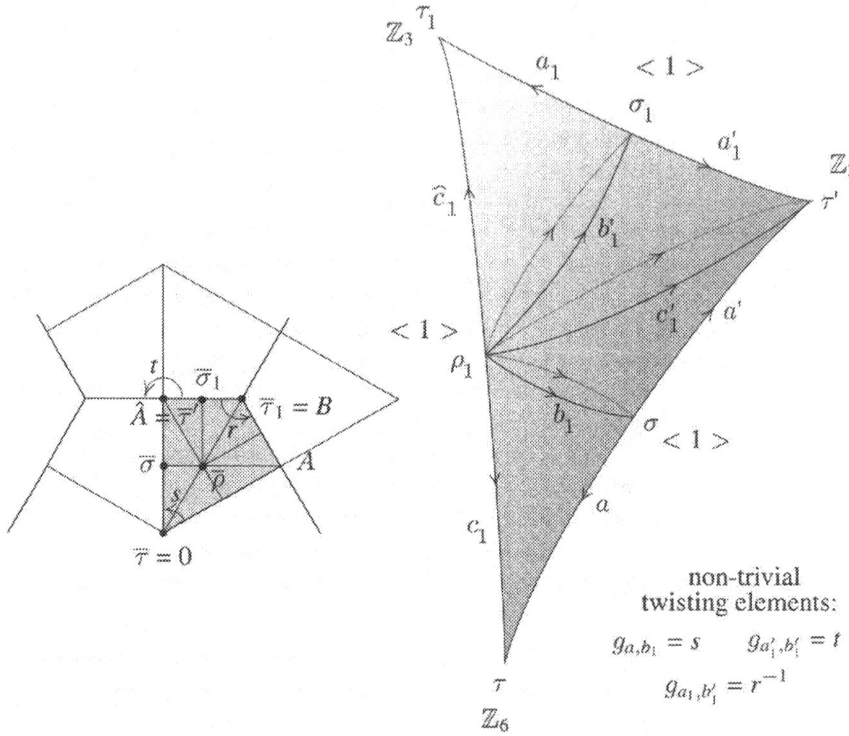

Fig. $\mathcal{C}$.12 The complex of groups associated to the action of figure $\mathcal{C}$.6

(2.9(3)) and then pass to the universal covering of this component as in example 2.10(2).

The Basic Construction

2.13 Theorem (The Basic Construction). *Let $G(\mathcal{Y}) = (G_\sigma, \psi_a, g_{a,b})$ be a complex of groups over a scwol $\mathcal{Y}$.*

(1) *Let G be a group. Canonically associated to each morphism $\phi : G(\mathcal{Y}) \to G$ there is an action of G on a scwol $D(\mathcal{Y}, \phi)$ with quotient $\mathcal{Y}$. ($D(\mathcal{Y}, \phi)$ is called the development of $\mathcal{Y}$ with respect to ϕ.) If ϕ is injective on the local groups, then $G(\mathcal{Y})$ is the complex of groups (with respect to canonical choices) associated to this action and $G(\mathcal{Y}) \to G$ is the associated morphism.*

(2) *If $G(\mathcal{Y})$ is the complex of groups associated to an action of a group G on a scwol $\mathcal{X}$ (with respect to some choices) and if $\phi : G(\mathcal{Y}) \to G$ is the associated morphism, then there is a G-equivariant isomorphism $D(\mathcal{Y}, \phi) \to \mathcal{X}$ that projects to the identity of $\mathcal{Y}$.*

Proof. For (1) we define the scwol $D(\mathcal{Y}, \phi)$ as follows:

$$V(D(\mathcal{Y}, \phi)) = \{(g\phi_\sigma(G_\sigma), \sigma) | \ \sigma \in V(\mathcal{Y}),\ g\phi_\sigma(G_\sigma) \in G/\phi_\sigma(G_\sigma)\}$$
$$E(D(\mathcal{Y}, \phi)) = \{(g\phi_{i(a)}(G_{i(a)}), a) | \ a \in E(\mathcal{Y}),\ g \in g\phi_{i(a)}(G_{i(a)}) \in G/\phi_{i(a)}(G_{i(a)})\};$$

the maps $i, t : E(D(\mathcal{Y}, \phi)) \to V(D(\mathcal{Y}, \phi))$ are

$$i((g\phi_{i(a)}(G_{i(a)})), a) = (g\phi_{i(a)}(G_{i(a)}), i(a)),$$
$$t((g\phi_{i(a)}(G_{i(a)}), a)) = (g\phi(a)^{-1}\phi_{t(a)}(G_{t(a)}), t(a));$$

and the composition is

$$(g\phi_{i(a)}(G_{i(a}), a)(h\phi_{i(b)}(G_{i(b))}, b) = (h\phi_{i(b)}(G_{i(b)}), ab),$$

where $(a, b) \in E^{(2)}(\mathcal{Y}),\ g, h \in G$ and $g\phi_{i(a)}(G_{i(a)}) = h\phi(b)^{-1}\phi_{i(a)}(G_{i(a)})$.

The map t is well-defined because if $x \in G_{i(a)}$, then by (2.5)

$$\phi_{i(a)}(x)\phi(a)^{-1} = \phi(a)^{-1}\phi_{t(a)}(\psi_a(x)) \equiv \phi(a)^{-1} \text{ modulo } \phi_{t(a)}(G_{t(a)}).$$

It is straightforward to check that $\mathcal{X}$ is a scwol. For instance to see that

$$t((h\phi_{i(b)}(G_{i(b}), ab)) = t((g\phi_{i(a)}(G_{i(a)}), a)),$$

we need to check that $h\phi(ab)^{-1}\phi_{t(a)}(G_{t(a)}) = g\phi(a)^{-1}\phi_{t(a)}(G_{t(a)})$, provided that $g = h\phi(b)^{-1}\phi_{i(a)}(x)$ for some $x \in G_{i(a)}$. Using 2.5 we get:

$$\begin{aligned} g\phi(a)^{-1} &= h\phi(b)^{-1}\phi_{i(a)}(x)\phi(a)^{-1} = h\phi(b)^{-1}\phi(a)^{-1}\phi(a)\phi_{i(a)}(x)\phi(a)^{-1} \\ &= h\phi(ab)^{-1}\phi_{t(a)}(g_{a,b}^{-1}\psi_a(x)) \equiv h\phi(ab)^{-1} \mod \phi_{t(a)}(G_{t(a)}). \end{aligned}$$

The group G acts naturally on $\mathcal{X}$: given $g, h \in G$ and $\alpha \in \mathcal{Y}$, define $h.(g\phi_{i(\alpha)}(G_{i(\alpha)}), \alpha) := (hg\phi_{i(\alpha)}(G_{i(\alpha)}), \alpha)$. It is clear that $\mathcal{Y} = G\backslash\mathcal{X}$. Consider the complex of groups $\overline{G}(\mathcal{Y})$ associated to this action and the (natural) choices $\overline{\sigma} = (\sigma, G_\sigma)$ and $h_a = \phi(a)$ is $(\overline{G}_\sigma, \overline{\psi}_a, \overline{g}_{a,b})$, where $\overline{G}_\sigma = \phi_\sigma(G_\sigma)$, the injection $\overline{\psi}_a$ is the restriction of $\mathrm{Ad}(\phi(a))$ to $\phi_{i(a)}(G_{i(a)})$, and $\overline{g}_{a,b} = \phi(a)\phi(b)\phi(ab)^{-1} \in \phi_{t(a)}(G_{t(a)})$. If each ϕ_σ is injective and we identify G_σ with its image $\phi_\sigma(G_\sigma)$, then $\overline{G}(\mathcal{Y}) = G(\mathcal{Y})$.

(2) If $(gG_{i(\alpha)}, \alpha) \in D(\mathcal{Y}, \phi)$, then $\alpha \in \mathcal{Y} = G\backslash\mathcal{X}$. Using the notation of 2.9(1), let $\overline{\alpha}$ be the unique element (edge or vertex) of $\mathcal{X}$ such that $p(\overline{\alpha}) = \alpha$ and $i(\overline{\alpha}) = \overline{i(\alpha)}$. It is straightforward to check that $(gG_{i(\alpha)}, \alpha) \mapsto g.\overline{\alpha}$ defines a G-equivariant isomorphism $D(\mathcal{Y}, \phi) \to \mathcal{X}$. □

2.14 Remark. Given a complex of groups $G(\mathcal{Y}) = (G_\sigma, \psi_a, g_{a,b})$ and a morphism $\phi : G(\mathcal{Y}) \to G$ to a group (where ϕ is not necessarily injective on the local groups), one can consider the complex of groups $\overline{G}(\mathcal{Y}) = (\overline{G}_\sigma, \overline{\psi}_a, \overline{g}_{a,b})$ (where $\overline{G}_\sigma = \phi_\sigma(G_\sigma)$, etc.) and the induced morphism $\overline{\phi} = (\overline{\phi}_\sigma, \overline{\phi}(a)) : \overline{G}(\mathcal{Y}) \to G$, where $\overline{\phi}_\sigma : \overline{G}_\sigma \to G$ is simply the inclusion and $\overline{\phi}(a) = \phi(a)$. It is clear that $D(\mathcal{Y}, \phi) = D(\mathcal{Y}, \overline{\phi})$, and by 2.13(1) the complex of groups canonically associated to the action of G on $D(\mathcal{Y}, \phi)$ is $\overline{G}(\mathcal{Y})$.

2.15 Corollary. *A complex of groups $G(\mathcal{Y})$ is developable if and only if there exists a morphism ϕ from $G(\mathcal{Y})$ to some group G which is injective on the local groups.*

2.16 Remark. Let $\mathcal{Y}$ be the scwol associated to a poset $\mathcal{Q}$ (see 1.2(1)). Let $G(\mathcal{Y})$ be a simple complex of groups and let $\phi : G(\mathcal{Y}) \to G$ be a morphism that is injective on local groups and simple in the sense that $\phi(a) = 1$ for all $a \in E(\mathcal{Y})$. Then $D(\mathcal{Y}, \phi)$ is the scwol associated to the poset yielded by the basic construction of II.12.18.

As mentioned in remark 2.6, if $\mathcal{Y}$ is simply connected then any morphism from $G(\mathcal{Y})$ to a group G is homotopic to a simple one (see 3.10(3)). Therefore a simple complex of groups over $\mathcal{Q}$ is strictly developable if and only if the corresponding complex of groups over the associated scwol is developable. In particular, for simplices or n-gons of groups the notions of developability and strict developability are equivalent. See parts (5) and (6) of II.12.17 for examples of complexes of groups which are not developable.

2.17 Corollary (Bass-Serre). *Any complex of groups $G(\mathcal{Y})$ over a scwol $\mathcal{Y}$ of dimension one is developable.*

Moreover, if $\mathcal{Y}$ is finite and all of the local groups are finite, there is an action of a finite group on a finite scwol $\mathcal{X}$ (which is connected if $\mathcal{Y}$ is connected) such that the associated complex of groups is isomorphic to $G(\mathcal{Y})$.

Proof. Before beginning the proof properly, we remark that whenever one has two free actions of a group H on a set Z such that the sets of orbits have the same cardinality, one can find a permutation of Z that conjugates one of the actions to the other. To see this, we write G to denote the group of permutations of Z and write the actions as homomorphisms $\phi, \phi' : H \to G$. In each orbit of both actions we choose a representative: let $Z_0 \subseteq Z$ and $Z_0' \subseteq Z$ be the sets of chosen representatives. By hypothesis, there is a bijection $f_0 : Z_0 \to Z_0'$. As the actions are free, for each $z \in Z$ there is a unique $z_0 \in Z_0$ and a unique $h_0 \in H$ such that $z = \phi(h_0)(z_0)$. Thus we may define a map $f : Z \to Z$ by sending z to $\phi'(h_0)(f_0(z_0))$. This map is a bijection, and by construction $f\phi(h) = \phi'(h)f$ for each $h \in H$.

We now turn to the proof of the corollary. Consider first the case where all the local groups G_σ have a bounded order. Let Z be a finite set whose cardinality is divisible by the cardinality of each G_σ, and let G be the group of permutations of Z. For each $\sigma \in V(\mathcal{Y})$ we can construct an injective homomorphism $\phi_\sigma : G_\sigma \to G$ whose image acts freely on Z by choosing a partition of Z into subsets whose cardinality is the order of G_σ and taking an arbitrary free action of G_σ on each subset.

For each $a \in E(\mathcal{Y})$, the images of $G_{i(a)}$ under the injective homomorphisms $\phi_{i(a)}$ and $\phi_{t(a)}\psi_a$ act freely on Z and in each case the action has $|Z|/|G_i(a)|$ orbits. Therefore our opening remark yields an element $\phi(a) \in G$ such that $\phi_{t(a)}\psi_a = \mathrm{Ad}(\phi(a))\phi_{i(a)}$. In this way we obtain a morphism $G(\mathcal{Y}) \to G$, denoted $\phi = (\phi_\sigma, \phi(a))$, that is injective on the local groups. If $\mathcal{Y}$ is finite, then the development $D(\mathcal{Y}, \phi)$ is a finite scwol on which G acts and the associated complex of groups is $G(\mathcal{Y})$. If $\mathcal{Y}$ is connected, we can consider a connected component of $D(\mathcal{Y}, \phi)$ as in 2.9(3) to prove the last assertion of the corollary.

In the general case, we let

$$Z = \prod_{\sigma \in V(\mathcal{Y})} G_\sigma,$$

and we again take G to be the group of permutations of Z. We let G_σ act by left translations on the factor G_σ and trivially on the other factors. For each $a \in E(\mathcal{Y})$ the homomorphisms $\phi_{i(a)}$ and $\phi_{t(a)}\psi_a$ are conjugate via a permutation $\phi(a)$ of Z because the cardinality of $G_{t(a)}$ is equal to the cardinality of $G_{i(a)} \times G_{t(a)}/\psi_a(G_{i(a)})$. Thus we may conclude as above. □

2.18 The Functorial Properties of the Basic Construction. *Let $G(\mathcal{Y}) = (G_\sigma, \psi_a, g_{a,b})$ and $G(\mathcal{Y}') = (G_{\sigma'}, \psi_{a'}, g_{a',b'})$ be complexes of groups over scwols $\mathcal{Y}$ and $\mathcal{Y}'$. Let $\phi : G(\mathcal{Y}) \to G$ and $\phi' : G(\mathcal{Y}') \to G'$ be morphisms and let $\Lambda : G \to G'$ be a group homomorphism.*

(1) *Let $\lambda : G(\mathcal{Y}) \to G(\mathcal{Y}')$ be a morphism over $l : \mathcal{Y} \to \mathcal{Y}'$. Functorially associated to each homotopy from $\Lambda\phi$ to $\phi'\lambda$ there is a Λ-equivariant morphism $L : D(\mathcal{Y}, \phi) \to D(\mathcal{Y}', \phi')$ that projects to $l : \mathcal{Y} \to \mathcal{Y}'$.*

(2) *If λ is an isomorphism and ϕ, ϕ' are injective on the local groups, then* $\ker \Lambda$ *acts simply transitively on the fibres of L, and L is surjective if and only if Λ is surjective.*

(3) *If ϕ and ϕ' are injective on the local groups then every Λ-equivariant morphism $L : D(\mathcal{Y}, \phi) \to D(\mathcal{Y}', \phi')$ arises as in (1).*

Proof. (1) A homotopy from $\Lambda\phi$ to $\phi'\lambda$ is determined by a family of elements $k_\sigma \in G'$ (notation of 2.4). It is straightforward to check that the map

$$(g\phi_{i(\alpha)}(G_{i(\alpha)}),\ \alpha) \mapsto (\Lambda(g)k_{i(\alpha)}^{-1}\phi'_{i(l(\alpha))}(G_{i(l(\alpha))}),\ l(\alpha))$$

is well-defined and that it is a Λ-equivariant morphism $L : D(\mathcal{Y}, \phi) \to D(\mathcal{Y}', \phi')$ which projects to l.

(2) If λ is an isomorphism and if ϕ and ϕ' are injective on local groups, then $\ker \Lambda$ acts freely on $D(\mathcal{Y}, \phi)$, because for each $\sigma \in V(\mathcal{Y})$ the equality $\phi'\lambda_\sigma = (\mathrm{Ad}k_\sigma)\Lambda\phi_\sigma$ implies that $\ker \Lambda \cap \phi_\sigma(G_\sigma) = 1$. Let us check that $\ker \Lambda$ acts transitively on the fibres of L. If two vertices of $D(\mathcal{Y}, \phi)$ are mapped by L to the same vertex, then they are of the form $(g_1\phi_\sigma(G_\sigma), \sigma)$ and $(g_2\phi_\sigma(G_\sigma), \sigma)$ with $\Lambda(g_1)k_\sigma^{-1} \equiv \Lambda(g_2)k_\sigma^{-1}$ modulo $\phi'_{l(\sigma)}(G_{l(\sigma)})$. As λ_σ is surjective, there exists $x \in G_\sigma$ such that $k_\sigma\Lambda(g_2^{-1}g_1)k_\sigma^{-1} = \phi'_{l(\sigma)}(\lambda_\sigma(x))$. This implies that $\Lambda(g_2^{-1}g_1) = \Lambda(\phi_\sigma(x))$, hence $h := g_2\phi_\sigma(x)g_1^{-1} \in \ker \Lambda$, and therefore $h.(g_1\phi_\sigma(G_\sigma), \sigma) = (g_2\phi_\sigma(G_\sigma), \sigma)$.

(3) This was already observed in 2.9(4). □

The case where Λ is the identity is already interesting:

2.19 Corollary. *Let $\phi, \phi' : G(\mathcal{Y}) \to G$ be two morphisms which are injective on the local groups. ϕ and ϕ' are homotopic if and only if there is a G-equivariant isomorphism $D(\mathcal{Y}, \phi) \to D(\mathcal{Y}, \phi')$ projecting to the identity on $\mathcal{Y}$.*

2.20 Exercise (The Mapping Cylinder). Let $\phi = (\phi_\sigma, \phi(a)) : G(\mathcal{Y}_0) \to G(\mathcal{Y}_1)$ be a morphism of complexes of groups over a non-degenerate morphism $f : \mathcal{Y}_0 \to \mathcal{Y}_1$ which is injective on the local groups.

(1) Define the mapping cylinder of f. This should be a scwol M_f that contains the disjoint union of $\mathcal{Y}_0$ and $\mathcal{Y}_1$ as subscwols and has extra edges u_α indexed by the elements $\alpha \in \mathcal{Y}_0$, with $i(u_\alpha) = i(\alpha)$ and $t(u_\alpha) = f(t(\alpha))$. For each pair of composable elements $\alpha, \beta \in \mathcal{Y}_0$, one should have compositions defined by the formulas $u_\alpha \beta = u_{\alpha\beta} = f(\alpha) u_\beta$. Show that M_f is indeed a scwol.

(2) Construct the mapping cylinder $G(M_f)$ of ϕ. This should be a complex of groups over M_f whose restriction to $\mathcal{Y}_i$ is $G(\mathcal{Y}_i)$, for $i = 0, 1$. The homomorphisms associated to the new edges should satisfy $\psi_{u_\alpha} = \phi_{t(\alpha)} \psi_\alpha$. Complete these data to obtain the desired complex of groups $G(M_f)$.

(3) Given $\sigma_0 \in V(\mathcal{Y}_0)$, show that the homomorphism

$$\pi_1(G(\mathcal{Y}_1), f(\sigma_0)) \to \pi_1(G(M_f), f(\sigma_0))$$

induced by the natural inclusion $G(\mathcal{Y}_1) \to G(M_f)$ (see 3.6) is an isomorphism.

3. The Fundamental Group of a Complex of Groups

In this section we define the fundamental group of a complex of groups $G(\mathcal{Y})$ and we give a presentation of this group in terms of presentations for the local groups and a choice of a maximal tree in $\mathcal{Y}$. For a developable complex of groups, we construct its universal covering explicitly. Our treatment is a natural generalization of the theory of graph of groups due to Bass and Serre [Ser77]. Before reading this section, the reader may wish to look at paragraphs A.10 to A.13 in the Appendix, where the corresponding notions for small categories are described.

The Universal Group $FG(\mathcal{Y})$

3.1 Definition. Let $G(\mathcal{Y}) = (G_\sigma, \psi_a, g_{a,b})$ be a complex of groups over the scwol $\mathcal{Y}$. The group $FG(\mathcal{Y})$ is the group given by the following presentation. It is generated by the set

$$\coprod_{\sigma \in V(\mathcal{Y})} G_\sigma \coprod E^{\pm}(\mathcal{Y})$$

subject to the relations

$$\left\{ \begin{array}{l} \text{the relations in the groups } G_\sigma, \\ (a^+)^{-1} = a^- \text{ and } (a^-)^{-1} = a^+, \\ a^+ b^+ = g_{a,b} (ab)^+, \quad \forall\, (a,b) \in E^{(2)}(\mathcal{Y}) \\ \psi_a(g) = a^+ g a^-, \quad \forall g \in G_{i(a)} \end{array} \right\}.$$

There is a natural morphism

$$\iota = (\iota_\sigma, \iota(a)) : G(\mathcal{Y}) \to FG(\mathcal{Y}),$$

where $\iota_\sigma : G_\sigma \to FG(\mathcal{Y})$ is the natural homomorphism mapping $g \in G_\sigma$ to the corresponding generator g of $FG(\mathcal{Y})$ and $\iota(a) = a^+$. (The relations defining $FG(\mathcal{Y})$ are the least set that one must impose in order to make ι a morphism.)

If one regards a group G as a complex of groups over a single vertex, then FG is canonically isomorphic to G.

3.2 The Universal Property and Functoriality of F

Given a morphism $\phi : G(\mathcal{Y}) \to G$, there is a unique group homomorphism $F\phi : FG(\mathcal{Y}) \to G$ such that $(F\phi) \circ \iota = \phi$.

More generally, if $\phi : G(\mathcal{Y}) \to G(\mathcal{Y}')$ is a morphism of complexes of groups over a morphism $f : \mathcal{Y} \to \mathcal{Y}'$, then associated to ϕ there is a homomorphism $F\phi : FG(\mathcal{Y}) \to FG(\mathcal{Y}')$ such that the following diagram commutes:

$$\begin{array}{ccc} G(\mathcal{Y}) & \xrightarrow{\iota} & FG(\mathcal{Y}) \\ {\scriptstyle\phi}\downarrow & & \downarrow{\scriptstyle F\phi} \\ G(\mathcal{Y}') & \xrightarrow{\iota'} & FG(\mathcal{Y}') \end{array}$$

Proof. Let $\phi = (\phi_\sigma, \phi(a)) : G(\mathcal{Y}) \to G(\mathcal{Y}')$. Then $F\phi$ maps the generator $g \in G_\sigma$ to $\phi_\sigma(g)$ and the generator a^+ to $\phi(a)f(a)^+$. It is straightforward to verify that $F\phi$ is well defined. □

3.3 $G(\mathcal{Y})$-Paths. Let $\mathcal{Y}$ be a scwol and let $G(\mathcal{Y}) = (G_\sigma, \psi_a, g_{a,b})$ be a complex of groups over $\mathcal{Y}$ as defined in (2.1). Given $\sigma_0 \in V(\mathcal{Y})$, a *$G(\mathcal{Y})$-path* issuing from σ_0 is a sequence $c = (g_0, e_1, g_1, \dots, e_k, g_k)$ where $(e_1, \dots, e_k)$ is an edge path in $\mathcal{Y}$ that issues from σ_0 (in the sense of 1.6), $g_0 \in G_{\sigma_0}$, and $g_i \in G_{t(e_i)}$ for $1 \le i \le k$. The vertex σ_0 is called the initial vertex $i(c)$ of c and $\sigma = t(e_k)$ is called the terminal vertex $t(c)$ of c; we say that c joins σ_0 to σ. A $G(\mathcal{Y})$-path joining σ_0 to σ_0 is called a *$G(\mathcal{Y})$-loop at σ_0*.

If $c' = (g_0', e_1', g_1', \dots, e_{k'}', g_{k'}')$ is a $G(\mathcal{Y})$-path issuing from the terminal vertex of c, then the *composition* or *concatenation* of c and c' is the $G(\mathcal{Y})$-path

$$c * c' = (g_0, e_1, g_1, \dots, e_k, g_k g_0', e_1', g_1', \dots, e_{k'}', g_{k'}').$$

This partially defined law of composition is associative: if c, c', c'' are $G(\mathcal{Y})$-paths such that $c * c'$ and $c' * c''$ are defined, then $(c * c') * c'' = c * (c' * c'')$; this $G(\mathcal{Y})$-path is denoted $c * c' * c''$. The inverse c^{-1} of c is the $G(\mathcal{Y})$-path $(g_0', e_1', g_1', \dots, e_k', g_k')$ where $g_i' = g_{k-i}^{-1}$ and $e_i' = e_{k-i+1}^{-1}$.

3.4 Homotopy of $G(\mathcal{Y})$-Paths. To each $G(\mathcal{Y})$-path $c = (g_0, e_1, g_1, \dots, e_k, g_k)$ we associate the element $\pi(c)$ of $FG(\mathcal{Y})$ represented by the word $g_0 e_1 g_1 \dots e_k g_k$. By definition, two $G(\mathcal{Y})$-paths c and c' joining σ to τ are *homotopic* if $\pi(c) = \pi(c')$. The homotopy class of c is denoted $[c]$ and the set of homotopy classes of $G(\mathcal{Y})$-paths joining σ_0 to σ is denoted $\pi_1(G(\mathcal{Y}), \sigma_0, \sigma)$.

Note that the homotopy class of the composition of two $G(\mathcal{Y})$-paths depends only on the homotopy class of those paths.

The Fundamental Group $\pi_1(G(\mathcal{Y}), \sigma_0)$

3.5 Definition. The operation $[c][c'] := [c * c']$ defines a group structure on the set of homotopy classes of $G(\mathcal{Y})$-loops at σ_0. The resulting group is denoted $\pi_1(G(\mathcal{Y}), \sigma_0)$ and is called the *fundamental group* of $G(\mathcal{Y})$ at σ_0.

Note that, by definition, the map π identifies the group $\pi_1(G(\mathcal{Y}), \sigma_0)$ to $\pi_1(G(\mathcal{Y}), \sigma_0, \sigma_0)$, which is a subgroup of $FG(\mathcal{Y})$. If c is a $G(\mathcal{Y})$-path joining σ_0 to σ, then the map associating to each $G(\mathcal{Y})$-loop ℓ at σ_0 the $G(\mathcal{Y})$-loop $(c^{-1} * \ell * c)$ at σ induces an isomorphism $\pi_1(G(\mathcal{Y}), \sigma_0) \to \pi_1(G(\mathcal{Y}), \sigma)$. If we identify these groups to the corresponding subgroups of $FG(\mathcal{Y})$, then this isomorphism is the restriction of the inner automorphism $\mathrm{Ad}(\pi(c))$.

If $G(\mathcal{Y})$ is the trivial complex of groups over $\mathcal{Y}$ (i.e. all of the G_σ are trivial), then $\pi_1(G(\mathcal{Y}), \sigma_0) = \pi_1(\mathcal{Y}, \sigma_0)$. More generally, the map associating to each $G(\mathcal{Y})$-loop $c = (g_0, e_1, g_1, \dots, e_k, g_k)$ at σ_0 the edge loop $(e_1, \dots, e_k)$ induces a surjective homomorphism $\pi_1(G(\mathcal{Y}), \sigma_0) \to \pi_1(\mathcal{Y}, \sigma_0)$.

Any morphism $\phi = (\phi_\sigma, \phi(a))$ from $G(\mathcal{Y})$ to a group G induces a homomorphism $\pi_1(\phi, \sigma_0) : \pi_1(G(\mathcal{Y}), \sigma_0) \to G$ mapping the homotopy class of a $G(\mathcal{Y})$-loop c to the element $(F\phi)(\pi(c)) \in G$.

More generally, we have:

3.6 Proposition. *Every morphism $\phi = (\phi_\sigma, \phi(a))$ of complexes of groups $G(\mathcal{Y}) \to G(\mathcal{Y}')$ over a morphism $f : \mathcal{Y} \to \mathcal{Y}'$ induces a natural homomorphism*

$$\pi_1(\phi, \sigma_0) : \pi_1(G(\mathcal{Y}), \sigma_0) \to \pi_1(G(\mathcal{Y}'), f(\sigma_0)),$$

namely the restriction of $F\phi : FG(\mathcal{Y}) \to FG(\mathcal{Y}')$.

Proof. We only need to check that $F\phi$ maps the subset $\pi_1(G(\mathcal{Y}), \sigma_0, \sigma) \subseteq FG(\mathcal{Y})$ to the subset $\pi_1(G(\mathcal{Y}'), f(\sigma_0), f(\sigma)) \subseteq FG(\mathcal{Y}')$.

For each $e \in E(\mathcal{Y})^{\pm}$, we shall define a $G(\mathcal{Y}')$-path denoted $\phi(e)$ such that $\phi(e^{-1}) = \phi(e)^{-1}$. Assume $e = a^+$. If $f(a) \in E(\mathcal{Y}')$, then $\phi(e)$ is the $G(\mathcal{Y}')$-path $(\phi(a), f(a)^+, 1_{f(i(a))})$; if $f(a)$ is a unit $1_{\sigma'}$, then $\phi(e)$ is the $G(\mathcal{Y}')$-path $(1_{\sigma'})$. For each element $g \in G_\sigma$, we define $\phi(g)$ to be the $G(\mathcal{Y}')$-path $(\phi_\sigma(g))$. For each $G(\mathcal{Y})$-path $c = (g_0, e_1, \dots, g_k)$, we define $\phi(c)$ to be the concatenation of the $G(\mathcal{Y}')$-paths $\phi(g_0) * \phi(e_1) * \dots * \phi(g_k)$. It is clear that $F\phi$ maps the homotopy class of c to the homotopy class of $F\phi$. □

A Presentation of $\pi_1(G(\mathcal{Y}), \sigma_0)$

Consider the graph which is the 1-skeleton $|\mathcal{Y}|^{(1)}$ of the geometric realization of $\mathcal{Y}$: its set of vertices is $V(\mathcal{Y})$ and its set of 1-cells is $E(\mathcal{Y})$. We assume that $\mathcal{Y}$ is connected (equivalently, that this graph is connected). Let T be a maximal tree in this graph, i.e. a subgraph which is a tree containing all the vertices. Such a maximal tree is not unique in general: any subgraph T' of $|\mathcal{Y}|^{(1)}$ which is a tree can be extended to a maximal tree.

3.7 Theorem. *The fundamental group $\pi_1(G(\mathcal{Y}), \sigma_0)$ is isomorphic to the abstract group $\pi_1(G(\mathcal{Y}), T)$ generated by the set*

$$\coprod_{\sigma \in V(\mathcal{Y})} G_\sigma \coprod E^{\pm}(\mathcal{Y})$$

subject to the relations

$$\left\{\begin{array}{l} \textit{the relations in the groups } G_\sigma, \\ (a^+)^{-1} = a^- \textit{ and } (a^-)^{-1} = a^+, \\ a^+b^+ = g_{a,b}(ab)^+, \quad \forall\,(a,b) \in E^{(2)}(\mathcal{Y}) \\ \psi_a(g) = a^+ g a^-, \quad \forall g \in G_{i(a)}, \\ a^+ = 1, \ \forall\, a \in T. \end{array}\right\}$$

Proof. Identify $\pi_1(G(\mathcal{Y}), \sigma_0)$ with $\pi_1(G(\mathcal{Y}), \sigma_0, \sigma_0)$ and let $\Psi : \pi_1(G(\mathcal{Y}), \sigma_0) \to \pi_1(G(\mathcal{Y}), T)$ be the homomorphism obtained by restricting the natural projection $FG(\mathcal{Y}) \to \pi_1(G(\mathcal{Y}), T)$. For each vertex $\sigma \in V(\mathcal{Y})$, let $c_\sigma = (e_1, \dots, e_k)$ be the unique edge path in $\mathcal{Y}$ joining σ_0 to σ such that no two consecutive edges are inverse to each other and each e_i is contained in T (i.e. $e_i = a_i^{\pm}$, with $a_i \in T$). Let π_σ be the corresponding element $e_1 \dots e_k \in \pi_1(G(\mathcal{Y}), \sigma_0, \sigma) \subseteq FG(\mathcal{Y})$. Note that if $a \in T$, then either $\pi_{i(a)} = \pi_{t(a)} a^+$ or $\pi_{t(a)} = \pi_{i(a)} a^-$. Note also that the image of π_σ in $\pi_1(G(\mathcal{Y}), T)$ is trivial. Let $\Theta : \pi_1(G(\mathcal{Y}), T) \to \pi_1(G(\mathcal{Y}), \sigma_0)$ be the homomorphism mapping the generator $g \in G_\sigma$ to $\pi_\sigma g \pi_\sigma^{-1}$ and the generator a^+ to the element $\pi_{t(a)} a^+ \pi_{i(a)}^{-1}$. This homomorphism is well-defined because the relations are satisfied; in particular $\Theta(a^+) = 1$ if $a \in T$.

The elements of the form $\pi_\sigma g \pi_\sigma^{-1}$, where $g \in G_\sigma$, together with those of the form $\pi_{t(a)} a^+ \pi_{i(a)}^{-1}$, where $a \in E(\mathcal{Y})$, generate $\pi_1(G(\mathcal{Y}), \sigma_0)$. Their images under Ψ are, respectively, g and a^+. This shows that Ψ and Θ are inverse to each other. $\square$

The fundamental groups of the complexes of groups of dimension one shown in figure $\mathcal{C}$.9 are, respectively, the amalgamated product $G_{\tau_1} *_{G_\sigma} G_{\tau_2}$, an *HNN*-extension $G_\tau *_{G_\sigma}$, and the amalgamation of the groups G_{τ_i} along their isomorphic subgroup $\psi(G_\sigma)$. The morphism of complexes of groups described in figure $\mathcal{C}$.10 induces an isomorphism on the fundamental groups.

3.8 Corollary. *If all of the groups G_σ are finitely generated (resp. finitely presented) and if $\mathcal{Y}$ is finite, then $\pi_1(G(\mathcal{Y}), \sigma)$ is finitely generated (resp. finitely presented).*

The criterion of developability given in 2.15 can be reformulated as follows.

3.9 Proposition. *A complex of groups $G(\mathcal{Y})$ over a connected scwol $\mathcal{Y}$ is developable if and only if each of the natural homomorphisms $G_\sigma \to \pi_1(G(\mathcal{Y}), T)$ (equivalently $G_\sigma \to FG(\mathcal{Y})$) is injective.*

Proof. According to 2.15, $G(\mathcal{Y})$ is developable if and only if there is a morphism ϕ from $G(\mathcal{Y})$ to a group G which is injective on the local groups G_σ. The universality of $FG(\mathcal{Y})$ (3.2) shows that such a morphism exists if and only if the canonical morphism $\iota : G(\mathcal{Y}) \to FG(\mathcal{Y})$ is injective on the local groups. Moreover the quotient map $FG(\mathcal{Y}) \to \pi_1(G(\mathcal{Y}), T)$ is injective on the subgroups $\iota_\sigma(G_\sigma)$. Indeed, by choosing $\sigma_0 = \sigma$ in the proof of 3.7, we see that the map $\Theta \circ \Psi$ restricted to $i_\sigma(G_\sigma)$ is the identity. □

3.10 Proposition. *Let $G(\mathcal{Y})$ be a complex of groups over a connected scwol $\mathcal{Y}$, let T be a maximal tree in $|\mathcal{Y}|^{(1)}$ and let G be a group.*

(1) *Given a morphism $\phi = (\phi_\sigma, \phi(a)) : G(\mathcal{Y}) \to G$, define*

$$\pi_1(\phi, T) : \pi_1(G(\mathcal{Y}), T) \to G$$

by composing the natural isomorphism $\Theta : \pi_1(G(\mathcal{Y}), T) \to \pi_1(G(\mathcal{Y}), \sigma_0)$ with the homomorphism $\pi_1(\phi, \sigma_0) : \pi_1(G(\mathcal{Y}), \sigma_0) \to G$ defined in (3.6). If ϕ is such that $\phi(a) = 1$ for each $a \in T$, then $\pi_1(\phi, T)$ maps each generator $g \in G_\sigma$ to $\phi_\sigma(g)$ and each generator a^+ to $\phi(a)$.

(2) *Every morphism $\phi' : G(\mathcal{Y}) \to G$ is homotopic to a morphism $\phi : G(Y) \to G$ such that $\phi(a) = 1$ for each $a \in T$. Moreover ϕ is unique up to conjugation by an element of G. Therefore the map associating to a morphism $G(\mathcal{Y}) \to G$ the induced homomorphism $\pi_1(G(\mathcal{Y}), \sigma_0) \to G$ gives a bijection*

$$H^1(G(\mathcal{Y}), G) \to H^1(\pi_1(G(\mathcal{Y}), \sigma_0), G),$$

where $H^1(G(\mathcal{Y}), G)$ is the set of homotopy classes of morphisms of $G(\mathcal{Y})$ to G, and $H^1(\pi_1(G(\mathcal{Y}), \sigma_0), G)$ is the set of homomorphisms from $\pi_1(G(\mathcal{Y}), \sigma_0)$ to G up to conjugation by an element of G.

(3) *If $G(Y)$ is a simple complex of groups over a simply connected scwol $\mathcal{Y}$, then every morphism $G(\mathcal{Y}) \to G$ is homotopic to a simple morphism.*

Proof. We use the notations established in the proof of 3.7. Part (1) follows immediately from the observation that $\pi_1(\phi, \sigma_0)$ maps each π_σ to $1 \in G$.

To prove (2), consider the image k_σ of π_σ under $\pi_1(\phi', \sigma_0)$. Let $\phi = (\phi_\sigma, \phi(a))$ be the morphism, homotopic to ϕ', defined by $\phi_\sigma = \mathrm{Ad}(k_\sigma)\phi'_\sigma$ and $\phi(a) = k_{t(a)}\phi'(a)k_{i(a)}^{-1}$. For each $a \in T$, we have $k_{i(a)} = k_{t(a)}\phi'(a)$ or $k_{t(a)} = k_{i(a)}\phi'(a)^{-1}$, therefore $\phi(a) = 1$.

It remains to show that ϕ is unique. Let $\overline{\phi} = (\overline{\phi}_\sigma, \overline{\phi}(a)) : G(\mathcal{Y}) \to G$ be another morphism such that $\overline{\phi}(a) = 1$ for $a \in T$. If $(k_\sigma)_{\sigma \in V(\mathcal{Y})}$ is a homotopy between ϕ and $\overline{\phi}$, the relation $\overline{\phi}(a) = k_{t(a)}\phi(a)k_{i(a)}^{-1}$ applied to the edges $a \in T$ implies that k_σ is an element $g \in G$ independent of σ, therefore $\overline{\phi}$ is the conjugate of ϕ by g.

To prove (3), we first observe that if $\phi' : G(\mathcal{Y}) \to G$ is a morphism, then $a \mapsto \phi'(a)$ gives a morphism from the trivial complex of groups over $\mathcal{Y}$ to G. By (2), since $\mathcal{Y}$ is simply connected, there is a family (k_σ) of elements of G such that $k_{t(a)}\phi'(a)k_{i(a)}^{-1} = 1$ for each $a \in E(\mathcal{Y})$. Therefore (k_σ) defines a homotopy between ϕ' and a simple morphism $\phi : G(\mathcal{Y}) \to G$. □

3.11 Examples

(1) Assume that $G(\mathcal{Y})$ is a *simple* complex of groups (so all the elements $g_{a,b}$ are trivial). Then the subgroup of $\pi_1(G(\mathcal{Y}), T)$ generated by $\{a^+ \mid a \in E(\mathcal{Y})\}$ is isomorphic to the fundamental group of $\mathcal{Y}$. It follows that *if $\mathcal{Y}$ is simply connected, then $\pi_1(G(\mathcal{Y}), \sigma_0) = \pi_1(G(\mathcal{Y}), T)$ is the direct limit (in the category of groups) of the family of subgroups G_σ and inclusions ψ_a*. Indeed all the generators of the form $a^+,\ a \in E(\mathcal{Y})$, represent the trivial element of $\pi_1(G(\mathcal{Y}), T)$.

In particular if $G(\mathcal{Q})$ is a simple complex of groups over a poset $\mathcal{Q}$, and the geometric realization of $\mathcal{Q}$ is simply connected, then the group $\varinjlim_\sigma G_\sigma$ (denoted $\widehat{G(\mathcal{Q})}$ in II.12.12) is the fundamental group of the corresponding complex of groups. It follows that $G(\mathcal{Q})$ is strictly developable (in the sense of II.12.15) if and only if it is developable.

(2) Let $\phi = (\phi_\sigma, \phi(a)) : G(\mathcal{Y}) \to G(\mathcal{Y}')$ be a morphism of complexes of groups over $f : \mathcal{Y} \to \mathcal{Y}'$ and suppose that one can find maximal trees T in $|\mathcal{Y}|^{(1)}$ and T' in $|\mathcal{Y}'|^{(1)}$ such that $f(T) \subseteq T'$ and $\phi(a) = 1$ for each $a \in T$. Then the homomorphism induced by ϕ on the fundamental groups $\pi_1(G(\mathcal{Y}), T) \to \pi_1(G(\mathcal{Y}'), T')$ (via the isomorphisms of 3.7) maps the generators $g \in G_\sigma$ (resp. a^+) to the generators $\phi_\sigma(g)$ (resp. $\phi(a)^+$).

(3) *A presentation of the fundamental group of the complex of groups $G(\mathcal{Y})$ in example 2.10(3).*

If we choose the maximal tree T in $|\mathcal{Y}|^{(1)}$ to be the union of the edges $a, a', a_1, a'_1, \hat{b}_1$, then we have $1 = a = a' = a_1 = a'_1 = \hat{b}_1$. As $g_{a',\hat{b}_1} = g_{a,\hat{b}_1} = g_{a'_1,\hat{b}'_1} = g_{a_1,\hat{b}'_1} = 1$, we see successively that $1 = \hat{c}'_1 = c_1 = \hat{b}'_1 = \hat{c}_1$.

Besides the relations $s^n = r^3 = t^2 = 1$, we are left with the four relations

$$g_{a,b_1}c_1^+ = a^+b_1^+,\ g_{a',b_1}c_1'^+ = a'^+b_1^+,\ g^+_{a'_1,b'_1c'_1} = a_1'^+b_1'^+,\ g_{a_1,b'_1}\hat{c}_1^+ = a_1^+b_1'^+.$$

As $g_{a,b_1} = s,\ g_{a',b_1} = 1,\ g_{a'_1,b'_1} = t,\ g_{a_1,b'_1} = r^{-1}$ and $1 = c_1 = a = a' = a_1 = a'_1 = \hat{c}_1$, these relations are equivalent to

$$s = b_1^+,\ c_1'^+ = b_1^+,\ tc_1'^+ = b_1'^+,\ r^{-1} = b_1'^+.$$

After eliminating $b_1^+, c_1'^+$ and $b_1'^+$, we get only one relation $rst = 1$. Eventually we get the presentation

$$\pi_1(G(\mathcal{Y}), T) = \langle t, s, r | t^2 = r^3 = s^n = rst = 1\rangle.$$

This is a presentation that we would have obtained directly by applying Poincaré's theorem (cf.[Rh71]) to the quadrilateral $C = OAB\hat{A}$ (see figure $\mathcal{C}$.6).

More generally, consider the scwol $\mathcal{Y}$ associated to the polyhedral complex obtained by identifying the sides OA, $O\hat{A}$ and the sides BA, $B\hat{A}$ of the quadrilateral C (in the notation of figure $\mathcal{C}$.6). Consider a complex of groups $G(\mathcal{Y})$ such that the local groups associated to the vertices ρ, σ, σ_1 are trivial. After modification by a coboundary, we can assume that the only twisting elements that might be non-trivial are $g_{a,b_1} \in G_\tau$, $g_{a'_1,b'_1} \in G_{\tau'}$ and $g_{a_1,b'_1} \in G_{\tau_1}$. The same calculation as above shows that $\pi_1(G(\mathcal{Y}), T)$ is isomorphic to the quotient of the free product $G_\tau * G_{\tau'} * G_{\tau_1}$ by the normal subgroup generated by $g_{a,b_1} g_{a'_1,b'_1} g^{-1}_{a_1,b'_1}$.

(4) On the scwol $\mathcal{Y}$ obtained by gluing two congruent n-gons along their boundary (see 1.4(3)), we consider a complex of groups $G(\mathcal{Y})$ such that the only non-trivial local groups are the groups $G_k = G_{\tau_k}$. After modification by a coboundary we may assume that the only non-trivial twisting elements are the n elements $g_{a_k,b_k} = s_k \in G_k$. We leave the reader to check that, as above, the fundamental group of $G(\mathcal{Y})$ is the quotient of the free product of the groups G_k by the normal subgroup generated by the element $s_1 \dots s_n$.

(5) *Seifert-van Kampen Theorem.* Let $G(\mathcal{Y})$ be a complex of groups over a scwol $\mathcal{Y}$. Assume that $\mathcal{Y}$ is the union of two connected subscwols $\mathcal{Y}_1$ and $\mathcal{Y}_2$ such that the intersection $\mathcal{Y}_0 = \mathcal{Y}_1 \cap \mathcal{Y}_2$ is connected. Let $\sigma_0 \in V(\mathcal{Y}_0)$ be a base vertex. Let $G(\mathcal{Y}_i)$ be the complex of groups which is the restriction of $G(\mathcal{Y})$ to the subscwol $\mathcal{Y}_i$, for $i = 0, 1, 2$. Then $\pi_1(G(\mathcal{Y}), \sigma_0)$ is the quotient of the free product $\pi_1(G(\mathcal{Y}_1), \sigma_0) * \pi_1(G(\mathcal{Y}_2), \sigma_0)$ by the normal subgroup generated by the elements $j_1(\gamma)j_2(\gamma)^{-1}$ for all $\gamma \in \pi_1(G(\mathcal{Y}_0), \sigma_0)$, where $j_i : \pi_1(G(\mathcal{Y}_0), \sigma_0) \to \pi_1(G(\mathcal{Y}_i)\sigma_0)$ is the homomorphism induced by the inclusion of $G(\mathcal{Y}_0)$ into $G(\mathcal{Y}_i)$, for $i = 1, 2$.

To see this, choose a maximal tree T_0 in $|\mathcal{Y}_0|^{(1)}$ and extend it to maximal trees T_1 and T_2 in $|\mathcal{Y}_1|^{(1)}$ and $|\mathcal{Y}_2|^{(1)}$. Then $T = T_1 \cup T_2$ is a maximal tree in $|\mathcal{Y}|^{(1)}$. It is clear from the presentation in (3.7) that $\pi_1(G(\mathcal{Y}), T)$ is the quotient of the free product of $\pi_1(G(\mathcal{Y}_1), T_1) * \pi_1(G(\mathcal{Y}_2), T_2)$ by the normal subgroup generated by the elements $j'_1(\gamma)j'_2(\gamma)^{-1}$ for all $\gamma \in \pi_1(G(\mathcal{Y}_0), T_0)$, where $j'_i : \pi_1(G(\mathcal{Y}_0), T_0) \to \pi_1(G(\mathcal{Y}_i), T_i)$ is the homomorphism induced by the inclusion of $G(\mathcal{Y}_0)$ to $G(\mathcal{Y}_i)$, for $i = 1, 2$. And we can identify $\pi_1(G(\mathcal{Y}_i), T_i)$ with $\pi_1(G(\mathcal{Y}_i), \sigma_0)$ (hence j'_i to j_i) as in (3.7).

3.12 Exercise (HNN-Extensions). Let $G(\mathcal{Y}_1)$ be a complex of groups over a connected scwol $\mathcal{Y}_1$ and let $G(\mathcal{Y}_0)$ be its restriction to a connected subscwol $\mathcal{Y}_0$. We assume that if $a \in E(\mathcal{Y}_1)$ has its terminal vertex in $\mathcal{Y}_0$ then $a \in E(\mathcal{Y}_0)$. Let $\phi : G(\mathcal{Y}_0) \to G(\mathcal{Y}_1)$ be a morphism over a non-degenerate morphism $f : \mathcal{Y}_0 \to \mathcal{Y}_1$ and assume that ϕ is injective on the local groups.

(1) Following (2.20), construct a new scwol $\mathcal{Y}$ containing $\mathcal{Y}_1$ as a subscwol, with the same set of vertices and new edges u_α indexed by the elements $\alpha \in \mathcal{Y}_0$. (Construct $\mathcal{Y}$ from the mapping cylinder M_f of f by identifying in M_f the subscwol $\mathcal{Y}_0 \subseteq \mathcal{Y}_1 \subseteq M_f$ to the subscwol $\mathcal{Y}_0 \subseteq M_f$.)

(2) Construct a complex of groups $G(\mathcal{Y})$ whose restriction to $\mathcal{Y}_1$ is $G(\mathcal{Y}_1)$, the homomorphisms associated to the new edges u_α being defined by $\psi_{u_\alpha} = \phi_{t(\alpha)}\psi_\alpha$.

(3) Choose $\sigma_0 \in V(\mathcal{Y}_0)$ and an edge path c in $\mathcal{Y}_1$ joining σ_0 to $f(\sigma_0)$. Let $\Phi_0 : \pi_1(G(\mathcal{Y}_0), \sigma_0) \to \pi_1(G(\mathcal{Y}_1), \sigma_0)$ be the homomorphism induced by the inclusion $\mathcal{Y}_0 \to \mathcal{Y}_1$ and let $\Phi_1 : \pi_1(G(\mathcal{Y}_0), \sigma_0) \to \pi_1(G(\mathcal{Y}_1), \sigma_0)$ be the homomorphism sending the homotopy class of each $G(\mathcal{Y}_0)$-loop l based at σ_0 to the homotopy class of the $G(\mathcal{Y}_1)$-loop $(c * \phi(l) * c^{-1})$. Show that $\pi_1(G(\mathcal{Y}), \sigma_0)$ is isomorphic to the quotient of the free product $\pi_1(G(\mathcal{Y}_1), \sigma_0) * \mathbb{Z}$ by the relations $\Phi_0(\gamma) = t\Phi_1(\gamma)t^{-1}$, for all $\gamma \in \pi_1(G(\mathcal{Y}_0, \sigma_0)$, where t is the generator of $\mathbb{Z}$. (If Φ_0 is an injection, this group is an HNN-extension.)

The Universal Covering of a Developable Complex of Groups

Let $G(\mathcal{Y})$ be a developable complex of groups over a connected scwol $\mathcal{Y}$. Choose a maximal tree T in the 1-skeleton of the geometric realization of $\mathcal{Y}$. Let

$$\iota_T : G(\mathcal{Y}) \to \pi_1(G(\mathcal{Y}), T)$$

be the morphism mapping each element of the local group G_σ to the corresponding generator of $\pi_1(G(\mathcal{Y}), T)$ and each edge a to the generator a^+. The developments associated to morphisms to groups were defined in (2.13).

3.13 Theorem. *The development $D(\mathcal{Y}, \iota_T)$ is connected and simply connected.*

Proof. We identify the groups G_σ to their images in $\pi_1(G(\mathcal{Y}), T)$. Thus the elements of $D(\mathcal{Y}, \iota_T)$ are pairs $(gG_{i(\alpha)}, \alpha)$ where $\alpha \in \mathcal{Y}$ and $g \in \pi_1(G(\mathcal{Y}), T)$. We prove first that $\mathcal{X} := D(\mathcal{Y}, \iota_T)$ is connected.

For each $\sigma \in V(\mathcal{Y})$ let $\overline{\sigma}$ be the vertex (G_σ, σ) of $D(Y, \iota_T)$, and for each $a \in E(\mathcal{Y})$ let $\overline{a}$ be the edge $(G_{i(a)}, a)$. Let W be the subset of the geometric realization of $D(\mathcal{Y}, \phi)$ consisting of the images of the edges $\{\overline{a} \mid a \in E(\mathcal{Y})\}$. The images of the $\overline{a}$ with $a \in T$ form a tree $\overline{T} \subseteq W$ that projects to T. Since $i(\overline{a}) = \overline{i(a)}$ for every $a \in E(\mathcal{Y})$, the tree $\overline{T}$ contains all of the vertices $\overline{\sigma}$ and hence W is connected. To prove that $D(\mathcal{Y}, \iota_T)$ is connected, it is sufficient to prove that, for each generator s of $\pi_1(G(\mathcal{Y}), T)$, the intersection $W \cap s.W$ is non-empty. But this is clear for both $s = g \in G_\sigma$ and $s = a^\pm$, because this intersection contains $\overline{\sigma}$ and either $\overline{t(a)}$ or $\overline{i(a)}$.

If $D(\mathcal{Y}, \iota_T)$ were not simply connected, we could consider as in 2.10 (2)) its universal covering $p : \tilde{\mathcal{X}} \to \mathcal{X} = D(\mathcal{Y}, \iota_T)$, which is equivariant with respect to the homomorphism $\Phi : \tilde{G} \to G := \pi_1(G(\mathcal{Y}), T)$ (in the notation of 2.10 (2)). Using the choices made in 2.10(2), the complex of groups associated to the action of $\tilde{G}$ on $\tilde{X}$ can be identified to $G(\mathcal{Y})$; the corresponding homomorphism $\tilde{\phi} : G(\mathcal{Y}) \to \tilde{G}$ is such that $\tilde{\phi}(a) = 1$ when $a \in T$. Therefore the map associating to each generator $g \in G_\sigma$ of $\pi_1(G(\mathcal{Y}), T)$ the element $\tilde{\phi}(g)$, and to each generator a^+ the element $\tilde{\phi}(a)$, extends to a homomorphism $\tilde{\Phi} : G \to \tilde{G}$. Moreover $\Phi \circ \tilde{\Phi}$ is the identity of G. The map

$\tilde{p} : X \to \tilde{X}$ sending $(gG_{i(\sigma)}, \alpha)$ to $\tilde{\Phi}(g).(\tilde{\alpha})$ is a morphism and $p \circ \tilde{p}$ is the identity of $\mathcal{X}$. As $\tilde{X}$ is connected, $\tilde{p}$ is a bijection and X is simply connected. □

Note that if $\mathcal{G}(\mathcal{Y})$ is a graph of groups, then $D(\mathcal{Y}, \iota_T)$ is called the associated Bass-Serre tree (see 11.18 and 11.21).

The following is a generalization of II.12.20.

3.14 Properties of the Development $D(\mathcal{Y}, \phi)$. *Let $\mathcal{Y}$ be a connected scwol and choose a base vertex $\sigma_0 \in V(\mathcal{Y})$. Let $\phi : G(\mathcal{Y}) \to G$ be a morphism which is injective on the local groups and let $D(\mathcal{Y}, \phi)$ be the corresponding development on which the group G acts. Let $\pi_1(\phi, \sigma_0) : \pi_1(G(\mathcal{Y}), \sigma_0) \to G$ be the homomorphism induced by ϕ.*

(1) *The connected components of $D(\mathcal{Y}, \phi)$ correspond bijectively to the elements of the cosets in G of the image of $\pi_1(\phi, \sigma_0)$).*

(2) *If $D(\mathcal{Y}, \phi)$ is connected (equivalently, $\pi_1(\phi, \sigma_0)$ is surjective), the fundamental group of $D(\mathcal{Y}, \phi)$ is the kernel of $\pi_1(\phi, \sigma_0)$. In particular $D(Y, \phi)$ is simply connected if and only if $\pi_1(\phi, \sigma_0)$ is an isomorphism.*

Proof. Let T be a maximal tree in the 1-skeleton of the geometric realization of $\mathcal{Y}$. As $D(\mathcal{Y}, \phi)$ depends up to isomorphism only on the homotopy class of ϕ (see 2.19), we may assume that $\phi(a) = 1$ for every $a \in T$, and hence replace $\pi_1(\phi, \sigma_0)$ by the homomorphism $\Lambda = \pi_1(\phi, T) : \pi_1(g(\mathcal{Y}), T) \to G$ that sends the generator $g \in G_\sigma$ to $\phi_\sigma(g)$ and the generator a^+ to $\phi(a)$. Let $G_0 \subseteq G$ be the image of this homomorphism. We have $\Lambda\iota_T = \phi$, and ϕ is equal to the composition of a morphism $\phi_0 : G(\mathcal{Y}) \to G_0$ with the inclusion $G_0 \to G$.

According to 2.18, the map $L : D(\mathcal{Y}, \iota_T) \to D(\mathcal{Y}, \phi_0)$ defined by

$$(\gamma G_{i(\alpha)}, \alpha) \mapsto (\Lambda(\gamma)\phi_{i(\alpha)}(G_{i(\alpha)}), \alpha),$$

where $\gamma \in \pi_1(G(\mathcal{Y}), T)$, is a Galois covering in the sense of 1.13 with Galois group the kernel of Λ. In particular L is surjective, $D(\mathcal{Y}, \phi_0)$ is connected and the fundamental group of $D(\mathcal{Y}, \phi)$ is isomorphic to the kernel of Λ. It is easy to see that $D(\mathcal{Y}, \phi_0)$ is a connected component of $D(\mathcal{Y}, \phi)$ and the connected components of $D(\mathcal{Y}, \phi)$ are in bijection with the elements of G/G_0. □

3.15 Corollary. *Let G be a group acting on a simply connected scwol $\mathcal{X}$ with quotient $\mathcal{Y} = G\backslash\mathcal{X}$ and let $G(\mathcal{Y})$ be the complex of groups associated to this action (with respect to some choices). Let T be a maximal tree in the 1-skeleton of the geometric realization of $\mathcal{Y}$. Then G is isomorphic to $\pi_1(G(\mathcal{Y}), T)$ and $\mathcal{X}$ is equivariantly isomorphic to $D(\mathcal{Y}, \iota_T)$.*

3.16 Example. This corollary can be used to give a presentation of such a group G (compare with II.12.22). For instance, one sees that the subgroup of the group of orientation preserving isometries of $\mathbb{S}^2$, $\mathbb{E}^2$ or $\mathbb{H}^2$ leaving invariant a tesselation by regular n-gons with vertex angle $2\pi/3$ is generated by three elements r, s, t subject to the relations $s^n = r^3 = t^2 = rst = 1$ (cf. 3.11(3)).

4. Local Developments of a Complex of Groups

In this section we show that complexes of groups are always developable locally. (Thus non-developability is a global phenomenon.) We construct local developments at the vertices (4.10, 4.21) and edges (4.13) of the underlying scwol and show that in the developable case these are faithful models for the local structure of the development (4.11). We shall also explain how the different developments are related to each other (4.14 and 4.22).

For the duration of this section we fix a scwol $\mathcal{Y}$.

The Local Structure of the Geometric Realization

Recall that the geometric realization $|\mathcal{Y}|$ of a scwol $\mathcal{Y}$ is the complex obtained as the quotient of the disjoint union of simplices

$$\bigcup_{n\geq 0,\ A\in E^{(n)}(\mathcal{Y})} \Delta^n \times \{A\}$$

by the equivalence relation generated by the identifications

$$(d_i(x), A) \sim (x, \partial_i(A))$$

where $(x, A) \in \Delta^{n-1} \times E^{(n)}(\mathcal{Y})$, and the maps $d_i : \Delta^{n-1} \to \Delta^n$ and $\partial_i : E^{(n)}(\mathcal{Y}) \to E^{(n-1)}(\mathcal{Y})$ are defined as in (1.3).

4.1 Definition of $\mathrm{st}(\sigma)$**.** Given $\sigma \in V(\mathcal{Y})$, we define the open star $\mathrm{st}(\sigma)$ of σ in $|\mathcal{Y}|$ to be the union of the interiors of the simplices containing σ. (This is an open subset of $|\mathcal{Y}|$.)

Given $\sigma \in V(\mathcal{Y})$ and $p, q \geq 0$, let $E^{(p,q)}(\sigma) \subseteq E^{(p+q)}(\mathcal{Y})$ be the subset consisting of sequences of composable edges

$$(z_1, \dots, z_p, c_1, \dots, c_q)$$

such that $i(z_p) = \sigma$ if $p > 0$, $t(c_1) = \sigma$ if $q > 0$; if $p = q = 0$, then by definition $E^{(0,0)}(\sigma) = \{\sigma\}$. The elements of $E^{(p,q)}(\sigma)$ label the simplices of $|\mathcal{Y}|$ containing σ. For $0 \leq i < p$ (resp. $p < i \leq p+q$), the map $\partial_i : E^{(p+q)}(\sigma) \to E^{(p+q-1)}(\sigma)$ restricts to a map $\partial_i : E^{(p,q)}(\sigma) \to E^{(p-1,q)}(\sigma)$ (resp. $E^{(p,q)}(\sigma) \to E^{(p,q-1)}(\sigma)$).

4.2 Definition of $\mathrm{St}(\sigma)$**.** Given $\sigma \in V(\mathcal{Y})$ we define $\mathrm{St}(\sigma)$ to be the complex obtained as the quotient of the disjoint union

$$\bigcup_{p,q,A} \Delta^{p+q} \times \{A\},$$

with $p \geq 0, q \geq 0, A \in E^{(p,q)}(\sigma)$, by the equivalence relation generated by the identifications

$$(d_i(x), A) \sim (x, \partial_i(A)),$$

where $(x, A) \in \Delta^{p+q-1} \times E^{(p,q)}(\sigma)$ and $i \neq p$. The vertex $\Delta^0 \times \{\sigma\})$ will also be denoted σ. There is a natural affine map $\mathrm{St}(\sigma) \to |\mathcal{Y}|$ which induces a bijection from the set of simplices containing $\sigma \in \mathrm{St}(\sigma)$ to the set of simplices containing $\sigma \in |\mathcal{Y}|$ and which restricts to an affine isomorphism from the open star of $\sigma \in \mathrm{St}(\sigma)$ onto the open star of $\sigma \in |\mathcal{Y}|$. We identify these two open sets $\mathrm{st}(\sigma)$ by this isomorphism.

4.3 Remark. The projection $\mathrm{St}(\sigma) \to |\mathcal{Y}|$ is not injective if there are at least two distinct edges with initial vertex σ (resp. with terminal vertex σ) and the same terminal vertex (resp. the same initial vertex).

We leave the reader the exercise of proving that $\mathrm{St}(\sigma)$ is the geometric realization of the scwol $\mathcal{Y}(\sigma)$ defined in (1.17).

4.4 Definition of $\mathrm{st}(a)$**.** Given $a \in E(\mathcal{Y})$, we define the open star $\mathrm{st}(a)$ of a to be the union of the interiors of those simplices in $|\mathcal{Y}|$ that contain the 1-simplex labelled a (which we also denote a).

Let $E^{(p,k,q)}(a) \subseteq E^{(p+k+q)}(\mathcal{Y})$ be the subset consisting of sequences

$$(z_1, \ldots, z_p, a_1, \ldots, a_k, c_1, \ldots, c_q)$$

such that $a_1 \ldots a_k = a$. The elements of $E^{(p,k,q)}(a)$ label the simplices of $|\mathcal{Y}|$ containing a.

4.5 Definition of $\mathrm{St}(a)$**.** Given $a \in E(\mathcal{Y})$, to construct the complex $\mathrm{St}(a)$ we start with the disjoint union of Euclidean simplices

$$\bigcup_{p,k,q,A} \Delta^{p+k+q} \times \{A\},$$

where $p \geq 0, k \geq 1, q \geq 0$ and $A \in E^{(p,k,q)}(a)$, and form the quotient by the equivalence relation generated by the identifications $(d_i(x), A) \sim (x, \partial_i(A))$, where $(x, A) \in \Delta^{p+k+q-1} \times E^{(p,k,q)}(a)$ and $i \neq p, p+k$. (We write a to denote the 1-simplex of $\mathrm{St}(a)$ labelled $(a) \in E^{(0,1,0)}(a)$.)

There is a natural map $\mathrm{St}(a) \to |\mathcal{Y}|$ whose restriction to the open star in $\mathrm{St}(a)$ of the 1-simplex a is an affine isomorphism onto $\mathrm{st}(a)$. As above, we identify these two sets by means of this isomorphism.

4.6 Exercise. Show that for all $\sigma, \tau \in V(\mathcal{Y})$, the intersection $\mathrm{st}(\sigma) \cap \mathrm{st}(\tau)$ is the disjoint union of the connected open sets $\mathrm{st}(a)$, where a is any edge such that its initial vertex is σ and its terminal vertex is τ or vice versa.

Let a, a' be two distinct edges of $\mathcal{Y}$ such that $t(a) = t(a')$. Show that $\mathrm{st}(a) \cap \mathrm{st}(a') \neq \emptyset$ if and only if there exists $b \in E(\mathcal{Y})$ such that either $a = a'b$ or $a' = ab$. In the latter case $\mathrm{st}(a) \cap \mathrm{st}(a')$ is contained in the disjoint union $\coprod_b \mathrm{st}(b)$, where $b \in E(\mathcal{Y})$ is such that $a' = ab$.

Consider the case where $i(a) = i(a')$ and $\mathrm{st}(a) \cap \mathrm{st}(a') \neq \emptyset$.

The Geometric Realization of the Local Development

Let $G(\mathcal{Y}) = (G_\sigma, \psi_a, g_{a,b})$ be a complex of groups over $\mathcal{Y}$. For each $\sigma \in V(\mathcal{Y})$ we shall construct a complex $\mathrm{St}(\widetilde{\sigma})$ with an action of the group G_σ such that $\mathrm{St}(\sigma)$ is the quotient of $\mathrm{St}(\widetilde{\sigma})$ by the action of G_σ. (The construction is motivated by Proposition 4.11.) First we need another definition.

4.7 Definition of $E^{(p,q)}(\widetilde{\sigma})$. The elements of $E^{(p,q)}(\widetilde{\sigma})$ are sequences

$$(g\psi_c(G_{i(\sigma)}), z_1, \dots, z_p, c_1, \dots, c_q),$$

where $(z_1, \dots, z_p, c_1, \dots, c_q) = A \in E^{(p,q)}(\sigma)$, $c = c_1, \dots c_q$ and $g\psi_c(G_{i(c)}) \in G_\sigma/\psi_c(G_{i(c)})$. If $q = 0$, by convention $\psi_c(G_{i(c)}) = G_\sigma$, so $E^{(p,0)}(\widetilde{\sigma})$ is in bijection with $E^{(p,0)}(\sigma)$.

We define

$$\partial_i : E^{(p,q)}(\widetilde{\sigma}) \to E^{(p-1,q)}(\widetilde{\sigma}) \ \text{ for } 0 \le i < p \quad \text{and}$$
$$\partial_i : E^{(p,q)}(\widetilde{\sigma}) \to E^{(p,q-1)}(\widetilde{\sigma}) \ \text{ for } p < i \le p+q$$

by sending $(g\psi_c(G_{i(c)}), A)$ to $(g\psi_c(G_{i(c)}), \partial_i A)$ if $i < p+q$ or $i = p+q$ and $0 \le q \le 1$, and if $i = p+q$, $q \ge 2$, then by definition

$$\partial_{p+q}(g\psi_c(G_{i(c)}), z_1, \dots, z_p, c_1, \dots, c_q) = (gg^{-1}_{c',c_q}\psi_{c'}(G_{i(c')}), z_1, \dots, z_p, c_1, \dots, c_{q-1}),$$

where $c' = c_1 \dots c_{q-1}$.

The group G_σ acts naturally on $E^{(p,q)}(\widetilde{\sigma})$ by left translation on the first component and the maps ∂_i are G_σ-equivariant. The quotient by this action is $E^{(p,q)}(\sigma)$.

4.8 Lemma. *If i and j satisfy $0 \le i < j < p$ or $p < i < j \le p+q$, then*

$$\partial_i \partial_j = \partial_{j+1} \partial_i.$$

Proof. Only the case $j = p+q$ and $i = p+q-1$ is not obvious. In that case, one must apply condition 2.1(ii). □

4.9 The Construction of $\mathrm{St}(\widetilde{\sigma})$ and $\mathrm{st}(\widetilde{\sigma})$. We start from the disjoint union of Euclidean simplices

$$\bigcup_{p,q,\widetilde{A}} \Delta^{p+q} \times \{\widetilde{A}\},$$

where $p \ge 0$, $q \ge 0$ and $\widetilde{A} \in E^{(p,q)}(\widetilde{\sigma})$. On this we consider the equivalence relation generated by

$$(d_i(x), \widetilde{A}) \sim (x, \partial_i(\widetilde{A})),$$

where $(x, \widetilde{A}) \in \Delta^{p+q-1} \times E^{(p,q)}(\widetilde{\sigma})$ and $0 \le i < p$ or $p < i \le p+q$.

The group G_σ acts naturally on this disjoint union by the identity on the first factor and as in (4.7) on the second factor. This action is compatible with the equivalence relation.

In the light of Lemma 4.8, we know that the quotient by this equivalence relation is a complex (with an action of G_σ); it is denoted $\mathrm{St}(\widetilde{\sigma})$. The $(p+q)$-simplex $(\Delta^{p+q}, \widetilde{A})$ is mapped bijectively onto a $(p+q)$-cell of $\mathrm{St}(\widetilde{\sigma})$ which we regard as being labelled $\widetilde{A} \in E^{(p,q)}(\widetilde{\sigma})$. For $p = q = 0$, there is only one such 0-simplex and it will be denoted $\widetilde{\sigma}$. The set of simplices of $\mathrm{St}(\widetilde{\sigma})$ containing $\widetilde{\sigma}$ are in bijection with the set of their labels.

If $\mathcal{Y}$ is the scwol associated to a poset $\mathcal{Q}$ and if $G(\mathcal{Y})$ is a simple complex of groups, then $\mathrm{St}(\widetilde{\sigma})$ is isomorphic to the complex constructed in II.12.24.

4.10 Definition. The open star of $\widetilde{\sigma}$ in $\mathrm{St}(\widetilde{\sigma})$ is denoted $\mathrm{st}(\widetilde{\sigma})$; it is invariant by the action of G_σ. We call $\mathrm{st}(\widetilde{\sigma})$, equipped with this action of G_σ, the *geometric realization of the local development of $G(\mathcal{Y})$ at σ*.

The natural projection $E^{(p,q)}(\widetilde{\sigma}) \to E^{(p,q)}(\sigma)$ induces a projection $\mathrm{St}(\widetilde{\sigma}) \to \mathrm{St}(\sigma)$ and an isomorphism $G_\sigma \backslash \mathrm{St}(\widetilde{\sigma}) \to \mathrm{St}(\sigma)$. The restriction of the projection to $\mathrm{st}(\widetilde{\sigma})$ is denoted

$$p_\sigma : \mathrm{st}(\widetilde{\sigma}) \to \mathrm{st}(\sigma) \subseteq |\mathcal{Y}|.$$

4.11 Proposition. *If the complex of groups $G(\mathcal{Y})$ is associated to an action of a group G on a scwol $\mathcal{X}$ (with respect to some choices, as in 2.9(1)), then there is a canonical G_σ-equivariant isomorphism*

$$f_\sigma : \mathrm{st}(\widetilde{\sigma}) \to \mathrm{st}(\overline{\sigma}),$$

where $\overline{\sigma}$ is the vertex of $|\mathcal{X}|$ chosen as the representative of σ.

Proof. We use the notations established in (2.9). As the projection $p : \mathcal{X} \to \mathcal{Y}$ is non-degenerate, given a simplex of $|\mathcal{Y}|$ that contains σ and is labelled $A = (z_1, \dots, z_p, c_1, \dots, c_q) \in E^{p,q}(\sigma)$, there is a unique simplex of $|\mathcal{X}|$ labelled $\overline{A}$ projecting to A with initial vertex the chosen representative of the initial vertex $i(c_q)$ of A. The simplices of $|\mathcal{X}|$ containing $\overline{\sigma}$ and projecting by $|p|$ to the simplex labelled $A = (z_1, \dots, z_p, c_1, \dots, c_q) \in E^{(p,q)}(\sigma)$ are of the form $(gh_c).\overline{A}$, where $g \in G_\sigma$, $c = c_1 \dots c_q$ and $h_c \in G$ is the chosen element of G mapping $t(\overline{c})$ to $\overline{\sigma}$. They are in bijection with the elements of $G_\sigma/\psi_c(G_{i(c)})$. Therefore the map

$$(g\psi_c(G_{i(c)}), A) \mapsto gh_c.\overline{A}$$

gives a bijection denoted $\lambda^{p,q} : E^{(p,q)}(\widetilde{\sigma}) \to E^{(p,q)}(\overline{\sigma})$, which is G_σ-equivariant.

We claim that these maps $\lambda^{p,q}$ commute with the maps ∂_i. It will then follow that the maps $(x, \widetilde{A}) \mapsto (x, \lambda^{p,q}(\widetilde{A}))$ from $\Delta^{p+q} \times E^{(p,q)}(\widetilde{\sigma})$ to $\Delta^{p+q} \times E^{(p,q)}(\overline{\sigma})$ induce a G_σ-equivariant isomorphism from $\mathrm{St}(\widetilde{\sigma})$ to $\mathrm{St}(\overline{\sigma})$.

To see this, first observe that, for $A = (z_1, \dots, z_p, c_1, \dots, c_q) \in E^{(p,q)}(\sigma)$, we have $\overline{\partial_i(A)} = \partial_i(\overline{A})$ for $i \neq p, p+q$ and $\overline{\partial_{p+q}(A)} = h_{c_q}.\partial_{p+q}(\overline{A})$. We check commutativity only in the non-trivial case $i = p+q$ and $q \geq 2$. In this case, $\lambda^{p,q-1}$ maps $\partial_{p+q}(g\psi_c(G_{i(c)}), A) = (gg_{c',c_q}^{-1}\psi_{c'}(G_{i(c')}), \partial_{p+q}A)$ to

$$(gg_{c',c_q}^{-1}h_{c'}).\overline{\partial_{p+q}(A)} = (gg_{c',c_q}h_{c'}h_{c_q}).\partial_{p+q}(\overline{A}) = (gh_c).\partial_{p+q}(\overline{A}),$$

which is equal to $\partial_{p+q}(\lambda^{p,q}(g\psi_c(G_{i(c)})), A)$. (Here we appealed to (2.9(1)) for the last equality.) □

The construction of $\mathrm{St}(\widetilde{\sigma})$ is functorial with respect to morphisms of complexes of groups.

4.12 Proposition. *If $\phi = (\phi_\sigma, \phi(a)) : G(\mathcal{Y}) \to G(\mathcal{Z})$ is a morphism of complexes of groups over a non-degenerate morphism $f : \mathcal{Y} \to \mathcal{Z}$, then it induces, for each vertex $\sigma \in V(\mathcal{Y})$, ϕ_σ-equivariant maps $\mathrm{St}(\widetilde{\sigma}) \to \mathrm{St}(\widetilde{f(\sigma)})$ and $\mathrm{st}(\widetilde{\sigma}) \to \mathrm{st}(\widetilde{f(\sigma)})$.*

Proof. For $p \geq 0, q \geq 0$, let $\phi^{p,q}(\widetilde{\sigma}) : E^{(p,q)}(\widetilde{\sigma}) \to E^{(p,q)}(\widetilde{f(\sigma)})$ be the ϕ_σ-equivariant map

$$(g\psi_c(G_{i(c)}), A) \mapsto (\phi_\sigma(g)\phi(c)\psi_{f(c)}(G_{i(f(c))}), f(A)),$$

where $A = (z_1, \dots, z_p, c_1, \dots, c_q) \in E^{(p,q)}(\sigma)$, $c = c_1, \dots c_q$ and

$$f(A) = (f(z_1), \dots, f(z_p), f(c_1), \dots, f(c_q)).$$

We claim that the maps $\phi^{p,q}(\widetilde{\sigma})$ commute with the maps ∂_i defined in (4.7). Again we check only the non trivial case where $i = p+q$ and $q \geq 2$. We have

$$(\phi^{p,q-1}(\widetilde{\sigma}) \circ \partial_{p+q})((g\psi_c(G_{i(c)}), A)) = (\phi_\sigma(gg_{c',c_q}^{-1})\phi(c')\psi_{f(c')}(G_{i(f(c'))}),\ f(\partial_{p+q}A)\ ),$$

where $c' = c_1, \dots c_{q-1}$. Using condition (ii) of (2.4), we see that this is equal to

$$(\partial_{p+q} \circ \phi^{p,q})(\widetilde{\sigma})((g\psi_c(G_{i(c)}), A)) = (\phi_\sigma(g)\phi(c)g_{f(c'),f(c_q)}^{-1}\psi_{f(c')}(G_{i(f(c'))}), f(\partial_{p+q}A)\ ).$$

Therefore the maps

$$(\mathrm{id}, \phi^{p,q}(\widetilde{\sigma})) : \Delta^{p+q} \times E^{(p,q)}(\widetilde{\sigma}) \to \Delta^{p+q} \times E^{(p,q)}(\widetilde{f(\sigma)})$$

induce a ϕ_σ-equivariant map $\mathrm{St}(\widetilde{\sigma}) \to \mathrm{St}(\widetilde{f(\sigma)})$, and by restriction a map $\mathrm{st}(\widetilde{\sigma}) \to \mathrm{st}(\widetilde{f(\sigma)})$. □

4.13 The Construction of St($\widetilde{a}$) and st($\widetilde{a}$). Let $a \in E(\mathcal{Y})$ with $i(a) = \sigma$ and $t(a) = \tau$. Let $E^{(p,k,q)}(\widetilde{a}) \subset E^{(p+k,q)}(\widetilde{\sigma})$ be the set of sequences

$$\widetilde{A} = (g\psi_c(G_{i(\sigma)}), z_1, \ldots, z_p, a_1, \ldots, a_k, c_1, \ldots, c_q),$$

where $a = a_1 \ldots a_k$, $c = c_1, \ldots c_q$ and $g\psi_c(G_{i(c)}) \in G/\psi_c(G_{i(c)})$. We have maps

$$\partial_i : E^{(p,k,q)}(\widetilde{a}) \to E^{(p-1,k,q)}(\widetilde{a}) \;\; 0 \leq i < p$$
$$\partial_i : E^{(p,k,q)}(\widetilde{a}) \to E^{(p,k-1,q)}(\widetilde{a}) \;\; p < i < p+k$$
$$\partial_i : E^{(p,k,q)}(\widetilde{a}) \to E^{(p,k,q-1)}(\widetilde{a}) \;\; p+k < i \leq p+q+k$$

which are the restriction to $E^{(p,k,q)}(\widetilde{a})$ of the maps ∂_i defined above in (4.7).

To construct the complex St($\widetilde{a}$) we start from the disjoint union

$$\bigcup_{p,k,q,\widetilde{A}} \Delta^{p+k+q} \times \{\widetilde{A}\},$$

where $p \geq 0, k \geq 1, q \geq 0$ and $\widetilde{A} \in E^{(p,k,q)}(\widetilde{a})$, and then we form the quotient by the equivalence relation generated by $(d_i(x), \widetilde{A}) \sim (x, \partial_i(\widetilde{A}))$, where $(x, \widetilde{A}) \in \Delta^{p+k+q} \times E^{(p,k,q)}(\widetilde{a})$ and $0 \leq i < p+q+k,\; i \neq p, p+k$.

The group G_σ acts naturally on this disjoint union and the action is compatible with the equivalence relation.

The quotient of this equivalence relation, equipped with the induced action of G_σ, is denoted St($\widetilde{a}$). The $(p+k+q)$-simplex $(\Delta^{p+k+q}, \widetilde{A})$ is mapped injectively to the quotient; its image is a simplex which we label $\widetilde{A}$. For $p = q = 0$ and $k = 1$, there is only one such 1-simplex and it will be denoted $\widetilde{a}$.

Finally, we define st($\widetilde{a}$) to be the union of the interiors of those simplices of St($\widetilde{a}$) that contain $\widetilde{a}$. Note that the natural projection St($\widetilde{a}$) $\to$ St(a) induces an affine isomorphism $G_\sigma \backslash \mathrm{St}(\widetilde{a}) \to \mathrm{St}(a)$.

Note also that there is a natural G_σ-equivariant map St($\widetilde{a}$) $\to$ St($\widetilde{\sigma}$) induced by the inclusion

$$E^{(p,k,q)}(\widetilde{a}) \to E^{(p+k,q)}(\widetilde{\sigma}).$$

This map sends st($\widetilde{a}$) by an affine isomorphism onto the union of the interiors of the simplices that contain the 1-simplex labelled (G_σ, a). This simplex will be also denoted $\widetilde{a}$. We identify st($\widetilde{a}$) to its image in st($\widetilde{\sigma}$).

4.14 Proposition. *Let $a \in E(\mathcal{Y})$ with $i(a) = \sigma$ and $t(a) = \tau$. There is a ψ_a-equivariant map*

$$f_a : \mathrm{st}(\widetilde{a}) \to \mathrm{st}(\widetilde{\tau})$$

sending st($\widetilde{a}$) *isomorphically onto an open set, and the induced map that one gets* $\mathrm{st}(a) = G_\sigma \backslash \mathrm{st}(\widetilde{a}) \to \mathrm{st}(\tau) = G_\tau \backslash \mathrm{st}(\widetilde{\tau})$ *is the natural inclusion. Moreover, for all* $(a, b) \in E^{(2)}(\mathcal{Y})$, *we have*

$$f_a(f_b(x)) = g_{a,b} . f_{ab}(x)$$

for each $x \in \mathrm{st}(\widetilde{a}) \cap f_b^{-1}(\mathrm{st}(\widetilde{b}))$.

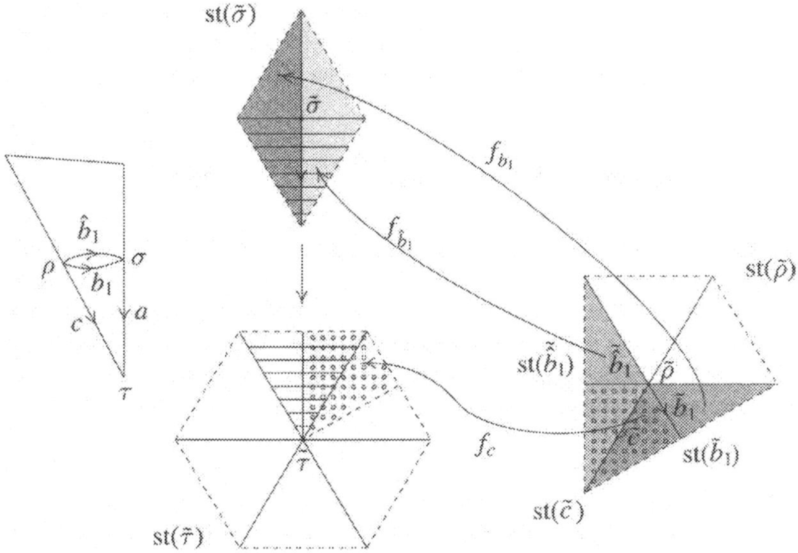

Fig. $\mathcal{C}$.**13** Local developments of the complex of groups $G(\mathcal{Y})$ of figure $\mathcal{C}$.12

Proof. There are ψ_a-equivariant maps

$$E^{(p,k,q)}(\widetilde{a}) \to E^{(p,k+q)}(\widetilde{\tau})$$

defined by

$$(g\psi_c(G_{i(c)}), A) \mapsto (gg_{a,c}\psi_{ac}(G_{i(ac)}), A),$$

where $A = (z_1, \ldots, z_p, a_1, \ldots, a_k, c_1, \ldots, c_q) \in E^{(p+k+q)}(\mathcal{Y})$ with $a = a_1 \ldots a_k$.

As above, one checks that these maps commute with the maps ∂_i and hence induce a ψ_a-equivariant map $\mathrm{St}(\widetilde{a}) \to \mathrm{St}(\widetilde{\tau})$. The restriction of this map

$$f_a : \mathrm{st}(\widetilde{a}) \to \mathrm{st}(\widetilde{\tau})$$

is an affine isomorphism onto the open star of the image of $\widetilde{a}$.

For $(a, b) \in E^{(2)}(\mathcal{Y})$, the open set $\mathrm{st}(\widetilde{a}) \cap f_b^{-1}(\mathrm{st}(\widetilde{b}))$ is the union of the interiors of those simplices of $\mathrm{St}(\widetilde{\sigma})$ labelled by

$$(g\psi_c(G_{i(c)}), z_1, \ldots, z_p, a_1, \ldots, a_k, b_1, \ldots, b_l, c_1, \ldots, c_q),$$

where $a = a_1, \ldots a_k$, $b = b_1 \ldots b_l$, and $c = c_1, \ldots c_q$.

The composition $f_a f_b$ maps the interior of the simplex labelled $(g\psi_c(G_{i(c)}), A)$ to the interior of the simplex labelled $(\psi_a(\psi_b(g))\psi_a(g_{b,c})g_{a,bc}\psi_{abc}(G_{i(c)}), A)$. The map f_{ab} maps the interior of this simplex to the interior of the simplex labelled $(\psi_{ab}(g)g_{ab,c}\psi_{abc}(G_{i(c)}), A)$. One checks that the conditions (i) and (ii) of (2.4) imply that

$$\psi_a(\psi_b(g))\psi_a(g_{b,c})g_{a,bc} = g_{a,b}\psi_{ab}(g_{ab,c}).$$

□

4.15 Remark. One can show that $\mathrm{St}(\widetilde{\sigma})$ and $\mathrm{St}(\widetilde{a})$ are the geometric realizations of scwols and that the maps defined above $\mathrm{St}(\widetilde{a}) \to \mathrm{St}(\widetilde{\sigma})$ and $\mathrm{St}(\widetilde{a}) \to \mathrm{St}(\widetilde{\tau})$ are geometric realizations of morphisms of scwols (see 4.21).

Local Development and Curvature

Assume that the geometric realization $|\mathcal{Y}|$ of the scwol $\mathcal{Y}$ is endowed with the structure of an M_κ-complex with a finite set of shapes in the sense of Chapter I.7 (cf. 1.3.3(2)). For each $\sigma \in V(\mathcal{Y})$, we can use the affine maps discussed above to endow $\mathrm{St}(\sigma)$ and $\mathrm{St}(\widetilde{\sigma})$ with an induced M_κ-complex structure such that $\mathrm{St}(\widetilde{\sigma}) \to \mathrm{St}(\sigma)$ restricted to each simplex is a local isometry. From this $\mathrm{st}(\widetilde{\sigma})$ inherits an induced length metric and the maps $f_a : \mathrm{st}(\widetilde{a}) \to \mathrm{st}(\widetilde{\sigma})$ constructed in (4.14) are local isometries.

4.16 Definition. Let $G(\mathcal{Y})$ be a complex of groups over a scwol $\mathcal{Y}$ such that $|\mathcal{Y}|$ is metrized as an M_κ-complex. We say that $G(\mathcal{Y})$ has curvature $\leq \kappa$ if $\mathrm{st}(\widetilde{\sigma})$, in the induced metric described above, has curvature $\leq \kappa$ for each $\sigma \in V(\mathcal{Y})$. If $\kappa \leq 0$ then we also say that $G(\mathcal{Y})$ is non-positively curved.

In the next chapter, we shall prove the following theorem which is a generalization of Gersten-Stallings theorem [St91] for triangles of groups. See also [Gro87], [Hae90, 91], [Spi92] and [Cors92], in the case $\dim\mathcal{Y} = 2$.

4.17 Theorem. *If the complex of groups $G(\mathcal{Y})$ is non-positively curved then it is developable.*

If $\mathcal{Y}$ is connected and $G(\mathcal{Y})$ is of curvature $\leq \kappa$, where $\kappa \leq 0$, then the geometric realization of the simply connected development of $\mathcal{Y}$ as constructed in Theorem 3.13 is an M_κ-polyhedral complex which is $\mathrm{CAT}(\kappa)$. It is equipped with an action of the fundamental group of $G(\mathcal{Y})$ by isometries, and $G(\mathcal{Y})$ is isomorphic to the complex of groups associated to this action.

4.18 Remark. Assume that $\mathcal{Y}$ is the scwol associated to an M_κ-polyhedral complex K. Then $G(\mathcal{Y})$ is of curvature $\leq \kappa$ in the sense of (4.16) if and only if, for each vertex τ of K, the geometric link of $\widetilde{\tau}$ in $\mathrm{st}(\widetilde{\tau})$, with the induced spherical structure, is CAT(1). To see this one simply applies the argument given in (II.5.2).

4.19 Examples

(1) In the notation of 1.4(3) and 3.11(4), if we assume that the two congruent n-gons in M_κ^2 have an angle $\geq \pi/n_k$ at the vertex τ_k, where n_k is the order of $s_k \in G_k$, then the complex of groups $G(\mathcal{Y})$ is of curvature $\leq \kappa$. If $\kappa \leq 0$, then $G(\mathcal{Y})$ is developable.

(2) The following is an alternative construction of an example due to Ballmann and Swiatkowski (see Theorem 2 and section 4 of [BaSw97]). We take m copies

$\mathcal{Y}_1, \dots \mathcal{Y}_m$ of the scwol $\mathcal{Y}$ described in figure $\mathcal{C}.6$ (see 1.14(2)), and glue them along the subscwol with edges a, a' and vertices σ, τ, τ'; the remaining vertices and edges in $\mathcal{Y}_k$ will be indexed by the subscript k (see figure $\mathcal{C}.14$). Let $\mathcal{Z}$ be the scwol obtained in this way; the set of vertices $V(\mathcal{Z})$ is $\sigma, \tau, \tau', \tau_k, \sigma'_k, \rho_k,\ k = 1, \dots, m$. We define a complex of groups $G(\mathcal{Z})$ over $\mathcal{Z}$ where the local groups associated to $\rho_k, \sigma_k, \sigma'_k$ are all trivial, $G_{\tau'} = \mathbb{Z}_2$ is generated by t, $G_{\tau_k} = \mathbb{Z}_3$ is generated by s_k and G_τ is a group H. We assume that the only non-trivial twisting elements are $g_{a,b_k} = s_k \in H,\ g_{a'_k,b'_k} = t$ and $g_{a_k,b'_k} = r_k$. We compute the fundamental group using 3.11(3); in $\mathcal{Y}_k$ we take the maximal tree considered in 3.11(3) and we take the union of these trees to be our maximal tree T in $\mathcal{Z}$. This yields the following presentation of $G := \pi_1(G(\mathcal{Z}), T)$: the generators are the elements of H and an extra element t, and these generators are subject to the relations in H and the extra relations $t^2 = 1,\ (s_k t)^3 = 1$ for $k = 1, \dots, m$.

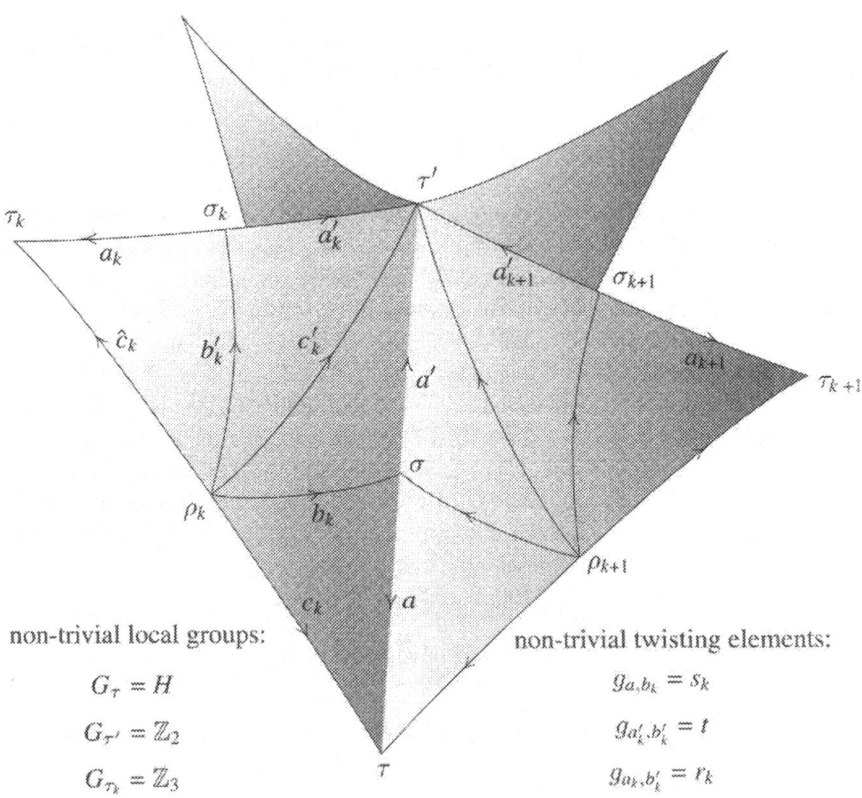

Fig. $\mathcal{C}$.14 Example 4.19(2)

We metrize $\mathcal{Z}$ as a piecewise Euclidean complex so that each $\mathcal{Y}_k$ is the union of two Euclidean triangles with angles $\pi/6$ at τ, $\pi/2$ at τ' and $\pi/3$ at τ_k. At the vertex

τ_k the local development is isometric to the interior of a hexagon in the Euclidean plane, and at the vertex τ' the local development is isometric to the interior of m Euclidean rectangles glued along a geodesic; in particular these local developments are of non-positive curvature. In the local development $\text{st}(\widetilde{\tau})$ at τ, the link of $\widetilde{\tau}$ is the following graph L: the vertices of L are the elements of H; the edges of L are labelled by the integers $1, \dots, m$ and each edge has length $\pi/3$; two vertices h, h' are joined by an edge labelled k if $h' = hs_k$.

The complex of groups $G(\mathcal{Z})$ has non-positive curvature if and only if the girth of L is at least 6. If this is the case, then $G(\mathcal{Y})$ is developable and the geometric realization of the simply connected development of $\mathcal{Y}$ as constructed in (3.13) is a piecewise Euclidean complex K whose 2-cells are equilateral Euclidean triangles; the link of each vertex will be isomorphic to L. The fundamental group G acts on K; the action is transitive on the set of vertices and the isotropy subgroup of a vertex is isomorphic to H. The isotropy subgroup of each 2-cell is the dihedral group with six elements. (The triangulation of K corresponding to the triangulation of $|\mathcal{Y}|$ is obtained by dividing each 2-cell of K into three congruent quadrilaterals and passing to the barycentric subdivision of this cell decomposition of K.)

The Local Development as a Scwol

Here we construct the local development purely in the framework of scwols. Specifically, given a complex of groups $G(\mathcal{Y}) = (G_\sigma, \psi_a, g_{a,b})$ over a scwol $\mathcal{Y}$, for each vertex $\sigma \in V(\mathcal{Y})$ we construct a scwol $\mathcal{Y}(\widetilde{\sigma})$ with an action of G_σ such that the geometric realization is equivariantly isomorphic to $\text{St}(\widetilde{\sigma})$.

4.20 The Construction of $\text{Lk}_{\widetilde{\sigma}}$. The upper link Lk_σ of $\sigma \in V(\mathcal{Y})$ was defined in (1.17). The natural morphism $\text{Lk}_\sigma \to \mathcal{Y}$ defined in (1.17) induces a complex of groups $G(\text{Lk}_\sigma)$ over Lk_σ from $G(\mathcal{Y})$ by the construction of (2.7). There is a natural morphism $\psi_\sigma : G(\text{Lk}_\sigma) \to G_\sigma$ mapping the local group $G_{i(a)}$ associated to $a \in V(\text{Lk}_\sigma) \subseteq E(\mathcal{Y})$ to the group G_σ by ψ_a and associating to each edge (a, b) of Lk_σ the element $g_{a,b} \in G_\sigma$. The development $D(\text{Lk}_\sigma, \psi_\sigma)$ of $G(\text{Lk}_\sigma)$ associated to this homomorphism shall be denoted $\text{Lk}_{\widetilde{\sigma}}$.

More explicitly, the scwol $\text{Lk}_{\widetilde{\sigma}}$ is defined as follows:

$$V(\text{Lk}_{\widetilde{\sigma}}) = \{(g\psi_a(G_{i(a)}), a) \mid a \in E(\mathcal{Y}),\ t(a) = \sigma;\ g\psi_a(G_{i(a)}) \in G_\sigma/\psi_a(G_{i(a)})\}$$

$$E(\text{Lk}_{\widetilde{\sigma}}) = \{(g\psi_{ab}(G_{i(b)}), a, b) \mid (a, b) \in E^{(2)}(\mathcal{Y}),\ t(a) = \sigma;\ g\psi_{ab}(G_{i(b)}) \in G_\sigma/\psi_{ab}(G_{i(b)})\}$$

The maps $i, t : E(\text{Lk}_{\widetilde{\sigma}}) \to V(\text{Lk}_{\widetilde{\sigma}})$ are defined by

$$i((g\psi_{ab}(G_{i(b)}), a, b)) = (g\psi_{ab}(G_{i(b)}), ab)$$
$$t((g\psi_a(G_{i(a)}), a, b)) = (gg_{a,b}^{-1}\psi_a(G_{i(a)}), a).$$

Composition is defined when $a' = ab$ and $g' = gg_{a',b'} \bmod \psi_{a'b'}(G_{i(b')})$ by the formula

$$(g\psi_{ab}(G_{i(b)}), a, b)(g'\psi_{a'b'}(G_{i(b')}), a', b') = (g'\psi_{abb'}(G_{i(b')}), a, bb').$$

The group G_σ acts naturally on $\mathrm{Lk}_{\tilde{\sigma}}$, namely $h \in G_\sigma$ acts on $(g\psi_{ab}(G_{i(b)}), a, b) \in E(\mathrm{Lk}_{\widetilde{\sigma}})$ by

$$h.(g\psi_{ab}(G_{i(b)}), a, b) = (hg\psi_{ab}(G_{i(b)}), a, b).$$

There is also a natural morphism $\mathrm{Lk}_{\tilde{\sigma}} \to \mathrm{Lk}_\sigma$ mapping $(g\psi_{ab}(G_{i(b)}), a, b)$ to (a, b); this induces an isomorphism $G_\sigma \backslash \mathrm{Lk}_{\tilde{\sigma}} \to \mathrm{Lk}_\sigma$.

4.21 Definition of the Local Development $\mathcal{Y}(\widetilde{\sigma})$. The local development of the complex of groups $G(\mathcal{Y})$ at the vertex $\sigma \in V(\mathcal{Y})$ is the scwol $\mathcal{Y}(\widetilde{\sigma}) := (\mathcal{Y}^\sigma * \mathrm{Lk}_{\tilde{\sigma}}) = (\mathrm{Lk}^\sigma * \sigma * \mathrm{Lk}_{\tilde{\sigma}})$ together with an action of G_σ on it: G_σ acts trivially on the first factor and as above on the second factor. We write $\mathcal{Y}_{\tilde{\sigma}}$ to denote the subscwol $(\sigma * \mathrm{Lk}_{\tilde{\sigma}}) \subseteq \widetilde{\mathcal{Y}}(\sigma)$.

We have a natural morphism

$$\mathcal{Y}(\widetilde{\sigma}) \to \mathcal{Y}(\sigma)$$

inducing an isomorphism $G_\sigma \backslash \mathcal{Y}(\widetilde{\sigma}) \to \mathcal{Y}(\sigma)$. This is simply the join of the identity on the first factor $\mathcal{Y}^\sigma$ and the morphism $\mathrm{Lk}_{\tilde{\sigma}} \to \mathrm{Lk}_\sigma$ on the second factor.

4.22 Proposition. $\mathrm{St}(\widetilde{\sigma})$ *is isomorphic to the geometric realization of* $\mathcal{Y}(\widetilde{\sigma})$.

Proof. We leave the proof as an exercise for the reader. We only point out the natural bijections

$$V(\mathrm{Lk}_{\tilde{\sigma}}) \leftrightarrow E^{(0,1)}(\widetilde{\sigma}), \quad \widetilde{\sigma} \leftrightarrow E^{(0,0)}(\widetilde{\sigma}), \quad V(\mathrm{Lk}^\sigma) \leftrightarrow E^{(1,0)}(\widetilde{\sigma}).$$

□

4.23 Proposition. *Let $\phi = (\phi_\sigma, \phi(a)) : G(\mathcal{Y}) \to G(\mathcal{Z})$ be a morphism of complexes of groups over a non-degenerate morphism of scwols $f : \mathcal{Y} \to \mathcal{Z}$, and let $\sigma \in V(\mathcal{Y})$. Then ϕ induces a ϕ_σ-equivariant morphism $\mathcal{Y}(\widetilde{\sigma}) \to \mathcal{Z}(\widetilde{f(\sigma)})$ of the local developments.*

Proof. The non-degenerate morphism $f : \mathcal{Y} \to \mathcal{Z}$ induces a morphism $\mathrm{Lk}^\sigma \to \mathrm{Lk}^{f(\sigma)}$ sending the vertex $a \in V(\mathrm{Lk}^\sigma)$ to the vertex $f(a) \in \mathrm{Lk}^{f(\sigma)}$ and the edge $(a, b) \in E(\mathrm{Lk}^\sigma)$ to the edge $(f(a), f(b)) \in \mathrm{Lk}^{f(\sigma)}$. The desired morphism $\mathcal{Y}(\widetilde{\sigma}) = (\mathrm{Lk}^\sigma * \sigma * \mathrm{Lk}_{\tilde{\sigma}}) \to \mathcal{Z}(\widetilde{f(\sigma)}) = (\mathrm{Lk}^\sigma * f(\sigma) * \mathrm{Lk}_{\widetilde{f(\sigma)}})$ is the join of the following morphisms of the three factors. On the first it is the morphism just described; on the second it maps σ to $f(\sigma)$; and the morphism on the third factor is described on the set of vertices (resp. edges) of $\mathrm{Lk}_{\tilde{\sigma}}$ by the following formulae (where $a \in E(\mathcal{Y})$ has $t(a) = \sigma$, $(a, b) \in E^{(2)}(\mathcal{Y})$ and $g \in G_\sigma$):

$$(g\psi_a(G_{i(a)}), a) \mapsto (\phi_\sigma(g)\phi(a)\psi_{f(a)}(G_{i(f(a))}), f(a))$$
$$(g\psi_{ab}(G_{i(ab)}), a, b) \mapsto (\phi_\sigma(g)\phi(ab)\psi_{f(ab)}(G_{i(f(a'b'))}), f(a), f(b)).$$

It is clear that this morphism is ϕ_σ-equivariant. (Compare with 2.18(1).) □

5. Coverings of Complexes of Groups

In this section we define coverings of complexes of groups and reformulate the definitions in several ways. We also define the fibres and monodromy of a covering. We show that a covering is determined up to isomorphism by its monodromy. The theory of coverings in the framework of small category theory is somewhat simpler and its relation with the theory of coverings of complexes of groups is indicated at the end of the appendix.

In the particular case of graphs of groups, our theory of coverings should be equivalent to the theory of Bass [Bass93], although it is not clear if his notion of morphism is the same as ours.

Definitions

5.1 Definition. Let $\phi : G'(\mathcal{Y}') \to G(\mathcal{Y})$ be a morphism of complexes of groups over a non-degenerate morphism of scwols $f : \mathcal{Y}' \to \mathcal{Y}$, where $\mathcal{Y}$ is connected. We say that ϕ is a *covering* if for each vertex $\sigma' \in V(\mathcal{Y}')$:

(i) the homomorphism $\phi_{\sigma'} : G_{\sigma'} \to G_{f(\sigma')}$ is injective, and

(ii) the $\phi_{\sigma'}$-equivariant map $\mathrm{st}(\widetilde{\sigma'}) \to \mathrm{st}(\widetilde{f(\sigma')})$ induced by ϕ (see 4.12) is a bijection.

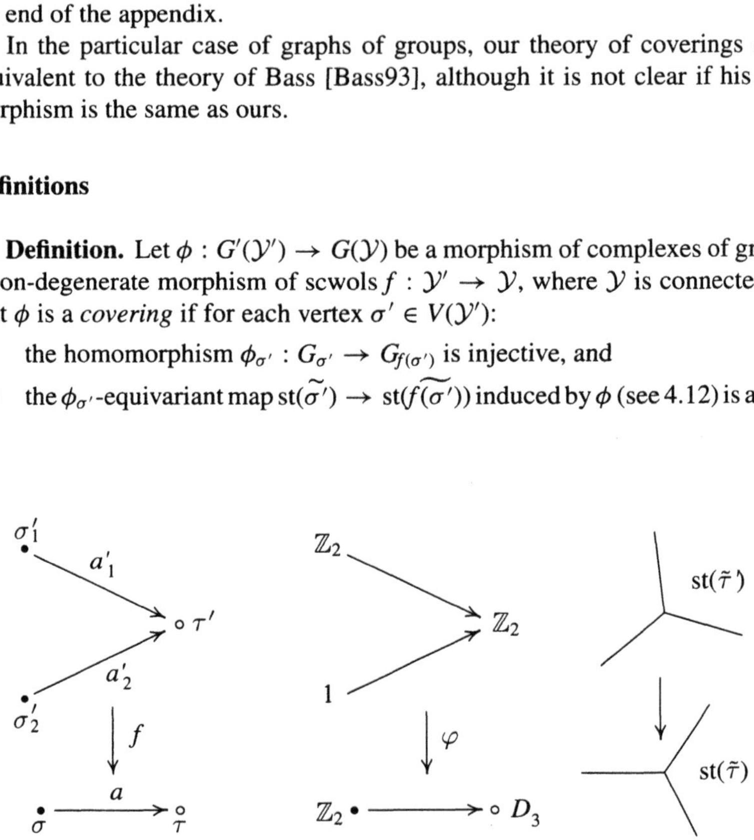

Fig. $\mathcal{C}$.15 A covering of a 1-dimensional complex of groups

5.2 Lemma. *Condition* (ii) *is equivalent to the condition:*

(ii)$'$ *For each $a \in E(\mathcal{Y})$ and $\sigma' \in V(\mathcal{Y}')$ with $t(a) = \sigma = f(\sigma')$, the map*

$$\coprod_{\substack{a' \in f^{-1}(a) \\ t(a')=\sigma'}} G_{\sigma'}/\psi_{a'}(G_{i(a')}) \to G_\sigma/\psi_a(G_{i(a)})$$

induced by $g \mapsto \phi_{\sigma'}(g)\phi(a')$ is bijective.

Proof. The proof is in terms of the maps $\phi^{p,q}(\widetilde{\sigma}')$ defined in the proof of (4.12). If (ii) is satisfied, then the map $\phi^{0,1}(\widetilde{\sigma}') : E^{(0,1)}(\widetilde{\sigma}') \to E^{(0,1)}(\widetilde{\sigma})$ is bijective. The bijectivity of this map is clearly equivalent to $\widetilde{\text{(ii)}}'$.

If $\phi^{0,1}(\widetilde{\sigma}')$ is bijective, then $\phi^{p,q}(\widetilde{\sigma}')$ is bijective for all p, q. This is obvious if $q = 0$ because f is assumed to be non-degenerate. Let us prove surjectivity for $q \geq 1$. Let $\widetilde{A} = (g\psi_c(G_{i(c)}), A)$, where $A = (z_1, \dots, z_p, c_1, \dots, c_q) \in E^{(p,q)}(\sigma)$ and $c = c_1 \dots c_q$. By hypothesis $\widetilde{\text{(ii)}}'$, there exists a unique element $(g'\psi_{c'}(G_{i(c')})), c') \in E^{(0,1)}(\widetilde{\sigma}')$ mapped by $\phi^{0,1}(\widetilde{\sigma}')$ to $(g\psi_c(G_{i(c)}), c)$. As f is non-degenerate, there is a unique $A' = (z'_1, \dots, z'_p, c'_1, \dots, c'_q) \in E^{(p,q)}(\sigma')$ with initial vertex $i(c') = i(c'_q)$ projecting by f to A. Therefore $(g'\psi_{c'}(G_{i(c')}), A')$ is mapped to $\widetilde{A}$. The proof of the injectivity is similar. □

5.3 Remark. Conditions (ii) or (ii)′ are equivalent also to the condition that the map $\mathrm{Lk}_{\widetilde{\sigma}'} \to \mathrm{Lk}_{\widetilde{\sigma}}$ induced by f, as defined in 4.23, is bijective. This in turn is equivalent to the condition that the map $\mathcal{Y}'(\widetilde{\sigma}') \to \mathcal{Y}(\widetilde{\sigma})$ is bijective.

5.4 Examples

(1) In the case where $G(\mathcal{Y}')$ and $G(\mathcal{Y})$ are trivial complexes of groups (i.e. all the local groups are trivial), definition (5.1) reduces to the definition of a covering of scwols (1.9). However if the local groups are not trivial, then the underlying morphism of scwols will not in general be a covering. (This is the case in figure $\mathcal{C}$.15.)

(2) In the notation of 2.9(1), consider an action of a group G on a scwol $\mathcal{X}$ and an associated complex of groups $G(\mathcal{Y})$ over the quotient $\mathcal{Y}$. We shall describe how to associate to this situation a covering $\lambda : \mathcal{X} \to G(\mathcal{Y})$ (where we consider $\mathcal{X}$ as the trivial complex of groups over $\mathcal{X}$) over the morphism $p : \mathcal{X} \to \mathcal{Y} = G\backslash\mathcal{X}$. For each $\widetilde{\sigma} \in V(\mathcal{X})$ we choose an element $g_{\widetilde{\sigma}} \in G$ such that $g_{\widetilde{\sigma}}.\overline{\sigma} = \widetilde{\sigma}$, where $\overline{\sigma}$ is the chosen representative in the orbit of $\widetilde{\sigma}$. We define λ to be the morphism that maps each edge $\widetilde{a} \in E(\mathcal{X})$ to the element

$$\lambda(\widetilde{a}) = g_{t(\widetilde{a})}^{-1} g_{i(\widetilde{a})} h_a^{-1} \in G_{p(t(\widetilde{a}))}.$$

It is straightforward to check that λ is a morphism, namely

$$\lambda(\widetilde{ab}) = \lambda(\widetilde{a})\psi_{p(\widetilde{a})}(\lambda(\widetilde{b}))g_{p(\widetilde{a}),p(\widetilde{b})}$$

for all $(\widetilde{a}, \widetilde{b}) \in E^{(2)}(\widetilde{\mathcal{X}})$. Moreover λ is a covering, because condition (ii)′ of 5.2 is satisfied. In this natural sense, the scwol $D(\mathcal{Y}, \iota_T)$ constructed in 3.13 can be considered as a simply connected covering of $G(\mathcal{Y})$.

We generalize the preceding construction. With the notation of 2.9(4), consider actions of groups G and G' on scwols $\mathcal{X}$ and $\mathcal{X}'$ respectively. Let $L : \mathcal{X} \to \mathcal{X}'$ be a morphism which is equivariant with respect to a homomorphism $\Lambda : G \to G'$. Let $\lambda : G(\mathcal{Y}) \to G'(\mathcal{Y}')$ be an associated morphism for the corresponding complex of groups as defined in 2.9(4). Assume that $\mathcal{X}$ is connected. Then λ is a covering if and only if both of the following conditions hold: L is a covering and the restriction of Λ to the isotropy subgroup of each vertex of $\mathcal{X}$ is injective.

The Fibres of a Covering

5.5 Definition. Let $\phi = (\phi_{\sigma'}, \phi(a')) : G(\mathcal{Y}') \to G(\mathcal{Y})$ be a morphism of complexes of groups over a non-degenerate morphism $f : \mathcal{Y}' \to \mathcal{Y}$. The *fibre* F_σ over a vertex $\sigma \in V(\mathcal{Y})$ is the set

$$F_\sigma = \coprod_{\sigma' \in f^{-1}(\sigma)} G_\sigma / \phi_{\sigma'}(G_{\sigma'})$$

with the natural action of G_σ by left multiplication.

For $a \in E(\mathcal{Y})$, let

$$F_a : F_{i(a)} \to F_{t(a)}$$

be the ψ_a-equivariant map which is the union of the maps defined by

$$F_a(g\phi_{i(a')}(G_{i(a')})) = \psi_a(g)\phi(a')^{-1}\phi_{t(a')}(G_{t(a')})),$$

where $a' \in f^{-1}(a)$ and $g \in G_{i(a)}$.

One calculates (using 2.1(i) and 2.4(ii)) that for composable edges $(a, b) \in E^{(2)}(\mathcal{Y})$ we have

$$F_a \circ F_b = g_{a,b}.F_{ab}.$$

5.6 Proposition. *Condition* (ii)$'$ *of 5.2 holds for every* $\sigma' \in f^{-1}(\sigma)$ *if and only if:*
(ii)$''$ $\forall a \in E(\mathcal{Y})$ *with* $t(a) = \sigma$, *the map* $F_a : F_{i(a)} \to F_{t(a)}$ *is a bijection.*

In the proof of this proposition we shall need the following general observation.

5.7 Lemma. *Let H and K be subgroups of a group G, and let $(g_j)_{j\in J}$ be a family of elements of G. We consider the action of H on G/K by left translations and similarly the action of K on G/H. For each $j \in J$, let H_j be a subgroup of H that fixes $g_jK \in G/K$. Note that the subgroup $K_j := g_j^{-1}Kg_j$ fixes $g_j^{-1}H \in G/H$. The following conditions are equivalent:*

(1) *The H-equivariant map*

$$\coprod_{j\in J} H/H_j \to G/K$$

given by $hH_j \mapsto hg_jK$ is a bijection.

(2)

$$H_j = H \cap g_jKg_j^{-1} \text{ and } G = \coprod_{j\in J} Hg_jK.$$

(This last condition means that $\{g_j\}_{j\in J}$ is a set of representatives for the double cosets $H\backslash G/K$, i.e. the set of subsets of G of the form HgK.)

(3) *The K-equivariant map*

$$\coprod_{j\in J} K/K_j \to G/H$$

given by $kK_j \mapsto kg_j^{-1}H$ is a bijection.

Proof. The injectivity of $H/H_j \to G/K$ is equivalent to the condition that H_j is the isotropy subgroup $H \cap g_j K g_j^{-1}$ of $g_j K$, and also to the condition that K_j is the isotropy subgroup of $g_j^{-1} H \in G/H$. The map associating to the H-orbit of $gK \in G/K$ the double coset HgK induces a bijection from the set of H-orbits in G/K to the set of orbits in G under the action $(h, k) : g \mapsto hgk^{-1}$ of $H \times K$ on G. Therefore $\{g_j K\}_{j \in J}$ is a set of representatives for the H-orbits in G/K, if and only if $\{g_j\}_{j \in J}$ is a set of representatives for the double cosets $H \backslash G/K$ in G. Interchanging the roles of H and K, we see that it is also equivalent to say that $(g_j^{-1} H)_{j \in J}$ is a set of representatives for the K-orbits in G/H. Thus (1), (2) and (3) are equivalent. □

Proof of the proposition 5.6. Condition (ii)″ is equivalent to the statement that for every $\sigma' \in f^{-1}(\sigma)$ and every $a \in E(\mathcal{Y})$ with $t(a) = \sigma$, the map

$$\coprod_{\substack{a' \in f^{-1}(a) \\ t(a') = \sigma'}} G_{i(a)}/\phi_{i(a')}(G_{i(a')}) \to G_\sigma/\phi_{\sigma'}(G_{\sigma'}),$$

which is the restriction of F_a, is bijective.

We apply the lemma with the following choices:

$$G = G_\sigma, \quad K = \phi_{\sigma'}(G_{\sigma'}), \quad H = \psi_a(G_{i(a)})$$

$$J = \{a' \in f^{-1}(a) \mid t(a') = \sigma'\},$$

$$g_{a'} = \phi(a')^{-1} \ \forall a' \in J \text{ and } H_{a'} = \psi_a \phi_{i(a')}(G_{i(a')}).$$

Condition (ii)″ corresponds to condition (1) of the lemma, while condition (ii)′ corresponds to condition (3). □

Our next goal is to establish a converse to 5.6: given an appropriate system of bijections F_a one can reconstruct an associated covering.

5.8 Proposition. *Let $G(\mathcal{Y})$ be a complex of groups over a connected scwol $\mathcal{Y}$. Suppose that for each $\sigma \in V(\mathcal{Y})$ a set F_σ is given with an action of G_σ, and for each edge $a \in E(\mathcal{Y})$ a ψ_a-equivariant bijection $F_a : F_{i(a)} \to F_{t(a)}$ is given such that $F_a \circ F_b = g_{a,b}.F_{ab}$ for all $(a, b) \in E^{(2)}(\mathcal{Y})$. Then one can construct a covering $\phi : G(\mathcal{Y}') \to G(\mathcal{Y})$ over a non-degenerate morphism $f : \mathcal{Y}' \to \mathcal{Y}$ such that the F_σ are naturally the fibres of ϕ. Moreover $G(\mathcal{Y}')$ is unique up to isomorphism and ϕ is unique up to homotopy.*

Proof. For each σ the subset $f^{-1}(\sigma) \subset V(\mathcal{Y}')$ will be a set of representatives for the G_σ-orbits in F_σ. For each $\sigma' \in f^{-1}(\sigma)$ there will be an edge $a' = (a, \sigma')$ in $\mathcal{Y}'$ that projects by f to a; the initial vertex $i(a')$ is σ' and the terminal vertex $t(a')$ will be the chosen representative in the $G_{t(a)}$-orbit of $F_a(\sigma')$. When the composition of two edges $(a, \sigma')(b, \rho')$ is defined, it is equal to (ab, ρ'). This information specifies the scwol $\mathcal{Y}'$ and the morphism $f : \mathcal{Y}' \to \mathcal{Y}$.

Define $G_{\sigma'}$ to be the subgroup of G_σ fixing $\sigma' \in F_\sigma$ and let $\phi_{\sigma'} : G_{\sigma'} \to G_\sigma$ be the natural inclusion. For each $a' \in V(\mathcal{Y}')$ with $f(a') = a$, choose an element $\phi(a') \in G_{t(a)}$

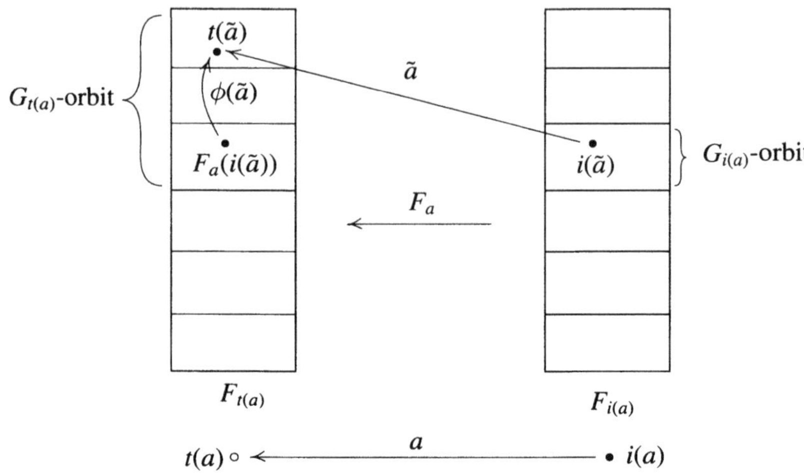

Fig. $\mathcal{C}$**.16** Construction of a covering from the monodromy

such that $\phi(a').F_a(i(a')) = t(a')$ and define a homomorphism $\psi_{a'} : G_{i(a')} \to G_{t(a')}$ by the formula

$$\psi_{a'}(g) = \phi(a')\psi_a(g)\phi(a')^{-1}.$$

For composable edges $(a', b') \in E^{(2)}(\mathcal{Y}')$, define

$$g_{a',b'} = \phi(a')\psi_a(\phi(b'))g_{a,b}\phi(a'b')^{-1} \in G_{t(a')}.$$

It is straightforward to check that the above data define a complex of groups $G(\mathcal{Y}')$ over $\mathcal{Y}'$ and a morphism $\phi : G(\mathcal{Y}') \to G(\mathcal{Y})$ over $f : \mathcal{Y}' \to \mathcal{Y}$ defined by the $\phi_{\sigma'}$ and the elements $\phi(a')$. The fibre of ϕ over σ is the disjoint union of the $G_\sigma/G_{\sigma'}$ for $\sigma' \in f^{-1}(\sigma)$; it can be identified to F_σ by the map sending the coset $gG_{\sigma'}$ to the point $g.\sigma'$. Under this identification, the map from the fibre over $i(a)$ to the fibre over $t(a)$, associated to a as in (5.5), is precisely F_a. Moreover, as these maps are bijections, (5.6) shows that the morphism ϕ is a covering.

It is clear that any other choice for the representatives σ' of the orbits, or for the elements $\phi(a')$, would lead to a homotopic morphism. □

5.9 Galois Covering. An important special case of the preceding construction arises naturally when one has a morphism $\Phi = (\Phi_\sigma, \Phi(a))$ from a complex of groups $G(\mathcal{Y})$ to a group G.

For each $\sigma \in V(\mathcal{Y})$ we define $F_\sigma = G$ with the action of G_σ given by right translations via Φ_σ (i.e. for $h \in G_\sigma,\ g \in G$, define $h.g = g\Phi_\sigma(h^{-1})$). If $F_a : G \to G$ is defined to be the right translation by $\Phi(a)^{-1}$, then F_a is ψ_a-equivariant and $(F_a \circ F_b)(g) = F_{ab}(g)\Phi_{t(a)}(g_{a,b}^{-1}) = g_{a,b}.F_{ab}(g)$. Following the construction of (5.8), we give an explicit description of the covering $\phi : G(\mathcal{Y}') \to G(\mathcal{Y})$ over the morphism $f : \mathcal{Y}' \to \mathcal{Y}$ in this special case.

$\mathcal{Y}'$ will be the development $D(\mathcal{Y}, \Phi)$ as constructed in (2.13) and $f : D(\mathcal{Y}, \Phi) \to \mathcal{Y}$ will be the natural projection. Thus $f^{-1}(\sigma)$ is the set of pairs $((g\Phi_\sigma(G_\sigma), \sigma) \in G/\Phi_\sigma(G_\sigma) \times \{\sigma\}$ and $f^{-1}(a)$, $a \in E(\mathcal{Y})$, is the set of pairs $a' = (g\Phi_{i(a)}((G_{i(a)}), a)$ with $i(a') = (g\Phi_{i(a)}(G_{i(a)}), i(a))$ and $t(a') = (g\Phi(a)\Phi_{t(a)}(G_{t(a)}), t(a))$. The group G acts on $\mathcal{Y}' = D(\mathcal{Y}, \Phi)$ and f induces an isomorphism $G\backslash\mathcal{Y}' \to \mathcal{Y}$.

For each $\sigma' \in f^{-1}(\sigma) \subseteq V(\mathcal{Y}')$, the group $G_{\sigma'}$ is the kernel of Φ_σ and $\phi_{\sigma'} : G_{\sigma'} \to G_\sigma$ is the inclusion. In order to define $\psi_{a'}$ and $g_{a',b'}$ we need to choose for each $\Phi_\sigma(G_\sigma)$-coset σ' a representative in G; we again denote this σ', thus identifying σ' to the coset $\sigma'\Phi_\sigma(G_\sigma)$. We also choose for each edge $a' \in f^{-1}(a) \subseteq E(\mathcal{Y}')$ an element $\phi(a') \in G_{t(a)}$ such that

$$i(a')\Phi(a)^{-1}\Phi_{t(a)}(\phi(a')^{-1}) = t(a').$$

We then define $\psi_{a'} : G_{i(a')} \to G_{t(a')}$ by $\psi_{a'} = \mathrm{Ad}\phi(a') \circ \psi_a$.

For composable edges $(a', b') \in E^{(2)}(\mathcal{Y}')$ with $a = f(a')$, $b = f(b')$, we define

$$g_{a',b'} = \phi(a')\psi_a(\phi(b'))g_{a,b}\phi(a'b')^{-1} \in \ker \Phi_{t(a)}.$$

The complex of groups $G(\mathcal{Y}')$ is defined to be $(G_{\sigma'}, \psi_{a'}, g_{a',b'})$ and the morphism $\phi : G(\mathcal{Y}') \to G(\mathcal{Y})$ is given by the homomorphisms $\phi_{\sigma'}$ and the elements $\phi(a')$.

5.10 Examples

(1) Even if $G(\mathcal{Y})$ is a simple complex of groups, in general a Galois covering $G(\mathcal{Y}')$ will not be isomorphic to a simple complex of groups, as the following example illustrates.

Let $\mathcal{Y}$ be the scwol associated to an n-gon P. The set of vertices of $G(\mathcal{Y})$ has n vertices τ_k corresponding to the vertices of P, has n vertices σ_k corresponding to the barycentres of the sides of P, where $k = 1, \ldots, n$, and has a vertex ρ corresponding to the barycentre of P. The set of edges has $2n$ elements a_k, a'_k with $i(a_k) = i(a'_k) = \sigma_k$, $t(a_k) = \tau_k$, $t(a'_k) = \tau_{k-1}$ (indices mod n), and a further $2n$ elements b_k, c_k with $i(b_k) = i(c_k) = \rho$, $t(b_k) = \sigma_k$, $t(c_k) = \tau_k$. Moreover $a_k b_k = c_k = a'_{k+1}b_{k+1}$. We consider on $\mathcal{Y}$ the complex of groups where G_ρ is trivial, G_{σ_k} is cyclic of order two generated by s_k, and G_{τ_k} is the dihedral group of order $2n_k$ generated by the images of s_k and s_{k+1}.

Let $\Phi : G(\mathcal{Y}) \to \mathbb{Z}_2$ be the morphism associating to each element s_k the generator t of $\mathbb{Z}_2$ and to each edge the trivial element. We construct the associated Galois covering $\phi : G(\mathcal{Y}') \to G(\mathcal{Y})$. The scwol $\mathcal{Y}'$ is associated to the complex obtained by gluing two copies of P along their boundary — in the notation of figure $\mathcal{C}$.4, the vertex ρ corresponds to the coset $\{t\}$ and ρ' corresponds to the coset $\{1\}$. The local groups associated to the vertices τ_k are the cyclic groups of order n_k generated by $s_k s_{k+1}$; the other local groups are trivial. To determine the twisting elements we choose coset representatives, these are trivial except for ρ, where we have the coset $\{t\}$. The elements $\phi(a')$ associated to the edges a' of $\mathcal{Y}'$ are trivial except $\phi(b_k) = s_k$ and $\phi(c_k) = s_{k+1}$. With these choices, all the twisting elements are trivial except $g_{a_k,b_k} = s_k s_{k+1}$ for $k = 1, \ldots, n$.

(2) We reconsider the triangle of groups $G_k(\mathcal{Y})$ studied in II.12.34(5) and maintain the notation established there. Thus $\mathcal{Y}$ is the scwol associated to a Euclidean triangle Δ with angles $(k-2)\pi/(2k, \pi/2, \pi/k$ at the vertices τ_0, $\tau_1 \tau_2$, respectively. Consider the dihedral group D_3 generated by the elements t_1 and t_2 of order 2. For k odd, there is a simple morphism $\Phi : G(\mathcal{Y}) \to D_3$ sending s_2 to t_2, both s_1 and s_3 to t_1, and s_0 to 1.

The corresponding Galois covering gives a simple complex of groups $G(\mathcal{Y}')$, with respect to suitable choices (cf. [Hag93]), where $\mathcal{Y}$ is the scwol associated to the polyhedral complex K obtained by gluing three isometric copies of Δ along the edge $[\tau_0, \tau_1]$. Let τ_2^i, $i = 1, 2, 3$, denote the remaining three vertices of K. The local group at τ_2^i is the dihedral group D_k generated by two involutions, t and x_i; the local group at τ_0 is the cyclic of order two generated by t; and the local group at τ_0 is isomorphic to $\mathbb{Z}_2 \times \mathbb{Z}_2$, the three non trivial elements being x_1, x_2, x_3 (the elements labelled by the same letter are identified by the edge monomorphisms). The groups associated to the barycentres of the 2-simplices and to the edge $[\tau_0, \tau_1]$ are all trivial, and the groups associated to the edges other than $[\tau_0, \tau_1]$ are cyclic of order two. We leave the reader to check that $G_k(\mathcal{Y}')$ is simply connected if $k = 3$.

The Monodromy

5.11 Construction of the Monodromy. Let $G(\mathcal{Y})$ be a complex of groups over a connected scwol $\mathcal{Y}$ and let the system of maps (F_σ, F_a) be as in (5.8). For each $g \in G_\sigma$, let $F_g : F_\sigma \to F_\sigma$ be the bijection $x \mapsto g.x$ given by the action of g on F_σ. For each oriented edge $e \in E(X)^\pm$ define $F_e : F_{t(e)} \to F_{i(e)}$ to be F_a if $e = a^+$ and F_a^{-1} if $e = a^-$. Given a $G(\mathcal{Y})$-path $c = (g_0, e_1, g_1, \ldots, e_k, g_k)$ joining σ to τ, let $F_c = F_{g_0} F_{e_1} F_{g_1} \ldots F_{e_k} F_{g_k} : F_\tau \to F_\sigma$. Note that if c and c' are homotopic, then $F_c = F_{c'}$. Also $F_{c^{-1}} = F_c^{-1}$, and if the concatenation $c * c'$ of two $G(\mathcal{Y})$-paths c and c' is defined, then $F_{c*c'} = F_c \circ F_{c'}$.

Fix a vertex σ_0. The *monodromy* associated to (F_σ, F_a) is the action of $\pi_1(G(\mathcal{Y}), \sigma_0)$ on F_{σ_0} where the homotopy class of a $G(\mathcal{Y})$-loop c at σ_0 acts as the bijection $F_c : F_{\sigma_0} \to F_{\sigma_0}$. When (F_σ, F_a) is the fibre system of a covering $\phi : G(\mathcal{Y}') \to G(\mathcal{Y})$, this action is called the *monodromy of the covering*.

We end this section by stating the classification of coverings in terms of monodromy.

5.12 Theorem. *Let $G(\mathcal{Y})$ be a complex of groups over a connected scwol $\mathcal{Y}$. Fix a vertex $\sigma_0 \in V(\mathcal{Y})$. Suppose that we are given an action of $\pi_1(G(Y), \sigma_0)$ on a set F_0. Then there is a covering $\phi : G(\mathcal{Y}') \to G(\mathcal{Y})$ whose fibre over σ_0 is F_0 and whose monodromy is the given action of $\pi_1(G(\mathcal{Y}), \sigma_0)$. This covering is connected if and only if the monodromy action is transitive.*

Let $\phi' : G(\mathcal{Y}') \to G(\mathcal{Y})$ over $f' : \mathcal{Y}' \to \mathcal{Y}$, and $\phi'' : G(\mathcal{Y}'') \to G(\mathcal{Y})$ over $f'' : \mathcal{Y}'' \to \mathcal{Y}$, be two coverings such that there is an equivariant bijection f_0 from the fibre F'_{σ_0} of ϕ over σ_0 to the fibre F''_{σ_0} of ϕ'' over σ_0 (equivariance is with respect *to the monodromy actions). Then there is an isomorphism* $\phi : G(\mathcal{Y}') \to G(\mathcal{Y}'')$ *over* $f : \mathcal{Y}' \to \mathcal{Y}''$ *such that* $\phi'' \circ \phi = \phi'$ *and* ϕ *induces the map* f_0 *on the fibres over* σ_0.

Proof. The proof of this result follows from the corresponding result proved in the more general framework of coverings of small categories in the Appendix. (The relationship between the general framework and coverings of complexes of groups is explained in A.24.) □

Appendix: Fundamental Groups and Coverings of Small Categories

As we indicated in the introduction to this chapter, the notational complexity that necessarily accompanies definitions concerning complexes of groups (morphisms, homotopy, fundamental groups, etc.) can obscure what are actually very natural and general constructions. In this appendix we strip away the specific notation associated to complexes of groups and explain these definitions in the general context of small categories. Thus the notions of morphisms of complexes of groups and homotopies between them are shown to correspond to functors and homotopies for the associated categories.

We give the definition of the fundamental group of a small category and give a presentation of this group using a maximal tree in the graph naturally associated to the category. The fundamental group of a complex of groups is the fundamental group of the associated category.

The theory of coverings for small categories is parallel to the theory of coverings for topological spaces. If $\phi : G(\mathcal{Y}') \to G(\mathcal{Y})$ is a morphism of complexes of groups which is a covering as defined in 5.1, then the corresponding functor $CG(\mathcal{Y}') \to CG(\mathcal{Y})$ for the associated categories is not a covering in general, but it is the composition of an equivalence $CG(\mathcal{Y}') \to \mathcal{C}'$ and a covering $\mathcal{C}' \to CG(\mathcal{Y})$. (The fact that the equivalence depends on some choices explains why the theory of coverings for complexes of groups is somewhat more complicated than the theory of coverings of categories.)

Basic Definitions

A.1 Category. A *(small) category*[71] $\mathcal{C}$ *consists of a set* $\mathcal{C}$, *an auxiliary set* $\mathcal{O}(\mathcal{C})$, *called the set of objects of* $\mathcal{C}$, *and two maps* $i : \mathcal{C} \to \mathcal{O}(\mathcal{C})$ *and* $t : \mathcal{C} \to \mathcal{O}(\mathcal{C})$. *The maps* i *and* t *associate to each element*[72] $\gamma \in \mathcal{C}$ *its initial object* $i(\gamma)$ *and its terminal object* $t(\gamma)$. *Given* $\sigma, \tau \in \mathcal{O}(\mathcal{C})$, *we let* $\mathcal{C}_{\tau,\sigma}$ *denote the set of* $\gamma \in \mathcal{C}$

[71] In this chapter we deal only with small categories, so for us "category" will always mean a small category.

[72] In the literature, elements of $\mathcal{C}$ are often called *morphisms*, but we shall not use this terminology. We interpret the elements of $\mathcal{C}$ as arrows γ with source $i(\gamma)$ and target $t(\gamma)$.

such that $i(\gamma) = \sigma$, $t(\gamma) = \tau$. *A law of composition (partially defined) is given in* $\mathcal{C}$*: two elements* $\gamma, \gamma' \in \mathcal{C}$ *can be composed if and only if* $i(\gamma) = t(\gamma')$*; their composition* $\gamma\gamma'$ *belongs to* $\mathcal{C}_{t(\gamma),i(\gamma')}$. *To each object* $\sigma \in \mathcal{O}(\mathcal{C})$ *is associated an element* $1_\sigma \in \mathcal{C}_{\sigma,\sigma}$ *which is a unit: if* $\gamma \in \mathcal{C}_{\tau,\sigma}$, *then* $\gamma 1_\sigma = \gamma = 1_\tau\gamma$. *The law of composition is associative: if* $\gamma\gamma'$ *and* $\gamma'\gamma''$ *are defined, then* $(\gamma\gamma')\gamma'' = \gamma(\gamma'\gamma'')$, *and this composition is denoted* $\gamma\gamma'\gamma''$.

We shall often identify $\mathcal{O}(\mathcal{C})$ to a subset of $\mathcal{C}$ by the inclusion $\sigma \mapsto 1_\sigma$. An element $\gamma \in \mathcal{C}_{\tau,\sigma}$ is called *invertible* if there is an element $\gamma^{-1} \in \mathcal{C}_{\sigma,\tau}$ such that $\gamma\gamma^{-1} = 1_\tau$ and $\gamma^{-1}\gamma = 1_\sigma$.

A *subcategory* $\mathcal{C}'$ of a category $\mathcal{C}$ consists of a subset $\mathcal{C}' \subseteq \mathcal{C}$ and a subset $\mathcal{O}(\mathcal{C}') \subseteq \mathcal{O}(\mathcal{C})$ such that if $\gamma \in \mathcal{C}'$ then $i(\gamma), t(\gamma) \in \mathcal{O}(\mathcal{C}')$; if $\gamma, \gamma' \in \mathcal{C}'$ are composable in $\mathcal{C}$ then $\gamma\gamma' \in \mathcal{C}'$; and if $\sigma \in \mathcal{O}(\mathcal{C}')$, then $1_\sigma \in \mathcal{C}'$.

On $\mathcal{O}(\mathcal{C})$ there is an equivalence relation generated by $[\sigma \sim \tau$ if $\mathcal{C}_{\sigma,\tau} \neq \emptyset]$. The category $\mathcal{C}$ is *connected* if there is only one equivalence class in $\mathcal{O}(\mathcal{C})$. In general, given an equivalence class, the union of the $\mathcal{C}_{\sigma,\tau}$ with σ and τ in that class is a subcategory of $\mathcal{C}$; it is called a *connected component*. $\mathcal{C}$ is the disjoint union of its connected components.

We write $\mathcal{C}^{(2)}$ to denote the set of pairs of composable elements, i.e. pairs $(\gamma_1, \gamma_2) \in \mathcal{C} \times \mathcal{C}$ with $i(\gamma_1) = t(\gamma_2)$. More generally, we write $\mathcal{C}^{(k)}$ to denote the set of k-tuples $(\gamma_1, \dots, \gamma_k)$ for which the composition $\gamma_1 \dots \gamma_k$ is defined. In particular $\mathcal{C}^{(1)} = \mathcal{C}$. By convention, $\mathcal{C}^{(0)}$ is the set of units of $\mathcal{C}$, often identified to $\mathcal{O}(\mathcal{C})$.

A.2 Examples

(0) A group can be considered as a category with one object where all elements of the category are invertible.

(1) Let G be a group acting on a set X. Let $\mathcal{C} = G \times X$, $\mathcal{O}(\mathcal{C}) = X$ and let $i, t : \mathcal{C} \to \mathcal{O}(\mathcal{C})$ be the maps defined for $\gamma = (g, x)$ by $i(\gamma) = x$, $t(\gamma) = g.x$. The composition $(g, x)(g', x')$ is defined iff $x = g'.x'$ and is equal to (gg', x'). For each object $x \in X$, we define $1_x = (1, x)$. These data define a category denoted $G \ltimes X$. The connected components of $G \ltimes X$ are the subcategories associated with the restriction of the action of G to the orbits.

There are two extreme cases. If X is a single point, then $G \ltimes X$ can be identified to the group G. If G is the trivial group, then $G \ltimes X$ is reduced to its set of units X.

(2) *Product of two categories.* Given two categories $\mathcal{C}$ and $\mathcal{C}'$ we can consider their product $\mathcal{C} \times \mathcal{C}'$ as a category: the set of objects is $\mathcal{O}(\mathcal{C}) \times \mathcal{O}(\mathcal{C}')$; the maps $i, t : \mathcal{C} \times \mathcal{C}' \to \mathcal{O}(\mathcal{C}) \times \mathcal{O}(\mathcal{C}')$ send (γ, γ') to $(i(\gamma), i(\gamma'))$ and $(t(\gamma), t(\gamma'))$ respectively; and the composition $(\gamma_1, \gamma_1')(\gamma_2, \gamma_2')$, whenever defined, is equal to $(\gamma_1\gamma_2, \gamma_1'\gamma_2')$.

(3) A small category $\mathcal{C}$ is a scwol (i.e a small category without loops) if and only if every invertible element of $\mathcal{C}$ is a unit and every element of $\mathcal{C}$ with the same initial and terminal vertex is a unit.

A.3 Functors and Equivalence. A *functor* ϕ from a category $\mathcal{C}$ to a category $\mathcal{C}'$ is a map $\phi : \mathcal{C} \to \mathcal{C}'$ together with a map $\mathcal{O}(\mathcal{C}) \to \mathcal{O}(\mathcal{C}')$ (also denoted ϕ) such that

(1) for $\gamma \in \mathcal{C}$, $\phi(i(\gamma)) = i(\phi(\gamma))$ and $\phi(t(\gamma)) = t(\phi(\gamma))$,

(2) for composable $\gamma, \gamma' \in \mathcal{C}$, $\phi(\gamma\gamma') = \phi(\gamma)\phi(\gamma')$,

(3) for $\sigma \in \mathcal{O}(\mathcal{C})$, $\phi(1_\sigma) = 1_{\phi(\sigma)}$.

A functor $\phi : \mathcal{C} \to \mathcal{C}'$ is said to be an *isomorphism* if there exists a functor $\phi^{-1} : \mathcal{C}' \to \mathcal{C}$ such that $\phi^{-1}\phi$ is the identity of $\mathcal{C}$ and $\phi\phi^{-1}$ is the identity of $\mathcal{C}'$.

Two functors $\phi_0, \phi_1 : \mathcal{C} \to \mathcal{C}'$ are *homotopic* if for each object $\sigma \in \mathcal{C}$ there is an invertible element $\kappa_\sigma \in \mathcal{C}'_{\phi_1(\sigma),\phi_0(\sigma)}$ such that for each $\gamma \in \mathcal{C}$ we have $\phi_1(\gamma) = \kappa_{t(\gamma)}\phi_0(\gamma)\kappa^{-1}_{i(\gamma)}$. If for some $\sigma_0 \in \mathcal{O}(\mathcal{C})$ we have $\phi_0(\sigma_0) = \phi_1(\sigma_0)$, and if κ_{σ_0} can be chosen to be $1_{\phi(\sigma_0)}$, we say ϕ_0 and ϕ_1 are homotopic relative to σ_0.

A functor $\phi : \mathcal{C} \to \mathcal{C}'$ is an *equivalence* if there is functor $\psi : \mathcal{C}' \to \mathcal{C}$ such that $\psi\phi$ is homotopic to the identity of $\mathcal{C}$ and $\phi\psi$ is homotopic to the identity of $\mathcal{C}'$. In that case we say that $\mathcal{C}$ is equivalent to $\mathcal{C}'$.

A.4 Proposition. *For each category $\mathcal{C}$, there is a subcategory $\mathcal{C}_0 \subseteq \mathcal{C}$ such that the inclusion $\iota : \mathcal{C}_0 \to \mathcal{C}$ is an equivalence and $\mathcal{C}_0$ is minimal in the sense that if $\gamma \in \mathcal{C}_0$ is invertible, then $i(\gamma) = t(\gamma)$. Any two such subcategories of $\mathcal{C}$ are isomorphic.*

Proof. In $\mathcal{O}(\mathcal{C})$ we consider the following equivalence relation: $\tau \sim \sigma$ if there is an invertible element γ such that $i(\gamma) = \sigma$ and $t(\gamma) = \tau$. For each object $\sigma \in \mathcal{C}$, choose a representative σ_0 in its equivalence class and an invertible element $\gamma_\sigma \in \mathcal{C}_{\sigma_0,\sigma}$; choose $\gamma_{\sigma_0} = 1_{\sigma_0}$. The set of objects of the subcategory $\mathcal{C}_0$ is the set of representatives of the equivalence classes, and for $\sigma_0, \tau_0 \in \mathcal{O}(\mathcal{C}_0)$ we have $(\mathcal{C}_0)_{\tau_0,\sigma_0} = \mathcal{C}_{\tau_0,\sigma_0}$. Let $\psi : \mathcal{C} \to \mathcal{C}_0$ be the functor mapping $\gamma \in \mathcal{C}$ to $\gamma_{t(\gamma)}\gamma\gamma^{-1}_{i(\gamma)} \in \mathcal{C}_0$. Then $\psi\iota$ is the identity of $\mathcal{C}_0$ and $\iota\psi$ is homotopic to the identity of $\mathcal{C}$, the homotopy being determined by the choices γ_σ. We leave to the reader to check that other choices would lead to isomorphic subcategories. $\square$

A.5 Example. Let G be a group acting on a scwol $\mathcal{X}$ as in (1.11). Associated to such an action is a category denoted $G \ltimes \mathcal{X}$ whose set of objects is $V(\mathcal{X})$ and whose set of morphisms consists of pairs $(g, \alpha) \in G \times \mathcal{X}$ with $i((g, \alpha)) = i(\alpha)$, $t((g, \alpha)) = g.t(\alpha)$; the composition $(g, \alpha)(h, \beta)$, whenever defined, is equal to $(gh, (h^{-1}.\alpha)\beta)$. Let $G(\mathcal{Y})$ be the complex of groups over the scwol $\mathcal{Y} = G\backslash\mathcal{X}$ associated to this action (as in 2.9(1)) with respect to some choices $\overline{\sigma}$ and h_a. Let $CG(\mathcal{Y})$ be the category associated to $G(\mathcal{Y})$ as in (2.8). Consider the map that sends $(g, \alpha) \in CG(\mathcal{Y})$ to the element $(g, \overline{\alpha}) \in G \ltimes \overline{X}$, where $\overline{\alpha}$ is the unique element of $\mathcal{X}$ such that $i(\overline{\alpha}) = \overline{i(\alpha)}$ and $p(\alpha) = \alpha$. This map gives an inclusion of $CG(\mathcal{Y})$ into $G \ltimes \mathcal{X}$, and in this way we identify $CG(\mathcal{Y})$ to a minimal equivalent subcategory of $G \ltimes \mathcal{X}$.

We now characterize categories associated to complexes of groups.

A.6 Proposition. *A small category $\mathcal{C}$ is associated to a complex of groups by the construction of (2.8) if and only if it satisfies the following conditions:*

(1) *For each object σ of $\mathcal{C}$, the subcategory $\mathcal{C}_{\sigma,\sigma}$ is a group (which we denote G_σ).*

(2) *For each pair of objects* (σ, τ), *if* $\gamma \in \mathcal{C}_{\tau,\sigma}$, $h \in G_\sigma$ *and* $g \in G_\tau$, *then the equality* $g\gamma = \gamma$ *implies* $g = 1_\tau$, *and the equality* $\gamma = \gamma h$ *implies* $h = 1_\sigma$.

(3) *For each* $\gamma \in \mathcal{C}_{\tau,\sigma}$ *and* $h \in G_\sigma$, *there is an element* $\psi_\gamma(h) \in G_\tau$ *such that* $\psi_\gamma(h)\gamma = \gamma h$. *(By (1) and (2),* $\psi_\gamma(h)$ *is unique and the map* $\psi_\gamma : G_\sigma \to G_\tau$ *is an injective homomorphism of groups.)*

(4) *If* $\gamma \in \mathcal{C}$ *is invertible, then* $\gamma \in G_{i(\gamma)}$.

Proof. These four properties are clearly satisfied if $\mathcal{C}$ is the category associated to a complex of groups. Conversely, assume that $\mathcal{C}$ satisfies properties (1) to (4). Such a category has a natural quotient $\mathcal{Y}$ which is a scwol. The set of objects $V(\mathcal{Y})$ of $\mathcal{Y}$ is equal to the set of objects of $\mathcal{C}$. The elements of $\mathcal{Y}$ are equivalence classes of elements of $\mathcal{C}$: two elements $\gamma, \gamma' \in \mathcal{C}$ are equivalent if $i(\gamma) = i(\gamma')$, $t(\gamma) = t(\gamma')$ and $\gamma' = g\gamma$ for some $g \in G_{t(\gamma)}$. Let $p : \mathcal{C} \to \mathcal{Y}$ be the map associating to an element γ its equivalence class. If we define the composition of two classes to be the class of the composition of elements in these classes, whenever defined, then p is a functor. Condition (4) implies that $\mathcal{Y}$ is a scwol.

For each edge $a \in E(\mathcal{Y})$, choose a representative $\overline{a}$ in the class a. Let $G(\mathcal{Y})$ be the complex of groups over $\mathcal{Y}$ given by the groups G_σ, the homomorphisms $\psi_a = \psi_{\overline{a}}$ and the elements $g_{a,b} \in G_{t(a)}$ that are uniquely defined by the equality $g_{a,b}\overline{ab} = \overline{a}\overline{b}$ in $\mathcal{C}$. Then $\mathcal{C}$ is clearly isomorphic to the category associated to $G(\mathcal{Y})$: the isomorphism sends (g, a) to $g\overline{a}$. Another choice of representatives would give a complex of groups deduced from $G(\mathcal{Y})$ by a coboundary (in the sense defined in (2.1)). $\square$

A.7 Definition of a Pre-Complex of Groups. A category $\mathcal{C}$ satisfying conditions (1), (2) and (3) of (A.6) is called a *pre-complex of groups*.

A.8 Example. Let G be a group acting on a category without loops $\mathcal{X}$. The category $G \ltimes \mathcal{X}$ defined in (A.5) is a pre-complex of groups.

A.9 Proposition. *If a category* $\mathcal{C}$ *is a pre-complex of groups, then it is equivalent to the category associated to a complex of groups, and this complex of groups is unique up to isomorphism.*

Proof. The minimal subcategory $\mathcal{C}_0$ of $\mathcal{C}$ constructed in (A.4) satisfies the conditions of (A.6), and is therefore the category associated to a complex of groups. $\square$

The Fundamental Group

A.10 $\mathcal{C}$-paths

We wish to define a notion of combinatorial path in a category $\mathcal{C}$. To this end, we associate two symbols γ^+ and γ^- to each element $\gamma \in \mathcal{C}$. The set of symbols γ^+, γ^- with $\gamma \in \mathcal{C}$ is denoted $\mathcal{C}^\pm$. Given $e \in \mathcal{C}^\pm$, we define its initial object $i(e)$ and its terminal object $t(e)$ by the formulae:

$$i(\gamma^+) = t(\gamma),\ t(\gamma^+) = i(\gamma) \text{ and } i(\gamma^-) = i(\gamma),\ t(\gamma^-) = t(\gamma).$$

For $e = \gamma^+$ (resp. γ^-), we define $e^{-1} = \gamma^-$ (resp. γ^+).

A *path in* $\mathcal{C}$ joining an object σ to an object τ is a sequence $c = (e_1, \ldots, e_k)$, where each $e_j \in \mathcal{C}^{\pm}$, $t(e_j) = i(e_{j+1})$ for $j = 1, \ldots, k-1$, and $i(e_1) = \sigma$, $t(e_k) = \tau$. The initial object $i(c)$ of c is σ, and its terminal object $t(c)$ is τ. If $\sigma = \tau$, then we also admit the constant path at σ (with $k = 0$). Note that $\mathcal{C}$ is connected if and only if for all $\sigma, \tau \in \mathcal{O}(\mathcal{C})$, there is a path joining σ to τ. If $c' = (e'_1, \ldots, e'_{k'})$ is a path in $\mathcal{C}$ joining $\sigma' = \tau$ to τ', then one can compose c and c' to obtain the path $c * c' = (e_1, \ldots, e_k, e'_1, \ldots, e'_k)$ joining σ to τ', called the *concatenation* of c and c'. Note that this composition is associative. The *inverse* of the path c is the path $c^{-1} = (e'_1, \ldots, e'_k)$ joining τ to σ, where $e'_j = e^{-1}_{k-j+1}$.

If $i(c) = t(c) = \sigma$, then c is called a *loop* at σ.

A.11 Homotopy of Paths

Let $c = (e_1, \ldots, e_k)$ be a path in $\mathcal{C}$ joining σ to τ. Consider the following three operations on c:

(1) Assume that for some $j < k$, we have $e_j = \gamma_j^+$ and $e_{j+1} = \gamma_{j+1}^+$ (resp. $e_j = \gamma_j^-$ and $e_{j+1} = \gamma_{j+1}^-$). Then the composition $\gamma_j\gamma_{j+1}$ (resp. $\gamma_{j+1}\gamma_j$) is defined and we get a new path c' in $\mathcal{C}$ by replacing the subsequence e_j, e_{j+1} of c by $(\gamma_j\gamma_{j+1})^+$ (resp. $(\gamma_{j+1}\gamma_j)^-$).

(2) Assume that for some $j < k$, we have $e_j = e_{j+1}^{-1}$. Then we get a new path c' by deleting from c the subsequence e_j, e_{j+1}.

(3) Assume that for some j, the edge e_j is associated to a unit (i.e. $e_j = 1_\rho^{\pm}$ for some object ρ). Then we get a new path c' by deleting e_j.

If c and c' are related as in (1),(2) or (3), then we say that they are obtained from each other by an elementary homotopy (thus we implicitly allow the inverses of operations (1) to (3)). Two paths joining σ to τ are defined to be *homotopic* if one can pass from the first to the second by a sequence of elementary homotopies. The set of homotopy classes of paths in $\mathcal{C}$ joining σ to τ is denoted $\pi_1(\mathcal{C}, \sigma, \tau)$. The set $\pi_1(\mathcal{C}, \sigma, \sigma)$ will also be denoted $\pi_1(\mathcal{C}, \sigma)$.

Note that if c joins σ to τ, then the path $c * c^{-1}$ is homotopic to the constant path at σ and that $c^{-1} * c$ is homotopic to the constant path at τ. The homotopy class of the concatenation $c * c'$ of two paths depends only on the homotopy classes of c and c'.

Another equivalent way of defining homotopy is to introduce the group $F\mathcal{C}$ with presentation $\langle \mathcal{C}^{\pm} \mid \mathcal{R} \rangle$ where the relations $\mathcal{R}$ are

$$(\gamma\gamma')^+ = \gamma^+\gamma'^+, \quad \forall\, (\gamma, \gamma') \in \mathcal{C}^{(2)}$$

$$(\gamma^+)^{-1} = \gamma^-, \quad \forall \gamma \in \mathcal{C}.$$

Two paths $c = (e_1, \ldots, e_k)$ and $c' = (e'_1, \ldots, e'_{k'})$ joining σ to τ are homotopic if and only if the elements of $F\mathcal{C}$ represented by the words $e_1 \ldots e_k$ and $e'_1 \ldots e'_{k'}$ are equal. In this way we identify $\pi_1(\mathcal{C}, \sigma, \tau)$ to a subset of $F\mathcal{C}$, in such a way that the homotopy class of the concatenation of two paths is the product in $F\mathcal{C}$ of the homotopy classes of those paths.

A.12 The Fundamental Group. The set $\pi_1(\mathcal{C}, \sigma_0)$ of homotopy classes of loops at σ_0, with the law of composition induced by concatenation of loops, is a group. It is called *the fundamental group of* $\mathcal{C}$ *at* σ_0.

If c is a path in $\mathcal{C}$ joining σ to τ, then the map associating to each loop l at σ the loop $c^{-1} * l * c$ induces an isomorphism of $\pi_1(\mathcal{C}, \sigma)$ onto $\pi_1(\mathcal{C}, \tau)$. A category $\mathcal{C}$ is *simply connected* if it is connected and $\pi_1(\mathcal{C}, \sigma)$ is the trivial group for some (hence all) $\sigma \in \mathcal{O}(\mathcal{C})$.

Let $\phi : \mathcal{C} \to \mathcal{C}'$ be a functor. For each $\gamma \in \mathcal{C}$, we define $\phi(\gamma^+) = \phi(\gamma)^+$ and $\phi(\gamma^-) = (\phi(\gamma))^-$. And if $c = (e_1, \dots, e_k)$ is a path in $\mathcal{C}$ joining σ to τ, its image under ϕ is the path $\phi(c) = (\phi(e_1), \dots, \phi(e_k))$ in $\mathcal{C}'$ joining $\phi(\sigma)$ to $\phi(\tau)$. The map $c \mapsto \phi(c)$ induces a homomorphism

$$\pi_1(\phi, \sigma_0) : \pi_1(\mathcal{C}, \sigma_0) \to \pi_1(\mathcal{C}', \phi(\sigma_0)).$$

In fact this homomorphism is the restriction of the homomorphism

$$F\phi : F\mathcal{C} \to F\mathcal{C}'$$

mapping each generator $\gamma^\pm$ of $F\mathcal{C}$ to the generator $\phi(\gamma)^\pm$ of $F\mathcal{C}'$.

If $\phi_0, \phi_1 : \mathcal{C} \to \mathcal{C}'$ are functors such that $\phi_0(\sigma_0) = \phi_1(\sigma_0)$ and if ϕ_0 and ϕ_1 are homotopic with respect to σ_0 (see A.3), then the homomorphisms induced by them on $\pi_1(\mathcal{C}, \sigma_0)$ are equal. Indeed if the homotopy is defined by the family of elements $(\kappa_\sigma)_{\sigma \in \mathcal{O}(\mathcal{C})}$ and we write $e_\sigma = \kappa_\sigma^+ \in \mathcal{C}'^\pm$, then for every $e \in \mathcal{C}^\pm$, the paths $(e_{i(e)}, \phi_0(e), e_{t(e)})$ and $(\phi_1(e))$ are homotopic, so in particular the images of each loop at σ_0 are homotopic. This shows in particular that the inclusion into $\mathcal{C}$ of a minimal equivalent subcategory $\mathcal{C}_0$ of $\mathcal{C}$ induces an isomorphism on fundamental groups.

If $\mathcal{C}$ is the category associated to a complex of groups $G(\mathcal{Y})$, then $\pi_1(G(\mathcal{Y}), \sigma)$ as defined in (3.5) is canonically isomorphic to $\pi_1(\mathcal{C}, \sigma)$. Indeed there is a natural isomorphism $FG(\mathcal{Y}) \to F\mathcal{C}$ that restricts to an isomorphism $\pi_1(G(\mathcal{Y}), \sigma) \to \pi_1(\mathcal{C}, \sigma)$; this is obtained by sending each $g \in G_\sigma$ to $(g, 1_\sigma)^+ \in \mathcal{C}^\pm$ and $a^\pm \in E(\mathcal{Y})^\pm$ to $(1_{t(a)}, a)^\pm$.

A.13 A Presentation of the Fundamental Group. *Assume that the category* $\mathcal{C}$ *is connected. Consider the graph* $|\mathcal{C}|^{(1)}$ *whose set of vertices is* $\mathcal{O}(\mathcal{C})$ *and whose set of 1-cells is* $\mathcal{C}$*; an element* $\gamma \in \mathcal{C}_{\tau,\sigma}$ *is considered as an edge joining the vertices* σ *and* τ*. Let* T *be a maximal tree in* $|\mathcal{C}|^{(1)}$.

Let $\pi_1(\mathcal{C}, T)$ *be the quotient of the free group on* $\mathcal{C}^\pm$ *by the relations:*

$$(\gamma\gamma')^+ = \gamma^+\gamma'^+, \quad (\gamma^+)^{-1} = \gamma^- \ \forall \gamma, \gamma' \in \mathcal{C} \text{ and } \gamma^+ = 1 \ \forall \gamma \in T.$$

Then $\pi_1(\mathcal{C}, T)$ *is canonically isomorphic to* $\pi_1(\mathcal{C}, \sigma_0)$.

Proof. The proof is the same as the proof of 3.7. □

A.14 Remark. For any category $\mathcal{C}$ one can construct a classifying space $B\mathcal{C}$ which is the geometric realization of the nerve of $\mathcal{C}$ (see [Qui73] and [Se68]). $B\mathcal{C}$ is a cell complex whose set of 0-cells is $\mathcal{O}(\mathcal{C})$ and whose 1-skeleton is the graph $|\mathcal{C}|^{(1)}$ defined above. The fundamental group $\pi_1(B\mathcal{C}, \sigma_0)$ is naturally isomorphic to $\pi_1(\mathcal{C}, \sigma_0)$. The covering spaces of $B\mathcal{C}$ are the geometric realizations of the nerves of the coverings of $\mathcal{C}$ as defined below.

Covering of a Category

A.15 Definition. Let $\mathcal{C}$ be a connected category. A functor $\phi : \mathcal{C}' \to \mathcal{C}$ from a category $\mathcal{C}'$ to a category $\mathcal{C}$ is a *covering* if for each object σ' of $\mathcal{C}'$ the restriction of ϕ to the subset of morphisms that have σ' as their terminal (resp. initial) object is a bijection onto the set of morphisms of $\mathcal{C}$ with terminal (resp. initial) object $\phi(\sigma')$.

It follows from A.17 that ϕ is surjective. As in remark 1.9(3), if when $\mathcal{C}$ is not connected, one could define a covering $\phi : \mathcal{C}' \to \mathcal{C}$ as a functor which is surjective and satisfies the above condition.

A.16 Examples

(1) For a connected category without loops $\mathcal{X}$, the notions of coverings defined in (1.9) and (A.15) are the same.

(2) A covering of the category $CG(\mathcal{Y})$ associated to a complex of groups $G(\mathcal{Y})$ is in general not associated to a complex of groups, but it is a pre-complex of groups (A.7).

(3) Let G be a group acting on a connected scwol $\mathcal{X}$ and let $\widetilde{G}$ be the extension of G by the fundamental group of $\mathcal{X}$ acting on the universal covering $\widetilde{\mathcal{X}}$ of $\mathcal{X}$ as in (1.15). Then the natural functor $\widetilde{G} \ltimes \widetilde{\mathcal{X}} \to G \ltimes \mathcal{X}$ is a covering.

A.17 Proposition (Path Lifting). *Let $\phi : \mathcal{C}' \to \mathcal{C}$ be a covering of the connected category $\mathcal{C}$. Let $\sigma \in \mathcal{O}(\mathcal{C})$ and $\sigma' \in \mathcal{O}(\mathcal{C}')$ be such that $\phi(\sigma') = \sigma$. Any path in $\mathcal{C}$ starting at σ can be lifted uniquely to a path in $\mathcal{C}'$ starting at σ'. Moreover, if two paths issuing from σ' project by ϕ to homotopic paths in $\mathcal{C}$, then the paths are also homotopic in $\mathcal{C}'$. Thus ϕ induces an injection of $\pi_1(\mathcal{C}', \sigma')$ into $\pi_1(\mathcal{C}, \sigma)$.*

Let $\mathcal{C}_0$ be a connected category and fix $\sigma_0 \in \mathcal{O}(\mathcal{C}_0)$. Let $\phi_1, \phi_2 : \mathcal{C}_0 \to \mathcal{C}'$ be two functors such that $\phi \circ \phi_1 = \phi \circ \phi_2$ and $\phi_1(\sigma_0) = \phi_2(\sigma_0)$. Then $\phi_1 = \phi_2$.

Proof. The first part follows directly from the definition of a covering and of homotopy, and the second part follows from the first. □

A.18 Definition. Let $\mathcal{C}$ be a connected category. A covering $\phi' : \mathcal{C}' \to \mathcal{C}$, where $\mathcal{C}'$ is connected, is a *universal covering* if it satisfies the following condition: for every covering $\phi'' : \mathcal{C}'' \to \mathcal{C}$ and every $\sigma' \in \mathcal{O}(\mathcal{C}')$ and $\sigma'' \in \mathcal{O}(\mathcal{C}')$ there is a functor $\psi : \mathcal{C}' \to \mathcal{C}''$ such that $\phi'' \circ \psi = \phi'$ and $\psi(\sigma') = \psi(\sigma'')$.

It follows from the above proposition that ψ is unique, and therefore the universal covering $\phi' : \mathcal{C}' \to \mathcal{C}$ is unique up to isomorphism.

A.19 Proposition. *A universal covering $\phi' : \mathcal{C}' \to \mathcal{C}$ of a connected category $\mathcal{C}$ always exists and it is necessarily simply connected. Conversely, any covering $\phi'' : \mathcal{C}'' \to \mathcal{C}$ such that $\mathcal{C}''$ is simply connected is universal.*

Proof. Choose $\sigma_0 \in \mathcal{O}(\mathcal{C})$. We define a category $\mathcal{C}'$ whose set of objects $\mathcal{O}(\mathcal{C}')$ is the set of homotopy classes $[c]$ of paths issuing from σ_0. The functor ϕ' will map $[c]$ to the terminal object of c. The set of elements of $\mathcal{C}'$ with initial object $[c]$ are pairs $(\gamma, [c])$, where $\gamma \in \mathcal{C}$ with $i(\gamma) = \phi'([c])$. The terminal object of $(\gamma, [c])$ is the homotopy class of the path $c * \gamma^-$. The object $[c]$ is identified to the pair $(1_{\phi'([c])}, [c])$. The composition $(\gamma', [c'])(\gamma, [c])$ is defined if $[c'] = t((\gamma, [c]))$ and is equal to $(\gamma'\gamma, [c])$. The functor ϕ' maps $(\gamma, [c])$ to γ. It is straightforward to check that $\mathcal{C}'$ is a connected category and that $\phi' : \mathcal{C}' \to \mathcal{C}$ is a covering. Let σ_0' denote the object of $\mathcal{C}'$ which is the homotopy class of the constant loop at σ_0.

We shall prove that ϕ' is a universal covering. Let $\phi'' : \mathcal{C}'' \to \mathcal{C}$ be a covering. Let $\sigma_0'' \in \mathcal{O}(\mathcal{C}'')$ be such that $\phi''(\sigma_0'') = \sigma_0$. Let $\psi : \mathcal{O}(\mathcal{C}') \to \mathcal{O}(\mathcal{C}'')$ be the map associating to each object $[c]$ the terminal object of the path in $\mathcal{C}''$ issuing from σ_0'' whose projection by ϕ'' is c. We extend this to a functor $\psi : \mathcal{C}' \to \mathcal{C}''$ by mapping $(\gamma, [c])$ to the unique element $\gamma'' \in \mathcal{C}''$ such that $i(\gamma'') = \phi''([c])$ and $\phi''(\gamma'') = \gamma$. This functor maps the homotopy class of the constant path at σ_0 to σ_0''. (If we want ψ to map a certain object τ' to a preferred object τ'' with $\phi'(\tau') = \phi''(\tau'') = \tau$, then it is sufficient to choose σ_0'' appropriately: choose a path c' in $\mathcal{C}'$ joining τ' to σ_0', then define σ_0'' to be the terminal object of the lifting of $\phi'(c')$ with initial object τ''.)

To see that $\mathcal{C}'$ is simply connected, note that the map associating to the homotopy class of a loop c at σ_0 the terminal point of the lifting of c with initial object σ_0' gives a bijection from $\pi_1(\mathcal{C}, \sigma_0)$ to $\phi^{-1}(\sigma_0)$, from which it follows (A.17) that every loop at σ_0' is homotopic to a constant loop.

If $\mathcal{C}''$ is simply connected, then it is easy to see that ψ is a bijection. Since $\psi^{-1}\psi(\sigma_0') = \sigma_0''$ and $\psi^{-1}(\psi(\sigma_0'')) = \sigma_0'$, it follows from (A.18) that ψ is an isomorphism. □

A.20 Examples

(1) Suppose that the category $\mathcal{C}$ is just a group G with a single object identified to the unit element of G. Then its universal covering is the category $\widetilde{\mathcal{C}}$ whose set of objects is G and whose set of elements is $G \times G$, where for each $(g, x) \in G \times G$ we define $i(g, x) = x$ and $t(g, x) = gx$; the composition $(g, x)(h, y)$, when defined, is equal to (gh, y).

(2) Let G be a group acting on a connected scwol $\mathcal{X}$. One can check that the fundamental group $\pi_1(G \ltimes \mathcal{X}, \sigma_0)$ of the associated category is isomorphic to the extension $\widetilde{G}$ of G by $\pi_1(\mathcal{X}, \sigma_0)$ described in (1.15).

A.21 The Monodromy of a Covering

Let $\phi : \mathcal{C}' \to \mathcal{C}$ be a covering. The *fibre* of ϕ over $\sigma \in \mathcal{O}(\mathcal{C})$ is the set $F_\sigma = \phi^{-1}(\sigma) \subseteq \mathcal{O}(\mathcal{C}')$. For $\gamma \in \mathcal{C}$ with $i(\gamma) = \sigma$, let $F_\gamma : F_{i(\gamma)} \to F_{t(\gamma)}$ be the map associating to each $\sigma' \in F_\sigma$ the terminal object of the unique element γ' of $\mathcal{C}'$ such that $\phi(\gamma') = \gamma$ and $i(\gamma') = \sigma'$. This map is a bijection because ϕ is a covering. Moreover, for composable morphisms $(\gamma_1, \gamma_2) \in \mathcal{C}^{(2)}$ we have $F_{\gamma_1\gamma_2} = F_{\gamma_1}F_{\gamma_2}$, and F_{1_σ} is the identity of F_σ. In other words F can be considered as a functor from the category $\mathcal{C}$ to the category whose elements are bijections of sets.

We can reconstruct $\mathcal{C}'$ and $\phi : \mathcal{C}' \to \mathcal{C}$ from the functor $\gamma \mapsto F_\gamma$, as follows. The set $\mathcal{O}(\mathcal{C}')$ is the disjoint union of the F_σ, $\sigma \in \mathcal{O}(\mathcal{C})$, and $\phi : \mathcal{O}(\mathcal{C}') \to \mathcal{O}(\mathcal{C})$ is the projection to the index set of the disjoint union. To recover the elements of $\mathcal{C}'$ we consider the set $\mathcal{C} \times_{\mathcal{O}(\mathcal{C})} \mathcal{O}(\mathcal{C}')$ of pairs $(\gamma, \sigma') \in \mathcal{C} \times \mathcal{O}(\mathcal{C}')$ such that $i(\gamma) = \phi(\sigma')$. This becomes a category with set of objects $\mathcal{O}(\mathcal{C}')$ if we define $i((\gamma, \sigma')) = \sigma'$ and $t((\gamma, \sigma')) = F_\gamma(\sigma')$; the composition is $(\gamma_1, \sigma_1')(\gamma_2, {\sigma'}_2) = (\gamma_1\gamma_2, {\sigma'}_2)$, when defined. The functor $\mathcal{C} \times_{\mathcal{O}(\mathcal{C})} \mathcal{O}(\mathcal{C}') \to \mathcal{C}$ sending (γ, σ') to γ is a covering which is equal to ϕ if we identify $\mathcal{C}'$ with $\mathcal{C} \times_{\mathcal{O}(\mathcal{C})} \mathcal{O}(\mathcal{C}')$ by the map sending γ' to the pair $(\phi(\gamma'), i(\gamma'))$.

The monodromy of the covering ϕ is defined as follows. For each $e \in \mathcal{C}^\pm$, define $F_e : F_{t(e)} \to F_{i(e)}$ to be F_γ if $e = \gamma^+$ and F_γ^{-1} if $e = \gamma^-$. For each path $c = (e_1, \dots, e_k)$ joining σ to τ, define $F_c = F_{e_1} \cdots F_{e_k} : F_\tau \to F_\sigma$. (Note that the unique path in $\mathcal{C}'$ that issues from $\sigma' \in F_\sigma$ and projects to c joins σ' to $F_c^{-1}(\sigma')$.)

If the concatenation $c * c'$ is defined, then $F_{c*c'} = F_cF_{c'}$ and $F_{c^{-1}} = F_c^{-1}$. If c and c' are homotopic paths, then $F_c = F_{c'}$. In particular the map associating to each loop c at σ_0 the map $F_c : F_{\sigma_0} \to F_{\sigma_0}$ gives a homomorphism M from $\pi_1(\mathcal{C}, \sigma_0)$ to the group of bijections of F_{σ_0}. This homomorphism is denoted M and is called the *monodromy representation* of ϕ at σ_0. The image $M(\pi_1(\mathcal{C}, \sigma_0))$ is called the *monodromy group of ϕ at σ_0.*

A.22 Proposition. *Let $\mathcal{C}$ be a connected category and let $\phi : \mathcal{C}' \to \mathcal{C}$ be a covering. Let $\sigma_0' \in \mathcal{C}'$ and let $\sigma_0 = \phi(\sigma_0')$.*

(1) *The homomorphism $\pi_1(\mathcal{C}', \sigma_0') \to \pi_1(\mathcal{C}, \sigma_0)$ described in (A.17) is injective and its image is precisely the subgroup of $\pi_1(\mathcal{C}, \sigma_0)$ that fixes σ_0' in the monodromy action M described above.*

(2) *$\mathcal{C}'$ is connected if and only if the fundamental group $\pi_1(\mathcal{C}, \sigma_0)$ acts transitively on the fibre F_{σ_0} via the monodromy of ϕ.*

Proof. The first part of the proposition follows directly from the definitions. Let us prove the second part. Suppose that $\mathcal{C}'$ is connected. Given $x, y \in F_{\sigma_0}$, there is a path c' in $\mathcal{C}'$ joining x to y whose projection by ϕ is a loop c at σ_0. The bijection F_c of F_{σ_0} associated to the homotopy class of c by the monodromy maps y to x. Conversely, if the monodromy group acts transitively on F_{σ_0}, then $\mathcal{C}'$ is connected: given two objects $x \in F_{\sigma_0}$ and $z \in F_\sigma$, let c be a path in $\mathcal{C}$ joining σ_0 to σ; the map $F_c : F_\sigma \to F_{\sigma_0}$ maps z to some $y \in F_{\sigma_0}$; by hypothesis there is a loop c' such that $F_{c'}$ maps y to x; therefore the path in $\mathcal{C}'$ that issues from x and has projection $c' * c$ joins x to z. □

Given two actions of a group G on sets X_1 and X_2, one says that a bijection $f : X_1 \to X_2$ *intertwines* the actions if $g.f(x) = f(g.x)\ \forall g \in G, \forall x \in X_1$.

A.23 Proposition. *Let $\mathcal{C}$ be a connected category and let $\phi' : \mathcal{C}' \to \mathcal{C}$ and $\phi'' : \mathcal{C}'' \to \mathcal{C}$ be two coverings of $\mathcal{C}$. Assume that there is a bijection f_0 from the fibre F'_{σ_0} of ϕ' over σ_0 to the fibre F''_{σ_0} of ϕ'' over σ_0 intertwining the monodromy representations M' and M'' of ϕ' and ϕ''. Then there is a unique isomorphism $\phi : \mathcal{C}' \to \mathcal{C}''$ with $\phi''\phi = \phi'$ such that the restriction of ϕ to F_{σ_0} is the given bijection f_0.*

Given an action of $\pi_1(\mathcal{C}, \sigma_0)$ on a set F_0, there is a covering $\phi' : \mathcal{C}' \to \mathcal{C}$ (unique up to isomorphism) whose fibre above σ_0 is equal to F_0 and whose monodromy at σ_0 is the given action.

Proof. For each object σ of $\mathcal{C}$, choose a path c_σ in $\mathcal{C}$ joining σ_0 to σ. Let F'_σ (resp. F''_σ) be the fibre of ϕ' (resp. ϕ'') above σ. Let $F'_{c_\sigma} : F'_\sigma \to F'_{\sigma_0}$ (resp. $F''_{c_\sigma} : F''_\sigma \to F''_{\sigma_0}$) be the bijection determined by c_σ. As in (A.21), we identify $\mathcal{C}'$ with $\mathcal{C} \times_{\mathcal{O}(\mathcal{C})} \mathcal{O}(\mathcal{C}')$ (resp. $\mathcal{C}''$ with $\mathcal{C} \times_{\mathcal{O}(\mathcal{C})} \mathcal{O}(\mathcal{C}'')$). Let

$$f_\sigma : {F''}^{-1}_{c_\sigma} f_0 F'_{c_\sigma} : F'_\sigma \to F''_\sigma .$$

Given $\gamma' = (\gamma, \sigma') \in \mathcal{C}'$, where $\gamma \in \mathcal{C}$, $i(\gamma) = \phi'(\sigma') = \sigma$ and $t(\gamma) = \tau$, we define

$$\phi(\gamma') = (\gamma, f_\sigma(\sigma')).$$

We claim that the map $\phi : \mathcal{C}' \to \mathcal{C}''$ is a functor. Clearly $i(\phi(\gamma')) = \phi(i(\gamma'))$. Let us check that $t(\phi(\gamma')) = \phi(t(\gamma'))$. By hypothesis, f_0 conjugates the monodromies around the loop $c_\tau * \gamma^+ * c_\sigma^{-1}$:

$$f_0 F'_{c_\tau} F'_\gamma {F'}^{-1}_{c_\sigma} = F''_{c_\tau} F''_\gamma {F''}^{-1}_{c_\sigma} f_0 .$$

Hence

$$t(\phi(\gamma')) = F''_\gamma(f_\sigma(\sigma')) = (F''_\gamma {F''}^{-1}_{c_\sigma} f_0 F'_{c_\sigma})(\sigma') =$$
$$({F''}^{-1}_{c_\tau} f_0 F'_{c_\tau} F'_\gamma)(\sigma') = (f_\tau F'_\gamma)(\sigma') = \phi(t(\gamma')).$$

It follows that if the composition $(\gamma_1, {\sigma'}_1)(\gamma_2, {\sigma'}_2) = (\gamma_1\gamma_2, {\sigma'}_2)$ is defined in $\mathcal{C}'$, then the composition $\phi((\gamma_1, {\sigma'}_2))\phi((\gamma_2, {\sigma'}_2))$ is also defined in $\mathcal{C}''$ and is equal to $(\gamma_1\gamma_2, f({\sigma'}_2)) = \phi((\gamma_1\gamma_2, {\sigma'}_2))$. It is clear that ϕ is bijective, therefore it is an isomorphism.

Given an action M of $\pi_1(\mathcal{C}, \sigma_0)$ on a set F_0, we construct an associated covering $\phi' : \mathcal{C}' \to \mathcal{C}$ as follows. The set $\mathcal{O}(\mathcal{C}')$ will be the product $\mathcal{O}(\mathcal{C}) \times F_0$ and $\phi : \mathcal{O}(\mathcal{C}') \to \mathcal{O}(\mathcal{C})$ will be the natural projection. The set $\mathcal{C}'$ is the set $\mathcal{C} \times F_0$. We define $i((\gamma, x)) = (i(\gamma), x)$ and $t((\gamma, x)) = (t(\gamma), M([c]).x)$, where c is the loop $c_{t(\gamma)} * \gamma^+ * c_{i(g)}^{-1}$. Given (γ_1, x_1) and (γ_2, x_2) in $\mathcal{C}'$ such that $i((\gamma_1, x_1)) = t((\gamma_2, x_2))$, the composition is defined to be equal to $(\gamma_1\gamma_2, x_2)$. Note that $t((\gamma_1\gamma_2, x_2)) = t((\gamma_1, x_1))$, because the loop $c_{t(\gamma_1\gamma_2)} * (\gamma_1\gamma_2)^+ * c_{i(\gamma_1\gamma_2)}^{-1}$ is homotopic to the loop $c_{t(\gamma_1)} * \gamma_1^+ * c_{i(\gamma_1)}^{-1} * c_{t(\gamma_2)} * \gamma_2^+ * c_{i(\gamma_2)}^{-1}$. This shows that $\mathcal{C}'$ with this partially defined law of composition is a category and that ϕ is a covering with the prescribed monodromy (if we identify the fibre above σ_0 to F_0 by the map $(\sigma_0, x) \mapsto x$). □

The Relationship with Coverings of Complexes of Groups

A.24 Proposition. *Let $G(\mathcal{Y})$ be a complex of groups over a connected scwol $\mathcal{Y}$ and let $\mathcal{C} = CG(\mathcal{Y})$ be the associated category.*

(1) *Let $\phi : G(\mathcal{Y}') \to G(\mathcal{Y})$ be a covering of $G(\mathcal{Y})$ in the sense of (5.1). There is a canonical covering $\phi' : \mathcal{C}' \to \mathcal{C} = CG(\mathcal{Y})$ and an inclusion $\lambda : CG(\mathcal{Y}') \to \mathcal{C}'$ which is an equivalence, such that $\phi = \phi'\lambda$.*

(2) *Conversely, if $\phi' : \mathcal{C}' \to \mathcal{C} = CG(\mathcal{Y})$ is a covering, then $\mathcal{C}'$ is a pre-complex of groups, and for any inclusion $\lambda : CG(\mathcal{Y}') \to \mathcal{C}'$ which is an equivalence, the composition $\phi'\lambda : CG(\mathcal{Y}') \to CG(\mathcal{Y})$ is associated to a covering $\phi : G(\mathcal{Y}') \to G(\mathcal{Y})$.*

Proof. (1) Let $\phi = (\phi_{\sigma'}, \phi(a')) : G(\mathcal{Y}') \to G(\mathcal{Y})$ be a covering over a morphism $f : \mathcal{Y}' \to \mathcal{Y}$. Let $\phi : CG(\mathcal{Y}') \to CG(\mathcal{Y})$ be the associated functor. For each $\sigma \in V(\mathcal{Y})$, let F_σ be the fibre over σ with the action of G_σ as in 5.5. Every element of $CG(\mathcal{Y})$ is of the form $\gamma = (g, \alpha)$, where $\alpha \in \mathcal{Y}$ and $g \in G_{t(\alpha)}$; define $F_\gamma = F_g F_\alpha : F_{i(\alpha)} \to F_{t(\alpha)}$ using the notations of 5.11 (if $\alpha = 1_\sigma$, then F_α is the identity of F_σ). For composable elements γ, γ' of $CG(\mathcal{Y})$, we have $F_{\gamma\gamma'} = F_\gamma F_{\gamma'}$. Let $\phi' : \mathcal{C}' \to \mathcal{C}$ be the associated covering as in A.23: the set of objects $\mathcal{O}(\mathcal{C}')$ is the disjoint union of the fibres F_σ and $\mathcal{C}'$ is the set of pairs $(\gamma, x) \in CG(\mathcal{Y}) \times_{\mathcal{O}(\mathcal{C})} \mathcal{O}(\mathcal{C}')$ with $x \in F_{i(\gamma)}$; the functor ϕ' maps (γ, x) to γ.

We now define the inclusion $\lambda : CG(\mathcal{Y}') \to \mathcal{C}'$. Let $\gamma' = (g', \alpha') \in CG(\mathcal{Y}')$, where $\alpha' \in \mathcal{Y}'$ with $i(\alpha') = \sigma'$, $t(\alpha') = \tau'$ and $g' \in G_{\sigma'}$. Then $\lambda(\gamma') = (\phi(\gamma'), \lambda(\sigma'))$, where $\lambda(\sigma') = \phi_{\sigma'}(G_{\sigma'}) \in G_\sigma/\phi_{\sigma'}(G_{\sigma'}) \subseteq F_{f(\sigma')}$. To show that λ is a functor, the main point to check is that $t(\lambda(\gamma')) = \lambda(t(\gamma'))$. We have

$$\begin{aligned} t(\lambda(\gamma')) &= F_{\phi(\gamma')}(\lambda(\sigma')) = \phi_{\tau'}(g')\phi(\alpha')F_{\alpha'}(\phi_{\sigma'}(G_{\sigma'}) \\ &= \phi_{\tau'}(g')\phi_{\tau'}(G_{\tau'}) = \phi_{\tau'}(G_{\tau'}) = \lambda(t(\gamma')). \end{aligned}$$

Thus if $\gamma_1', \gamma_2' \in CG(\mathcal{Y}')$ are composable, then $\lambda(\gamma_1')$ and $\lambda(\gamma_2')$ are composable and $\lambda(\gamma_1'\gamma_2') = (\phi(\gamma_1'\gamma_2'), \lambda(i(\gamma_2'))) = (\phi(\gamma_1')\phi(\gamma_2'), \lambda(i(\gamma_2'))) = \lambda(\gamma_1')\lambda(\gamma_2')$.

Part (2) essentially follows from the proof of (5.8). Indeed let $\phi' : \mathcal{C}' \to \mathcal{C} = CG(\mathcal{Y})$ be a covering; for $\sigma \in V(\mathcal{Y})$ let $F_\sigma = \phi^{-1}(\sigma)$, and for $\gamma \in \mathcal{C}$ let $F_\gamma : F_{i(\gamma)} \to F_{t(\gamma)}$ be as defined in A.21. The elements γ such that $i(\gamma) = t(\gamma) = \sigma$ form a group isomorphic to G_σ, and in this way we get an action of G_σ on F_σ. If $\gamma = (1_{t(a)}, a)$, we put $F_\gamma = F_a$; for $(a, b) \in E^{(2)}(\mathcal{Y})$, we have $F_a F_b = g_{a,b} F_{ab}$. Let $\phi : G(\mathcal{Y}') \to G(\mathcal{Y})$ be the covering over $f : \mathcal{Y}' \to \mathcal{Y}$ constructed in (5.8). The functor $\lambda : CG(\mathcal{Y}') \to \mathcal{C}'$ maps γ' to the pair $(\phi(\gamma'), 1_{\sigma'}) \in CG(\mathcal{Y}) \times_{\mathcal{O}(\mathcal{C})} \mathcal{O}(\mathcal{C}') = \mathcal{C}'$, where σ' is the chosen representative of $i(\gamma')$ in the $G_{f(\sigma')}$-orbit. $\square$

Chapter III.$\mathcal{G}$ Groupoids of Local Isometries

The purpose of this chapter is to prove a general result concerning the developability of groupoids of local isometries. We shall show that if such a groupoid $\mathcal{G}$ is Hausdorff and complete (in a suitable sense, 2.10), and if the metric on the space of units of $\mathcal{G}$ is locally convex, then $\mathcal{G}$ is equivalent to the groupoid associated to the proper action of a group of isometries on a complete geodesic space whose metric is (globally) convex in the sense of (II.1.3). This result unifies and extends several earlier developability theorems, as we shall now explain.

Groupoids of local isometries appear naturally in several contexts, for instance as holonomy groupoids of Riemannian foliations and as the groupoids associated to complexes of groups over M_κ-polyhedral complexes ($\mathcal{C}$.4.16 and 4.17). In the first case, the developability theorem which we prove (Theorem 2.15) generalizes a result of Hebda [Heb86], and in the second case it generalizes results of Gromov [Gro87, p.127-128], Gersten-Stallings [St91], Haefliger [Hae90,91], Corson[73] [Cors92] and Spieler [Spi92]. In its outline, our proof of Theorem 2.15 follows the Alexander-Bishop proof of the Cartan-Hadamard Theorem, which was explained in Chapter II.4. The proof of our main lemma (4.3) is also adapted from their proof.

In order to smooth the reader's introduction to the concepts of covering, equivalence and developability in the general setting of étale groupoids, we shall dedicate section 1 of this chapter to a discussion of these concepts in the more familiar context of differentiable orbifolds. Such an orbifold is said to be developable if it arises as the "quotient" of a differentiable manifold M by the faithful, proper action of a group of diffeomorphisms. We shall use the notion of covering to prove that an orbifold with a geometric structure (for instance an orbifold of constant curvature) is always developable; this was first proved by Thurston [Thu79].

The developability of *complete* Riemannian orbifolds of non-positive curvature is a deeper theorem; it is closely related to the Cartan-Hadamard theorem and was first discovered by Gromov. This result is subsumed by our main theorem (2.15), and although we do not give a separate proof in the special case of orbifolds, we advise the reader to keep this special case in mind while reading the proof.

In section 2 we define the basic objects of study in this chapter: étale groupoids, equivalence, and developability. We also define Hausdorff separability and completeness for groupoids of local isometries. By the time that we reach the end of section 2,

[73] The results of Gersten-Stallings [St91] and Corson [Cors92] also apply to certain non-metric 2-complexes where Theorem 2.15 does not apply.

the reader will be able to understand a precise statement of the developability theorem (2.15).

In section 3 we introduce the notions of fundamental group and covering for étale groupoids. Our approach to the fundamental group of a topological groupoid $\mathcal{G}$ is parallel to the usual description of the fundamental group of topological spaces in terms of homotopy classes of continuous paths. We first define the notion of an *equivalence class of $\mathcal{G}$-paths* (a notion which appears already in [Hae58]), these classes play the role that continuous paths do in the topological case. We then define the notion of homotopy of $\mathcal{G}$-paths, and define the fundamental group to be the group of homotopy classes of $\mathcal{G}$-loops at a base point. In both sections 2 and 3 we illustrate the main ideas with exercises and examples.

The proof of the main theorem is given in section 4. In order to follow the proof, the reader will need to understand the various definitions in sections 2 and 3, and the basic facts established in (2.14), (3.14) and (3.19).

1. Orbifolds

The notion of orbifold was introduced by Satake ([Sat56] and [Sat57]) under the name of V-manifold. It was rediscovered in the seventies by Thurston [Thu79] who introduced the term "orbifold". Thurston also defined the notion of covering of orbifolds and derived from it the notion of fundamental group of an orbifold. In this section we shall consider coverings of orbifolds, but postpone discussion of the fundamental group to section 3 where we work in the more general framework of étale topological groupoids. (As indicated above, our treatment of the fundamental group, unlike Thurston's, is based on an appropriate notion of homotopy of paths.)

Basic Definitions

1.1 Definitions. A (differentiable) orbifold structure[74] of dimension n on a Hausdorff topological space Q is given by the following data:

(i) An open cover $(V_i)_{i\in I}$ of Q indexed by a set I.

(ii) For each $i \in I$, a finite subgroup Γ_i of the group of diffeomorphisms of a simply connected n-manifold X_i and a continuous map $q_i : X_i \to V_i$, called a *uniformizing chart*, such that q_i induces a homeomorphism from $\Gamma_i\backslash X_i$ onto V_i.

(iii) For all $x_i \in X_i$ and $x_j \in X_j$ such that $q_i(x_i) = q_j(x_j)$, there is a diffeomorphism h from an open connected neighbourhood W of x_i to a neighbourhood of x_j such that $q_j \circ h = q_i|_W$. Such a map h is called a *change of chart*; it is well defined up to composition with an element of Γ_j (see exercise 1.5(1)). In particular if $i = j$ then h is the restriction of an element of Γ_i.

[74] To simplify, we shall consider only differentiable orbifolds in this section, because the analogue of exercise 1.5(1) is more difficult to prove in the topological case (cf. remark 1.6(2)).

The family $(X_i, q_i)_{i \in I}$ is called an atlas (of uniformizing charts) for the orbifold structure on Q.

By definition, two such atlases of uniformizing charts $(X_i, q_i)_{i \in I_1}$ and $(X_i, q_i)_{i \in I_2}$ define the same orbifold structure on Q if $(X_i, q_i)_{I_1 \cup I_2}$ satisfies the compatibility condition (iii).

Note that if all of the groups Γ_i are trivial (or more generally if they act freely on X_i), then Q is simply a differentiable manifold.

The orbifold structure on Q is said to be *Riemannian (complex analytic, etc)* if each X_i is a Riemannian (complex analytic, etc.) manifold and if the changes of charts are accordingly Riemannian isometries (complex analytic, etc.).

If the orbifold structure on Q is Riemannian, there is a natural pseudometric on Q, namely the quotient of the Riemannian length metric on the disjoint union of the X_i. The assumption that Q is Hausdorff implies that this pseudometric is actually a metric and induces the given topology on Q. This metric will be called the *quotient metric* on Q.

1.2 The Pseudogroup of Changes of Charts. Let X be the disjoint union of the X_i. We identify each X_i to a connected component of X and we call q the *union of the maps* q_i. Any diffeomorphism h of an open subset U of X to an open subset of X such that $q = qh$ on U will be called a change of charts. The change of charts form a pseudogroup[75] $\mathcal{H}$ of local diffeomorphisms of X called the *pseudogroup of changes of charts of the orbifold* Q (with respect to the atlas of uniformizing charts $(X_i, q_i)_{i \in I}$). It contains in particular all the elements of the groups Γ_i. If $h : U \to V$ is a change of charts such that U and V are contained in the same X_i and if U is connected, then h is the restriction to U of an element of Γ_i (see 1.5(1)).

Two points $x, x' \in X$ are said to be in the same orbit of $\mathcal{H}$ if and only if there is an element $h \in \mathcal{H}$ such that $h(x) = x'$. This defines an equivalence relation on X whose classes are called the orbits of $\mathcal{H}$. The quotient of X by this equivalence relation (with the quotient topology) will be denoted $\mathcal{H}\backslash X$. The map $q : X \to Q$ induces a homeomorphism from $\mathcal{H}\backslash X$ to Q.

1.3 Developable Orbifolds

Let Γ be a subgroup of the group of diffeomorphisms of a manifold M and suppose that the action of Γ on M is proper. Let $f : M \to Q$ be a continuous map that induces a homeomorphism from $\Gamma\backslash M$ with the quotient topology to Q. Note that Q is Hausdorff (I.8.4(2)).

---

[75] A pseudogroup $\mathcal{H}$ of local homeomorphisms of a topological space X is a collection $\mathcal{H}$ of homeomorphisms $h : U \to V$ of open sets of X such that:

(1) if $h : U \to V$ and $h' : U' \to V'$ belong to $\mathcal{H}$, then their composition $hh' : h'^{-1}(U \cap V') \to h(U \cap V')$ belongs to $\mathcal{H}$, as does h^{-1},

(2) the restriction of h to any open set of X belongs to $\mathcal{H}$,

(3) the identity of X belongs to $\mathcal{H}$,

(4) if a homeomorphism from an open set of X to an open set of X is the union of elements of $\mathcal{H}$, then it too belongs to $\mathcal{H}$.

The definition of a proper action ensures that one can find a collection of open balls $X_i \subseteq M$ such that:

(i) the subgroup $\Gamma_i = \{\gamma \in \Gamma \mid \gamma.X_i = X_i\}$ is finite and if $\gamma.X_i \cap X_i \neq \emptyset$, then $\gamma \in \Gamma_i$,

(ii) the open sets $V_i = f(X_i)$ cover Q.

Let q_i be the restriction of f to X_i. Then the atlas (X_i, q_i) defines an orbifold structure on Q. This structure depends only on the action of Γ and not on the choice of the open sets X_i; it is called the *orbifold quotient of M by the proper action of* Γ. We shall say that an orbifold structure on Q is *developable* (*good* in the terminology of Thurston [Thu79]) if it arises from an action in this way.

1.4 Examples

(1) *Examples of Orbifold Structures on the 2-Sphere* S^2. Identify S^2 with $\mathbb{C} \cup \{\infty\}$; let $V_0 = \mathbb{C} \subset S^2$ and let $V_\infty = S^2 \setminus \{0\}$. Given two positive integers m, n, we define an orbifold structure on S^2 using the two uniformizing charts $q_0 : \mathbb{C} \to V_0$ and $q_\infty : \mathbb{C} \to V_\infty$ defined by $q_0(z) = z^m$ and $q_\infty(w) = 1/w^n$. These two charts define a complex orbifold structure denoted $S^2_{(m,n)}$ on S^2: the group Γ_0 (resp. Γ_∞) consists of all the rotations of order m (resp. n) of $\mathbb{C}$ fixing 0; the changes from the chart q_0 to the chart q_∞ are local determinations of the multivalued holomorphic function $z \mapsto w = (1/z)^{m/n}$.

This orbifold is developable if and only if $m = n$. Indeed if $m = n$, then the map $z \mapsto z^m$ gives an isomorphism from the quotient of $\mathbb{C} \cup \{\infty\}$ by the action of the group generated by $z \mapsto e^{i\pi/m}z$ to the orbifold $S^2_{(n,n)}$. Conversely, if $S^2_{(m,n)}$ is developable, then there exists a connected manifold M, a subgroup Γ of the group of diffeomorphisms of M that acts properly, and a Γ-invariant continuous map $f : M \to S^2 = \mathbb{C} \cup \{\infty\}$ inducing the orbifold structure $S^2_{(m,n)}$ on S^2. The restriction of f to $M' = M \setminus f^{-1}(\{0, \infty\})$ is a connected covering of $\mathbb{C}^* = \mathbb{C} \setminus \{0\}$ whose number of sheets must be equal to both m and n, hence $m = n$.

(2) *Construction of 2-Dimensional Hyperbolic Orbifolds*. Consider a convex polygon P in the hyperbolic plane $\mathbb{H}^2$ such that the angle at each vertex v_i of P is π/n_i, where $n_i > 0$ is an integer. One can define an orbifold structure on P such that the uniformizing charts are defined on open subsets of $\mathbb{H}^2$ and the changes of charts are restrictions of hyperbolic isometries. One of these charts is the inclusion q_0 of the interior of P into P; and for each vertex v_i there is a uniformizing chart q_i defined on an open ball $B(v_i, \varepsilon)$ of small radius. The group Γ_i associated to this latter chart is generated by the reflections of $\mathbb{H}^2$ in the sides of P adjacent to v_i, and q_i maps $x \in B(v_i, \varepsilon)$ to the unique point of P which is in its Γ_i-orbit. Similarly, if y is a point of P in the interior of a side, a uniformizing chart is defined over a ball $B(y, \varepsilon)$ and maps $x \in B(y, \varepsilon)$ to itself if $x \in P$ and otherwise to its image under the reflection fixing the side containing it. A classical theorem of Poincaré (see for instance [Rh71], [Har91], [Rat94]) asserts that the subgroup of $\mathrm{Isom}(\mathbb{H}^2)$ generated by the reflections fixing the sides of P acts properly on $\mathbb{H}^2$ with P as a strict fundamental domain; it follows that the orbifold structure defined above on P is developable. This will also be a consequence of Theorem 1.13.

The same considerations apply to convex polygons P in $\mathbb{E}^2$ and $\mathbb{S}^2$ with vertex angles π/n_i. Note that in these cases, however, there are very few possibilities for such a polygon. (Exercise: List them.)

Let Q be the 2-sphere obtained by gluing two copies of P along their common boundaries. There is an obvious way to define a hyperbolic (resp. Euclidean or spherical) orbifold structure on Q with conical points corresponding to the vertices of P. (Exercise: Describe an atlas of uniformizing charts giving this structure.)

1.5 Exercises

(1) Let Γ be a finite subgroup of the group of diffeomorphisms of a connected, paracompact, differentiable manifold M. Let $q : M \to \Gamma\backslash M$ be the natural projection. Let $f : V \to W$ be a diffeomorphism between connected open subsets of M. Show that if $q \circ f = q|_V$ then f is the restriction of an element $g \in \Gamma$.

(Hint: Start from any Riemannian metric on M and by an averaging process prove the existence of a Γ-invariant Riemannian metric. Using the local exponential maps for this metric, show that one can introduce local coordinates around each point $x \in M$ such that, with respect to these coordinates, the action of the isotropy subgroup of x is orthogonal. Show that the open set $M_0 \subseteq M$ on which the action of Γ is free is everywhere dense. The restriction of f to $V \cap M_0$ is a Riemannian isometry, hence by continuity f itself is a Riemannian isometry. Given a point $x \in U \cap M_0$, there is a unique $g \in \Gamma$ such that $g.x = f(x)$. Then f is equal to g on a neighbourhood of x, hence the two isometries f and $g_{|U}$ are equal.)

(2) Let Γ be a group acting properly by diffeomorphisms on a smooth paracompact manifold M. Using a suitable partition of unity on M, show the existence of a Γ-invariant Riemannian metric on M.

(3) Show that every differentiable orbifold structure on a paracompact space Q carries a compatible Riemannian orbifold structure.

1.6 Remarks

(1) In the definition (1.1) of an orbifold, it is common to assume that the sources X_i of the uniformizing charts are open balls in $\mathbb{R}^n$. This leads to the same definition as ours. Equivalently we could have assumed that the sources X_i of the uniformizing charts $q_i : X_i \to V_i$ are differentiable manifolds on which the groups Γ_i (not supposed to be finite) act effectively and properly so that the maps q_i induce homeomorphisms $\Gamma_i\backslash X_i \to V_i$. And one could say that a connected orbifold is developable if and only if it can be defined by an atlas consisting of a single uniformizing chart.

(2) Topological orbifolds are defined in the same way as differentiable orbifolds: in definition (1.1) one replaces diffeomorphism by homeomorphism everywhere. It is still true that the changes of charts defined on a connected open set are well defined up to composition with an element of a finite group (see 1.1(iii) and 1.5(1)), but this is harder to prove than in the differentiable case. A proof of this fact , communicated to us by Bob Edwards, is based on the following basic theorem of M. H. A. Newman: if a finite group acts effectively by homeomorphisms of a connected manifold, then the set of points with trivial isotropy is open and dense ([New31] and [Dre69]).

Coverings of Orbifolds

1.7 Definition of a Covering of an Orbifold. Consider an orbifold structure on a space Q defined by an atlas $\{q_i : X_i \to V_i\}$ of uniformizing charts inducing homeomorphisms $\Gamma_i \backslash X_i \to V_i$. We assume that the X_i are disjoint; let X be the union of the X_i and let $q : X \to Q$ be the union of the q_i. Let $\mathcal{H}$ be the pseudogroup of changes of charts (1.2).

Let $p : \hat{X} \to X$ be a covering of X. On $\hat{X}$ there is a unique differentiable structure such that p is locally a diffeomorphism. We assume that for each change of charts $h : U \to U'$ in $\mathcal{H}$, there is a diffeomorphism $\hat{h} : p^{-1}(U) \to p^{-1}(U')$ commuting with p which is such that, for all $h, h' \in \mathcal{H}$,

$$\widehat{hh'} = \hat{h}\hat{h'} \text{ and } \widehat{h^{-1}} = \hat{h}^{-1}.$$

In particular this gives an action of each group Γ_i on $p^{-1}(X_i)$ that projects to the given action of Γ_i on X_i.

Let $\widehat{\mathcal{H}}$ be the pseudogroup generated by the diffeomorphisms $\hat{h}$. (More precisely the elements of $\widehat{\mathcal{H}}$ are the restrictions of the elements $\hat{h}$ to the open subsets of $\hat{X}$ and all unions of such restrictions.) Then $\widehat{\mathcal{H}}$ is the pseudogroup of changes of charts of an orbifold structure on the quotient $\hat{Q}$ of $\hat{X}$ by the equivalence relation whose classes are the orbits of $\widehat{\mathcal{H}}$. We shall now define this orbifold structure. Let $\hat{q} : \hat{X} \to \hat{Q}$ be the natural projection. There is a continuous map $\hat{p} : \hat{Q} \to Q$ which makes the diagram

$$\begin{array}{ccc} X & \xleftarrow{\ \ p\ \ } & \hat{X} \\ {\scriptstyle q}\downarrow & & \downarrow{\scriptstyle \hat{q}} \\ Q \cong \mathcal{H}\backslash X & \xleftarrow{\ \ \hat{p}\ \ } & \hat{Q} = \widehat{\mathcal{H}}\backslash \hat{X} \end{array} \tag{1.7-1}$$

commutative.

The uniformizing charts of $\widehat{Q}$ are the restrictions of $\hat{q}$ to the connected components of $p^{-1}(X_i)$. The group acting on such a component is the group formed by the restriction of those $\hat{\gamma}$, with $\gamma \in \Gamma_i$, that leave this component invariant. The only point to check is that $\hat{Q}$ with the quotient topology is Hausdorff. This follows easily from the hypothesis that Q is Hausdorff and the fact that the quotient of the open set $p^{-1}(X_i)$ by the action of the finite group Γ_i is Hausdorff.

A *covering of the orbifold* Q is a commutative diagram like (1.7-1). The orbifold $\hat{Q}$ is called a covering orbifold of the orbifold Q and the map $\hat{p} : \hat{Q} \to Q$ is the corresponding covering map.

1.8 Exercise. Show that, up to natural isomorphism, the notion of coverings for orbifolds is independent of the choice of atlas for the orbifold structure.

1.9 Galois Coverings. We maintain the notation of 1.7. Let G be a group acting on $\hat{X}$ so that the action is simply transitive on each fibre of p (in other words p is a Galois covering with Galois group G), and assume that the liftings $\hat{h}$ commute with the action of G. Then we say that $\hat{Q}$ is a *Galois covering* of Q with Galois group G. The action of G on $\hat{X}$ gives an action of G on $\hat{Q}$ and $Q = G\backslash\hat{Q}$. We shall see shortly that this action is proper.

As the sources X_i of the uniformizing charts are assumed to be simply connected, the covering p is trivial. After choosing a section $X_i \to p^{-1}(X_i)$, we can identify $p^{-1}(X_i)$ with $G \times X_i$; thus p becomes the projection on the second factor, and the action of G is by left translations on the first factor and is trivial on the second factor. The lifting $\hat{\gamma}$ of $\gamma \in \Gamma_i$ is of the form $(g, x) \mapsto (g\varphi_i(\gamma)^{-1}, \gamma.x)$, because $\hat{\gamma}$ commutes with the action of G. Moreover $\gamma \mapsto \varphi_i(\gamma)$ gives a homomorphism $\varphi_i : \Gamma_i \to G$.

Let $\hat{X}_i := \{1\} \times X_i \subseteq p^{-1}(X_i)$ and let $\hat{V}_i := \hat{q}(\hat{X}_i)$. The connected components of $\hat{X}$ are the open sets $g.\hat{X}_i$, indexed by $(g, i) \in G \times I$, and the uniformizing charts for $\hat{Q}$ are the maps $g.\hat{X}_i \to g.\hat{V}_i$ which are the restrictions of $\hat{q}$. Let $\Gamma_{(g,i)}$ be the group of diffeomorphisms of $g.\hat{X}_i$ formed by the restrictions to $g.\hat{X}_i$ of the elements $\hat{\gamma}$, where $\gamma \in \ker\varphi_i$ (in this way $\Gamma_{(g,i)}$ is isomorphic to $\ker\varphi_i$). The uniformizing charts $g.\hat{X}_i \to g.\hat{V}_i$ induce homeomorphisms $\Gamma_{(g,i)}\backslash g.\hat{X}_i \to g.\hat{V}_i$.

We have $\hat{Q} = \bigcup_{(g,i)} g.\hat{V}_i$. The subgroup $\varphi_i(\Gamma_i) \subseteq G$ leaves $\hat{V}_i$ invariant. Moreover $g.\hat{V}_i \cap \hat{V}_i \neq \emptyset$ if and only if $g \in \varphi_i(\Gamma_i)$. Indeed if $g.\hat{V}_i \cap \hat{V}_i \neq \emptyset$, then there exists $\hat{x} \in \hat{V}_i$ and $\gamma \in \Gamma_i$ such that $\hat{\gamma}(\hat{x}) \in \gamma.\hat{V}_i$, which implies that $g = \varphi_i^{-1}(\gamma)$. This shows that the action of G on $\hat{Q}$ is proper.

1.10 Lemma. *With the notations above, if the homomorphisms $\varphi_i : \Gamma_i \to G$ associated to the Galois covering are injective, then $\hat{Q}$ is naturally a differentiable manifold on which G acts properly by diffeomorphisms and the orbifold structure on Q is the quotient of $\hat{Q}$ by this action.*

Proof. The restrictions of $\hat{q}$ to the open sets $g.\hat{X}_i$ are homeomorphisms $g.\hat{X}_i \to g.\hat{V}_i$ which define on $\hat{Q}$ the structure of a differentiable manifold (because the changes of charts are the elements of $\widehat{\mathcal{H}}$ and therefore are differentiable). Clearly G acts on $\hat{Q}$ by diffeomorphisms. The restrictions $\hat{p}_i : \hat{V}_i \to V_i$ of $\hat{p}$ to the open sets $\hat{V}_i$ are the uniformizing charts for the quotient orbifold structure on Q. For each $i \in I$, let $f_i : X_i \to \hat{V}_i$ be the diffeomorphism sending x to the image under $\hat{q}$ of the point $(1, x) \in \hat{X}_i$. This map is a diffeomorphism which is φ_i-equivariant, i.e. $f_i(\gamma.x) = \varphi_i(\gamma).f_i(x)$ for all $\gamma \in \Gamma_i$. Indeed,

$$f_i(\gamma.x) = \hat{q}(\widehat{\gamma^{-1}}.(1, \gamma.x)) = \hat{q}((\varphi_i(\gamma), x)) = \varphi_i(\gamma).f_i(\gamma).$$

Moreover $\hat{p}_i(f_i(x)) = q_i(x)$. Therefore, via the isomorphisms f_i, the atlases (X_i, q_i) and $(\hat{V}_i, \hat{p}_i)$ are the same. □

Orbifolds with Geometric Structures

1.11 (G, Y)-Geometric Structure. Let G be a group acting by diffeomorphisms on a simply connected differentiable manifold Y. We assume that the action is *quasi-analytic*; this means that if the restriction of an element $g \in G$ to an open set U of Y is the identity of U, then g is the identity element of G.

A *(G, Y)-geometric structure* (or simply a (G, Y)-structure) on a differentiable manifold X is defined by a maximal atlas $\mathcal{A}$ made up of charts which are diffeomorphisms ψ from open sets of X to open sets of Y such that:

(i) the sources of the charts cover X,

(ii) for all $\psi, \psi' \in \mathcal{A}$, the change of charts $\psi'\psi^{-1}$, where it is defined, is locally the restriction of an element of G to an open set of Y.

Given (G, Y)-structures on X_1 and X_2, with maximal atlases $\mathcal{A}_1$ and $\mathcal{A}_2$, a diffeomorphism of an open subset of X_1 to an open subset of X_2 is called an isomorphism of (G, Y)-structures if the composition of h with any chart of $\mathcal{A}_2$ is a chart of $\mathcal{A}_1$.

For instance, a Euclidean, spherical or hyperbolic manifold of dimension n is an n-manifold with a (G, Y)-structure, where Y is respectively $\mathbb{E}^n$, $\mathbb{S}^n$, $\mathbb{H}^n$ and G is the full group of isometries of Y. An n-manifold with an affine structure is, by definition, a manifold with a (G, Y)-structure, where Y is the Euclidean space $\mathbb{R}^n$ and G is the group of affine transformations of $\mathbb{R}^n$. Interesting examples of (G, Y)-structures, where Y is the 2-sphere $\mathbb{S}^2$ and G the group of Möbius transformations, can be found in [SuTh83] (see also [Sco83]).

1.12 Orbifolds with a (G, Y)-Geometric Structure. Consider an orbifold structure on a space Q given by an atlas of uniformizing charts $q_i : X_i \to V_i$. A *(G, Y)-geometric structure on this orbifold* is given by a (G, Y)-geometric structure on each X_i which is left invariant by the changes of uniformizing charts (in particular by the groups Γ_i).

If Y is a Riemannian manifold and if G is a subgroup of the group of isometries of Y, then an orbifold with a (G, Y)-structure inherits a compatible Riemannian metric for which the charts of the (G, Y)-structure are Riemannian isometries.

Typical examples of such orbifolds are obtained by taking the quotient of Y by a subgroup of G acting properly on Y, or by taking the quotient of an open subset $Y_0 \subset Y$ by a subgroup of G that leaves Y_0 invariant and acts properly on Y_0. For interesting examples of hyperbolic orbifolds, see [Thu79, chapter 11] and [Rat94].

1.13 Theorem. *Let Q be a connected orbifold with a (G, Y)-geometric structure. Then*

(1) *Q is developable. More precisely, there is a subgroup Γ of G acting properly on a connected manifold M endowed with a (G, Y)-structure such that Q with its (G, Y)-structure is naturally the quotient of M by this action of Γ. Moreover there is a map $D : M \to Y$ which is Γ-equivariant and which is locally an*

isomorphism of (G, Y)-structures. The map D is called the development of Q and Γ is called the holonomy group; both D and Γ are well defined up to conjugation by an element of G.

(2) *Assume that Y is a Riemannian manifold and that G is a subgroup of the group of isometries of Y (so the orbifold structure on Q is Riemannian). Assume also that Q is complete with respect to the quotient metric (this is the case if Q is compact). Then there is a natural identification of M with Y under which D becomes the identity map.*

Before giving the proof of this theorem, which is very similar to the proof of (I.3.32), we follow a digression concerning the spaces of germs of continuous maps.

1.14 Spaces of Germs. Let X and Y be topological spaces. Consider the set of pairs (f, x), where f is a continuous map from an open subset $U \subseteq X$ to Y and $x \in U$. We introduce an equivalence relation on this set: $(f, x) \sim (f', x')$ if and only if $x = x'$ and f is equal to f' on some neighbourhood of x. The equivalence class of (f, x) is called the *germ of f at x*. The point x (resp. $f(x)$) is called the source (resp. the target) of the germ of f at x.

Let $\mathcal{M}(Y, X)$ be the set of germs of continuous maps from open sets of X to Y and let $\alpha : \mathcal{M}(Y, X) \to X$ (resp. $\omega : \mathcal{M}(Y, X) \to Y$) be the map associating to a germ its source (resp. its target). On $\mathcal{M}(Y, X)$ there is a natural topology, a basis of which consists of the subsets U_f which are the unions of the germs of continuous maps $f : U \to Y$ at the various points of U. The projections α and ω are continuous maps and α is an *étale map*, i.e. α maps open sets to open sets and its restriction to any sufficiently small open set is a homeomorphism onto its image. Note that in general $\mathcal{M}(Y, X)$ is not a Hausdorff space.

Consider a topological space Z, continuous maps $f : U \to Y$ and $f' : U' \to Z$, where U and U' are respectively open sets of X and Y, and a point $x \in U$ such that $f(x) \in U'$. The germ of $f'f : f^{-1}(U') \to Z$ at x depends only on the germ g of f at x and of the germ g' of f' at $f(x)$ and is called the *composition* $g'g$ of those germs. Let $\mathcal{M}(Z, Y) \times_Y \mathcal{M}(Y, X)$ be the subspace of $\mathcal{M}(Z, Y) \times \mathcal{M}(Y, X)$ formed by pairs (g', g) such that $\alpha(g') = \omega(g)$. The composition of germs gives a continuous map $\mathcal{M}(Z, Y) \times_Y \mathcal{M}(Y, X) \to \mathcal{M}(Z, X)$.

Proof of Theorem 1.13. To a (G, Y)-structure on an orbifold Q we shall associate a Galois covering with Galois group G satisfying the hypothesis of Lemma 1.10.

We can assume that the sources X_i of the uniformizing charts $q_i : X_i \to V_i$ are disjoint and are the sources of charts $\psi_i : X_i \to \psi_i(X_i) \subset Y$ for the given (G, Y)-geometric structure. Let X be the union of the X_i and let $q : X \to Q$ be the union of the uniformizing charts. Let $\hat{X}$ be the space of germs of all the charts from X to Y defining the (G, Y)-structure on X (this is an open subspace of the space of all germs of continuous maps from open sets of X to Y). The group G acts by diffeomorphisms on $\hat{X}$: if $g \in G$ and $\hat{x}$ is the germ of a chart ψ at x, then $g.\hat{x}$ is the germ of the chart $g \circ \psi$ at x.

We have two projections $p : \hat{X} \to X$ and $\Delta : \hat{X} \to Y$ associating to each germ, respectively, its source and target. Clearly p is G-invariant and Δ is both G-equivariant and locally a diffeomorphism.

We first claim that $p : \hat{X} \to X$ is a Galois covering with Galois group G, i.e. G acts simply transitively on the fibres of p. To see this, let $\hat{X}_i$ be the union of the germs of ψ_i at the points $x \in X_i$. It follows from condition (ii) in (1.11) that $p^{-1}(X_i)$ is the union of the open sets $g.\hat{X}_i$, where $g \in G$, and the condition of quasi-analyticity on the action of G on Y implies that these open sets are disjoint.

Let $h : U \to V$ be an element of the pseudogroup $\mathcal{H}$ of changes of uniformizing charts. Such a map has a canonical lifting $\hat{h} : p^{-1}(U) \to U \cap p^{-1}(U')$: if $\hat{x}$ is the germ at $x \in U$ of a chart ψ for the (G, Y)-structure on X, then $\hat{h}(\hat{x})$ is the germ at $h(x)$ of the chart ψh^{-1}. We have $\widehat{h^{-1}} = \hat{h}^{-1}$, and if $h' : U' \to V'$ is another orbifold change of charts, then $\hat{h}\hat{h'} = \widehat{hh'}$ on $p^{-1}(h^{-1}(U')$. Note that each $\hat{h}$ commutes with the action of G.

Let $\widehat{\mathcal{H}}$ be the pseudogroup of local diffeomorphisms of $\hat{X}$ generated by the elements $\hat{h}$. Let $\hat{Q}$ be the quotient of $\hat{X}$ by the equivalence relation whose classes are the orbits of $\widehat{\mathcal{H}}$. Let $\hat{q}$ be the natural projection $\hat{X} \to \hat{Q}$ and let $\hat{p} : \hat{Q} \to Q$ be the map such that $qp = \hat{p}\hat{q}$.

According to (1.8), these data define an orbifold structure on $\hat{Q}$ which can be considered as a covering of Q with covering projection $\hat{p} : \hat{Q} \to Q$. Moreover, as the elements $\hat{h}$ commute with the action of G on $\hat{X}$, we can consider $\hat{Q}$ as a Galois covering of Q (see 1.9); the group G acts properly on $\hat{Q}$ with quotient Q.

As $\Delta : \hat{X} \to Y$ obviously commutes with the action of $\widehat{\mathcal{H}}$, we get a continuous map $\hat{D} : \hat{Q} \to Y$ such that $\Delta = \hat{D} \circ \hat{q}$. Therefore we have the commutative diagram

$$\begin{array}{ccccc}
X & \xleftarrow[p]{\quad} & \hat{X} & \xrightarrow{\ \Delta\ } & Y \\
{\scriptstyle q}\big\downarrow & & {\scriptstyle \hat{q}}\big\downarrow & & {\scriptstyle id}\big\downarrow \\
Q & \xleftarrow[\hat{p}]{\quad} & \hat{Q} = \widehat{\mathcal{H}}\backslash\hat{X} & \xrightarrow{\ \hat{D}\ } & Y
\end{array}$$

It follows that the restriction of $\hat{q}$ to each of the open sets $\hat{X}_i$ is a homeomorphism onto its image (denoted $\hat{V}_i \subset \hat{Q}$) because Δ is a local isomorphism. Also the restriction of $\hat{q}$ to $g.\hat{X}_i$ is a homeomorphism onto $g.\hat{V}_i$. Since $\hat{X}$ is locally diffeomorphic to Y, this implies that $\hat{Q}$ is a manifold. It also implies that the homomorphisms $\varphi_i : \Gamma_i \to G$ described in (1.9) are injective. According to lemma 1.10, the orbifold Q is the quotient of the manifold $\hat{Q}$ by the proper action of G. Moreover $\hat{Q}$ is endowed with the (G, Y)-structure for which $\hat{D}$ is locally an isomorphism of (G, Y)-structures.

The connected components of $\hat{Q}$ are permuted by the elements of G. Fix a connected component M of $\hat{Q}$. Let Γ be the subgroup of G leaving M invariant and let D be the restriction of $\hat{D}$ to M. As Q is connected, it is the orbifold quotient of M by the action of Γ. This proves (1).

Suppose that G leaves invariant a Riemannian metric on Y. Then there is on X a unique Riemannian metric such that all of the maps ψ_i are isometries. This metric is invariant by the pseudogroup $\mathcal{H}$ of changes of uniformizing charts. $\hat{X}$ also has a unique Riemannian metric such that Δ is locally an isometry; it is invariant by the action of G and the pseudogroup $\widehat{\mathcal{H}}$. Assume that the metric on Q which is the quotient of the metric on X associated to this Riemannian metric is complete. This metric is the same as the quotient metric on Q associated to the action of Γ by isometries of the manifold M. As the action of Γ is proper, it follows that the metric on M is also complete. According to (I.3.28), this implies that $D : M \to Y$ is a covering. And as Y is assumed to be simply connected, this implies that D is an isometry. This proves (2). □

2. Étale Groupoids, Homomorphisms and Equivalences

Étale Groupoids

2.1 Definition of an Étale Groupoid. A *groupoid* $(\mathcal{G}, X)$ is a small category $\mathcal{G}$ (see $\mathcal{C}$.A.1) whose elements are all invertible; X is the set of its objects, identified to the set of units of $\mathcal{G}$ by the map associating to $x \in X$ the unit $1_x \in \mathcal{G}$. The projection $\mathcal{G} \to X$ that associates to each element of $\mathcal{G}$ its initial object (resp. its terminal object) will be denoted α (resp. ω) (these maps were denoted i and t in $\mathcal{C}$). Given $g \in \mathcal{G}$, the object $\alpha(g)$ (resp. $\omega(g)$) will be called its source (resp. its target).

Given $x \in X$, the set $\mathcal{G}_x = \{g \in \mathcal{G} \mid \alpha(g) = \omega(g) = x\}$ is a group called the *isotropy group* of x. The subset $\mathcal{G}.x = \{y \in X \mid \exists g \in \mathcal{G}, \alpha(g) = x, \omega(g) = y\}$ is the *$\mathcal{G}$-orbit* or simply the orbit of x. The $\mathcal{G}$-orbits form a partition of X and the set of $\mathcal{G}$-orbits will be denoted $\mathcal{G}\backslash X$.

A *topological groupoid* $(\mathcal{G}, X)$ is a groupoid with a topology on $\mathcal{G}$ and X such that the natural inclusion $X \to \mathcal{G}$ is a homeomorphism onto its image, and the projections α, ω, as well as composition and passage to inverses, are continuous. The space of orbits $\mathcal{G}\backslash X$ is endowed with the quotient topology.

We say that a topological groupoid is *étale* if the maps α and ω are étale maps, i.e. are locally homeomorphisms.

Associated to each étale groupoid $(\mathcal{G}, X)$ *there is a pseudogroup of local homeomorphisms of* X. The elements of this pseudogroup are the homeomorphisms $h : U \to V$ of the form $h = \omega \circ s$, where $s : U \to \mathcal{G}$ is a continuous section of α over U, i.e. $\alpha(s(x)) = x$, $\forall x \in U$. If X is a differentiable manifold and if the elements of the associated pseudogroup are diffeomorphisms, then $(\mathcal{G}, X)$ is called *a differentiable étale groupoid.*

2.2 The Groupoid Associated to a Group Action. Let Γ be a group acting by homeomorphisms on a set X. The étale groupoid $(\Gamma \ltimes X, X)$ associated to this action is the category whose space of objects is the space X and whose space of elements is $\mathcal{G} = \Gamma \times X$, where Γ is endowed with the discrete topology. The projections α and ω are defined by $\alpha(\gamma, x) = x$ and $\omega(\gamma, x) = \gamma.x$. The composition $(\gamma, x)(\gamma', x')$ is defined if $x = \gamma'.x'$ and is equal to $(\gamma\gamma', x')$. The inverse of (γ, x) is $(\gamma^{-1}, \gamma.x)$.

We describe a series of other basic examples.

2.3 Examples
(1) Among the étale groupoids, we have two extreme cases, namely:

(a) X is a single point and $\mathcal{G}$ is just a discrete group;

(b) X is a topological space and all the elements of $\mathcal{G}$ are units (equivalently, the inclusion $X \to \mathcal{G}$ is a bijection). We say that this is the trivial groupoid X.

(2) *The Étale Groupoid Associated to a Pseudogroup of Local Homeomorphisms.* Let $\mathcal{H}$ be a pseudogroup of local homeomorphisms of a topological space X. Its associated étale groupoid $(\mathcal{G}, X)$ is the groupoid of all the germs of the elements of $\mathcal{H}$, with the germ topology. (This is an open subspace of the space of all germs of continuous maps from open sets of X to X.) The natural inclusion of X into $\mathcal{G}$ associates to $x \in X$ the germ 1_x at x of the identity map of X. The projections α and ω associate to a germ its source and target, and the composition is the composition of germs. From $\mathcal{G}$ one can reconstruct $\mathcal{H}$ as the pseudogroup associated to $\mathcal{G}$.

In general, if $\mathcal{H}$ is the pseudogroup associated to an étale groupoid $(\mathcal{G}, X)$ (see 2.1), then the étale groupoid of germs associated to $\mathcal{H}$ is a quotient of $(\mathcal{G}, X)$. (Consider for instance example 2.3(1a).)

(3) *The Étale Groupoid Associated to an Atlas of Uniformizing Charts for an Orbifold.* This is the étale groupoid $(\mathcal{G}, X)$ associated to the pseudogroup of changes of charts (cf. 1.2); it determines the orbifold structure on Q if one identifies Q with the quotient $\mathcal{G}\backslash X$ of X by the equivalence relation whose classes are the orbits of $\mathcal{G}$.

(4) **The Étale Groupoid Associated to a Complex of Groups.** Let $G(\mathcal{Y}) = (G_\sigma, \psi_a, g_{a,b})$ be a complex of groups over a scwol $\mathcal{Y}$ as defined in ($\mathcal{C}$.2.1). We construct an étale groupoid $(\mathcal{G}(\mathcal{Y}), \tilde{Y})$ canonically associated to $G(\mathcal{Y})$ as follows (see fig. $\mathcal{C}$.13). We use the notations of ($\mathcal{C}$.4.9, 4.13, 4.14). The space of units $\tilde{Y}$ will be the disjoint union of the spaces $\mathrm{st}(\tilde{\sigma})$, where $\sigma \in V(\mathcal{Y})$. The groupoid $\mathcal{G}(\mathcal{Y})$ will be the disjoint union of the subspaces $\mathcal{G}_{\tau,\sigma}$ of elements with sources in $\mathrm{st}(\tilde{\sigma})$ and targets in $\mathrm{st}(\tilde{\tau})$. There are three cases to consider.

First case: $\sigma = \tau$. In this case $(\mathcal{G}_{\sigma,\sigma}, \mathrm{st}(\tilde{\sigma}))$ will be the groupoid

$$(G_\sigma \ltimes \mathrm{st}(\tilde{\sigma}), \mathrm{st}(\tilde{\sigma}))$$

associated to the action of G_σ on $\mathrm{st}(\tilde{\sigma})$.

Second case: There is an edge in $E(\mathcal{Y})$ with initial vertex σ and terminal vertex τ. In this case

$$\mathcal{G}_{\tau,\sigma} = \bigcup_a G_\tau \times \mathrm{st}(\tilde{a}),$$

where $a \in E(\mathcal{Y})$ with $i(a) = \sigma$ and $t(a) = \tau$. Recall that the subspaces $\mathrm{st}(\tilde{a})$ are disjoint in $\mathrm{st}(\tilde{\sigma})$ (exercise $\mathcal{C}$.4.6). The projection α (resp. ω) maps $(g, x) \in G_\tau \times \mathrm{st}(\tilde{a})$ to x (resp. $g.f_a(x)$), where f_a is as defined in ($\mathcal{C}$.4.14)). For this reason, it will be more suggestive to use the notation (g, f_a, x) for the element $(g, x) \in G_\tau \times \mathrm{st}(\tilde{a})$.

If $(h, y) \in \mathcal{G}_{\sigma,\sigma}$ is such that $x = h.y$, the composition $(g, f_a, x)(h, y)$ is defined to be $(g\psi_a(h), f_a, y)$. If $(g', x') \in \mathcal{G}_{\tau,\tau}$ with $x' = g.f_a(x)$, the composition $(g', x')(g, f_a, x)$

is equal to $(g'g, f_a, x)$. If $(a, b) \in E^{(2)}(\mathcal{Y})$ with $i(b) = \rho$ and if $(g, f_a, x) \in \mathcal{G}_{\tau,\sigma}$ and $(h, f_b, y) \in \mathcal{G}_{\sigma,\rho}$ with $x = h.f_b(y)$, the composition $(g, f_a, x)(h, f_b, y)$ is equal to $(g\psi_a(h)g_{a,b}, f_{ab}, y)$.

As a set $\mathcal{G}_{\sigma,\tau}$ is equal to $\mathcal{G}_{\tau,\sigma}$ but the projections α and ω are exchanged, thus its elements are (a priori formal) inverses of the elements of $\mathcal{G}_{\tau,\sigma}$. Thus to each element $(g, f_a, x) \in \mathcal{G}_{\tau,\sigma}$ is associated a symbol $(g, f_a, x)^{-1} \in \mathcal{G}_{\sigma,\tau}$ such that $\alpha((g, f_a, x)^{-1})) = g.f_a(x)$ and $\omega((g, f_a, x)) = x$. The composition is defined accordingly. For instance, if $(g', y') \in \mathcal{G}_{\tau,\tau}$ with $g'.y' = g.f_a(x)$, then $(g, f_a, x)^{-1}(g', y') = (g'^{-1}g, f_a, x)^{-1}$, and if $(h, x) \in \mathcal{G}_{\sigma,\sigma}$, then $(h, x)(g, f_a, x)^{-1} = (g\psi_a(h^{-1}), f_a, h.x)^{-1}$. Now assume that a, a' are two distinct elements of $E(\mathcal{Y})$ with $t(a) = t(a') = \tau$ and $i(a) = \sigma,\ i(a') = \sigma'$. Let $(g, f_a, x) \in \mathcal{G}_{\tau,\sigma}$ and $(g', f_{a'}, x') \in \mathcal{G}_{\tau,\sigma'}$ be such that $g.f_a(x) = g'.f_{a'}(x')$. Then $(g', f_{a'}, x')^{-1}(g, f_a, x)$ and $(g, f_a, x)^{-1}(g', f_{a'}, x')$ should be defined. The natural projections from $\mathrm{st}(\tilde{\sigma})$ and $\mathrm{st}(\tilde{\sigma}')$ to $|\mathcal{Y}|$ map x and x' to a point in $\mathrm{st}(a) \cap \mathrm{st}(a')$; it follows (see exercise $\mathcal{C}$.4.6) that there is a unique $b \in E(\mathcal{Y})$ such that either $a' = ab$ or $a = a'b$; exchanging the roles of a and a' we can assume that $a = a'b$. We then define the composition $(g', f_{a'}, x')^{-1}(g, f_a, x)$ as the unique element $(h, f_b, x) \in \mathcal{G}_{\sigma',\sigma}$ such that $(g, f_a, x) = (g', f_{a'}, x')(h, f_b, x)$, where h is defined by the equation $g'\psi_{a'}(h)g_{a',b} = g$. The composition $(g, f_a, x)^{-1}(g', f_{a'}, x')$ is defined to be $(h, f_b, x)^{-1}$. One proceeds similarly to define, for instance, the composition $(g', f_{a'}, x)(g, f_a.x)^{-1}$ when $i(a) = i(a')$.

Third case: σ and τ are distinct and not joined by an edge. In this case $\mathcal{G}_{\tau,\sigma}$ is empty.

We leave the reader to check that the law of composition defined above is associative (one needs to refer to ($\mathcal{C}$.2.1) and ($\mathcal{C}$.4.14)).

(5) *The Holonomy Groupoid of a Foliation.* A foliation $\mathcal{F}$ of codimension n on a manifold M of dimension m can be defined by a family of maps $f_i : U_i \to \mathbb{R}^n$, where $\{U_i\}_{i \in I}$ is an open cover of M. The maps f_i, called local projections, are submersions from U_i onto open sets V_i of $\mathbb{R}^n$ with connected fibres (i.e. $f_i^{-1}(x)$ is a connected submanifold of U_i for each $x \in V_i$). The following compatibility condition must be satisfied: for each $y \in U_i \cap U_j$ there is an open neighbourhood $U_{ji}^y \subseteq U_i \cap U_j$ and a homeomorphism $h_{ji}^y : f_i(U_{ji}^y) \to f_j(U_{ji}^y)$ such that $f_j = h_{ji}^y \circ f_i$ on $U_{i,j}^y$. Two such families of submersions define the same foliation $\mathcal{F}$ if their union still satisfies the analogous compatibility conditions. The leaves of $\mathcal{F}$ are the $(m-n)$-dimensional submanifolds of M which are the connected components of M endowed with the topology whose basis is the set of open sets in the fibres of the submersions f_i.

Let X be the disjoint union of the open sets V_i. The *holonomy pseudogroup* of $\mathcal{F}$, defined by the family of submersions $\{f_i : U_i \to V_i \subset X\}$, is the pseudogroup of local homeomorphisms of X generated by the germs of the h_{ji}^y. The *holonomy groupoid* of $\mathcal{F}$ is the groupoid of germs associated to its holonomy pseudogroup. The holonomy groupoid associated to a different family of compatible submersions that defines the same foliation will be equivalent to the previous one in the sense of the definition that follows.

Equivalence and Developability

2.4 Homomorphisms and Equivalences of Groupoids. A *homomorphism* (ϕ, f) from an étale groupoid $(\mathcal{G}, X)$ to an étale groupoid $(\mathcal{G}', X')$ consists of a continuous functor $\varphi : \mathcal{G} \to \mathcal{G}'$ inducing a continuous map $f : X \to X'$. We say that (φ, f) is an *étale homomorphism* if f is an étale map.

(φ, f) is an *equivalence* if f is an étale map and if the functor φ is an equivalence in the sense of ($\mathcal{C}$.A.3), i.e.

(1) for each $x \in X$, φ induces an isomorphism from $\mathcal{G}_x$ onto $\mathcal{G}'_{f(x)}$,

(2) $f : X \to X'$ induces a bijection $\mathcal{G}\backslash X \to \mathcal{G}'\backslash X'$ of the orbits sets.

In this case we say that the étale groupoids $(\mathcal{G}, X)$ and $(\mathcal{G}', X')$ are equivalent. This *generates* an equivalence relation among étale groupoids.

If $(\mathcal{G}, X)$ and $(\mathcal{G}', X')$ are differentiable étale groupoids (see 2.1), a homomorphism $(\varphi, f) : (\mathcal{G}, X) \to (\mathcal{G}', X')$ is called a *differentiable equivalence* if it is an equivalence and f is locally a diffeomorphism. The equivalence generated by this relation is called *differentiable equivalence*.

2.5 *Remark.* One can check (see exercise 2.8(1)) that two étale groupoids $(\mathcal{G}, X)$ and $(\mathcal{G}', X')$ are related by the above equivalence relation if and only if there is an étale groupoid $(\mathcal{G}'', X'')$ and étale homomorphisms $(\varphi, f) : (\mathcal{G}'', X'') \to (\mathcal{G}, X)$ and $(\varphi', f') : (\mathcal{G}'', X'') \to (\mathcal{G}', X')$ which are equivalences.

There is a more general way of describing directly an equivalence from $(\mathcal{G}, X)$ to $(\mathcal{G}', X')$. We shall explain this through a series of exercises (see 2.8(3)).

2.6 Definition of Developability. An étale groupoid $(\mathcal{G}, X)$ is *developable* if it is equivalent to the groupoid $(\Gamma \times \tilde{X}, \tilde{X})$ associated to an action of a group Γ by homeomorphisms of a space $\tilde{X}$ (see 1.3).

2.7 Examples

(1) Let $(\mathcal{G}, X)$ be an étale groupoid and let $U \subseteq X$ be an open subset meeting all the $\mathcal{G}$-orbits. Let $(\mathcal{G}|_U, U)$ be the restriction of $(\mathcal{G}, X)$ to U, i.e. the subgroupoid formed by the elements with source and target in U. Then the natural inclusion $(\mathcal{G}|_U, U) \to (\mathcal{G}, X)$ is an equivalence.

(2) Let Γ be a group acting freely by homeomorphisms on a topological space $\tilde{X}$ so that the natural projection $p : \tilde{X} \to \Gamma\backslash\tilde{X} = X$ is a covering map. Let $\pi : \Gamma \times \tilde{X} \to X$ be the map sending $(\gamma, \tilde{x})$ to $p(\tilde{x}) = p(\gamma.\tilde{x})$. Then the étale homomorphism (π, p) from the étale groupoid associated to the action of Γ on $\tilde{X}$ to the trivial groupoid X is an equivalence.

More generally, if $(\mathcal{G}, X)$ is an etale groupoid such that the natural projection of X onto $\mathcal{G}\backslash X$ with the quotient topology is étale, and if $\mathcal{G}_x$ is trivial for all $x \in X$, then $(\mathcal{G}, X)$ is equivalent to the trivial groupoid $\mathcal{G}\backslash X$. For instance the groupoid of germs of all the changes of charts of an atlas defining a given structure on a manifold M is equivalent to the trivial groupoid M.

(3) The étale groupoid of germs of changes of charts of an orbifold is developable if and only if the orbifold structure is developable.

An orbifold structure on a topological space Q could be defined as a differentiable equivalence class of differentiable étale groupoids $(\mathcal{G}, X)$ together with a homeomorphism from $\mathcal{G}\backslash X$ onto Q, such that:

(i) $\mathcal{G}\backslash X$ is Hausdorff, and

(ii) each point of X has a neighbourhood U such that the restriction of $(\mathcal{G}, X)$ to U is the groupoid associated to an effective action of a finite group on U.

(4) **The Developability of Complexes of Groups.** A complex of groups $G(\mathcal{Y})$ over a scwol $\mathcal{Y}$ is developable in the sense of ($\mathcal{C}$.2.11) if and only if its associated groupoid $(\mathcal{G}(\mathcal{Y}), \tilde{Y})$ is developable.

To see this, let us first assume that $G(\mathcal{Y}) = (G_\sigma, \psi_a, g_{a,b})$ is developable: thus there is a group G acting on a scwol $\mathcal{X}$ such that $\mathcal{Y} = G\backslash\mathcal{X}$ and $G(\mathcal{Y})$ is the associated complex of groups with respect to some choices (we use the notations of $\mathcal{C}$.2.9(1)). Let X be the geometric realization of $\mathcal{X}$ and let $G \times X \to X$ be the corresponding action. For each $\sigma \in V(\mathcal{Y})$, there is a canonical isomorphism $f_\sigma : \mathrm{st}(\tilde{\sigma}) \to \mathrm{st}(\overline{\sigma})$ which is G_σ-equivariant (see ($\mathcal{C}$.4.11)). There is a unique étale homomorphism (φ, f) : $(\mathcal{G}(\mathcal{Y}), \tilde{Y}) \to (G \ltimes X, X)$, where $f : \tilde{Y} \to X$ is the union of the maps f_σ, and φ is defined by:

$$\varphi(g, x) = (g, f_\sigma(x)) \ \text{ if } \ (g, x) \in G_\sigma \times \mathrm{st}(\tilde{\sigma}) = \mathcal{G}_{\sigma,\sigma}$$
$$\varphi(g, x) = (gh_a, f_{i(a)}(x)) \ \text{ if } \ (g, x) \in G_{t(a)} \times \mathrm{st}(\tilde{a}) \subseteq \mathcal{G}_{t(a),i(a)}.$$

These formulae define a functor $\varphi : \mathcal{G}(\mathcal{Y}) \to G \ltimes X$. To check that this is the case, note that for each $x \in \mathrm{st}(\tilde{a})$, we have $h_a.f_{i(a)}(x) = f_{t(a)}(f_a(x))$ and we use $\mathcal{C}$.2.9(1) and $\mathcal{C}$.4.14. It is clear that (φ, f) is an equivalence.

Conversely, assume that $(\mathcal{G}(\mathcal{Y}), \tilde{Y})$ is equivalent to the étale groupoid associated to the action of a group G acting by homeomorphisms on a topological space X. Let Y be the geometric realization of $\mathcal{Y}$. We can identify Y to $\mathcal{G}(\mathcal{Y})\backslash\tilde{Y}$ and to $G\backslash X$.

By exercise 2.8(3)(e), there is an étale homomorphism $(\varphi, f) : (\mathcal{G}(\mathcal{Y}), \tilde{Y}) \to (G \ltimes X, X)$ which is an equivalence (because each connected component of $\tilde{Y}$ is simply connected), and $f : \tilde{Y} \to X$ induces the identity on Y. Therefore, for each $\sigma \in V(\mathcal{Y})$, the homomorphism φ maps $(g, x) \in G_\sigma \times \mathrm{st}(\tilde{\sigma})$ to an element of the form $(\varphi_\sigma(g), f(x)) \in G \times X$, and the map $g \mapsto \varphi_\sigma(g)$ is an isomorphism onto the isotropy subgroup of $\overline{\sigma} = f(\sigma)$. If $a \in E(\mathcal{Y})$ and $x \in \mathrm{st}(\tilde{a})$, then $\varphi(1_{t(a)}, x) \in G \times X$ is of the form $(\varphi(a), f(x))$. It is easy to check that the injective homomorphisms φ_σ and the elements $\varphi(a) \in G$ define a morphism $G(\mathcal{Y}) \to G$, hence $G(\mathcal{Y})$ is developable by ($\mathcal{C}$.2.15). It is also easy to see that X is G-equivariantly homeomorphic to the geometric realization of the development of $\mathcal{Y}$ associated to this morphism (cf. $\mathcal{C}$.2.13).

(5) It is a remarkable fact that any connected étale groupoid of germs associated to a pseudogroup of analytic local diffeomorphisms of an analytic one-dimensional manifold is always developable. Indeed such a groupoid is equivalent to the groupoid associated to a quasi-analytic action of a group Γ on a simply connected analytic

manifold of dimension one (this manifold is not Hausdorff in general). This applies in particular to the groupoid (classically denoted $\Gamma_1^{(\omega)}$) of all germs of analytic local diffeomorphisms of $\mathbb{R}$, or to the holonomy groupoid of an analytic foliation of codimension one. This follows from the main lemma in [Hae58], its interpretation in [Hae70] and the criterion for developability given in exercise 3.20(2). The corresponding statement is not true in general if "analytic" is replaced by "differentiable".

(6) Let $\mathcal{F}$ be a foliation on a connected manifold M defined by a family of submersions $f_i : U_i \to V_i \subseteq \mathbb{R}^n$ and let $(\mathcal{G}, X)$ be the associated holonomy groupoid of $\mathcal{F}$ (see 2.3(5)). The induced foliation $p^{-1}(\mathcal{F})$ on a covering $p : \hat{M} \to M$ of M can be defined by the family of submersions obtained by composing f_i with the restrictions of p to the connected components of $p^{-1}(U_i)$.

The developability of the holonomy groupoid $(\mathcal{G}, X)$ of $\mathcal{F}$ can be interpreted in terms of foliation theory. One can show that $(\mathcal{G}, X)$ is equivalent to the groupoid associated to an action of a group Γ on a connected manifold[76] $\hat{X}$ if and only if there is a Galois covering $p : \hat{M} \to M$ with Galois group Γ and a Γ-equivariant submersion $\hat{f} : \hat{M} \to \hat{X}$ with connected fibres which are the leaves of the foliation $p^{-1}(\mathcal{F})$.

2.8 Exercises

(1) Let $(\mathcal{G}_1, X_1)$ and $(\mathcal{G}_2, X_2)$ be two equivalent étale groupoids. Show that there is a groupoid $(\mathcal{G}, X)$ and two étale homomorphisms $(\varphi_i, f_i) : (\mathcal{G}, X) \to (\mathcal{G}_i, X_i)$, $i = 1, 2$, which are equivalences.

(Hint: If $(\psi_i, h_i) : (\mathcal{G}_i, X_i) \to (\mathcal{G}', X')$ are two étale homomorphisms which are equivalences, let $\mathcal{G} = \{(g_1, g_2) \mid g_i \in \mathcal{G}_i, \psi_1(g_1) = \psi_2(g_2)\}$ and let $X = \{(x_1, x_2) \mid x_i \in X_i, h_1(x_1) = h_2(x_2)\}$. Show that $(\mathcal{G}, X)$ is naturally an étale groupoid and that the natural projections $(\mathcal{G}, X) \to (\mathcal{G}_i, X_i)$ are equivalences.)

(2) Given an étale groupoid $(\mathcal{G}, X)$ and an open cover $\mathcal{U} = \{U_i\}_{i \in I}$ of X indexed by a set I, let U be the disjoint union of the U_i, namely the set of pairs (i, x), with $x \in U_i$. For $i, j \in I$, let $\mathcal{G}_{ij}$ be the set of triples (i, g, j) where $g \in \mathcal{G}$ and $\alpha(g) \in U_j$, $\omega(g) \in U_i$. Let $\mathcal{G}_{\mathcal{U}}$ be the disjoint union of the $\mathcal{G}_{ij}$. Define on $(\mathcal{G}_{\mathcal{U}}, U)$ the structure of an étale groupoid such that the natural projection $(i, g, j) \mapsto g$ gives an étale equivalence to $(\mathcal{G}, X)$. We say that $(\mathcal{G}_{\mathcal{U}}, U)$ is the groupoid obtained from $(\mathcal{G}, X)$ by localization on the cover $\mathcal{U}$.

Show that two étale groupoids $(\mathcal{G}, X)$ and $(\mathcal{G}', X')$ are equivalent if and only if there exist open covers $\mathcal{U}$ and $\mathcal{U}'$ of X and X' such that the localizations $(\mathcal{G}_{\mathcal{U}}, U)$ and $(\mathcal{G}'_{\mathcal{U}'}, U')$ are isomorphic.

(3) In this series of exercises, we present a more direct definition of equivalence between two topological groupoids $(\mathcal{G}_1, X_1)$ and $(\mathcal{G}_2, X_2)$. This definition can be found in Jean Renault (see [Ren82]); it is equivalent to the definition given in (2.4).

Consider a topological space $\mathcal{E}$ with two surjective étale maps $\alpha : \mathcal{E} \to X_1$ and $\omega : \mathcal{E} \to X_2$. A right action of $\mathcal{G}_1$ on $\mathcal{E}$, over $\alpha : \mathcal{E} \to X_1$, is a continuous

[76] This manifold is not Hausdorff in general but the action is always quasi-analytic.

map $\mathcal{E} \times_{X_1} \mathcal{G}_1 \to \mathcal{E}$, written $(e, g_1) \mapsto e.g_1$, where $\mathcal{E} \times_{X_1} \mathcal{G}_1$ is the subspace of $\mathcal{E} \times \mathcal{G}_1$ consisting of pairs (e, g_1) such that $\alpha(e) = \omega(g_1)$. This action must satisfy the following conditions: if $e \in \mathcal{E}, g_1, g_1' \in \mathcal{G}_1$ are such that $\alpha(e) = \omega(g_1)$ and $\alpha(g_1) = \omega(g_1')$, then $(e.g_1).g_1' = e.(g_1 g_1')$; and $e.1_{\alpha(e)} = e$.

This action is said to be *simply transitive* with respect to $\omega : \mathcal{E} \to X_2$ if each point of X_2 has an open neighbourhood U with a continuous section $s : U \to \mathcal{E}$ of ω over U such that the map $(u, g_1) \mapsto s(u).g_1$ of $U \times_{X_1} \mathcal{G}_1$ to $\omega^{-1}(U)$ is a homeomorphism, where $U \times_{X_1} \mathcal{G}_1$ is the subspace of $U \times \mathcal{G}_1$ consisting of pairs (u, g_1) with $\alpha(s(u)) = \omega(g_1)$.

Similarly, a left action of $\mathcal{G}_2$ on $\mathcal{E}$, over $\omega : \mathcal{E} \to X_2$, is a continuous map $(g_2, e) \mapsto g_2.e$ of $\mathcal{G}_2 \times_{X_2} \mathcal{E}$ to $\mathcal{E}$, where $\alpha(g_2) = \omega(e)$, and, if $g_2, g_2' \in \mathcal{G}_2,\ e \in j$ are such that $\alpha(g_2) = \omega(g_2')$ and $\alpha(g_2') = \omega(e)$, then one requires $g_2.(g_2'.e) = (g_2 g_2').e$ and $1_{\omega(e)}.e = e$.

An *equivalence from* $(\mathcal{G}_1, X_1)$ *to* $(\mathcal{G}_2, X_2)$ is a space $\mathcal{E}$, with two étale maps $\alpha : \mathcal{E} \to X_1$ and $\omega : \mathcal{E} \to X_2$, a right action of $\mathcal{G}_1$ on $\mathcal{E}$ over $\alpha : \mathcal{E} \to X_1$ and a left action of $\mathcal{G}_2$ on $\mathcal{E}$ over $\omega : \mathcal{E} \to X_2$ such that:

(i) the action of $\mathcal{G}_1$ commutes with the action of $\mathcal{G}_2$,

(ii) the action of $\mathcal{G}_1$ (resp. $\mathcal{G}_2$) is simply transitive with respect to $\omega : \mathcal{E} \to X_2$ (resp. $\alpha : \mathcal{E} \to X_1$).

A typical example to bear in mind is the following. Let $\mathcal{H}_1$ and $\mathcal{H}_2$ be the pseudogroups of changes of uniformizing charts of two atlas $\mathcal{A}_1$ and $\mathcal{A}_2$ defining the same orbifold structure on a space Q. Let $(\mathcal{G}_1, X_1)$ and $(\mathcal{G}_2, X_2)$ be the associated groupoids of germs. Let $\mathcal{E}$ be the space of germs at the points of X_1 for the changes of charts from the atlas $\mathcal{A}_1$ to the atlas $\mathcal{A}_2$, with the source (resp. target) projections $\alpha : \mathcal{E} \to X_1$ (resp. $\omega : \mathcal{E} \to X_2$). The groupoid $\mathcal{G}_1$ acts on the right on $\mathcal{E}$ over α by composition of germs, and similarly $\mathcal{G}_2$ acts on the left over ω. Clearly $\mathcal{E}$ gives an equivalence from $(\mathcal{G}_1, X_1)$ to $(\mathcal{G}_2, X_2)$.

(a) Let $\mathcal{E}_{2,1}$ be an equivalence from the étale groupoid $(\mathcal{G}_1, X_1)$ to the étale groupoid $(\mathcal{G}_2, X_2)$. Show there is an equivalence $\mathcal{E}_{1,2}$ from $(\mathcal{G}_2, X_2)$ to $(\mathcal{G}_1, X_1)$. (Hint: take $\mathcal{E}_{1,2} = \mathcal{E}_{2,1}$ and reverse the roles of α and ω.)

(b) Let $\mathcal{E}_{3,2}$ be an equivalence from $(\mathcal{G}_2, X_2)$ to the étale groupoid $(\mathcal{G}_3, X_3)$. Construct an equivalence $\mathcal{E}_{3,1}$ from $(\mathcal{G}_1, X_1)$ to $(\mathcal{G}_3, X_3)$.

(Hint: Consider the subspace $\mathcal{E}_{3,2} \times_{X_2} \mathcal{E}_{2,1}$ of $\mathcal{E}_{3,2} \times \mathcal{E}_{2,1}$ consisting of pairs (e, e') with $\alpha(e) = \omega(e')$, and define $\mathcal{E}_{3,1}$ as the quotient of $\mathcal{E}_{3,2} \times_{X_2} \mathcal{E}_{2,1}$ by the equivalence relation which identifies $(e.g_2, e')$ to $(e, g_2.e')$, where $g_2 \in \mathcal{G}_2$.)

(c) Let $(\varphi, f) : (\mathcal{G}_1, X_1) \to (\mathcal{G}_2, X_2)$ be a homomorphism which is an equivalence in the sense of (2.4). Let $\mathcal{E}_{2,1} = \mathcal{G}_2 \times_{X_2} X_1$ be the subspace of $\mathcal{G}_2 \times X_1$ formed by the pairs (g_2, x) such that $\alpha(g_2) = f(x)$. Define $\alpha(g_2, x) = x$ and $\omega(g_2, x) = \omega(g_2)$. The map $((g_2, x), g_1) \mapsto (g_2\varphi(g_1), \alpha(g_1))$, where $g_1 \in \mathcal{G}_1$ with $\omega(g_1) = x$, defines a right action of $\mathcal{G}_1$ on $\mathcal{E}_{2,1}$ over α. Similarly the formula $(g_2', (g_2, x)) \mapsto (g_2' g_2, x)$ defines a left action of $\mathcal{G}_2$ over ω. Show that $\mathcal{E}_{2,1}$ defines an equivalence from $(\mathcal{G}_1, X_1)$ to $(\mathcal{G}_2, X_2)$.

The map $s : X_1 \to \mathcal{E}_{2,1}$ sending x to $(1_{f(x)}, x)$ is a section of α. Show that conversely an equivalence $\mathcal{E}_{2,1}$ is associated to a homomorphism (φ, f) if and only there is a continuous section $s : X_1 \to \mathcal{E}_{2,1}$ of α.

(d) Using exercise (1), show that $(\mathcal{G}_1, X_1)$ is equivalent to $(\mathcal{G}_2, X_2)$ in the sense of (2.4) if and only if there is an equivalence $\mathcal{E}_{2,1}$ from $(\mathcal{G}_1, X_1)$ to $(\mathcal{G}_2, X_2)$.

(e) Assume that $(\mathcal{G}_2, X_2)$ is the étale groupoid $(\Gamma \ltimes X_2, X_2)$ associated to an action of the group Γ by homeomorphisms of X_2. Let $\mathcal{E}_{2,1}$ be an equivalence from $(\mathcal{G}_1, X_1)$ to $(\mathcal{G}_2, X_2)$. Show that $\alpha : \mathcal{E}_{2,1} \to X_1$ is a Galois covering with group Γ.

Using (c) and (d), show that this implies the following: let $(\mathcal{G}_1, X_1)$ be an étale groupoid which is equivalent to the groupoid associated to an action of a group Γ on a space X_2; if each connected component of X_1 is simply connected, there is an étale homomorphism $(\varphi, f) : (\mathcal{G}_1, X_1) \to (\Gamma \ltimes X_2, X_2)$ which is an equivalence.

Groupoids of Local Isometries

2.9 Definition. A *groupoid* $(\mathcal{G}, X)$ *of local isometries* is an étale topological groupoid with a length metric on its space of units X that induces the given topology on X and is such that the elements of the associated pseudogroup (see 2.1) are local isometries of X.

Equivalence among groupoids of local isometries is the equivalence relation generated by étale homomorphisms $(\varphi, f) : (\mathcal{G}, X) \to (\mathcal{G}', X')$ which are equivalences and are such that $f : X \to X'$ is locally an isometry.

2.10 Hausdorff Separability and Completeness. A groupoid $(\mathcal{G}, X)$ of local isometries is said to be *Hausdorff* if $\mathcal{G}$ is Hausdorff as a topological space and for every continuous map $c : (0, 1] \to \mathcal{G}$, if $\lim_{t \to 0} \alpha \circ c$ and $\lim_{t \to 0} \omega \circ c$ exist, then $\lim_{t \to 0} c(t)$ exists.

A groupoid $(\mathcal{G}, X)$ of local isometries of X is said to be *complete* if X is locally complete (i.e. each point of X has a complete neighbourhood) and if the space of orbits $\mathcal{G}\backslash X$ with the quotient pseudometric is complete.

2.11 Remark. To justify the second condition in the definition of Hausdorff separability, consider the following example. Suppose that the metric space X is just the Euclidean space $\mathbb{E}^n$ and that $\mathcal{G}$ is generated by an isometry f from an non-empty open subset U of $\mathbb{E}^n$ to a disjoint open subset V of $\mathbb{E}^n$ (i.e. the elements of $\mathcal{G}$ are the germs of the identity map of $\mathbb{E}^n$ and the germs of f and its inverse at the points of U and V respectively). Then $(\mathcal{G}, X)$ is a groupoid of local isometries which is complete but not Hausdorff: the second condition in the definition of Hausdorff separability is not satisfied. Note that the quotient X' of $\mathbb{E}^n$ by the equivalence relation which identifies $x \in U$ to $f(x) \in V$ is a non-Hausdorff space and that $(\mathcal{G}, X)$ is equivalent to the trivial groupoid X'.

In general, if $(\mathcal{G}, X)$ is Hausdorff, then the space of orbits $\mathcal{G}\backslash X$ need not be Hausdorff in general (see the first example below).

2.12 Examples

(1) If a group Γ acts by isometries on a length space X, then the associated groupoid $(\Gamma \ltimes X, X)$ is a groupoid of local isometries, and it is Hausdorff. It is complete if and only if X is a complete metric space.

(2) The groupoid $(\mathcal{G}, X)$ of germs of changes of uniformizing charts of a Riemannian orbifold is a groupoid of local isometries, and it is Hausdorff (in this case Hausdorff being equivalent to the condition that the orbit space Q is Hausdorff). The quotient pseudometric on Q is always a metric, and $(\mathcal{G}, X)$ is complete if and only if Q is complete.

(3) Let $G(\mathcal{Y})$ be a complex of groups over a scwol $\mathcal{Y}$ whose geometric realization $|\mathcal{Y}|$ is a M_κ-simplicial complex with finitely many shapes (see $\mathcal{C}$.4.16). Then the associated étale groupoid $(\mathcal{G}(\mathcal{Y}), \tilde{Y})$ is a groupoid of local isometries; it is always Hausdorff and complete. Indeed the quotient pseudometric on the space of orbits $|\mathcal{Y}|$ is the length metric associated to the given M_κ-structure, which is always complete by (I.7.13).

(4) Assume that X is a Riemannian manifold and that $(\mathcal{G}, X)$ is an étale groupoid such that the elements of the associated pseudogroup are Riemannian isometries. Then $(\mathcal{G}, X)$ is a groupoid of local isometries, called a *groupoid of local Riemannian isometries*.

The holonomy groupoid of a Riemannian foliation (see [Mol88]) is a typical example of a groupoid of local Riemannian isometries. A Riemannian foliation $\mathcal{F}$ on a Riemannian manifold M is by definition a foliation such that the leaves are locally at constant distance; equivalently, if $\mathcal{F}$ is given by local submersions $f_i : U_i \to V_i \subset \mathbb{R}^n$, there should exist on each V_i a Riemannian metric such that f_i is Riemannian submersion (meaning that, for each point $x \in U_i$, the restriction of the differential of f to the subspace of T_xM orthogonal to the tangent space of the leaf through x is an isometry onto the tangent space of V_i at $f_i(x)$). It follows that the associated holonomy groupoid is a groupoid of local Riemannian isometries. If the Riemannian manifold M is complete, then the holonomy groupoid is Hausdorff and complete.

For a systematic study of closures of orbits in Hausdorff and complete groupoids of local Riemannian isometries, see [Hae88], where examples of such groupoids that have a single orbit but are non-developable are exhibited.

2.13 $\mathcal{G}$-Connectedness. An étale groupoid $(\mathcal{G}, X)$ is said to be *connected* (equivalently X is "*$\mathcal{G}$-connected*") if for any two points $x, y \in X$ there is a sequence of points $(x_1, y_1, \dots, x_k, y_k)$ such that $x_1 = x,\ y_k = y$, the point x_i is in the $\mathcal{G}$-orbit of y_{i+1} for $i = 1, \dots, k-1$ and there is a path joining x_i to y_i for each $i = 1, \dots, k$.

To illustrate the notion of Hausdorff separability and completeness, we prove the following lemma, which will be useful later.

2.14 Lemma. *Let $(\mathcal{G}, X)$ be a groupoid of local isometries which is Hausdorff. Let $x \in X$ and $\varepsilon > 0$ be such that the closed ball $\overline{B}(x, \varepsilon)$ is complete. Given a path*

$c : [a, b] \to X$ of length $< \varepsilon$ and an element $g \in \mathcal{G}$ with $c(a) = \alpha(g)$ and $\omega(g) = x$, there is a unique continuous map $t \mapsto g(t)$ from $[a, b]$ to $\mathcal{G}$ such that $g(a) = g$ and $\alpha(g(t)) = c(t)$, for all $t \in [a, b]$.

Thus, if a $\mathcal{G}$-orbit is a pseudodistance less than ε from the $\mathcal{G}$-orbit of x, then it meets $B(x, \varepsilon)$.

Proof. Let I be the maximal subinterval of $[a, b]$ containing a on which $g(t)$ can be defined; note that $g(t)$ is unique because $\mathcal{G}$ is a Hausdorff space. Because $\mathcal{G}$ is étale, the interval I is open in $[a, b]$. The length of the path $t \mapsto \omega(g(t))$ is equal to the length of the path c restricted to I, hence is smaller than ε. Using the hypothesis that $\mathcal{G}$ is Hausdorff and $\overline{B}(x, \varepsilon)$ is complete, we see that I is also closed, hence equal to $[a, b]$.

To prove the second assertion of the lemma, note that if in $\mathcal{G}\backslash X$ the orbit of y is within a pseudodistance ε of the orbit of x, then there is a sequence $(x_1, y_1, \dots, x_k, y_k)$ of points of X such that $x_1 = x$, $y_k = y$, the points x_i and y_{i+1} are in the same orbit and there are paths c_i joining x_i to y_i such that the sum of the lengths of the paths c_i is smaller than ε. Thus we can apply the first part of the lemma successively to the paths c_i. □

Statement of the Main Theorem

The aim of this chapter is to prove the following result.

2.15 Developability Theorem. *Let $(\mathcal{G}, X)$ be a connected groupoid of local isometries which is Hausdorff and complete. If the metric on X is locally convex*[77]*, then $(\mathcal{G}, X)$ is developable.*

The conclusion of the theorem is that $(\mathcal{G}, X)$ is equivalent to the groupoid associated to an action of a group Γ by isometries on a metric space $\tilde{X}$. The group Γ will be the fundamental group of $(\mathcal{G}, X)$ and the space $\tilde{X}$ will be the space of equivalence classes of $\mathcal{G}$-geodesics issuing from a base point $x_0 \in X$ with free endpoint. (All of these terms will be defined in the next two sections.) The space $\tilde{X}$ has a natural length metric making it locally isometric to X. As $\tilde{X}$ is complete and simply connected, its metric is globally convex by the Cartan-Hadamard theorem (II.4.1). If X is non-positively curved, then $\tilde{X}$ is a CAT(0) space.

As a special case of 2.15 we have:

2.16 Corollary (Gromov). *Every complete Riemannian orbifold of non-positive curvature is developable.*

2.17 Corollary. *Every complex of groups of non-positive curvature (see C.4.16) is developable.*

---

[77] in the sense of Busemann, see II.1.18 and II.4.1.

This follows from (2.15) because non-positive curvature implies that the metric is locally convex and the developability of a complex of groups is equivalent to the developability of its associated étale groupoid (see 2.7(4)).

By definition, a Riemannian foliation $\mathcal{F}$ is transversally of non-positive curvature if it is defined by local Riemannian submersions $f_i : U_i \to V_i$, where the Riemannian metric on V_i is of non-positive curvature. The proof of the following corollary is left as an exercise (see 2.7(6)).

2.18 Corollary (Hebda). *Let $\mathcal{F}$ be a Riemannian foliation on a complete Riemannian manifold M which is transversally of non-positive curvature. Then the foliation induced on a suitable Galois covering $p : \hat{M} \to M$ is defined by a Riemannian submersion with connected fibres $f : \hat{M} \to X$, where X is a Riemannian manifold of non-positive curvature.*

3. The Fundamental Group and Coverings of Étale Groupoids

Equivalence and Homotopy of $\mathcal{G}$-Paths

3.1 $\mathcal{G}$-Paths. Let $(\mathcal{G}, X)$ be an étale groupoid. A $\mathcal{G}$-path $c = (g_0, c_1, g_1, \dots, c_k, g_k)$ over a subdivision $a = t_0 \leq \dots \leq t_k = b$ of the interval $[a, b] \subset \mathbb{R}$ consists of:

(1) continuous maps $c_i : [t_{i-1}, t_i] \to X$,

(2) elements $g_i \in \mathcal{G}$ such that $\alpha(g_i) = c_{i+1}(t_i)$ for $i = 0, 1, \dots, k-1$ and $\omega(g_i) = c_i(t_i)$ for $i = 1, \dots, k$.

The *initial point* of c is $x := \omega(g_0)$ and $y := \alpha(g_k)$ is its *terminal* point. We say that c joins x to y. If $g_0 = 1_x$ and $g_k = 1_y$ are units, they can be dropped in the notation for c. (If $c : [a, b] \to X$ is a path joining x to y, it can be considered as the $\mathcal{G}$-path $(1_x, c, 1_y)$.)

If $(\mathcal{G}, X)$ is a groupoid of local isometries, then the *length* $l(c)$ of the $\mathcal{G}$-path $c = (g_0, c_1, g_1, \dots, c_k, g_k)$ is the sum of the lengths of the paths c_i. The *pseudodistance* between the $\mathcal{G}$-orbits of x and y is the infimum of the lengths of $\mathcal{G}$-paths joining x to y.

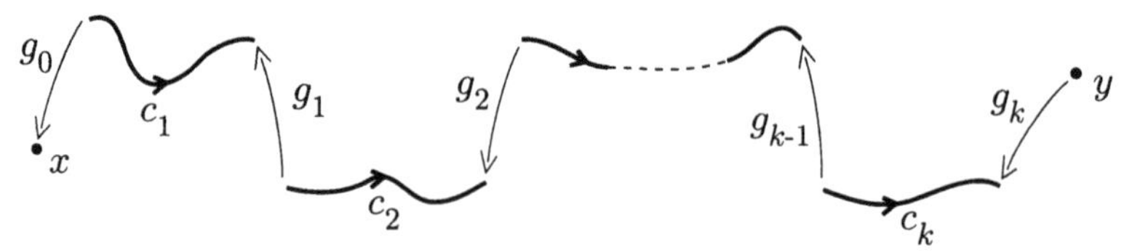

Fig. $\mathcal{G}$.1 A $\mathcal{G}$-path

3.2 Equivalence of $\mathcal{G}$-Paths. Among $\mathcal{G}$-paths parameterized by the same interval $[a, b]$ we define an equivalence relation generated by the following two operations.

(i) *Subdivision*: Given a $\mathcal{G}$- path $c = (g_0, c_1, \dots, g_k)$ over the subdivision $a = t_0 \le t_1 \le \cdots \le t_k = b$, we can add a new subdivision point $t'_i \in [t_{i-1}, t_i]$ together with the unit element $g'_i = 1_{c_i(t'_i)}$ to get a new sequence, replacing c_i by c'_i, g'_i, c''_i, where c'_i and c''_i are the restrictions of c_i to $[t_{i-1}, t'_i]$ and $[t'_i, t_i]$.

(ii) Replace the $\mathcal{G}$-path c by a new path $c' = (g'_0, c_1, g_1, \dots, c'_k, g'_k)$ over the same subdivision as follows: for each $i = 1, \dots, k$, choose a continuous map $h_i : [t_{i-1}, t_i] \to \mathcal{G}$ such that $\alpha(h_i(t)) = c_i(t)$ and define $c'_i : t \mapsto \omega(h_i(t))$, $g'_i = h_i(t_i)\, g_i h_{i+1}^{-1}(t_i)$ for $i = 1, \dots, k-1$, $g'_0 = g_0 h_1^{-1}(t_0)$ and $g'_k = h_k(t_k) g_k$.

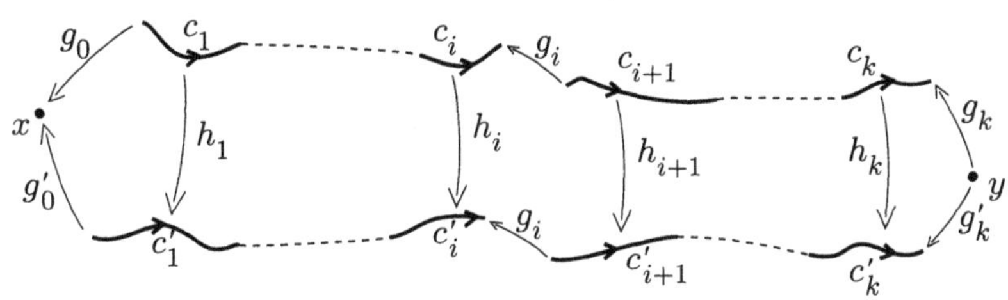

Fig. $\mathcal{G}$.2 Equivalent $\mathcal{G}$-paths

Note that if two $\mathcal{G}$-paths are equivalent, then under a suitable common subdivision one can pass from the first to the second by an operation of type (ii). Note that two equivalent $\mathcal{G}$-paths have the same initial and terminal points. We also note that if $(\mathcal{G}, X)$ is Hausdorff, then the continuous maps h_i in (ii) above are uniquely defined by c and c'.

If a $\mathcal{G}$-path c joining x to y is such that all the c_i are constant maps, then the equivalence class of c is completely characterized by an element $g \in \mathcal{G}$ with $\alpha(g) = y$ and $\omega(g) = x$.

3.3 Examples

(1) Let M be a differentiable manifold whose differentiable structure is given by an atlas of charts $q_i : X_i \to V_i$, where the V_i cover M and the X_i are disjoint. Let X be the union of the X_i and let q be the union of the q_i. Let $(\mathcal{G}, X)$ be the étale groupoid of germs of all the changes of charts. Then the set of equivalence classes of $\mathcal{G}$-paths joining x to y corresponds bijectively to the set of paths in M joining $q(x)$ to $q(y)$.

More generally, if $(\varphi, f) : (\mathcal{G}, X) \to (\mathcal{G}', X')$ is an equivalence of étale groupoids, then φ induces a bijection from the set of equivalences classes of $\mathcal{G}$-paths joining x to y to the set of equivalence classes of $\mathcal{G}'$-paths joining $f(x)$ to $f(y)$.

(2) Let X be $\mathbb{R}^2$ with rectangular coordinates (x, y). Let Q be the half-plane defined by $y \ge 0$ and let $q : X \to Q$ be the projection $(x, y) \mapsto (x, |y|)$. Let Γ be the cyclic group of order two generated by the symmetry $\sigma : (x, y) \mapsto (x, -y)$. The map q induces a homeomorphism $\Gamma\backslash X \to Q$ and can be considered as a uniformizing chart defining an orbifold structure on Q. The pseudogroup $\mathcal{H}$ of changes of charts is generated by σ; let $\mathcal{G}$ be the étale groupoid of germs of elements of $\mathcal{H}$.

Consider the $\mathcal{G}$-paths $c = (c_1, g_1, c_2)$ and $c' = (c_1, g_1', c_2)$ parameterized by $[0, 1]$ joining $(0, 1)$ to itself, where c_1 maps $t \in [0, 1/2]$ to $(0, 1 - 2t)$, c_2 maps $t \in [1/2, 1]$ to $(0, 2t - 1)$ and g_1 (resp. g_1') is the germ at $(0, 0)$ of the identity (resp. of σ). The images under q of these paths look the same, but they are not equivalent (not even homotopic in the sense of (3.5)).

3.4 Inverse $\mathcal{G}$-Paths and Composition. From now on, unless otherwise specified, all $\mathcal{G}$-paths will be parameterized by $[0, 1]$. Let $c = (g_0, c_1, \dots, g_k)$ be a $\mathcal{G}$-path over the subdivision $0 = t_0 \leq \dots, \leq t_k = 1$. The inverse of c is the $\mathcal{G}$-path $c^{-1} = (g_0', c_1', \dots, g_k')$ over the subdivision $0 = t_0' \leq \dots \leq t_k' = 1$, where $t_i' = 1 - t_i$, $g_i' = g_{k-i}^{-1}$ and $c_i'(t) = c_{k-i+1}(1 - t)$. Note that the terminal point of c is the initial point of c^{-1} and vice versa. The inverses of two equivalent paths are equivalent.

Given two $\mathcal{G}$-paths, $c = (g_0, c_1, \dots, g_k)$ over the subdivision $0 = t_0 \leq \dots \leq t_k = 1$ and $c' = (g_0', c_1', \dots, g_k')$ over the subdivision $0 = t_0' \leq \dots \leq t_{k'} = 1$, such that the terminal point of c is equal to the initial point of c', their *composition* $c * c'$ (or *concatenation*) is the $\mathcal{G}$-path $c'' = (g_0'', c_1'', \dots, g_{k+k'}'')$ defined over the subdivision $t_0'' \leq \dots \leq t_{k+k'}''$, where

$$t_i'' = t_i/2 \text{ if } 0 \leq i \leq k, \quad t_i'' = 1/2 + t_{i-k}'/2 \text{ if } k \leq i \leq k + k',$$

$$c_i''(t) = c_i(t/2) \text{ for } 1 \leq i \leq k; \quad c_i''(t) = c_{i-k}'(2t - 1) \text{ for } k < i \leq k + k',$$

$$g_i'' = g_i \text{ for } 0 \leq i < k, \ g_k'' = g_k g_0', \ g_i'' = g_{i-k}' \text{ for } k < i \leq k + k'.$$

Note that if c is equivalent to $\bar{c}$, and c' is equivalent to $\bar{c}'$, then $c * c'$ is equivalent to $\bar{c} * \bar{c}'$.

3.5 Homotopies of $\mathcal{G}$-Paths. Two $\mathcal{G}$-paths c and c' (parameterized by the interval $[0, 1]$) are homotopic[78]) if one can pass from the first to the second by a finite sequence of the following operations:

(1) equivalence of $\mathcal{G}$-paths,

(2) *elementary homotopies*: an elementary homotopy between two $\mathcal{G}$-paths c and c' is a family, parameterized by $s \in [s_0, s_1]$, of $\mathcal{G}$-paths $c^s = (g_0^s, c_1^s, \dots, g_k^s)$ over the subdivisions $0 = t_0^s \leq t_1^s \leq \dots \leq t_k^s = 1$, where t_i^s, c_i^s and g_i^s depend continuously on the parameter s, the elements g_0^s and g_k^s are independent of s and $c^{s_0} = c$, $c^{s_1} = c'$.

The homotopy class of a $\mathcal{G}$-path c will be denoted $[c]$. Note that two equivalent $\mathcal{G}$-paths are homotopic. If c and c' are composable $\mathcal{G}$-paths, the homotopy class of $c * c'$ depends only on the homotopy classes of c and c' and will be denoted $[c * c'] = [c] * [c']$. If c, c' and c'' are three $\mathcal{G}$-paths which are composable, then $[c * c'] * [c''] = [c] * [c' * c'']$. Note that although $(c * c') * c''$ is not equivalent to $c' * (c' * c'')$, the homotopy class of $(c * c') * c''$ is equal to the homotopy class of

[78] Always, implicitly, relative to their endpoints.

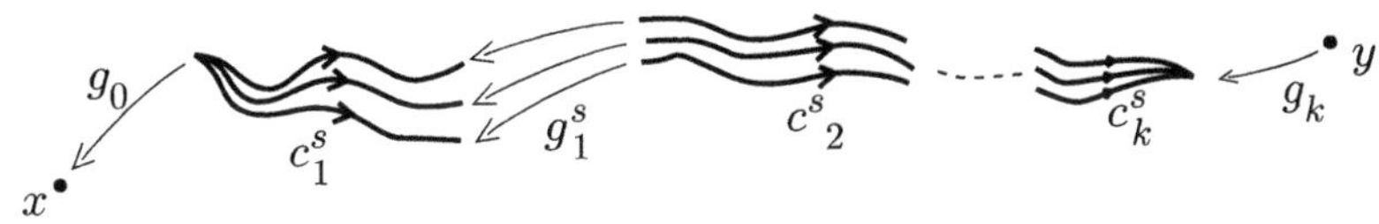

Fig. $\mathcal{G}$.3 An elementary homotopy between $\mathcal{G}$-paths

$c * (c' * c'')$; it will be denoted $[c] * [c'] * [c'']$. All these properties are proved as in the usual case of paths in topological spaces.

In the notation of example 3.3(1), there is an obvious bijection between the homotopy classes of $\mathcal{G}$-paths joining two points $x, y \in X$ and the homotopy classes of paths in M joining $q(x)$ to $q(y)$.

The Fundamental Group $\pi_1((\mathcal{G}, X), x_0)$

3.6 Definition. Let $(\mathcal{G}, X)$ be an étale groupoid. A *$\mathcal{G}$-loop based at* $x_0 \in X$ is a path parameterized by $[0, 1]$ joining x_0 to x_0. With the operation of composition, the homotopy classes of $\mathcal{G}$-loops form a group called the *fundamental group* $\pi_1((\mathcal{G}, X), x_0)$ *of* $(\mathcal{G}, X)$ *based at* x_0.

Let $(\varphi, f) : (\mathcal{G}, X) \to (\mathcal{G}', X')$ be a continuous homomorphism of étale groupoids. Given a $\mathcal{G}$-path $c = (g_0, c_1, \ldots, g_k)$ over a subdivision $t_0 \le \cdots \le t_k$, its image under (φ, f) is the $\mathcal{G}'$-path $\varphi(c) = (\varphi(g_0), f \circ c_1, \ldots, \varphi(g_k))$ over the same subdivision. It is clear that if two $\mathcal{G}$-paths are equivalent (resp. homotopic), then their images under any continuous homomorphism are equivalent (resp. homotopic). In particular (φ, f) induces a homomorphism

$$\pi_1((\mathcal{G}, X), x_0) \to \pi_1((\mathcal{G}', X'), f(x_0)).$$

The first part of the following proposition is proved as in the special case of topological spaces. The second part is left as an exercise.

3.7 Proposition. *Let $(\mathcal{G}, X)$ be an étale groupoid and let $x_0 \in X$ be a base point. Let a be a $\mathcal{G}$-path joining x_0 to a point x_1. Then the map that associates to each $\mathcal{G}$-loop c at x_0 the $\mathcal{G}$-loop $(a^{-1} * c * a)$ at x_1 induces an isomorphism from $\pi_1((\mathcal{G}, X), x_0)$ to $\pi_1((\mathcal{G}, X), x_1)$.*

Let $(\varphi, f) : (\mathcal{G}, X) \to (\mathcal{G}', X')$ be a continuous étale homomorphism of étale groupoids. If (φ, f) is an equivalence, then the induced homomorphism on the fundamental groups is an isomorphism.

3.8 Definition. Recall that an étale groupoid $(\mathcal{G}, X)$ is connected (equivalently X is $\mathcal{G}$-connected) if any two points of X can be joined by a $\mathcal{G}$-path. It is *simply connected* if it is connected and if $\pi_1((\mathcal{G}, X), x_0)$ is the trivial group.

3.9 Examples

(1) Assume that $(\mathcal{G}, X)$ is the étale groupoid associated to an action of a group Γ by homeomorphisms on X. Then any $\mathcal{G}$-path $(g_0, c_1, g_1, \dots, c_k, g_k)$ joining x to y defined over the subdivision $0 = t_0 \leq \dots \leq t_k = 1$ is equivalent to a unique $\mathcal{G}$-path of the form (c, g), where $c : [0, 1] \to X$ is a continuous path with $c(0) = x$ and $g = (\gamma, y)$ with $\gamma.y = c(1)$. Indeed if $g_i = (\gamma_i, x_i) \in \Gamma \times X$, and we choose c to be the path mapping $t \in [t_{i-1}, t_i]$ to $\gamma_0 \dots \gamma_{i-1}.c_i(t)$ and define $\gamma = \gamma_0 \dots \gamma_k$, then the given $\mathcal{G}$-path is equivalent to the $\mathcal{G}$-path (c, g), where $g = (\gamma, x_k)$.

We describe the group $\pi_1((\mathcal{G}, X), x_0)$ in this special case. Let X_0 be the path component of x_0 and let Γ_0 be the subgroup formed by the elements $\gamma \in \Gamma$ such that $\gamma.x_0 \in X_0$. Every $\mathcal{G}$-loop at x_0 is equivalent to a unique $\mathcal{G}$-loop of the form (c, g) where $c : [0, 1] \to X_0$ is a continuous path with $c(0) = x_0$ and $g = (\gamma, x_0)$, with $\gamma.x_0 = c(1)$; therefore $\gamma \in \Gamma_0$. Another such $\mathcal{G}$-loop (c', g') with $g' = (\gamma', x_0)$ represents the same element of $\pi_1((\mathcal{G}, X), x_0)$ if and only if $\gamma = \gamma'$ and c is homotopic to c'. Hence we have the exact sequence (compare with $\mathcal{C}$.1.15):

$$1 \to \pi_1(X_0, x_0) \to \pi_1((\mathcal{G}, X), x_0) \to \Gamma_0 \to 1.$$

(2) Consider an orbifold structure defined on a space Q by an atlas of uniformizing charts, and let $(\mathcal{G}, X)$ be the étale groupoid of germs of changes of charts. The orbifold fundamental group of Q based at $x_0 \in X$ is by definition the group $\pi_1((\mathcal{G}, X), x_0)$. If Q is connected, then up to isomorphism this group is independent of the choice of base point and the choice of a compatible atlas of uniformizing charts (see (3.6) and (3.7)).

3.10 Exercises

(0) Show that an equivalence $(\varphi, f) : (\mathcal{G}, X) \to (\mathcal{G}', X')$ of étale groupoids induces a bijection from the set of equivalence classes of $\mathcal{G}$-paths joining x to y to the set of equivalence classes of $\mathcal{G}'$-paths connecting $f(x)$ to $f(y)$. Show that (φ, f) induces an isomorphism $\pi_1((\mathcal{G}, X), x_0)) \to \pi_1((\mathcal{G}', X'), f(x_0))$.

(1) What are the fundamental groups of the trivial examples 2.3(1)? Show that the fundamental group of the orbifold $S^2_{m,n}$ described in 1.4(1) is a cyclic group whose order is the greatest common divisor of m and n.

(2) Let $(\mathcal{G}, \mathbb{R})$ be the étale groupoid of the germs (at all points of $\mathbb{R}$) of the elements of the group of diffeomorphisms of $\mathbb{R}$ that is generated by the diffeomorphisms $h_1, \dots, h_n$, where each h_i is the identity on some non-empty open set U_i. Show that $(\mathcal{G}, \mathbb{R})$ is simply connected. (Consider first the case where $n = 1$.)

(3) Let $G(\mathcal{Y})$ be a complex of groups over a connected scwol $\mathcal{Y}$ and let $(\mathcal{G}(\mathcal{Y}), \tilde{Y})$ be the associated étale groupoid (see 2.3(4)). Show that the fundamental group of $G(\mathcal{Y})$ as defined in $\mathcal{C}$.3.6 is isomorphic to the fundamental group of $(\mathcal{G}(\mathcal{Y}), \tilde{Y})$.

(4) *The Seifert-van Kampen Theorem.* Let $(\mathcal{G}, X)$ be an étale groupoid and let X_1 and X_2 be two open subsets of X such that $X = X_1 \cup X_2$. Let $X_0 = X_1 \cap X_2$. We assume that the restriction $(\mathcal{G}_i, X_i)$ of $(\mathcal{G}, X)$ to X_i is connected for $i = 0, 1, 2$. Let $x_0 \in X_0$ be a base point. Show that $\pi_1((\mathcal{G}, X), x_0)$ is isomorphic to the quotient of

the free product $\pi_1((\mathcal{G}_1, X_1), x_0) * \pi_1((\mathcal{G}_2, X_2), x_0)$ by the normal subgroup generated by the elements of the form $j_1(\gamma)j_2(\gamma)^{-1}$, where $\gamma \in \pi_1((\mathcal{G}_0, X_0), x_0)$ and j_i is the homomorphism induced on the fundamental groups by the inclusion from $(\mathcal{G}_0, X_0)$ into $(\mathcal{G}_i, X_i)$, $i = 1, 2$.

(5) Let $\mathcal{F}$ be a foliation on a connected manifold M, and let $(\mathcal{G}, X)$ be its holonomy groupoid as described in 2.3(5). Show there is a surjective homomorphism from the fundamental group of M to the fundamental group of $(\mathcal{G}, X)$. (Thus $(\mathcal{G}, X)$ is always simply connected if M is simply connected.)

(Hint: In the notations of example 2.3(5), let $c : [0, 1] \to M$ be a loop in M with $c(0) = c(1) = y_0 \in U_{j_0}$. Let $0 = t_0 < t_1 < \cdots < t_k = 1$ be a partition such that each $c([t_{i-1}, t_i])$ is contained in some U_{j_i} for $i = 1, \ldots, k-1$; let $y_i = c(t_i)$ and $x_i = f_{j_i}(y_i)$, $x_0 = f_{j_0}(y_0)$; let $\tilde{c} = (g_0, c_1, \ldots, c_k, g_k)$ be the $\mathcal{G}$-loop at x_0 defined as follows: $c_i : [t_{i-1}, t_i] \to V_{j_i}$ maps t to $f_{j_i}(c(t))$, g_i is the germ of $h^{y_i}_{j_i j_{i+1}}$ at $f_{j_{i+1}}(y_i)$ and g_k the germ at x_0 of $h^{y_0}_{j_k j_0}$. Show that the map associating to the homotopy class of c the homotopy class of $\tilde{c}$ is a surjective homomorphism from $\pi_1(M, y_0)$ to $\pi_1((\mathcal{G}, X), x_0)$.)

Coverings

3.11 The Action of a Groupoid on a Space. Let $(\mathcal{G}, X)$ be an étale groupoid, let $\hat{X}$ be a topological space and let $p : \hat{X} \to X$ be a continuous map. Let $\hat{\mathcal{G}} = \hat{X} \times_X \mathcal{G}$ be the subspace of $\hat{X} \times \mathcal{G}$ consisting of pairs $(\hat{x}, g)$ with $p(\hat{x}) = \omega(g)$. A *right action* of $(\mathcal{G}, X)$ on $\hat{X}$ over p is a continuous map $(\hat{x}, g) \mapsto \hat{x}.g$ of $\hat{\mathcal{G}}$ to $\hat{X}$, such that:

(1) $p(\hat{x}.g) = \alpha(g)$,

(2) if $\alpha(g) = \omega(g')$, then $\hat{x}.(gg') = (\hat{x}.g).g'$,

(3) if $\hat{x} \in p^{-1}(x)$, then $\hat{x}.1_x = \hat{x}$.

To such an action is associated the étale groupoid $(\hat{\mathcal{G}}, \hat{X})$ where $\omega((\hat{x}, g)) = \hat{x}$ and $\alpha((\hat{x}, g)) = \alpha(g) = \hat{x}.g$; the composition $(\hat{x}, g)(\hat{x}', g')$ is defined if $\hat{x}' = \hat{x}.g$ and is equal to $(\hat{x}, gg')$; the inverse of $(\hat{x}, g)$ is $(\hat{x}.g, g^{-1})$. The map $(\hat{x}, g) \mapsto (p(\hat{x}), g)$ is a continuous homomorphism denoted $(\pi, p) : (\hat{\mathcal{G}}, \hat{X}) \to (\mathcal{G}, X)$.

Note that any right action of $\mathcal{G}$ on $\hat{X}$ over $\hat{p}$ can be converted into a left action of $\mathcal{G}$ over p by defining $g.\hat{x} = \hat{x}.g^{-1}$, where $\alpha(g) = p(\hat{x})$.

3.12 Definition of a Covering. In the preceding definition, if $p : \hat{X} \to X$ is a covering, we say that $(\pi, p) : (\hat{\mathcal{G}}, \hat{X}) \to (\mathcal{G}, X)$ is a *covering* of étale groupoids.

Moreover, if $p : \hat{X} \to X$ is a Galois covering such that the action of its Galois group Γ on $\hat{X}$ commutes with the action of $\mathcal{G}$, then we say that (π, p) is a *Galois covering* with Galois group Γ.

3.13 Examples

(1) *Coverings of a Groupoid Associated to an Action of a Group.* Let Γ be a group acting by homeomorphisms on a simply connected topological space X. Let Γ_0 be a subgroup of Γ. Let $\hat{X} = \Gamma/\Gamma_0 \times X$, where Γ/Γ_0 has the discrete topology. The group Γ acts naturally on $\hat{X}$ by the rule $\gamma.(\gamma'\Gamma_0, x) = (\gamma\gamma'\Gamma_0, \gamma.x)$. Let $p : \Gamma/\Gamma_0 \times X \to X$

be the natural projection. The functor $\pi : \Gamma \ltimes \hat{X} \to \Gamma \ltimes X$ mapping $(\gamma, \hat{x})$ to $(\gamma, p(\hat{x}))$ gives a morphism $(\pi, p) : (\Gamma \ltimes \hat{X}, \hat{X}) \to (\Gamma \ltimes X, X)$ which can be considered as a covering. The natural inclusion $\Gamma_0 \times X \to \Gamma \times \hat{X}$ sending (γ_0, x) to $(\gamma_0, (\Gamma_0, x))$ defines an étale homomorphism $(\Gamma_0 \ltimes X, X) \to (\Gamma \ltimes \hat{X}, \hat{X})$ which is an equivalence.

In fact the equivalence classes of connected coverings of the groupoid $(\Gamma \ltimes X, X)$ are in bijection with the conjugacy classes of subgroups of Γ.

(2) *Coverings of an Orbifold.* Consider an orbifold structure on a space Q defined by an atlas of uniformizing charts and let $(\mathcal{G}, X)$ be the associated groupoid of germs of changes of charts. Then any covering $(\hat{\mathcal{G}}, \hat{X})$ of $(\mathcal{G}, X)$ is the groupoid of changes of charts of an orbifold structure on the space of orbits $\hat{Q} = \hat{\mathcal{G}} \backslash \hat{X}$, and $\hat{Q}$ is a covering orbifold of the orbifold Q, in the sense of (1.7).

(3) *The Orientation Covering.* Let $(\mathcal{G}, X)$ be the étale groupoid of germs of a pseudogroup of local homeomorphisms of a manifold X. Let $p : \hat{X} \to X$ be the 2-fold orientation covering of X: the points of the fibre $p^{-1}(x)$ are the two possible local orientations of X at x. The groupoid $(\mathcal{G}, X)$ acts naturally on $\hat{X}$. The corresponding étale covering is called the orientation covering of $(\mathcal{G}, X)$. The groupoid is defined to be orientable if there is a continuous section $s : X \to \hat{X}$ which is invariant by $\mathcal{G}$.

We considered the following lemma earlier in the special case of orbifolds (1.10).

3.14 Lemma. *With the notations of 3.12, let $(\pi, p) : (\hat{\mathcal{G}}, \hat{X}) \to (\mathcal{G}, X)$ be a Galois covering with Galois group Γ. If the natural projection $\hat{q} : \hat{X} \to \tilde{X} := \hat{\mathcal{G}} \backslash \hat{X} = \hat{X}/\mathcal{G}$ is étale and if $\hat{\mathcal{G}}_{\hat{x}} = 1_{\hat{x}}$ for each $\hat{x} \in \hat{X}$, then $(\mathcal{G}, X)$ is equivalent to the groupoid $(\Gamma \ltimes \tilde{X}, \tilde{X})$ associated to the natural action of Γ on $\tilde{X}$.*

Proof. Let $(\Gamma \times \hat{\mathcal{G}}, \hat{X})$ be the étale groupoid defined as follows. The space $\Gamma \times \hat{\mathcal{G}}$ is the subspace of the product $\Gamma \times \hat{X} \times \mathcal{G}$ (where the topology on Γ is discrete) consisting of the triples $(\gamma, \hat{x}, g)$ with $p(\hat{x}) = \omega(g)$. The source and target projections α and ω are $\alpha((\gamma, \hat{x}, g)) = \hat{x}.g$ and $\omega((\gamma, \hat{x}, g)) = \gamma.\hat{x}$. The composition $(\gamma, \hat{x}, g)(\gamma', \hat{x}', g')$ is defined if $\hat{x}.g = \gamma'.\hat{x}'$ and is equal to $(\gamma\gamma', \gamma'^{-1}.\hat{x}, gg')$, and the inverse of $(\gamma, \hat{x}, g)$ is $(\gamma^{-1}, \gamma.\hat{x}.g, g^{-1})$.

The lemma follows from the following two assertions:

(1) The étale homomorphism $(\Gamma \times \hat{\mathcal{G}}, \hat{X}) \to (\mathcal{G}, X)$ sending $(\gamma, \hat{x}, g)$ to g and $\hat{x}$ to $p(\hat{x})$ is an equivalence.

(2) The étale homomorphism $(\Gamma \times \hat{\mathcal{G}}, \hat{X}) \to (\Gamma \ltimes \tilde{X}, \tilde{X})$ sending $(\gamma, \hat{x}, g)$ to $(\gamma, \hat{q}(\hat{x}))$ and $\hat{x}$ to $\hat{q}(\hat{x})$ is an equivalence.

To prove (1), we only use the hypothesis that (π, p) is a Galois covering. The group Γ acts transitively on each fibre of p, hence the homomorphism induces a bijection on the spaces of orbits. The isotropy group of a point $\hat{x} \in \hat{X}$ with $x = p(\hat{x})$ consists of the triples $(\gamma, \hat{x}', g)$ such that $g \in \mathcal{G}_x$ and $\hat{x} = \hat{x}'.g = \gamma.\hat{x}'$; equivalently it consists of the pairs (γ, g) with $\gamma \in \Gamma$, $g \in \mathcal{G}_x$ and $\gamma.\hat{x} = \hat{x}.g$. Given $g \in \mathcal{G}_x$, there is a unique γ satisfying this equality, hence the homomorphism induces an isomorphism from the isotropy group of $\hat{x}$ onto $\mathcal{G}_x$.

To prove (2), we observe that the étale homomorphism $(\Gamma \times \hat{\mathcal{G}}, \hat{X}) \to (\Gamma \ltimes \tilde{X}, \tilde{X})$ induces a bijection $(\Gamma \times \hat{G})\backslash\hat{X} \to \Gamma\backslash\tilde{X}$. The isotropy subgroup of $\hat{q}(\hat{x})$ is the set of $\gamma \in \Gamma$ such that $\gamma.\hat{x} = \hat{x}.g$ for some $g \in \mathcal{G}_x$. The hypothesis that $\hat{\mathcal{G}}_{\hat{x}}$ is trivial implies that this element g is uniquely determined by γ. Therefore the homomorphism induces a bijection of the isotropy subgroups. □

The theory of coverings for étale groupoids is strictly parallel to the theory of coverings for topological spaces, as the following sequence of results illustrates.

3.15 Proposition (Liftings of $\mathcal{G}$-Paths). *Let $(\pi, p) : (\hat{\mathcal{G}}, \hat{X}) \to (\mathcal{G}, X)$ be a covering of étale groupoids, let $\hat{x}_0 \in \hat{X}$ be a base point and let $x_0 = p(\hat{x}_0)$. For every $\mathcal{G}$-path $c = (g_0, c_1, \dots, g_k)$ issuing from x_0, there is a unique $\hat{\mathcal{G}}$-path $\hat{c}$ issuing from $\hat{x}_0$ such that $\pi(\hat{c}) = c$. If c' is a $\mathcal{G}$-path homotopic to c, then its lifting $\hat{c}'$ issuing from $\hat{x}_0$ is homotopic to $\hat{c}$.*

Proof. Let $c = (g_0, c_1, \dots, g_k)$ be defined over the subdivision $0 = t_0 \leq \cdots \leq t_k = 1$. The lifting $\hat{c} = (\hat{g}_0, \hat{c}_1, \dots, \hat{g}_k)$ of c is constructed by induction as follows: $\hat{g}_0 = (\hat{x}_0, g_0)$, $\hat{c}_i : [t_{i-1}, t_i] \to \hat{X}$ is the unique continuous path such that $c_i = p\hat{c}_i$ and $\hat{c}_i(t_{i-1}) = \alpha(\hat{g}_{i-1})$, $\hat{g}_i = (\hat{c}_i(t_i), g_i)$. Each stage of the construction is uniquely determined by the previous stage and $\hat{g}_0$ is uniquely determined. It is clear that the liftings issuing from $\hat{x}_0$ of two equivalent $\mathcal{G}$-paths are equivalent $\hat{\mathcal{G}}$-paths, and the lifting of an elementary homotopy is an elementary homotopy because p is a covering projection. This proves the proposition. □

3.16 Corollary. *The homomorphism $\pi_1((\hat{\mathcal{G}}, \hat{X}), \hat{x}_0) \to \pi_1((\mathcal{G}, X), x_0)$ induced by (π, p) is injective.*

3.17 Proposition. *Let $(\varphi, f) : (\mathcal{G}', X') \to (\mathcal{G}, X)$ be a continuous homomorphism of étale groupoids and let $(\pi, p) : (\hat{\mathcal{G}}, \hat{X}) \to (\mathcal{G}, X)$ be a covering. Assume that X' is $\mathcal{G}'$-connected and locally arcwise connected. Suppose that $(\mathcal{G}', X')$ is simply connected and fix base points $x'_0 \in X'$ and $\hat{x}_0 \in \hat{X}$ such that $f(x'_0) = p(\hat{x}_0) = x_0$. Then there is a unique continuous homomorphism $(\hat{\varphi}, \hat{f}) : (\mathcal{G}', X') \to (\hat{\mathcal{G}}, \hat{X})$ such that $f = p \circ \hat{f}$, $\varphi = \hat{\pi} \circ \hat{\varphi}$ and $\hat{f}(x'_0) = \hat{x}_0$.*

Proof. Let c be a $\mathcal{G}'$-path issuing from x'_0. If $\hat{\varphi}$ exists, then $\hat{\varphi}(c)$ is a $\hat{\mathcal{G}}$-path $\hat{c}$ issuing from $\hat{x}_0$ such that $\varphi(c) = \pi(\hat{c})$. As such a path is unique (by (3.15)), this shows the uniqueness of $\hat{\varphi}$.

Given a point $x' \in X'$, there is a $\mathcal{G}'$-path c joining x'_0 to x', because X' is assumed to be $\mathcal{G}'$-connected. Moreover, as $\mathcal{G}'$ is simply connected, the homotopy class of such a path is unique. Let $\hat{c}$ be the $\hat{\mathcal{G}}$-path issuing from $\hat{x}_0$ whose projection by π is $\varphi(c)$. Let $\hat{f} : X' \to \hat{X}$ be the map associating to x' the end point of $\hat{c}$. To check that this map is continuous at x', choose a neighbourhood $\hat{U}$ of $\hat{f}(x')$ that is mapped by p homeomorphically onto an open set $U \subset X$; as f is continuous and X' is locally arcwise connected, $f^{-1}(U)$ contains an arcwise connected neighbourhood V of x';

we claim that $\hat{f}^{-1}(\hat{U})$ contains V. Indeed for each point $z \in V$, we may choose a continuous path c_z in V joining x' to z; its image under f can be lifted uniquely as a path $\hat{c}_z$ in $\hat{U}$ issuing from $\hat{f}(x')$; considering c_z as a $\mathcal{G}'$-path and $\hat{c}_z$ as a $\hat{\mathcal{G}}$-path, the lifting of $f(c * c_z)$ is $\hat{c} * \hat{c}_z$, hence $\hat{f}^{-1}(\hat{U})$ contains V. □

3.18 Corollary. *Let $(\pi, p) : (\hat{\mathcal{G}}, \hat{X}) \to (\mathcal{G}, X)$ and $(\pi', p') : (\hat{\mathcal{G}}', \hat{X}') \to (\mathcal{G}, X)$ be two simply connected coverings of $(\mathcal{G}, X)$, and choose basepoints $\hat{x}_0 \in \hat{X}$ and $\hat{x}'_0 \in \hat{X}'$ such that $p(\hat{x}_0) = p'(\hat{x}'_0)$. Then there is a unique isomorphism $(\hat{\varphi}, \hat{f}) : (\hat{\mathcal{G}}, \hat{X}) \to (\hat{\mathcal{G}}', \hat{X}')$ such that $\pi = \pi' \circ \hat{\varphi}$ and $\hat{f}(\hat{x}_0) = \hat{x}'_0$.*

By definition, an *automorphism* γ of a covering $(\pi, p) : (\hat{\mathcal{G}}, \hat{X}) \to (\mathcal{G}, X)$ is a homeomorphism $\gamma : \hat{X} \to \hat{X}$ that commutes with the projection p and the right action of $\mathcal{G}$, namely $p \circ \gamma = p$ and for all $g \in \mathcal{G}$, $\hat{x} \in \hat{X}$ with $p(\hat{x}) = \omega(g)$ we have $\gamma.(\hat{x}.g) = (\gamma.\hat{x}).g$.

3.19 Corollary. *Let $(\mathcal{G}, X)$ be an étale groupoid and assume that X is locally arcwise connected and $\mathcal{G}$-connected. Any simply connected covering $(\pi, f) : (\hat{\mathcal{G}}, \hat{X}) \to (\mathcal{G}, X)$ is naturally a Galois covering, with Galois group $\Gamma \cong \pi_1((\mathcal{G}, X), x_0)$.*

Proof. Fix a base point $x_0 \in X$. By Corollary 3.18, the group Γ of automorphisms of (π, p) acts simply transitively on the fibre $p^{-1}(x_0)$. Let $\hat{x}_0 \in p^{-1}(x_0)$. Given $\gamma \in \Gamma$, let $\hat{c}$ be a $\hat{\mathcal{G}}$-path joining $\gamma.\hat{x}_0$ to $\hat{x}_0$. Let $\pi(\gamma)$ be the homotopy class of the $\mathcal{G}$-loop $p(\hat{c})$ at x_0; it is independent of the choice of $\hat{c}$ because $(\hat{\mathcal{G}}, \hat{X})$ is simply connected. As in the case of topological spaces, one shows that the map $\Phi : \Gamma \to \pi_1((\mathcal{G}, X), x_0)$ is an isomorphism; we leave the details as an (instructive) exercise. □

3.20 Exercises

(1) *Construction of the Universal Covering.* Let $(\mathcal{G}, X)$ be an arcwise connected étale groupoid such that X is locally simply connected (this means that each point $x \in X$ has a fundamental system of neighbourhoods which are simply connected). Construct a simply connected covering $(\pi, p) : (\hat{\mathcal{G}}, \hat{X}) \to (\mathcal{G}, X)$ using an analogue of the classical construction of the universal covering of a locally simply connected space.

(Hints: $\hat{X}$ will be the set of homotopy classes $[c]$ of $\mathcal{G}$-paths c issuing from a base point $x_0 \in X$ (all the paths are parameterized by $[0, 1]$); the projection $p : \hat{X} \to X$ will map $[c]$ to the endpoint of c. Define a topology on $\hat{X}$ as follows: given a simply connected open set U in X and $[c] \in \hat{X}$ with $x = \hat{p}([c]) \in U$, there is a canonical lifting $U_{[c]}$ of U in $\hat{X}$; namely to $y \in U$ one associates the homotopy class of the composition of c with a path in U connecting x to y (this homotopy class is well defined because U is simply connected). Show, that the subsets of the form $U_{[c]}$ constitute a basis for a topology on $\hat{X}$ and that $p : \hat{X} \to X$ is a covering.

Consider the map that associates to each $([c], g) \in \hat{X} \times \mathcal{G}$ with $\omega(g) = p([c])$ the element $[c * g] \in \hat{X}$. Show that this map defines a right action of $(\mathcal{G}, X)$ on $\hat{X}$ over p. Let $[c'] \in \pi_1((\mathcal{G}, X), x_0)$. Check that $([c'], [c]) \mapsto [c' * c]$ defines a left action of

$\Gamma = \pi_1((\mathcal{G}, X), x_0)$ on $\hat{X}$ which is simply transitive on each fibre of $\hat{p}$ and commutes with the right action of $(\mathcal{G}, X)$.

Let $(\hat{\mathcal{G}}, \hat{X})$ be the étale groupoid associated to the action of $(\mathcal{G}, X)$ on $\hat{X}$ over p, and let $\varphi : \hat{\mathcal{G}} \to \hat{X}$ be the map sending $([c], g)$ to $[c].g$. Show that $(\varphi, p) : (\hat{\mathcal{G}}, \hat{X}) \to (\mathcal{G}, X)$ is a simply connected covering.)

(2) *A Criterion for Developability*. Let $(\mathcal{G}, X)$ be a connected étale groupoid such that X is locally simply connected. Show that the following conditions are equivalent.

(i) $(\mathcal{G}, X)$ is developable, i.e. equivalent to the étale groupoid associated to an action of $\pi_1((\mathcal{G}, X), x_0)$ on a simply connected space $\tilde{X}$.

(ii) Each point of X has a simply connected open neighbourhood U such that: if $g \in \mathcal{G}$ is not a unit and $\alpha(g),\ \omega(g) \in U$, and if $c : [0, 1) \to U$ is a continuous path with $c(1) = \omega(g)$ and $c(0) = \alpha(g)$, then the homotopy class of the $\mathcal{G}$-loop (c, g) is non-trivial.

(Hint for (ii) $\Longrightarrow$ (i): Using the notations of the preceding exercise, let $\tilde{X}$ be the quotient space of $\hat{X}$ by the equivalence relation which identifies $[c] \in \hat{\mathcal{G}}$ with $[c].g$, where $g \in \mathcal{G}$ and $\omega(g) = \hat{p}([c])$. Let $\hat{q} : \hat{X} \to \tilde{X}$ be the map associating to $[c]$ its equivalence class. Check that if condition (2) is satisfied, then $\hat{q}$ is an étale map. (It is sufficient to check that the restriction of $\tilde{p}$ to an open subset of the form $U_{[c]}$ is injective.) Then apply lemma (3.14).)

4. Proof of the Main Theorem

Outline of the Proof

We first outline the proof, which is adapted from the proof of the Cartan-Hadamard Theorem given by S. Alexander and R. L. Bishop (see II.4). Given a groupoid $(\mathcal{G}, X)$ of local isometries, we first define the notion of $\mathcal{G}$-geodesic. Assuming that $\mathcal{G}$ is Hausdorff and that the metric on X is locally complete and locally convex, we introduce a topology (see 4.4) on the set $\hat{X}$ of equivalence classes of $\mathcal{G}$-geodesics issuing from a base point $x_0 \in X$; for this topology the projection $p : \hat{X} \to X$ sending each $\mathcal{G}$-geodesic to its endpoint is étale.

The groupoid $\mathcal{G}$ acts naturally on the right on $\hat{X}$ over p. Let $\tilde{X} := \hat{X}/\mathcal{G}$ be the quotient of $\hat{X}$ by this action. The natural projection $\hat{q} : \hat{X} \to \tilde{X}$ is étale (4.5) and the groupoid $(\hat{\mathcal{G}}, \hat{X})$ associated to the action is equivalent to the trivial groupoid $\tilde{X}$. Moreover the space $\tilde{X}$ is contractible, in particular it is simply connected. Hence the equivalent groupoid $(\hat{\mathcal{G}}, \hat{X})$ associated to the action of $\mathcal{G}$ on $\hat{X}$ is also simply connected.

If $(\mathcal{G}, X)$ is complete, then $\hat{p} : \hat{X} \to X$ is a covering (4.7). Therefore $(\hat{\mathcal{G}}, \hat{X}) \to (\mathcal{G}, X)$ is a Galois covering with Galois group $\Gamma = \pi_1((\mathcal{G}, X), x_0)$. The action of Γ on $\hat{X}$ by covering automorphisms commutes with the right action of $\mathcal{G}$, hence it also acts on $\tilde{X}$. It then follows from lemma 3.14 that $(\mathcal{G}, X)$ is equivalent to the groupoid associated to the action of Γ on $\tilde{X}$.

$\mathcal{G}$-Geodesics

4.1 $\mathcal{G}$-Geodesics. Let $(\mathcal{G}, X)$ be a groupoid of local isometries. A $\mathcal{G}$-geodesic is a $\mathcal{G}$-path $c = (g_0, c_1, g_1, \dots, c_k, g_k)$ over the subdivision $0 = t_0 < \dots < t_k = 1$ that satisfies the following conditions for $i = 1, \dots, k$.

(1) $c_i : [t_{i-1}, t_i] \to X$ is a constant speed local geodesic.

(2) If $\tilde{g}_i : U \to \mathcal{G}$ is a local section of α defined on an open neighbourhood of $\alpha(g_i)$ such that $\tilde{g}_i(\alpha(g_i)) = g_i$, then for $\varepsilon > 0$ small enough, the concatenation of $c_i|_{[t_i-\varepsilon, t_i]}$ and $\omega \circ \tilde{g}_i \circ c_{i+1}|_{[t_i, t_i+\varepsilon]}$ is a constant speed local geodesic.

Note that any $\mathcal{G}$-path which is equivalent to a $\mathcal{G}$-geodesic is a $\mathcal{G}$-geodesic. If $(\mathcal{G}, X)$ is the groupoid associated to an action of a group Γ by isometries on a length space X, then any $\mathcal{G}$-geodesic joining x to y is equivalent to a unique $\mathcal{G}$-geodesic of the form $c = (c_1, g_1)$ where $c_1 : [0, 1] \to X$ is a constant speed local geodesic issuing from x and $g_1 = (\gamma, y)$ with $\gamma.y = c(1)$ (see 3.9(1)).

4.2 Lemma. *Let $(\mathcal{G}, X)$ be a groupoid of local isometries which is Hausdorff. Fix $g \in \mathcal{G}$ with $x = \alpha(g)$, $y = \omega(g)$ and let $r > 0$ be such that the closed balls $\overline{B}(x, r)$ and $\overline{B}(y, r)$ are complete and the metric restricted to these balls is convex (see II.1.3). Then there is a unique continuous map $\tilde{g} : B(x, r) \to \mathcal{G}$ such that $\tilde{g}(x) = g$ and $\alpha \circ \tilde{g}$ is the identity of $B(x, r)$; moreover $h := \omega \circ \tilde{g}$ is an isometry from $B(x, r)$ onto $B(y, r)$.*

Proof. Let $c_{x'} : [0, 1] \to B(x, r)$ be the unique constant speed geodesic segment joining x to $x' \in B(x, r)$. As $(\mathcal{G}, X)$ is Hausdorff and $\overline{B}(y, r)$ is complete, there is a unique continuous lifting $\tilde{c}_{x'} : [0, 1] \to \mathcal{G}$ such that $\alpha(\tilde{c}_{x'}(t)) = c_{x'}(t)$ and $\tilde{c}_{x'}(0) = g$ (see 2.14). We define $\tilde{g}(x') = \tilde{c}_{x'}(1)$; this gives a map $\tilde{g} : B(x, r) \to \mathcal{G}$ which is continuous (because geodesics vary continuously with their endpoints in $B(x, r)$ where the metric is convex) and $\omega \circ \tilde{g}$ is a local isometry from $B(x, r)$ to $B(y, r)$. The same argument applied to the inverse of g shows that this map has an inverse which is also a local isometry, Hence $\omega \circ \tilde{g}$ is an isometry from $B(x, r)$ onto $B(y, r)$. □

The next lemma is the analogue of lemma II.4.3.

4.3 Main Lemma. *Let $(\mathcal{G}, X)$ be a groupoid of local isometries which is Hausdorff. Assume that the metric on X is locally complete and locally convex. Let $c = (g_0, c_1, g_1, \dots, c_k, g_k)$ be a $\mathcal{G}$-geodesic joining x to y over the subdivision $0 = t_0 < \dots < t_k = 1$. Then there is an $\varepsilon > 0$ such that:*

(i) *for each i there are unique continuous maps $\tilde{g}_i : B(\alpha(g_i), \varepsilon) \to \mathcal{G}$ which are sections of α, and are such that $\tilde{g}_i(\alpha(g_i)) = g_i$;*

(ii) *for each pair of points $\overline{x}, \overline{y} \in X$ with $d(x, \overline{x}) < \varepsilon$, $d(y, \overline{y}) < \varepsilon$, there is a unique $\mathcal{G}$-geodesic $\overline{c} = (\overline{g}_0, \overline{c}_1, \overline{g}_1, \dots, \overline{c}_k, \overline{g}_k)$ that is defined over the same subdivision as c, joins $\overline{x}$ to $\overline{y}$, and is such that $d(c_i(t), \overline{c}_i(t)) < \varepsilon$ and $\overline{g}_i = \tilde{g}_i(\alpha(\overline{g}_i))$ for $i = 0, \dots, k$ and $t \in [0, 1]$. Moreover*

$$l(\overline{c}) \leq l(c) + d(x, \overline{x}) + d(y, \overline{y}).$$

Proof. The proof is an adaptation of the argument of Alexander-Bishop [AB90] used in lemma II.4.3. In the proof of II.4.3, instead of dividing the interval $[a, b]$ into three equal parts, one can use any fixed partition of the interval.

Choose $r > 0$ small enough so that, for each $i = 1, \dots, k$ and each $t \in [t_{i-1}, t_i]$, the closed balls $\overline{B}(c_i(t), r)$, as well as the closed balls $\overline{B}(x, r)$ and $\overline{B}(y, r)$, are complete and the distance function is convex on these balls.

It follows from (4.2) that, for each $i = 0, \dots, k$, there are unique continuous maps $\tilde{g}_i : B(\alpha(g_i), r) \to \mathcal{G}$ which are sections of α such that $\tilde{g}_i(\alpha(g_i)) = g_i$ and $h_i = \omega \circ \tilde{g}_i$ is an isometry from $B(\alpha(g_i), r)$ onto $B(\omega(g_i), r)$.

We first assume the existence assertion in (ii) and deduce from it the uniqueness assertion in (ii) and the inequality for the lengths. Consider two $\mathcal{G}$-geodesics $\overline{c}$ and $\overline{c}'$ defined over the subdivision $0 = t_0 < \cdots < t_k = 1$, as in (ii), with endpoints $\overline{x}, \overline{y}$ and $\overline{x}', \overline{y}'$. Assume $\varepsilon < r$. The function mapping $t \in [t_{i-1}, t_i]$ to $d(\overline{c}(t), \overline{c}'(t))$ is continuous on the interval $[0, 1]$ and locally convex, hence convex. This shows that if $\overline{c}$ and $\overline{c}'$ have the same endpoints, then they are equal. The stated inequality concerning lengths is proved like in lemma II.4.3. Also, provided $\varepsilon < r$, any Cauchy sequences $(\overline{x}_n)$ in $B(x, \varepsilon/2)$ and $(\overline{y}_n)$ in $B(y, \varepsilon/2)$ will converge to points $\overline{x} \in \overline{B}(x, \varepsilon/2)$ and $\overline{y} \in \overline{B}(y, \varepsilon/2)$, and the above convexity argument shows that the sequence of unique $\mathcal{G}$-geodesics joining $\overline{x}_n$ to $\overline{y}_n$ will converge (uniformly) to the unique $\mathcal{G}$-geodesic joining $\overline{x}$ to $\overline{y}$.

To prove the existence of $\overline{c}$ for ε small enough, we shall argue by induction on k. The case $k = 1$ follows from (II.4.3).

We may assume by induction that the analogue of the lemma is true for the $\mathcal{G}$-geodesic $c' = (g_0, \dots, c_{k-1}, g_{k-1})$ obtained by restricting the $\mathcal{G}$-geodesic c of the lemma to the subdivision $0 = t_0 < \cdots < t_{k-1}$ of the interval $[0, t_{k-1}]$ and for some ε' smaller than r. Let $\delta < 0$ be small enough so that $\delta < \varepsilon'$ and $t_{k-1} - \delta > t_{k-2}$. Let $c'' : [t_{k-1} - \delta, t_k] \to X$ be the local geodesic defined by $c''(t) = c_k(t)$ for $t \in [t_{k-1}, t_k]$ and $h_{k-1}(c''(t)) = c_{k-1}(t)$ for $t \in [t_{k-1} - \delta, t_{k-1}]$. Let $z'_0 = c''(t_{k-1})$ and $z''_0 = c''(t_{k-1} - \delta)$. Inductively, we define sequences of points z'_n and z''_n in $B(c_k(t_{k-1}), \varepsilon')$. Assume z'_{n-1} and z''_{n-1} are already defined. The first paragraph of the proof shows that our inductive hypothesis on c' gives a unique $\mathcal{G}$-geodesic c'_n over the subdivision $t_0 < \cdots < t_{k-1}$ joining $\overline{x}$ to z'_{n-1}. Let $c''_n : [t_{k-1} - \delta, t_k] \to X$ be the unique local geodesic joining z''_{n-1} to $h_k(\overline{y})$ which is r-close to c''(see II.4.3). Define $z'_n = c''_n(t_{k-1})$ and $z''_n = h^{-1}_{k-1}(c'_n(t_{k-1} - \delta))$.

By convexity we have

$$d(z''_n, z''_{n+1}) \leq \frac{t_{k-1} - \delta}{t_{k-1}} \, d(z'_{n-1}, z'_n) \quad \text{for } n \geq 1,$$

and

$$d(z''_0, z''_1) \leq \frac{\delta}{t_{k-1}} \, d(x, \overline{x}).$$

Similarly,

$$d(z'_n, z'_{n+1}) \leq \frac{t_k - t_{k-1}}{t_k - t_{k-1} + \delta} \, d(z''_{n-1}, z''_n) \quad \text{for } n \geq 1,$$

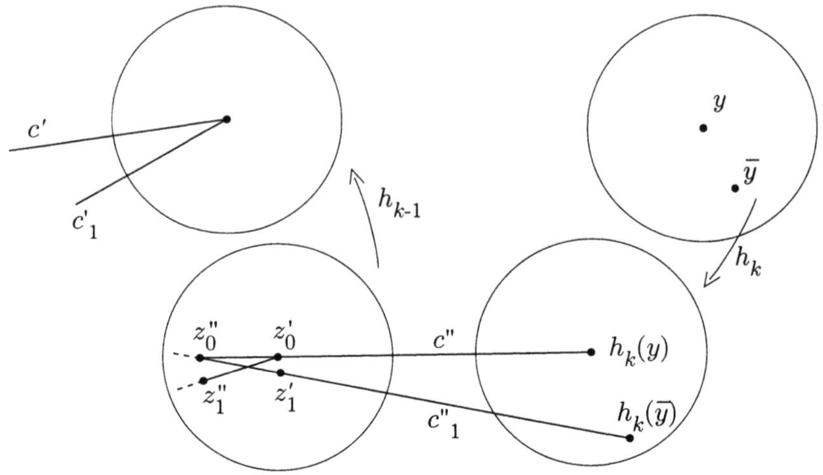

Fig. $\mathcal{G}$.4 The proof of the main lemma

and

$$d(z_0', z_1') \leq \frac{\delta}{t_k - t_{k-1} + \delta}\, d(y, \overline{y}).$$

Choose a number $\lambda \geq 1/2$ such that

$$\max\left\{\frac{t_{k-1} - \delta}{t_{k-1}}, \frac{\delta}{t_{k-1}}, \frac{t_k - t_{k-1}}{t_k - t_{k-1} + \delta}, \frac{\delta}{t_k - t_{k-1} + \delta}\right\} < \lambda < 1.$$

Assuming $d(x, \overline{x}) < \varepsilon'$ and $d(y, \overline{y}) < \varepsilon'$, these inequalities imply that $d(z_n', z_{n+1}') < \lambda^{n+1}\varepsilon'$ and $d(z_n'', z_{n+1}'') < \lambda^{n+1}\varepsilon'$, hence $d(z_0', z_n')$ and $d(z_0'', z_n'')$ are both smaller than $\frac{\lambda}{1-\lambda}\,\varepsilon'$. Therefore, if we assume that $\varepsilon < \frac{1-\lambda}{\lambda}\varepsilon'$, then the sequences (z_n') and (z_n'') that we have defined inductively, are Cauchy sequences. As the closed balls $\overline{B}(c(t), \varepsilon)$ are complete and the metric is locally convex, the sequence of $\mathcal{G}$-geodesics c_n' converges uniformly to a $\mathcal{G}$-geodesic $\overline{c}' = (\overline{g}_0, \overline{c}_1, \dots, \overline{g}_{k-1})$ over the subdivision $0 = t_0 < \cdots < t_{k-1}$; this $\mathcal{G}$-geodesic joins $\overline{x}$ to $z' = \lim z_n'$, and $\overline{g}_i = \tilde{g}_i(\overline{c}_i(t_{i-1}))$. Also, the sequence c_n'' converges to a local geodesic $\overline{c}'' : [t_{k-1} - \delta, t_k] \to X$ joining $z'' = \lim z_n''$ to $h_k(\overline{y})$. Moreover $h_{k-1}(\overline{c}''(t)) = \overline{c}_{k-1}(t)$ for $t \in [t_{k-1} - \delta, t_{k-1}]$. Therefore, if we define $\overline{c}_k = \overline{c}''|_{[t_{k-1}, t_k]}$ and $\overline{g}_k = \tilde{g}_k(\overline{y})$, then $\overline{c} = (\overline{g}_0, \dots, \overline{c}_{k-1}, \overline{g}_{k-1}, \overline{c}_k, \overline{g}_k)$ is the $\mathcal{G}$-geodesic joining x to y that we were looking for. □

The Space $\hat{X}$ of $\mathcal{G}$-Geodesics Issuing from a Base Point

Let $\hat{X}$ be the set of equivalence classes of $\mathcal{G}$-geodesics parameterized by $[0, 1]$ that issue from a base point $x_0 \in X$. Let $p : \hat{X} \to X$ be the map associating to each equivalence class its terminal point. Let $\hat{x}_0$ denote the point of $\hat{X}$ represented by the constant $\mathcal{G}$-geodesic at x_0, namely (g_0, c_1, g_1), where $g_0 = g_1 = 1_{x_0}$ and $c_1(t) = x_0$ for each $t \in [0, 1]$.

Given a $\mathcal{G}$-geodesic $c = (g_0, , c_1, \dots, g_k)$ joining x_0 to a point y over the subdivision $0 = t_0 < \cdots < t_k = 1$, we shall write X_c^ε to denote the set of $\mathcal{G}$-geodesics given by (4.3) that join x_0 to the various points of $B(y, \varepsilon)$, and we write $\hat{X}_c^\varepsilon$ to denote its image in $\hat{X}$. The projection p maps $\hat{X}_c^\varepsilon$ bijectively onto $B(y, \varepsilon)$. This shows that the map $X_c^\varepsilon \to \hat{X}_c^\varepsilon$ sending each $\mathcal{G}$-path to its equivalence class is injective.

Note that if we refine the subdivision of c we don't change $\hat{X}_c^\varepsilon$. Therefore there is no loss of generality if we assume in what follows that the $\mathcal{G}$-geodesic $c = (g_0, c_1, \dots, c_k, g_k)$ satisfies the following condition: there exists $r > 0$ such that the closed balls of radius r centred at the sources and targets of the g_i, and at each point $c_i(t)$, are complete geodesic spaces on which the distance function is convex; we also assume that $c_i([t_{i-1}, t_i]) \subset B(c_i(t_{i-1}), r/3)$ (in particular each c_i is a constant speed geodesic segment). We shall consider sets X_c^ε with $\varepsilon < r$.

4.4 Proposition. *With the assumptions of lemma 4.3, the subsets $\hat{X}_c^\varepsilon \subseteq \hat{X}$ form a basis for a Hausdorff topology on $\hat{X}$, which respect to which p is étale.*

Proof. Consider two sets $\hat{X}_c^\varepsilon$ and $\hat{X}_{c'}^{\varepsilon'}$ which intersect. Without loss of generality, we can assume that c and c' are defined over the same subdivision $0 = t_0 < \cdots < t_k = 1$ and satisfy the condition preceding the statement of the proposition. By construction $p(\hat{X}_c^\varepsilon)$ and $p(\hat{X}_{c'}^{\varepsilon'})$ are balls $B(y, \varepsilon)$ and $B(y', \varepsilon')$ in X. Assume that $\overline{c} = (\overline{g}_0, \overline{c}_1, \dots, \overline{g}_k) \in X_c^\varepsilon$ and $\overline{c}' = (\overline{g}'_0, \overline{c}'_1, \dots, \overline{g}'_k) \in X_{c'}^{\varepsilon'}$ are two equivalent $\mathcal{G}$-geodesics. The image under p of the equivalence classes of $\overline{c}$ and $\overline{c}'$ is a point $z \in B(y, \varepsilon) \cap B(y', \varepsilon')$.

The proposition is a consequence of the following assertion.

Claim: If δ is such that $B(z, \delta) \subset B(y, \varepsilon) \cap B(y', \varepsilon')$, then $B(z, \delta) \subset p(\hat{X}_c^\varepsilon \cap \hat{X}_{c'}^{\varepsilon'})$.

By definition, there are continuous maps $h_i : [t_{i-1}, t_i] \to \mathcal{G}$ such that $\alpha(h_i(t)) = \overline{c}_i(t)$, $\omega(h_i(t)) = \overline{c}'_i(t)$ and $h_i(t_i)\overline{g}'_i = \overline{g}_i h_{i+1}(t_i)$ for $i = 1, \dots, k-1$, also $\overline{g}_0 = \overline{g}'_0 h_1(0)$ and $\overline{g}'_k = h_k(1)\overline{g}_k$. Using lemma 4.2, we can find liftings $\tilde{h}_i : B(\overline{c}_i(t_{i-1}), \delta) \to \mathcal{G}$ of α such that $\tilde{h}_i(\overline{c}(t_{i-1})) = h_i(t_{i-1})$. As the projection α is étale and $\mathcal{G}$ is Hausdorff, for each $i = 1, \dots, k-1$ and $u \in B(c_i(t_i), \delta)$ we have

$$\tilde{h}_i(\omega(\tilde{g}_i(u)))\, \tilde{g}_i(u) = \tilde{g}'_i(\omega(\tilde{h}_{i+1}(u)))\, \tilde{h}_{i+1}(u),$$

also $\tilde{g}_0(u) = \tilde{g}'_0(\omega(\tilde{h}_1(u))\, \tilde{h}_1(u)$ and $\tilde{h}_k(\omega(\tilde{g}_k(u))\tilde{g}_k(u) = \tilde{g}'_k(u)$ for each u in, respectively, $B(\overline{c}_1(0), \delta)$ and $B(\alpha(\overline{g}_k), \delta)$, because these two last equalities are true when u is the centre of the ball. This implies that the $\tilde{h}_i$ induce an equivalence from elements of $X_{\overline{c}}^\delta \subset X_{\overline{c}}^\varepsilon$ to elements of $X_{\overline{c}'}^\delta \subset X_{\overline{c}}^{\varepsilon'}$. Thus $\hat{X}_{\overline{c}}^\delta = \hat{X}_{\overline{c}'}^\delta$. This proves the claim. □

The Space $\tilde{X} = \hat{X}/\mathcal{G}$

Given an element $\hat{x} \in \hat{X}$ represented by a $\mathcal{G}$-geodesic $c = (g_0, c_1, \dots, g_k)$ and an element $g \in \mathcal{G}$ with $\omega(g) = p(\hat{x})$, we write $\hat{x}.g$ for the element of $\hat{X}$ represented by the $\mathcal{G}$-geodesic $c.g$ obtained from c by replacing its last entry g_k with $g_k g$. This defines a continuous right action of $\mathcal{G}$ on $\hat{X}$ over $p : \hat{X} \to X$.

Let $(\hat{\mathcal{G}}, \hat{X})$ be the étale groupoid associated to this action (see 3.11). The following lemma is analogous to (II.4.5).

4.5 Lemma. *Let $\tilde{X}$ be the quotient of $\hat{X}$ by the right action of $\mathcal{G}$ described above, with the quotient topology. ($\tilde{X}$ is also the space of orbits of the groupoid $\hat{\mathcal{G}}$.) Then $\tilde{X}$ is Hausdorff and contractible. The natural projection $\hat{q} : \hat{X} \to \tilde{X}$ is étale and induces an equivalence from $(\hat{\mathcal{G}}, \hat{X})$ to the trivial groupoid $\tilde{X}$. In particular the groupoid $(\hat{\mathcal{G}}, \hat{X})$ is simply-connected.*

Proof. We first note that if $g \in \mathcal{G}$ is not a unit and if $\omega(g)$ is the terminal point of c, then $c.g$ is not equivalent to c, because $\mathcal{G}$ is Hausdorff (see 3.2). More generally, if $\tilde{g} : B(\alpha(g), \varepsilon) \to B(\omega(g), \varepsilon)$ is a lifting of α such that $\tilde{g}(\alpha(g)) = g$, then $\hat{X}_c^\varepsilon \cap \hat{X}_{c.g}^\varepsilon = \emptyset$. This shows that $\hat{q} : \hat{X} \to \tilde{X}$ is étale, that $\tilde{X}$ is Hausdorff, and that the map $(\hat{x}, g) \mapsto \hat{q}(\hat{x})$ gives an equivalence from the groupoid $(\hat{\mathcal{G}}, \hat{X})$ to the trivial groupoid $\hat{X}$.

It remains to check that $\tilde{X}$ is contractible. Given $s \in (0, 1]$ and $\tilde{x} \in \tilde{X}$ represented by the $\mathcal{G}$-geodesic $c = (g_0, c_1, \dots, g_k)$ over the subdivision $0 = t_1 < \dots < t_k = 1$, let $r_s(\tilde{x}) \in \tilde{X}$ be the element of $\tilde{X}$ represented by the $\mathcal{G}$-geodesic obtained from c by restricting the parameter t to the interval $[0, s]$ and reparameterizing by $[0, 1]$. (Note that if $s = t_j$, then the $\mathcal{G}$-geodesics $(g_0, c_1, \dots, c_i)$ and $(g_0, c_1, \dots c_i, g_i)$ over the subdivision $0 = t_0 < \dots t_i = s$ of $[0, s]$ (reparameterized by $[0, 1]$) represent the same point of $\tilde{X}$.) Then $(s, \tilde{x}) \mapsto r_s(\tilde{x})$ is a continuous map $[0, 1] \times \tilde{X} \to \tilde{X}$ defining a homotopy that connects the constant map from $\tilde{X}$ to $\tilde{x}_0$ to the identity map of $\tilde{X}$. □

The Covering $p : \hat{X} \to X$

4.6 Lemma. *With the assumptions of Theorem 2.15, given $\hat{x} \in \hat{X}$ and a rectifiable path $s \mapsto y^s$ in X parameterized by $s \in [0, 1]$, with $y^0 = p(\hat{x})$, there is continuous path $s \mapsto \hat{y}^s$ in $\hat{X}$ such that $p(\hat{y}^s) = y^s$ and $\hat{y}^0 = \hat{x}$.*

Proof. We assume that $s \mapsto y^s$ is a constant speed parameterization. As p is étale, the maximal interval containing 0 on which the lifting $s \mapsto \hat{y}^s$ can be defined is open. So it will be sufficient to prove that if a lifting $s \mapsto \hat{y}^s$ is defined on $[0, a)$, then $\lim_{s \to a} \hat{y}^s$ exists. Let ℓ be the sum of the length of the path $s \mapsto y^s$ and the length of a $\mathcal{G}$-geodesic representing y^0.

Let $q : X \to \mathcal{G}\backslash X$ be the natural projection. We endow $\mathcal{G}\backslash X$ with the quotient pseudometric and its associated topology. For $s \in [0, a)$, if $\hat{y}^s$ is represented by the $\mathcal{G}$-geodesic $c^s = (g_0^s, c_1^s, \dots, g_n^s)$ over the subdivision $0 = t_0^s < \dots < t_r^s = 1$, then the map $C : [0, 1] \times [0, a) \to \mathcal{G}\backslash X$ sending (t, s), with $t \in [t_{i-1}, t_i]$, to $q(c_i^s(t))$ is well-defined. For $t_1, t_2 \in [0, 1]$ and $s_1, s_2 \in [0, a)$, by (4.3) we have

$$d(C(t_1, s_1), C(t_2, s_2)) \leq \ell(|t_1 - t_2| + |s_1 - s_2|).$$

Therefore the map C extends to a continuous map $[0, 1] \times [0, a] \to \mathcal{G}\backslash X$, still denoted C. By the compactness of the interval $[0, 1]$, we can find a subdivision

$0 = t_0 < \cdots < t_k = 1$, a number $\varepsilon > 0$ and points $x_0, x_1, \ldots, x_k$ in X such that $q(x_i) = C(t_i, a)$, $x_k = y(a)$, the balls $\overline{B}(x_i, 3\varepsilon)$ are complete and geodesic with convex distance function, and each $(t_i - t_{i-1})$ is smaller than ε/ℓ.

Let $s_1 \in [0, a)$ be such that the length of the curve $s \mapsto y^s$ restricted to $[s_1, a]$ is smaller than ε and $d(C(t, s_1), C(t, a)) < \varepsilon$ for all $t \in [0, 1]$. Using (2.14) we can find points $z_i \in B(x_i, \varepsilon)$ for $i = 1, \ldots, k-1$ such that $q(z_i) = C(t_i, s_1)$; let $z_0 = x_0$ and $z_k = y^{s_1}$. Using (2.14) again, we can find a representative of $\hat{y}^{s_1}$ of the form $c^{s_1} = (g_0^{s_1}, c_1^{s_1}, \ldots, g_k^{s_1})$ over the subdivision $0 = t_0 < \cdots < t_k = 1$, where $c_i^{s_1}(t_{i-1}) = z_{i-1}$ and $g_0^{s_1}$ is the unit 1_{x_0}. We claim that we can also find, for each $s \in [s_1, a)$, a unique representative $c^s = (g_0^s, c_1^s, \ldots, g_k^s)$ of $\hat{y}^s$ defined over the same subdivision $0 = t_0 < \ldots t_k = 1$, where g_i^s and c_i^s vary continuously with s. By lemma 4.3 this is possible for s close to s_1. Suppose that c^s can be defined on the interval $[s^1, s^2)$; the length of each path c_i^s is smaller than ε, as is the length of each path $s \mapsto c^s(t_i)$ (because its length is no greater than that of the corresponding path $s \mapsto y(s)$, by lemma 4.3); therefore each path c_i^s is contained in $B(x_{i-1}, 3\varepsilon)$; as the closure of these balls is assumed to be complete, we can define c^{s_2} as the limit of c^s as $s \to s_2$, and hence c^s can be defined on the interval $[s_1, s_2]$. If $s_2 < 1$, we can extend the construction of c^s to a larger interval using (4.2), and eventually to the whole interval $[s_1, a]$. □

4.7 Corollary. *If $(\mathcal{G}, X)$ is a groupoid of local isometries which is connected, Hausdorff, complete and such that the metric on X is locally convex, then $p : \hat{X} \to X$ is a covering of X.*

Proof. Let $c = (g_0, c_1, \ldots, g_n)$ be a rectifiable $\mathcal{G}$-path defined over a subdivision $0 = s_0 < \cdots < s_n = 1$ such that $\omega(g_0) = x_0$. Using the preceding lemma, we can construct, by induction on n, a $\hat{\mathcal{G}}$-path $\hat{c} = (\hat{g}_0, \hat{c}_1, \ldots, \hat{g}_n)$ defined over the same subdivision such that $\hat{g}_0 = (\hat{x}_0, g_0)$, and $\hat{c}_i : [s_{i-1}, s_i] \to \hat{X}$ is the unique path in $\hat{X}$ such that $p(s) = c_i(s)$ and $\hat{c}_i(s_{i-1}) = \hat{c}_{i-1}(s_{i-1}).g_{i-1}$. In particular p maps the terminal point of $\hat{c}$ to the terminal point of c. As X is $\mathcal{G}$-connected, this shows that p is surjective. The preceding lemma implies that p is a covering over each connected component of X. □

The End of the Proof of the Main Theorem. Let $\pi : \hat{\mathcal{G}} \to \mathcal{G}$ be the map $(\hat{x}, g) \mapsto g$. We have just seen that $(\pi, p) : (\hat{\mathcal{G}}, \hat{X}) \to (\mathcal{G}, X)$ is a covering. By (4.5), $(\hat{\mathcal{G}}, \hat{X})$ is simply connected, therefore (π, p) is a Galois covering with Galois group $\Gamma = \pi_1((\mathcal{G}, X), x_0)$ (see (3.18). Thus the theorem follows from 3.14. □

References

[AB90] S. B. Alexander and R. L. Bishop, *The Hadamard–Cartan theorem in locally convex spaces*, Enseign. Math. **36** (1990), 309–320

[AB96] S. B. Alexander and R. L. Bishop, *Comparison theorems for curves of bounded geodesic curvature in metric spaces of curvature bounded above*, Differential Geom. Appl. **6** (1996), 67–86

[AB97] S. B. Alexander and R. L. Bishop, *Racetracks and extremal problems for curves of bounded curvature in meric spaces*, Geom. Dedicata **66** (1997), 331–341

[ABB87] S. B. Alexander, I. D. Berg and R. L. Bishop, *The Riemannian obstacle problem*, Illinois J. Math. **31** (1987),167–184

[ABB93a] S. B. Alexander, I. D. Berg and R. L. Bishop, *Cut loci, minimizers, and wavefronts in Riemannian manifolds with boundary*, Michigan Math. J. **40** (1993), 229–237

[ABB93b] S. B. Alexander, I. D. Berg and R. L. Bishop, *Geometric curvature bounds in Riemannian Manifolds with Boundary*, Trans. Amer. Math. Soc. **339** (1993), 703–716

[AiR90] I. R. Aitchison and J. H. Rubinstein, *An introduction to polyhedral metrics of nonpositive curvature on* 3*-manifolds*, Geometry of low-dimensional manifolds II, Durham 1989, (S. K. Donaldson, C. B. Thomas, ed), LMS Lecture Note Series **151**, Cambridge Univ. Press, 1990, 127–161

[Ale51] A. D. Alexandrov, *A theorem on triangles in a metric space and some of its applications*, Trudy Mat. Inst. Steklov **38** (1951), 5–23

[Ale55] A. D. Alexandrov, Die innere Geometrie der konvexen Flächen, Akadamie Verlag, Berlin, 1955

[Ale57a] A. D. Alexandrov, *Über eine Verallgemeinerung der Riemannschen Geometrie*, Schriftreihe des Forschinstituts für Mathematik **1** (1957), Berlin, 33–84

[Ale57b] A. D. Alexandrov, *Ruled surfaces in metric spaces*, Vestnik Leningrad Univ. **12** (1957), 5–26

[AleBN86] A. D. Alexandrov, V. N. Berestovskii and I. G. Nikolaev, *Generalized Riemannian spaces*, Russian Math. Surveys **41:3** (1986), 1–54

[AleZ67] A. D. Alexandrov and V. A. Zalgaller, Intrinsic Geometry of Surfaces, Transl. Math. Monog. **15**, Amer. Math. Soc., Providence RI, 1967

[Alm66] F. Almgren Jr., Plateau's Problem, Benjamin, New York NY, 1966

[Alo90] J. M. Alonso, *Inégalités isopérimétriques et quasi-isométries*, C. R. Acad. Sci. Paris, **311** (1990), 761–764

[AloB95] J. M. Alonso and M. R. Bridson, *Semihyperbolic groups*, Proc. London Math. Soc. (3) **70** (1995), 56–114

[Alo+98] J. M. Alonso, W. A. Bogley, S. J. Pride and X. Wang, *Second order Dehn functions of groups*, Quart. J. Math. Oxford (2) **49** (1998), 1–30

[AnDG97] F. D. Ancel, M. W. Davis and C. R. Guilbault, *CAT*(0) *reflection groups*, Geometric Topology (Athens, GA, 1993) (W. H. Kazez, ed), AMS/IP, Stud. Adv. Math., Amer. Math. Soc., Providence RI, 1997, Part 1, 441–445

[AnG97] F. D. Ancel and C. R. Guilbault, *Interiors of compact contractible n-manifolds are hyperbolic* ($n \geq 5$), J. Differential Geom. **45** (1997), 1–32

[Ba85] W. Ballmann, *Nonpositively curved manifolds of higher rank*, Ann. of Math. (2) **122** (1985), 597–609

[Ba90] W. Ballmann, *Singular spaces of non-positive curvature*, Sur les Groupes Hyperboliques d'après Mikhael Gromov (E. Ghys, P. de la Harpe, ed), Chapter 10, Progr. Math. vol. 83, Birkhäuser, Boston MA, 1990

[Ba95] W. Ballmann, Lectures on Spaces of Nonpositive Curvature, DMV Seminar, Band 25, Birkhäuser, Basel, 1995

[BaBr94] W. Ballmann and M. Brin, *Polygonal complexes and combinatorial group theory*, Geom. Dedicata **50** (1994), 164–191

[BaBr95] W. Ballmann and M. Brin, *Orbihedra of nonpositive curvature*, Inst. Hautes Études Sci. Publ. Math. **82** (1995), 169–209

[BaBu96] W. Ballmann and S. Buyalo, *Nonpositively curved metrics on 2-polyhedra*, Math. Z. **222** (1996), 97-134

[BaGS85] W. Ballmann, M. Gromov, V. Schroeder, Manifolds of Nonpositive Curvature, Progr. Math. vol. 61, Birkhäuser, Boston MA, 1985

[BaKu90] H. Bass and R. Kulkarni, *Uniform tree lattices*, J. Amer. Math. Soc. **3** (1990), 845–902

[BanS91] V. Bangert and V. Schroeder, *Existence of flat tori in analytic manifolds of nonpositive curvature*, Ann. Sci. École Norm. Sup. **24** (1991), 605–634

[Bana32] S. Banach, Théorie des Opérations Linéaires, Hafner, New York, 1932

[Bar93] C. Baribaud, Travail de diplôme, Univ. de Genève, 1993

[Barr95] S. Barré, *Polyèdres finis de dimension 2 à courbure* ≤ 0 *et de rang 2*, Ann. Inst. Fourier (Grenoble) **45** (1995), 1037–1059

[Barr97] S. Barré, *Sur les polyèdres de rang* 2, Sémin. Theor. Spectr. Geom. **15** (1996-7), Université de Grenoble I, Saint-Martin d'Hères

[Bass72] H. Bass, *The degree of polynomial growth of finitely generated nilpotent groups*, Proc. London Math. Soc. (3) **25** (1972), 603–614

[Bass93] H. Bass, *Covering theory for graph of groups*, J. Pure Appl. Algebra **89** (1993), 3–47

[BaSw97] W. Ballmann and J. Świątkowski, *On L^2-cohomology and property (T) for automorphism groups of polyhedral cell complexes*, Geom. Func. Anal.**7** (1997), 615–645

[BBMS97] G. Baumslag, M. R. Bridson, C. F. Miller III and H. Short, *Finitely presented subgroups of automatic groups and their isoperimetric functions*, J. London Math. Soc. (2) **56** (1997), 292–304

[Bear83] A. Beardon, The Geometry of Discrete Groups, Grad. Texts in Math., vol. 91, Springer Verlag, New York, 1983

[Ben91a] Nadia Benakli, *Polyèdres à géométrie locale donnée*, C. R. Acad. Sci. Paris **313** (1991), 561–564

[Ben91b] Nadia Benakli, *Polyèdres hyperboliques à groupe d'automorphismes non discret*, C. R. Acad. Sci. Paris **313** (1991), 667–669

[Ben92] Nadia Benakli, Polyèdres Hyperboliques, Passage du Local au Global, Thèse, Univ. de Paris Sud, 1992

[Ber83] V. N. Berestovskii, *Borsuk's problem on metrization of a polyhedron*, Dokl. Akad. Nauk. SSSR **27:1** (1983), 56–59

[Ber86] V. N. Berestovskii, *Spaces with bounded curvature and distance geometry*, Siberian Math. J. **27** (1986), 8–19

[Berg77] M. Berger, Géométrie, vols. 1–5, Cedric/F. Nathan, Paris, 1977-78

[BerN93] V. N. Berestovskii and I. G. Nikolaev, *Multidimensional generalized Riemannian spaces*, Geometry IV (Yu. G. Reshetnyak, ed), Encyclopedia of Math. Sci., vol. 70, Springer Verlag, Berlin, Heidelberg, New York, 1993, 165–243

[Bes88] M. Bestvina, *Degenerations of hyperbolic space*, Duke Math. J. **56** (1988), 143–161

[Bes93] M. Bestvina, *The virtual cohomological dimension of Coxeter groups*, LMS Lecture Note Series **181** (G. A. Niblo and M. A. Roller, ed), Cambridge Univ. Press, 1993, 19–23

[Bes96] M. Bestvina, *Local homological properties of boundaries of groups*, Michigan Math. J. **43** (1996), 123–139

[BesB97] M. Bestvina and N. Brady, *Morse theory and finitenesss properties of groups*, Invent. Math **129** (1997), 445–470

[BesF92] M. Bestvina and M. Feighn, *A combination theorem for negatively curved groups*, J. Differential Geom. **35** (1992), 85–101

[BesF95] M. Bestvina and M. Feighn, *Stable actions of groups on real trees*, Invent. Math. **121** (1995), 287–321

[BesM91] M. Bestvina and G. Mess, *The boundary of a negatively curved group*, J. Amer. Math. Soc. **4** (1991), 469–481

[BGSS9] G. Baumslag, S. M. Gersten, M. Shapiro and H. Short, *Automatic groups and amalgams*, J. Pure Appl. Algebra **76** (1991), 229–316

[BGP92] Y. Burago, M. Gromov and G. Perel'man, *A. D. Alexandrov spaces with curvature bounded below*, Russian Mathematical Surveys **47** (1992), 3–51

[Bi76a] R. Bieri, *Normal subgroups in duality groups and in groups of cohomological dimension 2*, J. Pure Appl. Algebra **7** (1976), 32–52

[Bi76b] R. Bieri, Homological Dimension of Discrete Groups, Queen Mary College Mathematics Notes, Queen Mary College, London, 1976

[Bi98] R. Bieri, *Finiteness length and connectivity*, Geometric Group Theory Down Under, Canberra 1996 (J. Cossey, C. F. Miller III, W. D. Neumann, M. Shapiro, ed), de Gruyter, Berlin, 1998

[Bing59] R.L. Bing, *The cartesian product of a certain non-manifold with a line is* E^4, Ann. of Math. (2) **70** (1959), 399–412

[BiNS87] R. Bieri, W. D. Neumann and R. Strebel, *A geometric invariant of discrete groups*, Invent. Math. **90** (1987), 451–477

[Bir76] J. Birman, Braids, Links and Mapping Class Groups, Ann. of Math. Stud., vol. 82, Princeton Univ. Press, Princeton NJ, 1976

[BisC66] R. L. Bishop and R. J. Crittenden, Geometry of manifolds, Academic Press, New York, 1966

[BisO69] R. L. Bishop and B. O'Neill, *Manifolds of negative curvature*, Trans. Amer. Math. Soc. **145** (1969), 1–49

[BlW97] J. Block and S. Weinberger, *Large scale homology theories and geometry*, Geometric Topology (Athens, GA, 1993) (W. H. Kazez, ed), AMS/IP, Stud. Adv. Math., Amer. Math. Soc., Providence RI, 1997, Part 1, 522–569

[BMS93] G. Baumslag, C. F. Miller III and H. Short, *Unsolvable problems about small cancellation and word hyperbolic groups*, Bull. London Math. Soc. **23** (1993), 97–101

[Bo63] A. Borel, *Compact Clifford-Klein forms of symmetric spaces*, Topology **2** (1963), 111–122

[BoHC62] A. Borel and Harish Chandra, *Arithmetic subgroups of algebraic groups*, Ann. of Math. **75** (1962), 485–535

[BoG95] O. V. Bogopolskii and V. N. Gerasimov, *Finite subgroups of hyperbolic groups*, Algebra Logic **34** (1995), 343–345

[Bor66] K. Borsuk, Theory of Retracts, PWN, Warsaw, 1966

[BoRu96a] P. Bowers and K. Ruane, *Fixed points in boundaries of negatively curved groups*, Proc. Amer. Math. Soc. **124** (1996), 1311–1313

[BoRu96b] P. Bowers and K. Ruane, *Boundaries of nonpositively curved groups of the form* $G \times Z^n$, Glasgow Math. J. **38** (1996) 177–189

[Bou93] M. Bourdon, Action Quasi-Convexes d'un Groupe Hyperbolique et le Flot Géodésique, Thèse, Univ. Paris Sud, 1993

[Bou95a] M. Bourdon, *Au bord de certains polyèdres hyperboliques*, Ann. Inst. Fourier (Grenoble) **45** (1995), 119–141

[Bou95b] M. Bourdon, *Structure conforme au bord et flot géodésiques d'un* CAT(-1) *espace*, Enseign. Math. **40** (1995), 267–289

[Bou96] M. Bourdon, *Sur le birapport au bord des CAT(-1)-espaces*, Inst. Hautes Études Sci. Publ. Math. **83** (1996), 95–104

[Bou97a] M. Bourdon, *Immeubles hyperboliques, dimension conforme et rigidité de Mostow*, Geom. Func. Anal. **7**(1997), 245–268

[Bou97b] M. Bourdon, *Sur la dimension de Hausdorff de l'ensemble limite de sous-groupes convexes co-compacts*, C.R. Acad. Sci. Paris, Série I, **325** (1997), 1097–1100

[Bour53] N. Bourbaki, Topologie Genérale, 1953, Hermann, Paris

[Bour59] N. Bourbaki, *Formes sesquilinéaires et formes quadratiques*, Chap IX, Livre II, Algèbre, Hermann, Paris, 1959

[Bour81] N. Bourbaki, Groupes et Algèbres de Lie, Chapitres 4–6, Hermann, Paris, 1981

[Bow91] B. H. Bowditch, *Notes on Gromov's hyperbolicity criterion for path-metric spaces*, Group Theory from a Geometrical Viewpoint (E. Ghys, A. Haefliger, A. Verjovsky, ed), Proc. ICTP Trieste 1990, World Scientific, Singapore, 1991, 373–464

[Bow95a] B. H. Bowditch, *A short proof that a subquadratic isoperimetric inequality implies a linear one*, Michigan J. Math. **42** (1995), 103–107

[Bow95b] B. H. Bowditch, *Minkowskian subspaces of non-positively curved metric spaces*, Bull. London Math. Soc. **27** (1995), 575–584

[Bow95c] B. H. Bowditch, *Notes on locally* CAT*(1)-spaces*, Geometric Group Theory (R. Charney, M. Davis, M. Shapiro, ed), de Gruyter, Berlin-New York, 1995, 1–48

[Bow98a] B. H. Bowditch *A topological characterization of hyperbolic groups*, J. Amer. Math. Soc. **11** (1998), 643–677

[Bow98b] B. H. Bowditch, *Convergence groups and configuration spaces*, Geometric Group Theory Down Under, Canberra 1996 (J. Cossey, C. F. Miller III, W. D. Neumann, M. Shapiro, ed), de Gruyter, Berlin, 1998

[Bow98c] B. H. Bowditch, *Cut points and canonical splittings of hyperbolic groups*, Acta Math. **180** (1998), 145–186

[BR84] G. Baumslag and J.E. Roseblade, *Subgroups of the direct product of two free groups*, J. London Math. Soc. (2) **30** (1984), 44–52

[Bra95] T. Brady, *Complexes of nonpositive curvature for extensions of F_2 by $\mathbb{Z}$*, Topology Appl. **63** (1995), 267–275

[Bra99] N. Brady, *Branched coverings of cubical complexes and subgroups of hyperbolic groups*, J. London Math. Soc., in press

[Bri91] M. R. Bridson, *Geodesics and curvature in metric simplicial complexes*, Group Theory from a Geometrical Viewpoint (E. Ghys, A. Haefliger, A. Verjovsky, ed), Proc. ICTP Trieste 1990, World Scientific, Singapore, 1991, 373–464

[Bri93a] M. R. Bridson, *On the geometry of normal forms in discrete groups*, Proc. London Math. Soc.(3) **67** (1993), 596–616

[Bri93b] M. R. Bridson, *Regular combings, non-positive curvature and the quasi-convexity of abelian subgroups*, J. Pure Appl. Algebra **88** (1993), 23–35

[Bri95a] M. R. Bridson, *On the existence of flat planes in spaces of nonpositive curvature*, Proc. Amer. Math. Soc. **123** (1995), 223–235.

[Bri95b] M. R. Bridson, *Optimal isoperimetric inequalities for abelian-by-free groups*, Topology **34** (1995), 547–564

[Bri98a] M. R. Bridson, *Controlled embeddings into groups that have no non-trivial finite quotients*, Geometry and Topology Monographs 1 (1998), The Epstein Birthday Schrift (I. Rivin, C. Rourke, C. Series, ed), International Press, 99–116

[Bri98b] M. R. Bridson, *Non-positive curvature in group theory*, Groups St. Andrews 1997 at Bath (C. M. Campbell, E. F. Robertson, N. Ruskuc, G. C. Smith, ed), LMS Lecture Note Series **260**, Cambridge Univ. Press, 1998, 124–175

[Bri99a] M. R. Bridson, *Asymptotic cones and polynomial isoperimetric inequalities*, Topology **38** (1999), 543–554

[Bri99b] M. R. Bridson, *Doubles, finiteness properties of groups and quadratic isoperimetric Inequalities*, J. Algebra **213** (1999) 652–667

[BriG96] M. R. Bridson and S. M. Gersten, *The optimal isoperimetric inequality for torus bundles over the circle*, Quart. J. Math. Oxford **47** (1996), 1–23

[BriS94] M. R. Bridson and G. A. Swarup, *On Hausdorff–Gromov convergence and a theorem of F. Paulin*, Enseign. Math. **40** (1994), 267–289

[BriV95] M. R. Bridson and K. Vogtmann, *On the geometry of the automorphism group of a free group*, Bull. London Math. Soc. **27** (1995), 544–552

[Brn83] A. Bronsted, An Introduction to Convex Polytopes, Springer Verlag, Berlin, Heidelberg, New York, 1983

[Bro82] K. S. Brown, Cohomology of Groups, Grad. Texts in Math., vol. 87, 1982, Springer Verlag, New York

[Bro87] K. S. Brown, *Finiteness properties of groups*, J. Pure Appl. Algebra **44** (1987), 45–75

[Bro88] K. S. Brown, Buildings, Springer Verlag, Berlin, Heidelberg, New York, 1988

[BruT72] F. Bruhat et J. Tits, *Groupes réductifs sur un corps local. I. Données radicielles valuées*, Inst. Hautes Études Sci. Publ. Math. **41** (1972), 5–251

[BSo62] G. Baumslag and D. Solitar, *Some two-generator one-relator non-Hopfian groups*, Bull. Amer. Math. Soc. **68** (1962), 199–201

[Bu95] M. Burger, *Rigidity Properties of group actions on* CAT(0)*-spaces*, Proceedings of the I.C.M., Zürich, 1994, Birkhäuser, Basel, 1995, 761–769

[BuM96] M. Burger and S. Mozes, CAT(-1)*-spaces, divergence groups and their commensurators*, J. Amer. Math. Soc. **9** (1996), 57–83

[BuM97] M. Burger and S. Mozes, *Finitely presented simple groups and products of trees*, C. R. Acad. Sci. Paris **324** (1997), 747–752

[Bus42] H. Busemann, Metric Methods in Finsler Spaces and in the Foundations of Geometry, Ann. of Math. Stud., vol. 8, Princeton Univ. Press, Princeton NJ, 1942

[Bus43] H. Busemann, *On spaces in which two points determine a geodesic*, Trans. Amer. Math. Soc. **54** (1943), 171–184

[Bus48] H. Busemann, *Spaces with non-positive curvature*, Acta Math. **80** (1948), 259–310

[Bus55] H. Busemann, Geometry of Geodesics, Academic Press, New York, 1955

[BusM41] H. Busemann and W. Mayer, *On the foundations of the calculus of variations*, Trans. Amer. Math. Soc. **49** (1941), 173–198

[Buy92] S. Buyalo, *Euclidean planes in open 3-dimensional manifolds of non-positive curvature*, St. Petersburg Math. J. **3** (1992), 83–96

[BuZi85] G. Burde and H. Zieschang, Knots, de Gruyter, Berlin, 1985

[Ca84] J. W. Cannon, *The combinatorial structure of cocompact discrete hyperbolic groups*, Geom. Dedicata **16** (1984), 123–148

[Ca91] J. W. Cannon, *The theory of negatively curved spaces and groups*, Ergodic Theory, Symbolic Dynamics and Hyperbolic Spaces (T. Bedford, M. Keane, C. Series, ed), Oxford Univ. Press, New York, 1991, 315–369

[Car26] E. Cartan, *Sur une classe remarquable d'espaces de Riemann*, Bull. Soc. Math. France **54** (1926), 214–264

[Car27] E. Cartan, *Sur une classe remarquable d'espaces de Riemann*, Bull. Soc. Math. France **55** (1927), 114–134

[Car28] E. Cartan, Leçons sur la Géométrie des Espaces de Riemann, Gauthier-Villars, Paris, 1928, 2nd edition 1951

[Car32] E. Cartan, *Les espaces riemanniens symétriques*, Verh. Int. Math. Kongr. Zürich I, 1932, 152–161

[CasB92] A. Casson and S. Bleiler, Automorphisms of Surfaces after Nielsen and Thurston, LMS Students Texts **9**, Cambridge Univ. Press, 1992

[CasJ94] A. Casson and D. Jungreis, *Convergence groups and Seifert fibered 3-manifolds*, Invent. Math. **118** (1994), 441–456

[Cay1878] A. Cayley, *On the theory of groups*, Proc. London Math. Soc. **9** (1878), 126–133

[CD93] R. Charney and M. Davis, *Singular metrics of nonpositive curvature on branched coverings of Riemannian manifolds*, Amer. J. Math **115** (1993), 929–1009

[CD95a] R. Charney and M. Davis, *The $K(\pi, 1)$-problem for hyperplane complements associated to infinite reflection groups*, J. Amer. Math. Soc. **8** (1995), 597–627

[CD95b] R. Charney and M. Davis, *Strict hyperbolization*, Topology **34** (1995), 329–350

[CD95c] R. Charney and M. Davis, *The Euler characteristic of a non-positively curved piecewise Euclidean manifold*, Pacific J. Math. **171** (1995), 117–137

[CD95d] R. Charney and M. Davis, *The polar dual of a convex polyhedral set in hyperbolic space*, Michigan Math. J. **42** (1995), 479–510

[CDM97] R. Charney, M. Davis and G. Moussong, *Nonpositively curved piecewise Euclidean structures on hyperbolic manifolds*, Michigan Math. J. **44** (1997), 201–208

[CDP90] M. Coornaert, T. Delzant and A. Papadopoulos, Notes sur les Groupes Hyperboliques de Gromov, Lecture Notes in Math. **1441**, Springer Verlag, New York, 1990

[Cham95] C. Champetier, *Propriétés statistiques des groupes de présentation finie*, Adv. Math. **116** (1995), 197–262

[ChaM82] B. Chandler and W. Magnus, The History of Combinatorial Group Theory, Spinger Verlag, New York, Heidelberg, Berlin, 1982

[ChEb75] J. Cheeger and B. Ebin, Comparison Theorems in Riemannian Geometry, North Holland, Amsterdam, 1975

[ChGr86] J. Cheeger and M. Gromov, *Collapsing Riemannian manifolds while keeping their curvature bounded, I and II*, J. Differential Geom. **23** (1986), 309–346, **32** (1990), 269–298

[Coh75] M. M. Cohen, Simple-Homotopy Theory, Springer Verlag, Berlin, Heidelberg, New York, Grad. Texts in Math., vol. 10, 1975

[CoMi77] D. J. Collins and C. F. Miller III, *The Conjugacy problem and subgroups of finite index*, Proc. London Math. Soc. **33** (1977), 535–556

[CoMi98] D. J. Collins and C. F. Miller III, *The word problem in groups of cohomological dimension 2*, Groups St. Andrews 1997 at Bath (C. Campbell, E. Robertson, N. Ruskuc, G. Smith, ed), LMS Lecture Note Series **260**, Cambridge Univ. Press, 1998, 211-218

[Con92] G. Conner, Metrics on Groups, Ph.D Thesis, University of Utah, 1992

[Cors92] J. M. Corson, *Complexes of groups*, Proc. London Math. Soc. **65** (1992), 199–224

[Cox34] H. S. M. Coxeter, *Discrete subgroups generated by reflections*, Ann. of Math. **34** (1934), 588–621

[CV35a] S. Cohn-Vossen, *Existenz Kürzester Wege*, Dokl. Akad. Naut. SSSR **8** (1935), 339–342

[CV35b] S. Cohn-Vossen, *Kürzeste Wege und Total Krümmung auf Flächen*, Compositio Math. **2** (1935), 69–133

[Da83] M. W. Davis, *Groups generated by reflections and aspherical manifolds not covered by Euclidean space*, Ann. of Math. (2) **117** (1983), 293–324

[Da87] M. W. Davis, *Some aspherical manifolds*, Duke Math. J. **55** (1987), 105–139

[Da98] M. W. Davis, *Buildings are* CAT(0), Geometry and Cohomology in Group Theory (P. H. Kropholler, G. A. Niblo and R. Stohr, ed), LMS Lecture Note Series **252**, Cambridge Univ. Press, Cambridge, 1988, 108–123

[Da99] M. W. Davis, *Nonpositive Curvature and Reflection Groups*, Handbook of Geometric Topology, Elsevier, Amsterdam, 1999

[DaJ91a] M. W. Davis and T. Januszkiewicz, *Hyperbolization of polyhedra*, J. Differential Geom. **34** (1991), 347–388

[DaJ91b] M. W. Davis and T. Januszkiewicz, *Convex polytopes, Coxeter orbifolds and torus actions*, Duke J. Math. **62** (1991), 417–451

[Dehn12a] M. Dehn, *Transformationen der Kurven auf zweiseitigen Flächen*, Math. Ann. **72** (1912), 413–421

[Dehn12b] M. Dehn, *Über unendliche diskontinuierliche Gruppen*, Math. Ann. **71** (1912), 116–144

[Dehn87] M. Dehn, Papers on Group Theory and Topology, translated and introduced by John Stillwell, Springer Verlag, Berlin, Heidelberg, New York, 1987

[Del91] T. Delzant, *Sous-groupes à deux générateurs des groupes hyperboliques*, Group Theory from a Geometrical Viewpoint (E. Ghys, A. Haefliger, A. Verjovsky, ed), Proc. ICTP Trieste 1990, World Scientific, Singapore, 1991, 177–189

[Del95] T. Delzant, *L'image d'un groupe dans un groupe hyperbolique*, Comment. Math. Helv. **70** (1995), 267–284

[Del96] T. Delzant, *Sous-groupes distingués et quotients des groupes hyperboliques*, Duke Math. J. **83** (1996), 661–682

[Dieu63] J. Dieudonné, Fondements de l'Analyse Moderne, Gauthier-Villars, Paris, 1963

[Dix60] J. Dixmier, *Opérateurs de rang fini dans les représentations unitaires*, Inst. Hautes Études Sci. Publ. Math. **6** (1960), 13–25

[Dow52] C. H. Dowker, *The topology of metric complexes*, Amer. J. Math. **74** (1952), 555–577

[Dr97] A. N. Dranishnikov, *On the virtual cohomological dimension of Coxeter groups*, Proc. Amer. Math. Soc. **125** (1997), 1885–1891

[Dre69] A. Dress, *Newman's theorem on transformation group*, Topology **8** (1969), 203–207

[Dru97] C. Drutu, *Nondistorsion des horosphères dans des immeubles euclidiens et dans des espaces symétriques*, Geom. Func. Anal. **7** (1997), 712–754

[DW84] L. van den Dries and A.J. Wilkie, *On Gromov's theorem concerning groups of polynomial growth and elementary logic*, J. Algebra **89** (1984), 49–374

[Eb72a] P. Eberlein, *Geodesic flow in certain manifolds without conjugate points*, Trans. Amer. Math. Soc. **167** (1972), 151–170

[Eb72b] P. Eberlein, *Geodesic flows on negatively curved manifolds, I*, Ann. of Math. (2) **95** (1972), 492–510

[Eb73] P. Eberlein, *Geodesic flows on negatively curved manifolds, II*, Trans. Amer. Math. Soc. **178** (1973), 57–82

[Eb96] P. Eberlein, Geometry of Non-Positively Curved Manifolds, Chicago Lectures in Math., U. Chicago Press, Chicago-London, 1996

[EbON73] P. Eberlein and B. O'Neill, *Visibility manifolds*, Pacific J. Math. **46** (1973), 45–109

[Ef53] V. A. Efromovich, *The proximity geometry of Riemannian manifolds*, Uspekhi Math. Nauk **8** (1953), 189–195

[Eg77] H. G. Eggleston, Convexity, Cambridge Univ. Press, 1977

[Ehr54] C. Ehresmann, *Structures locales*, Annali de Matematica, Serie IV, **36** (1954), 133–142

[Eps87] D. B. A. Epstein, *Complex hyperbolic geometry*, Analytical and Geometric Aspects of Hyperbolic Spaces, Coventry/Durham 1984 (D. B. A. Epstein, ed), LMS Lecture Note Series **111**, Cambridge Univ. Press, 1987, 93–111

[Ep+92] D. B. A. Epstein, J. W. Cannon, D. F. Holt, S. V. F. Levy, M. S. Paterson, W. P. Thurston, Word Processing in Groups, Jones and Bartlett, Boston MA, 1992

[Es98] A. Eskin, *Quasi-isometric rigiditiy for non-uniform lattices in higher rank*, J. Amer. Math. Soc. **11** (1998), 321–361

[EsF97] A. Eskin and B. Farb, *Quasi-flats and rigiditiy in higher rank symmetric spaces*, J. Amer. Math. Soc. **10** (1997), 653–692

[Fa97] B. Farb, *The quasi-isometry classification of lattices in semisimple Lie group*, Math. Res. Lett. **4** (1997), 705–718

[FaMo98] B. Farb and L. Mosher, *A rigidity theorem for the solvable Baumslag-Solitar groups*, Invent. Math. **131** (1998), 419–451

[Fe96] S. C. Ferry, *Topological finiteness theorems in Gromov-Hausdorff space*, Duke J. Math. **74**, (1996), 95–104

[Fe98] S. C. Ferry, *Limits of polyhedra in Gromov-Hausdorff space*, Topology **37** (1998), 1325-1338

[FJ89] T. Farrell and L. Jones, *A topological analogue of Mostow's rigidity theorem*, J. Amer. Math. Soc. **2** (1989), 257–369

[FJ93] T. Farrell and L. Jones, *Topological rigidity for compact non-positively curved manifolds*, Differential Geometry, Los Angeles 1990, Proc. Sympos. Pure Math. **54**, Amer. Math. Soc., Providence RI, 1993, 224–274

[FP97] W. Floyd and W. Parry, *The growth of non-positively curved triangles of groups*, Invent. Math. **129** (1997), 289–359

[Fr31] H. Freudenthal, *Über die Enden topologisher Räume und Gruppen*, Math. Z. **33** (1931), 692–713

[FrKl12] R. Fricke and F. Klein, Vorslesungen über die Theorie der automorphen Functionen, Vol 1 (1897) and Vol 2 (1912), Teubner, Leipzig-Berlin, reprinted by Academic Press, New York 1966

[Gab93] D. Gabai, *Convergence groups are Fuchsian groups*, Ann. of Math. (2) **136** (1993), 447–579

[Ger87] S. M. Gersten, *Reducible diagrams and equations over groups*, Essays in Group Theory (S. M. Gersten, ed), MSRI Publ. **8**, Springer Verlag, New York, 1987, 15–73

[Ger93a] S. M. Gersten, *Quasi-isometry invariance of cohomological dimension*, C. R. Acad. Sci. Paris **316** (1993), 411–416

[Ger93b] S. M. Gersten, *Isoperimetric and isodiametric functions of finite presentations*, Geometric Group Theory (G. Niblo and M. Roller, ed), vol. 1, LMS Lecture Note Series **181**, Cambridge Univ. Press, 1993, 79–96

[Ger94] S. M. Gersten, *The automorphism group of a free group is not a* CAT(0) *group*, Proc. Amer. Math. Soc. **121** (1994), 999–1002

[Ger96] S. M. Gersten, *Subgroups of word hyperbolic groups in dimension 2*, J. London Math. Soc. **54** (1996), 261–283

[GeS91] S. M. Gersten and H. Short, *Rational subgroups of biautomatic groups*, Ann. of Math. (2) **134** (1991), 125–158

[GhH90] E. Ghys and P. de la Harpe (ed), Sur les Groupes Hyperboliques d'après Mikhael Gromov, Progr. Math. vol. 83, Birkhäuser, Boston MA, 1990

[GHV91] E. Ghys, A. Haefliger and A. Verjovsky (editors), Group Theory from a Geometrical Viewpoint, Proc. ICTP Trieste 1990, World Scientific, Singapore, 1991

[GPW91] K. Grove, P. Peterson and J. Wu, *Geometric finiteness theorems via controlled topology*, Invent. Math. **99** (1990), 205–213

[GrH97] R. Grigorchuk and P. de la Harpe, *On problems related to growth, entropy and spectrum in group theory*, J. Dynam. Control Systems **3** (1997), 51–89

[Gri83] R. Grigorchuk, *On Milnor's problem of group growth*, Dokl. Akad. Nauk. SSSR **28** (1983), 23–26

[GrLP81] M. Gromov, J. Lafontaine and P. Pansu, Structures Métriques pour les Variétés Riemanniennes, Fernand Nathan, Paris, 1981

[Gro80] M. Gromov, *Synthetic geometry in Riemannian manifolds*, Proc. ICM, Helsinki 1978, Acad. Sc. Fennica, Helsinki, 1980, 415–419

[Gro81a] M. Gromov, *Hyperbolic manifolds, groups and actions*, Riemann Surfaces and Related Topics (Stony Brook, 1978), Ann. of Math. Stud., vol. 97, Princeton University Press, 1981, 183–213

[Gro81b] M. Gromov, *Groups of polynomial growth and expanding maps*, Inst. Hautes Études Sci. Publ. Math. **53** (1981), 53–78

[Gro84] M. Gromov, *Infinite groups as geometric objects*, Proc. ICM, Warsaw 1983, PWN, Warsaw 1984, 385–392

[Gro87] M. Gromov, *Hyperbolic groups*, Essays in group theory (S. M. Gersten, ed), Springer Verlag, MSRI Publ. **8** (1987), 75–263

[Gro93] M. Gromov, *Asymptotic invariants of infinite groups*, Geometric Group Theory (G. A. Niblo and M. A. Roller, ed), LMS Lecture Note Series **182**, Cambridge Univ. Press, 1993, 75–263

[GroP91] M. Gromov and P. Pansu, *Rigidity of lattices: an introduction*, Geometric Topology: Recent Developments (P. de Bartolomeis, F. Tricerri, ed), Lecture Notes in Math. **1504**, Springer Verlag, New York, 1991

[GrS92] M. Gromov and R. Schoen, *Harmonic maps into singular spaces and p-adic superrigidity for lattices in groups of rank one*, Inst. Hautes Études Sci. Publ. Math. **76** (1992), 165–256

[GrTh87] M. Gromov and W. P. Thurston, *Pinching constants for hyperbolic manifolds*, Invent. Math. **89** (1987), 1–12

[Grü67] B. Grünbaum, Convex Polytopes, Interscience, New York, 1967

[Gui70] Y. Guivarc'h, *Groupes de Lie à croissance polynomiale*, C.R. Acad. Sci. Paris, **271** (1970), 237–239

[Gui73] Y. Guivarc'h, *Croissance polynomiale et périodes des fonctions harmoniques*, Bull. Soc. Math. France, **101** (1973), 333–379

[GW71] D. Gromoll and J. Wolf, *Some relations between the metric structure and the algebraic structure of the fundamental group in manifolds of non-positive curvature*, Bull. Amer. Math. Soc. **77** (1971), 545–552

[Ha1898] J. Hadamard, *Les surfaces à courbure opposée et leurs lignes géodésiques*, J. Math. Pures Appl. **4** (1898), 27–73

[Hae58] A. Haefliger, *Structures feuilletées et cohomologie à valeur dans un faisceau de groupoides*, Comment. Math. Helv. **32** (1958), 248–329

[Hae70] A. Haefliger, *Homotopy and Integrability*, Manifolds, Amsterdam, Lecture Notes in Math. **197**, Springer Verlag, New York, 1970, 133–163

[Hae84] A. Haefliger, *Groupoides d'holonomie et classifiants*, Structure Transverse des Feuilletages, Toulouse 1982, Astérisque **116** (1984), 70–97

[Hae88] A. Haefliger, *Leaf closures in Riemannian foliations*, A Fête of Topology, Academic Press, Boston MA, 1988, 3–32

[Hae90] A. Haefliger, *Orbi-espaces*, Sur les Groupes Hyperboliques d'après Mikhael Gromov (E. Ghys et P. de la Harpe, ed), Chapter 11, Progr. Math. vol. 83, Birkhäuser, Boston MA, 1990

[Hae91] A. Haefliger, *Complexes of groups and orbihedra*, Group Theory from a Geometrical Viewpoint (E. Ghys, A. Haefliger, A. Verjovsky, ed), Proc. ICTP Trieste 1990, World Scientific, Singapore, 1991, 504–540

[Hae92] A. Haefliger, *Extension of complexes of group*, Ann. Inst. Fourier (Grenoble) **42** (1992), 275–311

[HaeQ84] A. Haefliger et Quach Ngc Du, *Appendice: une présentation du groupe fondamental d'une orbifold*, Structure transverse des feuilletages,Toulouse 1982, Astérisque **116** (1984), 98–107

[Hag91] F. Haglund, *Les polyèdres de Gromov*, C. R. Acad. Sc. Paris, Série I, **313** (1991), 603–606

[Hag93] F. Haglund, *Les polyèdres de Gromov*, Thèse ENS Lyon, 1993

[Har83] P. de la Harpe, *Free groups in linear groups*, Enseign. Math. **29** (1983), 129-144

[Har91] P. de la Harpe, *An invitation to Coxeter groups*, Group Theory from a Geometrical Viewpoint (E. Ghys, A. Haefliger, A. Verjovsky, ed), Proc. ICTP Trieste 1990, World Scientific, Singapore, 1991, 193–253

[Heb86] J. Hebda, *Curvature and focal points in Riemannian foliations*, Indiana Univ. Math. J. **35** (1986), 321–331

[Hebe87] J. Heber, *Hyperbolische geodätische Raume*, Diplomarbeit, Univ. Bonn, 1987

[Hel62] S. Helgason, Differential Geometry and Symmetric Spaces, Academic Press, New York, 1962

[Hel78] S. Helgason, Differential Geometry, Lie groups and Symmetric Spaces, Academic Press, New York, 1978

[Hem76] J. Hempel, 3-manifolds, Ann. of Math. Stud., vol. 86, Princeton Univ. Press, Princeton NJ, 1976.

[Hi1900] D. Hilbert, *Über das Dirichletsche Prinzip*, Jber. Dtsch. Math. Verein. **8** (1900), 184–188

[HiCV52] D. Hilbert and S. Cohn-Vossen, Geometry and the Imagination, Chelsea, New York, 1952

[Hid97] C. Hidber, *Isoperimetric functions of finitely generated nilpotent groups and their amalgams*, Thesis, Diss. ETH No. 12166, Zürich, 1997

[Hig51] G. Higman, *A finitely generated infinite simple group*, J. London Math. Soc. **26** (1951), 61–64

[HNN49] G. Higman, B. H. Neumann, H. Neumann, *Embedding theorems for groups*, J. London Math. Soc. **24** (1949), 247–254

[Hopf30] H. Hopf, *Zur Algebra der Abbildungen von Mannigfaltigkeiten*, J. Reine Angew. Math. **163** (1930), 71–88

[Hopf31] H. Hopf, *Beiträger zur Theorie der Flächenabbildungen*, J. Reine Angew. Math. **165** (1931), 225–236

[Hopf43] H. Hopf, *Enden offener Räume und unendliche diskontinuirliche Gruppen*, Comment. Math. Helv. **16** (1943-44), 81–100

[HoRi32] H. Hopf und W. Rinow, *Über den Begriff der vollständigen differentialgeometrischen Fläche*, Comment. Math. Helv. **3** (1932), 209–225

[How81] J. Howie, *Pairs of 2-complexes and equations over groups*, J. Reine Angew. Math. **324** (1981), 165–174

[HsW97] T. Hsu and D. T. Wise, *Embedding theorems for non-positively curved polygons of finite groups*, J. Pure Appl. Algebra (1997), 201–221

[Hu(B)93] B. Hu, *Whitehead groups of non-positively curved polyhedra*, J. Differential Geom. **38** (1993), 501–517

[Hu(S)65] S. Hu, Theory of Retracts, Wayne State University Press, Detroit MI, 1965

[Im79] H. C. Im Hof, *Die Geometrie der Weylkammern in symmetrischen Raumen vom nichtkompakten Typ*, Habilitationschrift, Bonn, 1979

[Iv92] B. Iversen, Hyperbolic Geometry, LMS Student Texts **25**, Cambridge Univ. Press, 1992

[IwMa65] N. Iwahori and H. Matsumoto, *On some Bruhat decompositions and of the structure of the Hecke ring of a* ($\mathfrak{p}$)*-adic Chevalley group*, Inst. Hautes Études Sci. Publ. Math. **25** (1965), 5–48

[Jam90] I. M. James, An Introduction to Uniform Spaces, LMS Lecture Note Series **144**, Cambridge Univ. Press, 1990

[JaS79] W. Jaco and P. Shalen, *Seifert fibered spaces in 3-manifold*, Mem. Amer. Math. Soc. **220**, 1979

[Jo79] K. Johannson, Homotopy Equivalence of 3-manifolds with Boundaries, Lecture Notes in Math. **761**, Springer Verlag, New York,1979

[JvN35] P. Jordan and J. von Neumann, *On inner products in linear metric spaces*. Ann. of Math. **36** (1935), 719–723

[K33a] E. R. van Kampen, *On the connection between the fundamental groups of some related spaces*, Amer. J. Math. **55** (1933), 261–267

[K33b] E. R. van Kampen, *On some lemmas in the theory of groups*, Amer. J. Math. **55** (1933), 268–273

[KaL95] M. Kapovich and B. Leeb, *Asymptotic cones and quasi-isometry classes of fundamental groups of 3-manifolds*, Geom. Func. Anal. **5** (1995), 582–603

[KaL96] M. Kapovich and B. Leeb, *Actions of discrete groups on nonpositively curved spaces*, Math. Ann. **306** (1996), 341–352

[Kap97] I. Kapovich, *Quasiconvexity and amalgams*, **7** (1997), 771–811

[Kar65] F. I. Karpelevic, *The geometry of geodesics and the eigenfunctions of the Beltrami-Laplace operator on symmetric spaces*, Trans. Moscow Math. Soc. (Amer. Math. Soc. Translations) **14** (1965), 51–199

[Ker83] S. P. Kerckhoff, *The Nielsen Realization Problem*, Ann. of Math. (2) **117** (1983), 235–265

[KKL97] M. Kapovich, B. Kleiner and B. Leeb, *Quasi-isometries and the de Rham decomposition*, Topology **37** (1998), 1193–1212

[KlC96] B. Kleiner and C. B. Croke, *Conjugacy and rigidity for non-positively curved manifolds of higher rank*, Topology **35** (1996), 273–286

[KlL97] B. Kleiner and B. Leeb, *Rigidity of quasi-isometries for symmetric spaces and Euclidean buildings*, Inst. Hautes Études Sci. Publ. Math. **86** (1997), 115–197

[KS93] N. J. Korevaar, R. M. Schoen, *Sobolev spaces and harmonic maps for metric space targets*, Comm. Anal. Geom. **1** (1993), 561–659

[La69] S. Lang, Analysis II, Addison Wesley, Reading MA,1969

[La80] S. Lang, Algebra, Addison Wesley, Reading MA, 1980

[Lan96] U. Lang, *Quasigeodesics outside horoballs*, Geom. Dedicata **63** (1996), 205-215

[LaY72] H. B. Lawson and S. T. Yau, *Compact manifolds of nonpositive curvature*, J. Differential Geom. **7** (1972), 211–228

[Le95] B. Leeb, *3-manifolds with(out) metrics of non-positive curvature*, Invent. Math. **122** (1995), 277–289

[Le97] B. Leeb, *A characterisation of irreducible symmetric spaces and Euclidean buildings of higher rank by their asymptotic geometry*, Habilitationsschrift, Bonn, May 1997

[Leu92] E. Leuzinger, *On the trigonometry of symmetric spaces*, Comment. Math. Helv. **67** (1992), 252–286

[Leu95] E. Leuzinger, *An exhaustion of locally symmetric spaces by compact submanifolds*, Invent. Math. **121** (1995), 389–410

[Leu96] E. Leuzinger, *On the Gauss-Bonnet formula for locally symmetric spaces of non compact type*, Enseign. Math. **42** (1996), 201-214

[LeuP96] E. Leuzinger and C. Pittet, *Isoperimetric inequalities for lattices in semisimple Lie groups of rank* 2, Geom. Func. Anal. **6** (1996), 389–410

[LS97a] U. Lang and V. Schroeder, *Jung's theorem for Alexandrov spaces of curvature bounded above*, Ann. Global Anal. Geom. **15** (1997), 263–275

[LS97b] U. Lang and V. Schroeder, *Kirszbraun's theorem and metric spaces of bounded curvature*, Geom. Func. Anal. **7** (1997), 535–560

[LS97c] U. Lang and V. Schroeder, *Quasiflats in Hadamard spaces*, Ann. Sci. École Norm. Sup. **30** (1997), 339–352

[Ly66] R. Lyndon, *On Dehn's algorithm*, Math. Ann. **166** (1966), 208–228

[LyS77] R. C. Lyndon and P. E. Schupp, Combinatorial Group Theory, Ergeb. Math. Grenzgeb., Springer Verlag, Berlin, Heidelberg, New York, 1977

[Lys90] I. G. Lysenok, *On some algorithmic properties of hyperbolic groups*, Math. USSR Izv. **35** (1990), 145–163

[Mac64] A. M. Macbeath, *Groups of homeomorphisms of a simply connected space*, Ann. of Math. (2) **79** (1964), 473–488

[Mal40] A. I. Mal'cev, *On faithful representations of infinite groups of matrices*, Mat. Sb. **8** (1940), 405–422, Amer. Math. Soc. Translations **45** (1965), 1–18

[Mar74] G. A. Margulis, *Discrete groups of motions of manifolds of non-positive curvature*, Proc. ICM, Vancouver 1974, Amer. Math. Soc. Translations **109** (1977), 33–45

[Mar90] G. A. Margulis, Discrete subgroups of Lie groups, Ergeb. Math. Grenzgeb., Springer Verlag, Berlin, Heidelberg, New York, 1990

[Mark58] A. A. Markov, *Insolubility of the problem of homeomorphy*, Proc. ICM Cambridge 1958, Cambridge Univ. Press, 1960, 300-306

[Mass91] W. Massey, A Basic Course in Algebraic Topology, Grad. Texts in Math., vol. 127, Springer Verlag, New York, 1991

[MaUl32] S. Mazur et S. Ulam, *Toute isométrie d'un espace normé qui préserve l'origine est linéaire*, C. R. Acad. Sci. Paris **194** (1932), 946–948

[Maz61] B. Mazur, *A note on some contractible 4-manifolds*, Ann. of Math. **73** (1961), 221–228

[Me82] D. Meier, *Non-Hopfian groups*, J. London Math. Soc. (2) **26** (1982), 265–270

[Mei94] J. Meier, *Automorphisms of negatively-curved polygonal amalgams*, Michigan Math. J. **41** (1994), 121–134

[Mei96] J. Meier, *When is the graph product of hyperbolic groups hyperbolic?*, Geom. Dedicata **61** (1996), 29–41

[MeiMV98] J. Meier, H. Meinert and L. van Wyck, *Higher generation subgroup sets and the Σ-invariants of graph groups*, Comment. Math. Helv. **73** (1998), 22–44

[MeiW95] J. Meier and L. van Wyk, *The Bieri-Neumann-Strebel invariants for graph groups*, Proc. London Math. Soc. **71** (1995), 263–280

[Men52] K. Menger, *The formative years of Abraham Wald and his work in geometry*, Ann. Math. Statistics **23** (1952), 14–20

[Mih58] K. A. Mihailova, *The occurrence problem for direct products of groups*, Dokl. Akad. Nauk SSSR **119** (1958), 1103–1105

[Mil63] J. Milnor, Morse Theory, Annals of Math. Stud., vol 51, Princeton Univ. Press, Princeton NJ, 1963

[Mil68] J. Milnor, *A note on curvature and the fundamental group*, J. Differential Geom. **2** (1968), 1–7

[Mil82] J. Milnor, *Hyperbolic geometry, the first 150 years*, Bull. Amer. Math. Soc. **6** (1982), 9–24

[Mill71] C. F. Miller III, On Group Theoretic Decision Problems and their Classification, Annals of Math. Stud., vol. 68, Princeton. Univ. Press, Princeton NJ, 1971

[Mill92] C. F. Miller III, *Decision problems for groups: survey and reflections*, Algorithms and Classification in Combinatorial Group Theory (G. Baumslag and C. F. Miller III, ed), MSRI Publ. **23**, Springer-Verlag, New York, 1992, 1–59

[Min05] H. Minkowski, *Diskontinuitätsbereich für arithmetische Äquivalenz*, J. Reine Angew. Math. **129** (1905), 220–274

[Mo95a] L. Mosher, *Mapping class groups are automatic*, Ann. of Math. (2) **142** (1995), 303–384

[Mo95b] L. Mosher, *Geometry of cubulated 3-manifolds*, Topology **34** (1995), 789–814

[Mol88] P. Molino, Riemannian Foliations, Progress in Math. **73**, 1988, Birkhäuser, Boston MA

[Mor84] J. Morgan, *Thurston's uniformization theorem*, Smith Conjecture (J. Morgan, H. Bass, ed), Academic Press, New York, 1984, 37–126

[Mos55] G. D. Mostow, *Some new decompositions of semi-simple groups*, Lie Algebras and Lie Groups, Mem. Amer. Math. Soc. **14** (1955), 31–34

[Mos73] G. D. Mostow, Strong rigidity of locally symmetric spaces, Ann. of Math. Stud., vol. 78, Princeton Univ. Press, Princeton NJ, 1973

[Mou88] G. Moussong, *Hyperbolic Coxeter groups*, Ph.D Thesis, The Ohio State University, 1988

[MoZ55] D. Montgomery and L. Zippin, Topological transformations groups, Interscience publishers, Inc., New-York, 1955

[NeR97] W. D. Neumann and L. Reeves, *Central extensions of word hyperbolic groups*, Ann. of Math. (2) **145** (1997), 183–192

[NeRe92] W. D. Neumann and A. W. Reid, *Arithmetic of hyperbolic manifolds*, Topology '90 (Apanasov et al., ed), Ohio State Univ. Math. Res. Inst. Publ. 1, 273–310, de Gruyter, Berlin, 1992

[New31] M. H. A. Newman, *A theorem on periodic transformations of spaces*, Quart. J. Math. Oxford **2** (1931), 1–9

[Ni95] I. G. Nikolaev, *The tangent cone of an Alexandrov Space of curvature* $\leq K$, Manuscripta Math. **86** (1995), 683-689

[NR97] G. Niblo and L. Reeves, *Groups acting on* CAT(0) *cube complexes*, Geometry and Topology **1** (1997), 1–7

[NR98] G. Niblo and L. Reeves, *The geometry of cube complexes and the complexity of their fundamental groups*, Topology **37** (1998), 621–633

[Ol91] A. Yu. Ol'shanskii, *Hyperbolicity of groups with subquadratic isoperimetric inequalities*, Internat. J. Algebra Comput. **1** (1991), 281–289

[Ol92] A. Yu. Ol'shanskii, *Almost every group is hyperbolic*, Internat. J. Algebra Comput. **2** (1992), 1–17

[Ol93] A. Yu. Ol'shanskii, *On residualising homomorphisms and G-subgroups of hyperbolic groups*, Internat. J. Algebra Comput. **3** (1993), 365–409

[Ol95] A. Yu. Ol'shanskii, *SQ-universality of hyperbolic groups*, Mat. Sb. **186** (1995), 119–132

[Ol96] A. Yu. Ol'shanskii, *Hyperbolic groups and their quotients of bounded exponents*, Trans. Amer. Math. Soc. **348** (1996), 2091–2138

[Ot92] J-P. Otal, *Sur la géométrie symplectique de l'espace des géodésiques d'une variété à courbure négative*, Rev. Mat. Iberoamericana **8** (1992), 441–456

[Ot96] J-P. Otal, Le Théorème d'Hyperbolisation pour les Variétés Fibrées de Dimension 3, Astérisque **235**, 1996

[Pan83] P. Pansu, *Croissance des boules et des géodesiques fermées dans les nilvariétés*, Ergod. Th. Dynam. Sys. **3** (1983), 415–445

[Pan89] P. Pansu, *Métriques de Carnot-Carathéodory et quasiisométries des espaces symétriques de rang un*, Ann. of Math. (2) **129** (1989), 1–60

[Pan90] P. Pansu, *Dimension conforme et sphère à l'infini des variétés à courbure négative*, Ann. Acad. Sci. Fenn. Math., Series A.I, **14** (1990), 177–212

[Pan95] P. Pansu, *Sous-groupes discrets des groupes de Lie: rigidité, arithméticité*, Sém. Bourbaki No. 778 (1993/94), Astérique **227** (1995), 69–105

[Papa57] C. D. Papakyriakopolous, *On Dehn's lemma and the asphericity of knots*, Ann. of Math. **66** (1957), 1–26

[Paps95a] P. Papasoglou, *On the sub-quadratic isoperimetric inequality*, Geometric Group Theory (R. Charney, M. Davis, M. Shapiro, ed), de Gruyter, Berlin-New York, 1995, 149–158

[Paps95b] P. Papasoglou, *Strongly geodesically automatic groups are hyperbolic*, Invent. Math. **121** (1995), 322–334

[Paps96] P. Papasoglou, *Asymptotic cones and the quadratic isoperimetric inequality*, J. Differential Geom. **44** (1996), 789–806

[Pau88] F. Paulin, *Topologie de Gromov équivariante, structures hyperboliques et arbres réels*, Invent. Math. **94** (1988), 53–80

[Pau89] F. Paulin, *The Gromov topology on* $\mathbb{R}$*-trees*, Topology Appl. **32** (1989), 197–221

[Pau91a] F. Paulin, *Construction of hyperbolic groups via hyperbolization of polyhedra*, Group Theory from a Geometrical Viewpoint (E. Ghys, A. Haefliger, A. Verjovsky, ed), Proc. ICTP Trieste 1990, World Scientific, Singapore, 1991, 313–373

[Pau91b] F. Paulin, *Outer automorphisms of hyperbolic groups and small actions on* $\mathbb{R}$*-trees*, Arboreal Group Theory (R. Alperin, ed), MSRI series, vol 19, Springer-Verlag 1991, 331–343

[Pau95] F. Paulin, *Actions de groupes sure les arbres*, Sém. Bourbaki, no. 808, November 1995

[Pau96] F. Paulin, *Un groupe hyperbolique est déterminé par son bord*, J. London Math. Soc. **54** (1996), 50–74

[Pet91] P. Peterson, Riemannian Geometry, Grad. Texts in Math., vol. 171, Springer Verlag, New York, 1991

[Po1895] H. Poincaré, *Analysis Situs*, J. de l'École Polyt. **1** (1895), 1–121

[Po04] H. Poincaré, *Cinquième complement à l'analysis situs*, Rend. Circ. Mat. Palermo **18** (1904), 45–110

[Poe60] V. Poenaru, *La décomposition de l'hypercube en produit topologique*, Bull. Soc. Math. France **88** (1960), 113-129

[Pr88] S. J. Pride, *Star complexes and the dependence problem for hyperbolic complexes*, Glasgow Math. J. **30** (1988), 155–170

[Qui73] D. Quillen, *Higher algebraic theory:I*, Algebraic K-theory I, Lecture Notes in Math. **341**, Springer Verlag, Berlin, 1973, 77–139

[Rag72] M. Raghunathan, Discrete Subgroups of Lie Groups, Ergeb. Math. Grenzgeb., Springer Verlag, Berlin, Heidelberg, New York, 1972

[Rat94] J. Ratcliffe, Foundations of Hyperbolic Manifolds, Grad. Texts in Math., vol. 149, Springer Verlag, New York, 1994

[Rau51] H. E. Rauch, *A contribution to Riemannian geometry in the large*, Ann. of Math. (2) **54** (1951), 38-55

[Ren82] Jean Renault, C^*-algebras of groupoids and foliations, Operator Algebras and Applications, Proc. of Symposia in Pure Mathematics, **38**, part I, A.M.S. Providence (1982), 339–350

[Resh60] Yu. G. Reshetnyak, *On the theory of spaces of curvature not greater than* K, Mat. Sb. **52** (1960), 789–798

[Resh68] Yu. G. Reshetnyak, *Non-expansive maps in a space of curvature no greater than* K, Sib. Mat. J. **9** (1968), 683–689

[Rh52] G. de Rham, *Sur la réductibilité d'un espace de Riemann*, Comment. Math. Helv. **26** (1952), 328–344

[Rh71] G. de Rham, *Sur les polygones générateurs des groupes fuchsiens*, Enseign. Math. **17** (1971), 49–61

[Rin61] W. Rinow, Die innere Geometrie der metrischen Raüme, Grund. Math. Wiss., vol. 105, Springer-Verlag, Berlin, Göttingen, Heidelberg, 1961

[Rips82] E. Rips, *Subgroups of small cancellation groups*, Bull. London Math. Soc. **14** (1982), 45–47

[RiS94] E. Rips and Z. Sela, *Structure and rigidity in hyperbolic groups*, Geom. Func. Anal. **4** (1994), 337–371

[RiS95] E. Rips and Z. Sela, *Canonical representatives and equations in hyperbolic groups*, Invent. Math. **120** (1995), 489–512

[Rob82] A. Robert, *Meilleure approximation linéaire et espaces euclidiens*, Enseign. Math. **28** (1982), 281–293

[Roe93] J. Roe, Elementary Geometry, Oxford Univ. Press, 1993

[Roe96] J. Roe, Index theory, Coarse Geometry and the Topology of Manifolds, CBMS Series, Amer. Math. Soc., Providence RI, 1996

[Rol68] D. Rolfsen *Strongly Convex Metric in Cells*, Bull. Amer. Math. Soc. **74** (1968), 171–175

[Rol70] D. Rolfsen, *Characterizing the 3-cell by its metric*, Fund. Math. **68** (1970), 215-223

[Rol90] D. Rolfsen, Knots and Links, second edition, Publish or Perish, Houston TX, 1990

[Ron89] M. A. Ronan, Lectures on Buildings, Academic Press, San Diego, 1989

[Ron92] M. A. Ronan, *Buildings: Main Ideas and Applications, I and II*, Bull. London Math. Soc. **24** (1992), 1–51 and 97–126

[Rot95] J. J. Rotman, An introduction to the theory of groups, Fourth Edition, Grad. Texts in Math., vol. 148, Springer-Verlag, Berlin-Heidelberg-New York, 1995

[Rua96] K. Ruane, *Boundaries of groups*, Ph.D. Thesis, Florida State University, 1996

[Rua99] K. Ruane, *Boundaries of* CAT(0) *groups of the form* $\Gamma = G \times H$, Topology Appl. **92** (1999), 131–152

[Sag95] M. Sageev, *Ends of group pairs and non-positively curved cube complexes*, Proc. London Math. Soc. **71** (1995), 585–617

[Sat56] I. Satake, *On a generalization of the notion of manifolds*, Proc. Nat. Acad. Sci., USA, **42** (1956), 359–363

[Sat57] I. Satake, *The Gauss-Bonnet theorem for V-manifolds*, J. Math. Soc. Japan **9** (1957), 464 –492

[Sch85] V. Schroeder, *A splitting theorem for spaces of nonpositive curvature*, Invent. Math. **79** (1985), 323–327

[Sch89] V. Schroeder, *A cusp closing theorem*, Proc. Amer. Math. Soc. **106** (1989), 797–802

[Sch90] V. Schroeder, *Codimension one tori in manifolds of nonpositive curvature*, Geom. Dedicata **33** (1990), 251–263

[Sch88] V. Schroeder, *Existence of immersed tori in manifolds of nonpositive curvature*, J. Reine Angew. Math. **390** (1988), 32–46

[Schw95] R. E. Schwartz, *The quasi-isometry classification of rank one lattices*, Inst. Hautes Études Sci. Publ. Math. **82** (1995), 133–168

[Schw96] R. E. Schwartz, *Quasi-isometric rigidity and diophantine approximation*, Acta Math. **177** (1996), 75–112

[Sco73a] G. P. Scott, *Finitely generated 3-manifold groups are finitely presented*, J. London Math. Soc. **6** (1973), 437–440

[Sco73b] G. P. Scott, *Compact submanifolds of 3-manifolds*, J. London Math. Soc. **7** (1973), 246–250

[Sco83] G. P. Scott, *The geometry of 3-manifolds*, Bull. London Math. Soc. **15** (1983), 401–487

[ScoW79] G. P. Scott and C. T. C. Wall, *Topological methods in group theory*, Homological Group Theory, LMS Lecture Note Series **36**, Cambridge Univ. Press, 1979, 137–203

[Se68] G. Segal, *Classifying spaces and spectral sequences*, Inst. Hautes Études Sci. Publ. Math. **134** (1968), 105–112

[Seg85] D. Segal, Polycyclic Groups, Cambridge Univ. Press, 1985

[Sei31] W. Seifert, *Konstruktion dreidimensionaler geshlossener Räume*, Ber. Sächs. Akad. Wiss. **83** (1931), 26-66

[Sel95] Z. Sela, *The isomorphism problem for hyperbolic groups I*, Ann. of Math. (2) **141** (1995), 217–283

[Sel96] Z. Sela, *The conjugacy problem for knot groups*, Topology **32**, (1996), 363–369

[Sel97] Z. Sela, *Structure and rigidity in (Gromov) hyperbolic groups and discrete groups in rank 1 Lie groups II*, Geom. Func. Anal. **7** (1997), 561–593

[Sel99] Z. Sela, *Endomorphisms of hyperbolic groups I: The Hopf property*, Topology **38**, 301–322

[Selb60] A. Selberg, *On discontinuous groups in higher dimensional symmetric spaces*, Contributions to Function Theory, Tata Institute for Fundamental Research, Bombay, 1960, 147–164

[Ser77] J-P. Serre, Trees, Springer Verlag, 1980, Translation of "Arbres, Amalgames, SL_2", Astérisque **46**, 1977

[Sha91] P. B. Shalen, *Dendrology and its applications*, Group Theory from a Geometrical Viewpoint (E. Ghys, A. Haefliger, A. Verjovsky, ed), Proc. ICTP Trieste 1990, World Scientific, Singapore, 1991, 543–617

[Sho91] H. Short (editor), *Notes on word hyperbolic groups*, Group Theory from a Geometrical Viewpoint (E. Ghys, A. Haefliger, A. Verjovsky, ed), Proc. ICTP Trieste 1990, World Scientific, Singapore, 1991, 3–64

[Sol69] L. Solomon, *The Steinberg character of a finite group with BN-pair*, Theory of Finite Groups, Benjamin, New York, 1969, 213–221

[Spa66] E. H. Spanier, Algebraic Topology, McGraw-Hill, New York, 1966

[Spi92] B. Spieler, *Developability of nonpositively curved orbihedra*, Ph.D. Thesis, The Ohio State University, 1992

[St63] J.R. Stallings, *A finitely presented group whose 3-dimensional integral homology is not finitely generated*, Amer. J. Math. **85** (1963), 541–543

[St68] J .R. Stallings, *On torsion-free groups with infinitely many ends*, Ann. of Math. (2) **88** (1968), 312–334

[St91] J .R. Stallings, *Non-positively curved triangles of groups*, Group Theory from a Geometrical Viewpoint (E. Ghys, A. Haefliger, A. Verjovsky, ed), Proc. ICTP Trieste 1990, World Scientific, Singapore, 1991, 491–503

[SuTh83] D. Sullivan and W. P. Thurston, *Manifolds with canonical coordinate charts, some examples*, Enseign. Math. **29** (1983), 15–25

[Sv55] A.S. Švarc, *Volume invariants of coverings*, Dokl. Akad. Nauk. SSSR **105** (1955), 32–34

[Sw98] J. Świątkowski, *Trivalent polygonal complexes of non-positive curvature and Platonic symmetry*, Geom. Dedicata **70** (1998), 87–110

[Swa97] G. A. Swarup, *On the cut point conjecture*, Electron. Res. Announc. Amer. Math. Soc. **2** (1996), 99–100

[Swe93] E. L. Swenson, *Negatively curved groups and related topics*, Ph.D. Thesis, Brigham Young University, 1993

[Th96a] P. Thurston, CAT(0) *4-manifolds possessing a single tame point are Euclidean*, J. Geom. Anal. **6** (1996), 475–494

[Th96b] P. Thurston, *4-dimensional Busemann G-spaces are 4-manifolds*, Differential Geom. Appl. **6** (1996), 245–270

[Thu79] W. P. Thurston, *The geometry and topology of 3-manifolds*, Lecture notes, Princeton University, 1978–79

[Thu82] W. P. Thurston, *Three dimensional manifolds, Kleinian groups and hyperbolic geometry*, Bull. Amer. Math. Soc. **6** (1982), 357-381

[Thu86] W. P. Thurston, *Hyperbolic structures on 3-manifolds I: Deformation of acylindrical manifolds*, Ann. of Math. (2) **124** (1986), 203–246

[Thu97] W. P. Thurston, Three-dimensional geometry and topology (edited by Silvio Levy), Princeton Univ. Press, Princeton Math Series, 35, 1997

[Tits74] J. Tits, *Buildings of spherical type and finite BN-pairs*, Lecture Notes in Mathematics **386**, Springer Verlag, Berline, 1974

[Tits75] J. Tits, *On buildings and their applications*, Proc. ICM, Vancouver 1974 **1**, Canad. Math. Congress, Montreal, 1975, 209–220

[Tits81] J. Tits, *A local approach to buildings*, The Geometric Vein (C. Davis, B. Grünbaum, F. A. Scherk, ed), Springer-Verlag, New York, 1981, 519–549

[Tits86a] J. Tits, *Buildings and group amalgamations*, LMS Lecture Note Series **121**, Proceedings of Groups St.Andrews 1985, Cambridge Univ. Press, 1986, 110–127

[Tits86b] J. Tits, *Ensemble ordonnés, immeubles et sommes amalgamées*, Bull. Soc. Math. Belgique, Ser.A **38** (1986), 367–387

[Vin85] E. B. Vinberg, *Hyperbolic reflection groups*, Russian Math. Surveys **40** (1985), 29–66

[VinS93] E. B. Vinberg and O. Shvartsman, *Discrete groups of motions of spaces of constant curvature*, Geometry II: Spaces of constant curvature (E. Vinberg, ed), Encyclopedia of Math. Sci., vol. 29, Springer Verlag, Berlin, Heidelberg, New York, 1993, 139-248

[Wa36] A. Wald, *Begründung einer koordinatenlosen Differentialgeometrie der Flächen*, Ergebnisse eines Math. Kolloquiums **7** (1936), 24–46

[We71] C. M. Weinbaum, *The word and conjugacy problems for the knot group of any tame prime alternating knot*, Proc. Amer. Math. Soc. **22** (1971), 22–26

[West77] J. E. West, *Mapping Hilbert cube manifolds to ANR's: a solution to a conjecture of Borsuk*, Ann. of Math. (2) **106** (1977), 1–18

[Wh36] J. H. C. Whitehead, *On equivalent sets of elements in a free group*, Ann. of Math. (2) **37** (1936), 782–800

[Wi96a] D. T. Wise, Non-positively curved squared complexes, aperiodic tilings, and non-residually finite groups, Ph.D. Thesis, Princeton Univ., 1996

[Wi96b] D. T. Wise, *An automatic group that is not Hopfian*, J. Algebra **126** (1996), 845–847

[Wi98a] D. T. Wise, *Incoherent negatively curved groups*, Proc. Amer. Math. Soc. **126** (1998), 957–964

[Wo64] J. Wolf, *Homogeneity and bounded isometries in manifolds of negative curvature*, Illinois J. Math. **8** (1964), 14–18

[Wo67] J. Wolf, Spaces of constant curvature, McGraw-Hill, 1967

[Wo68] J. Wolf, *Growth of finitely generated solvable groups and curvature of Riemannian manifolds*, J. Differential Geom. **2** (1968), 421–446

[Yau71] S-T Yau, *On the fundamental group of compact manifolds of nonpositive curvature*, Ann. of Math. (2) **93** (1971), 579–585

[Zim84] R. Zimmer, Ergodic Theory and Semisimple Groups, Monogr. Math. **81**, Birkhäuser, Boston MA, 1984

[Zuk96] A.Zuk, *La propriété (T) de Kazhdan pour les groupes agissant sur les polyèdres*, C.R. Acad. Sci. Paris **323** (1996), 453–458

Index

Grundlehren der mathematischen Wissenschaften

A Series of Comprehensive Studies in Mathematics

A Selection

219. Duvaut/Lions: Inequalities in Mechanics and Physics
220. Kirillov: Elements of the Theory of Representations
221. Mumford: Algebraic Geometry I: Complex Projective Varieties
222. Lang: Introduction to Modular Forms
223. Bergh/Löfström: Interpolation Spaces. An Introduction
224. Gilbarg/Trudinger: Elliptic Partial Differential Equations of Second Order
225. Schütte: Proof Theory
226. Karoubi: K-Theory. An Introduction
227. Grauert/Remmert: Theorie der Steinschen Räume
228. Segal/Kunze: Integrals and Operators
229. Hasse: Number Theory
230. Klingenberg: Lectures on Closed Geodesics
231. Lang: Elliptic Curves. Diophantine Analysis
232. Gihman/Skorohod: The Theory of Stochastic Processes III
233. Stroock/Varadhan: Multidimensional Diffusion Processes
234. Aigner: Combinatorial Theory
235. Dynkin/Yushkevich: Controlled Markov Processes
236. Grauert/Remmert: Theory of Stein Spaces
237. Köthe: Topological Vector Spaces II
238. Graham/McGehee: Essays in Commutative Harmonic Analysis
239. Elliott: Probabilistic Number Theory I
240. Elliott: Probabilistic Number Theory II
241. Rudin: Function Theory in the Unit Ball of C^n
242. Huppert/Blackburn: Finite Groups II
243. Huppert/Blackburn: Finite Groups III
244. Kubert/Lang: Modular Units
245. Cornfeld/Fomin/Sinai: Ergodic Theory
246. Naimark/Stern: Theory of Group Representations
247. Suzuki: Group Theory I
248. Suzuki: Group Theory II
249. Chung: Lectures from Markov Processes to Brownian Motion
250. Arnold: Geometrical Methods in the Theory of Ordinary Differential Equations
251. Chow/Hale: Methods of Bifurcation Theory
252. Aubin: Nonlinear Analysis on Manifolds. Monge-Ampère Equations
253. Dwork: Lectures on ρ-adic Differential Equations
254. Freitag: Siegelsche Modulfunktionen
255. Lang: Complex Multiplication
256. Hörmander: The Analysis of Linear Partial Differential Operators I
257. Hörmander: The Analysis of Linear Partial Differential Operators II
258. Smoller: Shock Waves and Reaction-Diffusion Equations
259. Duren: Univalent Functions
260. Freidlin/Wentzell: Random Perturbations of Dynamical Systems
261. Bosch/Güntzer/Remmert: Non Archimedian Analysis – A System Approach to Rigid Analytic Geometry
262. Doob: Classical Potential Theory and Its Probabilistic Counterpart
263. Krasnosel'skiĭ/Zabreĭko: Geometrical Methods of Nonlinear Analysis
264. Aubin/Cellina: Differential Inclusions
265. Grauert/Remmert: Coherent Analytic Sheaves
266. de Rham: Differentiable Manifolds
267. Arbarello/Cornalba/Griffiths/Harris: Geometry of Algebraic Curves, Vol. I
268. Arbarello/Cornalba/Griffiths/Harris: Geometry of Algebraic Curves, Vol. II

269. Schapira: Microdifferential Systems in the Complex Domain
270. Scharlau: Quadratic and Hermitian Forms
271. Ellis: Entropy, Large Deviations, and Statistical Mechanics
272. Elliott: Arithmetic Functions and Integer Products
273. Nikol'skiĭ: Treatise on the Shift Operator
274. Hörmander: The Analysis of Linear Partial Differential Operators III
275. Hörmander: The Analysis of Linear Partial Differential Operators IV
276. Liggett: Interacting Particle Systems
277. Fulton/Lang: Riemann-Roch Algebra
278. Barr/Wells: Toposes, Triples and Theories
279. Bishop/Bridges: Constructive Analysis
280. Neukirch: Class Field Theory
281. Chandrasekharan: Elliptic Functions
282. Lelong/Gruman: Entire Functions of Several Complex Variables
283. Kodaira: Complex Manifolds and Deformation of Complex Structures
284. Finn: Equilibrium Capillary Surfaces
285. Burago/Zalgaller: Geometric Inequalities
286. Andrianaov: Quadratic Forms and Hecke Operators
287. Maskit: Kleinian Groups
288. Jacod/Shiryaev: Limit Theorems for Stochastic Processes
289. Manin: Gauge Field Theory and Complex Geometry
290. Conway/Sloane: Sphere Packings, Lattices and Groups
291. Hahn/O'Meara: The Classical Groups and K-Theory
292. Kashiwara/Schapira: Sheaves on Manifolds
293. Revuz/Yor: Continuous Martingales and Brownian Motion
294. Knus: Quadratic and Hermitian Forms over Rings
295. Dierkes/Hildebrandt/Küster/Wohlrab: Minimal Surfaces I
296. Dierkes/Hildebrandt/Küster/Wohlrab: Minimal Surfaces II
297. Pastur/Figotin: Spectra of Random and Almost-Periodic Operators
298. Berline/Getzler/Vergne: Heat Kernels and Dirac Operators
299. Pommerenke: Boundary Behaviour of Conformal Maps
300. Orlik/Terao: Arrangements of Hyperplanes
301. Loday: Cyclic Homology
302. Lange/Birkenhake: Complex Abelian Varieties
303. DeVore/Lorentz: Constructive Approximation
304. Lorentz/v. Golitschek/Makovoz: Construcitve Approximation. Advanced Problems
305. Hiriart-Urruty/Lemaréchal: Convex Analysis and Minimization Algorithms I. Fundamentals
306. Hiriart-Urruty/Lemaréchal: Convex Analysis and Minimization Algorithms II. Advanced Theory and Bundle Methods
307. Schwarz: Quantum Field Theory and Topology
308. Schwarz: Topology for Physicists
309. Adem/Milgram: Cohomology of Finite Groups
310. Giaquinta/Hildebrandt: Calculus of Variations I: The Lagrangian Formalism
311. Giaquinta/Hildebrandt: Calculus of Variations II: The Hamiltonian Formalism
312. Chung/Zhao: From Brownian Motion to Schrödinger's Equation
313. Malliavin: Stochastic Analysis
314. Adams/Hedberg: Function Spaces and Potential Theory
315. Bürgisser/Clausen/Shokrollahi: Algebraic Complexity Theory
316. Saff/Totik: Logarithmic Potentials with External Fields
317. Rockafellar/Wets: Variational Analysis
318. Kobayashi: Hyperbolic Complex Spaces
319. Bridson/Haefliger: Metric Spaces of Non-Positive Curvature
320. Kipnis/Landim: Scaling Limits of Interacting Particle Systems
321. Grimmett: Percolation
322. Neukirch: Algebraic Number Theory
323. Neukirch/Schmidt/Wingberg: Cohomology of Number Fields
324. Liggett: Stochastic Interacting Systems: Contact, Voter and Exclusion Processes

Printing (computer to plate): Mercedes-Druck, Berlin
Binding: Stürtz AG, Würzburg

GPSR Compliance
The European Union's (EU) General Product Safety Regulation (GPSR) is a set of rules that requires consumer products to be safe and our obligations to ensure this.

If you have any concerns about our products, you can contact us on

ProductSafety@springernature.com

In case Publisher is established outside the EU, the EU authorized representative is:

Springer Nature Customer Service Center GmbH
Europaplatz 3
69115 Heidelberg, Germany

www.ingramcontent.com/pod-product-compliance
Ingram Content Group UK Ltd.
Pitfield, Milton Keynes, MK11 3LW, UK
UKHW021914280526
471597UK00003B/169
9783540643241